T0177218

Quantum Field Theory and the Standard Model

Providing a comprehensive introduction to quantum field theory, this textbook covers the development of particle physics from its foundations to the discovery of the Higgs boson. Its combination of clear physical explanations, with direct connections to experimental data, and mathematical rigor make the subject accessible to students with a wide variety of backgrounds and interests. Assuming only an undergraduate-level understanding of quantum mechanics, the book steadily develops the Standard Model and state-of-the art calculation techniques. It includes multiple derivations of many important results, with modern methods such as effective field theory and the renormalization group playing a prominent role. Numerous worked examples and end-of-chapter problems enable students to reproduce classic results and to master quantum field theory as it is used today. Based on a course taught by the author over many years, this book is ideal for an introductory to advanced quantum field theory sequence or for independent study.

MATTHEW D. SCHWARTZ is a Professor of Physics at Harvard University. He is one of the world's leading experts on quantum field theory and its applications to the Standard Model.

Quantum Field Theory and the Standard Model

MATTHEW D. SCHWARTZ

Harvard University

CAMBRIDGE
UNIVERSITY PRESS

University Printing House, Cambridge CB2 8BS, United Kingdom

Published in the United States of America by Cambridge University Press, New York

Cambridge University Press is part of the University of Cambridge.

It furthers the University's mission by disseminating knowledge in the pursuit of
education, learning, and research at the highest international levels of excellence.

www.cambridge.org
Information on this title: www.cambridge.org/9781107034730
DOI: 10.1017/9781139540940

First published 2014 (version 14, August 2024)

Printed in the United States of America by Books International, Virginia

A catalog record for this publication is available from the British Library

Library of Congress Cataloging in Publication data
Schwartz, Matthew Dean, 1976–
Quantum field theory and the standard model / Matthew D. Schwartz.
pages   cm
ISBN 978-1-107-03473-0 (hardback)
1. Quantum field theory – Textbooks.   2. Particles (Nuclear physics) – Textbooks.   I. Title.
QC174.45.S329   2014
530.14'3–dc23

2013016195

ISBN 978-1-107-03473-0 Hardback

To my mother,
and to Carolyn, Eve and Alec

Contents

Preface

Quantum field theory (QFT) provides an extremely powerful set of computational methods that have yet to find any fundamental limitations. It has led to the most fantastic agreement between theoretical predictions and experimental data in the history of science. It provides deep and profound insights into the nature of our universe, and into the nature of other possible self-consistent universes. On the other hand, the subject is a mess. Its foundations are flimsy, it can be absurdly complicated, and it is most likely incomplete. There are often many ways to solve the same problem and sometimes none of them are particularly satisfying. This leaves a formidable challenge for the design and presentation of an introduction to the subject.

This book is based on a course I have been teaching at Harvard for a number of years. I like to start my first class by flipping the light switch and pointing out to the students that, despite their comprehensive understanding of classical and quantum physics, they still cannot explain what is happening. Where does the light come from? The emission and absorption of photons is a quantum process for which particle number is not conserved; it is an everyday phenomenon which cannot be explained without quantum field theory. I then proceed to explain (with fewer theatrics) what is essentially Chapter 1 of this book. As the course progresses, I continue to build up QFT, as it was built up historically, as the logical generalization of the quantum theory of creation and annihilation of photons to the quantum theory of creation and annihilation of any particle. This book is based on lecture notes for that class, plus additional material.

The main guiding principle of this book is that QFT is primarily a theory of physics, not of mathematics, and not of philosophy. QFT provides, first and foremost, a set of tools for performing practical calculations. These calculations take as input measured numbers and predict, sometimes to absurdly high accuracy, numbers that can be measured in other experiments. Whenever possible, I motivate and validate the methods we develop as explaining natural (or at least in principle observable) phenomena. Partly, this is because I think having tangible goals, such as explaining measured numbers, makes it easier for students to understand the material. Partly, it is because the connection to data has been critical in the historical development of QFT.

The historical connection between theory and experiment weaves through this entire book. The great success of the Dirac equation from 1928 was that it explained the magnetic dipole moment of the electron (Chapter 10). Measurements of the Lamb shift in the late 1940s helped vindicate the program of renormalization (Chapters 15 to 21). Measurements of inelastic electron–proton scattering experiments in the 1960s (Chapter 32) showed that QFT could also address the strong force. Ironically, this last triumph occurred only a few years after Geoffrey Chew famously wrote that QFT "is sterile with respect to strong inter-actions and that, like an old soldier, it is destined not to die but just to fade away." [Chew, 1961, p. 2]. Once asymptotic freedom (Chapter 26) and the renormalizability of the Standard Model (Chapter 21 and Part IV) were understood in the 1970s, it was clear that QFT was capable of precision calculations to match the precision experiments that were being performed. Our ability to perform such calculations has been steadily improving ever since, for example through increasingly sophisticated effective field theories (Chapters 22, 28, 31, 33, 35 and 36), renormalization group methods (Chapter 23 and onward), and on-shell approaches (Chapters 24 and 27). The agreement of QFT and the Standard Model with data over the past half century has been truly astounding.

Beyond the connection to experiment, I have tried to present QFT as a set of related tools guided by certain symmetry principles. For example, Lorentz invariance, the symmetry group associated with special relativity, plays an essential role. QFT is the theory of the creation and destruction of particles, which is possible due to the most famous equation of special relativity $E = mc^2$. Lorentz invariance guides the definition of particle (Chapter 8), is critical to the spin-statistics theorem (Chapter 12), and strongly constrains properties of the main objects of interest in this book: scattering or S-matrix elements (Chapter 6 and onward). On the other hand, QFT is useful in space-times for which Lorentz invariance is not an exact symmetry (such as our own universe, which since 1998 has been known to have a positive cosmological constant), and in non-relativistic settings, where Lorentz invariance is irrelevant. Thus, I am reluctant to present Lorentz invariance as an axiom of QFT (I personally feel that as QFT is a work in progress, an axiomatic approach is premature). Another important symmetry is unitarity, which implies that probabilities should add up to 1. Chapter 24 is entirely dedicated to the implications of unitarity, with reverberations throughout Parts IV and V. Unitarity is closely related to other appealing features of our description of fundamental physics, such as causality, locality, analyticity and the cluster decomposition principle. While unitarity and its avatars are persistent themes within the book, I am cautious of giving them too much of a primary role. For example, it is not clear how well cluster decomposition has been tested experimentally.

I very much believe that QFT is not a finished product, but rather a work in progress. It has developed historically, it continues to be simplified, clarified, expanded and applied through the hard work of physicists who see QFT from different angles. While I do present QFT in a more or less linear fashion, I attempt to provide multiple viewpoints whenever possible. For example, I derive the Feynman rules in five different ways: in classical field theory (Chapter 3), in old-fashioned perturbation theory (Chapter 4), through a Lagrangian approach (Chapter 7), through a Hamiltonian approach (also Chapter 7), and through the Feynman path integral (Chapter 14). While the path-integral derivation is the quickest, it is also the furthest removed from the type of perturbation theory to which the reader might already be familiar. The Lagrangian approach illustrates in a transparent way how tree-level diagrams are just classical field theory. The old-fashioned perturbation theory derivation connects immediately to perturbation theory in quantum mechanics, and motivates the distinct advantage of thinking off-shell, so that Lorentz invariance can be kept manifest at all stages of the calculation. On the other hand, there are some instances where an on-shell approach is advantageous (see Chapters 24 and 27).

Other examples of multiple derivations include the four explanations of the spin-statistics theorem I give in Chapter 12 (direct calculation, causality, stability and Lorentz invariance of the S-matrix), the three ways I prove the path integral and canonical formulations of quantum field theory equivalent in Chapter 14 (through the traditional Hamiltonian derivation, perturbatively through the Feynman rules, and non-perturbatively through the Schwinger–Dyson equations), and the three ways in which I derive effective actions in Chapter 33 (matching, with Schwinger proper time, and with Feynman path integrals). As different students learn in different ways, providing multiple derivations is one way in which I have tried to make QFT accessible to a wide audience.

This textbook is written assuming that the reader has a solid understanding of quantum mechanics, such as what would be covered in a year-long undergraduate class. I have found that students coming in generally do not know much classical field theory, and must relearn special relativity, so these topics are covered in Chapters 2 and 3. At Harvard, much of the material in this book is covered in three semesters. The first semester covers Chapters 1 to 22. Including both QED and renormalization in a single semester makes the coursework rather intense. On the other hand, from surveying the students, especially the ones who only have space for a single semester of QFT, I have found that they are universally glad that renormalization is covered. Chapter 22, on non-renormalizable theories, is a great place to end a semester. It provides a qualitative overview of the four forces in the Standard Model through the lens of renormalization and predictivity.

The course on which this textbook is based has a venerable history, dominated by the thirty or so years it was taught by the great physicist Sidney Coleman. Sidney provides an evocative description of the period from 1966 to 1979 when theory and experiment collaborated to firmly establish the Standard Model [Coleman, 1985, p. xiii]:

> This was a great time to be a high-energy theorist, the period of the famous triumph of quantum field theory. And what a triumph it was, in the old sense of the word: a glorious victory parade, full of wonderful things brought back from far places to make the spectator gasp with awe and laugh with joy.

Sidney was able to capture some of that awe and joy in his course, and in his famous Erice Lectures from which this quote is taken. Over the past 35 years, the parade has continued. I hope that this book may give you a sense of what all the fuss is about.

Acknowledgements

The book would not have been possible without the perpetual encouragement and enthusiasm of the many students who took the course on which this book is based. Without these students, this book would not have been written. The presentation of most of the material in this book, particularly the foundational material in Parts I to III, arose from an iterative process. These iterations were promoted in no small part by the excellent questions the students posed to me both in and out of class. In ruminating on those questions, and discussing them with my colleagues, the notes steadily improved.

I have to thank in particular the various unbelievable teaching assistants I had for the course, particularly David Simmons-Duffin, Clay Cordova, Ilya Feige and Prahar Mitra for their essential contributions to improving the course material. The material in this book was refined and improved due to critical conversations that I had with many people. In particular, I would like to thank Frederik Denef, Ami Katz, Aneesh Manohar, Yasunori Nomura, Michael Peskin, Logan Ramalingam, Matthew Reece, Subir Sachdev, Iain Stewart, Matthew Strassler, and Xi Yin for valuable conversations. More generally, my approach to physics, by which this book is organized, has been influenced by three people most of all: Lisa Randall, Nima Arkani-Hamed and Howard Georgi. From them I learned to respect non-renormalizable field theories and to beware of smoke and mirrors in theoretical physics.

I am indebted to Anders Andreassen, David Farhi, William Frost, Andrew Marantan and Prahar Mitra for helping me convert a set of decent lecture notes into a coherent and comprehensive textbook. I also thank Ilya Feige, Yang-Ting Chien, Yale Fan, Thomas Becher, Zoltan Ligeti and Marat Freytsis for critical comments on the advanced chapters of the book.

Some of the material in the book is original, and some comes from primary literature. However, the vast majority of what I write is a rephrasing of results presented in the existing vast library of fantastic quantum field theory texts. The textbooks by Peskin and Schroeder and by Weinberg were especially influential. For example, Peskin and Schroeder's nearly perfect Chapter 5 guides my Chapter 13. Weinberg's comprehensive two volumes of *The Quantum Theory of Fields* are unequalled in their rigor and generality. Less general versions of many of Weinberg's explanations have been incorporated into my Chapters 8, 9, 14 and 24.

I have also taken some material from Srednicki's book (such as the derivation of the LSZ reduction formula in my Chapter 6), from Muta's book on quantum chromodynamics (parts of my Chapters 13, 26 and 32), from Banks' dense and deep *Concise Introduction* (particularly his emphasis on the Schwinger–Dyson equations which affected my Chapters 7 and 14), from Halzen and Martin's very physical book *Quarks and Leptons* (my Chapter 32), Rick Field's book *Applications of Perturbative QCD* (my Chapter 20). I have always found Manohar and Wise's monograph *Heavy Quark Physics* (on which my Chapter 35 is based) to be a valuable reference, in particular its spectacularly efficient first chapter. Zee's *Quantum Field Theory in a Nutshell* also had much influence on me (and on my Chapter 15). In addition, a few historical accounts come from Pais' *Inward Bound*, which I recommend any serious student of quantum field theory to devour.

Finally, I would like to thank my wife Carolyn, for her patience, love and support as this book was being written, as well as for some editorial assistance.

Matthew Dean Schwartz

Cambridge, Massachusetts
November 2013

PART I

FIELD THEORY

Microscopic theory of radiation   1

On October 19, 1900, Max Planck proposed an explanation of the blackbody radiation spectrum involving a new fundamental constant of nature, $h = 6.626 \times 10^{-34}$ J s [Planck, 1901]. Although Planck's result precipitated the development of quantum mechanics (i.e. the quantum mechanics of electrons), his original observation was about the quantum nature of light, which is a topic for quantum field theory. Thus, radiation is a great motivation for the development of a quantum theory of fields. This introductory topic involves a little history, a little statistical mechanics, a little quantum mechanics, and a little quantum field theory. It provides background and motivation for the systematic presentation of quantum field theory that begins in Chapter 2.

1.1 Blackbody radiation

In 1900, no one had developed a clear explanation for the spectrum of radiation from hot objects. A logical approach at the time was to apply the equipartition theorem, which implies that a body in thermal equilibrium should have energy equally distributed among all possible modes. For a hot gas, the theorem predicts the Maxwell–Boltzmann distribution of thermal velocities, which is in excellent agreement with data. When applied to the spectrum of light from a hot object, the equipartition theorem leads to a bizarre result.

A **blackbody** is an object at fixed temperature whose internal structure we do not care about. It can be treated as a hot box of light (or Jeans cube) in thermal equilibrium. Classically, a box of size L (with periodic boundary conditions for simplicity) supports standing electromagnetic waves with angular frequencies

$$\omega_n = \frac{2\pi}{L} |\vec{n}| c \tag{1.1}$$

for integer 3-vectors $\vec{n}$, with c being the speed of light. Before 1900, physicists believed you could have as much or as little energy in each mode as you want. By the (classical) equipartition theorem, blackbodies should emit light equally in all modes with the intensity growing as the differential volume of phase space:

$$I(\omega) \equiv \frac{1}{V} \frac{d}{d\omega} E(\omega) = \text{const} \times c^{-3} \omega^2 k_B T \quad \text{(classical)}. \tag{1.2}$$

More simply, this classical result follows from dimensional analysis: it is the only quantity with units of energy $\times$ time $\times$ distance^{-3} that can be constructed out of ω, $k_B T$ and

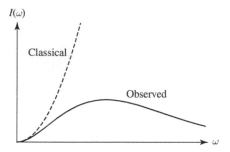

$I(\omega)$

Classical

Observed

ω

Fig. 1.1 The ultraviolet catastrophe. The classical prediction for the intensity of radiation coming from a blackbody disagrees with experimental observation at large frequencies.

c. We will set $c = 1$ from now on, since it can be restored by dimensional analysis (see Appendix A).

The classical spectrum implies that the amount of radiation emitted per unit frequency should increase with frequency, a result called the **ultraviolet catastrophe**. Experimentally, the distribution looks more like a Maxwell–Boltzmann distribution, peaked at some finite ω, as shown in Figure 1.1. Clearly the equipartition theorem does not work for blackbody radiation.

The incompatibility of observations with the classical prediction led Planck to postulate that the energy of each electromagnetic mode in the cavity is quantized in units of frequency:[1]

$$E_n = \hbar\omega_n = \frac{2\pi}{L}\hbar|\vec{n}| = |\vec{p}_n|, \tag{1.3}$$

where h is the Planck constant and $\hbar \equiv \frac{h}{2\pi}$. Albert Einstein later interpreted this as implying that light is made up of particles (later called **photons**, by the chemist Gilbert Lewis). Note that if the excitations are particles, then they are massless:

$$m_n^2 = E_n^2 - |\vec{p}_n|^2 = 0. \tag{1.4}$$

If Planck and Einstein are right, then light is really a collection of massless photons. As we will see, there are a number of simple and direct experimental consequences of this hypothesis: quantizing light resolves the blackbody paradox; light having energy leads to the photoelectric effect; and light having momentum leads to Compton scattering. Most importantly for us, the energy hypothesis was the key insight that led to the development of quantum field theory.

With Planck's energy hypothesis, the thermal distribution is easy to compute. Each mode of frequency ω_n can be excited an integer number j times, giving energy $jE_n = j(\hbar\omega_n)$

---

[1] Planck was not particularly worried about the ultraviolet catastrophe, since there was no strong argument why the equipartition theorem should hold universally; instead, he was trying to explain the observed spectrum. He first came up with a mathematical curve that fit data, generalizing previous work of Wilhelm Wien and Lord Rayleigh, then wrote down a toy model that generated this curve. The interpretation of his model as referring to photons and the proper statistical mechanics derivation of the blackbody spectrum did not come until years later.

in that mode. The probability of finding that much energy in the mode is the same as the probability of finding energy in *anything*, proportional to the Boltzmann weight $\exp(-\text{energy}/k_B T)$. Thus, the expectation value of energy in each mode is

$$\langle E_n \rangle = \frac{\sum_{j=0}^{\infty}(jE_n)e^{-jE_n\beta}}{\sum_{j=0}^{\infty}e^{-jE_n\beta}} = \frac{-\frac{d}{d\beta}\frac{1}{1-e^{-\hbar\omega_n\beta}}}{\frac{1}{1-e^{-\hbar\omega_n\beta}}} = \frac{\hbar\omega_n}{e^{\hbar\omega_n\beta}-1}, \tag{1.5}$$

where $\beta = 1/k_B T$. (This simple derivation is due to Peter Debye. The more modern one, using ensembles and statistical mechanics, was first given by Satyendra Nath Bose in 1924.)

Now let us take the continuum limit, $L \to \infty$. In this limit, the sums turn into integrals and the average total energy up to frequency ω in the blackbody is

$$E(\omega) = \int^{\omega} d^3\vec{n} \frac{\hbar\omega_n}{e^{\hbar\omega_n\beta}-1} = \int_{-1}^{1} d\cos\theta \int_{0}^{2\pi} d\phi \int_{0}^{\frac{\omega L}{2\pi}} d|\vec{n}| \frac{|\vec{n}|^2\hbar\omega_n}{e^{\hbar\omega_n\beta}-1}$$

$$= 4\pi\hbar \frac{L^3}{8\pi^3} \int_{0}^{\omega} d\omega' \frac{\omega'^3}{e^{\hbar\omega'\beta}-1}. \tag{1.6}$$

Thus, the intensity of light as a function of frequency is (adding a factor of 2 for the two polarizations of light)

$$I(\omega) = \frac{1}{V}\frac{dE(\omega)}{d\omega} = \frac{\hbar}{\pi^2}\frac{\omega^3}{e^{\hbar\omega\beta}-1}. \tag{1.7}$$

It is this functional form that Planck showed in 1900 correctly matches experiment.

What does this have to do with quantum field theory? In order for this derivation, which used equilibrium statistical mechanics, to make sense, light has to be able to equilibrate. For example, if we heat up a box with monochromatic light, eventually all frequencies must be excited. However, if different frequencies are different particles, equilibration must involve one kind of particle turning into another kind of particle. So, particles must be created and destroyed. Quantum field theory tells us how that happens.

1.2 Einstein coefficients

A straightforward way to quantify the creation of light is through the coefficient of spontaneous emission. This is the rate at which an excited atom emits light. Even by 1900, this phenomenon had been observed in chemical reactions, and as a form of radioactivity, but at that time it was only understood statistically. In 1916, Einstein came up with a simple proof of the relation between emission and absorption based on the existence of thermal equilibrium. In addition to being relevant to chemical phenomenology, his relation made explicit why a first principles quantum theory of fields was needed.

Einstein's argument is as follows. Suppose we have a cavity full of atoms with energy levels E_1 and E_2. Assume there are n_1 of the E_1 atoms and n_2 of the E_2 atoms and let $\hbar\omega = E_2 - E_1$. The probability for an E_2 atom to emit a photon of frequency ω and transition to state E_1 is called the **coefficient for spontaneous emission** A. The probability for a photon of frequency ω to induce a transition from 2 to 1 is proportional to the **coefficient of stimulated emission** B and to the number of photons of frequency ω in the cavity, that is, the intensity $I(\omega)$. These contribute to a change in n_2 of the form

$$dn_2 = -\left[A + BI(\omega)\right]n_2 + \cdots . \tag{1.8}$$

The probability for a photon to induce a transition from 1 to 2 is called the **coefficient of absorption** B'. Absorption decreases n_1 and increases n_2 by $B'I(\omega)n_1$. Since the total number of atoms is conserved in this two-state system, $dn_1 + dn_2 = 0$. Therefore,

$$dn_2 = -dn_1 = -\left[A + BI(\omega)\right]n_2 + B'I(\omega)n_1. \tag{1.9}$$

Even though we computed $I(\omega)$ above for the equilibrium blackbody situation, these equations should hold for any $I(\omega)$. For example, $I(\omega)$ could be the intensity of a laser beam we shine at some atoms in the lab.

At this point, Einstein assumes the gas is in equilibrium. In equilibrium, the number densities are constant, $dn_1 = dn_2 = 0$, and determined by Boltzmann distributions:

$$n_1 = Ne^{-\beta E_1}, \qquad n_2 = Ne^{-\beta E_2}, \tag{1.10}$$

where N is some normalization factor. Then

$$\left[B'e^{-\beta E_1} - Be^{-\beta E_2}\right]I(\omega) = Ae^{-\beta E_2} \tag{1.11}$$

and so

$$I(\omega) = \frac{A}{B'e^{\hbar\beta\omega} - B}. \tag{1.12}$$

However, we already know that in equilibrium

$$I(\omega) = \frac{\hbar}{\pi^2}\frac{\omega^3}{e^{\hbar\beta\omega} - 1} \tag{1.13}$$

from Eq. (1.7). Since equilibrium must be satisfied at any temperature, i.e. for any β, we must have

$$B' = B \tag{1.14}$$

and

$$\frac{A}{B} = \frac{\hbar}{\pi^2}\omega^3. \tag{1.15}$$

These are simple but profound results. The first, $B = B'$, says that the coefficient of absorption must be the same as the coefficient for stimulated emission. The coefficients B and B' can be computed in quantum mechanics (not quantum field theory!) using time-dependent perturbation theory with an external electromagnetic field. Then Eq. (1.15) determines A. Thus, all the Einstein coefficients A, B and B' can be computed without using quantum field theory.

You might have noticed something odd in the derivation of Eqs. (1.14) and (1.15). We, and Einstein, needed to use an equilibrium result about the blackbody spectrum to derive the A/B relation. Does spontaneous emission from an atom have anything to do with equilibrium of a gas? It seems like it shouldn't, that spontaneous emission should occur, at a calculable rate, even for an isolated atom. The calculation of A/B from first principles was not performed until 10 years after Einstein's calculation; it had to wait until the invention of quantum field theory.

1.3 Quantum field theory

The basic idea behind the calculation of the spontaneous emission coefficient in quantum field theory is to treat photons of each energy as separate particles, and then to study the system with multi-particle quantum mechanics. The following treatment comes from a paper of Paul Dirac from 1927 [Dirac, 1927], which introduced the idea of second quantization. This paper is often credited for initiating quantum field theory.

Start by looking at just a single-frequency (energy) mode of a photon, say of energy Δ. This mode can be excited n times. Each excitation adds energy Δ to the system. So, the energy eigenstates have energies $\Delta, 2\Delta, 3\Delta, \ldots$. There is a quantum mechanical system with this property that you may remember from your quantum mechanics course: the simple harmonic oscillator (reviewed in Section 2.2.1 and Problem 2.7).

The easiest way to study a quantum harmonic oscillator is with creation and annihilation operators, $a^\dagger$ and a. These satisfy

$$[a, a^\dagger] = 1. \tag{1.16}$$

There is also the number operator $\hat{N} = a^\dagger a$, which counts modes:

$$\hat{N}|n\rangle = n|n\rangle. \tag{1.17}$$

Then,

$$\hat{N}a^\dagger|n\rangle = a^\dagger a a^\dagger|n\rangle = a^\dagger|n\rangle + a^\dagger a^\dagger a|n\rangle = (n+1)a^\dagger|n\rangle. \tag{1.18}$$

Thus, $a^\dagger|n\rangle = C|n+1\rangle$ for some constant C, which can be chosen real. We can determine C from the normalization $\langle n|n\rangle = 1$:

$$C^2 = \langle n+1|C^2|n+1\rangle = \langle n|aa^\dagger|n\rangle = \langle n|(a^\dagger a + 1)|n\rangle = n+1, \tag{1.19}$$

so $C = \sqrt{n+1}$. Similarly, $a|n\rangle = C'|n-1\rangle$ and

$$C'^2 = \langle n-1|C'^2|n-1\rangle = \langle n|a^\dagger a|n\rangle = n, \tag{1.20}$$

so $C' = \sqrt{n}$. The result is that

$$a^\dagger|n\rangle = \sqrt{n+1}|n+1\rangle, \qquad a|n\rangle = \sqrt{n}|n-1\rangle. \tag{1.21}$$

While these normalization factors are simple to derive, they have important implications.

Now, you may recall from quantum mechanics that transition rates can be computed using Fermi's golden rule. Fermi's golden rule says that the transition rate between two states is proportional to the matrix element squared:

$$\Gamma \sim |\mathcal{M}|^2 \delta(E_f - E_i), \tag{1.22}$$

where the δ-function serves to enforce energy conservation. (We will derive a similar formula for the transition rate in quantum field theory in Chapter 5. For now, we just want to use quantum mechanics.) The matrix element $\mathcal{M}$ in this formula is the projection of the initial and final states on the interaction Hamiltonian:

$$\mathcal{M} = \langle f|H_{\text{int}}|i\rangle. \tag{1.23}$$

In this case, we do not need to know exactly what the interaction Hamiltonian H_{int} is. All we need to know is that H_{int} must have some creation operator or annihilation operator to create the photon. H_{int} also must be Hermitian. Thus it must look like[2]

$$H_{\text{int}} = H_I^\dagger a^\dagger + H_I a, \tag{1.24}$$

with H_I having non-zero matrix elements between initial and final atomic states.

For the $2 \to 1$ transition, the initial state is an excited atom we call atom$_2$ with n_ω photons of frequency $\omega = \Delta/\hbar$:

$$|i\rangle = |\text{atom}_2; n_\omega\rangle. \tag{1.25}$$

The final state is a lower energy atom we call atom$_1$ with $n_\omega + 1$ photons of energy Δ:

$$\langle f| = \langle \text{atom}_1; n_\omega + 1|. \tag{1.26}$$

So,

$$\begin{aligned}
\mathcal{M}_{2\to1} &= \langle \text{atom}_1; n_\omega + 1|(H_I^\dagger a^\dagger + H_I a)|\text{atom}_2; n_\omega\rangle \\
&= \langle \text{atom}_1|H_I^\dagger|\text{atom}_2\rangle\langle n_\omega + 1|a^\dagger|n_\omega\rangle + \langle \text{atom}_1|H_I|\text{atom}_2\rangle\langle n_\omega + 1|a|n_\omega\rangle \\
&= \mathcal{M}_0^\dagger\langle n_\omega + 1|n_\omega + 1\rangle\sqrt{n_\omega + 1} + 0 \\
&= \mathcal{M}_0^\dagger\sqrt{n_\omega + 1}
\end{aligned} \tag{1.27}$$

where $\mathcal{M}_0^\dagger = \langle \text{atom}_1|H_I^\dagger|\text{atom}_2\rangle$. Thus,

$$|\mathcal{M}_{2\to1}|^2 = |\mathcal{M}_0|^2(n_\omega + 1). \tag{1.28}$$

If instead we are exciting an atom, then the initial state has an unexcited atom and n_ω photons:

$$|i\rangle = |\text{atom}_1; n_\omega\rangle \tag{1.29}$$

---

[2] Dirac derived H_I from the canonical introduction of the vector potential into the Hamiltonian: $H = \frac{1}{2m}\vec{p}^2 \to \frac{1}{2m}(\vec{p} + e\vec{A})^2$. This leads to $H_{\text{int}} \sim \frac{e}{m}\vec{A} \cdot \vec{p}$ representing the photon interacting with the atom's electric dipole moment. In our coarse approximation, the photon field $\vec{A}$ is represented by a and so H_I must be related to the momentum operator $\vec{p}$. Fortunately, all that is needed to derive the Einstein relations is that H_I is something with non-zero matrix elements between different atomic states; thus, we can be vague about its precise definition. For more details consult [Dirac, 1927] or [Dirac, 1930, Sections 61–64].

and the final state has an excited atom and $n_\omega - 1$ photons:

$$\langle f| = \langle \text{atom}_2; n_\omega - 1|. \tag{1.30}$$

This leads to

$$\begin{aligned}
\mathcal{M}_{1\to 2} &= \langle \text{atom}_2; n_\omega - 1|H_I^\dagger a^\dagger + H_I a|\text{atom}_1; n_\omega\rangle \\
&= \langle \text{atom}_2|H_I|\text{atom}_1\rangle\langle n_\omega - 1|a|n_\omega\rangle \\
&= \mathcal{M}_0\sqrt{n_\omega} \tag{1.31}
\end{aligned}$$

and therefore,

$$dn_2 = -dn_1 = -|\mathcal{M}_{2\to 1}|^2 n_2 + |\mathcal{M}_{1\to 2}|^2 n_1 = -|\mathcal{M}_0|^2(n_\omega + 1)n_2 + |\mathcal{M}_0|^2(n_\omega)n_1. \tag{1.32}$$

This is pretty close to Einstein's equation, Eq. (1.9):

$$dn_2 = -dn_1 = -[A + BI(\omega)]\, n_2 + B'I(\omega)n_1. \tag{1.33}$$

To get them to match exactly, we just need to relate the number of photon modes of frequency ω to the intensity $I(\omega)$. Since the energies are quantized by $\Delta = \hbar\omega = \hbar\frac{2\pi}{L}|\vec{n}|$, the total energy is

$$E(\omega) = \int^\omega d^3\vec{n}(\hbar\omega)n_\omega = (4\pi)\hbar L^3 \int_0^\omega \frac{d\omega}{(2\pi)^3}\omega^3 n_\omega. \tag{1.34}$$

We should multiply this by 2 for the two polarizations of light. (Dirac actually missed this factor in his 1927 paper, since polarization was not understood at the time.) Including the factor of 2, the intensity is

$$I(\omega) = \frac{1}{L^3}\frac{dE}{d\omega} = \frac{\hbar\omega^3}{\pi^2}n_\omega. \tag{1.35}$$

This equation is a standard statistical mechanical relation, independent of what n_ω actually is; its derivation required no mention of temperature or of equilibrium, just a phase space integral.

So now we have

$$dn_2 = -dn_1 = -|\mathcal{M}_0|^2\left[1 + \frac{\pi^2}{\hbar\omega^3}I(\omega)\right]n_2 + |\mathcal{M}_0|^2\left[\frac{\pi^2}{\hbar\omega^3}I(\omega)\right]n_1 \tag{1.36}$$

and can read off Einstein's relations,

$$B' = B, \qquad \frac{A}{B} = \frac{\hbar}{\pi^2}\omega^3, \tag{1.37}$$

without ever having to assume thermal equilibrium. This beautiful derivation was one of the first ever results in quantum field theory.

2 Lorentz invariance and second quantization

In the previous chapter, we saw that by treating each mode of electromagnetic radiation in a cavity as a simple harmonic oscillator, we can derive Einstein's relation between the coefficients of induced and spontaneous emission without resorting to statistical mechanics. This was our first calculation in quantum electrodynamics (QED). It is not a coincidence that the harmonic oscillator played an important role. After all, electromagnetic waves oscillate harmonically. In this chapter we will review special relativity and the simple harmonic oscillator and show how they are connected. This leads naturally to the notion of **second quantization**, which is a poorly chosen phrase used to describe the canonical quantization of relativistic fields.

It is worth mentioning at this point that there are two ways commonly used to quantize a field theory, both of which are covered in depth in this book. The first is canonical quantization. This is historically how quantum field theory was understood, and closely follows what you learned in quantum mechanics. The second way is called the *Feynman path integral*. Path integrals are more concise, more general, and certainly more formal, but when using path integrals it is sometimes hard to understand physically what you are calculating. It really is necessary to understand both ways. Some calculations, such as the LSZ formula which relates scattering amplitudes to correlation function (see Chapter 6), require the canonical approach, while other calculations, such as non-perturbative quantum chromodynamics (see Chapter 25), require path integrals. There are other ways to perform quantum field theory calculations, for example using old-fashioned perturbation theory (Chapter 4), or using Schwinger proper time (Chapter 33). Learning all of these approaches will give you a comprehensive picture of how and why quantum field theory works. We start with canonical quantization, as it provides the gentlest introduction to quantum field theory.

From now on we will set $\hbar = c = 1$. This gives all quantities dimensions of mass to some power (see Appendix A).

2.1 Lorentz invariance

Quantum field theory is the result of combining quantum mechanics with special relativity. Special relativity is relevant when velocities are a reasonable fraction of the speed of light, $v \sim 1$. In this limit, a new symmetry emerges: Lorentz invariance. A system is Lorentz invariant if it is symmetric under the Lorentz group, which is the generalization of the rotation group to include both rotations and boosts.

Normally, the more symmetric a system, the easier it is to solve problems. For example, solving the Schrödinger equation with a spherically symmetric potential (as in the hydrogen atom) is much easier than solving it with a cylindrically symmetric potential (such as for the hydrogen molecule). So why is quantum field theory so much harder than quantum mechanics? The answer, as Sidney Coleman put it, is because $E = mc^2$. This famous relation holds for particles at rest. When particles move relativistically, their kinetic energy is comparable to or exceeds their rest mass, $E_{kin} \gtrsim m$, which is only a factor of 2 away from the threshold for producing two particles. Thus, there is no regime in which the relativistic corrections of order v/c are relevant, but the effect from producing new particles is not.

2.1.1 Rotations

Lorentz invariance is symmetry under rotations and boosts. If you get confused, focus on perfecting your understanding of rotations alone. Then, consider boosts as a generalization.

Rotations should be extremely familiar to you, and they are certainly more intuitive than boosts. Under two-dimensional (2D) rotations, a vector (x, y) transforms as

$$x \to x \cos\theta + y \sin\theta, \tag{2.1}$$

$$y \to -x \sin\theta + y \cos\theta. \tag{2.2}$$

We can write this as

$$\begin{pmatrix} x \\ y \end{pmatrix} \to \begin{pmatrix} x\cos\theta + y\sin\theta \\ -x\sin\theta + y\cos\theta \end{pmatrix} = \begin{pmatrix} \cos\theta & \sin\theta \\ -\sin\theta & \cos\theta \end{pmatrix} \begin{pmatrix} x \\ y \end{pmatrix}, \tag{2.3}$$

or as

$$x_i \to R_{ij} x_j, \quad x_i = \begin{pmatrix} x \\ y \end{pmatrix}, \quad i = 1, 2. \tag{2.4}$$

When an index appears twice, as in $R_{ij}x_j$, that index should be summed over (the **Einstein summation convention**), so $R_{ij}x_j = R_{i1}x_1 + R_{i2}x_2$. This is known as a **contraction**.

Technically, we should write $x_i = R_i{}^j x_j$. However, having upper and lower indices on the same object makes expressions difficult to read, so we will often just lower or raise all the indices. We will be careful about the index position if it is ever ambiguous. For the row vector,

$$x^i = \begin{pmatrix} x & y \end{pmatrix} \to \begin{pmatrix} x & y \end{pmatrix} \begin{pmatrix} \cos\theta & -\sin\theta \\ \sin\theta & \cos\theta \end{pmatrix} = x^j \left(R^T \right)^{ji}. \tag{2.5}$$

Note that $R^T = R^{-1}$. That is,

$$\left(R^T \right)_{ik} R_{kj} = \delta_{ij} = \begin{pmatrix} 1 & 0 \\ 0 & 1 \end{pmatrix}_{ij} = \mathbb{1}_{ij} \tag{2.6}$$

or equivalently,

$$R^T R = \mathbb{1}. \tag{2.7}$$

This property (orthogonality) along with R preserving orientation ($\det R = 1$) is enough to characterize R as a rotation. This algebraic characterization in Eq. (2.7) is a much more useful definition of the group than the explicit form of the rotation matrices as a function of θ. The group of 2D rotations is also called the **special orthogonal group** SO(2). The group of 3D rotations is called SO(3).

If we contract the upper and lower indices of a vector, we find

$$x^i x_i = (x, y) \begin{pmatrix} x \\ y \end{pmatrix} = x^2 + y^2. \tag{2.8}$$

This is just the norm of the vector x_i and is invariant under rotations. To see that, note that under a rotation

$$x^i x_i \rightarrow \left(x^i R^T_{ij}\right)\left(R_{jk} x_k\right) = x^i \delta_{ik} x_k = x^i x_i, \tag{2.9}$$

since $R^T = R^{-1}$. In fact, another way to define the rotation group is as the set of linear transformations on $\mathbb{R}^n$ preserving the inner product $x^i x_i = \delta_{ij} x^i x^j$:

$$R_{ki} R_{lj} \delta_{kl} = [(R^T)\mathbb{1}(R)]_{ij} = (R^T R)_{ij} = \delta_{ij}, \tag{2.10}$$

which you can check explicitly using Eq. (2.3).

2.1.2 Lorentz transformations

Lorentz transformations work exactly like rotations, except with some minus signs here and there. Instead of preserving $r^2 = x^2 + y^2 + z^2$ they preserve $s^2 \equiv t^2 - x^2 - y^2 - z^2$. Instead of 3-vectors $v^i = (x, y, z)$ we use 4-vectors $x^\mu = (t, x, y, z)$. We generally use Greek indices for 4-vectors and Latin indices for 3-vectors. We write x^0 for the time component of a 4-vector.

Lorentz transformations acting on 4-vectors are matrices Λ satisfying

$$\Lambda^T g \Lambda = g = \begin{pmatrix} 1 & & & \\ & -1 & & \\ & & -1 & \\ & & & -1 \end{pmatrix}. \tag{2.11}$$

In this and future matrices, empty entries are 0. $g_{\mu\nu}$ is known as the **Minkowski metric**. Sometimes we write $\eta_{\mu\nu}$ for this metric, with $g_{\mu\nu}$ reserved for a general metric, as in general relativity. But outside of quantum gravity contexts, which will be clear when we encounter them, taking $g_{\mu\nu} = \eta_{\mu\nu}$ will cause no confusion in quantum field theory. Equation (2.11) says that Lorentz transformations preserve the Minkowskian inner product:

$$x^\mu x_\mu = g_{\mu\nu} x^\mu x^\nu = t^2 - x^2 - y^2 - z^2. \tag{2.12}$$

A rotation around the z axis leaves $x^2 + y^2$ invariant while a boost in the z direction leaves $t^2 - z^2$ invariant. So, instead of being sines and cosines, which satisfy $\cos^2\theta + \sin^2\theta = 1$, boosts are made from hyperbolic sines and cosines, which satisfy $\cosh^2\beta - \sinh^2\beta = 1$.

The Lorentz group is the most general set of transformations preserving the Minkowski metric. Up to some possible discrete transformations (see Section 2.1.3 below), a general Lorentz transformation can be written as a product of rotations around the x, y or z axes:

$$\begin{pmatrix} 1 & & & \\ & 1 & & \\ & & \cos\theta_x & \sin\theta_x \\ & & -\sin\theta_x & \cos\theta_x \end{pmatrix}, \quad \begin{pmatrix} 1 & & & \\ & \cos\theta_y & & -\sin\theta_y \\ & & 1 & \\ & \sin\theta_y & & \cos\theta_y \end{pmatrix}, \quad \begin{pmatrix} 1 & & & \\ & \cos\theta_z & \sin\theta_z & \\ & -\sin\theta_z & \cos\theta_z & \\ & & & 1 \end{pmatrix}$$

$$(2.13)$$

and boosts in the x, y or z direction:

$$\begin{pmatrix} \cosh\beta_x & \sinh\beta_x & & \\ \sinh\beta_x & \cosh\beta_x & & \\ & & 1 & \\ & & & 1 \end{pmatrix}, \quad \begin{pmatrix} \cosh\beta_y & & \sinh\beta_y & \\ & 1 & & \\ \sinh\beta_y & & \cosh\beta_y & \\ & & & 1 \end{pmatrix}, \quad \begin{pmatrix} \cosh\beta_z & & & \sinh\beta_z \\ & 1 & & \\ & & 1 & \\ \sinh\beta_z & & & \cosh\beta_z \end{pmatrix}.$$

$$(2.14)$$

The θ_i are ordinary rotation angles around the i axis, with $0 \le \theta_i < 2\pi$, and the β_i are hyperbolic angles sometimes called **rapidities**, with $-\infty < \beta_i < \infty$. Note that these matrices do not commute, so the order in which we do the rotations and boosts is important. We will rarely need an actual matrix representation of the group elements like this, but it is helpful to see.

To relate the β_i to something useful, such as velocity, recall that for velocities $v \ll 1$ well below the speed of light, a boost should reduce to a Galilean transformation $x \to x + vt$. The unique transformations that preserve $t^2 - x^2$ and reduce to the Galilean transformations at small v are

$$x \to \frac{x + vt}{\sqrt{1 - v^2}}, \quad t \to \frac{t + vx}{\sqrt{1 - v^2}}. \qquad (2.15)$$

Thus we can identify

$$\cosh\beta_x = \frac{1}{\sqrt{1 - v^2}}, \quad \sinh\beta_x = \frac{v}{\sqrt{1 - v^2}}. \qquad (2.16)$$

These equations relate boosts to ordinary velocity. In particular, $\beta_x = v$ to leading order in v.

Scalar fields are functions of space-time that are Lorentz invariant. That is, under an arbitrary Lorentz transformation the field does not change:

$$\phi(x) \to \phi(x). \qquad (2.17)$$

Sometimes the notation $\phi(x^\mu) \to \phi\big((\Lambda^{-1})^\mu_{\ \nu} x^\nu\big)$ is used, which makes it seem like the scalar field is changing in some way. It is not. While our definitions of x^μ change in different frames $x^\mu \to \Lambda^\mu_\nu x^\nu$, the space-time point labeled by x^μ is fixed. That equations are invariant under relabeling of coordinates tells us absolutely nothing about nature. The physical content of Lorentz invariance is that nature has a symmetry under which scalar fields do not transform. Take, for example, the temperature of a fluid, which can vary from point to point. If we change reference frames, the labels for the points change, but the temperature at each point stays the same. A **scalar** (not scalar field) is just a number. For example, $\hbar$ and 7 and the electric charge e are scalars.

Under Lorentz transformations Λ^μ_ν, **4-vectors** V_μ transform as

$$V^\mu \to \Lambda^\mu_\nu V^\nu. \qquad (2.18)$$

This transformation law is the defining property of a 4-vector. If V^μ is not just a number but depends on x, we write $V^\mu(x)$ and call it a **vector field**. Under Lorentz transformations, vector fields transform just like 4-vectors. For a vector field, as for a scalar field, the coordinates of x transform but the space-time point to which they refer is invariant. The difference from a scalar field is that the components of a vector field at the point x transform into each other as well. If you need a concrete example, think about how the components of the electric field $\vec{E}(\vec{x})$ rotate into each other under 3D rotations, while a scalar potential $\phi(\vec{x})$ for which $\vec{E}(\vec{x}) = \vec{\nabla}\phi(\vec{x})$ is rotationally invariant.

A vector field $V_\mu(x)$ is a set of four functions of space-time. A Lorentz-invariant theory constructed with vector fields has a symmetry: the result of calculations will be the same if the four functions are mixed up according to Eq. (2.18). For example, $g^{\mu\nu}\partial_\mu V_\nu(x)$ is Lorentz invariant at each space-time point x if and only if $V_\mu(x)$ transforms as a vector field under Lorentz transformations. If $V_\mu(x)$ were just a collection of four scalar fields, $g^{\mu\nu}\partial_\mu V_\nu(x)$ would be frame-dependent.

Some important 4-vectors are position:

$$x^\mu = (t, x, y, z), \tag{2.19}$$

derivatives with respect to x^μ:

$$\partial_\mu = \frac{\partial}{\partial x^\mu} = (\partial_t, \partial_x, \partial_y, \partial_z), \tag{2.20}$$

and momentum:

$$p^\mu = (E, p_x, p_y, p_z). \tag{2.21}$$

Tensors transform as

$$T^{\mu\nu} \rightarrow \Lambda^\mu_\alpha \Lambda^\nu_\beta T^{\alpha\beta}. \tag{2.22}$$

Tensor fields are functions of space-time, such as the energy-momentum tensor $T^{\mu\nu}(x)$ or the metric $g^{\mu\nu}(x)$ in general relativity. If you add more indices, such as $Z^{\mu\nu\alpha\beta}$, we still call it a tensor. The number of indices is the **rank** of a tensor, so $T^{\mu\nu}$ is rank 2, $Z^{\mu\nu\alpha\beta}$ is rank 4, etc.

When the same index appears twice, it is contracted, just as for rotations. Contractions implicitly involve the Minkowski metric and are Lorentz invariant. For example:

$$V^\mu W_\mu = V_\mu g^{\mu\nu} W_\nu = V_0 W_0 - V_1 W_1 - V_2 W_2 - V_3 W_3. \tag{2.23}$$

Such a contraction is Lorentz invariant and transforms like a scalar (just as the dot product of two 3-vectors $\vec{V} \cdot \vec{W}$, which is a contraction with δ_{ij}, is rotationally invariant). So, under a Lorentz transformation,

$$V^\mu W_\mu = VgW \rightarrow (V\Lambda^T)g(\Lambda W) = VgW = V^\mu W_\mu. \tag{2.24}$$

When writing contractions this way, you can usually just pretend g is the identity matrix. You will only need to distinguish g from δ when you write out components. This is one of the reasons the 4-vector notation is very powerful. Contracting indices is just a notational convention, not a deep property of mathematics.

It is worth adding a few more words about raising and lowering indices in field theory. In general relativity, it is important to be careful about distinguishing vectors with lower indices (**covariant** vectors) and vectors with upper indices (**contravariant** vectors). When an index appears twice (in a contraction) the technically correct approach is for one index to be upper and one to be lower. However, that can make the notation very cumbersome. For example, if the indices are ordered, you must write $V^\mu(x) \to \Lambda^\mu{}_\nu V^\nu(x)$, which is different from $V^\mu(x) \to \Lambda_\nu{}^\mu V^\nu(x)$. It is easier just to write $V^\mu \to \Lambda^{\mu\nu} V^\nu$ where the index order is clear. In special relativity, we always contract with the Minkowski metric $g_{\mu\nu} = \eta_{\mu\nu}$. So, we will often forget about which indices are upper and which are lower and just use the modern contraction convention for which all contractions are equivalent:

$$V_\mu W^\mu = V^\mu W_\mu = V_\mu W_\mu = V^\mu W^\mu. \tag{2.25}$$

Index position is important only when we plug in explicit vectors or matrices.

Although the index position is not important for us, the actual indices are. You should never have anything such as

$$V_\mu W_\mu X_\mu \tag{2.26}$$

with three (or more) of the same indices. To avoid this, be very careful about relabeling. For example, do not write

$$(V^2)(W^2) = V_\mu V_\mu W_\mu W_\mu; \tag{2.27}$$

instead write

$$(V^2)(W^2) = V_\mu^2 W_\nu^2 = V_\mu V_\mu W_\nu W_\nu = g_{\mu\alpha} g_{\nu\beta} V_\mu V_\alpha W_\nu W_\beta. \tag{2.28}$$

You will quickly get the hang of all this contracting.

The simplest Lorentz-invariant operator that we can write down involving derivatives is the **d'Alembertian**:

$$\Box = \partial_\mu^2 = \partial_t^2 - \partial_x^2 - \partial_y^2 - \partial_z^2. \tag{2.29}$$

This is the relativistic generalization of the **Laplacian**:

$$\triangle = \vec\nabla^2 = \partial_x^2 + \partial_y^2 + \partial_z^2. \tag{2.30}$$

Finally, it is worth keeping the terminology straight. We say that objects such as

$$V^2 = V_\mu V^\mu, \quad \phi, \quad 1, \quad \partial_\mu V^\mu \tag{2.31}$$

are **Lorentz invariant**, meaning they do not depend on our Lorentz frame at all, while objects such as

$$V_\mu, \quad F_{\mu\nu}, \quad \partial_\mu, \quad x_\mu \tag{2.32}$$

are **Lorentz covariant**, meaning they *do* change in different frames, but precisely as the Lorentz transformation dictates. Something such as energy density is neither Lorentz invariant nor Lorentz covariant; it is instead the 00 component of a Lorentz tensor $T_{\mu\nu}$.

2.1.3 Discrete transformations

Lorentz transformations are defined to be those that preserve the Minkowski metric:

$$\Lambda^T g \Lambda = g. \tag{2.33}$$

Equivalently, they are those that leave inner products such as

$$V_\mu W^\mu = V_0 W_0 - V_1 W_1 - V_2 W_2 - V_3 W_3 \tag{2.34}$$

invariant. By this definition, the transformations

$$P : (t, x, y, z) \to (t, -x, -y, -z) \tag{2.35}$$

known as **parity** and

$$T : (t, x, y, z) \to (-t, x, y, z) \tag{2.36}$$

known as **time reversal** are also Lorentz transformations. They can be written as

$$P = \begin{pmatrix} 1 & & & \\ & -1 & & \\ & & -1 & \\ & & & -1 \end{pmatrix}, \quad T = \begin{pmatrix} -1 & & & \\ & 1 & & \\ & & 1 & \\ & & & 1 \end{pmatrix}. \tag{2.37}$$

Parity and time reversal are special because they cannot be written as the product of rotations and boosts, Eqs. (2.13) and (2.14). Discrete transformations play an important role in quantum field theory (see Chapter 11).

We say that a vector is **timelike** when

$$V^\mu V_\mu > 0 \quad \text{(timelike)} \tag{2.38}$$

and **spacelike** when

$$V^\mu V_\mu < 0 \quad \text{(spacelike)}. \tag{2.39}$$

Naturally, time $= (t, 0, 0, 0)$ is timelike and space $= (0, x, 0, 0)$ is spacelike. Whether something is timelike or spacelike is preserved under Lorentz transformations since the norm is preserved. If a vector has zero norm we say it is **lightlike**:

$$V^\mu V_\mu = 0 \quad \text{(lightlike)}. \tag{2.40}$$

If p^μ is a 4-momentum, then (since $p^2 = m^2$) it is lightlike if and only if it is massless. Photons are massless, which is the origin of the term *lightlike*.

Many more details of the mathematical structure of the Lorentz group (such as its unitary representations) will be covered in Chapters 8 and 10.

2.1.4 Solving problems with Lorentz invariance

Special relativity in quantum field theory is much easier than the special relativity you learned in your introductory physics course. We never need to talk about putting long cars in small garages or engineers with flashlights on trains. These situations are all designed to make your non-relativistic intuition mislead you. In quantum field theory, other than the perhaps unintuitive notion that energy can turn into matter through $E = mc^2$, your non-relativistic intuition will serve you perfectly well.

For field theory, all you really need from special relativity is the one equation that defines Lorentz transformations:

$$\Lambda^T g \Lambda = g. \tag{2.41}$$

This implies that contractions such as $p^2 \equiv p^\mu p_\mu$ are Lorentz invariant. For problems that involve changing frames, usually you know everything in one frame and are interested in some quantity in another frame. For example, you may know momenta p_1^μ and p_2^μ of two incoming particles that collide and are interested in the energy of an outgoing particle E_3 in the center-of-mass frame (the center-of-mass frame is defined as the frame in which the total 3-momenta, $\vec{p}_{\text{tot}} = 0$). For such problems, it is best to first calculate a Lorentz-invariant quantity such as $p_{\text{tot}}^2 = (p_1^\mu + p_2^\mu)^2$ in the first frame, then go to the second frame, and solve for the unknown quantity. Since p_{tot}^2 is Lorentz invariant, it has the same value in both frames. Usually, when you input everything you know about the second frame (e.g. $\vec{p}_{\text{tot}} = 0$ if it is the center-of-mass frame), you can solve for the remaining unknowns. If you find yourself plugging in explicit boost and rotation matrices, you are probably solving the problem the hard way. This trick is especially useful for situations in which there are many particles, say $p_1^\mu, \ldots, p_5^\mu$, and therefore many Lorentz-invariant quantities, such as $p_1^\mu p_{4\mu}$ or $(p_5^\mu + p_4^\mu)^2$.

2.2 Classical plane waves as oscillators

We next review the simple harmonic oscillator and discuss the connection to special relativity.

2.2.1 Simple harmonic oscillator

Anything with a linear restoring force (any conservative force is linear close enough to equilibrium), such as a spring, or a string with tension, or a wave, is a harmonic oscillator. For example, a spring has

$$m\frac{d^2x}{dt^2} + kx = 0, \tag{2.42}$$

which is satisfied by $x(t) = \cos\left(\sqrt{\frac{k}{m}}t\right)$, so it oscillates with frequency

$$\omega = \sqrt{\frac{k}{m}}. \tag{2.43}$$

A more general solution is

$$x(t) = c_1 e^{i\omega t} + c_2 e^{-i\omega t}. \tag{2.44}$$

The classical Hamiltonian for this system is the sum of kinetic and potential energies:

$$H = \frac{1}{2}\frac{p^2}{m} + \frac{1}{2}m\omega^2 x^2. \tag{2.45}$$

To quantize the harmonic oscillator, we promote x and p to operators and impose the canonical commutation relations

$$[x, p] = i. \tag{2.46}$$

Analysis of the harmonic oscillator spectrum is simplest if we change variables to

$$a = \sqrt{\frac{m\omega}{2}}\left(x + \frac{ip}{m\omega}\right), \quad a^\dagger = \sqrt{\frac{m\omega}{2}}\left(x - \frac{ip}{m\omega}\right), \tag{2.47}$$

which satisfy

$$[a, a^\dagger] = 1, \tag{2.48}$$

so that

$$H = \omega\left(a^\dagger a + \frac{1}{2}\right). \tag{2.49}$$

Thus, energy eigenstates are eigenstates of the number operator

$$\hat{N} = a^\dagger a, \tag{2.50}$$

which is Hermitian. The results we derived in Section 1.3:

$$\hat{N}|n\rangle = n|n\rangle, \tag{2.51}$$

$$a^\dagger|n\rangle = \sqrt{n+1}|n+1\rangle, \tag{2.52}$$

$$a|n\rangle = \sqrt{n}|n-1\rangle, \tag{2.53}$$

follow from these definitions. We can also calculate how the operators evolve in time (in the Heisenberg picture):

$$i\frac{d}{dt}a = [a, H] = \left[a, \omega\left(a^\dagger a + \frac{1}{2}\right)\right] = \omega(aa^\dagger a - a^\dagger aa) = \omega[a, a^\dagger]a = \omega a. \tag{2.54}$$

This equation is solved by

$$a(t) = e^{-i\omega t}a(0). \tag{2.55}$$

2.2.2 Connection to special relativity

To connect special relativity to the simple harmonic oscillator we note that the simplest possible Lorentz-invariant equation of motion that a field can satisfy is $\Box\phi = 0$. That is,

$$\Box\phi = (\partial_t^2 - \vec{\nabla}^2)\phi = 0. \tag{2.56}$$

The classical solutions are plane waves. For example, one solution is

$$\phi(x) = a_p(t)e^{i\vec{p}\cdot\vec{x}}, \tag{2.57}$$

where

$$(\partial_t^2 + \vec{p}\cdot\vec{p})a_p(t) = 0. \tag{2.58}$$

This is exactly the equation of motion of a harmonic oscillator. A general solution is

$$\phi(x,t) = \int \frac{d^3 p}{(2\pi)^3} \left[a_p(t) e^{i\vec{p}\cdot\vec{x}} + a_p^\star(t) e^{-i\vec{p}\cdot\vec{x}} \right],\tag{2.59}$$

with $(\partial_t^2 + \vec{p}\cdot\vec{p})a_p(t) = 0$, which is just a Fourier decomposition of the field into plane waves. Or more simply

$$\phi(x,t) = \int \frac{d^3 p}{(2\pi)^3} \left(a_p e^{-ipx} + a_p^\star e^{ipx} \right),\tag{2.60}$$

with a_p and $a_p^\star$ now just numbers and $p_\mu \equiv (\omega_p, \vec{p})$ with $\omega_p \equiv |\vec{p}|$. To be extra clear about notation, px contains an implicit 4-vector contraction: $px = p^\mu x_\mu = \omega_p x_0 - \vec{p}\cdot\vec{x}$.

Not only is $\Box\phi = 0$ the simplest Lorentz-invariant field equation possible, it is one of the equations that free massless fields will always satisfy (up to some exotic exceptions). For example, recall that there is a nice Lorentz-covariant treatment of electromagnetism using

$$F_{\mu\nu} \equiv \begin{pmatrix} 0 & E_x & E_y & E_z \\ -E_x & 0 & -B_z & B_y \\ -E_y & B_z & 0 & -B_x \\ -E_z & -B_y & B_x & 0 \end{pmatrix}.\tag{2.61}$$

This $F_{\mu\nu}$ transforms covariantly as a tensor under Lorentz transformations and thus concisely encodes how $\vec{E}$ and $\vec{B}$ rotate into each other under boosts. In terms of $F_{\mu\nu}$, Maxwell's equations in empty space have the simple forms

$$\partial_\mu F_{\mu\nu} - 0, \qquad \partial_\mu F_{\nu\rho} + \partial_\nu F_{\rho\mu} + \partial_\rho F_{\mu\nu} = 0.\tag{2.62}$$

Any field satisfying these equations can be written as

$$F_{\mu\nu} = \partial_\mu A_\nu - \partial_\nu A_\mu.\tag{2.63}$$

Although not necessary, we can also require $\partial_\mu A_\mu = 0$, which is a gauge choice (**Lorenz gauge**). We will discuss gauge invariance in great detail in Chapters 8 and 25. For now, it is enough to know that the physical $\vec{E}$ and $\vec{B}$ fields *can* be combined into an antisymmetric tensor $F_{\mu\nu}$, which is determined by a 4-vector A_μ satisfying $\partial_\mu A_\mu = 0$. In Lorenz gauge, Maxwell's equations reduce to

$$\partial_\mu F_{\mu\nu} = \Box A_\nu - \partial_\nu(\partial_\mu A_\mu) = \Box A_\nu = 0.\tag{2.64}$$

Thus, each component of A_ν satisfies the minimal Lorentz-invariant equation of motion.

That $\Box\phi = 0$ for a scalar field and $\Box A_\mu = 0$ for a vector field have the same form is not a coincidence. The electromagnetic field is made up of particles of spin 1 called photons. The polarizations of the field are encoded in the four fields $A_\nu(x)$. In fact, massless particles of *any* spin will satisfy $\Box\chi_i = 0$ where the different fields, indexed by i, encode different polarizations of that particle. This is not obvious, and we are not ready to prove it, so let us focus simply on the electromagnetic field. For simplicity, we will ignore polarizations for now and just treat A_ν as a scalar field ϕ (such approximations were used in some of the earliest QED papers, e.g. [Born *et al.*, 1926]). A general solution to Maxwell's equations in Lorenz gauge is therefore given by Eq. (2.60) for each polarization (polarizations will be explained in Chapter 8). Such a solution simply represents the Fourier decomposition of electromagnetic fields into plane waves. The oscillation of the waves is the same as the oscillation of a harmonic oscillator for each value of $\vec{p}$.

2.3 Second quantization

Since the modes of an electromagnetic field satisfy the same classical equations as a simple harmonic oscillator, we can quantize them in the same way. We introduce an annihilation operator a_p and its conjugate creation operator $a_p^\dagger$ for each wavenumber $\vec{p}$ and integrate over them to get the Hamiltonian for the free theory:

$$H_0 = \int \frac{d^3 p}{(2\pi)^3} \omega_p \left(a_p^\dagger a_p + \frac{1}{2} V \right), \qquad (2.65)$$

with V the volume (the volume term is infinite, but we can ignore it until Chapter 15) and

$$\omega_p = |\vec{p}|. \qquad (2.66)$$

This is known as **second quantization**. At the risk of oversimplifying things a little, that is all there is to quantum field theory. The rest is just quantum mechanics.

First quantization refers to the discrete modes, for example, of a particle in a box. Second quantization refers to the integer numbers of excitations of each of these modes. However, this is somewhat misleading – the fact that there are discrete modes is a classical phenomenon. The two steps really are (1) interpret these modes as having energy $E = \hbar\omega$ and (2) quantize each mode as a harmonic oscillator. In that sense we are only quantizing once. Whether second quantization is a good name for this procedure is semantics, not physics.

There are two new features in second quantization:

1. We have many quantum mechanical systems – one for each $\vec{p}$ – all at the same time.
2. We interpret the nth excitation of the $\vec{p}$ harmonic oscillator as having n **particles**.

Let us take a moment to appreciate this second point. Recall the old simple harmonic oscillator: the electron in a quadratic potential. We would never interpret the states $|n\rangle$ of this system as having n electrons. The fact that a pointlike electron in a quadratic potential has analogous equations of motion to a Fourier component of the electromagnetic field is just a coincidence. Do not let it confuse you. Both are just the simplest possible dynamical systems, with linear restoring forces.[1]

In second quantization, the Hilbert space is promoted to a **Fock space**, which is defined at each time as a direct sum,

$$\mathcal{F} = \oplus_n \mathcal{H}_n, \qquad (2.67)$$

---

[1] To set up a proper analogy we need to first treat the electron as a classical field (we do not know how to do that yet), and find a set of solutions (such as the discrete frequencies of the electromagnetic waves). Then we would quantize each of those solutions, allowing $|n\rangle$ excitations. However, if we did this, electrons would have Bose–Einstein statistics. Instead, they must have Fermi–Dirac statistics, so we would have to restrict n to 0 or 1. The second quantization of electrons will be discussed in Chapters 10 through 12, and the interpretation of an electron as a classical field, which requires Grassmann numbers, in Chapter 14.

of Hilbert spaces, $\mathcal{H}_n$, of physical n-particle states. If there is one particle type, states in $\mathcal{H}_n$ are linear combinations of states $\{|p_1^\mu, \ldots, p_n^\mu\rangle\}$ of all possible momenta satisfying $p_i^2 = m^2$ with $p_i^0 > 0$. If there are many different particle types, the Fock space is the direct sum of the Hilbert spaces associated with each particle. The Fock space is the same at all times, by time-translation invariance, and in any frame, by Lorentz invariance. Note that the Fock space is *not* a sum over Hilbert spaces defined with arbitrary 4-vectors, since the energy for a physical state is determined by its 3-momentum $\vec{p}_i$ and its mass m_i as $p_i^0 = \sqrt{\vec{p}_i^2 + m_i^2}$. We thus write $|\vec{p}\rangle$, $|p^\mu\rangle$ and $|p\rangle$ interchangeably.

2.3.1 Field expansion

Now let us get a little more precise about what the Hamiltonian in Eq. (2.65) means. The natural generalizations of

$$[a, a^\dagger] = 1 \tag{2.68}$$

are the equal-time commutation relations

$$[a_k, a_p^\dagger] = (2\pi)^3 \delta^3(\vec{p} - \vec{k}). \tag{2.69}$$

The factors of 2π are a convention, stemming from our convention for Fourier transforms (see Appendix A). These $a_p^\dagger$ operators create particles with momentum p:

$$a_p^\dagger |0\rangle = \frac{1}{\sqrt{2\omega_p}} |\vec{p}\rangle, \tag{2.70}$$

where $|\vec{p}\rangle$ is a state with a single particle of momentum $\vec{p}$. This factor of $\sqrt{2\omega_p}$ is just another convention, but it will make some calculations easier. Its nice Lorentz transformation properties are studied in Problem 2.6.

To compute the normalization of one-particle states, we start with

$$\langle 0|0\rangle = 1, \tag{2.71}$$

which leads to

$$\langle \vec{p}|\vec{k}\rangle = 2\sqrt{\omega_p \omega_k} \langle 0|a_p a_k^\dagger|0\rangle = 2\omega_p (2\pi)^3 \delta^3(\vec{p} - \vec{k}). \tag{2.72}$$

The identity operator for one-particle states is

$$\mathbb{1} = \int \frac{d^3 p}{(2\pi)^3} \frac{1}{2\omega_p} |\vec{p}\rangle\langle\vec{p}|, \tag{2.73}$$

which we can check with

$$|\vec{k}\rangle = \int \frac{d^3 p}{(2\pi)^3} \frac{1}{2\omega_p} |\vec{p}\rangle\langle\vec{p}|\vec{k}\rangle = \int \frac{d^3 p}{(2\pi)^3} \frac{1}{2\omega_p} 2\omega_p (2\pi)^3 \delta^3(\vec{p} - \vec{k})|\vec{p}\rangle = |\vec{k}\rangle. \tag{2.74}$$

We then define quantum fields as integrals over creation and annihilation operators for each momentum:

$$\phi_0(\vec{x}) = \int \frac{d^3 p}{(2\pi)^3} \frac{1}{\sqrt{2\omega_p}} \left(a_p e^{i\vec{p}\vec{x}} + a_p^\dagger e^{-i\vec{p}\vec{x}}\right), \tag{2.75}$$

where the subscript 0 indicates this is a free field. The factor of $\sqrt{2\omega_p}$ is included for later convenience.

This equation looks just like the classical free-particle solutions, Eq. (2.59), to Maxwell's equations (ignoring polarizations) but instead of a_p and $a_p^\dagger$ being *functions*, they are now the annihilation and creation *operators* for that mode. Sometimes we say the classical a_p is *c*-number valued and the quantum one is *q*-number valued. The connection with Eq. (2.59) is only suggestive. The quantum equation, Eq. (2.75), should be taken as the definition of a field operator $\phi_0(\vec{x})$ constructed from the creation and annihilation operators a_p and $a_p^\dagger$.

To get a sense of what the operator ϕ_0 does, we can act with it on the vacuum and project out a momentum component:

$$
\begin{aligned}
\langle \vec{p} | \phi_0(\vec{x}) | 0 \rangle &= \langle 0 | \sqrt{2\omega_p} a_p \int \frac{d^3 k}{(2\pi)^3} \frac{1}{\sqrt{2\omega_k}} \left(a_k e^{i\vec{k}\vec{x}} + a_k^\dagger e^{-i\vec{k}\vec{x}} \right) | 0 \rangle \\
&= \int \frac{d^3 k}{(2\pi)^3} \sqrt{\frac{\omega_p}{\omega_k}} \left[e^{i\vec{k}\vec{x}} \langle 0 | a_p a_k | 0 \rangle + e^{-i\vec{k}\vec{x}} \langle 0 | a_p a_k^\dagger | 0 \rangle \right] \\
&= e^{-i\vec{p}\vec{x}}.
\end{aligned} \tag{2.76}
$$

This is the same thing as the projection of a position state on a momentum state in one-particle quantum mechanics:

$$
\langle \vec{p} | \vec{x} \rangle = e^{-i\vec{p}\vec{x}}. \tag{2.77}
$$

So, $\phi_0(\vec{x})|0\rangle = |\vec{x}\rangle$, that is, $\phi_0(\vec{x})$ creates a particle at position $\vec{x}$. This should not be surprising, since $\phi_0(x)$ in Eq. (2.75) is very similar to $x = a + a^\dagger$ in the simple harmonic oscillator. Since ϕ_0 is Hermitian, $\langle \vec{x} | = \langle 0 | \phi_0(\vec{x})$ as well.

By the way, there are many states $|\psi\rangle$ in the Fock space that satisfy $\langle \vec{p} | \psi \rangle = e^{-i\vec{p}\vec{x}}$. Since $\langle \vec{p} |$ only has non-zero matrix elements with one-particle states, adding to $|x\rangle$ a two- or zero-particle state, as in $\phi_0^2(\vec{x})|0\rangle$, has no effect on $\langle \vec{p}|\vec{x}\rangle$. That is, $|\psi\rangle = \left(\phi_0(\vec{x}) + \phi_0^2(\vec{x}) \right)|0\rangle$ also satisfies $\langle \vec{p} | \psi \rangle = e^{-i\vec{p}\vec{x}}$. The state $|\vec{x}\rangle \equiv \phi_0(\vec{x})|0\rangle$ is the unique *one-particle* state with $\langle \vec{p} | \psi \rangle = e^{-i\vec{p}\vec{x}}$.

2.3.2 Time dependence

In quantum field theory, we generally work in the Heisenberg picture, where all the time dependence is in operators such as ϕ and a_p. For free fields, the creation and annihilation operators for each momentum $\vec{p}$ in the quantum field are just those of a simple harmonic oscillator. These operators should satisfy Eq. (2.55), $a_p(t) = e^{-i\omega_p t} a_p$, and its conjugate $a_p^\dagger(t) = e^{i\omega_p t} a_p^\dagger$, where a_p and $a_p^\dagger$ (without an argument) are time independent. Then, we can *define* a quantum scalar field as

$$
\phi_0(\vec{x}, t) = \int \frac{d^3 p}{(2\pi)^3} \frac{1}{\sqrt{2\omega_p}} \left(a_p e^{-ipx} + a_p^\dagger e^{ipx} \right), \tag{2.78}
$$

with $p^\mu \equiv (\omega_p, \vec{p})$ and $\omega_p = |\vec{p}|$ as in Eq. (2.60). The 0 subscript still indicates that these are free fields.

To be clear, there is no physical content in Eq. (2.78). It is just a definition. The physical content is in the algebra of a_p and $a_p^\dagger$ and in the Hamiltonian H_0. Nevertheless, we will see that collections of a_p and $a_p^\dagger$ in the form of Eq. (2.78) are very useful in quantum field theory. For example, you may note that while the integral is over only three components of p_μ, the phases have combined into a manifestly Lorentz-invariant form. This field now automatically satisfies $\Box\phi(x) = 0$. If a scalar field had mass m, we could still write it in exactly the same way but with a massive dispersion relation: $\omega_p \equiv \sqrt{\vec{p}^2 + m^2}$. Then the quantum field still satisfies the classical equation of motion: $(\Box + m^2)\phi(x) = 0$.

Let us check that our free Hamiltonian is consistent with the expectation for time evolution. Commuting the free fields with H_0 we find

$$
\begin{aligned}
[H_0, \phi_0(\vec{x}, t)] &= \int \frac{d^3p}{(2\pi)^3} \int \frac{d^3k}{(2\pi)^3\sqrt{2\omega_k}} \left[\omega_p \left(a_p^\dagger a_p + \frac{1}{2}V \right), a_k e^{-ikx} + a_k^\dagger e^{ikx} \right] \\
&= \int \frac{d^3p}{(2\pi)^3} \frac{1}{\sqrt{2\omega_p}} \left[-\omega_p a_p\, e^{-ipx} + \omega_p a_p^\dagger e^{ipx} \right] \\
&= -i\partial_t \phi_0(\vec{x}, t),
\end{aligned}
\tag{2.79}
$$

which is exactly the expected result.

For any Hamiltonian, quantum fields satisfy the Heisenberg equations of motion:

$$
i\partial_t \phi(x) = [\phi, H].
\tag{2.80}
$$

In a free theory, $H = H_0$, and this is consistent with Eq. (2.78). In an interacting theory, that is, one whose Hamiltonian H differs from the free Hamiltonian H_0, the Heisenberg equations of motion are still satisfied, but we will rarely be able to solve them exactly. To study interacting theories, it is often useful to use the same notation for interacting fields as for free fields:

$$
\phi(\vec{x}, t) = \int \frac{d^3p}{(2\pi)^3} \frac{1}{\sqrt{2\omega_p}} \left[a_p(t) e^{-ipx} + a_p^\dagger(t) e^{ipx} \right].
\tag{2.81}
$$

At any *fixed time*, the full interacting creation and annihilation operators $a_p^\dagger(t)$ and $a_p(t)$ satisfy the same algebra as in the free theory – the Fock space is the same at every time, due to time-translation invariance. We can therefore define the exact creation operators $a_p^\dagger(t)$ to be equal to the free creation operators $a_p^\dagger$ at any given fixed time, $a_p^\dagger(t_0) = a_p^\dagger$ and so $\phi(\vec{x}, t_0) = \phi_0(\vec{x}, t_0)$. However, the operators that create particular momentum states $|p\rangle$ in the interacting theory mix with each other as time evolves. We generally will not be able to solve the dynamics of an interacting theory exactly. Instead, we will expand $H = H_0 + H_{\text{int}}$ and calculate amplitudes using time-dependent perturbation theory with H_{int}, just as in quantum mechanics. In Chapter 7, we use this approach to derive the Feynman rules.

The first-quantized (quantum mechanics) limit of the second-quantized theory (quantum field theory) comes from restricting to the one-particle states, which is appropriate in the non-relativistic limit. A basis of these states is given by the vectors $\langle x| = \langle \vec{x}, t|$:

$$
\langle x| = \langle 0| \, \phi(\vec{x}, t).
\tag{2.82}
$$

Then, a Schrödinger picture wavefunction is

$$
\psi(x) = \langle x|\psi\rangle,
\tag{2.83}
$$

which satisfies

$$i\partial_t \psi(x) = i\partial_t \langle 0|\phi(\vec{x},t)|\psi\rangle = i\,\langle 0|\partial_t\phi(\vec{x},t)|\psi\rangle \,. \tag{2.84}$$

In the massive case, the free quantum field $\phi_0(x)$ satisfies $\partial_t^2\phi_0 = \left(\vec{\nabla}^2 - m^2\right)\phi_0$ and we have from Eq. (2.79) (with the massive dispersion relation $\omega_p = \sqrt{\vec{p}^2 + m^2}$):

$$i\langle 0|\partial_t\phi_0(\vec{x},t)|\psi\rangle = \langle 0|\int \frac{d^3p}{(2\pi)^3}\frac{\sqrt{\vec{p}^2+m^2}}{\sqrt{2\omega_p}}\left(a_p e^{-ipx} - a_p^\dagger e^{ipx}\right)|\psi\rangle$$

$$= \langle 0|\sqrt{m^2 - \vec{\nabla}^2}\phi_0(\vec{x},t)|\psi\rangle. \tag{2.85}$$

So,

$$i\partial_t\psi(x) = \sqrt{m^2 - \vec{\nabla}^2}\psi(x) = \left(m - \frac{\vec{\nabla}^2}{2m} + \mathcal{O}\left(\frac{1}{m^2}\right)\right)\psi(x). \tag{2.86}$$

The final form is the low-energy (large-mass) expansion. We can then define the non-relativistic Hamiltonian by subtracting off the mc^2 contribution to the energy, which is irrelevant in the non-relativistic limit. This gives

$$i\partial_t\psi(x) = -\frac{\vec{\nabla}^2}{2m}\psi(x), \tag{2.87}$$

which is the non-relativistic Schrödinger equation for a free theory. Another way to derive the quantum mechanics limit of quantum field theory is discussed in Section 33.6.2.

2.3.3 Commutation relations

We will occasionally need to use the equal-time commutation relations of the second-quantized field and its time derivative. The commutator of a field at two different points is

$$[\phi(\vec{x}),\phi(\vec{y})] = \int \frac{d^3p}{(2\pi)^3}\int\frac{d^3q}{(2\pi)^3}\frac{1}{\sqrt{2\omega_p 2\omega_q}}\left[\left(a_p e^{i\vec{p}\vec{x}} + a_p^\dagger e^{-i\vec{p}\vec{x}}\right),\left(a_q e^{i\vec{q}\vec{y}} + a_q^\dagger e^{-i\vec{q}\vec{y}}\right)\right]$$

$$= \int \frac{d^3p}{(2\pi)^3}\int\frac{d^3q}{(2\pi)^3}\frac{1}{\sqrt{2\omega_p 2\omega_q}}\left(e^{i\vec{p}\vec{x}}e^{-i\vec{q}\vec{y}}\left[a_p,a_q^\dagger\right] + e^{-i\vec{p}\vec{x}}e^{i\vec{q}\vec{y}}\left[a_p^\dagger,a_q\right]\right). \tag{2.88}$$

Using Eq. (2.69), $[a_k,a_p^\dagger] = (2\pi)^3\delta^3(\vec{p}-\vec{k})$, this becomes

$$[\phi(\vec{x}),\phi(\vec{y})] = \int \frac{d^3p}{(2\pi)^3}\int\frac{d^3q}{(2\pi)^3}\frac{1}{\sqrt{2\omega_p 2\omega_q}}\left[e^{i\vec{p}\vec{x}}e^{-i\vec{q}\vec{y}} - e^{-i\vec{p}\vec{x}}e^{i\vec{q}\vec{y}}\right](2\pi)^3\delta^3(\vec{p}-\vec{q})$$

$$= \int \frac{d^3p}{(2\pi)^3}\frac{1}{2\omega_p}\left[e^{i\vec{p}(\vec{x}-\vec{y})} - e^{-i\vec{p}(\vec{x}-\vec{y})}\right]. \tag{2.89}$$

Since the integral measure and $\omega_p = \sqrt{\vec{p}^2 + m^2}$ are symmetric under $\vec{p} \to -\vec{p}$ we can flip the sign on the exponent of one of the terms to see that the commutator vanishes:

$$[\phi(\vec{x}), \phi(\vec{y})] = 0. \tag{2.90}$$

The equivalent calculation at different times is much more subtle (we discuss the general result in Section 12.6 in the context of the spin-statistics theorem).

Next, we note that the time derivative of the free field, at $t = 0$, has the form

$$\pi(\vec{x}) \equiv \left. \partial_t \phi(x) \right|_{t=0} = -i \int \frac{d^3 p}{(2\pi)^3} \sqrt{\frac{\omega_p}{2}} \left(a_p e^{i\vec{p}\vec{x}} - a_p^\dagger e^{-i\vec{p}\vec{x}} \right), \tag{2.91}$$

where π is the operator canonically conjugate to ϕ. As $\phi(\vec{x})$ is the second-quantized analog of the $\hat{x}$ operator, $\pi(\vec{x})$ is the analog of the $\hat{p}$ operator. Note that $\pi(\vec{x})$ has nothing to do with the physical momentum of states in the Hilbert space: $\pi(\vec{x}) |0\rangle$ is not a state of given momentum. Instead, it is a state also at position $\vec{x}$ created by the time derivative of $\phi(\vec{x})$.

Now we compute

$$[\phi(\vec{x}), \pi(\vec{y})] = -i \int \frac{d^3 p}{(2\pi)^3} \int \frac{d^3 q}{(2\pi)^3} \sqrt{\frac{\omega_p}{2}} \frac{1}{\sqrt{2\omega_q}} \left(e^{i\vec{p}\vec{y}} e^{-i\vec{q}\vec{x}} [a_q^\dagger, a_p] - e^{i\vec{q}\vec{x}} e^{-i\vec{p}\vec{y}} [a_q, a_p^\dagger] \right)$$

$$= \frac{i}{2} \int \frac{d^3 p}{(2\pi)^3} \left[e^{i\vec{p}(\vec{x}-\vec{y})} + e^{-i\vec{p}(\vec{x}-\vec{y})} \right]. \tag{2.92}$$

Both of these integrals give $\delta^3(\vec{x} - \vec{y})$, so we find

$$[\phi(\vec{x}), \pi(\vec{y})] = i\delta^3(\vec{x} - \vec{y}), \tag{2.93}$$

which is the analog of $[\hat{x}, \hat{p}] = i$ in quantum mechanics. It encapsulates the field theory version of the uncertainty principle: you cannot know the properties of the field and its rate of change at the same place at the same time.

In a general interacting theory, at any fixed time, $\phi(\vec{x})$ and $\pi(\vec{x})$ have expressions in terms of creation and annihilation operators whose algebra is identical to that of the free theory. Therefore, they satisfy the commutation relations in Eqs. (2.90) and (2.93) as well as $[\pi(\vec{x}), \pi(\vec{y})] = 0$. The Hamiltonian in an interacting theory should be expressed as a functional of the operators $\phi(\vec{x})$ and $\pi(\vec{x})$ with time evolution given by $\partial_t \mathcal{O} = i[H, \mathcal{O}]$. Any such Hamiltonian can then be expressed entirely in terms of creation and annihilation operators using Eqs. (2.75) and (2.91); thus it has a well-defined action on the associated Fock space. Conversely, it is sometimes more convenient (especially for non-relativistic or condensed matter applications) to derive the form of the Hamiltonian in terms of a_p and $a_p^\dagger$. We can then express a_p and $a_p^\dagger$ in terms of $\phi(\vec{x})$ and $\pi(\vec{x})$ by inverting Eqs. (2.75) and (2.91) for a_p and $a_p^\dagger$ (the solution is the field theory equivalent of Eq. (2.47)).

In summary, all we have done to quantize the electromagnetic field is to treat it as an infinite set of simple harmonic oscillators, one for each wavenumber $\vec{p}$. More generally:

Quantum field theory is just quantum mechanics with an infinite number of harmonic oscillators.

2.3.4 Einstein coefficients revisited

In quantum mechanics we usually study a single electron in a background potential $V(x)$. In quantum field theory, the background (e.g. the electromagnetic system) is dynamical, so all kinds of new phenomena can be explained. We already saw one example in Chapter 1. We can now be a little more explicit about what the relevant Hamiltonian should be for Dirac's calculation of the Einstein coefficients.

We can always write a Hamiltonian as

$$H = H_0 + H_{\text{int}}, \tag{2.94}$$

where H_0 describes some system that we can solve exactly. In the case of the two-state system discussed in Chapter 1, we can take H_0 to be the sum of the Hamiltonians for the atom and the photons:

$$H_0 = H_{\text{atom}} + H_{\text{photon}}. \tag{2.95}$$

The eigenstates of H_{atom} are the energy eigenstates $|\psi_n\rangle$ of the hydrogen atom, with energies E_n. H_{photon} is the Hamiltonian in Eq. (2.65) above:

$$H_{\text{photon}} = \int \frac{d^3 k}{(2\pi)^3} \omega_k \left(a_k^\dagger a_k + \frac{1}{2} V \right). \tag{2.96}$$

The remaining H_{int} is hopefully small enough to let us use perturbation theory.

Fermi's golden rule from quantum mechanics says the rate for transitions between two states is proportional to the square of the matrix element of the interaction between the two states:

$$\Gamma \propto |\langle f|H_{\text{int}}|i\rangle|^2 \delta(E_f - E_i), \tag{2.97}$$

and we can treat the interaction semi-classically:

$$H_{\text{int}} = \phi H_I. \tag{2.98}$$

As mentioned in Footnote 2 in Chapter 1, H_I can be derived from the $\frac{e}{m}\vec{p} \cdot \vec{A}$ interaction of the minimally coupled non-relativistic Hamiltonian, $H = \frac{1}{2m}(\vec{p} + e\vec{A})^2$. Since we are ignoring spin, it does not pay to be too precise about H_I; the important point being only that H_{int} has a quantum field ϕ in it, representing the photon, and H_I has non-zero matrix elements between different atomic states.

According to Fermi's golden rule, the transition probability is proportional to the matrix element of the interaction squared. Then,

$$\mathcal{M}_{1\to 2} = \langle \text{atom}^\star; n_k - 1|H_{\text{int}}|\text{atom}; n_k\rangle \propto \langle \text{atom}^\star|H_I|\text{atom}\rangle \sqrt{n_k}, \tag{2.99}$$

$$\mathcal{M}_{2\to 1} = \langle \text{atom}; n_k + 1|H_{\text{int}}|\text{atom}^\star; n_k\rangle \propto \langle \text{atom}|H_I|\text{atom}^\star\rangle \sqrt{n_k + 1}, \tag{2.100}$$

where we have used

$$\langle n_k - 1|\phi|n_k\rangle = \int \frac{d^3 p}{(2\pi)^3} \frac{1}{\sqrt{2\omega_p}} \langle n_k - 1|a_p|n_k\rangle \propto \sqrt{n_k}, \tag{2.101}$$

$$\langle n_k + 1|\phi|n_k\rangle = \int \frac{d^3p}{(2\pi)^3} \frac{1}{\sqrt{2\omega_p}} \langle n_k + 1|a_p^\dagger|n_k\rangle \propto \sqrt{n_k + 1}. \tag{2.102}$$

Thus, $\mathcal{M}_{1\to2}$ and $\mathcal{M}_{2\to1}$ agree with what we used in Chapter 1 to reproduce Dirac's calculation of the Einstein coefficients. Note that we only used one photon mode, of momentum k, so this was really just quantum mechanics. Quantum field theory just gave us a δ-function from the d^3p integration.

Problems

2.1 Derive the transformations $x \to \frac{x+vt}{\sqrt{1-v^2}}$ and $t \to \frac{t+vx}{\sqrt{1-v^2}}$ in perturbation theory. Start with the Galilean transformation $x \to x + v_g t$. Add a transformation $t \to t + \delta t$ and solve for δt assuming it is linear in x and t and preserves $t^2 - x^2$ to $\mathcal{O}\left(v^2\right)$ with $v_g = v$. Repeat for δt and δx to second order in v and show that the result agrees with the second-order expansion of the full transformations for some function $v(v_g)$.

2.2 Special relativity and colliders.
  (a) The Large Hadron Collider was designed to collide protons together at 14 TeV center-of-mass energy. How many kilometers per hour less than the speed of light are the protons moving?
  (b) How fast is one proton moving with respect to the other?

2.3 The GZK bound. In 1966 Greisen, Zatsepin and Kuzmin argued that we should not see cosmic rays (high-energy protons hitting the atmosphere from outer space) above a certain energy, due to interactions of these rays with the cosmic microwave background.
  (a) The universe is a blackbody at 2.73 K. What is the average energy of the photons in outer space (in electronvolts)?
  (b) How much energy would a proton (p^+) need to collide with a photon (γ) in outer space to convert it to a 135 MeV pion (π^0)? That is, what is the energy threshold for $p^+ + \gamma \to p^+ + \pi^0$?
  (c) How much energy does the outgoing proton have after this reaction?
  This GZK bound was finally confirmed experimentally 40 years after it was conjectured [Abbasi *et al.*, 2008].

2.4 Is the transformation $Y : (t,x,y,z) \to (t,x,-y,z)$ a Lorentz transformation? If so, why is it not considered with P and T as a discrete Lorentz transformation? If not, why not?

2.5 Compton scattering. Suppose we scatter an X-ray off an electron in a crystal, but we cannot measure the electron's momentum, just the reflected X-ray momentum.
  (a) Why is it OK to treat the electrons as free?
  (b) Calculate the frequency dependence of the reflected X-ray on the scattering angle. Draw a rough plot.
  (c) What happens to the distribution as you take the electron mass to zero?

(d) If you did not believe in quantized photon momenta, what kind of distribution might you have expected? [Hint: see [Compton, 1923].]

2.6 Lorentz invariance.

(a) Show that

$$\int_{-\infty}^{\infty} dk^0 \delta(k^2 - m^2)\theta(k^0) = \frac{1}{2\omega_k}, \qquad (2.103)$$

where $\theta(x)$ is the unit step function and $\omega_k \equiv \sqrt{\vec{k}^2 + m^2}$.

(b) Show that the integration measure d^4k is Lorentz invariant.

(c) Finally, show that

$$\int \frac{d^3k}{2\omega_k} \qquad (2.104)$$

is Lorentz invariant.

2.7 Coherent states of the simple harmonic oscillator.

(a) Calculate $\partial_z(e^{-za^\dagger} a e^{za^\dagger})$ where z is a complex number.

(b) Show that $|z\rangle = e^{za^\dagger}|0\rangle$ is an eigenstate of a. What is its eigenvalue?

(c) Calculate $\langle n|z\rangle$.

(d) Show that these "coherent states" are minimally dispersive: $\Delta p \Delta x = \frac{1}{2}$, where $\Delta x^2 = \langle x^2 \rangle - \langle x \rangle^2$ and $\Delta p^2 = \langle p^2 \rangle - \langle p \rangle^2$, where $\langle x \rangle = \frac{\langle z|x|z\rangle}{\langle z|z\rangle}$ and $\langle p \rangle = \frac{\langle z|p|z\rangle}{\langle z|z\rangle}$.

(e) Why can you not make an eigenstate of $a^\dagger$?

Classical field theory 3

We have now seen how quantum field theory is just quantum mechanics with an infinite number of oscillators. We already saw that it can do some remarkable things, such as explain spontaneous emission. But it also seems to lead to absurdities, such as an infinite shift in the energy levels of the hydrogen atom (see Chapter 4). To show that quantum field theory is not absurd, but extremely predictive, we will have to be very careful about how we do calculations. We will begin by going through carefully some of the predictions that the theory gets right without infinities. These are called the tree-level processes, which means they are leading order in an expansion in $\hbar$. Since taking $\hbar \to 0$ gives the classical limit, tree-level calculations are closely related to calculations in classical field theory, which is the subject of this chapter.

3.1 Hamiltonians and Lagrangians

A classical field theory is just a mechanical system with a continuous set of degrees of freedom. Think about the density of a fluid $\rho(x)$ as a function of position, or the electric field $\vec{E}(x)$. Field theories can be defined in terms of either a Hamiltonian or a Lagrangian, which we often write as integrals over all space of Hamiltonian or Lagrangian densities:

$$H = \int d^3x \mathcal{H}, \quad L = \int d^3x \mathcal{L}. \tag{3.1}$$

We will use a calligraphic script for densities and an italic script for integrated quantities. The word "density" is almost always omitted.

Formally, the **Hamiltonian** (density) is a functional of fields and their conjugate momenta $\mathcal{H}[\phi, \pi]$. The **Lagrangian** (density) is the Legendre transform of the Hamiltonian (density). Formally, it is defined as

$$\mathcal{L}[\phi, \dot{\phi}] = \pi[\phi, \dot{\phi}] \, \dot{\phi} - \mathcal{H}[\phi, \pi[\phi, \dot{\phi}]], \tag{3.2}$$

where $\dot{\phi} = \partial_t \phi$ and $\pi[\phi, \dot{\phi}]$ is implicitly defined by $\frac{\partial \mathcal{H}[\phi, \pi]}{\partial \pi} = \dot{\phi}$. The inverse transform is

$$\mathcal{H}[\phi, \pi] = \pi \, \dot{\phi}[\phi, \pi] - \mathcal{L}[\phi, \dot{\phi}[\phi, \pi]], \tag{3.3}$$

where $\dot{\phi}[\phi, \pi]$ is implicitly defined by $\frac{\partial \mathcal{L}[\phi, \dot{\phi}]}{\partial \dot{\phi}} = \pi$.

To make this more concrete, consider this example:

$$\mathcal{L} = \frac{1}{2}(\partial_\mu \phi)(\partial_\mu \phi) - \mathcal{V}[\phi] = \frac{1}{2}\dot{\phi}^2 - \frac{1}{2}(\vec{\nabla}\phi)^2 - \mathcal{V}[\phi], \tag{3.4}$$

where $\mathcal{V}[\phi]$ is called the potential (density). Then $\pi = \frac{\partial \mathcal{L}}{\partial \dot{\phi}} = \dot{\phi}$, which is easy to solve for $\dot{\phi}$: $\dot{\phi}[\phi, \pi] = \pi$. Plugging in to Eq. (3.3) we find

$$\mathcal{H} = \pi \dot{\phi}[\phi, \pi] - \mathcal{L}[\phi, \dot{\phi}[\phi, \pi]] = \frac{1}{2}\pi^2 + \frac{1}{2}(\vec{\nabla}\phi)^2 + \mathcal{V}[\phi]. \tag{3.5}$$

We often just write $\mathcal{H} = \frac{1}{2}\dot{\phi}^2 + \frac{1}{2}(\nabla\phi)^2 + \mathcal{V}[\phi]$ so that we do not have to deal with the π fields. For a more complicated Lagrangian it may not be possible to produce a closed-form expression for $\dot{\phi}[\phi, \pi]$. For example, $\mathcal{L} = \phi^2\dot{\phi}^2 + \phi\dot{\phi}^3$ would imply $\pi = 2\phi^2\dot{\phi} + 3\phi\dot{\phi}^2$ from which $\dot{\phi}[\phi, \pi]$ is a mess. There are also situations where the Legendre transform may not exist, so that a Hamiltonian does not have a corresponding Lagrangian, or vice versa.[1]

Equations (3.4) and (3.5) inspire the identification of the Hamiltonian with the sum of the kinetic and potential energies of a system:

$$\mathcal{H} = \mathcal{K} + \mathcal{V}, \tag{3.6}$$

while the Lagrangian is their difference:

$$\mathcal{L} = \mathcal{K} - \mathcal{V}. \tag{3.7}$$

Matching onto Eqs. (3.4) and (3.5), the kinetic energy is the part with time derivatives, $\mathcal{K} = \frac{1}{2}\dot{\phi}^2$, and the potential energy is the rest, $\mathcal{V} = \frac{1}{2}(\vec{\nabla}\phi)^2 + \mathcal{V}[\phi]$.

The Hamiltonian corresponds to a conserved quantity – the total energy of a system – while the Lagrangian does not. The problem with Hamiltonians, however, is that they are not Lorentz invariant. The Hamiltonian picks out energy, which is not a Lorentz scalar; rather, it is the 0 component of a Lorentz vector: $P^\mu = (H, \vec{P})$. The Hamiltonian density is the 00 component of a Lorentz tensor, the energy-momentum tensor $\mathcal{T}_{\mu\nu}$. Hamiltonians are great for non-relativistic systems, but for relativistic systems we will almost exclusively use Lagrangians.

We do not usually talk about kinetic and potential energy in quantum field theory. Instead we talk about *kinetic terms* and then about *interactions*, for reasons that will become clear after we have done a few calculations. **Kinetic terms** are **bilinear**, meaning they have exactly two fields. So kinetic terms are

$$\mathcal{L}_K \supset \frac{1}{2}\phi\Box\phi, \quad \bar{\psi}\partial\!\!\!/\psi, \quad \frac{1}{4}F_{\mu\nu}^2, \quad \frac{1}{2}m^2\phi^2, \quad \frac{1}{2}\phi_1\Box\phi_2, \quad \phi_1\partial_\mu A_\mu, \quad \ldots \tag{3.8}$$

where

$$F_{\mu\nu} = \partial_\mu A_\nu - \partial_\nu A_\mu. \tag{3.9}$$

---

[1] The Legendre transform is just trading velocity, $\dot{\phi}$, for a new variable called π, which corresponds to momentum in simple cases. It does this trade at each value of ϕ, so ϕ just goes along for the ride in the Legendre transform. So let us hold ϕ fixed and write $\mathcal{L}[\dot{\phi}]$. No information is lost in writing $\dot{\phi}$ as π as long as $\pi = \mathcal{L}'[\dot{\phi}]$ and $\dot{\phi}$ are in one-to-one correspondence. For a function $f(x)$, x and $f'(x)$ are in one-to-one correspondence as long as $f''(x) > 0$ or $f''(x) < 0$ for all x, that is, if the function is convex. Therefore, one can go back and forth between the Hamiltonian and the Lagrangian as long as $\mathcal{L}[\phi, \dot{\phi}]$ is a convex function of $\dot{\phi}$ at each value of ϕ and $\mathcal{H}[\phi, \pi]$ is a convex function of π. For multiple fields, ϕ_n and π_n, the requirement is that $M_{ij} = \partial\mathcal{H}[\phi_n, \pi_n]/\partial\pi_i\partial\pi_j$ be an invertible matrix.

It is standard to use the letters ϕ or π for scalar fields, ψ, ξ, χ for fermions, A_μ, J_μ, V_μ for vectors and $h_{\mu\nu}, T_{\mu\nu}$ for tensors.

Anything with just two fields of the same or different type can be called a kinetic term. The kinetic terms tell you about the free (non-interacting) behavior. Fields with kinetic terms are said to be *dynamical* or **propagating**. More precisely, a field should have time derivatives in its kinetic term to be dynamical. It is also sometimes useful to think of a **mass term**, such as $m^2\phi^2$, as an interaction rather than a kinetic term (see Problem 7.4).

Interactions have three or more fields:

$$\mathcal{L}_{\text{int}} \supset \lambda\phi^3, \quad g\bar{\psi}A\!\!\!/\,\psi, \quad g\partial_\mu\phi A_\mu\phi^\star, \quad g^2 A_\mu^2 A_\nu^2, \quad \frac{1}{M_{\text{Pl}}}\partial_\mu h_{\mu\nu}\partial_\nu h_{\alpha\beta}h_{\alpha\beta}, \quad \ldots \quad (3.10)$$

Since the interactions are everything but the kinetic terms, we also sometimes write

$$\mathcal{L}_{\text{int}} = -\mathcal{V} = -\mathcal{H}_{\text{int}}. \qquad (3.11)$$

It is helpful if the coefficients of the interaction terms are small in some sense, so that the fields are **weakly interacting** and we can do perturbation theory.

3.2 The Euler–Lagrange equations

In quantum field theory, we will almost exclusively use Lagrangians. The simplest reason for this is that Lagrangians are manifestly Lorentz invariant. Dynamics for a Lagrangian system are determined by the principle of least action. The **action** is the integral over time of the Lagrangian:

$$S = \int dt\, L = \int d^4x\, \mathcal{L}(x). \qquad (3.12)$$

Say we have a Lagrangian $\mathcal{L}[\phi, \partial_\mu\phi]$ that is a functional only of a field ϕ and its first derivatives. Now imagine varying $\phi \to \phi + \delta\phi$ where $\delta\phi$ can be any field. Then,

$$\delta S = \int d^4x \left[\frac{\partial \mathcal{L}}{\partial \phi}\delta\phi + \frac{\partial \mathcal{L}}{\partial(\partial_\mu\phi)}\delta(\partial_\mu\phi) \right]$$

$$= \int d^4x \left\{ \left[\frac{\partial \mathcal{L}}{\partial \phi} - \partial_\mu \frac{\partial \mathcal{L}}{\partial(\partial_\mu\phi)} \right] \delta\phi + \partial_\mu \left[\frac{\partial \mathcal{L}}{\partial(\partial_\mu\phi)}\delta\phi \right] \right\}. \qquad (3.13)$$

The last term is a total derivative and therefore its integral only depends on the field values at spatial and temporal infinity. We will always make the physical assumption that our fields vanish on these asymptotic boundaries, which lets us drop such total derivatives from Lagrangians. In other words, it lets us integrate by parts within Lagrangians, without consequence. That is

$$A\partial_\mu B = -(\partial_\mu A)B \qquad (3.14)$$

in a Lagrangian. We will use this identity constantly in both classical and quantum field theory.

In classical field theory, just as in classical mechanics, the equations of motion are determined by the principle of least action: when the action is evaluated on fields that satisfy the equations of motion, it should be insensitive to small variations of those fields, $\frac{\delta S}{\delta \phi} = 0$. If this holds for all variations, then Eq. (3.13) implies

$$\frac{\partial \mathcal{L}}{\partial \phi} - \partial_\mu \frac{\partial \mathcal{L}}{\partial(\partial_\mu \phi)} = 0. \tag{3.15}$$

These are the celebrated **Euler–Lagrange equations**. They give the **equations of motion** following from a Lagrangian.

For example, if our action is

$$\mathcal{S} = \int d^4x \left[\frac{1}{2}(\partial_\mu \phi)(\partial_\mu \phi) - \mathcal{V}[\phi] \right], \tag{3.16}$$

then the equations of motion are

$$- \mathcal{V}'[\phi] - \partial_\mu(\partial_\mu \phi) = 0. \tag{3.17}$$

Or, more simply, $\Box \phi + \mathcal{V}'[\phi] = 0$, recalling the d'Alembertian $\Box \equiv \partial_\mu^2$. In particular, if $\mathcal{L} = \frac{1}{2}(\partial_\mu \phi)(\partial_\mu \phi) - \frac{1}{2}m^2\phi^2$, the equations of motion are

$$(\Box + m^2)\phi = 0. \tag{3.18}$$

This is known as the **Klein–Gordon equation**. The Klein–Gordon equation describes the equations of motion for a free scalar field.

Why do we restrict to Lagrangians of the form $\mathcal{L}[\phi, \partial_\mu \phi]$? First of all, this is the form that all "classical" Lagrangians had. If only first derivatives are involved, boundary conditions can be specified by initial positions and velocities only, in accordance with Newton's laws. In the quantum theory, if kinetic terms have too many derivatives, for example $\mathcal{L} = \phi \Box^2 \phi$, there will generally be disastrous consequences. For example, there may be states with negative energy or negative norm, permitting the vacuum to decay (see Chapters 8 and 24). But interactions with multiple derivatives may occur. Actually, they *must* occur due to quantum effects in all but the simplest *renormalizable* field theories; for example, they are generic in all *effective field theories*, which are introduced in Chapter 22 and are the subject of much of Part IV. You can derive the equations of motion for general Lagrangians of the form $\mathcal{L}[\phi, \partial_\mu \phi, \partial_\nu \partial_\mu \phi, \ldots]$ in Problem 3.1.

3.3 Noether's theorem

It may happen that a Lagrangian is invariant under some special type of variation $\phi \to \phi + \delta\phi$. For example, a Lagrangian for a complex field ϕ is

$$\mathcal{L} = |\partial_\mu \phi|^2 - m^2|\phi|^2. \tag{3.19}$$

This Lagrangian is invariant under $\phi \to e^{-i\alpha}\phi$ for any $\alpha \in \mathbb{R}$. This transformation is a **symmetry** of the Lagrangian. There are two independent real degrees of freedom in a complex field ϕ, which we can take as $\phi = \phi_1 + i\phi_2$ or more conveniently ϕ and $\phi^\star$. Then the Lagrangian is

$$\mathcal{L} = (\partial_\mu \phi)(\partial_\mu \phi^\star) - m^2 \phi \phi^\star, \tag{3.20}$$

and the symmetry transformations are

$$\phi \to e^{-i\alpha}\phi, \quad \phi^\star \to e^{i\alpha}\phi^\star. \tag{3.21}$$

You should check that the equations of motion following from this Lagrangian are $(\Box + m^2)\phi = 0$ and $(\Box + m^2)\phi^\star = 0$.

When there is such a symmetry that depends on some parameter α that can be taken small (that is, the symmetry is **continuous**), we find, similar to Eq. (3.13), that

$$0 = \frac{\delta\mathcal{L}}{\delta\alpha} = \sum_n \left\{ \left[\frac{\partial\mathcal{L}}{\partial\phi_n} - \partial_\mu \frac{\partial\mathcal{L}}{\partial(\partial_\mu\phi_n)} \right] \frac{\delta\phi_n}{\delta\alpha} + \partial_\mu \left[\frac{\partial\mathcal{L}}{\partial(\partial_\mu\phi_n)} \frac{\delta\phi_n}{\delta\alpha} \right] \right\}, \tag{3.22}$$

where ϕ_n may be ϕ and $\phi^\star$ or whatever set of fields the Lagrangian depends on. In contrast to Eq. (3.13), this equation holds even for field configurations ϕ_n for which the action is not extremal (i.e. for ϕ_n that do not satisfy the equations of motion), since the variation corresponds to a symmetry.

When the equations of motion *are* satisfied, then Eq. (3.22) reduces to $\partial_\mu J_\mu = 0$, where

$$J_\mu = \sum_n \frac{\partial\mathcal{L}}{\partial(\partial_\mu\phi_n)} \frac{\delta\phi_n}{\delta\alpha}. \tag{3.23}$$

This is known as a **Noether current**.

For example, with the Lagrangian in Eq. (3.19),

$$\frac{\delta\phi}{\delta\alpha} = -i\phi, \quad \frac{\delta\phi^\star}{\delta\alpha} = i\phi^\star, \tag{3.24}$$

so that

$$J_\mu = \frac{\partial\mathcal{L}}{\partial(\partial_\mu\phi)} \frac{\delta\phi}{\delta\alpha} + \frac{\partial\mathcal{L}}{\partial(\partial_\mu\phi^\star)} \frac{\delta\phi^\star}{\delta\alpha} = -i\left(\phi\partial_\mu\phi^\star - \phi^\star\partial_\mu\phi\right). \tag{3.25}$$

Note that the symmetry is continuous so that we can take small variations. We can check that

$$\partial_\mu J_\mu = -i\left(\phi\Box\phi^\star - \phi^\star\Box\phi\right), \tag{3.26}$$

which vanishes when the equations of motion $\Box\phi = -m^2\phi$ and $\Box\phi^\star = -m^2\phi^\star$ are satisfied.

A vector field J_μ that satisfies $\partial_\mu J_\mu = 0$ is called a **conserved current**. It is called *conserved* because the total charge Q, defined as

$$Q = \int d^3x\, J_0, \tag{3.27}$$

satisfies

$$\partial_t Q = \int d^3x\, \partial_t J_0 = \int d^3x\, \vec{\nabla}\cdot\vec{J} = 0. \tag{3.28}$$

In the last step we have assumed $\vec{J}$ vanishes at the spatial boundary, since, by assumption, nothing is leaving our experiment. Thus, the total charge does not change with time, and is conserved.

We have just proved a very general and important theorem known as **Noether's theorem**.

Box 3.1	Noether's theorem

If a Lagrangian has a continuous symmetry then there exists a current associated with that symmetry that is conserved when the equations of motion are satisfied.

Recall that we needed to assume the symmetry was continuous so that small variations $\frac{\delta\mathcal{L}}{\delta\alpha}$ could be taken. So, Noether's theorem does not apply to discrete symmetries, such as the symmetry under $\phi \to -\phi$ of $\mathcal{L} = \frac{1}{2}\phi\Box\phi - m^2\phi^2 - \lambda\phi^4$ with ϕ real.

Important points about this theorem are:

- The symmetry must be continuous, otherwise $\delta\alpha$ has no meaning.
- The current is conserved *on-shell*, that is, when the equations of motion are satisfied.
- It works for *global symmetries*, parametrized by numbers α, not only for *local (gauge) symmetries* parametrized by functions $\alpha(x)$.

This final point is an important one, although it cannot be fully appreciated with what we have covered so far. Gauge symmetries will be discussed in Chapter 8, where we will see that they are required for Lagrangian descriptions of massless spin-1 particles. Gauge symmetries imply global symmetries, but the existence of conserved currents holds whether or not there is a gauge symmetry or an associated massless spin-1 particle.

3.3.1 Energy-momentum tensor

There is a very important case of Noether's theorem that applies to a global symmetry of the action, not the Lagrangian. This is the symmetry under (global) space-time translations. In general relativity this symmetry is promoted to a local symmetry – diffeomorphism invariance – but all one needs to get a conserved current is a global symmetry. The current in this case is the energy-momentum tensor, $\mathcal{T}_{\mu\nu}$.

Space-time translation invariance says that physics at a point x should be the same as physics at any other point y. We have to be careful distinguishing this symmetry which acts on fields from a trivial symmetry under relabeling our coordinates. Acting on fields, it says that if we replace the value of the field $\phi(x)$ with its value at a different point $\phi(y)$, we will not be able to tell the difference. To turn this into mathematics, we consider cases where the new points y are related to the old points by a simple shift: $y^\nu = x^\nu - \xi^\nu$ with ξ^ν a constant 4-vector. Scalar fields then transform as $\phi(x) \to \phi(x + \xi)$. For infinitesimal ξ^μ, this is

$$\phi(x) \to \phi(x + \xi) = \phi(x) + \xi^\nu \partial_\nu \phi(x) + \cdots, \tag{3.29}$$

where the $\cdots$ are higher order in the infinitesimal transformation ξ^ν. To be clear, we are considering variations where we replace the field $\phi(x)$ with a linear combination of the field and its derivatives evaluated at the same point x. The point x does not change. Our coordinates do not change. A theory with a global translation symmetry is invariant under this replacement.

This transformation law,

$$\frac{\delta \phi}{\delta \xi^\nu} = \partial_\nu \phi, \tag{3.30}$$

applies for any field, whether tensor or spinor or anything else. It is also applies to the Lagrangian itself, which is a scalar:

$$\frac{\delta \mathcal{L}}{\delta \xi^\nu} = \partial_\nu \mathcal{L}. \tag{3.31}$$

Since this is a total derivative, $\delta S = \int d^4x \, \delta \mathcal{L} = \xi^\nu \int d^4x \, \partial_\nu \mathcal{L} = 0$, which is why we sometimes say this is a symmetry of the action, not the Lagrangian.

Proceeding as before, using the equations of motion, the variation of the Lagrangian is

$$\frac{\delta \mathcal{L}[\phi_n, \partial_\mu \phi_n]}{\delta \xi^\nu} = \partial_\mu \left(\sum_n \frac{\partial \mathcal{L}}{\partial(\partial_\mu \phi_n)} \frac{\delta \phi_n}{\delta \xi^\nu} \right). \tag{3.32}$$

Equating this with Eq. (3.31) and using Eq. (3.30) we find

$$\partial_\nu \mathcal{L} = \partial_\mu \left(\sum_n \frac{\partial \mathcal{L}}{\partial(\partial_\mu \phi_n)} \partial_\nu \phi_n \right) \tag{3.33}$$

or equivalently

$$\partial_\mu \left(\sum_n \frac{\partial \mathcal{L}}{\partial(\partial_\mu \phi_n)} \partial_\nu \phi_n - g_{\mu\nu} \mathcal{L} \right) = 0. \tag{3.34}$$

The four symmetries have produced four Noether currents, one for each ν:

$$\mathcal{T}_{\mu\nu} = \sum_n \frac{\partial \mathcal{L}}{\partial(\partial_\mu \phi_n)} \partial_\nu \phi_n - g_{\mu\nu} \mathcal{L}, \tag{3.35}$$

all of which are conserved: $\partial_\mu \mathcal{T}_{\mu\nu} = 0$. The four conserved quantities are energy and momentum. $\mathcal{T}_{\mu\nu}$ is called the **energy-momentum tensor**.

An important component of the energy-momentum tensor is the energy density:

$$\mathcal{E} = \mathcal{T}_{00} = \sum_n \frac{\partial \mathcal{L}}{\partial \dot{\phi}_n} \dot{\phi}_n - \mathcal{L}, \tag{3.36}$$

where $\dot{\phi}_n = \partial_t \phi_n$. Observe that this energy density is identical to the Legendre transform of the Lagrangian, Eq. (3.3), so that the energy density and the Hamiltonian density are identical.

The conserved charges corresponding to the energy-momentum tensor are $Q_\nu = \int d^3x \, \mathcal{T}_{0\nu}$. The components of Q_ν are the total energy and momentum of the system, which are time independent since $\partial_t Q_\nu = 0$ following from $\partial_\mu \mathcal{T}_{\mu\nu} = 0$. This symmetry (invariance of the theory under space-time translations) means that physics is independent of where in the universe you conduct your experiment. Noether's theorem tells us that this symmetry is *why* energy and momentum are conserved.

By the way, the energy-momentum tensor defined this way is not necessarily symmetric. There is another way to derive the energy-momentum tensor, in general relativity. There, the metric $g_{\mu\nu}$ is a field, and we can expand it as $g_{\mu\nu} = \eta_{\mu\nu} + \sqrt{G_N} h_{\mu\nu}$. If you insert this expansion in a general relativistic action, the terms linear in $h_{\mu\nu}$ that couple to matter will have the form $h_{\mu\nu} \mathcal{T}_{\mu\nu}$. This $\mathcal{T}_{\mu\nu}$ is the energy-momentum tensor for matter, and is conserved. The energy-momentum tensor defined by Eq. (3.35) is often called the **canonical energy-momentum tensor**.

3.3.2 Currents

Both the conserved vector J_μ associated with a global symmetry and the energy-momentum tensor $\mathcal{T}_{\mu\nu}$ are types of currents. The concept of a **current** is extremely useful for field theory. Currents are used in many ways. For example:

1. Currents can be Noether currents associated with a symmetry.
2. Currents can refer to *external* currents. These are given background configurations, such as electrons flowing through a wire. For example, a charge density $\rho(x)$ with velocity $v_i(x)$ has the current

$$J_\mu(x) : \quad \left\{ \begin{array}{l} J_0(x) = \rho(x), \\ J_i(x) = \rho(x) v_i(x). \end{array} \right. \tag{3.37}$$

3. Currents can be used as sources for fields, appearing in the Lagrangian as

$$\mathcal{L}(x) = \cdots - A_\mu(x) J_\mu(x). \tag{3.38}$$

This current can be the Noether current, an explicit external current such as the charge current above, or just a formal place-holder. The current is never a dynamical field; that is, it never has its own kinetic terms. We may include time dependence in $J_\mu(\vec{x}, t)$, but we will not generally try to solve for the dynamics of J_μ at the same time as solving for the dynamics of real propagating fields such as A_μ.

4. Currents can be place-holders for certain terms in a Lagrangian. For example, if our Lagrangian is

$$\mathcal{L} = -\frac{1}{4} F_{\mu\nu}^2 - \phi^\star \Box \phi - ieA_\mu(\phi^\star \partial_\mu \phi - \phi \partial_\mu \phi^\star), \tag{3.39}$$

we could write it as

$$\mathcal{L} = -\frac{1}{4} F_{\mu\nu}^2 - \phi^\star \Box \phi - A_\mu J_\mu \tag{3.40}$$

with $J_\mu = ie(\phi^\star \partial_\mu \phi - \phi \partial_\mu \phi^\star)$. The point of this use of currents is that it is independent of the type of interaction. For example, $A_\mu J_\mu$ could mean $A_\mu \bar{\psi} \gamma^\mu \psi$, in which case we would have $J_\mu = \bar{\psi} \gamma^\mu \psi$. This notation is particularly useful when we are only interested in the field A_μ itself, not in whether it was created by ϕ or ψ. Using currents helps separate the problem into two halves: how the field ϕ or ψ produces the field A_μ and then how A_μ affects other fields. Often we are interested in only half of the problem.

3.4 Coulomb's law

The best way to understand classical field theory is by doing some calculations. In this section we derive Coulomb's law using classical field theory.

Start with a charge of strength e at the origin. This can be represented with an external current:

$$J_\mu(x): \quad \begin{cases} J_0(x) = \rho(x) = e\delta^3(x), \\ J_i(x) = 0. \end{cases} \tag{3.41}$$

The Lagrangian is

$$\mathcal{L} = -\frac{1}{4}F_{\mu\nu}^2 - A_\mu J_\mu. \tag{3.42}$$

To calculate the equations of motion, we first expand

$$\mathcal{L} = -\frac{1}{4}(\partial_\mu A_\nu - \partial_\nu A_\mu)^2 - A_\mu J_\mu = -\frac{1}{2}(\partial_\mu A_\nu)^2 + \frac{1}{2}(\partial_\mu A_\mu)^2 - A_\mu J_\mu, \tag{3.43}$$

and note that

$$\partial_\mu \frac{\partial(\partial_\alpha A_\alpha)^2}{\partial(\partial_\mu A_\nu)} = \partial_\mu\left[2(\partial_\alpha A_\alpha)\frac{\partial(\partial_\beta A_\gamma)}{\partial(\partial_\mu A_\nu)}g_{\beta\gamma}\right] = \partial_\mu[2(\partial_\alpha A_\alpha)g_{\beta\mu}g_{\gamma\nu}g_{\beta\gamma}] = 2\partial_\nu(\partial_\alpha A_\alpha). \tag{3.44}$$

Then, the Euler–Lagrange equations $\frac{\partial\mathcal{L}}{\partial A_\nu} - \partial_\mu\frac{\partial\mathcal{L}}{\partial(\partial_\mu A_\nu)} = 0$ imply

$$- J_\nu - \partial_\mu(-\partial_\mu A_\nu) - \partial_\nu(\partial_\mu A_\mu) = 0, \tag{3.45}$$

which gives

$$\partial_\mu F_{\mu\nu} = J_\nu. \tag{3.46}$$

These are just Maxwell's equations in the presence of a source.

Expanding out $F_{\mu\nu}$ we find

$$J_\nu = \partial_\mu(\partial_\mu A_\nu - \partial_\nu A_\mu) = \Box A_\nu - \partial_\nu(\partial_\mu A_\mu). \tag{3.47}$$

Now choose Lorenz gauge, $\partial_\mu A_\mu = 0$. Then,

$$\Box A_\nu(x) = J_\nu(x), \tag{3.48}$$

which has a formal solution

$$A_\nu(x) = \frac{1}{\Box}J_\nu(x), \tag{3.49}$$

where $\frac{1}{\Box}$ just means the inverse of $\Box$, which we will define more precisely soon. This type of expression comes about in almost every calculation in quantum field theory. It says that the A_ν field is determined by the source J_ν after it propagates with the **propagator**

$$\Pi_A = \frac{1}{\Box}. \tag{3.50}$$

We will understand these propagators in great detail as we go along.

For the particular source we are interested in, the point charge at the origin, Eq. (3.41), the equations of motion are

$$A_i = 0, \tag{3.51}$$

$$A_0(x) = \frac{e}{\Box} \delta^3(x). \tag{3.52}$$

There are also homogeneous solutions for which $\Box A_\mu = 0$. These are electromagnetic waves that do not have anything to do with our source, so we will ignore them for now.

3.4.1 Fourier transform interlude

Continuing with the Coulomb calculation, we next take the Fourier transform. Recall that the Fourier transform of a δ-function is just 1: $\tilde{\delta}(k) = 1$. That is

$$\delta^3(\vec{x}) = \int \frac{d^3k}{(2\pi)^3} e^{i\vec{k}\vec{x}}. \tag{3.53}$$

Since the Laplacian is $\triangle = \partial_{\vec{x}}^2$, we have

$$\triangle^n \delta^3(\vec{x}) = \int \frac{d^3k}{(2\pi)^3} \triangle^n e^{i\vec{k}\vec{x}} = \int \frac{d^3k}{(2\pi)^3} (-\vec{k}^2)^n e^{i\vec{k}\vec{x}}. \tag{3.54}$$

Thus, we identify

$$\widetilde{[\triangle^n \delta]}(\vec{k}) = (-\vec{k}^2)^n. \tag{3.55}$$

This also works for Lorentz-invariant quantities:

$$\delta^4(x) = \int \frac{d^4k}{(2\pi)^4} e^{ik_\mu x_\mu}, \tag{3.56}$$

$$\Box^n \delta^4(x) = \int \frac{d^4k}{(2\pi)^4} \Box^n e^{ik_\mu x_\mu} = \int \frac{d^4k}{(2\pi)^4} (-k^2)^n e^{ik_\mu x_\mu}. \tag{3.57}$$

More generally,

$$\Box^n f(x) = \int \frac{d^4k}{(2\pi)^4} \Box^n \tilde{f}(k) e^{ik_\mu x_\mu} = \int \frac{d^4k}{(2\pi)^4} (-k^2)^n \tilde{f}(k) e^{ik_\mu x_\mu}. \tag{3.58}$$

So,

$$\widetilde{[\Box^n f]}(k) = (-k^2)^n \tilde{f}(k). \tag{3.59}$$

Thus, in general,

$$\triangle \leftrightarrow -\vec{k}^2 \quad \text{and} \quad \Box \leftrightarrow -k^2. \tag{3.60}$$

We will use this implicitly all the time. For a field theorist, *box* means "$-k^2$".

3.4.2 Coulomb potential

Since $\delta^3(\vec{x})$ is time independent, our scalar potential simplifies to

$$A_0(x) = \frac{e}{\Box}\delta^3(\vec{x}) = -\frac{e}{\triangle}\delta^3(\vec{x}). \qquad (3.61)$$

We can solve this equation in Fourier space:

$$
\begin{aligned}
A_0(x) &= \int \frac{d^3k}{(2\pi)^3}\frac{e}{\vec{k}^2}e^{i\vec{k}\vec{x}}\\
&= \frac{e}{(2\pi)^3}\int_0^\infty k^2 dk \int_{-1}^1 d\cos\theta \int_0^{2\pi} d\phi \frac{1}{k^2}e^{ikr\cos\theta}\\
&= \frac{e}{(2\pi)^2}\int_0^\infty dk \frac{e^{ikr}-e^{-ikr}}{ikr}\\
&= \frac{e}{8\pi^2}\frac{1}{ir}\int_{-\infty}^\infty dk \frac{e^{ikr}-e^{-ikr}}{k}. \qquad (3.62)
\end{aligned}
$$

Note that the integrand does not blow up as $k \to 0$. Thus, it should be insensitive to a small shift in the denominator, and we can simplify it with

$$\int_{-\infty}^\infty dk\frac{e^{ikr}-e^{-ikr}}{k} = \lim_{\delta\to 0}\left[\int_{-\infty}^\infty dk\frac{e^{ikr}-e^{-ikr}}{k+i\delta}\right]. \qquad (3.63)$$

If $\delta > 0$ then the pole at $k = -i\delta$ lies on the negative imaginary axis. For e^{ikr} we must close the contour up to get exponential decay at large k. This misses the pole, so this term gives zero. For e^{-ikr} we close the contour down and get

$$\int_{-\infty}^\infty dk\frac{-e^{-ikr}}{k+i\delta} = -(2\pi i)(-e^{-\delta r}) = 2\pi i e^{-\delta r}. \qquad (3.64)$$

Thus,

$$A_0(x) = \frac{e}{4\pi}\frac{1}{r}. \qquad (3.65)$$

This result can also be derived through the $m \to 0$ limit of the potential for a massive vector boson, as in Problem 3.6.

3.5 Green's functions

The important point is that we found the Coulomb potential by using

$$A_\mu = \frac{1}{\Box}J_\mu. \qquad (3.66)$$

Even if J_μ were much more complicated, producing all kinds of crazy-looking electromagnetic fields, we could still use this equation.

For example, consider the Lagrangian

$$\mathcal{L} = -\frac{1}{4}F_{\mu\nu}^2 - \phi^\star\Box\phi - ieA_\mu(\phi^\star\partial_\mu\phi - \phi\partial_\mu\phi^\star), \tag{3.67}$$

where ϕ represents a charged object that radiates the A field. Now A's equation of motion is (in Lorenz gauge)

$$\Box A_\mu = ie\left(\phi^\star\partial_\mu\phi - \phi\partial_\mu\phi^\star\right). \tag{3.68}$$

This is just what we had before but with $J_\mu = ie\left(\phi^\star\partial_\mu\phi - \phi\partial_\mu\phi^\star\right)$. And again we will have $A_\mu = \frac{1}{\Box}J_\mu$.

Using propagators is a very useful way to solve these types of equations, and quite general. For example, let us suppose our Lagrangian had an interaction term such as A^3 in it. The Lagrangian for the electromagnetic field does not have such a term (electromagnetism is *linear*), but there are plenty of self-interacting fields in nature. The gluon is one. Another is the graviton. The Lagrangian for the graviton is heuristically

$$\mathcal{L} = -\frac{1}{2}h\Box h + \frac{1}{3}\lambda h^3 + Jh, \tag{3.69}$$

where h represents the gravitational potential, as A_0 represents the Coulomb potential. We are ignoring spin and treating gravity as a simple scalar field theory. The h^3 term represents a graviton self-interaction, which is present in general relativity and so $\lambda \sim \sqrt{G_N}$. The equations of motion are

$$\Box h - \lambda h^2 - J = 0. \tag{3.70}$$

Now we solve perturbatively in λ. For $\lambda = 0$,

$$h_0 = \frac{1}{\Box}J. \tag{3.71}$$

This is what we had before. Then we plug in

$$h = h_0 + h_1 \tag{3.72}$$

with $h_1 = \mathcal{O}(\lambda^1)$. Then

$$\Box(h_0 + h_1) - \lambda(h_0 + h_1)^2 - J = 0, \tag{3.73}$$

which implies

$$\Box h_1 = \lambda h_0^2 + \mathcal{O}(\lambda^2), \tag{3.74}$$

so that

$$h_1 = \lambda\frac{1}{\Box}(h_0 h_0) = \lambda\frac{1}{\Box}\left[\left(\frac{1}{\Box}J\right)\left(\frac{1}{\Box}J\right)\right]. \tag{3.75}$$

Thus, the solution to order λ is

$$h = \frac{1}{\Box}J + \lambda\frac{1}{\Box}\left[\left(\frac{1}{\Box}J\right)\left(\frac{1}{\Box}J\right)\right] + \mathcal{O}(\lambda^2). \tag{3.76}$$

We can keep this up, resulting in a nice expansion for h.

This is known as the Green's function method. The object

$$\Pi = -\frac{1}{\Box} \tag{3.77}$$

is known as a 2-point Green's function or propagator. Propagators are integral parts of quantum field theory. Classically, they tell us how a field propagates through space when it is sourced by a current $J(x)$. Note that the propagator has nothing to do with the source. In fact it is entirely determined by the kinetic terms for a field.

It is not hard to be more precise about this expansion. We can define $\Pi = -\frac{1}{\Box}$ as the solution to

$$\Box_x \Pi(x, y) = -\delta^4(x - y) , \tag{3.78}$$

where $\Box_x = g^{\mu\nu} \frac{\partial}{\partial x^\mu} \frac{\partial}{\partial x^\nu}$. Up to some subtleties with boundary conditions, which will be addressed in future chapters, the solution is

$$\Pi(x, y) = \int \frac{d^4 k}{(2\pi)^4} e^{ik(x-y)} \frac{1}{k^2} , \tag{3.79}$$

which is easy to check:

$$\Box_x \Pi(x, y) = - \int \frac{d^4 k}{(2\pi)^4} e^{ik(x-y)} = -\delta^4(x - y) . \tag{3.80}$$

Note that $\Pi(x, y) = \Pi(y, x)$.

Using $\Box_y \Pi(x, y) = -\delta^4(x - y)$ we can then write a field as

$$h(x) = \int d^4 y \, \delta^4(x - y) \, h(y) = - \int d^4 y \, [\Box_y \Pi(x, y)] \, h(y) = - \int d^4 y \, \Pi(x, y) \, \Box_y h(y), \tag{3.81}$$

where we have integrated by parts in the last step. This lets us solve the free equation $\Box_y h_0(y) = J(y)$ by inserting it on the right-hand side of this identity, to give

$$h_0(x) = - \int d^4 y \, \Pi(x, y) \, J(y). \tag{3.82}$$

The next term in the expansion is Eq. (3.74), whose more precise form is

$$\Box_w h_1(w) = \lambda h_0^2(w) = \lambda \int d^4 y \, \Pi(w, y) J(y) \int d^4 z \, \Pi(w, z) \, J(z). \tag{3.83}$$

Substituting again into Eq. (3.81) and combining with the leading-order result, we find

$$h(x) = - \int d^4 y \, \Pi(x, y) J(y)$$
$$- \lambda \int d^4 w \int d^4 y \int d^4 z \, \Pi(x, w) \Pi(w, y) \Pi(w, z) J(y) J(z) + \mathcal{O}(\lambda^2), \tag{3.84}$$

which is what was meant by Eq. (3.76).

There is a nice pictorial representation of this solution:

$$h(x) = \quad \cdots \tag{3.85}$$

These are called **Feynman diagrams**. The rules for matching equations such as Eq. (3.84) to pictures like this are called **Feynman rules**. The Feynman rules for this classical field theory example are:

1. Draw a point x and a line from x to a new point x_i.
2. Either truncate a line at a source J or let the line branch into two lines adding a new point and a factor of λ.
3. Repeat previous step.
4. The final value for $h(x)$ is given by graphs up to some order in λ with the ends capped by currents $J(x_i)$, the lines replaced by propagators $\Pi(x_i, x_j)$, and all internal points integrated over.

As we will see in Chapter 7, the Feynman rules for quantum field theory are almost identical, except that for $\hbar \neq 0$ lines can close in on themselves.

Returning to our concrete example of classical gravity, these diagrams describe the way the Sun affects Mercury. The double wavy lines represent gravitons and the blobs on the right represent the source, which in this case is the Sun. Mercury, on the left, is also drawn as a blob, since it is classical. The first diagram represents Newton's potential, while the second diagram has the self-interaction in it, proportional to $\lambda \sim \sqrt{G_N}$. You can use this pictorial representation to immediately write down the additional terms. Drawing the next-order picture translates immediately into an integral expression representing the next term in the perturbative solution for $h(x)$. In this way, one can solve the equations of motion for a classical field by drawing pictures.

Problems

3.1 Find the generalization of the Euler–Lagrange equations for general Lagrangians, of the form $\mathcal{L}[\phi, \partial_\mu \phi, \partial_\nu \partial_\mu \phi, \ldots]$.

3.2 Lorentz currents.

    (a) Calculate the conserved currents $K_{\mu\nu\alpha}$ associated with (global) Lorentz transformations $x_\mu \to \Lambda_{\mu\nu} x_\nu$. Express the currents in terms of the energy-momentum tensor.

    (b) Evaluate the currents for $\mathcal{L} = -\frac{1}{2}\phi(\Box + m^2)\phi$. Check that these currents satisfy $\partial_\alpha K_{\mu\nu\alpha} = 0$ on the equations of motion.

    (c) What is the physical interpretation of the conserved quantities $Q_i = \int d^3x K_{0i0}$ associated with boosts?

    (d) Show that $\frac{dQ_i}{dt} = 0$ can still be consistent with $i\frac{\partial Q_i}{\partial t} = [Q_i, H]$. Thus, although these charges are conserved, they do not provide invariants for the equations of motion. This is one way to understand why particles have spin, corresponding to representations of the rotation group, and not additional quantum numbers associated with boosts.

3.3 Ambiguities in the energy-momentum tensor.

(a) If you add a total derivative to the Lagrangian $\mathcal{L} \to \mathcal{L} + \partial_\mu X^\mu$, how does the energy-momentum tensor change?

(b) Show that the total energy $Q = \int T_{00}\, d^3x$ is invariant under such changes.

(c) Show that $T_{\mu\nu} \neq T_{\nu\mu}$ is not symmetric for $\mathcal{L} = -\frac{1}{4}F_{\mu\nu}^2$. Can you find an X_μ so that $T_{\mu\nu}$ is symmetric in this case?

3.4 Write down the next-order diagrams in Eq. (3.85) and their corresponding integral expressions using Feynman rules. Check that your answer is correct by using the Green's function method.

3.5 Spontaneous symmetry breaking is an important subject, to be discussed in depth in Chapter 28. A simple classical example that demonstrates spontaneous symmetry breaking is described by the Lagrangian for a scalar with a *negative* mass term:

$$\mathcal{L} = -\frac{1}{2}\phi\Box\phi + \frac{1}{2}m^2\phi^2 - \frac{\lambda}{4!}\phi^4. \tag{3.86}$$

(a) How many constants c can you find for which $\phi(x) = c$ is a solution to the equations of motion? Which solution has the lowest energy (the ground state)?

(b) The Lagrangian has a symmetry under $\phi \to -\phi$. Show that this symmetry is not respected by the ground state. We say the vacuum expectation value of ϕ is c, and write $\langle\phi\rangle = c$. In this vacuum, the $\mathbb{Z}_2$ symmetry $\phi \to -\phi$ is spontaneously broken.

(c) Write $\phi(x) = c + \pi(x)$ and substitute back into the Lagrangian. Show that now $\pi = 0$ *is* a solution to the equations of motion. How does π transform under the $\mathbb{Z}_2$ symmetry $\phi \to -\phi$? Show that this is a symmetry of π's Lagrangian.

3.6 Yukawa potential.

(a) Calculate the equations of motion for a massive vector A_μ from the Lagrangian

$$\mathcal{L} = -\frac{1}{4}F_{\mu\nu}^2 + \frac{1}{2}m^2 A_\mu^2 - A_\mu J_\mu, \tag{3.87}$$

where $F_{\mu\nu} = \partial_\mu A_\nu - \partial_\nu A_\mu$. Assuming $\partial_\mu J_\mu = 0$, use the equations to find a constraint on A_μ.

(b) For J_μ the current of a point charge, show that the equation of motion for A_0 reduces to

$$A_0(r) = \frac{e}{4\pi^2 ir} \int_{-\infty}^\infty \frac{k\, dk}{k^2 + m^2} e^{ikr}. \tag{3.88}$$

(c) Evaluate this integral with contour integration to get an explicit form for $A_0(r)$.

(d) Show that as $m \to 0$ you reproduce the Coulomb potential.

(e) In 1935 Yukawa speculated that this potential might explain what holds protons together in the nucleus. What qualitative features does this Yukawa potential have, compared to a Coulomb potential, that make it a good candidate for the force between protons? What value for m might be appropriate (in MeV)?

(f) Plug the constraint on A_μ that you found in part (a) back into the Lagrangian, simplify, then *rederive* the equations of motion. Can you still find the constraint? What is acting as a Lagrange multiplier in Eq. (3.87)?

3.7 Nonlinear gravity as a classical field theory. In this problem, you will calculate the perihelion shift of Mercury simply by dimensional analysis.

(a) The interactions in gravity have

$$\mathcal{L} = M_{\text{Pl}}^2 \left(-\frac{1}{2} h_{\mu\nu} \Box h_{\mu\nu} + (\partial_\alpha h_{\mu\nu})(\partial_\beta h_{\mu\alpha}) h_{\nu\beta} + \cdots \right) - h_{\mu\nu} T_{\mu\nu}, \quad (3.89)$$

where $M_{\text{Pl}} = \frac{1}{\sqrt{G_N}}$ is the Planck scale. Rescaling h, and dropping indices and numbers of order 1, this simplifies to

$$\mathcal{L} = -\frac{1}{2} h \Box h + (M_{\text{Pl}})^a h^2 \Box h - (M_{\text{Pl}})^b h T. \quad (3.90)$$

What are a and b (i.e. what are the dimensions of these terms)?

(b) The equations of motion following from this Lagrangian are (roughly)

$$\Box h = (M_{\text{Pl}})^a \Box(h^2) - (M_{\text{Pl}})^b T. \quad (3.91)$$

For a point source $T = m\delta^{(3)}(x)$, solve Eq. (3.91) for h to *second* order in the source T (or equivalently to third order in M_{Pl}^{-1}). You may use the Coulomb solution we already derived.

(c) To first order, h is just the Newtonian potential. This causes Mercury to orbit. What is Mercury's orbital frequency, $\omega = \frac{2\pi}{T}$? How does it depend on m_{Mercury}, m_{Sun}, M_{Pl} and the distance R between Mercury and the Sun?

(d) To second order, there is a correction that causes a small shift in Mercury's orbit. Estimate the order of magnitude of the correction to ω in arcseconds/century using your second-order solution.

(e) Estimate how big the effect is of other planets on Mercury's orbital frequency. (Dimensional analysis will do – just get the right powers of masses and distances.)

(f) Do you think the shifts from either the second-order correction or from the other planets should be observable for Mercury? What about for Venus?

(g) If you derive Eq. (3.91) from Eq. (3.90), what additional terms do you get? Why is it OK to use Eq. (3.91) without these terms?

3.8 How does the blackbody paradox argument show that the electromagnetic field cannot be classical while electrons and atoms are quantum mechanical? Should the same arguments apply to treating gravity classically and electrons quantum mechanically?

3.9 Photon polarizations (this problem follows the approach in [Feynman *et al.*, 1996]).

(a) Starting with $\mathcal{L} = -\frac{1}{4} F_{\mu\nu}^2 + J_\mu A_\mu$, substitute in A_μ's equations of motion. This is called *integrating out* A_μ. In momentum space, you should get something like $J_\mu \frac{1}{k^2} J_\mu$.

(b) Choose $k_\mu = (\omega, \kappa, 0, 0)$. Use current conservation ($\partial_\mu J_\mu = 0$) to formally solve for J_1 in terms of J_0, ω and κ in this coordinate system.

(c) Rewrite the interaction $J_\mu \frac{1}{k^2} J_\mu$ in terms of J_0, J_2, J_3, ω and κ.

(d) In what way is a term without time derivatives instantaneous (non-causal)? How many causally propagating degrees of freedom are there?

(e) How do we know that the instantaneous term(s) do not imply that you can communicate faster than the speed of light?

3.10 Graviton polarizations. We will treat the graviton as a symmetric 2-index tensor field. It couples to a current $T_{\mu\nu}$ also symmetric in its two indices, which satisfies the conservation law $\partial_\mu T_{\mu\nu} = 0$.

(a) Assume the Lagrangian is $\mathcal{L} = -\frac{1}{2}h_{\mu\nu}\Box h_{\mu\nu} + \frac{1}{M_{\rm Pl}}h_{\mu\nu}T_{\mu\nu}$. Solve $h_{\mu\nu}$'s equations of motion, and substitute back to find an interaction like $T_{\mu\nu}\frac{1}{k^2}T_{\mu\nu}$.

(b) Write out the 10 terms in the interaction $T_{\mu\nu}\frac{1}{k^2}T_{\mu\nu}$ explicitly in terms of T_{00}, T_{01}, etc.

(c) Use current conservation to solve for $T_{\mu 1}$ in terms of $T_{\mu 0}$, ω and κ. Substitute in to simplify the interaction. How many causally propagating degrees of freedom are there?

(d) Add to the interaction another term of the form $cT_{\mu\mu}\frac{1}{k^2}T_{\nu\nu}$. What value of c can reduce the number of propagating modes? How many are there now?

4 Old-fashioned perturbation theory

The slickest way to perform a perturbation expansion in quantum field theory is with Feynman diagrams. These diagrams will be the main tool we will use in this book, and we will derive the diagrammatic expansion in Chapter 7. Feynman diagrams, while having advantages such as producing manifestly Lorentz-invariant results, can give a very unintuitive picture of what is going on. For example, they seem to imply that particles that cannot exist can appear from nowhere. Technically, Feynman diagrams introduce the idea that a particle can be **off-shell**, meaning not satisfying its classical equations of motion, for example, with $p^2 \neq m^2$. They trade on-shellness for exact 4-momentum conservation. This conceptual shift was critical in allowing the efficient calculation of amplitudes in quantum field theory. In this chapter, we explain where off-shellness comes from, why you do not need it, but why you want it anyway.

To motivate the introduction of the concept of off-shellness, we begin by using our second-quantized formalism to compute amplitudes in perturbation theory, just as in quantum mechanics. Since we have seen that quantum field theory is just quantum mechanics with an infinite number of harmonic oscillators, the tools of quantum mechanics such as perturbation theory (time-dependent or time-independent) should not have changed. We will just have to be careful about using integrals instead of sums and the Dirac δ-function instead of the Kronecker δ. So, we will begin by reviewing these tools and applying them to our second-quantized photon. This is called old-fashioned perturbation theory (OFPT).

As a historical note, OFPT was still a popular way of doing calculations through at least the 1960s. Some physicists, such as Schwinger, never embraced Feynman diagrams and continued to use OFPT. It was not until the 1950s through the work of Dyson and others that it was shown that OFPT and Feynman diagrams gave the same results. Despite the prevalence of Feynman's approach in modern calculations, and the efficient encapsulation by the path integral formalism (Chapters 14 and onward), OFPT is still worth understanding. It provides complementary physical insight into quantum field theory. For example, a souped-up version of OFPT is given by Schwinger's proper-time formalism (Chapter 33), which is still the best way to do certain effective-action calculations. Also, OFPT is closely related to the reduction of loop amplitudes into sums over on-shell states using unitarity (see Section 24.1.2).

This chapter can be skipped without losing continuity with the rest of the text.

4.1 Lippmann–Schwinger equation

Just as in quantum mechanics, perturbation theory in quantum field theory works by splitting the Hamiltonian up into two parts:

$$H = H_0 + V, \tag{4.1}$$

where the eigenstates of H_0 are known exactly, and the potential V gives corrections that are small in some sense. The difference from quantum mechanics is that in quantum field theory the states often have a continuous range of energies. For example, in a hydrogen atom coupled to an electromagnetic field, the associated photon energies, $E = \omega_k = |\vec{k}|$, can take any values. Because of the infinite number of states, the methods look a little different, but we will just be applying the natural continuum generalization of perturbation theory in quantum mechanics.

We are often interested in a situation where we know the state of a system at early times and would like to know the state at late times. Say the state has a fixed energy E at early and late times (of course, it is the same E). There will be some eigenstate of H_0 with energy E, call it $|\phi\rangle$. So,

$$H_0|\phi\rangle = E|\phi\rangle . \tag{4.2}$$

If the energies E are continuous, we should be able to find an eigenstate $|\psi\rangle$ of the full Hamiltonian with the same eigenvalue:

$$H|\psi\rangle = E|\psi\rangle, \tag{4.3}$$

and we can formally write

$$|\psi\rangle = |\phi\rangle + \frac{1}{E - H_0}V|\psi\rangle, \tag{4.4}$$

which is trivial to verify by multiplying both sides by $E - H_0$. This is called the **Lippmann–Schwinger equation**.[1] The inverted object appearing in the Lippmann–Schwinger equation is a kind of Green's function known as the **Lippmann–Schwinger kernel**:

$$\Pi_{\mathrm{LS}} = \frac{1}{E - H_0}. \tag{4.6}$$

The Lippmann–Schwinger equation is useful in scattering theory (see Chapter 5). In scattering calculations the potential acts at intermediate times to induce transitions among states $|\phi\rangle$ that are assumed to be free (non-interacting) at early and late times. It says the full wavefunction $|\psi\rangle$ is given by the free wavefunction $|\phi\rangle$ plus a scattering term.

---

[1] Formally, the inverse of $E - H_0$ is not well defined. Since E is an eigenvalue of H_0, $\det(E - H_0) = 0$ and $(E - H_0)^{-1}$ is singular. To regulate this singularity, we can add an infinitesimal imaginary factor $i\varepsilon$, leading to

$$|\psi\rangle = |\phi\rangle + \frac{1}{E - H_0 + i\varepsilon}V|\psi\rangle, \tag{4.5}$$

with the understanding that ε should be taken to zero at the end of the calculation.

What we would really like to do is express $|\psi\rangle$ entirely in terms of $|\phi\rangle$. Thus, we define an operator T by

$$V|\psi\rangle = T|\phi\rangle \, , \tag{4.7}$$

where T is known as the **transfer matrix**. Inserting this definition turns the Lippmann–Schwinger equation into

$$|\psi\rangle = |\phi\rangle + \frac{1}{E - H_0} T|\phi\rangle, \tag{4.8}$$

which formally gives $|\psi\rangle$ in terms of $|\phi\rangle$. Multiplying this by V and demanding the two sides be equal when contracted with any state $\langle\phi_j|$ gives an operator equation for T:

$$T = V + V \frac{1}{E - H_0} T. \tag{4.9}$$

We can then solve perturbatively in V to get

$$\begin{aligned}
T &= V + V \frac{1}{E - H_0} V + V \frac{1}{E - H_0} V \frac{1}{E - H_0} V + \cdots \\
&= V + V \, \Pi_{\mathrm{LS}} \, V + V \, \Pi_{\mathrm{LS}} \, V \, \Pi_{\mathrm{LS}} \, V + \cdots .
\end{aligned} \tag{4.10}$$

If we insert the complete set $\sum_j |\phi_j\rangle\langle\phi_j|$ of eigenstates $|\phi_j\rangle$ of H_0, the matrix elements become (leaving the summation symbol implicit):

$$\langle\phi_f|T|\phi_i\rangle = \langle\phi_f|V|\phi_i\rangle + \langle\phi_f|V \frac{1}{E - H_0}|\phi_j\rangle\langle\phi_j|V|\phi_i\rangle + \cdots . \tag{4.11}$$

Writing $T_{fi} = \langle\phi_f|T|\phi_i\rangle$ and $V_{ij} = \langle\phi_i|V|\phi_j\rangle$, this becomes

$$T_{fi} = V_{fi} + V_{fj}\Pi_{\mathrm{LS}}(j)V_{ji} + V_{fj}\Pi_{\mathrm{LS}}(j)V_{jk}\Pi_{\mathrm{LS}}(k)V_{ki} + \cdots , \tag{4.12}$$

where $\Pi_{\mathrm{LS}}(k) = \frac{1}{E - E_k}$. Again, $E = E_i = E_f$ is the energy of the initial and final state we are interested in. This expansion is **old-fashioned perturbation theory**.

Equation (4.12) describes how a transition rate can be calculated in perturbation theory as a sum of terms. In each term the potential creates an intermediate state $|\phi_j\rangle$ which propagates with the propagator $\Pi_{\mathrm{LS}}(j)$ until it hits another potential, where it creates a new field $|\phi_k\rangle$ which then propagates and so on, until they hit the final potential factor, which transitions it to the final state. There is a nice diagrammatic way of drawing this series, called Feynman graphs, which we will see through an example in a moment and in more detail in upcoming chapters. The first term V_{fi} gives the **Born approximation** (or first Born approximation), the second term, the second Born approximation and so on. See, for example, [Sakurai, 1993] for applications of the Lippmann–Schwinger equation in non-relativistic quantum mechanics.

4.1.1 Coulomb's law revisited

The example we will compute is one we will revisit many times: an electron scattering off another electron. The transition matrix element for this process is given by the Lippmann–Schwinger equation as

$$T_{fi} = V_{fi} + \sum_n V_{fn} \frac{1}{E_i - E_n} V_{ni} + \cdots . \tag{4.13}$$

Here, E_i is the initial energy (which is the same as the final energy), and E_n is the energy of the intermediate state. The initial and final states each have two electrons $|i\rangle = |\psi_e^1\psi_e^2\rangle$ and $\langle f| = \langle\psi_e^3\psi_e^4|$, where the superscripts label the momenta, i.e. ψ_e^1 has $\vec{p}_1$, etc. The intermediate state can be anything in the whole Fock space, but only certain intermediate states will have non-vanishing matrix elements with V.

In relativistic field theory, the instantaneous action-at-a-distance of Coulomb's law is replaced by a process where two electrons exchange a photon that travels only at the speed of light. Thus, there should be a photon in the intermediate state. Ignoring the spin of the electrons and the photon, the interaction of the electrons with the photon field can be written as

$$V = \frac{1}{2}e \int d^3x \, \psi_e(x)\phi(x)\psi_e(x). \tag{4.14}$$

This interaction is local, since the fields $\psi_e(x)$ and $\phi(x)$, corresponding to the electrons and photon, are all evaluated at the same point. The factor of $\frac{1}{2}$ comes from ignoring spin and treating all fields as representing real scalar particles.

This interaction can turn a state with an electron of momentum $\vec{p}_1$ into a state with an electron of momentum $\vec{p}_3$ and a photon of momentum $\vec{p}_\gamma$. Since initial and final states both have two electrons and no photons, the leading term in Eq. (4.13) vanishes, $V_{fi} = 0$.

To get a non-zero matrix element, we need an intermediate state $|n\rangle$ with a photon. There are two intermediate states that can contribute. In the first, the photon is emitted from the first electron and the intermediate state is before that photon hits the second electron. We can draw a picture representing this process:

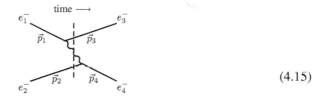

$$\tag{4.15}$$

The vertical dashed line indicates the time at which the intermediate state is evaluated. The second electron feels the effect of a photon that the source, the first electron, emitted at an earlier time. We say that the electron states interact in this case through a **retarded propagator**. For this retarded case, $|n\rangle = |\psi_e^3\phi^\gamma\psi_e^2\rangle$ where $|\phi^\gamma\rangle$ is a photon state of momentum p_γ. Then,

$$V_{ni}^{(R)} = \langle\psi_e^3\phi^\gamma\psi_e^2|V|\psi_e^1\psi_e^2\rangle = \langle\psi_e^3\phi^\gamma|V|\psi_e^1\rangle\langle\psi_e^2|\psi_e^2\rangle = \langle\psi_e^3\phi^\gamma|V|\psi_e^1\rangle. \tag{4.16}$$

The other possibility is that the photon is emitted from the second electron, corresponding to

$$\tag{4.17}$$

which requires $V_{ni}^{(A)} = \langle \psi_e^4 \phi^\gamma | V | \psi_e^2 \rangle$. In this case, from the second electron's point of view, the effect is felt before the source, the first electron, emitted the photon. The photon propagator in this case is called an **advanced propagator**. Obviously, which diagram is advanced or retarded depends on what we call the source, but either way there are two intermediate states, one with a retarded and the other with an advanced propagator.

To find an expression for these matrix elements, we insert our field operators:

$$V_{ni}^{(R)} = \langle \psi_e^3 \phi^\gamma | V | \psi_e^1 \rangle = \frac{e}{2} \int d^3x \langle \psi_e^3 \phi^\gamma | \psi_e(x) \phi(x) \psi_e(x) | \psi_e^1 \rangle. \tag{4.18}$$

To evaluate this, recall from Eq. (2.75) that the second-quantized fields are

$$\phi(\vec{x}) = \int \frac{d^3p}{(2\pi)^3} \frac{1}{\sqrt{2\omega_p}} \left(a_p e^{i\vec{p}\vec{x}} + a_p^\dagger e^{-i\vec{p}\vec{x}} \right), \tag{4.19}$$

with a similar form for the electron, and that

$$\langle \phi^\gamma | \phi(x) | 0 \rangle = e^{-i\vec{p}_\gamma \cdot \vec{x}} \tag{4.20}$$

and similarly for other matrix elements. The interaction $V(x)$ is a product of three fields, and one can pair either of the electron fields in $V(x)$ with either of the electron states in evaluating $\langle \psi_e^3 \phi^\gamma | V | \psi_e^1 \rangle$, so we pick up a factor of 2 in the matrix element. We then find

$$V_{ni}^{(R)} = e \int d^3x \, e^{i(\vec{p}_1 - \vec{p}_3 - \vec{p}_\gamma)\vec{x}} = e(2\pi)^3 \, \delta^3(\vec{p}_1 - \vec{p}_3 - \vec{p}_\gamma). \tag{4.21}$$

The other matrix elements, $V_{ni}^{(A)}$, and those involving the final state are similar, which you can verify.

Thus, we have at first non-vanishing order:

$$T_{fi} = \int d^3\vec{p}_\gamma \, (2\pi)^3 \, \delta^3(\vec{p}_1 - \vec{p}_3 - \vec{p}_\gamma) \, (2\pi)^3 \, \delta^3(\vec{p}_2 - \vec{p}_4 + \vec{p}_\gamma) \, \frac{e^2}{E_i - E_n}. \tag{4.22}$$

These δ-functions tell us that 3-momentum is conserved in the local interactions between the photon and the electrons. Note that nothing tells us that energy is conserved; if it were, then $E_n = E_i$ and this matrix element would blow up. This should not surprise you; the energy of intermediate states has always been different from the energy of the initial and final states in quantum mechanics – due to the uncertainty principle, energy can be not conserved for short times.

To find a form for E_n, let us first denote the intermediate photon energy as E_γ, the incoming electron energies E_1 and E_2, and the outgoing electron energies E_3 and E_4. The momenta of the electrons are $\vec{p}_1, \vec{p}_2, \vec{p}_3, \vec{p}_4$ as above. By conservation of momentum, the photon momentum must be $\vec{p}_\gamma = \vec{p}_1 - \vec{p}_3$. The photon energy is whatever it needs to be to put the photon on-shell: $0 = p_\gamma^2 = E_\gamma^2 - \vec{p}_\gamma^2$, so $E_\gamma = |\vec{p}_\gamma|$. That is,

$$p_\gamma^\mu = (E_\gamma, \vec{p}_\gamma) = (|\vec{p}_1 - \vec{p}_3|, \vec{p}_1 - \vec{p}_3). \tag{4.23}$$

For Eq. (4.22), we need the intermediate state energy, which is different for retarded and advanced cases.

In the retarded case the first electron emits the photon and we look at the state before the photon hits the second electron, as shown in the figure in Eq. (4.15). In this case, at the intermediate time the first electron is already in its final state, with energy E_3. So the *total* intermediate state energy is

$$E_n^{(R)} = E_3 + E_2 + E_\gamma. \tag{4.24}$$

Dropping the 2π factors and the overall momentum-conserving δ-functions for clarity (we will give a detailed derivation of these factors and their connection to scattering cross sections in the relativistic theory in Chapter 5), we then find

$$T_{fi}^{(R)} = \frac{e^2}{E_i - E_n^{(R)}} = \frac{e^2}{(E_1 + E_2) - (E_3 + E_2 + E_\gamma)} = \frac{e^2}{(E_1 - E_3) - E_\gamma}. \tag{4.25}$$

In the advanced case, the second electron emits the photon and we look at the intermediate state before the photon hits the first electron, as in the diagram in Eq. (4.17). Then the energy is

$$E_n^{(A)} = E_4 + E_1 + E_\gamma \tag{4.26}$$

and

$$T_{fi}^{(A)} = \frac{e^2}{E_i - E_n^{(A)}} = \frac{e^2}{(E_1 + E_2) - (E_4 + E_1 + E_\gamma)} = \frac{e^2}{(E_2 - E_4) - E_\gamma}. \tag{4.27}$$

Finally, we have to add the advanced and retarded contributions, since they are both valid intermediate states. Overall energy conservation says $E_1 + E_2 = E_3 + E_4$, so $E_1 - E_3 = E_4 - E_2 \equiv \Delta E$. So the sum is

$$T_{fi}^{(R)} + T_{fi}^{(A)} = \frac{e^2}{E_i - E_n^{(R)}} + \frac{e^2}{E_i - E_n^{(A)}} = \frac{e^2}{\Delta E - E_\gamma} + \frac{e^2}{-\Delta E - E_\gamma} = \frac{2e^2 E_\gamma}{(\Delta E)^2 - (E_\gamma)^2}. \tag{4.28}$$

To simplify this answer, let us *define* a 4-vector k^μ by

$$k^\mu \equiv p_3^\mu - p_1^\mu = (\Delta E, \vec{p}_\gamma). \tag{4.29}$$

Note, this is *not* the photon momentum in Eq. (4.23) since $E_\gamma = |\vec{p}_\gamma| \neq \Delta E$, or more simply, since $k^2 \neq 0$. But k^μ *is* a Lorentz 4-vector, since it comprises an energy and a 3-momentum. The norm of k^μ is

$$k^2 = (\Delta E)^2 - (E_\gamma)^2. \tag{4.30}$$

This is convenient, since it lets us write the transition matrix simply as

$$T_{fi} = T_{fi}^{(R)} + T_{fi}^{(A)} = 2E_\gamma \left(\frac{e^2}{k^2} \right). \tag{4.31}$$

The $2E_\gamma$ is related to normalization, which, along with the 2π and δ-function factors, will be properly accounted for in the relativistic treatment of the transfer matrix in the next chapter.

The remarkable feature of T_{fi} is that it contains a Lorentz-invariant factor of $\frac{1}{k^2}$. This $\frac{1}{k^2} = -\frac{1}{\Box}$ is the Green's function for a Lorentz-invariant theory. If one of the electrons were at rest, we would sum the appropriate combination of momentum eigenstates, which would amount to Fourier transforming $\frac{1}{k^2}$ to reproduce the Coulomb potential, as in Section 3.4.

4.1.2 Feynman rules for OFPT

Let us summarize some ingredients that went into the scattering calculation above:

- All states are *physical*, that is, they are on-shell at all times.
- Matrix elements V_{ij} will vanish unless 3-momentum is conserved at each vertex.
- Energy is *not* conserved at each vertex.

As mentioned in the introduction to this chapter, **on-shell** means that the state satisfies its free-field equations of motion. For example, a scalar field satisfying $(\Box + m^2)\phi = 0$ would have $p^2 = m^2$. It is called on-shell since $\vec{p}^{\,2} = E^2 - m^2$ at fixed E and m is the equation for the surface of a sphere. So on-shell particles live on the shell of the sphere.[2]

Despite the fact that the intermediate states in OFPT are on-shell, we saw that it was helpful to write the answer in terms of a Lorentz 4-vector k^μ with $k^2 \neq 0$ representing the momentum of an unphysical, off-shell photon. We were led to k^μ by combining two diagrams with different temporal orderings, which we called advanced and retarded.

It would be nice if we could get k^μ with just one diagram, where 4-momentum is conserved at vertices and so propagators can be Lorentz invariant from the start. In fact we can! That is what we will be doing in the rest of the book. As we will see, there is just one propagator in this approach, the Feynman propagator, which combines the advanced and retarded propagators into one in a beautifully efficient way. So we will not have to keep track of what happens first. This new formalism will give us a much more cleanly organized framework to address the confusing infinities that plague quantum field theory calculations. Before finishing OFPT, as additional motivation and for its important historical relevance, we will heuristically review one such infinity.

4.2 Early infinities

Historically, one of the first confusions about the second-quantized photon field was that the Hamiltonian

$$H = \int \frac{d^3k}{(2\pi)^3} \omega_k \left(a_k^\dagger a_k + \frac{1}{2} \right) \tag{4.32}$$

with $\omega_k = |\vec{k}|$ seemed to imply that the vacuum has infinite energy,

$$E_0 = \langle 0|H|0\rangle = \frac{1}{2} \int \frac{d^3k}{(2\pi)^3} |\vec{k}| = \infty. \tag{4.33}$$

Fortunately, there is an easy way out of this paradoxical infinity: How do you measure the energy of the vacuum? You do not! Only energy differences are measurable, and in these differences the **zero-point energy**, the energy of the ground state, drops out. This is the basic idea behind renormalization – infinities can appear in intermediate calculations,

---

[2]   For completeness, the Feynman rules for OFPT with relativistic normalization are: a factor of $2\pi\delta(E_i - E_f)$ for overall energy conservation, a propagator factor $1/(E_i - E_k + i\varepsilon)$ for each intermediate state k where E_k is the energy of the entire intermediate state, a factor of $(2\pi)^3\delta^3(\vec{p}_{\text{in}} - \vec{p}_{\text{out}})$ for each vertex and an integral $\int \frac{d^3 p_j}{(2\pi)^3 2E_j}$ for each internal line j. See [Sterman, 1993, Section 9.5] for their derivation.

but they must drop out of physical observables. This zero-point energy does have consequences, such as the Casimir effect (Chapter 15), which comes from the difference in zero-point energies in different size boxes, and the cosmological constant problem, which comes from the fact that energy gravitates. We will come to understand these two examples in detail in Part III, but it makes more sense to start with some less exotic physics.

In 1930, Oppenheimer thought to use perturbation theory to compute the shift of the energy of the hydrogen atom due to the photons [Oppenheimer, 1930]. He got infinity and concluded that QED was wrong. In fact, the result is not infinite but a finite calculable quantity known as the Lamb shift, which agrees perfectly with data. However, it is instructive to understand Oppenheimer's argument.

4.2.1 Oppenheimer and the Lamb shift

Using OFPT we would calculate the energy shift using

$$\Delta E_n = \langle \psi_n | H_{\text{int}} | \psi_n \rangle + \sum_{m \neq n} \frac{| \langle \psi_n | H_{\text{int}} | \psi_m \rangle |^2}{E_n - E_m}. \tag{4.34}$$

This is the standard formula from time-independent perturbation theory. The basic problem is that we have to sum over all possible intermediate states $|\psi_m\rangle$, including ones that have nothing much to do with the system of interest (for example, free plane waves). It is still true in field theory that there are only a finite number of states below any given energy level E, so that as $E \to \infty$, $\frac{1}{E - E_n} \to 0$. The catch is that there are an infinite number of states, and their phase space density goes as $\int d^3k \sim E^3$, so that you get $\frac{E^3}{E - E_n} \to \infty$ and perturbation theory breaks down. This is exactly what Oppenheimer found.

First, take something where the calculation makes sense, such as a fixed non-dynamical background field. Say there is an electric field in the $\hat{z}$ direction. Then the potential energy is proportional to the electric field:

$$H_{\text{int}} = e\vec{E} \cdot \vec{x} = e|E|z. \tag{4.35}$$

This interaction produces the **linear Stark effect**, which is a straightforward application of time-independent perturbation theory in quantum mechanics. Our discussion of the Stark effect here will be limited to a quick demonstration that it is finite, and a representation of the result in terms of diagrams.

Since an atom has no electric dipole moment, the first-order correction is zero:

$$\langle \psi_n | H_{\text{int}} | \psi_n \rangle = 0. \tag{4.36}$$

At second order:

$$\Delta E_0 = \sum_{m>0} \frac{| \langle \psi_0 | H_{\text{int}} | \psi_m \rangle |^2}{E_0 - E_m} = \tag{4.37}$$

The picture on the right side of this equation is the corresponding Feynman diagram: the ≡ symbols represent the electric field which sources photons that interact with the electron (more general background-field calculations will be discussed in Chapters 33 and 34); the electron is represented as the solid line on the bottom; the points where the photon meets the electron correspond to matrix elements of H_{int}; finally, the line between the two photon insertions is the electron propagator, the $\frac{1}{E_0 - E_m}$ factor in the second-order expression for ΔE_0.

To show that ΔE_0 is finite, we assume that $E_0 < 0$ without loss of generality and that $E_m > E_1 > E_0$ so that E_0 is the ground state. Since $\Delta E_0 < 0$, by Eq. (4.37), we need to show that ΔE_0 is bounded from below. Using the completeness relation

$$\mathbb{1} = \sum_{m \geq 0} |\psi_m\rangle\langle\psi_m| = |\psi_0\rangle\langle\psi_0| + \sum_{m>0} |\psi_m\rangle\langle\psi_m|, \tag{4.38}$$

we have

$$-\Delta E_0 \leq \frac{1}{E_1 - E_0} \sum_{m>0} \langle\psi_0|H_{\text{int}}|\psi_m\rangle \langle\psi_m|H_{\text{int}}|\psi_0\rangle$$

$$= \frac{1}{E_1 - E_0} \left[\langle\psi_0|H_{\text{int}}^2|\psi_0\rangle - \langle\psi_0|H_{\text{int}}|\psi_0\rangle^2 \right]. \tag{4.39}$$

The right-hand side of this equation is a positive number, thus ΔE_0 is bounded from below and above (by 0) and hence the energy correction to the ground state is finite. While it is not hard to calculate ΔE_0 exactly for a given system, such as the hydrogen atom, the only thing we want to observe here is that ΔE_0 is finite.

Now, instead of an external electric field, what would happen if this field were produced by the electron itself? Then we need to allow for the creation of photons by the electron and their annihilation back into the electron, which can be described with our second-quantized photon field. The starting Hamiltonian, for which we know the exact eigenstates, now has two parts:

$$H_0 = H_0^{\text{atom}} + H_0^{\text{photon}}, \tag{4.40}$$

with energy eigenstates given by electron wavefunctions associated with a set of photons, so

$$H_0|\psi_n; \{n_k\}\rangle = \left(E_n + \sum_k n_k \omega_k \right) |\psi_n; \{n_k\}\rangle, \tag{4.41}$$

where we allow for any number of excitations n_k of the photons of any momenta $\vec{k}$.

At second order in perturbation theory, only one photon can be created and destroyed, but we have to integrate over this photon's momentum. We are interested in the integration region where the photon has a very large momentum. By momentum conservation in OFPT, since the ground state only has support for small momentum, the excited state of the atom must have large momentum roughly backwards to that of the photon, $\vec{p} \sim -\vec{k}$. Thus, the excited state wavefunction will approach that of a free plane wave. The excited state energy is $E \approx |\vec{p}| + |\vec{k}|$ and so at large k the integral will be

$$\Delta E_0 \sim \int \frac{d^3p}{(2\pi)^3} \int \frac{d^3k}{(2\pi)^3} \int d^3x \frac{e^{i(\vec{k}-\vec{p})\cdot\vec{x}}}{E_0 - (|\vec{p}| + |\vec{k}|)}. \tag{4.42}$$

After evaluating the x integral to get $\delta^3(\vec{p} - \vec{k})$ and then the $\vec{p}$ integral, we find

$$\Delta E_0 \sim \int \frac{d^3k}{(2\pi)^3} \frac{1}{|\vec{k}|} = \frac{1}{2\pi^2} \int k \, dk = \infty. \tag{4.43}$$

This means that there should be an infinite shift in the energy levels of the hydrogen atom. Oppenheimer also showed that if you take the difference between two levels, relevant for the shift in spectral lines, the result is also divergent. He concluded, "It appears improbable that the difficulties discussed in this work will be soluble without an adequate theory of the masses of the electron and proton; nor is it certain that such a theory will be possible on the basis of the special theory of relativity" [Oppenheimer, 1930, p. 477].

What went wrong? In the Stark effect calculation we only had to sum over excited electron states, through $\sum_{m>0} |\psi_m\rangle\langle\psi_m|$ in Eq. (4.39), which was finite. For the Lamb shift calculation, the sum was also over photon states, which was divergent. It diverged because the phase space for photons, d^3k, is larger than the suppression, $\frac{1}{|\vec{k}|}$, due to the energies of the intermediate excited states. In terms of Feynman diagrams, the difference is that in the latter case we do not consider interactions with a fixed external field, but integrate over dynamical fields, corresponding to intermediate state photons. Since the photons relevant to the $\langle\psi_0|H_{\text{int}}|\psi_m; 1_k\rangle$ matrix element are the same as the photons relevant to the second, $\langle\psi_m; 1_k|H_{\text{int}}|\psi_0\rangle$ matrix element, the photon lines represent the same state and should be represented by a single line. Thus the diagram contracts,

$$\Delta E_0 = \quad\underset{\text{\textasciitilde\textasciitilde}}{}\quad \rightarrow \quad , \qquad (4.44)$$

and the Stark effect diagram becomes a loop diagram for the Lamb shift. These pictures are just shorthand for the perturbation expansion. The loop means that there is an unknown momentum, $\vec{k}$, over which we have to integrate. Momentum must be conserved, but it can split between the atom and the photon in an infinite number of ways.

There was actually nothing wrong with Oppenheimer's calculation. He did get the answer that OFPT predicts. What he missed was that there are other infinities that eventually cancel this infinity (for example, the electron mass is infinite too, so in fact his conclusion was on the right track). This discussion was really just meant as a preview to demonstrate the complexities we will be up against. To sort out all these infinities, it will be really helpful, but not strictly necessary, to have a formalism that keeps the symmetries, in particular Lorentz invariance, manifest along the way. Although Schwinger was able to tame the infinities using OFPT, his techniques were not for everyone. In his own words, "Like the silicon chips of more recent years, the Feynman diagram was bringing computation to the masses" [Brown and Hoddesdon, 1984, p. 329].

Problems

4.1 Calculate the transition matrix element T_{ij} for the process $e^+e^- \to \gamma \to \mu^+\mu^-$.
  (a) Write down the $\frac{1}{E_i - E_0}$ terms for the two possible intermediate states, from the two possible time slicings.
  (b) Show that they add up to $\frac{2E_\gamma}{k^2}$, where k_μ is now the 4-momentum of the virtual off-shell photon.

5   Cross sections and decay rates

The twentieth century witnessed the invention and development of collider physics as an efficient way to determine which particles exist in nature, their properties, and how they interact. In early experiments, such as Rutherford's discovery of the nucleus in 1911 using α-particles or Anderson's discovery of the positron in 1932 from cosmic rays, the colliding particles came from nature. Around 1931, E. O. Lawrence showed that particles could be accelerated to relativistic velocities in the lab, first through a 4-inch cyclotron, which gave protons 80 000 electronvolts of kinetic energy, soon to go up to around 1 million electronvolts. The Large Hadron Collider can collide beams of protons with 7 trillion electronvolts of energy. Colliders provide a great way to study fundamental interactions because they begin with initial states of essentially fixed momenta, i.e. plane waves, and end up with final states, which also have fixed momenta. By carefully measuring the mapping from initial state momenta to final state momenta, one can then compare to theoretical models, such as those of quantum field theory.

Quantum mechanics consists of an elaborate collection of rules for manipulating states in a Hilbert space. The experimentally measurable quantities that are predicted in quantum mechanics are differential *probabilities*. These probabilities are given by the modulus squared of inner products of states. We can write such inner products as $\langle \chi; t_2 | \psi; t_1 \rangle$, where $|\psi; t_1\rangle$ is the initial state we start with at time t_1 and $\langle \chi; t_2|$ is the state we are interested in at some later time t_2. Since quantum field theory is just quantum mechanics with many degrees of freedom, the experimental quantities we will be able to predict are also of the form $|\langle \chi; t_2 | \psi; t_1 \rangle|^2$.

The notation $\langle \chi; t_2 | \psi; t_1 \rangle$ refers to the Schrödinger picture representation, where the states evolve in time and the operators are static. In the Heisenberg picture, which will be the default picture for quantum field theory, we move the time evolution into the operators. We are often interested in scattering experiments, where the relevant states are momentum eigenstates at $t = \pm\infty$, called **asymptotic states** . These asymptotic states are produced by the creation operators $a_p^\dagger$ at asymptotically early or late times and we denote them here by $|i\rangle$ and $\langle f|$ respectively. The projection of one on the other gives the elements of the scattering or **S-matrix** :

$$S_{fi} = \langle f|S|i\rangle \tag{5.1}$$

In this equation the operator S in $\langle f|S|i\rangle$ is the Heisenberg-picture time-evolution operator from $-\infty$ to $+\infty$ which allows us to project $|i\rangle$ onto $\langle f|$. S-matrix elements tell us how freely-moving momentum-eigenstates at early times evolve into different freely-moving momentum eigenstates at late times.

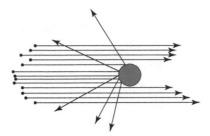

The number of particles scattered classically is proportional to the cross-sectional area of the scattering object.

Fig. 5.1

S-matrix elements are the primary objects of interest for high-energy physics. In this chapter, we will relate S-matrix elements to scattering cross sections, which are directly measured in collider experiments. We will also derive an expression for decay rates, which are also straightforward to measure experimentally. Quantum field theory is capable of calculating other quantities besides S-matrix elements, such as thermodynamic properties of condensed matter systems. However, since the tools we develop for S-matrix calculations, such as Feynman rules, are also relevant for these applications, it is logical to focus on S-matrix elements for concreteness.

5.1 Cross sections

A cross section is a natural quantity to measure experimentally. For example, Rutherford was interested in the size r of an atomic nucleus. By colliding α-particles with gold foil and measuring how many α-particles were scattered, he could determine the cross-sectional area, $\sigma = \pi r^2$, of the nucleus. Imagine there is just a single nucleus. Then the cross-sectional area is given by

$$\sigma = \frac{\text{number of particles scattered}}{\text{time} \times \text{number density in beam} \times \text{velocity of beam}} = \frac{1}{T}\frac{1}{\Phi}N, \qquad (5.2)$$

where T is the time for the experiment and Φ is the incoming flux ($\Phi = $ number density $\times$ velocity of beam) and N is the number of particles scattered. This is shown in Figure 5.1.

In a real gold foil experiment, we would also have to include additional factors for the number density of protons in the foil and the cross-sectional area of the beam if it is smaller than the size of the foil. These factors, like the flux and time factors in Eq. (5.2), depend on the details of how the experiment is actually performed. In contrast, the cross section σ is a property of the things being scattered independent of the way the experiment is done.

It is also natural to measure the differential cross section, $d\sigma/d\Omega$, which gives the number of scattered particles in a certain solid angle $d\Omega$. Classically, this gives us information about the shape of the object or form of the potential off of which the α-particles are scattered.

In quantum mechanics, we generalize the notion of cross-sectional area to a **cross section**, which still has units of area, but has a more abstract meaning as a measure of the

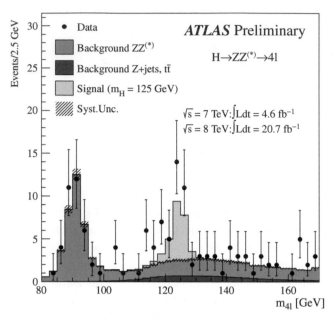

Fig. 5.2 ATLAS four lepton invariant mass measurement showing evidence for the Higgs boson [Atlas Collaboration, 2013]. Solid curves are the predictions from the Standard Model. ATLAS Experiment ©2013 CERN.

interaction strength. While classically an α-particle either scatters off the nucleus or it does not scatter, quantum mechanically it has a probability for scattering. The classical differential probability is $P = \frac{N}{N_{\text{inc}}}$, where N is the number of particles scattering into a given area and N_{inc} is the number of incident particles. So the quantum mechanical cross section is then naturally

$$d\sigma = \frac{1}{T}\frac{1}{\Phi}dP, \tag{5.3}$$

where Φ is the flux, now normalized as if the beam has just one particle, and P is now the quantum mechanical probability of scattering. The differential quantities $d\sigma$ and dP are differential in kinematical variables, such as the angles and energies of the final state particles. The number of scattering events measured in a collider experiment is

$$N = \int d\sigma \times \int \mathcal{L}(t)dt, \tag{5.4}$$

where $\mathcal{L}(t)$ is the **instantaneous luminosity**, defined by this equation.

In practice, experimental data are presented as the number of events seen in a given kinematic region for a given integrated luminosity. For example, Figure 5.2 shows the cross section for final states with four leptons (more precisely, four muons, four electrons, or two muons and two electrons) from colliding proton initial states, as measured by the ATLAS collaboration at the Large Hadron Collider. The cross section shown is differential in the invariant mass of the four leptons ($m_{4l} = \sqrt{(p_1 + p_2 + p_3 + p_4)^2}$). Each point on the plot shows the number of events where the measured mass fell inside the given 2.5 GeV interval. As indicated on the figure, the data plotted correspond to an integrated luminosity of $L_{\text{int}} = \int \mathcal{L}(t)\, dt = 25.3\,\text{fb}^{-1}$ combined from a 7 TeV and an 8 TeV run. To compare to these data, one would calculate $\frac{d\sigma}{dm_{4l}}$ using quantum field theory at the two energies, multiply by the appropriate luminosities, and add the resulting distributions. This final state can come from Z-boson pair production, top-quark pair production, Z-boson plus jet production or Higgs-boson production, as the solid histograms show. The sum of the contributions agrees very well with the data if the Higgs boson is included.

Now let us relate the formula for the differential cross section to S-matrix elements. From a practical point of view it is impossible to collide more than two particles at a time, thus we can focus on the special case of S-matrix elements where $|i\rangle$ is a two-particle state. So, we are interested in the differential cross section for the $2 \to n$ process:

$$p_1 + p_2 \to \{p_j\}. \tag{5.5}$$

In the rest frame of one of the colliding particles, the flux is just the magnitude of the velocity of the incoming particle divided by the total volume: $\Phi = |\vec{v}|/V$. In a different frame, such as the center-of-mass frame, beams of particles come in from both sides, and the flux is then determined by the difference between the particles' velocities. So, $\Phi = |\vec{v}_1 - \vec{v}_2|/V$. This should be familiar from classical scattering. Thus,

$$d\sigma = \frac{V}{T} \frac{1}{|\vec{v}_1 - \vec{v}_2|} dP. \tag{5.6}$$

From quantum mechanics we know that probabilities are given by the square of amplitudes. Since quantum field theory is just quantum mechanics with many degrees of freedom, the normalized differential probability is

$$dP = \frac{|\langle f|S|i\rangle|^2}{\langle f|f\rangle \langle i|i\rangle} d\Pi. \tag{5.7}$$

Here, $d\Pi$ is the region of final state momenta at which we are looking. It is proportional to the product of the differential momentum, $d^3 p_j$, of each final state and must integrate to 1. So

$$d\Pi = \prod_j \frac{V}{(2\pi)^3} d^3 p_j. \tag{5.8}$$

This has $\int d\Pi = 1$, since $\int \frac{dp}{2\pi} = \frac{1}{L}$ (by dimensional analysis and our 2π convention).[1]

The $\langle f|f\rangle$ and $\langle i|i\rangle$ in the denominator of Eq. (5.7) come from the fact that the one-particle states, defined at fixed time, may not be normalized to $\langle f|f\rangle = \langle i|i\rangle = 1$. In fact, such a convention would not be Lorentz invariant. Instead, in Chapter 2 we defined

$$a_k^\dagger |0\rangle = \frac{1}{\sqrt{2\omega_k}} |k\rangle \tag{5.9}$$

---

[1] This normalization is the natural continuum limit of having discrete points $x_i = \frac{i}{N} L$ and wavenumbers $p_i = \frac{2\pi}{L} \frac{i}{N}$ with $i = 1, \ldots, N$.

and $[a_p, a_q^\dagger] = (2\pi)^3 \delta^3(p - q)$, so that

$$\langle p|p \rangle = (2\pi)^3 (2\omega_p) \delta^3(0). \tag{5.10}$$

This $\delta^3(0)$ is formally infinite, but is regulated by the finite volume. It can be understood by using the relation

$$(2\pi)^3 \, \delta^3(p) = \int d^3x \, e^{i\vec{p}\vec{x}}. \tag{5.11}$$

So,

$$\delta^3(0) = \frac{1}{(2\pi)^3} \int d^3x = \frac{V}{(2\pi)^3}. \tag{5.12}$$

Similarly

$$\delta^4(0) = \frac{TV}{(2\pi)^4}, \tag{5.13}$$

where T is the total time for the process, which we will eventually take to ∞. Thus,

$$\langle p|p \rangle = 2\omega_p V = 2E_p V \tag{5.14}$$

and, using $|i\rangle = |p_1\rangle |p_2\rangle$ and $|f\rangle = \prod_j |p_j\rangle$,

$$\langle i|i \rangle = (2E_1 V)(2E_2 V), \quad \langle f|f \rangle = \prod_j (2E_j V). \tag{5.15}$$

We will see that all these V factors conveniently drop out of the final answer.

Now let us turn to the S-matrix element $\langle f|S|i\rangle$ in the numerator of Eq. (5.7). We usually calculate S-matrix elements perturbatively. In a free theory, where there are no interactions, the S-matrix is simply the identity matrix $\mathbb{1}$. We can therefore write

$$S = \mathbb{1} + i\mathcal{T}, \tag{5.16}$$

where $\mathcal{T}$ is called the **transfer matrix** and describes deviations from the free theory.[2] Since the S-matrix should vanish unless the initial and final states have the same total 4-momentum, it is helpful to factor an overall momentum-conserving δ-function:

$$\mathcal{T} = (2\pi)^4 \delta^4(\Sigma p) \mathcal{M}. \tag{5.17}$$

Here, $\delta^4(\Sigma p)$ is shorthand for $\delta^4\left(\Sigma p_i^\mu - \Sigma p_f^\mu\right)$, where p_i^μ are the initial particles' momenta and p_f^μ are the final particles' momenta. In this way, we can focus on computing the non-trivial part of the S-matrix, $\mathcal{M}$. In quantum field theory, "matrix elements" usually means $\langle f|\mathcal{M}|i\rangle$. Thus we have

$$\langle f|S - \mathbb{1}|i\rangle = i(2\pi)^4 \delta^4(\Sigma p) \langle f|\mathcal{M}|i\rangle. \tag{5.18}$$

Now, it might seem worrisome at first that we need to take the square of a quantity with a δ-function. However, this is actually simple to deal with. When integrated over, one of

---

[2] The i in this definition is just a convention, motivated by $S \approx e^{i\hat{T}}$, which makes $\hat{T}$ Hermitian if S is unitary. Note that $\mathcal{T}$ defined by Eq. (5.16) is not exactly $\hat{T}$ and does not have to be Hermitian. Hermiticity of $\hat{T}$ will play an important role in implications of unitarity, discussed in Chapter 24.

the δ-functions in the square is sufficient to enforce the desired condition; the remaining δ-function will always be non-zero and formally infinite, but with our finite time and volume will give $\delta^4(0) = \frac{TV}{(2\pi)^4}$. For $|f\rangle \neq |i\rangle$ (the case $|f\rangle = |i\rangle$, for which nothing happens, is special),

$$
\begin{aligned}
|\langle f|S|i\rangle|^2 &= \delta^4(0)\delta^4(\Sigma p)(2\pi)^8|\langle f|\mathcal{M}|i\rangle|^2 \\
&= \delta^4(\Sigma p)TV(2\pi)^4|\mathcal{M}|^2,
\end{aligned}
\tag{5.19}
$$

where Eq. (5.13) was used and $|\mathcal{M}|^2 \equiv |\langle f|\mathcal{M}|i\rangle|^2$.

So,

$$
\begin{aligned}
dP &= \frac{\delta^4(\Sigma p)TV(2\pi)^4}{(2E_1 V)(2E_2 V)} \frac{1}{\prod_j(2E_j V)}|\mathcal{M}|^2 \prod_j \frac{V}{(2\pi)^3} d^3 p_j \\
&= \frac{T}{V} \frac{1}{(2E_1)(2E_2)}|\mathcal{M}|^2 d\Pi_{\text{LIPS}},
\end{aligned}
\tag{5.20}
$$

where

$$
d\Pi_{\text{LIPS}} \equiv (2\pi)^4 \delta^4(\Sigma p) \times \prod_{\text{final states } j} \frac{d^3 p_j}{(2\pi)^3} \frac{1}{2E_{p_j}}
\tag{5.21}
$$

is called the **Lorentz-invariant phase space** (LIPS). You are encouraged to verify that $d\Pi_{\text{LIPS}}$ is Lorentz invariant in Problem 5.2.

Putting everything together, we have

$$
d\sigma = \frac{1}{(2E_1)(2E_2)|\vec{v}_1 - \vec{v}_2|}|\mathcal{M}|^2 d\Pi_{\text{LIPS}}.
\tag{5.22}
$$

All the factors of V and T have dropped out, so now it is trivial to take $V \to \infty$ and $T \to \infty$. Recall also that velocity is related to momentum by $\vec{v} = \vec{p}/p_0$.

5.1.1 Decay rates

A differential decay rate is the probability that a one-particle state with momentum p_1 turns into a multi-particle state with momenta $\{p_j\}$ over a time T:

$$
d\Gamma = \frac{1}{T}dP.
\tag{5.23}
$$

Of course, it is impossible for the incoming particle to be an asymptotic state at $-\infty$ if it is to decay, and so we should not be able to use the S-matrix to describe decays. The reason this is not a problem is that we calculate the decay rate in perturbation theory assuming the interactions happen only over a finite time T. Thus, a decay is really just like a $1 \to n$ scattering process.

Following the same steps as for the differential cross section, the decay rate can be written as

$$dΓ = \frac{1}{2E_1}|\mathcal{M}|^2 d\Pi_{\text{LIPS}}. \tag{5.24}$$

When a particle is moving at relativistic velocities, its energy is larger than in its rest frame, so it will decay more slowly. This is the effect of time dilation. The rate in a boosted frame can be calculated from the rest-frame decay rate using special relativity.

5.1.2 Special cases

For $2 \to 2$ scattering in the center-of-mass frame

$$p_1 + p_2 \longrightarrow p_3 + p_4, \tag{5.25}$$

with $\vec{p}_1 = -\vec{p}_2$ and $\vec{p}_3 = -\vec{p}_4$ and $E_1 + E_2 = E_3 + E_4 = E_{\text{CM}}$, where E_{CM} is the total energy in the center-of-mass frame. Then

$$d\Pi_{\text{LIPS}} = (2\pi)^4 \delta^4(\Sigma p) \frac{d^3 p_3}{(2\pi)^3} \frac{1}{2E_3} \frac{d^3 p_4}{(2\pi)^3} \frac{1}{2E_4}. \tag{5.26}$$

We can now integrate over $\vec{p}_4$ using the δ-function to give

$$d\Pi_{\text{LIPS}} = \frac{1}{16\pi^2} d\Omega \int dp_f \frac{p_f^2}{E_3} \frac{1}{E_4} \delta(E_3 + E_4 - E_{\text{CM}}), \tag{5.27}$$

where $p_f = |\vec{p}_3| = |\vec{p}_4|$ and $E_3 = \sqrt{m_3^2 + p_f^2}$ and $E_4 = \sqrt{m_4^2 + p_f^2}$. We now change variables from p_f to $x(p_f) = E_3(p_f) + E_4(p_f) - E_{\text{CM}}$. The Jacobian is

$$\frac{dx}{dp_f} = \frac{d}{dp_f}(E_3 + E_4 - E_{\text{CM}}) = \frac{p_f}{E_3} + \frac{p_f}{E_4} = \frac{E_3 + E_4}{E_3 E_4} p_f \tag{5.28}$$

and therefore, using $E_3 + E_4 = E_{\text{CM}}$ because of the δ-function, we get

$$d\Pi_{\text{LIPS}} = \frac{1}{16\pi^2} d\Omega \int_{m_3 + m_4 - E_{\text{CM}}}^{\infty} dx \frac{p_f}{E_{\text{CM}}} \delta(x)$$

$$= \frac{1}{16\pi^2} d\Omega \frac{p_f}{E_{\text{CM}}} \theta(E_{\text{CM}} - m_3 - m_4), \tag{5.29}$$

where θ is the **unit-step function** or **Heaviside function**: $\theta(x) = 1$ if $x > 0$ and 0 otherwise.

Plugging this into Eq. (5.22), we find

$$d\sigma = \frac{1}{(2E_1)(2E_2)|\vec{v}_1 - \vec{v}_2|} \frac{1}{16\pi^2} d\Omega \frac{p_f}{E_{\text{CM}}} |\mathcal{M}|^2 \theta(E_{\text{CM}} - m_3 - m_4). \tag{5.30}$$

After using

$$|\vec{v}_1 - \vec{v}_2| = \left| \frac{|\vec{p}_1|}{E_1} + \frac{|\vec{p}_2|}{E_2} \right| = p_i \frac{E_{\text{CM}}}{E_1 E_2} \tag{5.31}$$

with $p_i = |\vec{p}_1| = |\vec{p}_2|$, we end up with the fairly simple formula

$$\left(\frac{d\sigma}{d\Omega}\right)_{\text{CM}} = \frac{1}{64\pi^2 E_{\text{CM}}^2}\frac{p_f}{p_i}|\mathcal{M}|^2\theta(E_{\text{CM}} - m_3 - m_4), \qquad (5.32)$$

with the CM subscript reminding us that this formula holds only in the center-of-mass frame.

If all the masses are equal then $p_f = p_i$ and this formula simplifies further:

$$\left(\frac{d\sigma}{d\Omega}\right)_{\text{CM}} = \frac{1}{64\pi^2 E_{\text{CM}}^2}|\mathcal{M}|^2 \quad \text{(masses equal)}. \qquad (5.33)$$

5.2 Non-relativistic limit

In the non-relativistic limit, our formula for the cross section should reduce to the usual formula from non-relativistic quantum mechanics. To see this, consider the case where an electron ϕ_e of mass m_e scatters off a proton ϕ_p of mass m_p. From non-relativistic quantum mechanics, the cross section should be given by the Born approximation:

$$\left(\frac{d\sigma}{d\Omega}\right)_{\text{Born}} = \frac{m_e^2}{4\pi^2}|\widetilde{V}(\vec{k})|^2, \qquad (5.34)$$

where the Fourier transform of the potential is given by

$$\widetilde{V}(\vec{k}) = \int d^3x\, e^{-i\vec{k}\vec{x}}V(\vec{x}) \qquad (5.35)$$

and $\vec{k}$ is the difference in the electron momentum before and after scattering, sometimes called the **momentum transfer**. For example, if this is a Coulomb potential, $V(x) = \frac{e^2}{4\pi|\vec{x}|}$, then $\widetilde{V}(\vec{k}) = \frac{e^2}{\vec{k}^2}$ so

$$\left(\frac{d\sigma}{d\Omega}\right)_{\text{Born}} = \frac{m_e^2}{4\pi^2}\left(\frac{e^2}{\vec{k}^2}\right)^2. \qquad (5.36)$$

Let us check the mass dimensions in these formulas (see Appendix A). $[V(x)] = 1$, so $[\widetilde{V}(k)] = -2$ and then $[(\frac{d\sigma}{d\Omega})_{\text{Born}}] = -2$, which is the correct dimension for a cross section.

For the field theory version, the center-of-mass frame is the proton rest frame to a good approximation and $E_{\text{CM}} = m_p$. Also, the scattering is elastic, so $p_f = p_i$. Then, the prediction is

$$\left(\frac{d\sigma}{d\Omega}\right)_{\text{CM}} = \frac{1}{64\pi^2 m_p^2}|\mathcal{M}|^2. \qquad (5.37)$$

What dimension should $\mathcal{M}$ have? Since $\left[\frac{d\sigma}{d\Omega}\right] = -2$ and $\left[m_p^{-2}\right] = -2$, it follows that $\mathcal{M}$ should be dimensionless.

If we ignore spin, we will see in Chapter 9, Eq. (9.11), that the Lagrangian describing the interaction between the electron, proton and photon has the form

$$\mathcal{L} = -\frac{1}{4}F_{\mu\nu}^2 - \phi_e^\star(\Box + m_e^2)\phi_e - \phi_p^\star(\Box + m_p^2)\phi_p$$
$$- ieA_\mu(\phi_e^\star\partial_\mu\phi_e - \phi_e\partial_\mu\phi_e^\star) + ieA_\mu(\phi_p^\star\partial_\mu\phi_p - \phi_p\partial_\mu\phi_p^\star) + \mathcal{O}(e^2), \qquad (5.38)$$

with ϕ_e and ϕ_p representing the electron and proton respectively. (This is the Lagrangian for scalar QED.)

In the non-relativistic limit, the momentum $p^\mu = (E, \vec{p})$ is close to being at rest $(m, 0)$. So, $E \sim m$, that is, $\partial_t \phi \sim im\phi$ and $|\vec{p}| \ll m$. Let us use this to factorize out the leading-order time dependence, $\phi_e \to \phi_e e^{im_e t}$ and $\phi_p \to \phi_p e^{im_p t}$. Then the Lagrangian becomes

$$\mathcal{L} = -\frac{1}{4}F_{\mu\nu}^2 + \phi_e^\star\vec{\nabla}^2\phi_e + 2em_e A_0\phi_e^\star\phi_e + \phi_p^\star\vec{\nabla}^2\phi_p - 2em_p A_0\phi_p^\star\phi_p + \cdots, \qquad (5.39)$$

with $\cdots$ higher order in $\frac{\vec{\nabla}^2}{m^2}$. We have removed all the time dependence, which is appropriate because we are trying to calculate a static potential.

Although we do not know exactly how to calculate the matrix element, by now we are capable of guessing the kinds of ingredients that go into the calculation. The matrix element must have a piece proportional to $-2em_p$ from the interaction between the proton and the photon, a factor of the propagator $\frac{1}{\vec{k}^2}$ from the photon kinetic term, and a piece proportional to $2em_e$ from the photon interacting with the electron. Thus,

$$\mathcal{M} \sim (-2em_p)\frac{1}{\vec{k}^2}(2em_e). \qquad (5.40)$$

Then, from Eq. (5.37),

$$\left(\frac{d\sigma}{d\Omega}\right)_{\text{CM}} = \frac{1}{64\pi^2 m_p^2}|\mathcal{M}|^2 \sim \frac{e^4 m_e^2}{4\pi^2}\frac{1}{\vec{k}^4}, \qquad (5.41)$$

which agrees with Eq. (5.36) and so the non-relativistic limit works. We will perform this calculation again carefully and completely, without asking you to accept anything without proof, once we derive the perturbation expansion and Feynman rules. The answer will be the same.

The factors of m in the interaction terms are unconventional. It is more standard to rescale $\phi \to \frac{1}{\sqrt{2m}}\phi$ so that

$$\mathcal{L} = -\frac{1}{4}F_{\mu\nu}^2 + \frac{1}{2m_e}\phi_e^\star\vec{\nabla}^2\phi_e + \frac{1}{2m_p}\phi_p^\star\vec{\nabla}^2\phi_p + eA_0\phi_e^\star\phi_e - eA_0\phi_p^\star\phi_p, \qquad (5.42)$$

which has the usual $\frac{p^2}{2m}$ for the kinetic term and an interaction with just a charge e and no mass m in the coupling. Of course, the final result is independent of the normalization, but it is still helpful to see how relativistic and non-relativistic normalization conventions are related. Note that, since in the non-relativistic limit $\frac{1}{\sqrt{2m}} = \frac{1}{\sqrt{2E}}$, this rescaling is closely related to the normalization factors $\frac{1}{\sqrt{2\omega_p}}$ we added by convention to the definition of the quantum field.

5.3 $e^+e^- \rightarrow \mu^+\mu^-$ with spin

So far, we have approximated everything, electrons and protons, as being spinless. This is a good first approximation, as the basic $\frac{1}{r}$ form of Coulomb's law does not involve spin – it follows from flux conservation (Gauss's law) or, more simply, from dimensional analysis. In Chapter 10, we will understand the spin of the electron and proton using the Dirac equation and *spinors*. While spinors are an extremely efficient way to encode spin information in a relativistic setting, it is also important to realize that relativistic spin can be understood the same way as for non-relativistic scattering.

In this section we will do a simple example of calculating a matrix element with spin. Consider the process of electron–positron annihilation into muon–antimuon pairs (this process will be considered in more detail in Section 13.3 and Chapter 20). The electron does not interact with the muon directly, only through the electromagnetic force (and the weak force). The leading-order contribution should then come from a process represented by

$$(5.43)$$

This diagram has a precise meaning, as we will see in Chapter 7, but for now just think of it as a pictorial drawing of the process: the e^+e^- annihilate into a virtual photon, which propagates along, then decays into a $\mu^+\mu^-$ pair.

Let us get the dimensional part out of the way first. The propagator we saw in Chapters 3 and 4 (see Eqs. (3.79) and (4.31)) gives $\frac{1}{k^2}$, where $k^\mu = p_1^\mu + p_2^\mu = p_3^\mu + p_4^\mu$ is the off-shell photon momentum. For a scattering process, such as $e^-p^+ \rightarrow e^-p^+$, this propagator $\frac{1}{k^2}$ gives the scattering potential. For this annihilation process, it is much simpler; in the center-of-mass frame $\frac{1}{k^2} = \frac{1}{E_{\rm CM}^2}$, which is constant (if $E_{\rm CM}$ is constant). By dimensional analysis, $\mathcal{M}$ should be dimensionless. The $\frac{1}{E_{\rm CM}^2}$ is in fact canceled by factors of $\sqrt{2E_1} = \sqrt{2E_2} = \sqrt{2E_3} = \sqrt{2E_4} = \sqrt{E_{\rm CM}}$, which come from the (natural, non-relativistic) normalization of the electron and muon states. Thus, all these $E_{\rm CM}$ factors cancel and $\mathcal{M}$ is just a dimensionless number, given by the appropriate spin projections.

So, the only remaining part of $\mathcal{M}$ is given by projections of initial spins onto the intermediate photon polarizations, and then onto final spins. We can write

$$\mathcal{M}(s_1 s_2 \rightarrow s_3 s_4) = \sum_\epsilon \langle s_3 s_4 | \epsilon \rangle \langle \epsilon | s_1 s_2 \rangle, \qquad (5.44)$$

where s_1 and s_2 are the spins of the incoming states, s_3 and s_4 the spins of the outgoing states, and ϵ is the polarization of the intermediate photon.

Let us now try to guess the form of these spin projections by using angular momentum. This is easiest to do in the center-of-mass frame. At ultra-relativistic energies, we can neglect the electron and muon masses. Then, with some choice of direction $\hat{z}$, the incoming e^- and e^+ momenta are

$$p_1^\mu = (E, 0, 0, E), \qquad p_2^\mu = (E, 0, 0, -E). \tag{5.45}$$

Next, we will use that the electron is spin $\frac{1}{2}$. In the non-relativistic limit, we usually think of the spin states as being up and down. In the relativistic limit, it is better to think of the electron as being polarized, just like a photon. Polarizations for spin-$\frac{1}{2}$ particles are usually called helicities (helicity will be defined precisely in Section 11.1). We can use either a basis of circular polarization (called left and right helicity) or a basis of linear polarizations. Linearly polarized electrons are like linearly polarized light, and the polarizations must be transverse to the direction of motion. So the electron moving in the z direction can either be polarized in the x direction or in the y direction. So there are four possible initial states:

$$|s_1 s_2\rangle = |\leftrightarrow\leftrightarrow\rangle, \quad |s_1 s_2\rangle = |\updownarrow\updownarrow\rangle, \quad |s_1 s_2\rangle = |\updownarrow\leftrightarrow\rangle, \quad |s_1 s_2\rangle = |\leftrightarrow\updownarrow\rangle, \tag{5.46}$$

Next, we use that the photon has spin 1 and two polarizations (this will be derived in Chapter 8). To get spin 1 from spin $\frac{1}{2}$ and spin $\frac{1}{2}$, the electron and positron have to be polarized in the same direction. Thus, only the first two initial states could possibly annihilate into a photon. Since the electron polarization is perpendicular to its momentum, the photon polarization will be in either the x or y direction as well. The two possible resulting photon polarizations are

$$\epsilon^1 = (0, 1, 0, 0), \quad \epsilon^2 = (0, 0, 1, 0). \tag{5.47}$$

Both of these polarizations are produced by the $e^+ e^-$ annihilation: $|\leftrightarrow\leftrightarrow\rangle$ produces ϵ^1 and $|\updownarrow\updownarrow\rangle$ produces ϵ^2.

Next, the muon and antimuon are also spin $\frac{1}{2}$ so the final state has four possible spin states too. Similarly, only two of them can have non-zero overlap with the spin-1 photon. However, the μ^+ and μ^- are not going in the z direction. Their momenta can be written as

$$p_3^\mu = E(1, 0, \sin\theta, \cos\theta), \quad p_4^\mu = E(1, 0, -\sin\theta, -\cos\theta), \tag{5.48}$$

where θ is the angle to the $e^+ e^-$ axis. There is also an azimuthal angle ϕ about the z axis, which we have set to 0 since the problem has cylindrical symmetry. So in this case the two possible directions for the photon polarization are

$$\bar{\epsilon}^1 = (0, 1, 0, 0), \quad \bar{\epsilon}^2 = (0, 0, \cos\theta, -\sin\theta). \tag{5.49}$$

You can check that these are orthogonal to p_3^μ and p_4^μ.

The matrix element is given by summing over all possible intermediate photon polarizations for given electron and muon polarizations. There are only two non-vanishing possibilities:

$$\mathcal{M}_1 = \mathcal{M}\Big(|\leftrightarrow\leftrightarrow\rangle \to |\epsilon^1\rangle \to |\bar{\epsilon}_1\rangle \to |f\rangle\Big) = \epsilon^1 \bar{\epsilon}_1 = -1, \tag{5.50}$$

$$\mathcal{M}_2 = \mathcal{M}\Big(|\updownarrow\updownarrow\rangle \to |\epsilon^2\rangle \to |\bar{\epsilon}_2\rangle \to |f\rangle\Big) = \epsilon^2 \bar{\epsilon}_2 = -\cos\theta. \tag{5.51}$$

Then,

$$\sum_{\text{states}} |\mathcal{M}|^2 = |\mathcal{M}_1|^2 + |\mathcal{M}_2|^2 = 1 + \cos^2\theta, \tag{5.52}$$

and so

$$\frac{d\sigma}{d\Omega} = \frac{e^4}{64\pi^2 E_{\text{CM}}^2} \left(1 + \cos^2\theta\right). \tag{5.53}$$

This is the correct cross section for $e^+e^- \to \mu^+\mu^-$. We will re-derive this using the full machinery of quantum field theory: spinors, Feynman rules, etc., in Section 13.3.[3]

Problems

5.1 Show that the differential cross section for $2 \to 2$ scattering with $p_i^\mu + p_A^\mu \to p_f^\mu + p_B^\mu$ in the rest frame of particle A can be written as

$$\frac{d\sigma}{d\Omega} = \frac{1}{64\pi^2 m_A} \left[E_B + E_f \left(1 - \frac{|\vec{p}_i|}{|\vec{p}_f|}\cos\theta\right) \right]^{-1} \frac{|\vec{p}_f|}{|\vec{p}_i|} |\mathcal{M}|^2, \tag{5.54}$$

where θ is the angle between $\vec{p}_i$ and $\vec{p}_f$, $E_B = \sqrt{(\vec{p}_f - \vec{p}_i)^2 + m_B^2}$ and $E_f = \sqrt{\vec{p}_f^2 + m_f^2}$.

5.2 Show that $d\Pi_{\text{LIPS}}$ is Lorentz invariant and verify Eq. (5.21).

5.3 A muon decays to an electron, an electron anti-neutrino and a muon neutrino, $\mu^- \to e^- \nu_\mu \bar{\nu}_e$. The matrix element for this process, ignoring the electron and neutrino masses, is given by $|\mathcal{M}|^2 = 32G_F^2(m^2 - 2mE)mE$, where m is the mass of the muon and E is the energy of the outgoing ν_e. $G_F = 1.166 \times 10^{-5}\,\text{GeV}^{-2}$ is the Fermi constant.

(a) Perform the integral over $d\Pi_{\text{LIPS}}$ to show that the decay rate is

$$\Gamma = \frac{G_F^2 m^5}{192\pi^3}. \tag{5.55}$$

(b) Compare your result to the observed values $m = 106\,\text{MeV}$ and $\tau = \Gamma^{-1} = 2.20\,\mu\text{s}$. How big is the discrepancy as a percentage? What might account for the discrepancy?

5.4 Repeat the $e^+e^- \to \mu^+\mu^-$ calculation in Section 5.3 using circular polarizations.

5.5 One of the most important scattering experiments ever was Rutherford's gold foil experiment. Rutherford scattering is $\alpha N \to \alpha N$, where N is some atomic nucleus and α is an α-particle (helium nucleus). It is an almost identical process to Coulomb scattering ($e^- p^+ \to e^- p^+$).

(a) Look up or calculate the *classical* Rutherford scattering cross section. What assumptions go into its derivation?

(b) We showed that the quantum mechanical cross section for Coulomb scattering in Eq. (5.41) follows either from the Born approximation or from quantum field theory. Start from the formula for Coulomb scattering and make the appropriate replacements for αN scattering.

---

[3] We were a little sloppy with factors of 2. We should add a factor of $\frac{1}{4}$ for averaging over initial states, but this cancels a factor of 4 from the relativistic normalization of the linearly-polarized states. All the factors of 2 are handled carefully in Chapter 13.

(c) Draw the Feynman diagram for Rutherford scattering. What is the momentum of the virtual photon, k_μ, in terms of the scattering angle and the energy of the incoming α-particle?

(d) Substitute in for k^4 and rewrite the cross section in terms of the kinetic energy of the α-particle. Show that Rutherford's classical formula is reproduced.

(e) Why are the classical and quantum answers the same? Could you have known this ahead of time?

(f) Would the cross section for $e^- e^- \to e^- e^-$ also be given by the Coulomb scattering cross section?

5.6 In Section 5.3 we found that the $e^+ e^- \to \mu^+ \mu^-$ cross section had the form $\frac{d\sigma}{d\Omega} = \frac{e^4}{64\pi^2 E_{\mathrm{CM}}^2}(1 + \cos^2\theta)$ in the center-of-mass frame.

(a) Work out the Lorentz-invariant quantities $s = (p_{e^+} + p_{e^-})^2$, $t = (p_{\mu^-} - p_{e^-})^2$ and $u = (p_{\mu^+} - p_{e^-})^2$ in terms of E_{CM} and $\cos\theta$ (still assuming $m_\mu = m_e = 0$).

(b) Derive a relationship between s, t and u.

(c) Rewrite $\frac{d\sigma}{d\Omega}$ in terms of s, t and u.

(d) Now assume m_μ and m_e are non-zero. Derive a relationship between s, t and u and the masses.

The *S*-matrix and time-ordered products

As discussed in Chapter 5, scattering experiments have been a fruitful and efficient way to determine the particles that exist in nature and how they interact. In a typical collider experiment, two particles, generally in approximate momentum eigenstates at $t = -\infty$, are collided with each other and we measure the probability of finding particular outgoing momentum eigenstates at $t = +\infty$. All of the interesting interacting physics is encoded in how often given initial states produce given final states, that is, in the *S*-matrix.

The working assumption in scattering calculations is that all of the interactions happen in some finite time $-T < t < T$. This is certainly true in real collider scattering experiments. But more importantly, it lets us make the problem well defined; if there were always interactions, it would not be possible to set up our initial states at $t = -\infty$ or find the desired final states at $t = +\infty$. Without interactions at asymptotic times, the states we scatter can be defined as on-shell one-particle states of given momenta, known as **asymptotic states**. In this chapter, we derive an expression for the *S*-matrix using only that the system is free at asymptotic times. In Chapter 7 we will work out the Feynman rules, which make it easy to perform a perturbation expansion for the interacting theory.

The main result of this chapter is a derivation of the **LSZ (Lehmann–Symanzik–Zimmermann) reduction formula**, which relates *S*-matrix elements $\langle f|S|i \rangle$ for n asymptotic momentum eigenstates to an expression involving the quantum fields $\phi(x)$:

$$\langle f|S|i \rangle = \left[i \int d^4x_1\, e^{-ip_1x_1}(\Box_1 + m^2) \right] \cdots \left[i \int d^4x_n\, e^{ip_nx_n}(\Box_n + m^2) \right]$$
$$\times \langle \Omega|T\{\phi(x_1)\,\phi(x_2)\,\phi(x_3)\cdots\phi(x_n)\}|\Omega \rangle\,, \quad (6.1)$$

with the $-i$ in the exponent applying for initial states and the $+i$ for final states, and $\Box_j \equiv g^{\mu\nu}\frac{\partial}{\partial x_j^\mu}\frac{\partial}{\partial x_j^\nu}$. In this formula, $T\{\cdots\}$ refers to a time-ordered product, to be defined below, and $|\Omega \rangle$ is the ground state or vacuum of the interacting theory, which in general may be different from the vacuum in a free theory.

The time-ordered correlation function in this formula can be very complicated and encodes a tremendous amount of information besides *S*-matrix elements. The factors of $\Box + m^2$ project onto the *S*-matrix: $\Box + m^2$ becomes $-p^2 + m^2$ in Fourier space, which vanishes for the asymptotic states. These factors will therefore remove all terms in the time-ordered product except those with poles of the form $\frac{1}{p^2-m^2}$, corresponding to propagators of on-shell particles. Only the terms with poles for each factor of $p^2 - m^2$ will survive, and the S-matrix is given by the residue of these poles. Thus, the physical content of the LSZ formula is that the S-matrix projects out one-particle asymptotic states from the time-ordered product of fields.

6.1 The LSZ reduction formula

In Chapter 5, we derived a formula for the differential cross section for $2 \to n$ scattering of asymptotic states, Eq. (5.22):

$$d\sigma = \frac{1}{(2E_1)(2E_2)|\vec{v}_1 - \vec{v}_2|} |\mathcal{M}|^2 d\Pi_{\text{LIPS}}, \qquad (6.2)$$

where $d\Pi_{\text{LIPS}}$ is the Lorentz-invariant phase space, and $\mathcal{M}$, which is shorthand for $\langle f | \mathcal{M} | i \rangle$, is the S-matrix element with an overall momentum-conserving δ-function factored out:

$$\langle f | S - \mathbb{1} | i \rangle = i(2\pi)^4 \delta^4(\Sigma p) \mathcal{M}. \qquad (6.3)$$

The state $|i\rangle$ is the initial state at $t = -\infty$, and $\langle f|$ is the final state at $t = +\infty$. More precisely, using the operators $a_p^\dagger(t)$, which create particles with momentum p at time t, these states are

$$|i\rangle = \sqrt{2\omega_1} \sqrt{2\omega_2} \; a_{p_1}^\dagger(-\infty) \, a_{p_2}^\dagger(-\infty) |\Omega\rangle, \qquad (6.4)$$

where $|\Omega\rangle$ is the ground state, with no particles, and

$$|f\rangle = \sqrt{2\omega_3} \cdots \sqrt{2\omega_n} \, a_{p_3}^\dagger(\infty) \cdots a_{p_n}^\dagger(\infty) |\Omega\rangle. \qquad (6.5)$$

We are generally interested in the case where some scattering actually happens, so let us assume $|f\rangle \neq |i\rangle$, in which case the $\mathbb{1}$ does not contribute. Then the S-matrix is

$$\langle f | S | i \rangle = 2^{n/2} \sqrt{\omega_1 \omega_2 \omega_3 \cdots \omega_n} \langle \Omega | a_{p_3}(\infty) \cdots a_{p_n}(\infty) \, a_{p_1}^\dagger(-\infty) \, a_{p_2}^\dagger(-\infty) |\Omega\rangle. \qquad (6.6)$$

This expression is not terribly useful as is. We would like to relate it to something we can compute with our Lorentz-invariant quantum fields $\phi(x)$.

Recall that we defined the fields as a sum over creation and annihilation operators:

$$\phi(x) = \phi(\vec{x}, t) = \int \frac{d^3p}{(2\pi)^3} \frac{1}{\sqrt{2\omega_p}} \left[a_p(t) e^{-ipx} + a_p^\dagger(t) e^{ipx} \right], \qquad (6.7)$$

where $\omega_p = \sqrt{\vec{p}^2 + m^2}$. We also start to use the notation $\phi(x) = \phi(\vec{x}, t)$ as well, for simplicity. These are Heisenberg picture operators which create states at some particular time. However, the creation and annihilation operators at time t are in general different from those at some other time t'. An interacting Hamiltonian will rotate the basis of creation and annihilation operators, which encodes all the interesting dynamics. For example, if H is time independent, $a_p(t) = e^{iH(t-t_0)} a_p(t_0) e^{-iH(t-t_0)}$, just as $\phi(x) = e^{iH(t-t_0)} \phi(\vec{x}, t_0) e^{-iH(t-t_0)}$, where t_0 is some arbitrary reference time where we have matched the interacting fields onto the free fields.

The key to proving LSZ is the algebraic relation

$$i \int d^4x\, e^{ipx}(\Box + m^2)\phi(x) = \sqrt{2\omega_p}[a_p(\infty) - a_p(-\infty)], \qquad (6.8)$$

where $p^\mu = (\omega_p, \vec{p})$. To derive this, we only need to assume that all the interesting dynamics happens in some finite time interval, so that the theory is free at $t \to \pm\infty$.[1]

To prove Eq. (6.8), we will obviously have to be careful about boundary conditions at $t = \pm\infty$. However, we can safely assume that the fields die off at $\vec{x} = \pm\infty$, allowing us to integrate by parts in $\vec{x}$. Then,

$$i \int d^4x\, e^{ipx}(\Box + m^2)\phi(x) = i \int d^4x\, e^{ipx}(\partial_t^2 - \vec{\partial}_x^2 + m^2)\phi(x)$$

$$= i \int d^4x\, e^{ipx}(\partial_t^2 + \vec{p}^2 + m^2)\phi(x)$$

$$= i \int d^4x\, e^{ipx}(\partial_t^2 + \omega_p^2)\phi(x). \qquad (6.9)$$

Note this is true for any kind of $\phi(x)$, whether classical field or operator. Also,

$$\partial_t\left[e^{ipx}(i\partial_t + \omega_p)\phi(x)\right] = \left[i\omega_p e^{ipx}(i\partial_t + \omega_p) + e^{ipx}(i\partial_t^2 + \omega_p\partial_t)\right]\phi(x)$$
$$= ie^{ipx}(\partial_t^2 + \omega_p^2)\phi(x), \qquad (6.10)$$

which holds independently of boundary conditions. So,

$$i \int d^4x\, e^{ipx}(\Box + m^2)\phi(x) = \int d^4x\, \partial_t\left[e^{ipx}(i\partial_t + \omega_p)\phi(x)\right]$$

$$= \int dt\, \partial_t\left[e^{i\omega_p t}\int d^3x\, e^{-i\vec{p}\vec{x}}(i\partial_t + \omega_p)\phi(x)\right]. \qquad (6.11)$$

Again, this is true for whatever kind of crazy interacting field $\phi(x)$ might be.

This integrand is a total derivative in time, so it only depends on the fields at the boundary $t = \pm\infty$. By assumption, our $a_p(t)$ and $a_p^\dagger(t)$ operators are time independent at late and early times. For the particular case of $\phi(x)$ being a quantum field, Eq. (6.7), we can do the $\vec{x}$ integral:

$$\int d^3x\, e^{-i\vec{p}\vec{x}}(i\partial_t + \omega_p)\phi(x)$$

$$= \int d^3x\, e^{-i\vec{p}\vec{x}}(i\partial_t + \omega_p)\int \frac{d^3k}{(2\pi)^3}\frac{1}{\sqrt{2\omega_k}}\left(a_k(t)e^{-ikx} + a_k^\dagger(t)e^{ikx}\right)$$

$$= \int \frac{d^3k}{(2\pi)^3}\int d^3x\left[\left(\frac{\omega_k + \omega_p}{\sqrt{2\omega_k}}\right)a_k(t)e^{-ikx}e^{-i\vec{p}\vec{x}} + \left(\frac{-\omega_k + \omega_p}{\sqrt{2\omega_k}}\right)a_k^\dagger(t)e^{ikx}e^{-i\vec{p}\vec{x}}\right]. \qquad (6.12)$$

Here we used $\partial_t a_k(t) = 0$, which is not true in general, but true at $t = \pm\infty$ where the fields are free, which is the only region relevant to Eq. (6.11). The $\vec{x}$ integral gives a $\delta^3(\vec{p} - \vec{k})$

[1] To achieve this, in the rigorous proof of the LSZ reduction theorem, the asymptotic states are constructed as the limit of wavepackets with exponentially small overlap at asymptotic times. Unfortunately, the rigorous proof fails if there are massless particles in the theory, as in QED. We choose therefore simply to impose the physical assumption that particles are approximately free at asymptotic times, which seems to hold even for QED.

in the first term and a $\delta^3(\vec{p} + \vec{k})$ in the second term. Either way, it forces $\omega_k = \omega_p$ and so we get

$$\int d^3x \, e^{-i\vec{p}\vec{x}}(i\partial_t + \omega_p)\phi(x) = \sqrt{2\omega_p}a_p(t)e^{-i\omega_p t}. \qquad (6.13)$$

Thus,

$$i\int d^4x \, e^{ipx}(\Box + m^2)\phi(x) = \int dt \, \partial_t[(e^{i\omega_p t})\left(\sqrt{2\omega_p}a_p(t)e^{-i\omega_p t}\right)]$$

$$= \sqrt{2\omega_p}\left[a_p(\infty) - a_p(-\infty)\right], \qquad (6.14)$$

which is what we wanted. Similarly (by taking the Hermitian conjugate),

$$\sqrt{2\omega_p}[a_p^\dagger(\infty) - a_p^\dagger(-\infty)] = -i\int d^4x \, e^{-ipx}(\Box + m^2)\phi(x). \qquad (6.15)$$

Now we are almost done. We wanted to compute

$$\langle f|S|i\rangle = \sqrt{2^n\omega_1\cdots\omega_n}\langle\Omega|a_{p_3}(\infty)\cdots a_{p_n}(\infty)\, a_{p_1}^\dagger(-\infty)a_{p_2}^\dagger(-\infty)|\Omega\rangle \qquad (6.16)$$

and we have an expression for $a_p(\infty) - a_p(-\infty)$. Note that all the initial states have $a_p^\dagger$ operators and $-\infty$, and the final states have a_p operators and $+\infty$, so the operators are already in time order:

$$\langle f|S|i\rangle = \sqrt{2^n\omega_1\cdots\omega_n}\langle\Omega|T\{a_{p_3}(\infty)\cdots a_{p_n}(\infty)\, a_{p_1}^\dagger(-\infty)a_{p_2}^\dagger(-\infty)\}|\Omega\rangle, \qquad (6.17)$$

where the **time-ordering operation** $T\{\cdots\}$ indicates that all the operators should be ordered so that those at later times are always to the left of those at earlier times. That is, $T\{\cdots\}$ just manhandles the operators within the brackets, placing them in order regardless of whether they commute or not.

Time ordering lets us write the S-matrix element as

$$\langle f|S|i\rangle = \sqrt{2^n\omega_1\cdots\omega_n}\langle\Omega|T\{[a_{p_3}(\infty) - a_{p_3}(-\infty)]\cdots[a_{p_n}(\infty) - a_{p_n}(-\infty)]$$

$$\times [a_{p_1}^\dagger(-\infty) - a_{p_1}^\dagger(\infty)][a_{p_2}^\dagger(-\infty) - a_{p_2}^\dagger(\infty)]\}|\Omega\rangle. \qquad (6.18)$$

The time ordering migrates all the unwanted $a^\dagger(\infty)$ operators associated with the initial states to the left, where they annihilate $\langle\Omega|$, and all the unwanted $a(-\infty)$ operators to the right, where they annihilate $|\Omega\rangle$. Then there is no ambiguity in commuting the $a_{p_i}^\dagger(\infty)$ past the $a_{p_j}(\infty)$ and everything we do not want drops out of this expression.[2]

The result is then

$$\langle p_3\cdots p_n\,|S|\,p_1 p_2\rangle = \left[i\int d^4x_1 e^{-ip_1 x_1}(\Box_1 + m^2)\right]\cdots\left[i\int d^4x_n e^{ip_n x_n}(\Box_n + m^2)\right]$$

$$\times \langle\Omega|T\{\phi(x_1)\phi(x_2)\phi(x_3)\cdots\phi(x_n)\}|\Omega\rangle, \qquad (6.19)$$

where $\Box_i = (\frac{\partial}{\partial x_i^\mu})^2$, which agrees with Eq. (6.1).[3] This is the LSZ reduction formula.

---

[2] The only subtlety is when some momenta are identical. We will define the S-matrix elements for such configurations by analytic continuation of the case when all momenta are different.

[3] Pulling the $\Box$ factors through the time-ordering operator is technically not allowed. However, as we will see in the next chapter, the effect of doing this is to introduce contact terms that do not contribute to the connected part of the S-matrix. See also [Itzykson and Zuber, 1980] for a more detailed treatment.

6.1.1 Discussion

The LSZ reduction says that to calculate an S-matrix element, multiply the time-ordered product of fields by some $\Box + m^2$ factors and Fourier transform. If the fields $\phi(x)$ were free fields, they would satisfy $(\Box + m^2)\phi(x,t) = 0$ and so the $(\Box_i + m^2)$ terms would give zero. However, as we will see, when calculating amplitudes, there will be factors of propagators $\frac{1}{\Box+m^2}$ for the one-particle states. These blow up as $(\Box + m^2) \to 0$. The LSZ formula guarantees that the zeros and infinities in these terms cancel, leaving a finite non-zero result. Moreover, the $\Box + m^2$ terms will kill anything that does not have a divergence, which will be anything *but* the exact initial and final state we want.[4] That is the whole point of the LSZ formula: it isolates the asymptotic states by adding a carefully constructed zero to cancel everything that does not correspond to the state we want.

It is easy to think that LSZ is totally trivial, but it is not. The projections are the only thing that tells us what the initial states are (the things created from the vacuum at $t = -\infty$) and what the final states are (the things that annihilate into the vacuum at $t = +\infty$). Initial and final states are distinguished by the $\pm i$ in the phase factors. The time ordering is totally physical: all the creation of the initial states happens before the annihilation of the final states. In fact, because this is true not just for free fields, all the crazy stuff that happens at intermediate times in an interacting theory must be time-ordered too. But the great thing is that we do not need to know which are the initial states and which are the final states anymore when we do the hard part of the computation. We just have to calculate time-ordered products, and the LSZ formula sorts out what is being scattered for us.

6.1.2 LSZ for operators

For perturbation theory in the Standard Model, which is mostly what we will study in this book, the LSZ formula in the above form is all that is needed. However, the LSZ formula is more powerful than it seems and applies even if we do not know what the particles are.

If you go back through the derivation, you will see that we never needed an explicit form for the full field $\phi(x)$ and its creation operators $a_p^\dagger(t)$, which did not necessarily evolve like creation operators in the free theory. In fact, all we used was that the field $\phi(x)$ creates free particle states at asymptotic times. So the LSZ reduction actually implies

$$\langle p_3 \cdots p_n | S | p_1 p_2 \rangle = \left[i \int d^4 x_1 e^{-ip_1 x_1} (\Box_1 + m^2) \right] \cdots \left[i \int d^4 x_n e^{ip_n x_n} (\Box_n + m^2) \right]$$
$$\times \langle \Omega | T\{\mathcal{O}_1(x_1)\mathcal{O}_2(x_2)\mathcal{O}_3(x_3) \cdots \mathcal{O}_n(x_n)\} | \Omega \rangle , \tag{6.20}$$

where the $\mathcal{O}_i(x)$ are any operators that can create one-particle states. By this we mean that

$$\langle p | \mathcal{O}(x) | \Omega \rangle = W e^{ipx} \tag{6.21}$$

---

[4] It should not be obvious at this point that there cannot be higher-order poles, such as $\frac{1}{(p^2-m^2)^2}$, coming out of time-ordered products. Such terms would signal the appearance of unphysical states known as ghosts, which violate unitarity. The fastest a correlation function can decay at large p^2 in a unitary theory is as p^{-2}, a result we will prove in Section 24.2.

for some number W. LSZ does not distinguish **elementary** particles, which we define to mean particles that have corresponding fields appearing in the Lagrangian, from any other type of particle. Anything that overlaps with one-particle states will produce an appropriate pole to be canceled by the $\Box + m^2$ factors giving a non-zero S-matrix element. Therefore, particles in the Hilbert space can be produced whether or not we have elementary fields for them.

It is probably worth saying a little more about what these operators $\mathcal{O}_n(x)$ are. The operators can be defined as they would be in quantum mechanics, by their matrix elements in a basis of states ψ_n of the theory $C_{nm} = \langle \psi_n | \mathcal{O} | \psi_m \rangle$. Any such operator can be written as a sum over creation and annihilation operators:

$$\mathcal{O} = \sum_{n,m} \int dq_1 \cdots dq_n dp_1 \cdots dp_m a_{q_1}^\dagger \cdots a_{q_n}^\dagger a_{p_m} \cdots a_{p_1} C_{nm}(q_1, \ldots, p_m). \quad (6.22)$$

It is not hard to prove that the C_{nm} are in one-to-one correspondence with the matrix elements of $\mathcal{O}$ in n and m particle states (see Problem 6.3). One can turn the operator into a functional of fields, using Eq. (6.13) and its conjugate. The most important operators in relativistic quantum field theory are Lorentz-covariant composite operators constructed out of elementary fields, e.g. $\mathcal{O}(x) = \phi(x)\partial_\mu \phi(x)\partial_\mu \phi(x)$. However, some operators, such as the Hamiltonian, are not Lorentz invariant. Other operators, such as Wilson lines (see Section 25.2), are non-local. Also, non-Lorentz-invariant operators are essential for many condensed-matter applications.

As an example of this generalized form of LSZ, suppose there is a bound state in our theory, such as positronium. We will derive the Lagrangian for quantum electrodynamics in Chapter 13. We will find, as you might imagine, that it is a functional of only the electron, positron and photon fields. Positronium is a composite state, composed of an electron, a positron and lots of photons binding them together. It has the same quantum numbers as the operator $\mathcal{O}_P(x) = \bar{\psi}_e(x)\psi_e(x)$, where $\bar{\psi}(x)$ and $\psi(x)$ are the fields for the positron and electron. Thus, $\mathcal{O}_P(x)$ should have some non-zero overlap with positronium, and we can insert it into the time-ordered product to calculate the S-matrix for positronium scattering or production. This is an important conceptual fact: *there do not have to be fundamental fields associated with every asymptotic state in the theory to calculate the S-matrix.*

Conversely, even if we do not know what the elementary particles actually are in the theory, we can introduce fields corresponding to them in the Lagrangian to calculate S-matrix elements in perturbation theory. For example, in studying the proton or other nucleons, we can treat them as elementary particles. As long as we are interested in questions that do not probe the substructure of the proton, such as non-relativistic scattering, nothing will go wrong. This is a general and very useful technique, known generally as **effective field theory**, which will play an important role in this book. Thus, quantum field theory is very flexible: it works if you use fields that do not correspond to elementary particles (effective field theories) or if you scatter particles that do not have corresponding fields (such as bound states). It even can provide a predictive description of unstable composite particles, such as the neutron, which neither have elementary fields nor are proper asymptotic states.

6.2 The Feynman propagator

To recap, our immediate goal, as motived in Chapter 5, is to calculate cross sections, which are determined by S-matrix elements. We now have an expression, the LSZ reduction formula, for S-matrix elements in terms of time-ordered products of fields. Next, we need to figure out how to compute those time-ordered products. As an example, we will now calculate a time-ordered product in the free theory. In Chapter 7, we will derive a method for computing time-ordered products in interacting theories using perturbation theory.

We start with the free-field operator:

$$\phi_0(x,t) = \int \frac{d^3k}{(2\pi)^3} \frac{1}{\sqrt{2\omega_k}} \left(a_k e^{-ikx} + a_k^\dagger e^{ikx}\right), \tag{6.23}$$

where $k_0 = \omega_k = \sqrt{m^2 + \vec{k}^2}$ and a_k and $a_k^\dagger$ are time independent (all time dependence is in the phase). Then, using $|0\rangle$ instead of $|\Omega\rangle$ to denote the vacuum in the free theory,

$$\langle 0|\phi_0(x_1)\phi_0(x_2)|0\rangle = \int \frac{d^3k_1}{(2\pi)^3} \int \frac{d^3k_2}{(2\pi)^3} \frac{1}{\sqrt{2\omega_{k_1}}} \frac{1}{\sqrt{2\omega_{k_2}}} \langle 0|a_{k_1} a_{k_2}^\dagger|0\rangle e^{i(k_2 x_2 - k_1 x_1)}. \tag{6.24}$$

The $\langle 0|a_{k_1} a_{k_2}^\dagger|0\rangle$ gives $(2\pi)^3 \delta^3(\vec{k}_1 - \vec{k}_2)$ so that

$$\langle 0|\phi_0(x_1)\phi_0(x_2)|0\rangle = \int \frac{d^3k}{(2\pi)^3} \frac{1}{2\omega_k} e^{ik(x_2 - x_1)}. \tag{6.25}$$

Now, we are interested in $\langle 0|T\{\phi(x_1)\phi(x_2)\}|0\rangle$. Recalling that time ordering puts the later field on the left, we get

$$\begin{aligned}
\langle 0|T\{\phi_0(x_1)\phi_0(x_2)\}|0\rangle &= \langle 0|\phi_0(x_1)\phi_0(x_2)|0\rangle\, \theta(t_1 - t_2) + \langle 0|\phi_0(x_2)\phi_0(x_1)|0\rangle\, \theta(t_2 - t_1) \\
&= \int \frac{d^3k}{(2\pi)^3} \frac{1}{2\omega_k} \left[e^{ik(x_2-x_1)}\theta(t_1 - t_2) + e^{ik(x_1-x_2)}\theta(t_2 - t_1) \right] \\
&= \int \frac{d^3k}{(2\pi)^3} \frac{1}{2\omega_k} \left[e^{i\vec{k}(\vec{x}_1 - \vec{x}_2)} e^{-i\omega_k \tau}\theta(\tau) + e^{-i\vec{k}(\vec{x}_1 - \vec{x}_2)} e^{i\omega_k \tau}\theta(-\tau) \right],
\end{aligned} \tag{6.26}$$

where $\tau = t_1 - t_2$. Taking $k \to -k$ in the first term leaves the volume integral $\int d^3k$ invariant and gives

$$\langle 0|T\{\phi_0(x_1)\phi_0(x_2)\}|0\rangle = \int \frac{d^3k}{(2\pi)^3} \frac{1}{2\omega_k} e^{-i\vec{k}(\vec{x}_1 - \vec{x}_2)} \left[e^{i\omega_k \tau}\theta(-\tau) + e^{-i\omega_k \tau}\theta(\tau) \right]. \tag{6.27}$$

The two terms in this sum are the advanced and retarded propagators that we saw were relevant in relativistic calculations using old-fashioned perturbation theory.

The next step is to simplify the right-hand side using the mathematical identity

$$e^{-i\omega_k \tau}\theta(\tau) + e^{i\omega_k \tau}\theta(-\tau) = \lim_{\varepsilon \to 0} \frac{-2\omega_k}{2\pi i} \int_{-\infty}^{\infty} \frac{d\omega}{\omega^2 - \omega_k^2 + i\varepsilon} e^{i\omega \tau}. \tag{6.28}$$

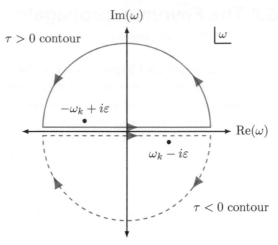

Fig. 6.1 Contour integral for the Feynman propagator. Poles are at $\omega = \pm\omega_k \mp i\varepsilon$. For $\tau > 0$ we close the contour upward, picking up the left pole, for $\tau < 0$ we close the contour downward, picking up the right pole.

To derive this identity, first separate out the poles with partial fractions:

$$
\frac{1}{\omega^2 - \omega_k^2 + i\varepsilon} = \frac{1}{[\omega - (\omega_k - i\varepsilon)]\,[\omega - (-\omega_k + i\varepsilon)]}
$$

$$
= \frac{1}{2\omega_k}\left[\frac{1}{\omega - (\omega_k - i\varepsilon)} - \frac{1}{\omega - (-\omega_k + i\varepsilon)}\right]. \tag{6.29}
$$

Here, we dropped terms of order ε^2 and wrote $2\varepsilon\omega_k = \varepsilon$, which is fine since we will take $\varepsilon \to 0$ in the end. The location of the two poles in the complex plane is shown in Figure 6.1.

Including an $e^{i\omega\tau}$ factor, as on the right-hand side of Eq. (6.28), we can integrate from $-\infty < \omega < \infty$ by closing the contour upwards when $\tau > 0$ and downwards when $\tau < 0$. The first fraction in Eq. (6.29) then picks up the pole only if $\tau < 0$, giving 0 otherwise. That is,

$$
\int_{-\infty}^{\infty} \frac{d\omega}{\omega - (\omega_k - i\varepsilon)} e^{i\omega\tau} = -2\pi i e^{i\omega_k\tau}\theta(-\tau) + O(\varepsilon), \tag{6.30}
$$

with the extra minus sign coming from doing the contour integration clockwise. For the second fraction,

$$
\int_{-\infty}^{\infty} \frac{d\omega}{\omega - (-\omega_k + i\varepsilon)} e^{i\omega\tau} = 2\pi i e^{-i\omega_k\tau}\theta(\tau) + O(\varepsilon). \tag{6.31}
$$

Thus,

$$
\lim_{\varepsilon \to 0}\int_{-\infty}^{\infty} \frac{d\omega}{\omega^2 - \omega_k^2 + i\varepsilon} e^{i\omega\tau} = -\frac{2\pi i}{2\omega_k}\left[e^{i\omega_k\tau}\theta(-\tau) + e^{-i\omega_k\tau}\theta(\tau)\right] \tag{6.32}
$$

as desired.

Putting it together, we find

$$
\langle 0|T\{\phi_0(x_1)\phi_0(x_2)\}|0\rangle = \lim_{\varepsilon \to 0}\int \frac{d^3k}{(2\pi)^3}\frac{1}{2\omega_k}e^{-i\vec{k}(\vec{x}_1 - \vec{x}_2)}\int d\omega \frac{-2\omega_k}{2\pi i}\frac{1}{\omega^2 - \omega_k^2 + i\varepsilon}e^{i\omega\tau}. \tag{6.33}
$$

Letting the limit be implicit, this is

$$D_F(x_1, x_2) = \langle 0|T\{\phi_0(x_1)\phi_0(x_2)\}|0\rangle = \int \frac{d^4k}{(2\pi)^4} \frac{i}{k^2 - m^2 + i\varepsilon} e^{ik(x_1 - x_2)}. \quad (6.34)$$

This beautiful Lorentz-invariant object is called the **Feynman propagator**. It has a pole at $k^2 = m^2$, exactly to be canceled by prefactors in the LSZ reduction formula in the projection onto one-particle states.

Points to keep in mind:

- $k_0 \neq \sqrt{\vec{k}^2 + m^2}$ anymore. It is a separate integration variable. The propagating field can be off-shell!
- The i comes from a contour integral. We will always get factors of i in 2-point functions of real fields.
- The ε is just a trick for representing the time ordering in a simple way. We will always take $\varepsilon \to 0$ at the end, and often leave it implicit. You always need a $+i\varepsilon$ for time-ordered products, but it is really just shorthand for a pole prescription in the contour integral, which is exactly equivalent to adding various $\theta(t)$ factors.
- For $\varepsilon = 0$ the Feynman propagator looks just like a classical Green's function for the Klein–Gordon equation $(\Box + m^2) D_F(x, y) = -i\delta^4(x)$ with certain boundary conditions. That is because it is. We are just computing classical propagation in a really complicated way.

As we saw in Chapter 4 using old-fashioned perturbation theory, when using physical intermediate states there are contributions from advanced and retarded propagators, both of which are also Green's functions for the Klein–Gordon equation. The Lorentz-invariant Feynman propagator encodes both of these contributions, with its boundary condition represented by the $i\varepsilon$ in the denominator. The advanced and retarded propagators have more complicated integral representations, as you can explore in Problem 6.2.

Problems

6.1 Calculate the Feynman propagator in position space. To get the pole structure correct, you may find it helpful to use Schwinger parameters (see Appendix B). Take the $m \to 0$ limit of your result to find

$$\langle 0|T\{\phi_0(x_1)\phi_0(x_2)\}|0\rangle = -\frac{1}{4\pi^2}\frac{1}{(x_1 - x_2)^2 - i\varepsilon}. \quad (6.35)$$

6.2 Find expressions for the advanced and retarded propagators as d^4k integrals.

6.3 Prove that any operator can be put in the form of Eq. (6.22).

Feynman rules

In the previous chapter, we saw that scattering cross sections are naturally expressed in terms of time-ordered products of fields. The S-matrix has the form

$$\langle f|S|i\rangle \sim \langle \Omega|T\left\{\phi(x_1)\cdots\phi(x_n)\right\}|\Omega\rangle, \tag{7.1}$$

where $|\Omega\rangle$ is the ground state/vacuum in the interacting theory. In this expression the fields $\phi(x)$ are not free but are the full interacting quantum fields. We also saw that in the free theory, the time-ordered product of two fields is given by the Feynman propagator:

$$D_F(x,y) \equiv \langle 0|T\left\{\phi_0(x)\phi_0(y)\right\}|0\rangle = \lim_{\varepsilon\to 0}\int\frac{d^4k}{(2\pi)^4}\frac{i}{k^2-m^2+i\varepsilon}e^{ik(x-y)}, \tag{7.2}$$

where $|0\rangle$ is the ground state in the free theory.

In this chapter, we will develop a method of calculating time-ordered products in perturbation theory in terms of integrals over various Feynman propagators. There is a beautiful pictorial representation of the perturbation expansion using Feynman diagrams and an associated set of Feynman rules. There are position-space Feynman rules, for calculating time-ordered products, and also momentum-space Feynman rules, for calculating S-matrix elements. The momentum-space Feynman rules are by far the more important – they provide an extremely efficient way to set up calculations of physical results in quantum field theory. The momentum-space Feynman rules are the main result of Part I.

We will first derive the Feynman rules using a Lagrangian formulation of time evolution and quantization. This is the quickest way to connect Feynman diagrams to classical field theory. We will then derive the Feynman rules again using time-dependent perturbation theory, based on an expansion of the full interacting Hamiltonian around the free Hamiltonian. This calculation much more closely parallels the way we do perturbation theory in quantum mechanics. While the Hamiltonian-based calculation is significantly more involved, it has the distinct advantage of connecting time evolution directly to a Hermitian Hamiltonian, so time evolution is guaranteed to be unitary. The Feynman rules resulting from both approaches agree, confirming that the approaches are equivalent (at least in the case of the theory of a real scalar field, which is all we have seen so far). As we progress in our understanding of field theory and encounter particles of different spin and more complicated interactions, unitarity and the requirement of a Hermitian Hamiltonian will play a more important role (see in particular Chapter 24). A third independent way to derive the Feynman rules is through the path integral (Chapter 14).

7.1 Lagrangian derivation

In Section 2.3 we showed that free quantum fields satisfy

$$[\phi(\vec{x}, t), \phi(\vec{x}', t)] = 0, \tag{7.3}$$

$$[\phi(\vec{x}, t), \partial_t \phi(\vec{x}', t)] = i\hbar \delta^3(\vec{x} - \vec{x}') \tag{7.4}$$

(we have temporarily reinstated $\hbar$ to clarify the classical limit). We also showed that free quantum fields satisfy the free scalar field Euler–Lagrange equation $(\Box + m^2)\phi = 0$. In an arbitrary interacting theory, we must generalize these equations to specify how the dynamics is determined. In quantum mechanics, this is done with the Hamiltonian. So, one natural approach is to assume that $i\partial_t \phi(x) = [\phi, H]$ for an interacting quantum field theory, which leads to the Hamiltonian derivation of the Feynman rules in Section 7.2. In this section we discuss the simpler Lagrangian approach based on the Schwinger–Dyson equations, which has the advantage of being manifestly Lorentz invariant from start to finish.

In the Lagrangian approach, Hamilton's equations are replaced by the Euler–Lagrange equations. We therefore assume that our interacting fields satisfy the Euler–Lagrange equations derived from a Lagrangian $\mathcal{L}$ (the generalization of $(\Box + m^2)\phi = 0$), just like classical fields. We will also assume Eqs. (7.3) and (7.4) are still satisfied. This is a natural assumption, since at any given time the Hilbert space for the interacting theory is the same as that of a free theory. Equation (7.3) is a necessary condition for causality: at the same time but at different points in space, all operators, in particular fields, should be simultaneously observable and commute (otherwise there could be faster-than-light communication). This causality requirement will be discussed more in the context of the spin-statistics theorem in Section 12.6. Equation (7.4) is the equivalent of the canonical commutation relation from quantum mechanics: $[\hat{x}, \hat{p}] = i\hbar$. It indicates that a quantity and its time derivative are not simultaneously observable – the hallmark of the uncertainty principle.

At this point we only know how to calculate $\langle 0|T\{\phi(x)\phi(x')\}|0\rangle$ in the free theory. To calculate this quantity in an interacting theory, it is helpful to have the intermediate result

$$(\Box + m^2)\langle\Omega|T\{\phi(x)\phi(x')\}|\Omega\rangle = \langle\Omega|T\{(\Box + m^2)\phi(x)\phi(x')\}|\Omega\rangle - i\hbar\delta^4(x - x'), \tag{7.5}$$

where $|\Omega\rangle$ is the vacuum in the interacting theory. The $\delta^4(x - x')$ on the right side of this equation is critically important. It signifies the difference between the classical and quantum theories in a way that will be clear shortly.

To derive Eq. (7.5) we calculate

$$\partial_t \langle\Omega|T\{\phi(x)\phi(x')\}|\Omega\rangle = \partial_t[\langle\Omega|\phi(x)\phi(x')|\Omega\rangle\theta(t - t') + \langle\Omega|\phi(x')\phi(x)|\Omega\rangle\theta(t' - t)]$$
$$= \langle\Omega|T\{\partial_t\phi(x)\phi(x')\}|\Omega\rangle + \langle\Omega|\phi(x)\phi(x')|\Omega\rangle\partial_t\theta(t - t') + \langle\Omega|\phi(x')\phi(x)|\Omega\rangle\partial_t\theta(t' - t)$$
$$= \langle\Omega|T\{\partial_t\phi(x)\phi(x')\}|\Omega\rangle + \delta(t - t')\langle\Omega|[\phi(x), \phi(x')]|\Omega\rangle, \tag{7.6}$$

where we have used $\partial_x \theta(x) = \delta(x)$ in the last line. The second term on the last line vanishes, since $\delta(t - t')$ forces $t = t'$ and $[\phi(x), \phi(x')] = 0$ at equal times. Taking a second time derivative then gives

$$\partial_t^2 \langle \Omega | T\{\phi(x)\phi(x')\} | \Omega \rangle = \langle \Omega | T\{\partial_t^2 \phi(x)\phi(x')\} | \Omega \rangle + \delta(t - t')\langle \Omega | [\partial_t \phi(x), \phi(x')] | \Omega \rangle.$$
$$(7.7)$$

Here again $\delta(t - t')$ forces the two times to be equal, in which case $[\partial_t \phi(x), \phi(x')] = -i\hbar\delta^3(\vec{x} - \vec{x}')$ as in Eq. (7.4). Thus,

$$\partial_t^2 \langle \Omega | T\{\phi(x)\phi(x')\} | \Omega \rangle = \langle \Omega | T\{\partial_t^2 \phi(x)\phi(x')\} | \Omega \rangle - i\hbar\delta^4(x - x') \qquad (7.8)$$

and Eq. (7.5) follows.

For example, in the free theory, $(\Box + m^2)\phi_0(x) = 0$. Then Eq. (7.5) implies

$$(\Box + m^2) D_F(x, x') = -i\delta^4(x - x'), \qquad (7.9)$$

which is easy to verify from Eq. (7.2) (the $\hbar$ cancels because $\langle 0 | T\{\phi_0(x)\phi_0(y)\} | 0 \rangle = \hbar D_F(x, y)$).

Introducing the notation $\langle \cdots \rangle = \langle \Omega | T\{\cdots\} | \Omega \rangle$ for time-ordered correlation functions in the interacting theory, Eq. (7.5) can be written as

$$(\Box + m^2)\langle \phi(x)\phi(x') \rangle = \langle (\Box + m^2)\phi(x)\phi(x') \rangle - i\hbar\delta^4(x - x'). \qquad (7.10)$$

It is not hard to see that similar equations hold for commutators involving more fields. We will get $[\partial_t \phi(x), \phi(x_j)]$ terms from the time derivatives acting on the time-ordering operator giving δ-functions. The result is that

$$\Box_x \langle \phi(x)\phi(x_1) \cdots \phi(x_n) \rangle = \langle \Box_x \phi(x)\phi(x_1) \cdots \phi(x_n) \rangle$$
$$- i\hbar \sum_j \delta^4(x - x_j)\langle \phi(x_1) \cdots \phi(x_{j-1})\phi(x_{j+1}) \cdots \phi(x_n) \rangle. \qquad (7.11)$$

You should check this generalization by calculating $\Box_x \langle \phi(x)\phi(x_1)\phi(x_2) \rangle$ on your own.

Now we use the fact that the quantum field satisfies the equations of motion as the classical field, by assumption. In particular, if the Lagrangian has the form $\mathcal{L} = -\frac{1}{2}\phi(\Box + m^2)\phi + \mathcal{L}_{\text{int}}[\phi]$ then the (quantum) field satisfies $(\Box + m^2)\phi - \mathcal{L}'_{\text{int}}[\phi] = 0$, where $\mathcal{L}'_{\text{int}}[\phi] = \frac{d}{d\phi}\mathcal{L}_{\text{int}}[\phi]$, giving

$$(\Box_x + m^2)\langle \phi_x \phi_1 \cdots \phi_n \rangle = \langle \mathcal{L}'_{\text{int}}[\phi_x] \phi_1 \cdots \phi_n \rangle$$
$$- i\hbar \sum_j \delta^4(x - x_j)\langle \phi_1 \cdots \phi_{j-1}\phi_{j+1} \cdots \phi_n \rangle, \qquad (7.12)$$

where $\phi_x \equiv \phi(x)$ and $\phi_j \equiv \phi(x_j)$. These are known as the **Schwinger–Dyson equations**.

The Schwinger–Dyson equations encode the difference between the classical and quantum theories. Note that their derivation did not require any specification of the dynamics of the theory, only that the canonical commutation relations in Eq. (7.4) are satisfied.

In particular, in a classical theory, $[\phi(\vec{x}', t), \partial_t \phi(\vec{x}, t)] = 0$ and therefore classical time-ordered correlation functions would satisfy a similar equation but without the $\delta^4(x - x_j)$ terms (i.e. $\hbar = 0$). That is, in a classical theory, correlation functions satisfy the same differential equations as the fields within the correlation functions. In a quantum theory, that is true only up to δ-functions, which in this context are also called **contact interactions**. These contact interactions allow virtual particles to be created and destroyed, which permits closed loops to form in the Feynman diagrammatic expansion, as we will now see.

7.1.1 Position-space Feynman rules

The Schwinger–Dyson equations specify a completely non-perturbative relationship among correlation functions in the fully interacting theory. Some non-perturbative implications will be discussed in later chapters (in particular Sections 14.8 and 19.5). In this section, we will solve the Schwinger–Dyson equations in perturbation theory.

For efficiency, we write $\delta_{xi} = \delta^4(x - x_i)$ and $D_{ij} = D_{ji} = D_F(x_i, x_j)$. We will also set $m = 0$ for simplicity (the $m \neq 0$ case is a trivial generalization), and $\hbar = 1$. With this notation, the Green's function equation for the Feynman propagator can be written concisely as

$$\Box_x D_{x1} = -i\delta_{x1}. \tag{7.13}$$

This relation can be used to rewrite correlation functions in a suggestive form. For example, the 2-point function can be written as

$$\langle \phi_1 \phi_2 \rangle = \int d^4x \, \delta_{x1} \langle \phi_x \phi_2 \rangle = i \int d^4x \, (\Box_x D_{x1}) \langle \phi_x \phi_2 \rangle = i \int d^4x \, D_{x1} \Box_x \langle \phi_x \phi_2 \rangle, \tag{7.14}$$

where we have integrated by parts in the last step. This is suggestive because $\Box_x$ acting on a correlator can be simplified with the Schwinger–Dyson equations.

Now first suppose we are in the free theory where $\mathcal{L}_{\text{int}} = 0$. Then the 2-point function can be evaluated using the Schwinger–Dyson equation, $\Box_x \langle \phi_x \phi_y \rangle = -i\delta_{xy}$, to give

$$\langle \phi_1 \phi_2 \rangle = \int d^4x \, D_{x1} \delta_{x2} = D_{12}, \tag{7.15}$$

as expected. For a 4-point function, the expansion is similar:

$$\langle \phi_1 \phi_2 \phi_3 \phi_4 \rangle = i \int d^4x \, D_{x1} \Box_x \langle \phi_x \phi_2 \phi_3 \phi_4 \rangle$$
$$= \int d^4x \, D_{x1} \{ \delta_{x2} \langle \phi_3 \phi_4 \rangle + \delta_{x3} \langle \phi_2 \phi_4 \rangle + \delta_{x4} \langle \phi_2 \phi_3 \rangle \}. \tag{7.16}$$

Collapsing the δ-functions and using Eq. (7.15), this becomes

$$\langle \phi_1 \phi_2 \phi_3 \phi_4 \rangle = D_{12} D_{34} + D_{13} D_{24} + D_{14} D_{23}$$

$$\tag{7.17}$$

Each of these terms is drawn as a diagram. In the diagrams, the points $x_1 \ldots x_4$ correspond to points where the correlation function is evaluated and the lines connecting these points correspond to propagators.

Next, we will add interactions. Consider for example the 2-point function again with Lagrangian $\mathcal{L} = -\frac{1}{2}\phi\Box\phi + \frac{g}{3!}\phi^3$ (the 3! is a convention that will be justified shortly). Up to Eq. (7.14) things are the same as before. But now an application of the Schwinger–Dyson equations involves $\mathcal{L}'_{\text{int}}[\phi] = \frac{g}{2}\phi^2$, so we get

$$\langle \phi_1 \phi_2 \rangle = i \int d^4x \, D_{1x} \left(\frac{g}{2} \langle \phi_x^2 \phi_2 \rangle - i\delta_{x2} \right). \tag{7.18}$$

To simplify this, we introduce another integral, use $\delta_{2y} = i\Box_y D_{y2}$, and integrate by parts again to give

$$\langle \phi_1 \phi_2 \rangle = D_{12} - \frac{g}{2} \int d^4x \, d^4y \, D_{x1} D_{y2} \Box_y \langle \phi_x^2 \phi_y \rangle$$

$$= D_{12} - \frac{g^2}{4} \int d^4x \, d^4y \, D_{x1} D_{2y} \langle \phi_x^2 \phi_y^2 \rangle + ig \int d^4x \, D_{1x} D_{2x} \langle \phi_x \rangle. \tag{7.19}$$

If we are only interested in order g^2, the $\langle \phi_x^2 \phi_y^2 \rangle$ term can then be simplified using the free field Schwinger–Dyson result, Eq. (7.17),

$$\langle \phi_x^2 \phi_y^2 \rangle = 2D_{xy}^2 + D_{xx} D_{yy} + \mathcal{O}(g). \tag{7.20}$$

The $\langle \phi_x \rangle$ term in Eq. (7.19) can be expanded using the Schwinger–Dyson equations again:

$$\langle \phi_x \rangle = i \int d^4y \, D_{xy} \Box_y \langle \phi_y \rangle = i\frac{g}{2} \int d^4y \, D_{xy} \langle \phi_y^2 \rangle = i\frac{g}{2} \int d^4y \, D_{xy} D_{yy} + \mathcal{O}(g^2). \tag{7.21}$$

Thus the final result is

$$\langle \phi_1 \phi_2 \rangle = D_{12} - g^2 \int d^4x \, d^4y \left(\frac{1}{2} D_{1x} D_{xy}^2 D_{y2} + \frac{1}{4} D_{1x} D_{xx} D_{yy} D_{y2} \right.$$

$$\left. + \frac{1}{2} D_{1x} D_{2x} D_{xy} D_{yy} \right). \tag{7.22}$$

The three new terms correspond to the diagrams

$$\tag{7.23}$$

These diagrams now have new points, labeled x and y, which are integrated over.

From these examples, and looking at the pictures, it is easy to infer the way the perturbative expansion will work for higher-order terms or more general interactions.

1. Start with (external) points x_i for each position at which fields in the correlation function are evaluated. Draw a line from each point.
2. A line can then either contract to an existing line, giving a Feynman propagator connecting the endpoints of the two lines, or it can split, due to an interaction. A split gives a new (internal) vertex proportional to the coefficient of $\mathcal{L}'_{\text{int}}[\phi]$ times i and new lines corresponding to the fields in $\mathcal{L}'_{\text{int}}[\phi]$.

3. At a given order in the perturbative couplings, the result is the sum of all diagrams with all the lines contracted, integrated over the positions of internal vertices.

These are known as the **position-space Feynman rules**. The result is a set of diagrams. The original time-ordered product is given by a sum over integrals represented by the diagrams with an appropriate numerical factor. To determine the numerical factor, it is conventional to write interactions normalized by the number of permutations of identical fields, for example

$$\mathcal{L}_{\text{int}} = \frac{\lambda}{4!}\phi^4, \quad \frac{g}{3!}\phi^3, \quad \frac{\kappa}{5!3!2!}\phi_1^5\phi_2^3\phi_3^2, \quad \cdots \tag{7.24}$$

Thus, when the derivative is taken to turn the interaction into a vertex, the prefactor becomes $\frac{1}{(n-1)!}$. This $(n-1)!$ is then canceled by the number of permutations of the lines coming out of the vertex, not including the line coming in, which we already fixed. In this way, the $n!$ factors all cancel. The diagram is therefore associated with just the prefactor λ, g, κ, etc. from the interaction.

In some cases, such as theories with real scalar fields, some of the permutations give the same amplitude. For example, if a line connects back to itself, then permuting the two legs gives the same integral. In this case, a factor of $\frac{1}{2}$ in the normalization is not canceled, so we must divide by 2 to get the prefactor for a diagram. That is why the third diagram in Eq. (7.23) has a $\frac{1}{2}$ and the second diagram has a $\frac{1}{4}$. For the first diagram, the factor of $\frac{1}{2}$ comes from exchanging the two lines connecting x and y. So there is one more rule:

4. Drop all the $n!$ factors in the coefficient of the interaction, but then divide by the geometrical symmetry factor for each diagram.

Symmetries are ways that a graph can be deformed so that it looks the same with the *external points, labeled x_i, held fixed*. Thus, while there are symmetry factors for the graphs in Eq. (7.23), a graph such as

$$\tag{7.25}$$

has no symmetry factor, since the graph cannot be brought back to itself without tangling up the external lines. The safest way to determine the symmetry factor is simply to write down all the diagrams using the Feynman rules and see which give the same integrals. In practice, diagrams almost never have geometric symmetry factors; occasionally in theories with scalars there are factors of 2.

As mentioned in the introduction, an advantage of this approach is that it provides an intuitive way to connect and contrast the classical and quantum theories. In a classical theory, as noted above, the contact interactions are absent. It was these contact interactions that allowed us to *contract* two fields within a correlation function to produce a term in the expansion with fewer fields. For example, $\Box\langle\phi_1\phi_2\phi_3\phi_4\rangle = i\delta_{12}\langle\phi_3\phi_4\rangle + \cdots$. In the classical theory, all that can happen is that the fields will proliferate. Thus, we can have diagrams such as

$$(7.26)$$

The first process may represent general relativistic corrections to Mercury's orbit (see Eq. (3.85)), which can be calculated entirely with classical field theory. The external points in this case are all given by external sources, such as Mercury or the Sun, which are illustrated with the stars. The second process represents an electron in an external electromagnetic field (see Eq. (4.37)). This is a semi-classical process in which a single field is quantized (the electron) and does not get classical-source insertions on the end of its lines. But since quantum mechanics is first-quantized, particles cannot be created or destroyed and no closed loops can form. Thus, neither of these first two diagrams involve virtual pair creation. The third describes a process that can only be described with quantum field theory (or, with difficulty, with old-fashioned perturbation theory as in Eq. (4.44)). It is a Feynman diagram for the electron self-energy, which will be calculated properly using quantum field theory in Chapter 18.

7.2 Hamiltonian derivation

In this section, we reproduce the position-space Feynman rules using time-dependent perturbation theory. Instead of assuming that the quantum field satisfies the Euler–Lagrange equations, we instead assume its dynamics is determined by a Hamiltonian H through the Heisenberg equation of motion $i\partial_t \phi(x) = [\phi(x), H]$. This is the dynamical equation in the **Heisenberg picture** where all the time dependence is in operators. States including the vacuum state $|\Omega\rangle$ in the Heisenberg picture are time-independent.

The first step in time-dependent perturbation theory is to write the Hamiltonian as

$$H = H_0 + V, \qquad (7.27)$$

where the time evolution induced by H_0 can be solved exactly and V is small in some sense. For example, H_0 could be the free Hamiltonian and V might be a ϕ^3 interaction:

$$V(t) = \int d^3x \frac{g}{3!} \phi(\vec{x}, t)^3. \qquad (7.28)$$

Here the time-dependence in $V(t)$ is induced from the time dependence of the fields $\phi(\vec{x}, t)$, according to the equations of motion.

For time-dependent perturbation theory, a powerful tool is in the **interaction picture**, where the leading time-evolution, that generated by the free Hamiltonian H_0, is factored out. The interaction picture fields are what we had been calling (and will continue to call) the free fields $\phi_0(x)$; they evolve in time according to H_0 alone: $i\partial_t \phi_0(x, t) = [\phi_0(x), H_0]$. Thus we can write

$$\phi_0(\vec{x}, t) = e^{iH_0(t-t_0)} \phi(\vec{x}, t_0) e^{-iH_0(t-t_0)} = \int \frac{d^3p}{(2\pi)^3} \frac{1}{\sqrt{2\omega_p}} (a_p e^{-ipx} + a_p^\dagger e^{ipx}). \quad (7.29)$$

Here, t_0 is the time when the interaction-picture fields $\phi_0(\vec{x}, t)$ are matched onto the Heisenberg picture fields $\phi(\vec{x}, t)$.

In the interaction picture, the free Hamiltonian H_0 is time-independent, $i\partial_t H_0 = [H_0, H_0] = 0$, but the full Hamiltonian may not be. Allowing for possible time-dependence in the Hamiltonian, the Heisenberg-picture fields $\phi(x, t)$ evolve according to $i\partial_t \phi = [\phi(x), H(t)]$. The formal solution of this equation is

$$\phi(\vec{x}, t) = S(t, t_0)^\dagger \phi(\vec{x}, t_0) S(t, t_0), \tag{7.30}$$

where $S(t, t_0)$ satisfies

$$i\partial_t S(t, t_0) = S(t, t_0) H(t). \tag{7.31}$$

Using that the fields are matched at t_0, the Heisenberg picture fields are related to the free (interaction-picture) fields at time t by

$$\phi(\vec{x}, t) = S^\dagger(t, t_0)\, e^{-iH_0(t-t_0)} \phi_0(\vec{x}, t)\, e^{iH_0(t-t_0)} S(t, t_0)$$
$$= U^\dagger(t, t_0) \phi_0(\vec{x}, t)\, U(t, t_0). \tag{7.32}$$

The operator $U(t, t_0) \equiv e^{iH_0(t-t_0)} S(t, t_0)$ therefore relates the full Heisenberg picture fields to the free fields at the *same time t*. The evolution begins from the time t_0 where the fields in the two pictures are equal.

We can find a differential equation for $U(t, t_0)$ using Eq. (7.31):

$$i\partial_t U(t, t_0) = i\left(\partial_t e^{iH_0(t-t_0)}\right) S(t, t_0) + e^{iH_0(t-t_0)} i\partial_t S(t, t_0)$$
$$= -e^{iH_0(t-t_0)} H_0 S(t, t_0) + e^{iH_0(t-t_0)} H(t_0) S(t, t_0)$$
$$= e^{iH_0(t-t_0)} [-H_0 + H(t_0)]\, e^{-iH_0(t-t_0)} e^{iH_0(t-t_0)} S(t, t_0)$$
$$= V_I(t) U(t, t_0), \tag{7.33}$$

where $V_I(t) \equiv e^{iH_0(t-t_0)} V(t_0) e^{-iH_0(t-t_0)}$ is the potential $V(t)$ from Eq. (7.27), now expressed in the interaction picture.

If everything commuted, the solution to Eq. (7.33) would be $U(t, t_0) = \exp(-i\int_{t_0}^t V_I(t')dt')$. But $V_I(t_1)$ does not necessarily commute with $V_I(t_2)$, so this is not the right answer. It turns out that the right answer is very similar:

$$U(t, t_0) = T\left\{\exp\left[-i\int_{t_0}^t dt' V_I(t')\right]\right\}, \tag{7.34}$$

where $T\{\}$ is the time-ordering operator, introduced in Chapter 6. This solution works because time ordering effectively makes everything inside commute:

$$T\{A \cdots B \cdots\} = T\{B \cdots A \cdots\}. \tag{7.35}$$

Taking the derivative, you can see immediately that Eq. (7.34) satisfies Eq. (7.33). Since it has the right boundary conditions, namely $U(t, t) = 1$, this solution is unique.

Time ordering of an exponential is defined in the obvious way through its expansion:

$$U(t, t_0) = 1 - i\int_{t_0}^t dt' V_I(t') - \frac{1}{2}\int_{t_0}^t dt' \int_{t_0}^t dt'' T\{V_I(t') V_I(t'')\} + \cdots. \tag{7.36}$$

This is known as a **Dyson series**. Dyson defined the time-ordered product and this series in his classic paper [Dyson, 1949]. In that paper he showed the equivalence of old-fashioned perturbation theory or, more exactly, the interaction picture method developed by Schwinger and Tomonaga based on time-dependent perturbation theory, and Feynman's method, involving space-time diagrams.

7.2.1 Perturbative solution for the Dyson series

We guessed and checked the solution to Eq. (7.33), which is often the easiest way to solve a differential equation. We can also solve it using perturbation theory.

Removing the subscript on V for simplicity, the differential equation we want to solve is

$$i\partial_t U(t, t_0) = V(t)U(t, t_0). \tag{7.37}$$

Integrating this equation lets us write it in an equivalent form:

$$U(t, t_0) = 1 - i \int_{t_0}^t dt' V(t')U(t', t_0), \tag{7.38}$$

where 1 is the appropriate integration constant so that $U(t_0, t_0) = 1$.

Now we will solve the integral equation order-by-order in V. At zeroth order in V,

$$U(t, t_0) = 1. \tag{7.39}$$

To first order in V we find

$$U(t, t_0) = 1 - i \int_{t_0}^t dt' V(t') + \cdots . \tag{7.40}$$

To second order,

$$U(t, t_0) = 1 - i \int_{t_0}^t dt' V(t') \left[1 - i \int_{t_0}^{t'} dt'' V(t'') + \cdots \right]$$

$$= 1 - i \int_{t_0}^t dt' V(t') + (-i)^2 \int_{t_0}^t dt' \int_{t_0}^{t'} dt'' V(t')V(t'') + \cdots . \tag{7.41}$$

The second integral has $t_0 < t'' < t' < t$, which is the same as $t_0 < t'' < t$ and $t'' < t' < t$. So it can also be written as

$$\int_{t_0}^t dt' \int_{t_0}^{t'} dt'' V(t')V(t'') = \int_{t_0}^t dt'' \int_{t''}^t dt' V(t')V(t'') = \int_{t'}^t dt'' \int_{t_0}^t dt' V(t'')V(t'), \tag{7.42}$$

where we have relabeled $t'' \leftrightarrow t'$ and swapped the order of the integrals to get the third form. Averaging the first and third form gives

$$\int_{t_0}^t dt' \int_{t_0}^{t'} dt'' V(t')V(t'') = \frac{1}{2} \int_{t_0}^t dt' \left[\int_{t_0}^{t'} dt'' V(t')V(t'') + \int_{t'}^t dt'' V(t'')V(t') \right]$$

$$= \frac{1}{2} \int_{t_0}^t dt' \int_{t_0}^t dt'' T\left\{ V(t')V(t'') \right\}. \tag{7.43}$$

Thus,

$$U(t, t_0) = 1 - i \int_{t_0}^{t} dt' V(t') + \frac{(-i)^2}{2} \int_{t_0}^{t} dt' \int_{t_0}^{t} dt'' T\{V(t')V(t'')\} + \cdots . \quad (7.44)$$

Continuing this way, we find, restoring the subscript on V, that

$$U(t, t_0) = T\left\{\exp\left[-i \int_{t_0}^{t} dt' V_I(t')\right]\right\} . \quad (7.45)$$

7.2.2 U relations

It is convenient to abbreviate U with

$$U_{21} \equiv U(t_2, t_1) = T\left\{\exp\left[-i \int_{t_1}^{t_2} dt' V_I(t')\right]\right\} . \quad (7.46)$$

Remember that in field theory we always have later times on the left. It follows that

$$U_{21} U_{12} = 1 , \quad (7.47)$$

$$U_{21}^{-1} = U_{21}^{\dagger} = U_{12} \quad (7.48)$$

and for $t_1 < t_2 < t_3$

$$U_{32} U_{21} = U_{31}. \quad (7.49)$$

Multiplying this by U_{12} on the right, we find

$$U_{31} U_{12} = U_{32}, \quad (7.50)$$

which is the same identity with $2 \leftrightarrow 1$. Multiplying Eq. (7.49) by U_{23} on the left gives the same identity with $3 \leftrightarrow 1$. Therefore, this identity holds for any time ordering.

Finally, our defining relation, Eq. (7.32),

$$\phi(\vec{x}, t) = U^{\dagger}(t, t_0) \phi_0(\vec{x}, t) U(t, t_0) \quad (7.51)$$

lets us write

$$\phi(x_1) = \phi(\vec{x}_1, t_1) = U_{10}^{\dagger} \phi_0(\vec{x}_1, t_1) U_{10} = U_{01} \phi_0(x_1) U_{10}. \quad (7.52)$$

7.2.3 Vacuum matrix elements

In deriving the LSZ reduction theorem, we used that the vacuum state $|\Omega\rangle$ was annihilated by the fully interacting Heisenberg-picture operators $a_p(t)$ at a time $t = -\infty$. We need to know how this relates to the vacuum $|0\rangle$ in the interaction picture which is annihilated by the interaction-picture (free) creation operators contained in the free field $\phi_0(x, t)$. At $t = -\infty$, our working assumption is that the interactions can be neglected. So we should be able to match $|0\rangle$, which evolves with H_0 through $e^{iH_0 t}$, onto $|\Omega\rangle$ which evolves with $H(t)$ through $S(t, t_0)$, as $t \to -\infty$. That is, $e^{iH_0(t-t_0)}|0\rangle \to S(t, t_0)|\Omega\rangle$ as $t \to \infty$. Noting that states in quantum field theory are the same even if they differ by normalization, we can write the relation between the interacting theory vacuum and free theory vacuum as

$$|\Omega\rangle = \mathcal{N}_i \lim_{t \to -\infty} S^\dagger(t, t_0) e^{-iH_0(t-t_0)} |0\rangle = \mathcal{N}_i U_{0-\infty} |0\rangle \tag{7.53}$$

for some number $\mathcal{N}_i$. Similarly, $\langle\Omega| = \mathcal{N}_f \langle 0| U_{\infty 0}$ for some number $\mathcal{N}_f$.[1]

Now let us see what happens when we rewrite correlation functions in the interaction picture. We are interested in time-ordered products $\langle\Omega|T\{\phi(x_1)\cdots\phi(x_n)\}|\Omega\rangle$. Since all the $\phi(x_i)$ are within a time-ordered product, we can write them in any order we want. So let us put them in time order, or equivalently we assume $t_1 > \cdots > t_n$ without loss of generality. Then,

$$\langle\Omega|T\{\phi(x_1)\cdots\phi(x_n)\}|\Omega\rangle = \langle\Omega|\phi(x_1)\cdots\phi(x_n)|\Omega\rangle$$
$$= \mathcal{N}_i\mathcal{N}_f\langle 0|U_{\infty 0}U_{01}\phi_0(x_1)U_{10}U_{02}\phi_0(x_2)U_{20}\cdots U_{0n}\phi_0(x_n)U_{n0}U_{0-\infty}|0\rangle$$
$$= \mathcal{N}_i\mathcal{N}_f\langle 0|U_{\infty 1}\phi_0(x_1)U_{12}\phi_0(x_2)U_{23}\cdots U_{(n-1)n}\phi_0(x_n)U_{n-\infty}|0\rangle. \tag{7.54}$$

Now, since the t_i are in time order and the U_{ij} are themselves time-ordered products involving times between t_i and t_j, everything in this expression is in time order. Thus

$$\langle\Omega|T\{\phi(x_1)\cdots\phi(x_n)\}|\Omega\rangle$$
$$= \mathcal{N}_i\mathcal{N}_f\langle 0|T\{U_{\infty 1}\phi_0(x_1)U_{12}\phi_0(x_2)U_{23}\cdots\phi_0(x_n)U_{n-\infty}\}|0\rangle$$
$$= \mathcal{N}_i\mathcal{N}_f\langle 0|T\{\phi_0(x_1)\cdots\phi_0(x_n)U_{\infty,-\infty}\}|0\rangle. \tag{7.55}$$

The normalization should be set so that $\langle\Omega|\Omega\rangle = 1$, just as $\langle 0|0\rangle = 1$ in the free theory. This implies $\mathcal{N}_i\mathcal{N}_f = \langle 0|U_{\infty-\infty}|0\rangle^{-1}$ and therefore

$$\langle\Omega|T\{\phi(x_1)\cdots\phi(x_n)\}|\Omega\rangle = \frac{\langle 0|T\{\phi_0(x_1)\cdots\phi_0(x_n)U_{\infty,-\infty}\}|0\rangle}{\langle 0|U_{\infty,-\infty}|0\rangle}. \tag{7.56}$$

Substituting in Eq. (7.46) we then get

$$\langle\Omega|T\{\phi(x_1)\cdots\phi(x_n)\}|\Omega\rangle = \frac{\langle 0|T\left\{\phi_0(x_1)\cdots\phi_0(x_n)\exp[-i\int_{-\infty}^{\infty}dtV_I(t)]\right\}|0\rangle}{\langle 0|T\left\{\exp[-i\int_{-\infty}^{\infty}dtV_I(t)]\right\}|0\rangle}. \tag{7.57}$$

7.2.4 Interaction potential

The only thing left to understand is what $V_I(t)$ is. We have defined the time t_0 as when the interacting fields are the same as the free fields. For example, a cubic interaction would be

$$V(t_0) = \int d^3x \frac{g}{3!}\phi(\vec{x}, t_0)^3 = \int d^3x \frac{g}{3!}\phi_0(\vec{x}, t_0)^3 = \int d^3x \frac{g}{3!}\phi_0(\vec{x})^3, \tag{7.58}$$

[1] It may be worth emphasizing that $|0\rangle$ and $|\Omega\rangle$ are *not* the same state at the reference time t_0. After evolving from $-\infty$ to t_0, $|\Omega\rangle$ becomes a complicated superposition of multiparticle states.

Recall that the time dependence of the free fields is determined by the free Hamiltonian,

$$\phi_0(\vec{x}, t) = e^{iH_0(t-t_0)} \phi_0(\vec{x}) e^{-iH_0(t-t_0)}, \tag{7.59}$$

and therefore

$$V_I = e^{iH_0(t-t_0)} \left[\int d^3x \frac{g}{3!} \phi_0(\vec{x})^3 \right] e^{-iH_0(t-t_0)} = \int d^3x \frac{g}{3!} \phi_0(\vec{x}, t)^3. \tag{7.60}$$

So the interaction picture potential is expressed in terms of the free fields at all times.

Now we will make our final transition away from non-Lorentz-invariant Hamiltonians to Lorentz-invariant Lagrangians, leaving old-fashioned perturbation theory for good. Recall that the potential is related to the Lagrangian by $V_I = -\int d^3x \, \mathcal{L}_{\text{int}}[\phi_0]$, where $\mathcal{L}_{\text{int}}$ is the interacting part of the Lagrangian density. Then,

$$U_{\infty,-\infty} = T\left\{ \exp\left[-i \int_{-\infty}^{\infty} dt \, V_I(t) \right] \right\} = T\left\{ \exp\left[i \int_{-\infty}^{\infty} d^4x \, \mathcal{L}_{\text{int}}[\phi_0] \right] \right\}. \tag{7.61}$$

The $\int_{-\infty}^{\infty} dt$ combines with the $\int d^3x$ to give a Lorentz-invariant integral.

In summary, matrix elements of interacting fields in the interacting vacuum are given by

$$\langle \Omega | \phi(x_1) \cdots \phi(x_n) | \Omega \rangle = \frac{\langle 0 | U_{\infty 1} \phi_0(x_1) U_{12} \phi_0(x_2) U_{23} \cdots \phi_0(x_n) U_{n,-\infty} | 0 \rangle}{\langle 0 | U_{\infty,-\infty} | 0 \rangle}, \tag{7.62}$$

where $|\Omega\rangle$ is the ground state in the interacting theory and

$$U_{ij} = T\left\{ \exp\left[i \int_{t_j}^{t_i} d^4x \, \mathcal{L}_{\text{int}}[\phi_0] \right] \right\}, \tag{7.63}$$

with $\mathcal{L}_{\text{int}}[\phi] = \mathcal{L}[\phi] - \mathcal{L}_0[\phi]$, where $\mathcal{L}_0[\phi]$ is the free Lagrangian. The free Lagrangian is defined as whatever goes into the free-field evolution, usually taken to be just kinetic terms.

For the special case of time-ordered products, such as what we need for S-matrix elements, this simplifies to

$$\langle \Omega | T\{\phi(x_1) \cdots \phi(x_n)\} | \Omega \rangle = \frac{\langle 0 | T\left\{ \phi_0(x_1) \cdots \phi_0(x_n) e^{i \int d^4x \mathcal{L}_{\text{int}}[\phi_0]} \right\} | 0 \rangle}{\langle 0 | T\left\{ e^{i \int d^4x \mathcal{L}_{\text{int}}[\phi_0]} \right\} | 0 \rangle}, \tag{7.64}$$

which is a remarkably simple and manifestly Lorentz-invariant result.

7.2.5 Time-ordered products and contractions

We will now see that the expansion of Eq. (7.64) produces the same position-space Feynman rules as those coming from the Lagrangian approach described in Section 7.1. To see that, let us take as an example our favorite ϕ^3 theory with interaction Lagrangian

$$\mathcal{L}_{\text{int}}[\phi] = \frac{g}{3!} \phi^3, \tag{7.65}$$

and consider $\langle \Omega | T\{\phi(x_1)\phi(x_2)\} | \Omega \rangle$.

The numerator of Eq. (7.64) can be expanded perturbatively in g as

$$\langle 0| T\big\{\phi_0(x_1)\phi_0(x_2)e^{i\int d^4x \mathcal{L}_{\text{int}}[\phi_0]}\big\} |0\rangle = \langle 0|T\{\phi_0(x_1)\phi_0(x_2)\}|0\rangle$$

$$+ \frac{ig}{3!}\int d^4x \langle 0| T\{\phi_0(x_1)\phi_0(x_2)\phi_0(x)^3\} |0\rangle$$

$$+ \left(\frac{ig}{3!}\right)^2 \frac{1}{2}\int d^4x \int d^4y \langle 0| T\{\phi_0(x_1)\phi_0(x_2)\phi_0(x)^3\phi_0(y)^3\} |0\rangle + \cdots . \quad (7.66)$$

A similar expansion would result from any time-ordered product of interacting fields. Thus, we now only need to evaluate correlation functions of products of free fields.

To do so, it is helpful to write $\phi_0(x) = \phi_+(x) + \phi_-(x)$, where

$$\phi_+(x) = \int \frac{d^3p}{(2\pi)^3}\frac{1}{\sqrt{2\omega_p}}a_p^\dagger e^{ipx}, \quad \phi_-(x) = \int \frac{d^3p}{(2\pi)^3}\frac{1}{\sqrt{2\omega_p}}a_p e^{-ipx}, \quad (7.67)$$

with ϕ_+ containing only creation operators and ϕ_- only annihilation operators. Then products of ϕ_0 fields at different points become sums of products of ϕ_+ and ϕ_- fields at different points. For example,

$$\langle 0| T\big\{\phi_0(x_1)\phi_0(x_2)\phi_0(x)^3\phi_0(y)^3\big\} |0\rangle$$

$$= \langle 0| T\big\{[\phi_+(x_1)+\phi_-(x_1)][\phi_+(x_2)+\phi_-(x_2)][\phi_+(x)+\phi_-(x)]^3[\phi_+(y)+\phi_-(y)]^3\big\} |0\rangle$$

$$= \langle 0| T\big\{\phi_+(x_1)\phi_+(x_2)\phi_+(x)^3\phi_+(y)^3\big\} |0\rangle$$

$$+ 3\langle 0| T\big\{\phi_+(x_2)\phi_+(x_1)\phi_+(x)^3\phi_+(y)^2\phi_-(y)\big\} |0\rangle + \cdots . \quad (7.68)$$

The last line indicates that the result is the sum of a set of products of ϕ_+ and ϕ_- operators evaluated at different points. In each element of this sum, a ϕ_+ would create a particle that, to give a non-zero result, must then be annihilated by some ϕ_- operator. The matrix element can only be non-zero if every particle that is created is destroyed, so every term must have four ϕ_+ operators and four ϕ_- operators. Each pairing of ϕ_+ with ϕ_- to get a Feynman propagator is called a **contraction** (not to be confused with a Lorentz contraction). The result is then the sum of all possible contractions.

Each contraction represents the creation and then annihilation of a particle, with the creation happening earlier than the annihilation. Each contraction gives a factor of the Feynman propagator:

$$\langle 0|T\{\phi_0(x)\phi_0(y)\}|0\rangle = \int \frac{d^4k}{(2\pi)^4}\frac{i}{k^2-m^2+i\varepsilon}e^{ik(x-y)} \equiv D_F(x,y). \quad (7.69)$$

A time-ordered correlation function of free fields is given by a sum over all possible ways in which all of the fields in the product can be contracted with each other. This is a result known as **Wick's theorem**. Wick's theorem is given in Box 7.3 and proven in the appendix to this chapter.

To see how Wick's theorem works, let us return to our example and use the notation $D_{ij} \equiv D_F(x_i, x_j)$. The first term in the expansion of $\langle \Omega|T\{\phi(x_1)\phi(x_2)\}|\Omega\rangle$ is $\langle 0|T\{\phi_0(x_1)\phi_0(x_2)\}|0\rangle$, from Eq. (7.66). There is only one contraction here, which gives the propagator $D_F(x_1, x_2) = D_{12}$. The second term in Eq. (7.66) has an odd number of

ϕ fields, and therefore cannot be completely contracted and must vanish. The third term in Eq. (7.66) involves eight fields, and there are multiple possible contractions:

$$\langle 0|T\{\phi_0(x_1)\phi_0(x_2)\phi_0(x)\phi_0(x)\phi_0(x)\phi_0(y)\phi_0(y)\phi_0(y)\}|0\rangle$$
$$= 9D_{12}D_{xx}D_{xy}D_{yy} + 6D_{12}D_{xy}^3$$
$$+ 18D_{1x}D_{2x}D_{xy}D_{yy} + 9D_{1x}D_{2y}D_{xx}D_{yy} + 18D_{1x}D_{2y}D_{xy}^2$$
$$+ 18D_{1y}D_{2y}D_{xy}D_{xx} + 9D_{1y}D_{2x}D_{xx}D_{yy} + 18D_{1y}D_{2x}D_{xy}^2. \tag{7.70}$$

As in Eq. (7.66), we have to integrate over x and y. Thus, many of these terms (those on the last line) give the same contributions as other terms. We find, to order g^2,

$$\langle\Omega|T\{\phi(x_1)\phi(x_2)\}|\Omega\rangle = \frac{1}{\langle 0|T\left\{e^{i\int\mathcal{L}_{\rm int}}\right\}|0\rangle}\left\{D_{12}\right.$$
$$- g^2\int d^4x\int d^4y\left[\frac{1}{8}D_{12}D_{xx}D_{xy}D_{yy} + \frac{1}{12}D_{12}D_{xy}^3 + \frac{1}{2}D_{1x}D_{2x}D_{xy}D_{yy}\right.$$
$$\left.\left. + \frac{1}{4}D_{1x}D_{xx}D_{yy}D_{y2} + \frac{1}{2}D_{1x}D_{xy}^2D_{y2}\right] + \cdots\right\} \tag{7.71}$$

The position-space Feynman rules that connect this expansion to diagrams are the same as those coming from the Lagrangian approach in Section 7.1. Comparing to Eq. (7.22) we see that the sum of terms is exactly the same, including combinatoric factors, with two exceptions: the $\langle 0|T\{e^{i\int\mathcal{L}_{\rm int}}\}|0\rangle$ factor and the first two terms on the second line. The two new terms correspond to diagrams

$$\text{and} \tag{7.72}$$

To see the cancellation, note that the extra diagrams both include **bubbles**. That is, they have connected subgraphs not involving any external point. The bubbles are exactly what are in $\langle 0|T\{e^{i\int\mathcal{L}_{\rm int}}\}|0\rangle$. To see this, note that Wick's theorem also applies to the denominator of Eq. (7.64). Up to order g^2, Wick's theorem implies

$$\langle 0|T\left\{e^{i\int d^4x\mathcal{L}_{\rm int}[\phi_0]}\right\}|0\rangle = \langle 0|0\rangle + \left(\frac{ig}{3!}\right)^2\frac{1}{2}\int d^4x\int d^4y\langle 0|T\left\{\phi_0(x)^3\phi_0(y)^3\right\}|0\rangle + \cdots . \tag{7.73}$$

We have dropped the $\mathcal{O}(g)$ term since it involves an odd number of fields and therefore vanishes by Wick's theorem. Performing a similar expansion as above, we find

$$\langle 0|T\left\{e^{i\int d^4x\mathcal{L}_{\rm int}[\phi_0]}\right\}|0\rangle = 1 + \left(\frac{ig}{3!}\right)^2\frac{1}{2}\int d^4x\int d^4y\left[9D_{xx}D_{xy}D_{yy} + 6D_{xy}^3\right] + \mathcal{O}(g^3). \tag{7.74}$$

These diagrams are the bubbles and . Expanding Eq. (7.71) including terms up to $\mathcal{O}(g^2)$ in the numerator and denominator, we find

$$\frac{\langle 0|T\left\{\phi_0(x_1)\phi_0(x_2)e^{i\int\mathcal{L}_{\rm int}}\right\}|0\rangle}{\langle 0|T\{e^{i\int\mathcal{L}_{\rm int}}\}|0\rangle} = \frac{D_{12} - g^2\int\left[\frac{1}{8}D_{12}D_{xx}D_{xy}D_{yy} + \frac{1}{12}D_{12}D_{xy}^3 + \cdots\right]}{1 - g^2\int\left[\frac{1}{8}D_{xx}D_{xy}D_{yy} + \frac{1}{12}D_{xy}^3\right]}. \tag{7.75}$$

Since $\frac{1}{1+g^2x} = 1 - g^2x + \mathcal{O}(g^4)$, we can invert the denominator in perturbation theory to see that the bubbles exactly cancel.

More generally, the bubbles will always cancel. Since the integrals in the expansion of the numerator corresponding to the bubbles never involve any external point, they just factor out. The sum over all graphs, in the numerator, is then the sum over all graphs with no bubbles multiplying the sum over the bubbles. In pictures,

$$
\tag{7.76}
$$

The sum over bubbles is exactly $\langle 0| T\left\{ e^{i\int \mathcal{L}_{\text{int}}}\right\} |0\rangle$. So,

$$\langle\Omega| T\{\phi(x_1)\phi(x_2)\} |\Omega\rangle = \langle 0| T\left\{\phi_0(x_1)\phi_0(x_2)e^{i\int \mathcal{L}_{\text{int}}}\right\} |0\rangle_{\text{no bubbles}}, \tag{7.77}$$

where "no bubbles" means that every connected subgraph involves an external point.

7.2.6 Position-space Feynman rules

We have shown that the same sets of diagrams appear in the Hamiltonian and the Lagrangian approaches: each point x_i in the original n-point function $\langle\Omega|T\{\phi(x_1)\cdots \phi(x_n)\}|\Omega\rangle$ gets an external point and each interaction gives a new vertex whose position is integrated over and whose coefficient is given by the coefficient in the Lagrangian.

As long as the vertices are normalized with appropriate permutation factors, as in Eq. (7.24), the combinatoric factors will work out the same, as we saw in the example. In the Lagrangian approach, we saw that the coefficient of the diagram will be given by the coefficient of the interaction multiplied by the geometrical symmetry factor of the diagram. To see that this is also true for the Hamiltonian, we have to count the various combinatoric factors:

- There is a factor of $\frac{1}{m!}$ from the expansion of $\exp(i\mathcal{L}_{\text{int}}) = \sum \frac{1}{m!}(i\mathcal{L}_{\text{int}})^m$. If we expand to order m there will be m identical vertices in the same diagram. We can also swap these vertices around, leaving the diagram looking the same. If we only include the diagram once in our final sum, the $m!$ from permuting the diagrams will cancel the $\frac{1}{m!}$ from the exponential. Neither of these factors were present in the Lagrangian approach, since internal vertices came out of the splitting of lines associated with external vertices, which was unambiguous, and there was no exponential to begin with.

- If interactions are normalized as in Eq. (7.24), then there will be a $\frac{1}{j!}$ for each interaction with j identical particles. This factor is canceled by the $j!$ ways of permuting the j identical lines coming out of the same internal vertex. In the Lagrangian approach, one of the lines was already chosen so the factor was $(j-1)!$, with the missing j coming from using $\mathcal{L}'_{\text{int}}[\phi]$ instead of $\mathcal{L}_{\text{int}}[\phi]$.

The result is the same Feynman rules as were derived in the Lagrangian approach. In both cases, symmetry factors must be added if there is some geometric symmetry (there rarely is in theories with complex fields, such as QED). In neither case do any of the diagrams include bubbles (subdiagrams that do not connect with any external vertex).

7.3 Momentum-space Feynman rules

The position-space Feynman rules derived in either of the previous two sections give a recipe for computing time-ordered products in perturbation theory. Now we will see how those time-ordered products simplify when all the phase-space integrals over the propagators are performed to turn them into S-matrix elements. This will produce the momentum-space Feynman rules.

Consider the diagram

$$\mathcal{T}_1 = \quad \overset{x_1 \quad x}{\bullet\text{---}\bullet} \bigcirc \overset{y \quad x_2}{\text{---}\bullet} \quad = -\frac{g^2}{2} \int d^4x \int d^4y \, D_{1x} D_{xy}^2 D_{y2}. \qquad (7.78)$$

To evaluate this diagram, first write every propagator in momentum space (taking $m = 0$ for simplicity):

$$D_{xy} = \int \frac{d^4p}{(2\pi)^4} \frac{i}{p^2 + i\varepsilon} e^{ip(x-y)}. \qquad (7.79)$$

Then there will be four d^4p integrals from the four propagators and all the positions will appear only in exponentials. So,

$$\mathcal{T}_1 = -\frac{g^2}{2} \int d^4x \int d^4y \int \frac{d^4p_1}{(2\pi)^4} \int \frac{d^4p_2}{(2\pi)^4} \int \frac{d^4p_3}{(2\pi)^4} \int \frac{d^4p_4}{(2\pi)^4}$$
$$\times \, e^{ip_1(x_1-x)} e^{ip_2(y-x_2)} e^{ip_3(x-y)} e^{ip_4(x-y)} \frac{i}{p_1^2 + i\varepsilon} \frac{i}{p_2^2 + i\varepsilon} \frac{i}{p_3^2 + i\varepsilon} \frac{i}{p_4^2 + i\varepsilon}. \qquad (7.80)$$

Now we can do the x and y integrals, which produce $\delta^4(-p_1+p_3+p_4)$ and $\delta^4(p_2-p_3-p_4)$ respectively, corresponding to momentum being conserved at the vertices labeled x and y in the Feynman diagram. If we integrate over p_3 using the first δ-function then we can replace $p_3 = p_1 - p_4$ and the second δ-function becomes $\delta^4(p_1 - p_2)$. Then we have, relabeling $p_4 = k$,

$$T_1 = -\frac{g^2}{2} \int \frac{d^4 k}{(2\pi)^4} \int \frac{d^4 p_1}{(2\pi)^4} \int \frac{d^4 p_2}{(2\pi)^4} e^{ip_1 x_1} e^{-ip_2 x_2}$$

$$\times \frac{i}{p_1^2 + i\varepsilon} \frac{i}{p_2^2 + i\varepsilon} \frac{i}{(p_1 - k)^2 + i\varepsilon} \frac{i}{k^2 + i\varepsilon} (2\pi)^4 \delta^4(p_1 - p_2). \quad (7.81)$$

Next, we use the LSZ formula to convert this to a contribution to the S-matrix:

$$\langle f|S|i\rangle = \left[-i \int d^4 x_1 e^{-ip_i x_1} (p_i^2) \right] \left[-i \int d^4 x_2 e^{ip_f x_2} (p_f^2) \right] \langle \Omega| T \{\phi(x_1)\phi(x_2)\} |\Omega\rangle ,$$
$$(7.82)$$

where p_i^μ and p_f^μ are the initial state and final state momenta. So the contribution of this diagram gives

$$\langle f|S|i\rangle = -\int d^4 x_1 e^{-ip_i x_1} (p_i)^2 \int d^4 x_2 e^{ip_f x_2} (p_f^2) T_1 + \cdots . \quad (7.83)$$

Now we note that the x_1 integral gives $(2\pi)^4 \delta^4(p_1 - p_i)$ and the x_2 integral gives $(2\pi)^4 \delta^4(p_2 - p_f)$. So we can now do the p_1 and p_2 integrals, giving

$$\langle f|S|i\rangle = -\frac{g^2}{2} \int \frac{d^4 k}{(2\pi)^4} \frac{i}{(p_i - k)^2 + i\varepsilon} \frac{i}{k^2 + i\varepsilon} (2\pi)^4 \delta^4(p_i - p_f) + \cdots . \quad (7.84)$$

Note how the two propagator factors in the beginning get canceled. This always happens for external legs – remember the point of LSZ was to force the external lines to be on-shell one-particle states. By the way, this integral is infinite; Part III of this book is devoted to making sense out of these infinities.

Finally, the $\delta^4(p_i - p_f)$ term in the answer forces overall momentum conservation, and will always be present in any calculation. But we will always factor it out, as we did when we related differential scattering amplitudes to S-matrix elements. Recalling that

$$S = \mathbb{1} + (2\pi)^4 \delta^4(\Sigma p_i) i\mathcal{M}, \quad (7.85)$$

we have

$$i\mathcal{M} = -\frac{g^2}{2} \int \frac{d^4 k}{(2\pi)^4} \frac{i}{(p_i - k)^2 + i\varepsilon} \frac{i}{k^2 + i\varepsilon} + \cdots . \quad (7.86)$$

We can summarize this procedure with the **momentum-space Feynman rules**. These Feynman rules, given in Box 7.1, tell us how to directly calculate $i\mathcal{M}$ from pictures. With these rules, you can forget about practically anything else we have covered so far.

A couple of notes about the rules. The combinatoric factor for the diagram, as contributing to the momentum-space Feynman rules, is given only by the geometric symmetry factor of the diagram. Identical particles are already taken care of in Wick's theorem; moving around the a_p's and $a_p^\dagger$'s has the algebra of identical particles in them. The only time identical particles need extra consideration is when we cannot distinguish the particles we are scattering. This only happens for final states, since we distinguish our initial states by the setup of the experiment. Thus, when n of the same particles are produced, we have to divide the cross section by $n!$.

| Momentum-space Feynman rules | Box 7.1 |

- Internal lines (those not connected to external points) get propagators $\frac{i}{p^2-m^2+i\varepsilon}$.
- Vertices come from interactions in the Lagrangian. They get factors of the coupling constant times i.
- Lines connected to external points do not get propagators (their propagators are canceled by terms from the LSZ reduction formula).
- Momentum is conserved at each vertex.
- Integrate over undetermined 4-momenta.
- Sum over all possible diagrams.

7.3.1 Signs of momenta

There is unfortunately no standard convention about how to choose the direction in which the momenta are going. For external momenta it makes sense to assign them their physical values, which should have positive energy. Then momentum conservation becomes

$$\sum p_i = \sum p_f, \qquad (7.87)$$

which appears in δ-functions as $\delta^4(\sum p_i - \sum p_f)$.

For internal lines, we integrate over the momenta, so it does not matter if we use k_μ or $-k_\mu$. Still, it is important to keep track of which way the momentum is going so that all the δ-functions at the vertices are also $\sum(p_{\text{in}} - p_{\text{out}})$. We draw arrows next to the lines to indicate the flow of momentum:

$$(7.88)$$

We also sometimes draw arrows superimposed on lines, as ——▶——. These arrows point in the direction of momentum for particles and opposite to the direction of momentum for antiparticles. We will discuss these *particle-flow arrows* more when we introduce antiparticles in Chapter 9.

You should be warned that sometimes Feynman diagrams are drawn with time going upwards, particularly in describing hadronic collisions.

7.3.2 Disconnected graphs

A lot of the contractions will result in diagrams where some subset of the external vertices connect to each other without interacting with the other subsets. What do we do with graphs where subsets are independently connected, such as the contribution to the 8-point function shown on the left in Figure 7.1? Diagrams like this have physical effects. For example, at

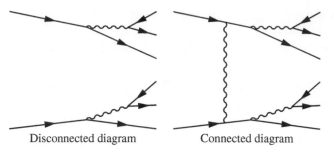

Disconnected diagram   Connected diagram

Fig. 7.1 Disconnected graphs like the one on the left have important physical effects. However, they have a different singularity structure and therefore zero interference with connected graphs, like the one on the right.

a muon collider, there would be a contribution to the S-matrix from situations where the muons just decay independently, somewhat close to the interaction region, which look like the left graph, in addition to the contribution where the muons scatter off each other, which might look like the right graph in Figure 7.1.

Clearly, both processes need to be incorporated for an accurate description of the collision. However, the disconnected decay process can be computed from the S-matrix for $1 \to 3$ scattering (as in either half of the left diagram). The probability for the $2 \to 6$ process from the disconnected diagram is then just the product of the two $1 \to 3$ probabilities. More generally, the S-matrix (with bubbles removed) factorizes into a product of sums of connected diagrams, just as the bubbles factorized out of the full S-matrix (see Eq. (7.76)).

The only possible complication is if there could be interference between the disconnected diagrams and the connected ones. However, this cannot happen: there is zero interference. To see why, recall that the definition of the matrix element that these time-ordered calculations produce has only a single δ-function:

$$S = \mathbb{1} + i\delta^4(\Sigma p)\mathcal{M}. \tag{7.89}$$

Disconnected matrix elements will have extra δ-functions $\mathcal{M}_{\text{disconnected}} = \delta^4(\Sigma_{\text{subset}}p)(\cdots)$. Connected matrix elements are just integrals over propagators, as given by the Feynman rules. Such integrals can only have poles or possibly branch cuts, but are analytic functions of the external momenta away from these. They can never produce singularities as strong as δ-functions. (The same decoherence is also relevant for meta-stable particles produced in collisions, where it leads to the narrow-width approximation, to be discussed in Section 24.1.4.) Therefore, the disconnected amplitudes are always infinitely larger than the connected ones, and the interference vanishes. You can check this in Problem 7.2.

More profoundly, the fact that there can never be more than a single δ-function coming out of connected amplitudes is related to a general principle called **cluster decomposition**, which is sometimes considered an axiom of quantum field theory [Weinberg, 1995]. The cluster decomposition principle says that experiments well-separated in space cannot influence each other. More precisely, as positions in one subset become well-separated from positions in the other subsets, the connected S-matrix should vanish. If there were an extra δ-function, one could asymptotically separate some of the points in such a way that the

S-matrix went to a constant, violating cluster decomposition. Constructing local theories out of *fields* made from creation and annihilation operators guarantees cluster decomposition, as we have seen. However, it is not known whether the logic is invertible, that is, if the only possible theories that satisfy cluster decomposition are local field theories constructed out of creation and annihilation operators. It is also not clear how well cluster decomposition has been tested experimentally.

Technicalities of cluster decomposition aside, the practical result of this section is that the only thing we ever need to compute for scattering processes is

$$\langle 0|T\{\phi_0(x_1)\cdots\phi_0(x_n)\}|0\rangle_{\text{connected}}, \tag{7.90}$$

where "connected" means every external vertex connects to every other external vertex through the graph somehow. Everything else is factored out or normalized away. Bubbles come up occasionally in discussions of vacuum energy; disconnected diagrams are never important.

7.4 Examples

The Feynman rules will all make a lot more sense after we do some examples. Let us start with the Lagrangian,

$$\mathcal{L} = -\frac{1}{2}\phi\Box\phi - \frac{1}{2}m^2\phi^2 + \frac{g}{3!}\phi^3, \tag{7.91}$$

and consider the differential cross section for $\phi\phi \to \phi\phi$ scattering. In the center-of-mass frame, the cross section is related to the matrix element by Eq. (5.32),

$$\frac{d\sigma}{d\Omega}(\phi\phi \to \phi\phi) = \frac{1}{64\pi^2 E_{\text{CM}}^2}|\mathcal{M}|^2. \tag{7.92}$$

Let the incoming momenta be p_1^μ and p_2^μ and the outgoing momenta be p_3^μ and p_4^μ.

There are three diagrams. The first gives

$$i\mathcal{M}_s = \quad = (ig)\frac{i}{(p_1+p_2)^2 - m^2 + i\varepsilon}(ig) = \frac{-ig^2}{s - m^2 + i\varepsilon}, \tag{7.93}$$

where $s \equiv (p_1 + p_2)^2$. The second gives

$$i\mathcal{M}_t = \quad = (ig)\frac{i}{(p_1-p_3)^2 - m^2 + i\varepsilon}(ig) = \frac{-ig^2}{t - m^2 + i\varepsilon}, \tag{7.94}$$

where $t \equiv (p_1 - p_3)^2$. The final diagram evaluates to

$$i\mathcal{M}_3 = \quad = (ig)\frac{i}{(p_1 - p_4)^2 - m^2 + i\varepsilon}(ig) = \frac{-ig^2}{u - m^2 + i\varepsilon}, \quad (7.95)$$

where $u \equiv (p_1 - p_4)^2$. The sum is

$$\frac{d\sigma}{d\Omega}(\phi\phi \to \phi\phi) = \frac{g^4}{64\pi^2 E_{\mathrm{CM}}^2}\left[\frac{1}{s - m^2} + \frac{1}{t - m^2} + \frac{1}{u - m^2}\right]^2, \quad (7.96)$$

We have dropped the $i\varepsilon$, which is fine as long as s, t and u are not equal to m^2. (For that to happen, the intermediate scalar would have to go on-shell in one of the diagrams, which is a degenerate situation, usually contributing only to $\mathbb{1}$ in the S-matrix. The $i\varepsilon$'s will be necessary for loops, but in tree-level diagrams you can pretty much ignore them.)

7.4.1 Mandelstam variables

The variables s, t and u are called **Mandelstam variables**. They are a great shorthand, used almost exclusively in $2 \to 2$ scattering and in $1 \to 3$ decays, although there are generalizations for more momenta. For $2 \to 2$ scattering, with initial momenta p_1 and p_2 and final momenta p_3 and p_4, they are defined by

$$s \equiv (p_1 + p_2)^2 = (p_3 + p_4)^2, \quad (7.97)$$
$$t \equiv (p_1 - p_3)^2 = (p_2 - p_4)^2, \quad (7.98)$$
$$u \equiv (p_1 - p_4)^2 = (p_2 - p_3)^2. \quad (7.99)$$

These satisfy

$$s + t + u = \sum m_j^2, \quad (7.100)$$

where m_j are the invariant masses of the particles.

As we saw in the previous example, s, t and u correspond to particular diagrams where momentum in the propagator has invariant $p_\mu^2 = s$, t or u. This correspondence is summarized in Box 7.2. The s-channel is an annihilation process. In the s-channel, the

Box 7.2 $2 \to 2$ **scattering channels.**

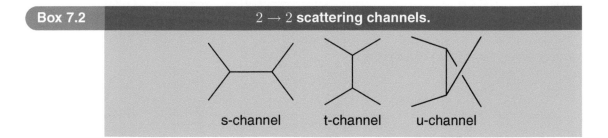

s-channel   t-channel   u-channel

intermediate state has $p_\mu^2 = s \geq 0$. The t- and u-channels are scattering diagrams and have $t \leq 0$ and $u \leq 0$. s, t and u are great because they are Lorentz invariant. So we compute $\mathcal{M}(s, t, u)$ in the center-of-mass frame, and then we can easily find out what it is in any other frame, for example the frame of the lab in which we are doing the experiment. We will use s, t and u a lot.

7.4.2 Derivative couplings

Suppose we have an interaction with derivatives in it, such as

$$\mathcal{L}_{\text{int}} = \lambda \phi_1 (\partial_\mu \phi_2)(\partial_\mu \phi_3), \tag{7.101}$$

where three different scalar fields are included for clarity. In momentum space, these ∂_μ's give factors of momenta. But now remember that

$$\phi(x) = \int \frac{d^3 p}{(2\pi)^3} \frac{1}{\sqrt{2\omega_p}} \left(a_p e^{-ipx} + a_p^\dagger e^{ipx} \right). \tag{7.102}$$

So, if the particle is being created (emerging from a vertex) it gets a factor of ip_μ, and if it is being destroyed (entering a vertex) it gets a factor of $-ip_\mu$. So, we get a minus for incoming momentum and a plus for outgoing momentum. In this case, it is quite important to keep track of whether momentum is flowing into or out of the vertex.

For example, take the diagram

(7.103)

Label the initial momenta p_1^μ and p_2^μ and the final momenta $p_1'^\mu$ and $p_2'^\mu$. The exchanged momentum is $k^\mu = p_1^\mu + p_2^\mu = p_1'^\mu + p_2'^\mu$. Then this diagram gives

$$i\mathcal{M} = (i\lambda)^2 \left(-ip_2^\mu\right)(ik^\mu) \frac{i}{k^2} (ip_2''^\nu)(-ik^\nu) = -i\lambda^2 \frac{[p_2 \cdot p_1 + (p_2)^2][p_2' \cdot p_1' + (p_2')^2]}{(p_1 + p_2)^2}. \tag{7.104}$$

As a cross check, we should get the same answer if we use a different Lagrangian related to the one we used by integration by parts:

$$\mathcal{L}_{\text{int}} = -\lambda \phi_3 [(\partial_\mu \phi_1)(\partial_\mu \phi_2) + \phi_1 \Box \phi_2]. \tag{7.105}$$

Now our one diagram becomes four diagrams, from the two types of vertices on the two sides, all of which look like Eq. (7.103). It is easiest to add up the contributions to the vertices before multiplying, which gives

$$i\mathcal{M} = (i\lambda)^2 \left[(-ip_2^\mu)(-ip_1^\mu) + (-ip_2)^2\right] \frac{i}{k^2} \left[(ip_2''^\nu)(ip_1''^\nu) + (ip_2')^2\right]$$

$$= -i\lambda^2 \frac{[p_2 \cdot p_1 + (p_2)^2][p_2' \cdot p_1' + (p_2')^2]}{(p_1 + p_2)^2}, \tag{7.106}$$

which is exactly what we had above. So, integrating by parts does not affect the matrix elements, as expected. Thus the Feynman rules passed our cross check.

To see more generally that integrating by parts does not affect matrix elements, it is enough to show that total derivatives do not contribute to matrix elements. Suppose we have a term

$$\mathcal{L}_{\text{int}} = \partial_\mu(\phi_1 \cdots \phi_n), \tag{7.107}$$

where there are any number of fields in this term. This would give a contribution from the derivative acting on each field, with a factor of that field's momenta. So if the vertex would have given V without the derivative, adding the derivative makes it

$$\Big(\underbrace{\sum p_\mu^i}_{\text{incoming}} - \underbrace{\sum p_\mu^j}_{\text{outgoing}} \Big) V. \tag{7.108}$$

Since the sum of incoming momenta is equal to the sum of outgoing momenta, because momentum is conserved at each vertex, we conclude that total derivatives do not contribute to matrix elements

To be precise, total derivatives do not contribute to matrix elements *in perturbation theory*. The term

$$\varepsilon^{\mu\nu\alpha\beta} F_{\mu\nu} F_{\alpha\beta} = 4\partial_\mu(\varepsilon^{\mu\nu\alpha\beta} A_\alpha \partial_\beta A_\nu) \tag{7.109}$$

is a total derivative. If we add a term $\theta\, \varepsilon^{\mu\nu\alpha\beta} F_{\mu\nu} F_{\alpha\beta}$ to the Lagrangian, indeed nothing happens in perturbation theory. It turns out that there are effects of this term that will never show up in Feynman diagrams, but are perfectly real. They have physical consequences. For example, if this term appeared in the Lagrangian with anything but an exponentially small coefficient, it would lead to an observable electric dipole moment for the neutron. That no such moment has been seen is known as the strong CP problem (see Section 29.5.3). A closely related effect from such a total derivative is the mass of the η' meson, which is larger than it could possibly be without total-derivative terms (see Section 30.5.2). In both cases the physical effect comes from the strong interactions which are non-perturbative.

7.A Normal ordering and Wick's theorem

In this appendix we prove that the vacuum matrix element of a time-ordered product of free fields is given by the sum of all possible full contractions, a result known as Wick's theorem. This theorem is necessary for the derivation of the Feynman rules in the Hamiltonian approach.

7.A.1 Normal ordering

To prove Wick's theorem, we will manipulate expressions with creation and annihilation operators into the form of a c-number expression plus terms that vanish when acting on the vacuum. This is always possible since we can commute the annihilation operators past the creation operators until they are all on the right, at which point they give zero when acting on the vacuum.

For example, we can write

$$(a_p^\dagger + a_p)(a_k^\dagger + a_k) = [a_p, a_k^\dagger] + a_k^\dagger a_p + a_p^\dagger a_k + a_p a_k + a_p^\dagger a_k^\dagger$$
$$= (2\pi)^3 \delta^3(p-k) + a_k^\dagger a_p + a_p^\dagger a_k + a_p a_k + a_p^\dagger a_k^\dagger. \quad (7.A.110)$$

Then, since the terms with annihilation operators on the right vanish, as do the terms with creation operators on the left, we get

$$\langle 0|(a_p^\dagger + a_p)(a_k^\dagger + a_k)|0\rangle = (2\pi)^3 \delta^3(p-k). \quad (7.A.111)$$

We call terms with all annihilation operators on the right *normal ordered.*

Normal ordered: all the $a_p^\dagger$ operators are on the left of all the a_p operators.

We represent normal ordering with colons. So,

$$:(a_p^\dagger + a_p)(a_k^\dagger + a_k): = a_k^\dagger a_p + a_p^\dagger a_k + a_p a_k + a_p^\dagger a_k^\dagger. \quad (7.A.112)$$

When you normal order something, you just pick up the operators and move them. Just manhandle them over, without any commuting, just as you manhandled the operators within a time-ordered product. Thus the $\delta(p-k)$ from Eq. (7.A.110) does not appear in Eq. (7.A.112).

The point of normal ordering is that vacuum matrix elements of normal-ordered products of free fields vanish:

$$\langle 0|:\phi_0(x_1)\cdots\phi_0(x_n):|0\rangle = 0. \quad (7.A.113)$$

The only normal-ordered expressions that do not vanish in the vacuum are *c*-number functions. Such a function f satisfies

$$\langle 0| :f: |0\rangle = f. \quad (7.A.114)$$

The nice thing about normal ordering is that we can use it to specify operator relations. For example,

$$T\{\phi_0(x)\phi_0(y)\} = :\phi_0(x)\phi_0(y) + D_F(x,y):. \quad (7.A.115)$$

This is obviously true in vacuum matrix elements, since $D_F(x,y) = \langle 0|T\{\phi_0(x)\phi_0(y)\}|0\rangle$ and vacuum matrix elements of normal-ordered products vanish. But it is also true at the level of the operators, as we show below. The point is that by normal ordering expressions we can read off immediately what will happen when we take vacuum matrix elements, but no information is thrown out.

7.A.2 Wick's theorem

Wick's theorem relates time-ordered products of fields to normal-ordered products of fields and contractions. It is given in Box 7.3. A **contraction** means taking two fields $\phi_0(x_i)$ and

Box 7.3   Wick's theorem

$$T\left\{\phi_0(x_1)\cdots\phi_0(x_n)\right\} = \, :\phi_0(x_1)\cdots\phi_0(x_n) + \begin{smallmatrix}\text{all possible}\\\text{contractions}\end{smallmatrix}:$$

$\phi_0(x_j)$ from anywhere in the series and replacing them with a factor of $D_F(x_i, x_j)$ for each pair of fields. "All possible contractions" includes one contraction, two contractions, etc., involving any of the fields. But each field can only be contracted once. Since normal-ordered products vanish unless all the fields are contracted, this implies that the time-ordered product is the sum of all the full contractions, which is what we will actually use to generate Feynman rules.

Wick's theorem is easiest to prove first by breaking the field up into creation and annihilation parts, $\phi_0(x) = \phi_+(x) + \phi_-(x)$, where

$$\phi_+(x) = \int \frac{d^3p}{(2\pi)^3} \frac{1}{\sqrt{2\omega_p}} a_p^\dagger e^{ipx}, \quad \phi_-(x) = \int \frac{d^3p}{(2\pi)^3} \frac{1}{\sqrt{2\omega_p}} a_p e^{-ipx}. \quad (7.A.116)$$

Since $[a_k, a_p^\dagger] = (2\pi)^3\delta^3(\vec{p} - \vec{k})$, commutators of these operators are just functions. In fact, the Feynman propagator can be written as

$$\begin{aligned}D_F(x_1, x_2) &= \langle 0|T\{\phi_0(x_1)\,\phi_0(x_2)\}|\,0\rangle \\ &= [\phi_-(x_1)\,,\phi_+(x_2)]\,\theta(t_1 - t_2) + [\phi_-(x_2)\,,\phi_+(x_1)]\,\theta(t_2 - t_1)\,. \quad (7.A.117)\end{aligned}$$

This particular combination represents a contraction.

Let us verify Wick's theorem for two fields. For $t_1 > t_2$

$$T\{\phi_0(x_1)\phi_0(x_2)\} = \phi_+(x_1)\,\phi_+(x_2) + \phi_+(x_1)\,\phi_-(x_2) + \phi_-(x_1)\,\phi_+(x_2) + \phi_-(x_1)\,\phi_-(x_2)\,. \quad (7.A.118)$$

All terms in this expression are normal ordered except for $\phi_-(x_1)\,\phi_+(x_2)$. So,

$$T\{\phi_0(x_1)\phi_0(x_2)\} = \,:\phi_0(x_1)\phi_0(x_2): + \,[\phi_-(x_1)\,,\phi_+(x_2)]\,, \quad t_1 > t_2. \quad (7.A.119)$$

For $t_2 > t_1$, the expression is the same with $x_1 \leftrightarrow x_2$. Thus,

$$T\{\phi_0(x_1)\phi_0(x_2)\} = \,:\phi_0(x_1)\phi_0(x_2): + \,D_F(x_1, x_2)\,, \quad (7.A.120)$$

exactly as Wick's theorem requires.

The full proof is straightforward to do by mathematical induction. We have shown that it works for two fields. Assume it holds for $n - 1$ fields. Without loss of generality, let t_1 be the latest time for all n fields. Then,

$$T\{\phi_0(x_1)\phi_0(x_2)\cdots\phi_0(x_n)\} = [\phi_+(x_1) + \phi_-(x_1)]\,:\phi_0(x_2)\cdots\phi_0(x_n) + \begin{smallmatrix}\text{all possible}\\\text{contractions}\end{smallmatrix}:. \quad (7.A.121)$$

Since $\phi_+(x_1)$ is on the left and contains $a_p^\dagger$ operators, we can just move it into the normal-ordering. The $\phi_-(x_1)$ must be moved through to the right. Each time it passes a $\phi_+(x_i)$ field in the normal-ordered product, a contraction results. The result is the sum over the normal-ordered product of n fields and all possible contractions of $\phi_-(x_1)$ with any of the $\phi_+(x_i)$ in any of the terms in the normal-ordered product in Eq. (7.A.121). That is exactly what all possible contractions of the fields $\phi_0(x_2)$ to $\phi_0(x_n)$ means. Thus, Wick's theorem is proven.

The result of Wick's theorem is that time-ordered products are given by a bunch of contractions plus normal-ordered products. Since the normal-ordered products vanish in vacuum matrix elements, all that remains for vacuum matrix elements of time-ordered products are the Feynman propagators.

Problems

7.1 Consider the Lagrangian for ϕ^3 theory,

$$\mathcal{L} = -\frac{1}{2}\phi(\Box + m^2)\phi + \frac{g}{3!}\phi^3. \qquad (7.122)$$

(a) Draw a tree-level Feynman diagram for the decay $\phi \to \phi\phi$. Write down the corresponding amplitude using the Feynman rules.
(b) Now consider the one-loop correction, given by

$$(7.123)$$

Write down the corresponding amplitude using the Feynman rules.
(c) Now start over and write down the diagram from part (b) in position space, in terms of integrals over the intermediate points and Wick contractions, represented with factors of D_F.
(d) Show that after you apply LSZ, what you got in (c) reduces to what you got in (b), by integrating the phases into δ-functions, and integrating over those δ-functions.

7.2 Calculate the contribution to $2 \to 4$ scattering from the Lagrangian $\mathcal{L} = -\frac{1}{2}\phi\Box\phi + \frac{g}{3!}\phi^3 + \frac{1}{6!}\lambda\phi^6$ from both the connected diagram, with the 6-point vertex, and the disconnected diagram with the 3-point vertex. Show that there is no interference between the two diagrams. (There are of course many connected diagrams with the 3-point vertex that you can ignore.)

7.3 Non-relativistic Møller scattering: $e^-e^- \to e^-e^-$. If the electron and photon were spinless, we could write the Lagrangian as

$$\mathcal{L} = -\frac{1}{2}\phi_e(\Box + m_e^2)\phi_e - \frac{1}{2}A_0\Box A_0 + em_e A_0\phi_e\phi_e, \qquad (7.124)$$

where A_0 is the scalar potential and the factor of m_e comes from the non-relativistic limit as in Section 5.2 (or by dimensional analysis!).

(a) Draw the three tree-level $e^- e^- \to e^- e^-$ diagrams following from this Lagrangian.

(b) Which one of the diagrams would be forbidden in real QED?

(c) Evaluate the other two diagrams, and express the answers in terms of s, t and u. Give the diagrams an extra relative minus sign, because electrons are fermions.

(d) Now let us put back the spin. In the non-relativistic limit, the electron spin is conserved. This should be true at each vertex, since the photon is too soft to carry off any spin angular momentum. Thus, a vertex can only allow for $|\uparrow\rangle \to |\uparrow; \gamma\rangle$ or $|\downarrow\rangle \to |\downarrow; \gamma\rangle$. This forbids, for example, $|\uparrow\downarrow\rangle \to |\uparrow\uparrow\rangle$ from occurring. For each of the 16 possible sets of spins for the four electrons (for example $|\uparrow\downarrow\rangle \to |\uparrow\uparrow\rangle$), which processes are forbidden, and which get contributions from the s-, t- or u-channels?

(e) It is difficult to measure electron spins. Thus, assume the beams are unpolarized, meaning that they have an equal fraction of spin-up and spin-down electrons, and that you do not measure the final electron spins, only the scattering angle θ. What is the total rate $\frac{d\sigma}{d\cos\theta}$ you would measure? Express the answer in terms of E_{CM} and θ. Sketch the angular distribution.

7.4 We made a distinction between kinetic terms, which are bilinear in fields, and interactions, which have three or more fields. Time evolution with the kinetic terms is solved exactly as part of the free Hamiltonian. Suppose, instead, we only put the derivative terms in the free Hamiltonian and treated the mass as an interaction. So,

$$H_0 = \frac{1}{2}\phi\Box\phi, \quad H_{\mathrm{int}} = \frac{1}{2}m^2\phi^2. \tag{7.125}$$

(a) Draw the (somewhat degenerate looking) Feynman graphs that contribute to the 2-point function $\langle 0|T\{\phi(x)\phi(y)\}|0\rangle$ using only this interaction, up to order m^6.

(b) Evaluate the graphs.

(c) Sum the series to all orders in m^2 and show you reproduce the propagator that would have come from taking $H_0 = \frac{1}{2}\phi\Box\phi + \frac{1}{2}m^2\phi^2$.

(d) Repeat the exercise classically: Solve for the massless propagator using an external current, perturb with the mass, sum the series, and show that you get the same answer as if you included the mass to begin with.

7.5 Show in general that integrating by parts does not affect matrix elements.

7.6 Use the Lagrangian

$$\mathcal{L} = -\frac{1}{2}\phi_1\Box\phi_1 - \frac{1}{2}\phi_2\Box\phi_2 + \frac{\lambda}{2}\phi_1(\partial_\mu\phi_2)(\partial_\mu\phi_2) + \frac{g}{2}\phi_1^2\phi_2 \tag{7.126}$$

to calculate the differential cross section

$$\frac{d\sigma}{d\Omega}(\phi_1\phi_2 \to \phi_1\phi_2) \tag{7.127}$$

at tree level.

7.7 Consider a Feynman diagram that looks like a regular tetrahedron, with the external lines coming out of the four corners. This can contribute to $2 \to 2$ scattering in a scalar field theory with interaction $\frac{\lambda}{4!}\phi^4$. You can take ϕ to be massless.

(a) Write down the corresponding amplitude including the appropriate symmetry factor.

(b) What would the symmetry factor be for the same diagram in ϕ^3 theory without the external lines?

7.8 Radioactive decay. The muon decays to an electron and two neutrinos through an intermediate massive particle called the W^- boson. The muon, electron and W^- all have charge -1.

(a) Write down a Lagrangian that would allow for $\mu^- \to e^- \bar{\nu}_e \nu_\mu$. Assume the W and other particles are all scalars, and the e^-, ν_e and ν_μ are massless. Call the coupling g.

(b) Calculate $|\mathcal{M}|^2$ for this decay in the limit that the W mass, m_W, is large.

(c) The decay rate Γ $(= \frac{1}{\text{lifetime}})$ is proportional to $|\mathcal{M}|^2$. The coupling g should be dimensionless (like the coupling e for the photon), but appears dimensionful because we ignore spin. If the W spin were included, you would get extra factors of p^μ, which would turn into a factor of $\sqrt{s} = m_\mu$ in $|\mathcal{M}|^2$. Use dimensional analysis to figure out what power of m_μ should be there. Also, throw in a $\frac{1}{192\pi^3}$ for the three-body phase space, as in Eq. (5.55) from Problem 5.3.

(d) Let's guess that the coupling g is close to the coupling for electromagnetism, so $\frac{g^2}{4\pi} \approx \frac{1}{137}$. Use this and the muon mass $(m_\mu = 105 \text{ MeV})$ and lifetime $(2.2 \times 10^{-6} \text{ s})$ to estimate the W mass.

(e) The tauon, τ, decays to $e^- \bar{\nu}_e \nu_\tau$. Use the τ lifetime $\Gamma_{\text{tot}}^{-1} = 2.9 \times 10^{-13}$ s and previous parts to estimate the τ mass. Which of m_W, g, m_μ, the muon lifetime, or the $192\pi^3$ we threw in does your prediction depend on?

(f) In reality, the tauon only decays as $\tau \to e^- \bar{\nu}_e \nu_\tau$ 17.8% of the time. Use this fact to refine your τ mass estimate.

(g) How could you measure g and M_W separately using precise measurements of the μ and τ decay distributions? What % precision would you need?

7.9 Unstable particles. Unstable particles pick up imaginary parts that generate a width Γ in their resonance line shape. This problem will develop an understanding of what is meant by the terms *width* and *pick up*.

(a) What would the cross section be for s-channel scattering if the intermediate propagator were $\frac{i}{p^2 - m^2 + im\Gamma}$, where $\Gamma > 0$? This is called the Breit–Wigner distribution.

(b) Sketch the cross section as a function of $x = \frac{s}{m^2}$ for $\frac{\Gamma}{m}$ small and for $\frac{\Gamma}{m}$ large.

(c) Show that a propagator only has an imaginary part if it goes on-shell. Explicitly, show that $\text{Im}(\mathcal{M}) = -\pi\delta(p^2 - m^2)$, when $i\mathcal{M} = \frac{i}{p^2 - m^2 + i\varepsilon}$.

(d) Loops of particles can produce effective interactions that have imaginary parts. Suppose we have another particle ψ and an interaction $\phi\psi\psi$ in the Lagrangian. Loops of ψ will have imaginary parts if and only if ψ is lighter than half of ϕ, that is, if $\phi \to \psi\psi$ is allowed kinematically. Draw a series of loop corrections to the ϕ propagator. Show that, if these give an imaginary number, you can sum the graphs to reproduce the propagator in part (a).

(e) What is the connection between parts (c) and (d)? Can you see why the width is related to the decay rate?

PART II

QUANTUM ELECTRODYNAMICS

Spin 1 and gauge invariance

8

Up until now, we have dealt with general features of quantum field theories. For example, we have seen how to calculate scattering amplitudes starting from a general Lagrangian. Now we will begin to explore what the Lagrangian of the real world could possibly be. In Part IV we will discuss what it actually is, or at least what is known about it so far.

A good way to start understanding the consistency requirements of the physical universe is with a discussion of spin. There is a deep connection between spin and Lorentz invariance that is obscure in non-relativistic quantum mechanics. For example, well before quantum field theory, it was known from atomic spectroscopy that the electron had two spin states. It was also known that light had two polarizations. The polarizations of light are easy to understand at the classical level since light is a field, but how can an individual photon be polarized? For the electron, we can at least think of it as a spinning top, so there is a classical analogy, but photons are massless and structureless, so what exactly is spinning? The answers to these questions follow from an understanding of Lorentz invariance and the requirements of a consistent quantum field theory.

Our discussion of spin and the Lorentz group is divided into a discussion of integer spin particles (tensor representations) in this chapter and half-integer spin particles (spinor representations) in Chapter 10.

8.1 Unitary representations of the Poincaré group

Our universe has a number of apparent symmetries that we would like our quantum field theory to respect. One symmetry is that no place in space-time seems any different from any other place. Thus, our theory should be translation invariant: if we take all our fields $\psi(x)$ and replace them by $\psi(x + a)$ for any constant 4-vector a^ν, the observables should look the same. Another symmetry is Lorentz invariance: physics should look the same whether we point our measurement apparatus to the left or to the right, or put it on a train. The group of translations and Lorentz transformations is called the **Poincaré group**, ISO(1,3) (the *iso*metry group of Minkowski space).

Our universe also has a bunch of different types of particles in it. Particles have mass and spin and all kinds of other quantum numbers. They also have momentum and the value of spin projected on some axis. If we rotate or boost to change frame, only the momenta and the spin projection change, as determined by the Poincaré group, but the other quantum numbers do not. So a **particle** can be defined as a set of states that mix only among themselves under Poincaré transformations.

Generically, we can write that our states transform as

$$|\psi\rangle \to \mathcal{P}|\psi\rangle \tag{8.1}$$

under a Poincaré transformation $\mathcal{P}$. A set of objects ψ that mix under a transformation group is called a **representation** of the group. For example, scalar fields $\phi(x)$ at all points x form a representation of translations, since $\phi(x) \to \phi(x+a)$. Quite generally, in a given representation there should be a basis for the states $|\psi\rangle$, call it $\{|\psi_i\rangle\}$, where i is a discrete or continuous index, so that

$$|\psi_i\rangle \to \mathcal{P}_{ij}|\psi_j\rangle, \tag{8.2}$$

where the transformed states are expressible in the original basis. If no subset of states transform only among themselves, the representation is **irreducible**.

In addition, we want **unitary** representations. The reason for this is that the things we compute in field theory are matrix elements,

$$\mathcal{M} = \langle\psi_1|\psi_2\rangle, \tag{8.3}$$

which should be Poincaré **invariant**. If $\mathcal{M}$ is Poincaré invariant, and $|\psi_1\rangle$ and $|\psi_2\rangle$ transform covariantly under a Poincaré transformation $\mathcal{P}$, we find

$$\mathcal{M} = \langle\psi_1|\mathcal{P}^\dagger\mathcal{P}|\psi_2\rangle. \tag{8.4}$$

So we need $\mathcal{P}^\dagger\mathcal{P} = 1$, which is the definition of unitarity. The unitary representations of the Poincaré group are only a small subset of all the representations of the Poincaré group. For example, as we will discuss, the 4-vector representation, A_μ, is not unitary. But the unitary ones are the only ones from which we will be able to compute Poincaré-invariant matrix elements, so we have to understand how to find them. Thus,

> Particles transform under irreducible unitary representations of the Poincaré group.

This statement can even be interpreted as the *definition* of what a particle is. Of course, many particles can transform under the same representation of the Poincaré group. What makes two particles identical is discussed in Section 12.1.

By the way, there is an even stronger requirement on physical theories: the S-matrix must be unitary. Requiring a unitary S-matrix constrains the dynamics of the theory, while demanding unitary representations of the Poincaré group is just a statement about free-particle states. Implications of unitarity of the S-matrix is the subject of Chapter 24.

One way to think of the allocation into irreducible representations is that our universe is clearly filled with different kinds of particles in different states. By doing things such as putting an electron in a magnetic field, or sending a photon through a polarizer, we manipulate the momenta and spins. Some states will mix with each other under these manipulations and some will not. We look at the irreducible representations because those are the building blocks with which we can construct the most general description of nature.

We already know some representations of the Poincaré group: the constant tensors, ϕ, V_μ, $T_{\mu\nu}$, …. These are finite-dimensional representations, with $1, 4, 16, \ldots$ elements. They transform under rotations and boosts as discussed in Section 2.1, and are invariant under translations. Unfortunately, these are not unitary representations, as we will see below. In fact, there are *no finite-dimensional unitary representations of the Poincaré group*.

The unitary irreducible representations of the Poincaré group were classified by Eugene Wigner in 1939 [Wigner, 1939]. They are all infinite dimensional and naturally described by fields. As you might imagine, before Wigner people did not really know what the rules were for constructing physical theories, and by trial and error they were coming across all kinds of problems. Wigner showed that irreducible unitary representations are uniquely classified by mass m and spin J, where m is a non-negative real number and spin is a non-negative half integer, $J = 0, \frac{1}{2}, 1, \frac{3}{2}, \ldots$ Moreover, Wigner showed that, if $J > 0$, for each value of the momentum with $p^2 = m^2$ there are $2J + 1$ independent states in the representation if $m > 0$ and exactly 2 states for $m = 0$.[1] These states correspond to linearly independent polarizations of particles with spin J. If $J = 0$ there is only one state for any m. You can find the proof of Wigner's theorem in [Weinberg, 1995]. We are not going to reproduce the proof. Instead, we will do some examples that will make the ingredients that go into the proof clear.

Knowing what the representations of the Poincaré group are is a great start, but we still have to figure out how to construct a unitary interacting theory of particles in these representations. To do that, we would like to embed the irreducible representations into objects with space-time indices. That is, we want to squeeze states of spin $0, \frac{1}{2}, 1, \frac{3}{2}, 2$ etc. into scalar fields $\phi(x)$, vector fields $V_\mu(x)$, tensor fields $T_{\mu\nu}(x)$, spinor fields $\psi(x)$ etc. That way we can write down simple-looking Lagrangians and develop general methods for doing calculations. We see an immediate complication: tensors have $1, 4, 16, 64, \ldots, 4^n$ elements, but spin states have $1, 3, 5, 7, \ldots, 2J + 1$ physical degrees of freedom. The embedding of the $2J + 1$ states for a unitary representation in the 4^n-dimensional tensors is tricky, and leads to things such as gauge invariance, as we will see in this chapter.[2]

8.1.1 Unitarity versus Lorentz invariance

We do not need fancy mathematics to see the conflict between unitarity and Lorentz invariance. In non-relativistic quantum mechanics, you have an electron with spin up $|\uparrow\rangle$ or spin down $|\downarrow\rangle$. This is your basis, and you can have a state which is any linear combination of these two:

$$|\psi\rangle = c_1|\uparrow\rangle + c_2|\downarrow\rangle. \tag{8.5}$$

---

[1] To be accurate, there are also tachyon representations with $m^2 < 0$, and continuous spin representations for $m = 0$. These exotic representations seem not to be realized in nature and we will not discuss them further.

[2] If we did not care to write down local Lagrangians, we could avoid introducing gauge invariance altogether. Alternate approaches based on using on-shell physical states only are discussed in Chapters 24 and 27. However, quantum field theory with gauge invariance remains the most complete method for studying massless spin-1 particles.

The norm of such a state is

$$\langle\psi|\psi\rangle = |c_1|^2 + |c_2|^2 > 0. \tag{8.6}$$

This norm is invariant under rotations, which send

$$|\uparrow\rangle \to \cos\theta|\uparrow\rangle + \sin\theta|\downarrow\rangle, \qquad |\downarrow\rangle \to -\sin\theta|\uparrow\rangle + \cos\theta|\downarrow\rangle. \tag{8.7}$$

(In fact, the norm is invariant under the larger group SU(2), which you can see using the Pauli matrices, but that is not important right now.)

Say we wanted to do the same thing with a basis of four states $|V_\mu\rangle$ which transform as a 4-vector. Then an arbitrary linear combination would be

$$|\psi\rangle = c_0|V_0\rangle + c_1|V_1\rangle + c_2|V_2\rangle + c_3|V_3\rangle. \tag{8.8}$$

The norm of this state would be

$$\langle\psi|\psi\rangle = |c_0|^2 + |c_1|^2 + |c_2|^2 + |c_3|^2 > 0. \tag{8.9}$$

This is the norm for any basis and it is always positive, which is one of the postulates of quantum mechanics. However, the norm is not Lorentz invariant. For example, suppose we start with $|\psi\rangle = |V_0\rangle$, which has norm $\langle\psi|\psi\rangle = 1$. Then we boost in the 1 direction, so we get $|\psi'\rangle = \cosh\beta|V_0\rangle + \sinh\beta|V_1\rangle$. Now the norm is

$$\langle\psi'|\psi'\rangle = \cosh^2\beta + \sinh^2\beta \neq 1 = \langle\psi|\psi\rangle. \tag{8.10}$$

Thus, the probability of finding that a state is itself depends on what frame we are in! We see that the norm is not invariant under the boost. In terms of matrices, the boost matrix

$$\Lambda = \begin{pmatrix} \cosh\beta & \sinh\beta \\ \sinh\beta & \cosh\beta \end{pmatrix} \tag{8.11}$$

is not unitary: $\Lambda^\dagger \neq \Lambda^{-1}$.

One way out, you might suppose, could be to modify the norm to be

$$\langle\psi|\psi\rangle = |c_0|^2 - |c_1|^2 - |c_2|^2 - |c_3|^2. \tag{8.12}$$

This is Lorentz invariant, but not positive definite. That is not automatically a problem, since inner products in quantum mechanics are in general complex numbers. In fact, even with this norm the probability $P = |\langle\psi|\psi\rangle|^2 \geq 0$ for any state. However, the probabilities will no longer be ≤ 1. For example, suppose $|\psi\rangle = |V_0\rangle$ so that $\langle\psi|\psi\rangle = 1$ as before. Any state related to this one by a boost such as $|\psi'\rangle = \cosh\beta|V_0\rangle + \sinh\beta|V_1\rangle$ must also be in the Hilbert space, by Lorentz invariance. And $\langle\psi'|\psi'\rangle = 1$, by construction. However, the probability of finding $|\psi'\rangle$ in the state $|\psi\rangle = |V_0\rangle$ is $|\langle V_0|\psi'\rangle|^2 = \cosh^2\beta$. Since for $\beta \neq 0, \cosh\beta > 1$, there is no way to interpret this projection as a probability. Thus, because Lorentz transformations can mix positive norm and negative norm states, the probabilities are not bounded. In Problem 8.1, you can show that having a probability interpretation, with $0 \leq P \leq 1$, requires us to have only positive (or only negative) norm states. So unitarity, with a positive definite norm, is critical to have any physical interpretation of quantum mechanics.

In summary, there is a conflict between having a Hilbert space with a positive norm, which is a physical requirement leading to the $\delta^{\mu\nu}$ inner product preserved under unitary transformations, and the requirement of Lorentz invariance, which needs the $g^{\mu\nu}$ inner product preserved under Lorentz transformations. When we study general representations of the Lorentz group in Chapter 10, we will be able to trace this conflict to the Lorentz group being non-compact and the boosts having anti-Hermitian generators.

What do we do about the conflict? Well, there are two things we need to fix. First of all, note that $V_\mu^2 = V_0^2 - V_1^2 - V_2^2 - V_3^2$ has one positive term and three negative terms. In fact, the vector representation of the Lorentz group V_μ, which is four-dimensional, is the direct sum of two irreducible representations: a spin-0 representation, which is one-dimensional, and a spin-1 representation, which is three-dimensional. If we could somehow project the spin-1 (or spin-0) representation out of the reducible tensor representations (V_μ or $h_{\mu\nu}$), then we might be able write down Lorentz-invariant Lagrangians for a theory with positive norm.

The second thing is that, while there are in fact no non-trivial finite-dimensional irreducible unitary representations of the Poincaré group, there are some infinite-dimensional ones. We will see that instead of constant basis vectors, such as $(1,0,0,0)$, $(0,1,0,0)$ etc., we will need a basis $\epsilon_\mu(p)$ that depends on the momentum of the particle. So the plan is to first see how to embed the right number of degrees of freedom for a particular mass and spin (irreducible representation of the Poincaré group) into tensors such as A_μ. Then we will see how the infinite dimensionality of the representation comes about.

8.2 Embedding particles into fields

In this section we explore how to construct Lagrangians for fields that contain only particles of single spins. We will start with the classical theory, where we cannot ask for unitarity (there is no classical norm) but we can ask for the energy to be positive definite, or more generally, bounded from below. Having both positive and negative energy states classically heralds disaster after quantization. For example, if photons could have positive and negative energy, the vacuum could decay into pairs of photons with $p_1^\mu + p_2^\mu = 0$. This process does not violate energy or momentum conservation; it is normally only forbidden by photons having positive energies. An alternative criterion for determining whether a classical theory would be non-unitary when quantized is discussed in Section 8.7.

The classical energy density $\mathcal{E}$ is given by the 00 component of the energy-momentum tensor, which was calculated in Section 3.3.1, Eq. (3.36), to be

$$\mathcal{E} = \mathcal{T}_{00} = \sum_n \frac{\partial \mathcal{L}}{\partial \dot\phi_n}\dot\phi_n - \mathcal{L}. \tag{8.13}$$

The energy is $E = \int d^3x\, \mathcal{E}$.

8.2.1 Spin 0

For spin 0, the embedding is easy, we just put the one degree of freedom into a $J = 0$ scalar field $\phi(x)$. The Lagrangian is

$$\mathcal{L}(x) = \frac{1}{2}\partial_\mu\phi(x)\partial_\mu\phi(x) - \frac{1}{2}m^2\phi(x)^2, \tag{8.14}$$

which is Lorentz invariant and transforms covariantly under translations. The equation of motion is

$$(\Box + m^2)\phi = 0, \tag{8.15}$$

which has solutions $\phi = e^{\pm ipx}$ with $p^2 = m^2$. So this field has mass m. The Lagrangian is unique up to an overall constant for which the conventional normalization is given.

The energy density corresponding to this Lagrangian is given by

$$\mathcal{E} = \frac{\partial \mathcal{L}}{\partial \dot{\phi}}\dot{\phi} - \mathcal{L} = \frac{1}{2}\Big[(\partial_t\phi)^2 + (\vec{\nabla}\phi)^2 + m^2\phi^2\Big]. \tag{8.16}$$

This is a positive definite quantity and bounded from below by 0. Thus the overall sign in the scalar Lagrangian is consistent with positive energy.

8.2.2 Massive spin 1

For spin 1, there are three degrees of freedom if $m > 0$. This is a mathematical result, which we will not derive formally, but we will see how it works in practice. The smallest tensor field we could possibly embed these three degrees of freedom in is a vector field A_μ which has four components. Sometimes we write $4 = 3 \oplus 1$ to indicate that the four-dimensional representation of the Lorentz group is the direct sum of three-dimensional (spin-1) and one-dimensional (spin-0) representations of the rotation group $SO(3)$. A complete mathematical classification of the representations of the Lorentz group will be given in Chapter 10. In this chapter we will take the more physical approach of trying to engineer a Lagrangian that engenders a positive definite energy density, which we will see requires removing the spin-0 degree of freedom.

A natural guess for the Lagrangian for a massive spin-1 field is

$$\mathcal{L} = -\frac{1}{2}(\partial_\nu A_\mu)(\partial_\nu A_\mu) + \frac{1}{2}m^2 A_\mu^2, \tag{8.17}$$

where $A_\mu^2 = A_\mu A^\mu$. Then the equations of motion are

$$(\Box + m^2)A_\mu = 0, \tag{8.18}$$

which has four propagating modes. In fact, this Lagrangian is not the Lagrangian for a massive spin-1 field, but the Lagrangian for four massive scalar fields, A_0, A_1, A_2 and A_3. That is, we have reduced $4 = 1 \oplus 1 \oplus 1 \oplus 1$, which is not what we wanted. The energy density in this case is

$$\mathcal{E} = \frac{\partial \mathcal{L}}{\partial(\partial_t A_\mu)}\partial_t A_\mu - \mathcal{L}$$

$$= -\frac{1}{2}\left[(\partial_t A_0)^2 + (\vec{\nabla} A_0)^2 + m^2 A_0^2\right] + \frac{1}{2}\left[(\partial_t \vec{A})^2 + (\nabla_i A_j)^2 + m^2 \vec{A}^2\right], \quad (8.19)$$

which has a negative sign for the A_0 field and a positive sign for the $\vec{A}$ fields. If we switched the overall sign, we would still have some fields with negative energy. So this Lagrangian will not produce a physical theory.

By the way, you may wonder how we know if A_μ transforms as a vector or as four scalars, since the Lagrangian is invariant under both transformations. That is, why did we get four scalars when we wanted a vector? As a very general statement, we do not get to impose symmetries on a theory. We just pick the Lagrangian, then we let the theory go. If there are symmetries, and the Lagrangian is constructed correctly to preserve them, the symmetries will hold up in matrix elements in the full interacting theory. This is true even if we never figured out that the symmetries were there. For example, Maxwell's equations are Lorentz invariant. They work the same way if you have $\vec{E}$ and $\vec{B}$ instead of A_μ. The Lorentz invariance is then obscure, but it still works. In fact, a very important tool in making progress in physics has been to observe symmetries in a physical result, such as a matrix element, then to go back and figure out why they are there at a deeper level, which leads to generalizations. That happened with Maxwell for electromagnetism, with Einstein for special and general relativity, with Fermi, Feynman, Glashow, Weinberg and Salam for the $V - A$ theory of weak interactions, with Gell-Mann for the quark model, and in many other cases.

Back to massive spin 1. There is one more Lorentz-invariant two-derivative kinetic term we can write down with the same dimension,[3] $A_\mu \partial_\mu \partial_\nu A_\nu$. Allowing arbitrary coefficients for the different possible terms, the most general free Lagrangian is

$$\mathcal{L} = \frac{a}{2}A_\mu \Box A_\mu + \frac{b}{2}A_\mu \partial_\mu \partial_\nu A_\nu + \frac{1}{2}m^2 A_\mu^2, \quad (8.20)$$

where a and b are numbers. As long as b is non-zero, the $\partial_\mu A_\mu$ contraction forces A_μ to transform as a 4-vector; if A_μ transformed as four scalars, $\partial_\mu A_\mu$ would not be Lorentz invariant. Thus, we should now have $4 = 3 \oplus 1$ instead of $4 = 1 \oplus 1 \oplus 1 \oplus 1$ and have a chance to get rid of the one degree of freedom corresponding to spin 0, isolating the three degrees of freedom for a spin-1 particle.

The equations of motion are

$$a\Box A_\mu + b\partial_\mu \partial_\nu A_\nu + m^2 A_\mu = 0. \quad (8.21)$$

Taking ∂_μ of this equation gives

$$\left[(a + b)\Box + m^2\right](\partial_\mu A_\mu) = 0. \quad (8.22)$$

---

[3] Terms with more derivatives such as $A_\mu \Box^2 A_\mu$ can also be considered, but they will always lead to negative energy. A simple explanation of this fact is given in Section 8.7 and a complete proof is given in Section 24.2.

If $a = -b$ and $m \neq 0$, this reduces to $\partial_\mu A_\mu = 0$, which removes one degree of freedom. Since $\partial_\mu A_\mu = 0$ is a Lorentz-invariant condition, it has to remove a complete representation, which with one degree of freedom can only be the spin-0 component. Taking $a = 1$ and $b = -1$, we find

$$
\begin{aligned}
\mathcal{L} &= \frac{1}{2} A_\mu \Box A_\mu - \frac{1}{2} A_\mu \partial_\mu \partial_\nu A_\nu + \frac{1}{2} m^2 A_\mu^2 \\
&= -\frac{1}{4} F_{\mu\nu}^2 + \frac{1}{2} m^2 A_\mu^2,
\end{aligned} \tag{8.23}
$$

where the Maxwell tensor is $F_{\mu\nu} = \partial_\mu A_\nu - \partial_\nu A_\mu$. This is sometimes called the **Proca Lagrangian**. Note that we did not say anything here about gauge invariance or electromagnetism, we just derived that $F_{\mu\nu}$ appears based on constructing a Lagrangian that generates a constraint to propagate only the spin-1 field by removing the spin-0 field. The equations of motion now imply $\left(\Box + m^2\right) A_\mu = 0$ and $\partial_\mu A_\mu = 0$.

The energy-momentum tensor for the Proca Lagrangian is

$$
\mathcal{T}_{\mu\nu} = \frac{\partial \mathcal{L}}{\partial(\partial_\mu A_\alpha)} \partial_\nu A_\alpha - g_{\mu\nu} \mathcal{L} = -F_{\mu\alpha} \partial_\nu A_\alpha + g_{\mu\nu} \left(\frac{1}{4} F_{\alpha\beta}^2 - \frac{1}{2} m^2 A_\alpha^2 \right). \tag{8.24}
$$

To simplify this, we will use the classical result (which you are encouraged to check) that the Maxwell action can be written as

$$
-\frac{1}{4} F_{\mu\nu}^2 = \frac{1}{2} \left(\vec{E}^2 - \vec{B}^2 \right), \tag{8.25}
$$

where $\vec{E} = \partial_t \vec{A} - \vec{\nabla} A_0$ and $\vec{B} = \vec{\nabla} \times \vec{A}$. Then,

$$
\begin{aligned}
\mathcal{E} = \mathcal{T}_{00} &= -(\partial_t A_\alpha - \partial_\alpha A_0) \partial_t A_\alpha + \frac{1}{2} \vec{B}^2 - \frac{1}{2} \vec{E}^2 - \frac{1}{2} m^2 A_\alpha A_\alpha \\
&= \frac{1}{2} \left(\vec{B}^2 + \vec{E}^2 \right) + \partial_i A_0 (\partial_t A_i - \partial_i A_0) - \frac{1}{2} m^2 A_0^2 + \frac{1}{2} m^2 \vec{A}^2.
\end{aligned} \tag{8.26}
$$

This looks like it has negative energy components. However, we can rewrite this energy density in the suggestive form

$$
\begin{aligned}
\mathcal{E} = \frac{1}{2} \left(\vec{E}^2 + \vec{B}^2 \right) + \frac{1}{2} m^2 \left(A_0^2 + \vec{A}^2 \right) \\
+ A_0 \partial_t (\partial_\mu A_\mu) - A_0 \left(\Box + m^2 \right) A_0 + \partial_i (A_0 F_{0i}).
\end{aligned} \tag{8.27}
$$

The second line is the sum of three terms. The first two vanish on the equations of motion $\partial_\mu A_\mu = 0$ and $\left(\Box + m^2\right) A_0 = 0$. Since the equations of motion were already used in the derivation of the energy-momentum tensor in Noether's theorem, we can use them again here. The final term is a total spatial derivative. Thus, while it contributes to the energy density, it makes no contribution to the total energy. Therefore, the total energy of the fields in the Proca Lagrangian is positive definite, as desired.

Let us now find explicit solutions to the equations of motion. We start by Fourier transforming our (classical) fields. Since $\left(\Box + m^2\right) A_\mu = 0$, we can write any solution as

$$
A_\mu(x) = \sum_i \int \frac{d^3 \vec{p}}{(2\pi)^3} \tilde{a}_i(\vec{p}) \epsilon_\mu^i(p) e^{ipx}, \quad p_0 = \omega_p = \sqrt{\vec{p}^2 + m^2}, \tag{8.28}
$$

for some basis vectors $\epsilon_\mu^i(p)$. For example, we could trivially take $i = 1 \dots 4$ and use four vectors $\epsilon_\mu^i(p) = \delta_\mu^i$ in this decomposition. Instead, we want a basis that forces $A_\mu(x)$ to automatically satisfy also its equation of motion $\partial_\mu A_\mu = 0$. This will happen if $p_\mu \epsilon_\mu^i(p) = 0$. For any fixed 4-momentum p_μ with $p^2 = m^2$, there are three independent solutions to this equation given by three 4-vectors $\epsilon_\mu^i(p)$, necessarily p_μ-dependent, which we call **polarization vectors**. Thus, we only have to sum over $i = 1 \dots 3$ in Eq. (8.28). We conventionally normalize the polarizations by $\epsilon_\mu^\star \epsilon_\mu = -1$.

To be explicit, let us choose a canonical basis. Take p^μ to point in the z direction,

$$p^\mu = (E, 0, 0, p_z), \quad E^2 - p_z^2 = m^2, \tag{8.29}$$

then two obvious vectors satisfying $p_\mu \epsilon_\mu = 0$ and $\epsilon_\mu^2 = -1$ are

$$\epsilon_1^\mu = (0, 1, 0, 0), \quad \epsilon_2^\mu = (0, 0, 1, 0). \tag{8.30}$$

These are the **transverse polarizations**. The other one is

$$\epsilon_L^\mu = (\frac{p_z}{m}, 0, 0, \frac{E}{m}). \tag{8.31}$$

This is the **longitudinal polarization**. It is easy to check that $(\epsilon_L^\mu)^2 = -1$ and $p_\mu \epsilon_\mu^L = 0$. These three polarization vectors $\epsilon_i^\mu(p)$ generate the irreducible representation. The basis vectors depend on p^μ, and since there are an infinite number of possible momenta, it is an infinite-dimensional representation. The vector space generated by integrating these basis vectors against arbitrary Fourier components $\tilde{a}_i(p)$ in Eq. (8.28) is the space of *fields* satisfying the equations of motion, which form an infinite-dimensional unitary representation of the Poincaré group.

By the way, massive spin-1 fields are not a purely theoretical concept: they exist! There is one called the ρ meson, which is lighter than the proton, but unstable, so we do not often see it. More importantly, there are really heavy ones, the W and Z bosons, which mediate the weak force and radioactivity. We will study them in great detail, particularly in Chapter 29. But there is an important feature of these heavy bosons that is easy to see already. At high energy, $E \gg m$, the longitudinal polarization becomes

$$\epsilon_L^\mu \sim \frac{E}{m}(1, 0, 0, 1). \tag{8.32}$$

If we scatter these modes, we might have a cross section whose high-energy behavior gets a contribution like $d\sigma \sim g^2 \epsilon_L^0 \epsilon_L^3 \sim g^2 \frac{E^2}{m^2}$, where g is the coupling constant (an explicit example where this really happens is the theory of weak interactions described in Chapter 29). Then, no matter how small g is, if we go to high enough energies, this cross section blows up. However, cross sections cannot be arbitrarily big. After all, they are probabilities, which are bounded by 1. So, at some scale, what we are calculating becomes not a very good representation of what is really going on. In other words, our perturbation theory is breaking down. We can see already that this happens at $E \sim \frac{m}{g}$. If $m \sim 100\,\text{GeV}$ and $g \sim 0.1$, corresponding to the mass and coupling strength of the W and Z bosons (which are massive spin-1 particles) we find $E \sim 1\,\text{TeV}$. That is why the TeV scale has been the focus of the Tevatron and Large Hadron Colliders. A longer discussion of perturbative unitary violation is given in Sections 24.1.5 and 29.2.

Also, the fact that there is a spin-1 particle in this Lagrangian follows completely from the Lagrangian itself – we never have to impose any additional constraints. In fact, we did not have to talk about spin, or Lorentz invariance at all – all the properties associated with that spin would just have fallen out when we tried to calculate physical quantities. That is part of the beauty of symmetries: they work even if you do not know about them! It would be fine to think of A_μ as four scalar fields that happen to conspire so that when you compute something in one frame, certain ones contribute, and when you compute in a different frame, other ones contribute, but the final answer is frame independent. Obviously it is a lot easier to do the calculation if we know this ahead of time, so we can choose a nice frame, but in no way is it required.

8.2.3 Massless spin 1

The easiest way to come up with a theory of massless spin-1 is to simply take the $m \to 0$ limit of the massive spin-1 theory. Then the Lagrangian becomes

$$\mathcal{L} = -\frac{1}{4}F_{\mu\nu}^2, \tag{8.33}$$

which is the Lagrangian for electrodynamics, confirming that we are on the right track. Unfortunately, the massless limit is not quite as smooth as we would like. First of all, the constraint equation $m^2(\partial_\mu A_\mu) = 0$ is automatically satisfied for $m = 0$, so we no longer automatically have $\partial_\mu A_\mu = 0$. Thus, it seems the spin-0 mode we removed should now be back. Another problem with the massless limit is that as $m \to 0$ the longitudinal polarization blows up:

$$\epsilon_L^\mu = \left(\frac{p_z}{m}, 0, 0, \frac{E}{m}\right) \to \infty. \tag{8.34}$$

Partly, this is due to normalization. In the massless limit, $p_z \to E$ and the momentum becomes lightlike, that is,

$$p^\mu \to (E, 0, 0, E), \tag{8.35}$$

so a more invariant statement is that $\epsilon_L^\mu \to p^\mu$ up to normalization. Finally, we expect from representation theory that there should only be two polarizations for a massless spin-1 particle, so the spin-0 and the longitudinal mode should somehow decouple from the physical system.

Instead of trying to analyze what happens to the massive modes, let us just postulate the Lagrangian and start over with analyzing the degrees of freedom. So we start with

$$\mathcal{L} = -\frac{1}{4}F_{\mu\nu}^2, \quad F_{\mu\nu} = \partial_\mu A_\nu - \partial_\nu A_\mu. \tag{8.36}$$

This Lagrangian has an important property that the massive Lagrangian did not have: **gauge invariance**. It is invariant under the transformation

$$A_\mu(x) \to A_\mu(x) + \partial_\mu \alpha(x) \tag{8.37}$$

for any function $\alpha(x)$. Thus, two fields A_μ that differ by the derivative of a scalar are physically equivalent.

The equations of motion following from the Lagrangian are

$$\Box A_\mu - \partial_\mu(\partial_\nu A_\nu) = 0. \tag{8.38}$$

This is really four equations and it is helpful to separate out the 0 and i components:

$$-\partial_j^2 A_0 + \partial_t \partial_j A_j = 0, \tag{8.39}$$

$$\Box A_i - \partial_i(\partial_t A_0 - \partial_j A_j) = 0. \tag{8.40}$$

To count the physical degrees of freedom, let us use the freedom of transforming the fields in Eq. (8.37) to impose constraints on A_μ, a procedure known as **gauge-fixing**. Since $\partial_j A_j \to \partial_j A_j + \partial_i^2 \alpha$, unless $\partial_j A_j$ is singular we can choose α so that $\partial_j A_j = 0$, known as **Coulomb gauge**. Then the A_0 equation of motion becomes

$$\partial_j^2 A_0 = 0, \tag{8.41}$$

which has no time derivative. Now, under gauge transformations $\partial_i A_i \to \partial_i A_i + \partial_i^2 \alpha$, so Coulomb gauge is preserved under $A_\mu \to A_\mu + \partial_\mu \alpha$ for any α satisfying $\partial_i^2 \alpha = 0$. Since $A_0 \to A_0 + \partial_t \alpha$ and A_0 also satisfies $\partial_i^2 A_0 = 0$ we have exactly the residual symmetry we need to set $A_0 = 0$. Thus, we have eliminated one degree of freedom from A_μ completely, and we are down to three. One more to go!

In Coulomb gauge, the other equations reduce to

$$\Box A_i = 0, \tag{8.42}$$

which seem to propagate three modes. But do not forget that A_i is constrained by $\partial_i A_i = 0$. In Fourier space

$$A_\mu(x) = \int \frac{d^4 p}{(2\pi)^4} \epsilon_\mu(p) e^{ipx} \tag{8.43}$$

and the equations become $p^2 = 0$ (equations of motion), $p_i \epsilon_i = 0$ (gauge choice), and $\epsilon_0 = 0$ (gauge choice). Choosing a frame, we can write the momentum as $p_\mu = (E, 0, 0, E)$. Then these equations have two solutions,

$$\epsilon_1^\mu = (0, 1, 0, 0), \quad \epsilon_2^\mu = (0, 0, 1, 0), \tag{8.44}$$

which represent linearly polarized light. Thus, we have constructed a theory propagating only two degrees of freedom, as is appropriate for irreducible unitary representations of a massless spin-1 particle.

Another common basis for the transverse polarizations of light is

$$\epsilon_R^\mu = \frac{1}{\sqrt{2}}(0, 1, i, 0), \quad \epsilon_L^\mu = \frac{1}{\sqrt{2}}(0, 1, -i, 0). \tag{8.45}$$

These polarizations correspond to circularly polarized light and are called helicity eigenstates.

We could also have used Lorenz gauge ($\partial_\mu A_\mu = 0$), in which case we would have found that three vectors satisfy $p_\mu \epsilon_\mu = 0$:

$$\epsilon_1^\mu = (0, 1, 0, 0), \quad \epsilon_2^\mu = (0, 0, 1, 0), \quad \epsilon_f^\mu = (1, 0, 0, 1). \tag{8.46}$$

The first two modes are the physical transverse polarizations. The third apparent solution denoted ϵ_f^μ is called the **forward polarization**. It does not correspond to a physical state. One way to see this is to note that ϵ_f^μ is not normalizable ($\epsilon_f^{\mu\star}\epsilon_f^\mu = 0$). Another way is to note that $\epsilon_f^\mu \propto p^\mu$, which corresponds to $A_\mu = \partial_\mu\phi$ for some ϕ. This field configuration is gauge-equivalent to $A_\mu = 0$ (choose $\alpha = -\phi$ in Eq. (8.37)). Thus, the forward polarization corresponds to a field configuration that is pure gauge. Similarly, if we had not imposed the second Coulomb gauge condition, $\epsilon^0 = 0$, we would have found another polarization satisfying $p_i\epsilon_i = 0$ is $\epsilon_t^\mu = (1,0,0,0)$. This **timelike polarization** cannot be normalized so $\epsilon_i^{\mu\star}\epsilon_j^\mu = -\delta_{ij}$, since ϵ_t^μ is timelike, and is therefore unphysical.

8.2.4 Summary

To summarize, for massive spin 1, we chose the kinetic term to be $-\frac{1}{4}F_{\mu\nu}^2 + \frac{1}{2}m^2A_\mu^2$ in order to enforce $\partial_\mu A_\mu = 0$, which eliminated one degree of freedom from A_μ, leaving the three for massive spin-1. We found that the energy density is positive definite if and only if the Lagrangian has this form, up to an overall normalization. The Lagrangian for a massive spin-1 particle does not have gauge invariance, but we still need $F_{\mu\nu}^2$.

For the massless case, having $F_{\mu\nu}^2$ gives us gauge invariance. This allows us to remove an additional polarization, leaving two, which is the correct number for a massless spin-1 representation of the Poincaré group.

For both massive and massless spin 1, we found a basis of polarization vectors $\epsilon_\mu^i(p)$, with $i = 1, 2, 3$ for $m > 0$ and $i = 1, 2$ for $m = 0$. The fact that the polarizations depend on p^μ make these infinite-dimensional representations. The representation of the full Poincaré group is **induced** by a representation of the subgroup of the Poincaré group that holds p^μ fixed, called the **little group**. The little group has finite-dimensional representations. For the massive case, the little group, holding for example $p^\mu = (m, 0, 0, 0)$ fixed (or any other 4-vector of mass m), is just the group of three-dimensional rotations, SO(3). SO(3) has finite-dimensional irreducible representations of spin J with $2J + 1$ degrees of freedom. For the massless case, the group that holds a massless 4-vector such as $(E, 0, 0, E)$ fixed is the group ISO(2) (the *iso*metry group of the two-dimensional Euclidean plane), which has representations of spin J with two degrees of freedom for each J. Studying representations of the little group is the easiest way to prove Wigner's classification. Rather than work through the mathematics, we will understand the little group and induced representations through example, particularly in Section 8.4 below. The little group is revisited in Chapters 10 and 27.

8.3 Covariant derivatives

In order not to affect our counting of degrees of freedom, the interactions in the Lagrangian must respect gauge invariance. For example, you might try to add an interaction

$$\mathcal{L} = \cdots + A_\mu\phi\partial_\mu\phi, \tag{8.47}$$

but this is not invariant. Under the gauge transformation

$$A_\mu \phi \partial_\mu \phi \to A_\mu \phi \partial_\mu \phi + (\partial_\mu \alpha) \phi \partial_\mu \phi. \tag{8.48}$$

In fact, it is impossible to couple A_μ to any field with only one degree of freedom, such as the scalar field ϕ. We must be able to make ϕ transform to compensate for the gauge transformation of A_μ, in order to cancel the $\partial_\mu \alpha$ term. But if there is only one field ϕ, it has nothing to mix with so it cannot transform.

Thus, we need at least two fields ϕ_1 and ϕ_2. It is easiest to deal with such a doublet by putting them together into a complex field $\phi = \phi_1 + i\phi_2$, and then to work with ϕ and $\phi^\star$. Under a gauge transformation, ϕ can transform as

$$\phi(x) \to e^{-i\alpha(x)} \phi(x), \tag{8.49}$$

which makes $m^2 \phi^\star \phi$ gauge invariant. But what about the derivatives? $|\partial_\mu \phi|^2$ is not invariant.

We can in fact make the kinetic term gauge invariant using something we call a covariant derivative. Adding a conventional constant e to the transformation of A_μ, so $A_\mu \to A_\mu + \frac{1}{e} \partial_\mu \alpha$, we find

$$(\partial_\mu + ieA_\mu)\phi \to (\partial_\mu + ieA_\mu + i\partial_\mu \alpha)e^{-i\alpha(x)}\phi = e^{-i\alpha(x)}(\partial_\mu + ieA_\mu)\phi. \tag{8.50}$$

This leads us to define the **covariant derivative** as

$$D_\mu \phi \equiv (\partial_\mu + ieA_\mu)\phi \to e^{-i\alpha(x)} D_\mu \phi, \tag{8.51}$$

which transforms just like the field does. Thus

$$\mathcal{L} = -\frac{1}{4} F_{\mu\nu}^2 + (D_\mu \phi)^\star (D_\mu \phi) - m^2 \phi^\star \phi \tag{8.52}$$

is gauge invariant. This is the Lagrangian for **scalar QED**.

More generally, different fields ϕ_n can have different charges Q_n and transform as

$$\phi_n \to e^{Q_n i\alpha(x)} \phi_n. \tag{8.53}$$

Then the covariant derivative is $D_\mu \phi_n = (\partial_\mu - ieQ_n A_\mu) \phi_n$, where in Eq. (8.51) we have taken $Q = -1$ for ϕ, thinking of it as an electron with charge -1. Thus, we write Q for the charges of the fields, and e is the strength of the electric charge, normalized so that $Q = -1$ for the electron, whence $\frac{e^2}{4\pi} \approx \frac{1}{137}$ is the normal fine-structure constant.[4] Until we deal with quarks (for which $Q = \frac{2}{3}$ or $Q = -\frac{1}{3}$), we will not write Q explicitly, and we will just take $D_\mu = \partial_\mu + ieA_\mu$.

By the way, there is also a beautiful geometric way to understand covariant derivatives, similar to how they are understood in general relativity. Since the phase of ϕ is unobservable, one can pick different phase conventions in different regions without consequence. Thus $\phi(x) - \phi(y)$ or even $|\phi(x) - \phi(y)|$ is not well defined. The gauge field records the change in our phase convention from point to point, with a gauge transformation representing a change in this convention. Turning these words into mathematics leads to the notion of Wilson lines, which will play an important role in non-Abelian gauge theories.

---

[4] It is interesting to note that the electric charge itself is $e \approx 0.3 \approx \frac{1}{3}$, which is not actually that small. Doing an expansion in $\frac{1}{3}$ is also popular in QCD, where $3 = N_c$ is the number of colors.

Thus, we postpone the detailed discussion of this interpretation of covariant derivatives until Chapter 25.

8.3.1 Gauge symmetries and conserved currents

Symmetries parametrized by a function such as $\alpha(x)$ are called **gauge** or **local symmetries**, while if they are only symmetries for constant α they are called **global symmetries**. For gauge symmetries, we can pick a separate transformation at each point in space-time. A gauge symmetry automatically implies a global symmetry. Global symmetries imply conserved currents by Noether's theorem. For example, the Lagrangian $\mathcal{L} = -\phi^\star \Box \phi$ of a free complex scalar field is not gauge invariant, but it does have a symmetry under which $\phi \to e^{-i\alpha}\phi$ for a constant α and it does have an associated Noether current.

Let us see how the Noether current changes when the gauge field is included. Expanding out the scalar QED Lagrangian, Eq. (8.52), gives

$$\mathcal{L} = -\frac{1}{4}F_{\mu\nu}^2 + \partial_\mu\phi^\star \partial_\mu\phi + ieA_\mu(\phi\partial_\mu\phi^\star - \phi^\star\partial_\mu\phi) + e^2 A_\mu^2 \phi^\star\phi - m^2\phi^\star\phi. \quad (8.54)$$

The equations of motion are

$$\left(\Box + m^2\right)\phi = -2ieA_\mu\partial_\mu\phi + e^2 A_\mu^2\phi, \quad (8.55)$$

$$\left(\Box + m^2\right)\phi^\star = 2ieA_\mu\partial_\mu\phi^\star + e^2 A_\mu^2\phi^\star. \quad (8.56)$$

The Noether current associated with the global symmetry for which $\frac{\delta\phi}{\delta\alpha} = -i\phi$ and $\frac{\delta\phi^\star}{\delta\alpha} = i\phi^\star$ is (using Eq. (3.23))

$$J_\mu = \sum_n \frac{\partial\mathcal{L}}{\partial(\partial_\mu\phi_n)}\frac{\delta\phi_n}{\delta\alpha} = -i(\phi\partial_\mu\phi^\star - \phi^\star\partial_\mu\phi) - 2eA_\mu\phi^\star\phi. \quad (8.57)$$

The first term on the right-hand side is the Noether current in the free theory ($e = 0$). You should check this full current is also conserved on the equations of motion.

By the way, you might have noticed that the term in the scalar QED Lagrangian linear in A_μ is just $-eA_\mu J_\mu$. There is a quick way to see why this will happen in general. Define $\mathcal{L}_0$ as the limit of a gauge-invariant Lagrangian when $A_\mu = 0$ (or equivalently $e = 0$). $\mathcal{L}_0$ will still be invariant under the global symmetry for which A_μ is the gauge field, since A_μ does not transform when α is constant. If we then let α be a function of x, the transformed $\mathcal{L}_0$ can only depend on $\partial_\mu\alpha$. Thus, for infinitesimal $\alpha(x)$,

$$\delta\mathcal{L}_0 = (\partial_\mu\alpha)J_\mu + \mathcal{O}(\alpha^2) \quad (8.58)$$

for some J_μ. For example, in scalar QED with $A_\mu = 0$, $\mathcal{L}_0 = (\partial_\mu\phi)^\star(\partial_\mu\phi) - m^2\phi^\star\phi$ and

$$\delta\mathcal{L}_0 = (\partial_\mu\alpha)J_\mu + (\partial_\mu\alpha)^2\phi^\star\phi, \quad (8.59)$$

with J_μ given by Eq. (8.57). Returning to the general theory, after integration by parts the term linear in α is $\delta\mathcal{L}_0 = \alpha\partial_\mu J_\mu$. Since the variation of the Lagrangian vanishes on the equations of motion for any transformation, including this one parametrized by α, we must have $\partial_\mu J_\mu = 0$ implying that J_μ is conserved. In fact, J_μ is the Noether current, since we have just rederived Noether's theorem a different way. To make the Lagrangian invariant without using the equations of motion, we can add a field A_μ with $\delta A_\mu = \partial_\mu \alpha$ and define $\mathcal{L} = \mathcal{L}_0 - A_\mu J_\mu$ so that

$$\delta\mathcal{L} = \delta\mathcal{L}_0 - \delta A_\mu J_\mu = (\partial_\mu \alpha) J_\mu - (\partial_\mu \alpha) J_\mu = 0. \tag{8.60}$$

Hence, the coupling $A_\mu J_\mu$ between a gauge field and a Noether current is generic and universal. In scalar field theory there is also a term quadratic in A_μ required to cancel the $(\partial_\mu \alpha)^2$ term in Eq. (8.59). In spinor QED, as we will see, there is just the linear term.

8.4 Quantization and the Ward identity

To quantize fields with multiple degrees of freedom, we simply need creation and annihilation operators for each degree separately. For example, if we have two spin-0 fields, we can write

$$\phi_1(x) = \int \frac{d^3p}{(2\pi)^3} \frac{1}{\sqrt{2\omega_p}} (a_{p,1} e^{-ipx} + a_{p,1}^\dagger e^{ipx}), \tag{8.61}$$

$$\phi_2(x) = \int \frac{d^3p}{(2\pi)^3} \frac{1}{\sqrt{2\omega_p}} (a_{p,2} e^{-ipx} + a_{p,2}^\dagger e^{ipx}). \tag{8.62}$$

Then the complex field $\phi = \phi_1 + i\phi_2$ can be written in the suggestive form as a real doublet:

$$\vec{\phi}(x) = \begin{pmatrix} \phi_1 \\ \phi_2 \end{pmatrix} = \int \frac{d^3p}{(2\pi)^3} \frac{1}{\sqrt{2\omega_p}} \left[\begin{pmatrix} a_{p,1} \\ a_{p,2} \end{pmatrix} e^{-ipx} + \begin{pmatrix} a_{p,1}^\dagger \\ a_{p,2}^\dagger \end{pmatrix} e^{ipx} \right]$$

$$= \int \frac{d^3p}{(2\pi)^3} \frac{1}{\sqrt{2\omega_p}} \sum_{j=1}^{2} \left(\vec{\epsilon}_j a_{p,j} e^{-ipx} + \vec{\epsilon}_j a_{p,j}^\dagger e^{ipx} \right), \tag{8.63}$$

with $\vec{\epsilon}_1 = \begin{pmatrix} 1 \\ 0 \end{pmatrix}$ and $\vec{\epsilon}_2 = \begin{pmatrix} 0 \\ 1 \end{pmatrix}$. In this notation you can think of $\vec{\epsilon}_j$ as the polarization vectors of the complex scalar field. To quantize spin-1 fields, we will just allow for the polarizations to be in a basis that has four components instead of two and can depend on momentum $\vec{\epsilon}_j \to \epsilon_j^\mu(p)$.

8.4.1 Massive spin 1

The quantum field operator for massive spin 1 is

$$A_\mu(x) = \int \frac{d^3p}{(2\pi)^3} \frac{1}{\sqrt{2\omega_p}} \sum_{j=1}^{3} \left[\epsilon_\mu^j(p) a_{p,j} e^{-ipx} + \epsilon_\mu^{j\star}(p) a_{p,j}^\dagger e^{ipx} \right]. \tag{8.64}$$

There are separate creation and annihilation operators for each of the polarizations, and we sum over them. $\epsilon_\mu^j(p)$ represents a canonical set of basis vectors.

The creation and annihilation operators have polarization indices. To specify our asymptotic states we will now need to give both the momentum and the polarization. So

$$a_{p,j}^\dagger |0\rangle = \frac{1}{\sqrt{2\omega_p}} |p, \epsilon^j\rangle \qquad (8.65)$$

up to normalization. Thus

$$\langle 0|A_\mu(x)|p, \epsilon^j\rangle = \epsilon_\mu^j e^{-ipx}, \qquad (8.66)$$

so our field creates a particle at position x whose polarization can be projected out with the appropriate contraction.

Recall that the basis has to depend on p^μ because there are no finite-dimensional unitary representations of the Lorentz group. To see it again, let us suppose instead that we tried to pick constants for our basis vectors. Say, $\epsilon_1^\mu = (0,1,0,0)$, $\epsilon_2^\mu = (0,0,1,0)$ and $\epsilon_3^\mu = (0,0,0,1)$. The immediate problem is that this basis is not complete, because under Lorentz transformations

$$\epsilon_i^\mu \to \Lambda_{\mu\nu}\epsilon_i^\nu, \qquad (8.67)$$

so that for boosts these will mix with the timelike polarization $\epsilon_t^\mu = (1,0,0,0)$.

We saw from solving the classical equations of motion that we can choose a momentum-dependent basis $\epsilon_1^\mu(p)$, $\epsilon_2^\mu(p)$ and $\epsilon_3^\mu(p)$. For example, for the massive case, for p^μ pointing in the z direction,

$$p^\mu = (E,0,0,p_z), \qquad (8.68)$$

we can use the basis

$$\epsilon_1^\mu(p) = (0,1,0,0), \quad \epsilon_2^\mu(p) = (0,0,1,0), \quad \epsilon_L^\mu(p) = \left(\frac{p_z}{m},0,0,\frac{E}{m}\right), \qquad (8.69)$$

which all satisfy $\epsilon_i^\mu \epsilon_i^{\mu\star} = -1$ and $\epsilon_i^\mu p^\mu = 0$.

What happened to the fourth degree of freedom in the vector representation? The vector orthogonal to these is $\epsilon_S^\mu(p) = \frac{1}{m}p^\mu = \left(\frac{E}{m},0,0,\frac{p_z}{m}\right)$. In position space, this is $\epsilon_\mu^S = \frac{1}{m}\partial_\mu \alpha(x)$ for some scalar function $\alpha(x)$. So we do not want to include this spin-0 polarization $\epsilon_S^\mu(p)$ in the sum in Eq. (8.64). To see that the polarization based on the scalar $\alpha(x)$ does not mix with the other three is easy: if something is the derivative of a function $\alpha(x)$, under a Lorentz transformation it will still be the derivative of the same function, just in a different frame. So the polarizations in the spin-1 representation (the ϵ_i^μ's) do not mix with the polarization in the spin-0 representation, ϵ_S^μ.

Now, you may wonder, if we are redefining our basis with every boost, so that $\epsilon_1^\mu(p) \to \epsilon_1^\mu(p')$, $\epsilon_2^\mu(p) \to \epsilon_2^\mu(p')$, and $\epsilon_L^\mu(p) \to \epsilon_L^\mu(p')$, when do the polarization vectors ever mix? Have we gone too far and just made four separate one-dimensional representations? The answer is that there are Lorentz transformations that leave p^μ alone, and therefore leave our basis alone. These are, by definition, the elements of the *little group*. For little-group transformations, we need to check that our basis vectors rotate into each other and form a complete representation. For example, suppose we go to the frame

$$q^\mu = (m,0,0,0). \qquad (8.70)$$

Then we can choose our polarization basis vectors as

$$\epsilon_1^\mu(q) = (0,1,0,0), \quad \epsilon_2^\mu(q) = (0,0,1,0), \quad \epsilon_3^\mu(q) = (0,0,0,1) \tag{8.71}$$

and $\epsilon_S^\mu = (1,0,0,0)$. The little group which preserves q_μ in this case is simply the 3D rotation group SO(3). It is then easy to see that, under 3D rotations, the three ϵ_i^μ polarizations will mix among each other, and $\epsilon_S^\mu = (1,0,0,0)$ stays fixed. If we boost, it looks like the ϵ_i^μ will mix with ϵ_S^μ. However, we have to be careful, because the basis vectors will also change, for example to $\epsilon_1^\mu, \epsilon_2^\mu$ and ϵ_L^μ, above. The group that fixes $p^\mu = (E,0,0,p_z)$ is also SO(3), although it is harder to see. And these SO(3) rotations will also only mix $\epsilon_1^\mu, \epsilon_2^\mu$ and ϵ_L^μ, leaving ϵ_S^μ fixed. So everything works. The non-trivial effect of Lorentz transformations is to mix up the polarization vectors at fixed p^μ. So the spin-1 representation is characterized by this smaller group, the little group, which is the subgroup of the Lorentz group that leaves p^μ unchanged. This method of studying representations of the Lorentz group is called the *method of induced representations*.

The little group also helps resolve the conflict between Lorentz invariance and unitarity discussed in Section 8.1.1. In quantum mechanics, we can expand any polarization in this basis. Let us fix the momentum q^μ. Then, any physical polarization vector ϵ^μ can be written as

$$\epsilon^\mu = c_j \epsilon_j^\mu(q), \tag{8.72}$$

corresponding to the state $|\epsilon\rangle = c_j |j\rangle$. Since the basis states all have $\langle j|j\rangle = 1$, we find

$$\langle \epsilon|\epsilon\rangle = |c_1|^2 + |c_2|^2 + |c_3|^2. \tag{8.73}$$

This inner product is rotation invariant, by the defining property of rotations, and boost invariant in a trivial way: under boosts the c_j's do not change because the basis vectors ϵ_j^μ do. Thus, $\langle \epsilon|\epsilon\rangle$ is positive definite and Lorentz invariant.

In quantum field theory, we will be calculating matrix elements with the field A_μ. These matrix elements will depend on the polarization vector and must have the form

$$\mathcal{M} = \epsilon^\mu M_\mu, \tag{8.74}$$

where M_μ transforms as a 4-vector. Here, ϵ^μ is the polarization vector, which can be any of the ϵ_j^μ or any linear combination of them. For example, say we start with ϵ_μ^1. Now change frames, so $M_\mu \to M_\mu' = \Lambda_{\mu\nu} M_\nu$. Then the matrix element is invariant:

$$\mathcal{M} = \epsilon^{\mu'}(p') M_\mu', \tag{8.75}$$

where $\epsilon^{\mu'}(p') = \Lambda_{\mu\nu} \epsilon^\nu(\Lambda_{\alpha\beta} p_\beta)$. This new polarization $\epsilon^{\mu'}(p')$ is still a physical state in our Hilbert space, since the basis is closed under the Lorentz group. More simply, we can say that $\epsilon^\mu M_\mu$ is Lorentz invariant on the restricted space of 4-vectors $\epsilon^\mu(p) = c_j \epsilon_j^\mu(p)$. This sounds pretty obvious, but having understood the massive case in this language will greatly facilitate understanding the massless case, which is much more subtle.

8.4.2 Massless spin 1

We quantize massless spin 1 exactly like massive spin 1, but summing over two polarizations instead of three:

$$A_\mu(x) = \int \frac{d^3p}{(2\pi)^3} \frac{1}{\sqrt{2\omega_p}} \sum_{i=1}^{2} \left[\epsilon^i_\mu(p)a_{p,i}e^{-ipx} + \epsilon^{i\star}_\mu(p)a^\dagger_{p,i}e^{ipx} \right]. \tag{8.76}$$

A sample basis is, for p^μ in the z direction,

$$p^\mu = (E,0,0,E), \tag{8.77}$$

$$\epsilon^\mu_1(p) = (0,1,0,0), \quad \epsilon^\mu_2(p) = (0,0,1,0). \tag{8.78}$$

These satisfy $\left(\epsilon^i_\mu\right)^2 = -1$ and $\epsilon^i_\mu p_\mu = 0$. The two orthogonal polarizations are

$$\epsilon^\mu_f(p) = (1,0,0,1), \quad \epsilon^\mu_b(p) = (1,0,0,-1), \tag{8.79}$$

where f and b stand for forward and backward.

But now we have a problem. Even though there is an irreducible unitary representation of massless spin-1 particles involving two polarizations, it is *impossible* to embed these polarizations in vector fields like ϵ^μ. To see the problem, recall that in the massive case ϵ^μ_1 and ϵ^μ_2 mixed not only with each other under the little group SO(3), that is, Lorentz transformations preserving $p^\mu = (E,0,0,p_z)$, but they also mixed with the longitudinal mode $\epsilon^\mu_L(p) = (\frac{p_z}{m},0,0,\frac{E}{m})$. We saw this because $|c_1|^2 + |c_2|^2 + |c_L|^2$ is invariant under this SO(3), but $|c_1|^2+|c_2|^2$ itself would not be. There is nothing particularly discontinuous about the $m \to 0$ limit. The momentum goes to $p^\mu = (E,0,0,E)$ and the longitudinal mode becomes the same as our forward-polarized photon, up to normalization

$$\lim_{m\to 0} \epsilon^\mu_L(p) = \epsilon^\mu_f(p) \propto p^\mu. \tag{8.80}$$

The little group goes to ISO(2) in the massless case. There are still little-group members that mix ϵ^1_μ and ϵ^μ_2 with the other polarization, $\epsilon^\mu_f(p) \propto p^\mu$. In general

$$\epsilon^\mu_1(p) \to c_{11}(\Lambda)\epsilon^\mu_1(p) + c_{12}(\Lambda)\epsilon^\mu_2(p) + c_{13}(\Lambda)p^\mu, \tag{8.81}$$

$$\epsilon^\mu_2(p) \to c_{21}(\Lambda)\epsilon^\mu_1(p) + c_{22}(\Lambda)\epsilon^\mu_2(p) + c_{13}(\Lambda)p^\mu, \tag{8.82}$$

where the c_{ij} are numbers.

To be really explicit, consider the Lorentz transformation

$$\Lambda^\mu_\nu = \begin{pmatrix} \frac{3}{2} & 1 & 0 & -\frac{1}{2} \\ 1 & 1 & 0 & -1 \\ 0 & 0 & 1 & 0 \\ \frac{1}{2} & 1 & 0 & \frac{1}{2} \end{pmatrix}. \tag{8.83}$$

This satisfies $\Lambda^T g\Lambda = g$, so it is a Lorentz transformation. It also has $\Lambda^\mu_\nu p^\nu = p^\mu$ so it preserves the momentum $p^\mu = (E,0,0,E)$. Thus, this Λ is an honest member of the little group. However,

$$\Lambda^\mu_\nu \epsilon^\nu_1 = (1,1,0,1) = \epsilon^\mu_1 + \frac{1}{E}p^\mu, \tag{8.84}$$

so it mixes the physical polarization with the momentum. This is in contrast to the case for massive spin 1, where the basis vectors ϵ_i^μ only mix with themselves.

Now, consider the kind of matrix element we would get from scattering a photon using the field A_μ. It would, just like the massive case, be

$$\mathcal{M} = \epsilon_\mu M_\mu, \tag{8.85}$$

where now ϵ^μ is some linear combination of the two physical polarizations ϵ_1^μ and ϵ_2^μ. Then, under a Lorentz transformation,

$$\mathcal{M} \to \epsilon^{\mu\prime} M_\mu' + c(\Lambda) p^\mu M_\mu' \tag{8.86}$$

for some $c(\Lambda)$, where $M_\mu' = \Lambda_{\mu\nu} M_\nu$ and ϵ_μ' is a linear combination of ϵ_μ^1 and ϵ_μ^2, but p_μ is not. For example, under the explicit Lorentz transformation above,

$$\mathcal{M} = \epsilon_1^\mu M_\mu \to \left(\epsilon_1^\mu + \frac{1}{E} p^\mu \right) M_\mu'. \tag{8.87}$$

So we have a problem. The state with polarization $\epsilon_1^\mu + \frac{1}{E} p^\mu$ is not in our Hilbert space! Thus, there is no physical polarization for which the matrix element is the same in the new frame as it was in the old frame. There is only one way out – if $p^\mu M_\mu = 0$. Then there is a physical polarization that gives the same matrix element and $\mathcal{M}$ is invariant. Thus, to have a Lorentz-invariant theory with a massless spin-1 particle, we must have $p^\mu M_\mu = 0$.

This is extremely important and worth repeating. We have found that under Lorentz transformations the massless polarizations transform as

$$\epsilon^\mu \to c_1 \epsilon_1^\mu + c_2 \epsilon_2^\mu + c_3 p^\mu. \tag{8.88}$$

Generally, this transformed polarization is not physical and not in our Hilbert space because of the p^μ term. The best we can do is transform it into $\epsilon^{\mu\prime} = c_1 \epsilon_1^\mu + c_2 \epsilon_2^\mu$. When we calculate something in QED we will get matrix elements

$$\mathcal{M} = \epsilon^\mu M_\mu \tag{8.89}$$

for some M_μ transforming like a Lorentz vector. If we Lorentz transform this expression we will get

$$\mathcal{M} \to (a_1 \epsilon_1^\mu + a_2 \epsilon_2^\mu + a_3 p^\mu) M_\mu'. \tag{8.90}$$

It is therefore only possible for $\mathcal{M}$ to be Lorentz invariant if $\mathcal{M} = \epsilon^{\mu\prime} M_\mu'$, which happens only if

$$p^\mu M_\mu = 0. \tag{8.91}$$

This is known as the **Ward identity**. The Ward identity must hold by *Lorentz invariance* and the fact that *unitary* representations for *massless spin-1 particles* have *two polarizations*. We did *not* show that it holds, only that it *must* hold in a reasonable physical theory. That it holds in QED is complicated to show in perturbation theory, but we will sketch the ingredients in the next chapter. We will eventually prove it non-perturbatively using path integrals in Chapter 14. The Ward identity is closely related to gauge invariance. Since the Lagrangian is invariant under $A_\mu \to A_\mu + \partial_\mu \alpha$, in momentum space this should directly imply that $\epsilon^\mu \to \epsilon^\mu + p^\mu$ is a symmetry of the theory, which is the Ward identity.

8.5 The photon propagator

In order to calculate anything with a photon, we are going to need to know its propagator $\Pi^{\mu\nu}$, defined by

$$\langle 0|T\{A^\mu(x)A^\nu(y)\}|0\rangle = i \int \frac{d^4p}{(2\pi)^4} e^{ip(x-y)}\Pi^{\mu\nu}(p), \qquad (8.92)$$

evaluated in the free theory. The easiest way to calculate the propagator is to solve for the classical Green's function and then add the time ordering with the $i\varepsilon$ prescription, as for a scalar.

Let us first try to calculate the classical Green's function by using the equations of motion, without choosing a gauge. In the presence of a current, the equations of motion following from $\mathcal{L} = -\frac{1}{4}F_{\mu\nu}^2 - A_\mu J_\mu$ are

$$\partial_\mu F_{\mu\nu} = J_\nu, \qquad (8.93)$$

so

$$\partial_\mu \partial_\mu A_\nu - \partial_\mu \partial_\nu A_\mu = J_\nu, \qquad (8.94)$$

or in momentum space,

$$(-p^2 g_{\mu\nu} + p_\mu p_\nu)A_\mu = J_\nu. \qquad (8.95)$$

We would like to write $A_\mu = \Pi_{\mu\nu}J_\nu$, so that $(-p^2 g_{\mu\nu} + p_\mu p_\nu)\Pi_{\nu\alpha} = g_{\mu\alpha}$. That is, we want to invert the kinetic term. The problem is that

$$\det(-p^2 g_{\mu\nu} + p_\mu p_\nu) = 0, \qquad (8.96)$$

which follows since $-p^2 g_{\mu\nu} + p_\nu p_\mu$ has a zero eigenvalue, with eigenvector p_μ. Because it has a zero eigenvalue, the kinetic term cannot be invertible, just as for a finite-dimensional linear operator. The non-invertibility is a manifestation of gauge invariance: A_μ is not uniquely determined by J_μ; different gauges will give different values for A_μ from the same J_μ.

So what do we do? We could try to just choose a gauge, for example $\partial_\mu A_\mu = 0$. This would reduce the Lagrangian to

$$-\frac{1}{4}F_{\mu\nu}^2 \to \frac{1}{2}A_\mu \Box A_\mu. \qquad (8.97)$$

However, now it seems there are four propagating degrees of freedom in A_μ instead of two. In fact, you can do this, but you have to keep track of the gauge constraint $\partial_\mu A_\mu = 0$ all along. A cleaner solution, which more easily generalizes to non-Abelian theories, is to add a new term directly to the Lagrangian. In this way, the constraints are generated from the equations of motion, rather than imposed on the theory. A standard modification is to introduce a parameter ξ and write

$$\mathcal{L} = -\frac{1}{4}F_{\mu\nu}^2 - \frac{1}{2\xi}(\partial_\mu A_\mu)^2 - J_\mu A_\mu. \tag{8.98}$$

In the limit of very small ξ there is a tremendous pressure on the Lagrangian to have $\partial_\mu A_\mu = 0$ to stay near the minimum. Thus $\xi = 0$ will correspond to Lorenz gauge. Conveniently, however, this modification can be applied for any value of ξ and $\xi = 1$ (Feynman-'t Hooft gauge) will prove particularly convenient.

With the ξ term, the equations of motion for A_μ are

$$\left[-p^2 g_{\mu\nu} + \left(1 - \frac{1}{\xi}\right)p_\mu p_\nu\right]A_\nu = J_\mu. \tag{8.99}$$

Although not obvious, but easy to check, the inverse of the operator in brackets is

$$\Pi_{\mu\nu} = -\frac{g_{\mu\nu} - (1-\xi)\frac{p_\mu p_\nu}{p^2}}{p^2}. \tag{8.100}$$

To check, we calculate

$$\left[-p^2 g_{\mu\alpha} + \left(1 - \frac{1}{\xi}\right)p_\mu p_\alpha\right]\Pi_{\alpha\nu} = \left[p^2 g_{\mu\alpha} - \left(1 - \frac{1}{\xi}\right)p_\mu p_\alpha\right]\left[p^2 g_{\alpha\nu} - (1-\xi)p_\alpha p_\nu\right]\frac{1}{p^4}$$

$$= g_{\mu\nu} + \left[-\left(1 - \frac{1}{\xi}\right) - (1-\xi) + \left(1 - \frac{1}{\xi}\right)(1-\xi)\right]\frac{p_\mu p_\nu}{p^2}$$

$$= g_{\mu\nu}. \tag{8.101}$$

This confirms that we have inverted the operator correctly.

The time-ordered Feynman propagator for a photon is then

$$i\Pi^{\mu\nu}(p) = \frac{-i}{p^2 + i\varepsilon}\left[g^{\mu\nu} - (1-\xi)\frac{p^\mu p^\nu}{p^2}\right]. \tag{8.102}$$

This is the photon propagator in **covariant** or **R_ξ-gauge**. Although one can (with some effort) derive this Feynman propagator using the canonical picture [Greiner and Reinhardt, 1996], it follows trivially from Eq. (8.100) when using path-integrals (see Chapter 14).

As with the scalar propagator, the $i\varepsilon$ is a quick way to combine the advanced and retarded propagators into the time-ordered propagator. The sign for the numerator can be remembered using

$$-ig^{\mu\nu} = \begin{pmatrix} -i & & & \\ & i & & \\ & & i & \\ & & & i \end{pmatrix}. \tag{8.103}$$

Since it is the spatial components A_i of the vector field that propagate, they should have the same form as the scalar propagator, $i\Pi_S = \frac{i}{p^2 + i\varepsilon}$, confirming the $-ig^{\mu\nu}$.

8.5.1 Covariant gauges

In the covariant gauges, each choice of ξ gives a different Lorentz-invariant gauge. Some useful gauges are:

- Feynman–'t Hooft gauge $\xi = 1$:

$$i\Pi^{\mu\nu}(p) = \frac{-ig^{\mu\nu}}{p^2 + i\varepsilon}.$$

(8.104)

This is the gauge we will use for most calculations.
- Lorenz gauge $\xi = 0$:

$$i\Pi^{\mu\nu}(p) = -i\frac{g^{\mu\nu} - \frac{p^\mu p^\nu}{p^2}}{p^2 + i\varepsilon}.$$

(8.105)

We saw that $\xi \to 0$ forces $\partial_\mu A_\mu = 0$. Note that we could not set $\xi = 0$ and then invert the kinetic term, but we can invert and then set $\xi = 0$.
- Unitary gauge $\xi \to \infty$. This gauge is useless for QED, since the propagator blows up. But it is extremely useful for the gauge theory of the weak interactions.

Other non-covariant gauges are occasionally useful. Lightcone gauge, with $n_\mu A_\mu = 0$ for some fixed lightlike 4-vector n_μ is occasionally handy if there is a preferred direction. For example, in situations with multiple collinear fields, such as the quarks inside a fast-moving proton, lightcone gauge is useful (see Section 32.5 and Chapter 36). Coulomb gauge, $\nabla \cdot A = 0$, and radial or Fock–Schwinger gauge, $x_\mu A_\mu(x) = 0$, also facilitate some calculations. For QED we will stick to covariant gauges.

The final answer for any Lorentz-invariant quantity had better be gauge invariant. In covariant gauges,

$$i\Pi^{\mu\nu}(p) = \frac{-i}{p^2 + i\varepsilon}\left[g^{\mu\nu} - (1 - \xi)\frac{p^\mu p^\nu}{p^2}\right].$$

(8.106)

This means the final answer should be independent of ξ. Thus, whatever we contract $\Pi_{\mu\nu}$ with should give 0 if $\Pi_{\mu\nu} \propto p_\mu p_\nu$. This is very similar to the requirement of the Ward identities, which say that the matrix elements vanish if the physical external polarization is replaced by $\epsilon_\mu \to p_\mu$. We will sketch a diagrammatic proof of gauge invariance in the next chapter, and give a full non-perturbative proof of both gauge invariance and the Ward identity in Chapter 14 on path integrals.

8.6 Is gauge invariance real?

Gauge invariance is not physical. It is not observable and is not a symmetry of nature. Global symmetries are physical, since they have physical consequences, namely conservation of charge. That is, we measure the total charge in a region, and if nothing leaves that region, whenever we measure it again the total charge will be exactly the same. There is no such thing that you can actually measure associated with gauge invariance. We introduce gauge invariance to have a local description of massless spin-1 particles. The existence of these particles, with only two polarizations, is physical, but the gauge invariance is merely a redundancy of description we introduce to be able to describe the theory with a local Lagrangian.

A few examples may help drive this point home. First of all, an easy way to see that gauge invariance is not physical is that we can choose any gauge, and the physics is going to be exactly the same. In fact, we *have to* choose a gauge to do any computations. Therefore, there cannot be any physics associated with this artificial symmetry. But note that even though we gauge-fix by modifying the kinetic terms, this is a very particular breaking of gauge symmetry. The interactions of the gauge field with matter are still gauge invariant, as would be the interactions of the gauge field with itself in gravity or in Yang–Mills theories (Chapters 25 and 26). The controlled way that gauge invariance is broken, particularly by the introduction of covariant gauges, is critical to proving renormalizability of gauge theories, as we will see in Chapter 21.

A useful toy model that may help distinguish gauge invariance (artificial) from the physical spectrum (real) is

$$\mathcal{L} = -\frac{1}{4}F_{\mu\nu}^2 + \frac{1}{2}m^2(A_\mu + \partial_\mu\pi)^2. \tag{8.107}$$

This has a gauge invariance under which $A_\mu(x) \to A_\mu(x) + \partial_\mu\alpha(x)$ and $\pi(x) \to \pi(x) - \alpha(x)$. However, we can use that symmetry to set $\pi = 0$ everywhere. Then the Lagrangian reduces to that of a massive gauge boson. So the physics is that of three polarizations of a massive spin-1 particle. When π is included there are still three degrees of freedom, but now these are two polarizations in A_μ and one in π, with the third polarization of A_μ unphysical because of the exact gauge invariance.

We could do something even more crazy with this Lagrangian: integrate out π. By setting π equal to its equations of motion and substituting back into the Lagrangian, it becomes

$$\mathcal{L} = -\frac{1}{4}F_{\mu\nu}\left(1 + \frac{m^2}{\Box}\right)F_{\mu\nu}. \tag{8.108}$$

This Lagrangian is also manifestly gauge invariant, but it is very strange. In reasonable field theories, the effects of a field are **local**, meaning that they decrease with distance. For example, a massive scalar field generates a Yukawa potential $V(r) = \frac{1}{4\pi r}e^{-mr}$ so that its effects are confined to within the correlation length $\xi \sim \frac{1}{m}$. In contrast, at distances $r \gg \xi$, the $\frac{m^2}{\Box} \sim r^2m^2$ term in Eq. (8.108) becomes increasingly important. Thus Eq. (8.108) appears to describe a non-local theory.

In quantum field theory, non-local theories have S-matrices that can have poles not associated with particles in the Hilbert space. If there are poles without particles, the theory is not unitary (as we will show explicitly in Section 24.3). So non-locality and unitarity are intimately tied together. In this case, the Lagrangian looks like a Lagrangian for a massless spin-1 field with two degrees of freedom. However, the missing particle, which would correspond to the extra pole in the S-matrix, is precisely the longitudinal mode of A_μ, which we can call either π or the third polarization of a massive spin-1 particle.

In practice, local symmetries make it much easier to do computations. You might wonder why we even bother introducing this field A_μ, which has this huge redundancy to it. Instead, why not just quantize the electric and magnetic fields, that is $F_{\mu\nu}$ itself? Well, you could do that, but it turns out to be more complicated than using A_μ. To see why, first note that $F_{\mu\nu}$ as a field does not propagate with the Lagrangian $\mathcal{L} = -\frac{1}{4}F_{\mu\nu}^2$. All the dynamics will be moved to the interactions. Moreover, if we include interactions, either with a simple current $A_\mu J_\mu$ or with a scalar field $\phi^\star A_\mu \partial_\mu \phi$ or with a fermion $\bar{\psi}\gamma_\mu A_\mu \psi$, we see that they naturally involve A_μ. If we want to write these in terms of $F_{\mu\nu}$ we have to solve for A_μ in terms of $F_{\mu\nu}$ and we will get some non-local thing such as $A_\mu = \frac{1}{\Box}\partial_\nu F_{\mu\nu}$. Then we would have to spend all our time showing that the theory is actually local and causal. It turns out to be much easier to deal with a little redundancy so that we do not have to check locality all the time.

Another reason is that all of the physics of the electromagnetic field is, in fact, not *entirely* contained in $F_{\mu\nu}$. There are global topological properties of A_μ that are not contained in $F_{\mu\nu}$ but have physical consequences. An example is the Aharonov–Bohm effect, which you might remember from quantum mechanics. Other examples that come up in field theory are instantons and sphalerons, which are relevant for the U(1) problem and baryogenesis respectively, to be discussed in Section 30.5. There are more general gauge-invariant objects than $F_{\mu\nu}$ that can encode these effects. In particular, Wilson loops (see Section 25.2) are gauge invariant, but they are non-local. An approach to reformulating gauge theories entirely in terms of Wilson lines achieved some limited success in the 1980s, but remains a longshot approach to reformulating quantum field theory completely.

In summary, although gauge invariance is merely a redundancy of description, it makes it a lot easier to study field theory. The physical content is what we saw in the previous section with the Lorentz transformation properties of spin-1 fields: *massless spin-1 fields have two polarizations*. If there were a way to compute S-matrix elements without a local Lagrangian (and to some extent there is, for example, using recursion relations, as we will see in Chapter 27), we might be able to do without this redundancy altogether.

By the way, the word **gauge** means size; the original symmetry of this type was conceived by Hermann Weyl as an invariance under scale transformations, now known as Weyl or scale invariance. The Lagrangian $\mathcal{L} = -\frac{1}{4}F_{\mu\nu} + |D_\mu\phi|^2$ is classically scale invariant. However, at the quantum level, scale invariance is broken (see Chapters 16 and 23). Effectively, the coupling constant becomes dependent on the characteristic energy of the process. A classical symmetry broken by quantum effects is said to be **anomalous**. The gauge symmetry associated with the photon, or other gauge fields, cannot be anomalous or else the Ward identity would be violated and the theory would be non-unitary. Anomalies are the subject of Chapter 30.

8.7 Higher-spin fields

This section, which can be skipped without losing continuity with the rest of the book, generalizes the discussion of spin 1 to particles of higher integer spin. In particular, we will construct the Lagrangian for spin 2 from the bottom up. A spin-2 particle has five polarizations if it is massive or two polarizations if it is massless. The smallest tensor in which five polarizations would fit is a 2-index tensor $h_{\mu\nu}$. To determine the Lagrangian, rather than looking for positive energy as a sign of unitarity, we will look for the absence of ghosts. A **ghost** is a state with negative norm, or wrong-sign kinetic term, such as the A_0 component of a vector field if the Lagrangian is $\mathcal{L} = \frac{1}{2} A_\mu (\Box + m^2) A_\mu$. To decide if there are ghosts we will separate out the longitudinal and transverse modes. This method was developed by Ernst Stueckelberg for the Abelian case [Stueckelberg, 1938], Sidney Coleman *et al.* [Coleman *et al.*, 1969; Callan *et al.*, 1969] for the non-Abelian case, and Arkani-Hamed *et al.* [Arkani-Hamed *et al.*, 2003] for gravity. The bottom-up construction of the Lagrangian for general relativity was discussed by Feynman *et al.* [Feynman *et al.*, 1996].

8.7.1 Longitudinal fields and spin 1

Before turning to spin 2, let us re-analyze spin 1 in a way that makes it easier to see the ghosts. Any vector field can be written as

$$A_\mu(x) = A_\mu^T(x) + \partial_\mu \pi(x) \tag{8.109}$$

with

$$\partial_\mu A_\mu^T = 0. \tag{8.110}$$

To see this, observe that this decomposition is invariant under shifts $A_\mu^T \to A_\mu^T + \partial_\mu \alpha$ and $\pi \to \pi - \alpha$. Thus, there are an infinite number of ways to split a generic A_μ up this way. But from the point of view of A_μ^T, this is just a gauge transformation, and we already know that we can pick α so that the field is in Lorenz gauge where $\partial_\mu A_\mu^T = 0$.

The beauty of this decomposition is that it lets us see whether the non-transverse polarizations are physical simply by looking at the Lagrangian. Start with the most general Lorentz-invariant Lagrangian for a vector field A_μ:

$$\mathcal{L} = a A_\mu \Box A_\mu + b A_\mu \partial_\mu \partial_\nu A_\nu + m^2 A_\mu^2. \tag{8.111}$$

Performing our substitution and using Eq. (8.110) gives

$$\mathcal{L} = a A_\mu^T \Box A_\mu^T + m^2 (A_\mu^T)^2 - (a + b)\pi \Box^2 \pi - m^2 \pi \Box \pi. \tag{8.112}$$

We will now show that for $a + b \neq 0$, there are ghosts and the theory cannot be unitary.

An easy way to see this is from π's propagator. In momentum space it is

$$\Pi_\pi = \frac{-1}{2(a+b)k^4 - 2m^2 k^2} = \frac{1}{2m^2}\left[\frac{1}{k^2} - \frac{(a+b)}{(a+b)k^2 - m^2}\right]. \tag{8.113}$$

Thus, π really represents two fields, one of which has negative norm for generic a and b and therefore represents a ghost. If we choose $a = -b$ however, the propagator is just $\frac{1}{m^2 k^2}$, which can represent unitary propagation. More generally, a kinetic term with more than two derivatives always indicates that a theory is not unitary. We will show in Section 24.2 using the spectral decomposition that, in a unitary theory, the fastest that propagators can die off at large p^2 is $\Pi \sim \frac{1}{p^2}$. With a $\Box^2$ kinetic term, the propagator would die off as $\frac{1}{p^4}$.

We can only remove the dangerous 4-derivative kinetic terms by choosing $a = -b$, leading to the unique physical Lagrangian for a massive spin-1 field we derived before. Taking $a = -b = \frac{1}{2}$ and rescaling $m^2 \to \frac{1}{2}m^2$ we get

$$\mathcal{L} = \frac{1}{2}A_\mu \Box A_\mu - \frac{1}{2}A_\mu \partial_\mu \partial_\nu A_\nu + \frac{1}{2}m^2 A_\mu^2 = -\frac{1}{4}F_{\mu\nu}^2 + \frac{1}{2}m^2 A_\mu^2. \qquad (8.114)$$

In this case, we see that the longitudinal modes get a kinetic term from the mass term, as expected.

In the massless limit, there is no kinetic term at all for the longitudinal mode. If a mode has interactions but no kinetic terms, the theory is also sick. One way to see that is to take the limit that the kinetic term goes to zero. Suppose we had

$$\mathcal{L} = Z\pi\Box\pi + \lambda\pi^3. \qquad (8.115)$$

If we rescale the fields to their canonical normalization, $\pi_c = \sqrt{Z}\pi$, we get

$$\mathcal{L} = \pi_c \Box \pi_c + \frac{\lambda}{Z^{3/2}}\pi_c^3. \qquad (8.116)$$

So $Z \to 0$ indicates infinitely strong interactions. Thus, we have to make sure that π never appears when we substitute $A_\mu \to A_\mu + \partial_\mu\pi$, which is just the statement that the theory must be gauge invariant. For interactions, this will be true if

$$\mathcal{L} = \cdots + A_\mu J_\mu \qquad (8.117)$$

with $\partial_\mu J_\mu = 0$.

We can use the same method to determine the interactions. Start with a real field ϕ. The simplest Lorentz-invariant interaction we can write down involving A_μ and ϕ is

$$\mathcal{L}_{\text{int}} = A_\mu \phi \partial_\mu \phi. \qquad (8.118)$$

This is not gauge invariant. Nor is there any way to transform ϕ so that it is gauge invariant. For a complex field, the unique real interaction term we can write is

$$-iA_\mu(\phi^\star \partial_\mu \phi - \phi \partial_\mu \phi^\star) \to -iA_\mu(\phi^\star \partial_\mu \phi - \phi \partial_\mu \phi^\star) + i\pi(\phi^\star \Box \phi - \phi \Box \phi^\star), \quad (8.119)$$

where the substituted part has been integrated by parts. This is not zero. However, if we allow ϕ to transform as

$$\phi \to \phi - i\pi\phi, \qquad (8.120)$$

then ϕ's kinetic term $(\partial_\mu \phi^\star)(\partial_\mu \phi)$ will transform as

$$(\partial_\mu \phi^\star)(\partial_\mu \phi) \to (\partial_\mu \phi^\star)(\partial_\mu \phi) - i\left[(\partial_\mu \phi^\star)(\partial_\mu \pi)\phi - (\partial_\mu \phi)(\partial_\mu \pi)\phi^\star\right] - (\pi\phi^\star)\,\Box\,(\pi\phi)$$
$$= (\partial_\mu \phi^\star)(\partial_\mu \phi) - i\pi(\phi^\star \Box \phi - \phi \Box \phi^\star) - (\pi\phi^\star)\,\Box\,(\pi\phi), \qquad (8.121)$$

which exactly cancels the unwanted piece. However, now there are terms quadratic in π that do not cancel. These can be compensated by adding terms second order in the field,

$$\phi \to \phi - i\pi\phi - \frac{1}{2}\pi^2\phi \qquad (8.122)$$

and adding another term to the Lagrangian:

$$\mathcal{L} = -\frac{1}{4}F_{\mu\nu}^2 + (\partial_\mu\phi^\star)(\partial_\mu\phi) - iA_\mu(\phi^\star\partial_\mu\phi - \phi\partial_\mu\phi^\star) + A_\mu^2\phi^\star\phi. \tag{8.123}$$

This is invariant up to terms of order π^3, but it is exactly invariant under $A_\mu \to A_\mu + \partial_\mu\pi$ and $\phi \to e^{-i\pi}\phi$. In fact, it is just the scalar QED Lagrangian that we have derived from the bottom up.

8.7.2 Spin 2

For spin-1 fields, the procedure in the previous section is a bit tedious, since we already know about gauge invariance. The power of this technique becomes clear when we generalize to spin 2. Let us start with a symmetric tensor $h_{\mu\nu}$ (we can force it to be symmetric by introducing only 10 independent elements in its Lagrangian). Then we can replace

$$h_{\mu\nu} = h_{\mu\nu}^T + \partial_\mu\pi_\nu + \partial_\nu\pi_\mu \tag{8.124}$$

with $\partial_\mu h_{\mu\nu}^T = 0$. A massive spin-2 field should have five polarizations, two in the transverse components and three in the longitudinal components, π_ν. To make sure that π_ν has three physical components, we can further transform by

$$\pi_\mu = \pi_\mu^T + \partial_\mu\pi^L \tag{8.125}$$

with $\partial_\mu\pi_\mu^T = 0$.

The most general kinetic terms of dimension 4 we can write down are

$$\mathcal{L} = ah_{\mu\nu}\Box h_{\mu\nu} + bh_{\mu\nu}\partial_\mu\partial_\alpha h_{\nu\alpha} + ch\Box h + dh\partial_\mu\partial_\nu h_{\mu\nu} + m^2(xh_{\mu\nu}^2 + yh^2), \tag{8.126}$$

where $h = h_{\alpha\alpha}$ is the trace of the tensor. Let us start by looking at the mass term. After inserting the replacements in Eqs. (8.124) and (8.125), we find

$$m^2(xh_{\mu\nu}^2 + yh^2) = 4m^2(x + y)\pi_L\Box^2\pi_L + \cdots, \tag{8.127}$$

which says that a component of the field has a dangerous 4-derivative kinetic term. We can eliminate the 4-derivative kinetic term uniquely by taking $x = -y$. A similar analysis for the rest of the Lagrangian leads to

$$\mathcal{L} = \frac{1}{2}h_{\mu\nu}\Box h_{\mu\nu} - h_{\mu\nu}\partial_\mu\partial_\alpha h_{\nu\alpha} + h\partial_\mu\partial_\nu h_{\mu\nu} - \frac{1}{2}h\Box h + \frac{1}{2}m^2(h_{\mu\nu}^2 - h^2). \tag{8.128}$$

This is the unique Lagrangian for a massive spin-2 field. It was first derived by Markus Fierz and Wolfgang Pauli in 1939 [Fierz and Pauli, 1939].

In the massless limit,

$$\mathcal{L}_{\text{kin}} = \frac{1}{4}h_{\mu\nu}\Box h_{\mu\nu} - \frac{1}{2}h_{\mu\nu}\partial_\mu\partial_\alpha h_{\nu\alpha} + \frac{1}{2}h\partial_\mu\partial_\nu h_{\mu\nu} - \frac{1}{4}h\Box h. \tag{8.129}$$

This happens to be the leading terms in the expansion of the Einstein–Hilbert Lagrangian (see Eq. (8.146) below).

For a massless spin-2 field, as with a massless spin-1 field, the π_μ should never appear in the interactions or the theory will be sick. A generic interaction would be

$$\mathcal{L} = \cdots + h_{\mu\nu}T_{\mu\nu}. \tag{8.130}$$

Then, having π_ν decouple when $h_{\mu\nu} \to h_{\mu\nu} + \partial_\mu \pi_\nu + \partial_\nu \pi_\mu$ forces $\partial_\mu T_{\mu\nu} = 0$. Thus, a massless spin-2 field must couple to a conserved tensor current. The simplest interaction would be

$$\mathcal{L}_1 = \frac{1}{2} h\phi. \tag{8.131}$$

Under

$$h_{\mu\nu} \to h_{\mu\nu} + \partial_\mu \pi_\nu + \partial_\nu \pi_\mu \tag{8.132}$$

this gives

$$\mathcal{L}_1 \to \mathcal{L}_1 + \partial_\nu \pi_\nu \phi, \tag{8.133}$$

which does not vanish.

The way out is, as in the spin-1 case, that we are allowed to perform replacements on ϕ at the same time. However, with only this one interaction, any change in $\phi \to \phi'[\phi, \pi]$ would automatically have a term with a π in it. We can make progress, however, if we modify our interaction Lagrangian to

$$\mathcal{L}_2 = \phi + \frac{1}{2} h\phi \tag{8.134}$$

and allow for ϕ to transform as

$$\phi \to \phi + \pi_\nu \partial_\nu \phi. \tag{8.135}$$

That works to cancel the term in Eq. (8.133). The Lagrangian $\mathcal{L}_2$ is now invariant up to terms with three or more fields:

$$\mathcal{L}_2 \to \mathcal{L}_2 + \frac{1}{2} h\pi_\nu(\partial_\nu \phi) + (\partial_\nu \pi_\nu)(\pi_\alpha \partial_\alpha \phi). \tag{8.136}$$

Let us focus on the term linear in π, since small π represents an infinitesimal transformation. The $\frac{1}{2} h\pi_\nu(\partial_\nu \phi)$ can be canceled if we generalize the transformation of h to

$$h_{\mu\nu} \to h_{\mu\nu} + \partial_\mu \pi_\nu + \partial_\nu \pi_\mu + \pi_\alpha \partial_\alpha h_{\mu\nu} \tag{8.137}$$

and add another term to our Lagrangian

$$\mathcal{L}_3 = \phi + \frac{1}{2} h\phi + \frac{1}{8} h^2\phi, \tag{8.138}$$

so that

$$\mathcal{L}_3 \to \mathcal{L}_3 + \frac{1}{2} h\pi_\nu(\partial_\nu \phi) + (\partial_\nu \pi_\nu)(\pi_\alpha \partial_\alpha \phi) + \frac{1}{2} \pi_\alpha(\partial_\alpha h)\phi + \frac{1}{2}(\partial_\alpha \pi_\alpha)h\phi$$
$$+ \frac{1}{2}(\partial_\alpha \pi_\alpha)(\partial_\nu \pi_\nu)\phi + \cdots, \tag{8.139}$$

where the $\cdots$ contain terms with four or more fields. Integrating by parts, all the terms with one factor of π cancel. This process can continue as a perturbation expansion in the number of fields for terms with one factor of π.

Continuing in this way, we are led to transformations

$$h_{\mu\nu} \to h_{\mu\nu} + \partial_\mu \pi_\nu + \partial_\nu \pi_\mu + \pi^\alpha \partial_\alpha h_{\mu\nu} + (\partial_\mu \pi^\alpha)h_{\alpha\nu} + (\partial_\nu \pi^\alpha)h_{\mu\alpha} \tag{8.140}$$

and

$$\phi \to \phi + \pi^\alpha \partial_\alpha \phi, \tag{8.141}$$

with a Lagrangian

$$\mathcal{L} = \left(1 + \frac{1}{2}h + \frac{1}{8}h^2 + \cdots\right)\phi. \tag{8.142}$$

This Lagrangian will be independent of π to linear order in π.

To make something invariant to all orders in π, the complete transformation can be written as

$$\phi \to \phi(x^\alpha + \pi^\alpha), \tag{8.143}$$

$$h_{\mu\nu} \to (\eta_{\alpha\mu} + \partial_\alpha \pi_\mu)(\eta_{\beta\nu} + \partial_\beta \pi_\nu)\left[\eta_{\alpha\beta} + h_{\alpha\beta}(x^\gamma + \pi^\gamma)\right] - \eta_{\mu\nu}, \tag{8.144}$$

where $\phi(x^\alpha + \pi^\alpha)$ and $h_{\alpha\beta}(x^\gamma + \pi^\gamma)$ are to be understood as Taylor expansions in π. Here, $\eta_{\mu\nu}$ is the Minkowski metric, which we usually call $g_{\mu\nu}$. A reader familiar with general relativity will recognize this as a general coordinate transformation, and the Lagrangian as

$$\mathcal{L} = \sqrt{-\det(\eta_{\mu\nu} + \frac{1}{M_{\mathrm{Pl}}}h_{\mu\nu})}\phi, \tag{8.145}$$

where $M_{\mathrm{Pl}} = G_N^{-1/2} \approx 10^{19}$ GeV is the Planck scale, which has been introduced to make $h_{\mu\nu}$ have mass dimension 1.

In the same way, the kinetic terms for $h_{\mu\nu}$ become

$$\mathcal{L} = M_{\mathrm{Pl}}^2 \sqrt{-\det\left(\eta_{\mu\nu} + \frac{1}{M_{\mathrm{Pl}}}h_{\mu\nu}\right)} R\left[\eta_{\mu\nu} + \frac{1}{M_{\mathrm{Pl}}}h_{\mu\nu}\right], \tag{8.146}$$

where R is the Ricci scalar. We have rescaled $h_{\mu\nu}$ by a factor of $\frac{1}{M_{\mathrm{Pl}}}$ by dimensional analysis to give $h_{\mu\nu}$ canonical dimension, since R has dimension 2 (every term in R has two derivatives). Thus, the Lagrangian for general relativity is given uniquely as the only Lagrangian that can couple a massless spin-2 particle to matter.

This is obviously a very inefficient way to deduce the Lagrangian for a massless spin-2 field. It is much nicer to use symmetry arguments, general coordinate invariance, the equivalence principle, etc. The one thing those arguments do not tell you is why that theory is unique. For example, in general relativity, at some point you have to assume the connection is torsion free and compatible with the metric. What happened to the torsion tensor? If all you know about is general coordinate invariance, you have not yet constructed a physical theory. Constructing it this way you see that you generate the curvature tensor, not the torsion tensor. More precisely, it might be the torsion tensor for a different geometric construction, but the expansion in terms of $h_{\mu\nu}$ will be identical. The simple fact that there is a *unique* theory of a massless spin-2 particle coupled to matter is an important consequence of this approach.

Of course, there are huge advantages to general relativity. In particular, this approach is based on Lorentz invariance and is entirely perturbative. In contrast, general relativity is background independent and non-perturbative. The Schwarzschild solution, from this language, is a coherent background of gravitons. That is not a productive language for doing calculations in general, although it is useful for certain calculations. For example, the perihelion shift of Mercury can be computed perturbatively this way (Problem 3.7).

8.7.3 Spin greater than 2

One can continue this procedure for integer spin greater than 2. There exist spin-3 particles in nature, for example the ω_3 with mass of 1670 MeV, as well as spin 4, spin 5, etc. These particles are all massive. One can construct free Lagrangians for them using the same trick. An interesting and profound result is that it is impossible to have an *interacting* theory of massless particles with spin greater than 2. The required gauge invariance would be so restrictive that nothing could satisfy it. We will prove this in the next chapter. Constructing the kinetic term for a spin-3 particle is done in Problem 8.8.

Problems

8.1 Show that having a probability interpretation, with $0 \leq P \leq 1$, requires us to have only positive (or only negative) norm states.

8.2 Calculate the energy-momentum tensor corresponding to the Lagrangian $\mathcal{L} = -\frac{1}{4}F_{\mu\nu}^2$. Show that the energy density is positive definite, up to a total spatial divergence $\mathcal{E} - \partial_i X_i > 0$ for some X_i.

8.3 Calculate the classical propagator for a massive spin-1 particle by inverting the equations of motion to the form $A_\mu = \Pi_{\mu\nu}J_\nu$.

8.4 Calculate the propagator for a photon in axial gauge, where $A_0 = 0$.

8.5 Vector polarization sums. In this problem you can build some intuition for the way in which the numerator of a spin-1 particle propagator represents an outer product of physical polarizations $|\epsilon\rangle\langle\epsilon|$. Calculate the 4×4 matrix outer product $|\epsilon\rangle\langle\epsilon| \equiv \sum_j \epsilon_\mu^j \epsilon_\nu^j$ by the following:

(a) Sum over the physical polarizations for a massive spin-1 particle in some frame. Re-express your answer in a Lorentz-covariant way, in terms of m, $k_\mu k_\nu$ and $g_{\mu\nu}$.

(b) Show that the numerator of the massive vector propagator (Problem 8.3) is the same as the polarization sum. Why should this be true?

(c) Sum over the two physical polarizations for a massless vector. A helpful basis for these polarizations comes from choosing them orthogonal to both momentum, $\epsilon^i \cdot p = 0$ and an arbitrary reference vector r^μ: $\epsilon^i \cdot r = 0$. Find explicit forms for the two polarizations, do the sum, and then express your answer in a Lorentz-covariant way (i.e. in terms of p^μ and r^μ).

(d) Write down a Lagrangian so that the photon propagator derived from it has the numerator you found in part (c).

(e) Compare the numerator from part (c) to the numerator of the photon propagator in the R_ξ gauges. What might be an advantage of using the numerator from (c) rather than Feynman gauge? What might be a disadvantage?

8.6 Tensor polarization sums. A spin-2 particle can be embedded in a 2-index tensor $h_{\mu\nu}$. Therefore, its polarizations are tensors too, $\epsilon^i_{\mu\nu}$. These should be orthonormal, $\epsilon^i_{\mu\nu}\epsilon^{*j}_{\mu\nu} = \delta^{ij}$, where the sum is over μ and ν contracted with the Minkowski metric.

(a) The polarizations should be transverse, $k_\mu \epsilon^i_{\mu\nu} = 0$, and symmetric, $\epsilon^i_{\mu\nu} = \epsilon^i_{\nu\mu}$. How many degrees of freedom do these conditions remove?

(b) For a massive spin-2 particle, choose a frame in which the momentum k_μ is simple. How many orthonormal $\epsilon^i_{\mu\nu}$ can you find? Write your basis out explicitly, as 4×4 matrices.

(c) Guess which of these correspond to spin 0, spin 1 or spin 2. What kind of Lorentz-invariant condition can you impose so that you just get the spin-2 polarizations?

(d) If you use the same conditions but take k_μ to be the momentum of a massless tensor, what are the polarizations? Do you get the right number?

(e) What would you embed a massive spin-3 field in? What conditions could you impose to get the right number of degrees of freedom?

8.7 Using the method of Section 8.7.2 construct the set of cubic interactions of a massless spin-2 field embedded in $h_{\mu\nu}$. There are many terms, all with two derivatives, but their coefficients are precisely fixed. You can also check that this is the same thing you get from expanding $M_{\mathrm{Pl}}^2 \sqrt{\eta_{\mu\nu} + \frac{1}{M_{\mathrm{Pl}}}h_{\mu\nu}} R\left[\eta_{\mu\nu} + \frac{1}{M_{\mathrm{Pl}}}h_{\mu\nu}\right]$ to cubic order in $h_{\mu\nu}$.

It should be clear that the same method will produce the terms fourth order in $h_{\mu\nu}$, however, these are suppressed by $\frac{1}{M_{\mathrm{Pl}}^2}$. Most tests of general relativity probe only that it is described by a minimally coupled spin-2 field (e.g. bending of light, gravitational waves, frame dragging). Some precision tests assay the cubic interactions (e.g. the perihelion shift of Mercury). No experiment has yet tested the quartic interactions.

8.8 Construct the free kinetic Lagrangian for a massive spin-3 particle by embedding it in a tensor $Z_{\mu\nu\alpha}$.

8.9 Show that it is impossible to write down a Lorentz-invariant Lagrangian for a single scalar field with 4-derivative kinetic terms (e.g. $\mathcal{L} = -\phi\Box^2\phi$) that generates a non-negative energy density.

9 Scalar quantum electrodynamics

Now that we have Feynman rules and we know how to quantize the photon, we are very close to QED. All we need is the electron, which is a spinor. Before we get into spinors, however, it is useful to explore a theory that is an approximation to QED in which the spin of the electron can be neglected. This is called scalar QED. The Lagrangian is

$$\mathcal{L} = -\frac{1}{4}F_{\mu\nu}^2 + |D_\mu\phi|^2 - m^2|\phi|^2, \tag{9.1}$$

with

$$D_\mu\phi = \partial_\mu\phi + ieA_\mu\phi, \tag{9.2}$$

$$D_\mu\phi^\star = \partial_\mu\phi^\star - ieA_\mu\phi^\star. \tag{9.3}$$

Although there actually do exist charged scalar fields in nature which this Lagrangian describes, for example the charged pions, that is not the reason we are introducing scalar QED before spinor QED. Spinors are somewhat complicated, so starting with this simplified Lagrangian will let us understand some elements of QED without having to deal with spinor algebra.

9.1 Quantizing complex scalar fields

We saw that for a scalar field to couple to A_μ it has to be complex. This is because the charge is associated with a continuous global symmetry under which

$$\phi \rightarrow e^{-i\alpha}\phi. \tag{9.4}$$

Such phase rotations only make sense for complex fields. The first thing to notice is that the classical equations of motion for ϕ and $\phi^\star$ are[1]

$$\left(\Box + m^2\right)\phi = i(-eA_\mu)\,\partial_\mu\phi + i\partial_\mu(-eA_\mu\phi) + (-eA_\mu)^2\phi, \tag{9.5}$$

$$\left(\Box + m^2\right)\phi^\star = i(eA_\mu)\,\partial_\mu\phi^\star + i\partial_\mu(eA_\mu\phi^\star) + (eA_\mu)^2\phi^\star. \tag{9.6}$$

So we see that ϕ and $\phi^\star$ couple to the electromagnetic field with opposite charge, but have the same mass. Of course, something having an equation does not mean we can produce

---

[1] We are treating ϕ and $\phi^\star$ as separate real degrees of freedom. If you find this confusing you can always write $\phi = \phi_1 + i\phi_2$ and study the physics of the two independent fields ϕ_1 and ϕ_2, but the ϕ and $\phi^\star$ notation is much more efficient.

it. However, in a second-quantized relativistic theory, the radiation process, $\phi \rightarrow \phi\gamma$, automatically implies that $\gamma \rightarrow \phi\phi^\star$ is also possible (as we will see). Thus, we must be able to produce these $\phi^\star$ particles. In other words, in a relativistic theory with a massless spin-1 field, **antiparticles** must exist and we know how to produce them!

To see antiparticles in the quantum theory, first recall that a quantized real scalar field is

$$\phi(x) = \int \frac{d^3p}{(2\pi)^3} \frac{1}{\sqrt{2\omega_p}} \left(a_p e^{-ipx} + a_p^\dagger e^{ipx} \right). \tag{9.7}$$

Since a complex scalar field must be different from its conjugate by definition, we have to allow for a more general form. We can do this by introducing two sets of creation and annihilation operators and writing

$$\phi(x) = \int \frac{d^3p}{(2\pi)^3} \frac{1}{\sqrt{2\omega_p}} \left(a_p e^{-ipx} + b_p^\dagger e^{ipx} \right). \tag{9.8}$$

Then, by complex conjugation

$$\phi^\star(x) = \int \frac{d^3p}{(2\pi)^3} \frac{1}{\sqrt{2\omega_p}} \left(a_p^\dagger e^{ipx} + b_p e^{-ipx} \right). \tag{9.9}$$

Thus, we can conclude that b_p annihilates particles of opposite charge and the same mass to what a_p annihilates. That is, b_p annihilates the antiparticles. Note that in both cases $\omega_p = \sqrt{\vec{p}^2 + m^2} > 0$.

All we used was the fact that the field was complex. Clearly $a_p^\dagger \neq b_p^\dagger$ as these operators create particles of opposite charge. So a global symmetry under phase rotations implies charge, which implies complex fields, which implies antiparticles. That is,

Matter coupled to massless spin-1 particles automatically implies the existence of antiparticles, which are particles of identical mass and opposite charge.

This profound conclusion is an inevitable consequence of relativity and quantum mechanics.

To recap, we saw that to have a consistent theory with a massless spin-1 particle we needed gauge invariance. This required a conserved current, which in turn required that charge be conserved. To couple the photon to matter, we needed more than one degree of freedom so we were led to ϕ and $\phi^\star$. Upon quantization, complex scalar fields imply antiparticles. Thus, there are many profound consequences of consistent theories of massless spin-1 particles.

9.1.1 Historical note: holes

Historically, it was the Dirac equation that led to antiparticles. In fact, in 1931 Dirac predicted there should be a particle exactly like the electron except with opposite charge. In 1932 the positron was discovered by Anderson, beautifully confirming Dirac's prediction and inspiring generations of physicists.

Actually, Dirac had an interpretation of antiparticles that sounds funny in retrospect, but was much more logical to him for historical reasons. Suppose we had written

$$\phi(x) = \int \frac{d^3p}{(2\pi)^3} \frac{1}{\sqrt{2\omega_p}} \left(a_p^\dagger e^{ipx} + c_p^\dagger e^{-ipx}\right), \tag{9.10}$$

where both $a_p^\dagger$ and $c_p^\dagger$ are creation operators. Then $c_p^\dagger$ seems to be creating states of negative frequency, or equivalently negative energy. This made sense to Dirac at the time, since there are classical solutions to the Klein–Gordon equation, $E^2 - p^2 = m^2$, with negative energy, so something should create these solutions. Dirac interpreted these negative energy creation operators as removing something of positive energy, and creating an energy hole. But an energy hole in what? His answer was that the universe is a sea full of positive energy states. Then $c_p^\dagger$ creates a hole in this sea, which moves around like an independent excitation.

Then why does the sea stay full, and not collapse to the lower-energy configuration? Dirac's explanation for this was to invoke the Pauli exclusion principle. The sea is like the orbitals of an atom. When an atom loses an electron it becomes ionized, but it looks like it gained a positive charge. So positive charges can be interpreted as the absence of negative charges, as long as all the orbitals are filled. Dirac argued that the universe might be almost full of particles, so that the negative energy states are the absences of those particles [Dirac, 1930].

It is not hard to see that this is total nonsense. For example, it should work only for fermions, not our scalar field, which is a boson. As we have seen, it is much easier to write the creation operator $c_p^\dagger$ as an annihilation operator to begin with, $c_p^\dagger = b_p$, which cleans everything up immediately. Then the negative energy solutions correspond to the absence of antiparticles, which does not require a sea.

9.2 Feynman rules for scalar QED

Expanding out the scalar QED Lagrangian we find

$$\mathcal{L} = -\frac{1}{4}F_{\mu\nu}^2 - \phi^\star(\Box + m^2)\phi - ieA_\mu[\phi^\star(\partial_\mu\phi) - (\partial_\mu\phi^\star)\phi] + e^2A_\mu^2|\phi|^2. \tag{9.11}$$

We can read off the Feynman rules from the Lagrangian. The complex scalar propagator is

$$= \frac{i}{p^2 - m^2 + i\varepsilon}. \tag{9.12}$$

This propagator is the Fourier transform of $\langle 0|\phi^\star(x)\phi(0)|0\rangle$ in the free theory. It propagates both ϕ and $\phi^\star$, that is both particles and antiparticles at the same time – they cannot be disentangled.

The photon propagator was calculated in Section 8.5:

$$= \frac{-i}{p^2 + i\varepsilon}\left[g_{\mu\nu} - (1-\xi)\frac{p_\mu p_\nu}{p^2}\right], \tag{9.13}$$

where ξ parametrizes a set of covariant gauges.

Some of the interactions that connect A_μ to ϕ and $\phi^\star$ have derivatives in them, which will give momentum factors in the Feynman rules. To see which momentum factors we get, look back at the quantized fields:

$$\phi(x) = \int \frac{d^3p}{(2\pi)^3} \frac{1}{\sqrt{2\omega_p}} \left(a_p e^{-ipx} + b_p^\dagger e^{ipx} \right), \tag{9.14}$$

$$\phi^\star(x) = \int \frac{d^3p}{(2\pi)^3} \frac{1}{\sqrt{2\omega_p}} \left(a_p^\dagger e^{ipx} + b_p e^{-ipx} \right). \tag{9.15}$$

A ϕ in the interaction implies the creation of an antiparticle or the annihilation of a particle at position x. A $\phi^\star$ implies the creation of a particle or the annihilation of an antiparticle. When a derivative acts on these fields, we will pull down a factor of $\pm ip^\mu$ which enters the vertex Feynman rule.

Since the interaction has the form

$$- ieA_\mu[\phi^\star(\partial_\mu\phi) - (\partial_\mu\phi^\star)\phi], \tag{9.16}$$

it always has one ϕ and one $\phi^\star$. Each p^μ comes with an i, and there is another i from the expansion of $\exp(i\mathcal{L}_{\text{int}})$, so we always get an overall $(-ie)\, i^2 = ie$ multiplying whichever $\pm p^\mu$ comes from the derivative. There are four possibilities, each one getting a contribution from $A_\mu\phi^\star(\partial_\mu\phi)$ and $-A_\mu\phi(\partial_\mu\phi^\star)$. Calling what a_p annihilates an e^- and what b_p annihilates an e^+ the possibilities are:

- Annihilate e^- and create e^- – particle scattering

$$\text{(diagram)} = ie(-p_\mu^1 - p_\mu^2). \tag{9.17}$$

Here, the term $\phi^\star(\partial_\mu\phi)$ gives a $-p_\mu^1$ because the e^- is annihilated by ϕ and the $-\phi(\partial_\mu\phi^\star)$ gives a $-(+p_\mu^2)$ because an e^- is being created by $\phi^\star$. We will come back to the arrows in a moment.

- Annihilate e^+ and create e^+ – antiparticle scattering

$$\text{(diagram)} = ie(p_\mu^1 + p_\mu^2). \tag{9.18}$$

Here, $\phi^\star(\partial_\mu\phi)$ creates the e^+ giving p_μ^2 and $-(\partial_\mu\phi^\star)\phi$ annihilates an e^+ giving $-(-p_\mu^1)$. The next two you can do yourself.

- Annihilate e^- and annihilate e^+ – pair annihilation

$$\text{(diagram)} = ie(-p_\mu^1 + p_\mu^2). \tag{9.19}$$

- Create e^- and create e^+ – pair creation

$$\vcenter{\hbox{\includegraphics{diagram}}} = ie(-p_\mu^1 + p_\mu^2). \qquad (9.20)$$

First of all, we see that there are only four types of vertices. It is impossible for a vertex to create two particles of the same charge. That is, the Feynman rules guarantee that charge is conserved.

Now let us explain the arrows. In the above vertices, the arrows outside the scalar lines are *momentum-flow arrows*, indicating the direction that momentum is flowing. We conventionally draw momentum flowing from left to right. The arrows superimposed on the lines in the diagram are *particle-flow arrows*. These arrows point in the direction of momentum for particles (e^-) but opposite to the direction of momentum for antiparticles (e^+). If you look at all the vertices, you will see that if the particle-flow arrow points to the right, the vertex gives $-iep_\mu$, if the particle-flow arrow points to the left, the vertex gives $+iep_\mu$. So the particle-flow arrows make the scalar QED Feynman rule easy to remember:

> A scalar QED vertex gives $-ie$ times the sum of the momentum of the particles whose particle-flow arrows point to the right minus the momentum of the particles whose arrows point to the left.

The four cases in Eqs. (9.17) to (9.20) are reproduced by this single rule.

Particle-flow arrows should always make a connected path through the Feynman diagram. For internal lines and loops, whether your arrows point left or right is arbitrary; as long as the direction of the arrows is consistent with particle flow the answer will be the same. If your diagram represents a physical process, external line particle-flow arrows should always point right for particles and to the left for antiparticles.

For loops it is impossible to always have the momentum going to the right. It is conventional in loops to have the momentum flow in the same direction as the charge-flow arrows. For uncharged particles, such as photons or real scalars, you can pick any directions for the loop momenta you want, as long as momentum is conserved at each vertex. Some examples are

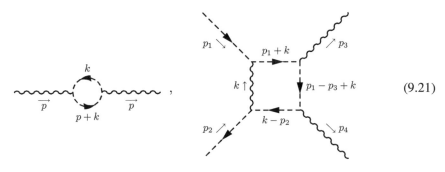

$$(9.21)$$

For antiparticles, momentum is flowing backwards to the direction of the arrow. Thus, if particles go forwards in time, antiparticles must be going backwards in time. This idea was proposed by Stueckelberg in 1941 and independently by Feynman at the famous Poconos conference in 1948 as an interpretation of his Feynman diagrams. The Feynman–Stuckelberg interpretation gives a funny picture of the universe with electrons flying around, bouncing off photons and going back in time, etc. You can have fun thinking about this, but the picture does not seem to have much practical application.

Finally, we cannot forget that there is another vertex in scalar QED:

$$\mathcal{L}_{\text{int}} = e^2 A_\mu^2 |\phi|^2. \tag{9.22}$$

This vertex comes from $|D_\mu \phi|^2$, so it is forced by gauge invariance. Its Feynman rule is

$$= 2ie^2 g_{\mu\nu}. \tag{9.23}$$

This is sometimes called a *seagull* vertex, perhaps due to its vague resemblence to the head-on view of a bird. The 2 comes from the symmetry factor for the two A fields. There would not have been a 2 if we had written $\frac{1}{2} e^2 A_\mu^2 |\phi|^2$, but this is not what the Lagrangian gives us. The i comes from the expansion of $\exp(i\mathcal{L}_{\text{int}})$ which we always have for Feynman rules.

9.2.1 External states

Now we know the vertex factors and propagators for the photon and the complex scalar field. The only thing left in the Feynman rules is how to handle external states. For a scalar field, this is easy – we just get a factor 1. That is because a complex scalar field is just two real scalar fields, so we just take the real scalar field result. The only thing left is external photons.

For external photons, recall that the photon field is

$$A_\mu(x) = \int \frac{d^3 k}{(2\pi)^3} \frac{1}{\sqrt{2\omega_k}} \sum_{i=1}^{2} \left(\epsilon_\mu^i(k) a_{k,i} e^{-ikx} + \epsilon_\mu^{i\star}(k) a_{k,i}^\dagger e^{ikx} \right). \tag{9.24}$$

As far as free states are concerned, which is all we need for S-matrix elements, the photon is just a bunch of scalar fields integrated against some polarization vectors $\epsilon_\mu^i(k)$. Recall that external states with photons have momenta and polarizations, $|k, \epsilon\rangle$, so that $\langle 0 | A_\mu(x) | k, \epsilon_i \rangle = \epsilon_\mu^i(k) e^{-ikx}$. This leads to LSZ being modified only by adding a factor of the photon polarization for each external state: ϵ_μ if it is incoming and $\epsilon_\mu^\star$ if it is outgoing.

For example, consider the following diagram:

$$
i\mathcal{M} = \quad\text{[diagram]}\quad = (-ie)\epsilon_\mu^1(p_2^\mu + k^\mu)\frac{i}{k^2 - m^2 + i\varepsilon}(-ie)(p_3^\nu + k^\nu)\epsilon_\nu^{\star 4},
$$

$$(9.25)$$

where $k^\mu = p_1^\mu + p_2^\mu$. The first polarization ϵ_μ^1 is the polarization of the photon labeled with p_μ^1. It gets contracted with the momenta $p_2^\mu + k^\mu$ which come from the $-ieA_\mu[\phi^\star(\partial_\mu\phi) -(\partial_\mu\phi^\star)\phi]$ vertex. The other polarization, ϵ_μ^4, is the polarization of the photon labeled with p_μ^4 and contracts with the second vertex.

9.3 Scattering in scalar QED

As a first application, let us calculate the cross section for Møller scattering, $e^- e^- \rightarrow e^- e^-$, in scalar QED. There are two diagrams. The t-channel diagram (recall the Mandelstam variables s, t and u from Section 7.4.1) gives

$$
i\mathcal{M}_t = \quad\text{[diagram]}\quad = (-ie)(p_1^\mu + p_3^\mu)\frac{-i\left[g_{\mu\nu} - (1-\xi)\frac{k_\mu k_\nu}{k^2}\right]}{k^2}(-ie)(p_2^\nu + p_4^\nu),
$$

$$(9.26)$$

with $k^\mu = p_3^\mu - p_1^\mu$. But note that

$$
k^\mu(p_1^\mu + p_3^\mu) = (p_3^\mu - p_1^\mu)(p_3^\mu + p_1^\mu) = p_3^2 - p_1^2 = m^2 - m^2 = 0. \tag{9.27}
$$

So this simplifies to

$$
\mathcal{M}_t = e^2\frac{(p_1^\mu + p_3^\mu)(p_2^\mu + p_4^\mu)}{t} \tag{9.28}
$$

and the ξ dependence has vanished. We expected this to happen, by gauge invariance, and now we have seen that it does indeed happen.

The u-channel gives

$$
i\mathcal{M}_u = \quad\text{[diagram]}\quad = (-ie)(p_1^\mu + p_4^\mu)\frac{-i\left[g_{\mu\nu} - (1-\xi)\frac{k_\mu k_\nu}{k^2}\right]}{k^2}(-ie)(p_2^\nu + p_3^\nu),
$$

$$(9.29)$$

where $k^\mu = p_4^\mu - p_1^\mu$. In this case,

$$k^\mu(p_1^\mu + p_4^\mu) = p_4^2 - p_1^2 = 0 \tag{9.30}$$

so that

$$\mathcal{M}_u = e^2 \frac{(p_1^\mu + p_4^\mu)(p_2^\mu + p_3^\mu)}{u}. \tag{9.31}$$

Thus, the cross section for scalar Møller scattering is

$$\begin{aligned}
\frac{d\sigma(e^-e^- \to e^-e^-)}{d\Omega} &= \frac{e^4}{64\pi^2 E_{\mathrm{CM}}^2}\left[\frac{(p_1^\mu + p_3^\mu)(p_2^\mu + p_4^\mu)}{t} + \frac{(p_1^\mu + p_4^\mu)(p_2^\mu + p_3^\mu)}{u}\right]^2 \\
&= \frac{\alpha^2}{4s}\left[\frac{s-u}{t} + \frac{s-t}{u}\right]^2,
\end{aligned} \tag{9.32}$$

where $\alpha = \frac{e^2}{4\pi}$ is the fine-structure constant.

9.4 Ward identity and gauge invariance

We saw in the previous example that the matrix elements for a particular amplitude in scalar QED were independent of the gauge parameter ξ. The photon propagator is

$$i\Pi_{\mu\nu} = \frac{-i\left[g_{\mu\nu} - (1-\xi)\frac{p_\mu p_\nu}{p^2}\right]}{p^2 + i\varepsilon}. \tag{9.33}$$

A general matrix element involving an internal photon will be $M_{\mu\nu}\Pi_{\mu\nu}$ for some $M_{\mu\nu}$. So gauge invariance, which in this context means ξ independence, requires $M_{\mu\nu}p_\mu p_\nu = 0$. Gauge invariance in this sense is closely related to the Ward identity, which required $p_\mu M_\mu = 0$ if the matrix element involving an on-shell photon is $\epsilon_\mu M_\mu$. Both gauge invariance and the Ward identity hold for any amplitude in scalar QED. However, it is somewhat tedious to prove this in perturbation theory. In this section, we will give a couple of examples illustrating what goes into the proof, with the complete non-perturbative proof postponed until Section 14.8 after path integrals are introduced.

As a non-trivial example where the Ward identity can be checked, consider the process $e^+e^- \to \gamma\gamma$. A diagram contributing to this is

$$i\mathcal{M}_t = \quad = (-ie)^2 \frac{i(2p_1^\mu - p_3^\mu)(p_4^\nu - 2p_2^\nu)}{(p_1 - p_3)^2 - m^2}\epsilon_3^{\star\mu}\epsilon_4^{\star\nu}. \tag{9.34}$$

Using only that the electron is on shell (not assuming $p_3^2 = p_4^2 = p_3 \cdot \epsilon_3 = p_4 \cdot \epsilon_4$), this simplifies slightly to

$$M_t = e^2 \frac{(p_3 \cdot \epsilon_3^\star - 2p_1 \cdot \epsilon_3^\star)(p_4 \cdot \epsilon_4^\star - 2p_2 \cdot \epsilon_4^\star)}{p_3^2 - 2p_3 \cdot p_1}. \tag{9.35}$$

The crossed diagram gives the same thing with $1 \leftrightarrow 2$ (or equivalently $3 \leftrightarrow 4$)

$$iM_u = \qquad\qquad = ie^2 \frac{(p_3 \cdot \epsilon_3^\star - 2p_2 \cdot \epsilon_3^\star)(p_4 \cdot \epsilon_4^\star - 2p_1 \cdot \epsilon_4^\star)}{p_3^2 - 2p_3 \cdot p_2}. \tag{9.36}$$

To check whether the Ward identity is satisfied with just these two diagrams, we replace $\epsilon_3^{\star\nu}$ with p_3^μ giving

$$\mathcal{M}_t + \mathcal{M}_u = e^2[p_4 \cdot \epsilon_4^\star - 2p_2 \cdot \epsilon_4^\star + p_4 \cdot \epsilon_4^\star - 2p_1 \cdot \epsilon_4^\star] = 2e^2\epsilon_4^{\star\mu}(p_4^\mu - p_2^\mu - p_1^\mu), \tag{9.37}$$

which is in general non-zero. The resolution is the missing diagram involving the 4-point vertex:

$$iM_4 = \qquad\qquad = 2ie^2 g_{\mu\nu}\epsilon_3^{\star\mu}\epsilon_4^{\star\nu}. \tag{9.38}$$

Thus, replacing $\epsilon_3^{\star\mu}$ with p_3^μ and summing all the diagrams, we have

$$\mathcal{M}_t + \mathcal{M}_u + \mathcal{M}_4 = 2e^2\epsilon_4^{\star\mu}(p_4^\mu - p_2^\mu - p_1^\mu + p_3^\mu) = 0, \tag{9.39}$$

and the Ward identity is satisfied.

The above derivation did not require us to use that the photons are on-shell or massless. That is, we did not apply any of $p_3^2 = p_4^2 = \epsilon_3^\star \cdot p_3 = \epsilon_4^\star \cdot p_4 = 0$. Thus, the Ward identity would be satisfied even if the external photon states were not physical; for example, if they were in a loop. In fact, that is exactly what we need for gauge invariance, so the same calculation can be used to prove ξ independence.

To prove gauge invariance, we need to consider internal photon propagators, for example in a diagram such as

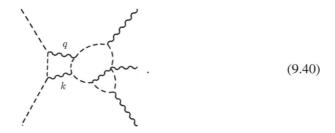

$$(9.40)$$

Let us focus on showing ξ independence for the propagators labeled q and k. For this purpose, the entire right side of the diagram (or the left side) can be replaced by a generic tensor $X_{\alpha\beta}$ depending only on the virtual momenta of the photons entering it. The index α will contract with the q photon propagator, $\Pi_{\mu\alpha}(q)$, and β with the k photon propagator, $\Pi_{\nu\beta}(k)$. Diagrammatically, this means

$$i\mathcal{M}_t = $$

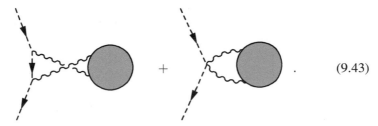

$$(9.41)$$

which is very closely related to the t-channel diagram above, Eq. (9.34). The integral can be written in the form

$$\mathcal{M}_t = \int \frac{d^4 q}{(2\pi)^4} \frac{d^4 k}{(2\pi)^4} (2\pi)^4 \delta^4(p_1 + p_2 - k - q) e^2$$
$$\times \frac{(q^\mu - 2p_1^\mu)(k^\nu - 2p_2^\nu)}{q^2 - 2q \cdot p_1} \Pi_{\mu\alpha}(q)\Pi_{\nu\beta}(k)X_{\alpha\beta}(q, k), \qquad (9.42)$$

where we have inserted an extra integral over momentum and an extra δ-function to keep the amplitude symmetric in q and k. Comparing with Eq. (9.34), the polarization vectors ϵ_3^μ and ϵ_4^ν have been replaced by contractions with the photon propagators and $p_3 \to q$ and $p_4 \to k$. Replacing $\Pi_{\mu\alpha}(q)$ by $\xi q_\mu q_\alpha$ we see that the result does not vanish, implying that this diagram alone is not gauge invariant.

To see gauge invariance, we need to include all the diagrams that contribute at the same order. This includes the u-channel diagrams and the one involving the 4-point vertex:

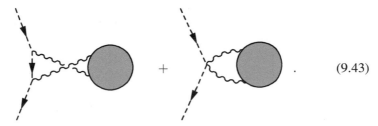

$$(9.43)$$

Adding these graphs, we get same sum as before:

$$\mathcal{M}_t + \mathcal{M}_u + \mathcal{M}_4 = e^2 \int \frac{d^4q}{(2\pi)^4} \frac{d^4k}{(2\pi)^4} (2\pi)^4 \delta^4(p_1 + p_2 - k - q)$$

$$\times \left[\frac{(q^\mu - 2p_1^\mu)(k^\nu - 2p_2^\nu)}{q^2 - 2q \cdot p_1} + \frac{(q^\mu - 2p_2^\mu)(k^\nu - 2p_1^\nu)}{q^2 - 2q \cdot p_2} + 2g^{\mu\nu} \right] \Pi_{\mu\alpha}(q) \Pi_{\nu\beta}(k) X_{\alpha\beta}(q, k).$$

$$(9.44)$$

Now if we replace $\Pi_{\mu\alpha}(q) \to \xi q_\mu q_\alpha$ we find

$$\mathcal{M}_t + \mathcal{M}_u + \mathcal{M}_4 \to 2\xi e^2 \int \frac{d^4q}{(2\pi)^4} \frac{d^4k}{(2\pi)^4} (2\pi)^4 \delta^4(p_1 + p_2 - k - q)$$

$$\times (k^\nu - p_2^\nu - p_1^\nu + q^\nu) q^\alpha \Pi_{\nu\beta}(k) X_{\alpha\beta}(q, k), \qquad (9.45)$$

which exactly vanishes. Thus, gauge invariance holds in this case. The case of a photon attaching to a closed scalar loop is similar and you can explore it in Problem 9.2.

A general diagrammatic proof involves arguments like this, generalized to an arbitrary number of photons and possible loops. The only challenging part is keeping track of the combinatorics associated with the different diagrams. Some examples can be found in [Zee, 2003] and in [Peskin and Schroeder, 1995]. The complete diagrammatic proof is actually easier in genuine QED (with spinors) than in scalar QED, since there is no 4-point vertex in QED. As mentioned above, we will give a complete non-perturbative proof of both gauge invariance and the Ward identity in Section 14.8.

9.5 Lorentz invariance and charge conservation

There is a beautiful and direct connection between Lorentz invariance and charge conservation that bypasses gauge invariance completely. What we will now show is that a theory with a massless spin-1 particle automatically has an associated conserved charge. This profound result, due to Steven Weinberg, does not require a Lagrangian description: it only uses little-group invariance and the fact that for a massless field one can take the soft limit [Weinberg, 1964].

Imagine we have some diagram with lots of external legs and loops and things. Say the matrix element for this process is $\mathcal{M}_0$. Now tack an outgoing photon of momentum q_μ and polarization ϵ_μ onto an external leg. For simplicity, we take ϵ_μ real to avoid writing $\epsilon_\mu^\star$ everywhere. Let us first tack the photon onto leg i, which we take to be an incoming e^-:

$$i\mathcal{M}_0(p_i) = \qquad\qquad\qquad \longrightarrow \qquad\qquad\qquad = \mathcal{M}_i(p_i, q). \quad (9.46)$$

This modifies the amplitude to

$$\mathcal{M}_i(p_i, q) = (-ie)\frac{i\,[p_i^\mu + (p_i^\mu - q^\mu)]}{(p_i - q)^2 - m^2}\epsilon^\mu \mathcal{M}_0(p_i - q)\,. \qquad (9.47)$$

We can simplify this using $p_i^2 = m^2$ and $q^2 = 0$ in the denominator, since the electron and photon are on-shell, and $q_\mu\epsilon_\mu = 0$ in the numerator, since the polarizations of physical photons are transverse to their own momenta. Then we get

$$\mathcal{M}_i(p_i, q) = -e\frac{p_i \cdot \epsilon}{p_i \cdot q}\mathcal{M}_0(p_i - q)\,. \qquad (9.48)$$

Now take the soft limit. By soft we mean that $|q \cdot p_i| \ll |p_j \cdot p_k|$ for all the external momenta p_i, not just the one we modified. Then $\mathcal{M}_0(p_i - q) \approx \mathcal{M}_0(p_i)$, where $\approx$ indicates the soft limit. Note that photons attached to loop momenta in the blob in $\mathcal{M}_0$ are subdominant to photons attached to external legs, since the loop momenta are off-shell and hence the associated propagators are not singular as $q \to 0$. That is, photons coming off loops cannot give $\frac{1}{p_i \cdot q}$ factors. Thus, in the soft limit, the dominant effect comes only from diagrams where photons are attached to external legs. We must sum over all such diagrams.

If the leg is an incoming e^+, we would get

$$\mathcal{M}_i(p_i, q) \approx e\frac{p_i \cdot \epsilon}{p_i \cdot q}\mathcal{M}_0(p_i)\,, \qquad (9.49)$$

where the sign flip comes from the charge of the e^+. If the leg is an outgoing electron, it is a little different. The photon is still outgoing, so we have

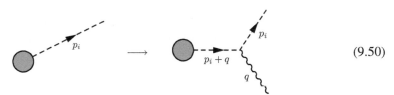

$$(9.50)$$

and the amplitude is modified to

$$\mathcal{M}_i(p_i, q) = (-ie)\frac{i\,[p_i^\mu + (p_i^\mu + q^\mu)]}{(p_i + q)^2 - m^2}\epsilon_\mu \mathcal{M}_0(p_i + q) \approx e\frac{p_i \cdot \epsilon}{p_i \cdot q}\mathcal{M}_0(p_i)\,. \qquad (9.51)$$

Similarly for an outgoing positron, we would get another sign flip and

$$\mathcal{M}_i(p_i, q) \approx -e\frac{p_i \cdot \epsilon}{p_i \cdot q}\mathcal{M}_0(p_i)\,. \qquad (9.52)$$

If we had many different particles with different charges, these formulas would be the same but the charge Q_i would appear instead of ± 1.

Summing over all the particles we get

$$\mathcal{M} \approx e\mathcal{M}_0\left[\sum_{\text{incoming}} Q_i\frac{p_i \cdot \epsilon}{p_i \cdot q} - \sum_{\text{outgoing}} Q_i\frac{p_i \cdot \epsilon}{p_i \cdot q}\right]\,, \qquad (9.53)$$

where Q_i is the charge of particle i.

Here comes the punchline. Under a Lorentz transformation, $\mathcal{M}(p_i, \epsilon) \to \mathcal{M}(p'_i, \epsilon')$, where p'_i and ϵ' are the momenta and polarization in the new frame. Since $\mathcal{M}$ must be Lorentz invariant, the transformed $\mathcal{M}$ must be the same. However, polarization vectors do not transform exactly like 4-vectors. As we showed explicitly in Section 8.4.2, there are certain Lorentz transformations for which q_μ is invariant and

$$\epsilon_\mu \to \epsilon_\mu + q_\mu. \tag{9.54}$$

These transformations are members of the little group, so the basis of polarization vectors does not change. Since there is no polarization proportional to q_μ, there does not exist a physical polarization ϵ'_μ in the new frame that is equal to the transformed ϵ_μ. Therefore, $\mathcal{M}$ has to change, violating Lorentz invariance. The only way out is if the q_μ term does not contribute. In terms of $\mathcal{M}$, the little-group transformation effects

$$\mathcal{M} \to \mathcal{M} + e\mathcal{M}_0 \left[\sum_{\text{incoming}} Q_i - \sum_{\text{outgoing}} Q_i \right], \tag{9.55}$$

and therefore the only way for $\mathcal{M}$ to be Lorentz invariant is if

$$\sum_{\text{incoming}} Q_i = \sum_{\text{outgoing}} Q_i, \tag{9.56}$$

which says that charge is conserved. This is a sum over all of the particles in the original $\mathcal{M}_0$ diagram, without the soft photon. Since this process was arbitrary, we conclude that charge must always be conserved.

Although we used the form of the interaction in scalar QED to derive the above result, it turns out this result is completely general. For example, suppose the photon had an arbitrary interaction with ϕ. Then the Feynman rule for the vertex could have arbitrary dependence on momenta:

$$= -ie\Gamma_\mu (p, q). \tag{9.57}$$

The vertex must have a μ index to contract with the polarization, by Lorentz invariance. Furthermore, also by Lorentz invariance, since the only 4-vectors available are p^μ and q^μ, we must be able to write $\Gamma^\mu = 2p^\mu F(p^2, q^2, p \cdot q) + q^\mu G(p^2, q^2, p \cdot q)$. Functions such as F and G are sometimes called **form factors**. In scalar QED, $F = G = 1$. Since $q^\mu \epsilon_\mu = 0$ we can discard G. Moreover, since $p^2 = m^2$ and $q^2 = 0$, the remaining form factor can only be a function of $\frac{p \cdot q}{m^2}$ by dimensional analysis, so we write $\Gamma^\mu = 2p^\mu F\left(\frac{p \cdot q}{m^2}\right)$. Now we put this general form into the above argument, so that

$$\approx -eF_i(0)\frac{p_i \cdot \epsilon}{p_i \cdot q}\mathcal{M}_0(p_i). \tag{9.58}$$

$F_i(0)$ is the only relevant value of $F_i(x)$ in the soft limit. We have added a subscript i on F since F_i can be different for different particles i. Although $F_i(x)$ does not have to be an analytic function, its limit as $x \to 0$ should be finite or else the matrix element for emitting a soft photon would diverge. Then Eq. (9.55) becomes

$$\mathcal{M} \to \mathcal{M} - e\mathcal{M}_0 \left[\sum_{\text{incoming}} F_i(0) - \sum_{\text{outgoing}} F_i(0) \right]. \qquad (9.59)$$

Thus, we get the same result as before, and moreover produce a general definition of the charge $Q_i = -F_i(0)$. (This definition will re-emerge in the context of renormalization, in Section 19.3.)

Thus, the connection between a massless spin-1 particle and conservation of charge is completely general. In fact, the same result holds for charged particles of any spin. The $\frac{p \cdot \epsilon}{p \cdot q}$ form of the interaction between light and matter in the soft limit is universal and spin independent. (It is called an **eikonal interaction**. The soft limit of gauge theories is discussed in more detail in Section 36.3.) The conclusion is:

> Massless spin-1 particles imply conservation of charge.

Note that masslessness of the photon was important in two places: that there are only two physical polarizations, and that we can take the soft limit with the photon on-shell.

"What's the big deal?" you say, "we know that already." But in the derivation from the previous chapter, we had to use gauge invariance, gauge-fix, isolate the conserved current, etc. Those steps were all artifacts of trying to write down a nice simple Lagrangian. The result we just derived does not require Lagrangians or gauge invariance at all. It just uses that a massless particle of spin-1 has two polarizations and the soft limit. Little-group scaling was important, but only to the extent that the final answer had to be a Lorentz-invariant function of the polarizations and momenta 4-vectors. The final conclusion, that charge is conserved, does not care that we embedded the two polarizations in a 4-vector ϵ_μ. It would be true even if we only used on-shell helicity amplitudes (an alternative proof without polarization vectors is given in Chapter 27).

To repeat, this is a non-perturbative statement about the physical universe, not a statement about our way of doing computations, like gauge invariance and the Ward identities are. Proofs like this are rare and very powerful. In Problem 9.3 you can show in a similar way that, when multiple massless spin-1 particles are involved, the soft limit forces them to transform in the adjoint representation of a Lie group. We now turn to the implications of the soft limit for massless particles of integer spin greater than 1.

9.5.1 Lorentz invariance for spin 2 and higher

A massless spin-2 field has two polarizations $\epsilon^i_{\mu\nu}$, which rotate into each other under Lorentz transformations, and also into $q_\mu q_\nu$. There are little-group transformations that send

$$\epsilon_{\mu\nu} \to \epsilon_{\mu\nu} + \Lambda_\mu q_\nu + \Lambda_\nu q_\mu + \Lambda q_\mu q_\nu, \qquad (9.60)$$

where these Λ_μ vectors have to do with the explicit way the Lorentz group acts, which we do not care about so much. Thus, any theory involving a massless spin-2 field should satisfy a Ward identity: if we replace even one index of the polarization tensor by q_μ the matrix elements must vanish. The spin-2 polarizations can be projected out of $\epsilon_{\mu\nu}$ as the transverse-traceless modes: $q_\mu \epsilon_{\mu\nu} = \epsilon_{\mu\mu} = 0$.

What do the interactions look like? As in the scalar case, they do not actually matter, and we can write a general interaction as

$$
\text{} = -i\Gamma^{\mu\nu}(p,q) = -2ip^\mu p^\nu \tilde{F}\left(\frac{p \cdot q}{m^2}\right), \qquad (9.61)
$$

where $\tilde{F}(x)$ is some function, different in general from the spin-1 form factor $F(x)$. The μ and ν indices on $\Gamma^{\mu\nu}$ will contract with the indices of the spin-2 polarization vector $\epsilon_{\mu\nu}$.

Taking the soft limit and adding up diagrams where all the incoming and outgoing particles emit the spin-2 particle, we find

$$
\mathcal{M} = \mathcal{M}_0 \left[\sum_{\text{incoming}} \tilde{F}_i(0) \frac{p_i^\mu}{p_i \cdot q} \epsilon_{\mu\nu} p_i^\nu - \sum_{\text{outgoing}} \tilde{F}_i(0) \frac{p_i^\mu}{p_i \cdot q} \epsilon_{\mu\nu} p_i^\nu \right], \qquad (9.62)
$$

which is similar to what we had for spin 1, but with an extra factor of p_ν^i in each sum.

By Lorentz invariance, little-group transformations such as those in Eq. (9.60) imply that this should vanish if $\epsilon_{\mu\nu} = q_\mu \Lambda_\nu$ for any Λ_ν. So, writing $\kappa_i \equiv \tilde{F}_i(0)$, which is just a number for each particle, we find

$$
\mathcal{M}_0 \Lambda_\nu \left[\sum_{\text{incoming}} \kappa_i p_i^\nu - \sum_{\text{outgoing}} \kappa_i p_i^\nu \right] = 0, \qquad (9.63)
$$

which implies

$$
\sum_{\text{incoming}} \kappa_i p_i^\nu = \sum_{\text{outgoing}} \kappa_i p_i^\nu. \qquad (9.64)
$$

In other words, the sum of $\kappa_i p_i^\nu$ is conserved. But we already know, by momentum conservation, that the sum of p_i^μ is conserved. So, for example, we can solve for p_1^μ in terms of the others. If we add another constraint on the p_i^μ then there would be a different solution for p_1^μ, which is impossible unless all the p_i^μ are zero. The only way we can have non-trivial scattering is for all the charges to be the same:

$$
\kappa_i = \kappa \quad \text{for all } i. \qquad (9.65)
$$

But that is exactly what gravity does! All particles gravitate with the same strength, $\kappa_i \equiv \frac{1}{M_{\text{Pl}}} = \sqrt{G_n}$. In other words, gravity is universal. So,

> Massless spin-2 particles imply gravity is universal.

We can keep going. For massless spin 3 we would need

$$\sum_{\text{incoming}} \beta_i p_\nu^i p_\mu^i = \sum_{\text{outgoing}} \beta_i p_\nu^i p_\mu^i, \tag{9.66}$$

where $\beta_i = \hat{F}_i(0)$ for some generic spin-3 form factor $\hat{F}_i\left(\frac{p \cdot q}{m^2}\right)$. For example, the $\mu = \nu = 0$ component of this says

$$\sum_{\text{incoming}} \beta_i E_i^2 = \sum_{\text{outgoing}} \beta_i E_i^2, \tag{9.67}$$

that is, the sum of the squares of the energies times some charges are conserved. That is way too constraining. The only way out is if all the charges are 0, which is a boring, non-interacting theory of free massless spin-3 field. So,

> There are no interacting theories of massless particles of spin greater than 2.

And in fact, no massless particles with spin > 2 have ever been seen. (Massive particles of spin > 2 are plentiful [Particle Data Group (Beringer *et al.*), 2012].)

Problems

9.1 Compton scattering in scalar QED.
  (a) Calculate the tree-level matrix elements for $(\gamma\phi \rightarrow \gamma\phi)$. Show that the Ward identity is satisfied.
  (b) Calculate the cross section $\frac{d\sigma}{d\cos\theta}$ for this process as a function of the incoming and outgoing polarizations, ϵ_μ^{in} and $\epsilon_\mu^{\text{out}}$, in the center-of-mass frame.
  (c) Evaluate $\frac{d\sigma}{d\cos\theta}$ for ϵ_μ^{in} polarized in the plane of the scattering, for each $\epsilon_\mu^{\text{out}}$.
  (d) Evaluate $\frac{d\sigma}{d\cos\theta}$ for ϵ_μ^{in} polarized transverse to the plane of the scattering, for each $\epsilon_\mu^{\text{out}}$.
  (e) Show that when you sum (c) and (d) you get the same thing as having replaced $(\epsilon_\mu^{\text{in}})^\star \epsilon_\nu^{\text{in}}$ with $-g_{\mu\nu}$ and $(\epsilon_\mu^{\text{out}})^\star \epsilon_\nu^{\text{out}}$ with $-g_{\mu\nu}$.
  (f) Should this replacement work for any scattering calculation?
9.2 Consider the following 3-loop diagram for light-by-light scattering:

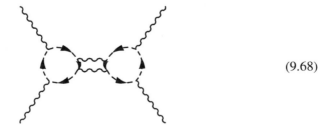

$$\tag{9.68}$$

  (a) Approximately how many other diagrams contribute at the same order in perturbation theory? [Hint: you do not need to draw the diagrams.]

(b) This diagram is not gauge invariant (independent of ξ) by itself. What is the minimal set of diagrams you need to add to this one for the sum to be gauge invariant? Why should the other diagrams cancel on their own?

9.3 In this problem you will prove the uniqueness of non-Abelian gauge theories by considering the soft limit when there are multiple scalar fields ϕ_i. Suppose these fields have a mass matrix M (i.e. the mass term in the Lagrangian is $\mathcal{L} = M_{ij}\phi_i^\star \phi_j$) and there are N massless spin-1 particles A_μ^a, $a = 1 \ldots N$ we will call *gluons*. Then the generic interaction between A_μ^a, ϕ_i and ϕ_j can be written as $\Gamma_{ij}^{a\mu}(p,q)$ as in Eq. (9.57).

(a) Show that in the soft limit, $q \ll p$, the charges are now described by a matrix $T^a = T_{ij}^a$.

(b) For $N = 1$, show that only if $[M,T] = 0$ can the theory be consistent. Conclude that gluons (or the photon) can only couple between particles of the same mass.

(c) Consider Compton scattering, $\phi_i(p)A_\mu^a(q^a) \rightarrow \phi_j(p')A_\nu^b(q^b)$, in the soft limit $q^a, q^b \ll p, p'$. Evaluate the two diagrams for this process and then show that, by setting $\epsilon_\mu^a = q_\mu^a$ and $\epsilon_\nu^b = q_\nu^b$, the interactions are consistent with Lorentz invariance only if $[T^a, T^b] = 0$, assuming nothing else is added.

(d) Show that one can modify this theory with a contact interaction involving $\phi_i\phi_j A_\mu^a A_\nu^b$ of the generic form $\Gamma_{ij}^{ab\mu\nu}(p,q^a,q^b)$ so that Lorentz invariance is preserved in the soft limit. How must $\Gamma_{ij}^{ab\mu\nu}$ relate to T_{ij}^a? Show also that $\Gamma_{ij}^{ab\mu\nu}$ must have a pole, for example as $(q^a + q^b)^2 \rightarrow 0$.

(e) Such a pole indicates a massless particle being exchanged, naturally identified as a gluon. In this case, the $\Gamma_{ij}^{ab\mu\nu}$ interaction in part (d) can be resolved into a 3-point interaction among gluons, of the form $\Gamma_{\mu\nu\alpha}^{abc}(q^a,q^b,q^c)$ and the $\Gamma_{ij}^{a\mu}(p,q)$ vertex. Show that if $\Gamma_{\mu\nu\alpha}^{abc}$ itself has no poles, then in the soft limit it can be written uniquely as $\Gamma_{\mu\nu\alpha}^{abc}(q^a,q^b,q^c) = f^{abc}(g^{\mu\nu}q_\alpha^\alpha + \cdots)$ for some constants f^{abc} and work out the $\cdots$. Show that if and only if $[T^a, T^b] = if^{abc}T^c$ can the Compton scattering amplitude be Lorentz invariant in the soft limit. This implies that the gluons transform in the adjoint representation of a Lie group, as will be discussed in Chapter 25.

9.4 The soft limit also implies that massless spin-2 particles (gravitons) must have self-interactions.

(a) To warm up, consider the soft limit of massless spin-1 particles coupled to scalars (as in scalar QED). Just assuming generic interactions (not the scalar QED Lagrangian), show that there must be an $AA\phi^\star\phi$ interaction for Compton scattering to be Lorentz invariant.

(b) Now consider Compton scattering of gravitons h off scalars. Show that there must be an $hh\phi\phi$ interaction. Then show that unlike the massless spin-1 case the new interaction must have a pole at $(q_1 + q_2)^2 = 0$. This pole should be resolved into a graviton exchange graph. Derive a relationship between the form of the graviton self-coupling and the $h\phi\phi$ coupling.

Spinors

The structure of the periodic table is due largely to the electron having spin $\frac{1}{2}$. In non-relativistic quantum mechanics you learned that the spin $+\frac{1}{2}$ and spin $-\frac{1}{2}$ states of the electron projected along a particular direction are efficiently described by a complex doublet:

$$|\psi\rangle = \left| \begin{array}{c} \uparrow \\ \downarrow \end{array} \right\rangle. \tag{10.1}$$

You probably also learned that the dynamics of this doublet, in the non-relativistic limit, is governed by the **Schrödinger–Pauli equation**:

$$i\partial_t |\psi\rangle = \left\{ \frac{1}{2m_e}(i\vec{\nabla} - e\vec{A})^2 \begin{pmatrix} 1 & 0 \\ 0 & 1 \end{pmatrix} - eA_0 \begin{pmatrix} 1 & 0 \\ 0 & 1 \end{pmatrix} + \mu_B \begin{pmatrix} B_z & B_x - iB_y \\ B_x + iB_y & -B_z \end{pmatrix} \right\} |\psi\rangle, \tag{10.2}$$

where $\vec{A}$ and A_0 are the vector and scalar potentials, $\vec{B} = \vec{\nabla} \times \vec{A}$ and $\mu_B = \frac{e}{2m_e}$ is the **Bohr magneton**, which characterizes the strength of the electron's magnetic dipole moment. The last term in this equation is responsible for the Stern–Gerlach effect.

You may also have learned of a shorthand notation for this involving the Pauli matrices:

$$\sigma_1 = \begin{pmatrix} 0 & 1 \\ 1 & 0 \end{pmatrix}, \quad \sigma_2 = \begin{pmatrix} 0 & -i \\ i & 0 \end{pmatrix}, \quad \sigma_3 = \begin{pmatrix} 1 & 0 \\ 0 & -1 \end{pmatrix}, \tag{10.3}$$

which let us write the Schrödinger–Pauli equation more concisely:

$$i\partial_t \psi = \left[\left(\frac{1}{2m}\left(i\vec{\nabla} - e\vec{A}\right)^2 - eA_0 \right) \mathbb{1}_{2\times 2} + \mu_B \vec{B} \cdot \vec{\sigma} \right] \psi, \tag{10.4}$$

where $\psi(x) = \langle x | \psi \rangle$ as usual. This equation is written with the Pauli matrices combined into a vector $\vec{\sigma} = (\sigma_1, \sigma_2, \sigma_3)$ so that rotationally invariant quantities such as $(\vec{\sigma} \cdot \vec{B})\psi$ are easy to write. That $(\vec{\sigma} \cdot \vec{B})\psi$ is rotationally invariant is non-trivial, and only works because

$$[\sigma_i, \sigma_j] = 2i\varepsilon_{ijk}\sigma_k, \tag{10.5}$$

which are the same algebraic relations satisfied by infinitesimal rotations (we will review this shortly). Keep in mind that σ_i do not change under rotations – they are always given by Eq. (10.3) in any frame. ψ is changing and B_i is changing, and these changes cancel in $(\vec{\sigma} \cdot \vec{B})\psi$.

We could also have written down a rotationally invariant equation of motion for ψ:

$$\mathbb{1}\partial_t \psi - \partial_i \sigma_i \psi = 0. \tag{10.6}$$

Since ∂_i transforms as a 3-vector and so does $\sigma_i \psi$, this equation is rotationally invariant. It turns out it is Lorentz invariant too. In fact, this is just the Dirac equation! If we write

$$\sigma^\mu = (\mathbb{1}_{2\times 2}, \sigma_1, \sigma_2, \sigma_3), \tag{10.7}$$

then Eq. (10.6) becomes

$$\sigma^\mu \partial_\mu \psi = 0, \tag{10.8}$$

which is nice and simple looking. (Actually, this is the Dirac equation for a Weyl spinor, which is not exactly the same as the equation commonly called the Dirac equation.)

Unfortunately, it does not follow that this equation is Lorentz invariant just because we have written it as $\sigma^\mu \partial_\mu$. For example,

$$(\sigma^\mu \partial_\mu + m)\psi = 0 \tag{10.9}$$

is *not* Lorentz invariant. To understand these enigmatic transformation properties, we have to know how to relate the Lorentz group to the Pauli matrices. It turns out that the Pauli matrices naturally come out of the mathematical analysis of the representations of the Lorentz group. By studying these representations, we will find spin-$\frac{1}{2}$ particles, which transform in *spinor* representations. The Dirac equation and its non-relativistic limit, the Schrödinger–Pauli equation, will immediately follow.

10.1 Representations of the Lorentz group

In Chapter 8, we identified particles with unitary representations of the Poincaré group. Due to Wigner's theorem, these representations are characterized by two quantum numbers: mass m and spin j. Recall where these quantum numbers come from. Mass is Lorentz invariant, so it is an obvious quantum number. Momentum is also conserved, but it is Lorentz *covariant*; that is, momentum is not a good quantum number for characterizing particles since it is frame dependent. If we choose a frame in which the momentum has some canonical form, for example $p^\mu = (m, 0, 0, 0)$ for $m > 0$, then the particles are characterized by the group that holds this momentum fixed, known as the *little group*. For example, the little group for $p^\mu = (m, 0, 0, 0)$ is the group of 3D rotations, SO(3). The little group representations provide the second quantum number, j. The way the states transform under the full Poincaré group is then induced by the transformations under the little group and the way the momentum transforms under boosts.

There are no finite-dimensional non-trivial unitary representations of the Poincaré group, but there are infinite-dimensional ones. We have seen how these can be embedded into fields, such as $V_\mu(x)$, $\phi(x)$ or $T_{\mu\nu}(x)$. As we saw for spin 1, a lot of trouble comes from having to embed particles of fixed mass and spin into these fields. The problem is that, except for $\phi(x)$, these fields describe reducible and non-unitary representations. For example, $V_\mu(x)$ has four degrees of freedom, which describes spin 0 and spin 1. We found that we could construct a unitary theory for massive spin 1 by carefully choosing the Lagrangian so that the physical theory never excites the spin-0 component. For massless spin 1, we could also choose a Lagrangian that only propagated the spin-1 component, but only by introducing gauge invariance. This led directly to charge conservation.

The next logical step to make these embeddings more systematic is to determine all possible Lorentz-invariant fields we can write down. This will reveal the existence of the spin-$\frac{1}{2}$ states, and help us characterize their embeddings into fields.

10.1.1 Group theory

A **group** is a set of elements $\{g_i\}$ and a rule $g_i \times g_j = g_k$ which tells how each pair of elements is multiplied to get a third. The rule defines the group, independent of any particular way to write the group elements down as matrices. More precisely, the mathematical definition requires the rule to be associative $(g_i \times g_j) \times g_k = g_i \times (g_j \times g_k)$, there to be an identity element for which $\mathbb{1} \times g_i = g_i \times \mathbb{1} = g_i$, and for the group elements to have inverses, $g_i^{-1} \times g_i = \mathbb{1}$. A **representation** is a particular embedding of these g_i into operators that act on a vector space. For finite-dimensional representations, this means an embedding of the g_i into matrices. Often we talk about the vectors on which the matrices act as being the representation, but technically the matrix embedding is the representation. Any group has the trivial representation $r : g_i \to \mathbb{1}$. A representation in which each group element gets its own matrix is called a **faithful representation**.

Recall that the Lorentz group is the set of rotations and boosts that preserve the Minkowski metric: $\Lambda^T g \Lambda = g$. The Λ matrices in this equation are in the 4-vector representation under which

$$X_\mu \to \Lambda_{\mu\nu} X_\nu. \tag{10.10}$$

Examples of Lorentz transformations are rotations around the x, y or z axes:

$$\begin{pmatrix} 1 & & & \\ & 1 & & \\ & & \cos\theta_x & \sin\theta_x \\ & & -\sin\theta_x & \cos\theta_x \end{pmatrix}, \begin{pmatrix} 1 & & & \\ & \cos\theta_y & & -\sin\theta_y \\ & & 1 & \\ & \sin\theta_y & & \cos\theta_y \end{pmatrix}, \begin{pmatrix} 1 & & & \\ & \cos\theta_z & \sin\theta_z & \\ & -\sin\theta_z & \cos\theta_z & \\ & & & 1 \end{pmatrix}$$

and boosts in the x, y or z directions:

$$\begin{pmatrix} \cosh\beta_x & \sinh\beta_x & & \\ \sinh\beta_x & \cosh\beta_x & & \\ & & 1 & \\ & & & 1 \end{pmatrix}, \begin{pmatrix} \cosh\beta_y & & \sinh\beta_y & \\ & 1 & & \\ \sinh\beta_y & & \cosh\beta_y & \\ & & & 1 \end{pmatrix}, \begin{pmatrix} \cosh\beta_z & & & \sinh\beta_z \\ & 1 & & \\ & & 1 & \\ \sinh\beta_z & & & \cosh\beta_z \end{pmatrix}.$$

These matrices give an embedding of elements of the Lorentz group into a set of matrices. That is, they describe one particular representation of the Lorentz group (the 4-vector representation). We would now like to find all the representations.

The Lorentz group itself is a mathematical object independent of any particular representation. To extract the group away from its representations, it is easiest to look at infinitesimal transformations. In the 4-vector representation, an infinitesimal Lorentz transformation can be written in terms of six infinitesimal angles θ_i and β_i as

$$\delta X_0 = \beta_i X_i, \tag{10.11}$$

$$\delta X_i = \beta_i X_0 - \varepsilon_{ijk}\theta_j X_k, \tag{10.12}$$

where the **Levi-Civita** or **totally antisymmetric tensor** ε_{ijk} is defined by $\varepsilon_{123} = 1$ and the rule that the sign flips when you swap any two indices.

Alternatively, we can write the infinitesimal transformations as

$$\delta X_\mu = i \sum_{i=1}^{3} \left[\theta_i (J_i)_{\mu\nu} + \beta_i (K_i)_{\mu\nu} \right] X_\nu, \tag{10.13}$$

where

$$J_1 = i \begin{pmatrix} 0 & & & \\ & 0 & & \\ & & 0 & -1 \\ & & 1 & 0 \end{pmatrix}, \; J_2 = i \begin{pmatrix} 0 & & & \\ & 0 & & 1 \\ & & 0 & \\ & -1 & & 0 \end{pmatrix}, \; J_3 = i \begin{pmatrix} 0 & & & \\ & 0 & -1 & \\ & 1 & 0 & \\ & & & 0 \end{pmatrix}, \tag{10.14}$$

$$K_1 = i \begin{pmatrix} 0 & -1 & & \\ -1 & 0 & & \\ & & 0 & \\ & & & 0 \end{pmatrix}, \; K_2 = i \begin{pmatrix} 0 & & -1 & \\ & 0 & & \\ -1 & & 0 & \\ & & & 0 \end{pmatrix}, \; K_3 = i \begin{pmatrix} 0 & & & -1 \\ & 0 & & \\ & & 0 & \\ -1 & & & 0 \end{pmatrix}. \tag{10.15}$$

These matrices are known as the **generators** of the Lorentz group in the 4-vector basis. They generate the group in the sense that any element of the group can be written uniquely as

$$\Lambda = \exp(i\theta_i J_i + i\beta_i K_i) \tag{10.16}$$

up to some discrete transformations. The advantage of writing the group elements this way is that it is completely general. In any finite-dimensional representation the group elements can be written as an exponential of matrices.

For any group G, some group elements $g \in G$ can be written as $g = \exp(ic_i^g \lambda_i)$, where c_i^g are numbers and λ_i are group generators. The generators are in an **algebra**, because you can add and multiply them, while the group elements are in a **group**, because you can only multiply them. For example, the real numbers form an algebra (there is a rule for addition and a rule for multiplication) but rotations are a group (there is only one rule, multiplication). **Lie groups** are a class of groups, including the Lorentz group, with an infinite number of elements but a finite number of generators. The generators of the Lie group form an algebra called its **Lie algebra**. Lie groups are critical to understanding the Standard Model, since QED is described by the unitary group $U(1)$, the weak force by the special unitary group $SU(2)$ and the strong force by the group $SU(3)$. The Lorentz group is sometimes called $O(1, 3)$. This is an orthogonal (preserves a metric) group corresponding to a metric with $(1, 3)$ signature (i.e. $g_{\mu\nu} = \text{diag}\,(1, -1, -1, -1)$).

Lie groups also have the structure of a differentiable manifold. For most applications of quantum field theory, the manifold is totally irrelevant, but it is occasionally important. For example, topological properties of the 3D rotation group $SO(3)$ will help us understand the spin-statistics theorem. We sometimes distinguish the proper orthochronous Lorentz group, which is the elements of the Lorentz group continuously connected to the identity, from the full Lorentz group, which includes time reversal (T) and parity reversal (P).

In a Lie algebra the multiplication rule is defined as the **Lie bracket**. With matrix representations, this Lie bracket is just an ordinary commutator. Since any element of a Lie algebra can be written as a linear combination of the generators, a Lie algebra is fixed by the commutation relations of its generators. For the Lorentz group, these commutation relations can be calculated using any representation, for example the 4-vector representation with generators in Eq. (10.14). We find

$$[J_i, J_j] = i\epsilon_{ijk}J_k, \tag{10.17}$$

$$[J_i, K_j] = i\epsilon_{ijk}K_k, \tag{10.18}$$

$$[K_i, K_j] = -i\epsilon_{ijk}J_k. \tag{10.19}$$

These commutation relations define the **Lorentz algebra**, so$(1,3)$. You might recognize that $[J_i, J_j] = i\epsilon_{ijk}J_k$ is the algebra for rotations, SO(3), and in fact the J_i generate the 3D rotation subgroup of the Lorentz group.

These commutation relations define the Lie algebra of the Lorentz group. Although these commutation relations were derived using Eq. (10.14) they must hold for any representation; for example in the rank-2 tensor representation J_i and K_i can be written as 16-dimensional matrices. It is sometimes useful to use a different form for these commutation relations. We can index the generators by $V^{\mu\nu}$ instead of J_i and K_i:

$$V^{\mu\nu} = \begin{pmatrix} 0 & K_1 & K_2 & K_3 \\ -K_1 & 0 & J_3 & -J_2 \\ -K_2 & -J_3 & 0 & J_1 \\ -K_3 & J_2 & -J_1 & 0 \end{pmatrix}. \tag{10.20}$$

Here each $V^{\mu\nu}$ is itself a 4×4 matrix, for example, $V^{23} = J_1$. A Lorentz transformation can be written in terms of $V^{\mu\nu}$ as $\Lambda_V = \exp(i\theta_{\mu\nu}V^{\mu\nu})$ for six numbers $\theta_{\mu\nu}$. These $V^{\mu\nu}$ satisfy

$$[V^{\mu\nu}, V^{\rho\sigma}] = i(g^{\nu\rho}V^{\mu\sigma} - g^{\mu\rho}V^{\nu\sigma} - g^{\nu\sigma}V^{\mu\rho} + g^{\mu\sigma}V^{\nu\rho}). \tag{10.21}$$

By definition, the generators in any other representation must satisfy these same relations. For example, another representation of the Lorentz group is given by

$$L^{\mu\nu} = i(x^\mu \partial^\nu - x^\nu \partial^\mu). \tag{10.22}$$

This is an infinite-dimensional representation which acts on functions rather than a finite-dimensional vector space. These are the classical generators of angular momentum generalized to include time. You can check that $L^{\mu\nu}$ satisfy the commutation relations of the Lorentz algebra.

By the way, not all the elements of the Lorentz group can be written as $\exp(ic_i^g \lambda_i)$ for some c_i^g. The generators of the Lorentz algebra so$(1,3)$ only generate the part of the Lorentz group connected to the identity, known as the proper orthochronous Lorentz group SO$^+(1,3)$. It is possible for two different groups to have the same algebra. For example, the proper orthochronous Lorentz group and the full Lorentz group have the same algebra, but the full Lorentz group has in addition time reversal and parity. The group generated by o$(1,3)$ and P is called the **orthochronous** Lorentz group, denoted O$^+(1,3)$. The **proper** Lorentz group is the special orthogonal group SO$(1,3)$, which contains only the elements with determinant 1, so it excludes T and P. Sometimes SO$(1,3)$ is taken to include only P with SO$^+(1,3)$ excluding also T. These notations are more general than we need: in odd space-time dimensions, parity has determinant 1 and is therefore a special orthogonal transformation. Rather than worry about group naming conventions, we will simply talk about the Lorentz group with or without T and P.

10.1.2 General representations of the Lorentz group

The irreducible representations of the Lorentz group can be constructed from irreducible representations of SU(2). To see how this works, we start with the rotation generators J_i and the boost generators K_j. You can think of them as the matrices in Eq. (10.14), which is a particular representation, but the algebraic properties in Eqs. (10.17) to (10.19) are representation independent.

Now take the linear combinations

$$J_i^+ \equiv \frac{1}{2}(J_i + iK_i), \quad J_i^- \equiv \frac{1}{2}(J_i - iK_i), \tag{10.23}$$

which satisfy

$$[J_i^+, J_j^+] = i\epsilon_{ijk}J_k^+, \tag{10.24}$$

$$[J_i^-, J_j^-] = i\epsilon_{ijk}J_k^-, \tag{10.25}$$

$$[J_i^+, J_j^-] = 0. \tag{10.26}$$

These commutation relations indicate that the Lie algebra for the Lorentz group has two commuting subalgebras. The algebra generated by J_i^+ (or J_i^-) is the 3D rotation algebra, which has multiple names, $so(3) = sl(2, \mathbb{R}) = so(2, 1) = su(2)$, due to multiple Lie groups having the same algebra. So we have shown that

$$so(1, 3) = su(2) \oplus su(2). \tag{10.27}$$

Thus, representations of $su(2) \oplus su(2)$ will determine representations of the Lorentz group.

The decomposition $so(1, 3) = su(2) \oplus su(2)$ makes studying the irreducible representations very easy. We already know from quantum mechanics what the representations of $su(2)$ are, since $su(2) = so(3)$ is the algebra of Pauli matrices, which generates the 3D rotation group SO(3). Each irreducible representation of $su(2)$ is characterized by a half-integer j. The representation acts on a vector space with $2j+1$ basis elements (see Problem 10.2). It follows that irreducible representations of the Lorentz group are characterized by two half-integers: A and B. The (A, B) representation has $(2A + 1)(2B + 1)$ degrees of freedom.

The regular rotation generators are $\vec{J} = \vec{J}^+ + \vec{J}^-$, where we use the vector superscript to call attention to the fact that the spins must be added vectorially, as you might remember from studying Clebsch–Gordan coefficients. Since the 3D rotation group SO(3) is a subgroup of the Lorentz group, every representation of the Lorentz group will also be a representation of SO(3). In fact, finite-dimensional irreducible representations of the Lorentz algebra, which are characterized by two half-integers (A, B), generate many representations of SO(3): with spins $j = A+B, A+B-1, \ldots, |A-B|$, as shown in Table 10.1. For

Representation of su(2) ⊕ su(2)	$(0,0)$	$(\frac{1}{2},0)$	$(0,\frac{1}{2})$	$(\frac{1}{2},\frac{1}{2})$	$(1,0)$	$(1,1)$
Table 10.1 Decomposition of irreducible representations of the Lorentz algebra su(2) ⊕ su(2) into irreducible representations of its so(3) subalgebra describing spin.						
Representations of so(3)	0	$\frac{1}{2}$	$\frac{1}{2}$	$1 \oplus 0$	1	$2 \oplus 1 \oplus 0$

example, the general tensor representations $T_{\mu_1 \cdots \mu_n}$ correspond to the $(\frac{n}{2}, \frac{n}{2})$ representations of the Lorentz algebra. These are each *irreducible* representations of the Lorentz algebra, but *reducible* representations of the su(2) subalgebra corresponding to spin.

The relevance of the decomposition in Table 10.1 for particle physics is that Lagrangians are constructed out of fields, $V_\mu(x)$ and $\psi(x)$, which transform under the Lorentz group. However, particles transform under irreducible unitary representations of the Poincaré group, which have spins associated with the little group (as discussed in Chapter 8). So, the decomposition of Lorentz representations as in Table 10.1 determines the spins of particles that might be described by given fields. For example, the Lorentz representation acting on real 4-vectors $A_\mu(x)$ is the $(\frac{1}{2}, \frac{1}{2})$ representation (containing four degrees of freedom). It can describe spin-1 or -0 representations of $SO(3)$, with three and one degrees of freedom, respectively. We saw in Section 8.2.2 how the Lagrangian for a massive vector field could be chosen so that only the spin-1 particle propagates.

By the way, the group generated by exponentiating the Lie algebra of a given group is known as the **universal cover** of the given group. For example, exponentiating su(2) gives SU(2). Since SU(2) and SO(3) have the same Lie algebra, SU(2) is the universal cover of SO(3). The Lie algebra su(2) ⊕ su(2) generates SL(2, ℂ), which is therefore the universal cover of the Lorentz group. We will revisit the distinction between SL(2, ℂ) and the Lorentz group more in Section 10.5.1. For now, we will simply study su(2) ⊕ su(2). Group theory is discussed further in the context of Yang–Mills theories in Chapter 25.

10.2 Spinor representations

So far we have only considered the tensor representations, $T_{\mu_1 \cdots \mu_n}$, that have only integer spins. We will now discuss representations with half-integer spins.

There exist two complex $J = \frac{1}{2}$ representations, $(\frac{1}{2}, 0)$ and $(0, \frac{1}{2})$. What do these representations actually look like? The vector spaces they act on have $2J + 1 = 2$ degrees of freedom. Thus we need to find 2×2 matrices that satisfy

$$[J_i^+, J_j^+] = i\varepsilon_{ijk} J_k^+, \tag{10.28}$$

$$[J_i^-, J_j^-] = i\varepsilon_{ijk} J_k^-, \tag{10.29}$$

$$[J_i^+, J_j^-] = 0. \tag{10.30}$$

But we already know such matrices: the Pauli matrices. They satisfy Eq. (10.5): $[\sigma_i, \sigma_j] = 2i\varepsilon_{ijk}\sigma_k$. Rescaling, we find

$$\left[\frac{\sigma_i}{2}, \frac{\sigma_j}{2}\right] = i\varepsilon_{ijk}\frac{\sigma_k}{2}, \tag{10.31}$$

which is the SO(3) algebra. Another useful fact is that

$$\{\sigma_i, \sigma_j\} = \sigma_i\sigma_j + \sigma_j\sigma_i = 2\delta_{ij}, \tag{10.32}$$

where the anticommutator is defined by

$$\left\{A, B\right\} \equiv AB + BA. \tag{10.33}$$

Thus, we can set $J_i^- = \frac{1}{2}\sigma_i$, which generates the "$\frac{1}{2}$" in $(\frac{1}{2}, 0)$. What about J_i^+? This should be the "0" in $(\frac{1}{2}, 0)$. The obvious thing to do is just take the trivial representation $J_i^+ = 0$. So the $(\frac{1}{2}, 0)$ representation is

$$\left(\frac{1}{2}, 0\right): \quad \vec{J}^- = \frac{1}{2}\vec{\sigma}, \quad \vec{J}^+ = 0. \tag{10.34}$$

Similarly, the $(0, \frac{1}{2})$ representation is

$$\left(0, \frac{1}{2}\right): \quad \vec{J}^- = 0, \quad \vec{J}^+ = \frac{1}{2}\vec{\sigma}. \tag{10.35}$$

What does this mean for actual Lorentz transformations? Well, the rotations are $\vec{J} = \vec{J}^- + \vec{J}^+$ and the boosts are $\vec{K} = i(\vec{J}^- - \vec{J}^+)$ so

$$\left(\frac{1}{2}, 0\right): \quad \vec{J} = \frac{1}{2}\vec{\sigma}, \quad \vec{K} = \frac{i}{2}\vec{\sigma}, \tag{10.36}$$

$$\left(0, \frac{1}{2}\right): \quad \vec{J} = \frac{1}{2}\vec{\sigma}, \quad \vec{K} = -\frac{i}{2}\vec{\sigma}. \tag{10.37}$$

Since the Pauli matrices are Hermitian, $\vec{\sigma}^\dagger = \vec{\sigma}$, the rotations are Hermitian and the boosts are anti-Hermitian ($\vec{K}^\dagger = -\vec{K}$). Also notice that the group generators in the $(\frac{1}{2}, 0)$ and $(0, \frac{1}{2})$ representations are adjoints of each other. So we sometimes say these are complex-conjugate representations.

Elements of the vector space on which the spin-$\frac{1}{2}$ representations act are known as **spinors**. The $(0, \frac{1}{2})$ spinors are called **right-handed Weyl spinors** and often denoted ψ_R. Under rotation angles θ_j and boost angles β_j

$$\psi_R \rightarrow e^{\frac{1}{2}(i\theta_j\sigma_j + \beta_j\sigma_j)}\psi_R = \left(1 + \frac{i}{2}\theta_j\sigma_j + \frac{1}{2}\beta_j\sigma_j + \cdots\right)\psi_R, \tag{10.38}$$

where $\cdots$ are higher order in the expansion of the exponential. Similarly, the $(\frac{1}{2}, 0)$ representation acts on **left-handed Weyl spinors**, ψ_L,

$$\psi_L \rightarrow e^{\frac{1}{2}(i\theta_j\sigma_j - \beta_j\sigma_j)}\psi_L = \left(1 + \frac{i}{2}\theta_j\sigma_j - \frac{1}{2}\beta_j\sigma_j + \cdots\right)\psi_L. \tag{10.39}$$

Infinitesimally,

$$\delta\psi_R = \frac{1}{2}(i\theta_j + \beta_j)\sigma_j\psi_R, \qquad (10.40)$$

$$\delta\psi_L = \frac{1}{2}(i\theta_j - \beta_j)\sigma_j\psi_L. \qquad (10.41)$$

Note again that the angles θ_j and β_j are real numbers. Although we mapped $\vec{J}^-$ or $\vec{J}^+$ to 0, we still have non-trivial action of all the Lorentz generators. So these are *faithful* irreducible representations of the Lorentz group. Similarly,

$$\delta\psi_R^\dagger = \frac{1}{2}(-i\theta_j + \beta_j)\psi_R^\dagger\sigma_j, \qquad (10.42)$$

$$\delta\psi_L^\dagger = \frac{1}{2}(-i\theta_j - \beta_j)\psi_L^\dagger\sigma_j. \qquad (10.43)$$

10.2.1 Unitary representations

We have just constructed two 2D representations of the Lorentz group. But these representations are not *unitary*. Unitarity means $\Lambda^\dagger\Lambda = 1$, which is necessary to have Lorentz-invariant matrix elements:

$$\langle\psi|\psi\rangle \rightarrow \langle\psi|\Lambda^\dagger\Lambda|\psi\rangle. \qquad (10.44)$$

Since a group element is the exponential of a generator $\Lambda = e^{i\lambda}$, unitarity requires that $\lambda^\dagger = \lambda$, that is, that λ be Hermitian. We saw that the boost generators in the spinor representations are instead anti-Hermitian.

It is not hard to see that any representation of $SU(2)\times SU(2)$ constructed as above (which are all the finite-dimensional representations) will not be unitary. Since su(2) is the special unitary algebra, all of its representations are unitary. So, the generators for the su(2)⊕su(2) decomposition $\vec{J}_\pm = \frac{1}{2}\left(\vec{J}\pm i\vec{K}\right)$ are Hermitian. Thus $\exp(i\theta_+^j J_+^j + i\theta_-^j J_-^j)$ is unitary, for *real* θ_+^j and θ_-^j. But this does not mean that the corresponding representations of the Lorentz group are unitary. A Lorentz group element is

$$\Lambda = \exp(i\theta_j J_j + i\beta_j K_j), \qquad (10.45)$$

where the θ_j are the rotation angles and β_j the boost "angles." These are *real* numbers. They are related to the angles for the $\vec{J}_\pm$ generators of su(2) ⊕ su(2) by $\theta_+^j = \theta_j - i\beta_j$ and $\theta_-^j = \theta_j + i\beta_j$. So for a boost, the $\vec{J}_+$ and $\vec{J}_-$ generators get multiplied by imaginary angles, which makes the transformation anti-unitary. Thus, none of the representations of the Lorentz group generated this way will be unitary and therefore there are *no finite-dimensional unitary representations of the Lorentz group*.

To construct a unitary field theory, we need unitary representations of the Poincaré group, which are infinite dimensional; the corresponding representations of the Lorentz subgroup of the Poincaré group are also infinite dimensional. To construct these representations, we will use the same trick we used for spin 1 in Chapter 8. We will construct an infinite-dimensional representation by having the basis depend on the momentum p^μ. For fixed momentum, say $p^\mu = (m, 0, 0, 0)$ in the massive case, or $p^\mu = (E, 0, 0, E)$ in the massless case, the group reduces to the appropriate little group, $SO(3)$ or $ISO(2)$ respectively. These little groups *do* have unitary representations. Implementing this procedure for spin 1, we were led uniquely to Lagrangians with kinetic terms of the form $-\frac{1}{4}F_{\mu\nu}^2$, and gauge invariance and charge conservation if $m = 0$. We will now see how to construct Lorentz-invariant Lagrangians that describe unitary theories with spinors.

10.2.2 Lorentz-invariant Lagrangians

Having seen that we need infinite-dimensional representations, we are now ready to talk about fields. These fields are spinor-valued functions of space-time, which we write as $\psi_R(x) = \begin{pmatrix} \psi_1(x) \\ \psi_2(x) \end{pmatrix}$ for the $(0, \frac{1}{2})$ representation, or $\psi_L(x) = \begin{pmatrix} \psi_1(x) \\ \psi_2(x) \end{pmatrix}$ for the $(\frac{1}{2}, 0)$ representation.

As in the spin-1 case, we would like first to write down a Lorentz-invariant Lagrangian for these fields with the right number of degrees of freedom (two). The simplest thing to do would be to write down a Lagrangian with terms such as

$$(\psi_R)^\dagger \Box \psi_R + m^2 (\psi_R)^\dagger \psi_R. \tag{10.46}$$

However, using the infinitesimal transformations Eqs. (10.40) and (10.42), it is easy to see that these terms are not Lorentz invariant:

$$\delta\left(\psi_R^\dagger \psi_R\right) = \frac{1}{2}\psi_R^\dagger[(i\theta_i + \beta_i)\sigma_i\psi_R] + \frac{1}{2}[\psi_R^\dagger(-i\theta_i + \beta_i)\sigma_i]\psi_R$$
$$= \beta_i \psi_R^\dagger \sigma_i \psi_R \neq 0. \tag{10.47}$$

This is just the manifestation of the fact that the representation is not unitary because the boost generators are anti-Hermitian.

If we allow ourselves two fields, ψ_R and ψ_L, we can write down terms such as $\psi_L^\dagger \psi_R$. Under infinitesimal Lorentz transformations,

$$\delta(\psi_L^\dagger \psi_R) = \left[\psi_L^\dagger \frac{1}{2}(-i\theta_i - \beta_i)\sigma_i^\dagger\right]\psi_R + \psi_L^\dagger\left[\frac{1}{2}(i\theta_i + \beta_i)\sigma_i\psi_R\right] = 0, \tag{10.48}$$

which is great. We need to add the Hermitian conjugate to get a term in a real Lagrangian. Thus, we find that

$$\mathcal{L}_{\text{Dirac mass}} = m\left(\psi_L^\dagger \psi_R + \psi_R^\dagger \psi_L\right) \tag{10.49}$$

is real and Lorentz invariant for any m. This combination is bilinear in the fields, but lacks derivatives, so it is a type of mass term known as a **Dirac mass**. A theory with only this term in its Lagrangian would have no dynamics.

What about kinetic terms? We could try

$$\mathcal{L} = \psi_L^\dagger \Box \psi_R + \psi_R^\dagger \Box \psi_L, \tag{10.50}$$

which is both Lorentz invariant and real. But this is actually not a very interesting Lagrangian. We can always split up our field into components $\psi_R = \begin{pmatrix} \psi_1 \\ \psi_2 \end{pmatrix}$, where ψ_1

and ψ_2 are just regular fields. Then we see that this is just the Lagrangian for a couple of scalars. So it is not enough to declare the Lorentz transformation properties of something, the Lagrangian has to *force* those transformation properties. In the same way, a vector field A_μ is just four scalars until we contract it with ∂_μ in the Lagrangian, as in the $(\partial_\mu A_\mu)^2$ part of $F_{\mu\nu}^2$.

To proceed, let us look at $\psi_R^\dagger \sigma_i \psi_R$. This transforms as

$$\delta(\psi_R^\dagger \sigma_i \psi_R) = \frac{1}{2}\psi_R^\dagger \sigma_i[(i\theta_j + \beta_j)\sigma_j \psi_R] + \frac{1}{2}[\psi_R^\dagger(-i\theta_j + \beta_j)\sigma_j]\sigma_i \psi_R$$
$$= \frac{\beta_j}{2}\psi_R^\dagger(\sigma_i\sigma_j + \sigma_j\sigma_i)\psi_R + \frac{i\theta_j}{2}\psi_R^\dagger(\sigma_i\sigma_j - \sigma_j\sigma_i)\psi_R$$
$$= \beta_i\psi_R^\dagger\psi_R - \theta_j\varepsilon_{ijk}\psi_R^\dagger\sigma_k\psi_R. \tag{10.51}$$

Thus, we have found that

$$\delta\left(\psi_R^\dagger\psi_R, \ \psi_R^\dagger\sigma_i\psi_R\right) = \left(\beta_i\psi_R^\dagger\sigma_i\psi_R, \ \beta_i\psi_R^\dagger\psi_R - \varepsilon_{ijk}\theta_j\psi_R^\dagger\sigma_k\psi_R\right), \tag{10.52}$$

which is exactly how a vector transforms:

$$\delta(V_0, V_i) = (\beta_i V_i, \ \beta_i V_0 - \varepsilon_{ijk}\theta_j V_k) \tag{10.53}$$

as in Eq. (10.12). So $V_R^\mu = (\psi_R^\dagger\psi_R, \ \psi_R^\dagger\vec{\sigma}\psi_R)$ is an honest-to-goodness Lorentz 4-vector. Therefore,

$$\psi_R^\dagger\partial_t\psi_R + \psi_R^\dagger\partial_j\sigma_j\psi_R \tag{10.54}$$

is Lorentz invariant. Note that $\partial_t(\psi_R^\dagger\psi_R) + \partial_j(\psi_R^\dagger\sigma_j\psi_R)$ is also Lorentz invariant, but not a viable candidate for the spinor Lagrangian since it is a total derivative. Similarly,

$$\delta\left(\psi_L^\dagger\psi_L, \ -\psi_L^\dagger\sigma_i\psi_L\right) = \left(-\beta_i\psi_L^\dagger\sigma_i\psi_L, \ \beta_i\psi_L^\dagger\psi_L + \varepsilon_{ijk}\theta_j\psi_L^\dagger\sigma_k\psi_L\right) \tag{10.55}$$

so $(\psi_L^\dagger\psi_L, \ -\psi_L^\dagger\sigma_i\psi_L)$ also transforms like a vector and the combination $\psi_L^\dagger\partial_t\psi_L - \psi_L^\dagger\partial_j\sigma_j\psi_L$ is Lorentz invariant.

Defining

$$\sigma^\mu \equiv (\mathbb{1}, \vec{\sigma}), \quad \bar{\sigma}^\mu \equiv (\mathbb{1}, -\vec{\sigma}), \tag{10.56}$$

we can write all the Lorentz-invariant terms we have found as

$$\mathcal{L} = i\psi_R^\dagger\sigma_\mu\partial_\mu\psi_R + i\psi_L^\dagger\bar{\sigma}_\mu\partial_\mu\psi_L - m(\psi_R^\dagger\psi_L + \psi_L^\dagger\psi_R). \tag{10.57}$$

We added a factor of i in the kinetic term to make the Lagrangian Hermitian:

$$(i\psi_R^\dagger\sigma_\mu\partial_\mu\psi_R)^\dagger = -i(\partial_\mu\psi_R^\dagger)\,\sigma_\mu\psi_R = i\psi_R^\dagger\sigma_\mu\partial_\mu\psi_R, \tag{10.58}$$

where we have used $\sigma_\mu^\dagger = \sigma_\mu$ and integrated by parts.

There is an even shorter-hand way to write this. Let us combine the two spinors into a four-component object known as a **Dirac spinor**:

$$\psi = \begin{pmatrix} \psi_L \\ \psi_R \end{pmatrix}. \tag{10.59}$$

If we also define

$$\bar{\psi} = \left(\begin{array}{cc} \psi_R^\dagger & \psi_L^\dagger \end{array} \right),$$ (10.60)

and use the 4×4 matrices

$$\gamma^\mu = \left(\begin{array}{cc} & \sigma^\mu \\ \bar{\sigma}^\mu & \end{array} \right),$$ (10.61)

known as **Dirac matrices** or **γ-matrices**, our Lagrangian becomes

$$\mathcal{L} = \bar{\psi}(i\gamma^\mu \partial_\mu - m)\psi.$$ (10.62)

which is the conventional form of the **Dirac Lagrangian**. The equations of motion that follow are

$$(i\gamma^\mu \partial_\mu - m)\psi = 0\,,$$ (10.63)

which is the **Dirac equation**.

10.3 Dirac matrices

Expanding them out, the Dirac matrices from Eq. (10.61) are

$$\gamma^0 = \left(\begin{array}{cc} & \mathbb{1} \\ \mathbb{1} & \end{array} \right), \quad \gamma^i = \left(\begin{array}{cc} 0 & \sigma_i \\ -\sigma_i & 0 \end{array} \right).$$ (10.64)

Or, even more explicitly,

$$\gamma^0 = \left(\begin{array}{cccc} & & 1 & 0 \\ & & 0 & 1 \\ 1 & 0 & & \\ 0 & 1 & & \end{array} \right), \quad \gamma^1 = \left(\begin{array}{cccc} & & 0 & 1 \\ & & 1 & 0 \\ 0 & -1 & & \\ -1 & 0 & & \end{array} \right),$$

$$\gamma^2 = \left(\begin{array}{cccc} & & 0 & -i \\ & & i & 0 \\ 0 & i & & \\ -i & 0 & & \end{array} \right), \quad \gamma^3 = \left(\begin{array}{cccc} & & 1 & 0 \\ & & 0 & -1 \\ -1 & 0 & & \\ 0 & 1 & & \end{array} \right).$$ (10.65)

They satisfy

$$\left\{ \gamma^\mu, \gamma^\nu \right\} = 2g^{\mu\nu}.$$ (10.66)

In the same way that the algebra of the Lorentz group is more fundamental than any particular representation, the algebra of the γ-matrices is more fundamental than any particular representation of them. We say the γ-matrices generate the Dirac algebra, which is a special case of a **Clifford algebra**. This particular form of the Dirac matrices is known as the **Weyl representation**.

Next we define a useful shorthand:

$$\sigma^{\mu\nu} \equiv \frac{i}{2}[\gamma^\mu, \gamma^\nu]. \tag{10.67}$$

The Lorentz generators when acting on Dirac spinors can be written as

$$S^{\mu\nu} = \frac{i}{4}[\gamma^\mu, \gamma^\nu] = \frac{1}{2}\sigma^{\mu\nu}, \tag{10.68}$$

which you can check by expanding in terms of σ-matrices. More generally, $S^{\mu\nu}$ will satisfy the Lorentz algebra when constructed from any γ-matrices satisfying the Clifford algebra. That is, you can derive from $\{\gamma^\mu, \gamma^\nu\} = 2g^{\mu\nu}$ that

$$[S^{\mu\nu}, S^{\rho\sigma}] = i(g^{\nu\rho}S^{\mu\sigma} - g^{\mu\rho}S^{\nu\sigma} - g^{\nu\sigma}S^{\mu\rho} + g^{\mu\sigma}S^{\nu\rho}). \tag{10.69}$$

It is important to appreciate that the matrices $S_{\mu\nu}$ are different from the matrices $V_{\mu\nu}$ corresponding to the Lorentz generators in the 4-vector representation. In particular, $S_{\mu\nu}$ are complex. So we have found two inequivalent four-dimensional representations. In each case, the group element is determined by six real angles $\theta_{\mu\nu}$ (three rotations and three boosts). The vector or $(\frac{1}{2}, \frac{1}{2})$ representation is irreducible, and has Lorentz group element

$$\Lambda_V = \exp(i\theta_{\mu\nu}V^{\mu\nu}), \tag{10.70}$$

while the Dirac or $(\frac{1}{2}, 0) \oplus (0, \frac{1}{2})$ representation is reducible and has Lorentz group elements

$$\Lambda_s = \exp(i\theta_{\mu\nu}S^{\mu\nu}). \tag{10.71}$$

There are actually a number of Dirac representations, depending on the form of the γ-matrices. We will consider two: the Weyl and Majorana representations.

In the Weyl representation, the Lorentz generators are

$$S^{ij} = \frac{1}{2}\varepsilon_{ijk}\begin{pmatrix} \sigma_k & \\ & \sigma_k \end{pmatrix}, \quad S^{0i} = -\frac{i}{2}\begin{pmatrix} \sigma_i & \\ & -\sigma_i \end{pmatrix}, \tag{10.72}$$

or, very explicitly,

$$S^{12} = \frac{1}{2}\begin{pmatrix} 1 & & & \\ & -1 & & \\ & & 1 & \\ & & & -1 \end{pmatrix}, \quad S^{13} = \frac{i}{2}\begin{pmatrix} 0 & 1 & & \\ -1 & 0 & & \\ & & 0 & 1 \\ & & -1 & 0 \end{pmatrix}, \quad S^{23} = \frac{1}{2}\begin{pmatrix} 0 & 1 & & \\ 1 & 0 & & \\ & & 0 & 1 \\ & & 1 & 0 \end{pmatrix},$$

$$S^{01} = \frac{i}{2}\begin{pmatrix} 0 & -1 & & \\ -1 & 0 & & \\ & & 0 & 1 \\ & & 1 & 0 \end{pmatrix}, \quad S^{02} = \frac{1}{2}\begin{pmatrix} 0 & -1 & & \\ 1 & 0 & & \\ & & 0 & 1 \\ & & -1 & 0 \end{pmatrix}, \quad S^{03} = \frac{i}{2}\begin{pmatrix} -1 & & & \\ & 1 & & \\ & & 1 & \\ & & & -1 \end{pmatrix}.$$
$$\tag{10.73}$$

These are block diagonal. These are the same generators we used for the $(\frac{1}{2},0)$ and $(0,\frac{1}{2})$ representations above. This makes it clear that the Dirac representation of the Lorentz group is reducible; it is the direct sum of a left-handed and a right-handed spinor representation.

Another representation is the **Majorana representation**:

$$\gamma^0 = \begin{pmatrix} 0 & \sigma^2 \\ \sigma^2 & 0 \end{pmatrix}, \gamma^1 = \begin{pmatrix} i\sigma^3 & 0 \\ 0 & i\sigma^3 \end{pmatrix}, \gamma^2 = \begin{pmatrix} 0 & -\sigma^2 \\ \sigma^2 & 0 \end{pmatrix}, \gamma^3 = \begin{pmatrix} -i\sigma^1 & 0 \\ 0 & -i\sigma^1 \end{pmatrix}.$$

(10.74)

In this basis the γ-matrices are purely imaginary. The Majorana is another $(\frac{1}{2},0) \oplus (0,\frac{1}{2})$ representation of the Lorentz group that is physically equivalent to the Weyl representation.

The Weyl spinors, ψ_L and ψ_R, are in a way more fundamental than Dirac spinors such as ψ because they correspond to irreducible representations of the Lorentz group. But the electron is a Dirac spinor. More importantly, QED is symmetric under $L \leftrightarrow R$. Thus, for QED the γ-matrices make calculations a lot easier than separating out the ψ_L and ψ_R components. In fact, we will develop such efficient machinery for manipulating the γ-matrices that even in theories which are not symmetric to $L \leftrightarrow R$, such as the theory of weak interactions (Chapter 29), it will be convenient to embed the Weyl spinors into Dirac spinors and add projectors to remove the unphysical components. These projections are discussed in Section 11.1.

10.3.1 Lorentz transformation properties

When using Dirac matrices and spinors, we often suppress spinor indices but leave vector indices explicit. So an equation such as $\{\gamma^\mu, \gamma^\nu\} = 2g^{\mu\nu}$ really means

$$\gamma^\mu_{\alpha\gamma}\gamma^\nu_{\gamma\beta} + \gamma^\nu_{\alpha\gamma}\gamma^\mu_{\gamma\beta} = 2g^{\mu\nu}\delta_{\alpha\beta},$$

(10.75)

and the equation $S^{\mu\nu} = \frac{i}{4}[\gamma^\mu, \gamma^\nu]$ means

$$S^{\mu\nu}_{\alpha\beta} = \frac{i}{4}\left(\gamma^\mu_{\alpha\gamma}\gamma^\nu_{\gamma\beta} - \gamma^\nu_{\alpha\gamma}\gamma^\mu_{\gamma\beta}\right).$$

(10.76)

For an expression such as

$$V^2 = V_\mu g^{\mu\nu} V_\nu = \frac{1}{2}V_\mu \{\gamma^\mu, \gamma^\nu\} V_\nu$$

(10.77)

to be invariant, the Lorentz transformations in the vector and Dirac representations must be related. Indeed, since $\bar{\psi}\gamma^\mu\psi$ transforms like a 4-vector we can deduce that

$$\Lambda_s^{-1}\gamma^\mu\Lambda_s = (\Lambda_V)^{\mu\nu}\gamma^\nu,$$

(10.78)

where the Λ_s are the Lorentz transformations acting on spinor indices and Λ_V are the Lorentz transformations in the vector representation. Writing out the matrix indices $\gamma^\mu_{\alpha\beta}$ this means

$$(\Lambda_s^{-1})_{\delta\alpha}\gamma^\mu_{\alpha\beta}(\Lambda_s)_{\beta\gamma} = (\Lambda_V)^{\mu\nu}\gamma^\nu_{\delta\gamma},$$

(10.79)

where μ refers to which γ-matrix, and α and β index the elements of that matrix. You can check this with the explicit forms for Λ_V and Λ_s in Eqs. (10.70) and (10.71) above.

It is useful to study the properties of the Lorentz generators from the Dirac algebra itself, without needing to choose a particular basis for the γ^μ. First note that

$$\{\gamma^\mu, \gamma^\nu\} = 2g^{\mu\nu} \quad \Rightarrow \quad (\gamma^0)^2 = \mathbb{1}, \quad (\gamma^i)^2 = -\mathbb{1}. \tag{10.80}$$

So the eigenvalues of γ^0 are ± 1 and the eigenvalues of γ^i are $\pm i$. Thus, if we diagonalize γ^0, we will see that it is Hermitian, and if we diagonalize γ^1, γ^2 or γ^3 we will see that they are anti-Hermitian. This is true for any representation of the γ-matrices:

$$\gamma^{0\dagger} = \gamma^0, \quad \gamma^{i\dagger} = -\gamma^i. \tag{10.81}$$

(Technically, we can also find representations that cannot be diagonalized with unitary transforms, but we do not need to consider those for physics.) Then,

$$(S^{\mu\nu})^\dagger = \left(\frac{i}{4}[\gamma^\mu, \gamma^\nu]\right)^\dagger = -\frac{i}{4}[\gamma^{\nu\dagger}, \gamma^{\mu\dagger}] = \frac{i}{4}[\gamma^{\mu\dagger}, \gamma^{\nu\dagger}], \tag{10.82}$$

which implies

$$S^{ij\dagger} = S^{ij}, \quad S^{0i\dagger} = -S^{0i}. \tag{10.83}$$

Again, we see that the rotations are unitary and the boosts are not. You can see this from the explicit representations in Eq (10.73). But because we showed it algebraically, using only the defining equation $\{\gamma^\mu, \gamma^\nu\} = 2g^{\mu\nu}$, it is true in *any* representation of the Dirac algebra.

Now, observe that *one* of the Dirac matrices is Hermitian, γ^0. Moreover,

$$\gamma^0 \gamma^i \gamma^0 = -\gamma^i = \gamma^{i\dagger}, \quad \gamma^0 \gamma^0 \gamma^0 = \gamma^0 = \gamma^{0\dagger}, \tag{10.84}$$

so $\gamma^{\mu\dagger} = \gamma^0 \gamma^\mu \gamma^0$. Then

$$\gamma^0 (S^{\mu\nu})^\dagger \gamma^0 = \gamma^0 \frac{i}{4}[\gamma^{\mu\dagger}, \gamma^{\nu\dagger}]\gamma^0 = \frac{i}{4}[\gamma^0\gamma^{\mu\dagger}\gamma^0, \gamma^0\gamma^{\nu\dagger}\gamma^0] = \frac{i}{4}[\gamma^\mu, \gamma^\nu] = S^{\mu\nu}, \tag{10.85}$$

and so

$$\left(\gamma^0 \Lambda_s \gamma^0\right)^\dagger = \gamma^0 \exp(i\theta_{\mu\nu} S^{\mu\nu})^\dagger \gamma^0 = \exp(-i\theta_{\mu\nu}\gamma^0 S^{\mu\nu\dagger}\gamma^0) = \exp(-i\theta_{\mu\nu}S^{\mu\nu}) = \Lambda_s^{-1}. \tag{10.86}$$

Then, finally,

$$\psi^\dagger \gamma^0 \psi \to (\psi^\dagger \Lambda_s^\dagger)\gamma^0(\Lambda_s \psi) = \left(\psi^\dagger \gamma^0 \Lambda_s^{-1}\Lambda_s \psi\right) = \psi^\dagger \gamma^0 \psi, \tag{10.87}$$

which is Lorentz invariant.

We have just been re-deriving from the Dirac algebra point of view what we found by hand from the Weyl point of view. We have seen that the natural conjugate for ψ out of which real Lorentz-invariant expressions are constructed is not $\psi^\dagger$ but

$$\bar\psi \equiv \psi^\dagger \gamma^0. \tag{10.88}$$

The point is that $\bar\psi$ transforms according to Λ_s^{-1}. Thus $\bar\psi\psi$ is Lorentz invariant. In contrast, $\psi^\dagger\psi$ is not Lorentz invariant, since $\psi^\dagger\psi \to (\psi^\dagger \Lambda_s^\dagger)(\Lambda_s\psi)$. For this to be invariant, we would need $\Lambda_s^\dagger = \Lambda_s^{-1}$, that is, for the representation of the Lorentz group to be unitary.

But the finite-dimensional spinor representation of the Lorentz group, like the 4-vector representation, is *not* unitary, because the boost generators are anti-Hermitian. As with vectors, for unitary representations we will need *fields* $\psi(x)$ that transform in infinite-dimensional representations of the Poincaré group.

We can also construct objects such as

$$\bar{\psi}\gamma_\mu\psi, \quad \bar{\psi}\gamma_\mu\gamma_\nu\psi, \quad \bar{\psi}\partial_\mu\psi; \tag{10.89}$$

all transform like tensors under the Lorentz group. Also

$$\mathcal{L} = \bar{\psi}(i\gamma^\mu\partial_\mu - m)\psi \tag{10.90}$$

is Lorentz invariant. We abbreviate this with

$$\mathcal{L} = \bar{\psi}(i\slashed{\partial} - m)\psi, \tag{10.91}$$

which is the Dirac Lagrangian.

The Dirac equation follows from this Lagrangian by the equations of motion:

$$(i\slashed{\partial} - m)\psi = 0. \tag{10.92}$$

To be explicit, this is shorthand for

$$(i\gamma^\mu_{\alpha\beta}\partial_\mu - m\delta_{\alpha\beta})\psi_\beta = 0. \tag{10.93}$$

After multiplying the Dirac equation by $(i\slashed{\partial} + m)$ we find

$$0 = (i\slashed{\partial} + m)(i\slashed{\partial} - m)\psi = \left(-\frac{1}{2}\partial_\mu\partial_\nu\{\gamma^\mu,\gamma^\nu\} - \frac{1}{2}\partial_\mu\partial_\nu[\gamma^\mu,\gamma^\nu] - m^2\right)\psi$$

$$= -(\partial^2 + m^2)\psi. \tag{10.94}$$

So ψ satisfies the Klein–Gordon equation:

$$(\Box + m^2)\psi = 0. \tag{10.95}$$

In Fourier space, this implies that on-shell spinor momenta satisfy the unique relativistic dispersion relation $p^2 = m^2$, just like scalars. Because spinors also satisfy an equation linear in derivatives, people sometimes say the Dirac equation is the "square root" of the Klein–Gordon equation.

We can integrate the Lagrangian by parts to derive the equations of motion for $\bar{\psi}$:

$$\mathcal{L} = \bar{\psi}i\slashed{\partial}\psi - m\bar{\psi}\psi = -i\left(\partial_\mu\bar{\psi}\right)\gamma^\mu\psi - m\bar{\psi}\psi. \tag{10.96}$$

So,

$$-i\partial_\mu\bar{\psi}\gamma^\mu - m\bar{\psi} = 0. \tag{10.97}$$

This γ^μ on the opposite side from ∂_μ is a little annoying, so we often write

$$\bar{\psi}(-i\overleftarrow{\slashed{\partial}} - m) = 0, \tag{10.98}$$

where the derivative acts to the left. This makes the conjugate equation look more like the original Dirac equation.

10.4 Coupling to the photon

Under a gauge transform ψ transforms just like a scalar. For a spinor with charge $Q = -1$, such as the electron,

$$\psi \to e^{-i\alpha}\psi. \tag{10.99}$$

Then we can use the same covariant derivative $\partial_\mu + ieA_\mu$ as for a scalar. So

$$D_\mu\psi = (\partial_\mu + ieA_\mu)\psi. \tag{10.100}$$

Then the Dirac equation becomes

$$(i\slashed{\partial} - e\slashed{A} - m)\psi = 0. \tag{10.101}$$

Something very interesting happens if we try to compare the Dirac equation to the Klein–Gordon equation for a scalar field ϕ coupled to A_μ:

$$\left[(i\partial_\mu - eA_\mu)^2 - m^2\right]\phi = 0. \tag{10.102}$$

Following the same route as before, we multiply 0 by $(i\slashed{\partial} - e\slashed{A} + m)$ giving

$$\begin{aligned}
0 &= (i\slashed{\partial} - e\slashed{A} + m)(i\slashed{\partial} - e\slashed{A} - m)\psi \\
&= \left[(i\partial_\mu - eA_\mu)(i\partial_\nu - eA_\nu)\gamma^\mu\gamma^\nu - m^2\right]\psi \\
&= \left(\frac{1}{4}\{i\partial_\mu - eA_\mu, i\partial_\nu - eA_\nu\}\{\gamma^\mu, \gamma^\nu\} + \frac{1}{4}[i\partial_\mu - eA_\mu, i\partial_\nu - eA_\nu][\gamma^\mu, \gamma^\nu] - m^2\right)\psi.
\end{aligned} \tag{10.103}$$

In Eq. (10.94), the antisymmetric combination dropped out, but now we find

$$[i\partial_\mu - eA_\mu, i\partial_\nu - eA_\nu] = -e[i\partial_\mu A_\nu - i\partial_\nu A_\mu] = -eiF_{\mu\nu}. \tag{10.104}$$

So we get

$$\left((i\partial_\mu - eA_\mu)^2 - \frac{e}{2}F_{\mu\nu}\sigma^{\mu\nu} - m^2\right)\psi = 0, \tag{10.105}$$

where $\sigma^{\mu\nu} = \frac{i}{2}[\gamma^\mu, \gamma^\nu]$ as in Eq. (10.67). This equation contains an extra term compared to the spin-0 equation, Eq. (10.102).

The above manipulation can be condensed into the useful identity

$$\slashed{D}^2 = D_\mu^2 + \frac{e}{2}F_{\mu\nu}\sigma^{\mu\nu}, \tag{10.106}$$

which concisely describes the difference between covariant derivatives on spinors and scalars.

What is this extra term? Well, recall that the Lorentz generators acting on Dirac spinors are $S^{\mu\nu} = \frac{1}{2}\sigma^{\mu\nu}$. These have the form (in the Weyl representation)

$$S_{ij} = \frac{1}{2}\varepsilon_{ijk}\begin{pmatrix} \sigma_k & \\ & \sigma_k \end{pmatrix}, \quad S_{0i} = -\frac{i}{2}\begin{pmatrix} \sigma_i & \\ & -\sigma_i \end{pmatrix}, \tag{10.107}$$

and since

$$F_{0i} = E_i, \quad F_{ij} = -\varepsilon_{ijk}B_k, \qquad (10.108)$$

we get

$$\left\{ (\partial_\mu + ieA_\mu)^2 + m^2 - e\begin{pmatrix} (\vec{B} + i\vec{E}) \cdot \vec{\sigma} & \\ & (\vec{B} - i\vec{E}) \cdot \vec{\sigma} \end{pmatrix} \right\} \psi = 0. \qquad (10.109)$$

This corresponds to a magnetic dipole moment. With conventional normalization, the size of the magnetic moment is $\mu_B = \frac{e}{2m_e}$. In the non-relativistic limit, as you can explore in Problem 10.1, the Schrödinger–Pauli equation, Eq. (10.2), is reproduced with correct magnetic moment. So the Dirac equation makes a testable prediction: charged fermions should have magnetic dipole moments with size given by $\mu_B = \frac{e}{2m_e}$. Experimentally, the moment is $\sim 1.002\mu_B$. The 0.002 will be calculated later.

To summarize, we found that while free spinors satisfy the equation of motion for a scalar field, when spinors are coupled to the photon, an additional interaction appears which corresponds to a magnetic dipole moment. The size of the electron's magnetic moment can be read off as the coefficient of this additional interaction. That the correct magnetic moment comes out of the Dirac equation is a remarkable *physical* prediction of Dirac's equation. Note that the coupling to the electric field in Eq. (10.109) is not an electric dipole moment – that would not have an i, but is simply the effect of a magnetic moment in a boosted frame. Electric dipole moments will be explored in Section 29.5.3 and in Problem 11.10.

Finally, note that the Noether current associated with the global symmetry $\psi \to e^{-i\alpha}\psi$ is

$$J_\mu = \bar{\psi}\gamma_\mu\psi. \qquad (10.110)$$

This, like any Noether current, is conserved on the equations of motion even if we set $A_\mu = 0$. The 0 component of this current gives the charge density

$$J_0 = \psi^\dagger\psi = \psi_L^\dagger\psi_L + \psi_R^\dagger\psi_R. \qquad (10.111)$$

We originally hoped this would be Lorentz invariant, which it is not. Now we see that it transforms as the 0 component of a conserved current. We can interpret this as the probability density for a fermion. The conserved charge $Q = \int d^3x J_0$ is electron number, which is the number of electrons minus the number of positrons. The spatial components of J_μ denote electron number flow. The electron number current J_μ is related to the charge current eJ_μ, which couples to A_μ, by a factor of the electric charge e.

10.5 What does spin $\frac{1}{2}$ mean?

To understand spin-$\frac{1}{2}$ particles, let us begin by looking at what happens when we rotate them by an angle θ around the z axis. For any representation, such a rotation is given by

$$\Lambda(\theta_z) = \exp(i\theta_z J_z), \qquad (10.112)$$

with J_z the generator in an appropriate representation. The easiest way to exponentiate a matrix is to first diagonalize it with a unitary transformation, then exponentiate the eigenvalues, then transform back. This unitary transformation is like choosing a (possibly complex) direction. If we are only ever rotating around one axis, we can simply use the diagonal basis for the exponentiation.

First, for the vector representation,

$$J_3 = V_{12} = i \begin{pmatrix} 0 & & \\ & \begin{matrix} 0 & -1 \\ 1 & 0 \end{matrix} & \\ & & 0 \end{pmatrix} = U^{-1} \begin{pmatrix} 0 & & \\ & \begin{matrix} -1 & \\ & 1 \end{matrix} & \\ & & 0 \end{pmatrix} U. \tag{10.113}$$

Note that the eigenvalues of J_3 are $-1, 0, 1$ and 0, which is what one expects from the $\left(\frac{1}{2}, \frac{1}{2}\right)$ representation of the Lorentz group describing spins 1 and 0, as in Table 10.1. So,

$$\Lambda_V(\theta_z) = \exp(i\theta_z V_{12}) = U^{-1} \begin{pmatrix} 1 & & & \\ & \exp(-i\theta_z) & & \\ & & \exp(i\theta_z) & \\ & & & 1 \end{pmatrix} U \tag{10.114}$$

and

$$\Lambda_V(2\pi) = \mathbb{1}. \tag{10.115}$$

That is, we rotate 360 degrees and we are back to where we started.

For the spinor representation

$$\Lambda_s(\theta_z) = \exp(i\theta_z S_{12}) \tag{10.116}$$

the 12 rotation is already diagonal:

$$S_{12} = \begin{pmatrix} \frac{1}{2} & & & \\ & -\frac{1}{2} & & \\ & & \frac{1}{2} & \\ & & & -\frac{1}{2} \end{pmatrix}. \tag{10.117}$$

Here the eigenvalues are $\frac{1}{2}, -\frac{1}{2}, \frac{1}{2}$ and $-\frac{1}{2}$, as one expects for the $\left(\frac{1}{2}, 0\right) \oplus \left(0, \frac{1}{2}\right)$ representation of the Lorentz group. So,

$$\Lambda_s(\theta_z) = \exp(i\theta_z S_{12}) = \begin{pmatrix} \exp(\frac{i}{2}\theta_z) & & & \\ & \exp(-\frac{i}{2}\theta_z) & & \\ & & \exp(\frac{i}{2}\theta_z) & \\ & & & \exp(-\frac{i}{2}\theta_z) \end{pmatrix} \tag{10.118}$$

and

$$\Lambda_s(2\pi) = \begin{pmatrix} -1 & & & \\ & -1 & & \\ & & -1 & \\ & & & -1 \end{pmatrix} = -\mathbb{1}. \tag{10.119}$$

Thus, a 2π rotation does not bring us back where we started! If we rotate by 4π it would, but with a 2π rotation we pick up a -1.

By the way, this odd factor of -1 is the origin of the connection between spin and statistics. As a quick way to see the connection, consider a state containing two identical fermions localized on opposite sides of the origin in the $\hat{x}$ direction. Let their spins both point in the $+\hat{z}$ direction. So the state is $|\psi_{12}\rangle = |\psi_\uparrow(\vec{x})\,\psi_\uparrow(-\vec{x})\rangle$. Now rotate the state around the z axis by π. Such a rotation interchanges the two particles, and does not affect the spins. It also induces a factor of $\Lambda_s(\pi) = i$ for each spinor. Thus, $|\psi_{12}\rangle \to -|\psi_{21}\rangle$ since the particles are identical, $|\psi_{12}\rangle = |\psi_{21}\rangle$. Thus, the wavefunction picks up a -1 when the particles are interchanged. That is, the spinors are fermions. This argument is repeated with somewhat more detail in Chapter 12, where additional implications of the spin-statistics theorem are discussed.

10.5.1 Projective representations

How can something go back to minus itself under a 2π rotation? This is not something that can happen in the Lorentz group. By definition, all representations of the Lorentz group map 2π rotations to the identity element of the group: $r[2\pi] = \mathbb{1}$. And, by definition, the identity group element sends objects to themselves. The problem is that by exponentiating elements of the Lie algebra for the group we generated a different group, $\mathrm{SL}(2,\mathbb{C})$, which is the universal cover of the Lorentz group, not the Lorentz group itself. So, technically, spinors transform as representations of $\mathrm{SL}(2,\mathbb{C})$. Why is this OK?

Recall that the Lorentz group is defined as the group preserving the Minkowski metric $\Lambda^T g \Lambda = g$. Observables, in particular the S-matrix, should be invariant under this symmetry. In quantum mechanics, we learned that states are identified with rays, so that $\|\psi\rangle$ and $\lambda|\psi\rangle$ for any complex number λ are the same state. In field theory, we have carefully normalized our fields (and we will carefully *re*normalize them), so we do not want that norm to change in different frames. However, we can still have the fields change by a phase without upsetting their norms. Thus, for physical purposes what we are looking for is not exactly representations of the Lorentz group, but **projective representations** of the Lorentz group, in which group elements can change the phase of a state. Projective representations can have

$$r[g_1]r[g_2] = e^{i\phi(g_1,g_2)}r[g_1g_2], \qquad (10.120)$$

which is a generalization of the normal requirement that $r[g_1g_2] = r[g_1]r[g_2]$ for a group representation. The projective representations of $\mathrm{O}(1,3)$ are the same as the representations of $\mathrm{SL}(2,\mathbb{C})$, which include the spinors.

Using objects that have properties that are not directly observable is not new. For example, in quantum mechanics we learned that wavefunctions are complex. There are plenty of implications of the complexity, but you do not actually measure complex things. In the same way, although we only measure Lorentz-invariant things (matrix elements), the most general theory can have objects, spinors, that are a little bit more complicated than the Lorentz group alone would naively suggest. Although spinors transform in representations of $\mathrm{SL}(2,\mathbb{C})$, the Poincaré group is still the symmetry group of observables.

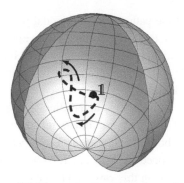 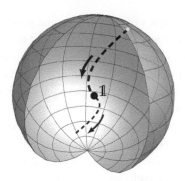

The group $SO(3)$ can be thought of as a ball of radius π with antipodal points identified. On the left is a contractible path through $SO(3)$ and on the right is a non-contractible path. **Fig. 10.1**

The existence of objects, spinors, that transform as $\psi \to -\psi$ under $\theta = 2\pi$ rotations is closely related to an interesting fact about the 3D rotation group that you might not be aware of: it is not **simply connected**. In a group that is not simply connected, there are closed paths through the group that are not contractible, that is, they cannot be smoothly deformed to a point. For example, the group $SO(2)$ of 2D rotations is specified by angles θ. Let us describe our path by a number t, with $0 \le t < 1$. Then the path $\theta(t) = 2\pi t$ is not smoothly deformable to $\theta(t) = 0$. That means there is no smooth function $\theta(t, u)$ for $0 \le u \le 1$, such that $\theta(t, 0) = \theta(t)$ and $\theta(t, 1) = 0$. In fact, none of the paths $\theta(t) = 2\pi n t$ for an integer n can be deformed into each other. We say the **fundamental group** for $SO(2)$ is $\mathbb{Z}$.

The group $SO(3)$ is not simply connected either. To see that, define rotations around the z axis by an angle θ_z, and consider the path $\theta_z(t) = 2\pi t$ corresponding to the group elements

$$R(t) = \begin{pmatrix} \cos 2\pi t & \sin 2\pi t & 0 \\ -\sin 2\pi t & \cos 2\pi t & 0 \\ 0 & 0 & 1 \end{pmatrix}. \tag{10.121}$$

This path cannot be smoothly deformed to the identity. Try it! Try to find $R(t, u)$ so that $R(t, 0) = R(t)$ and $R(t, 1) = \mathbb{1}$.

To see that $SO(3)$ is not simply connected, consider the geometric pictures shown in Figure 10.1. Any rotation in $SO(3)$ can be specified by an axis $\vec{v}$ and an angle $-\pi \le \theta < \pi$. If we think of the axis as a point on the surface of a ball of radius $r = \pi$, then the rotation can be specified by a point in the ball, with the distance from the origin being the angle θ. Thus, a path through the group is a path through the ball. The identity group element is the center of the ball. There is one catch, however: rotations about an axis $\vec{v}$ by $\theta = \pi$ are the same as rotations about the axis $\vec{v}$ by $\theta = -\pi$ (or the axis $-\vec{v}$ by $\theta = \pi$). Thus, we have to identify antipodal points on the sphere as the same group element in $SO(3)$.[1] Paths from $\mathbb{1}$ to $\mathbb{1}$ in $SO(3)$ go from the center of the ball back to the center. Figure 10.1

---

[1] Topologically, $SO(3)$ is a real projective space $\mathbb{RP}^3 = S^3/\mathbb{Z}_2$ where S^3 is the surface of a 3-sphere. The full 3-sphere is $SU(2)$, which is the universal cover of $SO(3)$, and so $\mathbb{Z}_2$ is the fundamental group of $SO(3)$. For the full Lorentz group, the cover is $SL(2, \mathbb{C})$ and the fundamental group is still $\mathbb{Z}_2$.

shows examples of contractible and non-contractible paths. Since antipodal points on the sphere are π and $-\pi$ rotations around the same axis, they are the same group element; that is why the second path cannot be deformed to the identity without breaking the line. If you imaging pulling the contractible path through the surface, it will have two points where it crosses. Such paths can be deformed back to the identity.

You can actually see the non-contractible paths without too much difficulty by just holding something (like a glass of water) in your hand with your arm outstretched and rotating your arm 360 degrees in a plane parallel to your body. Then your arm (the path) will be twisted, but the object in your hand will have mapped back to itself. You can untangle your arm (the path) with another 360 degree rotation, in this case in a plane parallel to the ground, which gives another $\mathbb{Z}_2$ undoing the twist. If you are careful, you will not even spill the water. Spinors maintain an imprint of how they have been rotated, which shows up as a minus sign after a 2π rotation, much like your arm would if it were an internal degree of freedom of the glass of water. This demonstration is sometimes called Dirac's belt trick, Feynman's plate trick, the Balinese cup trick or the quaternionic handshake.

10.6 Majorana and Weyl fermions

For QED, one only needs the electron, which is efficiently described in the reducible Dirac representation $\left(\frac{1}{2},0\right) \oplus \left(0,\frac{1}{2}\right)$ of the Lorentz group. In other theories, such as the Standard Model or supersymmetric theories, spinors that are not Dirac spinors are prevalent. In this section we discuss other Lorentz-invariant quantities that can be constructed using spinors that are not in the Dirac representation and introduce some efficient notation.

10.6.1 Majorana masses

There is one more way to get a Lorentz-invariant quantity out of ψ_L and ψ_R. Recall that we could not write down a mass term $\psi_R^\dagger \psi_R$ for just a right-handed spinor. The Lorentz transformations, in Eqs. (10.40) and (10.42),

$$\delta\psi_R = \frac{1}{2}(i\theta_j + \beta_j)\sigma_j\psi_R, \qquad \delta\psi_R^\dagger = \frac{1}{2}(-i\theta_j + \beta_j)\psi_R^\dagger\sigma_j, \qquad (10.122)$$

imply that the natural candidate mass term $m\psi_R^\dagger\psi_R$ is not boost invariant:

$$\delta\left(\psi_R^\dagger\psi_R\right) = \beta_j\psi_R^\dagger\sigma_j\psi_R \neq 0. \qquad (10.123)$$

It turns out that there is a different bilinear quantity that *is* Lorentz invariant:

$$\mathcal{L}_{\text{Maj}} = \psi_R^T\sigma_2\psi_R. \qquad (10.124)$$

This is known as a **Majorana mass**.

To see the Lorentz invariance, recall that for the Pauli matrices σ_1 and σ_3 are real, and σ_2 is imaginary:

$$\sigma_1 = \begin{pmatrix} 0 & 1 \\ 1 & 0 \end{pmatrix}, \quad \sigma_2 = \begin{pmatrix} 0 & -i \\ i & 0 \end{pmatrix}, \quad \sigma_3 = \begin{pmatrix} 1 & 0 \\ 0 & -1 \end{pmatrix}. \qquad (10.125)$$

So,

$$\sigma_1^\star = \sigma_1, \quad \sigma_2^\star = -\sigma_2, \quad \sigma_3^\star = \sigma_3, \tag{10.126}$$

$$\sigma_1^T = \sigma_1, \quad \sigma_2^T = -\sigma_2, \quad \sigma_3^T = \sigma_3. \tag{10.127}$$

This implies $\sigma_1^T \sigma_2 = \sigma_1 \sigma_2 = -\sigma_2 \sigma_1$, $\sigma_3^T \sigma_2 = \sigma_3 \sigma_2 = -\sigma_2 \sigma_3$ and $\sigma_2^T \sigma_2 = -\sigma_2 \sigma_2$. That is,

$$\sigma_j^T \sigma_2 = -\sigma_2 \sigma_j \tag{10.128}$$

and so

$$\delta(\psi_R^T \sigma_2) = \frac{1}{2}(i\theta_j + \beta_j)\psi_R^T \sigma_j^T \sigma_2 = \frac{1}{2}(-i\theta_j - \beta_j)\left(\psi_R^T \sigma_2\right)\sigma_j. \tag{10.129}$$

Combining this with the transformation of ψ_R in Eq. (10.122) we see that $\mathcal{L}_{\text{Maj}}$ in Eq. (10.124) is Lorentz invariant.

Since $\sigma_2 = \begin{pmatrix} & -i \\ i & \end{pmatrix}$ the Majorana mass can be expanded out to

$$\psi_R^T \sigma_2 \psi_R = \begin{pmatrix} \psi_1 & \psi_2 \end{pmatrix} \begin{pmatrix} & -i \\ i & \end{pmatrix} \begin{pmatrix} \psi_1 \\ \psi_2 \end{pmatrix} = -i(\psi_1 \psi_2 - \psi_2 \psi_1). \tag{10.130}$$

Thus, we have shown that $\psi_1 \psi_2 - \psi_2 \psi_1$ is Lorentz invariant. We often write this as

$$\psi_1 \psi_2 - \psi_2 \psi_1 = \psi^\alpha \psi^\beta \varepsilon_{\alpha\beta}, \quad \varepsilon_{\alpha\beta} = \begin{pmatrix} 0 & -1 \\ 1 & 0 \end{pmatrix}, \tag{10.131}$$

which avoids picking a σ_2.

There is only one problem: if the fermion components commute, $\psi_1 \psi_2 - \psi_2 \psi_1 = 0$! For Majorana masses to be non-trivial, fermion components cannot be regular numbers, they must be anticommuting numbers. Such things are called **Grassmann numbers** and satisfy a Grassmann algebra. Further explanation of why spinors must anticommute is given in Chapter 12, on the spin-statistics theorem. The mathematics of Grassmann numbers is discussed more in Section 14.6 on the path integral.

10.6.2 Notation for Weyl spinors

In QED, we will be mostly interested in Dirac spinors, such as the electron. But since Weyl spinors correspond to irreducible representations of the Lorentz group, it is sometimes helpful to have concise notation for constructing products and contractions of Weyl spinors only. This notation is useful in many contexts besides gauge theories, such as supersymmetry. It is also related to the spinor-helicity formalism we will discuss in Chapter 27. If you are just interested in QED, you can skip this part.

Let us write ψ for left-handed spinors and $\tilde{\psi}$ for right-handed spinors. Sometimes the notation $\bar{\psi}$ is used, especially in the contexts of supersymmetry, but this can be confused with the bar notation for a Dirac spinor, $\bar{\psi} \equiv \psi^\dagger \gamma_0$, so we will stick with $\tilde{\psi}$. We index the two components of left-handed Weyl spinors with Greek indices from the beginning of the alphabet, i.e. ψ_α. For right-handed spinors, we use dotted Greek indices, i.e. $\tilde{\psi}_{\dot{\alpha}}$. A Dirac spinor is

$$\psi = \begin{pmatrix} \psi^\alpha \\ \tilde{\psi}_{\dot{\beta}} \end{pmatrix}. \tag{10.132}$$

Conventionally, left-handed spinors (and right-handed antispinors) have upper undotted indices and right-handed spinors (and left-handed antispinors) have lower dotted indices.

Recall that we showed that a Majorana mass is Lorentz invariant. This mass has the form

$$\mathcal{L}_{\text{maj}} = \psi^T \sigma_2 \psi = -i\left(\psi_1 \psi_2 - \psi_2 \psi_1\right) = i\psi_\alpha \varepsilon^{\alpha\beta} \psi_\beta, \tag{10.133}$$

where $\varepsilon^{\alpha\beta} = -i\sigma_2^{\alpha\beta}$ is the totally antisymmetric 2×2 tensor

$$\varepsilon^{\alpha\beta} = -\varepsilon_{\alpha\beta} = \begin{pmatrix} 0 & 1 \\ -1 & 0 \end{pmatrix}. \tag{10.134}$$

That is, $\varepsilon^{12} = \varepsilon_{21} = 1$, which leads to $\varepsilon_{\alpha\beta}\varepsilon^{\beta\gamma} = \delta_\alpha^\gamma$. The ε tensor serves the function for Weyl spinors that $g_{\mu\nu}$ does for tensors – we can always contract spinors into Lorentz-invariant combinations with the ε tensor. However, we have to be careful raising and lowering indices, since

$$\psi^\alpha \chi_\alpha = \varepsilon^{\alpha\beta} \varepsilon_{\alpha\gamma} \psi_\beta \chi^\gamma = -\varepsilon^{\beta\alpha} \varepsilon_{\alpha\gamma} \psi_\beta \chi^\gamma = -\delta_\gamma^\beta \psi_\beta \chi^\gamma = -\psi_\beta \chi^\beta. \tag{10.135}$$

While it seems that the index position makes things messy, it actually makes things easier, since spinors anticommute. We can define the inner product between two left-handed spinors as

$$\psi\chi \equiv \psi_\alpha \chi^\alpha = \psi_\alpha \varepsilon^{\alpha\beta} \chi_\beta = -\chi_\beta \varepsilon^{\alpha\beta} \psi_\alpha = \chi_\beta \varepsilon^{\beta\alpha} \psi_\alpha = \chi_\alpha \psi^\alpha = \chi\psi, \tag{10.136}$$

so that the product is symmetric. Note that $\psi\psi \neq 0$ even though $\psi^\alpha \psi^\alpha = 0$. For right-handed spinors, we define

$$\tilde{\psi}\tilde{\chi} = \tilde{\psi}^{\dot{\alpha}} \tilde{\chi}_{\dot{\alpha}} = -\tilde{\chi}_{\dot{\alpha}} \tilde{\psi}^{\dot{\alpha}} = \tilde{\chi}^{\dot{\alpha}} \tilde{\psi}_{\dot{\alpha}} = \tilde{\chi}\tilde{\psi}, \tag{10.137}$$

which is also symmetric.[2]

For Weyl spinors, the σ-matrices $\sigma^\mu = (1, \vec{\sigma})$ and $\bar{\sigma}^\mu = (1, -\vec{\sigma})$ replace the Dirac γ-matrices. Recall that the kinetic term for a Dirac spinor $\psi = (\psi, \tilde{\chi})$ is

$$\mathcal{L}_{\text{kin}} = i\bar{\psi}\slashed{\partial}\psi = i\psi^\dagger \bar{\sigma}^\mu \partial_\mu \psi + i\tilde{\chi}^\dagger \sigma^\mu \partial_\mu \tilde{\chi}. \tag{10.138}$$

Each of these two terms is separately Lorentz invariant. With spinor indices, $\sigma^\mu = \sigma^\mu_{\alpha\dot{\alpha}}$, the contractions are

$$\tilde{\chi}^\dagger \sigma^\mu \tilde{\chi} = (\tilde{\chi}^\dagger)^\alpha \sigma^\mu_{\alpha\dot{\alpha}} \tilde{\chi}^{\dot{\alpha}} = \chi^\alpha \sigma^\mu_{\alpha\dot{\alpha}} \tilde{\chi}^{\dot{\alpha}}, \tag{10.139}$$

where we have defined a left-handed spinor $\chi \equiv \tilde{\chi}^\dagger$ so that we can drop the $\dagger$. You can think of χ as the particle and $\tilde{\chi}$ as the antiparticle for the same Weyl spinor. Similarly,

$$\psi^\dagger \bar{\sigma}^\mu \psi = (\psi^\dagger)^{\dot{\alpha}} \bar{\sigma}^\mu_{\dot{\alpha}\alpha} \psi^\alpha = \tilde{\psi}^{\dot{\alpha}} \bar{\sigma}^\mu_{\dot{\alpha}\alpha} \psi^\alpha \tag{10.140}$$

with $\tilde{\psi} \equiv \psi^\dagger$.

---

[2] These are opposite conventions to [Wess and Bagger, 1992], but consistent with what is used in spinor-helicity calculations (Chapter 27).

Two very useful relations are

$$g_{\mu\nu}\sigma^{\mu}_{\alpha\dot{\alpha}}\sigma^{\nu}_{\beta\dot{\beta}} = 2\varepsilon_{\alpha\beta}\varepsilon_{\dot{\alpha}\dot{\beta}} \tag{10.141}$$

and

$$\varepsilon_{\alpha\beta}\varepsilon_{\dot{\alpha}\dot{\beta}}\sigma^{\mu\beta\dot{\beta}} = \bar{\sigma}^{\mu}_{\dot{\alpha}\alpha}. \tag{10.142}$$

You can prove these relations in Problem 10.3.

Problems

10.1 We saw that the Dirac equation predicted that there is an interaction between the electron spin and the magnetic field, $\vec{S}\vec{B}$, with strength $\mu_B = \frac{\hbar e}{2m_e c}$. When the electron has angular momentum $\vec{L}$, such as in an atomic orbital, there is also a $\vec{B}\vec{L}$ interaction and a spin-orbit coupling $\vec{S}\vec{L}$. The Dirac equation (along with symmetry arguments) predicts the strength of all three interactions, as well as other corrections. To see the effect of these terms on the hydrogen atom, we have to take the non-relativistic limit.

(a) For the Schrödinger equation, we need the Hamiltonian, not the Lagrangian. Find the Dirac Hamiltonian by writing the Dirac equation as $i\partial_t \psi = H_D \psi$. Write the Hamiltonian in terms of momenta p_i rather than derivatives ∂_i.

(b) Calculate $(H_D - eA_0)^2$ in the Weyl representation for $\psi = (\psi_L \psi_R)$. Leave in terms of σ_i, p_i and A_i. Put back in the factors of c and $\hbar$, keeping the charge e dimensionless.

(c) Now take the square root of this result and expand in $\frac{1}{c}$, subtracting off the zero-point energy mc^2, i.e. compute $H = H_D - mc^2$ to order c^0. Looking at the σ_i term, how big are the electron's electric and magnetic dipole moments?

(d) The size of the terms in this Hamiltonian are only meaningful because the spin and angular momentum operators have the same normalization. Check the normalization of the angular momentum operators $L_i = \varepsilon_{ijk}x_j p_k$ and the spin operators $S_i = \frac{1}{2}\sigma_i$ by showing that they both satisfy the rotation algebra: $[J_i, J_j] = i\varepsilon_{ijk}J_k$.

(e) The gyromagnetic ratio, g_e (sometimes called the g-factor), is the relative size of the $\vec{S}\vec{B}$ and $\vec{L}\vec{B}$ interactions. Choose a constant magnetic field in the z direction, then isolate the $B_z L_z$ coupling in H. Extract the electron gyromagnetic ratio g_e by writing the entire coupling to the magnetic field in the Hamiltonian as $\mu_B B_z(L_z + g_e S_z) = B_z \mu_z$, with $\vec{\mu} \equiv \mu_B(\vec{L} + g_e \vec{S})$. How could you experimentally measure g_e (e.g. with spectroscopy of the hydrogen atom)?

(f) In spherical coordinates, the Schrödinger equation has an $\vec{L}^2$ term. With spin, you might expect that this becomes $\vec{L}\vec{\mu} = \mu_B(\vec{L}^2 + g_e \vec{L}\vec{S})$, making the $\vec{L}\vec{S}$ term proportional to the g-factor. This is wrong. It misses an important relativistic effect, Thomas precession. It is very hard to calculate directly, but easy to calculate using symmetries. With no magnetic field, the atom, with spin included, is still rotationally invariant. Which of $\vec{J} = \vec{L} + \vec{S}$ or $\vec{\mu} = \vec{L} + g_e\vec{S}$ is conserved (i.e. commutes with H)? Using this result, how does the spin-orbit coupling depend on g_e?

(g) There are additional relativistic effects coming from the Dirac equation. Expand the Dirac equation to next order in $\frac{1}{c^2}$, producing a term that scales as $\vec{p}^4$.

(h) Now let us do some dimensional analysis – there is only one scale m_e. Show that the electron's Compton wavelength, the classical electron radius, r_e, the Bohr radius, a_0, and the inverse-Rydberg constant, Ry^{-1}, are all m_e^{-1} times powers of α_e. Are the splittings due to the p^4 term fine structure ($\Delta E \sim \alpha_e^2 E$), hyperfine structure ($\Delta E \sim \alpha_e^4 E$) or something else? [Hint: write out a formula for the energy shift using time-independent perturbation theory, then see which of the above length scales appears.]

10.2 In this problem you will construct the finite-dimensional irreducible representations of $SU(2)$. By definition, such a representation is a set of three $n \times n$ matrices τ_1, τ_2 and τ_3 satisfying the algebra of the Pauli matrices $[\tau_i, \tau_j] = i\varepsilon_{ijk}\tau_k$. It is also helpful to define the linear combinations $\tau^{\pm} = \tau_1 \pm i\tau_2$.

(a) In any such representation we can diagonalize τ_3. Its eigenvectors are n complex vectors V_j with $\tau_3 V_j = \lambda_j V_j$. Show that $\tau^+ V_j$ and $\tau^- V_j$ either vanish or are eigenstates of τ_3 with eigenvalues $\lambda_j + 1$ and $\lambda_j - 1$ respectively.

(b) Prove that exactly one of the eigenstates V_{max} of τ_3 must satisfy $\tau^+ V_{\text{max}} = 0$. The eigenvalue $\lambda_{\text{max}} = j$ of V_{max} is known as the spin. Similarly, there will be an eigenvector V_{min} of τ_3 with $\tau^- V_{\text{min}} = 0$.

(c) Since there are a finite number of eigenvectors, $V_{\text{min}} = (\tau^-)^N V_{\text{max}}$ for some integer N. Prove that $N = 2J$ so that $n = 2J + 1$.

(d) Construct explicitly the five-dimensional representation of $SU(2)$.

10.3 Derive Eqs. (10.141) and (10.142):

(a) $g^{\mu\nu}\sigma_\mu^{\alpha\dot{\alpha}}\sigma_\nu^{\beta\dot{\beta}} = 2\varepsilon^{\alpha\beta}\varepsilon^{\dot{\alpha}\dot{\beta}}$,

(b) $\varepsilon_{\alpha\beta}\varepsilon_{\dot{\alpha}\dot{\beta}}\sigma^{\mu\beta\dot{\beta}} = \bar{\sigma}_{\dot{\alpha}\alpha}^\mu$.

10.4 Majorana representation.

(a) Write out the form of the Lorentz generators in the Majorana representation.

(b) Compute $\vec{J}^2$ in the Majorana representation, the left-handed Weyl representation and 4-vector representation. How do you interpret the eigenvalues of $\vec{J}^2$?

(c) Calculate $\gamma^5 = i\gamma^0\gamma^1\gamma^2\gamma^3$ in the Majorana representation.

10.5 Supersymmetry.

(a) Show that the Lagrangian

$$\mathcal{L} = \partial_\mu\phi^\star\partial^\mu\phi + \chi^\dagger i\bar{\sigma}\partial\chi + F^\star F + m\phi F + \frac{i}{2}m\chi^T\sigma^2\chi + h.c. \quad (10.143)$$

is invariant under

$$\delta\phi = -i\epsilon^T\sigma^2\chi, \quad (10.144)$$

$$\delta\chi = \epsilon F + \sigma^\mu\partial_\mu\phi\sigma^2\epsilon^\star, \quad (10.145)$$

$$\delta F = -i\epsilon^\dagger\bar{\sigma}^\mu\partial_\mu\chi, \quad (10.146)$$

where ϵ is a spinor, χ is a spinor, and F and ϕ are scalars. All spinors anticommute. σ^2 is the second Pauli spin matrix.

(b) The field F is an auxiliary field, since it has no kinetic term. A useful trick for dealing with auxiliary fields is to solve their equations of motion exactly and plug the result back into the Lagrangian. This is called *integrating out a field*. Integrate out F to show that ϕ and χ have the same mass.

(c) Auxiliary fields such as F act like Lagrange multipliers. One reason to keep the auxiliary fields in the Lagrangian is because they make symmetry transformations exact at the level of the Lagrangian. After the field has been integrated out, the symmetries are only guaranteed to hold if you use the equations of motion. Still using $\delta\phi = i\varepsilon^T \sigma^2 \chi$, what is the transformation of χ that makes the Lagrangian in (b) invariant, if you are allowed to use the equations of motion?

Spinor solutions and *CPT*

In the previous chapter, we cataloged the irreducible representations of the Lorentz group O(1, 3). We found that in addition to the obvious tensor representations, $\phi, A_\mu, h_{\mu\nu}$ etc., there are a whole set of spinor representations, such as Weyl spinors ψ_L, ψ_R. A Dirac spinor ψ transforms in the reducible $\left(\frac{1}{2}, 0\right) \oplus \left(0, \frac{1}{2}\right)$ representation. We also found Lorentz-invariant Lagrangians for spinor *fields*, $\psi(x)$. The next step towards quantizing a theory with spinors is to use these Lorentz group representations to generate irreducible unitary representations of the Poincaré group.

We discussed how unitary representations of the Poincaré group are induced from representations of its little group. The little group is the group that leaves a given momentum 4-vector p_μ invariant. When p_μ is massive, the little group is SO(3); when p_μ is massless, the little group is ISO(2). As a consequence, massive particles of spin J should have $2J + 1$ degrees of freedom and massless particles of any spin > 0 have two degrees of freedom. In the spin-1 case, we found that there were ambiguities in what the free Lagrangian was (it could have been $aA_\mu \Box A_\mu + bA_\mu \partial_\mu \partial_\nu A_\nu$ for any a and b), but we found that there was a unique Lagrangian that propagated the correct degrees of freedom. We then solved the free equations of motion for a fixed momentum p_μ generating two or three polarizations $\epsilon_\mu^i(p)$. These solutions, which were representations of the little group, if known for every value of p_μ, induce representations of the full Poincaré group.

For the spin-$\frac{1}{2}$ case, there is a unique free Lagrangian (up to Majorana masses) that automatically propagates the right degrees of freedom. In this sense, spin $\frac{1}{2}$ is easier than spin 1, since there are no unphysical degrees of freedom. The mass term couples left- and right-handed spinors, so it is natural to use the Dirac representation. As in the spin-1 case, we will solve the free equations of motion to find basis spinors, $u_s(p)$ and $v_s(p)$ (analogs of ϵ_μ^i), which we will use to define our quantum fields. As with complex scalars, we will naturally find both particles and antiparticles in the spectrum with the same mass and opposite charge: these properties fall out of the unique Lagrangian we can write down.

A spinor can also be its own antiparticle, in which case we call it a Majorana spinor. As we saw, since particles and antiparticles have opposite charges, Majorana spinors must be neutral. We will define the operation of charge conjugation C as taking particles to antiparticles, so Majorana spinors are invariant under C. After introducing C, it is natural to continue to discuss how the discrete symmetries parity, P, and time reversal, T, act on spinors.

11.1 Chirality, helicity and spin

In a relativistic theory, spin can be a confusing subject. There are actually three concepts associated with spin: spin, helicity and chirality. In this section we define and distinguish these different quantities.

Recall from Eq. (10.105) that the Dirac equation $(i\not{D} - m)\psi = 0$ implies

$$\left[(i\partial_\mu - eA_\mu)^2 - \frac{e}{2}F_{\mu\nu}\sigma^{\mu\nu} - m^2\right]\psi = 0, \tag{11.1}$$

and for the conjugate field $\bar{\psi} = \psi^\dagger\gamma_0$,

$$\bar{\psi}\left[(i\overleftarrow{\partial}_\mu + eA_\mu)^2 - \frac{e}{2}F_{\mu\nu}\sigma^{\mu\nu} - m^2\right] = 0. \tag{11.2}$$

Thus, $\bar{\psi}$ is a particle with mass m and charge opposite to ψ; that is, $\bar{\psi}$ is the antiparticle of ψ (the additional sign flip in the $F_{\mu\nu}\sigma^{\mu\nu}$ term indicates that particle and antiparticle spins are different). We will often call ψ an electron and $\bar{\psi}$ a positron, although the Dirac equation describes many particle–antiparticle pairs.

When we constructed the Dirac representation, we saw that it was the direct sum of two irreducible representations of the Lorentz group: $\left(\frac{1}{2}, 0\right) \oplus \left(0, \frac{1}{2}\right)$. Now we see that it describes two physically distinguishable particles: the electron and the positron. Irreducible unitary spin-$\frac{1}{2}$ representations of the Poincaré group, Weyl spinors, have two degrees of freedom. Dirac spinors have four. These are two spin states for the electron and two spin states for the positron. For charged spinors, there is no other way. Uncharged spinors can be their own antiparticles if they are Majorana spinors, as discussed in Section 11.3 below.

To understand the degrees of freedom within a four-component Dirac spinor, first recall that in the Weyl basis the γ-matrices have the form

$$\gamma_\mu = \begin{pmatrix} 0 & \sigma_\mu \\ \bar{\sigma}_\mu & 0 \end{pmatrix}, \tag{11.3}$$

and the Lorentz generators $S^{\mu\nu} = \frac{i}{4}[\gamma^\mu, \gamma^\nu]$ are block diagonal. Under an infinitesimal Lorentz transformation,

$$\psi \to \psi + \frac{1}{2}\begin{pmatrix} (i\theta_i - \beta_i)\sigma_i & \\ & (i\theta_i + \beta_i)\sigma_i \end{pmatrix}\psi. \tag{11.4}$$

In this basis, a Dirac spinor is a doublet of a left- and a right-handed Weyl spinor:

$$\psi = \begin{pmatrix} \psi_L \\ \psi_R \end{pmatrix}. \tag{11.5}$$

Here **left-handed** and **right-handed** refer to the $\left(\frac{1}{2}, 0\right)$ or $\left(0, \frac{1}{2}\right)$ representations of the Lorentz group. The handedness of a spinor is also known as its **chirality**.

It is helpful to be able to project out the left- or right-handed Weyl spinors from a Dirac spinor. We can do that with the γ_5-matrix:

$$\gamma^5 \equiv i\gamma^0\gamma^1\gamma^2\gamma^3. \tag{11.6}$$

In the Weyl representation

$$\gamma^5 = \begin{pmatrix} -\mathbb{1} & \\ & \mathbb{1} \end{pmatrix}, \tag{11.7}$$

so left- and right-handed spinors are eigenstates of γ_5 with eigenvalues ∓ 1. We can also define projection operators,

$$P_R = \frac{1+\gamma^5}{2} = \begin{pmatrix} 0 & \\ & \mathbb{1} \end{pmatrix}, \quad P_L = \frac{1-\gamma^5}{2} = \begin{pmatrix} \mathbb{1} & \\ & 0 \end{pmatrix}, \tag{11.8}$$

which satisfy $P_R^2 = P_R$ and $P_L^2 = P_L$ and

$$P_R \begin{pmatrix} \psi_L \\ \psi_R \end{pmatrix} = \begin{pmatrix} 0 \\ \psi_R \end{pmatrix}, \quad P_L \begin{pmatrix} \psi_L \\ \psi_R \end{pmatrix} = \begin{pmatrix} \psi_L \\ 0 \end{pmatrix}. \tag{11.9}$$

Writing projectors as $\frac{1 \pm \gamma^5}{2}$ is basis independent.

It is easy to check that $\left(\gamma^5\right)^2 = \mathbb{1}$ and $\{\gamma^5, \gamma^\mu\} = 0$. Thus γ^5 is like another γ-matrix, which is why we call it γ_5. This lets us formally extend the Clifford algebra to five generators, $\gamma^M = \gamma^0, \gamma^1, \gamma^2, \gamma^3, i\gamma^5$ so that $\{\gamma^M, \gamma^N\} = 2g^{MN}$ with $g^{MN} = \text{diag}(1, -1, -1, -1, -1)$. If we were looking at representations of the five-dimensional Lorentz group, we would use this extended Clifford algebra. See [Polchinski, 1998] for a discussion of spinors in various dimensions.

To understand the degrees of freedom in the spinor, let us focus on the free theory. In the Weyl basis, the Dirac equation is

$$\begin{pmatrix} -m & i\sigma^\mu \partial_\mu \\ i\bar{\sigma}^\mu \partial_\mu & -m \end{pmatrix} \begin{pmatrix} \psi_L \\ \psi_R \end{pmatrix} = 0. \tag{11.10}$$

In Fourier space, this implies

$$\sigma^\mu p_\mu \psi_R = (E - \vec{\sigma} \cdot \vec{p})\psi_R = m\psi_L, \tag{11.11}$$

$$\bar{\sigma}^\mu p_\mu \psi_L = (E + \vec{\sigma} \cdot \vec{p})\psi_L = m\psi_R. \tag{11.12}$$

So the mass mixes the left- and right-handed states.

In the absence of a mass, left- and right-handed states are eigenstates of the operator $\hat{h} = \frac{\vec{\sigma} \cdot \vec{p}}{|\vec{p}|}$ with opposite eigenvalue, since $E = |\vec{p}|$ for massless particles. This operator projects the spin on the momentum direction. Spin projected on the direction of motion is called the **helicity**, so the left- and right-handed states have opposite helicity in the massless theory.

When there is a mass, the left- and right-handed fields mix due to the equations of motion. However, since momentum and spin are good quantum numbers in the free theory, even with a mass, helicity is conserved as well. Therefore, helicity can still be a useful concept for the massive theory. The distinction is that, when there is a mass, helicity eigenstates are no longer the same as the chirality eigenstates ψ_L and ψ_R.

By the way, the independent solutions to the free equations of motion for massless particles of any spin are the helicity eigenstates. For any spin, we will always find $\vec{S} \cdot \vec{p}\, \Psi_s = \pm s|\vec{p}|\Psi_s$, where $\vec{S} = \vec{J}$ are the rotation generators in the Lorentz group for spin s. For spin $\frac{1}{2}$, $\vec{S} = \frac{1}{2}\vec{\sigma}$. For spin 1, the rotation generators are given in Eqs. (10.14). For example, J_3 has eigenvalues ± 1 with eigenstates $(0, -i, 1, 0)$ and $(0, i, 1, 0)$. These are the states of circularly polarized light in the z direction, which are helicity eigenstates. In general, the polarizations of massless particles with spin > 0 can always be taken to be helicity eigenstates. This is true for spin $\frac{1}{2}$ and spin 1, as we have seen; it is also true for gravitons (spin 2), Rarita–Schwinger fields (spin $\frac{3}{2}$) and spins $s > 2$ (although, as we proved in Section 9.5.1, it is impossible to have interacting theories with massless fields of spin $s > 2$).

We have seen that the left- and right-handed *chirality* states ψ_L and ψ_R

- do not mix under Lorentz transformations – they transform in separate irreducible representations.
- each have two components on which the $\vec{\sigma}$-matrices act. These are the two spin states of the electron; both left- and right-handed spinors have two spin states.
- are eigenstates of helicity in the massless limit.

We have now seen three different spin-related quantities:

Spin is used in two ways. We say a particle has spin s along the $\vec{v}$ axis if it is an eigenvector of $\vec{v} \cdot \vec{S}$ with eigenvalue s. Here, $\vec{S}$ is the spin operator (for example, $\vec{S} = \frac{\vec{\sigma}}{2}$ for a Weyl fermion). When we say a particle has spin s without referring to an axis we mean the particle has eigenvalue $s(s + 1)$ of the operator $\vec{S}^2$ (e.g. a Weyl fermion has spin $\frac{1}{2}$). The maximal value of the spin along any axis is the spin of the particle.

Helicity refers to the projection of spin on the direction of motion. Helicity eigenstates satisfy $\frac{\vec{S} \cdot \vec{p}}{s|\vec{p}|}\Psi = \pm\Psi$. Helicity eigenstates exist for any spin. The helicity eigenstates of the photon correspond to what we normally call circularly polarized light.

Chirality is a concept that only exists for spinors, or more precisely for (A, B) representations of the Lorentz group with $A \neq B$. You may remember the word chiral from chemistry: DNA is chiral, so is glucose and many organic molecules. These are not symmetric under reflection in a mirror. In field theory, a chiral theory is one that is not symmetric on interchange of the (A, B) representations with the (B, A) representations. Almost always, chirality means that a theory is not symmetric between left-handed Weyl spinors ψ_L and right-handed spinors ψ_R. These chiral spinors can also be written as Dirac spinors that are eigenstates of γ_5. By abuse of notation we also write ψ_L and ψ_R for Dirac spinors, with $\gamma_5\psi_L = -\psi_L$ and $\gamma_5\psi_R = \psi_R$. Whether a Weyl or Dirac spinor is meant by ψ_L and ψ_R will be clear from context. Chirality works for higher half-integer spins too. For example, a spin-$\frac{3}{2}$ field can be put in a Dirac spinor with a μ index, ψ_μ. Then $\gamma_5\psi_\mu = \pm\psi_\mu$ are the chirality eigenstates.

Whether spin, helicity or chirality is important depends on the physical question you are interested in. For free massless spinors, the spin eigenstates are also helicity eigenstates and chirality eigenstates. In other words, the Hamiltonian for the massless Dirac equation commutes with the operators for chirality, γ_5, helicity, $\frac{\vec{S} \cdot \vec{p}}{E}$, and the spin operators, $\vec{S}$. The QED interaction $\bar{\psi}\slashed{A}\psi = \bar{\psi}_L\slashed{A}\psi_L + \bar{\psi}_R\slashed{A}\psi_R$ is **non-chiral**, that is, it preserves chirality. Helicity, on the other hand, is not necessarily preserved by QED: if a left-handed spinor has its direction reversed by an electric field, its helicity flips. When particles are massless (or ultra-relativistic) they do not change direction so easily, but the helicity can flip due to an interaction.

In the massive case, it is also possible to take the non-relativistic limit. Then it is often better to talk about spin, the vector. Projecting on the direction of motion does not make so much sense when the particle is nearly at rest, or in a gas, say, when its direction of motion is constantly changing. The QED interactions do not preserve spin. However, only a strong magnetic field can flip an electron's spin, so as long as magnetic fields are weak, spin is a good quantum number. That is why spin is used in quantum mechanics.

In QED, we hardly ever talk about chirality. The word is basically reserved for chiral theories, which are theories that are not symmetric under $L \leftrightarrow R$, such as the theory of the weak interactions. We talk very often about helicity. In the high-energy limit, helicity is often used interchangeably with chirality. As a slight abuse of terminology, we say ψ_L and ψ_R are helicity eigenstates. In the non-relativistic limit, we use helicity for photons and spin (the vector) for spinors. Helicity eigenstates for photons are circularly polarized light.

11.2 Solving the Dirac equation

Now let us solve the free Dirac equation. Since spinors satisfy the Klein–Gordon equation, $(\Box + m^2)\psi = 0$ (in addition to the Dirac equation) they have plane-wave solutions:

$$\psi_s(x) = \int \frac{d^3p}{(2\pi)^3} u_s(p) e^{-ipx}, \tag{11.13}$$

with $p_0 = \sqrt{\vec{p}^2 + m^2} > 0$. These are like the solutions $A_\mu(x) = \int \frac{d^4p}{(2\pi)^4}\epsilon_\mu(p)e^{-ipx}$ for spin-1 plane waves. There are of course also solutions to $(\Box + m^2)\psi = 0$ with $p^0 < 0$. We will give these antiparticle interpretations, as in the complex scalar case (Chapter 9), and write

$$\chi_s(x) = \int \frac{d^3p}{(2\pi)^3} v_s(p) e^{ipx}, \tag{11.14}$$

also with $p_0 = \sqrt{\vec{p}^2 + m^2} > 0$. These are classical solutions, but the quantum versions will annihilate particles and create the appropriate positive-energy antiparticles. The spinors $u_s(p)$ and $\bar{v}_s(p)$ are the polarizations for particles and antiparticles, respectively. They transform under the Poincaré group through the transformation of p^μ and the little group that stabilizes p^μ. Thus, we only need to find explicit solutions for fixed p^μ, as we did for the spin-1 polarizations.

To find the spinor solutions, we use the Dirac equation in the Weyl basis:

$$\begin{pmatrix} -m & p\cdot\sigma \\ p\cdot\bar{\sigma} & -m \end{pmatrix} u_s(p) = \begin{pmatrix} -m & -p\cdot\sigma \\ -p\cdot\bar{\sigma} & -m \end{pmatrix} v_s(p) = 0. \tag{11.15}$$

In the rest frame, $p^\mu = (m, 0, 0, 0)$ and the equations of motion reduce to

$$\begin{pmatrix} -1 & 1 \\ 1 & -1 \end{pmatrix} u_s = \begin{pmatrix} -1 & -1 \\ -1 & -1 \end{pmatrix} v_s = 0. \tag{11.16}$$

So, solutions are constants:

$$u_s = \begin{pmatrix} \xi_s \\ \xi_s \end{pmatrix}, \quad v_s = \begin{pmatrix} \eta_s \\ -\eta_s \end{pmatrix}, \tag{11.17}$$

for any two-component spinors ξ_s and η_s. For example, four linearly independent solutions are

$$u_\uparrow = \sqrt{m} \begin{pmatrix} 1 \\ 0 \\ 1 \\ 0 \end{pmatrix}, \quad u_\downarrow = \sqrt{m} \begin{pmatrix} 0 \\ 1 \\ 0 \\ 1 \end{pmatrix}, \quad v_\uparrow = \sqrt{m} \begin{pmatrix} -1 \\ 0 \\ 1 \\ 0 \end{pmatrix}, \quad v_\downarrow = \sqrt{m} \begin{pmatrix} 0 \\ 1 \\ 0 \\ -1 \end{pmatrix}. \tag{11.18}$$

The Dirac spinor is a complex four-component object, with eight real degrees of freedom. The equations of motion reduce it to four degrees of freedom, which, as we will see, can be interpreted as spin up and spin down for particle and antiparticle.

To derive a more general expression, we can solve the equations again in the boosted frame and match the normalization. If $p^\mu = (E, 0, 0, p_z)$ then

$$p \cdot \sigma = \begin{pmatrix} E - p_z & 0 \\ 0 & E + p_z \end{pmatrix}, \quad p \cdot \bar\sigma = \begin{pmatrix} E + p_z & 0 \\ 0 & E - p_z \end{pmatrix}. \tag{11.19}$$

Let $a = \sqrt{E - p_z}$ and $b = \sqrt{E + p_z}$, then $m^2 = (E - p_z)(E + p_z) = a^2 b^2$ and Eq. (11.15) becomes

$$\begin{pmatrix} -ab & 0 & a^2 & 0 \\ 0 & -ab & 0 & b^2 \\ b^2 & 0 & -ab & 0 \\ 0 & a^2 & 0 & -ab \end{pmatrix} u_s(p) = 0. \tag{11.20}$$

The solutions are

$$u_s = \begin{pmatrix} \begin{pmatrix} a & 0 \\ 0 & b \end{pmatrix} \xi_s \\ \begin{pmatrix} b & 0 \\ 0 & a \end{pmatrix} \xi_s \end{pmatrix} \tag{11.21}$$

for any two-component spinor ξ_s. Note that in the rest frame $p_z = 0$, $a^2 = b^2 = m$, and these solutions reduce to Eq. (11.17) above. The solutions in the p_z frame are

$$u_s(p) = \begin{pmatrix} \begin{pmatrix} \sqrt{E - p_z} & 0 \\ 0 & \sqrt{E + p_z} \end{pmatrix} \xi_s \\ \begin{pmatrix} \sqrt{E + p_z} & 0 \\ 0 & \sqrt{E - p_z} \end{pmatrix} \xi_s \end{pmatrix}. \tag{11.22}$$

Similarly,

$$v_s(p) = \begin{pmatrix} \begin{pmatrix} \sqrt{E-p_z} & 0 \\ 0 & \sqrt{E+p_z} \end{pmatrix} \eta_s \\ \begin{pmatrix} -\sqrt{E+p_z} & 0 \\ 0 & -\sqrt{E-p_z} \end{pmatrix} \eta_s \end{pmatrix}. \tag{11.23}$$

Using

$$\sqrt{p \cdot \sigma} = \begin{pmatrix} \sqrt{E-p_z} & 0 \\ 0 & \sqrt{E+p_z} \end{pmatrix}, \quad \sqrt{p \cdot \bar{\sigma}} = \begin{pmatrix} \sqrt{E+p_z} & 0 \\ 0 & \sqrt{E-p_z} \end{pmatrix} \tag{11.24}$$

we can write more generally

$$u_s(p) = \begin{pmatrix} \sqrt{p \cdot \sigma}\, \xi_s \\ \sqrt{p \cdot \bar{\sigma}}\, \xi_s \end{pmatrix}, \quad v_s(p) = \begin{pmatrix} \sqrt{p \cdot \sigma}\, \eta_s \\ -\sqrt{p \cdot \bar{\sigma}}\, \eta_s \end{pmatrix}, \tag{11.25}$$

where the square root of a matrix can be defined by changing to the diagonal basis, taking the square root of the eigenvalues, then changing back to the original basis. In practice, we will usually pick p^μ along the z axis, so we do not need to know how to make sense of $\sqrt{p \cdot \sigma}$. Then the four solutions are

$$u_p^1 = \begin{pmatrix} \sqrt{E-p_z} \\ 0 \\ \sqrt{E+p_z} \\ 0 \end{pmatrix}, \quad u_p^2 = \begin{pmatrix} 0 \\ \sqrt{E+p_z} \\ 0 \\ \sqrt{E-p_z} \end{pmatrix}, \quad v_p^1 = \begin{pmatrix} \sqrt{E-p_z} \\ 0 \\ -\sqrt{E+p_z} \\ 0 \end{pmatrix}, \quad v_p^2 = \begin{pmatrix} 0 \\ \sqrt{E+p_z} \\ 0 \\ -\sqrt{E-p_z} \end{pmatrix}. \tag{11.26}$$

In any frame u^s are the positive frequency electrons, and the v^s are negative frequency electrons, or equivalently, positive frequency positrons.

For massless spinors, $p_z = \pm E$ and the explicit solutions in Eq. (11.26) are 4-vectors with one non-zero component describing spinors with fixed helicity. The spinor solutions for massless electrons are sometimes called polarizations, and are useful for computing polarized electron scattering amplitudes.

For Weyl spinors, there are only four real degrees of freedom off-shell and two real degrees of freedom on-shell. Recalling that the Dirac equation splits up into separate equations for ψ_L and ψ_R, the Dirac spinors with zeros in the bottom two rows will be ψ_L and those with zeros in the top two rows will be ψ_R. Since ψ_L and ψ_R have two degrees of freedom each, these must be particle and antiparticle for the same helicity. The embedding of Weyl spinors into fields this way induces irreducible unitary representations of the Poincaré group for $m = 0$.

11.2.1 Normalization and spin sums

To figure out what the normalization is that we have implicitly chosen, let us compute the inner product:

$$\bar{u}_s(p)u_{s'}(p) = u_s^\dagger(p)\gamma_0 u_{s'}(p) = \begin{pmatrix} \sqrt{p\cdot\sigma}\,\xi_s \\ \sqrt{p\cdot\bar{\sigma}}\,\xi_s \end{pmatrix}^\dagger \begin{pmatrix} 0 & 1 \\ 1 & 0 \end{pmatrix} \begin{pmatrix} \sqrt{p\cdot\sigma}\,\xi_{s'} \\ \sqrt{p\cdot\bar{\sigma}}\,\xi_{s'} \end{pmatrix}$$

$$= \begin{pmatrix} \xi_s \\ \xi_s \end{pmatrix}^\dagger \begin{pmatrix} \sqrt{(p\cdot\sigma)\,(p\cdot\bar{\sigma})} & \\ & \sqrt{(p\cdot\sigma)\,(p\cdot\bar{\sigma})} \end{pmatrix} \begin{pmatrix} \xi_{s'} \\ \xi_{s'} \end{pmatrix}$$

$$= 2m\delta_{ss'}. \tag{11.27}$$

Similarly, $\bar{v}_s(p)v_{s'}(p) = -2m\delta_{ss'}$. This is the (conventional) normalization for the **spinor inner product** for massive Dirac spinors. It is also easy to check that $\bar{v}_s(p)u_{s'}(p) = \bar{u}_s(p)v_{s'}(p) = 0$.

We can also calculate

$$u_s^\dagger(p)u_{s'}(p) = \begin{pmatrix} \sqrt{p\cdot\sigma}\,\xi_s \\ \sqrt{p\cdot\bar{\sigma}}\,\xi_s \end{pmatrix}^\dagger \begin{pmatrix} \sqrt{p\cdot\sigma}\,\xi_{s'} \\ \sqrt{p\cdot\bar{\sigma}}\,\xi_{s'} \end{pmatrix} = 2E\xi_s^\dagger\xi_{s'} = 2E\delta_{ss'}, \tag{11.28}$$

and similarly, $v_s^\dagger(p)v_{s'}(p) = 2E\delta_{ss'}$. This is the conventional normalization for massless Dirac spinors. Another useful relation is that, if we define $\bar{p}^\mu = (E_p, -\vec{p})$ as a momentum backwards to p^μ, then $v_s^\dagger(p)u_{s'}(\bar{p}) = u_s^\dagger(p)v_{s'}(\bar{p}) = 0$.

We can also compute the **spinor outer product**:

$$\sum_{s=1}^{2} u_s(p)\bar{u}_s(p) = \slashed{p} + m, \tag{11.29}$$

where the sum is over the spins. Both sides of this equation are matrices. It may help to think of this equation as $\sum_s |s\rangle\langle s|$. For the antiparticles,

$$\sum_{s=1}^{2} v_s(p)\bar{v}_s(p) = \slashed{p} - m. \tag{11.30}$$

You should verify these relations on your own (see Problem 11.2).

To keep straight the inner and outer products, it may be helpful to compare to spin-1 particles. We have found

$$\langle s|s'\rangle: \qquad \epsilon_\mu^{i\star}(p)\epsilon_\mu^j(p) = -\delta^{ij} \qquad\qquad \leftrightarrow \quad \bar{u}_s(p)u_{s'}(p) = 2m\delta_{ss'}, \tag{11.31}$$

$$\sum_s |s\rangle\langle s|: \quad \sum_{i=1}^{3} \epsilon_i^\mu(p)\epsilon_i^{\star\nu}(p) = -g^{\mu\nu} + \frac{p^\mu p^\nu}{m^2} \quad \leftrightarrow \quad \sum_{s=1}^{2} u_s(p)\bar{u}_s(p) = \slashed{p} + m. \tag{11.32}$$

So, when we sum over internal spin indices, we use an inner product and get a number. When we sum over polarizations/spins, we get a matrix.

$$u_\uparrow = \sqrt{m} \begin{pmatrix} 1 \\ 0 \\ 1 \\ 0 \end{pmatrix}, \quad u_\downarrow = \sqrt{m} \begin{pmatrix} 0 \\ 1 \\ 0 \\ 1 \end{pmatrix}, \quad v_\uparrow = \sqrt{m} \begin{pmatrix} -1 \\ 0 \\ 1 \\ 0 \end{pmatrix}, \quad v_\downarrow = \sqrt{m} \begin{pmatrix} 0 \\ 1 \\ 0 \\ -1 \end{pmatrix}.$$

$$(11.42)$$

Then

$$(u_\uparrow)^c = -i\gamma_2 \sqrt{m} \begin{pmatrix} 1 \\ 0 \\ 1 \\ 0 \end{pmatrix}^\star = -i\sqrt{m} \begin{pmatrix} 0 & 0 & 0 & -i \\ 0 & 0 & i & 0 \\ 0 & i & 0 & 0 \\ -i & 0 & 0 & 0 \end{pmatrix} \begin{pmatrix} 1 \\ 0 \\ 1 \\ 0 \end{pmatrix} = \sqrt{m} \begin{pmatrix} 0 \\ 1 \\ 0 \\ -1 \end{pmatrix} = v_\downarrow$$

$$(11.43)$$

and so on, giving

$$(u_\uparrow)^c = v_\downarrow, \quad (u_\downarrow)^c = v_\uparrow, \quad (v_\uparrow)^c = u_\downarrow, \quad (v_\downarrow)^c = u_\uparrow. \tag{11.44}$$

Thus, charge conjugation takes particles to antiparticles and flips the spin. In particular, invariance under C of a theory constrains how different spin states interact.

Charge conjugation may or may not be a symmetry of a particular Lagrangian. The operation of charge conjugation acts on spinors and their conjugates by

$$C: \quad \psi \to -i\gamma_2 \psi^\star. \tag{11.45}$$

In the Weyl basis, $\gamma_2^\star = -\gamma_2$ and $\gamma_2^T = \gamma_2$, so

$$C: \quad \psi^\star \to -i\gamma_2 \psi, \tag{11.46}$$

and in particular $C^2 = 1$, which is why C is called a *conjugation* operator. Then

$$C: \quad \bar{\psi}\psi \to (-i\gamma_2\psi)^T \gamma_0 (-i\gamma_2\psi^\star) = -\psi^T \gamma_2^T \gamma_0 \gamma_2 \psi^\star = -\psi^T \gamma_0 \psi^\star. \tag{11.47}$$

The transpose on a spinor is not really necessary. This last expression just means

$$-\psi^T \gamma_0 \psi^\star = -(\gamma_0)_{\alpha\beta} \psi_\alpha \psi_\beta^\star. \tag{11.48}$$

Now, anticommuting the spinors, relabeling $\alpha \leftrightarrow \beta$ and combining shows that

$$-(\gamma_0)_{\alpha\beta} \psi_\alpha \psi_\beta^\star = (\gamma_0)_{\alpha\beta} \psi_\beta^\star \psi_\alpha = (\gamma_0)_{\beta\alpha} \psi_\alpha^\star \psi_\beta = \psi^\dagger \gamma_0^T \psi = \bar{\psi}\psi. \tag{11.49}$$

Thus,

$$C: \quad \bar{\psi}\psi \to \bar{\psi}\psi. \tag{11.50}$$

Similarly,

$$C: \quad \bar{\psi}\slashed{\partial}\psi \to \bar{\psi}\slashed{\partial}\psi. \tag{11.51}$$

So the free Dirac Lagrangian is C invariant.

We can also check that

$$C: \quad \bar{\psi}\gamma^\mu\psi \to -\bar{\psi}\gamma^\mu\psi. \tag{11.52}$$

This implies that the interaction $eA_\mu\bar{\psi}\gamma^\mu\psi$ will only be C invariant if

$$C: \quad A_\mu \to -A_\mu. \tag{11.53}$$

Since the kinetic term $F_{\mu\nu}^2$ is invariant under $A_\mu \to \pm A_\mu$, the whole QED Lagrangian is therefore C invariant.

The transformation $A_\mu \to -A_\mu$ under C may seem strange, since a vector field is real, so it should not transform under an operation that switches particles with antiparticles. Since particles and antiparticles have opposite charge and A_μ couples proportionally to charge, this transformation is needed to compensate for the transformation of the charged fields.

There is an important lesson here: you *could* take $C : A_\mu \to A_\mu$, but then the Lagrangian would not be invariant. Thus, rather than trying to figure out how C acts, the right question is: How can we enlarge the action of the transformation C, which we know for Dirac spinors, to a full interacting theory so that the symmetry is preserved? Whether we interpret C with the words "takes particles to antiparticles," has no physical implications. In contrast, a symmetry of a theory does have physical implications: preservation of the symmetry gives a selection rule – certain transitions cannot happen. An important example is that C invariance forces matrix elements involving an odd number of photons to vanish, a result known as Furry's theorem (see Problem 14.2). Thus, cataloging the symmetries of a theory is important, whether or not we have interesting names or simple physical interpretations of those symmetries.

For future reference, it is also true that

$$C: \quad i\bar\psi\gamma^5\psi \to i\bar\psi\gamma^5\psi, \tag{11.54}$$

$$C: \quad i\bar\psi\gamma^5\gamma^\mu\psi \to i\bar\psi\gamma^5\gamma^\mu\psi, \tag{11.55}$$

$$C: \quad \bar\psi\sigma^{\mu\nu}\psi \to -\bar\psi\sigma^{\mu\nu}\psi, \tag{11.56}$$

which you can prove in Problem 11.5.

11.5 Parity

Recall that the full Lorentz group, $O(1, 3)$, is the group of 4×4 matrices Λ with $\Lambda^T g \Lambda = g$. In addition to the transformations smoothly connected to $\mathbb{1}$, this group also contains the transformations of parity and time reversal:

$$P: \quad (t, \vec{x}) \to (t, -\vec{x}), \tag{11.57}$$

$$T: \quad (t, \vec{x}) \to (-t, \vec{x}). \tag{11.58}$$

Just as with charge conjugation, we would like to know how to define these transformations acting on spinors, and other fields, so that they are symmetries of QED or whatever theory we are studying.

You might expect that the action of P and T should be determined from representation theory. However, recall that technically spinors do not actually transform under the Lorentz group, $O(1, 3)$, only its universal cover, $SL(2, \mathbb{C})$, so we are not guaranteed that T and P will act in any nice way on irreducible spinor representations. In fact they do not. Although we can *define* an action of T and P on spinors (and other fields), these definitions are only useful to the extent that they are symmetries of the theory we are interested in. For example, we will define P so that it is a good symmetry of QED, but there is no way to define it so that it is preserved under the weak interactions. In any representation, we should have $P^2 = T^2 = 1$.

11.5.1 Scalars and vectors

For real scalars, parity should be a symmetry of the kinetic terms $\mathcal{L} = -\frac{1}{2}\phi\Box\phi - \frac{1}{2}m^2\phi^2$ or we are dead in the water. Thus, $P^2 = 1$ (we do not need to use $P^2 = 1$ in the Lorentz group for this argument) and there are two choices:

$$P: \quad \phi(t,\vec{x}) \to \pm\phi(t,-\vec{x}). \tag{11.59}$$

The sign is known as the **intrinsic parity** of a particle. In nature, there are particles with even parity (scalars, such as the Higgs boson) and particles with odd parity (pseudoscalars, such as the π^0). Since the action integrates over all $\vec{x}$, we can change $\vec{x} \to -\vec{x}$ and the action will be invariant.

For complex scalars, the free theory has Lagrangian $\mathcal{L} = -\phi^\star\Box\phi - m^2\phi^\star\phi$, so the most general possibility is

$$P: \quad \phi(t,\vec{x}) \to \eta\,\phi(t,-\vec{x}), \tag{11.60}$$

where η is a pure phase. Note that $P^2 : \phi(t,\vec{x}) \to \eta^2\phi(t,\vec{x})$, so P^2 is an internal symmetry (not acting on spacetime). In principle, the parity phase η can be arbitrary for each particle.

Consider scalar QED with a bunch of particles of different charges. It has a continuous global symmetry under which $\phi \to e^{i\alpha Q}\phi$ for any $\alpha \in \mathbb{R}$ with Q the charge of ϕ. If P^2 is a subgroup of this global charge symmetry, then $\eta^2 = e^{i\beta Q}$. Thus the discrete symmetry $(P')^2 = P^2 e^{-i\beta Q} = \mathbb{1}$ and $P' = Pe^{-i\frac{\beta}{2}Q} : \phi(t,\vec{x}) \to \pm\phi(t,-\vec{x})$ for all particles. Thus, we might as well take the convention that P' is called parity rather than P, and therefore all particles in this theory have parity ± 1.

In the Standard Model, there are three continuous global symmetries: lepton number L (leptons have $L = 1$, everything else has $L = 0$), baryon number B (quarks have $B = \frac{1}{3}$, everything else has $B = 0$) and electric charge Q. If P^2 is a subgroup of the product of these, then $P^2 = e^{i(\alpha B + \beta L + \gamma Q)}$ for some α, β, and γ. We then define $P' = Pe^{-\frac{i}{2}(\alpha B + \beta L + \gamma Q)}$ so that all particles have parity phases of ± 1. Moreover, if the proton has parity -1, we can redefine again by $P'' = (-1)^B P'$ so that proton has parity $+1$. Similarly using up L and Q, we get to pick three parity phases in total, which conventionally are that the proton, neutron and electron all have parity $+1$. See [Weinberg, 1995, p.125] for more details of this argument.

From nuclear physics measurements, it was deduced that the pion, π^0, and its charged siblings, π^+ and π^-, all have parity -1. Then it was very strange to find that a particle called the kaon, K^+, decayed to both two pions and three pions. People thought for a while that the kaon was two particles, the θ^+ (with parity $+1$, which decayed to two pions) and the τ^+ (with parity -1, which decayed to three pions). Lee and Yang finally figured out, in 1956, that these were the same particle, and that parity was not conserved in kaon decays.

For vector fields, P acts as it does on 4-vectors. However, for the free vector theory to be invariant, we only require that

$$P: \quad V_0(t,\vec{x}) \to \pm V_0(t,-\vec{x}), \qquad V_i(t,\vec{x}) \to \mp V_i(t,-\vec{x}). \tag{11.61}$$

The notation is that if $P : V_i \to -V_i$, like $\vec{x}$, we say V_μ has parity -1 and call it a vector. If $P : V_i \to V_i$, we call it a pseudovector, with parity $+1$. For example, the ρ meson is a vector and the a_1 meson is a pseudovector. You have already seen pseudovectors in three dimensions: the electric field is a vector that flips sign under parity, while the magnetic field is a pseudovector that remains invariant under parity.

Massless vectors such as the photon have to have parity -1. To see this, just look at the coupling to a charged scalar. Under parity we would like

$$P: \quad A_\mu (\phi^\star \partial_\mu \phi - \phi \partial_\mu \phi^\star) \to A_\mu (\phi^\star \partial_\mu \phi - \phi \partial_\mu \phi^\star) , \tag{11.62}$$

which is only possible if A_μ transforms like ∂_μ. That is, A_μ is a vector:

$$P: \quad A_0(t, \vec{x}) \to A_0(t, -\vec{x}), \qquad A_i(t, \vec{x}) \to -A_i(t, -\vec{x}). \tag{11.63}$$

11.5.2 Spinors

Now let us turn to spinors. In the Lorentz group, P commutes with the rotations. Thus, P does not change the spin of a state embedded in a vector field. This should be true for spinors too. For massless spinors, recall that left- and right-handed spinors are eigenstates of the helicity operator, which projects spin onto the momentum axis:

$$\frac{\vec{\sigma} \cdot \vec{p}}{|\vec{p}|} \psi_R = \psi_R, \qquad \frac{\vec{\sigma} \cdot \vec{p}}{|\vec{p}|} \psi_L = -\psi_L. \tag{11.64}$$

Since parity commutes with spin, $\vec{\sigma}$, and energy but flips the momentum, it will take left-handed spinors to right-handed spinors. That is, it will map $\left(\frac{1}{2}, 0\right)$ representations to $\left(0, \frac{1}{2}\right)$. Therefore, P cannot be appended to either of the spin-$\frac{1}{2}$ irreducible representations alone.

For Dirac spinors, which comprise left- and right-handed spinors, we can see that parity just swaps left and right, keeping the spin invariant. In the Weyl basis, this transformation can be written in the simple form

$$P : \psi \to \gamma_0 \psi. \tag{11.65}$$

There is in principle a phase ambiguity here, as for charged scalars. But, as in that case, we can use invariance under global phase rotations, associated with charge conservation, to simply choose this phase to be 1, as we have done here. Despite this phase, a chiral theory (one with no symmetry under $L \leftrightarrow R$), such as the theory of weak interactions, cannot be invariant under parity.

Note that

$$P: \quad \bar{\psi}\psi(t, \vec{x}) \to \psi^\dagger \gamma_0 \gamma_0 \gamma_0 \psi(t, -\vec{x}) = \bar{\psi}\psi(t, -\vec{x}), \tag{11.66}$$

$$P: \quad \bar{\psi}\gamma_\mu \psi (t, \vec{x}) \to \psi^\dagger \gamma_0 \gamma_0 \gamma_\mu \gamma_0 \psi(t, -\vec{x}) = \bar{\psi}\gamma_\mu^\dagger \psi(t, -\vec{x}). \tag{11.67}$$

Recalling that $\gamma_0^\dagger = \gamma_0$ and $\gamma_i^\dagger = -\gamma_i$, we see that

$$P: \quad \bar{\psi}\gamma_0 \psi(t, \vec{x}) \to \bar{\psi}\gamma_0 \psi(t, -\vec{x}), \qquad \bar{\psi}\gamma_i \psi(t, \vec{x}) \to -\bar{\psi}\gamma_i \psi(t, -\vec{x}), \tag{11.68}$$

so that $\bar{\psi}\gamma_\mu\psi$ transforms exactly as a 4-vector and hence the Dirac Lagrangian is parity invariant. The parity transformations are opposite for bilinears with γ^5:

$$P: \quad \bar{\psi}\gamma_0\gamma^5\psi \to -\bar{\psi}\gamma_0\gamma^5\psi(t,-\vec{x}), \qquad \bar{\psi}\gamma_i\gamma^5\psi \to \bar{\psi}\gamma_i\gamma^5\psi(t,-\vec{x}), \qquad (11.69)$$

so that

$$P: \quad \bar{\psi}\slashed{A}\psi \to \bar{\psi}\slashed{A}\psi(t,-\vec{x}), \qquad\qquad\qquad (11.70)$$

$$P: \quad \bar{\psi}\slashed{A}\gamma^5\psi \to -\bar{\psi}\slashed{A}\gamma^5\psi(t,-\vec{x}). \qquad\qquad (11.71)$$

The currents contracted with A^μ in these terms are known as the **vector current**, $J_V^\mu = \bar{\psi}\gamma^\mu\psi$, and the **axial vector current**, $J_A^\mu = \bar{\psi}\gamma^\mu\gamma^5\psi$. These currents play a crucial role in the theory of weak interactions, which involves $J_V^\mu - J_A^\mu$, or the $V - A$ current.

11.6 Time reversal

Finally, let us turn to the most confusing of the discrete symmetries, time reversal. As a Lorentz transformation,

$$T: \quad (t,\vec{x}) \to (-t,\vec{x}). \qquad\qquad\qquad (11.72)$$

We are going to need a transformation of our spinor fields, ψ, such that (at least) the kinetic Lagrangian is invariant. To do this, we need $\bar{\psi}\gamma^\mu\psi$ to transform as a 4-vector under T, so that $i\bar{\psi}\slashed{\partial}\psi(t,\vec{x}) \to i\bar{\psi}\slashed{\partial}\psi(-t,x)$ and the action will be invariant. In particular, we need the 0-component, $\bar{\psi}\gamma^0\psi \to -\bar{\psi}\gamma^0\psi$, which implies $\psi^\dagger\psi \to -\psi^\dagger\psi$. But this last form of the requirement is very odd – it says we need to turn a positive definite quantity into a negative definite quantity. This is impossible for any linear transformation $\psi \to \Gamma\psi$. Thus, we need to think harder.

We will discuss two possibilities. One we will call "simple $\hat{T}$," and denote $\hat{T}$. It is the obvious parallel to parity. The other is the T symmetry, which is normally what is meant by T in the literature. This second T was invented by Wigner in 1932 and requires T to take $i \to -i$ in the whole Lagrangian in addition to acting on fields. While the simple $\hat{T}$ is the more natural generalization of the action of T on 4-vectors, it is also kind of trivial. Wigner's T has important physical implications.

11.6.1 The simple $\hat{T}$

Before doing anything drastic, the simplest thing besides $T: \psi \to \Gamma\psi$ would be $T: \psi \to \Gamma\psi^\star$, as with charge conjugation. We will call this transformation $\hat{T}$ to distinguish it from what is conventionally called T in the literature. So,

$$\hat{T}: \quad \psi \to \Gamma\psi^\star, \quad \psi^\dagger \to (\Gamma\psi^\star)^\dagger = \psi^T\Gamma^\dagger. \qquad (11.73)$$

That $\hat{T}$ should take particles to antiparticles is also understandable from the picture of antiparticles as particles moving backwards in time.

Then,

$$\psi^\dagger\psi \to \psi^T\Gamma^\dagger\Gamma\psi^\star = \Gamma^\dagger_{\alpha\beta}\Gamma_{\beta\gamma}\psi_\alpha\psi^\star_\gamma = -\psi^\star_\gamma\Gamma^T_{\gamma\beta}\Gamma^{\dagger T}_{\beta\alpha}\psi_\alpha = -\psi^\dagger(\Gamma^\dagger\Gamma)^T\psi, \quad (11.74)$$

so we need $\Gamma^\dagger\Gamma = \mathbb{1}$, which says that Γ is a unitary matrix. That is fine. But we also need $\bar\psi\gamma_i\psi$ and the mass term $\bar\psi\psi$ to be preserved. For the mass term,

$$\bar\psi\psi \to \psi^T\Gamma^\dagger\gamma_0\Gamma\psi^\star = -\bar\psi(\Gamma^\dagger\gamma_0\Gamma\gamma_0)^T\psi \quad (11.75)$$

This equals $\bar\psi\psi$ only if $\{\Gamma,\gamma_0\} = 0$. Next,

$$\bar\psi\gamma_i\psi \to \psi^T\Gamma^\dagger\gamma_0\gamma_i\Gamma\psi^\star = -\bar\psi(\Gamma^\dagger\gamma_0\gamma_i\Gamma\gamma_0)^T\psi = -\bar\psi(\Gamma^\dagger\gamma_i\Gamma)^T\psi, \quad (11.76)$$

which should be equal to $\bar\psi\gamma_i\psi$ for $i=1,2,3$. So $\gamma_i\Gamma + \Gamma\gamma_i^T = 0$, which implies $[\Gamma,\gamma_1] = 0$, $[\Gamma,\gamma_3] = 0$ and $\{\Gamma,\gamma_2\} = 0$. The unique (up to a constant) matrix that commutes with γ_1 and γ_3 and anticommutes with γ_2 and γ_0 is $\Gamma = \gamma_0\gamma_2$. Thus,

$$\psi(t,\vec x) \to \gamma_0\gamma_2\psi^\star(-t,\vec x), \quad \psi^\dagger(t,\vec x) \to -\psi^T\gamma_2\gamma_0(-t,\vec x). \quad (11.77)$$

Note that this is very similar to $P \cdot C$. On vectors, we should have

$$\hat T: \quad A_0(t,\vec x) \to -A_0(-t,\vec x), \quad A_i(t,\vec x) \to A_i(-t,\vec x), \quad (11.78)$$

so that the interaction $\bar\psi\slashed A\psi$ in the action is invariant.

Now consider the action of $C \cdot P \cdot \hat T$. This sends

$$C P\hat T: \quad \psi(t,\vec x) \to -i\gamma_2(\gamma_0[\gamma_0\gamma_2\psi^\star])^\star(-t,-\vec x) = -i\psi(-t,-\vec x) \quad (11.79)$$

and so

$$C P\hat T: \quad \bar\psi(t,\vec x)\gamma^\mu\psi(t,\vec x) \to \bar\psi(-t,-\vec x)\gamma^\mu\psi(-t,\quad\vec x), \quad (11.80)$$

$$C P\hat T: \quad A_\mu(t,\vec x) \to A_\mu(-t,-\vec x). \quad (11.81)$$

$C P\hat T$ also sends $\partial_\mu \to -\partial_\mu$. Thus, $\bar\psi\psi$, $i\bar\psi\gamma^\mu\partial_\mu\psi$ and $\bar\psi A_\mu\gamma^\mu\psi$ are all invariant in the action.

This time reversal is essentially defined to be $\hat T \sim (CP)^{-1}$, which makes $C P\hat T$ invariance trivial. The actual CPT theorem concerns a different T symmetry, which we will now discuss.

11.6.2 Wigner's T (i.e. what is normally called T)

What is normally called time reversal is a symmetry T that was described in a 1932 paper by Wigner, and shown to be an explanation of Kramer's degeneracy. To understand Kramer's degeneracy, consider the Schrödinger equation,

$$i\partial_t\psi(t,\vec x) = H\psi(t,\vec x), \quad (11.82)$$

where, for simplicity, let us say $H = \frac{p^2}{2m} + V(x)$, which is real and time independent. If we take the complex conjugate of this equation and also $t \to -t$, we find

$$i\partial_t \psi^\star(-t, \vec{x}) = H\psi^\star(-t, \vec{x}). \tag{11.83}$$

Thus, $\psi'(t, \vec{x}) = \psi^\star(-t, \vec{x})$ is another solution to the Schrödinger equation. If ψ is an energy eigenstate, then as long as $\psi \neq \xi\psi^\star$ for any complex number ξ, ψ' will be another state with the same energy. This doubling of states at each energy is known as **Kramer's degeneracy**. In particular, for the hydrogen atom, $\psi_{nlm}(\vec{x}) = R_n(r)Y_{lm}(\theta, \phi)$ are the energy eigenstates, so Kramer's degeneracy says that the states with m and $-m$ will be degenerate (which they are). The importance of this theorem is that it also holds for more complicated systems, and for systems in external electric fields, for which the exact eigenstates may not be known.

As we will soon see, this mapping, $\psi(t, \vec{x}) \to \psi^\star(-t, \vec{x})$, sends particles to particles (not antiparticles), unlike the simple $\hat{T}$ operator above. It has a nice interpretation: Suppose you made a movie of some physics process, then watched the movie backwards; time reversal implies you should not be able to tell which was "play" and which was "reverse."

The trick to Wigner's T is that we had to complex conjugate and *then* take $\psi' = \psi^\star$. This means in particular that the i in the Schrödinger equation goes to $-i$ as well as the field transforming. This is the key to finding a way out of the problem that $\psi^\dagger\psi$ needed to flip sign under T, which we discussed at the beginning of the section. The kinetic term for ψ is $i\bar{\psi}\gamma^0\partial_0\psi$; so if $i \to -i$ then, since $\partial_0 \to -\partial_0$, $\psi^\dagger\psi$ can be invariant. Thus we need

$$T: \quad i \to -i. \tag{11.84}$$

This makes T an **anti-linear** operator. What that means is that if we write any object on which T acts as a real plus an imaginary part $\psi = \psi_1 + i\psi_2$, with ψ_1 and ψ_2 real, then $T(\psi_1 + i\psi_2) = T\psi_1 - iT\psi_2$.

Since T changes all the factors of i in the Lagrangian to $-i$, it also affects the γ-matrices. In the Weyl basis, only γ_2 is imaginary, so

$$T: \quad \gamma_{0,1,3} \to \gamma_{0,1,3}, \quad \gamma_2 \to -\gamma_2. \tag{11.85}$$

For a real spinor, T is simply linear, so we can write its action as

$$T: \quad \psi(t, \vec{x}) \to \tilde{\Gamma}\psi(-t, \vec{x}), \tag{11.86}$$

with $\tilde{\Gamma}$ a Dirac matrix. Then, for $i\bar{\psi}\gamma^\mu\partial_\mu\psi$ to be invariant, we need $\bar{\psi}\gamma^0\psi$ to be invariant and $\bar{\psi}\gamma^i\psi \to -\bar{\psi}\gamma^i\psi$. Thus,

$$[\tilde{\Gamma}, \gamma_0] = \{\tilde{\Gamma}, \gamma_1\} = [\tilde{\Gamma}, \gamma_2] = \{\tilde{\Gamma}, \gamma_3\} = 0. \tag{11.87}$$

The only element of the Dirac algebra that satisfies these constraints is $\tilde{\Gamma} = \gamma_1\gamma_3$, up to a constant. Thus, we take

$$T: \quad \psi(t, \vec{x}) \to \gamma_1\gamma_3\psi(-t, \vec{x}) = \begin{pmatrix} 0 & 1 & & \\ -1 & 0 & & \\ & & 0 & 1 \\ & & -1 & 0 \end{pmatrix} \psi(-t, \vec{x}). \tag{11.88}$$

Thus, T flips the spins of particles, but does not turn particles into antiparticles, as expected. T does not have a well-defined action on Weyl spinors, which have one spin state. T also reverses the momenta, $\vec{p} = i\vec{\nabla}$, because of the i. Thus, T makes it look like things are going forwards in time, but with their momenta and spins flipped.

Similarly, for $\bar{\psi}\not{D}\psi$ to be invariant, we need A_μ to transform as $i\partial_\mu$, which is

$$T: \quad A_0(t,\vec{x}) \to A_0(-t,\vec{x}), \quad A_i(t,\vec{x}) \to -A_i(-t,\vec{x}). \tag{11.89}$$

It is straightforward to check now that the Dirac Lagrangian is invariant under T.

Next, consider the combined operation of CPT. This sends particles into antiparticles moving as if you watched them in reverse in a mirror. On Dirac spinors, it acts as

$$C \cdot P \cdot T: \quad \psi(x) \to -i\gamma_2\gamma_0\gamma_1\gamma_3\psi^\star(-x) = -\gamma_5\psi^\star(-x). \tag{11.90}$$

It also sends $A_\mu(x) \to -A_\mu(-x)$, $\phi \to \phi^\star(-x)$ and of course $i \to -i$.

You can check (Problem 11.7) that any terms you could possibly write down, for example,

$$\mathcal{L} = \bar{\psi}\psi, \quad i\bar{\psi}\not{\partial}\psi, \quad \bar{\psi}\not{A}\psi, \quad \bar{\psi}\gamma^\mu\gamma^5\psi W_\mu, \quad \bar{\psi}\sigma^{\mu\nu}\psi F_{\mu\nu} \tag{11.91}$$

and so on, are all CPT invariant. The **CPT theorem** says that this is a consequence of Lorentz invariance and unitarity. A rigorous mathematical proof of the CPT theorem can be found in [Streater and Wightman, 1989]. It is not hard to check that any term you could write down in a local Lagrangian is CPT invariant; however, the rigorous proof does not require a Lagrangian description. Some examples of how unitarity can be used without a Lagrangian are given in Chapter 24.

Problems

11.1 In practice, we only rarely use explicit representations of the Dirac matrices. Most calculations can be done using algebraic identities that depend only on $\{\gamma^\mu, \gamma^\nu\} = 2g^{\mu\nu}$. Derive algebraically (without using an explicit representation):

(a) $(\gamma^5)^2 = \mathbb{1}$

(b) $\gamma_\mu \not{p} \gamma^\mu = -2\not{p}$

(c) $\gamma_\mu \not{p}\not{q}\not{p}\gamma^\mu = -2\not{p}\not{q}\not{p}$

(d) $\{\gamma^5, \gamma^\mu\} = 0$

(e) $\mathrm{Tr}[\gamma^\alpha\gamma^\mu\gamma^\beta\gamma^\nu] = 4(g^{\alpha\mu}g^{\beta\nu} - g^{\alpha\beta}g^{\mu\nu} + g^{\alpha\nu}g^{\mu\beta})$

11.2 Spinor identities.

(a) Show that $\sum_s u_s(p)\bar{u}_s(p) = \not{p} + m$ and $\sum_s v_s(p)\bar{v}_s(p) = \not{p} - m$.

(b) Show that $\bar{u}_\sigma(p)\gamma^\mu u_{\sigma'}(p) = 2\delta_{\sigma\sigma'}p^\mu$.

11.3 Prove that massless spin-1 particles coupled to spin-0 or spin-$\frac{1}{2}$ particles imply a conserved charge. You may use results from Section 9.5.

11.4 Show that for on-shell spinors

$$\bar{u}(q)\gamma^\mu u(p) = \bar{u}(q)\left[\frac{q^\mu + p^\mu}{2m} + i\frac{\sigma^{\mu\nu}(q_\nu - p_\nu)}{2m}\right]u(p), \tag{11.92}$$

where $\sigma_{\mu\nu} = \frac{i}{2}[\gamma_\mu, \gamma_\nu]$. This is known as the **Gordon identity**. We will use this when we calculate the 1-loop correction to the electron's magnetic dipole moment.

11.5 Derive the charge-conjugation properties of the spinor bilinears in Eqs. (11.54) to (11.56).

11.6 The physics of spin, helicity and chirality.

(a) Use the left and right chirality projection operators to show that the QED vertex $\bar\psi\gamma^\mu\psi$ vanishes unless ψ and $\bar\psi$ are both left-handed or both right-handed.

(b) For the non-relativistic limit, choose explicit spinors for a spinor at rest. Show that $\bar\psi_s\gamma^\mu\psi_{s'}$ vanishes unless $s = s'$.

(c) Use the Schrödinger equation to show that in the non-relativistic limit the electric field cannot flip an electron's spin, only the magnetic field can.

(d) Suppose we take a spin-up electron going in the $+z$ direction and turn it around carefully with electric fields so that now it goes in the $-z$ direction but is still spin up. Has its helicity or chirality flipped (or both)? How is your answer consistent with part (a)?

(e) How can you measure the spin of a slow electron?

(f) Suppose you have a radioactive source, such as cobalt-60, which undergoes β-decay $^{60}_{27}\text{Co} \rightarrow {}^{60}_{28}\text{Ni} + e^- + \bar\nu$. How could you (in principle) find out if those electrons coming out are polarized; that is, if they all have the same helicity? Do you think they would be polarized? If so, which polarization do you expect more of?

11.7 Show that the most general Lagrangian term you can write down in terms of Dirac spinors, γ-matrices, and the photon field A_μ is automatically invariant under CPT. To warm up, consider first the terms in Eq. (11.91).

11.8 Fierz rearrangement formulas (Fierz identities). It is often useful to rewrite spinor contractions in other forms to simplify formulas. Show that

(a) $\left(\bar\psi_1\gamma^\mu P_L\psi_2\right)\left(\bar\psi_3\gamma^\mu P_L\psi_4\right) = -\left(\bar\psi_1\gamma^\mu P_L\psi_4\right)\left(\bar\psi_3\gamma^\mu P_L\psi_2\right)$

(b) $\left(\bar\psi_1\gamma^\mu\gamma^\alpha\gamma^\beta P_L\psi_2\right)\left(\bar\psi_3\gamma^\mu\gamma^\alpha\gamma^\beta P_L\psi_4\right) = -16\left(\bar\psi_1\gamma^\mu P_L\psi_4\right)\left(\bar\psi_3\gamma^\mu P_L\psi_2\right)$

(c) $\text{Tr}\left[\Gamma^M\Gamma^N\right] = 4\delta^{MN}$, with $\Gamma^M \in \{\mathbb{1}, \gamma^\mu, \sigma^{\mu\nu}, \gamma_5\gamma^\mu, \gamma_5\}$

(d) $\left(\bar\psi_1\Gamma^M\psi_2\right)\left(\bar\psi_3\Gamma^N\psi_4\right) = \sum_{PQ}\frac{1}{16}\text{Tr}\left[\Gamma^P\Gamma^M\Gamma^Q\Gamma^N\right]\left(\bar\psi_1\Gamma^P\psi_4\right)\left(\bar\psi_3\Gamma^Q\psi_2\right)$

where $P_L = \frac{1-\gamma_5}{2}$ projects out the left-handed spinor from a Dirac fermion. The identities with P_L play an important role in the theory of weak interactions, which only involves left-handed spinors (see Chapter 29).

11.9 The electron neutrino is a nearly massless neutral particle. Its interactions violate parity: only the left-handed neutrino couples to the W and Z bosons.

(a) The Z is a vector boson, like the photon but heavier, and has an associated U(1) gauge invariance (it is actually broken in nature, but that is not relevant for this problem). If there is only a left-handed neutrino ν_L, the only possible mass term of dimension four is a Majorana mass, of the form $iM\nu_L^T\sigma_2\nu_L$. Show that this mass is forbidden by the U(1) symmetry.

This motivates the introduction of a right-handed neutrino ν_R. The most general kinetic Lagrangian involving ν_L and ν_R is

$$\mathcal{L}_{\text{kin}} = \nu_L^\dagger \bar{\sigma}^\mu \partial_\mu \nu_L + \nu_R^\dagger \sigma^\mu \partial_\mu \nu_R + m(\nu_L^\dagger \nu_R + \nu_R^\dagger \nu_L)$$
$$+ iM\left(\nu_R^T \sigma_2 \nu_R - \nu_R^\dagger \sigma_2 \nu_R^\star\right), \tag{11.93}$$

where ν_L is a left-handed $(\frac{1}{2}, 0)$ two-component Weyl spinor and ν_R is a right-handed $(0, \frac{1}{2})$ Weyl spinor. Note that there are two mass terms: a Dirac mass m, as for the electron, and a Majorana mass, M.

(b) We want to figure out what the mass eigenstates are, but as written the Lagrangian is mixing everything up. First, show that $\chi_L \equiv i\sigma_2 \nu_R^\star$ transforms as a left-handed spinor under the Lorentz group, so that it can mix with ν_L. Then rewrite the mass terms in terms of ν_L and χ_L.

(c) Next, rewrite the Lagrangian in terms of a doublet $\vec{\Theta} \equiv (\nu_L, \chi_L)$. This is not a Dirac spinor, but a doublet of left-handed Weyl spinors. Using $\mathcal{L}_{\text{kin}}$, show that this doublet satisfies the Klein–Gordon equation. What are the mass eigenstates for the neutrinos? How many particles are there?

(d) Suppose $M \gg m$. For example, $M = 10^{16}$ GeV and $m = 100$ GeV. What are the masses of the physical particles? The fact that as M goes up, the physical masses go down, inspired the name **see-saw mechanism** for this neutrino mass arrangement. What other choice of M and m would give the same spectrum of observed particles (i.e. particles less than $\sim$1 TeV)?

(e) The left-handed neutrino couples to the Z boson and also to the electron through the W boson. The W boson also couples the neutron and proton. The relevant part for the weak-force Lagrangian is

$$\mathcal{L}_{\text{weak}} = g_W(\nu_L^\dagger \slashed{W} e_L + e_L^\dagger \slashed{W} \nu_L) + g_Z(\nu_L^\dagger \slashed{Z} \nu_L) + g_W(n\slashed{W}\bar{p} + \bar{n}\slashed{W}p). \tag{11.94}$$

Using these interactions, draw a Feynman diagram for neutrinoless double β-decay, in which two neutrons decay to two protons and two electrons.

(f) Which of the terms in $\mathcal{L}_{\text{kin}}$ and $\mathcal{L}_{\text{weak}}$ respect a global symmetry (lepton number) under which $\nu_L \to e^{i\theta}\nu_L$, $\nu_R \to e^{i\theta}\nu_R$ and $e_L \to e^{i\theta}e_L$? Define arrows on the e and ν lines to respect lepton number flow. Show that you cannot connect the arrows on your diagram without violating lepton number. Does this imply that neutrinoless double β-decay can tell if the neutrino has a Majorana mass?

11.10 In Section 10.4, we showed that the electron has a magnetic dipole moment, of order $\mu_B = \frac{e}{2m_e}$, by squaring the Dirac equation. An additional magnetic moment could come from an interaction of the form $\mathcal{B} = iF_{\mu\nu}\bar{\psi}[\gamma^\mu, \gamma^\nu]\psi$ in the Lagrangian. An electric dipole moment (EDM) corresponds to a term of the form $\mathcal{E} = F_{\mu\nu}\bar{\psi}\gamma_5[\gamma^\mu, \gamma^\nu]\psi$.

(a) Expand the contribution of the electric dipole term to the Dirac equation in terms of electric and magnetic fields to show that it does in fact give an EDM.

(b) Which of the symmetries C, P or T are respected by the magnetic dipole moment operator, $\mathcal{B}$, and the EDM operator, $\mathcal{E}$?

(c) It turns out that C, P and T are all separately violated in the Standard Model, even though they are preserved in QED (and QCD). P is violated by the weak interactions, but T (and CP) is only very weakly violated. Thus we expect, unless there is a new source of CP violation beyond the Standard Model, the electron, the neutron, the proton, the deuteron etc., all should have unmeasurably small (but non-zero) EDMs. Why is it OK for a molecule (such as H_2O) or a battery to have an EDM but not the neutron (which is made up of quarks with different charges)?

Spin and statistics 12

One of the most profound consequences of merging special relativity with quantum mechanics is the spin-statistics theorem: states with identical particles of integer spin are symmetric under the interchange of the particles, while states with identical particles of half-integer spin are antisymmetric under the interchange of the particles. This is equivalent to the statement that the creation and annihilation operators for integer spin particles satisfy canonical commutation relations, while creation and annihilation operators for half-integer spin particles satisfy canonical anticommutation relations. Particles quantized with canonical commutation relations are called **bosons**, and satisfy Bose–Einstein statistics, and particles quantized with canonical anticommutation relations are called **fermions**, and satisfy Fermi–Dirac statistics.

The simplest way to see the connection between spin and statistics, mentioned in Chapter 10, is as follows. One way to interchange two particles is to rotate them around their midpoint by π. For a particle of spin s, this rotation will introduce a phase factor of $e^{i\pi s}$. Thus, a two-particle state with identical particles both of spin s will pick up a factor of $e^{2\pi i s}$. For s a half-integer, this will give a factor of -1; for s an integer, it will give a factor of $+1$. This argument is made more precise in Section 12.2.

Traditionally, the spin-statistics theorem is derived by pointing out that things go terribly awry if the wrong statistics are applied. For example, the spin-statistics theorem follows from **Lorentz invariance of the S-matrix**. Since the S-matrix is constructed from Lorentz-covariant fields, Lorentz invariance is almost obvious. The only possible catch is the explicit time-ordering that appears in the definition of the S-matrix: $S \sim T\{\phi_1(x_1)\cdots\phi_n(x_n)\}$. If you choose commutation relations for particles of half-integer spin, this time-ordered product will not be Lorentz invariant. If you choose anticommutation relations, it will be. The relevant calculations are given in Section 12.4. An important result of this section is the propagator for a Dirac spinor.

Another criterion that can be used to prove the spin-statistics theorem is that the total energy of a system should be bounded from below. When applied to free particles, we call this the **stability** requirement (instabilities due to interactions are a different story; see, for example, Chapter 28). For free particles, if the wrong statistics are used, antiparticles will have arbitrarily negative energy. This would allow kinematical processes, such as an electron decaying into a muon, $e^- \rightarrow \mu^- \nu_e \bar{\nu}_\mu$, which is normally forbidden by energy conservation (*not* momentum conservation). We take it for granted that light particles cannot decay to heavier particles, but this is actually a non-trivial consequence of the spin-statistics theorem. In studying the stability requirement, in Section 12.5, we will investigate the Hamiltonian and energy-momentum tensor, which provide more generally useful results.

One does not have to *postulate* stability for free particles, since it follows from spin-statistics, which follows from Lorentz invariance. However, requiring stability is a necessary and sufficient condition for the spin-statistics theorem. This is important in contexts such as condensed matter systems in which Lorentz invariance is irrelevant. There, you could study representations of whatever the appropriate group is, say the Galilean group, and you would still find spinors, but you would not be interested in the S-matrix or causality. In this case, spinors would still have to be fermions to ensure stability of the system you are studying.

There are other ways to see the connection between spin and statistics. A very important requirement historically was that operators corresponding to observables that are constructed out of fields should commute at spacelike separation:

$$[\mathcal{O}_1(x), \mathcal{O}_2(y)] = 0, \quad (x - y)^2 < 0. \tag{12.1}$$

We call this the **causality** criterion. Note that it is pretty crazy to imagine that a theory which involves generally smooth functions could produce objects that vanish in a compact region but do not vanish everywhere. This would be mathematically impossible if $[\mathcal{O}_1(x), \mathcal{O}_2(y)]$ were an analytic function of x and y. Quantum field theory can get away with this because operator products give distributions, not functions. In fact, as we will show in Section 12.6, they give distributions with precisely the property of Eq. (12.1).

The causality criterion was first proposed by Pauli in his seminal paper on spin-statistics from 1940 [Pauli, 1940]. The idea behind this requirement comes from quantum mechanics: When two operators commute, they are simultaneously observable; they cannot influence each other. If they could influence each other at spacelike separations, one could use them to communicate faster than the speed of light. This is a weaker requirement than Lorentz invariance of the S-matrix. Unfortunately, causality can only be used to prove that integer spin particles commute, but not that half-integer spin particles anticommute. The reason is that observables are bilinear in spinors, and hence have integer spin (can you think of an observable linear in a spinor?).

Causality actually follows directly from Lorentz invariance of the S-matrix: time ordering is only Lorentz invariant for timelike separations. That is, the inequality $t_i < t_j$ is Lorentz invariant as long as $x_i^\mu - x_j^\mu$ is timelike. If two points are spacelike separated, then one can boost to a frame where $t_j < t_i$. Thus, for spacelike separation, time ordering of a pair of fields is an ambiguous operation unless the fields commute. So causality follows from Lorentz invariance of the S-matrix. The converse is not true: Eq. (12.1) is a necessary condition, but not sufficient, for Lorentz invariance of the S-matrix.

12.1 Identical particles

To talk about spin-statistics, we first need to talk about identical particles. The universe is full of many types of particles: photons, electrons, muons, quarks, etc. Each particle has a momentum, $\vec{p}_i$, a spin, s_i, and a bunch of additional quantum numbers, n_i, which say what type of particle it is. For each type of particle we have a set of creation and annihilation operators, $a^\dagger_{\vec{p}_i s_i n_i}$ and $a_{\vec{p}_i s_i n_i}$. Particles transforming in the same representation of the Poincaré group and having the same additional quantum numbers, n_i, are said to be **identical particles**.

If we act with creation operators for identical particles on the vacuum, we get a multiparticle state:

$$|\cdots s_1\vec{p}_1 n \cdots s_2\vec{p}_2 n \cdots\rangle = \cdots \sqrt{2\omega_1} a^\dagger_{\vec{p}_1 s_1 n} \cdots \sqrt{2\omega_2} a^\dagger_{\vec{p}_2 s_2 n} \cdots |0\rangle. \qquad (12.2)$$

Recall that the multi-particle states are normalized so that

$$\langle s_1\vec{p}_1 n_1 \cdots | s'_1\vec{p}'_1 n'_1 \cdots \rangle = \prod_i \delta_{n_i n'_i} \delta_{s_i s'_i} 2\omega_i (2\pi)^3 \delta^3(\vec{p}_i - \vec{p}'_i). \qquad (12.3)$$

We could have also acted with the creation operators in a different order, giving

$$|\cdots s_2\vec{p}_2 n \cdots s_1\vec{p}_1 n \cdots\rangle = \cdots \sqrt{2\omega_2} a^\dagger_{\vec{p}_2 s_2 n} \cdots \sqrt{2\omega_1} a^\dagger_{\vec{p}_1 s_1 n} \cdots |0\rangle. \qquad (12.4)$$

Since the particles are identical, this must be the same physical state, so it can only differ by normalization. Since we have fixed the normalization, it can only differ by a phase:

$$|\cdots s_1\vec{p}_1 n \cdots s_2\vec{p}_2 n \cdots\rangle = \alpha|\cdots s_2\vec{p}_2 n \cdots s_1\vec{p}_1 n \cdots\rangle, \qquad (12.5)$$

where $\alpha = e^{i\phi}$ for some real ϕ.

What can α depend on? Since it is just a number, it cannot depend on the momenta $\vec{p}_i$ or the spins s_i, as there are no non-trivial one-dimensional representations of the (proper) Lorentz group. It could possibly depend on a Lorentz-invariant characterization of the path by which the particles are interchanged. However, in $3 + 1$ dimensions, there are no such invariants (we derive this in the next section). Thus, α can only depend on n, the species of particle. So let us write $\alpha = \alpha_n$.

Now we can swap the particles back, giving

$$|\cdots s_1\vec{p}_1 n \cdots s_2\vec{p}_2 n \cdots\rangle = \alpha_n^2|\cdots s_1\vec{p}_1 n \cdots s_2\vec{p}_2 n \cdots\rangle. \qquad (12.6)$$

Thus $\alpha_n = \pm 1$. We call $\alpha_n = 1$ **bosons**, which we say satisfy **Bose–Einstein statistics**, and we call $\alpha_n = -1$ **fermions**, which we say satisfy **Fermi–Dirac statistics**. So every particle is either a fermion or a boson. The boson case implies that

$$a^\dagger_{\vec{p}_1 s_1 n} a^\dagger_{\vec{p}_2 s_2 n} |\psi\rangle = a^\dagger_{\vec{p}_2 s_2 n} a^\dagger_{\vec{p}_1 s_1 n} |\psi\rangle \qquad (12.7)$$

for all $|\psi\rangle$ and therefore

$$[a^\dagger_{\vec{p}_1 s_1 n}, a^\dagger_{\vec{p}_2 s_2 n}] = [a_{\vec{p}_1 s_1 n}, a_{\vec{p}_2 s_2 n}] = 0 \quad \text{(bosons)}. \qquad (12.8)$$

Also, since $\langle \vec{p}_1 | \vec{p}_2 \rangle = 2\omega_1 (2\pi)^3 \delta^3(\vec{p}_1 - \vec{p}_2)$, we can use the same argument to show that

$$[a_{\vec{p}_1 s_1 n}, a^\dagger_{\vec{p}_2 s_2 n}] = (2\pi)^3 \delta^3(\vec{p}_1 - \vec{p}_2)\delta_{s_1,s_2}. \qquad (12.9)$$

For the fermion case, the same logic implies

$$\{a_{\vec{p}_1 s_1 n}^\dagger, a_{\vec{p}_2 s_2 n}^\dagger\} = \{a_{\vec{p}_1 s_1 n}, a_{\vec{p}_2 s_2 n}\} = 0 \quad \text{(fermions)}, \tag{12.10}$$

$$\{a_{\vec{p}_1 s_1 n}^\dagger, a_{\vec{p}_2 s_2 n}\} = (2\pi)^3 \delta^3(\vec{p}_1 - \vec{p}_2)\delta_{s_1, s_2}. \tag{12.11}$$

The physics of fermions is very different from the physics of bosons. With bosons, such as the photon, we can have multiple particles of the same momentum in the same state. In fact, thinking about multi-particle excitations in Chapter 1 led to the connection with the simple harmonic oscillator and second quantization in Chapter 2. Now consider what happens if we try to construct a state with two identical fermionic particles with the same momenta (a two-particle state). We find

$$a_{\vec{p}}^\dagger a_{\vec{p}}^\dagger |0\rangle = -a_{\vec{p}}^\dagger a_{\vec{p}}^\dagger |0\rangle = 0. \tag{12.12}$$

This is the Pauli exclusion principle, and it follows directly from the anticommutation relations.

By the way, that identical particles must exist is an automatic consequence of using creation and annihilation operators in quantum field theory. You might wonder why we have to consider states produced with creation operators at all. If we demand that all operators are constructed out of creation and annihilation operators we are guaranteed that the **cluster decomposition principle** holds. The cluster decomposition principle requires that when you separate two measurements asymptotically far apart, they cannot influence each other. Technically, it says that the S-matrix should factorize into clusters of interactions. Many other methods for calculating S-matrix elements have been considered over the years, but the quantum field theory approach, based on creation and annihilation operators, remains the most efficient way to guarantee cluster decomposition.

12.2 Spin-statistics from path dependence

Rather than simply relating two states $a_1^\dagger a_2^\dagger |0\rangle$ and $a_2^\dagger a_1^\dagger |0\rangle$, we can consider actually interchanging the particles physically. This will let us connect statistics directly to spin and representations of the Lorentz group. Suppose we have two particles at positions x_1 and x_2 at time $t = 0$. Then, at some later time, we find them also at x_1 and x_2. Since they are identical, we could have had the particle at x_1 move back to x_1 and the particle at x_2 move back to x_2, or the particles could have switched places. We could also have had the particles spin around each other many times. There is a well-defined way to characterize the transformation, by the angle ϕ by which one particle rotated around the other. This angle is frame independent and a topological property associated with the path. In Figure 12.1, pictures of these exchanges are shown.

In general, it is possible for the two-particle state to pick up a phase proportional to this angle ϕ, as in Eq. (12.5). So we can define an operator S that switches the particles. Then the most general possibility for what would happen when we switch the particles is that

$$S|\phi_1(x_1)\phi_2(x_2)\rangle = e^{i\phi\kappa}|\phi_2(x_1)\phi_1(x_2)\rangle \tag{12.13}$$

 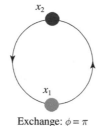 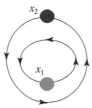

No exchange: $\phi = 0$ Exchange: $\phi = \pi$ No exchange: $\phi = 2\pi$

The angle that one particle travels around another before coming back to its own or the other's position is a Lorentz-invariant characterization of the path.

Fig. 12.1

for some number κ characteristic of the particle type.

Now, with three spatial dimensions, the angle ϕ can only be defined up to 2π. For example, the diagram in the third figure can be unwrapped by pulling the x_2 loop out of the page so that the particles do not go around each other. Thus, the action of $\mathcal{S}$ on the states is not Lorentz invariant unless it gives the same answer for ϕ and $\phi + 2\pi$. Thus $\kappa \in \mathbb{Z}$. This implies that, for an interchange with $\phi = \pi$, we can only have

$$\mathcal{S}|\phi_1(x_1)\phi_2(x_2)\rangle = \pm |\phi_2(x_1)\phi_1(x_2)\rangle . \tag{12.14}$$

So only fermionic and bosonic statistics are possible. In other words, in three dimensions, there is no way to characterize the path other than that the particles were swapped ($\phi = \pi$) or not ($\phi = 0$).

Thus, there are only two possibilities, given by the first two paths in Figure 12.1. Consider the second path. In a free-field theory, we can actually perform the exchange by acting with Poincaré generators on the fields. One way would be to translate by the distance between x_1 and x_2, then to rotate the whole system by π so that particle 2 is back at x_1, as shown in Figure 12.2.

Under the translation, nothing interesting happens. Under the rotation, what happens depends on the spin. For scalars, there is no spin, so our transformation takes

$$\mathcal{S}|\phi_1(x_1)\phi_2(x_2)\rangle = |\phi_2(x_1)\phi_1(x_2)\rangle . \tag{12.15}$$

On the other hand, for spinors there is a non-trivial transformation. In fact, we worked it out explicitly in Section 10.5: for Dirac spinors, a rotation by an angle θ_z is represented by

Particles' positions can be interchanged by first translating the pair by $x_2 - x_1$, then rotating the pair around x_2. Alternatively, we could have just rotated the two particles around their midpoint.

Fig. 12.2

Eq. (10.118):

$$\Lambda_s(\theta_z) = \begin{pmatrix} \exp(\tfrac{i}{2}\theta_z) & & & \\ & \exp(-\tfrac{i}{2}\theta_z) & & \\ & & \exp(\tfrac{i}{2}\theta_z) & \\ & & & \exp(-\tfrac{i}{2}\theta_z) \end{pmatrix}, \qquad (12.16)$$

which for $\theta_z = \pi$ is the matrix with i and $-i$ in the diagonal. So, suppose the particles were both spin out-of-the page (spin into-the-page is the same, but for spins in the x or y direction, this manipulation will not take the particles back to themselves and one needs to consider a different route). Then,

$$\psi_1 = \begin{pmatrix} 1 \\ 0 \\ 1 \\ 0 \end{pmatrix} \rightarrow \begin{pmatrix} i \\ 0 \\ i \\ 0 \end{pmatrix} = i\psi_1. \qquad (12.17)$$

So the two-particle state with identical spins has

$$\mathcal{S}|\psi_1(x_1)\psi_2(x_2)\rangle = -|\psi_2(x_1)\psi_1(x_2)\rangle, \qquad (12.18)$$

which is to say that the spinors pick up a minus sign under the interchange.

This derivation works for particles of any half-integer or integer spin. The only thing we need is that under a 2π rotation half-integer spin particles go to minus themselves, while integer spin particles go to themselves. This is practically the definition of spin. This derivation is appealing because it is directly related to spinors transforming in representations of the universal cover of the Lorentz group, $\mathrm{SL}(2, \mathbb{C})$, which is simply connected, while the Lorentz group itself is doubly connected. If you like this argument, then you can skip the rest of this chapter (except for the calculation of the Feynman propagator for Dirac spinors, which we will use later).

In $2 + 1$ dimensions, the situation is more interesting. There, paths with ϕ and $\phi + 2\pi n$ are distinguishable – we cannot unwrap the third path in Figure 12.1 into the first path anymore by pulling it out of the page. So in this case,

$$\mathcal{S}|\phi_1(x_1)\phi_2(x_2)\rangle = e^{i\phi\kappa}|\phi_2(x_1)\phi_1(x_2)\rangle, \qquad (12.19)$$

and κ can be an arbitrary number. Particles with $\kappa \notin \mathbb{Z}$ are called **anyons**.

We can understand anyons also from the representations of the 3D Lorentz group, $\mathrm{SO}(2, 1)$. Recall that for four dimensions the little group of the Poincaré group, which determined its irreducible representations, was $\mathrm{SO}(3)$ (for the massive case). For $\mathrm{SO}(3)$, we found that there were paths through the group, from $\mathbb{1}$ to $\mathbb{1}$, that were not smoothly deformable to the trivial path. For example, rotations by 2π around any axis have this property. However, any 4π rotation can be deformed to the trivial path. In other words, the fundamental group of $\mathrm{SO}(3)$ is Z_2. With two spatial dimensions, the little group is $\mathrm{SO}(2)$.

Here a 2π rotation, as a path through the group, also cannot be deformed to the trivial path. Moreover, rotations $e^{in\kappa}$ with $0 \leq \kappa < 2\pi$ for any $n \in \mathbb{Z}$ make up separate paths. Thus, the fundamental group of SO(2) is $\mathbb{Z}$. Then, in the same way that spinors picked up a factor of -1 under 2π rotations in $3 + 1$ dimensions, there are representations that can pick up factors of e^{in} in $2 + 1$ dimensions. These are the anyons.

12.3 Quantizing spinors

The remaining connections between spin and statistics we want to explore involve quantum fields. So the first thing we must do is quantize our spinors. This is straightforward, up to the statistics issue.

Recall that for a complex scalar the field we had

$$\phi(x) = \int \frac{d^3 p}{(2\pi)^3} \frac{1}{\sqrt{2\omega_p}} \left(a_p e^{-ipx} + b_p^\dagger e^{ipx} \right), \tag{12.20}$$

$$\phi^\star(x) = \int \frac{d^3 p}{(2\pi)^3} \frac{1}{\sqrt{2\omega_p}} \left(a_p^\dagger e^{ipx} + b_p e^{-ipx} \right). \tag{12.21}$$

Remember, $a_p^\dagger$ creates particles and $b_p^\dagger$ creates antiparticles, which are particles of the opposite charge and same mass.

For the Dirac equation the Lagrangian is

$$\mathcal{L} = \bar{\psi}(i\slashed{D} - m)\psi \tag{12.22}$$

and the equations of motion are

$$(i\slashed{\partial} - e\slashed{A} - m)\psi = 0, \tag{12.23}$$

$$\bar{\psi}(-i\overleftarrow{\slashed{\partial}} - e\slashed{A} - m) = 0. \tag{12.24}$$

In Section 11.2, we saw that the free-field solutions can be written in terms of constant two-component spinors ξ_s and η_s, with $s = 1, 2$ in the concise notation:

$$u_s(p) = \begin{pmatrix} \sqrt{p \cdot \sigma}\, \xi_s \\ \sqrt{p \cdot \bar{\sigma}}\, \xi_s \end{pmatrix}, \quad v_s(p) = \begin{pmatrix} \sqrt{p \cdot \sigma}\, \eta_s \\ -\sqrt{p \cdot \bar{\sigma}}\, \eta_s \end{pmatrix}. \tag{12.25}$$

To be clear, $\xi_1 = \eta_1 = (1, 0)^T$ and $\xi_2 = \eta_2 = (0, 1)^T$ are constants, while $u^s(p)$ and $v^s(p)$ are the solutions to the Dirac equation with arbitrary momentum describing electrons and positrons respectively.

Thus, we take

$$\psi(x) = \sum_s \int \frac{d^3 p}{(2\pi)^3} \frac{1}{\sqrt{2\omega_p}} \left(a_p^s u_p^s e^{-ipx} + b_p^{s\dagger} v_p^s e^{ipx} \right), \tag{12.26}$$

$$\bar{\psi}(x) = \sum_s \int \frac{d^3 p}{(2\pi)^3} \frac{1}{\sqrt{2\omega_p}} \left(a_p^{s\dagger} \bar{u}_p^s e^{ipx} + b_p^s \bar{v}_p^s e^{-ipx} \right), \tag{12.27}$$

where, as always, the energy is positive and determined by the 3-momentum $\omega_p = \sqrt{\vec{p}^2 + m^2}$. So $\psi(x)$ annihilates incoming electrons and $\bar{\psi}(x)$ annihilates incoming positrons. The full Feynman rules for QED will be derived in the next chapter.

The next three sections will be devoted to deriving the spin-statistics theorem in three different ways.

12.4 Lorentz invariance of the *S*-matrix

As mentioned in the introduction to this chapter, Lorentz invariance of the S-matrix is a sufficient condition for the spin-statistics theorem. The S-matrix is constructed from time-ordered products, with the simplest non-trivial time-ordered product being the Feynman propagator.

Time ordering must be defined for bosons and fermions. Fermionic creation and annihilation operators anticommute at generic momenta and times. Therefore,

$$T\{a_p(t)a_q(t')\} = -T\{a_q(t')a_p(t)\}. \tag{12.28}$$

Thus we cannot just define time ordering as "take all the operators and put them in time order," or else this equation would imply the time-ordered product must vanish. So we have to define time ordering for fermions by anticommuting the operators past each other, keeping track of minus signs. Thus, for generic functions $\psi(x)$ of fermionic creation and annihilation operators, the only consistent definition of time ordering is

$$T\{\psi(x)\chi(y)\} = \psi(x)\chi(y)\theta(x_0 - y_0) - \chi(y)\psi(x)\theta(y_0 - x_0). \tag{12.29}$$

Now we can consider time-ordered products of fermionic fields.

12.4.1 Spin 0

Let us first review what happens with a complex scalar. The vacuum matrix element of a field and its conjugate is

$$\langle 0|\phi^\star(x)\phi(0)|0\rangle = \int \frac{d^3p}{(2\pi)^3} \int \frac{d^3q}{(2\pi)^3} \frac{1}{\sqrt{2\omega_p}} \frac{1}{\sqrt{2\omega_q}} \langle 0|(a_p^\dagger e^{ipx} + b_p e^{-ipx})(a_q + b_q^\dagger)|0\rangle$$

$$= \int \frac{d^3p}{(2\pi)^3} \frac{1}{2\omega_p} e^{-i\omega_p t + i\vec{p}\vec{x}}, \tag{12.30}$$

and similarly,

$$\langle 0|\phi(0)\phi^\star(x)|0\rangle = \int \frac{d^3p}{(2\pi)^3} \frac{1}{2\omega_p} e^{i\omega_p t - i\vec{p}\vec{x}}. \tag{12.31}$$

Combining these equations we get

$$\langle 0|T\{\phi^\star(x)\phi(0)\}|0\rangle = \int \frac{d^3p}{(2\pi)^3} \frac{1}{2\omega_p} \left[e^{i\vec{p}\vec{x} - i\omega_p t}\theta(t) + e^{-i\vec{p}\vec{x} + i\omega_p t}\theta(-t) \right]. \tag{12.32}$$

Now we take $\vec{p} \to -\vec{p}$ in the first term, giving

$$\langle 0|T\{\phi^\star(x)\phi(0)\}|0\rangle = \int \frac{d^3p}{(2\pi)^3} \frac{1}{2\omega_p} e^{-i\vec{p}\vec{x}} \left[e^{-i\omega_p t}\theta(t) + e^{i\omega_p t}\theta(-t) \right]. \tag{12.33}$$

Then recalling the identities from Eqs. (6.30) and (6.31):

$$e^{i\omega_p t}\theta(-t) = \frac{i}{2\pi} \int_{-\infty}^{\infty} \frac{d\omega}{\omega - (\omega_p - i\varepsilon)} e^{i\omega t},$$

$$e^{-i\omega_p t}\theta(t) = -\frac{i}{2\pi} \int_{-\infty}^{\infty} \frac{d\omega}{\omega - (-\omega_p + i\varepsilon)} e^{i\omega t}, \tag{12.34}$$

we arrive at

$$\langle 0|T\{\phi^\star(x)\phi(0)\}|0\rangle = \int \frac{d^4p}{(2\pi)^4} \frac{i}{2\omega_p} e^{ipx} \left(\frac{1}{\omega - (\omega_p - i\epsilon)} - \frac{1}{\omega - (-\omega_p + i\epsilon)} \right)$$

$$= \int \frac{d^4p}{(2\pi)^4} \frac{i}{p^2 - m^2 + i\varepsilon} e^{ipx}, \tag{12.35}$$

where $p_0 \equiv \omega$ is an integration variable. This is a beautiful manifestly Lorentz-invariant expression.

If we instead take anticommutation relations, we would need to use anti-time ordering. Then,

$$\langle 0|T\{\phi^\star(x)\phi(0)\}|0\rangle = \int \frac{d^3p}{(2\pi)^3} \frac{1}{2\omega_p} e^{-i\vec{p}\vec{x}} \left[-e^{i\omega_p t}\theta(-t) + e^{-i\omega_p t}\theta(t) \right]$$

$$= -\int \frac{d^4p}{(2\pi)^4} \frac{i}{2\omega_p} e^{ipx} \left(\frac{1}{\omega - (\omega_p - i\epsilon)} + \frac{1}{\omega - (-\omega_p + i\epsilon)} \right)$$

$$= \int \frac{d^4p}{(2\pi)^4} \frac{\omega}{\sqrt{\vec{p}^2 + m^2}} \frac{-i}{p^2 - m^2 + i\varepsilon} e^{ipx}. \tag{12.36}$$

This is not Lorentz invariant. Therefore the S-matrix for spin-0 particles is Lorentz invariant *if and only if* they are bosons.

12.4.2 Spinors

Now let us repeat the calculation with spinors. We start the same way:

$$\langle 0|\psi(0)\bar\psi(x)|0\rangle = \int \frac{d^3p}{(2\pi)^3} \int \frac{d^3q}{(2\pi)^3} \frac{1}{\sqrt{2\omega_p}} \frac{1}{\sqrt{2\omega_q}}$$

$$\times \sum_{s,s'} \langle 0|(a_p^s u_p^s + b_p^{s\dagger} v_p^s)(a_q^{s'\dagger} \bar{u}_q^{s'} e^{iqx} + b_q^{s'} \bar{v}_q^{s'} e^{-iqx})|0\rangle$$

$$= \int \frac{d^3p}{(2\pi)^3} \int \frac{d^3q}{(2\pi)^3} \frac{1}{\sqrt{2\omega_p}} \frac{1}{\sqrt{2\omega_q}} \sum_{s,s'} u_p^s \bar{u}_q^{s'} \langle 0|a_p^s a_q^{s'\dagger}|0\rangle e^{iqx}$$

$$= \int \frac{d^3p}{(2\pi)^3} \frac{1}{2\omega_p} \sum_s u_p^s \bar{u}_p^s e^{ipx}. \tag{12.37}$$

Note that $\psi(0)\bar{\psi}(x)$ refers to a matrix in spinor space: $\langle 0|\psi_\alpha(0)\bar{\psi}_\beta(x)|0\rangle \sim (u_p^s)_\alpha (\bar{u}_p^s)_\beta$. Now we sum over polarizations using the outer products from Eqs. (11.29) and (11.30):

$$\sum_{s=1}^{2} u_s(p)\bar{u}_s(p) = \not{p} + m, \quad \sum_{s=1}^{2} v_s(p)\bar{v}_s(p) = \not{p} - m, \tag{12.38}$$

giving

$$\langle 0|\psi(0)\bar{\psi}(x)|0\rangle = \int \frac{d^3p}{(2\pi)^3} \frac{1}{2\omega_p} e^{ipx} (\not{p} + m) = (-i\not{\partial} + m) \int \frac{d^3p}{(2\pi)^3} \frac{1}{2\omega_p} e^{ipx}. \tag{12.39}$$

Similarly,

$$\langle 0|\bar{\psi}(x)\psi(0)|0\rangle = \sum_{ss'} \int \frac{d^3p}{(2\pi)^3} \int \frac{d^3q}{(2\pi)^3} \frac{1}{\sqrt{2\omega_p}} \frac{1}{\sqrt{2\omega_q}}$$

$$\times \langle 0|(b_q^{s'}\bar{v}_q^{s'}e^{-iqx} + a_q^{s'\dagger}\bar{u}_q^{s'}e^{iqx})(a_p^s u_p^s + b_p^{s\dagger} v_p^s)|0\rangle$$

$$= \int \frac{d^3p}{(2\pi)^3} \frac{1}{2\omega_p} e^{-ipx} (\not{p} - m) = (i\not{\partial} - m) \int \frac{d^3p}{(2\pi)^3} \frac{1}{2\omega_p} e^{-ipx}. \tag{12.40}$$

This is also a matrix in spinor space: $\bar{\psi}(x)\psi(0)$ in this expression means $\bar{\psi}_\alpha(x)\psi_\beta(0)$. Note that this convention contrasts with when we write Lagrangian terms such as $\bar{\psi}\psi = \bar{\psi}_\alpha(x)\psi_\alpha(x) = \text{Tr}(\bar{\psi}_\alpha(x)\psi_\beta(x))$, which have no free indices. Whether there is a contraction of spinor indices will be clear from context or explicitly indicated.

In summary, we have found

$$\langle 0|\psi(0)\bar{\psi}(x)|0\rangle = (-i\not{\partial} + m) \int \frac{d^3p}{(2\pi)^3} \frac{1}{2\omega_p} e^{ipx}, \tag{12.41}$$

$$\langle 0|\bar{\psi}(x)\psi(0)|0\rangle = -(-i\not{\partial} + m) \int \frac{d^3p}{(2\pi)^3} \frac{1}{2\omega_p} e^{-ipx}. \tag{12.42}$$

These equations are independent of whether commutation or anticommutation relations are assumed.

Now let us first assume commutation relations. Then the time-ordered product is defined in the usual way: $T\{\psi(0)\bar{\psi}(x)\} = \psi(0)\bar{\psi}(x)\theta(-t) + \bar{\psi}(x)\psi(0)\theta(t)$. Then we get, recycling results from Section 12.4.1,

$$\langle 0|T\{\psi(0)\bar{\psi}(x)\}|0\rangle = (i\not{\partial} - m) \int \frac{d^3p}{(2\pi)^3} \frac{1}{2\omega_p} \left[e^{i\vec{p}\vec{x} - i\omega_p t}\theta(t) - e^{-i\vec{p}\vec{x} + i\omega_p t}\theta(-t) \right]$$

$$= (-i\not{\partial} + m) \int \frac{d^4p}{(2\pi)^4} \frac{\omega}{\sqrt{\vec{p}^2 + m^2}} \frac{i}{p^2 - m^2 + i\varepsilon} e^{ipx}, \tag{12.43}$$

which is not Lorentz invariant. If instead we assume anticommutation relations, and the fermionic time-ordered product, $T\{\psi(0)\bar{\psi}(x)\} = \psi(0)\bar{\psi}(x)\theta(-t) - \bar{\psi}(x)\psi(0)\theta(t)$, we find

$$\langle 0|T\{\psi(0)\bar{\psi}(x)\}|0\rangle = (-i\not{\partial} + m) \int \frac{d^4p}{(2\pi)^4} \frac{i}{\omega^2 - \omega_p^2 + i\varepsilon} e^{ipx}, \tag{12.44}$$

which is beautifully Lorentz invariant.

The Feynman propagator for Dirac spinors is more conventionally written as

$$\langle 0|T\{\psi(0)\bar{\psi}(x)\}|0\rangle = \int \frac{d^4p}{(2\pi)^4} \frac{i(\not{p}+m)}{p^2-m^2+i\varepsilon} e^{ipx}. \tag{12.45}$$

This is an extremely important result, used in practically every calculation in QED.

Let us trace back to what happened. We found for a scalar:

$$\langle 0|\phi^\star(x)\phi(0)|0\rangle = \int \frac{d^3p}{(2\pi)^3} \frac{1}{2\omega_p} e^{-ipx}, \tag{12.46}$$

$$\langle 0|\phi(0)\phi^\star(x)|0\rangle = \int \frac{d^3p}{(2\pi)^3} \frac{1}{2\omega_p} e^{ipx}, \tag{12.47}$$

as compared to (for $m=0$):

$$\langle 0|\bar{\psi}(x)\psi(0)|0\rangle = \int \frac{d^3p}{(2\pi)^3} \frac{\not{p}}{2\omega_p} e^{-ipx}, \tag{12.48}$$

$$\langle 0|\psi(0)\bar{\psi}(x)|0\rangle = \int \frac{d^3p}{(2\pi)^3} \frac{\not{p}}{2\omega_p} e^{ipx}. \tag{12.49}$$

Now we can see that the problem is that $\not{p}$ is odd under the rotation that takes $p \rightarrow -p$, so that an extra -1 is generated when we try to combine the time-ordered sum for the fermion. Rotating $p \rightarrow -p$ is a rotation by π. We saw that this gives a factor of i in the fermion case. So here we have two fermions, and we get a -1. So it is directly related to the spin $\frac{1}{2}$. This will happen for any half-integer spin, which gets an extra -1 in the rotation.

Another way to look at it is that the $\not{p}$ factor comes from the polarization sum, which in turn comes from the requirement that the free solutions satisfy the equations of motion, $\not{p}u_s(p) = \not{p}v_s(p) = 0$. In fact, we can now see directly that the same problem will happen for any particle of half-integer spin. A particle of spin $n + \frac{1}{2}$ for integer n will have a field with n vector indices and a spinor index, $\psi_{\mu_1 \cdots \mu_n}$. So the corresponding polarization sum must have a factor of γ_μ and the only thing around to contract γ_μ with is its momentum p_μ. Thus, we always get a $\not{p}$, plus possibly additional factors of p_μ^2, and the time-ordered product can never be Lorentz invariant unless the fields anticommute. These are fermions. They obey Fermi–Dirac statistics. In contrast, for integer spin there can only be an even number of p_μ^2 factors in the polarization sum. So these fields must commute to have Lorentz-invariant time-ordered products. These are bosons. They obey Bose–Einstein statistics.

12.5 Stability

One does not have to deal with time-ordered products to see the consequences of wrongly chosen statistics. In fact, a universe in which spinors commute would have disastrous consequences – particles with finite momentum could have negative energy. The particles would still be on-shell, $E^2 = \vec{p}^2 + m^2$, so this is not a problem with Lorentz invariance, but it would mean that all kinds of things such as $p^+ \to p^+ e^+ e^-$ would not be forbidden.

Recall from Eq. (8.13) that the energy density is given by the 00-component of the energy-momentum tensor:

$$\mathcal{E} = \mathcal{T}_{00} = \sum_n \frac{\partial \mathcal{L}}{\partial \dot{\phi}_n} \dot{\phi}_n - \mathcal{L}. \tag{12.50}$$

We derived this equation by identifying the energy-momentum tensor as the Noether current associated with space-time translations. We have already used this general definition of the energy density to constrain theories with integer spin particles in Chapter 8. Here, we will see how spin-statistics follows from having a positive-definite energy density. More precisely, we need the total energy given by

$$E = \int d^3x \mathcal{E} \tag{12.51}$$

to be bounded from below, since a constant shift has no physical consequences.

12.5.1 Free scalar fields

For a free complex scalar field,

$$\mathcal{L} = |\partial_\mu \phi|^2 - m^2 |\phi|^2, \tag{12.52}$$

the energy-momentum tensor is, starting from Eq. (3.35),

$$\mathcal{T}_{\mu\nu} = \sum_n \frac{\partial \mathcal{L}}{\partial(\partial_\mu \phi_n)} \partial_\nu \phi_n - g_{\mu\nu} \mathcal{L}$$
$$= \partial_\mu \phi^\star \partial_\nu \phi + \partial_\mu \phi \partial_\nu \phi^\star - g_{\mu\nu} \left[|\partial_\mu \phi|^2 - m^2 |\phi|^2 \right]. \tag{12.53}$$

The energy density is

$$\mathcal{E} = \mathcal{T}_{00} = (\partial_t \phi^\star)(\partial_t \phi) + (\vec{\nabla} \phi^\star) \cdot (\vec{\nabla} \phi) + m^2 \phi^\star \phi. \tag{12.54}$$

Classically, this would obviously be positive definite. It is not quite that simple in the quantum theory.

Using the free scalar fields, Eqs. (12.20) and (12.21), the total energy is

$$E = \int d^3x \mathcal{E} = \int d^3x \int \frac{d^3q}{(2\pi)^3} \frac{1}{\sqrt{2\omega_q}} \int \frac{d^3p}{(2\pi)^3} \frac{1}{\sqrt{2\omega_p}}$$
$$\times \left[-(\omega_q \omega_p + \vec{q} \cdot \vec{p})(a_q^\dagger e^{iqx} - b_q e^{-iqx})(-a_p e^{-ipx} + b_p^\dagger e^{ipx}) \right.$$
$$\left. + m^2 (a_q^\dagger e^{iqx} + b_q e^{-iqx})(a_p e^{-ipx} + b_p^\dagger e^{ipx}) \right]. \tag{12.55}$$

Doing the x integral first turns the phases into δ-functions. Using $\omega_p^2 = \vec{p}^2 + m^2$ then reduces the whole thing to

$$E = \int \frac{d^3p}{(2\pi)^3} \omega_p \left(a_p^\dagger a_p + b_p b_p^\dagger \right)$$

$$= \int \frac{d^3p}{(2\pi)^3} \omega_p \left[a_p^\dagger a_p + b_p^\dagger b_p + (2\pi)^3 \delta^3(0) \right]. \tag{12.56}$$

Using $\delta^3(0) = \frac{V}{(2\pi)^3}$, as in Eq. (5.12) and defining $\mathcal{E}_0 = \int \frac{d^3p}{(2\pi)^3} \omega_p$, this gives

$$E = \int \frac{d^3p}{(2\pi)^3} \omega_p \left[a_p^\dagger a_p + b_p^\dagger b_p \right] + V \mathcal{E}_0 \quad \text{(with commutators)}. \tag{12.57}$$

This $V\mathcal{E}_0$ term is an infinite contribution to the energy, which is independent of what state the system is in. It is just the zero-point energy for the sum of the particles and antiparticles in the Hilbert space (each of which gives $\frac{\omega_p}{2}$). Just as in classical mechanics, only differences in energy can have measurable effects (see Chapter 15). The important point for stability is that the energy difference between two states is

$$\Delta E = \int \frac{d^3p}{(2\pi)^3} \omega_p \left(\Delta\# \text{ particles} + \Delta\# \text{ antiparticles} \right). \tag{12.58}$$

States with more particles (or antiparticles) have more energy.

Now, suppose we had used anticommutation relations instead, then we would have had

$$E = \int \frac{d^3p}{(2\pi)^3} \omega_p \left(a_p^\dagger a_p - b_p^\dagger b_p \right) + V \mathcal{E}_0 \quad \text{(with anticommutators)}, \tag{12.59}$$

which would mean

$$\Delta E = \int \frac{d^3p}{(2\pi)^3} \omega_p \left(\Delta\# \text{ particles} - \Delta\# \text{ antiparticles} \right). \tag{12.60}$$

In particular, the energy can be lowered by producing antiparticles! Thus, the vacuum could spontaneously decay into particle–antiparticle pairs. Particles could spontaneously decay into particles and antiparticles. Nothing would be stable – this would be a huge disaster.

12.5.2 Free fermions

Now we will do the same computation for fermions. Here the Lagrangian is

$$\mathcal{L} = \bar{\psi}(i\slashed{\partial} - m)\psi \tag{12.61}$$

and the energy-momentum tensor is

$$\mathcal{T}_{\mu\nu} = i\bar{\psi}\gamma_\mu\partial_\nu\psi - g_{\mu\nu}\left[\bar{\psi}(i\slashed{\partial} - m)\psi\right]. \tag{12.62}$$

So the energy density is

$$\mathcal{E} = \mathcal{T}_{00} = \bar{\psi}(i\gamma^i\partial_i + m)\psi. \tag{12.63}$$

Using the equations of motion, this simplifies to

$$\mathcal{E} = i\bar{\psi}\gamma^0\partial_t\psi. \tag{12.64}$$

In the quantum theory

$$
E = \int d^3x\, \mathcal{E} = i \int d^3x \int \frac{d^3q}{(2\pi)^3} \frac{1}{\sqrt{2\omega_q}} \int \frac{d^3p}{(2\pi)^3} \frac{1}{\sqrt{2\omega_p}}
$$
$$
\times \sum_{s,s'} \big(e^{ipx} a_p^{s'\dagger} \bar{u}_p^{s'} + e^{-ipx} b_p^{s'} \bar{v}_p^{s'}\big) \gamma^0 \partial_t \big(e^{-iqx} a_q^s u_q^s + e^{iqx} b_q^{s\dagger} v_q^s\big). \quad (12.65)
$$

The x integral forces $\vec{q} = \vec{p}$ for the $\bar{u}u$ and $\bar{v}v$ terms, which then simplify using $\bar{u}_s(p)\gamma^0 u_{s'}(p) = u_s^\dagger(p) u_{s'}(p) = 2\omega_p \delta_{ss'}$. It also forces $\vec{q} = -\vec{p}$ for the $\bar{u}v$ and $\bar{v}u$ terms, which simplify with $u_s^\dagger(\vec{p}) v_{s'}(-\vec{p}) = v_s^\dagger(\vec{p}) u_{s'}(-\vec{p}) = 0$. The result is

$$
E = \sum_s \int \frac{d^3p}{(2\pi)^3} \omega_p \big(a_p^{s\dagger} a_p^s - b_p^s b_p^{s\dagger}\big). \quad (12.66)
$$

Now if we have anticommutators, this is just

$$
E = \sum_s \left[\int \frac{d^3p}{(2\pi)^3} \omega_p \big(a_p^{s\dagger} a_p^s + b_p^{s\dagger} b_p^s\big) - V\mathcal{E}_0 \right] \quad \text{(with anticommutators)}, \quad (12.67)
$$

which again counts the number of particles and antiparticles, weighted by the energy. Note that for fermions the zero-point energy is negative.

If we had commutators instead, we would have

$$
E = \sum_s \left[\int \frac{d^3p}{(2\pi)^3} \omega_p \big(a_p^{s\dagger} a_p^s - b_p^{s\dagger} b_p^s\big) - V\mathcal{E}_0 \right] \quad \text{(with commutators)}, \quad (12.68)
$$

which would have an energy unbounded from below.

So the stability requirement, that is, that the energy must grow when we add more particles or antiparticles, holds if and only if the spin-statistics theorem holds. Again, just postulating Lorentz invariance of the S-matrix implies spin-statistics, which implies stability.

12.5.3 General spins

The spinor calculation in this case was much easier than the scalar case. Nevertheless, we can track through and find that the terms that survived the scalar calculation came from

$$
E = \int d^3x \frac{d^3q}{(2\pi)^3} \frac{d^3p}{(2\pi)^3} \frac{1}{\sqrt{\omega_p \omega_q}} \left[(\partial_t e^{iqx})(\partial_t e^{-ipx}) a_q^\dagger a_p + b_q b_p^\dagger (\partial_t e^{-iqx})(\partial_t e^{ipx}) \right]
$$
$$
= \int \frac{d^3p}{(2\pi)^3} \omega_p \big(a_p^\dagger a_p + b_p b_p^\dagger\big). \quad (12.69)
$$

In contrast, the relevant terms for spin $\frac{1}{2}$ were (using commutation relations)

$$
E = \int d^3x \frac{d^3q}{(2\pi)^3} \frac{d^3p}{(2\pi)^3} \left[e^{iqx}(i\partial_t e^{-ipx}) a_q^\dagger a_p + b_q b_p^\dagger e^{-iqx}(i\partial_t e^{ipx}) \right]
$$
$$
= \int \frac{d^3p}{(2\pi)^3} \omega_p \big(a_p^\dagger a_p - b_p b_p^\dagger\big). \quad (12.70)
$$

The difference is that the spinor expression is linear in the time derivative, while the scalar is quadratic. This in turn comes from the fact that the Lagrangian for the scalar has two derivative kinetic terms, while the spinor has single derivatives.

More generally, every integer spin particle will be embedded in a tensor $(A_\mu, h_{\mu\nu}, Z_{\mu\nu\alpha}, \ldots)$. The terms quadratic in these fields will have an even number of indices to contract, forcing an even number of derivatives in the kinetic terms. In contrast, every half-integer spin particle will be embedded in a spinor field, with tensor indices $(\psi, \chi_\mu, \eta_{\mu\nu}, \ldots)$. They must be contracted with barred spinors $(\bar\psi, \overline\chi_\mu, \bar\eta_{\mu\nu}, \ldots)$, which transform in complex conjugate representations of the Lorentz group. To contract these, we must insert a γ_μ matrix, which must be contracted with a single ∂_μ. Thus, all kinetic terms for integer spin fields will have an even number of derivatives and kinetic terms for half-integer spin fields will have an odd number of derivatives. This will lead to the same minus signs in the derivation of the Hamiltonian. Thus, all integer (half-integer) spin particles must be bosons (fermions).

12.6 Causality

The other connection between spin and statistics that is often discussed comes from considerations of causality. Causality is a reasonable physical requirement. The precise condition is that the commutator of observables should vanish outside the lightcone, that is, at spacelike separation. For spin 0, the field itself is observable, so we require

$$[\phi(x), \phi(y)] = 0, \quad (x - y)^2 < 0. \tag{12.71}$$

What does this commutator have to do with physics? Remember, we are just doing quantum mechanics here. So when two operators commute they are simultaneously observable. Another way to say this is that if the operators commute they are uncorrelated and cannot influence each other. So, if we measure the field (remember ϕ measures the field) at $x = 0$ it should not influence the measurement at a distant point y at the same time. On the other hand, if we measure the field at $t = 0$ it might affect the field at a later time t at the same position x. This is a precise statement of causality.

What we are going to show below is that $[\bar\psi_\alpha(x), \psi_\beta(y)]$ does not vanish outside the lightcone. This would imply that if we could measure $\psi(x)$, then we would have a violation of causality. Unfortunately, spinors appear not to be observables. The only things we ever measure are numbers, which are constructed out of bilinears in spinors. Thus, the physical requirement is only that

$$[\bar\psi(x)\psi(x), \bar\psi(y)\psi(y)] = 0, \quad (x - y)^2 < 0. \tag{12.72}$$

This condition will be guaranteed if either $[\bar\psi_\alpha(x), \psi_\beta(y)] = 0$ *or* $\{\bar\psi_\alpha(x), \psi_\beta(y)\} = 0$ outside the lightcone. Thus, having the spinors anticommute (or commute) at spacelike separation would be a sufficient condition for causality, but it may not be necessary. In fact, one expects that perhaps the commutator of spinor bilinears will vanish outside the lightcone because two spinor fields at the same point transform like a combination of fields with integer spin.

Now let us compute this commutator, first for a scalar field, then for a spin-$\frac{1}{2}$ field:

$$\phi(x) = \int \frac{d^3q}{(2\pi)^3} \frac{1}{\sqrt{2\omega_q}} \left(a_q^\dagger e^{iqx} + a_q e^{-iqx}\right), \qquad (12.73)$$

so, using $[a_p, a_q^\dagger] = (2\pi)^3 \delta^3(p-q)$ and $[a_p, a_q] = [a_p^\dagger, a_q^\dagger] = 0$,

$$[\phi(x), \phi(y)] = \int \frac{d^3q}{(2\pi)^3} \frac{1}{\sqrt{2\omega_q}} \int \frac{d^3p}{(2\pi)^3} \frac{1}{\sqrt{2\omega_p}} \left(e^{iqx} e^{-ipy}[a_q^\dagger, a_p] + e^{-iqx} e^{ipy}[a_q, a_p^\dagger]\right)$$

$$= \int \frac{d^3q}{(2\pi)^3} \frac{1}{2\omega_q} \left(e^{-iq(x-y)} - e^{iq(x-y)}\right). \qquad (12.74)$$

Letting $t = x_0 - y_0$ and $\vec{r} = \vec{x} - \vec{y}$ we have

$$[\phi(x), \phi(y)] = \frac{1}{(2\pi)^2} \int \frac{q^2 dq}{2\omega_q} \int_{-1}^{1} d\cos\theta \left(e^{-i\omega_q t} e^{iqr\cos\theta} - e^{i\omega_q t} e^{-iqr\cos\theta}\right)$$

$$= \frac{-i}{2\pi^2} \int_0^\infty q^2 dq \frac{\sin(\sqrt{q^2+m^2}\,t)}{\sqrt{q^2+m^2}} \frac{\sin(qr)}{qr}$$

$$\equiv iD(t,r). \qquad (12.75)$$

This integral is tricky. We have to be careful since we expect it to be something non-analytic – that is the only way it can vanish everywhere outside the lightcone, but not vanish inside the lightcone.

The result is a function we call $D(t,r)$. For $m=0$ it is

$$D(t,r) = \frac{1}{4\pi r}\left[\delta(r+t) - \delta(r-t)\right], \qquad (12.76)$$

which has support only on the lightcone. For $m \neq 0$, we can find an exact expression for $D(t,r)$ in terms of the Bessel function $\mathcal{J}_0(x)$:

$$D(t,r) = \frac{1}{4\pi r}\frac{\partial}{\partial r} \begin{cases} \mathcal{J}_0(m\sqrt{t^2-r^2}), & t > r, \\ 0, & r > t > -r, \\ -\mathcal{J}_0(m\sqrt{t^2-r^2}), & t < -r. \end{cases} \qquad (12.77)$$

We see that $D(t,r)$ has support only in the future and past lightcones.

More generally, $D(t,\vec{r})$ is a Green's function for the Klein–Gordon equation with boundary conditions:

$$(\Box + m^2)D(t,\vec{r}) = 0, \quad D(0,\vec{r}) = 0, \quad \frac{\partial}{\partial t}D(t,\vec{r})\Big|_{t=0} = -\delta(\vec{r}). \qquad (12.78)$$

$D(t,\vec{r})$ satisfies

$$D(t,\vec{r}) = -D(-t,\vec{r}) \quad \text{and} \quad D(t,\vec{r}) = D(t,-\vec{r}). \qquad (12.79)$$

That is, it is odd under time reversal and even under parity. This can be seen from the explicit form, or from the boundary condition on the Green's function. So, with $x - y \equiv (t, \vec{r})$,

$$[\phi(x), \phi(y)] = iD(t, \vec{r}), \tag{12.80}$$

which has support only within the future and past lightcones, as desired.

If we had chosen anticommutation relations for the scalar, then

$$\{\phi(x), \phi(y)\} = \int \frac{d^3q}{(2\pi)^3} \frac{1}{2\omega_q} \left(e^{-iq(x-y)} + e^{iq(x-y)} \right)$$

$$= \frac{1}{2\pi^2} \int_0^\infty q^2 dq \frac{\cos(\sqrt{q^2 + m^2}t)}{\sqrt{q^2 + m^2}} \frac{\sin(qr)}{qr}$$

$$\equiv D_1(t, r). \tag{12.81}$$

For $m = 0$, the explicit form is

$$D_1(t, r) = \frac{1}{2\pi^2} \frac{1}{r^2 - t^2}. \tag{12.82}$$

For $m \neq 0$,

$$D_1(t, r) = -\frac{1}{4\pi r} \frac{\partial}{\partial r} \begin{cases} i\mathcal{Y}_0(m\sqrt{t^2 - r^2}), & t > r, \\ \mathcal{H}_0(im\sqrt{r^2 - t^2}), & r > t > -r, \\ i\mathcal{Y}_0(m\sqrt{t^2 - r^2}), & t < -r, \end{cases} \tag{12.83}$$

where $\mathcal{Y}_0(x)$ is a Bessel function of the second kind and $\mathcal{H}_0(x) = \mathcal{J}_0(x) + i\mathcal{Y}_0(x)$ is a Hankel function. This does not vanish outside the lightcone and therefore *spin-0 particles must be bosons*.

12.6.1 Spinor case

Now let us do the same calculation with quantized spinors. We start by assuming

$$[a_p^s, a_q^{s'\dagger}] = [b_p^s, b_q^{s'\dagger}] = (2\pi)^3 \delta^3(p - q)\delta_{ss'} \tag{12.84}$$

to see what goes wrong. Then,

$$[\psi(x), \bar{\psi}(y)] = \int \frac{d^3q}{(2\pi)^3} \frac{1}{\sqrt{2\omega_q}} \int \frac{d^3p}{(2\pi)^3} \frac{1}{\sqrt{2\omega_p}}$$

$$\times \sum_{s,s'} \left[\left(e^{-iqx} a_q^s u_q^s + e^{iqx} b_q^{s\dagger} v_q^s \right), \left(e^{ipy} a_p^{s'\dagger} \bar{u}_p^{s'} + e^{-ipy} b_p^{s'} \bar{v}_p^{s'} \right) \right]$$

$$= \sum_s \int \frac{d^3q}{(2\pi)^3} \frac{1}{2\omega_q} \left[u_q^s \bar{u}_q^s e^{-iq(x-y)} - v_q^s \bar{v}_q^s e^{iq(x-y)} \right]. \tag{12.85}$$

Now we sum over polarizations using the outer product to get

$$[\psi(x), \bar{\psi}(y)] = \int \frac{d^3q}{(2\pi)^3} \frac{1}{2\omega_q} \left[(\slashed{q} + m)e^{-iq(x-y)} - (\slashed{q} - m)e^{iq(x-y)} \right] \tag{12.86}$$

and

$$
\begin{aligned}
[\psi(x), \bar{\psi}(y)] &= \int \frac{d^3 q}{(2\pi)^3} \frac{1}{2\omega_q} \left[(i\partial\!\!\!/_x + m) e^{-iq(x-y)} - (-i\partial\!\!\!/_x - m) e^{iq(x-y)} \right] \\
&= (i\partial\!\!\!/_x + m) \int \frac{d^3 q}{(2\pi)^3} \frac{1}{2\omega_q} \left[e^{-iq(x-y)} + e^{iq(x-y)} \right] \\
&= (i\partial\!\!\!/_x + m) D_1(t, r).
\end{aligned}
\tag{12.87}
$$

Thus, we get the function that does not vanish outside the lightcone.

If, instead, we take anticommutation relations

$$
\{a_p^r, a_q^{s\dagger}\} = \{b_p^r, b_q^{s\dagger}\} = (2\pi)^3 \delta^{(3)}(p-q)\delta^{rs},
\tag{12.88}
$$

then

$$
\begin{aligned}
\{\psi(x), \bar{\psi}(y)\} &= \int \frac{d^3 q}{(2\pi)^3} \frac{1}{2\omega_q} \left[(q\!\!\!/ + m) e^{-iq(x-y)} + (q\!\!\!/ - m) e^{iq(x-y)} \right] \\
&= (i\partial\!\!\!/_x + m) \int \frac{d^3 q}{(2\pi)^3} \frac{1}{2\omega_q} \left[e^{-iq(x-y)} - e^{iq(x-y)} \right] \\
&= (i\partial\!\!\!/_x + m) D(t, r),
\end{aligned}
\tag{12.89}
$$

which vanishes outside the lightcone as desired.

The vanishing of anticommutators of spinors outside the lightcone is a sufficient but not necessary condition for causality; see Problem 12.1.

12.6.2 Higher spins

For spin 0 and spin $\frac{1}{2}$ we found

$$
[\phi(x), \phi(y)] = D(t, r), \qquad\qquad \{\phi(x), \phi(y)\} = D_1(r, t),
\tag{12.90}
$$

$$
[\psi(x), \bar{\psi}(y)] = (i\partial\!\!\!/_x + m) D_1(t, r), \quad \{\psi(x), \bar{\psi}(y)\} = (i\partial\!\!\!/_x + m) D(t, r),
\tag{12.91}
$$

Since $D(t, r)$ vanishes outside of the lightcone, but $D_1(t, r)$ does not, we concluded that we needed commutators for spin 0 and anticommutators for spin $\frac{1}{2}$. The prefactors are just spin sums – recall that $\sum_{\text{spins}} \bar{u}u = (p\!\!\!/ + m)$ for Dirac spinors and $\sum_{\text{spins}} = 1$ for a scalar. For higher spin fields, the canonical quantization will result in the same integrals, but with a different prefactor operator.

For higher spin fields, we will get the appropriate polarization sum. For massive spin 1, we would get

$$
[A_\mu(x), A_\nu(y)] = \left(g_{\mu\nu} + \frac{1}{m^2} \partial_\mu \partial_\nu \right) D(t, r),
$$

$$
\{A_\mu(x), A_\nu(y)\} = \left(g_{\mu\nu} + \frac{1}{m^2} \partial_\mu \partial_\nu \right) D_1(t, r),
\tag{12.92}
$$

and again we have to pick commutators.

For higher spin fields there will be more derivatives acting on either the function D or D_1. We can see whether D or D_1 appears by a simple symmetry argument (due to Pauli). Observe that under the combined time reversal and parity transformation, PT,

$$D(t, \vec{r}) = -D(-t, -\vec{r}) \quad \text{and} \quad D_1(t, \vec{r}) = D_1(-t, -\vec{r}). \tag{12.93}$$

This can be seen at once from Eqs. (12.75) and (12.81). Also, the commutator $[\phi(x), \phi(y)]$ is odd under $x \leftrightarrow y$ and the anticommutator is even. Derivatives are odd. Therefore,

$$[A_{\mu_1 \cdots \mu_n}(x), A_{\nu_1 \cdots \nu_n}(y)] = f(\Box, g_{\mu\nu}, \partial_\mu \partial_\nu) D(t, r), \tag{12.94}$$

$$\{A_{\mu_1 \cdots \mu_n}(x), A_{\nu_1 \cdots \nu_n}(y)\} = f(\Box, g_{\mu\nu}, \partial_\mu \partial_\nu) D_1(t, r), \tag{12.95}$$

since there must be an even number of derivatives in the function f by Lorentz invariance.

For half-integer spin, we will get an odd number of derivatives. In general, the quantity $[\bar{\psi}_{\mu_1 \cdots \mu_n}(x), \psi_{\nu_1 \cdots \nu_n}(y)]$ does not have definite quantum number under PT, since the fields are complex. But if we combine with the interchange of $a_p \leftrightarrow b_p$, charge conjugation, the CPT transformation properties determine that

$$[\bar{\psi}_{\mu_1 \cdots \mu_n}(x), \psi_{\nu_1 \cdots \nu_n}(y)] = f(\Box, g_{\mu\nu}, \partial_\mu \partial_\nu) \slashed{\partial} D_1(t, r), \tag{12.96}$$

$$\{\bar{\psi}_{\mu_1 \cdots \mu_n}(x), \psi_{\nu_1 \cdots \nu_n}(y)\} = f(\Box, g_{\mu\nu}, \partial_\mu \partial_\nu) \slashed{\partial} D(t, r), \tag{12.97}$$

showing that all integer spin fields must have commutation relations and all half-integer spin fields must have anticommutation relations.

In summary, for integer spins, which can be observables, causality is consistent only with commutation relations. For half-integer spins, anticommutation relations are a sufficient condition for causality. This method does not show that anticommutation relations for half-integer spins are a necessary condition for causality.

Problems

12.1 In a causal theory, commutators of observables should vanish outside the lightcone, $[\phi(x), \phi(y)] = 0$ for $(x - y)^2 < 0$. For spinors, we found that with anticommutation relations $\{\bar{\psi}(x), \psi(y)\} = 0$ outside the lightcone. This implies that integer spin quantities constructed out of spinors are automatically causal, e.g. $[\bar{\psi}\psi(x), \bar{\psi}\psi(y)] = 0$. However, this is not a proof that spinors *must* anticommute. What would happen to $[\bar{\psi}\psi(x), \bar{\psi}\psi(y)]$ outside the lightcone if we used *commutation* relations for spinors? For simplicity, you can just look at $\langle 0| [\bar{\psi}\psi(x), \bar{\psi}\psi(y)] |0\rangle$.

13 Quantum electrodynamics

Now we are ready to do calculations in QED. We have found that the Lagrangian for QED is

$$\mathcal{L} = -\frac{1}{4}F_{\mu\nu}^2 + i\bar{\psi}\slashed{D}\psi - m\bar{\psi}\psi, \tag{13.1}$$

with $D_\mu\psi = \partial_\mu\psi + ieA_\mu\psi$. We have also introduced quantized Dirac fields:

$$\psi(x) = \sum_s \int \frac{d^3p}{(2\pi)^3}\frac{1}{\sqrt{2\omega_p}}\left(a_p^s u_p^s e^{-ipx} + b_p^{s\dagger} v_p^s e^{ipx}\right), \tag{13.2}$$

$$\bar{\psi}(x) = \sum_s \int \frac{d^3p}{(2\pi)^3}\frac{1}{\sqrt{2\omega_p}}\left(b_p^s \bar{v}_p^s e^{-ipx} + a_p^{s\dagger} \bar{u}_p^s e^{ipx}\right). \tag{13.3}$$

The creation and annihilation operators for spinors must anticommute by the spin-statistics theorem:

$$\left\{a_p^{s\dagger}, a_q^{s'\dagger}\right\} = \left\{a_p^s, a_q^{s'}\right\} = \left\{b_p^{s\dagger}, b_q^{s'\dagger}\right\} = \left\{b_p^s, b_q^{s'}\right\} = 0 \tag{13.4}$$

and

$$\left\{a_p^s, a_q^{s'\dagger}\right\} = \left\{b_p^s, b_q^{s'\dagger}\right\} = \delta_{ss'}(2\pi)^3\delta^3(p-q). \tag{13.5}$$

A basis of spinors for each momentum p^μ can be written as

$$u_s(p) = \begin{pmatrix} \sqrt{p\cdot\sigma}\xi_s \\ \sqrt{p\cdot\bar{\sigma}}\xi_s \end{pmatrix}, \quad v_s(p) = \begin{pmatrix} \sqrt{p\cdot\sigma}\eta_s \\ -\sqrt{p\cdot\bar{\sigma}}\eta_s \end{pmatrix}, \tag{13.6}$$

with $\xi_1 = \eta_1 = \begin{pmatrix} 1 \\ 0 \end{pmatrix}$ and $\xi_2 = \eta_2 = \begin{pmatrix} 0 \\ 1 \end{pmatrix}$. These spinors satisfy

$$\sum_{s=1}^2 u_s(p)\bar{u}_s(p) = \slashed{p} + m, \tag{13.7}$$

$$\sum_{s=1}^2 v_s(p)\bar{v}_s(p) = \slashed{p} - m. \tag{13.8}$$

We have also calculated the Feynman propagator for a Dirac spinor:

$$\langle 0|T\{\psi(0)\bar{\psi}(x)\}|0\rangle = \int \frac{d^4p}{(2\pi)^4}\frac{i(\slashed{p}+m)}{p^2 - m^2 + i\varepsilon}e^{ipx}. \tag{13.9}$$

In this chapter we will derive the Feynman rules for QED and then perform some important calculations.

13.1 QED Feynman rules

The Feynman rules for QED can be read directly from the Lagrangian just as in scalar QED. The only subtlety is possible extra minus signs coming from anticommuting spinors within the time ordering. First, we write down the Feynman rules, then derive the supplementary minus sign rules.

A photon propagator is represented with a squiggly line:

$$= \frac{-i}{p^2 + i\varepsilon}\left[g_{\mu\nu} - (1-\xi)\frac{p_\mu p_\nu}{p^2}\right]. \tag{13.10}$$

Unless we are explicitly checking gauge invariance, we will usually work in Feynman gauge, $\xi = 1$, where the propagator is

$$= \frac{-ig_{\mu\nu}}{p^2 + i\varepsilon} \quad \text{(Feynman gauge)} . \tag{13.11}$$

A spinor propagator is a solid line with an arrow:

$$= \frac{i(\slashed{p} + m)}{p^2 - m^2 + i\varepsilon}. \tag{13.12}$$

The arrow points to the right for particles and to the left for antiparticles. For internal lines, the arrow points with momentum flow.

External photon lines get polarization vectors:

$$= \epsilon_\mu(p) \quad \text{(incoming)}, \tag{13.13}$$

$$= \epsilon_\mu^\star(p) \quad \text{(outgoing)}. \tag{13.14}$$

Here the blob means the rest of the diagram.

External fermion lines get spinors, with u spinors for electrons and v spinors for positrons.

$$= u^s(p), \tag{13.15}$$

$$= \bar{u}^s(p), \tag{13.16}$$

$$= \bar{v}^s(p), \tag{13.17}$$

$$= v^s(p). \tag{13.18}$$

External spinors are on-shell (they are forced to be on-shell by LSZ). So, for external spinors we can use the equations of motion:

$$(\slashed{p} - m)u^s(p) = \bar{u}^s(p)(\slashed{p} - m) = 0, \tag{13.19}$$

$$(\slashed{p} + m)v^s(p) = \bar{v}^s(p)(\slashed{p} + m) = 0, \tag{13.20}$$

which will simplify a number of calculations.

Expanding the Lagrangian,

$$\mathcal{L} = -\frac{1}{4}F_{\mu\nu}^2 + \bar{\psi}(i\gamma^\mu\partial_\mu - m)\psi - e\bar{\psi}\gamma^\mu\psi A_\mu, \tag{13.21}$$

we see that the interaction is $\mathcal{L}_{\text{int}} = -e\bar{\psi}\gamma^\mu\psi A_\mu$. Since there is no factor of momentum, the Feynman rule is the same for any combination of incoming or outgoing fields (unlike in scalar QED):

$$= -ie\gamma^\mu. \tag{13.22}$$

The μ on the γ^μ will get contracted with the μ of the photon, which will either be in the $g_{\mu\nu}$ of the photon propagator (if the photon is internal) or the ϵ_μ of a polarization vector (if the photon is external).

The $\gamma^\mu = \gamma^\mu_{\alpha\beta}$ as a matrix will always get sandwiched between spinors, as in

$$\bar{u}\gamma^\mu u = \bar{u}_\alpha\gamma^\mu_{\alpha\beta}u_\beta \tag{13.23}$$

for e^-e^- scattering, or $\bar{v}\gamma^\mu u$ for e^+e^- annihilation, etc. The barred spinor always goes on the left, since the interaction is $\bar{\psi}A_\mu\gamma^\mu\psi$. If there is an internal fermion line between the ends, the fermion propagator goes between the end spinors:

$$= (-ie)^2\bar{u}(p_3)\gamma^\mu\frac{i(\not{p}_2+m)}{p_2^2-m^2+i\varepsilon}\gamma^\nu u(p_1)\epsilon^2_\mu(q_2)\,\epsilon^1_\nu(q_1),$$

$$\tag{13.24}$$

where the photon momenta are $q_1^\mu = p_2^\mu - p_1^\mu$ and $q_2^\mu = p_3^\mu - p_2^\mu$. In this example, the three γ-matrices get multiplied and then sandwiched between the spinors. To see explicitly what is a matrix and what is a vector, we can add in the spinor indices:

$$\bar{u}(p_3)\,\gamma^\mu\frac{i(\not{p}_2+m)}{p_2^2-m^2+i\varepsilon}\gamma^\nu u(p_1) = \bar{u}_\alpha(p_3)\,\gamma^\mu_{\alpha\beta}\frac{i(\not{p}_2+m)_{\beta\gamma}}{p_2^2-m^2+i\varepsilon}\gamma^\nu_{\gamma\delta}u_\delta(p_1). \tag{13.25}$$

If we tie the ends of the diagram above together we get a loop:

$$\tag{13.26}$$

For fermion loops we use the same convention as for scalar loops that the loop momentum goes in the direction of the particle-flow arrow. In the loop, since any possible intermediate states are allowed, we must integrate over the momenta of the virtual spinors as well as sum over their possible spins. The $u_\delta\bar{u}_\alpha$ in Eq. (13.25) then gets replaced by a propagator that sums over all possible spins. This is done automatically since the numerator of the propagator is $(\not{p}_2+m)_{\delta\alpha} = \sum_s u_\alpha^s\bar{u}_\alpha^s$. We also must integrate over all possible momenta constrained by momentum conservation at each vertex. So the loop in Eq. (13.26) evaluates to

$$iM = -(-ie)^2 \int \frac{d^4p_1}{(2\pi)^4} \frac{d^4p_2}{(2\pi)^4} (2\pi)^4 \delta^4\left(p + p_2 - p_1\right) \epsilon_\mu^{2\star}(p)\epsilon_\nu^1(p)$$

$$\times \left[\gamma^\mu_{\alpha\beta} \frac{i(\not{p}_1 + m)_{\beta\gamma}}{p_1^2 - m^2 + i\varepsilon} \gamma^\nu_{\gamma\delta} \frac{i(\not{p}_2 + m)_{\delta\alpha}}{p_2^2 - m^2 + i\varepsilon} \right]. \quad (13.27)$$

The extra minus sign is due to spin-statistics, as will be explained shortly. Contracting all the spinor indices and replacing p_1^μ by $p^\mu + k^\mu$ and p_2^μ by k^μ:

$$= e^2 \epsilon_\mu^{2\star} \epsilon_\nu^1 \int \frac{d^4k}{(2\pi)^4} \text{Tr}\left[\gamma^\mu \frac{i(\not{p} + \not{k} + m)}{(p + k)^2 - m^2 + i\varepsilon} \gamma^\nu \frac{i(\not{k} + m)}{k^2 - m^2 + i\varepsilon} \right], \quad (13.28)$$

where the trace is a trace of spinor indices. Computing Feynman diagrams in QED will often involve taking the trace of products of γ-matrices.

A useful general rule is that the spinor matrices are always multiplied together in the direction opposite to the particle-flow arrow, which allows us to read off Eqs. (13.24) and (13.28) easily from the corresponding diagrams.

13.1.1 Signs

Recall from Eq. (12.29) that spinors anticommute within a time-ordered product:

$$T\{\cdots\psi(x)\psi(y)\cdots\} = -T\{\cdots\psi(y)\psi(x)\cdots\}. \quad (13.29)$$

Minus signs coming from such anticommutations appear in the Feynman rules. It is easiest to see when they should appear by example.

Consider Møller scattering ($e^-e^- \to e^-e^-$) at tree-level. There are two Feynman diagrams, for the t-channel (in Feynman gauge):

$$= \pm(-ie)\bar{u}(p_3)\gamma^\mu u(p_1)\frac{-ig_{\mu\nu}}{(p_1-p_3)^2}(-ie)\bar{u}(p_4)\gamma^\nu u(p_2),$$

$$(13.30)$$

and the u-channel:

$$= \pm(-ie)\bar{u}(p_3)\gamma^\mu u(p_2)\frac{-ig_{\mu\nu}}{(p_1-p_4)^2}(-ie)\bar{u}(p_4)\gamma^\nu u(p_1).$$

$$(13.31)$$

The question is: What sign should each diagram have?

To find out, recall that Feynman diagrams represent contributions to S-matrix elements. By the LSZ reduction theorem, these Feynman diagrams come from the Fourier transform of the Green's function:

$$G_4(x_1, x_2, x_3, x_4) = \langle \Omega | T\left\{ \psi(x_3)\bar{\psi}(x_1)\psi(x_4)\bar{\psi}(x_2) \right\} | \Omega \rangle, \qquad (13.32)$$

with external propagators removed and external spinors added. Here, we have picked a convention for the ordering of spinors in the definition of G_4. While a different convention might differ by an overall sign, the overall sign does not matter for cross sections; it is the relative sign for different perturbative contributions to G_4 that has physical consequences.

The first non-zero contribution to this Green's function in perturbation theory comes at order e^2 in an expansion of free fields:

$$G_4(x_1, x_2, x_3, x_4) = (-ie)^2 \int d^4x \int d^4y$$

$$\times \langle 0 | T\left\{ \psi(x_3)\bar{\psi}(x_1)\psi(x_4)\bar{\psi}(x_2)\left(\bar{\psi}(x)\slashed{A}(x)\psi(x)\right)\left(\bar{\psi}(y)\slashed{A}(y)\psi(y)\right) \right\} | 0 \rangle, \quad (13.33)$$

where the big $(\cdots)$ indicate that the spinors inside are contracted. More explicitly,

$$G_4(x_1, x_2, x_3, x_4) = (-ie)^2 \gamma^\mu_{\beta_1\beta_2} \gamma^\nu_{\beta_3\beta_4} \int d^4x \int d^4y$$

$$\times \langle 0 | T\left\{ \psi_{\alpha_3}(x_3)\bar{\psi}_{\alpha_1}(x_1)\psi_{\alpha_4}(x_4)\bar{\psi}_{\alpha_2}(x_2)\bar{\psi}_{\beta_1}(x)A^\mu(x)\psi_{\beta_2}(x)\bar{\psi}_{\beta_3}(y)A^\nu(y)\psi_{\beta_4}(y) \right\} | 0 \rangle.$$
$$(13.34)$$

In this form, we can anticommute the spinors within the time ordering before performing any contractions.

To get Feynman diagrams out of this Green's function, we have to perform contractions, which means creating fields from the vacuum and then annihilating them. To be absolutely certain about the sign coming from the contraction, it is easiest to anticommute the fields so that the fields that annihilate spinors are right next to the fields that create them. For example, in the t-channel diagram, the top electron line is created by $\bar{\psi}(x_1)$, then annihilated by $\psi(x)$, then created by $\bar{\psi}(x)$ and then annihilated by $\psi(x_3)$. Putting the fields so that the annihilation operators will occur right after the creation operators, we have

$$G_4(x_1, x_2, x_3, x_4) = (-ie)^2 \gamma^\mu_{\beta_1\beta_2} \gamma^\nu_{\beta_3\beta_4} \int d^4x \int d^4y$$

$$\times \langle 0 | T\left\{ \psi_{\alpha_3}(x_3)\bar{\psi}_{\beta_1}(x)A^\mu(x)\psi_{\beta_2}(x)\bar{\psi}_{\alpha_1}(x_1)\psi_{\alpha_4}(x_4)\bar{\psi}_{\beta_3}(y)A^\nu(y)\psi_{\beta_4}(y)\bar{\psi}_{\alpha_2}(x_2) \right\} | 0 \rangle.$$
$$(13.35)$$

Contractions of these spinors in this order gives the t-channel diagram in Eq. (13.30).

For the u-channel, ordering the fields so that the contractions are in order gives

$$G_4(x_1, x_2, x_3, x_4) = -(-ie)^2 \gamma^\mu_{\beta_1\beta_2} \gamma^\nu_{\beta_3\beta_4} \int d^4x \int d^4y$$

$$\times \langle 0 | T\left\{ \psi_{\alpha_4}(x_4)\bar{\psi}_{\beta_1}(x)A^\mu(x)\psi_{\beta_2}(x)\bar{\psi}_{\alpha_1}(x_1)\psi_{\alpha_3}(x_3)\bar{\psi}_{\beta_3}(y)A^\nu(y)\psi_{\beta_4}(y)\bar{\psi}_{\alpha_2}(x_2) \right\} | 0 \rangle,$$
$$(13.36)$$

so that the u-channel in Eq. (13.31) has a relative minus with respect to the t-channel. The result is that the matrix element for Møller scattering has the form

$$\mathcal{M} = \mathcal{M}_t + \mathcal{M}_u = e^2 \left\{ \frac{[\bar{u}(p_3)\gamma^\mu u(p_1)][\bar{u}(p_4)\gamma^\mu u(p_2)]}{(p_1 - p_3)^2} \right.$$
$$\left. - \frac{[\bar{u}(p_4)\gamma^\mu u(p_1)][\bar{u}(p_3)\gamma^\mu u(p_2)]}{(p_1 - p_4)^2} \right\}. \quad (13.37)$$

A shortcut to remembering the relative minus sign is simply to note that $G_4(x_1, x_2, x_3, x_4) = -G_4(x_1, x_2, x_4, x_3)$. A minus sign from interchanging the identical fermions at x_3 and x_4 is exactly what you would expect from Fermi–Dirac statistics. The overall sign of the sum of the matrix elements is an unphysical phase, but the relative sign of the t- and u-channels is important for the cross term in the $|\mathcal{M}|^2 = |\mathcal{M}_u + \mathcal{M}_t|^2$ and has observable effects.

One can do the same exercise for loops. For example, a loop such as

$$(13.38)$$

comes from a term in the perturbative expansion of the time-ordered product for two photon fields of the form (leaving all spinor indices implicit)

$$G_2 = (-ie)^2 (\gamma_{\alpha\beta} \cdots) \langle 0| T\{ A^\mu(x_1) A^\nu(x_2) \bar{\psi}(x) A^\alpha(x) \psi(x) \bar{\psi}(y) A^\beta(y) \psi(y) \} |0\rangle. \quad (13.39)$$

To get the spinors into the order where they are created and then immediately destroyed, we need to anticommute $\psi(y)$ from the right to the left. That is, we use

$$\bar{\psi}_\alpha(x) \psi_\beta(x) \bar{\psi}_\gamma(y) \psi_\delta(y) = -\psi_\delta(y) \bar{\psi}_\alpha(x) \psi_\beta(x) \bar{\psi}_\gamma(y). \quad (13.40)$$

Thus, the Feynman rule for this fermion loop should be supplemented with an additional minus sign. As an exercise, you should check that adding more photons to a fermionic loop does not change this overall minus sign.

In summary, the Feynman rules for fermions must be supplemented by a factor of

- -1 for interchange of external identical fermions. Diagrams such as s- and t-channel exchanges, which would be present even for non-identical particles, do not get an extra minus sign. The -1 is a relative minus sign between two diagrams that are related by interchanging two external identical particles.
- -1 for each fermion loop.

13.2 γ-matrix identities

Before beginning the QED calculations, let us derive some useful identities about γ-matrices. We will often need to take traces of products of γ-matrices. These can often be simplified using the cyclic property of the trace:

$$\text{Tr}[AB \cdots C] = \text{Tr}[B \cdots CA]. \quad (13.41)$$

We will often also use $\gamma_5 \equiv i\gamma_0\gamma_1\gamma_2\gamma_3$, which satisfies

$$\gamma_5^2 = 1, \quad \gamma_5\gamma_\mu = -\gamma_\mu\gamma_5. \tag{13.42}$$

To keep the spinor indices straight, we sometimes write $\mathbb{1}$ for the identity on spinor indices. So,

$$\{\gamma^\mu, \gamma^\nu\} = 2g^{\mu\nu}\mathbb{1} \tag{13.43}$$

and

$$\mathrm{Tr}[g^{\mu\nu}\mathbb{1}] = g^{\mu\nu}\mathrm{Tr}[\mathbb{1}] = 4g^{\mu\nu}. \tag{13.44}$$

The $g^{\mu\nu}$ are just numbers for each μ and ν and pull out of the trace.

Then

$$\mathrm{Tr}[\gamma^\mu] = \mathrm{Tr}[\gamma_5\gamma_5\gamma^\mu] = \mathrm{Tr}[\gamma_5\gamma^\mu\gamma_5] = -\mathrm{Tr}[\gamma_5\gamma_5\gamma^\mu] = -\mathrm{Tr}[\gamma^\mu], \tag{13.45}$$

where we have cycled the γ_5-matrix in the second step. Thus

$$\mathrm{Tr}[\gamma^\mu] = 0. \tag{13.46}$$

Similarly,

$$\mathrm{Tr}[\gamma^\mu\gamma^\nu] = \mathrm{Tr}[-\gamma^\nu\gamma^\mu + 2g^{\mu\nu}\mathbb{1}] = -\mathrm{Tr}[\gamma^\mu\gamma^\nu] + 8g^{\mu\nu}, \tag{13.47}$$

which leads to

$$\mathrm{Tr}[\gamma^\mu\gamma^\nu] = 4g^{\mu\nu}. \tag{13.48}$$

In a similar way, you can show

$$\mathrm{Tr}[\gamma^\alpha\gamma^\beta\gamma^\mu] = 0 \tag{13.49}$$

and more generally that the trace of an odd number of γ-matrices is zero. For four γ-matrices, the result is (Problem 11.1)

$$\mathrm{Tr}[\gamma^\alpha\gamma^\mu\gamma^\beta\gamma^\nu] = 4(g^{\alpha\mu}g^{\beta\nu} - g^{\alpha\beta}g^{\mu\nu} + g^{\alpha\nu}g^{\mu\beta}). \tag{13.50}$$

You will use this last one a lot! You can remember the signs because adjacent indices give plus and the other one gives minus.

A summary of important γ-matrix identities is given in Appendix A.

13.3 $e^+e^- \rightarrow \mu^+\mu^-$

The muon, μ^-, is a particle that is identical to the electron as far as QED is concerned, except heavier. Studying processes with muons therefore provides simple tests of QED. Indeed, the simplest tree-level scattering process in QED is $e^+e^- \rightarrow \mu^+\mu^-$, which we calculate at tree-level here and at 1-loop in Chapter 20. The leading-order contribution is

$$i\mathcal{M} = $$

$$= (-ie)\bar{v}_\alpha(p_2)\gamma^\mu_{\alpha\beta}u_\beta(p_1)\frac{-i\left[g_{\mu\nu} - (1-\xi)\frac{k_\mu k_\nu}{k^2}\right]}{k^2}(-ie)\bar{u}_\delta(p_3)\gamma^\nu_{\delta\iota}v_\iota(p_4), \quad (13.51)$$

where $k^\mu = p_1^\mu + p_2^\mu = p_3^\mu + p_4^\mu$. Each of these spinors has a spin, thus we should properly write $u_\alpha^{s_1}(p_1)$ and so on. It is conventional to leave these spin labels implicit. Since the spinors are on-shell, we can use the equations of motion $p\!\!\!/_1 u(p_1) = m\,u(p_1)$ and $\bar{v}(p_2)p\!\!\!/_2 = -m\bar{v}(p_2)$. Thus,

$$\bar{v}_\alpha(p_2)\gamma^\mu_{\alpha\beta}u_\beta(p_1)k_\mu = \bar{v}(p_2)p\!\!\!/_1 u(p_1) + \bar{v}(p_2)p\!\!\!/_2 u(p_1)$$
$$= m\,\bar{v}(p_2)\,u(p_1) - m\,\bar{v}(p_2)u(p_1) = 0, \quad (13.52)$$

implying that the $k^\mu k^\nu$ term does not contribute, as expected by gauge invariance. So,

$$\mathcal{M} = \frac{e^2}{s}\bar{v}(p_2)\gamma^\mu u(p_1)\bar{u}(p_3)\gamma_\mu v(p_4), \quad (13.53)$$

where $s = (p_1 + p_2)^2$ as usual.

To calculate $|\mathcal{M}|^2$ we need the conjugate amplitude. To get this, we first recall that

$$\gamma_\mu^\dagger\gamma_0 = \gamma_0\gamma_\mu \quad \text{and} \quad \gamma_0^\dagger = \gamma_0. \quad (13.54)$$

So,

$$(\bar{\psi}_1\gamma^\mu\psi_2)^\dagger = (\psi_1^\dagger\gamma_0\gamma^\mu\psi_2)^\dagger = \psi_2^\dagger\gamma^{\mu\dagger}\gamma_0^\dagger\psi_1 = \psi_2^\dagger\gamma_0\gamma^\mu\psi_1 = \bar{\psi}_2\gamma^\mu\psi_1. \quad (13.55)$$

This nice transformation property is another reason why using $\bar{\psi}$ instead of $\psi^\dagger$ is useful. Then,

$$\mathcal{M}^\dagger = \frac{e^2}{s}\bar{v}(p_4)\gamma^\mu u(p_3)\bar{u}(p_1)\gamma_\mu v(p_2) \quad (13.56)$$

and therefore

$$|\mathcal{M}|^2 = \frac{e^4}{s^2}\left[\bar{v}(p_2)\gamma^\mu u(p_1)\right]\left[\bar{u}(p_3)\gamma_\mu v(p_4)\right]\left[\bar{v}(p_4)\gamma^\nu u(p_3)\right]\left[\bar{u}(p_1)\gamma_\nu v(p_2)\right]. \quad (13.57)$$

The grouping is meant to emphasize that each term in brackets is just a number for each μ and each set of spins. Thus $|\mathcal{M}|^2$ is a product of these numbers. For example, we could also have written

$$|\mathcal{M}|^2 = \frac{e^4}{s^2}\left[\bar{v}(p_2)\gamma^\mu u(p_1)\right]\left[\bar{u}(p_1)\gamma^\nu v(p_2)\right]\left[\bar{u}(p_3)\gamma_\mu v(p_4)\right]\left[\bar{v}(p_4)\gamma_\nu u(p_3)\right], \quad (13.58)$$

which shows that $|\mathcal{M}|^2$ is the contraction of two tensors, one depending only on the initial state, and the other depending only on the final state.

13.3.1 Unpolarized scattering

The easiest thing to calculate from this is the cross section for scattering assuming spin is not measured. The spin sum can be performed with a Dirac trace. To see this, we will sum over the μ^+ spins using

$$\sum_s v_\alpha^s(p_4)\,\bar{v}_\beta^s(p_4) = \sum_s \bar{v}_\beta^s(p_4)\,v_\alpha^s(p_4) = \left(p\!\!\!/_4 - m_\mu \mathbb{1}\right)_{\alpha\beta}, \tag{13.59}$$

and over the μ^- spins using

$$\sum_s u_\alpha^s(p_3)\,\bar{u}_\beta^s(p_3) = \sum_s \bar{u}_\beta^s(p_3)\,u_\alpha^s(p_3) = \left(p\!\!\!/_3 + m_\mu \mathbb{1}\right)_{\alpha\beta}. \tag{13.60}$$

We have written each sum two ways to emphasize that these are sums over vectors of complex numbers corresponding to external spinors, not over fermion fields. Thus, no minus sign is induced from reversing the order of the sum: $u_\alpha^s \bar{u}_\beta^s = \bar{u}_\beta^s u_\alpha^s$.

Using these relations

$$\sum_{s'}\sum_s [\bar{u}^{s'}(p_3)\gamma^\mu v^s(p_4)][\bar{v}^s(p_4)\gamma^\nu u^{s'}(p_3)] = \sum_{s'}[\bar{u}_\beta^{s'}(p_3)\gamma_{\beta\delta}^\mu(p\!\!\!/_4 - m_\mu\mathbb{1})_{\delta\iota}\gamma_{\iota\alpha}^\nu u_\alpha^{s'}(p_3)]$$

$$= \left(p\!\!\!/_3 + m_\mu\mathbb{1}\right)_{\alpha\beta}\gamma_{\beta\delta}^\mu(p\!\!\!/_4 - m_\mu\mathbb{1})_{\delta\iota}\gamma_{\iota\alpha}^\nu$$

$$= \mathrm{Tr}[(p\!\!\!/_3 + m_\mu)\gamma^\mu(p\!\!\!/_4 - m_\mu)\gamma^\nu], \tag{13.61}$$

which is a simple expression we can evaluate using γ-matrix identities.

Let us also assume that we do not know the polarization of the initial states. Then, if we do the measurement many times, we will get the *average* over each polarization. This leads to contractions and traces of the initial state, with a factor of $\frac{1}{4}$ ($\frac{1}{2}$ each for the incoming e^+ and incoming e^-) to average over our ignorance. Thus we need

$$\frac{1}{4}\sum_{\text{spins}}|\mathcal{M}|^2 = \frac{e^4}{4s^2}\mathrm{Tr}[(p\!\!\!/_1 + m_e)\gamma^\nu(p\!\!\!/_2 - m_e)\gamma^\mu]\mathrm{Tr}[(p\!\!\!/_3 + m_\mu)\gamma^\mu(p\!\!\!/_4 - m_\mu)\gamma^\nu]. \tag{13.62}$$

These traces simplify using trace identities:

$$\mathrm{Tr}[(p\!\!\!/_3 + m_\mu)\gamma^\alpha(p\!\!\!/_4 - m_\mu)\gamma^\beta] = p_3^\rho p_4^\sigma \mathrm{Tr}[\gamma^\rho\gamma^\alpha\gamma^\sigma\gamma^\beta] - m_\mu^2 \mathrm{Tr}[\gamma^\alpha\gamma^\beta]$$

$$= 4(p_3^\alpha p_4^\beta + p_4^\alpha p_3^\beta - p_3^\rho p_4^\rho g^{\alpha\beta}) - 4m_\mu^2 g^{\alpha\beta}. \tag{13.63}$$

So,

$$\frac{1}{4}\sum_{\text{spins}}|\mathcal{M}|^2 = \frac{4e^4}{s^2}\left(p_1^\alpha p_2^\beta + p_2^\alpha p_1^\beta - (p_1^\rho p_2^\rho + m_e^2)g^{\alpha\beta}\right)$$

$$\times \left(p_3^\alpha p_4^\beta + p_4^\alpha p_3^\beta - (p_3^\sigma p_4^\sigma + m_\mu^2)g^{\alpha\beta}\right)$$

$$= \frac{8e^4}{s^2}\left(p_{13}p_{24} + p_{14}p_{23} + m_\mu^2 p_{12} + m_e^2 p_{34} + 2m_e^2 m_\mu^2\right), \tag{13.64}$$

with $p_{ij} \equiv p_i \cdot p_j$. We can simplify this further with Mandelstam variables:

$$s = (p_1 + p_2)^2 = (p_3 + p_4)^2 = 2m_e^2 + 2p_{12} = 2m_\mu^2 + 2p_{34}, \tag{13.65}$$

$$t = (p_1 - p_3)^2 = (p_2 - p_4)^2 = m_e^2 + m_\mu^2 - 2p_{13} = m_e^2 + m_\mu^2 - 2p_{24}, \tag{13.66}$$

$$u = (p_1 - p_4)^2 = (p_2 - p_3)^2 = m_e^2 + m_\mu^2 - 2p_{14} = m_e^2 + m_\mu^2 - 2p_{23}. \tag{13.67}$$

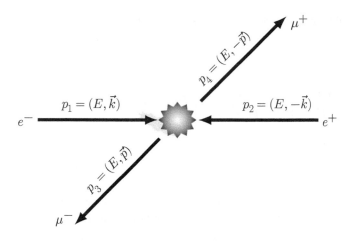

Kinematics of $e^-e^+ \rightarrow \mu^-\mu^+$ in the center-of-mass frame. Since the particles are all on-shell, $|\vec{k}| = \sqrt{E^2 - m_e^2}$ and $|\vec{p}| = \sqrt{E^2 - m_\mu^2}$.

Fig. 13.1

After some algebra, the result is

$$\frac{1}{4}\sum_{\text{spins}}|\mathcal{M}|^2 = \frac{2e^4}{s^2}\left[t^2 + u^2 + 4s(m_e^2 + m_\mu^2) - 2\left(m_e^2 + m_\mu^2\right)^2\right]. \qquad (13.68)$$

13.3.2 Differential cross section

For $2 \rightarrow 2$ scattering of particles of different mass, the cross section in the center-of-mass frame can be computed from the matrix element with Eq. (5.32):

$$\left(\frac{d\sigma}{d\Omega}\right)_{\text{CM}} = \frac{1}{64\pi^2 E_{\text{CM}}^2}\frac{|\vec{p}_f|}{|\vec{p}_i|}|\mathcal{M}|^2. \qquad (13.69)$$

There are only two variables on which the cross section depends: E_{CM} and the scattering angle between the incoming electron and the outgoing muon. In the center-of-mass frame, the kinematics are as shown in Figure 13.1.

With this choice of momenta, we find

$$s = (p_1 + p_2)^2 = 4E^2 = E_{\text{CM}}^2, \qquad (13.70)$$

$$t = (p_1 - p_3)^2 = m_e^2 + m_\mu^2 - 2E^2 + 2\vec{k}\cdot\vec{p}, \qquad (13.71)$$

$$u = -(\vec{k} + \vec{p})^2 = m_e^2 + m_\mu^2 - 2E^2 - 2\vec{k}\cdot\vec{p}, \qquad (13.72)$$

and so

$$\begin{aligned}\frac{d\sigma}{d\Omega} &= \frac{e^4}{32\pi^2 E_{\text{CM}}^2 s^2}\frac{|\vec{p}|}{|\vec{k}|}\left[t^2 + u^2 + 4s(m_e^2 + m_\mu^2) - 2(m_e^4 + 2m_e^2 m_\mu^2 + m_\mu^4)\right]\\ &= \frac{\alpha^2}{16E^6}\frac{|\vec{p}|}{|\vec{k}|}\left(E^4 + (\vec{k}\cdot\vec{p})^2 + E^2(m_e^2 + m_\mu^2)\right).\end{aligned} \qquad (13.73)$$

The only angular dependence comes from the $\vec{k} \cdot \vec{p}$ term:

$$\vec{k} \cdot \vec{p} = |\vec{k}||\vec{p}| \cos \theta. \tag{13.74}$$

So,

$$\frac{d\sigma}{d\Omega} = \frac{\alpha^2}{16E^6} \frac{|\vec{p}|}{|\vec{k}|} \left(E^4 + |\vec{k}|^2 |\vec{p}|^2 \cos^2\theta + E^2(m_e^2 + m_\mu^2) \right), \tag{13.75}$$

where $\alpha = \frac{e^2}{4\pi}$ and

$$|\vec{k}| = \sqrt{E^2 - m_e^2}, \quad |\vec{p}| = \sqrt{E^2 - m_\mu^2}, \tag{13.76}$$

which is the general result for the $e^+ e^- \to \mu^+ \mu^-$ rate in the center-of-mass frame.

Taking $m_e = 0$ for simplicity gives $|\vec{k}| = E$ and this reduces to

$$\frac{d\sigma}{d\Omega} = \frac{\alpha^2}{4E_{\text{CM}}^2} \sqrt{1 - \frac{m_\mu^2}{E^2}} \left(1 + \frac{m_\mu^2}{E^2} + \left(1 - \frac{m_\mu^2}{E^2}\right) \cos^2\theta \right). \tag{13.77}$$

If, in addition, we take $m_\mu = 0$, which is the ultra-high-energy limit, we find

$$\frac{d\sigma}{d\Omega} = \frac{\alpha^2}{4E_{\text{CM}}^2}(1 + \cos^2\theta), \tag{13.78}$$

which is the same thing we had from the naive sum over spin states back in Eq. (5.53). Recall that scattering with spins transverse to the plane gave $\mathcal{M} \propto 1$ and scattering with spins in the plane gave $\mathcal{M} \propto \cos^2\theta$, so this agrees with our previous analysis. You can check explicitly by choosing explicit spinors that our intuition with spin scattering agrees with QED even for the polarized cross section. Integrating the differential cross section over θ gives $\sigma_0 = \frac{4\pi\alpha^2}{3E_{\text{CM}}^2}$ for the total cross section at tree-level. The 1-loop correction to the total cross section will be calculated in Chapter 20.

13.4 Rutherford scattering $e^- p^+ \to e^- p^+$

Now let us go back to the problem we considered long ago, scattering of an electron by a Coulomb potential. Recall the classical Rutherford scattering formula,

$$\frac{d\sigma}{d\Omega} = \frac{m_e^2 e^4}{64\pi^4 p^4 \sin^4\frac{\theta}{2}}, \tag{13.79}$$

where $p = |\vec{p}_i| = |\vec{p}_f|$ is the magnitude of the incoming electron momentum, which is the same as the magnitude of the outgoing electron momentum for elastic scattering. Rutherford calculated this using classical mechanics to describe how an electron would get deflected in a central potential, as from an atomic nucleus. We recalled in Section 5.2 that Rutherford's formula is reproduced in quantum mechanics through the Born approximation, which relates the cross section to the Fourier transform of the Coulomb potential $V(r) = \frac{e^2}{4\pi r}$:

$$\left(\frac{d\sigma}{d\Omega}\right)_{\text{Born}} = \frac{m_e^2}{4\pi^2}\tilde{V}(k)^2 = \frac{m_e^2}{4\pi^2}\left(\int d^3x\, e^{-i\vec{k}\cdot\vec{x}}\frac{e^2}{4\pi\,|\vec{x}|}\right)^2 = \frac{m_e^2}{4\pi^2}\left(\frac{e^2}{|\vec{k}|^2}\right)^2 = \frac{m_e^2 e^4}{64\pi^4 p^4 \sin^4\frac{\theta}{2}},$$

(13.80)

where $\vec{k} = \vec{p}_i - \vec{p}_f$ is the momentum transfer satisfying $|\vec{k}| = 2p\sin\frac{\theta}{2}$.

We also reproduced these results from field theory, taking the non-relativistic limit before doing the calculation. We found that the amplitude is given by a t-channel diagram:

$$\left(\frac{d\sigma}{d\Omega}\right)_{\text{QFT}} = \frac{e^4}{64 p^4 \pi^2 E_{\text{CM}}^2}\frac{(2m_e)^2(2m_p)^2}{t^2},$$

(13.81)

where the $2m_e^2$ and $2m_p^2$ factors come from the non-relativistic normalization of the electron and proton states. Since $t = (p_3 - p_1)^2 = -2p^2(1 - \cos\theta) = -4p^2\sin^2\frac{\theta}{2}$ and $E_{\text{CM}} = m_p$ in the center-of-mass frame, we reproduce the Rutherford formula.

We will now do the calculation in QED. This will allow us to reproduce the above equation, but it will also give us the relativistic corrections. In this whole section, we neglect any internal structure of the proton, treating it, like the muon, as a pointlike particle. A discussion of what actually happens at extremely high energy, $E_{\text{CM}} \gg m_p$, is given in Chapter 32.

13.4.1 QED amplitude

As far as QED is concerned, a proton and a muon are the same thing, up to the sign of the charge, which gets squared anyway, and the mass. So let us start with $e^-\mu^- \to e^-\mu^-$. The amplitude is given by a t-channel diagram:

$$i\mathcal{M} = \quad = (-ie)\bar{u}(p_3)\gamma^\mu u(p_1)\frac{-i\left[g_{\mu\nu} - (1-\xi)\frac{k_\mu k_\nu}{k^2}\right]}{(p_1 - p_3)^2}(-ie)\bar{u}(p_4)\gamma^\nu u(p_2),$$

(13.82)

with $k^\mu = p_1^\mu - p_3^\mu$. As in $e^+e^- \to \mu^+\mu^-$, the $k^\mu k^\nu$ term drops out for on-shell spinors, as expected by gauge invariance. So this matrix element simplifies to

$$\mathcal{M} = \frac{e^2}{t}\bar{u}(p_3)\gamma^\mu u(p_1)\bar{u}(p_4)\gamma_\mu u(p_2),$$

(13.83)

with $t = (p_1 - p_3)^2$. Summing over final states and averaging over initial states,

$$\frac{1}{4}\sum_{\text{spins}}|\mathcal{M}|^2 = \frac{e^4}{4t^2}\text{Tr}[(\not{p}_1 + m_e)\gamma_\nu(\not{p}_3 + m_e)\gamma^\mu]\text{Tr}[(\not{p}_4 + m_\mu)\gamma_\mu(\not{p}_2 + m_\mu)\gamma^\nu].$$ (13.84)

This is remarkably similar to what we had for $e^+e^- \to \mu^+\mu^-$:

$$\frac{1}{4}\sum_{\text{spins}}|\mathcal{M}|^2 = \frac{e^4}{4s^2}\text{Tr}[(\slashed{p}_1 + m_e)\gamma_\nu(\slashed{p}_2 - m_e)\gamma^\mu]\text{Tr}[(\slashed{p}_3 + m_\mu)\gamma_\mu(\slashed{p}_4 - m_\mu)\gamma^\nu]. \quad (13.85)$$

In fact, the two are identical if we take the $e^+e^- \to \mu^+\mu^-$ formula and replace

$$(p_1, p_2, p_3, p_4) \to (p_1, -p_3, p_4, -p_2). \quad (13.86)$$

These changes send $s \to t$, or more generally,

$$s = (p_1 + p_2)^2 \to (p_1 - p_3)^2 = t, \quad (13.87)$$

$$t = (p_1 - p_3)^2 \to (p_1 - p_4)^2 = u, \quad (13.88)$$

$$u = (p_1 - p_4)^2 \to (p_1 + p_2)^2 = s. \quad (13.89)$$

These replacements are not physical, since $p_2 \to -p_3$ produces a momentum with negative energy, which cannot be an external particle. It is just a trick, called a **crossing relation**, that lets us recycle tedious algebraic manipulations. You can prove crossing symmetries in general, even for polarized cross sections with particular spins, and there exist general crossing rules. However, rather than derive and apply these rules, it is often easier simply to write down the amplitude that you want and inspect it to find the right transformation.

With the crossing symmetry we can just skip to the final answer. For $e^+e^- \to \mu^+\mu^-$ we had

$$\frac{1}{4}\sum_{\text{spins}}|\mathcal{M}|^2 = \frac{2e^4}{s^2}\left[t^2 + u^2 + 4s(m_e^2 + m_\mu^2) - 2(m_e^2 + m_\mu^2)^2\right]. \quad (13.90)$$

Therefore, for $e^-p^+ \to e^-p^+$ we get

$$\frac{1}{4}\sum_{\text{spins}}|\mathcal{M}|^2 = \frac{2e^4}{t^2}\left[u^2 + s^2 + 4t(m_e^2 + m_p^2) - 2(m_e^2 + m_p^2)^2\right]. \quad (13.91)$$

13.4.2 Corrections to Rutherford's formula

Now let us take the limit $m_p \gg m_e$ to get the relativistic corrections to Rutherford's formula. In this limit we can treat the proton mass as effectively infinite, but we have to treat the electron mass as finite to go from the non-relativistic to the relativistic limit. As the proton mass goes to infinity, the momenta are

$$p_1^\mu = (E, \vec{p}_i), \quad p_2^\mu = (m_p, 0), \quad p_3^\mu = (E, \vec{p}_f), \quad p_4^\mu = (m_p, 0). \quad (13.92)$$

The scattering angle is defined by

$$\vec{p}_i \cdot \vec{p}_f = p^2 \cos\theta = v^2 E^2 \cos\theta, \quad (13.93)$$

where $p = |\vec{p}_i| = |\vec{p}_f|$ and

$$v = \frac{p}{E} = \sqrt{1 - \frac{m_e^2}{E^2}} \quad (13.94)$$

is the electron's relativistic velocity. Then, to leading order in m_e/m_p,

$$p_{13} = E^2(1 - v^2 \cos \theta), \tag{13.95}$$

$$p_{12} = p_{23} = p_{34} = p_{14} = E m_p, \tag{13.96}$$

$$p_{24} = m_p^2, \tag{13.97}$$

where $p_{ij} \equiv p_i^\mu p_j^\mu$ and

$$t = (p_1 - p_3)^2 = -(\vec{p}_i - \vec{p}_f)^2 = -2p^2(1 - \cos \theta), \tag{13.98}$$

so that

$$\frac{1}{4} \sum_{\text{spins}} |\mathcal{M}|^2 = \frac{8e^4}{t^2} \left[p_{14}p_{23} + p_{12}p_{34} - m_p^2 p_{13} - m_e^2 p_{24} + 2 m_e^2 m_p^2 \right]$$

$$= \frac{8e^4}{4 v^4 E^4 (1 - \cos \theta)^2} \left[E^2 m_p^2 + E^2 m_p^2 v^2 \cos \theta + m_e^2 m_p^2 \right]$$

$$= \frac{2 e^4 m_p^2}{v^4 E^2 (1 - \cos \theta)^2} \left[2 - v^2 (1 - \cos \theta) \right]$$

$$= \frac{e^4 m_p^2}{v^4 E^2 \sin^4 \frac{\theta}{2}} \left[1 - v^2 \sin^2 \frac{\theta}{2} \right]. \tag{13.99}$$

Note that each term in the top line of this equation scales as m_p^2, as does the final answer, so dropping subleading terms in Eq. (13.92) is justified. Since the center-of-mass frame is essentially the lab frame, the differential cross section is given by $\frac{d\sigma}{d\Omega} = \frac{1}{64\pi^2 E_{\text{CM}}^2} |\mathcal{M}|^2$:

$$\frac{d\sigma}{d\Omega} = \frac{e^4}{64\pi^2 v^2 p^2 \sin^4 \frac{\theta}{2}} \left(1 - v^2 \sin^2 \frac{\theta}{2} \right), \qquad E \ll m_p. \tag{13.100}$$

This is known as the **Mott formula**. Note that the limit $m_p \to \infty$ exists: there is no dependence on the proton mass. For slow velocities we can use $v \ll 1$ and $p \ll E \sim m_e$ so $v \sim \frac{p}{m_e}$. Thus,

$$\frac{d\sigma}{d\Omega} = \frac{e^4 m_e^2}{64\pi^2 p^4 \sin^4 \frac{\theta}{2}}, \qquad v \ll 1 \text{ and } E \ll m_p \tag{13.101}$$

which is the Rutherford formula. In particular, note that the normalization factors, m_e^2, worked out correctly.

In the very high energy limit, $E \gg m_e$, one can no longer assume that the final state proton is also at rest. However, one can now neglect the electron mass, so that $v = 1$. Then the momenta are

$$p_1^\mu = (E, \vec{p}_i), \quad p_2^\mu = (m_p, 0), \quad p_3^\mu = (E', \vec{p}_f), \quad p_4^\mu = p_1^\mu + p_2^\mu - p_3^\mu, \tag{13.102}$$

with $|\vec{p}_i| = E$ and $|\vec{p}_f| = E'$. For $m_e = 0$ in the proton rest frame, following the same steps as above, we find the cross section:

$$\frac{d\sigma}{d\Omega} = \frac{e^4}{64\pi^2 E^2 \sin^4 \frac{\theta}{2}} \frac{E'}{E} \left(\cos^2 \frac{\theta}{2} + \frac{E - E'}{m_p} \sin^2 \frac{\theta}{2} \right), \qquad m_e \ll E, \qquad (13.103)$$

where E' is the final state electron's energy. As $m_p \to \infty$, $E \to E'$ and this reduces to the $v \to 1$ limit of the Mott formula, Eq. (13.100).

These formulas characterize the scattering of pointlike particles from other pointlike particles. Note that the final forms in which we have written these cross sections depend only on properties of the initial and final electrons. Thus, they are suited to experimental situations in which electrons are scattered by hydrogen gas and the final state proton momenta are not measured. Such experiments were carried out in the 1950s, notably at Stanford, and deviations of the measured cross section from the form of Eq. (13.103) led to the conclusion that the proton must have substructure. More shockingly, at very high energy, this pointlike scattering form was once again observed, indicating the presence of pointlike constituents within the proton, now known as quarks. We will discuss these important $e^- p^+$ scattering experiments and their theoretical interpretation in great detail in Chapter 32.

13.5 Compton scattering

The next process worth studying is the QED prediction for Compton scattering, $\gamma e^- \to \gamma e^-$. By simple relativistic kinematics, Compton was able to predict the shift in wavelength of the scattered light as a function of angle,

$$\Delta \lambda = \frac{1}{m}(1 - \cos \theta), \qquad (13.104)$$

but he could not predict the intensity of radiation at each angle.

In the classical limit, for scattering of soft radiation against electrons, J. J. Thomson had derived the formula

$$\frac{d\sigma}{d\cos\theta} = \pi r_e^2 \left(1 + \cos^2\theta\right) = \frac{\pi\alpha^2}{m^2}(1 + \cos^2\theta), \qquad (13.105)$$

where r_e is the classical electron radius, $r_e = \frac{\alpha}{m}$, defined so that if the electron were a disk of radius r, the cross section would be πr^2. The 1 comes from radiation polarized out of the plane of scattering and the $\cos^2\theta$ from polarization in the plane, just as we saw for $e^+ e^- \to \mu^+ \mu^-$ in Section 5.3. From QED we should be able to reproduce this formula, plus the relativistic corrections.

There are two diagrams:

$$i\mathcal{M}_s = \quad = (-ie)^2 \epsilon_1^\mu \epsilon_4^{\star\nu} \bar{u}(p_3) \gamma^\nu \frac{i(\not{p}_1 + \not{p}_2 + m)}{(p_1 + p_2)^2 - m^2} \gamma^\mu u(p_2),$$

$$(13.106)$$

$$i\mathcal{M}_t = \quad\text{[diagram]}\quad = (-ie)^2 \epsilon_1^\mu \epsilon_4^{\star\nu} \bar{u}(p_3)\gamma^\mu \frac{i(\not{p}_2 - \not{p}_4 + m)}{(p_2 - p_4)^2 - m^2}\gamma^\nu u(p_2), \quad (13.107)$$

so the sum is

$$\mathcal{M} = -e^2 \epsilon_1^\mu \epsilon_4^{\star\nu} \bar{u}(p_3) \left[\frac{\gamma^\nu \left(\not{p}_1 + \not{p}_2 + m \right)\gamma^\mu}{s - m^2} + \frac{\gamma^\mu \left(\not{p}_2 - \not{p}_4 + m \right)\gamma^\nu}{t - m^2} \right] u(p_2). \quad (13.108)$$

We would next like to calculate the unpolarized cross section.

13.5.1 Photon polarization sums

To square this and sum over on-shell physical polarizations, it is helpful to employ a trick for the photon polarization sum. There is no way to write the sum over transverse modes in a Lorentz-invariant way, since the only available dimensionless tensors are $g_{\mu\nu}$ and $\frac{p_\mu p_\nu}{p^2}$, but on-shell $p^2 = 0$ so $\frac{p^\mu p^\nu}{p^2}$ is undefined.

Physical polarizations can be defined as orthogonal to p^μ and orthogonal to any other lightlike reference vector r^μ as long as r^μ is not proportional to p^μ. For example, if $p^\mu = (E, 0, 0, E)$, then the canonical polarizations $\epsilon_1^\mu = (0, 1, 0, 0)$ and $\epsilon_2^\mu = (0, 0, 1, 0)$ are orthogonal to $\bar{p}^\mu = (E, 0, 0, -E)$. More generally, if $p^\mu = (E, \vec{p})$, then choosing the reference vector as $r^\mu = \bar{p}^\mu$, where

$$\bar{p}^\mu = (E, -\vec{p}), \quad (13.109)$$

will uniquely determine the two transverse polarizations. Other choices of reference vector r^μ lead to transverse polarizations that are related to the canonical transverse polarizations by little-group transformations (Lorentz transformations that hold p^μ fixed). For example, with $p^\mu = (E, 0, 0, E)$ choosing $r^\mu = (1, 0, 1, 0)$ leads to $\hat{\epsilon}_1^\mu = (0, 1, 0, 0)$ and $\hat{\epsilon}_2^\mu = (1, 0, 1, 1)$. Since $\hat{\epsilon}_2^\mu = \epsilon_2^\mu + \frac{1}{E}p^\mu$, there will be no difference in matrix elements calculated using these different polarization sets by the Ward identity. In fact, invariance under change of reference vector provides an important constraint on the form that matrix elements can have. This constraint will be efficiently exploited in the calculation of amplitudes using helicity spinors in Chapter 27.

With the choice $r^\mu = \bar{p}^\mu$, you should verify that (see Problem 8.5)

$$\sum_{i=1}^{2} \epsilon_\mu^{i\star} \epsilon_\nu^i = -g_{\mu\nu} + \frac{1}{2E^2}\left(p_\mu \bar{p}_\nu + \bar{p}_\mu p_\nu \right). \quad (13.110)$$

Now, suppose we have an amplitude involving a photon. Writing $\mathcal{M} = \epsilon_\mu M_\mu$, we find

$$\sum_{\text{pols. } i} |\mathcal{M}|^2 = \epsilon_\mu^{\star i} M_\mu^\star M_\nu \epsilon_\nu^i = -M_\mu^\star M_\mu + \frac{1}{2E^2}\left(p_\mu M_\mu^\star M_\nu \bar{p}_\nu + \bar{p}_\mu M_\mu^\star M_\nu p_\nu \right). \quad (13.111)$$

By the Ward identity, $p^\mu M_\mu = 0$, and therefore we can simply replace

$$\sum_{\text{pols. } i} \epsilon_\mu^{i\star} \epsilon_\nu^i \quad \rightarrow \quad -g_{\mu\nu} \tag{13.112}$$

in any physical matrix element. Note that this replacement only works for the sum of all relevant diagrams – individual diagrams are not gauge invariant, as you can explore in Problem 13.7.

13.5.2 Matrix element

Returning to the Compton scattering process, we are now ready to evaluate $|\mathcal{M}|^2$ summed over spins and polarizations. $|\mathcal{M}|^2$ includes terms from the t-channel and s-channel diagrams squared as well as their cross terms ($\mathcal{M}_t^\star \mathcal{M}_s + \mathcal{M}_s^\star \mathcal{M}_t$). To see what is involved, let us just evaluate one piece in the high-energy limit where we can set $m = 0$. In this limit

$$\mathcal{M}_t = -\frac{e^2}{t} \epsilon_\mu^1 \epsilon_\nu^{4\star} \bar{u}(p_3) \gamma^\mu (\not{p}_2 - \not{p}_4) \gamma^\nu u(p_2), \tag{13.113}$$

and so, using Eqs. (13.8) and (13.112),

$$\sum_{\text{spins/pols.}} |\mathcal{M}_t|^2 = \frac{e^4}{t^2} \text{Tr}[\not{p}_3 \gamma^\mu (\not{p}_2 - \not{p}_4) \gamma^\nu \not{p}_2 \gamma_\nu (\not{p}_2 - \not{p}_4) \gamma_\mu]. \tag{13.114}$$

Now use $\gamma^\nu \not{p} \gamma_\nu = -2\not{p}$ and $q^\mu = p_2^\mu - p_4^\mu = p_3^\mu - p_1^\mu$ to get

$$\sum_{\text{spins/pols.}} |\mathcal{M}_t|^2 = 4\frac{e^4}{t^2} \text{Tr}[\not{p}_3 \not{q} \not{p}_2 \not{q}] = 16\frac{e^4}{t^2} \left(2(p_3 \cdot q)(p_2 \cdot q) - p_{23}q^2\right). \tag{13.115}$$

Using $p_3^2 = p_2^2 = 0$, we can simplify this to

$$\sum_{\text{spins/pols.}} |\mathcal{M}_t|^2 = 16\frac{e^4}{t^2} \left(2p_{13}p_{24} + 2p_{23}p_{13}\right) = 8\frac{e^4}{t^2} \left(t^2 + ut\right) = -8e^4 \frac{s}{t} = 8e^4 \frac{p_{12}}{p_{24}}. \tag{13.116}$$

Note that one of the factors of t canceled, so the divergence at $t = 0$ is not $\frac{1}{t^2}$ but simply $\frac{1}{t}$.

Including all the terms gives

$$\mathcal{M} = -e^2 \epsilon_\mu^1 \epsilon_\nu^{4\star} \bar{u}(p_3) \left[\frac{\gamma^\nu (\not{p}_1 + \not{p}_2 + m)\gamma^\mu}{s - m^2} + \frac{\gamma^\mu (\not{p}_2 - \not{p}_4 + m)\gamma^\nu}{t - m^2} \right] u(p_2). \tag{13.117}$$

Then, summing/averaging over spins and polarizations we find

$$\frac{1}{4} \sum_{\text{spins/pols.}} |\mathcal{M}|^2 = \frac{1}{4} e^4 \text{Tr} \left\{ (\not{p}_3 + m) \left[\frac{\gamma^\nu (\not{p}_1 + \not{p}_2 + m)\gamma^\mu}{s - m^2} + \frac{\gamma^\mu (\not{p}_2 - \not{p}_4 + m)\gamma^\nu}{t - m^2} \right] \right.$$
$$\left. \times (\not{p}_2 + m) \left[\frac{\gamma^\mu (\not{p}_1 + \not{p}_2 + m)\gamma^\nu}{s - m^2} + \frac{\gamma^\nu (\not{p}_2 - \not{p}_4 + m)\gamma^\mu}{t - m^2} \right] \right\}. \tag{13.118}$$

This is a bit of a mess, but after some algebra the result is rather simple:

$$\frac{1}{4} \sum_{\text{spins/pols.}} |\mathcal{M}|^2 = 2e^4 \left[\frac{p_{24}}{p_{12}} + \frac{p_{12}}{p_{24}} + 2m^2 \left(\frac{1}{p_{12}} - \frac{1}{p_{24}} \right) + m^4 \left(\frac{1}{p_{12}} - \frac{1}{p_{24}} \right)^2 \right].$$

(13.119)

13.5.3 Klein–Nishina formula

Let us start with the low-energy limit, $\omega \ll m$, where it makes sense to work in the lab frame. Then

$$p_1 = (\omega, 0, 0, \omega), \qquad\qquad p_2 = (m, 0, 0, 0),$$
$$p_4 = (\omega', \omega' \sin\theta, 0, \omega' \cos\theta), \quad p_3 = p_1 + p_2 - p_4 = (E', p'). \qquad (13.120)$$

Note that the on-shell condition $p_3^2 = m^2$ implies

$$0 = p_{12} - p_{14} - p_{24} = \omega m - \omega\omega'(1 - \cos\theta) - m\omega', \qquad (13.121)$$

so

$$\omega' = \frac{\omega}{1 + \frac{\omega}{m}(1 - \cos\theta)}, \qquad (13.122)$$

which is the formula for the shifted frequency as a function of angle. There is no QED in this relation – it is just momentum conservation and is the same as Compton's formula for the wavelength shift:

$$\Delta\lambda = \frac{1}{\omega'} - \frac{1}{\omega} = \frac{1}{m}(1 - \cos\theta), \qquad (13.123)$$

but it is still a very important relation!

Then, since $p_{12} = \omega m$ and $p_{24} = \omega' m$, we get a simple formula for $|\mathcal{M}|^2$:

$$\frac{1}{4} \sum_{\text{pols.}} |\mathcal{M}|^2 = 2e^4 \left[\frac{\omega'}{\omega} + \frac{\omega}{\omega'} - 2(1 - \cos\theta) + (1 - \cos\theta)^2 \right]$$
$$= 2e^4 \left[\frac{\omega'}{\omega} + \frac{\omega}{\omega'} - \sin^2\theta \right]. \qquad (13.124)$$

Now we need to deal with the phase space. In the lab frame, we have to go back to our general formula, Eq. (5.22),

$$d\sigma = \frac{1}{(2E_1)(2E_2)|\vec{v}_1 - \vec{v}_2|} |\mathcal{M}|^2 d\Pi_{\text{LIPS}} = \frac{1}{4\omega m} |\mathcal{M}|^2 d\Pi_{\text{LIPS}} \qquad (13.125)$$

and

$$\int d\Pi_{\text{LIPS}} = \int \frac{d^3 p_3}{(2\pi)^3} \frac{1}{2E'} \int \frac{d^3 p_4}{(2\pi)^3} \frac{1}{2\omega'} \left[(2\pi)^4 \delta^4(p_1^\mu + p_2^\mu - p_3^\mu - p_4^\mu) \right]. \qquad (13.126)$$

The δ-function fixes the 3-momenta when we integrate over d^3p_4, leaving the energy constraint

$$\int d\Pi_{\text{LIPS}} = \frac{1}{4(2\pi)^2} \int \omega'^2 d\Omega \, d\omega' \frac{1}{\omega' E'} \delta\left(\sum E\right)$$

$$= \frac{1}{8\pi} \int d\cos\theta \, d\omega' \frac{\omega'}{E'} \delta\left(\sum E\right). \qquad (13.127)$$

Now we want to integrate over ω' to enforce the energy constraint $E' + \omega' = m + \omega$. But we have to be a little careful because E' and ω' are already constrained by the electron's on-shell condition:

$$E'^2 = m^2 + p'^2 = m^2 + (\omega' \sin\theta)^2 + (\omega' \cos\theta - \omega)^2$$

$$= m^2 + \omega'^2 + \omega^2 - 2\omega\omega' \cos\theta. \qquad (13.128)$$

So,

$$E' \frac{dE'}{d\omega'} = \omega' - \omega \cos\theta \qquad (13.129)$$

and thus

$$\int d\Pi_{\text{LIPS}} = \frac{1}{8\pi} \int d\cos\theta \, d\omega' \frac{\omega'}{E'} \delta(\omega' + E'(\omega') - m - \omega)$$

$$= \frac{1}{8\pi} \int d\cos\theta \frac{\omega'}{E'} \left(1 + \frac{dE'}{d\omega'}\right)^{-1}$$

$$= \frac{1}{8\pi} \int d\cos\theta \frac{\omega'}{E'} \left(1 + \frac{\omega' - \omega\cos\theta}{E'}\right)^{-1}$$

$$= \frac{1}{8\pi} \int d\cos\theta \frac{(\omega')^2}{\omega m}, \qquad (13.130)$$

where ω' now refers to Eq. (13.122), not the integration variable. This leads to

$$\frac{d\sigma}{d\cos\theta} = \frac{1}{4\omega m} \frac{1}{8\pi} \frac{(\omega')^2}{\omega m} 2e^4 \left[\frac{\omega'}{\omega} + \frac{\omega}{\omega'} - \sin^2\theta\right], \qquad (13.131)$$

or more simply,

$$\frac{d\sigma}{d\cos\theta} = \frac{\pi\alpha^2}{m^2} \left(\frac{\omega'}{\omega}\right)^2 \left[\frac{\omega'}{\omega} + \frac{\omega}{\omega'} - \sin^2\theta\right]. \qquad (13.132)$$

This is the **Klein–Nishina formula**. It was first calculated by Klein and Nishina in 1929 and was one of the first tests of QED.

Substituting in for ω' using Eq. (13.122),

$$\frac{d\sigma}{d\cos\theta} = \frac{\pi\alpha^2}{m^2} \left[1 + \cos^2\theta - \frac{2\omega}{m}(1 + \cos^2\theta)(1 - \cos\theta) + \mathcal{O}\left(\frac{\omega^2}{m^2}\right)\right]. \qquad (13.133)$$

Note that, in the limit $m \to \infty$,

$$\frac{d\sigma}{d\cos\theta} = \frac{\pi\alpha^2}{m^2} \left[1 + \cos^2\theta\right]. \qquad (13.134)$$

This is the Thomson scattering cross section for classical electromagnetic radiation by a free electron. We have calculated the full relativistic corrections.

13.5.4 High-energy behavior

Next, consider the opposite limit, $\omega \gg m$. In this limit, we will be able to understand some of the physics of Compton scattering, in particular, the spin and polarization dependence and the origin of an apparent singularity for exactly backwards scattering, $\theta = \pi$.

At high energy, the center-of-mass frame makes the most sense. Then

$$p_1 = (\omega, 0, 0, \omega), \qquad\qquad p_2 = (E, 0, 0, -\omega),$$
$$p_3 = (E, -\omega \sin\theta, 0, -\omega \cos\theta), \quad p_4 = (\omega, \omega \sin\theta, 0, \omega \cos\theta), \qquad (13.135)$$

so that

$$p_{12} = \omega(E + \omega), \qquad (13.136)$$
$$p_{24} = \omega(E + \omega \cos\theta), \qquad (13.137)$$

and

$$\frac{1}{4}\sum_{\text{pols.}} |\mathcal{M}|^2 \approx 2e^4 \left[\frac{p_{24}}{p_{12}} + \frac{p_{12}}{p_{24}} \right] = 2e^4 \left[\frac{E + \omega\cos\theta}{E + \omega} + \frac{E + \omega}{E + \omega\cos\theta} \right]. \qquad (13.138)$$

For $\omega \gg m$, $E = \sqrt{m^2 + \omega^2} \approx \omega \left(1 + \frac{m^2}{2\omega^2} \right)$ and

$$\frac{1}{4}\sum_{\text{pols.}} |\mathcal{M}|^2 \approx 4e^4 \left[\frac{1 + \cos\theta}{4} + \frac{1}{\frac{m^2}{2\omega^2} + 1 + \cos\theta} \right]. \qquad (13.139)$$

We have only kept the factors of m required to cut off the singularity at $\cos\theta = -1$. The cross section for $\omega \gg m$ is

$$\frac{d\sigma}{d\cos\theta} \approx \frac{2\pi}{64\pi^2(2\omega)^2} \left(\frac{1}{4}\sum_{\text{pols.}} |\mathcal{M}|^2 \right) \approx \frac{\pi\alpha^2}{2\omega^2} \left[\frac{1 + \cos\theta}{4} + \frac{1}{\frac{m^2}{2\omega^2} + 1 + \cos\theta} \right]. \qquad (13.140)$$

Near $\theta = \pi$, as $\omega \gg m$, we see that the cross section becomes very large (but still finite). In this region of phase space, the photon and electron bounce off each other and go back the way they came. Or, in more Lorentz-invariant language, the direction of the outgoing photon momentum is the same as the direction of incoming electron momentum. Let us now try to understand the origin of the $\theta = \pi$ singularity.

Since the matrix element can be written in the massless limit as

$$\frac{1}{4}\sum_{\text{pols.}} |\mathcal{M}|^2 = 2e^4 \left[\frac{p_{24}}{p_{12}} + \frac{p_{12}}{p_{24}} \right] \approx -2e^4 \left[\frac{t}{s} + \frac{s}{t} \right] \qquad (13.141)$$

for $\omega \gg m$ and

$$t \approx -2p_{24} = -2\omega^2(1 + \cos\theta), \qquad (13.142)$$

we see that the origin of the pole at $\theta = \pi$ is due to the t-channel exchange. Looking back, the t-channel matrix element is

$$\mathcal{M}_t = -\frac{e^2}{t}\epsilon_\mu^1 \epsilon_\nu^{4\star} \bar{u}(p_3)\gamma^\mu(\not{p}_2 - \not{p}_4)\gamma^\nu u(p_2). \tag{13.143}$$

Since this scales as $\frac{1}{t}$ we might expect the cross section to diverge as $\frac{1}{t^2} \sim \frac{1}{(1+\cos\theta)^2}$. In fact this would happen in a scalar field theory, such as one with interaction $g\phi^3$ for which

$$\mathcal{M} \sim \frac{g^2}{t}, \quad |\mathcal{M}|^2 \sim \frac{g^4}{t^2}, \tag{13.144}$$

which has a strong t^2 pole. In QED, we calculated the t-channel diagram in the massless limit and found

$$\frac{1}{4}\sum |\mathcal{M}_t|^2 = -2e^4\frac{s}{t} = 4e^4\frac{1}{1 + \cos\theta}. \tag{13.145}$$

This gives the entire $\frac{1}{t}$ pole, so we do not have to worry about interference for the purpose of understanding the singularity. Where did the other factor of t come from to cancel the pole?

For $\theta = \pi - \phi$ with $\phi \sim 0$, the momenta become

$$p_1 = (\omega, 0, 0, \omega), \qquad p_2 = (\omega, 0, 0, -\omega),$$
$$p_3 = (\omega, -\omega\phi, 0, \omega), \quad p_4 = (\omega, \omega\phi, 0, -\omega), \tag{13.146}$$

and then

$$t = -\omega^2\phi^2. \tag{13.147}$$

So a $\frac{1}{t}$ pole goes as $\frac{1}{\phi^2}$, but $\frac{1}{t^2}$ goes as $\frac{1}{\phi^4}$. But notice that the momentum factor in the matrix element also vanishes as $p_2 \to p_4$:

$$\not{p}_2 - \not{p}_4 = -\omega\phi\not{k}, \quad k^\mu = (0, 1, 0, 0). \tag{13.148}$$

So,

$$\mathcal{M}_t = \frac{e^2}{\omega^2\phi^2}\bar{u}(p_3)\not{\epsilon}_1(\not{p}_2 - \not{p}_4)\not{\epsilon}_4^\star u(p_2) = -\frac{e^2}{\omega\phi}\bar{u}(p_3)\not{\epsilon}_1\not{k}\not{\epsilon}_4^\star u(p_2). \tag{13.149}$$

Thus, one factor of ϕ is canceling. This factor came from the spinors, and is an effect of angular momentum conservation.

If we include the electron mass, as in Eq. (13.140), we would have found $\frac{\omega\phi}{\omega^2\phi^2+m_e^2}$ instead of $\frac{1}{\omega\phi}$, which is finite even for exactly backwards scattering. So there is not really a divergence. Still, the cross section becomes very large for nearly backwards scattering. More discussion of these types of *infrared divergences* is given in Chapter 20.

Let us further explore the singular $t \to 0$ region by looking at the helicity structure. Recall that the left- and right-handed spinors, ψ_L and ψ_R, satisfy

$$\frac{1-\gamma_5}{2}\psi_L = \psi_L, \quad \frac{1+\gamma_5}{2}\psi_L = 0, \quad \bar{\psi}_L\frac{1+\gamma_5}{2} = \bar{\psi}_L, \quad \bar{\psi}_L\frac{1-\gamma_5}{2} = 0, \tag{13.150}$$

$$\frac{1+\gamma_5}{2}\psi_R = \psi_R, \quad \frac{1-\gamma_5}{2}\psi_R = 0, \quad \bar{\psi}_R\frac{1-\gamma_5}{2} = \bar{\psi}_R, \quad \bar{\psi}_R\frac{1+\gamma_5}{2} = 0. \tag{13.151}$$

Since

$$1 = \frac{1+\gamma_5}{2} + \frac{1-\gamma_5}{2}, \tag{13.152}$$

we can write $\psi = \psi_L + \psi_R$. Then we use $\gamma_\mu(1+\gamma_5) = (1-\gamma_5)\gamma_\mu$ to see that each γ-matrix flips L to R. This lets us derive that

$$\bar\psi\psi = \bar\psi_L\psi_R + \bar\psi_R\psi_L, \tag{13.153}$$

$$\bar\psi\gamma^\mu\psi = \bar\psi_L\gamma^\mu\psi_L + \bar\psi_R\gamma^\mu\psi_R, \tag{13.154}$$

$$\bar\psi\gamma^\mu\gamma^\nu\psi = \bar\psi_L\gamma^\mu\gamma^\nu\psi_R + \bar\psi_R\gamma^\mu\gamma^\nu\psi_L, \tag{13.155}$$

$$\bar\psi\gamma^\mu\gamma^\alpha\gamma^\beta\psi = \bar\psi_L\gamma^\mu\gamma^\alpha\gamma^\beta\psi_L + \bar\psi_R\gamma^\mu\gamma^\alpha\gamma^\beta\psi_R, \tag{13.156}$$

and so on. The general rule is that an odd number of γ-matrices couples RR and LL while an even number couples LR and RL.

In particular, our interaction $\mathcal{M}_t$ has three γ-matrices, so it couples RR and LL. Thus,

$$\mathcal{M}_t = -\frac{e^2}{\omega\phi}\left[\bar u_L(p_3)\epsilon\!\!\!/_1\,k\!\!\!/\,\epsilon\!\!\!/_4^*\,u_L(p_2) + \bar u_R(p_3)\epsilon\!\!\!/_1\,k\!\!\!/\,\epsilon\!\!\!/_4^*\,u_R(p_2)\right], \tag{13.157}$$

which is helicity conserving. So, $u(p_2)$ and $\bar u(p_3)$ should be either both right-handed or both left-handed. This is consistent with a general property of QED, that in the limit of a massless electron, the left- and right-handed states completely decouple.

Now recall our explicit electron polarizations in the massless limit:

$$u_R = \sqrt{2E}\begin{pmatrix} 0 \\ 0 \\ 1 \\ 0 \end{pmatrix}, \quad u_L = \sqrt{2E}\begin{pmatrix} 0 \\ 1 \\ 0 \\ 0 \end{pmatrix}. \tag{13.158}$$

For the photons, we need to use the helicity eigenstates:[1]

$$\epsilon_R^\mu = \frac{1}{\sqrt{2}}(0,1,i,0), \quad \epsilon_L^\mu = \frac{1}{\sqrt{2}}(0,1,-i,0). \tag{13.159}$$

Note that

$$\sqrt{2}\epsilon\!\!\!/_R = -\gamma_1 - i\gamma_2 = \begin{pmatrix} & & 0 & -2 \\ & & 0 & 0 \\ 0 & 2 & & \\ 0 & 0 & & \end{pmatrix}, \quad \sqrt{2}\epsilon\!\!\!/_L = -\gamma_1 + i\gamma_2 = \begin{pmatrix} & & 0 & 0 \\ & & -2 & 0 \\ 0 & 0 & & \\ 2 & 0 & & \end{pmatrix},$$

$$\tag{13.160}$$

---

[1] To see that the convention for "left" and "right" is being used consistently, it is easy to check that $\frac{\vec{S}\cdot\vec{p}}{E}\epsilon_L = -\epsilon_L$ and $\frac{\vec{S}\cdot\vec{p}}{E}u_L = -\frac{1}{2}u_L$ using $p^\mu = (E,0,0,p_z)$ and $S_z = S_{12}$ or $S_z = V_{12}$ in Eqs. (10.117) and (10.113) respectively.

so

$$\epsilon_L^\star u_R = \epsilon_R^\star u_L = \bar{u}_R \epsilon_L = \bar{u}_L \epsilon_R = 0. \qquad (13.161)$$

Thus, everything is right-handed or everything is left-handed.

This has an important physical implication. Consider shooting a laser beam at a high-energy beam of electrons. Lasers are polarized. Suppose the laser produces left circularly polarized light. Such a beam will dominantly back-scatter only left-handed electrons. This is a useful way to polarize your electron beam. It also directly connects helicity for spinors to helicity for spin-1 particles.

13.6 Historical note

Considering only $2 \rightarrow 2$ scattering involving electrons, positrons, muons, antimuons and photons, there are quite a number of historically important processes in QED. Some examples are

- $\gamma e^- \rightarrow \gamma e^-$: Compton scattering. Observed in 1923 by American physicist Arthur Holly Compton [Compton, 1923]. The differential scattering formula was calculated by Oskar Klein and Yoshio Nishina in 1929 [Klein and Nishina, 1929]. This was one of the first results obtained from QED, and was crucial in convincing us of the correctness of Dirac's equation. Before this, all that was known was the classical Thomson scattering formula, which was already in disagreement with experiment by the early 1920s. The Klein–Nishina formula agreed perfectly with available experiments in the late 1920s. However, at higher energies, above $2\,\mathrm{MeV}$ or so, it looked wrong. It was not until many years later that the discrepancy was shown to be due to the production of $e^- e^+$ pairs, with the positron annihilating into some other electron, and to Bremsstrahlung.
- $e^- e^- \rightarrow e^- e^-$: Møller scattering. First calculated in the ultra-relativistic regime by Danish physicist Christian Møller [Møller, 1932]. In the non-relativistic regime it is called Coulomb scattering or Rutherford scattering. Møller calculated the cross section based on some guesses and consistency requirements, not using QED. The cross section was calculated in QED soon after by Bethe and Fermi [Bethe and Fermi, 1932]. Møller scattering was not measured until 1950 by Canadian physicist Lorne Albert Page [Page, 1950]. This was partly because researchers did not consider it interesting until renormalization was understood and radiative corrections could be measured.
- $e^+ e^- \rightarrow e^+ e^-$: Bhabha scattering. First calculated by Indian physicist Homi Jehengir Bhabha in 1936 [Bhabha, 1936]. The positron was not discovered until 1932, so it was a while before the differential cross section that Bhabha predicted could be measured in the lab. However, the total cross section for $e^+ e^- \rightarrow e^+ e^-$ was important for cosmic-ray physics from the 1930s onward.
- $\gamma\gamma \rightarrow \gamma\gamma$: Light-by-light scattering. In 1933, German physicist Otto Halpern realized that QED predicted that light could scatter off light [Halpern, 1933]. There is no tree-level contribution to this process in QED. The first contribution comes from a box

diagram at 1-loop. Heisenberg and his students Hans Euler and Bernhard Kockel [Euler and Kockel, 1935; Euler and Heisenberg, 1936] were able to show that this box diagram was finite. They expressed the result in terms of an effective Lagrangian now known as the Euler–Heisenberg Lagrangian (see Chapter 33). The closest experiment has come to observing light-by-light scattering is through light scattering off of the Coulomb field of a nucleus (Delbrück scattering) [Schumacher, 1975], In going beyond the box diagram, Euler and Heisenberg encountered divergences in the loop graphs, concluding that "QED must be considered provisional" [Schweber, 1994, p. 119].

Although QED had great successes at tree-level, that is at leading-order in the fine-structure constant α, it appeared in the 1930s incapable of making quantitative predictions at higher orders. For example, the infinite contribution of the Coulomb potential to the electron mass in the classical theory was still infinite in QED; and QED could not be used to compute corrections to the energy levels of the hydrogen atom. By the late 1930s, the experts generally believed that QED was incomplete, if not wrong.

One should keep in mind that QED was being developed not long after quantum mechanics itself was discovered. Physicists were still coming to terms with the violations of classical causality inherent in the quantum theory, and some, including Bohr and Dirac, suspected that the difficulties of QED might be related to an incomplete understanding of causality. Bohr, with Kramers and Slater, had proposed in 1924 a version of quantum mechanics in which energy was not conserved microscopically, only statistically [Bohr *et al.*, 1924]. Although experiments in the late 1920s confirmed that energy was indeed conserved microscopically, an experiment by Shankland in 1936 implied that perhaps it was not [Shankland, 1936]. Dirac immediately jumped at this opportunity to disown QED, claiming, "because of its extreme complexity, most physicists will be glad to see the end of it" [Dirac, 1936, p. 299]. Bohr, as late as 1938, ruminated that perhaps the violations of causality in quantum mehanics were just the beginning and a more "radical departure" from classical theory would be necessary [Bohr, 1938, p. 29]. He nevertheless was sufficiently impressed with QED and its "still more complex abstractions" that he argued it "entails the greatest encouragement to proceed on such lines." [Bohr, 1938, p. 17]. It turns out that the resolution of the difficulties of QED are not related to causality (although they do involve more complex abstractions). As we will see in Part III, the key to performing calculations in QED beyond leading order in α is to carefully relate observable quantities to other observable quantities.

It was not until 1947, at the famous Shelter Island conference, that experiments finally showed that there were finite effects subleading in α, which gave theorists something precise to calculate. The next year, Schwinger came out with his celebrated calculation of the leading radiative correction to the electron magnetic moment: $g - 2 = \frac{\alpha}{\pi}$ (Chapter 17). That, and the agreement between Willis Lamb's measurement of the splitting between the $2S_{1/2}$ and $2P_{1/2}$ levels of the hydrogen atom (the hyperfine structure) and Hans Bethe's calculation of that splitting firmly established QED as predictive and essentially correct.

For additional information about the history of QED, there are a number of excellent accounts. Abraham Pais' *Inward Bound* [Pais, 1986] is classic; Mehra and Milton's scientific biography of Schwinger [Mehra *et al.*, 2000] and Schweber's book [Schweber, 1994] are also highly recommended.

Problems

13.1 Of the tree-level processes in QED, Møller scattering ($e^-e^- \to e^-e^-$) is especially interesting because it involves identical particles.

(a) Calculate the spin-averaged differential cross section for Møller scattering, $e^-e^- \to e^-e^-$. Express your answer in terms of s, t, u and m_e.

(b) Show that in the non-relativistic limit you get what we guessed by spin-conservation arguments in Problem 7.3:

$$\frac{d\sigma}{d\Omega} = \frac{m_e^4 \alpha^2}{E_{\rm CM}^2 p^4}\left(\frac{1 + 3\cos^2\theta}{\sin^4\theta}\right), \quad p^2 = \left(\frac{E_{\rm CM}}{2}\right)^2 - m_e^2. \tag{13.162}$$

(c) Simplify the Møller scattering formula in the ultra-relativistic limit ($m_e \to 0$). [Hint: you should get something proportional to $(3 + \cos^2\theta)^2$.]

13.2 Derive Eq. (13.103). It may be helpful to use the formula for scattering in the target rest frame derived in Problem 5.1.

13.3 Particle decays. Recall that the decay rate is given by the general formula

$$d\Gamma = \frac{1}{2E_1}|\mathcal{M}|^2 \frac{d^3p_2}{(2\pi)^3}\frac{1}{2E_2}\cdots\frac{d^3p_n}{(2\pi)^3}\frac{1}{2E_n}(2\pi)^4\delta^4(p_1 - p_2 - \cdots - p_n). \tag{13.163}$$

(a) Evaluate the phase-space integrals for $1 \to 2$ decays. Show that the total rate is

$$\Gamma(\phi \to e^+ + e^-) = \frac{\sqrt{1 - 4x^2}}{16\pi m_\phi}|\mathcal{M}|^2, \quad x = \frac{m_e}{m_\phi}. \tag{13.164}$$

(b) Evaluate Γ for a particle ϕ of mass m_ϕ decaying to e^+e^- of mass m_e if
  1. ϕ is a *scalar*, with interaction $g_S\phi\bar\psi\psi$;
  2. ϕ is a *pseudoscalar*, with interaction $ig_P\phi\bar\psi\gamma_5\psi$;
  3. ϕ is a *vector*, with interaction $g_V\phi_\mu\bar\psi\gamma^\mu\psi$;
  4. ϕ is an *axial vector*, with interaction $ig_A\phi_\mu\bar\psi\gamma^\mu\gamma_5\psi$.

(c) Breaking news! A collider experiment reports evidence of a new particle that decays only to leptons (τ, μ and e) whose mass is around 4 GeV. About 25% of the time it decays to $\tau^+\tau^-$. What spin and parity might this particle have?

13.4 Show that you always get a factor of -1 in the Feynman rules for each fermionic loop.

13.5 Consider the following diagram for $e^+e^- \to \mu^+\mu^-$ in QED:

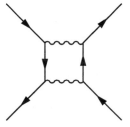

(a) How many diagrams contribute at the same order in perturbation theory?

(b) What is the minimal set of diagrams you need to add to this one for the sum to be gauge invariant (independent of ξ)?

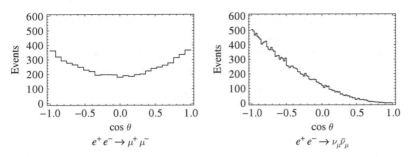

Angular distributions in e^+e^- annihilation produced with a Monte-Carlo simulation.

Fig. 13.2

(c) Show explicitly that the sum of diagrams in part (b) is gauge invariant.

13.6 Parity violation. We calculated that $e^+e^- \to \mu^+\mu^-$ has a $1 + \cos^2\theta$ angular dependence (see Eq. (13.78)), where θ is the angle between the e^- and μ^- directions. This agrees with experiment, as the simulated data on the left side of Figure 13.2 show. The angular distribution for scattering into muon neutrinos, $e^+e^- \to \nu_\mu\bar\nu_\mu$, is very different, as shown on the right side of Figure 13.2, where now θ is the angle between the e^- and $\bar\nu_\mu$ directions.

(a) At *low* energy, the total cross section, σ_{tot}, for $e^+e^- \to \nu_\mu\bar\nu_\mu$ scattering grows with energy, in contrast to the total $e^+e^- \to \mu^+\mu^-$ cross section. Show that this is consistent with neutrino scattering being mediated by a massive vector boson, the Z. Deduce how σ_{tot} should depend on E_{CM} for the two processes.

(b) Place the neutrino in a Dirac spinor ψ_ν. There are two possible couplings we could write down for the ν to the new massive gauge boson: $g_V \bar\psi_\nu \not{Z} \psi_\nu + g_A \bar\psi_\nu \not{Z} \gamma^5 \psi_\nu$. These are called vector and axial-vector couplings, respectively. Assume the Z couples to the electron in the same way as it couples to neutrinos. Calculate the full angular dependence for $e^+e^- \to \nu_\mu\bar\nu_\mu$ as a function of g_V and g_A (you can drop masses).

(c) What values of g_V and g_A reproduce Figure 13.2? Show that this choice is equivalent to the Z boson having chiral couplings: it only interacts with left-handed fields. Argue that this is evidence of parity violation, where the parity operator P is reflection in a mirror: $\vec{x} \to -\vec{x}$.

(d) An easier way to see parity violation is in β-decay. This is mediated by charged gauge bosons, the $W^\pm$, that are "unified" with the Z. Assuming they have the same chiral couplings as the Z, draw a diagram to show that the electron coming out of $^{60}_{27}\text{Co} \to {}^{60}_{28}\text{Ni} + e^- + \bar\nu$ will always be left-handed, independent of the spin of the cobalt nucleus. What handedness would the positron be in anti-cobalt decay: $^{60}_{27}\overline{\text{Co}} \to {}^{60}_{28}\overline{\text{Ni}} + e^+ + \nu$?

(e) If you are talking to aliens on the telephone (i.e. with light only), tell them how to use nuclear β-decay to tell clockwise from counterclockwise. For this, you will need to figure out how to relate the L in ψ_L to "left" in the real world. You are allowed to assume that all the materials on Earth are available to them, including things such as cobalt, and lasers.

(f) If you meet those aliens, and put out your right hand to greet them, but they put out their left hand, why should you not shake? (This scenario is due to Feynman.)

(g) Now forget about neutrinos. Could you have the aliens distinguish right from left by actually sending them circularly polarized light, for example using polarized radio waves for your intergalactic telephone?

13.7 One should be very careful with polarization sums and in giving physical interpretations to individual Feynman diagrams. This problem illustrates some of the dangers.

(a) We saw that the t-channel diagram for Compton scattering scales as $\mathcal{M}_t \sim \frac{1}{t}$. Calculate $|\mathcal{M}_t|^2$ summed over spins and polarizations. Be sure to sum over *physical* transverse polarizations only.

(b) Calculate $|\mathcal{M}_t|^2$ summed over spins and polarizations, but do the sum by replacing $\epsilon_\mu \epsilon_\nu^\star$ by $-g_{\mu\nu}$. Show that you get a different answer from part (a). Why is the answer different?

(c) Show that when you sum over all the diagrams you get the same answer whether you sum over physical polarizations or use the $\epsilon_\mu \epsilon_\nu^\star \to -g_{\mu\nu}$ replacement. Why is the answer the same?

(d) Repeat this exercise for scalar QED.

Path integrals   14

So far, we have studied quantum field theory using the canonical quantization approach, which is based on creation and annihilation operators. There is a completely different way to do quantum field theory called the path integral formulation. It says

$$\langle \Omega | T \{ \phi(x_1) \cdots \phi(x_n) \} | \Omega \rangle = \frac{\int \mathcal{D}\phi \, \phi(x_1) \cdots \phi(x_n) e^{iS[\phi]}}{\int \mathcal{D}\phi \, e^{iS[\phi]}}. \tag{14.1}$$

The left-hand side is exactly the kind of time-ordered product we use to calculate S-matrix elements. The $\mathcal{D}\phi$ on the right-hand side means integrate over all possible *classical* field configurations $\phi(\vec{x}, t)$ with a phase given by the *classical* action evaluated in that field configuration.

14.1 Introduction

The intuition for the path integral comes from a simple thought experiment you can do in quantum mechanics. Recall the double-slit experiment: the amplitude for a field to propagate from a source through a screen with two slits to a detector is the sum of the amplitudes to propagate through each slit separately. We add up the two amplitudes and then square to get the probability. If instead we had three slits and three screens, the amplitude would come from the sum of all possible paths through the slits and screens. And so on, for four slits, five slits, etc. Taking the continuum limit, we can keep slitting until the screen is gone. The result is that the final amplitude is the sum of all possible different paths. That is all the path integral is calculating. This is illustrated in Figure 14.1.

There is something very similar in classical physics called **Huygens' principle**. Huygens proposed in 1678 that to calculate the propagation of waves you can treat each point in the wavefront as the center of a fresh disturbance and a new source for the waves. A very intuitive example is surface waves in a region with obstructions, as shown in Figure 14.2. As the wave goes through a gap between barriers, a new wave starts from the gap and keeps going. This is useful, for example, in thinking about diffraction, where you can propagate the plane wave along to the slits, and then start waves propagating anew from each slit. Actually, it was not until 1816 that Fresnel realized that you could add amplitudes for the waves weighted by a phase given by the distance divided by the wavelength to explain

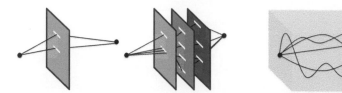

Fig. 14.1 The classic double slit allows for two paths between the initial and final points. Adding more screens and more slits allows for more diverse paths. An infinite number of screens and slits makes the amplitude the sum over all possible paths, as encapsulated in the path integral.

interference and diffraction. Thus, the principle is sometimes called the Huygens–Fresnel principle. The path integral is an implementation of this principle for quantum mechanical waves, with the phase determined by $\frac{1}{\hbar}$ times the action. Huygens' principle follows from the path integral since, as you take $\hbar \to 0$, this phase is dominated by the minimum of the action which is the classical action evaluated along the classical path. For $\hbar \neq 0$, there is a contribution from non-minimal action configurations that provide the quantum corrections.

There are a number of amazing things about path integrals. For example, they imply that by dealing with only classical field configurations you can get the quantum amplitude. This is really crazy if you think about it – these classical fields all commute, so you are also getting the non-commutativity for free somehow. Time ordering also just seems to pop out. And where are the particles? What happened to second quantization?

One way to think about path integrals is that they take the wave nature of matter to be primary, in contrast to the canonical method which is all about particles. Path integral quantization is in many ways simpler than canonical quantization, but it obscures some of the physics. Nowadays, people often just start with the path integral, using it to *define* the quantum theory. Path integrals are particularly useful to quantify non-perturbative effects. Examples include lattice QCD, instantons, black holes, etc. On the other hand, for calculations of discrete quantities such as energy eigenstates, and for many non-relativistic problems, the canonical formalism is much more practical.

Another important contrast between path integrals and the canonical approach is which symmetries they take to be primary. In the canonical approach, with the Hilbert space defined on spatial slices, matrix elements came out Lorentz invariant almost magically. With path integrals, Lorentz invariance is manifest the whole way through and Feynman diagrams appear very natural, as we will see. On the other hand, the Hamiltonian and Hilbert space are obscure in the path integral. That the Hamiltonian should be Hermitian and have positive definite eigenvalues on the Hilbert space (implying unitarity) is very hard to see with path integrals. So manifest unitarity is traded for manifest Lorentz invariance. Implications of unitarity for a general quantum field theory are discussed more in Chapter 24.

In this chapter, we will first derive the path integral from the canonical approach in the traditional way. Then we will perform two alternate derivations: we will show that we

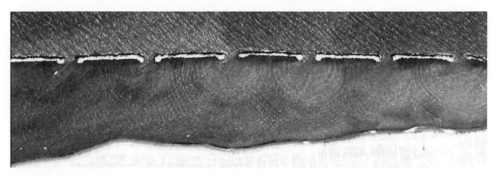

Ocean waves near Rimini, Italy (440 05 15.02 N , 120 32 26.07 E) illustrate Huygens' principle [Logiurato, 2012]. Image ©2013 Google Earth and ©2013 DigitalGlobe.

Fig. 14.2

reproduce the same perturbation series for time-ordered products (Feynman rules), and also show that the Schwinger–Dyson equations are satisfied. As applications, we will demonstrate the power of the path integral by proving gauge invariance and the Ward identity non-perturbatively in QED.

14.1.1 Historical note

Before around 1950, most QED calculations were done simply with old-fashioned perturbation theory. Schwinger (and Tomonaga around the same time) figured out how to do the calculations systematically using the interaction picture and applied the theory to radiative corrections. In particular, this method was used in the seminal calculations of the Lamb shift and magnetic moment of the electron in 1947/8. There were no diagrams. The diagrams, with loops, and Feynman propagators came from Feynman's vision of particles going forwards and backwards in time, and from his path integral. For example, Feynman knew that you could sum the retarded and advanced propagators together into one object (the Feynman propagator), while Schwinger and Tomonaga would add them separately.

Actually, Feynman did not know at the time how to prove that what he was calculating was what he wanted; he only had his intuition and some checks that he was correct. One of the ways Feynman could validate his approach was by showing that his tree-level calculations matched all the known results of QED. He then just drew the next picture and calculated the radiative correction. He could check his answers, eventually, by comparing to Schwinger and Tomonaga and, of course, to data, which were not available before 1947. He also knew his method was Lorentz covariant, which made the answers simple – another check. But what he was doing was not understood mathematically until Freeman Dyson cleaned things up in two papers in 1949 [Dyson, 1949]. Dyson's papers went a long way to convincing skeptics that QED was consistent.

There is a great story that Feynman recounted about the famous Poconos conference of 1948, where he and Schwinger both presented their calculations of the Lamb shift. Schwinger's presentation was polished and beautiful (but unintelligible, even to the experts such as Dirac and Pauli in the audience). Feynman got up and started drawing his pictures, but not knowing exactly how it worked, was unable to convince the bewildered audience. Feynman recounted [Mehra *et al.*, 2000, p. 233]:

> Already in the beginning I had said that I'll deal with single electrons, and I was going to describe this idea about a positron being an electron going backward in time, and Dirac asked, "Is it unitary?" I said, "Let me try to explain how it works, and you can tell me whether it is unitary or not!" I didn't even know then what "unitary" meant. So I proceeded further a bit, and Dirac repeated his question: "Is it unitary?" So I finally said: "Is what unitary?" Dirac said: "The matrix which carries you from the present to the future position." I said, "I haven't got any matrix which carries me from the present to the future position. I go forwards and backwards in time, so I do not know what the answer to your question is."

Teller was asking about the exclusion principle for virtual electrons; Bohr was asking about the uncertainty principle. Feynman did not have answers for any of these questions, he just knew his method worked. He concluded, "I'll just have to write it all down and publish it, so that they can read it and study it, because I know it is right! That's all there is to it." And so he did.

14.2 The path integral

The easiest way to derive the path integral is to start with non-relativistic quantum mechanics. Before deriving it, we will work out a simple mathematical formula for Gaussian integrals that is used in practically every path integral calculation. We then reproduce the derivation of the path integral in non-relativistic quantum mechanics, which you have probably already seen. The quantum field theory derivation is then a more-or-less straightforward generalization to the continuum.

14.2.1 Gaussian integrals

A general one-dimensional Gaussian integral is defined as

$$\mathcal{I} = \int_{-\infty}^{\infty} dp\, e^{-\frac{1}{2}ap^2 + Jp}. \tag{14.2}$$

To compute this integral, we first complete the square

$$\mathcal{I} = \int_{-\infty}^{\infty} dp\, e^{-\frac{1}{2}a\left(p - \frac{J}{a}\right)^2 + \frac{J^2}{2a}}, \tag{14.3}$$

then shift $p \to p + \frac{J}{a}$. The measure does not change under this shift, implying

$$\mathcal{I} = e^{\frac{J^2}{2a}} \int_{-\infty}^{\infty} dp\, e^{-\frac{1}{2}ap^2} = \frac{1}{\sqrt{a}} e^{\frac{J^2}{2a}} \int dp\, e^{-\frac{1}{2}p^2}. \qquad (14.4)$$

Now we use a trick to compute this:

$$\left[\int dp\, e^{-\frac{1}{2}p^2} \right]^2 = \int dx \int dy\, e^{-\frac{1}{2}x^2} e^{-\frac{1}{2}y^2}$$

$$= 2\pi \int_0^{\infty} r\, dr\, e^{-\frac{1}{2}r^2} = \pi \int_0^{\infty} dr^2 e^{-\frac{1}{2}r^2} = 2\pi, \qquad (14.5)$$

so,

$$\int_{-\infty}^{\infty} dp\, e^{-\frac{1}{2}ap^2 + Jp} = \sqrt{\frac{2\pi}{a}}\, e^{\frac{J^2}{2a}}. \qquad (14.6)$$

For multi-dimensional integrals, we need only generalize to many p_i, which may be complex. Then $ap^2 \to p_i^\star a_{ij} p_j = \vec{p}^\dagger \mathbf{A} \vec{p}$, with $\mathbf{A}$ a matrix. After diagonalizing $\mathbf{A}$ the integral becomes just a product of integrals over the p_i, and the result is the product of one-dimensional Gaussian integrals, with a being replaced by an eigenvalue of $\mathbf{A}$. That is,

$$\int_{-\infty}^{\infty} d\vec{p}\ e^{-\frac{1}{2}\vec{p}^\dagger \mathbf{A} \vec{p} + \vec{J}^\dagger \vec{p}} = \sqrt{\frac{(2\pi)^n}{\det \mathbf{A}}}\, e^{\frac{1}{2}\vec{J}^\dagger \mathbf{A}^{-1} \vec{J}}, \qquad (14.7)$$

where $\det \mathbf{A}$ comes from the product of the eigenvalues in the diagonal basis and n is the dimension of $\vec{p}$.

14.2.2 Path integral in quantum mechanics

Consider one-dimensional non-relativistic quantum mechanics with the Hamiltonian given by

$$\hat{H}(t) = \frac{\hat{p}^2}{2m} + V(\hat{x}, t). \qquad (14.8)$$

Here $\hat{H}$, $\hat{p}$ and $\hat{x}$ are operators acting on the Hilbert space, and t is just a number.

Suppose our initial state $|i\rangle = |x_i\rangle$ is localized at x_i at time t_i and we want to project it onto the final state $\langle f| = \langle x_f|$ localized at x_f at time t_f. If $\hat{H}$ did *not* depend on t, then we could just solve for the matrix element as

$$\langle f|i\rangle = \langle x_f | e^{-i(t_f - t_i)\hat{H}} | x_i \rangle. \qquad (14.9)$$

If instead we only assume $\hat{H}(t)$ is a smooth function of t, then we can only solve for the matrix element this way for infinitesimal time intervals. So, let us break this down into n small time intervals δt and define $t_j = t_i + j\delta t$. Then,

$$\langle f|i\rangle = \int dx_n \cdots dx_1 \langle x_f | e^{-iH(t_n)\delta t} | x_n \rangle \langle x_n | \cdots | x_2 \rangle \langle x_2 | e^{-iH(t_1)\delta t} | x_1 \rangle \langle x_1 | e^{-iH(t_i)\delta t} | x_i \rangle.$$

$$(14.10)$$

Each matrix element can be evaluated by inserting a complete set of momentum eigenstates and using $\langle p|x \rangle = e^{-ipx}$:

$$\langle x_{j+1}|e^{-iH(t_j)\delta t}|x_j \rangle = \int \frac{dp}{2\pi} \langle x_{j+1}|p \rangle \langle p|e^{-i\left[\frac{\hat{p}^2}{2m} + V(\hat{x}_j, t_j)\right]\delta t}|x_j \rangle$$

$$= e^{-iV(x_j, t_j)\delta t} \int \frac{dp}{2\pi} e^{-i\frac{p^2}{2m}\delta t} e^{ip(x_{j+1} - x_j)}. \qquad (14.11)$$

Now we can use the Gaussian integral in Eq. (14.6), $\int dp \exp\left(-\frac{1}{2}ap^2 + Jp\right) = \sqrt{\frac{2\pi}{a}} \exp\left(\frac{J^2}{2a}\right)$, with $a = i\frac{\delta t}{m}$ and $J = i(x_{j+1} - x_j)$ to get

$$\langle x_{j+1}|e^{-iH\delta t}|x_j \rangle = N e^{-iV(x_j, t_j)\delta t} e^{i\frac{m}{2}\delta t \frac{(x_{j+1} - x_j)^2}{(\delta t)^2}} = N e^{iL(x, \dot{x})\delta t}, \qquad (14.12)$$

where N is an x- and t-independent normalization constant, which we will justify ignoring later, and

$$L(x, \dot{x}) = \frac{1}{2}m\dot{x}^2 - V(x, t) \qquad (14.13)$$

is the Lagrangian. We see that the Gaussian integral effected a Legendre transform to go from $H(x, p)$ to $L(x, \dot{x})$.

Using Eq. (14.12), each term in Eq. (14.10) becomes just a number and the product reduces to

$$\langle f|i \rangle = N^n \int dx_n \cdots dx_1 e^{iL(x_n, \dot{x}_n)\delta t} \cdots e^{iL(x_1, \dot{x}_1)\delta t}. \qquad (14.14)$$

Finally, taking the limit $\delta t \to 0$, the exponentials combine into an integral over dt and we get

$$\langle f|i \rangle = N \int_{x(t_i)=x_i}^{x(t_f)=x_f} \mathcal{D}x(t) e^{iS[x]}, \qquad (14.15)$$

where $\mathcal{D}x$ means sum over all paths $x(t)$ with the correct boundary conditions and the action is $S[x] = \int dt\, \mathcal{L}[x(t), \dot{x}(t)]$. Note that N has been redefined and is now formally infinite, but it will drop out of any physical quantities, as we will see in the path integral case.

14.2.3 Path integral in quantum field theory

The field theory derivation is very similar, but the set of intermediate states is more complicated. We will start by calculating the vacuum matrix element $\langle 0; t_f|0; t_i \rangle$. In quantum mechanics we broke the amplitude down into integrals over $|x\rangle\langle x|$ for intermediate times where the states $|x\rangle$ are eigenstates of the $\hat{x}$ operator. In field theory, the equivalents of $\hat{x}$ are the Schrödinger picture fields $\hat{\phi}(\vec{x})$, which at any time t can be written as

$$\hat{\phi}(\vec{x}) = \int \frac{d^3p}{(2\pi)^3} \frac{1}{\sqrt{2\omega_p}} \left(a_p e^{i\vec{p}\vec{x}} + a_p^\dagger e^{-i\vec{p}\vec{x}}\right). \qquad (14.16)$$

Each field comprises an infinite number of operators, one at each point $\vec{x}$. We put the hat on ϕ to remind you that it is an operator.

Up to this point, we have been treating the Hamiltonian and Lagrangian as functionals of fields and their derivatives. Technically, the Hamiltonian should not have time derivatives in it, since it is supposed to be *generating* time translation. Instead of $\partial_t \phi$ the Hamiltonian should depend on canonical conjugate operators, which we introduced in Section 2.3.3 as

$$\hat{\pi}(\vec{x}) \equiv -i \int \frac{d^3 p}{(2\pi)^3} \sqrt{\frac{\omega_p}{2}} \left(a_p e^{i\vec{p}\vec{x}} - a_p^\dagger e^{-i\vec{p}\vec{x}} \right), \tag{14.17}$$

and satisfy

$$\left[\hat{\phi}(\vec{x}), \hat{\pi}(\vec{y}) \right] = i \delta^3(\vec{x} - \vec{y}). \tag{14.18}$$

These canonical commutation relations and the Hamiltonian that generates time translation define the quantum theory.

The equivalent of $|x\rangle$ is a complete set of eigenstates of $\hat{\phi}$:

$$\hat{\phi}(\vec{x})|\Phi\rangle = \Phi(\vec{x})|\Phi\rangle. \tag{14.19}$$

The eigenvalues are functions of space $\Phi(\vec{x})$.[1] The equivalents of $|p\rangle$ are the eigenstates of $\hat{\pi}(\vec{x})$ that satisfy

$$\hat{\pi}(\vec{x})|\Pi\rangle = \Pi(\vec{x})|\Pi\rangle. \tag{14.20}$$

The $|\Pi\rangle$ states are conjugate to the $|\Phi\rangle$ states, and satisfy

$$\langle \Pi | \Phi \rangle = \exp\left(-i \int d^3 x \, \Pi(\vec{x}) \Phi(\vec{x}) \right), \tag{14.21}$$

which is the equivalent of $\langle \vec{p} | \vec{x} \rangle = e^{-i\vec{p}\vec{x}}$. The inner product of two $|\Phi\rangle$ states is

$$\langle \Phi' | \Phi \rangle = \int \mathcal{D}\Pi \langle \Phi' | \Pi \rangle \langle \Pi | \Phi \rangle = \int \mathcal{D}\Pi \exp\left(-i \int d^3 x \, \Pi(\vec{x}) \left[\Phi(\vec{x}) - \Phi'(\vec{x}) \right] \right), \tag{14.22}$$

which is the generalization of $\langle \vec{x}' | \vec{x} \rangle = \delta(\vec{x} - \vec{x}') = \int \frac{dp}{2\pi} \exp(-i\vec{p}(\vec{x} - \vec{x}'))$. You can construct these states explicitly and check these inner products in Problem 14.4.

Using $\hat{\phi}$ and $\hat{\pi}$ one can rewrite the Hamiltonian so as not to include any time derivatives. We found in Eq. (8.16) that the energy density for a real scalar field was given by

$$\mathcal{E} = \frac{1}{2}(\partial_t \phi)^2 + \frac{1}{2}(\vec{\nabla}\phi)^2 + \frac{1}{2}m^2\phi^2. \tag{14.23}$$

This is the same as the Hamiltonian density

$$\widehat{\mathcal{H}} = \frac{1}{2}\hat{\pi}^2 + \frac{1}{2}(\nabla\hat{\phi})^2 + \frac{1}{2}m^2\hat{\phi}^2 \tag{14.24}$$

---

[1] In some field theory texts the path integral is constructed using eigenstates not of $\hat{\phi}$ but of the part of $\hat{\phi}$ that involves only annihilation operators, $\hat{\phi}_-$. Writing $\hat{\phi} = \hat{\phi}_- + \hat{\phi}_+$, these eigenstates are $|\Phi\rangle = \exp\left(\int d^3 y \, \hat{\phi}_+ (\vec{y}) \, \Phi(\vec{y}) \right) |0\rangle$. These satisfy $\hat{\phi}_-(\vec{x}) |\Phi\rangle = \Phi(\vec{x}) |\Phi\rangle$ and are the field theory version of coherent states for a single harmonic oscillator. See, for example [Altland and Simons, 2010; Brown, 1992; Itzykson and Zuber, 1980].

after the replacement of $\frac{\partial \mathcal{L}}{\partial(\partial_t \hat{\phi})} = \partial_t \phi$ by $\hat{\pi}$ (as in a Legendre transform). More generally, let us write

$$\hat{\mathcal{H}} = \frac{1}{2}\hat{\pi}^2 + \mathcal{V}(\hat{\phi}), \qquad (14.25)$$

where $\mathcal{V}(\hat{\phi})$ can include interactions. One can consider more general Hamiltonians, as long as they are Hermitian and positive definite, but we stick to ones of this form for simplicity. We will also write $\hat{H}(t) = \int d^3x\, \hat{\mathcal{H}}$ with the t dependence of $\hat{H}(t)$ coming from how the field operators change with time in the full interacting theory.[2]

Now we calculate the vacuum matrix element by inserting complete sets of intermediate states, as in quantum mechanics:

$$\langle 0; t_f | 0; t_i \rangle = \int \mathcal{D}\Phi_1(x) \cdots \mathcal{D}\Phi_n(x) \langle 0 | e^{-i\delta t \hat{H}(t_n)} | \Phi_n \rangle \langle \Phi_n | \cdots | \Phi_1 \rangle \langle \Phi_1 | e^{-i\delta t \hat{H}(t_0)} | 0 \rangle. \qquad (14.26)$$

Each of these pieces becomes

$$\langle \Phi_{j+1} | e^{-i\delta t \hat{H}(t_j)} | \Phi_j \rangle = \int \mathcal{D}\Pi_j \langle \Phi_{j+1} | \Pi_j \rangle \langle \Pi_j | \exp\left[-i\delta t \int d^3x \left(\frac{1}{2}\hat{\pi}^2 + \mathcal{V}(\hat{\phi}) \right) \right] | \Phi_j \rangle$$

$$= \int \mathcal{D}\Pi_j \exp\left[i \int d^3x\, \Pi_j(\vec{x}) \left(\Phi_{j+1}(\vec{x}) - \Phi_j(\vec{x}) \right) \right]$$

$$\times \exp\left[-i\delta t \int d^3x \left(\frac{1}{2}\Pi_j^2(\vec{x}) + \mathcal{V}(\Phi_j) \right) \right]. \qquad (14.27)$$

Now we perform the Gaussian integral over Π_j to give

$$\langle \Phi_{j+1} | e^{-i\delta t \hat{H}(t_j)} | \Phi_j \rangle = N \exp\left(-i\delta t \int d^3x \left[\mathcal{V}[\Phi_j] - \frac{1}{2}\left(\frac{\Phi_{j+1}(\vec{x}) - \Phi_j(\vec{x})}{\delta t} \right)^2 \right] \right)$$

$$= N \exp\left(i\delta t \int d^3x\, \mathcal{L}[\Phi_j, \partial_t \Phi_j] \right), \qquad (14.28)$$

where

$$\mathcal{L}[\Phi_j, \partial_t \Phi_j] = \frac{1}{2}(\partial_t \Phi_j)^2 - \mathcal{V}[\Phi_j]. \qquad (14.29)$$

Collapsing up the pieces of Eq. (14.26) gives

$$\langle 0; t_f | 0; t_i \rangle = N \int \mathcal{D}\Phi(\vec{x}, t) e^{iS[\Phi]}, \qquad (14.30)$$

where $S[\Phi] = \int d^4x\, \mathcal{L}[\Phi]$ with the time integral going from t_i to t_f. For S-matrix elements, we take $t_i = -\infty$ and $t_f = +\infty$, in which case the integral in $S[\Phi] = \int d^4x\, \mathcal{L}[\Phi]$ is over all space-time.

So the path integral tells us to integrate over all *classical* field configurations Φ. Note that Φ does not just consist of the one-particle states, it can have two-particle states, etc. We can remember this by drawing pictures for the paths – including disconnected bubbles – as

---

[2] If $\mathcal{V}$ depended on $\hat{\pi}$ and $\hat{\phi}$, there might be an ordering ambiguity; this is no different than in the non-relativistic case and it is conventional to define the Hamiltonian to be **Weyl ordered** with the $\hat{\pi}$ operators all to the left of the $\hat{\phi}$ operators.

we would using Feynman rules. Actually, we really sum over all kinds of discontinuous, disconnected random fluctuations. In perturbation theory, only paths corresponding to sums of states of fixed particle number contribute. Non-perturbatively, for example with bound states or situations where multiple soft photons are relevant, particle number may not be a useful concept. The path integral allows us to perform calculations in non-perturbative regimes.

14.2.4 Classical limit

As a first check on the path integral, we can take the classical limit. To do that, we need to put back $\hbar$, which can be done by dimensional analysis. Since $\hbar$ has dimensions of action, it appears as

$$\langle 0; t_f | 0; t_i \rangle = N \int \mathcal{D}\Phi(\vec{x}, t) e^{\frac{i}{\hbar} S[\Phi]}. \tag{14.31}$$

Using the method of stationary phase we see that, in the limit $\hbar \to 0$, this integral is dominated by the value of Φ for which $S[\Phi]$ has an extremum. But $\delta S = 0$ is precisely the condition that determines the Euler–Lagrange equations which a classical field satisfies. Therefore, the only configuration that contributes in the classical limit is the classical solution to the equations of motion.

In case you are not familiar with the **method of stationary phase** (also known as the **method of steepest descent**), it is easy to understand. The quickest way is to start with the same integral without the i:

$$\int \mathcal{D}\Phi(\vec{x}, t) e^{-\frac{1}{\hbar} S[\Phi]}. \tag{14.32}$$

In this case, the integral would clearly be dominated by the Φ_0 where $S[\Phi]$ has a minimum; everything else would give a bigger $S[\Phi]$ and be infinitely more suppressed as $\hbar \to 0$. Now, when we put the i back in, the same thing happens, not because the non-minimal terms are zero, but because away from the minimum you have to sum over phases swirling around infinitely fast. When you sum infinitely swirling phases, you also get something that goes to zero when compared to something with a constant phase. Another way to see it is to use the more intuitive case with $e^{-\frac{i}{\hbar} S[\Phi]}$. Since we expect the answer to be well defined, it should be an analytic function of Φ_0. So we can take $\hbar \to 0$ in the imaginary direction, showing that the integral is still dominated by $S[\Phi_0]$.

14.2.5 Time-ordered products

Suppose we insert a field at fixed position and time into the path integral:

$$\mathcal{I} = \int \mathcal{D}\Phi e^{iS[\Phi]} \Phi(\vec{x}_j, t_j). \tag{14.33}$$

What does this represent?

Going back through our derivation, this integral can be written as

$$\mathcal{I} = \int \mathcal{D}\Phi_1(\vec{x}) \cdots \mathcal{D}\Phi_n(\vec{x})$$

$$\times \langle 0|e^{-iH(t_n)\delta t}|\Phi_n\rangle \cdots \langle\Phi_2|e^{-iH(t_2)\delta t}|\Phi_1\rangle\langle\Phi_1|e^{-iH(t_1)\delta t}|0\rangle \Phi_j(\vec{x}_j), \quad (14.34)$$

with $\Phi(\vec{x}_j, t_j)$ getting replaced by $\Phi_j(\vec{x}_j)$ since the j subscript on $\Phi_j(\vec{x})$ refers to the time. Now we want to replace $\Phi_j(\vec{x}_j)$ by an operator. Since the subscript on Φ is just its point in time, we have

$$\int \mathcal{D}\Phi_j(\vec{x}) \left\{ e^{-iH(t_j)\delta t}|\Phi_j\rangle \Phi_j(\vec{x}_j)\langle\Phi_j| \right\} = \hat{\phi}(x_j) \int \mathcal{D}\Phi_j(\vec{x}) e^{-iH(t_j)\delta t}|\Phi_j\rangle\langle\Phi_j|.$$

$$(14.35)$$

So we get to replace $\Phi(x_j)$ by the operator $\hat{\phi}(x_j)$ put in at the time t_j. Then we can collapse up all the integrals to give

$$N \int \mathcal{D}\Phi(\vec{x}, t) e^{iS[\Phi]} \Phi(\vec{x}_j, t_j) = \langle 0|\hat{\phi}(\vec{x}_j, t_j)|0\rangle. \quad (14.36)$$

If you find the collapsing-up-the-integrals confusing, just think about the derivation backwards. An insertion of $\hat{\phi}(\vec{x}_j, t_j)$ will end up by $|\Phi_j\rangle\langle\Phi_j|$, producing the eigenvalue $\Phi(\vec{x}_j, t_j)$.

Now say we insert two fields:

$$\int \mathcal{D}\Phi(\vec{x}, t) e^{iS[\Phi]} \Phi(\vec{x}_1, t_1)\Phi(\vec{x}_2, t_2). \quad (14.37)$$

The fields will be inserted in the appropriate matrix element. In particular, the earlier field will always come out on the right of the later field. So we get

$$N \int \mathcal{D}\Phi(x) e^{iS[\Phi]} \Phi(x_1)\Phi(x_2) = \langle 0|T\{\hat{\phi}(x_1)\hat{\phi}(x_2)\}|0\rangle. \quad (14.38)$$

In general,

$$N \int \mathcal{D}\Phi(x) e^{iS[\Phi]} \Phi(x_1) \cdots \Phi(x_n) = \langle 0|T\{\hat{\phi}(x_1) \cdots \hat{\phi}(x_n)\}|0\rangle. \quad (14.39)$$

Thus, we get time ordering for free in the path integral!

Why does this work? As a quick cross check, suppose the answer were

$$N \int \mathcal{D}\Phi(x) e^{iS[\Phi]} \Phi(x_1)\Phi(x_2) = \langle 0|\hat{\phi}(x_1)\hat{\phi}(x_2)|0\rangle \quad (14.40)$$

without the time ordering. The left-hand side does not care whether we write $\Phi(x_1)\Phi(x_2)$ or $\Phi(x_2)\Phi(x_1)$, since these are classical fields, but the right-hand side does distinguish $\hat{\phi}(x_1)\hat{\phi}(x_2)$ from $\hat{\phi}(x_2)\hat{\phi}(x_1)$, since the fields do not commute (at timelike separation). Thus, Eq. (14.40) cannot be correct. The only possible equivalent of the left-hand side would be something in which the operators effectively commute, such as the time-ordering operation.

We are also generally interested in interacting theories. For an interacting theory, one has to be able to go between the Hamiltonian and the Lagrangian to derive the path integral.

This is rarely done explicitly, and for theories such as non-Abelian gauge theories, it may not even be possible. Fortunately, we can simply define the quantum theory through the path integral expressed in terms of an action. In the interacting case, we must normalize so that the interacting vacuum remains the vacuum, $\langle \Omega | \Omega \rangle = 1$. This fixes the normalization and leads to

$$\langle \Omega | T\{\hat{\phi}(x_1)\hat{\phi}(x_2)\} | \Omega \rangle = \frac{\int \mathcal{D}\Phi(x) e^{iS[\Phi]} \Phi(x_1)\Phi(x_2)}{\int \mathcal{D}\Phi(x) e^{iS[\Phi]}}, \qquad (14.41)$$

from which the constant N drops out. The generalization to arbitrary Green's functions is given in Eq. (14.1).

Unless there is any ambiguity, from now on we will use the standard notation $\phi(x)$ instead of $\Phi(x)$ for the classical fields being integrated over in the path integral.

14.3 Generating functionals

There is a great way to calculate path integrals using currents. Consider the action in the presence of an external classical source $J(x)$. The vacuum amplitude in the presence of this source is then a functional called the **generating functional** and is denoted by $Z[J]$:

$$Z[J] = \int \mathcal{D}\phi \exp\left\{ iS[\phi] + i\int d^4x\, J(x)\phi(x) \right\}. \qquad (14.42)$$

At $J = 0$, this reduces to the vacuum amplitude without the source:

$$Z[0] = \int \mathcal{D}\phi\, e^{i\int d^4x \mathcal{L}[\phi]}. \qquad (14.43)$$

We next introduce the variational partial derivative. Since $J(y) = \int d^4x\, \delta(x-y) J(x)$ it is natural to define

$$\frac{\partial J(x)}{\partial J(y)} = \delta^4(x-y). \qquad (14.44)$$

This partial derivative can be thought of as varying the value of J at y, holding all other values of J fixed. This equation implies that

$$\frac{\partial}{\partial J(x_1)} \int d^4x\, J(x)\, \phi(x) = \phi(x_1). \qquad (14.45)$$

Then,

$$-i\frac{\partial Z}{\partial J(x_1)} = \int \mathcal{D}\phi \exp\left\{ iS[\phi] + i\int d^4x\, J(x)\,\phi(x) \right\} \phi(x_1), \qquad (14.46)$$

and thus,

$$-i\frac{1}{Z[0]} \frac{\partial Z}{\partial J(x_1)}\bigg|_{J=0} = \frac{\int \mathcal{D}\phi \exp\{iS[\phi]\}\,\phi(x_1)}{\int \mathcal{D}\phi\, e^{i\int d^4x \mathcal{L}[\phi]}} = \langle \Omega|\hat{\phi}(x_1)|\Omega\rangle. \qquad (14.47)$$

Similarly,

$$(-i)^n \frac{1}{Z[0]} \left. \frac{\partial^n Z}{\partial J(x_1)\cdots\partial J(x_n)}\right|_{J=0} = \langle\Omega|T\{\hat\phi(x_1)\cdots\hat\phi(x_n)\}|\Omega\rangle. \qquad (14.48)$$

So this is a nice way of calculating time-ordered products – we can calculate $Z[J]$ once and for all, and then to get time-ordered products all we have to do is take derivatives.

The generating functional is the quantum field theory analog of the partition function in statistical mechanics – it tells us everything we could possibly want to know about a system. The generating functional is the holy grail of any particular field theory: if you have an exact closed-form expression for $Z[J]$ for a particular theory, you have solved it completely.

14.3.1 Solving the free theory

In the free theory, we can calculate the generating functional exactly. For a real scalar field, the Lagrangian is

$$\mathcal{L} = -\frac{1}{2}\phi(\Box + m^2)\phi. \qquad (14.49)$$

Then, using the notation $Z_0[J]$ for the generating functional in the free theory,

$$Z_0[J] = \int \mathcal{D}\phi \exp\left\{i \int d^4x \left(-\frac{1}{2}\phi(\Box+m^2)\phi + J(x)\phi(x)\right)\right\}. \qquad (14.50)$$

We can solve this exactly since it is quadratic in the fields. We just need to use our relation

$$\int_{-\infty}^{\infty} d\vec{p}\, e^{-\frac{1}{2}\vec{p}A\vec{p}+\vec{J}\vec{p}} = \sqrt{\frac{(2\pi)^n}{\det A}}\, e^{\frac{1}{2}\vec{J}A^{-1}\vec{J}} \qquad (14.51)$$

with $A = i(\Box + m^2)$. To compute A^{-1} we need to take the inverse of $-(\Box+m^2)$ and multiply by i. This inverse is a function $\Pi(x-y)$ satisfying

$$(\Box_x + m^2)\Pi(x-y) = -\delta(x-y). \qquad (14.52)$$

As we know, this equation is solved by the propagator

$$\Pi(x-y) = \int \frac{d^4p}{(2\pi)^4}\frac{1}{p^2-m^2}e^{ip(x-y)} \qquad (14.53)$$

up to boundary conditions. Thus,

$$Z_0[J] = N\exp\left\{-i\int d^4x \int d^4y \frac{1}{2}J(x)\Pi(x-y)J(y)\right\} \qquad (14.54)$$

and so,

$$\begin{aligned}
\langle 0|T\{\hat\phi_0(x)\hat\phi_0(y)\}|0\rangle &= (-i)^2 \frac{1}{Z_0[0]}\left.\frac{\partial^2 Z_0[J]}{\partial J(x)\partial J(y)}\right|_{J=0}\\
&= i\Pi(x-y)\\
&= \int \frac{d^4p}{(2\pi)^4}\frac{i}{p^2-m^2}e^{ip(x-y)},
\end{aligned} \qquad (14.55)$$

where $|0\rangle$ is used instead of $|\Omega\rangle$ for the free vacuum and $\hat{\phi}_0(x)$ are the free quantum fields. This agrees with the Feynman propagator that we calculated using creation and annihilation operators, up to the factor of $i\varepsilon$, which will be discussed in Section 14.4.

14.3.2 Four-point function

We can also compute higher-order products:

$$
\langle 0| T\left\{\hat{\phi}_0(x_1)\,\hat{\phi}_0(x_2)\,\hat{\phi}_0(x_3)\,\hat{\phi}_0(x_4)\right\} |0\rangle = (-i)^4 \frac{1}{Z_0[0]} \frac{\partial^4 Z_0}{\partial J(x_1)\cdots\partial J(x_4)}\bigg|_{J=0}
$$

$$
= \frac{\partial^4}{\partial J(x_1)\cdots\partial J(x_4)} e^{-\frac{1}{2}\int d^4x \int d^4y\, J(x) D_F(x-y) J(y)}\bigg|_{J=0}
$$

$$
= \frac{\partial^3}{\partial J(x_1)\,\partial J(x_2)\,\partial J(x_3)}
$$

$$
\times \left(-\int d^4z\, D_F(x_4 - z) J(z)\right) e^{-\frac{1}{2}\int d^4x \int d^4y\, J(x) D_F(x-y) J(y)}\bigg|_{J=0}.
$$

(14.56)

Before we continue, let us simplify the notation by replacing arguments by subscripts. Then

$$
\frac{\partial^4}{\partial J_1 \partial J_2 \partial J_3 \partial J_4} e^{-\frac{1}{2}J_x D_{xy} J_y}\bigg|_{J=0} = \frac{\partial^3}{\partial J_1 \partial J_2 \partial J_3}\left(-J_z D_{z4}\right) e^{-\frac{1}{2}J_x D_{xy} J_y}\bigg|_{J=0}
$$

$$
= \frac{\partial^2}{\partial J_1 \partial J_2}\left(-D_{34} + J_z D_{z3} J_w D_{w4}\right) e^{-\frac{1}{2}J_x D_{xy} J_y}\bigg|_{J=0}
$$

$$
= \frac{\partial}{\partial J_1}\left(D_{34} J_z D_{z2} + D_{23} J_w D_{w4} + J_z D_{z3} D_{24} - J_z D_{z3} J_w D_{w4} J_r D_{r2}\right) e^{-\frac{1}{2}J_x D_{xy} J_y}\bigg|_{J=0}
$$

$$
= D_{34} D_{12} + D_{23} D_{14} + D_{13} D_{24}.
$$

(14.57)

Thus,

$$
\langle 0| T\left\{\hat{\phi}_0(x_1)\,\hat{\phi}_0(x_2)\,\hat{\phi}_0(x_3)\,\hat{\phi}_0(x_4)\right\} |0\rangle =
$$

(14.58)

These are the same three contractions we found in the canonical approach in Chapter 7 (cf. Eq. (7.17)). More generally, each derivative can either kill a J factor or pull a J factor down from the exponential. At the end, we set $J = 0$ so the kills must be paired up with the pull-downs. The $Z_0[0]$ factor gives the vacuum bubbles that drop out of the connected part of the S-matrix, as they did in the Hamiltonian derivation of the Feynman rules presented in Section 7.2.

14.3.3 Interactions

Now suppose we have interactions

$$\mathcal{L} = -\frac{1}{2}\phi(\Box + m^2)\phi + \frac{g}{3!}\phi^3. \tag{14.59}$$

Then, we can write

$$
\begin{aligned}
Z[J] &= \int \mathcal{D}\phi\, e^{i\int d^4x\left[\frac{1}{2}\phi(-\Box - m^2)\phi + J(x)\phi(x) + \frac{g}{3!}\phi^3\right]} \\
&= \int \mathcal{D}\phi\, e^{i\int d^4x\left[\frac{1}{2}\phi(-\Box - m^2)\phi + J(x)\phi(x)\right]} e^{i\int d^4x\frac{g}{3!}\phi^3} \\
&= \int \mathcal{D}\phi\, e^{i\int d^4x\left[\frac{1}{2}\phi(-\Box - m^2)\phi + J(x)\phi(x)\right]} \\
&\quad \times \left[1 + \frac{ig}{3!}\int d^4z\,\phi^3(z) + \left(\frac{ig}{3!}\right)^2 \frac{1}{2}\int d^4z \int d^4w\,\phi^3(z)\,\phi^3(w) + \cdots \right].
\end{aligned}
\tag{14.60}
$$

Each term in this expansion is a path integral in the free theory. Thus we can write

$$
\begin{aligned}
Z[J] &= Z_0[J] + \frac{ig}{3!}\int d^4z(-i)^3 \frac{\partial^3 Z_0[J]}{(\partial J(z))^3} \\
&\quad + \left(\frac{ig}{3!}\right)^2 \frac{1}{2}\int d^4z \int d^4w(-i)^6 \frac{\partial^6 Z_0[J]}{(\partial J(z))^3 (\partial J(w))^3} + \cdots,
\end{aligned}
\tag{14.61}
$$

where $Z_0[J]$ is the generating functional in the free theory.

This expansion reproduces the Feynman rules we calculated in the canonical picture. For example, taking two derivatives to form the 2-point function and normalizing by $Z[0]$ we find

$$
\begin{aligned}
\langle \Omega | T\{\hat{\phi}(x_1)\,\hat{\phi}(x_2)\} | \Omega \rangle &= \frac{Z_0[0]}{Z[0]} \langle 0 | T\{\hat{\phi}_0(x_1)\,\hat{\phi}_0(x_2)\} | 0 \rangle \\
&\quad + \frac{ig}{3!}\frac{Z_0[0]}{Z[0]}\int d^4z \langle 0 | T\{\hat{\phi}_0(x_1)\,\hat{\phi}_0(x_2)\,\hat{\phi}_0(z)^3\} | 0 \rangle + \cdots \\
&= \frac{\langle 0 | T\{\hat{\phi}_0(x_1)\,\hat{\phi}_0(x_2)\,e^{i\int d^4z\frac{g}{3!}\hat{\phi}_0(z)^3}\} | 0 \rangle}{\langle 0 | T\{e^{i\int d^4z\frac{g}{3!}\hat{\phi}_0(z)^3}\} | 0 \rangle},
\end{aligned}
\tag{14.62}
$$

which agrees with Eq. (7.64) from Chapter 7. So we have reproduced the Feynman rules from the path integral.

14.4 Where is the $i\varepsilon$?

In the derivation of the path integral, propagators seemed to come out as $\frac{1}{p^2 - m^2}$ without the $i\varepsilon$. What happened to the $i\varepsilon$, which was supposed to tell us about time ordering? Without the $i\varepsilon$ the path integral is actually undefined, both physically (for example, not specifying whether the propagator is advanced, retarded, Feynman or something else) and mathematically (it is not convergent). From the physical point of view, we have so far only been talking about correlation functions, not S-matrix elements. As in the canonical approach, the emergence of time ordering as the relevant boundary condition is connected to the importance of causal processes, such as scattering, where the initial state is *before* the final state. In the path integral, the $i\varepsilon$ can be derived by including the appropriate boundary conditions on the path integral for S-matrix calculations, as we will now show.

14.4.1 *S*-matrix

In using the path integral to calculate S-matrix elements, the fields being integrated over must match onto the free fields at $t = \pm\infty$. We can write the S-matrix in terms of the path integral as

$$\langle f|S|i\rangle = \int_{\phi(t=\pm\infty)\ \text{constrained}} \mathcal{D}\phi\, e^{iS[\phi]}. \tag{14.63}$$

This notation matches how boundary conditions are imposed in the non-relativistic path integral, where one integrates over $x(t)$ constrained so that the path satisfies $x(t_i) = x_i$ and $x(t_f) = x_f$. In the path integral, the requirement is that the functions $\phi(x)$ that are being integrated over match onto the free fields at $t = \pm\infty$. To make this more precise, we can write the constraints as projections on the states for which $\phi(\vec{x})$ are eigenvalues:

$$\langle f|S|i\rangle = \int \mathcal{D}\phi\, e^{iS[\phi]} \langle f|\phi\,(t=+\infty)\rangle\langle\phi\,(t=-\infty)|i\rangle. \tag{14.64}$$

Here, we have reinstated the notation from Section 14.2.3 that $|\phi\rangle$ is the eigenstate of the field operator $\hat{\phi}(\vec{x})$, as in Eq. (14.19): $\hat{\phi}(\vec{x})\,|\phi\rangle = \phi(\vec{x})\,|\phi\rangle$. Equation (14.64) says that the path integral is restricted to an integral over field configurations with the right boundary conditions for a scattering problem.

Let us consider the free theory, and restrict to the case where $|f\rangle = |i\rangle = |0\rangle$, which is enough to derive the $i\varepsilon$. For the vacuum amplitude, we need to evaluate $\langle\Phi|0\rangle$ with $|0\rangle$ defined by $a_p|0\rangle = 0$. For a single harmonic oscillator, the vacuum is replaced by the ground state and, as you may recall, the ground state's wavefunction is $\phi(x) = \langle x|0\rangle = e^{-\frac{1}{2}x^2}$, up to some constants. The free-field theory version is also a Gaussian:

$$\langle 0|\Phi\rangle = \mathcal{N}\exp\left(-\frac{1}{2}\int d^3\vec{x}\, d^3\vec{y}\,\mathcal{E}(\vec{x},\vec{y})\,\phi(\vec{x})\,\phi(\vec{y})\right), \tag{14.65}$$

where $\mathcal{N}$ is some constant and

$$\mathcal{E}(\vec{x},\vec{y}) = \int \frac{d^3p}{(2\pi)^3} e^{i\vec{p}(\vec{x}-\vec{y})}\omega_p. \tag{14.66}$$

In Problem 14.3 you can derive this, and also find an explicit expression for $\mathcal{E}(\vec{x},\vec{y})$ in terms of Hankel functions. We give neither the derivation nor the explicit form since neither is relevant for the final answer.

At this point, we have

$$\langle 0|\Phi(t=+\infty)\rangle\langle\Phi(t=-\infty)|0\rangle$$
$$= |\mathcal{N}|^2 \exp\left(-\frac{1}{2}\int d^3\vec{x}\, d^3\vec{y}\, [\phi(\vec{x},\infty)\,\phi(\vec{y},\infty) + \phi(\vec{x},-\infty)\,\phi(\vec{y},-\infty)]\,\mathcal{E}(\vec{x},\vec{y})\right).$$
$$(14.67)$$

To massage this into a form that looks more like a local interaction in the path integral, we need to insert a dt integral. We can do that with the identity (see Problem 14.4)

$$f(\infty) + f(-\infty) = \lim_{\varepsilon\to 0^+}\varepsilon\int_{-\infty}^{\infty} dt\, f(t) e^{-\varepsilon|t|}, \qquad (14.68)$$

which holds for any smooth function $f(\tau)$ (here, $\varepsilon\to 0^+$ means ε is taken to zero from above). Then

$$\langle\Phi(-\infty)|0\rangle\langle 0|\Phi(+\infty)\rangle$$
$$= \lim_{\varepsilon\to 0^+}|\mathcal{N}|^2\exp\left(-\frac{1}{2}\varepsilon\int dt\frac{d^3p}{(2\pi)^3}\int d^3\vec{x}\, d^3\vec{y}\,\phi(\vec{x},t)\,\phi(\vec{y},t)\,e^{i\vec{p}(\vec{x}-\vec{y})}\omega_p\right), \quad (14.69)$$

where we set $e^{-\varepsilon|t|}=1$ since we only care about the leading term as $\varepsilon\to 0$.

The vacuum amplitude is then

$$\langle 0|0\rangle = \lim_{\varepsilon\to 0^+}|\mathcal{N}|^2\int\mathcal{D}\phi$$
$$\times\exp\left(\frac{-i}{2}\int d^4x\int d^3y\int\frac{d^3p}{(2\pi)^3}e^{i\vec{p}(\vec{x}-\vec{y})}\phi(\vec{y},t)(\Box+m^2-i\varepsilon\omega_p)\,\phi(\vec{x},t)\right).$$
$$(14.70)$$

For $\varepsilon\to 0$ the $i\varepsilon\omega_p$ can be replaced with $i\varepsilon$ giving

$$\langle 0|0\rangle = \lim_{\varepsilon\to 0^+}|\mathcal{N}|^2\int\mathcal{D}\phi\exp\left(\frac{-i}{2}\int d^4x\,\phi(x)(\Box+m^2-i\varepsilon)\,\phi(x)\right). \qquad (14.71)$$

The derivation with fields inserted into the correlation function is identical. So we derive that the free propagator is

$$\langle 0|T\{\hat\phi_0(x)\,\hat\phi_0(y)\}|0\rangle = \lim_{\varepsilon\to 0^+}\int\frac{d^4p}{(2\pi)^4}\frac{i}{p^2-m^2+i\varepsilon}e^{ip(x-y)}, \qquad (14.72)$$

which is the normal Feynman propagator. For more details, see [Weinberg, 1995, Section 9.1].

14.4.2 Reflection positivity

Mathematical physicists will tell you that the $i\varepsilon$ is required by the condition of **reflection positivity**. This is the requirement that under time-reversal, fields should have positive energy. More precisely, the restricted Hilbert space of physical fields, $\phi(\vec{x},t)$ with positive energy, generates another Hilbert space of positive-energy fields when reflected in time $\phi(\vec{x}, -t)$ (this restriction avoids fields such as $\phi(\vec{x}, t) - \phi(\vec{x}, -t)$, which will have eigenvalue -1 under the reflection). Reflection positivity is a succinct encapsulation of the requirement for a positive definite Hamiltonian and a unitary theory. The derivation of the $i\varepsilon$ starts by defining reflection positivity in Euclidean space, then analytically continuing to Minkowski space, where the $i\varepsilon$ comes from the contour close to the real t axis.

A quick way to see how consistency affects the path integral is that without the $i\varepsilon$ the path integral is not convergent. To make it convergent, we can make a slight deformation of order ε, defining

$$Z_0[J] = \int \mathcal{D}\phi \exp\left\{i \int d^4x\left[-\frac{1}{2}\phi(\Box + m^2)\phi + J(x)\,\phi(x)\right]\right\} \exp\left\{-\frac{\varepsilon}{2}\int d^4x\,\phi^2\right\}$$
$$= \int \mathcal{D}\phi \exp\left\{i \int d^4x\left[\frac{1}{2}\phi\left(-\Box - m^2 + i\varepsilon\right)\phi + J(x)\,\phi(x)\right]\right\}. \qquad (14.73)$$

Although this is the quickest way to justify the $i\varepsilon$ factor, it does not explain why $i\varepsilon$ appears and not $-i\varepsilon$, which would be anti-time ordering. In fact, both $\pm i\varepsilon$ are equally valid path integrals, although only $+i\varepsilon$ leads to causal scattering ($-i\varepsilon$ gives anti-time-ordered products).

One problem with the mathematical physics arguments is that even with reflection positivity and with the $i\varepsilon$ factor, the path integral still is not completely well defined. In fact, the path integral has only been shown to exist for a few cases. As of the time of this writing, the path integral (and field theories more generally) is only known to exist (i.e. have a precise mathematical definition) for free theories, and for ϕ^4 theory in two or three dimensions. ϕ^4 theory in five dimensions is known not to exist. In four dimensions, we do not know much, exactly. We do not know if QED exists, or if scalar ϕ^4 exists, or even if asymptotically free or conformal field theories exist. In fact, we do not know if any field theory exists, in a mathematically precise way, in four dimensions.

14.5 Gauge invariance

One of the key things that makes path integrals useful is that we can do field redefinitions. Here we will use field redefinitions to prove gauge invariance, by which we mean independence of the covariant-gauge parameter ξ. To do so, we will explicitly separate out the gauge degrees of freedom by rewriting $A_\mu = A_\mu + \partial_\mu \pi$ and then factor out the path integral over π. The following is a simplified version of a general method introduced by Faddeev and Popov, which is covered in Sections 25.4 and 28.4.

Recall that the Lagrangian for a massless spin-1 particle is $\mathcal{L} = -\frac{1}{4}F_{\mu\nu}^2 + J_\mu A_\mu$, which leads to the momentum space equations of motion:

$$(k^2 g_{\mu\nu} - k_\mu k_\nu)A_\nu = J_\mu. \qquad (14.74)$$

These equations are not invertible because the operator $k^2 g_{\mu\nu} - k_\mu k_\nu$ has zero determinant (it has an eigenvector k_μ with eigenvalue 0). The physical reason it is not invertible is because we cannot uniquely solve for A_μ in terms of J_μ because of gauge invariance:

$$A_\mu \to A_\mu + \partial_\mu \alpha(x). \tag{14.75}$$

In other words, many vector fields correspond to the same current. Our previous solution was to gauge-fix by adding the term $\frac{1}{2\xi}(\partial_\mu A_\mu)^2$ to the Lagrangian. Now we will justify that prescription, and prove gauge-invariance: any matrix element of gauge-invariant operators will be independent of ξ. More precisely, with a general set of fields ϕ_i and interactions we will show that correlation functions

$$\langle \Omega \,|\, T\{\mathcal{O}(x_1 \cdots x_n)\}|\Omega\rangle = \frac{1}{Z[0]} \int \mathcal{D}A_\mu \mathcal{D}\phi_i \mathcal{D}\phi_i^\star e^{i \int d^4x \mathcal{L}[A,\phi_i]} \mathcal{O}(x_1 \cdots x_n) \tag{14.76}$$

are ξ independent, where $\mathcal{O}(x_1 \cdots x_n)$ refers to any gauge-invariant collection of fields.

Recall that we can always go to a gauge where $\partial_\mu A_\mu = 0$. Since under a gauge transformation $\partial_\mu A_\mu \to \partial_\mu A_\mu + \Box \alpha$, we can always find a function α such that $\Box \alpha = \partial_\mu A_\mu$. We will write this function as $\alpha = \frac{1}{\Box}\partial_\mu A_\mu$. Now consider the following function:

$$f(\xi) = \int \mathcal{D}\pi e^{-i \int d^4x \frac{1}{2\xi}(\Box \pi)^2}, \tag{14.77}$$

which is just some function of ξ, probably infinite. As we will show, this represents the path integral over gauge orbits which will factor out of the full path integral. To see that, shift the field by

$$\pi(x) \to \pi(x) - \alpha(x) = \pi(x) - \frac{1}{\Box}\partial_\mu A_\mu. \tag{14.78}$$

This is just a shift, so the integration measure does not change. Then,

$$f(\xi) = \int \mathcal{D}\pi e^{-i \int d^4x \frac{1}{2\xi}(\Box \pi - \partial_\mu A_\mu)^2}. \tag{14.79}$$

This is still just the same function of ξ, which despite appearances is independent of A_μ. We can multiply and divide Eq. (14.76) by $f(\xi)$ in the two different forms, giving

$$\langle \Omega | T\{\mathcal{O}(x_1 \cdots x_n)\}|\Omega\rangle = \frac{1}{Z[0]} \frac{1}{f(\xi)} \int \mathcal{D}\pi \mathcal{D}A_\mu \mathcal{D}\phi_i \mathcal{D}\phi_i^\star$$
$$\times e^{i \int d^4x [\mathcal{L}[A,\phi_i] - \frac{1}{2\xi}(\Box \pi - \partial_\mu A_\mu)^2]} \mathcal{O}(x_1 \cdots x_n). \tag{14.80}$$

Now let us do the "Stueckelberg trick" and perform a gauge transformation shift, with $\pi(x)$ as our gauge parameter:

$$A_\mu = A'_\mu + \partial_\mu \pi, \qquad \phi_i = e^{i\pi}\phi'_i. \tag{14.81}$$

Again, the measure $\mathcal{D}\pi \mathcal{D}A_\mu \mathcal{D}\phi_i$, the action $\mathcal{L}[A, \phi_i]$, and the operator $\mathcal{O}$, which is gauge-invariant by assumption, do not change. We conclude that the path integral is the same as the gauge-fixed version up to normalization:

$$\langle\Omega|T\left\{\mathcal{O}(x_1\cdots x_n)\right\}|\Omega\rangle = \frac{1}{Z[0]}\left[\frac{1}{f(\xi)}\int\mathcal{D}\pi\right]\int\mathcal{D}A_\mu\mathcal{D}\phi_i\mathcal{D}\phi_i^\star$$
$$\times\, e^{i\int d^4x\left[\mathcal{L}[A,\phi_i]-\frac{1}{2\xi}(\partial_\mu A_\mu)^2\right]}\mathcal{O}(x_1\cdots x_n). \qquad (14.82)$$

Conveniently, the same normalization appears when we perform the same manipulations to $Z[0]$:

$$Z[0] = \left[\frac{1}{f(\xi)}\int\mathcal{D}\pi\right]\int\mathcal{D}A_\mu\mathcal{D}\phi_i\mathcal{D}\phi_i^\star e^{i\int d^4x\left[\mathcal{L}[A,\phi_i]-\frac{1}{2\xi}(\partial_\mu A_\mu)^2\right]}. \qquad (14.83)$$

Thus, the normalization drops out and we find that

$$\langle\Omega|T\left\{\mathcal{O}(x_1\cdots x_n)\right\}|\Omega\rangle = \frac{\int\mathcal{D}A_\mu\mathcal{D}\phi_i\mathcal{D}\phi_i^\star e^{i\int d^4x\mathcal{L}[A,\phi_i]}\mathcal{O}(x_i)}{\int\mathcal{D}A_\mu\mathcal{D}\phi_i\mathcal{D}\phi_i^\star e^{i\int d^4x\mathcal{L}[A,\phi_i]}}$$
$$= \frac{\int\mathcal{D}A_\mu\mathcal{D}\phi_i\mathcal{D}\phi_i^\star e^{i\int d^4x\left[\mathcal{L}[A,\phi_i]-\frac{1}{2\xi}(\partial_\mu A_\mu)^2\right]}\mathcal{O}(x_i)}{\int\mathcal{D}A_\mu\mathcal{D}\phi_i\mathcal{D}\phi_i^\star e^{i\int d^4x\left[\mathcal{L}[A,\phi_i]-\frac{1}{2\xi}(\partial_\mu A_\mu)^2\right]}}. \qquad (14.84)$$

That is, correlation functions calculated with the gauge-fixed Lagrangian will give the same results as correlation functions calculated with the gauge-invariant Lagrangian. In other words, $\langle\Omega|T\{\mathcal{O}(x_1\cdots x_n)\}|\Omega\rangle$ calculated with the gauge-fixed Lagrangian is completely independent of ξ.

Unfortunately, the above argument does not apply to correlation functions of fields that are gauge covariant. For example, $\langle\Omega|\bar{\psi}(x_1)\psi(x_2)|\Omega\rangle$ in general *will* depend on ξ. A simple example is $\langle\Omega|A_\mu(x)A_\nu(y)|\Omega\rangle$, which (at leading order) is just the ξ-dependent photon propagator. That the S-matrix is gauge invariant, a fact that was understood in perturbation theory in Section 9.4, requires additional insight. A proof valid to all orders in perturbation theory using a different approach is discussed in Section 14.8.4.

14.6 Fermionic path integral

A path integral over fermions is basically the same as for bosons, but we have to allow for the fact that the fermions anticommute. At the end of the day, all you really need to use is that classical fermion fields satisfy $\{\psi(x),\chi(y)\} = 0$. This section gives some of the mathematics behind anticommuting classical numbers.

A **Grassmann algebra** is a set of objects $\mathcal{G}$ that are generated by a basis $\{\theta_i\}$. These θ_i are **Grassmann numbers**, which anticommute with each other, $\theta_i\theta_j = -\theta_j\theta_i$, add commutatively, $\theta_i + \theta_j = \theta_j + \theta_i$, and can be multiplied by complex numbers, $a\theta \in \mathcal{G}$ for $\theta \in \mathcal{G}$ and $a \in \mathbb{C}$. The algebra must also have an element 0 so that $\theta_i + 0 = \theta_i$.

For one θ, the most general element of the algebra is

$$g = a + b\theta, \quad a,b \in \mathbb{C}, \qquad (14.85)$$

since $\theta^2 = 0$. For two θ's, the most general element is

$$g = A + B\theta_1 + C\theta_2 + F\theta_1\theta_2, \qquad (14.86)$$

and so on. Elements of the algebra that have an even number of θ_i commute with all elements of the algebra, so they compose the **even-graded** or bosonic subalgebra. Similarly, the **odd-graded** or fermionic subalgebra has an odd number of θ_i and anticommutes within itself (but commutes with the bosonic subalgebra). The fermionic subalgebra is not closed, since $\theta_1\theta_2$ is bosonic.

Sometimes it is helpful to compare what we will do with fermions to an example of a Grassmann algebra that you might already be familiar with: the exterior algebra of differential forms. Two forms A and B form a Grassmann algebra with the product usually denoted with a wedge, so that $A \wedge B = -B \wedge A$. So, for example, $\mathbf{dx}$ and $\mathbf{dy}$ would generate a two-dimensional Grassmann algebra.

In physics our Grassmann numbers will be $\theta_1 = \psi(x_1), \theta_2 = \psi(x_2), \ldots$, so we will have an infinite number of them. Then quantities such as the Lagrangian are (bosonic) elements of $\mathcal{G}$. To get regular numbers out, we need to integrate over $\mathcal{D}\psi$. So we need to figure out a consistent way to define such integrals.

To begin, we want integration to be linear, so that

$$\int d\theta_1 \cdots d\theta_n(sX+tY) = s \int d\theta_1 \cdots d\theta_n X + t \int d\theta_1 \cdots d\theta_n Y, \quad s,t \in \mathbb{C}, \quad X,Y \in \mathcal{G}.$$
(14.87)

We do not put limits of integration on the integrals since there is only one Grassmann number in each direction. These are the analogs of the definite integrals, $\int_{-\infty}^{\infty} dx \, f(x)$, in the bosonic case.

Next, we want the integrals to be like sums so that $d\theta$, like θ, is an anticommuting object, and so is $\int d\theta$. First consider one θ. The most general integral is

$$\int d\theta(a + b\theta) = a \int d\theta + b \int d\theta \, \theta.$$
(14.88)

Since the integral is supposed to be a map from $\mathcal{G}$ to $\mathbb{C}$, the first term must vanish. We conventionally define $\int d\theta \, \theta = 1$ and so

$$\int d\theta(a + b\theta) = b.$$
(14.89)

Note that the obvious definition for derivatives is

$$\frac{d}{d\theta}(a + b\theta) = b,$$
(14.90)

so integration and differentiation do the same thing on Grassmann numbers.

For more θ_i we define

$$\int d\theta_1 \cdots d\theta_n X = \frac{\partial}{\partial\theta_1} \cdots \frac{\partial}{\partial\theta_n} X,$$
(14.91)

so that

$$\int d\theta_1 \cdots d\theta_n \theta_n \cdots \theta_1 = 1.$$
(14.92)

Note that we evaluate these nested integrals from the inside out. That is,

$$\int d\theta_1 d\theta_2 \theta_2 \theta_1 = - \int d\theta_1 d\theta_2 \theta_1 \theta_2 = 1.$$
(14.93)

This is consistent with the order in which derivatives usually act. That is all there is to it. This is a consistent definition of integration and differentiation.

One important feature of these integrals is that they have the same kind of shift symmetry as the bosonic case. In the bosonic case $\int_{-\infty}^{\infty} dx f(x) = \int_{-\infty}^{\infty} dx\, f(x + a)$, where a is independent of x. That is, $\partial_x a = 0$. The analog here would be

$$\int d\theta (A + B\theta) = \int d\theta (A + B(\theta + X)), \qquad (14.94)$$

where X is any element of $\mathcal{G}$ that is constant with respect to θ: $\frac{\partial}{\partial \theta} X = 0$. This equality then holds by definition of integration.

For the path integral, we need Gaussian integrals. For two θ_i, we have

$$\int d\theta_1 d\theta_2 e^{-\theta_1 A_{12} \theta_2} = \int d\theta_1 d\theta_2 (1 - A_{12}\theta_1\theta_2) = A_{12}, \qquad (14.95)$$

where we have Taylor expanded the exponential. One does not need to think of θ as small in any way to do this. Rather, the exponential is *defined* by its Taylor expansion, as it is for other anticommuting things, such as Lie group generators.[3]

Now say we have n θ_i and n other independent θ_i that we will call $\bar{\theta}_i$. Then consider an integral that is an exponential of something quadratic in them:

$$\int d\bar{\theta}_1 d\theta_1 \cdots d\bar{\theta}_n d\theta_n e^{-\bar{\theta}_i A_{ij}\theta_j} = \int d\bar{\theta}_1 d\theta_1 \cdots d\bar{\theta}_n d\theta_n$$
$$\times \left(1 - \bar{\theta}_i A_{ij}\theta_j + \frac{1}{2}(\bar{\theta}_i A_{ij}\theta_j)(\bar{\theta}_k A_{kl}\theta_l) + \cdots \right). \quad (14.96)$$

The only term in this expansion that will survive is the one with all n θ_i and all n $\bar{\theta}_i$. This will give

$$\int d\bar{\theta}_1 d\theta_1 \cdots d\bar{\theta}_n d\theta_n e^{-\bar{\theta}_i A_{ij}\theta_j} = \frac{1}{n!} \sum_{\text{permutations}\{i_n, j_n\}} \pm A_{i_1 j_1} \cdots A_{i_n j_n}. \qquad (14.97)$$

If we think of A_{ij} as a matrix, this is a sum over all elements $\{i, j\}$ where we choose each row and column once, with the sign from the ordering. This is exactly how you compute a determinant. The $n!$ for the number of permutations cancels the $\frac{1}{n!}$ in front. So

$$\int d\bar{\theta}_1 d\theta_1 \cdots d\bar{\theta}_n d\theta_n e^{-\bar{\theta}_i A_{ij}\theta_j} = \det(\mathbf{A}). \qquad (14.98)$$

Note that this is different from what we found for ordinary numbers:

$$\int dx_1 \cdots dx_n e^{-\frac{1}{2}x_i A_{ij}x_j} = \sqrt{\frac{(2\pi)^n}{\det(\mathbf{A})}}. \qquad (14.99)$$

---

[3] In the literature, authors often talk about general functions $f(\theta_1, \theta_2, \ldots)$, which are defined from their Taylor series. This notation does not mean f is a function in the usual sense, but rather that f is an element of the algebra generated by the θ_i. This general notation is not particularly useful, and in the same way trying to decipher general functions $f(\mathbf{dx}, \mathbf{dy})$ is usually unnecessary.

Whether the determinant is in the numerator or denominator is occasionally important (but not for QED). With external currents η_i and $\bar{\eta}_i$,

$$\int d\bar{\theta}_1 d\theta_1 \cdots d\bar{\theta}_n d\theta_n e^{-\bar{\theta}_i A_{ij}\theta_j + \bar{\eta}_i\theta_i + \bar{\theta}_i\eta_i} = e^{\vec{\bar{\eta}}\mathbf{A}^{-1}\vec{\eta}} \int d\vec{\bar{\theta}}d\vec{\theta} e^{-(\vec{\bar{\theta}} - \vec{\bar{\eta}}\mathbf{A}^{-1})\mathbf{A}(\vec{\theta} - \mathbf{A}^{-1}\vec{\eta})}$$

$$= \det(\mathbf{A})e^{\vec{\bar{\eta}}\mathbf{A}^{-1}\vec{\eta}}, \tag{14.100}$$

which is all we need for the path integral.

Now let us take the continuum limit, replacing the index i by a continuous variable x, and θ_i by $\psi(x)$ and $\bar{\theta}_i$ by $\bar{\psi}(x)$. Then functions of θ_i and $\bar{\theta}_i$ become functionals of $\psi(x)$ and $\bar{\psi}(x)$. The fermionic path integral is over all such fields:

$$Z[\bar{\eta}, \eta] = \int \mathcal{D}[\bar{\psi}(x)]\mathcal{D}[\psi(x)]e^{i\int d^4x[\bar{\psi}(i\not{\partial} - m)\psi + \bar{\eta}\psi + \bar{\psi}\eta + i\varepsilon\bar{\psi}\psi]}. \tag{14.101}$$

As in the bosonic case, the $i\varepsilon$ comes from the boundary condition at $t = \pm\infty$. Then we have $\mathbf{A} = -i(i\not{\partial} - m + i\varepsilon)$ and so

$$Z[\bar{\eta}, \eta] = \mathcal{N}e^{i\int d^4x \int d^4y \bar{\eta}(y)(i\not{\partial} - m + i\varepsilon)^{-1}\eta(x)}, \tag{14.102}$$

where $\mathcal{N} = \det(i\not{\partial} - m)$ is some infinite constant.

The 2-point function in the free theory is

$$\langle 0|T\{\psi(x)\bar{\psi}(y)\}|0\rangle = \frac{1}{Z[0]}\frac{\partial^2}{\partial\bar{\eta}(x)\,\partial\eta(y)}Z[\bar{\eta}, \eta]\Big|_{\eta=0} = \frac{i}{i\not{\partial} - m + i\varepsilon}\delta^4(x - y)$$

$$= \int \frac{d^4p}{(2\pi)^4}\frac{i}{\not{p} - m + i\varepsilon}e^{-ip(x-y)}. \tag{14.103}$$

This simplifies using $(\not{p} - m)(\not{p} + m) = p^2 - m^2$, which implies

$$\frac{1}{\not{p} - m + i\varepsilon} = \frac{\not{p} + m}{p^2 - m^2 + i\varepsilon}. \tag{14.104}$$

So,

$$\langle 0|T\{\psi(x)\bar{\psi}(y)\}|0\rangle = \int \frac{d^4p}{(2\pi)^4}\frac{i(\not{p} + m)}{p^2 - m^2 + i\varepsilon}e^{-ip(x-y)}, \tag{14.105}$$

which is the Dirac propagator.

Fermionic path integrals may seem really hard and confusing, but in the end they are quite simple, and you can usually forget about the fact that there is a lot of weird mathematics going into them.

14.7 Schwinger–Dyson equations

One odd thing about the path integral is that it only involves classical fields. Where is the quantum mechanics? Where is the non-commutativity? We saw in Section 7.1 that an efficient way to see the difference between the classical and quantum theories was through the Schwinger–Dyson equations:

$$(\Box_x + m^2)\langle \hat{\phi}(x)\hat{\phi}(x_1)\cdots\hat{\phi}(x_n)\rangle = \langle \mathcal{L}'_{\text{int}}\left[\hat{\phi}(x)\right]\hat{\phi}(x_1)\cdots\hat{\phi}(x_n)\rangle$$
$$- i\sum_i \delta^4(x - x_i)\langle \hat{\phi}(x_1)\cdots\hat{\phi}(x_{i-1})\,\hat{\phi}(x_{i+1})\cdots\hat{\phi}(x_n)\rangle. \quad (14.106)$$

Here $\mathcal{L}'_{\text{int}}[\phi] = \frac{\partial}{\partial\phi}\mathcal{L}_{\text{int}}[\phi]$ is the variational derivative of the interaction Lagrangian, and we are using $\langle\cdots\rangle$ as an abbreviation for $\langle\Omega|T\{\cdots\}|\Omega\rangle$ for time-ordered matrix elements in the interacting vacuum to avoid clutter. Recall also from Section 7.1 that the derivation of these equations in the canonical quantization approach required that the interacting quantum fields satisfy the Euler–Lagrange equations $(\Box + m^2)\phi = \mathcal{L}'_{\text{int}}[\phi]$ and that the canonical commutation relations $[\hat{\phi}(x), \partial_t\hat{\phi}(y)] = i\delta^3(x - y)$ be satisfied.

The Schwinger–Dyson equations assert that vacuum matrix elements of time-ordered products satisfy the classical equations of motion up to contact terms. They specify non-perturbative relations among correlation functions. In fact, as we will see in this section, they are enough to completely specify the quantum theory. We will also show that these equations follow from the path integral and therefore they can be used to prove that the canonical and path integral approaches agree. Keep in mind that the classical fields in the path integral are not classical in the sense that they satisfy the classical equations of motion. In the path integral, one just integrates over all field configurations, whether or not they satisfy the equations of motion.

14.7.1 Contact terms

Since the contact terms in the Schwinger–Dyson equations indicate how the quantum field theory deviates from the corresponding classical field theory, it is natural to suspect that they are related to how the principle of least action is modified. In classical field theory, the Euler–Lagrange equations are derived by requiring that the action be stationary under variations $\phi(x) \to \phi(x) + \varepsilon(x)$, where $\varepsilon(x)$ is an arbitrary function. Let us now investigate how the derivation is modified in the quantum theory. In this section, we take $m = 0$ for simplicity.

Consider first the 1-point function:

$$\langle\hat{\phi}(x)\rangle = -i\frac{1}{Z[0]}\frac{\partial Z[J]}{\partial J(x)}\bigg|_{J=0} = \frac{1}{Z[0]}\int \mathcal{D}\phi\, e^{i\int d^4y\left(-\frac{1}{2}\phi\Box_y\phi\right)}\phi(x). \quad (14.107)$$

Now replace $\phi(x) \to \phi(x) + \varepsilon(x)$ in the path integral. This is just a field redefinition, and since the path integral integrates over all configurations, the same answer must result. Since this is a linear shift, the measure is invariant, so

$$\langle\hat{\phi}(x)\rangle = \frac{1}{Z[0]}\int \mathcal{D}\phi\, e^{i\int d^4y\left(-\frac{1}{2}(\phi+\varepsilon)\Box(\phi+\varepsilon)\right)}[\phi(x) + \varepsilon(x)]. \quad (14.108)$$

Expanding to first order in ε,

$$\langle\hat{\phi}(x)\rangle = \frac{1}{Z[0]}\int \mathcal{D}\phi\, e^{i\int d^4y\left(-\frac{1}{2}\phi\Box_y\phi\right)}\left\{\phi(x) + \varepsilon(x) - i\phi(x)\int d^4z\,\varepsilon(z)\Box_z\phi(z)\right\}, \tag{14.109}$$

where we have integrated by parts to combine the $\varepsilon\Box\phi$ and $\phi\Box\varepsilon$ terms. Comparing with Eq. (14.107), the $\phi(x)$ term already saturates the equality, so the remaining terms must add to zero. Thus,

$$\int d^4z\left[\varepsilon(z)\right]\int \mathcal{D}\phi\, e^{i\int d^4y\left(-\frac{1}{2}\phi\Box_y\phi\right)}\left[\phi(x)\Box_z\phi(z) + i\delta^4(z-x)\right] = 0. \tag{14.110}$$

Since the path integral does not depend on z except through the field insertion, the $\Box_z$ can be pulled outside of the integral. For the equality to hold for any $\varepsilon(z)$, we must have

$$(-i)^2\left(\Box_z\frac{\partial^2 Z[J]}{\partial J(z)\partial J(x)}\right)\Bigg|_{J=0} = -i\delta^4(z-x)\,Z[0]. \tag{14.111}$$

In terms of correlation functions, this is

$$\Box_z\langle\hat{\phi}(z)\hat{\phi}(x)\rangle = -i\delta^4(z-x)\,, \tag{14.112}$$

which is of course nothing but the Green's function equation for the Feynman propagator. It is also the Schwinger–Dyson equation for the 2-point function in a free scalar field theory.

For an interacting theory, let us add a potential so that $\mathcal{L} = -\frac{1}{2}\phi\Box\phi + \mathcal{L}_{\text{int}}[\phi]$. Then the classical equations of motion are $\Box\phi = \mathcal{L}'_{\text{int}}[\phi]$. In the path integral, the addition of the potential contributes a term $i\int d^4z\,\varepsilon(z)\mathcal{L}'_{\text{int}}[\phi(z)]$ to the $\{\}$ in Eq. (14.109) and Eq.(14.110) is modified to

$$\int d^4z\,\varepsilon(z)\left\{\Box_z\int \mathcal{D}\phi\left[e^{iS}\phi(z)\phi(x)\right]\right.$$
$$\left. - \int \mathcal{D}\phi\, e^{iS}\phi(x)\mathcal{L}'_{\text{int}}[\phi(z)] + i\delta^4(z-x)\int \mathcal{D}\phi\, e^{iS}\right\} = 0, \tag{14.113}$$

This can be written as a statement about correlation functions in the canonical picture:

$$\Box_z\langle\hat{\phi}(z)\,\hat{\phi}(x)\rangle = \langle\mathcal{L}'_{\text{int}}\left[\hat{\phi}(z)\right]\hat{\phi}(x)\rangle - i\delta^4(z-x)\,, \tag{14.114}$$

which is the Schwinger–Dyson equation for the 2-point function in the presence of interactions.

If we have more field insertions, the Schwinger–Dyson equations add contact interactions, contracting the field on which the operator acts with all the other fields in the correlator. For example, with three fields:

$$\Box_x\langle\hat{\phi}(x)\,\hat{\phi}(y)\,\hat{\phi}(z)\rangle = \langle\mathcal{L}'_{\text{int}}\left[\hat{\phi}(x)\right]\hat{\phi}(y)\hat{\phi}(z)\rangle - i\delta^4(x-z)\langle\hat{\phi}(y)\rangle - i\delta^4(x-y)\langle\hat{\phi}(z)\rangle \tag{14.115}$$

and so on. In this way, the complete set of Schwinger–Dyson equations can be derived.

Similar equations hold for theories with spinors or gauge bosons. For example, write the QED Lagrangian as

$$\mathcal{L} = \frac{1}{2} A_\mu \Box^{\mu\nu} A_\nu + \bar{\psi}(i\slashed{\partial} - m)\psi - e A_\mu \bar{\psi}\gamma^\mu\psi, \qquad (14.116)$$

with $\Box^{\mu\nu} = \Box g^{\mu\nu} - (1 - \frac{1}{\xi})\partial^\mu\partial^\nu$ in covariant gauges. The classical equations of motion for A^ν are $\Box_{\mu\nu} A^\nu = ej^\mu = e\bar{\psi}\gamma^\mu\psi$. By varying $A_\mu(x) \to A_\mu(x) + \varepsilon_\mu(x)$ and considering the correlation function $\langle A^\alpha \bar{\psi}\psi \rangle$ we would find

$$\Box^x_{\mu\nu} \langle A^\nu(x) A^\alpha(y) \bar{\psi}(z_1)\,\psi(z_2)\rangle$$
$$= e\langle j_\mu(x) A^\alpha(y)\bar{\psi}(z_1)\,\psi(z_2)\rangle - i\delta^4(x-y)\,\delta^\alpha_\mu \langle\bar{\psi}(z_1)\,\psi(z_2)\rangle. \qquad (14.117)$$

Another Schwinger–Dyson equation, for QED, is

$$\left(i\gamma^\mu_{\kappa\rho}\partial_\mu + m\delta_{\kappa\rho}\right)\langle\bar{\psi}_\kappa(x)\,\psi_\alpha(y)\bar{\psi}_\beta(z)\psi_\gamma(w)\rangle = -e\langle\bar{\psi}_\kappa(x)\slashed{A}_{\kappa\rho}\psi_\alpha(y)\,\bar{\psi}_\beta(z)\psi_\gamma(w)\rangle$$
$$i\delta_{\gamma\rho}\delta^4(x-w)\langle\psi_\alpha(y)\,\bar{\psi}_\beta(z)\rangle - i\delta_{\alpha\rho}\delta^4(x-y)\,\langle\psi_\gamma(w)\bar{\psi}_\beta(z)\rangle, \quad (14.118)$$

with the minus sign coming from anticommuting $\bar{\psi}_\beta(z)$ past $\psi_\gamma(w)$ in the last term.

14.7.2 Schwinger–Dyson differential equation

One has to be *very* careful going back and forth between the time-ordered products and path integrals. For example, the Schwinger–Dyson equation in Eq. (14.114) does not imply

$$\Box_z \int \mathcal{D}\phi \left[e^{iS}\phi(z)\phi(x)\right] - \int \mathcal{D}\phi \left[c^{iS}\Box_z\phi(z)\phi(x)\right] = -i\delta^4(z-x)\int \mathcal{D}\phi e^{iS}. \quad (14.119)$$

In fact, the left-hand side of this equation is zero, since $\Box_z$ only acts on $\phi(z)$. The correct relationship is Eq. (14.113). To avoid confusion, it is safest not to go back and forth between the pictures, but rather to express the Schwinger–Dyson equations as expressions relating observables, which can then be compared. The natural way to codify the observables is through the generating functional, which can be defined in both pictures.

Let us then repeat the path integral derivation of the Schwinger–Dyson equation above for the generating functional based on the scalar Lagrangian $\mathcal{L}[\phi] = -\frac{1}{2}\phi\Box\phi + \mathcal{L}_{\text{int}}[\phi]$. Shifting $\phi(y) \to \phi(y) + \varepsilon(y)$ we find

$$Z[J] = \int \mathcal{D}\phi\, e^{i\int d^4 y\left(-\frac{1}{2}(\phi+\varepsilon)\Box(\phi+\varepsilon)+\mathcal{L}_{\text{int}}[\phi+\varepsilon]+J\phi+J\varepsilon\right)}$$
$$= \int \mathcal{D}\phi\, e^{i\int d^4 y \mathcal{L}[\phi]+J\phi}\left[1 + i\int d^4 x\, \varepsilon(x)\left(-\Box_x\phi(x) + \frac{\partial\mathcal{L}_{\text{int}}[\phi]}{\partial\phi[x]} + J(x)\right) + \mathcal{O}(\varepsilon^2)\right].$$
$$(14.120)$$

As before, this should equal $Z[J]$ for any $\varepsilon(z)$. Thus,

$$\Box_x \int \mathcal{D}\phi\, e^{i\int d^4 y\mathcal{L}[\phi]+J\phi}\phi(x) = \int \mathcal{D}\phi\, e^{i\int d^4 y\mathcal{L}[\phi]+J\phi}\frac{\partial\mathcal{L}_{\text{int}}[\phi]}{\partial\phi[x]} + \int \mathcal{D}\phi\, e^{i\int d^4 y\mathcal{L}[\phi]+J\phi}J(x).$$
$$(14.121)$$

Or equivalently,

$$-i\Box_x \frac{\partial Z[J]}{\partial J(x)} = \left\{ \mathcal{L}'_{\text{int}}\left[-i\frac{\partial}{\partial J(x)}\right] + J(x) \right\} Z[J], \qquad (14.122)$$

which is the **Schwinger–Dyson differential equation**. The slick notation $\mathcal{L}'_{\text{int}}\left[\frac{-i\partial}{\partial J(x)}\right]$, which means the functional $\mathcal{L}'[X]$ taking $X = \frac{-i\partial}{\partial J(x)}$ as an argument, will be clarified below.

An amazing thing about the Schwinger–Dyson differential equation is that, since it encodes the difference between the classical and quantum theories, it can be used to *define* the quantum theory. Therefore, it can be used to prove that the path integral and canonical approaches are equivalent. In particular, it can be used to define the generating functional: $Z[J]$ is the unique solution to this differential equation (with appropriate boundary conditions). Since $Z[J]$ defines all of the correlation functions, which define the theory, the Schwinger–Dyson differential equation also defines the theory.

To show that the Schwinger–Dyson equation holds in the canonical theory, we first define a generating function $\hat{Z}[J]$ by

$$\hat{Z}[J] = \langle e^{i\int \hat{\phi}J}\rangle. \qquad (14.123)$$

Here, $J(x)$ is an arbitrary classical current, but now $\hat{\phi}(x)$ is the quantum operator. This generates the correlation functions as well:

$$\langle \hat{\phi}(x_1)\cdots\hat{\phi}(x_n)\rangle = \frac{1}{\hat{Z}[J]}(-i)^n \left.\frac{\partial^n \hat{Z}}{\partial J(x_1)\cdots\partial J(x_n)}\right|_{J=0}, \qquad (14.124)$$

exactly as $Z[J]$. Thus, if we show that Eq. (14.122) holds for $Z[J]$ and for $\hat{Z}[J]$, we have shown that $Z[J] = \hat{Z}[J]$, which shows that the path integral and canonical definitions agree.

To demonstrate that Eq. (14.122) holds in the canonical theory, start with the Schwinger–Dyson equations in Eq. (14.106) and insert factors of J at the same points as the field insertions. This gives

$$\Box_x \langle \hat{\phi}(x)\hat{\phi}(y_1)\cdots\hat{\phi}(y_n)J(y_1)\cdots J(y_n)\rangle = \langle \mathcal{L}'_{\text{int}}\left[\hat{\phi}(x)\right]\hat{\phi}(y_1)\cdots\hat{\phi}(y_n)J(y_1)\cdots J(y_n)\rangle$$

$$-i\sum_j \delta^4(x-y_j)\langle\hat{\phi}(y_1)\cdots\hat{\phi}(y_{j-1})\hat{\phi}(y_{j+1})\cdots\hat{\phi}(y_n)J(y_1)\cdots J(y_n)\rangle. \qquad (14.125)$$

What we will show is that each term in this expression is in one-to-one correspondence with the Taylor expansion of Eq. (14.122). To show this, we need the expansion of the generating functional:

$$\hat{Z}[J] = \langle 1 + i\int_y \hat{\phi}(y)J(y) + \frac{i^2}{2}\int_y\int_z \hat{\phi}(y)J(y)\hat{\phi}(z)J(z) + \cdots\rangle, \qquad (14.126)$$

where $\int_y$ means $\int d^4y$. This expansion can be used for either the path integral or the canonical definition of the generating functional.

The Taylor expansion of the left-hand side of Eq. (14.122) gives

$$- i\Box_x \frac{\partial \hat{Z}[J]}{\partial J(x)} = \Box_x \langle \hat{\phi}(x) + i\hat{\phi}(x) \int_y \hat{\phi}(y)J(y) + \cdots \rangle. \qquad (14.127)$$

This is the sum of all possible terms on the left-hand side of Eq. (14.125).

For the $\mathcal{L}'_{\text{int}}$ term in Eq. (14.122), Schwinger's slick notation $\mathcal{L}'_{\text{int}}\left[\frac{-i\partial}{\partial J(x)}\right]$ can be best understood with an example. Suppose $\mathcal{L}_{\text{int}}[\phi] = \frac{g}{3!}\phi^3$, then $\mathcal{L}'_{\text{int}}[\phi] = \frac{g}{2}\phi^2$ and so

$$\mathcal{L}'_{\text{int}}\left[\frac{-i\partial}{\partial J(x)}\right] \hat{Z}[J]$$
$$= \frac{g}{2!}\left(\frac{-i\partial}{\partial J(x)}\right)^2 \langle \cdots + \frac{i^3}{3!} \int_y \int_z \int_w \hat{\phi}(y)J(y)\hat{\phi}(z)J(z)\hat{\phi}(w)J(w) + \cdots \rangle, \quad (14.128)$$

where only one term is shown. Applying the $\frac{\partial}{\partial J}$ this becomes

$$\mathcal{L}_{\text{int}}'\left[\frac{-i\partial}{\partial J(x)}\right] \hat{Z}[J] = \langle \cdots + i\frac{g}{2!}\hat{\phi}^2(x) \int_w \hat{\phi}(w)J(w) + \cdots \rangle. \qquad (14.129)$$

Then, since the full interacting quantum operator satisfies $\Box\hat{\phi} = \mathcal{L}_{\text{int}}'[\hat{\phi}]$, the expression simplifies to

$$\mathcal{L}_{\text{int}}'\left[\frac{-i\partial}{\partial J(x)}\right] \hat{Z}[J] = \langle \cdots + i\Box_x\hat{\phi}(x) \int_w \hat{\phi}(w)J(w) + \cdots \rangle, \qquad (14.130)$$

which is a sum of terms given by the first term on the right-hand side of Eq. (14.125). Finally,

$$J(x)\hat{Z}[J] = \langle J(x) + iJ(x) \int_y \hat{\phi}(y)J(y) + \cdots \rangle$$
$$= \langle \int_w \delta(w-x)J(w) + i \int_w \int_y \delta(w-x)J(w)\hat{\phi}(y)J(y) + \cdots \rangle, \quad (14.131)$$

which has all the terms on the second line of Eq. (14.125). So each term in the expansion of Eq. (14.122) is verified and therefore Eq. (14.122) holds.

Since the Schwinger–Dyson differential equation holds for both the path integral $Z[J]$ and the canonically defined $\hat{Z}[J]$, Eq. (14.123), the two generating functionals must be identical. Thus, the path integral and canonical quantization are equivalent.

By the way, you often hear that the canonical approach is purely perturbative. That is not true, since $\hat{Z}[J]$ is identical to $Z[J]$. Although non-perturbative statements can be made with the canonical approach, they are generally easier to make with path integrals, which is a practical distinction, not one of principle.

14.8 Ward–Takahashi identity

Recall that in the derivation of Noether's theorem, in Section 3.3, we performed a variation of the field that was also a global symmetry of the Lagrangian. This led to the existence of a classically conserved current. Performing a similar variation on the path integral and following the steps that led to the Schwinger–Dyson equations will produce a general and powerful relation among correlation functions known as the Ward–Takahashi identity. The Ward–Takahashi identity not only implies the usual Ward identity and gauge invariance, but since it is non-perturbative it will also play an important role in the renormalization of QED.

14.8.1 Schwinger–Dyson equations for a global symmetry

Consider the correlation function of $\psi(x_1)\bar{\psi}(x_2)$ in a theory with a global symmetry under $\psi \to e^{i\alpha}\psi$:

$$I_{12} = \langle \psi(x_1)\bar{\psi}(x_2) \rangle = \int \mathcal{D}\psi \mathcal{D}\bar{\psi} \exp\left(i \int d^4x \left[\bar{\psi}(i\slashed{\partial} - m)\psi + \cdots \right] \right) \psi(x_1)\bar{\psi}(x_2),$$
(14.132)

where the $\cdots$ represent any locally symmetric additional terms. We do not need the photon, but you can add it if you like. Under a field redefinition which is a local transformation,

$$\psi(x) \to e^{-i\alpha(x)}\psi(x), \quad \bar{\psi}(x) \to e^{i\alpha(x)}\bar{\psi}(x),$$
(14.133)

the measure is invariant. The Lagrangian is not invariant, since we have not transformed A_μ (or even included it). Instead,

$$i\bar{\psi}(x)\slashed{\partial}\psi(x) \to i\bar{\psi}(x)\slashed{\partial}\psi(x) + \bar{\psi}(x)\gamma^\mu\psi(x)\partial_\mu\alpha(x)$$
(14.134)

and

$$\psi(x_1)\,\bar{\psi}(x_2) \to e^{-i\alpha(x_1)}e^{i\alpha(x_2)}\psi(x_1)\,\bar{\psi}(x_2).$$
(14.135)

Since the path integral is an integral over all field configurations ψ and $\bar{\psi}$, it is invariant under *any* redefinition, including Eq. (14.133) (up to a Jacobian factor, which in this case is just 1). Thus, expanding to first order in α, as in the derivation of the Schwinger–Dyson equations for a scalar field,

$$0 = \int \mathcal{D}\psi \, \mathcal{D}\bar{\psi} \, e^{iS} \left[i \int d^4x \, \bar{\psi}(x)\gamma^\mu \, \psi(x) \, \partial_\mu\alpha(x) \right] \psi(x_1)\bar{\psi}(x_2)$$

$$+ \int \mathcal{D}\psi \, \mathcal{D}\bar{\psi} \, e^{iS} \left[-i\alpha(x_1)\,\psi(x_1)\,\bar{\psi}(x_2) + i\alpha(x_2)\,\psi(x_1)\,\bar{\psi}(x_2) \right],$$
(14.136)

which implies

$$\int d^4x\, \alpha(x) i\partial_\mu \int \mathcal{D}\psi\, \mathcal{D}\bar{\psi}\, e^{iS}\, \bar{\psi}(x)\, \gamma^\mu\, \psi(x)\psi(x_1)\, \bar{\psi}(x_2)$$

$$= \int d^4x\, \alpha(x) \left[-i\delta(x - x_1) + i\delta(x - x_2) \right] \int \mathcal{D}\psi\, \mathcal{D}\bar{\psi}\, e^{iS}\psi(x_1)\, \bar{\psi}(x_2).$$
(14.137)

That this equality must hold for arbitrary $\alpha(x)$ implies

$$\partial_\mu \langle j^\mu(x)\psi(x_1)\, \bar{\psi}(x_2) \rangle = -\delta(x - x_1)\langle \psi(x_1)\, \bar{\psi}(x_2) \rangle + \delta(x - x_2)\langle \psi(x_1)\, \bar{\psi}(x_2) \rangle,$$
(14.138)

where $j^\mu(x) = \bar\psi(x)\,\gamma^\mu\,\psi(x)$ is the QED current. This is the Schwinger–Dyson equation associated with charge conservation. It is a non-perturbative relation between correlation functions. It has the same qualitative content as the other Schwinger–Dyson equations; the classical equations of motion, in this case $\partial_\mu j^\mu = 0$, hold within time-ordered correlation functions up to contact interactions.

The generalization of this to higher-order correlation functions has one δ-function for each field ψ_i of charge Q_i in the correlation function that $j^\mu(x)$ could contract with:

$$\partial_\mu \langle j^\mu(x)\,\psi_1(x_1)\,\bar\psi_2(x_2) A^\nu(x_3)\bar\psi_4(x_4)\cdots\rangle$$
$$= (Q_1\delta(x-x_1) - Q_2\delta(x-x_2) - Q_4\delta(x-x_4) + \cdots)\langle \psi_1(x_1)\,\bar\psi_2(x_2)\,A^\nu(x_3)\,\bar\psi_4(x_4)\cdots\rangle.$$
$$(14.139)$$

Photon fields A^ν have no effect since they are not charged and the interaction $A_\mu\bar\psi\gamma^\mu\psi$ is invariant under Eq. (14.133). More importantly, the kinetic term for the photon also has no effect, thus these equations are independent of gauge-fixing.

14.8.2 Ward–Takahashi identity

To better understand the implications of Eq. (14.138), it is helpful to Fourier transform. We first define a function $M^\mu(p, q_1, q_2)$ by the Fourier transform of the matrix element of the current with fields:

$$M^\mu(p, q_1, q_2) = \int d^4x\, d^4x_1\, d^4x_2\, e^{ipx} e^{iq_1 x_1} e^{-iq_2 x_2}\, \langle j^\mu(x)\,\psi(x_1)\,\bar\psi(x_2)\rangle. \quad (14.140)$$

We have chosen signs so that the momenta represent $j(p) + e^-(q_1) \to e^-(q_2)$. We also define

$$M_0(q_1, q_2) = \int d^4x_1\, d^4x_2\, e^{iq_1 x_1} e^{-iq_2 x_2}\, \langle \psi(x_1)\,\bar\psi(x_2)\rangle, \quad (14.141)$$

with signs to represent $e^-(q_1) \to e^-(q_2)$ so that

$$M_0(q_1 + p, q_2) = \int d^4x\, d^4x_1\, d^4x_2\, e^{ipx} e^{iq_1 x_1} e^{-iq_2 x_2} \delta^4(x - x_1)\, \langle \psi(x_1)\bar\psi(x_2)\rangle, \quad (14.142)$$

which is the Fourier transform of the first term on the right of Eq. (14.138). The second term is similar, and therefore Eq. (14.138) implies

$$ip_\mu M^\mu(p, q_1, q_2) = M_0(q_1 + p, q_2) - M_0(q_1, q_2 - p). \quad (14.143)$$

This is known as a **Ward–Takahashi** identity. It has important implications. In Section 19.5, we will show that it implies that charge conservation survives renormalization, which is highly non-trivial. The reason it is so powerful is that it applies not just to S-matrix elements, but to general correlation functions. It also implies the regular Ward identity, as we will show below.

One can give a diagrammatic interpretation of this Ward–Takahashi identity:

$$p_\mu \left(\begin{array}{c} {\scriptstyle p\downarrow} \\ {\scriptstyle q_1} \,\otimes\, {\scriptstyle q_2} \end{array} \right) = \underset{q_1+p \quad q_2}{\longrightarrow\bullet\longrightarrow} - \underset{q_1 \quad q_2-p}{\longrightarrow\bullet\longrightarrow} \; . \qquad (14.144)$$

Here, the $\otimes$ represents the insertion of momentum through the current. Note that these are not Feynman diagrams for S-matrix elements since the momenta are not on-shell. Instead, they are Feynman diagrams for correlation functions, also sometimes called **off-shell S-matrix** elements. The associated Feynman rules are the Fourier transforms of the position-space Feynman rules. Equivalently, the rules are the usual momentum space Feynman rules with the *addition* of propagators for external lines and *without* the external polarizations (that is, without removing the stuff that the LSZ formula removes). Momentum is not necessarily conserved, which is why we can have q_1+p coming in with q_2 going out for general q_1, p and q_2.

For correlation functions with f fermions and b currents, the matrix element can be defined as

$$M^{\mu\nu_1\cdots\nu_b}(p, p_1\cdots p_b, q_1\cdots q_f)$$
$$= \int d^4x d^4x_1 d^4y_1 \cdots e^{ipx} e^{ip_1x_1} e^{-iq_1y_1} \cdots \langle j^\mu(x) j^{\nu_1}(x_1) \cdots \bar\psi(y_1) \cdots \rangle \quad (14.145)$$

and the contractions as

$$M^{\nu_1\cdots\nu_b}(p_1\cdots p_b, q_1\cdots q_f)$$
$$= \int d^4x d^4x_1 d^4y_1 \cdots e^{ip_1x_1} e^{-iq_1y_1} \cdots \langle j^{\nu_1}(x_1) \cdots \bar\psi(y_1) \cdots \rangle . \quad (14.146)$$

Then, the generalized Ward–Takahashi identity is

$$ip_\mu M^{\mu\nu_1\cdots\nu_b}(p, p_1\cdots p_b, q_1\cdots q_f) = \sum_{\text{outgoing}} Q_i M^{\nu_1\cdots\nu_b}(p_1,\ldots,q_i-p,\ldots,q_f)$$
$$- \sum_{\text{incoming}} Q_i M^{\nu_1\cdots\nu_b}(p_1,\ldots,q_i+p,\ldots,q_f). \qquad (14.147)$$

This sum is over all places where the momentum of the current can be inserted into one of the fermion lines. There are no terms where the momentum of the current goes out through another current, since currents $j^\mu = \bar\psi(x)\gamma^\mu\psi(x)$ are gauge invariant and do not contribute to the Schwinger–Dyson equation.

14.8.3 Ward identity

Now let us connect the Ward–Takahashi identity to the normal Ward identity. Recall that the Ward identity is the requirement that if we replace ϵ_μ by p_μ in an S-matrix element with an external photon, we get 0. The basic idea behind the proof is that the S-matrix involves objects such as $\epsilon_\mu \Box \langle A_\mu \cdots \rangle$. By the Schwinger–Dyson equations, we can use $\Box A_\mu = J_\mu$ up to contact terms to write $\epsilon_\mu \Box \langle A_\mu \cdots \rangle = \epsilon_\mu \langle J_\mu \cdots \rangle$; then replacing $\epsilon_\mu \to p_\mu$ gives zero since $\partial_\mu \langle J_\mu \cdots \rangle = 0$ on-shell, by the Ward–Takahashi identity. The tricky part of the proof is showing that all the contact terms in the Schwinger–Dyson equations and Ward–Takahashi identity do not contribute.

From the LSZ reduction formula the S-matrix element with two polarizations ϵ and ϵ_k explicit is

$$\langle \epsilon \cdots \epsilon^k \cdots |S| \cdots \rangle$$
$$= \epsilon_\mu \epsilon_\alpha^k \left[i^n \int d^4 x e^{ipx} \Box_{\mu\nu} \int d^4 x_k e^{ip_k x_k} \Box_{\alpha\beta}^k \int \cdots \right] \langle A_\nu(x) \cdots A_\beta(x_k) \cdots \rangle,$$
$$\tag{14.148}$$

where the $\cdots$ are for the other particles involved in the scattering; $\Box_{\mu\nu}$ here is shorthand for the photon kinetic terms. For example, in covariant gauges

$$\Box_{\mu\nu} = \Box g_{\mu\nu} - (1 - \frac{1}{\xi})\partial_\mu \partial_\nu. \tag{14.149}$$

Whether the photon is gauge-fixed or not will not affect the following argument.

To simplify Eq. (14.148) we next use the Schwinger–Dyson equation for the photon:

$$\Box_{\alpha\beta}^k \Box_{\mu\nu} \langle A_\nu(x) \cdots A_\beta(x_k) \cdots \rangle = \Box_{\alpha\beta}^k \left[\langle j_\mu(x) \cdots A_\beta(x_k) \cdots \rangle - i\delta^4(x - x_k)g_{\mu\beta}\langle \cdots \rangle \right]$$
$$= \langle j_\mu(x) \cdots j_\alpha(x_k) \cdots \rangle + \Box_{\mu\alpha}^k \Box D_F(x, x_k)\langle \cdots \rangle, \tag{14.150}$$

where we have replaced $-i\delta^4(x - x_k) = \Box D_F(x, x_k)$ on the second line to connect to the perturbation expansion, as in Section 7.1. The first term represents the replacement of the photon fields by currents. The second term represents a contraction of two external photons with each other. In diagrams:

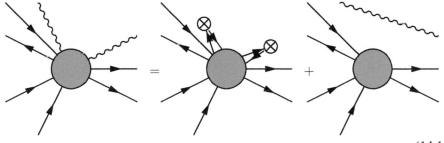

$$\tag{14.151}$$

where the $\otimes$ indicate current insertions. Since the contraction of two external photons gives a disconnected Feynman diagram, it does not contribute to the S-matrix. Thus,

$$\Box_{\alpha\beta}^k \Box_{\mu\nu} \langle A_\nu(x) \cdots A_\beta(x_k) \cdots \rangle = \langle j_\mu(x) \cdots j_\alpha(x_k) \cdots \rangle. \tag{14.152}$$

This result is a very general and useful property of diagrams involving photons:

S-matrix elements involving photons in QED with the external polarizations removed are equal to time-ordered products involving currents.

This is also true for S-matrix elements in which the external momenta p_i are not assumed to be on-shell.

If we then replace the polarization ϵ_μ in Eq. (14.148) by the associated photon's momentum p_μ, we find

$$\langle p\epsilon^1 \cdots |S| \cdots \rangle = \epsilon^1_{\alpha_1} \cdots \left[i^n \int d^4x\, e^{ipx} \int d^4x_k\, e^{ip_1 x_1} \int d^4y_1\, e^{iq_1 y_1} \left(i\partial\!\!\!/_{y_1} + m_1 \right) \cdots \right]$$
$$\times \partial_\mu \langle j_\mu(x) j_{\alpha_1}(x_1) \cdots \cdots \psi(y_1) \cdots \rangle$$
$$= i^n [\epsilon^1_{\alpha_1} \cdots][(q\!\!\!/_1 - m_1)\ldots] p_\mu M^{\mu\alpha_1\cdots\alpha_b}(p, p_1 \ldots p_b, q_1 \ldots q_f), \qquad (14.153)$$

where m_i are the masses of the fermions $q_i^2 = m_i^2$ and $M^{\mu\alpha_1\cdots\alpha_b}$ is given by Eq. (14.145).

Using the Ward–Takahashi identity, Eq. (14.147), this becomes

$$\langle p\epsilon^1_{\alpha_1} \cdots |S| \cdots \rangle = e[\epsilon^1_{\alpha_1} \cdots][(q\!\!\!/_1 - m_1)\ldots] \sum_j \pm Q_i M^{\alpha_1\cdots\alpha_b}(p_1, \ldots, q_j \pm p, \ldots, q_f)$$
$$(14.154)$$

In terms of diagrams, we have found

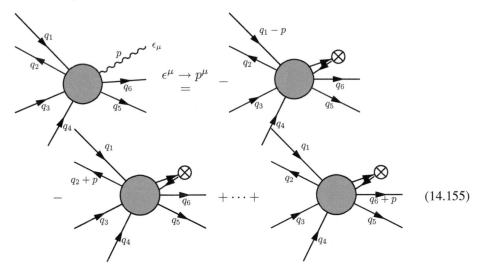

$$(14.155)$$

To get these diagrams, we first replace the external photons by currents, as in Eq. (14.151), and then remove the current associated with the photon with polarization ϵ_μ and feed its momentum p^μ into each of the possible external fermions, as dictated by Eq. (14.147).

Now, each term in the sum in Eq. (14.154) has a pole at $(q_i \pm p)^2 = m_i^2$, not at $q_i^2 = m_i^2$, and will vanish when multiplied by the prefactor $q\!\!\!/_i - m_i = \frac{q_i^2 - m_i^2}{q\!\!\!/_i + m_i}$ since q_i is on-shell. Therefore, the sum vanishes and the Ward identity is proven. Note that this proof is non-perturbative, and holds whether or not the external photons are assumed to have $p^2 = 0$.

By the way, the above derivation used that the photon interacted with the Noether current linearly. That is, that the interaction is $\mathcal{L}_{\text{int}} = ej^\mu A_\mu$. This is not true for scalar QED, where the interaction is $\mathcal{L}_{\text{int}} = ieA_\mu[\phi^\star(\partial_\mu\phi) - (\partial_\mu\phi^\star)\phi] + e^2 A_\mu^2|\phi|^2$ (cf. Eq. (9.11)). In scalar QED one can therefore have contractions of photons with other photons that do not only contribute to the disconnected part of the S-matrix. The Schwinger–Dyson equations in this case get additional pieces known as **Schwinger terms**. You can explore these terms in Problem 14.5.

14.8.4 Gauge invariance

Another consequence of the proof of the Ward identity in the previous section is that it lets us also prove gauge invariance in the sense of independence of the covariant gauge parameter ξ. Consider an arbitrary S-matrix element involving b external photons and f external fermions at order e^n in perturbation theory. All the diagrams contributing at this order will involve the same number of internal photons, namely $m = \frac{n-b}{2}$, since each external photon gives one factor of e and each internal photon gives two factors of e. Thus, the amplitude can be written as a sum over m propagators:

$$\mathcal{M} = e^n \epsilon_1^{\alpha_1} \cdots \epsilon_b^{\alpha_b} \int d^4 k_1 \cdots d^4 k_m \Pi_{\mu_1 \nu_1}(k_1) \cdots \Pi_{\mu_m \nu_m}(k_m)$$
$$\times \mathcal{M}^{\mu_1 \nu_1 \cdots \mu_m \nu_m \alpha_1 \cdots \alpha_b}(\cdots k_i \cdots q_i), \quad (14.156)$$

where q_i are all the external momenta and $\epsilon_i^{\alpha_i}$ the external photon polarizations. Here $\mathcal{M}^{\mu_1 \cdots}$ on the right-hand side can be written as an integral over matrix elements of time-ordered products of currents and evaluated at $e = 0$, that is, in the free theory.

By the Ward identity, which we saw does not require the photons to have $p^2 = 0$, $p_{\mu_1} \mathcal{M}^{\mu_1 \cdots} = 0$. Thus, if we replace any of the photon propagators by

$$\Pi_{\mu\nu}(k) \rightarrow \Pi_{\mu\nu}(k) + \xi k_\mu k_\nu, \quad (14.157)$$

the correction will vanish. Therefore, the matrix element is independent of ξ. This proof requires the external fermions to be on-shell, since otherwise there are contact interactions that give additional matrix elements on the right-hand side. It does not require the external photons to be on-shell.

Problems

14.1 Show that for complex scalar fields

$$\int \mathcal{D}\phi^* \, \mathcal{D}\phi \exp\left\{ i \int d^4x \, d^4y \left[\phi^*(x) M(x,y)\phi(y) \right] + i \int d^4x \left[J^*(x)\phi(x) + \phi^*(x) J(x) \right] \right\}$$
$$= \mathcal{N} \frac{1}{\det M} \exp\left\{ -i \int d^4x \, d^4y \, J^*(x) M^{-1}(x,y) J(y) \right\} \quad (14.158)$$

for some (infinite) constant $\mathcal{N}$.

14.2 Furry's theorem states that $\langle \Omega \, | T\{ A_{\mu_1}(q_1) \cdots A_{\mu_n}(q_n) \} | \, \Omega \rangle = 0$ if n is odd. It is a consequence of charge-conjugation C invariance.

(a) In scalar QED, charge conjugation swaps ϕ and ϕ^*. How must A_μ transform so that the Lagrangian is invariant?

(b) Prove Furry's theorem in scalar QED non-perturbatively using the path integral.

(c) Does Furry's theorem hold if the photons are off-shell or just on-shell?

(d) Prove Furry's theorem in QED.

(e) In the Standard Model, charge conjugation is violated by the weak interactions. Does your proof, for correlation functions of photons, still work in the Standard Model, or do you expect small violations of Furry's theorem?

14.3 In this problem, you will calculate $\langle \Phi | 0 \rangle$ to verify Eqs. (14.65) and (14.66).

(a) Invert the expansion of free fields in creation and annihilation operators (Eq. (2.78)) to solve for a_p in terms of $\hat{\phi}(x)$ and $\hat{\pi}(x) = \partial_t \hat{\phi}(x)$.

(b) Show that $\hat{\pi}$ acts on eigenstates of $\hat{\phi}$ as the variational derivative $-i \frac{\delta}{\delta \phi}$.

(c) Write a differential equation for $\langle \Phi | 0 \rangle$ using $a_p | 0 \rangle = 0$.

(d) Show that the solution is given by $\langle \Phi | 0 \rangle$ in Eqs. (14.65) and (14.66).

(e) Find a closed form for $\mathcal{E}(\vec{x}, \vec{y})$ in the massive and massless cases.

14.4 In this problem, you will construct all the states that satisfy Eq. (14.19), $\hat{\phi}(\vec{x})|\Phi\rangle = \Phi(\vec{x})|\Phi\rangle$, explicitly. This is one way to define the measure on the path integral.

(a) Write the eigenstates of $\hat{x} = c(a + a^\dagger)$ for a single harmonic oscillator in terms of creation operators acting on the vacuum. That is, find $f_z(a^\dagger)$ such that $\hat{x}|\psi\rangle = z|\psi\rangle$, where $|\psi\rangle = f_z(a^\dagger)|0\rangle$.

(b) Generalize the above construction to field theory, to find the eigenstates $|\Phi\rangle$ of $\hat{\phi}(\vec{x})$ that satisfy $\hat{\phi}(\vec{x})|\Phi\rangle = \Phi(\vec{x})|\Phi\rangle$.

(c) Prove that these eigenstates satisfy the orthogonality relation Eq. (14.22).

14.5 Schwinger terms.

(a) What are the Schwinger–Dyson equations for photons and charged scalar fields in scalar QED? That is, give an equation for $\Box^{\mu\nu}\langle A_\nu A_\alpha \phi^\star \phi \rangle = ?$

(b) How is the current-conservation Schwinger–Dyson equation different in QED and scalar QED?

14.6 Anticommutation.

(a) Since Grassmann numbers anticommute, $\theta_1 \theta_2 \theta_1 \theta_2 = 0$, why does a term in the Lagrangian such as $\bar{\psi}(x)\psi(x)\bar{\psi}(x)\psi(x)$ not automatically vanish? What about $(\bar{\psi}\psi)^5$? Would you get the same answer for $e^+ e^- \to 4 e^+ e^-$ pairs from a $(\bar{\psi}\psi)^5$ term in the Lagrangian in the canonical formalism and with the path integral?

(b) We showed that correlation functions of gauge-invariant operators come out the same if we add a term $-\frac{1}{2\xi}(\partial_\mu A_\mu)^2$ to the Lagrangian. Would they come out the same if we added a term of the form $-\frac{1}{2\xi}(\partial_\mu A_\mu)^4$? What about a term of the form ξA_μ^2?

14.7 To derive the Schwinger–Dyson equations for scalars in the canonical picture, we needed to use the equations $(\Box + m^2)\hat{\phi} = \mathcal{L}'_{\text{int}}[\hat{\phi}]$ and $[\hat{\phi}(\vec{x}, t), \partial_t \hat{\phi}(\vec{y}, t)] = i\delta^3(\vec{x} - \vec{y})$:

(a) What is the equivalent of these equations for Dirac spinors?

(b) Verify the Schwinger–Dyson equation in Eq. (14.118) using the canonical approach.

(c) Verify the Schwinger–Dyson equation in Eq. (14.118) using the path integral.

RENORMALIZATION

The Casimir effect

15

Now we come to the real heart of quantum field theory: loops. Loops generically are infinite. For example, the *vacuum polarization* diagram in scalar QED is

$$i\mathcal{M} = \quad = e^2 \int \frac{d^4 k}{(2\pi)^4} \frac{2k^\mu - p^\mu}{(k-p)^2 - m^2 + i\varepsilon} \frac{2k^\nu - p^\nu}{k^2 - m^2 + i\varepsilon}.$$

(15.1)

In the region of the integral at large, $k^\mu \gg p^\mu, m$, this is[1]

$$i\mathcal{M} \sim 4e^2 \int \frac{d^4 k}{(2\pi)^4} \frac{k^2}{k^4} \sim \int k \, dk = \infty.$$

(15.2)

This kind of divergent integral appears in almost every attempt to calculate matrix elements beyond leading order in perturbation theory: corrections to the electron mass, corrections to the hydrogen atom energy levels, etc. Even by the late 1930s, Dirac, Bohr, Oppenheimer and others were ready to give up on QED because of these divergent integrals.

So what are we supposed to do about these divergences? The basic answer is very simple: this loop is not by itself measurable. As long as we are always computing measurable quantities, the answer will come out finite. In practice, the way it works is a bit more complicated – instead of computing a physical observable all along, we deform the theory in such a way that the integrals come out finite, depending on some regulating parameter. When all the integrals are put together, the answer for the observable turns out to be independent of the regulator and the regulator can be removed. This is the program of renormalization. Why it is called "renormalization" will become clear in Chapter 18.

15.1 Casimir effect

Let us start with the simplest divergence, the one in the free Hamiltonian. Recall that for a free scalar field, which is just the sum of an infinite number of harmonic oscillators, the Hamiltonian is

$$H = \int \frac{d^3 k}{(2\pi)^3} \omega_k \left(a_k^\dagger a_k + \frac{1}{2} V \right),$$

(15.3)

---

[1] $k^\mu \gg p^\mu$ can be made precise by analytically continuing to Euclidean space, where it implies $|k_E^\mu| \gg |p_E^\mu|$. For scaling arguments, we will more simply treat all the components of k^μ as larger than all the components of p^μ when considering such limits.

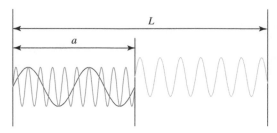

Fig. 15.1 A box of size a in a box of size L.

where $\omega_k = |\vec{k}|$. So the contribution to the vacuum energy of the photon zero modes is

$$E = \langle 0|H|0\rangle = V \int \frac{d^3 k}{(2\pi)^3} \frac{\omega_k}{2} = \frac{V}{4\pi^2} \int k^3 dk = \infty, \tag{15.4}$$

known as the **zero-point energy**.

While the zero-point energy is infinite, it is also not observable. As with potential energy in classical mechanics, only energy differences matter and the absolute energy is unphysical (with the exception of the cosmological constant, to be discussed in Section 22.7.1). To get physics out of the zero-point energy we must consider the free theory in some context other than just sitting there in the vacuum.

Consider the zero-point energy in a box of size a. If the energy changes with a, then we can calculate $F = -\frac{dE}{da}$, which will be a force on the walls of the box. In this case, we have a natural low-energy or infrared (IR) cutoff: $|\vec{k}| > \frac{1}{a}$. Of course, this does not cut off the high-energy or ultraviolet (UV) divergence at large k, but if we are careful there will be a finite residual dependence on a that will give an observable force, called the **Casimir force**.

Being careful, we realize immediately that if we change a then the energy inside *and* outside the box will change, which means we have to deal with all space again, complicating the problem. So let us put in a third wall on our box far away, at $L \gg a$. Then the zero-point energy completely outside the box is independent of a, so we can immediately drop it. The setup is shown in Figure 15.1.

We will work with a one-dimensional box for simplicity, and use a scalar field instead of the photon. In a one-dimensional box of size r the (classically) quantized frequencies are $\omega_n = \frac{\pi}{r} n$. Then the integral in the quantum Hamiltonian becomes a discrete (but still infinite) sum:[2]

$$E(r) = \langle 0|H|0\rangle = \sum_n \frac{\omega_n}{2}, \quad \omega_n = \frac{\pi}{r} n, \tag{15.5}$$

which represents the energy in a box of size r.

---

[2] Continuous modes are normalized as $[a_k, a_p^\dagger] = (2\pi)^3 \delta^3(p - k)$. For the Casimir force, the modes will be discrete, so $[a_k, a_p^\dagger] = \delta_{pk}$ is appropriate. Then the Hamiltonian, $H = \int \frac{d^3 p}{(2\pi)^3} \frac{\omega_p}{2} (a_p^\dagger a_p + a_p a_p^\dagger)$, reduces to Eq. (15.3).

The total energy is the sum of the energy on the right side, $r = (L - a)$, plus the energy on the left side, $r = a$:

$$E_{\text{tot}}(a) = E(a) + E(L - a) = \left(\frac{1}{a} + \frac{1}{L - a}\right) \frac{\pi}{2} \sum_{n=1}^{\infty} n. \tag{15.6}$$

We do not expect the total energy to be finite, but we *do* hope to find a finite value for the force:

$$F(a) = -\frac{dE_{\text{tot}}}{da} = \left(\frac{1}{a^2} - \frac{1}{(L - a)^2}\right) \frac{\pi}{2} \sum_{n=1}^{\infty} n. \tag{15.7}$$

For $L \to \infty$ this becomes

$$F(a) = \frac{\pi}{2} \frac{1}{a^2}(1 + 2 + 3 + \cdots) = \infty. \tag{15.8}$$

So the plates are infinitely repulsive. Needless to say, our prediction at this point does not agree with experiment.

What are we missing? Physics! These boundaries at $0, a$ and L are forcing the electromagnetic waves to be quantized due to the interactions between the photons and the boundary plates. These plates are made of atoms. Now think about the super-high-energy radiation modes, with super-small wavelengths. They are going to just plow through the walls of the box. Since we are only interested in the modes that are affected by the walls, these ultra-high-frequency modes should be irrelevant. The free theory is a little too idealized: without interactions, nothing can ever be measured.

15.2 Hard cutoff

Instead of putting in the detailed physics of the plates, it is easier to employ effective approximations. As we will see, all approximations that take into account certain gross properties of the interactions will be equivalent, providing a valuable lesson about renormalization.

Say we put in a high-frequency cutoff Λ so that $\omega < \pi\Lambda$. We can think of Λ as $\frac{1}{\text{atomic size}}$, or some other natural scale that limits the high-frequency light. Then

$$n_{\max}(r) = \Lambda r. \tag{15.9}$$

So,

$$E(r) = \frac{1}{r} \frac{\pi}{2} \sum_{n=1}^{n_{\max}} n = \frac{\pi}{2r} \frac{n_{\max}(n_{\max} + 1)}{2} = \frac{\pi}{4r}(\Lambda r)(\Lambda r + 1) = \frac{\pi}{4}(\Lambda^2 r + \Lambda). \tag{15.10}$$

Then

$$E_{\text{tot}} = E(L - a) + E(a) = \frac{\pi}{4}\left(\Lambda^2 L + 2\Lambda\right). \tag{15.11}$$

So, we get some infinite constant, but one that is independent of a. Thus $F(a) = -\frac{dE_{\text{tot}}}{da} = 0$. Now the force is no longer infinite, but vanishes. Is that the right answer?

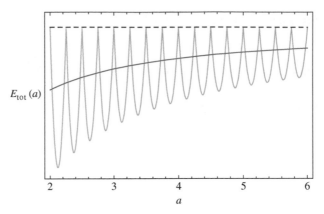

$E_{\text{tot}}(a)$

2   3   4   5   6

a

Fig. 15.2 The total energy with a floor-function cutoff does depend on a. The smooth line is the average of the oscillations, with $-\frac{1}{2}x(1-x) \to -\frac{1}{12}$, as explained in the text. The dashed line on top is the large L limit of the hard cutoff energy, $\frac{\pi}{4}\Lambda^2 L$. The values $\Lambda = 4$ and $L = 1000$ have been used.

Yes, to leading order. But we were a little too quick with this calculation. The hard cutoff means a mode is either included or not. Thus, even though we change r continuously, $n_{\max}$ can only change by discrete amounts. We can write this mathematically with a floor function

$$n_{\max}(r) = \lfloor \Lambda r \rfloor, \tag{15.12}$$

where $\lfloor x \rfloor$ means the greatest integer less than x. Then the sum is

$$E(r) = \frac{\pi}{4r} \lfloor \Lambda r \rfloor (\lfloor \Lambda r \rfloor + 1). \tag{15.13}$$

Now the total energy, $E_{\text{tot}}(a) = E(L - a) + E(a)$, oscillates with a, as shown in Figure 15.2, and we see that total energy is not a smooth function of a.

To deal with this oscillation, define a number x as

$$x = \Lambda a - \lfloor \Lambda a \rfloor \in [0, 1), \tag{15.14}$$

which gives

$$E(a) = \frac{\pi}{4} \left[\Lambda^2 a + \Lambda - 2\Lambda x - \frac{x(1-x)}{a} \right]. \tag{15.15}$$

We will also take ΛL to be an integer, which is allowed because L was some arbitrary fixed size that does not change when we move the wall at a. Then $\lfloor \Lambda L - \Lambda a \rfloor = \Lambda L - \lceil \Lambda a \rceil$. For simplicity, let us also assume Λa is not an integer, which lets us use $\lceil \Lambda a \rceil = \lfloor \Lambda a \rfloor + 1$. Then,

$$
\begin{aligned}
E(L - a) &= \frac{\pi}{4} \left[\frac{(\Lambda L - \lceil \Lambda a \rceil)(\Lambda L - \lceil \Lambda a \rceil + 1)}{L - a} \right] \\
&= \frac{\pi}{4} \left[\Lambda^2 (L - a) - \Lambda + 2\Lambda x - \frac{x(1-x)}{L - a} \right]
\end{aligned}
\tag{15.16}
$$

and

$$E_{\text{tot}}(a) = E(L-a) + E(a) = \frac{\pi}{4}\left[\Lambda^2 L - \frac{x(1-x)}{a} - \frac{x(1-x)}{L-a}\right]. \qquad (15.17)$$

The $\Lambda^2 L$ piece is the extrinsic energy of the whole system, which does not contribute to the force, and a part that oscillates as x goes between 0 and 1. Keeping only terms up to order L^0, the total energy is

$$E_{\text{tot}}(a) = \frac{\pi}{4}L\Lambda^2 - \frac{\pi}{4a}x(1-x). \qquad (15.18)$$

The $\frac{\pi}{4}L\Lambda^2$ term is the extrinsic energy, which does not contribute to the force, and a part that oscillates as $0 \le x < 1$.

Since $x = \Lambda a - \lfloor \Lambda a \rfloor$, as $\Lambda \to \infty$ at fixed a, there are more and more oscillations. In the continuum limit ($\Lambda \to \infty$), the plate will only experience the average force. Thus, we can average x between 0 and 1, using $\int x(1-x) = \frac{1}{6}$. So,

$$E_{\text{tot}}(a) \approx \frac{\pi}{4}L\Lambda^2 - \frac{\pi}{24a}. \qquad (15.19)$$

This average is shown as the smooth line in Figure 15.2.

The result is a non-zero and finite result for the force:

$$F(a) = -\frac{dE_{\text{tot}}}{da} = -\frac{\pi}{24a^2}. \qquad (15.20)$$

Putting back in the $\hbar$ and c, we find that the Casimir force in one dimension is

$$F(a) = -\frac{\pi\hbar c}{24a^2}. \qquad (15.21)$$

This is an attractive force. We can see that the force is purely quantum mechanical because it is proportional to $\hbar$.

In three dimensions, remembering to account for the two photon polarizations, the answer is

$$F(a) = -\frac{\pi^2 \hbar c}{240a^4}A, \qquad (15.22)$$

where A is the area of the walls of the box. Although predicted by Casimir in 1948 [Casimir, 1948], the force was not conclusively for another 50 years [Lamoreaux, 1997; Bressi, 2002].

15.3 Regulator independence

You should find the calculation of the Casimir effect incredibly disconcerting. We found the force to be independent of Λ, but we needed to use a crazy model of the walls where the discreteness of the hard cutoff played an important role. What if we took a different model besides the hard cutoff for regulating the UV modes? It seems obvious that we should get a different answer with each model.

However, it turns out we will not. We get the same answer no matter what, as long as the cutoff satisfies some basic requirements. That is a pretty amazing fact. We will first try a few more regulators, then we will present the precise requirements and a proof of regulator independence.

15.3.1 Heat-kernel regularization

Another reasonable physical assumption besides a hard cutoff would be that there is some penetration depth of the modes into the walls, with high-frequency modes getting further. This means that the contribution of high-frequency modes to the relevant energy sum is exponentially suppressed. Thus we can try

$$E(r) = \frac{1}{2} \sum_n \omega_n e^{-\omega_n/(\pi \Lambda)}. \tag{15.23}$$

This is called heat-kernel regularization.

Expanding with $\omega_n = \frac{\pi}{r} n$:

$$E(r) = \frac{1}{r} \frac{\pi}{2} \sum_{n=1}^{\infty} n e^{-n/(\Lambda r)} = \frac{1}{r} \frac{\pi}{2} \sum_{n=1}^{\infty} n e^{-\varepsilon n}, \qquad \varepsilon = \frac{1}{\Lambda r} \ll 1. \tag{15.24}$$

Now we can calculate

$$\sum_{n=1}^{\infty} n e^{-\varepsilon n} = -\partial_\varepsilon \sum_{n=1}^{\infty} e^{-\varepsilon n} = -\partial_\varepsilon \frac{1}{e^\varepsilon - 1} = \frac{e^{-\varepsilon}}{(1 - e^{-\varepsilon})^2} = \frac{1}{\varepsilon^2} - \frac{1}{12} + \frac{\varepsilon^2}{240} + \cdots. \tag{15.25}$$

Already, we see the factor $-\frac{1}{12}$ appearing.

So,

$$E(r) = \frac{1}{r} \frac{\pi}{2} \left[\Lambda^2 r^2 - \frac{1}{12} + \frac{1}{240 r^2 \Lambda^2} \cdots \right] = \frac{\pi}{2} r \Lambda^2 - \frac{\pi}{24 r} + \cdots, \tag{15.26}$$

and then

$$F(a) = -\frac{d}{da} [E(L - a) + E(a)] = -\frac{d}{da} \left[\frac{\pi}{2} L \Lambda^2 - \frac{\pi}{24} \left(\frac{1}{L - a} + \frac{1}{a} \right) + \cdots \right]$$

$$= \frac{\pi}{24} \left(\frac{1}{(L - a)^2} - \frac{1}{a^2} \right) + \cdots. \tag{15.27}$$

Now take $L \to \infty$ and we get

$$F(a) = -\frac{\pi \hbar c}{24 a^2}, \tag{15.28}$$

which is the same thing we found with the floor-function cutoff. Note, however, that the extrinsic energy term was $\frac{\pi}{4} L \Lambda^2$ in the previous case and is $\frac{\pi}{2} L \Lambda^2$ in this case.

15.3.2 Other regulators

What else can we try? We can use a Gaussian regulator:

$$E(r) = \frac{1}{2} \sum_n \omega_n e^{-\left(\frac{\omega_n}{\pi\Lambda}\right)^2}, \tag{15.29}$$

or a ζ-function regulator:

$$E(r) = \frac{1}{2} \sum_n \omega_n \left(\frac{\omega_n}{\mu}\right)^{-s}, \tag{15.30}$$

where we take $s \to 0$ instead of $\omega_{\max} \to \infty$ and have added an arbitrary scale μ to keep the dimensions correct. μ does not have to be large – it should drop out for any μ.

Let us work out the ζ-function case. Substituting in for ω_n we get

$$E(r) = \frac{1}{2} \left(\frac{\pi}{r}\right)^{1-s} \mu^s \sum_n n^{1-s}. \tag{15.31}$$

This sum is the definition of the Riemann ζ-function:

$$\sum n^{1-s} = \zeta(s-1) = -\frac{1}{12} - 0.165s + \cdots. \tag{15.32}$$

So we get

$$E(r) = \frac{1}{r}\frac{\pi}{2}\zeta(s-1) = \frac{1}{r}\frac{\pi}{2}\left[-\frac{1}{12} + O(s)\cdots\right], \tag{15.33}$$

and the energy comes out as

$$E(r) = -\frac{\pi}{24r} + \cdots. \tag{15.34}$$

This is the same as what the heat-kernel and floor-function regularization gave, although now note that the extrinsic energy term is absent.

All four of these regulators agree:

$$E(r) = \frac{1}{2} \sum_n \omega_n \theta(\pi\Lambda - \omega_n) \quad \text{(hard cutoff)}, \tag{15.35}$$

$$E(r) = \frac{1}{2} \sum_n \omega_n e^{-\frac{\omega_n}{\pi\Lambda}} \qquad \text{(heat kernel)}, \tag{15.36}$$

$$E(r) = \frac{1}{2} \sum_n \omega_n e^{-\left(\frac{\omega_n}{\pi\Lambda}\right)^2} \qquad \text{(Gaussian)}, \tag{15.37}$$

$$E(r) = \frac{1}{2} \sum_n \omega_n \left(\frac{\omega_n}{\mu}\right)^{-s} \qquad (\zeta\text{-function}). \tag{15.38}$$

That these regulators all agree is reassuring, but still somewhat mysterious.

15.3.3 Regulator-independent derivation

Casimir showed in his original paper [Casimir, 1948] a way to calculate the force in a regulator-independent way. Define the energy as

$$E(a) = \frac{\pi}{2} \sum_n \frac{n}{a} f\left(\frac{n}{a\Lambda}\right), \tag{15.39}$$

where $f(x)$ is some function whose properties we will determine shortly.

With this definition, the energy of the $L - a$ side of the box is

$$E(L - a) = \frac{\pi}{2}(L - a)\Lambda^2 \sum_n \frac{n}{(L-a)^2\Lambda^2} f\left(\frac{n}{(L-a)\Lambda}\right). \tag{15.40}$$

We can take the continuum limit of this ($L \to \infty$) with $x = \frac{n}{(L-a)\Lambda}$. Then,

$$E(L - a) = \frac{\pi}{2}L\Lambda^2 \int x\, dx\, f(x) - \frac{\pi}{2}a\Lambda^2 \int x\, dx\, f(x). \tag{15.41}$$

The first integral is just the extrinsic vacuum energy, with energy density

$$\rho = \frac{\pi}{2}\Lambda^2 \int x\, dx\, f(x). \tag{15.42}$$

The second integral simplifies with the change of variables $x = \frac{n}{a\Lambda}$. Adding the discrete sum, for the a side, with the continuum limit of the $L - a$ side, gives

$$E_{\text{tot}} = E(a) + E(L - a) = \rho L + \frac{\pi}{2a}\left[\sum_n nf\left(\frac{n}{a\Lambda}\right) - \int n\, dn\, f\left(\frac{n}{a\Lambda}\right)\right]. \tag{15.43}$$

This contains the difference between an infinite sum and an infinite integral. Such a difference is given by the Euler–Maclaurin series:

$$\sum_{n=1}^N F(n) - \int_0^N F(n)dn$$
$$= \frac{F(0) + F(N)}{2} + \frac{F'(N) - F'(0)}{12} + \cdots + B_j \frac{F^{(j-1)}(N) - F^{(j-1)}(0)}{j!} + \cdots , \tag{15.44}$$

where $F^{(j)}(N) = \frac{d^j F(N)}{dN^j}$ and B_j are the Bernoulli numbers. In particular, $B_2 = \frac{1}{6}$ and B_j for odd $j > 1$ happen to vanish.

In our case, $F(n) = nf(\frac{n}{a\Lambda})$. So, assuming that $f(x)$ dies sufficiently fast,

$$\lim_{x \to \infty} xf^{(j)}(x) = 0, \tag{15.45}$$

then

$$E_{\text{tot}} = \rho L - \frac{\pi f(0)}{24a} - \frac{B_4}{4!}\frac{3\pi}{2a^3\Lambda^2}f''(0) + \cdots . \tag{15.46}$$

For example, if $f(x) = e^{-x}$, then

$$E_{\text{tot}} = \frac{\pi}{2}\Lambda^2 L - \frac{\pi}{24a} + \mathcal{O}\left(\frac{1}{a^2\Lambda}\right), \tag{15.47}$$

which gives the correct Casimir force.

In fact, it is now clear that any regulator will give this force as long as

$$\lim_{x \to \infty} x f^{(j)}(x) = 0 \quad \text{and} \quad f(0) = 1. \tag{15.48}$$

You can see that all four of the regulators in Eq. (15.38) satisfy these requirements. The first requirement, that $f(x)$ die fast enough at high energy, means that UV modes go right through the box. It is this requirement that makes the force finite. The second requirement, that $f(0) = 1$, means that the regulator does not affect the spectrum in the IR. On physical grounds, only modes of size $\frac{1}{a}$ can reach both walls of the box to transmit the force, thus our deformation should not affect those modes.

We have two conclusions from this analysis:

> The Casimir force is independent of any regulator.
> The Casimir force is an infrared effect.

15.3.4 Counterterms

In the above analysis, we not only took $\omega_{\text{max}} \to \infty$ but also $L \to \infty$. Why did we need this third boundary at $r = L$ at all? Let us suppose we did not have it, and just calculated the energy inside the box of size r. Then we would get (with the heat-kernel regulator)

$$E(r) = \frac{\pi}{2} r \Lambda^2 - \frac{\pi}{24r} + \cdots . \tag{15.49}$$

This first term is the extrinsic energy, which is linear in the volume and is regulator dependent. It can be interpreted as saying there is some finite energy density $\rho = \frac{E_0}{r} = \frac{\pi}{2} \Lambda^2$. Now suppose that instead of just the free-field Lagrangian $\mathcal{L}$ we used to calculate the ground-state energy, we took

$$\mathcal{L} = \mathcal{L}' + \rho_c, \tag{15.50}$$

where ρ_c is constant. This new term gives an infinite contribution of $\int dx \, \rho_c$ in the action. Now if we choose $\rho_c = -\frac{\pi}{2}\Lambda^2$, the new term exactly cancels the $\frac{\pi}{2} r \Lambda^2$ term we found using the heat-kernel regulator. In the ζ-function regulator, where no divergent terms come out of the calculation, we could take $\rho_c = 0$.

The point is that since ρ_c is unmeasurable we can choose it to be whatever is convenient. ρ_c is called a **counterterm**. Counterterms give purely infinite contributions to *intermediate steps* in calculations, but when we compute *physical quantities* they drop out. Counterterms are an important tool in renormalization in quantum field theory.

15.3.5 String theory aside

A terse way to summarize the Casimir force calculation is that it amounts to the replacement

$$\frac{\pi}{2r} \sum_{n=1}^{\infty} n \quad \to \quad -\frac{\pi}{24r}, \tag{15.51}$$

or equivalently

$$1 + 2 + 3 + \cdots = -\frac{1}{12}. \tag{15.52}$$

This bizarre identity has an important use in string theory. In string theory, the mass of particles is determined by the string tension α':

$$m^2 = \frac{1}{\alpha'}j + E_0, \tag{15.53}$$

where j is the excitation number (the string harmonic) and E_0 is the Casimir energy of the string, which is independent of j. So there is a whole tower of particles with different masses. In d dimensions, the Casimir energy is

$$E_0 = \frac{1}{\alpha'}\left(\frac{d-2}{2}\right)\left(-\frac{1}{12}\right), \tag{15.54}$$

where the $-\frac{1}{12}$ comes from the same series we have just summed. Now, you can show in string theory that the $j = 1$ excitations comprise spin-1 particles with two polarizations. So they must be massless. Then, solving for $m = 0$ you find $d = 26$. That is why string theory takes place in 26 dimensions. If you do the same calculation for the superstring, you find $d = 10$.

15.4 Scalar field theory example

Before we do any physical calculations, let us get an overview of the way things are going to work out in quantum field theory. Consider the theory of a massless real scalar field with Lagrangian

$$\mathcal{L} = -\frac{1}{2}\phi \Box \phi - \frac{\lambda}{4!}\phi^4, \tag{15.55}$$

where λ is a dimensionless coupling constant. At tree-level, $\phi\phi \to \phi\phi$ scattering is given by the simple cross diagram:

$$i\mathcal{M}_1 = \qquad\qquad = -i\lambda. \tag{15.56}$$

The leading correction comes from loops, such as this s-channel one:

$$i\mathcal{M}_2 = \qquad\qquad = (-i\lambda)^2 \int \frac{d^4k}{(2\pi)^4}(\cdots) \tag{15.57}$$

There are also t- and u-channel diagrams, but let us forget about them (for example, if we had three fields with a $\frac{\lambda}{4}(\phi_1^2\phi_2^2 + \phi_2^2\phi_3^2)$ interaction, there would only be an s-channel contribution to $\phi_1\phi_1 \to \phi_3\phi_3$).

Let $p = p_1 + p_2 = p_3 + p_4$, then $k_1 + k_2 = p$ so we can set $k_1 = k$ and $k_2 = p - k$ and integrate over k. The diagram is then

$$i\mathcal{M}_2 = \frac{(-i\lambda)^2}{2} \int \frac{d^4 k}{(2\pi)^4} \frac{i}{k^2} \frac{i}{(p-k)^2}, \qquad (15.58)$$

where the $\frac{1}{2}$ is a symmetry factor. This is a Lorentz-invariant quantity, so it can only depend on $s = p^2$. It is also dimensionless and diverges as $\int \frac{dk}{k}$. So we expect $\mathcal{M}_2 \sim \log \frac{s}{\Lambda^2}$, where Λ is some cutoff parameter with dimensions of mass.

As a quick and dirty way to get the answer, take the derivative with respect to s:

$$\frac{\partial}{\partial s} \mathcal{M}_2(s) = \frac{p^\mu}{2s} \frac{\partial}{\partial p^\mu} \mathcal{M}_2(s) = \frac{i\lambda^2}{2s} \int \frac{d^4 k}{(2\pi)^4} \frac{1}{k^2} \frac{(p^2 - p \cdot k)}{(p-k)^4}. \qquad (15.59)$$

Now the integral is convergent. It is not too hard to work out this integral, and we will do some examples like this soon. But for now, we will just quote the answer:

$$\frac{\partial}{\partial s} \mathcal{M}_2(s) = -\frac{\lambda^2}{32\pi^2} \frac{1}{s}. \qquad (15.60)$$

This means that

$$\mathcal{M}_2 = -\frac{\lambda^2}{32\pi^2} \ln s + c, \qquad (15.61)$$

where c is an integration constant. Since the integral in Eq. (15.58) is divergent, c will have to be infinite. Also, since s has dimensions of mass squared, it is nice to write the constant as $c = \frac{\lambda^2}{32\pi^2} \ln \Lambda^2$, where Λ has dimensions of mass. Then we have

$$\mathcal{M}_2 = -\frac{\lambda^2}{32\pi^2} \ln \frac{s}{\Lambda^2}. \qquad (15.62)$$

So the total matrix element is

$$\mathcal{M}(s) = -\lambda - \frac{\lambda^2}{32\pi^2} \ln \frac{s}{\Lambda^2}. \qquad (15.63)$$

This is an analog of the Casimir energy, $E_{\text{tot}}(a) = c\Lambda^2 L - \frac{\pi}{24a}$, which has Λ dependence and dependence on the physical scale ($\sqrt{s}$ or a). We now need the analog of the observable, the force on the plates in the Casimir calculation.

15.4.1 Renormalization of λ

First of all, notice that, while $\mathcal{M}(s)$ is infinite, the difference between $\mathcal{M}(s_1)$ and $\mathcal{M}(s_2)$ at two different scales is finite:

$$\mathcal{M}(s_1) - \mathcal{M}(s_2) = \frac{\lambda^2}{32\pi^2} \ln \frac{s_2}{s_1}. \qquad (15.64)$$

Should we also expect that $\mathcal{M}(s)$ itself be finite? After all, $\mathcal{M}^2$ is supposed to be a physical cross section.

To answer this, let us think more about λ. It should be characterizing the strength of the ϕ^4 interaction. So to measure λ we would simply measure the cross section for $\phi\phi \to \phi\phi$ scattering, or equivalently, $\mathcal{M}$. But this matrix element is not just proportional to λ but also has the λ^2 correction above. Thus, it is impossible to simply extract λ from this scattering process. Instead, let us just *define* a **renormalized coupling** λ_R as the value of the matrix element at a particular $s = s_0$.

So,

$$\lambda_R \equiv -\mathcal{M}(s_0) = \lambda + \frac{\lambda^2}{32\pi^2} \ln \frac{s_0}{\Lambda^2} + \cdots . \tag{15.65}$$

This equation relates the parameter λ of the Lagrangian to the value of the observed scattering amplitude λ_R at a particular center-of-mass energy s_0. We can also conclude that since λ_R is observable and hence finite, λ must be infinite, to cancel the infinity from $\ln \Lambda^2$.

Next, we will solve for λ in terms of λ_R in perturbation theory by writing

$$\lambda = \lambda_R + a\lambda_R^2 + \cdots \tag{15.66}$$

and solving for a. Substituting into Eq. (15.65) we find

$$\lambda_R = \left(\lambda_R + a\lambda_R^2 + \cdots\right) + \frac{\left(\lambda_R + a\lambda_R^2 + \cdots\right)^2}{32\pi^2} \ln \frac{s_0}{\Lambda^2} + \cdots$$

$$= \lambda_R + a\lambda_R^2 + \frac{\lambda_R^2}{32\pi^2} \ln \frac{s_0}{\Lambda^2} + \cdots . \tag{15.67}$$

So, $a = -\frac{1}{32\pi^2} \ln \frac{s_0}{\Lambda^2}$ and

$$\lambda = \lambda_R - \frac{\lambda_R^2}{32\pi^2} \ln \frac{s_0}{\Lambda^2} + \cdots . \tag{15.68}$$

Although the second term may be larger than the first as $\Lambda \to \infty$, this should be thought of as a formal solution as a power series in λ_R.

Now, suppose we measure the cross section at a different center-of-mass energy s. Then

$$\mathcal{M}(s) = -\lambda - \frac{\lambda^2}{32\pi^2} \ln \frac{s}{\Lambda^2}$$

$$= -\left[\lambda_R - \frac{\lambda_R^2}{32\pi^2} \ln \frac{s_0}{\Lambda^2}\right] - \frac{\lambda_R^2}{32\pi^2} \ln \frac{s}{\Lambda^2} + \cdots$$

$$= -\lambda_R - \frac{\lambda_R^2}{32\pi^2} \ln \frac{s}{s_0} + \cdots . \tag{15.69}$$

This equation gives us an expression for $\mathcal{M}(s)$ for any s that is finite order-by-order in perturbation theory. More importantly it gives us a *physical prediction*. The ϕ^4 cross section with $s = s_1$ differs from the cross section with $s = s_0$ by logarithmic terms. Remember, by definition λ_R is observable: it is the *exact* cross section at the scale s_0. So we are predicting one observable (cross section at s) in terms of another (cross section at s_0). By the way, the logarithmic behavior is a characteristic of loop effects – tree-level graphs only give you rational polynomials in momenta and couplings, never logarithms. This will play an important role in proofs of renormalizability (Chapter 21) and in making predictions in non-renormalizable theories (Chapter 22).

15.4.2 Interpretation of counterterms

Another way of getting the same result is to add a counterterm to the Lagrangian. That means adding another interaction that is just like the first, but infinite. So we take as our Lagrangian

$$\mathcal{L} = -\frac{1}{2}\phi\Box\phi - \frac{\lambda_R}{4!}\phi^4 - \frac{\delta_\lambda}{4!}\phi^4, \tag{15.70}$$

where the counterterm δ_λ is infinite, but formally of order λ_R^2. Then, working to order λ_R^2, the amplitude is

$$\mathcal{M}(s) = -\lambda_R - \delta_\lambda - \frac{\lambda_R^2}{32\pi^2}\ln\frac{s}{\Lambda^2} + \mathcal{O}(\lambda_R^4). \tag{15.71}$$

Now we can choose δ_λ to be whatever we want. If we take it to be

$$\delta_\lambda = -\frac{\lambda_R^2}{32\pi^2}\ln\frac{s_0}{\Lambda^2}, \tag{15.72}$$

then

$$\mathcal{M}(s) = -\lambda_R - \frac{\lambda_R^2}{32\pi^2}\ln\frac{s}{s_0}, \tag{15.73}$$

which is finite. In particular, this choice of δ_λ makes $\mathcal{M}(s_0) = -\lambda_R$, which was our definition of λ_R above.

Doing things this way, with counterterms but as a perturbative expansion in the physical coupling λ_R, is known as *renormalized perturbation theory*. The previous way, where we compute physical quantities such as $\mathcal{M}(s_1) - \mathcal{M}(s_2)$ directly, is sometimes called *physical* or *on-shell perturbation theory*. The two are equivalent, but for complicated calculations, renormalized perturbation theory is often much easier.

Problems

15.1 Evaluate the Casimir force using the Gaussian regulator in Eq. (15.29).

15.2 Show that the Casimir force from the vacuum energy of fermions has the opposite sign than from bosons.

15.3 It has been proposed that geckos use the Casimir force to climb walls. It is known that geckos do not use suction (like salamanders) or capillary adhesion (like some frogs). A gecko's foot is covered in a million tiny hairs called setae, which terminate in spatula-shaped structures around 0.5 μm wide. Use dimensional analysis and the form of the Casimir force to decide whether you think this could be possible.

15.4 The vacuum energy of massive particles also contributes to the Casimir force. Before doing the calculation, how do you expect the Casimir force to depend on mass? Now do the calculation and see if you are correct (use any approximations you want – this problem is challenging).

16   Vacuum polarization

In the previous chapter, we found that although the energy of a system involving two plates is infinite, the force between the plates (the Casimir force), which is what is actually observable, is finite. At an intermediate step in the calculation, we needed to model the inability of the plates to restrict ultra-high-frequency radiation. We found that the force was independent of the model and only determined by radiation with wavelengths of the plate separation, exactly as physical intuition would suggest. More precisely, we proved the force was independent of how we modeled the interactions of the fields with the plates as long as the very short wavelength modes were effectively removed and the longest wavelength modes were not affected. Some of our models were inspired by physical arguments, as in a step-function cutoff representing an atomic spacing; others, such as the ζ-function regulator, were not. That the calculated force is independent of the model is very satisfying: macroscopic physics (the force) is independent of microscopic physics (the atoms). Indeed, for the Casimir calculation, it does not matter if the plates are made of atoms, aether, phlogiston or little green aliens.

The program of systematically making testable predictions about long-distance physics in spite of formally infinite short-distance fluctuations is known as **renormalization**. Because physics at short and long distance decouples, we can deform the theory at short distance any way we like to get finite answers – we are unconstrained by physically justifiable models. In fact, our most calculationally efficient deformation will be analytic continuation to $d = 4 - \varepsilon$ dimensions with $\varepsilon \to 0$. The beauty of renormalization is that the existence of a physical cutoff is totally irrelevant: quantitative predictions about long-distance physics do not care what the short-distance cutoff really is, or even whether or not it exists.

The core idea behind renormalization in quantum field theory is:

Observables are finite and in-principle calculable functions of other observables.

One can think of general correlation functions $\langle \Omega | T \{ \phi(x_1) \cdots \phi(x_n) \} | \Omega \rangle$ as a useful proxy for observables. Most of the conceptual confusion, both historically and among students learning the subject, stems from trying to express observables in terms of non-observable quantities, such as coupling constants in a Lagrangian. In practice:

- Infinite results associated with high-energy divergences may appear in intermediate steps of calculations, such as in loop graphs.
- Infinities are tamed by a deformation procedure called *regularization*. The regulator dependence must drop out of physical predictions.

- Coefficients of terms in the Lagrangian, such as coupling constants, are *not* observable. They can be solved for in terms of the regulator and will drop out of physical predictions.

We will find that loops can produce behavior different from anything possible at tree-level. In particular,

- Non-analytic behavior, such as $\ln\frac{s}{s_0}$, is characteristic of loop effects.

Tree-level amplitudes are always rational polynomials in external momenta and never involve logarithms. In many cases, the non-analytic behavior will comprise the entire physical prediction associated with the loop.

In Section 15.4, we gave an example of renormalization in ϕ^4 theory. We found that the expression for a correlation function in terms of the coupling constant λ was infinite: $\langle\phi^4\rangle_s = -\lambda - \frac{\lambda^2}{16\pi^2}\ln\frac{s}{\Lambda^2} + \cdots = \infty$. However, expressing the correlation function at the scale s in terms of the correlation function at a different scale s_0 gave a finite prediction: $\langle\phi^4\rangle_s = \langle\phi^4\rangle_{s_0} - \frac{1}{16\pi^2}\langle\phi^4\rangle_{s_0}^2 \ln\frac{s}{s_0} + \cdots$. Although ϕ^4 theory was just a toy example, renormalization in QED, which we begin in this chapter, is conceptually identical.

Recall that the Coulomb potential $V(r) = \frac{e^2}{4\pi r}$ is given by the exchange of a single photon:

$$\otimes\!\!\sim\!\!\sim\!\!\sim\!\!\otimes \;=\; \frac{e^2}{p^2}. \qquad (16.1)$$

Indeed, $\frac{1}{4\pi r}$ is just the Fourier transform of the propagator (cf. Section 3.4). A 1-loop correction to Coulomb's law comes from an e^+e^- loop inside the photon line:

$$\otimes\!\!\sim\!\!\bigcirc\!\!\sim\!\!\otimes \;. \qquad (16.2)$$

This will give us a correction to $V(r)$ proportional to e^4. We will show that while the charge e is infinite it can be replaced by a finite renormalized charge order-by-order in perturbation theory. The physical effect will be a measurable correction to Coulomb's law predicted by quantum field theory with logarithmic scale dependence, as in the ϕ^4 toy model.

The process represented by the Feynman diagram in Eq. (16.2) is known as **vacuum polarization**. The diagram shows the creation of virtual e^+e^- pairs, which act like a virtual dipole. In the same way that a dielectric material such as water would become polarized if we put it in a electric field, this vacuum polarization tells us how the vacuum itself is polarized when it interacts with electromagnetic radiation.

Since the renormalization of the graph is no different than it was in ϕ^4 theory, the only difficult part of calculating vacuum polarization is in the evaluation of the loop. Indeed, the loop in Eq. (16.2) is complicated, involving photons and spinors, but we can evaluate it by exploiting some tricks developed through the hard work of our predecessors. Our approach will be to build up the QED vacuum polarization graph in pieces, starting with ϕ^3 theory, then scalar QED, and finally real QED. For convenience, some of the more mathematical aspects of regularization are combined into one place in Appendix B, which is meant to provide a general reference. In the following, we assume familiarity with the results from that appendix.

16.1 Scalar ϕ^3 theory

As a warm-up for the vacuum polarization calculation in QED, we will start with scalar ϕ^3 theory with Lagrangian

$$\mathcal{L} = -\frac{1}{2}\phi(\Box + m^2)\phi + \frac{g}{3!}\phi^3. \tag{16.3}$$

Now we want to compute

$$= \frac{1}{2}(ig)^2 \int \frac{d^4k}{(2\pi)^4} \frac{i}{(k-p)^2 - m^2 + i\varepsilon} \frac{i}{k^2 - m^2 + i\varepsilon}, \tag{16.4}$$

which will tell us the 1-loop correction to the Yukawa potential. We will allow the initial and final line to be off-shell: $p^2 \neq m^2$, since we are calculating the correction to the initial ϕ propagator, $\frac{i}{p^2 - m^2}$, which also must have $p^2 \neq m^2$ to make any sense, and since we will be embedding this graph into a correction to Coulomb's law (see Eq. (16.14) below).

First, we can use the Feynman parameter trick from Appendix B:

$$\frac{1}{AB} = \int_0^1 dx \frac{1}{[A + (B - A)x]^2}, \tag{16.5}$$

with $A = (p - k)^2 - m^2 + i\varepsilon$ and $B = k^2 - m^2 + i\varepsilon$. Then we complete the square

$$\begin{aligned} A + [B - A]\,x &= (p-k)^2 - m^2 + i\varepsilon + \left[k^2 - (p-k)^2\right]x \\ &= [k - p(1-x)]^2 + p^2 x(1-x) - m^2 + i\varepsilon, \end{aligned} \tag{16.6}$$

which gives

$$i\mathcal{M}_{\text{loop}}(p) = \frac{g^2}{2} \int \frac{d^4k}{(2\pi)^4} \int_0^1 dx \frac{1}{[(k - p(1-x))^2 + p^2 x(1-x) - m^2 + i\varepsilon]^2}. \tag{16.7}$$

Now shift $k^\mu \to k^\mu + p^\mu(1-x)$ in the integral. The measure is unchanged, and we get

$$i\mathcal{M}_{\text{loop}}(p) = \frac{g^2}{2} \int \frac{d^4k}{(2\pi)^4} \int_0^1 dx \frac{1}{[k^2 - (m^2 - p^2 x(1-x)) + i\varepsilon]^2}. \tag{16.8}$$

At this point, we need to introduce a regulator. We will use Pauli–Villars regularization (see Appendix B), which adds a fictitious scalar of mass Λ with fermionic statistics. This particle is an unphysical *ghost* particle. We can use the Pauli–Villars formula from Appendix B:

$$\int \frac{d^4k}{(2\pi)^4} \frac{1}{(k^2 - \Delta + i\varepsilon)^2} = -\frac{i}{16\pi^2}\ln\frac{\Delta}{\Lambda^2}. \tag{16.9}$$

Comparing to Eq. (16.8), we have $\Delta = m^2 - p^2 x(1-x)$ so that

$$i\mathcal{M}_{\text{loop}}(p) = -\frac{ig^2}{32\pi^2} \int_0^1 dx \ln\left(\frac{m^2 - p^2 x(1-x)}{\Lambda^2}\right). \tag{16.10}$$

This integral can be done – the integrand is perfectly well behaved between $x = 0$ and $x = 1$. For $m = 0$ it has the simple form

$$\mathcal{M}_{\text{loop}}(p) = \frac{g^2}{32\pi^2}\left[2 - \ln\frac{-p^2}{\Lambda^2}\right]. \tag{16.11}$$

Note that the 2 cannot be physical, because we can remove it by redefining $\Lambda^2 \to \Lambda^2 e^{-2}$. Also note that when this diagram contributes to the Coulomb potential (as in Eq. (16.14) below), the virtual momentum p^μ is spacelike ($p^2 < 0$), so $\ln\frac{-p^2}{\Lambda^2}$ is real. Then,

$$\mathcal{M}_{\text{loop}}(p) = -\frac{g^2}{32\pi^2}\ln\frac{Q^2}{\Lambda^2}. \tag{16.12}$$

An important point is that the regulator scale Λ has to be just a number, independent of any external momenta. With the Pauli–Villars regulator we are using here, Λ is the mass of some heavy fictitious particle. It corresponds to a deformation of the theory at very high energies/short distances, like the modeling of the wall in the Casimir force. On the other hand, Q is a physical scale, like the plate separation in the Casimir force. Thus, Λ cannot depend on Q. In particular, the $\ln Q^2$ dependence cannot be removed by a redefinition of Λ like the 2 in Eq. (16.11) was. This point is so important it is worth repeating: the short-distance deformation (Λ) cannot depend on long-distance physical quantities (Q). This separation of scales is critical to being able to take $\Lambda \to \infty$ to make predictions by relating observables at different long-distance scales such as Q and Q_0. The coefficient of $\ln Q^2$ is in fact regulator independent and will generate the physical prediction from the loop.

16.1.1 Renormalization

The diagram we computed is a correction to the tree-level ϕ propagator. To see this, observe that the propagator is essentially the same as the t-channel scattering diagram:

$$i\mathcal{M}^0(p) = \qquad = (ig)^2\frac{i}{p^2}. \tag{16.13}$$

If we insert our scalar bubble in the middle, we get

$$i\mathcal{M}^1(p) = \qquad = (ig)^2\frac{i}{p^2}i\mathcal{M}_{\text{loop}}(p)\frac{i}{p^2} = ig^2\frac{1}{p^2}\left[-\frac{g^2}{32\pi^2}\ln\frac{-p^2}{\Lambda^2}\right]\frac{1}{p^2}. \tag{16.14}$$

Since $p^2 < 0$, let us write $Q^2 = -p^2$ with $Q > 0$. Then,

$$\mathcal{M}(Q) = \mathcal{M}^0(Q) + \mathcal{M}^1(Q) = \frac{g^2}{Q^2}\left(1 - \frac{1}{32\pi^2}\frac{g^2}{Q^2}\ln\frac{Q^2}{\Lambda^2} + \mathcal{O}(g^4)\right). \tag{16.15}$$

Note that g is not a number in ϕ^3 theory but has dimensions of mass. This actually makes ϕ^3 a little more confusing than QED, but not insurmountably so. Let us substitute for g a new Q-dependent variable $\tilde{g}^2 \equiv \frac{g^2}{Q^2}$, which is dimensionless. Then,

$$\mathcal{M}(Q) = \tilde{g}^2 - \frac{1}{32\pi^2}\tilde{g}^4 \ln\frac{Q^2}{\Lambda^2} + \mathcal{O}(\tilde{g}^6). \tag{16.16}$$

Then we can *define* a renormalized coupling $\tilde{g}_R$ at some fixed scale Q_0 by

$$\tilde{g}_R^2 \equiv \mathcal{M}(Q_0). \tag{16.17}$$

This is called a **renormalization condition**. It is a definition and, by definition, it holds to all orders in perturbation theory. The renormalization condition defines the coupling in terms of an observable. Therefore, *you can only have one renormalization condition for each parameter in the theory*. This is critical to the predictive power of quantum field theory.

It follows that $\tilde{g}_R^2$ is a formal power series in $\tilde{g}$:

$$\tilde{g}_R^2 = \mathcal{M}(Q_0) = \tilde{g}^2 - \frac{1}{32\pi^2}\tilde{g}^4 \ln\frac{Q_0^2}{\Lambda^2} + \mathcal{O}(\tilde{g}^6), \tag{16.18}$$

which can be inverted to give $\tilde{g}$ as a power series in $\tilde{g}_R$:

$$\tilde{g}^2 = \tilde{g}_R^2 + \frac{1}{32\pi^2}\tilde{g}_R^4 \ln\frac{Q_0^2}{\Lambda^2} + \mathcal{O}(\tilde{g}_R^6). \tag{16.19}$$

Substituting into Eq. (16.16) produces a prediction for the matrix element at the scale Q in terms of the matrix element at the scale Q_0:

$$\mathcal{M}(Q) = \tilde{g}^2 - \frac{1}{32\pi^2}\tilde{g}^4 \ln\frac{Q^2}{\Lambda^2} + \mathcal{O}(\tilde{g}^6) = \tilde{g}_R^2 + \frac{1}{32\pi^2}\tilde{g}_R^4 \ln\frac{Q_0^2}{Q^2} + \mathcal{O}(\tilde{g}_R^6). \tag{16.20}$$

Thus, we can measure $\mathcal{M}$ at one Q and then make a non-trivial prediction at another value of Q.

16.2 Vacuum polarization in QED

In ϕ^3 theory, we found

$$= -\frac{ig^2}{32\pi^2}\int_0^1 dx \ln\frac{m^2 - p^2 x(1-x)}{\Lambda^2}. \tag{16.21}$$

The integral in QED is quite similar. We will first evaluate the vacuum polarization graph in scalar QED, and then in spinor QED.

16.2.1 Scalar QED

In scalar QED the vacuum polarization diagram is

$$= (-ie)^2 \int \frac{d^4k}{(2\pi)^4} \frac{i(2k^\mu - p^\mu)}{(k-p)^2 - m^2 + i\varepsilon} \frac{i(2k^\nu - p^\nu)}{k^2 - m^2 + i\varepsilon}.$$

(16.22)

For external photons, we could contract the μ and ν indices with polarization vectors, but instead we keep them free so that this diagram can be embedded in a Coulomb exchange diagram as in Eq. (16.14). This integral is the same as in ϕ^3 theory, except for the numerator factors. In scalar QED there is another diagram:

$$= 2ie^2 g^{\mu\nu} \int \frac{d^4k}{(2\pi)^4} \frac{i}{k^2 - m^2 + i\varepsilon}.$$

(16.23)

Adding the diagrams gives

$$i\Pi_2^{\mu\nu} = -e^2 \int \frac{d^4k}{(2\pi)^4} \frac{-4k^\mu k^\nu + 2p^\mu k^\nu + 2p^\nu k^\mu - p^\mu p^\nu + 2g^{\mu\nu}\left[(p-k)^2 - m^2\right]}{[(p-k)^2 - m^2 + i\varepsilon][k^2 - m^2 + i\varepsilon]}.$$

(16.24)

Fortunately, we do not need to evaluate the entire integral. By looking at what possible form it could have, we can isolate the part that will contribute to a correction to Coulomb's law and just calculate that part. By Lorentz invariance, the most general form that $\Pi_2^{\mu\nu}$ could have is

$$\Pi_2^{\mu\nu} = \Delta_1(p^2, m^2)p^2 g^{\mu\nu} + \Delta_2(p^2, m^2)p^\mu p^\nu$$

(16.25)

for some form factors Δ_1 and Δ_2. Note that $\Pi_2^{\mu\nu}$ cannot depend on k^μ, since k^μ is integrated over.

As a correction to Coulomb's law, this vacuum polarization graph will contribute to the same process that the photon propagator does. Let us define the photon propagator in momentum space by

$$\langle\Omega|T\left\{A^\mu(x)A^\nu(y)\right\}|\Omega\rangle = \int \frac{d^4p}{(2\pi)^4} e^{ip(x-y)} iG^{\mu\nu}(p).$$

(16.26)

Note that this expression only depends on $x - y$ by translation invariance. This is the all-orders non-perturbative definition of the propagator $G^{\mu\nu}(p)$, which is sometimes called the **dressed propagator**. At leading order, in Feynman gauge, the dressed propagator reduces to the free propagator:

$$iG^{\mu\nu}(p) = \frac{-ig^{\mu\nu}}{p^2 + i\varepsilon} + \mathcal{O}(e^2).$$

(16.27)

Including the 1-loop correction, with the parametrization in Eq. (16.25), the propagator is (suppressing the $i\varepsilon$ pieces)

$$
\begin{aligned}
iG^{\mu\nu}(p) &= \frac{-ig^{\mu\nu}}{p^2} + \frac{-ig^{\mu\alpha}}{p^2} i\Pi^2_{\alpha\beta} \frac{-ig^{\beta\nu}}{p^2} + \mathcal{O}(e^4) \\
&= \frac{-ig^{\mu\nu}}{p^2} + \frac{-i}{p^2}\left(\Delta_1 g^{\mu\nu} + \Delta_2 \frac{p^\mu p^\nu}{p^2}\right) + \mathcal{O}(e^4) \\
&= -i\frac{(1+\Delta_1)g^{\mu\nu} + \Delta_2 \frac{p^\mu p^\nu}{p^2}}{p^2 + i\varepsilon}.
\end{aligned}
\tag{16.28}
$$

Note that we are calculating loop corrections to a Green's function, not an S-matrix element, so we do not truncate the external propagators and add polarization vectors. One point of using a dressed propagator is that once we calculate Δ_1 and Δ_2 we can just use $G^{\mu\nu}(p)$ instead of the tree-level propagator in QED calculations to include the loop effect.

Next note that the Δ_2 term is just a change of gauge – it gives a correction to the unphysical gauge parameter ξ in covariant gauges. Since ξ drops out of any physical process, by gauge invariance, so will Δ_2. Thus we only need to compute Δ_1. This means extracting the term proportional to $g^{\mu\nu}$ in $\Pi^{\mu\nu}$.

Most of the terms in the amplitude in Eq. (16.24) cannot give $g^{\mu\nu}$. For example, the $p^\mu p^\nu$ term must be proportional to $p^\mu p^\nu$ and can therefore only contribute to Δ_2, so we can ignore it. For the $p^\mu k^\nu$ term, we can pull p^μ out of the integral, so whatever the remaining integral gives, it must provide a p^ν by Lorentz invariance. So these terms can be ignored too. The $k^\mu k^\nu$ term is important – it may give a $p^\mu p^\nu$ piece, but may also give a $g^{\mu\nu}$ piece, which is what we are looking for. So we only need to consider

$$
\Pi_2^{\mu\nu} = ie^2 \int \frac{d^4k}{(2\pi)^4} \frac{-4k^\mu k^\nu + 2g^{\mu\nu}\left[(p-k)^2 - m^2\right]}{[(p-k)^2 - m^2 + i\varepsilon][k^2 - m^2 + i\varepsilon]}.
\tag{16.29}
$$

Now we need to compute the integral.

The denominator can be manipulated using Feynman parameters, just as with the ϕ^3 theory:

$$
\Pi_2^{\mu\nu} = ie^2 \int \frac{d^4k}{(2\pi)^4} \int_0^1 dx \frac{-4k^\mu k^\nu + 2g^{\mu\nu}\left[(p-k)^2 - m^2\right]}{[(k - p(1-x))^2 + p^2x(1-x) - m^2 + i\varepsilon]^2}.
\tag{16.30}
$$

However, now when we shift $k^\mu \to k^\mu + p^\mu(1-x)$ we get a correction to the numerator. We get

$$
\begin{aligned}
\Pi_2^{\mu\nu} = ie^2 &\int \frac{d^4k}{(2\pi)^4} \\
&\times \int_0^1 dx \frac{-4\left[k^\mu + p^\mu(1-x)\right]\left[k^\nu + p^\nu(1-x)\right] + 2g^{\mu\nu}\left[(xp-k)^2 - m^2\right]}{[k^2 + p^2x(1-x) - m^2 + i\varepsilon]^2}.
\end{aligned}
\tag{16.31}
$$

As we have said, we do not care about $p^\mu p^\nu$ pieces, or pieces linear in p^ν. Also, pieces such as $p \cdot k$ are odd under $k \to -k$ while the rest of the integrand, including the measure, is even. So these terms must give zero by symmetry. All that is left is

$$
\Pi_2^{\mu\nu} = 2ie^2 \int \frac{d^4k}{(2\pi)^4} \int_0^1 dx \frac{-2k^\mu k^\nu + g^{\mu\nu}(k^2 + x^2p^2 - m^2)}{[k^2 + p^2x(1-x) - m^2 + i\varepsilon]^2}.
\tag{16.32}
$$

It seems this integral is much more badly divergent than the ϕ^3 theory – it is now quadratically instead of logarithmically divergent. That is, if we cut off at $k = \Lambda$ we will get something proportional to Λ^2 due to the $k^\mu k^\nu$ and k^2 terms. Quadratic divergences are not technically a problem for renormalization. However, in Chapter 21 we will see, on very general grounds, that in gauge theories such as scalar QED, all divergences should be logarithmic. In this case, the quadratic divergence from the $k^\mu k^\nu$ term and the k^2 term precisely cancel due to gauge invariance. This cancellation can only be seen using a regulator that respects gauge invariance, such as dimensional regularization. In d dimensions (using $k^\mu k^\nu \to \frac{1}{d} k^2 g^{\mu\nu}$ from Appendix B), the integral becomes

$$\Pi_2^{\mu\nu} = 2ie^2 \mu^{4-d} g^{\mu\nu} \int_0^1 dx \int \frac{d^d k}{(2\pi)^d} \frac{(1 - \frac{2}{d})k^2 + x^2 p^2 - m^2}{[k^2 + p^2 x(1-x) - m^2 + i\varepsilon]^2}. \tag{16.33}$$

Using the formulas from Appendix B,

$$\int \frac{d^d k}{(2\pi)^d} \frac{k^2}{(k^2 - \Delta + i\varepsilon)^2} = -\frac{d}{2} \frac{i}{(4\pi)^{d/2}} \frac{1}{\Delta^{1-\frac{d}{2}}} \Gamma\left(\frac{2-d}{2}\right) \tag{16.34}$$

and

$$\int \frac{d^d k}{(2\pi)^d} \frac{1}{(k^2 - \Delta + i\varepsilon)^2} = \frac{i}{(4\pi)^{d/2}} \frac{1}{\Delta^{2-\frac{d}{2}}} \Gamma\left(\frac{4-d}{2}\right), \tag{16.35}$$

with $\Delta = m^2 - p^2 x(1-x)$, we find

$$\Pi_2^{\mu\nu} = -2\frac{e^2}{(4\pi)^{d/2}} g^{\mu\nu} \mu^{4-d} \int_0^1 dx \left[\left(1 - \frac{d}{2}\right) \Gamma\left(1 - \frac{d}{2}\right) \left(\frac{1}{\Delta}\right)^{1-\frac{d}{2}} \right.$$
$$\left. + (x^2 p^2 - m^2)\Gamma\left(2 - \frac{d}{2}\right)\left(\frac{1}{\Delta}\right)^{2-\frac{d}{2}} \right]. \tag{16.36}$$

Using $\Gamma\left(2 - \frac{d}{2}\right) = \left(1 - \frac{d}{2}\right)\Gamma\left(1 - \frac{d}{2}\right)$ this simplifies to

$$\Pi_2^{\mu\nu} = -2\frac{e^2}{(4\pi)^{d/2}} p^2 g^{\mu\nu} \Gamma\left(2 - \frac{d}{2}\right) \mu^{4-d} \int_0^1 dx\, x(2x - 1)\left(\frac{1}{\Delta}\right)^{2-\frac{d}{2}}. \tag{16.37}$$

For completeness, we also give the result including the $p^\mu p^\nu$ terms:

$$\Pi_2^{\mu\nu} = \frac{-2e^2}{(4\pi)^{d/2}} \left(p^2 g^{\mu\nu} - p^\mu p^\nu\right) \Gamma\left(2 - \frac{d}{2}\right) \mu^{4-d}$$
$$\times \int_0^1 dx\, x(2x - 1) \left(\frac{1}{m^2 - p^2 x(1-x)}\right)^{2-\frac{d}{2}}. \tag{16.38}$$

You should verify this through direct calculation (see Problem 16.1), but it is the unique result consistent with Eq. (16.37) that satisfies the Ward identity, $p_\mu \Pi_2^{\mu\nu} = 0$.

Expanding $d = 4 - \varepsilon$ we get, in the $\epsilon \to 0$ limit,

$$\Pi_2^{\mu\nu} = -\frac{e^2}{8\pi^2} \left(p^2 g^{\mu\nu} - p^\mu p^\nu\right) \int_0^1 dx\, x(2x - 1) \left[\frac{2}{\varepsilon} + \ln\left(\frac{4\pi e^{-\gamma_E} \mu^2}{m^2 - p^2 x(1-x)}\right) + \mathcal{O}(\varepsilon)\right]. \tag{16.39}$$

The $\frac{1}{\varepsilon}$ gives the infinite, regulator-dependent constant. It is also standard to define $\tilde{\mu}^2 = 4\pi e^{-\gamma_E}\mu^2$, which removes the $\ln(4\pi)$ and $e^{-\gamma_E}$ factors. For $Q^2 = -p^2 > 0$ and $m \ll Q$, the integral over x is easy to do and we find

$$\Pi_2^{\mu\nu} = -\frac{e^2}{48\pi^2}\left(p^2 g^{\mu\nu} - p^\mu p^\nu\right)\left(\frac{2}{\varepsilon} + \ln\frac{\tilde{\mu}^2}{Q^2} + \frac{8}{3}\right), \quad m \ll Q. \tag{16.40}$$

At this point, rather than continue with the scalar QED calculation, let us calculate the loop in QED, as it is almost exactly the same.

16.2.2 Spinor QED

In spinor QED, the loop is

$$= -(-ie)^2 \int \frac{d^4 k}{(2\pi)^4} \frac{i}{(p-k)^2 - m^2 + i\varepsilon} \frac{i}{k^2 - m^2 + i\varepsilon}$$

$$\times \mathrm{Tr}[\gamma^\mu(\slashed{k} - \slashed{p} + m)\gamma^\nu(\slashed{k} + m)], \tag{16.41}$$

where the -1 in front comes from the fermion loop. Note that there is only one diagram in this case.

Using our trace formulas (see Sections 13.2 or A.4), we find

$$\mathrm{Tr}[\gamma^\mu(\slashed{k} - \slashed{p} + m)\gamma^\nu(\slashed{k} + m)] = 4[-p^\mu k^\nu - k^\mu p^\nu + 2k^\mu k^\nu + g^{\mu\nu}(-k^2 + p\cdot k + m^2)]. \tag{16.42}$$

We can drop the p^μ and p^ν terms as before giving

$$i\Pi_2^{\mu\nu} = -4e^2 \int \frac{d^4 k}{(2\pi)^4} \frac{2k^\mu k^\nu + g^{\mu\nu}(-k^2 + p\cdot k + m^2)}{[(p-k)^2 - m^2 + i\varepsilon][k^2 - m^2 + i\varepsilon]} + \cdots . \tag{16.43}$$

Introducing Feynman parameters and changing $k^\mu \to k^\mu + p^\mu(1-x)$ and again dropping the p^μ and p^ν terms (and omitting the $\cdots$) we get

$$\Pi_2^{\mu\nu} = 4ie^2 \int \frac{d^4 k}{(2\pi)^4} \int_0^1 dx \frac{2k^\mu k^\nu - g^{\mu\nu}\left[k^2 - x(1-x)p^2 - m^2 + i\varepsilon\right]}{[k^2 + p^2 x(1-x) - m^2]^2}. \tag{16.44}$$

This integral is quite similar to the one for scalar QED, Eq. (16.32). The result is

$$\Pi_2^{\mu\nu} = -8p^2 g^{\mu\nu} \frac{e^2}{(4\pi)^{d/2}} \Gamma\left(2 - \frac{d}{2}\right)\mu^{4-d}\int_0^1 dx\, x(1-x)\left(\frac{1}{m^2 - p^2 x(1-x)}\right)^{2-\frac{d}{2}}$$

$$= -\frac{e^2}{2\pi^2}p^2 g^{\mu\nu}\int_0^1 dx\, x(1-x)\left[\frac{2}{\varepsilon} + \ln\left(\frac{\tilde{\mu}^2}{m^2 - p^2 x(1-x)}\right) + \mathcal{O}(\varepsilon)\right]. \tag{16.45}$$

So, we find (for large $Q^2 = -p^2 \gg m^2$)

$$\Pi_2^{\mu\nu} = -\frac{e^2}{12\pi^2}p^2 g^{\mu\nu}\left(\frac{2}{\varepsilon} + \ln\frac{\tilde{\mu}^2}{Q^2} + \frac{5}{3} + \mathcal{O}(\varepsilon)\right), \quad m \ll Q. \tag{16.46}$$

We see that the electron loop gives the same pole and $\ln\frac{\tilde{\mu}^2}{Q^2}$ terms as a scalar loop, multiplied by a factor of 4.

It is not hard to compute the $p^\mu p^\nu$ pieces as well (see Problem 16.1). The full result is

$$\Pi_2^{\mu\nu} = \frac{-8e^2}{(4\pi)^{d/2}} \left(p^2 g^{\mu\nu} - p^\mu p^\nu\right) \Gamma\left(2 - \frac{d}{2}\right) \mu^{4-d}$$

$$\times \int_0^1 dx\, x(1-x) \left(\frac{1}{m^2 - p^2 x(1-x)}\right)^{2-\frac{d}{2}}, \quad (16.47)$$

which, as in the scalar QED case, automatically satisfies the Ward identity.

16.3 Physics of vacuum polarization

We have found that the vacuum polarization loop in QED gives

$$i\Pi_2^{\mu\nu} = i\left(-p^2 g^{\mu\nu} + p^\mu p^\nu\right) e^2 \Pi_2(p^2), \quad (16.48)$$

where

$$\Pi_2(p^2) = \frac{1}{2\pi^2} \int_0^1 dx\, x(1-x) \left[\frac{2}{\varepsilon} + \ln\left(\frac{\tilde{\mu}^2}{m^2 - p^2 x(1-x)}\right)\right]. \quad (16.49)$$

Thus, the dressed photon propagator at 1-loop in Feynman gauge is

$$iG^{\mu\nu} = \quad \vphantom{x}$$

$$= -i\frac{g^{\mu\nu}}{p^2} + \frac{-i}{p^2} i\Pi_2^{\mu\nu} \frac{-i}{p^2} + p^\mu p^\nu \text{ terms}$$

$$= -i\frac{\left[1 - e^2 \Pi_2(p^2)\right] g^{\mu\nu}}{p^2} + p^\mu p^\nu \text{ terms}. \quad (16.50)$$

This directly gives the Fourier transform of the corrected Coulomb potential:

$$\tilde{V}(p) = e^2 \frac{1 - e^2 \Pi_2(p^2)}{p^2}. \quad (16.51)$$

Now we need to renormalize.

A natural renormalization condition is that the potential between two particles at some reference scale r_0 should be $V(r_0) \equiv -\frac{e_R^2}{4\pi r_0}$, which would define a renormalized e_R. It is easier to continue working in momentum space and to define the renormalized charge as $\tilde{V}(p_0) \equiv e_R^2 p_0^{-2}$ exactly. So

$$e_R^2 \equiv p_0^2 \tilde{V}(p_0) = e^2 - e^4 \Pi_2(p_0^2) + \cdots. \quad (16.52)$$

Solving for the bare coupling e as a function of e_R to order e_R^4 gives

$$e^2 = e_R^2 + e_R^4 \Pi_2(p_0^2) + \cdots. \quad (16.53)$$

Since $\Pi_2(p_0^2)$ is infinite, e is infinite as well, but that is OK since e is not observable.

The potential at another scale p, which *is* measurable, is

$$p^2 \widetilde{V}(p) = e^2 - e^4 \Pi_2(p^2) + \cdots = e_R^2 - e_R^4 \left[\Pi_2(p^2) - \Pi_2(p_0^2) \right] + \cdots . \quad (16.54)$$

For concreteness, let us take $p_0 = 0$, corresponding to $r = \infty$, so that the renormalized electric charge agrees with the macroscopically measured electric charge. Then

$$\Pi_2(p^2) - \Pi_2(0) = -\frac{1}{2\pi^2} \int_0^1 dx \, x(1-x) \ln\left[1 - \frac{p^2}{m^2} x(1-x) \right] . \quad (16.55)$$

Thus, we have

$$\widetilde{V}(p) = \frac{e_R^2}{p^2} \left\{ 1 + \frac{e_R^2}{2\pi^2} \int_0^1 dx \, x(1-x) \ln\left[1 - \frac{p^2}{m^2} x(1-x) \right] + \mathcal{O}(e_R^4) \right\}, \quad (16.56)$$

which is a totally finite correction to the Coulomb potential. It is also a well-defined perturbation expansion in terms of a small parameter e_R, which is also finite. We will now study some of the physical implications of this potential.

16.3.1 Small momentum: Lamb shift

First, let us look at the small-momentum, large-distance limit. For $|p^2| \ll m^2$,

$$\int_0^1 dx \, x(1-x) \ln\left[1 - \frac{p^2}{m^2} x(1-x) \right] \approx \int_0^1 dx \, x(1-x) \left[-\frac{p^2}{m^2} x(1-x) \right] = -\frac{p^2}{30m^2}, \quad (16.57)$$

implying

$$\widetilde{V}(p) = \frac{e_R^2}{p^2} - \frac{e_R^4}{60\pi^2 m^2} + \cdots . \quad (16.58)$$

The Fourier transform of a 1 is $\delta(r)$, so we find

$$V(r) = -\frac{e_R^2}{4\pi r} - \frac{e_R^4}{60\pi^2 m^2} \delta(r) + \cdots . \quad (16.59)$$

This agrees with the Coulomb potential up to a correction known as the **Uehling term**.

What is the physical effect of this extra term? One way to find out is to plug this potential into the Schrödinger equation and see how the states of the hydrogen atom change. Equivalently, we can evaluate the effect in time-independent perturbation theory by evaluating the leading-order energy shift $\Delta E = \langle \psi_i | \Delta V | \psi_i \rangle$ using $\Delta V = -\frac{e^4}{60\pi^4 m^2} \delta(r)$. Since only the $L = 0$ atomic orbitals have support at $r = 0$, this extra term will only affect the S states of the hydrogen atom. The energy is negative, so their energies will be lowered. You might recall that, at leading order, the energy spectrum of the hydrogen atom is determined only by the principal atomic number n, so the $2\mathrm{P}_{1/2}$ and $2\mathrm{S}_{1/2}$ levels (for example) are degenerate. Thus, the Uehling term contributes to the splitting of these levels, known as the *Lamb shift*. It changes the $2\mathrm{S}_{1/2}$ state by $-27\,\mathrm{MHz}$, which is a measurable contribution to the $1058\,\mathrm{MHz}$ Lamb shift.

More carefully, you can show in Problem 16.2 that the 1-loop potential is

$$V(r) = -\frac{e^2}{4\pi r}\left(1 + \frac{e^2}{6\pi^2}\int_1^\infty dx\, e^{-2mrx}\frac{2x^2+1}{2x^4}\sqrt{x^2-1}\right). \tag{16.60}$$

This is known as the **Uehling potential** [Uehling, 1935]. For $r \gg \frac{1}{m}$,

$$V(r) = -\frac{\alpha}{r}\left[1 + \frac{\alpha}{4\sqrt{\pi}}\frac{1}{(mr)^{3/2}}e^{-2mr}\right], \quad r \gg \frac{1}{m}. \tag{16.61}$$

This shows that the finite correction has extent $1/m = r_e$, the Compton wavelength of the electron. Since r_e is much smaller than the characteristic size of the L modes, the Bohr radius $a_0 \sim \frac{1}{m\alpha}$, our δ-function approximation is valid.

By the way, the measurement of the Lamb shift in 1947 by Wallis Lamb [Lamb and Retherford, 1947] was one of the key experiments that convinced people to take quantum field theory seriously. Measurements of the hyperfine splitting between the $2S_{1/2}$ and $2P_{1/2}$ states of the hydrogen atom had been attempted for many years, but it was only by using microwave technology developed during the Second World War that Lamb was able to provide an accurate measurement. He found $\Delta E \simeq 1000\,\text{MHz}$. Shortly after his measurement, Hans Bethe calculated the dominant theoretical contribution. His calculation was of the logarithmically-divergent difference between the free and bound electron's self-energy. Bethe cut off the divergence by hand at what he argued was a natural physical scale, the electron mass. His result was that $\Delta E = -\frac{Z^4\alpha^5 m_e}{12\pi}\ln(\alpha^4 Z^4) \approx -1000\,\text{MHz}$, in excellent agreement with Lamb's value. The next year, Feynman, Schwinger and Tomonaga all independently provided the complete calculation, including the Uehling term and the spin-orbit coupling. Due to a subtlety regarding gauge invariance, only Tomonaga got it right the first time. The full 1-loop result gives $E(2S_{1/2}) - E(2P_{1/2}) = 1051\,\text{MHz}$. The current best measurement of this shift is $1057.833 \pm 0.004\,\text{MHz}$.

16.3.2 Large momentum: logarithms of p

In the small distance limit, $r \ll \frac{1}{m}$, it is easier to consider the potential in momentum space. Then we have from Eq. (16.56)

$$\begin{aligned}
\widetilde{V}(p) &= \frac{e_R^2}{p^2} + \frac{e_R^4}{p^2}\frac{1}{2\pi^2}\int_0^1 dx\, x(1-x)\ln\left[1 - \frac{p^2}{m^2}x(1-x)\right] + \mathcal{O}(e_R^6)\\
&\approx \frac{e_R^2}{p^2} + \frac{e_R^4}{p^2}\frac{1}{2\pi^2}\ln\frac{-p^2}{m^2}\int_0^1 dx\, x(1-x) + \mathcal{O}(e_R^6)\\
&= \frac{e_R^2}{p^2}\left(1 + \frac{e_R^2}{12\pi^2}\ln\frac{-p^2}{m^2}\right) + \mathcal{O}(e_R^6).
\end{aligned} \tag{16.62}$$

Recall that for t-channel exchange, $Q^2 = -p^2 > 0$, so this logarithm is real.

If we compare the potential at two high-energy scales, $Q = \sqrt{-p^2} \gg m$ and $Q_0 = \sqrt{-p_0^2} \gg m$, we find

$$Q^2\widetilde{V}(p) - Q_0^2\widetilde{V}(p_0) = \frac{e_R^4}{12\pi^2}\ln\frac{Q_0^2}{Q^2}, \tag{16.63}$$

which is independent of m. Note, however, that setting $m = 0$ directly in Eq. (16.62) results in a divergence. This kind of divergence is known as a mass singularity, which is a type of IR divergence. In this case, the divergence is naturally regulated by $m \neq 0$. On other occasions we will have to introduce an artificial IR regulator (such as a photon mass) to produce finite answers. Infrared divergences are the subject of Chapter 20.

One way to write the radiative correction to the potential is

$$\tilde{V}(p) = \frac{e_{\text{eff}}^2(\sqrt{-p^2})}{p^2},\qquad(16.64)$$

where

$$e_{\text{eff}}^2(Q) = e_R^2 \left[1 + \frac{e_R^2}{12\pi^2} \ln \frac{Q^2}{m^2} \right].\qquad(16.65)$$

In this case, for simplicity, we have defined the renormalized charge, $e_R \equiv e_{\text{eff}}(m)$, at $Q = m$ rather than at $Q = 0$. (One could also define e_R at $Q = 0$, as with the Uehling potential; however, then one would need to include the full m dependence to regulate the $Q = 0$ singularity.)

Equation (16.65) is to be interpreted as an **effective charge** in QED that grows as the distance gets smaller (momentum gets larger). Near any particular fixed value of the momentum transfer p^μ, the potential looks like a Coulomb potential with a charge $e_{\text{eff}}(p^2)$ instead of e_R. This is a useful concept because the charge depends only weakly on p^2, through a logarithm. Thus, for small variations of p around a reference scale, the same effective charge can be used. Equation (16.65) only comes into play when one compares the charge at very different momentum transfers.

The sign of the coefficient of the $\ln \frac{Q}{m}$ term is very important; this sign implies that the effective charge gets larger at short distances. At large distances, the charge is increasingly screened by the virtual electron–positron dipole pairs. At smaller distances, there is less room for the screening and the effective charge increases. However, the effective charge only increases at small distances very slowly. In fact, taking $\alpha_R = \frac{e_R^2}{4\pi} = \frac{1}{137}$ so that $e_R = 0.303$, we get an effective fine-structure constant of the form

$$\alpha_{\text{eff}}(Q) = \frac{1}{137} \left[1 + 0.00077 \ln \frac{Q^2}{m^2} \right].\qquad(16.66)$$

Because the coefficient of the logarithm is numerically small, one has to measure the potential at extremely high energies to see its effect. In fact, only very few high-precision measurements are sensitive to this logarithm.

Despite the difficulty of probing extremely high energies in QED experimentally, one can at least ask what would happen if we attempted scattering at $Q \gg m$. From Eq. (16.66) we can see that at some extraordinarily high energies, $Q \sim 10^{286}$ eV, the loop correction, the logarithm, is as important as the tree-level value, the 1. Thus, perturbation theory is breaking down. At these scales, the 2-loop value will also be as large as the 1-loop and tree-level values, and so on. The scale where this happens is known as a **Landau pole**. So,

QED has a Landau pole: perturbation theory breaks down at short distances.

This means that QED is not a complete theory in the sense that it does not tell us how to compute scattering amplitudes at all energies.

16.3.3 Running coupling

It is not difficult to include certain higher-order corrections to the effective electric charge. Adding more loops in series, we can sum a set of graphs to all orders in the coupling constant:

$$(16.67)$$

These corrections to the propagator immediately translate into corrections to the momentum space potential:

$$
\tilde{V}(p) = \frac{e_R^2}{p^2} \left[1 + \frac{e_R^2}{12\pi^2} \ln \frac{-p^2}{m^2} + \left(\frac{e_R^2}{12\pi^2} \ln \frac{-p^2}{m^2} \right)^2 + \cdots \right]
$$

$$
= \frac{1}{p^2} \left[\frac{e_R^2}{1 - \frac{e_R^2}{12\pi^2} \ln \frac{-p^2}{m^2}} \right]. \tag{16.68}
$$

So now the momentum-dependent electric charge becomes

$$
e_{\text{eff}}^2(Q) = \frac{e_R^2}{1 - \frac{e_R^2}{12\pi^2} \ln \frac{Q^2}{m^2}}, \tag{16.69}
$$

which is known as a **running coupling**. Note that we have defined this running coupling to have the same renormalization condition as the 1-loop effective charge: $e_{\text{eff}} = e_R$ at $p^2 = -m^2$. Although the running coupling includes contributions from all orders in perturbation theory, it still has a Landau pole at $p \sim 10^{286}$ eV.

Running couplings will play an increasingly important role as we study more complicated problems in quantum field theory. They are best understood through the renormalization group. As a preview of how the renormalization group works, note that Eq. (16.69) can be written as

$$
\frac{1}{e_{\text{eff}}^2(Q)} = \frac{1}{e_R^2} - \frac{1}{12\pi^2} \ln \frac{Q^2}{m^2}. \tag{16.70}
$$

The renormalization group comes from the simple observation that there is nothing special about the renormalization point. Here we have defined $e_R = e_{\text{eff}}(m)$, but we could have renormalized at any other point μ^2 instead of m^2, and the results would be the same. Then we would have

$$
\frac{1}{e_{\text{eff}}^2(Q)} = \frac{1}{e_{\text{eff}}^2(\mu)} - \frac{1}{12\pi^2} \ln \frac{Q^2}{\mu^2}. \tag{16.71}
$$

The left-hand side is independent of μ. So, taking the μ derivative gives

$$0 = -\frac{2}{e_{\text{eff}}^3} \frac{d}{d\mu} e_{\text{eff}} + \frac{1}{12\pi^2} \frac{2}{\mu}, \tag{16.72}$$

or

$$\mu \frac{de_{\text{eff}}}{d\mu} = \frac{e_{\text{eff}}^3}{12\pi^2}. \tag{16.73}$$

This is known as a **renormalization group equation**. We even have a special name for the left-hand side of this particular equation, the β-**function**. In general,

$$\mu \frac{de}{d\mu} \equiv \beta(e) \tag{16.74}$$

and we have derived that $\beta(e) = \frac{e^3}{12\pi^2}$ at 1-loop. The renormalization group is the subject of Chapter 23.

Problems

16.1 Calculate the $p^\mu p^\nu$ pieces of the vacuum polarization graph in scalar QED and in spinor QED. Show that your result is consistent with the Ward identity.

16.2 Calculate the Uehling potential, Eq. (16.60), by Fourier transforming the effective potential.

16.3 The pions, $\pi^\pm$, are charged scalar quark–antiquark bound states (mesons) with masses of 139 MeV. The tauon is a lepton with mass 1770 MeV. Consider the contribution of the vacuum polarization amplitude to $\pi^+\pi^- \to \pi^+\pi^-$ through a virtual τ loop in QED. For simplicity, consider the s-channel contribution only.

    (a) Plot $|\mathcal{M}|^2$ as a function of s for forward scattering ($t = 0$). You should find a kink at $s = s_0$. What is s_0? What is going on physically when $s > s_0$?

    (b) Plot the real and imaginary parts of $\mathcal{M}$ separately. Calculate $\text{Im}(\mathcal{M})$ explicitly and show that it agrees with your plot.

    (c) Find a relationship between $\text{Im}(\mathcal{M})$ at $t = 0$ and the total rate for $\pi^+\pi^- \to \tau^+\tau^-$. This is a special case of a general and powerful result known as the optical theorem, which is discussed in detail in Chapter 24.

16.4 Where is the location of the Landau pole in QED if you include contributions from the electron, muon and tauon (all with charge $Q = -1$), from nine quarks (three colors times three flavors) with charge $Q = \frac{2}{3}$ and from nine quarks with charge $Q = -\frac{1}{3}$?

The anomalous magnetic moment   17

In the non-relativistic limit, the Dirac equation in the presence of an external magnetic field produces a Hamiltonian,

$$H = \frac{\vec{p}^2}{2m} + V(r) + \frac{e}{2m}\vec{B} \cdot (\vec{L} + g\vec{S}), \qquad (17.1)$$

acting on electron doublets $|\psi\rangle$, where $\vec{S} = \frac{1}{2}\vec{\sigma}$. This was derived in Problem 10.1. The coupling g is the g-**factor** of the electron, representing the relative strength of its intrinsic magnetic dipole moment to the strength of the spin-orbit coupling. From the point of view of the Schrödinger equation, g is a free parameter and could be anything. However, the Dirac equation implies that $g = 2$, which was a historically important postdiction in excellent agreement with data when Dirac presented his equation in 1932. A natural question is then: is $g = 2$ exactly, or does g receive quantum corrections? The answer should not be obvious. For example, the charge of the electron is *exactly* opposite to the charge of the proton, receiving no radiative corrections (we will prove this in Section 19.5), so perhaps the magnetic moment is exact as well. By the late 1940s there were experimental data that could be partially explained by the electron having an **anomalous magnetic moment**, that is, one different from 2. The calculation of this anomalous moment by Schwinger, Feynman and Tomonaga in 1948, and its agreement with data, was a triumph of quantum field theory.

17.1 Extracting the moment

We would like a way to extract the radiative corrections to g without having to take the non-relativistic limit. To see how to do this, recall from Section 10.4 how the electron's magnetic dipole moment was derived from the Dirac equation. Charged spinors satisfy $(i\not{D} - m)\psi = 0$. Multiplying this by $(i\not{D} + m)$ shows that charged spinors also satisfy $(\not{D}^2 + m^2)\psi = 0$. We then use the operator relation (cf. Eq. (10.106))

$$\not{D}^2 = D_\mu^2 + \frac{e}{2}F_{\mu\nu}\sigma^{\mu\nu}, \qquad (17.2)$$

where $\sigma_{\mu\nu} = \frac{i}{2}[\gamma_\mu, \gamma_\nu]$, to find $\left(D_\mu^2 + m^2 + \frac{e}{2}F_{\mu\nu}\sigma^{\mu\nu}\right)\psi = 0$. The $\frac{e}{2}F_{\mu\nu}\sigma^{\mu\nu}$ in this equation therefore encodes the difference between the way a scalar field, obeying

315

$\left(D_\mu^2 + m^2\right)\phi = 0$, and a spinor field interact with an electromagnetic field. In particular, in the Weyl representation,

$$\frac{e}{2}F_{\mu\nu}\sigma^{\mu\nu} = -e\begin{pmatrix}(\vec{B}+i\vec{E})\vec{\sigma} & \\ & (\vec{B}-i\vec{E})\vec{\sigma}\end{pmatrix}. \tag{17.3}$$

Going to momentum space, $(\slashed{D}^2 + m^2)\psi = 0$ implies (cf. Eq. (10.109))

$$\frac{(H - eA_0)^2}{2m}\psi = \left(\frac{m}{2} + \frac{(\vec{p}-e\vec{A})^2}{2m} - 2\frac{e}{2m}\vec{B}\cdot\vec{S} \pm i\frac{e}{m}\vec{E}\cdot\vec{S}\right)\psi, \tag{17.4}$$

which can be compared directly to Eq. (17.1) to read off the strength of the magnetic dipole interaction $g e\vec{B}\cdot\vec{S}$.[1] Since $\vec{S} = \frac{\vec{\sigma}}{2}$ for spin $\frac{1}{2}$, we find again that $g = 2$. If Eq. (17.2) had $g'\frac{e}{4}F_{\mu\nu}\sigma^{\mu\nu}$ in it, we would have found $g = g'$ instead. Thus, a general and relativistic way to extract corrections to g is to look for loops that have the same effect as an additional $F_{\mu\nu}\sigma^{\mu\nu}$ term.

A generally useful way to think about corrections to the way photons interact with spinors, such as corrections to g, is to consider *off-shell* S-matrix elements. The Feynman rules for off-shell S-matrix elements are the same as for on-shell S-matrix elements, except that $p_i^2 = m_i^2$ for the various external states is not enforced. In this case, the relevant process is $e^-(q_1)A_\mu(p) \to e^-(q_2)$, with polarization vector $\epsilon_\mu(p)$ and two spinor states $\bar{u}(q_2)$ and $u(q_1)$. At tree-level, the matrix element is just $\epsilon_\mu\mathcal{M}_0^\mu$, where

$$i\mathcal{M}_0^\mu = \quad\text{}\quad = -ie\bar{u}(q_2)\gamma^\mu u(q_1), \tag{17.5}$$

with the photon momentum constrained by momentum conservation to be $p^\mu = q_2^\mu - q_1^\mu$. This result actually contains $g = 2$ in it, although it is hard to see in this form. We expect something equivalent to an $F_{\mu\nu}\sigma^{\mu\nu}$ term, which should look like $\bar{u}(q_2)\,p_\nu\sigma^{\mu\nu}u(q_1)$ in momentum space. To see where $F_{\mu\nu}\sigma^{\mu\nu}$ is hiding, we need to massage the result a little.

For the magnetic moment, we only have to allow for the photon, which corresponds to an unconstrained external magnetic field, to be off-shell; the spinors can be on-shell, which helps simplify things. For example, we can use the **Gordon identity**, which you derived in Problem 11.4, and which holds for on-shell spinors:

$$\bar{u}(q_2)\left(q_1^\mu + q_2^\mu\right)u(q_1) = (2m)\,\bar{u}(q_2)\gamma^\mu u(q_1) + i\bar{u}(q_2)\,\sigma^{\mu\nu}(q_1^\nu - q_2^\nu)u(q_1). \tag{17.6}$$

Therefore

$$\mathcal{M}_0^\mu = -e\left(\frac{q_1^\mu + q_2^\mu}{2m}\right)\bar{u}(q_2)\,u(q_1) - \frac{e}{2m}i\bar{u}(q_2)\,p_\nu\sigma^{\mu\nu}u(q_1). \tag{17.7}$$

The first term is an interaction just like the scalar QED interaction: the photon couples to the momentum of the field, as in the D_μ^2 term in the Klein–Gordon equation. The q_μ^1 and q_μ^2 in this first term are just the momentum factors that appear in the scalar QED Feynman rule.

---

[1] The $\vec{E}\cdot\vec{S}$ term is not an electric dipole moment since it has an imaginary coefficient. Instead, it is the Lorentz-invariant completion of the magnetic moment.

The second term in Eq. (17.7) is spin dependent and gives the magnetic moment. So we can identify g as $\frac{4m}{e}$ times the coefficient of $ip_\nu \bar{u}\sigma^{\mu\nu}u$. Therefore, to calculate corrections to g we need to find how the coefficient of $i\bar{u}p_\nu\sigma^{\mu\nu}u$ is modified at loop level.

The correction to the magnetic moment must come from graphs involving the photon and the electron that contribute corrections to the process in Eq. (17.5). We can parametrize the most general possible result, at any-loop order, as

$$i\mathcal{M}^\mu = \qquad\qquad\qquad\qquad = \bar{u}(q_2)\left(f_1\gamma^\mu + f_2 p^\mu + f_3 q_1^\mu + f_4 q_2^\mu\right)u(q_1). \quad (17.8)$$

Here we have included all Lorentz vectors that might possibly appear, with the f_i their unknown Lorentz scalar coefficients. The f_i can depend in general on contractions of momenta, such as $p\cdot q$ or p^2, or on contractions with γ-matrices, such as $\slashed{p}$. (In more general theories, they could also depend on γ_5, but QED is parity invariant so γ_5 cannot appear.) For the magnetic moment application, we can assume the external spinors are on-shell, but the photon, representing an unconstrained external magnetic field, must still be off-shell. (Or, if you prefer, imagine this diagram is embedded in a larger Coulomb-scattering diagram with an off-shell intermediate photon and on-shell external spinors.)

The f_i are not all independent. Using momentum conservation, $p^\mu = q_2^\mu - q_1^\mu$, we can set $f_2 = 0$ and substitute away all the p^μ dependence. Then, if there are factors of $\slashed{q}_1$ or $\slashed{q}_2$ in the f_i, they can be removed by using the Dirac equation, $\slashed{q}_1 u(q_1) = m\,u(q_1)$, and $\bar{u}(q_2)\slashed{q}_2 = m\,\bar{u}(q_2)$. So, we can safely assume the f_i are real functions that can only depend on $q_1\cdot q_2$ and m, or more conventionally on $p^2 = 2m^2 - 2q_1\cdot q_2$ and m^2. Moreover, we can fix the relative dependence by dimensional analysis so the f_i are functions of $\frac{p^2}{m^2}$.

Next, the Ward identity (which we showed in Section 14.8 holds even if the photon is off-shell) implies

$$0 = p_\mu \bar{u}(f_1\gamma^\mu + f_3 q_1^\mu + f_4 q_2^\mu)u$$
$$= f_1 \bar{u}\slashed{p}u + (p\cdot q_1)f_3\bar{u}u + (p\cdot q_2)f_4\bar{u}u$$
$$= (p\cdot q_1)f_3\bar{u}u + (p\cdot q_2)f_4\bar{u}u. \quad (17.9)$$

We then use $p\cdot q_1 = q_2\cdot q_1 - m^2 = -p\cdot q_2$ to get $f_3 = f_4$. Thus, there are only two independent form factors. We can then use the Gordon identity, Eq. (17.6), to rewrite the q_1^μ and q_2^μ dependence in terms of $\sigma^{\mu\nu}$, leading to

$$i\mathcal{M}^\mu = (-ie)\bar{u}(q_2)\left[F_1\left(\frac{p^2}{m^2}\right)\gamma^\mu + \frac{i\sigma^{\mu\nu}}{2m}p_\nu F_2\left(\frac{p^2}{m^2}\right)\right]u(q_1), \quad (17.10)$$

which is our final form. This parametrization holds to all orders in perturbation theory. The functions F_1 and F_2 are known as **form factors**. The leading graph, Eq. (17.5), gives

$$F_1 = 1, \quad F_2 = 0. \quad (17.11)$$

Loops will give contributions to F_1 and F_2 at order α and higher.

Which of these two form factors could give an electron magnetic moment? F_1 modifies the original $eA_\mu\bar{\psi}\gamma^\mu\psi$ coupling. This renormalizes the electric charge, as we saw from the vacuum polarization diagram. In fact, the entire effect of this form factor is to give scale dependence to the electric charge, so no other effect, such as an anomalous magnetic moment, can come from it. F_2, on the other hand, has precisely the structure of a magnetic moment (which is, of course, why we put it in this form with the Gordon identity). Using that such a term without the F_2 factor gives $g = 2$, as in Eq. (17.7), we conclude that $F_2(\frac{p^2}{m^2})$ modifies the moment at the scale associated with p^2 by $g \to 2 + 2F_2(\frac{p^2}{m^2})$. Since the actual magnetic moment is measured at non-relativistic energies with $|\vec{p}| \ll m$, the moment that can be compared to data is

$$g = 2 + 2F_2(0). \tag{17.12}$$

Thus, we have reduced the problem to calculating $F_2(0)$.

17.2 Evaluating the graphs

There are four possible 1-loop graphs that could contribute to $\mathcal{M}^\mu$. Three of them,

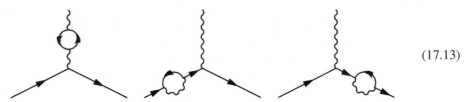

$$\tag{17.13}$$

can only give terms proportional to γ^μ. This is easy to see because these graphs just correct the propagators for the corresponding particles. Thus, these graphs can only contribute to F_1 and have no effect on the magnetic moment. The fourth graph is

$$i\mathcal{M}_2^\mu = \tag{17.14}$$

with $p^\mu = q_2^\mu - q_1^\mu$. This is the only graph we have to consider for $g - 2$.

Employing the Feynman rules, this graph is

$$i\mathcal{M}_2^\mu = (-ie)^3 \int \frac{d^4k}{(2\pi)^4} \frac{-ig^{\nu\alpha}}{(k-q_1)^2 + i\varepsilon} \bar{u}(q_2)\gamma^\nu$$
$$\times \frac{i(\slashed{p}+\slashed{k}+m)}{(p+k)^2 - m^2 + i\varepsilon}\gamma^\mu \frac{i(\slashed{k}+m)}{k^2 - m^2 + i\varepsilon}\gamma^\alpha u(q_1)$$
$$= -e^3\bar{u}(q_2)\int \frac{d^4k}{(2\pi)^4} \frac{\gamma^\nu(\slashed{p}+\slashed{k}+m)\gamma^\mu(\slashed{k}+m)\gamma_\nu}{[(k-q_1)^2 + i\varepsilon][(p+k)^2 - m^2 + i\varepsilon][k^2 - m^2 + i\varepsilon]}u(q_1). \tag{17.15}$$

To simplify this, we start by combining denominators and completing the square. The denominator has three terms and can be simplified with the identity

$$\frac{1}{ABC} = 2 \int_0^1 dx\, dy\, dz\, \delta(x+y+z-1)\frac{1}{[xA+yB+zC]^3}. \tag{17.16}$$

In this case

$$A = k^2 - m^2 + i\varepsilon, \tag{17.17}$$
$$B = (p+k)^2 - m^2 + i\varepsilon, \tag{17.18}$$
$$C = (k - q_1)^2 + i\varepsilon. \tag{17.19}$$

The new denominator is the cube of

$$xA + yB + zC = k^2 + 2k(yp - zq_1) + yp^2 + zq_1^2 - (x+y)m^2 + i\varepsilon$$
$$= (k^\mu + yp^\mu - zq_1^\mu)^2 - \Delta + i\varepsilon \tag{17.20}$$

with

$$\Delta = -xyp^2 + (1-z)^2 m^2. \tag{17.21}$$

Thus, we want to shift $k^\mu \to k^\mu - yp^\mu + zq_1^\mu$ to make the denominator $(k^2 - \Delta)^3$.

The numerator in Eq. (17.15) is

$$N^\mu = \bar{u}(q_2)\gamma^\nu(\slashed{p} + \slashed{k} + m)\gamma^\mu(\slashed{k} + m)\gamma_\nu u(q_1)$$
$$= -2\bar{u}(q_2)\left[\slashed{k}\gamma^\mu\slashed{p} + \slashed{k}\gamma^\mu\slashed{k} + m^2\gamma^\mu - 2m(2k^\mu + p^\mu)\right]u(q_1). \tag{17.22}$$

Shifting $k^\mu \to k^\mu - yp^\mu + zq_1^\mu$ then gives

$$-\frac{1}{2}N^\mu = \bar{u}(q_2)\left[(\slashed{k} - y\slashed{p} + z\slashed{q_1})\gamma^\mu\slashed{p} + (\slashed{k} - y\slashed{p} + z\slashed{q_1})\gamma^\mu(\slashed{k} - y\slashed{p} + z\slashed{q_1})\right]u(q_1)$$
$$+ \bar{u}(q_2)\left[m^2\gamma^\mu - 2m(2k^\mu - 2yp^\mu + 2zq_1^\mu + p^\mu)\right]u(q_1). \tag{17.23}$$

Using $k^\mu k^\nu = \frac{1}{4}g^{\mu\nu}k^2$ (see Appendix B), the Gordon identity, $x+y+z = 1$ and a fair amount of algebra, this simplifies to

$$-\frac{1}{2}N^\mu = \left[-\frac{1}{2}k^2 + (1-x)(1-y)p^2 + (1-4z+z^2)m^2\right]\bar{u}(q_2)\gamma^\mu u(q_1)$$
$$+ imz(1-z)p_\nu\bar{u}(q_2)\sigma^{\mu\nu}u(q_1)$$
$$+ m(z-2)(x-y)p^\mu\bar{u}(q_2)u(q_1). \tag{17.24}$$

We have found three independent terms instead of two since we have not used the Ward identity. Indeed, the Ward identity should fall out of the calculation automatically. To see that it does, note that the p^μ term gives a contribution to $\mathcal{M}_2^\mu$ of the form

$$i\mathcal{M}_2^\mu = \cdots + 4e^3 \int_0^1 dx\, dy\, dz\, \delta(x+y+z-1)m(z-2)(x-y)$$
$$\times \int \frac{d^4k}{(2\pi)^4}\frac{p^\mu}{(k^2 - \Delta + i\varepsilon)^3}\bar{u}(q_2)u(q_1). \tag{17.25}$$

Next, note that both Δ in Eq. (17.21) and the integral measure are symmetric in $x \leftrightarrow y$, but the integrand is antisymmetric. Thus this term is zero.

For the magnetic moment calculation we only need the $\sigma^{\mu\nu}$ term. Thus,

$$i\mathcal{M}_2^\mu = p_\nu \bar{u}(q_2)\sigma^{\mu\nu}u(q_1)\left[4ie^3 m \int_0^1 dx\,dy\,dz\,\delta(x+y+z-1)\right.$$
$$\left.\times \int \frac{d^4k}{(2\pi)^4}\frac{z(1-z)}{(k^2-\Delta+i\varepsilon)^3}\right]+\cdots, \quad (17.26)$$

where the $\cdots$ do not contribute to the moment. Recalling that $F_2(p^2)$ was defined as the coefficient of this operator, normalized by $\frac{2m}{e}$, we have

$$F_2(p^2) = \frac{2m}{e}\left(4ie^3 m\right)\int_0^1 dx\,dy\,dz\,\delta(x+y+z-1)\int \frac{d^4k}{(2\pi)^4}\frac{z(1-z)}{(k^2-\Delta+i\varepsilon)^3}+\mathcal{O}(e^4).$$
$$(17.27)$$

For completeness, the other form factor is $F_1(p^2) = 1 + f(p^2) + \mathcal{O}(e^4)$, where

$$f(p^2) = -2ie^2 \int_0^1 \frac{d^4k}{(2\pi)^4}dx\,dy\,dz\,\delta(x+y+z-1)$$
$$\times \frac{k^2 - 2(1-x)(1-y)p^2 - 2(1-4z+z^2)m^2}{[k^2-(m^2(1-z)^2-xyp^2)]^3}. \quad (17.28)$$

We will come back and evaluate $f(p^2)$ when we need to, in Section 19.3.

To evaluate F_2, we use the identity from Appendix B:

$$\int \frac{d^4k}{(2\pi)^4}\frac{1}{(k^2-\Delta+i\varepsilon)^3} = \frac{-i}{32\pi^2\Delta}, \quad (17.29)$$

to get that, up to terms of order α^2,

$$F_2(p^2) = \frac{\alpha}{\pi}m^2 \int_0^1 dx\,dy\,dz\,\delta(x+y+z-1)\frac{z(1-z)}{(1-z)^2m^2-xyp^2}. \quad (17.30)$$

At $p^2=0$ this integral is finite. Explicitly,

$$F_2(0) = \frac{\alpha}{\pi}\int_0^1 dz\int_0^1 dy\int_0^1 dx\,\delta(x+y+z-1)\frac{z}{(1-z)}$$
$$= \frac{\alpha}{\pi}\int_0^1 dz\int_0^{1-z} dy\frac{z}{(1-z)}$$
$$= \frac{\alpha}{2\pi}. \quad (17.31)$$

Thus

$$g = 2 + \frac{\alpha}{\pi} = 2.00232, \quad (17.32)$$

with the next correction of order α^2.

As a historical note, this result was first announced at the APS meeting in January 1948, by Schwinger. Feynman and Tomonaga had both calculated the same result independently at the same time. Schwinger actually found different values for $g-2$ for an electron bound in an atom and a free electron, while Feynman found they were the same. Feynman's result was the correct one, and it was relativistically invariant, while Schwinger's was not. The discrepancy was quickly resolved. Tomonaga was the first to correctly present the full 1-loop formula for the Lamb shift.

Unfortunately, it is not easy to measure g directly. Schwinger was able to check his calculation indirectly as giving part of the contribution to various hyperfine splittings in hydrogen, such as the Lamb shift. In order to make the comparison, he needed also to be able to get finite predictions out of the divergent integrals, such as the contributions to F_1 in addition to the finite $g - 2$ integral. The comparison with data really required a full understanding of all the 1-loop corrections in QED. For this reason, the simplicity of the finite $g - 2$ calculation we have just done was not immediately appreciated. Nevertheless, this calculation, and the Lamb shift calculation more generally, was critically important historically for convincing us that loops in quantum field theory had physical consequences.

The current best measurement is $g = 2.002\,319\,304\,3617 \pm (3 \times 10^{-13})$. The theory calculation has been performed up to 4-loop level. One cannot compare theory to experiment directly, since the theory is expressed as a function of α, which cannot be measured more precisely any other way. Therefore $g - 2$ is now used to define the renormalized value of the fine-structure constant, which comes out to $\alpha^{-1} = 137.035\,999\,070 \pm (9.8 \times 10^{-10})$.

Problems

17.1 In supersymmetry, each fermion has a scalar partner, and each gauge boson has a fermionic partner. For example, the partner of the electron is the selectron ($\tilde{e}$), the partner of the muon is the smuon ($\tilde{\mu}$), and the partner of the photon is the photino ($\tilde{A}$). The Lagrangian gets additional terms:

$$\mathcal{L}_{\text{SUSY}} - \mathcal{L}_{\text{SM}} + (\partial_\mu \tilde{e} + igA_\mu \tilde{e})^\star (\partial_\mu \tilde{e} + igA_\mu \tilde{e}) + m_{\tilde{e}}^2 |\tilde{e}|^2 + g\tilde{e}\bar{e}\tilde{A} + g\tilde{e}^\star \overline{\tilde{A}}e$$

$$+ \overline{\tilde{A}}(\partial\!\!\!/ + m_{\tilde{A}})\tilde{A} + (\partial_\mu \tilde{\mu} + igA_\mu \tilde{\mu})^\star (\partial_\mu \tilde{\mu} + igA_\mu \tilde{\mu}) + m_{\tilde{\mu}}^2 |\tilde{\mu}|^2 + g\tilde{\mu}\bar{\mu}\tilde{A} + g\tilde{\mu}^\star \overline{\tilde{A}}\mu.$$
(17.33)

The smuon and selectron have the same electric charge, -1 (here g denotes the electric charge, $\alpha_e = \frac{g^2}{4\pi} \sim \frac{1}{137}$). The size of the Yukawa couplings is fixed to be g as well, by gauge invariance and supersymmetry.

(a) Calculate the contribution of loops involving the smuon to the muon's magnetic dipole moment.

(b) The current best experimental value for $g - 2$ of the muon is $\frac{g_\mu - 2}{2} = 0.001\,165\,920\,8 \pm (6.3 \times 10^{-10})$. The current theory prediction (assuming the Standard Model only) is $\frac{g_\mu - 2}{2} = 0.001\,165\,918\,2 \pm (8.0 \times 10^{-10})$. What bound on $m_{\tilde{\mu}}$ do you get from this measurement?

(c) For other reasons, we expect $m_{\tilde{A}} \sim m_{\tilde{\mu}} \sim m_{\tilde{e}} \sim M_{\text{SUSY}}$. What bound on M_{SUSY} do you get from the muon $g - 2$?

Mass renormalization

In this chapter we will study the following 1-loop Feynman diagram:

which is known as the **electron self-energy graph**. You may recall we encountered this diagram way back in Chapter 4 in the context of Oppenheimer's Lamb shift calculation using old-fashioned perturbation theory. Indeed, this graph is important for the Lamb shift. However, rather than compute the Lamb shift (which is rather tedious and mostly of historical interest for us), we will use this graph to segue to a more general understanding of renormalization. You may also recall Oppenheimer's frustrated comment, quoted at the end of Chapter 4, where he suggested that the resolution of these infinities would require an "adequate theory of the masses of the electron and proton." In this chapter, we will provide such an adequate theory.

The electron self-energy graph corrects the electron propagator in the same way that the photon self-energy graph corrects the photon propagator. Recall from Chapter 16 that the photon self-energy graph could be interpreted as a vacuum polarization effect that generated a logarithmic strengthing of the Coulomb potential at short distances. Thus, by measuring $r_1 V(r_1) - r_2 V(r_2)$ with two different values of r one could measure vacuum polarization and compare it to the theoretical prediction. In particular, we were able to renormalize the divergent vacuum polarization graph by relating it to something (the potential) that can be directly connected to observables (e.g. the force between two currents or the energy levels of hydrogen).

Proceeding in the same way, the electron self-energy graph would correct the effect generated by the exchange of an electron. However, since the electron is a fermion, and charged, this exchange cannot be interpreted as generating a potential in any useful way. Thus, it is not clear what exactly one would measure to test whatever result we find by evaluating the self-energy diagram.

For the self-energy graph, and many other divergent graphs we will evaluate, it is helpful to navigate away from observables such as the Lamb shift or the Coulomb potential, which are particular to one type of correction, to thinking of general observables. Unfortunately, the question of what is observable and what is not is extremely subtle and has no precise definition in quantum field theory. For example, one might imagine that S-matrix elements are observable; in many cases they are actually infinite due to IR divergences, as we will see in Chapter 20. Luckily, one does not need a precise definition of an observable to understand renormalization, since even non-observable quantities can be renormalized. We will therefore consider the renormalization of general time-ordered correlation functions or **Green's functions**:

$$G(x_1, \ldots, x_n) = \langle \Omega | T\{\phi_1(x_1) \cdots \phi_n(x_n)\} | \Omega \rangle, \qquad (18.1)$$

where ϕ_i can be any type of field (scalars, electrons, photons, etc.). These Green's functions are in general *not* observable. In fact, they are in general not even gauge invariant. We will nevertheless show within a few chapters that all UV divergences can be removed from all Green's functions in *any* local quantum field theory through a systematic process of renormalization. Once the Green's functions are UV finite, S-matrix elements constructed from them using the LSZ reduction formula will also be UV finite. Infrared divergences and what can actually be observed are another matter.

One advantage of renormalizing general Green's functions rather than S-matrix elements is that the Green's functions can appear as internal subgraphs in many different S-matrix calculations. In particular, we will find that in QED, while there are an infinite number of divergent graphs contributing to the S-matrix, the divergences can be efficiently categorized and renormalized through the **one-particle irreducible** subgraphs (defined as graphs that cannot be cut in two by cutting a single propagator). As we will see, these one-particle irreducible graphs compose the minimal basis of Green's functions out of which any S-matrix can be built. Organizing the discussion in terms of Green's functions and one-particle irreducible diagrams will vastly simplify the proof of renormalizability in QED (in Chapter 21) and is critical to a general understanding of how renormalization works in various quantum field theories.

In this chapter, we abbreviate $\langle \Omega | T\{\cdots\} | \Omega \rangle$ with $\langle \cdots \rangle$ for simplicity.

18.1 Vacuum expectation values

We begin our consideration of the renormalization of general Green's functions by considering the simplest Green's functions, the 1-point functions:

$$\langle \phi(x) \rangle, \quad \langle \psi(x) \rangle, \quad \langle A_\mu(x) \rangle, \quad \ldots \qquad (18.2)$$

These give the expectation values of fields in the vacuum, also known as **vacuum expectation values**.

At tree-level, the vacuum expectation value of a field is the lowest energy configuration that satisfies the classical equations of motion. All Lagrangians we have considered so far begin at quadratic order in the fields, so that $\psi = A = \phi = 0$ are solutions to the equations of motion. Other solutions, such as plane waves in the free theory, contribute to the gradient terms in the energy density and thus have higher energy than the constant solution. Thus, $\psi = A = \phi = 0$ is the minimum energy solution and all the expectation values in Eq. (18.2) vanish at tree-level. More directly, we can see that $\langle \phi \rangle = \langle \psi \rangle = \langle A_\mu \rangle = 0$ at tree-level in the canonically quantized theory, since each quantum field has creation and annihilation operators that vanish in the vacuum.

At 1-loop, vacuum expectation values, for example for $\langle A_\mu \rangle$, could come from diagrams such as

This is called a **tadpole** diagram. It and all higher-loop contributions to $\langle A_\mu \rangle$ vanish identically in QED. This is easy to see since the graph must be proportional to the photon momentum (by Lorentz invariance) but the photon momentum must vanish (by momentum conservation). It is also true that $\langle \psi \rangle = 0$ to all orders in QED, simply because one cannot draw any diagrams.

A somewhat simpler proof that $\langle A_\mu \rangle$ or $\langle \psi \rangle$ must vanish is that non-zero values would violate Lorentz invariance, and Lorentz invariance is a symmetry of the QED Lagrangian. However, it may sometimes happen that the vacuum does not in fact satisfy every symmetry of the Lagrangian, in which case we say **spontaneous symmetry breaking** has occurred. Spontaneous symmetry breaking is covered in depth in Chapter 28. A familiar example is the spontaneous breaking of rotational invariance by a ferromagnet when cooled below its Curie temperature. At low temperature, the magnet has a preferred spin direction, which could equally well have pointed anywhere, but must point somewhere. Another example is the ground state of our universe, which has a preferred frame, the rest frame of the cosmic microwave background. In both cases space-time symmetries are symmetries of the Lagrangian but not of the ground state.

Spontaneous symmetry breaking can also apply to internal symmetries, such as global or gauge symmetries of a theory. For example, in the Bardeen–Cooper–Schrieffer (BCS) theory of superconductivity, the $U(1)$ symmetry of QED is spontaneously broken in type-II superconductors as they are cooled below their critical temperature. The attractive force between electrons due to phonon exchange becomes stronger than the repulsive Coulomb force and the vacuum becomes charged. Another important example is the Glashow–Weinberg–Salam theory of weak interactions (Chapter 29). This theory embeds the low-energy theory of weak interactions into a larger theory which has an exact $SU(2)$ symmetry that acts on the left-handed quarks and leptons.

Spontaneous symmetry breaking is an immensely important topic in quantum field theory, which we will systematically discuss beginning in Chapter 28, including more details of the above examples. Now, it is merely a distraction from our current task of understanding renormalization. Since $\langle A_\mu \rangle = \langle \psi \rangle = 0$ in QED to all orders in perturbation theory, there is nothing to renormalize and we can move on to 2-point functions.

18.2 Electron self-energy

There are a number of 2-point functions in QED. In Chapter 16, we discussed the renormalization of the photon propagator that corresponds to $\langle A_\mu A_\nu \rangle$. Two-point functions such as $\langle \psi A_\mu \rangle$ vanish identically in QED since there are simply no diagrams that could contribute to them. That leaves the fermion 2-point function $\langle \psi \bar{\psi} \rangle$.

As with the photon, it is helpful to study $\langle \psi \bar{\psi} \rangle$ in momentum space. We define the momentum space Green's function by

$$\langle \psi(x) \, \bar{\psi}(y) \rangle = \int \frac{d^4 p}{(2\pi)^4} e^{-ip(x-y)} iG(\not{p}). \tag{18.3}$$

This is possible since the left-hand side can only depend on $x - y$ by translation invariance.

At tree-level, $G(\not{p})$ is just the momentum space fermion propagator:

$$iG_0(\not{p}) \equiv \frac{i}{\not{p} - m}. \tag{18.4}$$

At 1-loop it gets a correction due to the self-energy graph:

$$iG_2(\not{p}) = \quad\quad\quad\quad\quad\quad\quad\quad\quad = iG_0(\not{p})[i\Sigma_2(\not{p})]\, iG_0(\not{p}), \tag{18.5}$$

where, in Feynman gauge,

$$i\Sigma_2(\not{p}) = (-ie)^2 \int \frac{d^4 k}{(2\pi)^4} \gamma^\mu \frac{i(\not{k} + m)}{k^2 - m^2 + i\varepsilon} \gamma^\mu \frac{-i}{(p-k)^2 + i\varepsilon}. \tag{18.6}$$

If this graph were contributing to an S-matrix element, rather than just a Green's function, we would remove the propagators from the external lines (the G_0 factors in Eq. (18.5)) and contract with external on-shell spinors. This $i\Sigma_2(\not{p})$ is what we would get from the normal Feynman rules without the external spinors.

Before evaluating this graph, we can observe an interesting feature that was not present in the photon case (the vacuum polarization graph). Including the self-energy graph, the *effective* electron propagator to 1-loop is

$$iG(\not{p}) = \quad\quad\quad\quad\quad\quad\quad = \quad\quad\quad\quad\quad + \quad\quad\quad\quad\quad + \cdots$$

$$= \frac{i}{\not{p} - m} + \frac{i}{\not{p} - m} i\Sigma_2(\not{p}) \frac{i}{\not{p} - m} + \mathcal{O}(e^4). \tag{18.7}$$

In an S-matrix element, this correction might appear on an external leg, such as

. In that case $G(\not{p})$ is contracted with an on-shell external spinor and the result multiplied by a factor of $\not{p} - m$ from the LSZ reduction formula. Now, there is no reason to expect that $\Sigma_2(m) = 0$ (and in fact it is not), so even after removing a single pole with $\not{p} - m$ we see from Eq. (18.7) that there will still be a pole left over. That is, the S-matrix will be singular. This problem did not come up for the photon propagator and vacuum polarization, where the corrected photon propagator had only a single pole to all orders in perturbation theory. The resolution of this apparently singular S-matrix for electron scattering is that the electron mass appearing in the LSZ formula does not necessarily have to match the electron mass appearing in the Lagrangian. In the photon case, they were equal, since both were zero. Once we evaluate the self-energy graph, we will then discuss how the electron mass is renormalized and why the S-matrix remains finite.

18.2.1 Self-energy loop graph

Evaluating the self-energy graph with Feynman parameters (see Appendix B) gives

$$
\begin{aligned}
i\Sigma_2(\not{p}) &= (-ie)^2 \int \frac{d^4k}{(2\pi)^4} \gamma^\mu \frac{i(\not{k}+m)}{k^2-m^2+i\varepsilon} \gamma^\mu \frac{-i}{(k-p)^2+i\varepsilon} \\
&= e^2 \int \frac{d^4k}{(2\pi)^4} \int_0^1 dx \frac{2\not{k}-4m}{\left[(k^2-m^2)(1-x)+(p-k)^2 x+i\varepsilon\right]^2}.
\end{aligned} \tag{18.8}
$$

Now we complete the square in the denominator and shift $k \to k + px$ to give

$$
i\Sigma_2(\not{p}) = 2e^2 \int_0^1 dx \int \frac{d^4k}{(2\pi)^4} \frac{x\not{p}-2m}{\left[k^2-\Delta+i\varepsilon\right]^2}, \tag{18.9}
$$

where $\Delta = (1-x)(m^2 - p^2 x)$ and we have dropped the term linear in k in the numerator since it is odd under $k \to -k$ and its integral therefore vanishes. This integrand scales as $\frac{d^4k}{k^4}$ and is therefore logarithmically divergent in the UV.

To regulate the UV divergence, we have to choose a regularization scheme. For pedagogical purposes we will evaluate this loop with both Pauli–Villars (PV) and dimensional regularization (DR). Recall (from Appendix B) that Pauli–Villars introduces heavy particles, of mass Λ with negative energy, for each physical particle in the theory. Pauli–Villars is nice because the scale Λ is clearly a UV deformation, with the Pauli–Villars ghosts having no effect on the low-energy theory as $\Lambda \to \infty$. In dimensional regularization, which analytically continues to $4 - \varepsilon$ dimensions, it is not clear that ε is a UV deformation in any sense. Dimensional regularization is much easier to use for more complicated theories than QED, so eventually we will use it exclusively. For now, it is helpful to use two regulators to see that results are regulator independent.

With a Pauli–Villars photon, the self-energy graph becomes

$$
\Sigma_2(\not{p}) = -2ie^2 \int_0^1 dx \, (x\not{p}-2m) \int \frac{d^4k}{(2\pi)^4} \left[\frac{1}{(k^2-\Delta)^2} - \frac{1}{(k^2-\Delta')^2}\right], \tag{18.10}
$$

with $\Delta' = (1-x)(m^2-p^2 x)+x\Lambda^2$. Since we take $\Lambda \to \infty$, we can more simply take $\Delta' = x\Lambda^2$. The regulated integral is now convergent and can be evaluated using formulas from Appendix B. The result is

$$
\begin{aligned}
\Sigma_2(\not{p}) &= -\frac{\alpha}{2\pi} \int_0^1 dx \, (2m - x\not{p}) \ln \frac{x\Lambda^2}{(1-x)(m^2-p^2 x)} \\
&= -\frac{\alpha}{\pi} \left(m \ln \Lambda^2 - \frac{1}{4}\not{p} \ln \Lambda^2 + \text{finite}\right) \quad \text{(PV)}.
\end{aligned} \tag{18.11}
$$

In dimensional regularization, in $d = 4 - \varepsilon$ dimensions, the loop is

$$\Sigma_2(\not{p}) = -ie^2\mu^{4-d}\int_0^1 dx\left[(d-2)x\not{p} - d\,m\right]\int\frac{d^dk}{(2\pi)^d}\frac{1}{(k^2 - \Delta + i\varepsilon)^2}$$

$$= \frac{\alpha}{4\pi}\int_0^1 dx\left[(2-\varepsilon)x\not{p} - (4-\varepsilon)m\right]\left[\frac{2}{\varepsilon} + \ln\frac{\tilde{\mu}^2}{(1-x)\,(m^2 - p^2x)}\right] \quad \text{(DR)},$$

$$(18.12)$$

where $\tilde{\mu}^2 \equiv 4\pi e^{-\gamma_E}\mu^2$. Extracting the divergent parts, the loop can be written as

$$\Sigma_2(\not{p}) = \frac{\alpha}{\pi}\left(\frac{\not{p} - 4m}{2\varepsilon} + \text{finite}\right). \quad (18.13)$$

Note that in both cases $\Sigma_2(m) \neq 0$, so there will be a double-pole in the 2-point function at 1-loop with the possibly dangerous consequences discussed below Eq. (18.7). Also note that both results have divergences proportional to both m and $\not{p}$. This implies that we need two quantities to renormalize, to remove both divergences.

18.2.2 Renormalization

As discussed in the introduction, we want the Green's function $G(\not{p})$ defined in Eq. (18.3) to be finite. Thus, the infinities from the $\mathcal{O}(e^2)$ contribution to this Green's function must be removed through renormalization.

As with the vacuum polarization, we need to figure out what parameters in the theory can be renormalized to cancel the infinities in the self-energy graph. To begin, let us write the Lagrangian as

$$\mathcal{L} = -\frac{1}{4}F_{\mu\nu}^2 + i\bar{\psi}\,\not{\partial}\psi - m_0\bar{\psi}\psi - e_0\bar{\psi}\not{A}\psi. \quad (18.14)$$

In the study of vacuum polarization in Chapter 16, we concluded that the charge in the Lagrangian, now written as e_0, called the **bare charge**, could be used to absorb an infinity. Recall that we defined a renormalized electric charge via

$$e_0^2 = e_R^2 + e_R^4\Pi_2(p_0^2) + \cdots = e_R^2\left(1 - \frac{e_R^2}{12\pi^2}\ln\frac{\Lambda^2}{-p_0^2} + \cdots\right), \quad (18.15)$$

where $\Pi_2(p_0^2)$ is formally infinite. Since e_0 has already been renormalized by vacuum polarization, we cannot renormalize it in a different way for the self-energy graph.

To make $G(\not{p})$ finite the obvious Lagrangian parameter that might absorb the infinity is the **bare electron mass**, m_0. Indeed, from Eq. (18.7),

$$iG_2(\not{p}) = \frac{i}{\not{p} - m_0} + \frac{i}{\not{p} - m_0}[i\Sigma_2(\not{p})]\frac{i}{\not{p} - m_0}, \quad (18.16)$$

we can see that an (infinite) redefinition of $m_0 = m + \Delta m$ with Δm of order e^2 could compensate for an infinity at order e^2 in $\Sigma_2(\not{p})$. Unfortunately, we saw in Eqs. (18.11) and (18.13) that $\Sigma_2(\not{p})$ has two types of infinities, one independent of $\not{p}$ and the other proportional to $\not{p}$. The mass renormalization can only remove one of these infinities. Thus, to progress further we need something else to renormalize. But what could it be? Our

Lagrangian only had two parameters, m and e, and we have already defined how e is renormalized.

In fact, there is another parameter: the normalization of the fermion wavefunction. Let us write the fermion field in terms of creation and annihilation operators that we have been using all along as the **bare free field**:

$$\psi^0(x) = \sum_s \int \frac{d^3p}{(2\pi)^3} \frac{1}{\sqrt{2\omega_p}} \left(a_p^s u_p^s e^{-ipx} + b_p^{s\dagger} v_p^s e^{ipx} \right). \qquad (18.17)$$

The bare free field is canonically normalized to give all the tree-level scattering results we have already calculated. We then define the renormalized field as

$$\psi^R(x) = \frac{1}{\sqrt{Z_2}} \sum_s \int \frac{d^3p}{(2\pi)^3} \frac{1}{\sqrt{2\omega_p}} \left(a_p^s u_p^s e^{-ipx} + b_p^{s\dagger} v_p^s e^{ipx} \right) \equiv \frac{1}{\sqrt{Z_2}} \psi^0 \qquad (18.18)$$

for some (formally infinite) number Z_2. This is the origin of the term *renormalization*. We index bare (infinite) fields and parameters with a 0 and physical finite renormalized fields and parameters with an R.

For the tree-level theory, $Z_2 = 1$ is required to be consistent with the normalization used in all our scattering formulas. So it is natural to account for radiative corrections by writing

$$Z_2 = 1 + \delta_2, \qquad (18.19)$$

where δ_2 is the **counterterm**, which has a formal Taylor series expansion in e starting at order e^2. We also write

$$m_0 = Z_m m_R \qquad (18.20)$$

and expand $Z_m = 1 + \delta_m$, with δ_m the mass counterterm.[1] Then

$$m_0 = m_R + m_R \delta_m. \qquad (18.21)$$

As we will see, particularly when we cover renormalized perturbation theory in Chapter 19, using counterterms rather than bare and renormalized quantities directly will be extremely efficient.

All the calculations we have done so far have been with fields with the conventional (bare) normalization. However, it is the Green's function of *renormalized* fields that should have finite physical values. So we define

$$\left\langle \psi^0(x)\, \bar\psi^0(y) \right\rangle = i \int \frac{d^4p}{(2\pi)^4} e^{-ip(x-y)} G^{\text{bare}}(\slashed{p}) \qquad (18.22)$$

and

$$\left\langle \psi^R(x)\, \bar\psi^R(y) \right\rangle = i \int \frac{d^4p}{(2\pi)^4} e^{-ip(x-y)} G^R(\slashed{p}) \qquad (18.23)$$

and expect $G^R(\slashed{p})$ to be finite. By definition,

$$G^R(\slashed{p}) = \frac{1}{Z_2} G^{\text{bare}}(\slashed{p}). \qquad (18.24)$$

---

[1] Another common convention is $Z_2 m_0 \equiv m_R + \delta_m$. Our convention is more commonly used in modern field theory calculations.

Now, since Z_2 is just a number, the tree-level propagator for the renormalized fields can be expressed in terms of the propagator of the bare fields as

$$iG^R(\slashed{p}) = \frac{1}{Z_2}\frac{i}{\slashed{p}-m_0} + \text{loops}$$

$$= \left(\frac{1}{1+\delta_2}\right)\left(\frac{i}{\slashed{p}-m_R-\delta_m m_R}\right) + \text{loops}$$

$$= \frac{i}{\slashed{p}-m_R} + \frac{i}{\slashed{p}-m_R}\left[i(\delta_2\slashed{p}-(\delta_2+\delta_m)m_R)\right]\frac{i}{\slashed{p}-m_R} + \text{loops} + \mathcal{O}(e^4). \tag{18.25}$$

Adding the 1-loop contribution, as in Eqs. (18.7) or (18.16), gives

$$iG^R(\slashed{p}) = \frac{i}{\slashed{p}-m_R} + \frac{i}{\slashed{p}-m_R}\left[i(\delta_2\slashed{p}-(\delta_2+\delta_m)m_R+\Sigma_2(\slashed{p}))\right]\frac{i}{\slashed{p}-m_R} + \mathcal{O}(e^4). \tag{18.26}$$

So now we can choose δ_2 and δ_m to remove all the infinities in the electron self-energy.

To be explicit, from Eq. (18.11) we see that choosing

$$\delta_2 = -\frac{\alpha}{4\pi}\ln\Lambda^2, \quad \delta_m = -\frac{3\alpha}{4\pi}\ln\Lambda^2 \quad \text{(PV)} \tag{18.27}$$

for Pauli–Villars or

$$\delta_2 = -\frac{\alpha}{4\pi}\frac{2}{\varepsilon}, \quad \delta_m = -\frac{3\alpha}{4\pi}\frac{2}{\varepsilon} \quad \text{(DR)} \tag{18.28}$$

for dimensional regularization will remove the infinities. With these choices, we will get finite answers for the 2-point function $G^R(\slashed{p})$ at any scale p.

We can choose different values for the counterterms which differ from these by finite numbers and $G^R(\slashed{p})$ will still be finite. Any prescription for choosing the finite parts of the counterterms is known as a **subtraction scheme**. Not only must observables in a renormalized theory be finite, but they also must be independent of the subtraction scheme, as we will see. Nevertheless, there are some smart choices for subtraction schemes and some not-so-smart choices.

The two subtraction schemes most often used in quantum field theory are the *on-shell* subtraction scheme and the *minimal subtraction* (MS) scheme. Minimal subtraction is by far the simplest scheme and the one used in almost all modern quantum field theory calculations. In minimal subtraction the counterterms are defined to have no finite parts at all, so that δ_2 and δ_m are given by Eqs. (18.27) and (18.28). More commonly, a slightly modified version of this prescription known as modified minimal subtraction $\overline{\text{MS}}$ is used, in which $\ln(4\pi)$ and γ_E finite parts in dimensionally regulated results are also subtracted off. $\overline{\text{MS}}$ just turns $\tilde{\mu}$ back into μ in dimensionally regularized amplitudes.

In on-shell subtraction, the renormalized mass m_R appearing in Green's functions is identified with the observed electron mass m_P which can be defined to all orders as the position of the pole in the S-matrix.[2] To see how this identification works in practice, it is helpful to look at the possible form of the higher-order corrections.

---

[2] Actually, there is no isolated pole in the S-matrix associated with the electron. Rather, the electron mass is the beginning of a cut in the complex plane. This will be discussed more in Chapter 24.

18.3 Pole mass

So far, we have only included one particular self-energy correction. The 2-point function $G(\not p)$ in fact gets corrections from an infinite number of graphs. One particular series of corrections, of the form

$$iG^{\text{bare}}(\not p) = \underline{\qquad} + \underline{\quad\curlywedge\quad} + \underline{\quad\curlywedge\quad\curlywedge\quad} + \cdots, \quad (18.29)$$

just produces a geometric series

$$iG^{\text{bare}}(\not p) = \frac{i}{\not p - m_0} + \frac{i}{\not p - m_0}\left(i\Sigma_2(\not p)\right)\frac{i}{\not p - m_0}$$
$$+ \frac{i}{\not p - m_0}\left(i\Sigma_2(\not p)\right)\frac{i}{\not p - m_0}\left(i\Sigma_2(\not p)\right)\frac{i}{\not p - m_0} + \cdots, \quad (18.30)$$

which is easy to sum. More generally, any possible graph contributing to this Green's function is part of *some* geometric series. Conversely, the entire Green's function can be written as the sum of a single geometric series constructed by sewing together graphs that cannot be cut in two by slicing a single propagator. We call such graphs **one-particle irreducible (1PI)**. For example,

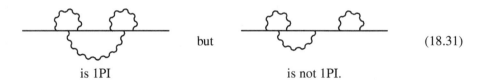

is 1PI   but   is not 1PI. $\quad (18.31)$

Thus,

$$iG(\not p) = \underline{\qquad} + \underline{\quad}\!\boxed{1\text{PI}}\!\underline{\quad} + \underline{\quad}\!\boxed{1\text{PI}}\!\boxed{1\text{PI}}\!\underline{\quad} + \cdots, \quad (18.32)$$

Defining $i\Sigma(\not p)$ as the sum of all of the 1PI graphs, we find

$$iG(\not p) = \frac{i}{\not p - m} + \frac{i}{\not p - m}\left(i\Sigma(\not p)\right)\frac{i}{\not p - m} + \frac{i}{\not p - m}\left(i\Sigma(\not p)\right)\frac{i}{\not p - m}\left(i\Sigma(\not p)\right)\frac{i}{\not p - m} + \cdots$$
$$= \frac{i}{\not p - m}\left[1 + \frac{-\Sigma(\not p)}{\not p - m} + \left(\frac{-\Sigma(\not p)}{\not p - m}\right)^2 + \cdots\right]$$
$$= \frac{i}{\not p - m}\frac{1}{1 + \frac{\Sigma(\not p)}{\not p - m}}$$
$$= \frac{i}{\not p - m + \Sigma(\not p)}. \quad (18.33)$$

This is just a general expression for a sum of Feynman diagrams, applying either $m = m_0$ or $m = m_R$. For the bare Green's function, there was just a single 1PI diagram at order e^2 and so $\Sigma(\not{p}) = \Sigma_2(\not{p}) + \mathcal{O}(e^4)$. Then we have

$$iG^{\text{bare}}(\not{p}) = \frac{i}{\not{p} - m_0 + \Sigma_2(\not{p}) + \cdots}. \tag{18.34}$$

This expression is the sum of the series in Eq. (18.29).

From the bare Green's function we can compute the renormalized Green's function as

$$
\begin{aligned}
iG^R(\not{p}) &= \frac{1}{1 + \delta_2} iG^{\text{bare}}(\not{p}) \\
&= \left(\frac{1}{1 + \delta_2}\right) \frac{i}{\not{p} - m_0 + \Sigma_2(\not{p}) + \cdots} \\
&= \frac{i}{\not{p} - m_0 + \delta_2\not{p} - m_0\delta_2 + \Sigma_2(\not{p}) + \cdots},
\end{aligned}
\tag{18.35}
$$

where the $\cdots$ are formally $\mathcal{O}(e^4)$ or higher. Then, using Eq. (18.21), $m_0 = m_R + m_R\delta_m$, this becomes

$$iG^R(\not{p}) = \frac{i}{\not{p} - m_R + \delta_2\not{p} - (\delta_2 + \delta_m)m_R + \Sigma_2(\not{p}) + \cdots}. \tag{18.36}$$

We will write this more conveniently as

$$iG^R(\not{p}) = \frac{i}{\not{p} - m_R + \Sigma_R(\not{p})}, \tag{18.37}$$

with $\Sigma_R(\not{p}) = \Sigma_2(\not{p}) + \delta_2\not{p} - (\delta_m + \delta_2)m_R + \mathcal{O}(e^4)$.

You may have noted that this result would follow easily from Eq. (18.26) if we could treat the counterterms as contributions to 1P1 graphs. To justify such treatment, all we have to do is rewrite the bare free Lagrangian in terms of renormalized fields:

$$\mathcal{L} = i\bar{\psi}^0\not{\partial}\psi^0 - m_0\bar{\psi}^0\psi^0 = iZ_2\bar{\psi}^R\not{\partial}\psi^R - Z_2 Z_m m_R\bar{\psi}^R\psi^R. \tag{18.38}$$

Using Eqs. (18.19) and (18.20) this becomes

$$\mathcal{L} = i\bar{\psi}^R\not{\partial}\psi^R - m^R\bar{\psi}^R\psi^R + i\delta_2\bar{\psi}^R\not{\partial}\psi^R - m_R(\delta_2 + \delta_m)\bar{\psi}^R\psi^R. \tag{18.39}$$

Thus, we can treat the counterterms, which start at order e^2, as interactions whose Feynman rules give contributions $\delta_2\not{p}$ and $-(\delta_2 + \delta_m)m_R$ to the 1PI graphs. Then Eq. (18.37) follows from the general form Eq. (18.33) with $m = m_R$ and $\Sigma = \Sigma_R$. Expanding the Lagrangian in terms of renormalized quantities leads to so-called **renormalized perturbation theory**. Renormalized perturbation theory will be discussed more completely, including interactions and the photon field, in the next chapter.

18.3.1 On-shell subtraction

Having summed all of the 1PI diagrams into the renormalized propagator, we can now identify the physical electron mass m_P as the location of its pole. More precisely, the renormalized propagator should have a single pole at $\not{p} = m_P$ with residue i. The location of the pole is a *definition* of mass, known as the **pole mass**. It is important to keep in mind that the pole mass is physical and independent of any subtraction scheme used to set the finite parts of the counterterms. In the on-shell subtraction scheme, the finite parts of the counterterms are chosen so that $m_R = m_P$. In minimal subtraction, $m_R \neq m_P$. In either case the 2-point Green's function still has a pole at m_P.

From Eq. (18.37), for $G^R(\not{p})$ to have a pole at $\not{p} = m_P$ the 1PI graphs must satisfy

$$\Sigma_R(m_P) = m_R - m_P. \tag{18.40}$$

This condition defines the pole mass, independent of the subtraction scheme.

In the **on-shell subtraction scheme**, the renormalized mass m_R is set equal to the pole mass m_P and the residue set to i. Having residue i implies

$$i = \lim_{\not{p} \to m_P} (\not{p} - m_P) \frac{i}{\not{p} - m_R + \Sigma_R(\not{p})} = \lim_{\not{p} \to m_P} \frac{i}{1 + \frac{d}{d\not{p}} \Sigma_R(\not{p})}, \tag{18.41}$$

where we have used L'Hôpital's rule. This implies

$$\left. \frac{d}{d\not{p}} \Sigma_R(\not{p}) \right|_{\not{p} = m_P} = 0. \tag{18.42}$$

Recalling $\Sigma_R(\not{p}) = \Sigma_2(\not{p}) + \delta_2 \not{p} - (\delta_m + \delta_2) m_R + \cdots$, these two on-shell conditions imply to order e^2

$$\delta_2 = - \left. \frac{d}{d\not{p}} \Sigma_2(\not{p}) \right|_{\not{p} = m_P} \tag{18.43}$$

and

$$\delta_m m_P = \Sigma_2(m_P), \tag{18.44}$$

which we can now evaluate in our different regulators.

With Pauli–Villars, Eq. (18.44) implies

$$\Sigma_2(m_P) = -\frac{\alpha}{2\pi} m_P \left(\frac{3}{2} \ln \frac{\Lambda^2}{m_P^2} + \frac{3}{4} \right) \qquad \text{(PV)}, \tag{18.45}$$

which is one of our conditions. Unfortunately, when we try to evaluate the derivative, we find

$$\left. \frac{d}{d\not{p}} \Sigma_2(\not{p}) \right|_{\not{p} = m_P} = \frac{\alpha}{2\pi} \left(\frac{1}{2} \ln \frac{\Lambda^2}{m_P^2} + \frac{5}{4} - \int_0^1 dx \frac{2x(2-x)}{1-x} \right) \qquad \text{(PV)}. \tag{18.46}$$

This last integral is divergent. This divergence is an **infrared divergence**, due to the integration region near $k^2 = 0$. In this case, the divergence does not come from the loop integral itself, but from our choice of subtraction scheme, which involved $\Sigma'(m_P)$. Nevertheless, IR divergences in renormalized Green's functions and S-matrix elements are unavoidable. We will see how they drop out of physical observables in Chapter 20.

For now, a quick way to sequester the IR divergence is to pretend that the photon has a tiny mass, m_γ. As with UV divergences, IR divergences will cancel in physical processes, so we will eventually be able to take $m_\gamma \to 0$. If you are skeptical about how this could happen, recall that in the vacuum polarization calculation at momentum transfers $-p^2 \gg m^2$, the corrections to the Coulomb potential were independent of m. In fact, the vacuum polarization graph would be IR divergent if we set $m = 0$ before evaluating the loop. Thus, at very short distances, the electron mass acts only as a regulator, just as m_γ does here.

The effect of a photon mass is to change Δ to $\Delta = (1-x)(m_P^2 - p^2 x) + x m_\gamma^2$, so that

$$\Sigma_2(\slashed{p}) = \frac{\alpha}{2\pi} \int_0^1 dx (x\slashed{p} - 2m_P) \ln \frac{x\Lambda^2}{(1-x)(m_P^2 - p^2 x) + x m_\gamma^2} \qquad \text{(PV)}. \qquad (18.47)$$

Then, keeping only the leading terms in m_γ,

$$\delta_2 = -\Sigma_2'(m_P) = \frac{\alpha}{2\pi} \left(-\frac{1}{2} \ln \frac{\Lambda^2}{m_P^2} - \frac{9}{4} - \ln \frac{m_\gamma^2}{m_P^2} \right) \qquad \text{(PV)}, \qquad (18.48)$$

which is now finite. Then,

$$\delta_m = \frac{1}{m_P} \Sigma_2(m_P) = \frac{\alpha}{2\pi} \left(-\frac{3}{2} \ln \frac{\Lambda^2}{m_P^2} - \frac{3}{4} \right) \qquad \text{(PV)}. \qquad (18.49)$$

In dimensional regularization, with the photon mass added, the loop gives

$$\Sigma_2(\slashed{p}) = \frac{\alpha}{4\pi} \int_0^1 dx \left[(2-\varepsilon)x\slashed{p} - (4-\varepsilon)m_P \right] \left[\frac{2}{\varepsilon} + \ln \frac{\tilde{\mu}^2}{(1-x)(m_P^2 - p^2 x) + x m_\gamma^2} \right] \qquad \text{(DR)}, \qquad (18.50)$$

leading to

$$\delta_2 = -\Sigma_2'(m_P) = -\frac{\alpha}{2\pi} \left(\frac{1}{\varepsilon} + \frac{1}{2} \ln \frac{\tilde{\mu}^2}{m_P^2} + 2 + \ln \frac{m_\gamma^2}{m_P^2} \right) \qquad \text{(DR)} \qquad (18.51)$$

and

$$\delta_m = \frac{1}{m_P} \Sigma_2(m_P) = \frac{\alpha}{2\pi} \left(-\frac{3}{\varepsilon} - \frac{3}{2} \ln \frac{\tilde{\mu}^2}{m_P^2} - 2 \right) \qquad \text{(DR)}. \qquad (18.52)$$

18.3.2 Amputation

Recall that the LSZ theorem converts Green's functions to S-matrix elements by adding external polarizations and factors of $\slashed{p} - m_0$ to project onto physical one-particle states. However, we have now seen that the location of the pole in the electron propagator is not the value of the mass m_0 appearing in the Lagrangian, but rather at some other location m_P. Moreover, we have found that only Green's functions of renormalized fields, such as $G^R \sim \langle \bar{\psi}^R \psi^R \rangle$, should be finite. Thus, it would be natural to modify the LSZ theorem to

$$\langle f|S|i\rangle \sim (p\!\!\!/_f - m_P) \cdots (p\!\!\!/_i - m_P)\langle \psi^R \cdots \psi^R\rangle. \tag{18.53}$$

This is almost correct.

The subtlety is that in the derivation of LSZ we had to assume that all the interactions happened during some finite time interval, and that as $t \to \pm\infty$ we could treat the theory as free. In the free theory, the pole would be at m_0. Thus, we really want the theory not to be entirely free at asymptotic times, but to include all of the corrections that move the pole from m_0 to m_P. Those corrections are precisely the series of 1PI insertions onto the electron propagator. Thus, in projecting onto the pole mass, with the $(p\!\!\!/ - m_P)$ factors, we must assume that all of the corrections to the on-shell external electron propagator have been included. For example, diagrams such as ⎯⎯●⎯⎯⎯ would only contribute to correcting the external electron propagator, which would then be removed by LSZ.

Thus, the LSZ theorem in a renormalized theory is

$$\langle f|S|i\rangle = (p\!\!\!/_f - m_P) \cdots (p\!\!\!/_i - m_P)\langle \psi^R \cdots \psi^R\rangle_{\text{amputated}}, \tag{18.54}$$

where **amputated** means the external lines are chopped off until they begin interacting with the other fields. Only amputated diagrams contribute to S-matrix elements.

Note that amputating diagrams does not mean that self-energy graphs are never important. When a self-energy bubble occurs on an internal line, as in ⟩⎯⎯⟨ , which provides a radiative correction to Compton scattering, it will have an important physical effect. All the renormalized LSZ theorem says is that you should not correct external lines for S-matrix elements since those corrections are already accounted for in the updated definition of asymptotic states.

18.4 Minimal subtraction

In minimal subtraction, the counterterms are fixed with no reference to the pole mass. The prescription is simply that the counterterms should have no finite parts. Thus, with Pauli–Villars, we get Eq. (18.27):

$$\delta_2 = -\frac{\alpha}{4\pi}\ln\Lambda^2, \quad \delta_m = -\frac{3\alpha}{4\pi}\ln\Lambda^2 \quad (\text{PV}), \tag{18.55}$$

and then $\Sigma_R(p\!\!\!/) = \Sigma_2(p\!\!\!/) + \delta_2 p\!\!\!/ - (\delta_m + \delta_2)m_R$ is

$$\Sigma_R(p\!\!\!/) = \frac{\alpha}{2\pi}\int_0^1 dx \,(x p\!\!\!/ - 2m_R)\ln\frac{x}{(1-x)(m_R^2 - p^2 x)}, \tag{18.56}$$

which is finite, but has nonsensical dimensions. Instead, we can modify the minimal subtraction for use with Pauli–Villars so that

$$\delta_2 = -\frac{\alpha}{4\pi}\ln\frac{\Lambda^2}{\mu^2}, \quad \delta_m = -\frac{3\alpha}{4\pi}\ln\frac{\Lambda^2}{\mu^2} \quad (\text{PV}), \tag{18.57}$$

with μ some arbitrary scale with dimensions of mass. μ should be thought of as a low-energy scale, say 1 GeV, which is not taken to infinity. Then,

$$\Sigma_R(\not{p}) = \frac{\alpha}{2\pi}\int_0^1 dx\,(x\not{p}-2m_R)\ln\frac{x\mu^2}{(1-x)(m_R^2-p^2x)}. \tag{18.58}$$

By introducing μ we have established a one-parameter family of subtraction schemes. Any physical observable must be independent of μ, but μ is *not* taken to infinity. μ is sometimes called the **subtraction point**.

The subtraction point already appeared in Chapter 16 on vacuum polarization, where it was set equal to the long-distance scale where the renormalized electric charge, e_R, was defined. As in that case, when one compares observables, such as combinations of the Coulomb potential $r_1 V(r_1) - r_2 V(r_2)$ measured at different scales, the subtraction point will drop out.

The subtraction point also appears as the parameter μ in dimensional regularization. Recall that in dimensional regularization μ is introduced by the rescaling $e^2 \to \mu^{4-d}e^2$, which lets the electric charge remain dimensionless in d dimensions. In dimensional regularization, minimal subtraction gives Eq. (18.28):

$$\delta_2 = -\frac{\alpha}{4\pi}\frac{2}{\varepsilon}, \quad \delta_m = -\frac{3\alpha}{4\pi}\frac{2}{\varepsilon} \quad (\text{DR, MS}). \tag{18.59}$$

In dimensional regularization, minimal subtraction is almost always upgraded to modified minimal subtraction ($\overline{\text{MS}}$), where the $\ln(4\pi)$ and γ_E factors are also removed. Expanding $\tilde{\mu}^2$ in Eq. (18.12):

$$\begin{aligned}
\Sigma_2(\not{p}) &= \frac{\alpha}{4\pi}\int_0^1 dx\left[(2-\varepsilon)x\not{p} - (4-\varepsilon)m_R\right]\left[\frac{2}{\varepsilon} + \ln\frac{4\pi e^{-\gamma_E}\mu^2}{(1-x)(m_R^2-p^2x)}\right] \\
&= \frac{\alpha}{2\pi}\left[\frac{1}{2}\not{p}\left(\frac{2}{\varepsilon}+\ln\!\left(4\pi e^{-\gamma_E}\right)\right) - 2m_R\left(\frac{2}{\varepsilon}+\ln\!\left(4\pi e^{-\gamma_E}\right)\right) + \text{finite}\right] \quad (\text{DR, MS}).
\end{aligned} \tag{18.60}$$

So in $\overline{\text{MS}}$,

$$\delta_2 = -\frac{\alpha}{4\pi}\left(\frac{2}{\varepsilon}+\ln\!\left(4\pi e^{-\gamma_E}\right)\right), \quad \delta_m = -\frac{3\alpha}{4\pi}\left(\frac{2}{\varepsilon}+\ln\!\left(4\pi e^{-\gamma_E}\right)\right) \quad (\text{DR } \overline{\text{MS}}), \tag{18.61}$$

and then

$$\Sigma_R(\not{p}) = \frac{\alpha}{2\pi}\int_0^1 dx\left\{(x\not{p}-2m_R)\left[\ln\frac{\mu^2}{(1-x)\,(m_R^2-p^2x)}\right] - x\not{p} + m_R\right\} \quad (\text{DR } \overline{\text{MS}}), \tag{18.62}$$

which is UV finite and has μ in it, not $\tilde{\mu}$. As with Pauli–Villars, there is a one-parameter family of renormalized 1PI corrections. In both cases, the subtraction point μ is an arbitrary scale which is *not* taken to infinity but will drop out of physical calculations. The $\ln(4\pi e^{-\gamma_E})$ terms in the counterterms are almost always left implicit in $\overline{\text{MS}}$, and μ and $\tilde{\mu}$ used interchangeably.

The value of m_R is finite in $\overline{\text{MS}}$ and known as the $\overline{\text{MS}}$ **mass**. The renormalized electron propagator will in general not have a pole at $\not{p}=m_R$. There is still a pole at $\not{p}=m_P$, but

$m_P \neq m_R$. Recalling the renormalized electron propagator from Eq. (18.37),

$$iG^R(\not{p}) = \frac{i}{\not{p} - m_R + \Sigma_R(\not{p})}, \tag{18.63}$$

we can now easily relate the pole mass and the $\overline{\text{MS}}$ mass. Requiring a pole in this propagator at $\not{p} = m_P$ gives

$$m_P - m_R + \Sigma_R(m_P) = 0. \tag{18.64}$$

Using $m_P = m_R$ at leading order, we then have

$$m_R = m_P + \Sigma_R(m_P) = m_P\left[1 - \frac{\alpha}{4\pi}\left(4 + 3\ln\frac{\mu^2}{m_P^2}\right) + \mathcal{O}(\alpha^2)\right] \quad (\text{DR}). \tag{18.65}$$

In particular, the $\overline{\text{MS}}$ mass depends on the arbitrary scale μ.

While your first instinct might be that this extra parameter μ in minimal subtraction adds an unnecessary complication, it is actually extremely useful. The fact that physical observables are independent of μ gives a powerful constraint. Indeed, demanding $\frac{d}{d\mu}\mathcal{O} = 0$, where $\mathcal{O}$ is some observable, is the *renormalization group equation*, to be discussed in Chapter 23.

18.5 Summary and discussion

In this chapter we saw that the electron self-energy graph contributes loop corrections to the electron propagator. This loop was divergent, but the divergence could be removed by renormalizing the electron's quantum field, $\psi^0 = \sqrt{Z_2}\psi^R$, and redefining the electron mass, $m_0 = Z_m m_R$. In these equations, ψ^0 and m^0 refer to bare quantities that are formally infinite, while ψ^R and m_R are finite renormalized quantities. The quantities δ_m and δ_2 defined by expanding the renormalization factors around the classical values, e.g. $Z_2 = 1 + \delta_2$, are known as *counterterms*. These counterterms can be chosen to cancel the infinite contribution of the electron self-energy graph to the renormalized electron propagator. While the cancellation fixes the infinite parts of the counterterms, the finite parts are arbitrary. Conventions for fixing the finite parts are known as *subtraction schemes*.

We saw that the general geometric series of loops correcting the propagator can be summed to all orders in α, leading to a renormalized propagator of the form

$$iG^R(\not{p}) = \frac{i}{\not{p} - m_R + \Sigma_R(\not{p})}. \tag{18.66}$$

Here, $\Sigma_R(\not{p})$ represents *one-particle irreducible Feynman diagrams* plus counterterm contributions. Up to order e^2, we found $\Sigma_R(\not{p}) = \Sigma_2(\not{p}) + \delta_2\not{p} - (\delta_m + \delta_2)m_R$. This renormalized propagator should have a pole at the physical electron mass, the *pole mass*, with residue i:

$$iG^R(\not{p}) = \frac{i}{\not{p} - m_P} + \text{terms regular at } \not{p} = m_P. \tag{18.67}$$

In terms of the bare propagator, $G^{\text{bare}}(\not{p}) = Z_2 G^R(\not{p})$, we can write

$$iG^{\text{bare}}(\not{p}) = \frac{iZ_2}{\not{p} - m_P} + \text{terms regular at } \not{p} = m_P. \qquad (18.68)$$

Sometimes this is used to interpret Z_2 as the residue of the pole. However, since both Z_2 and the bare propagator are formally infinite, this interpretation must be made with care.

Two subtraction schemes were discussed. The first, the *on-shell scheme*, was defined by equating the location of the pole of the propagator, m_P, with the renormalized mass, $m_R \equiv m_P$. This, along with a constraint on the residue of the pole, generated two equations:

$$\Sigma_R(m_P) = 0, \quad \frac{d}{d\not{p}}\Sigma_R(\not{p})\bigg|_{\not{p}=m_P} = 0. \qquad (18.69)$$

These equations, which apply to all orders in perturbation theory, fix the counterterms δ_2 and δ_m. They are known as the **on-shell renormalization conditions**.

The second scheme, known as *minimal subtraction*, simply sets the finite parts of δ_2 and δ_m to zero. *Modified minimal subtraction* also subtracts off $\ln(4\pi)$ and γ_E factors, which effectively replaces $\tilde{\mu}$ by μ in dimensionally regulated amplitudes. In minimal subtraction, the renormalized mass (written as m_R or often just m) is known as the $\overline{MS}$ mass. It is in general different from the pole mass. At 1-loop, we found

$$m_R = m_P + \Sigma_R(m_P) = m_P\left[1 - \frac{\alpha}{4\pi}\left(4 + 3\ln\frac{\mu^2}{m_P^2}\right)\right]. \qquad (18.70)$$

This expression depends on an arbitrary scale μ known as the *subtraction point*, which is not taken to ∞. While the extra parameter μ may seem superfluous, we will see in Chapter 23 that physical observables being independent of μ leads to an important constraint, the renormalization group equations. Even without using the renormalization group, μ independence order-by-order in perturbation theory gives an important check that an observable has been calculated correctly. We will provide a number of examples in the next two chapters.

You might wonder why on earth anyone would use an unphysical and arbitrary $\overline{MS}$ mass rather than the physical pole mass. The basic answer is that $\overline{MS}$ is a much simpler subtraction scheme than the on-shell scheme. It is often easier to compute loops in $\overline{MS}$ and then convert the masses back to the pole mass at the end rather than to do the computations in terms of the pole mass from the beginning. Numerically, the differences between pole masses and $\overline{MS}$ are often quite small for μ chosen of order m_P. One important exception is the top-quark mass, where $m_P \sim 175$ GeV but $m_R \sim 163$ GeV. This 5% difference is important for precision physics, to be discussed in Chapter 31. A more sophisticated answer is that the $\overline{MS}$ mass has an appealing property that it is free of ambiguities related to non-perturbative effects in quantum chromodynamics (so-called *renormalon ambiguities*). Indeed, for particles such as quarks, which can never be seen as asymptotic states, there is not actually a pole in the S-matrix, so the pole mass is not always a useful mass definition.

It is important to keep in mind that the physical electron mass, m_P, is the location of the pole in the electron propagator whether or not we identified this mass with m_R. In the on-shell scheme, we cannot ask about radiative corrections to the electron mass m_P

$$\langle \psi^R(x)\bar{\psi}^R(y) \rangle = \int \frac{d^4p}{(2\pi)^4} e^{-ip(x-y)} \left(\frac{i}{\not{p} - m_P} + \textbf{regular} \right), \qquad (19.2)$$

where the "regular" part is non-singular as $\not{p} \to m_P$. Here m_P is the pole mass, a finite, non-perturbative definition of the electron mass.

The renormalized fields are conventionally related to the bare fields appearing in the QED Lagrangian in Eq. (19.1) by **field strength renormalizations** Z_2 and Z_3 as

$$\psi^0 = \sqrt{Z_2}\psi^R, \quad A_\mu^0 = \sqrt{Z_3}A_\mu^R. \qquad (19.3)$$

The (infinite) bare mass m_0 is related to a (finite) renormalized mass m_R by a *mass renormalization* Z_m:

$$m_0 = Z_m m_R. \qquad (19.4)$$

The (infinite) bare electric charge e_0 is related to a (finite) renormalized electric charge e_R by a *charge renormalization* Z_e:

$$e_0 = Z_e e_R. \qquad (19.5)$$

In Chapter 16, the renormalized electric charge was defined so that the Coulomb potential was $V(r) = \frac{e_R^2}{4\pi r}$ at very large r; in Chapter 18 the renormalized electron mass was defined as the location of the pole in the exact electron 2-point function. For now, we do not need to know how e_R and m_R are defined, just that they can be taken finite.

After rescaling the fields in this, the QED Lagrangian becomes

$$\mathcal{L} = -\frac{1}{4}Z_3\big(\partial_\mu A_\nu^R - \partial_\nu A_\mu^R\big)^2$$
$$+ iZ_2\bar{\psi}_R\not{\partial}\psi_R - Z_2 Z_m m_R\overline{\psi}_R\psi_R - e_R Z_e Z_2\sqrt{Z_3}\bar{\psi}_R\not{A}_R\psi_R + \rho_0. \qquad (19.6)$$

We will from now on drop the subscript R on renormalized fields. Since we use ψ^0 and A_μ^0 for bare fields, this introduces no ambiguity. It is conventional also to define

$$Z_1 \equiv Z_e Z_2\sqrt{Z_3}. \qquad (19.7)$$

Then,

$$\mathcal{L} = -\frac{1}{4}Z_3 F_{\mu\nu}^2 + iZ_2\bar{\psi}\not{\partial}\psi - Z_2 Z_m m_R\bar{\psi}\psi - e_R Z_1\bar{\psi}\not{A}\psi + \rho_0. \qquad (19.8)$$

We will ignore ρ_0 unless otherwise stated from now on, as the vacuum energy density plays merely a spectator role in the renormalization of QED.

Next we want to expand around some classical tree-level values for these parameters. The field strengths are naturally expanded around $Z_2 = Z_3 = 1$; Z_1 should also be expanded around 1 so that e_R represents the classical electric charge. Finally, we expand m_0 around some renormalized mass m_R, which can be taken to be the pole mass or $\overline{\text{MS}}$ mass or any other convenient choice. It is not necessary to specify exactly how e_R and m_R are defined at this point. The expansions are conventionally written as

$$Z_1 \equiv 1 + \delta_1, \quad Z_2 \equiv 1 + \delta_2, \quad Z_3 \equiv 1 + \delta_3, \quad Z_m = 1 + \delta_m, \qquad (19.9)$$

with all the **counterterms** δ_i starting at order e_R^2. Sometimes we will also write

$$Z_e = 1 + \delta_e, \qquad (19.10)$$

where, following Eq. (19.7),

$$\delta_e = \delta_1 - \delta_2 - \frac{1}{2}\delta_3 + \mathcal{O}\left(e_R^4\right). \tag{19.11}$$

With these expansions the Lagrangian becomes

$$\mathcal{L} = -\frac{1}{4}F_{\mu\nu}^2 + i\bar{\psi}\slashed{\partial}\psi - m_R\bar{\psi}\psi - e_R\bar{\psi}\slashed{A}\psi$$
$$\qquad -\frac{1}{4}\delta_3 F_{\mu\nu}^2 + i\delta_2\bar{\psi}\,\slashed{\partial}\psi - (\delta_m + \delta_2)m_R\bar{\psi}\psi - e_R\delta_1\bar{\psi}\slashed{A}\psi. \tag{19.12}$$

This is the Lagrangian for **renormalized perturbation theory**.

In renormalized perturbation theory, the counterterms appear as interactions in the Lagrangian and can be used in Feynman diagrams, just like any other interactions. The Feynman rules are as follows:

$$\begin{array}{c}\text{———}\bigstar\text{———}\end{array} = i\left(\slashed{p}\delta_2 - (\delta_m + \delta_2)m_R\right). \tag{19.13}$$

The $\bigstar$ indicates a counterterm insertion as an interaction on an electron line. A counterterm on a photon line gives the vertex

$$\mu\;\rightsquigarrow\!\!\bigstar\!\!\rightsquigarrow\;\nu \;\; = -i\delta_3\left(p^2 g^{\mu\nu} - p^\mu p^\nu\right). \tag{19.14}$$

In a gauge-fixed Lagrangian, there is another term, like $\frac{1}{2\xi}(\partial_\mu A_\mu)^2$, which gives a new counterterm to renormalize ξ and modifies the $p^\mu p^\nu$ term in Eq. (19.14). In Feynman gauge, the Feynman rule for the photon line counterterm simplifies to

$$\mu\;\rightsquigarrow\!\!\bigstar\!\!\rightsquigarrow\;\nu = -i\delta_3 p^2 g^{\mu\nu} \quad \text{(Feynman gauge)}. \tag{19.15}$$

Finally, there is the vertex counterterm:

$$\begin{array}{c}\bigstar\end{array} \;\; = -ie_R\delta_1\gamma^\mu. \tag{19.16}$$

A virtue of renormalized perturbation theory is that even though the counterterms are all large numbers, proportional to some regulator cutoff such as $\frac{1}{\varepsilon}$, they are defined through their Taylor expansions in powers of e_R (starting at order e_R^2). In particular, the perturbation expansion can be justified since e_R is small, even if $\varepsilon \ll 1$ (that is, $e_R^2 < e_R$). In contrast, the way we had renormalized in previous chapters was through an expansion in the bare coupling, $e_0 \sim \frac{1}{\varepsilon}$, which is not small for $\varepsilon \ll 1$ (that is $e_0^2 > e_0$). Thus, in renormalized perturbation theory one has a more legitimate perturbation expansion.

It is important to keep in mind that the counterterms must be numbers (or functions of e_R and m_R) – they cannot depend on derivatives or momenta. For example, what would it mean if a field strength renormalization were $\delta_2 = \Box$? Then our quantum field would be $\psi_R = \sqrt{1 + \Box}\psi_0$, which would have completely different Feynman rules and interactions. As long as the counterterms are numbers, and finite numbers once the theory is regulated, the rules we have developed for quantum field theory are unchanged. Now it may happen (but not in QED) that there is an infinity in a Green's function that appears as if it could only be canceled if $\delta_2 = \frac{1}{\varepsilon}\Box$. In that case, we would need to have a term in the Lagrangian of the form $c_0 \bar{\psi}^0 \Box \psi^0$. Then, by renormalizing this term by expanding $c_0 = \frac{1}{\varepsilon}$, the infinity could be removed. The introduction of new terms in this way can be made systematic and underlies the renormalization of so-called *non-renormalizable field theories*. Such theories play an important role in modern quantum field theory. In QED, we will not need to introduce any new terms since it is *renormalizable*. Renormalizability is the subject of Chapter 21.

19.2 Two-point functions

As a warm-up, let us redo the electron self-energy calculation using renormalized perturbation theory. Recall our notation for the 2-point Green's function:

$$\langle \psi(x)\bar{\psi}(y) \rangle = \int \frac{d^4 p}{(2\pi)^4} e^{-ip(x-y)} iG(\not{p}). \tag{19.17}$$

In renormalized perturbation theory, the tree-level Feynman diagram for the 2-point Green's function is

$$iG_{\text{tree}}(\not{p}) = \underline{\qquad\longrightarrow\qquad} = \frac{i}{\not{p} - m_R}. \tag{19.18}$$

This is just the renormalized electron propagator. Note that we are calculating Green's functions in this chapter, not S-matrix elements, so the external lines are not truncated and external polarizations/spinors are not added.

At order e_R^2 there is the loop graph, involving the ordinary vertices, from Eq. (19.12),

$$\underline{\qquad\qquad\overbrace{}\qquad\qquad} = \frac{i}{\not{p} - m_R}[i\Sigma_2(\not{p})]\frac{i}{\not{p} - m_R}, \tag{19.19}$$

where $\Sigma_2(\not{p})$ was computed in Section 18.2; and there is also the counterterm graph,

$$\underline{\qquad\qquad\bigstar\qquad\qquad} = \frac{i}{\not{p} - m_R} i(\not{p}\delta_2 - (\delta_m + \delta_2)m_R)\frac{i}{\not{p} - m_R}. \tag{19.20}$$

Here, the counterterm is acting like a vertex, and since we are computing Green's functions not S-matrix elements, we do not amputate the external lines. So,

$$iG(\not{p}) = \frac{i}{\not{p} - m_R} + \frac{i}{\not{p} - m_R} i(\Sigma_2(\not{p}) + \not{p}\delta_2 - (\delta_m + \delta_2)m_R)\frac{i}{\not{p} - m_R} + \mathcal{O}(e_R^4), \tag{19.21}$$

which agrees with Eq. (18.26).

Now we see that the one-particle irreducible graphs (including counterterms) are $\Sigma(\not{p}) = \Sigma_2(\not{p}) + \not{p}\delta_2 - (\delta_m + \delta_2)m_R + \mathcal{O}(e_R^4)$. Summing them results in

$$iG(\not{p}) = \frac{i}{\not{p} - m_R + \Sigma(\not{p})}. \tag{19.22}$$

Then we can use the on-shell renormalization conditions

$$\Sigma(\not{p})\big|_{\not{p}=m_P} = 0, \qquad \frac{d}{d\not{p}}\Sigma(\not{p})\bigg|_{\not{p}=m_P} = 0, \qquad (19.23)$$

with $m_R = m_P$ to fix δ_2 and δ_m as

$$\delta_2 = -\frac{d}{d\not{p}}\Sigma_2(\not{p})\bigg|_{\not{p}=m_P}, \qquad \delta_m = \frac{1}{m_P}\Sigma_2(m_P), \qquad (19.24)$$

as in Eqs. (18.43) and (18.44).

Of particular interest to us in this chapter will be the value of the δ_2 counterterm in the on-shell scheme, which was calculated in Chapter 18 both in dimensional regularization,

$$\delta_2 = \frac{e_R^2}{8\pi^2}\left(-\frac{1}{\varepsilon} - \frac{1}{2}\ln\frac{\tilde{\mu}^2}{m_R^2} - 2 - \ln\frac{m_\gamma^2}{m_R^2}\right) \qquad \text{(DR)}, \qquad (19.25)$$

and with a Pauli–Villars regulator,

$$\delta_2 = \frac{e_R^2}{8\pi^2}\left(-\frac{1}{2}\ln\frac{\Lambda^2}{m_R^2} - \frac{9}{4} - \ln\frac{m_\gamma^2}{m_R^2}\right) \qquad \text{(PV)}. \qquad (19.26)$$

Next we will use a similar analysis for the photon self-energy to fix δ_3.

19.2.1 Photon self-energy

Proceeding as with the electron self-energy, we define the Fourier-transformed Green's function $G^{\mu\nu}(p)$ in terms of the exact 2-point function in the full interacting theory as

$$\langle A^\mu(x)A^\nu(y)\rangle = \int \frac{d^4p}{(2\pi)^4}e^{ip(x-y)}iG^{\mu\nu}(p). \qquad (19.27)$$

At order e_R^2 there is a contribution to $G^{\mu\nu}$ from the 1-loop graph using the ordinary Feynman rules in Eq. (19.12). The result was calculated in Section 16.2 and found to have the form

$= -i(p^2 g^{\mu\nu} - p^\mu p^\nu)e_R^2\Pi_2(p^2), \qquad (19.28)$

where

$$\Pi_2(p^2) = \frac{8}{(4\pi)^{d/2}}\Gamma\left(2 - \frac{d}{2}\right)\mu^{4-d}\int_0^1 dx\, x(1-x)\left(\frac{1}{m_R^2 - p^2 x(1-x)}\right)^{2-\frac{d}{2}}$$

$$= \frac{1}{2\pi^2}\int_0^1 dx\, x(1-x)\left[\frac{2}{\varepsilon} + \ln\left(\frac{\tilde{\mu}^2}{m_R^2 - p^2 x(1-x)}\right) + \mathcal{O}(\varepsilon)\right]. \qquad (19.29)$$

The other contribution at order e_R^2 in renormalized perturbation theory comes from the counterterm graph,

$= -i\delta_3\left(p^2 g^{\mu\nu} - p^\mu p^\nu\right). \qquad (19.30)$

These are the only two one-particle irreducible graphs contributing at order e_R^2.

leg corrections, which we have already calculated and rendered finite by the counterterms δ_2, δ_m and δ_3 in the renormalization of the 2-point functions. As before, we write

$$-ie_R\Gamma^\mu = \quad \text{(diagram)} \quad . \tag{19.42}$$

This is normalized so that at leading order $\Gamma^\mu = \gamma^\mu$. More generally, we showed in Chapter 17 that, by Lorentz invariance and the Ward identity (which holds for off-shell photons), arbitrary contributions to Γ^μ can be written in terms of two Lorentz-scalar form factors, F_1 and F_2:

$$\Gamma^\mu(p) = F_1(p^2)\gamma^\mu + \frac{i\sigma^{\mu\nu}}{2m_R}p_\nu F_2(p^2). \tag{19.43}$$

At leading order:

$$F_1(p^2) = 1, \quad F_2(p^2) = 0. \tag{19.44}$$

At next-to-leading order (order e_R^2), the form factors get contributions from a loop graph and from counterterms:

$$-ie_R\Gamma^\mu = \quad \text{(diagrams)} \quad + \cdots . \tag{19.45}$$

From Eq. (19.16) we see that the counterterm gives $\Gamma^\mu = \delta_1\gamma^\mu$, which contributes only to $F_1(p^2)$.

We calculated $F_2(p^2)$ at 1-loop when we considered corrections to the magnetic moment of the electron in Chapter 17. There we found a finite answer:

$$F_2(p^2) = \frac{e_R^2}{4\pi^2}\int_0^1 d^3x\, \delta(x+y+z-1)\frac{z(1-z)m_R^2}{(1-z)^2m_R^2 - xyp^2} + \mathcal{O}(e_R^4). \tag{19.46}$$

In particular, $F_2(0) = \frac{\alpha}{2\pi}$, which led to a prediction for the anomalous magnetic moment of the electron: $g - 2 = 2F_2(0) = \frac{\alpha}{\pi}$. Since this correction was finite, no counterterm was needed.

We also began the calculation of $F_1(p^2)$ at 1-loop. Appending the counterterm diagram to the expression for $F_1(p^2)$ in Chapter 17, we find

$$F_1(p^2) = 1 + f(p^2) + \delta_1 + \mathcal{O}(e_R^4), \tag{19.47}$$

where

$$f(p^2) = -2ie_R^2\int\frac{d^4k}{(2\pi)^4}\int dx\, dy\, dz\, \delta(x+y+z-1)$$
$$\times\frac{k^2 - 2(1-x)(1-y)p^2 - 2(1-4z+z^2)m_R^2}{[k^2 - (m_R^2(1-z)^2 - xyp^2)]^3}. \tag{19.48}$$

Before evaluating this integral, note that $F_1(0)$ gives the coefficient of the $e_R \bar{\psi} \slashed{A} \psi$ coupling in the Dirac equation. In particular, $F_1(0) = 1$ implies that e_R is the electric charge as measured by Coulomb's law at large distances. It is therefore natural to define the renormalized electric charge so that $F_1(0) = 1$ is true exactly. In other words:

$$\Gamma^\mu(0) = \gamma^\mu. \tag{19.49}$$

This is the final renormalization condition. It implies that the renormalized electric charge is what is measured by Coulomb's law at asymptotically large distances, and, *by definition*, does not get radiative corrections. This condition sets $\delta_1 = -f(0)$ at order e_R^2.

Now let us evaluate $f(p^2)$. The integral is both UV and IR divergent. We will regulate the UV divergence with dimensional regularization and the IR divergence with a photon mass, as we did the electron self-energy graph calculation in Section 18.2. In d dimensions and with a photon mass, you are encouraged to check that the integral is modified to

$$f(p^2) = -2ie_R^2 \mu^{4-d} \int \frac{d^d k}{(2\pi)^d} \int dx\, dy\, dz\, \delta(x + y + z - 1)$$

$$\times \frac{\frac{(d-2)^2}{d} k^2 - [(d-2)xy + 2z]p^2 + [2 + 2z^2 - d(1-z)^2]m_R^2}{(k^2 - \Delta + i\varepsilon)^3}, \tag{19.50}$$

where

$$\Delta = (1-z)^2 m_R^2 - xyp^2 + zm_\gamma^2. \tag{19.51}$$

Now the only UV-divergent term is the k^2 one, which can be evaluated with

$$\mu^{4-d} \int \frac{d^d k}{(2\pi)^d} \frac{\frac{(d-2)^2}{d} k^2}{(k^2 - \Delta + i\varepsilon)^3} = \mu^{4-d} \frac{i}{(4\pi)^{d/2}} \frac{(d-2)^2}{d} \frac{d}{4} \Delta^{\frac{d}{2}-2} \Gamma\left(\frac{4-d}{2}\right)$$

$$= \frac{i}{16\pi^2} \left(\frac{2}{\varepsilon} + \ln\frac{\tilde{\mu}^2}{\Delta} - 2\right). \tag{19.52}$$

The remaining terms are UV finite but IR divergent, so we can set $d = 4$ in them and use

$$\int \frac{d^4 k}{(2\pi)^4} \frac{-2(1-x)(1-y)p^2 - 2(1 - 4z + z^2)m_R^2}{(k^2 - \Delta + i\varepsilon)^3}$$

$$= i \frac{p^2(1-x)(1-y) + m_R^2(1 - 4z + z^2)}{16\pi^2 \Delta}. \tag{19.53}$$

Expanding in $d = 4 - \varepsilon$, we then get

$$f(p^2) = \frac{e_R^2}{8\pi^2} \left(\frac{1}{\varepsilon} - 1 + \int_0^1 dx\, dy\, dz\, \delta(x + y + z - 1)\right.$$

$$\left. \times \left[\frac{p^2(1-x)(1-y) + m_R^2(1 - 4z + z^2)}{\Delta} + \ln\frac{\tilde{\mu}^2}{\Delta}\right]\right). \tag{19.54}$$

At $p = 0$, this simplifies to

$$
\begin{aligned}
f(0) &= \frac{e_R^2}{8\pi^2} \left(\frac{1}{\varepsilon} - 1 + \int_0^1 dz (1 - z) \left[\frac{m_R^2 (1 - 4z + z^2)}{(1 - z)^2 m_R^2 + z m_\gamma^2} + \ln \frac{\tilde{\mu}^2}{(1 - z)^2 m_R^2 + z m_\gamma^2} \right] \right) \\
&= \frac{e_R^2}{8\pi^2} \left(\frac{1}{\varepsilon} + \frac{1}{2} \ln \frac{\tilde{\mu}^2}{m_R^2} + 2 + \ln \frac{m_\gamma^2}{m_R^2} \right).
\end{aligned}
\tag{19.55}
$$

Since $F_1(0) = 1 + f(0) + \delta_1 + \cdots$, at order e_R^2 the renormalization condition $F_1(0) = 1$ implies

$$
\delta_1 = -f(0) = \frac{e_R^2}{8\pi^2} \left(-\frac{1}{\varepsilon} - \frac{1}{2} \ln \frac{\tilde{\mu}^2}{m_R^2} - 2 - \ln \frac{m_\gamma^2}{m_R^2} \right) \qquad \text{(DR)}.
\tag{19.56}
$$

Comparing with Eq. (19.25), we find a surprise: $\delta_1 = \delta_2$ at order e_R^2.

An obvious test of whether this relationship could possibly be significant is to repeat the calculation with a different regulator. Using Pauli–Villars to cut off the UV divergences, we find

$$
f(p^2) = \frac{e_R^2}{8\pi^2} \int_0^1 dx \, dy \, dz \, \delta(x + y + z - 1) \left[\ln \frac{z \Lambda^2}{\Delta} + \frac{p^2 (1 - x)(1 - y) + m_R^2 (1 - 4z + z^2)}{\Delta} \right].
\tag{19.57}
$$

So,

$$
\begin{aligned}
f(0) &= \frac{e_R^2}{8\pi^2} \int_0^1 dz (1 - z) \left[\ln \frac{z \Lambda^2}{(1 - z)^2 m_R^2 + z m_\gamma^2} + \frac{(1 - 4z + z^2) m_R^2}{(1 - z)^2 m_R^2 + z m_\gamma^2} \right] \\
&= \frac{e_R^2}{8\pi^2} \left(\frac{1}{2} \ln \frac{\Lambda^2}{m_R^2} + \frac{9}{4} + \ln \frac{m_\gamma^2}{m_R^2} \right) \qquad \text{(PV)},
\end{aligned}
\tag{19.58}
$$

which gives

$$
\delta_1 = -\frac{e_R^2}{8\pi^2} \left(\frac{1}{2} \ln \frac{\Lambda^2}{m_R^2} + \frac{9}{4} + \ln \frac{m_\gamma^2}{m_R^2} \right) \qquad \text{(PV)},
\tag{19.59}
$$

which is exactly the same as what we found for δ_2 in Eq. (19.26).

Given that δ_1 and δ_2 came from entirely different loop calculations (the vertex correction and the electron self-energy graph), it appears almost magical that $\delta_1 = \delta_2$. So their equality, if not just a coincidence, would imply something highly non-trivial about QED. In fact, $\delta_1 = \delta_2$ exactly, as we will prove in Section 19.5. This result is equivalent to the QED charge current, $J^\mu = \bar{\psi} \gamma^\mu \psi$, not getting renormalized.

19.4 Renormalization conditions in QED

We have found a set of four renormalization conditions that fix the four counterterms δ_1, δ_2, δ_3 and δ_m in QED. In the on-shell scheme, the renormalized electron mass m_R is identified with the pole mass, $m_R = m_P$, and the conditions are

$$\Sigma(m_P) = 0, \tag{19.60}$$

$$\Sigma'(m_P) = 0, \tag{19.61}$$

$$\Gamma^\mu(0) = \gamma^\mu, \tag{19.62}$$

$$\Pi(0) = 0. \tag{19.63}$$

In these equations, $i\Sigma(\slashed{p})$ is the coefficient of $\frac{i}{\slashed{p} - m_R}$ in the sum of 1PI contributions to the electron 2-point function; $\Pi(p^2)$ is the coefficient of $\frac{i}{p^2}(-g^{\mu\nu} + \frac{p^\mu p^\nu}{p^2})$ in the sum of all 1PI contributions to the photon 2-point function in Lorenz gauge; and $-ie_R\Gamma^\mu(p)$ is the sum of all 1PI contributions to the 3-point function $\langle \bar\psi A^\mu \psi \rangle$ with p the photon momentum.

The first two conditions fix the electron propagator to

$$iG(\slashed{p}) = \frac{i}{\slashed{p} - m_P + i\varepsilon} + \text{regular at } \slashed{p} = m_P; \tag{19.64}$$

the third condition fixes the renormalized electric charge e_R to be what is measured by Coulomb's law at large distances. The final condition forces the photon propagator to be

$$iG^{\mu\nu}(p) = \frac{-ig^{\mu\nu}}{p^2 + i\varepsilon} + p^\mu p^\nu \text{ pieces} + \text{ regular at } \slashed{p} = m_P. \tag{19.65}$$

These four conditions give non-perturbative definitions for the four free parameters, e_0, Z_2, Z_3 and m_0, in the QED Lagrangian.

The four renormalization conditions listed above are not the only way to define the counterterms in QED. In fact, as discussed in Chapter 18, any definition for counterterms that differs from these by only finite parts will also remove all the infinities in these Green's functions. Different conventions for the finite parts of counterterms are known as different subtraction schemes. In minimal subtraction, the finite parts of the counterterms are set to zero. In modified minimal subtraction, which is used in conjunction with dimensional regularization, the only finite parts that are kept are the $\ln(4\pi)$ and γ_E factors, which effectively convert $\tilde\mu$ back to μ in unrenormalized amplitudes.

In dimensional regularization with minimal subtraction, the QED counterterms are

$$\delta_1 = \delta_2 = \frac{e_R^2}{16\pi^2}\left[-\frac{2}{\varepsilon}\right], \quad \delta_3 = \frac{e_R^2}{16\pi^2}\left[-\frac{8}{3\epsilon}\right], \quad \delta_m = \frac{e_R^2}{16\pi^2}\left[-\frac{6}{\varepsilon}\right]. \tag{19.66}$$

Thus, for example, $\Pi_2(p^2)$ becomes, in the $\overline{\text{MS}}$ scheme,

$$\Pi_2(p^2) = \frac{e_R^2}{2\pi^2} \int_0^1 dx\, x(1-x) \ln\left(\frac{\mu^2}{m_R^2 - p^2 x(1-x)}\right). \qquad (19.67)$$

This is a finite function, but depends on an arbitrary parameter μ.

An important point, which is often confused, is that there are two scales involved in any renormalization: the cutoff scale Λ, which is taken to infinity, and a finite low-energy scale μ, the **subtraction point**. Λ has to do with the way the theory is deformed in the UV to make it convergent, and we can always take the limit $\Lambda \to \infty$ (after renormalization). μ is related to the renormalization condition. In the on-shell scheme, μ is implicit. For example, in the on-shell scheme in the electron self-energy, m_R is set equal to the pole mass; this is effectively the choice $\mu = m_P$. For the photon, $\mu = 0$. Neither scale Λ nor μ can ever affect a physical calculation, but for different reasons. Λ can never matter, because it is entirely unphysical and we always take $\Lambda \to \infty$ after renormalization. μ can never matter because the subtraction point is arbitrary.

Let us recap the different quantities we have introduced related to renormalization:

- The renormalized mass m_R and electric charge e_R are parameters in the Lagrangian of renormalized perturbation theory used in calculations. They are finite, but only well-defined after a set of renormalization conditions or, equivalently, a subtraction scheme is introduced.
- The counterterms δ_1, δ_2, δ_3 and δ_m come from expanding the bare parameters in the un-renormalized QED Lagrangian around their tree-level values. The divergent parts of the counterterms depend on the regulator but not on the subtraction scheme. The finite parts of the counterterms depend on the subtraction scheme.
- The cutoff Λ or $\frac{1}{\varepsilon}$ is an unphysical scale used to make formally divergent quantities finite in a consistent way. The divergent part of the cutoff dependence cancels between loop graphs and counterterm graphs. After this cancellation, the cutoff can be taken to ∞.
- The subtraction point μ allows for a one-parameter family of renormalization conditions. Physical predictions that relate observables to other observables must be independent of μ.

19.5 $Z_1 = Z_2$: implications and proof

We found by explicit calculation that the two counterterms δ_1 and δ_2 were exactly equal at order e_R^2. This was true with the counterterms defined in the on-shell scheme, where δ_1 was fixed by $\Gamma^\mu(0) = \gamma^\mu$ and δ_2 was fixed by $\Sigma(m_P) = 0$, where m_P is the electron pole mass. The two loops required to determine δ_1 and δ_2 were the 1PI vertex correction and the 1PI electron self-energy graph. Now, we will understand why these seemingly unrelated calculations are in fact very closely connected.

First, note that $\delta_1 = \delta_2$ implies $Z_1 = Z_2$. Recalling Eq. (19.7), $Z_1 \equiv Z_e Z_2 \sqrt{Z_3}$, where $e_0 = Z_e e_R$, it follows that

$$e_R = \sqrt{Z_3} e_0. \tag{19.68}$$

Thus, the renormalization of the electric charge is determined completely by the renormalization of the photon field strength. This explains why we were able to calculate the renormalization of the electric charge from only the vacuum polarization graphs in Chapter 16.

There is an important physical implication of $Z_1 = Z_2$. Suppose we have a theory with two different kinds of particles: for example, a quark with charge $Q_q = \frac{2}{3}$ and an electron with charge $Q_e = -1$. The Lagrangian including both fields is

$$\mathcal{L} = -\frac{1}{4} Z_3 F_{\mu\nu}^2 + i Z_{2e} \bar{\psi}_e \slashed{\partial} \psi_e - e_R Z_{1e} \bar{\psi}_e \slashed{A} \psi_e + i Z_{2q} \bar{\psi}_q \slashed{\partial} \psi_q + \frac{2}{3} e_R Z_{1q} \bar{\psi}_q \slashed{A} \psi_q. \tag{19.69}$$

If $Z_{1e} = Z_{2e}$ and $Z_{1q} = Z_{2q}$ for both the electron and quark, then this Lagrangian is

$$\mathcal{L} = -\frac{1}{4} Z_3 F_{\mu\nu}^2 + Z_{2e} \bar{\psi}_e (i\slashed{\partial} - e_R \slashed{A}) \psi_e + Z_{2q} \bar{\psi}_q (i\slashed{\partial} + \frac{2}{3} e_R \slashed{A}) \psi_q. \tag{19.70}$$

Thus, $Z_1 = Z_2$ implies that the relationship between the coefficient of $i\slashed{\partial}$ and of $e_R \slashed{A}$ does not receive radiative corrections. In other words, the ratio of charges of the electron and the quark is the same in the quantum theory as they would be classically.

This is pretty remarkable. It explains why the observed charge of the proton and the charge of the electron can be exactly opposite, even in the presence of vastly different interactions of the two particles. A priori, we might have suspected that, because of strong interactions and virtual mesons surrounding the proton, the types of radiative corrections for the proton would be vastly more complicated than for the electron. But, as it turns out, this does not happen – the renormalization of the photon field strength rescales the electric charge, but the corrections to the relative charges of the proton and the electron cancel.

For a quick way to see that $Z_1 = Z_2$ to all orders, first rescale $A_\mu \to \frac{1}{e_R} A_\mu$. Then the Lagrangian becomes

$$\mathcal{L} = -\frac{1}{4 e_R^2} Z_3 F_{\mu\nu}^2 + Z_{2e} \bar{\psi}_e \left(i\slashed{\partial} - \frac{Z_{1e}}{Z_{2e}} \slashed{A} \right) \psi_e + Z_{2q} \bar{\psi}_q \left(i\slashed{\partial} + \frac{Z_{1q}}{Z_{2q}} \frac{2}{3} \slashed{A} \right) \psi_q. \tag{19.71}$$

At tree-level, with $Z_i = 1$, this Lagrangian is invariant under the gauge transformations

$$\psi_q \to e^{\frac{2}{3} i \alpha} \psi_q, \quad \psi_e \to e^{-i\alpha} \psi_e, \quad A_\mu \to A_\mu + \partial_\mu \alpha. \tag{19.72}$$

Note that the charges, $Q_i = -1, \frac{2}{3}$, appear in the transformation law but e_R does not. Second, observe that the transformation has nothing to do with perturbation theory. Since the Lagrangian is gauge invariant as long as the regulator preserves gauge invariance, the loop corrections will be gauge invariant, and the counterterms should respect the symmetry too. That is, since charge is conserved at each vertex, it will be conserved in all the loops.

Here, m can represent either the electron or muon mass. We also do not write m_R, since mass renormalization will not be relevant to the calculations in this chapter. We found that the second form factor at order e_R^2 was

$$F_2(p^2) = \frac{e_R^2}{4\pi^2} \int_0^1 dx\, dy\, dz\, \delta\left(x+y+z-1\right) \frac{z(1-z)m^2}{(1-z)^2m^2 - xyp^2} + \mathcal{O}(e_R^4). \quad (20.10)$$

In the high-energy limit, $\frac{p^2}{m^2} \to \infty$, this form factor vanishes, $F_2(p^2) \to 0$. This makes sense, since F_2 couples right- and left-handed spinors, which are uncoupled in massless QED.

The first form factor was both UV and IR divergent. Regulating the UV divergence in $F_1(p^2)$ with Pauli–Villars and the IR divergence with a photon mass, we found that

$$F_1(p^2) = 1 + f(p^2) + \delta_1 + \mathcal{O}(e_R^4), \quad (20.11)$$

where from Eq. (19.57)

$$f(p^2) = \frac{e_R^2}{8\pi^2} \int_0^1 dx\, dy\, dz\, \delta(x+y+z-1)\left[\ln\frac{z\Lambda^2}{\Delta} + \frac{p^2(1-x)(1-y)+m^2(1-4z+z^2)}{\Delta}\right] \quad (20.12)$$

with

$$\Delta = (1-z)^2m^2 - xyp^2 + zm_\gamma^2. \quad (20.13)$$

For $e^+e^- \to \mu^+\mu^-$ we need $p^2 = Q^2$ and we can take $m = 0$ for the high-energy limit ($Q \gg m$).

The counterterm is set by $F_1(0) = 1$, which normalizes the electric charge to what is measured at large distances. In Section 19.3. we calculated δ_1 for finite m. Now, with $m = 0$, we find

$$\delta_1 = -f(0) = -\frac{e_R^2}{8\pi^2}\left(\frac{1}{2}\ln\frac{\Lambda^2}{m_\gamma^2}\right). \quad (20.14)$$

Evaluating $f(Q^2)$ is more challenging. It has the form

$$f(Q^2) = \frac{e_R^2}{8\pi^2} \int_0^1 dx \int_0^{1-x} dy$$
$$\times \left[\ln\frac{(1-x-y)\Lambda^2}{-xyQ^2 + (1-x-y)m_\gamma^2} + \frac{Q^2(1-x)(1-y)}{-xyQ^2 + (1-x-y)m_\gamma^2}\right]. \quad (20.15)$$

The first term is IR finite and gives

$$\int_0^1 dx \int_0^{1-x} dy \left[\ln\frac{(1-x-y)\Lambda^2}{-xyQ^2 + (1-x-y)m_\gamma^2}\right] = \frac{3}{4} + \frac{1}{2}\ln\frac{\Lambda^2}{-Q^2} + \mathcal{O}(m_\gamma). \quad (20.16)$$

Note that the $\ln\Lambda^2$ has the right coefficient to be canceled by δ_1. More generally, the divergences in the vertex correction and δ_1 will always cancel for arbitrarily complicated processes involving a photon–fermion vertex. This is simply because the divergent part of the counterterm was determined by calculating the 1PI contributions to $\langle\Omega|T\{\psi A^\mu \bar\psi\}|\Omega\rangle$. In the divergent region of loop momentum, the external scales are irrelevant. Thus, the divergences for the 3-point function are the same whether or not it is embedded in a larger diagram, and therefore they will always be canceled by δ_1.

The second term in Eq. (20.15) is IR divergent but UV finite. Moreover, for real Q^2 there is a pole in the integration region. Fortunately, there is a small imaginary part in the denominator (due to the $i\varepsilon$ prescription) which makes the integral converge. Since x and y are positive we can perform the integral by taking $Q^2 \to Q^2 + i\varepsilon$, which gives

$$\int_0^1 dx \int_0^{1-x} dy \frac{Q^2(1-x)(1-y)}{-xyQ^2 + (1-x-y)m_\gamma^2}$$
$$= -\frac{1}{2}\ln^2 \frac{m_\gamma^2}{-Q^2 - i\varepsilon} - 2\ln \frac{m_\gamma^2}{-Q^2 - i\varepsilon} - \frac{\pi^2}{3} - \frac{5}{2} + \mathcal{O}(m_\gamma). \quad (20.17)$$

So that,

$$f(Q^2) + \delta_1 = \frac{e_R^2}{16\pi^2}\left\{ -\ln^2 \frac{m_\gamma^2}{-Q^2 - i\varepsilon} - 3\ln \frac{m_\gamma^2}{-Q^2 - i\varepsilon} - \frac{2\pi^2}{3} - \frac{7}{2} + \mathcal{O}(m_\gamma)\right\}. \quad (20.18)$$

Then we use

$$\lim_{\varepsilon \to 0} \ln(-Q^2 - i\varepsilon) = \ln Q^2 - i\pi \quad (20.19)$$

to write

$$f(Q^2) + \delta_1 = \frac{e_R^2}{16\pi^2}\left\{ -\ln^2 \frac{m_\gamma^2}{Q^2} - (3 + 2\pi i)\ln \frac{m_\gamma^2}{Q^2} + \frac{\pi^2}{3} - \frac{7}{2} - 3\pi i + \mathcal{O}(m_\gamma)\right\}. \quad (20.20)$$

Note that the $-\frac{2\pi^2}{3}$ has combined with the π^2 coming from the expansion of $-\ln^2 \frac{m_\gamma^2}{-Q^2}$ to give the $\frac{\pi^2}{3}$ term.

To evaluate the cross section at next-to-leading order, we need the first subleading term in $|\mathcal{M}_\Gamma + \mathcal{M}_0|^2$. The $\mathcal{O}(e_R^6)$ term in this comes from

$$\frac{1}{4}\sum_{\text{spins}} \left[\mathcal{M}_\Gamma^\dagger \mathcal{M}_0 + \mathcal{M}_0^\dagger \mathcal{M}_1\right] = \frac{e_R^4}{Q^4} \text{Tr}\left[\slashed{p}_2 \gamma_\mu \slashed{p}_1 \gamma_\nu\right] \text{Tr}[\slashed{p}_3 \Gamma_2^\mu \slashed{p}_4 \gamma^\nu] + \text{c.c.} \quad (20.21)$$

In the high-energy limit in which we are interested, the $\sigma^{\mu\nu}$ term in Γ^μ gives an odd number of γ-matrices in the second trace, forcing the contribution of the F_2 form factor to vanish. This is consistent with F_2 itself vanishing for $p^2 \gg m^2$. So we have simply

$$\frac{1}{4}\sum_{\text{spins}} \left[\mathcal{M}_\Gamma^\dagger \mathcal{M}_0 + \mathcal{M}_0^\dagger \mathcal{M}_\Gamma\right] = 2\text{Re}\left[f(Q^2) + \delta_1\right] \frac{1}{4}\sum_{\text{spins}} |\mathcal{M}_0|^2, \quad (20.22)$$

with $f(Q)$ just a number. Thus, the total loop (virtual) correction at order e_R^6 is given by

$$\sigma_V = 2\text{Re}\left[(f(Q^2) + \delta_1]\sigma_0 = \frac{e_R^2}{8\pi^2}\sigma_0 \left\{ -\ln^2 \frac{m_\gamma^2}{Q^2} - 3\ln \frac{m_\gamma^2}{Q^2} - \frac{7}{2} + \frac{\pi^2}{3}\right\}. \quad (20.23)$$

An important qualitative feature of this result is the $\ln^2 \frac{m_\gamma^2}{Q^2}$ term. This is known as a **Sudakov double logarithm**, and is characteristic of IR divergences. Sudakov logarithms play an important role in many areas of physics, such as the physics of *jets* and of *parton*

20.2 Jets

We have found that the sum of the $e^+e^- \to \mu^+\mu^-$ cross section σ_V, at order e_R^6 from the graphs $+$, and the $e^+e^- \to \mu^+\mu^-\gamma$ cross section σ_R also at order e_R^6 from the graphs $+$, was IR and UV finite. Photons emitted from final state particles, such as the muons in this case, are known as **final state radiation**. The explanation of why one has to include final state radiation to get a finite cross section is that it is impossible to tell whether the final state in a scattering process is just a muon or a muon plus an arbitrary number of soft or collinear photons. Trying to make this more precise leads naturally to the notion of jets.

For simplicity, we calculated only the total cross section for e^+e^- annihilation into states containing a muon and antimuon pair, inclusive over an additional photon. One could also calculate something less inclusive. For example, experimentally, a muon might be identified as a track in a cloud chamber or an energy deposition in a calorimeter. So one could calculate the cross section for the production of a track or energy deposition. This cross section gets contributions from different processes. Even with an amazing detector, there will be some lower limit E_{res} on the energy of photons that can be resolved. Even for energetic photons, if the photon is going in exactly the same direction as the muon there would be no way to resolve it and the muon separately. That is, there will be some lower limit θ_{res} on the angle that can be measured between either muon and the photon.

With these experimental parameters,

$$\sigma_{\text{tot}} = \sigma_{2\to2} + \sigma_{2\to3}, \tag{20.52}$$

where

$$\sigma_{2\to2} = \sigma\left(e^+e^- \to \mu^+\mu^-\right) + \sigma\left(e^+e^- \to \mu^+\mu^-\gamma\right)\bigg|_{E_\gamma < E_{\text{res}} \text{ or } \theta_{\gamma\mu} < \theta_{\text{res}}} \tag{20.53}$$

is the rate for producing something that looks just like a $\mu^+\mu^-$ pair and

$$\sigma_{2\to3} = \sigma\left(e^+e^- \to \mu^+\mu^-\gamma\right)\bigg|_{E_\gamma > E_{\text{res}} \text{ and } \theta_{\gamma\mu} > \theta_{\text{res}}} \tag{20.54}$$

is the rate for producing a muon pair in association with an observable photon.

The cross section for muons plus a hard photon is now IR finite due to the energy cutoff, even for $E_{\text{res}} \ll Q$ and $\theta_{\text{res}} \ll 1$. Unfortunately, the phase space integral within these cuts, even with $m_\gamma = 0$, is complicated enough to be unilluminating. The result, which we quote from [Ellis et al., 1996], is that the rate for producing all but a fraction $\frac{E_{\text{res}}}{Q}$ of the total energy in a pair of cones of half-angle θ_{res} is

$$\sigma_{2\to3} = \sigma_0 \frac{e_R^2}{8\pi^2} \left\{ \ln\frac{1}{\theta_{\text{res}}} \left[\ln\left(\frac{Q}{2E_{\text{res}}} - 1\right) - \frac{3}{4} + 3\frac{E_{\text{res}}}{Q} \right] \right.$$
$$\left. + \frac{\pi^2}{12} - \frac{7}{16} - \frac{E_{\text{res}}}{Q} + \frac{3}{2}\left(\frac{E_{\text{res}}}{Q}\right)^2 + \mathcal{O}\left(\theta_{\text{res}}\ln\frac{E_{\text{res}}}{Q}\right) \right\}. \tag{20.55}$$

To calculate $\sigma_{2 \to 2}$ one cannot take $m_\gamma = 0$ since the two contributions are separately IR divergent. Conveniently, since we have already calculated $\sigma_{\text{tot}} = \sigma_{2 \to 2} + \sigma_{2 \to 3}$, we can just read off that

$$\sigma_{2 \to 2} = \sigma_{\text{tot}} - \sigma_{2 \to 3} = \sigma_0 \left(1 - \frac{e_R^2}{8\pi^2} \left\{ \ln \frac{1}{\theta_{\text{res}}} \left[\ln \left(\frac{Q}{2E_{\text{res}}} - 1 \right) - \frac{3}{4} + 3 \frac{E_{\text{res}}}{Q} \right] + \cdots \right\} \right). \tag{20.56}$$

This result was first calculated by Sterman and Weinberg in 1977 [Sterman and Weinberg, 1977]. They interpreted $\sigma_{2 \to 2}$ as the rate for jet production, where a **jet** is defined as a two-body final state by the parameters θ_{res} and E_{res}. More precisely, these paramaters define a **Sterman–Weinberg jet**.

Sterman–Weinberg jets are not the most useful jet definition in practice. There are many other ways to define a jet. Any definition is acceptable as long as it allows a separation into finite cross sections for $\sigma_{2 \to 2}$ (the two-jet rate), $\sigma_{2 \to 3}$ (the three-jet rate), and $\sigma_{2 \to n}$ (the n-jet rate), which starts at higher order in perturbation theory. A jet definition simpler than Sterman–Weinberg simply puts a lower bound on the invariant mass of the photon–muon pair, $(p_\gamma + p_{\mu \pm})^2 > M_J^2$. This single parameter limits both the collinear and soft singularities. An invariant mass cutoff is sometimes known as a **JADE jet** after the JADE (Japan, Deutschland, England) experiment, which ran at DESY in Hamburg from 1979 to 1986.

Restricting $(p_\gamma + p_{\mu \pm})^2 > M_J^2$ implies $t > M_J^2$ and $u > M_J^2$ in the notation of Eqs. (20.38)–(20.40), or equivalently, $x_1 < 1 - \beta_J$ and $x_2 < 1 - \beta_J$, where $\beta_J = \frac{M_J^2}{Q^2}$. Then the cross section is

$$\sigma_{2 \to 3} = \frac{e_R^2}{8\pi^2} \sigma_0 \int_0^{1 - \beta_J} dx_1 \int_{1 - x_1}^{1 - \beta_J} dx_2 \frac{x_1^2 + x_2^2}{(1 - x_1)(1 - x_2)}$$

$$= \frac{e_R^2}{8\pi^2} \sigma_0 \left\{ 2 \ln^2 \frac{M_J^2}{Q^2} + 3 \ln \frac{M_J^2}{Q^2} - \frac{\pi^2}{3} + \frac{5}{2} + \mathcal{O}(\beta_J) \right\}, \tag{20.57}$$

where the $M_J \ll Q$ limit has been taken in the second line. One does not have to take this limit; however, the limit shows, as with Eq. (20.57), a general result:

- In physical cross sections, an experimental resolution parameter acts as an IR regulator.

In other words, we did not need to introduce m_γ. In practice, it is much easier to calculate the total cross section using m_γ than by using a more physical regulator associated with the details of an experiment.

An important qualitative feature of results such as the two- or three-jet rates is that for very small resolution parameters, $M_J \ll Q$, it can happen that $\frac{e_R^2}{8\pi^2} \ln^2 \frac{M_J^2}{Q^2} > 1$. In this limit, the perturbation expansion breaks down, since an order e_R^4 correction of the form $\left(\frac{e_R^2}{8\pi^2} \ln^2 \frac{M_J^2}{Q^2} \right)^2$ would be of the same order. Thus, to be able to compare to experiment, one should not take M_J too small. As a concrete example, the experiment BABAR at SLAC measured the decay of B mesons to kaons and photons ($B \to K\gamma$). This experiment was sensitive only to photons harder than $E_{\text{res}} = 1.8$ GeV. In other words, it could not distinguish a kaon in the final state from a kaon plus a photon softer than this energy.

To compare to theory, a calculation was needed of the rate for $B \to K\gamma$ with the γ energy integrated up to E_{res}. The rate has a term of the form $\ln^2 \frac{E_{\text{res}}}{m_B} \sim 1$ in it, which has a quantitatively important effect. Since the logarithm is large, higher orders in perturbation theory are also important. The summation of these Sudakov double logarithms to all orders in perturbation theory was an important impetus for the development of new powerful theoretical tools, in particular, Soft-Collinear Effective Theory (see Chapter 36) in the 2000s.

While these muon–photon packets are hard to see in QED, they are easy to see in QCD. In QCD, the muon is replaced by a quark and the photon replaced by a gluon. The quark itself and the additional soft gluons turn into separate observable particles, such as pions and kaons. Thus, a quark in QCD turns into a *jet* of *hadrons*. These jets are a very real and characteristic phenomenon of all high-energy collisions. We have explained their existence by studying the infrared singularity structure of Feynman diagrams in quantum field theory.

In modern collider physics, it is common to look not at the rate for jet production for a fixed resolution parameter, but instead to look at the distribution of jets themselves. To do this, one needs to define a jet through a *jet algorithm*. For example, one might cluster together any observed particles closer than some θ_{res}. The result would be a set of jets of angular size θ_{res}. Then one can look at the distribution of properties of those jets, such as $\frac{d\sigma}{dm_J}$, where m_J is the jet mass defined as the invariant mass of the sum of the 4-momenta of all the particles in the jet. It turns out that such distributions have a peak at some finite value of m_J. However, at any order in perturbation theory, one would just find results such as $\frac{d\sigma}{dm_J} = \frac{1}{m_J} \ln \frac{m_J}{Q}$, which grow arbitrarily large at small mass. Calculating the mass distribution of jets therefore requires tools beyond perturbation theory, some of which are discussed in Chapter 36.

20.3 Other loops

Now let us return to the other loops in Eq. (20.7). The box and crossed box diagrams

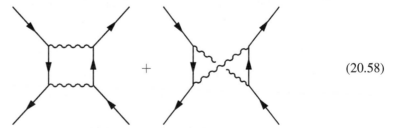

$$+ \tag{20.58}$$

are UV finite. To see this, note that the loop integrals for either graph will be of the form

$$\int \frac{d^4k}{(2\pi)^4} \frac{1}{k^2} \frac{1}{k^2} \frac{1}{\not{k}} \frac{1}{\not{k}} \sim \int \frac{d^4k}{k^6}, \tag{20.59}$$

where $k \gg p_i$ has been taken to isolate the UV-divergent region. These graphs are therefore UV finite, so no renormalization is necessary. The interference of these graphs with the tree-level graph contributes at order Q_e^3 and Q_μ^3 in the electron and muon charges, which is the same order as

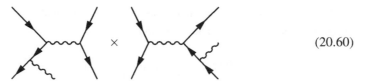

(20.60)

and similar cross terms. Besides the UV finiteness of the loops, there is nothing qualitatively new in these graphs. You can explore them in Problem 20.5.

20.3.1 Vacuum polarization correction

Next, we consider the vacuum polarization graph and its counterterm:

$$i\mathcal{M}_\beta = \quad + \quad . \tag{20.61}$$

The interference between the tree-level amplitude for $e^+ e^- \to \mu^+ \mu^-$ and these graphs gives a contribution to the cross section at order e_R^6. This contribution is proportional to the square of the charges of whatever particle is going around the loop. For a loop involving a generic charge, there are no corresponding real emission graphs of the same order in that charge; thus, any IR divergences must cancel between these graphs alone.

We evaluated these graphs in Section 16.2 (and in Section 19.2.1) for an off-shell photon. Copying over those results, the sum of the loop and its counterterms in this case gives an interference contribution $2\mathrm{Re}\,(\mathcal{M}_0 \mathcal{M}_\beta)$, which leads to a correction to the cross section of the form

$$\Delta\sigma_\beta = -2\mathrm{Re}[\Pi(Q^2)]\sigma_0, \tag{20.62}$$

with

$$\Pi(Q^2) = \frac{e_R^2}{2\pi^2} \sum_j Q_j^2 \int_0^1 dx\, x(1-x) \ln\left(\frac{m_j^2}{m_j^2 - Q^2 x(1-x)}\right). \tag{20.63}$$

For this physical application we have to sum over all particles j with masses m_j and charges Q_j that can go around the loop. This sum therefore includes electrons, muons, quarks, and everything else with electric charge in the Standard Model.

A more suggestive way to write the vacuum polarization contribution is through an effective charge. Recall that it was these same vacuum polarization graphs that contributed to the running of the Coulomb potential. In the Coulomb potential, the virtual photon is spacelike, with $-p^2 > 0$. In Chapter 16, we found that for $-p^2 \gg m^2$ the effective charge at 1-loop was (Eq. (16.65))

$$e_{\mathrm{eff}}^2(-p^2) = e_R^2 \left[1 + \frac{e_R^2}{12\pi^2} \ln \frac{-p^2}{m^2}\right], \tag{20.64}$$

with the convention that $e_R \equiv e_{\mathrm{eff}}(-m^2)$.

Now look at the correction to the cross section, with just one virtual fermion for simplicity and $Q^2 \gg m^2$. Then we can use

$$\Pi(Q^2) = \frac{e_R^2}{12\pi^2} \ln \frac{m^2}{-Q^2} + \text{regular as } \frac{m}{Q} \to 0. \tag{20.65}$$

Now recalling $\sigma_0(Q^2) = \frac{e_R^4}{12\pi Q^2}$ we find

$$\begin{aligned}
\sigma(Q^2) &= \frac{e_R^4}{12\pi Q^2} \left\{ 1 + 2\mathrm{Re}\left[\frac{e_R^2}{12\pi^2} \ln \frac{-Q^2}{m^2} \right] + \mathcal{O}(e_R^4) \right\} \\
&= \frac{1}{12\pi Q^2} \left| e_R^2 + \frac{e_R^4}{12\pi^2} \ln \frac{-Q^2}{m^2} \right|^2 + \mathcal{O}(e_R^8) \\
&= \frac{1}{12\pi Q^2} \left| e_{\text{eff}}(-Q^2) \right|^4 + \mathcal{O}(e_{\text{eff}}^8(-Q^2)).
\end{aligned} \tag{20.66}$$

Including the final state radiation and virtual correction from the muon vertex, we also have

$$\sigma = \frac{e_{\text{eff}}^4(Q^2)}{12\pi Q^2} \left(1 + \frac{3e_{\text{eff}}^2(Q^2)}{16\pi^2} \right) + \mathcal{O}(e_{\text{eff}}^8), \tag{20.67}$$

and thus

> The entire effect of the vacuum polarization graph is encapsulated in the scale-dependent effective charge.

This is true quite generally (as long as the electron mass can be neglected) and explains why an effective charge is such a useful concept.

You may have noticed that in the limit $m \to 0$ the effective charge in Eq. (20.67) appears to be IR divergent. However, since

$$e_{\text{eff}}^2(-Q_1^2) - e_{\text{eff}}^2(-Q_2^2) = \frac{e_R^4}{12\pi^2} \ln \frac{-Q_1^2}{-Q_2^2}, \tag{20.68}$$

as long as the effective charge measured at *some* scale is finite, the charge at any other scale will be finite. In particular, we can measure the charge before neglecting the electron mass, then *run* the charge up to high energy. Or more simply, measure the electric charge through the $e^+e^- \to \mu^+\mu^-$ cross section at some scale Q_1 and predict the effect at Q_2 (if we do this, however, the finite effect from the vertex correction and final state radiation contribution cannot be measured).

Although we only showed the agreement for a single virtual fermion, since the same vacuum polarization graphs correct Coulomb's law as correct the $e^+e^- \to \mu^+\mu^-$ cross section, the agreement will hold with arbitrary charged particles. If there are many particles, it is unlikely that Q will be much much larger than all their masses. Of course, if $Q \ll m_j$ for some mass, that particle has little effect (the logarithm in Eq. (20.63) goes to zero). But we may measure the cross section at various Q above and below some particle thresholds. In this case, the effective charge changes, sometimes even discontinuously. Physical observables (such as cross sections) are not discontinuous, since finite corrections to the cross section exactly cancel the discontinuities of the effective charge.[2]

Note that the way we have defined the effective charge, through the Coulomb potential where p^μ is spacelike, $e_{\text{eff}}(-p^2)$ is naturally evaluated at a positive argument. Here we see that to use the same charge for $e^+ e^- \to \mu^+ \mu^-$ it must be evaluated at a negative argument, $e_{\text{eff}}(-Q^2)$ with $Q^2 > 0$. In fact, it is natural for a process with a timelike intermediate state to have a factor such as $\ln \frac{-Q^2}{m^2}$ with a non-zero imaginary part. This imaginary part is actually required by unitarity, as will be discussed in Chapter 24. It also has a measurable effect, through terms such as the π^2 that contributed to the real part of the virtual amplitude in going from Eq. (20.18) to Eq. (20.20). This π^2 *does* contribute a non-zero amount to the cross section. In fact, since π^2 is not a small number, π^2 corrections can sometimes provide the dominant subleading contribution to a cross section. For example, they can be shown to account for a large part of the approximate doubling of the $pp \to e^+ e^-$ cross section at next-to-leading order [Magnea and Sterman, 1990].

20.3.2 Initial state radiation

Finally, we need to discuss the contributions to the $e^+ e^- \to \mu^+ \mu^- (+\gamma)$ cross section to third order in the electron charge and first order in the muon charge. In other words, the following diagrams:

$$ \tag{20.69} $$

In the same way that final state radiation was necessary to cancel the IR singularity of the vertex correction involving the photon, the sum of these diagrams will be finite. The radiation coming off the electrons in this process is known as **initial state radiation**. These real emission graphs are closely related to the real emission graphs with the photon coming off the muons, and their integrals over phase space have IR divergences. However, the IR-divergent region is a little different and the physical interpretation of the divergences is very different.

Let us suppose that the sum of the diagrams in Eq. (20.69) gives a finite total cross section for $e^+ e^- \to \mu^+ \mu^- (+\gamma)$ we call σ_{tot}. Then we should be able to calculate a more exclusive two-jet cross section, as in the previous section, for producing less than E_{res} of energy outside of cones of half-angle θ_{res} around the muons. In this case, however, there is no collinear singularity with the photons going collinearly to the muons. Instead, the IR divergences come from the intermediate electron propagator going on-shell. This propagator has a factor of

---

[2] The effective charge is regulator and subtraction scheme dependent. In the on-shell scheme, the effective charge is very difficult to calculate through particle thresholds. It is therefore more common to use dimensional regularization with minimal subtraction to define the effective charge. In particular, in QCD, where the thresholds are very important for the effective strong coupling constant α_s, $\overline{\text{MS}}$ is almost exclusively used, and there the effective charge is known to 4-loop order.

and positrons have variable energy in a typical beam, real or virtual soft photons were often emitted from the initial state to bring the Z to the resonance peak, a process called **radiative return**. The result was that you could just measure the decay of the Z and ignore the initial state completely. Thus, you only need the final state loops. The decay width is calculable, finite, and does not depend on whether it was e^+e^- or something else that produced the Z. In fact, $Z \to \mu^+\mu^-(+\gamma)$ gets precisely the $\frac{3e^2}{16\pi^2}$ correction in QED we calculated for the $\sigma_{\text{tot}}\left(e^+e^- \to \mu^+\mu^-\left(+\gamma\right)\right)$ rate in Eq. (20.51). Because the Z decays not just to muons but also to quarks, which have charges $\pm\frac{2}{3}$ or $\pm\frac{1}{3}$, this correction becomes $\frac{3e^2}{16\pi^2}Q_i^2$ and is therefore a way to test the Standard Model. In particular, the branching ratio for $Z \to b\bar{b}$ has proven a particularly powerful way to look for physics beyond the Standard Model, since it happens to be sensitive not just to loops involving electrons, but also to loops involving hypothetical particles (such as charged Higgses). On the other hand, if you want to calculate the line shape of the Z boson in the resonance region, then initial state radiation is important. Indeed, the importance of the large logarithms, as in Eq. (20.72) has been experimentally validated at LEP.

By the way, there is actually an interesting difference between initial state radiation in QED and QCD. In QED, there is an important theorem due to Bloch and Nordsieck [Bloch and Nordsieck, 1937], which says:

Box 20.1   **Bloch–Nordsieck theorem**

Infrared singularities will always cancel when summing over final state radiation in QED with a massive electron as long as there is a finite energy resolution.

In QCD, this is not true. At 2-loops, IR singularities in QCD with massive quarks will not cancel when summing over $2 \to n$ processes only; one also needs to sum over $3 \to n$ processes [Doria *et al.*, 1980]. The uncanceled singularity, however, vanishes as a power of the quark mass and therefore disappears as $\frac{m_q}{Q} \to 0$. Thus, in the high-energy limit of QCD, where the mass can be neglected, one can get an IR-finite answer summing only cross sections with two particles in the initial state. (This result has nothing to do with QCD being asymptotically free, and would hold even if there were enough flavors so that QCD were infrared free, like QED.)

A more general theorem, due to Kinoshita, Lee and Nauenberg (KLN) [Kinoshita, 1962; Lee and Nauenberg, 1964] is that

---

Box 20.2   **Kinoshita–Lee–Nauenberg (KLN) theorem**

Infrared divergences will cancel in any unitary theory when all possible final *and* initial states in a finite energy window are summed over.

The KLN theorem is mostly of formal interest, since we do not normally sum over initial states when computing cross sections. Proofs of the Bloch–Nordsieck and KLN theorems can be found in [Sterman, 1993].

20.A Dimensional regularization

The calculation of the total cross section for $e^+ e^- \to \mu^+ \mu^- \, (+\gamma)$ at next-to-leading order can also be done in dimensional regularization. Repeating the calculation this way helps illustrate regulator independence of physical quantities and will give us some practice with dimensional regularization.

20.A.1 $e^+ e^- \to \mu^+ \mu^-$

The first step is to calculate the tree-level cross section in $d = 4 - \varepsilon$ dimensions. It is of course non-singular as $\varepsilon \to 0$; however, we will need the $\mathcal{O}(\varepsilon)$ parts of the cross section for the virtual correction. We work in the limit $E_{\mathrm{CM}} = Q \gg m_e, m_\mu$ so that we can treat the fermions as massless. We first write an expression for a general $e^+ e^- \to \gamma^\star \to X$ process, then specialize to $e^+ e^- \to \mu^+ \mu^-$.

To calculate the cross section for $e^+ e^- \to \gamma^\star \to X$, we use the observation from Section 20.1.2 that the cross section factorizes into $e^+ e^- \to \gamma^\star$ and $\gamma^\star \to X$. In d dimensions we can still write

$$\sigma_R = \frac{1}{2Q^2} \int |\mathcal{M}|^2 d\Pi_{\mathrm{LIPS}} = \frac{e_R^4}{2Q^6} L^{\mu\nu} X_{\mu\nu}, \qquad (20.A.75)$$

with the electron tensor exactly as in Eq. (20.29):

$$L^{\mu\nu} = \frac{1}{4} \mathrm{Tr}\left[\slashed{p}_2 \gamma^\mu \slashed{p}_1 \gamma^\nu\right] = p_1^\mu p_2^\nu + p_1^\nu p_2^\mu - \frac{Q^2}{2} g^{\mu\nu}. \qquad (20.A.76)$$

The other tensor $X_{\mu\nu}$ is the matrix element squared for a generic $\gamma^\star \to X$ final state averaged over $\gamma^\star$ spins integrated over the associated Lorentz–invariant phase space. This definition makes the total decay rate have the form

$$\Gamma(\gamma^\star \to X) = -\frac{e_R^2}{2Q} g^{\mu\nu} X_{\mu\nu}, \qquad (20.A.77)$$

with the $-g_{\mu\nu}$ coming from a polarization sum over the $\gamma^\star$, assuming the Ward identity holds. Indeed, the Ward identity does hold in d dimensions, since dimensional regularization preserves gauge invariance, and so we can still write

$$X^{\mu\nu} = \left(p^\mu p^\nu - p^2 g^{\mu\nu}\right) X\left(p^2\right). \qquad (20.A.78)$$

However, in d dimensions, $X\left(Q^2\right) = -\frac{1}{(d-1)Q^2} g^{\mu\nu} X_{\mu\nu}$ and

$$L^{\mu\nu} X_{\mu\nu} = \frac{(d-2)Q^4}{2} X\left(Q^2\right) = -\frac{1}{2}\left(\frac{d-2}{d-1}\right) Q^2 g^{\mu\nu} X_{\mu\nu}, \qquad (20.A.79)$$

and therefore

$$\sigma\left(e^+ e^- \to X\right) = -\frac{e_R^4}{4Q^4} \mu^{2(4-d)} \left(\frac{d-2}{d-1}\right) g^{\mu\nu} X_{\mu\nu}. \qquad (20.A.80)$$

Using $\sigma_0\left(Q^2\right) = \frac{e_R^4}{12\pi Q^2}$ as before we can write this alternatively as

$$\sigma\left(e^+e^- \to X\right) = \sigma_0\mu^{2(4-d)}\frac{3\pi}{Q^2}\left(\frac{d-2}{d-1}\right)\left(-g^{\mu\nu}X_{\mu\nu}\right)$$

$$= \sigma_0\mu^{2(4-d)}\frac{6\pi}{Qe_R^2}\frac{d-2}{d-1}\Gamma(\gamma^\star \to X)\,, \qquad (20.A.81)$$

which reduces to Eq. (20.36) in four dimensions.

For the tree-level process, we need $\gamma^\star \to \mu^+\mu^-$ for which $X_{\mu\nu}$ is just like $L_{\mu\nu}$ but with the phase space tacked on. Then,

$$-g_{\mu\nu}X^{\mu\nu} = g_{\mu\nu}\left(2Q^2g^{\mu\nu} - 4p_3^\mu p_4^\nu - 4p_3^\nu p_4^\mu\right)\int d\Pi_{\mathrm{LIPS}} = 2(d-2)\,Q^2\int d\Pi_{\mathrm{LIPS}}\,. \qquad (20.A.82)$$

Since there is no angular dependence in the spin-summed $\gamma^\star \to \mu^+\mu^-$, this phase space is straightforward to evaluate:

$$\int d\Pi_{\mathrm{LIPS}} = (2\pi)^d\int\frac{d^{d-1}p_3}{(2\pi)^{d-1}}\frac{d^{d-1}p_4}{(2\pi)^{d-1}}\frac{1}{(2E_3)\,(2E_4)}\delta^d\left(p_3+p_4-p\right)\,. \qquad (20.A.83)$$

We first rescale the momenta by $p_i = \frac{Q}{2}\hat{p}_i$ to make them dimensionless. We also use $x_i = \frac{2}{Q}E_i$ as the energy components of the rescaled momenta. Then, evaluating the p_4 integral over the δ-function we get

$$\int d\Pi_{\mathrm{LIPS}} = (2\pi)^{2-d}\left(\frac{Q}{2}\right)^{d-2}\frac{1}{Q^2}\int\frac{d^{d-1}\hat{p}_3}{x_3 x_4}\delta\left(x_3+x_4-2\right)\,, \qquad (20.A.84)$$

where x_4 is an implicit function of $\hat{p}_3^\mu$ determined by spatial momentum conservation and the mass-shell conditions. Explicitly $x_4 = |\vec{\hat{p}}_3| = x_3$. So,

$$\int d\Pi_{\mathrm{LIPS}} = \left(\frac{Q}{4\pi}\right)^{d-2}\frac{1}{Q^2}\int\frac{d^{d-2}x_3}{x_3}\delta\left(2x_3-2\right)\int d\Omega_{d-1}$$

$$= \left(\frac{Q}{4\pi}\right)^{d-2}\frac{1}{2Q^2}\Omega_{d-1}$$

$$= \left(\frac{4\pi}{Q^2}\right)^{\frac{4-d}{2}}\frac{2^{-d}}{\sqrt{\pi}\,\Gamma\left(\frac{d-1}{2}\right)}\,. \qquad (20.A.85)$$

Combining this with Eqs. (20.A.81) and (20.A.82),

$$\sigma_0^d\left(e^+e^- \to \mu^+\mu^-\right) = \sigma_0\mu^{2(4-d)}\left(\frac{4\pi}{Q^2}\right)^{\frac{4-d}{2}}\frac{3\sqrt{\pi}\,(d-2)^2}{2^d\Gamma\left(\frac{d+1}{2}\right)}\,, \qquad (20.A.86)$$

which reduces to σ_0 in $d=4$.

20.A.2 Loops

Next, we will compute the loop amplitude in pure dimensional regularization. The easiest way to do the calculation is by evaluating the form factor, which corrects the $-ie_R\gamma^\mu$

vertex. Then we can use the result for the phase-space integral in d dimensions we have already calculated. To make sure we get all the factors of d correct, we will compute the loop from scratch.

The loop gives

$$-ie_R\mu^{\frac{4-d}{2}}\bar{u}\left(q_2\right)\Gamma_2^\mu v\left(q_1\right) =$$

$$= -\left(e_R\mu^{\frac{4-d}{2}}\right)^3 \int \frac{d^d k}{(2\pi)^d}\frac{\bar{u}(q_2)\gamma^\nu(\slashed{k}+\slashed{q_2})\gamma^\mu(\slashed{k}-\slashed{q_1})\gamma^\nu v(q_1)}{[(k+q_2)^2+i\varepsilon][(k-q_1)^2+i\varepsilon][k^2+i\varepsilon]}. \qquad (20.\text{A}.87)$$

We can simplify this using $\slashed{q_1}v(q_1)=\bar{u}(q_2)\slashed{q_2}=q_1^2=q_2^2=0$ and $q_1\cdot q_2=\frac{Q^2}{2}$. Using Feynman parameters for the three denominator factors we get

$$\bar{u}(q_2)\Gamma_2^\mu v(q_1)=-ie_R^2\mu^{4-d}\int_0^1 dx\int_0^{1-x}dy\int\frac{d^d k}{(2\pi)^d}$$

$$\times \frac{\bar{u}(q_2)N^\mu(k,q_1,q_2)v(q_1)}{[(k+xq_2-yq_1)^2+Q^2 xy+i\varepsilon]^3} \qquad (20.\text{A}.88)$$

with

$$N^\mu = 2\left[(d-2)k^2+4k\cdot q_2-4k\cdot q_1-2Q^2\right]\gamma^\mu-4\left[(d-2)k^\mu+2q_2^\mu-2q_1^\mu\right]\slashed{k}. \qquad (20.\text{A}.89)$$

Shifting $k^\mu\to k^\mu-xq_2^\mu+yq_1^\mu$ and dropping terms linear in k turns the numerator into

$$N^\mu = 2\left[(d-2)k^2+Q^2((2-d)xy+2x+2y-2)\right]\gamma^\mu-4(d-2)k^\mu\slashed{k}. \qquad (20.\text{A}.90)$$

Using $k^\mu k^\nu\to\frac{k^2}{d}g^{\mu\nu}$, as discussed in Appendix B, we can then replace $k^\mu\slashed{k}=\gamma^\alpha g^{\alpha\nu}k^\mu k^\nu\to\frac{k^2}{d}\gamma^\mu$ giving

$$\Gamma_2^\mu = -2i\gamma^\mu e_R^2\mu^{4-d}\int dx\,dy\frac{d^d k}{(2\pi)^d}\frac{\frac{(d-2)^2}{d}k^2+Q^2((2-d)xy+2x+2y-2)}{(k^2+Q^2 xy+i\varepsilon)^3}. \qquad (20.\text{A}.91)$$

This has two terms: the k^2 term is UV divergent, and the Q^2 term is IR divergent.

The k^2 term can be evaluated with $d<4$ using

$$\int\frac{d^d k}{(2\pi)^d}\frac{k^2}{(k^2-\Delta+i\varepsilon)^3}=i\frac{d/4}{(4\pi)^{d/2}}\frac{1}{\Delta^{2-\frac{d}{2}}}\Gamma\left(2-\frac{d}{2}\right), \qquad (20.\text{A}.92)$$

with $\Delta=-Q^2 xy$ to get

$$\int_0^1 dx\int_0^{1-x}dy\int\frac{d^d k}{(2\pi)^d}\frac{\frac{(d-2)^2}{d}k^2}{(k^2-(-Q^2 xy)+i\varepsilon)^3}=\frac{i}{16\pi^2}\left(\frac{4\pi}{-Q^2}\right)^{\frac{4-d}{2}}\frac{\Gamma\left(\frac{4-d}{2}\right)\Gamma\left(\frac{d}{2}\right)^2}{\Gamma(d-1)}$$

$$=\frac{i}{16\pi^2}\left(\frac{4\pi}{-Q^2}\right)^{\frac{4-d}{2}}\left[\frac{1}{\varepsilon_{\text{UV}}}-\frac{\gamma_E}{2}+\frac{1}{2}+\mathcal{O}(\varepsilon_{\text{UV}})\right]. \qquad (20.\text{A}.93)$$

Finally, adding in Eq. (20.A.102),

$$\sigma_V^d = \sigma_0 \frac{e_R^2}{\pi^2} \left(\frac{4\pi e^{-\gamma_E} \mu^2}{Q^2} \right)^{4-d} \left(-\frac{1}{\varepsilon^2} - \frac{13}{12\varepsilon} + \frac{5\pi^2}{24} - \frac{29}{18} + \mathcal{O}(\varepsilon) \right) \qquad (20.A.117)$$

gives a total cross section of

$$\sigma_R^d + \sigma_V^d = \sigma_0 \frac{3e_R^2}{16\pi^2} + \mathcal{O}(\varepsilon), \qquad (20.A.118)$$

which is finite as $\varepsilon \to 0$ and exactly the result we found with Pauli–Villars and a photon mass, Eq. (20.50).

Problems

20.1 Derive the phase space formula in Eq. (20.42).

20.2 Calculate the Sterman–Weinberg jet rates in Eqs. (20.55) and (20.56).

20.3 Calculate the total cross section for $e^+ e^- \to \mu^+ \mu^- (+\gamma)$ including the initial state radiation contribution.

20.4 Calculate the cross section for $e^+ e^- \to \mu^+ \mu^-$ directly in dimensional regularization, without factorizing into $e^+ e^- \to \gamma^\star$ and $\gamma^\star \to \mu^+ \mu^-$.

20.5 Calculate the box and crossed box loop graphs in Eq. (20.58). Are they IR divergent?

20.6 Calculate the splitting function for the QED function in Eq. (20.73).

At this point, we have calculated some 2-, 3- and 4-point functions in QED where we found three UV-divergent 1-loop graphs:

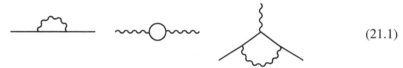

$$(21.1)$$

We saw that these UV divergences were artifacts of not computing something physical, since the UV-divergent answer was calculated using parameters in a Lagrangian that were not defined based on observables. More precisely, we saw that the normalizations of the electron and photon fields were not observable, and so these fields could be rescaled by wavefunction renormalization factors $Z_2 = 1 + \delta_2$ and $Z_3 = 1 + \delta_3$, with the counterterms δ_3 and δ_2 dependent on the UV regularization and subtraction scheme. We also saw that the bare electric charge parameter e_0 appearing in the Lagrangian and the bare Lagrangian electron mass parameter m_0 could be redefined keeping physical quantities (such as the charge measured by Coulomb's law at large distances and the location of the pole in the electron propagator) finite. This introduced two new counterterms, δ_1 and δ_m. We found that these same counterterms, and the four associated renormalization conditions that define them to all orders in perturbation theory, made all the 2-, 3- and 4-point functions we have so far considered finite.

The next question we will address is: Will this always be the case? Can these same four counterterms remove *all* of the infinities in QED? If so, QED is renormalizable. The general definition of renormalizable is

Renormalizable	**Box 21.1**

In a renormalizable theory, all UV divergences can be canceled with a finite number of counterterms.

It will not be hard to show that QED is renormalizable at 1-loop. The important observation is that UV divergences are the same whether or not the external legs are on-shell; they come from regions of loop momenta with $k \gg p_i$ for any external momentum p_i. In particular, the same counterterms will cancel the UV divergences of divergent graphs even when the 1-loop graphs are subgraphs in more complicated higher-order correlation functions. We saw this explicitly in Chapter 20 for $e^+ e^- \to \mu^+ \mu^-$: the counterterms we derived from 2- and 3-point functions removed the UV divergences in this 4-point case.

Recall that we introduced the notion of one-particle irreducibility when trying to deal with mass renormalization in Chapter 18. By summing 1PI graphs in external lines we justified using the exact renormalized propagator (with a pole at the physical mass) instead of the bare propagator. Now we see that we only need to look at 1PI graphs when trying to figure out what UV divergences are present. Our previous definition of 1PI was those graphs that could not be cut in two by slicing a single propagator. An equivalent definition, more useful for our present purposes, is

---

Box 21.2　　　　　　　　　**One-particle irreducible (1PI)**

A Feynman diagram is 1PI if all internal lines have some loop momentum going through them.

Any graph involved in the computation of any Green's function can be computed by sewing together 1PI graphs, with off-shell momenta, without doing any additional integrals. Thus, if the four QED counterterms cancel all the UV divergences in 1PI graphs, they will cancel the UV divergences in any Green's function. Keep in mind that for S-matrix elements we need to compute all (amputated) graphs, but for studying general properties of renormalizability it is enough to consider only the 1PI graphs.

It will be useful to consider a quantity D, the **superficial degree of divergence**, defined as the overall power of loop momenta k_i in the loop integrals, including the powers of k_i in the various $d^4 k_i$. For example, we say $\int d^4 k k^{-2}$ has $D = 2$ and $\int d^4 k_1 \int d^4 k_2 k_1^{-4} k_2^{-4}$ has $D = 0$. If we cut off all the components of all the k_μ^i at a single scale Λ, then a graph with degree of divergence D scales as Λ^D as we take $\Lambda \to \infty$ and as $\ln \Lambda$ for $D = 0$.

21.1 Renormalizability of QED

To approach renormalizability, we will continue our systematic study of removing infinities in Green's functions (which we began in Chapter 19), focusing on 1PI graphs. We have already shown that the QED counterterms cancel the UV divergences in all the 2- and 3-point functions in QED. So now we continue to 4-point and higher-point functions.

21.1.1 Four-point functions

Let us first consider the Green's function with four fermions, $\langle \Omega | T \{ \bar{\psi} \psi \bar{\psi} \psi \} | \Omega \rangle$. We evaluated this correlation function in Chapter 20 for $e^+ e^- \to \mu^+ \mu^-$ and found it to be UV finite. The only 1PI graph contributing to the scattering amplitude based on the 4-point function is

$$\langle \bar{\psi} \psi \bar{\psi} \psi \rangle \sim \quad \quad \sim \int \frac{d^4 k}{(2\pi)^4} \frac{1}{k^2} \frac{1}{k^2} \frac{1}{\not{k}} \frac{1}{\not{k}} \sim \frac{1}{\Lambda^2} \qquad (21.2)$$

or one of its various crossings. The notation $\langle \bar{\psi}\psi\bar{\psi}\psi \rangle$ means a (possibly) off-shell S-matrix element involving four external fermions and $\sim$ means expand the integrand in the limit that $k \gg p_i, m_i$ and then cutoff $|k^\mu| < \Lambda$. Since this amplitude scales as Λ^{-2} (its superficial degree of divergence is $D = -2$), it is not UV divergent. Therefore, no renormalization is required in the computation of this graph.

Note that, in the limit $k \gg p_i, m_i$, whether the lines are on-shell or off-shell is irrelevant. Also, because all propagators have some factor of loop momentum in them (by definition of 1PI), a 1-loop diagram can never be more divergent than its superficial degree of divergence. Thus, for 1-loop 1PI graphs, if $D < 0$ the graph is not UV divergent.

Next, the 1PI contribution to the two-fermion and two-photon Green's function is

$$\langle \bar{\psi}\psi AA \rangle \sim \quad \sim \int \frac{d^4 k}{(2\pi)^4} \frac{1}{k^2} \frac{1}{\not{k}} \frac{1}{\not{k}} \frac{1}{\not{k}} \sim \frac{1}{\Lambda}. \tag{21.3}$$

This has $D = -1$ and is also not UV divergent.

Finally, the last non-vanishing 4-point function is the four-photon function, which describes light-by-light scattering $\gamma\gamma \to \gamma\gamma$:

$$\mathcal{M} = \langle AAAA \rangle \sim \quad \sim \int \frac{d^4 k}{(2\pi)^4} \frac{1}{\not{k}} \frac{1}{\not{k}} \frac{1}{\not{k}} \frac{1}{\not{k}} \sim \Lambda^0. \tag{21.4}$$

This one has $D = 0$ and appears logarithmically divergent. However, we know that after regulating and performing the integrals, the result must be linear in the four photon polarizations and therefore have the form

$$\mathcal{M} = \epsilon_1^\mu \epsilon_2^\nu \epsilon_3^{\rho\star} \epsilon_4^{\sigma\star} M_{\mu\nu\rho\sigma}. \tag{21.5}$$

By Lorentz invariance, dimensional analysis, and symmetry under the interchange of the photons, $M_{\mu\nu\rho\sigma}$ must have the form

$$M_{\mu\nu\rho\sigma} = c \ln \Lambda^2 (g_{\mu\nu}g_{\rho\sigma} + g_{\mu\rho}g_{\nu\sigma} + g_{\mu\sigma}g_{\nu\rho}) + \text{finite} \tag{21.6}$$

for some constant c. We also know by the Ward identity that this must vanish when any one of the photons is replaced by its momentum. Say A_μ has momentum q_μ. Then

$$0 = q^\mu M_{\mu\nu\rho\sigma} = c \ln \Lambda^2 (q_\nu g_{\rho\sigma} + q_\rho g_{\nu\sigma} + q_\sigma g_{\nu\rho}) + q^\mu \cdot \text{finite}. \tag{21.7}$$

This must hold for all q^μ, which is impossible unless $c = 0$, and therefore this loop must be UV finite. (The loop is actually quite a mess to compute; the low-energy limit of the result will be computed using effective actions in Chapter 33.)

Renormalizability in QED means that all the UV divergences are canceled by the same four counterterms we introduced at 1-loop. These are fit by two numbers: the physical value of the electric charge e_R (measured in Coulomb's law at long distance) and the physical value of the electron mass, m_P. The other two counterterms are fixed by normalizing the electron and photon fields. This is actually a pretty amazing conclusion: QED is completely specified once e_R and m_P are measured;[1] after that, we can make an infinite number of arbitrarily precise predictions. The two initial measurements are needed to define even the classical theory. In the quantum theory, both logarithmic corrections can be calculated, such as to the scale-dependent effective charge (Chapter 16), as well as finite corrections, such as to the value of the anomalous magnetic moment (Chapter 17) or the $e^+e^- \to \mu^+\mu^-$ total cross section (Chapter 20).

Renormalizability played a very important role in the historical development of quantum field theory and gauge theories. In particular, 't Hooft's 1971 proof ['t Hooft, 1971] that spontaneously broken gauge theories were renormalizable made people take Weinberg's model of leptons seriously. Weinberg's model, which has now evolved into the Standard Model, had been proposed in 1967 and was subsequently ignored over concerns of renormalizability.

21.2 Non-renormalizable field theories

All else being equal, renormalizability is a desirable property for a theory to have: an infinite number of predictions follow from a finite number of measurements. Unfortunately, to make these predictions we have to be able to perform computations in the renormalizable theory. In practice, this is extremely challenging. Not only are loops difficult to evaluate, but perturbation theory in the coupling constants of a renormalizable theory often breaks down. For example, as we saw in Section 16.3.2, QED has a Landau pole. Thus, Coulomb scattering above $E = 10^{286}$ eV is a complete mystery in QED. In other words, *we cannot predict every observable just because we can cancel all the UV divergences*. Moreover, precisely because QED is renormalizable, low-energy measurements are totally insensitive to whatever completion QED might have above the Landau pole. That is, we have no way of probing the mysterious high-energy regime without building a 10^{286} eV collider. Other renormalizable theories are unpredictive in much more relevant regimes. For example, QCD does not make perturbative predictions below ~ 1 GeV. Or consider string theory, which is not only renormalizable but actually finite: it has no UV divergences. Despite its formal beauty, string theory has yet to relate any observable to any other observable at all.

A more modern view is that if one is interested in actually making physical predictions, renormalizability (or finiteness in the case of string theory) is somewhat irrelevant.

---

[1] In pure QED, only one measurement would actually be needed, since the electron mass m_P is dimensionful. This measurement would give $\frac{m_P}{\Lambda_{\text{QED}}}$, where Λ_{QED} is the location of the Landau pole, which is in one-to-one correspondence with e_R.

Table 21.1 Superficial degree of divergence $D_{f,b} = 4 - \frac{3}{2}f - b$ for a process with f fermions and b bosons.

$\langle AA \rangle$	$\int \frac{d^4k}{(2\pi)^4} \frac{1}{\not{k}} \frac{1}{\not{k}} \sim \Lambda^2$	$D_{0,2} = 2$
$\langle \bar\psi\psi \rangle$	$\int \frac{d^4k}{(2\pi)^4} \frac{1}{k^2} \frac{1}{\not{k}} \sim \Lambda^1$	$D_{2,0} = 1$
$\langle \bar\psi\psi A \rangle$	$\int \frac{d^4k}{(2\pi)^4} \frac{1}{k^2} \frac{1}{\not{k}} \frac{1}{\not{k}} \sim \Lambda^0$	$D_{2,1} = 0$
$\langle AAAA \rangle$	$\int \frac{d^4k}{(2\pi)^4} \frac{1}{\not{k}} \frac{1}{\not{k}} \frac{1}{\not{k}} \frac{1}{\not{k}} \sim \Lambda^0$	$D_{0,4} = 0$
$\langle \bar\psi\psi AA \rangle$	$\int \frac{d^4k}{(2\pi)^4} \frac{1}{k^2} \frac{1}{\not{k}} \frac{1}{\not{k}} \frac{1}{\not{k}} \sim \frac{1}{\Lambda}$	$D_{2,2} = -1$
$\langle \bar\psi\psi\bar\psi\psi \rangle$	$\int \frac{d^4k}{(2\pi)^4} \frac{1}{k^2} \frac{1}{k^2} \frac{1}{\not{k}} \frac{1}{\not{k}} \sim \frac{1}{\Lambda^2}$	$D_{4,0} = -2$

In many contexts, non-renormalizable theories are in fact much more useful than renormalizable ones, despite the fact that renormalizable theories have fewer parameters. To understand better the connection between renormalizability and predictability, we first have to examine non-renormalizable theories. We will study their UV divergence structure for the remainder of this chapter, and give a number of concrete examples in the next chapter. Non-renormalizable theories will play an increasingly important role as we progress through Parts IV and V as well.

21.2.1 Divergences in non-renormalizable theories

We saw that in QED there are a finite number of superficially divergent one-particle irreducible contributions to off-shell scattering amplitudes. The superficial degree of divergence of a scattering amplitude is well defined, because, as we showed in the previous section, inserting additional photon or fermion propagators into a loop does not change the degree of divergence. Call the superficial degree of divergence of a scattering amplitude with f fermions and b photons $D_{f,b}$. Some example amplitudes and values of $D_{f,b}$ are shown in Table 21.1.

It is not hard to work out the general formula:

$$D_{f,b} = 4 - \frac{3}{2}f - b. \tag{21.13}$$

With scalar external states, the generalization is

$$D_{f,b,s} = 4 - \frac{3}{2}f - b - s, \tag{21.14}$$

where s is the number of scalars being scattered. Divergent 1PI graphs can only possibly contribute to Green's functions with $D > 0$.

Besides counting loop momentum factors in integrals, another way to derive Eq. (21.13) is to recall that the LSZ reduction formula relates Green's functions to matrix elements by

$$\delta^4(\Sigma p)\, \mathcal{M} \sim \prod_{i=1}^{b} \int d^4 x_i \, e^{\pm i p_i x_i} \, \Box_i \cdots \prod_{j=1}^{f} \int d^4 y_j \, e^{\pm i q_j y_j} \, \partial_j \cdots$$

$$\times \langle \psi_1(y_1) \cdots \psi_f(y_f)\, A_1(x_1) \cdots A_b(x_b) \rangle. \quad (21.15)$$

Since fermions have dimension $\frac{3}{2}$ and photons dimension 1, the actual Green's function has dimension $\frac{3}{2}f + b$. The x_i and y_j integrals and prefactors have mass dimension $-2b - 3f$, and the δ-function has dimension -4. Thus, the dimension of $\mathcal{M}$ is $\frac{3}{2}f + b - 2b - 3f + 4 = 4 - \frac{3}{2}f - b$, as in Eq. (21.13).

QED is a special theory because it only has a single interaction vertex:

$$\mathcal{L}_{\text{QED}} = \mathcal{L}_{\text{kin}} - e\bar{\psi}A_\mu \gamma^\mu \psi. \quad (21.16)$$

The coefficient of this interaction is the dimensionless charge e. More generally, we might have a theory with couplings of arbitrary dimension. For example,

$$\mathcal{L} = -\frac{1}{2}\phi(\Box + m^2)\phi + g_1 \phi^3 + g_2 \phi^2 \Box \phi^3 + \cdots . \quad (21.17)$$

These additional couplings change the power counting.

Call the mass dimension of the coefficient of the ith interaction Δ_i. For example, the g_1 term above has $\Delta_1 = [g_1] = 1$ and g_2 has $\Delta_2 = [g_2] = -3$. Now consider a loop contribution to a Green's function with n_i insertions of the vertices with dimension Δ_i. For $k \gg p_i$, the only scales that can appear are k's and Δ's. So, by dimensional analysis, the superficial degree of divergence of the same integral changes as

$$\int k^D \to \left(\prod_i g_i^{n_i} \right) \int k^{D - \sum_i n_i \Delta_i}. \quad (21.18)$$

Thus,

$$D_{f,b,n_i} = 4 - \frac{3}{2}f - b - \sum_i n_i \Delta_i. \quad (21.19)$$

So, if there are interactions with $\Delta_i < 0$, then there can be an infinite number of values of n_i, and therefore an infinite number of values of f and b with $D_{f,b,n_i} > 0$. This means that there are an infinite number of Green's functions for which some 1PI graph has $D > 0$. Thus, we will need an infinite number of counterterms to cancel all the infinities. Such theories are called **non-renormalizable**.

We generalize this terminology also to describe individual interactions. We also sometimes describe interactions of dimension 0 as **marginal**, dimension >0 as **relevant**, and dimension <0 as **irrelevant**. These terms come from the Wilsonian renormalization group and will be discussed in Chapter 23.

Non-renormalizable interactions are those of mass dimension $\Delta_i < 0$. Having any non-renormalizable interaction term in the Lagrangian makes a theory non-renormalizable. On the other hand, if all the interactions have mass dimension $\Delta_i > 0$, then the theory is called **super-renormalizable** (for example, $\mathcal{L} = -\frac{1}{2}\phi\Box\phi + \frac{g}{3!}\phi^3$ describes a super-renormalizable theory).

It is worth pointing out that a theory can also be non-renormalizable due to the propagators generating new divergences. For example, consider the theory of a massive vector boson. Recall that the propagator for a massive spin-1 field is

$$i\Pi^{\mu\nu} \sim \frac{-i\left(g^{\mu\nu} - \frac{p^\mu p^\nu}{m^2}\right)}{p^2 - m^2 + i\varepsilon} \tag{21.20}$$

At high energy, $p \gg m$, this goes as $\frac{1}{m^2}$ not $\frac{1}{p^2}$. Thus, each loop contribution with a massive vector propagator contributes two more factors of k^2 than the corresponding loop with a photon. For example, adding a massive spin-1 particle to the light-by-light box diagram (the ~► indicates the massive spin-1 particle)

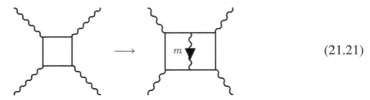

$$\tag{21.21}$$

turns a superficially logarithmically divergent integral into a quadratically divergent one:

$$\int \frac{d^4k}{(2\pi)^4} \frac{1}{\not{k}}\frac{1}{\not{k}}\frac{1}{\not{k}}\frac{1}{\not{k}} \sim \Lambda^0 \quad \to \quad \int \frac{d^4k}{(2\pi)^4} \frac{1}{\not{k}}\frac{1}{\not{k}}\frac{1}{\not{k}}\frac{1}{\not{k}} \int \frac{d^4k}{(2\pi)^4} \frac{1}{m^2}\frac{1}{\not{k}}\frac{1}{\not{k}} \sim \frac{\Lambda^2}{m^2}. \tag{21.22}$$

Thus, there are an infinite number of superficially divergent Feynman diagrams for a theory with a massive vector boson, and hence such theories are not renormalizable. That is not to say that they cannot be renormalized (they can!), but only that all of the UV divergences cannot be canceled with a finite number of counterterms.

21.2.2 Non-renormalizable theories are renormalizable

Although non-renormalizable theories have an infinite number of superficially divergent integrals, that does not mean that they give nonsense (infinities) for observables. Instead, non-renormalizable theories can be renormalized, but only by continually adding terms to the Lagrangian to provide counterterms to cancel divergences. While such a procedure seems like it would destroy the predictivity of a theory, in fact non-renormalizable theories are still extremely predictive.

As usual, let us start with an example. Consider the Lagrangian

$$\mathcal{L} = -\frac{1}{2}\phi(\Box + m^2)\phi + \frac{g}{4!}\phi^2\Box\phi^2, \tag{21.23}$$

where g has mass dimension -2. A 1-loop amplitude involving this vertex could generate a contribution to the 4-point amplitude:

$$\sim gp^2 + g^2(c_1\Lambda^4 + c_2\Lambda^2 p^2 + c_3 p^4 \ln\Lambda + \cdots), \tag{21.24}$$

where p refers generically to some external momentum and c_i are numbers. (The exact expression will have many terms with many different momenta involved.) If we had only g to renormalize, only the $c_2 \Lambda^2 p^2$ divergence could be removed. This follows because the tree-level contribution $g p^2$ has the same momentum dependence as the $c_2 \Lambda^2 p^2$ divergence. To remove the other divergences, we have to add more terms. So let us enlarge our Lagrangian to

$$\mathcal{L} = -\frac{1}{2}\phi(\Box + m^2)\phi + \lambda_R Z_\lambda \phi^4 + g_R Z_g \phi^2 \Box \phi^2 + \kappa_R Z_\kappa \phi^2 \Box^2 \phi^2 + \cdots, \quad (21.25)$$

expanding $Z_\lambda = 1 + \delta_\lambda$, $Z_g = 1 + \delta_g$ and $Z_\kappa = 1 + \delta_\kappa$, the counterterm contribution to the 4-point function is

$$\text{\Large✳} \quad \sim \lambda_R \delta_\lambda + g_R \delta_g p^2 + \kappa_R \delta_\kappa p^4. \quad (21.26)$$

Thus, we can choose

$$\lambda_R \delta_\lambda = -g_R^2 c_1 \Lambda^4, \quad \delta_g g_R = -g_R^2 c_2 \Lambda^2, \quad \delta_\kappa \kappa_R = -g_R^2 c_3 \ln \Lambda, \quad (21.27)$$

and all the infinities in Eq. (21.24) will cancel.

Of course, there will now be new infinities from loops involving λ_R and κ_R, but as long as we add every possible term consistent with the symmetries of the theory, we will always be able to remove all of the infinities at any given order. This will be possible as long as the divergences multiply functions that are polynomials in external momenta, such as could come from counterterms in a local Lagrangian. Now we will show that this always happens.

In the region of loop momentum for which $k \gg p$ for all external momenta p, the divergent integrals can always be written as sums of terms of the form

$$\mathcal{I}_{\text{div}} = (p_\mu^1 \cdots p_\nu^m) g_1 \cdots g_n \int \frac{dk}{k^j} \quad (21.28)$$

for some number m of the various external momenta p_μ^i. These integrals can produce logarithms of the regularization scale Λ, or powers of Λ:

$$\mathcal{I}_{\text{div}} = \sum_{m,n} g_1 \cdots g_n (p_\mu^1 \cdots p_\nu^m)[c_0^{m,n} \ln \Lambda + c_1^{m,n} \Lambda + c_2^{m,n} \Lambda^2 + c_3^{m,n} \Lambda^3 + \cdots], \quad (21.29)$$

where the sum is over all possible products of g_i and external momenta. It is very important that there can never be terms such as $\ln p^2$ coming from the *divergent* part of the integral; that is, nothing like $\Lambda^2 \ln p^2$ can appear. This is simply because integrands do not have any $\ln p^2$ terms to begin with and we can go to the divergent region of the integral by taking $k \gg p$ before integrating over anything that might give a logarithm.

More generally:

Divergences coming from loop integrals will always multiply polynomials in the external momenta.

A simple proof due to Weinberg of this important result is as follows [Weinberg, 1995]. A general divergent integral will have various momenta factors in it, such as

$$\mathcal{I}(p) = \int_0^\infty \frac{k\,dk}{k+p}. \tag{21.30}$$

Let us assume there is at least one denominator with a factor of p (if not, the loop trivially gives a polynomial in external momenta). If we differentiate the integral with respect to p enough times, the integral becomes convergent. For example,

$$\mathcal{I}''(p) = \int_0^\infty \frac{2\,k\,dk}{(k+p)^3} = \frac{1}{p}. \tag{21.31}$$

Then we can integrate over p to produce a polynomial, up to constants of integration we can call Λ and $c_1\Lambda$:

$$\mathcal{I}(p) = p\ln\frac{p}{\Lambda} - p + c_1\Lambda = p\ln p - p(\ln\Lambda + 1) + c_1\Lambda. \tag{21.32}$$

The constants of integration are in one-to-one correspondence with the divergences coming from any regulator. Moreover, multiple integrals of an integration constant over momenta can only ever produce a polynomial in momenta. Thus, the non-analytic terms must be independent of the integration constants or, equivalently, of the divergences (and the regulator). This proves the theorem.

Now, polynomials in external momenta are exactly what we get at tree-level from terms in the Lagrangian. So we can always introduce counterterms to cancel these divergences, as in the scalar field example above. In this way, all S-matrix elements can be made UV finite. In order to have a counterterm, we need the corresponding term to actually be in our Lagrangian. So the easiest thing to do is just to add every possible term with any number of derivatives acting on any fields. Symmetries often make certain terms unnecessary, but by adding all the possible terms we guarantee that counterterms can be chosen so that every S-matrix element will be finite.

21.2.3 Non-renormalizable theories are predictive

As we have seen, non-renormalizable theories require the addition of an infinite number of terms in the Lagrangian to guarantee that all infinities can be removed with counterterms. Despite the infinite number of free parameters, these theories are still very predictive. We will give a number of examples in the next chapter. Here we sketch a simple argument of why this is true.

The first observation is that, at tree-level, terms with more derivatives have weaker effects at low energy (long distances). For example, consider a theory with Lagrangian

$$\mathcal{L} = -\frac{1}{2}\phi(\Box + m^2)\phi + \lambda\phi^4 + \frac{g_1}{M^2}\phi^2\Box\phi^2 + \frac{g_2}{M^4}\phi^2\Box^2\phi^2 + \cdots, \tag{21.33}$$

where M is some scale added to make all the coupling constants dimensionless and the $\cdots$ represent operators with more derivatives or more fields (which have to be added to guarantee that the infinities can be canceled). Now consider some observable, such as the 4-point function $\langle\phi^4\rangle$, as a function of some energy scale E. To the extent that the energy dependence of this 4-point function is polynomial in E, we can fit the various renormalized couplings in the Lagrangian to the terms in its expansion around $E = 0$: $\langle \phi^4 \rangle = \lambda_R + g_1 \frac{E^2}{M^2} + g_2 \frac{E^4}{M^4} + \cdots$. As long as we are only interested in physics at low energy, only a finite number of terms in this series will be important. Thus, we can fit those terms with a few measurements and then predict the complete momentum dependence. In this way, the non-renormalizable theory is predictive even at tree-level.

A remarkable and important fact is that non-renormalizable theories are predictive not just at tree-level but also at loop-level, through calculable quantum corrections. The key to the predictivity of non-renormalizable theories is the result we proved in Section 21.2.2: UV divergences are always proportional to polynomials in momenta. Thus, the infinite number of terms required to renormalize a non-renormalizable theory are all polynomial in derivatives. Such terms lead to local, short-distance effects. In contrast, the non-divergent part of the loops in a non-renormalizable theory may have non-analytic momentum dependence, which can lead to long-distance interactions.

To see in what way analytic functions of momenta correspond to local effects, consider the effective potential, $V(r)$. By the Born approximation, $V(r)$ is given by the Fourier transform of the 2-point function (see Section 13.4). Thus, a term $\frac{1}{M^2} \phi \Box^2 \phi$ might contribute to this potential in perturbation theory as

$$\mathcal{M}(p^2) = \quad \longrightarrow\!\!\otimes\!\!\longrightarrow \quad = \frac{1}{p^2} \frac{p^4}{M^2} \frac{1}{p^2} = \frac{1}{M^2}. \tag{21.34}$$

Since the Fourier transform of a constant is $\delta(r)$, this term gives $V(r) \sim \frac{1}{M^2} \delta(r)$, which is short-ranged. This should be reminiscent of the Uehling potential calculation in Chapter 16. The $\delta(r)$ term in the potential is totally irrelevant at large distances. More insertions of this $\phi \Box^2 \phi$ operator, or contributions from other non-renormalizable operators, will give more positive powers of momentum. We can Fourier transform these contributions by noting that

$$V(r) \sim \int \frac{d^3\vec{p}}{(2\pi)^3} e^{-i\vec{p}\cdot\vec{x}} \left(\frac{\vec{p}^2}{M^2} \right)^n = \left(\frac{-\triangle}{M^2} \right)^n \int \frac{d^3\vec{p}}{(2\pi)^3} e^{-i\vec{p}\cdot\vec{x}} = \left(\frac{-\triangle}{M^2} \right)^n \delta^3(\vec{x}). \tag{21.35}$$

Thus, the tree-level contribution of any of the new terms we must add can have only short-ranged effects. In this sense, the terms we introduce in the Lagrangian for non-renormalizable theories are **local**.

In contrast, loops can give corrections that are non-analytic in momenta. For example, a loop may give $\ln p^2$. The Fourier transform is then

$$V(r) \sim \int \frac{d^3\vec{p}}{(2\pi)^3} e^{-i\vec{p}\cdot\vec{x}} \ln \vec{p}^2 = \frac{-1}{2\pi r^3}, \tag{21.36}$$

which completely dominates over the terms coming from polynomials in momentum. This dominance is beautifully exhibited in quantum gravity, discussed in the next chapter, where quantum corrections to Newton's potential completely dominate over corrections from higher-curvature terms in the Lagrangian for general relativity.

21.2.4 Summary

In summary:

- Renormalizable theories require only a *finite* number of counterterms.
- Non-renormalizable theories require an infinite number of counterterms.
- To renormalize non-renormalizable Lagrangians we must include every term not forbidden by symmetries.
- Non-renormalizable theories can be renormalized. After renormalization all Green's functions are UV finite.
- Non-renormalizable theories are predictive at loop level, particularly through non-analytic dependence on external momenta.

From a practical point of view, having a finite number of counterterms and renormalization conditions is a huge advantage. Nevertheless, non-renormalizable theories are still very predictive, often more so than renormalizable ones. We discuss these issues further in the next chapter through a number of Standard Model examples. Non-renormalizable theories play a central role in Part IV and especially Part V of this book.

Problems

21.1 Write down all the superficially divergent amplitudes in QED at 2-loops. Prove that all of the UV divergences can be removed with the same four counterterms required to remove the 1-loop divergences.

21.2 Calculate the contributions of $\frac{\vec{p}^4}{M^4}$, $\sqrt{\frac{\vec{p}^2}{M^2}}$ and $\ln^2 \frac{\vec{p}^2}{M^2}$ to a potential $V(r)$ by taking their Fourier transforms. Which gives the strongest contribution to the potential at large distances? Which gives the weakest contribution?

21.3 Write down all the renormalizable interactions for a field theory with a single scalar field $\phi(x)$ in two, three, four, five and six dimensions.

Thus, the Schrödinger equation, and its generalization in Eq. (22.2), describe a very predictive *quantum* theory. This theory is predictive despite it being non-renormalizable and having an infinite number of terms – the Schrödinger equation made quantum predictions many years before the Dirac equation was discovered.

It is also important to note that the Schrödinger equation is *not* predictive for momenta $|\vec{p}| \gtrsim m$, since all of the higher-order terms are then important. Thus, the Schrödinger equation is predictive at low energy, but also indicates the scale at which perturbation theory breaks down. If we can find a theory that reduces to the Schrödinger equation at low energy, but for which perturbation theory still works at high energy, it is called a **UV completion** of the Schrödinger equation. Thus, the Dirac equation is a UV completion of the Schrödinger equation. The Dirac equation (and QED) are predictive to much higher energies (but not at all energies, because of the Landau pole). The Klein–Gordon equation is a different UV completion of the Schrödinger equation.

22.2 The 4-Fermi theory

Weak decays were first modeled by Enrico Fermi in 1933. He observed that the easiest way to model β-decay, in which a proton decays into a neutron, positron and neutrino, is with an interaction of the form

$$\mathcal{L}_{\text{Fermi}} = G_F \bar{\psi}_p \psi_n \bar{\psi}_e \psi_\nu, \tag{22.4}$$

with maybe some γ-matrices thrown in between the spinors. This is known as a **4-Fermi interaction**, both because there are four fermions in it and because Fermi used it as a very successful model of radioactive decay. Similar 4-Fermi operators, such as $G_F \bar{\psi}_\mu \psi_{\nu_\mu} \bar{\psi}_e \psi_{\nu_e}$, also model the decay of the muon, $\mu^- \to e^- \nu_e \nu_\mu$. The Fermi constant is in fact best measured from the decay rate of the muon with the result

$$G_F = 1.166 \times 10^{-5} \, \text{GeV}^{-2} = \left(\frac{1}{292.9 \, \text{GeV}}\right)^2. \tag{22.5}$$

Since this was extracted from an actual experiment, it corresponds to the renormalized value of the coupling. It is not obvious that G_F in the muon 4-Fermi operator should be the same G_F in the nuclear β-decay operator; that they are the same implies a deeper structure and a symmetry governing these decays, now understood through the theory of weak interactions.

Since G_F has mass dimension -2, $\mathcal{L}_{\text{Fermi}}$ is a non-renormalizable interaction. Thus, there will be an infinite number of divergent one-particle irreducible graphs and an infinite number of counterterms are needed to cancel them. To prepare for this, we must add to $\mathcal{L}_{\text{Fermi}}$ all terms consistent with its symmetries (whatever those might be). For example, we may have terms such as

$$\mathcal{L} = G_F \bar{\psi}\psi\bar{\psi}\psi + a_1 G_F^2 \bar{\psi}\psi \Box \bar{\psi}\psi + a_2 G_F^3 \bar{\psi}\slashed{\partial}\psi \Box \bar{\psi}\slashed{\partial}\psi + \cdots, \tag{22.6}$$

where the a_i are numbers and the factors of G_F have been added by dimensional analysis. Derivatives can act anywhere and γ-matrices can be inserted anywhere; we are just showing some representative terms and dropping the fermion species labels for simplicity. Despite these additional terms with unknown coefficients, the 4-Fermi theory is very predictive, even at tree-level. One prediction from this interaction is that the rate for β-decay, $p^+ \to ne^+\nu$, will be related to the rate for $n \to p^+e^-\bar{\nu}$. The higher-order terms in $\mathcal{L}$ will affect the β-decay rate by factors of $(G_F E^2)^j$ for $j \geq 1$, where E is some energy in the process. Since the masses of the particles and the energies involved in β-decay are much less than $G_F^{-1/2}$, these higher-order terms will do practically nothing. The 4-Fermi theory also makes a prediction for the angular dependence and energy distribution of the decay products. In addition, the 4-Fermi theory can also be used to study parity violation, say, by comparing the predictions of $\bar{\psi}\psi\bar{\psi}\psi$ to those of $\bar{\psi}\gamma_5\gamma^\mu\psi\bar{\psi}\gamma_5\gamma^\mu\psi$. All of these predictions are for low-energy measurements and therefore almost totally independent of the a_i (assuming the a_i are not enormously large).

Besides tree-level predictions, one can also calculate loops in this non-renormalizable theory and derive physically testable predictions from those loops. For simplicity, let us imagine that all the fermions in Eq. (22.6) are identical. Then there will be both tree-level and loop contributions to the process $\psi\psi \to \psi\psi$. At tree-level, the Lagrangian generates S-matrix elements of the form

$$\mathcal{M}_{\text{tree}}(s) \sim G_F + a_1 G_F^2 s + a_2 G_F^3 s^2 + \cdots, \tag{22.7}$$

where s does not necessarily represent $s = (p_1 + p_2)^2$ but any kinematical Lorentz-invariant quantity of mass dimension 2, and we are ignoring the external spinors for simplicity. At low energies, $s \ll G_F^{-1}$, this scattering is dominated by the leading term, with subleading terms suppressed by powers of $sG_F \ll 1$. At 1-loop, there is a contribution of the form

$$\mathcal{M}_{\text{loop}}(s) = \text{\scriptsize(diagram)} \sim G_F^2 \int \frac{d^4k}{(2\pi)^4} \frac{1}{\slashed{k}}\frac{1}{\slashed{k}} \sim G_F^2\left(b_0\Lambda^2 + b_1 s + b_2 s \ln\frac{\Lambda^2}{s}\right). \tag{22.8}$$

On the right, we have parametrized the possible forms the result could take with three finite and calculable constants b_0, b_1, and b_2 and a regulator scale Λ. Without any symmetry arguments, there is no reason to expect that any of the constants b_i should vanish. Thus,

$$\mathcal{M}_{\text{tree}} + \mathcal{M}_{\text{loop}} \sim (G_F + b_0\Lambda^2 G_F^2) + sG_F^2(a_1 + b_1 + b_2 \ln\Lambda^2) \\ - b_2 G_F^2 s \ln s + a_2 G_F^3 s^2 + \cdots, \tag{22.9}$$

where we have grouped terms by their momentum dependence. The key term in this expression is the $s \ln s$ term, which has no analog coming from the classical Lagrangian, Eq. (22.6).

To make physical predictions, we have to renormalize. To do so, we introduce counterterms in the usual way. Equation (22.6) is treated as a bare Lagrangian and Z-factors are introduced:

$$\mathcal{L} = Z_F G_F \bar{\psi}\psi\bar{\psi}\psi + Z_1 a_1 G_F^2 \bar{\psi}\psi\Box\bar{\psi}\psi + Z_2 a_2 G_F^3 \bar{\psi}\slashed{\partial}\psi\Box\bar{\psi}\slashed{\partial}\psi + \cdots. \tag{22.10}$$

Then, we write $Z_F = 1 + \delta_F$, $Z_1 = 1 + \delta_1$, etc. (these are different Z_i from the QED renormalization factors with the same name). Then we find

$$\mathcal{M}_{\text{loop}} + \mathcal{M}_{\text{tree}} + \mathcal{M}_{\text{c.t.}} \sim \left(G_F + b_0\Lambda^2 G_F^2 + G_F\delta_F\right)$$
$$+ sG_F^2\left(a_1 + b_1 + b_2\ln\Lambda^2 + a_1\delta_1\right) - b_2 G_F^2 s\ln s + \cdots \quad (22.11)$$

and we can choose $\delta_F = -b_0\Lambda^2 G_F$ and $\delta_1 = -\frac{1}{a_1}\left(b_1 + b_2\ln\frac{\Lambda^2}{s_0}\right)$, with s_0 an arbitrary scale, to remove the infinities and reduce the leading two terms to the form of Eq. (22.7), where now G_F and a_i are the renormalized coefficients of these terms. This renormalization removes *almost* the entire result of the loop; however, one term remains. We find the renormalized matrix element is

$$\mathcal{M}(s) = \mathcal{M}_{\text{loop}} + \mathcal{M}_{\text{tree}} + \mathcal{M}_{\text{c.t.}} \sim G_F + sG_F^2\left(a_1 - b_2\ln\frac{s}{s_0}\right) + a_2 s^2 G_F^3 + \cdots . \quad (22.12)$$

At the scale $s = s_0$ this is identical to the tree-level prediction. If the s dependence of the distribution at low energies is well-enough measured, G_F, a_1, b_2, a_2, etc. can be extracted from data. Although the constants a_i are not calculable, the constant b_2 is. More precisely, one could plot

$$\frac{\mathcal{M}(s_1) - G_F}{s_1 G_F^2} - \frac{\mathcal{M}(s_2) - G_F}{s_2 G_F^2} \sim b_2\ln\frac{s_2}{s_1} + \mathcal{O}(G_F s_1), \quad (22.13)$$

and see whether the logarithmic scale dependence agrees with the theoretical calculation. Thus, b_2 is a genuine testable prediction from a loop calculation in a non-renormalizable theory.

The reason this works is because the $\ln s$ dependence can never come from a tree-level calculation. This is because tree-level calculations come from local Lagrangians that have only integer powers of derivatives, never terms such as $\bar{\psi}\psi\bar{\psi}\ln\Box\psi$. This is a general result:

Non-analytic energy dependence is a testable quantum prediction of non-renormalizable (or renormalizable) theories.

We will see phenomenologically relevant examples of these logarithmic corrections to the real 4-Fermi theory in Chapter 23 (on the renormalization group) and in Chapter 31 (on precision tests of the Standard Model).

22.2.1 UV completing the Fermi theory

Although the 4-Fermi theory is very predictive, its predictive power is confined to the low-energy regime. As energies approach $G_F^{-1/2} \sim 300$ GeV, each term in Eq. (22.7) becomes important and perturbation theory breaks down. Thus, the 4-Fermi theory calls out for a UV completion.

A UV completion of the 4-Fermi theory is a theory with massive vector bosons, the $W^\pm$ bosons (which are charged) and the Z boson (which is neutral). This UV completion actually combines the weak interactions with QED to form the electroweak theory, which is a gauge theory based on the Lie group $SU(2) \times U(1)$. We will discuss this theory in great detail in Chapter 29. Here, we skip the details of the gauge structure to concentrate only on the UV-completion aspect.

The Lagrangian for a fermion interacting with a massive vector boson W_μ has the form

$$\mathcal{L}_M = -\frac{1}{4}F_{\mu\nu}^2 + \frac{1}{2}M^2 W_\mu^2 + \bar{\psi}(i\slashed{\partial} + g\slashed{W})\psi, \tag{22.14}$$

where here $F_{\mu\nu} = \partial_\mu W_\nu - \partial_\nu W_\mu$ with g some gauge coupling. The actual Lagrangian for the W boson is more complicated (see Chapter 29); here we are approximating the electroweak gauge theory with a toy model with a single fermion and a single gauge boson. The matrix element for $\psi\psi \to \psi\psi$ in this theory in the s-channel is given by

$$iM = \quad \sim (ig)^2 \, \bar{v}_2 \gamma^\mu u_1 \frac{-i\left(g^{\mu\nu} - \frac{p^\mu p^\nu}{M^2}\right)}{s - M^2} \bar{u}_3 \gamma^\nu v_4. \tag{22.15}$$

In this $U(1)$ approximation, the $\frac{p^\mu p^\nu}{M^2}$ term in the numerator of the propagator does not contribute due to the Ward identity. At low energy, $s \ll M$, this matrix element is well approximated by

$$iM = \quad = -i\frac{g^2}{M^2}\bar{v}_2\gamma^\mu u_1 \bar{u}_3 \gamma^\mu v_4. \tag{22.16}$$

Physically, the W boson propagates over such short distances (of order M^{-1}) that at large distances one cannot see it, just as one cannot see the W propagator in the diagram in Eq. (22.16) since it has been contracted to a point.

Equation (22.16) is the same matrix element we would get from the 4-Fermi interaction $G_F \bar{\psi}\gamma^\mu\psi\bar{\psi}\gamma^\mu\psi$ if $G_F = \frac{g^2}{M^2}$. (The actual expression for the Fermi constant in terms of the weak coupling constant g_w and the W mass is $G_F = \frac{\sqrt{2}}{8}\frac{g_w^2}{m_W^2}$, where $m_W = 80.4\,\text{GeV}$ and $g_w = 0.65$, as discussed in Chapter 29.) Taylor expanding the propagator in Eq. (22.15) to higher orders in $\frac{s}{M^2}$ gives predictions for the higher-order terms in the non-renormalizable Lagrangian in Eq. (22.6). For example, the next term would be $M \sim g^2 \frac{s}{M^4}\bar{v}_1\gamma^\mu u_1 \bar{u}_2\gamma^\mu v_2$, which would correspond to a term $\frac{g^2}{M^2}\bar{\psi}\gamma^\mu\psi\frac{\Box}{M^2}\bar{\psi}\gamma^\mu\psi$. This exactly parallels how the expansion of the Dirac equation predicted the higher-order terms in the non-renormalizable theory it UV completed, the Schrödinger equation.

The actual electroweak theory involves four gauge bosons corresponding to the generators of a non-Abelian gauge group $SU(2) \times U(1)$. We will study these non-Abelian gauge theories in great detail in Part IV, but for now, we only need one important fact: the $\frac{p^\mu p^\nu}{M^2}$ terms in the numerator of the gauge boson propagator are no longer guaranteed to drop out. Thus, as discussed in the previous chapter, propagators can scale as $\frac{1}{M^2}$ instead of $\frac{1}{k^2}$

Expanding the Chiral Lagrangian out to quadratic order gives normal kinetic terms and photon interactions from scalar QED:

$$\mathcal{L}_{\text{kin}} = \frac{1}{2}\left(\partial_\mu \pi^0\right)\left(\partial_\mu \pi^0\right) + (D_\mu \pi^+)(D_\mu \pi^-). \tag{22.19}$$

Expanding to higher orders produces interactions such as

$$\mathcal{L}_{\text{int}} = \frac{1}{F_\pi^2}\left[-\frac{1}{3}\pi^0\pi^0\partial_\mu\pi^+\partial_\mu\pi^- + \cdots\right] + \frac{1}{F_\pi^4}\left[\frac{4}{45}(\pi^-\pi^+)^2\partial_\mu\pi^0\partial_\mu\pi^0 + \cdots\right] + \cdots. \tag{22.20}$$

Since F_π has dimensions of mass, the Chiral Lagrangian is non-renormalizable. The important point is that the interactions in the Chiral Lagrangian have a special form since they are constrained by the SU(2) symmetry. In particular, each term has two derivatives, so for example, a term such as $\frac{1}{F_\pi^2}\pi_0^6$ is forbidden. The coefficient of each term is also completely fixed.

Since this theory is non-renormalizable, we should also add more terms to absorb infinities from loops. Since $U^\dagger U = 1$ we cannot write down any non-trivial term without derivatives. There are only three terms you can write down with four derivatives:

$$\mathcal{L}_4 = L_1 \text{tr}[(D_\mu U)(D_\mu U)^\dagger]^2 + L_2 \text{tr}[(D_\mu U)(D_\nu U)^\dagger]^2$$
$$+ L_3 \text{tr}\left[(D_\mu U)(D_\mu U)^\dagger (D_\nu U)(D_\nu U)^\dagger\right]. \tag{22.21}$$

Thus, the Chiral Lagrangian admits a derivative expansion, with the leading term, $\mathcal{L}_\chi$ in Eq. (22.18), dominant and $\mathcal{L}_4$ being suppressed at low energies. One could add additional terms, which would have six or more derivatives, but these would be additionally suppressed, and unmeasurable from a practical point of view. The coefficients L_1, L_2 and L_3 have been fit from low-energy pion scattering experiments from which it has been found that $L_1 = 0.65$, $L_2 = 1.89$ and $L_3 = -3.06$. Additional interactions are suppressed by powers of momentum divided by the parameter $F_\pi = 92$ MeV.

As with the 4-Fermi theory, the quantum effects of the Chiral Lagrangian are calculable and measurable as well. They take the form of non-analytic logarithmic corrections to pion scattering cross sections and even have a name: **chiral logs** (see for example [Weinberg, 1979]).

As with any non-renormalizable theory, the Chiral Lagrangian points to its own demise – it becomes non-perturbative at a scale $\sqrt{s} \sim 4\pi F_\pi \approx 1200$ MeV. Above this scale, all the higher-order interactions become relevant and the theory is not predictive. A UV completion of the Chiral Lagrangian is QCD, the theory of quarks and gluons. This is a completely different type of UV completion than the electroweak theory which UV-completed the 4-Fermi theory or the Dirac equation which UV-completed the Schrödinger equation. For both of these theories, the fermions in the low-energy theory were present in the UV completion, but with different interactions. The theory of QCD does not have pions in it at all! Thus, one cannot ask about pion scattering at high energy in QCD. Instead, one must try to match the two theories indirectly, for example through correlation functions of external currents. The correlation functions can be measured by scattering photons or electrons off pions, but to calculate them we need a non-perturbative description of QCD, such as the lattice (see Section 25.5). So, although QCD is a renormalizable UV completion of the Chiral Lagrangian in the sense that it is well defined and perturbative up to arbitrarily high energies, it cannot answer the questions that the Chiral Lagrangian could not answer: What does $\pi\pi$ scattering look like for $s \gg F_\pi^2$? For low-energy pion scattering, the Chiral Lagrangian is much more useful than QCD.

22.4 Quantum gravity

The final non-renormalizable field theory we will discuss in this chapter is quantum gravity. This is the effective description of a massless spin-2 particle. We have already shown two important results about massless spin-2 particles. In Section 8.7, we embedded the spin-2 particles in a tensor field $h_{\mu\nu}$. The only consistent way to do this had a gauge symmetry under local space-time translations:

$$x^\alpha \to x^\alpha + \xi^\alpha(x), \tag{22.22}$$

also known as **general coordinate transformations**. The Noether current for this symmetry is the energy-momentum tensor $T_{\mu\alpha}$, which we derived in Section 3.3.1, whose conserved charges are energy and momentum. In Section 9.5, we bypassed the discussion of gauge invariance and showed, by considering the soft limit, that Lorentz invariance implies that massless spin-2 particles are associated with a conserved charge. It is, nevertheless, useful to describe massless spin-2 particles with a local Lagrangian, so we will review the results of Section 8.7, and continue to discuss quantum effects in this theory.

In Section 8.7 we showed that a massless spin-2 particle can be embedded in a tensor field $h_{\mu\nu}$ only if the Lagrangian for $h_{\mu\nu}$ is invariant under infinitesimal transformations parametrized by four functions ξ^α:

$$h_{\mu\nu} \to h_{\mu\nu} + \partial_\mu \xi_\nu + \partial_\nu \xi_\mu + (\partial_\mu \xi^\alpha)h_{\alpha\nu} + (\partial_\nu \xi^\alpha)h_{\mu\alpha} + \xi^\alpha \partial_\alpha h_{\mu\nu}. \tag{22.23}$$

The first two terms are the gauge part; they are the analog of $A_\mu \to A_\mu + \partial_\mu \alpha$ in QED but with four types of α, now called ξ^α. The last three terms are just the transformation properties of a tensor representation of the Poincaré group under infinitesimal general coordinate transformations. We also showed that the unique kinetic term for $h_{\mu\nu}$ was

$$\mathcal{L}_{\text{kin}} = \frac{1}{2} h_{\mu\nu} \Box h_{\mu\nu} - h_{\mu\nu} \partial_\mu \partial_\alpha h_{\nu\alpha} + h \partial_\mu \partial_\nu h_{\mu\nu} - \frac{1}{2} h \Box h. \tag{22.24}$$

The leading interactions have two derivatives and three factors of h and are therefore of the form $\mathcal{L}_{\text{int}} \sim \frac{1}{M_{\text{Pl}}} \Box h^3$ for some dimensional scale M_{Pl}. Thus, any interacting field theory of massless spin-2 particles is automatically non-renormalizable. Finally, it is possible to show [Feynman *et al.*, 1996] that the minimal set of interactions can be combined into the concise form

$$\mathcal{L}_{\text{EH}} = M_{\text{Pl}}^2 \sqrt{-\det\left(\eta_{\mu\nu} + \frac{1}{M_{\text{Pl}}} h_{\mu\nu}\right)} R\left[\eta_{\mu\nu} + \frac{1}{M_{\text{Pl}}} h_{\mu\nu}\right], \tag{22.25}$$

where $\eta_{\mu\nu}$ is the Minkowski metric, which we usually denote $g_{\mu\nu}$, and R is the scalar Ricci curvature. This Lagrangian, the Einstein–Hilbert Lagrangian, is more commonly written as

$$\mathcal{L}_{\text{EH}} = M_{\text{Pl}}^2 \sqrt{-\det(g)} R, \tag{22.26}$$

where $g_{\mu\nu} = \eta_{\mu\nu} + \frac{1}{M_{\text{Pl}}} h_{\mu\nu}$ and $M_{\text{Pl}} = G_N^{-1/2} \approx 10^{19}$ GeV is the Planck scale (alternative definitions with extra factors of 8π or $32\pi^2$ are sometimes used).

You can either review these results from the bottom-up approach of Section 8.7, derive them using the top-down approach of general relativity, or just take them as given. You do not need to know general relativity to follow the subsequent discussion of quantum corrections. The only thing you need to know is that there is a symmetry, general coordinate invariance, which vastly restricts the terms one can write down in a Lagrangian for a massless spin-2 particle.

22.4.1 Quantum predictions

General coordinate invariance implies that the Lagrangian must be a functional of $h_{\mu\nu}$ and the Riemann curvature tensor $R_{\mu\nu\alpha\beta}[h_{\mu\nu}]$. We also write

$$R_{\mu\nu} = g^{\alpha\beta} R_{\alpha\mu\beta\nu}, \quad R = g^{\mu\nu} R_{\mu\nu} \tag{22.27}$$

for the Ricci tensor and scalar.

The Riemann tensor can be thought of as

$$R_{\mu\nu\alpha\beta} \sim \partial_\mu \partial_\nu \exp\left(\frac{1}{M_{\text{Pl}}} h_{\alpha\beta}\right). \tag{22.28}$$

This heuristic notation, which is meant to mimic $U = \exp(\frac{i}{F_\pi} \sigma^a \pi^a)$ in the Chiral Lagrangian, encapsulates that all terms in the expansion of the curvature have two derivatives and an infinite number of factors $h_{\mu\nu}$. With this notation, $\mathcal{L}_{\text{EH}} \sim R \sim \text{Tr}[R_{\mu\nu}]$ becomes very similar to the form of the Chiral Lagrangian $\mathcal{L}_\chi = \text{Tr}[(D_\mu U)(D_\mu U)^\dagger]$.

Just like the Chiral Lagrangian, the Lagrangian for gravity is non-renormalizable but strongly constrained by symmetries. The higher-order terms we must add to be able to renormalize this non-renormalizable theory are all products of the metric and the Riemann tensor:

$$\mathcal{L} = \sqrt{-\det(g)}\left(M_{\text{Pl}}^2 R + L_1 R^2 + L_2 R_{\mu\nu} R^{\mu\nu} + L_3 R_{\mu\nu\rho\sigma} R^{\mu\nu\rho\sigma} + \cdots\right). \tag{22.29}$$

In this case, there are three terms, just as in the Chiral Lagrangian. Actually, one linear combination is a total derivative, called the Gauss–Bonnet term, which has no effect in perturbation theory, so we will set $L_3 = 0$. Since $R_{\mu\nu\alpha\beta}$ has two derivatives, the R^2 and $R_{\mu\nu}^2$ terms have four derivatives. Thus, the expansion of $\mathcal{L}$ becomes

$$\mathcal{L} \sim \left(\frac{1}{2} h \Box h + \frac{1}{M_{\text{Pl}}} \Box h^3 + \cdots\right) + L_i \left(\frac{1}{M_{\text{Pl}}^2} h \Box^2 h + \frac{1}{M_{\text{Pl}}^3} h \Box^2 h^2 + \cdots\right) + \cdots, \tag{22.30}$$

where we are only counting derivatives and factors of M_{Pl}.

The reason gravity is predictive is because $M_{\text{Pl}} \approx 10^{19}$ GeV, so $E \ll M_{\text{Pl}}$ for any reasonable experimentally accessible energy E. In fact, it is difficult to test even the terms in the Lagrangian cubic in h with two derivatives. These are terms such as $\frac{1}{M_{\text{Pl}}}\Box h^3$ coming from $M_{\text{Pl}}^2\sqrt{g}R$. To measure interactions in the Einstein–Hilbert Lagrangian at all, one either needs energies of order M_{Pl} or very large field values, $h \gtrsim M_{\text{Pl}}$. Such large field values are conveniently produced in nature, for example from the gravitational field around the Sun. There,

$$h(r) \sim \phi_{\text{Newton}} \sim \text{\ \ \ } \sim \frac{M_{\text{Sun}}}{M_{\text{Pl}}}\frac{1}{r}. \tag{22.31}$$

The corrections to this from the $\frac{1}{M_{\text{Pl}}}\Box h^3$ term are given by the classical field theory diagram:

$$\Delta h(r) \sim \text{\ \ \ } \sim \frac{1}{M_{\text{Pl}}}\left(\frac{M_{\text{Sun}}}{M_{\text{Pl}}}\frac{1}{r}\right)\left(\frac{M_{\text{Sun}}}{M_{\text{Pl}}}\frac{1}{r}\right), \tag{22.32}$$

with the $\frac{1}{M_{\text{Pl}}}$ coming from the vertex, the two factors of $\frac{M_{\text{Sun}}}{M_{\text{Pl}}}$ coming from the sources (the Sun) and the factors of r added by dimensional analysis. Using $\frac{M_{\text{Sun}}}{M_{\text{Pl}}} \sim 10^{38}$ and $M_{\text{Pl}}r \sim 10^{45}$ for r, the Mercury–Sun distance, we find $\frac{\Delta h}{h} \sim 10^{-7}$. This is the precision by which the orbit of Mercury would have to be measured to see the effect of this term.

The higher-order terms, like the ones proportional to L_1 and L_2, contribute corrections to Newton's potential as well. One can actually solve Einstein's equations exactly with L_1 and L_2. For L_1, the result is that at large distances the potential around the Sun has the form [Stelle, 1978]:

$$h(r) = \frac{M_{\text{Sun}}}{M_{\text{Pl}}}\frac{1}{r}\left[1 - \frac{1}{3}\exp\left(-\frac{rM_{\text{Pl}}}{\sqrt{96\pi L_1}}\right) + \cdots\right]. \tag{22.33}$$

Thus, the effects of the L_i terms are short-ranged, as expected from the general argument in Section 21.2.3. Expanding around $L_1 = 0$ and $L_2 = 0$, the leading term in the potential can be written as [Donoghue, 1994]

$$h(r) = \frac{M_{\text{Sun}}}{M_{\text{Pl}}}\left[\frac{1}{r} - 128\pi^2\frac{L_1 + L_2}{M_{\text{Pl}}^2}\delta^3(\vec{r}) + \cdots\right]. \tag{22.34}$$

This is consistent with what we expect from the Feynman diagram

$$\text{\ \ \ } \tag{22.35}$$

with the $\otimes$ representing an insertion of the $\frac{1}{M_{\text{Pl}}^2}h\Box^2 h$ term from Eq. (22.30). The result is that the higher-order terms in the gravity Lagrangian are unmeasurable.

Now let us consider loops. The simplest loop that contributes a correction to Newton's potential is a correction to the graviton propagator, which has the same general form as the vacuum polarization graph. Since the calculations are tedious, we will just summarize results. In harmonic gauge, $2\partial_\mu h_{\mu\nu} = \partial_\nu h_{\mu\mu}$, the graviton propagator is

$$\langle 0 \,|\, T\{h_{\mu\nu}(x)h_{\alpha\beta}(y)\}\,|\, 0\rangle = \int \frac{d^4p}{(2\pi)^4}e^{ipx}\frac{P_{\mu\nu,\alpha\beta}}{p^2 + i\varepsilon}, \tag{22.36}$$

with

$$P_{\mu\nu,\alpha\beta} = \frac{1}{2}\left(\eta_{\mu\alpha}\eta_{\nu\beta} + \eta_{\mu\beta}\eta_{\nu\alpha} - \eta_{\mu\nu}\eta_{\alpha\beta}\right). \tag{22.37}$$

The vacuum polarization graph gives a correction to this of the form

$$= \frac{1}{p^2} \left\{ \frac{p^4}{M_{\text{Pl}}^2} \left[\frac{21}{120} (\eta_{\mu\rho}\eta_{\nu\sigma} + \eta_{\mu\sigma}\eta_{\nu\rho}) + \frac{1}{120} \eta_{\mu\nu}\eta_{\rho\sigma} \right] \left[\frac{1}{\varepsilon} - \ln(-p^2) \right] \right\} \frac{1}{p^2} \quad (22.38)$$

up to $p_\mu p_\nu$ type terms, which have no physical effect due to gauge invariance. For the correction to Newton's law, p^2 is spacelike, $-p^2 > 0$, as with the vacuum polarization correction to Coulomb's law (see Chapter 16). To cancel the UV divergence in this graph, one needs a counterterm from L_1 or L_2 (or perhaps both). The important point is that counterterms and any possible additional finite contributions from the L_i terms cannot remove the $\ln(-p^2)$ contribution to the potential.

Fourier transforming the logarithmic term using Eq. (21.36) gives a contribution to the potential that scales as $\frac{1}{r^3}$. This correction is not short-ranged, like the tree-level contributions from the L_i terms. Combining all the contributions the result is

$$h(r) = \frac{M_{\text{Sun}}}{M_{\text{Pl}}} \frac{1}{r} \left[1 - \frac{M_{\text{Sun}}}{M_{\text{Pl}}^2 r} - \frac{127}{30\pi^2} \frac{1}{M_{\text{Pl}}^2 r^2} - 128\pi^2 \frac{L_1 + L_2}{M_{\text{Pl}}^2} \delta^3(\vec{r}) + \cdots \right], \quad (22.39)$$

corresponding to the Feynman diagrams

$$ \qquad \qquad + \qquad \qquad + \qquad \qquad + \qquad \qquad . \quad (22.40)$$

Thus, the radiative correction (the $\frac{127}{30\pi^2} \frac{1}{M_{\text{Pl}}^2 r^2}$ term) is a testable prediction that is parametrically more important than the L_i terms. For the perihelion shift of Mercury, the effect is one part in $(M_{\text{Pl}} r)^2 \sim 10^{90}$, which we are not going to measure any time soon. Nevertheless, it is a genuine prediction of quantum gravity. This prediction is entirely *independent* of the UV completion of the Einstein–Hilbert Lagrangian.

This calculation should make it clear that:

> There is nothing inconsistent about general relativity and quantum mechanics.

General relativity is the only consistent theory of an interacting massless spin-2 particle. It is a quantum theory, just as solid and calculable as the 4-Fermi theory. It is non-renormalizable, and therefore non-perturbative for energies $E \gtrsim M_{\text{Pl}}$, but it is not inconsistent. At distances $r \sim \frac{1}{M_{\text{Pl}}} \sim 10^{-33}$ cm (the **Planck length**), all of the quantum corrections and all of the higher-order terms in the Lagrangian become important.

So, if we want to use gravity at very short distances we need a UV completion. String theory is one such theory. It is capable of calculating the L_i terms in Eq. (22.29). If we could measure the L_i, then we could test string theory. However, as noted above, these L_i terms have exponentially suppressed effects at distances greater than the Planck length. In fact, we can now understand why it is so difficult to test string theory: long-distance physics is determined by symmetries in an effective quantum theory that is independent of the UV completion. The quantum prediction, the $\frac{127}{30\pi^2} \frac{1}{M_{\text{Pl}}^2 r^3}$ correction to Newton's potential, is determined only by the existence of a massless spin-2 particle. Assuming only that the long-distance description of gravity is a quantum field theory, its UV completion (which may not be a quantum field theory) must be screened at distances beyond the Planck length.

22.5 Summary of non-renormalizable theories

We have looked at four important non-renormalizable theories:

- The Schrödinger equation is perturbative for $E < m_e$. Its UV completion is the Dirac equation and QED, which is perturbative up to its Landau pole, $E \sim 10^{286}$ GeV.
- The Fermi theory of weak interactions is perturbative for $E < G_F^{-1/2} \sim 300$ GeV. Its UV completion is the electroweak theory with massive vector bosons W and Z. The electroweak theory is also non-renormalizable. Its UV completion contains a Higgs boson.
- The Chiral Lagrangian is the low-energy theory of pions. It is perturbative and very predictive for $E < 4\pi F_\pi \sim 1200$ MeV. Its UV completion is QCD. QCD is predictive at high energies. The fields in QCD, quarks and gluons, are related to pions and other hadrons (quark and gluon bound states) in a complicated, non-perturbative way. Thus, to study hadrons in QCD, we need non-perturbative methods, such as the lattice. In contrast, at low energy the Chiral Lagrangian is perturbative and therefore more useful than QCD for answering certain questions.
- General relativity is the low-energy theory of gravity. It is perturbative for $E < M_{\mathrm{Pl}} \sim 10^{19}$ GeV. It is extremely predictive at low energies, including predictive quantum corrections. One possible UV completion is string theory. Gravitational physics at distances larger than the Planck length, 10^{-33} cm, is independent of the UV completion, which explains why string theory (as a quantum theory of gravity) is so hard to test.

These four examples correspond to the four forces of nature: the electromagnetic force, the weak force, the strong force, and gravity. Notice that the UV completions are all qualitatively very different. In some cases, certainly for many physical applications, the non-renormalizable theory is more useful than the renormalizable one. Renormalizable just means there are a finite number of counterterms; it does *not* mean that you can calculate every observable perturbatively.

22.6 Mass terms and naturalness

Having discussed non-renormalizable interactions, which correspond to terms in a Lagrangian whose coefficients have negative mass dimension, we turn to terms whose coefficients have positive mass dimension. We begin with a discussion of renormalization of masses, with other possibilities discussed in Section 22.7.

space of models, with unobservable consequences. Thus, from the model-building point-of-view, naturalness is a statement about whether two different models predict the same values for renormalized couplings.

A possible explanation of fine-tuning in particle physics is that there may be patches of the universe probing different values of parameters in some finite theory (such as the various vacua of string theory). In this way, model space is explored cosmologically. Thus, if there are 10^{34} patches of the universe with different Higgs boson masses, it is then natural for us to live in the only one that can support life. One can then argue that life requires $m_H \ll M_{\text{Pl}}$, which eliminates the fine-tuning problem. This line of reasoning, known generally as the **anthropic principle**, has been increasing in popularity since the 1990s. The scientific merit of the anthropic principle is often debated. At this point, there are no testable predictions of the anthropic principle.

22.6.3 Fermion and gauge boson masses

Other coefficients of positive mass dimension are fermion and gauge boson masses. Consider first radiative corrections to fermion masses. For example, we already calculated the self-energy graph of the electron in QED in Chapter 18. With dimensional regularization, the result was (Eq. (18.50))

$$i\Sigma_2(\slashed{p}) = \underbrace{}_{p \qquad\qquad p}$$

$$= i\frac{e^2}{16\pi^2} \int_0^1 dx \left[(2-\varepsilon)x\slashed{p} - (4-\varepsilon)m_e \right] \left[\frac{2}{\varepsilon} + \ln \frac{\tilde{\mu}^2}{(1-x)(m_e^2 - p^2 x) + x m_\gamma^2} \right],$$
(22.56)

which can be compared to Eq. (22.45). The difference between the pole and $\overline{\text{MS}}$ mass for the electron in QED was also calculated in Chapter 18, in Eq. (18.65):

$$m_P - m_{\overline{\text{MS}}} = \frac{e^2}{16\pi^2} m_P \left(4 + 3\ln \frac{\mu^2}{m_P^2} \right),$$
(22.57)

which can be compared to Eq. (22.54).

Although not apparent in the expansion around $d = 4$, the full result has no pole in $d = 2$ or $d = 3$ and is therefore not quadratically or linearly divergent. That is a non-trivial fact. In non-relativistic quantum mechanics, you do get a linearly divergent shift. This can be seen from a simple integral over the classical electron self-energy. In the non-relativistic limit, the energy density of the electromagnetic field is $\rho \propto |\vec{E}|^2 + |\vec{B}|^2$. So

$$\Delta m \sim \int d^3r\, \rho(r) \sim \int d^3r \left(\frac{e}{r^2} \right)^2 \sim \alpha \int_{\Lambda^{-1}}^\infty \frac{r^2\, dr}{r^4} \sim \alpha\Lambda.$$
(22.58)

In a relativistic theory, there is only a logarithmically divergent self-energy.

Next, note that in QED the self-energy correction at $\slashed{p} = m_e$ is proportional to the electron mass, not any other mass scale in the problem. In this case, the other mass is a fictitious photon mass, but the result implies that if the photon in the loop were replaced by a real heavy gauge boson, such as the Z, the correction would still be proportional to m_e not m_Z. For another example, consider the Yukawa theory in Eq. (22.41). There the self-energy graph is

$$
i\Sigma_2(\slashed{p}) =
$$

$$
= \lambda^2 \int \frac{d^4 k}{(2\pi)^4} \frac{\slashed{p} - \slashed{k} + M}{[(p-k)^2 - M^2][k^2 - m^2]}
$$

$$
= i \frac{\lambda^2}{16\pi^2} \left[\frac{\slashed{p} + 2M}{\varepsilon} - \int_0^1 dx \left[(1-x)\slashed{p} + M \right] \ln \frac{M^2 x + (1-x)(m^2 - p^2 x)}{\tilde{\mu}^2} \right].
$$
(22.59)

There is no correction proportional to the scalar mass m, only to the fermion mass M. This graph is also not linearly divergent.

What if we throw in some more fermions or a couple more scalars, or look at 6-loops? It turns out that the mass shift will always be proportional to the fermion mass. The reason this happens is because the electron mass is protected by a **custodial chiral symmetry**.

A chiral symmetry is a global symmetry under which the left- and right-handed electron have opposite charge: $\psi_L \to e^{-i\alpha}\psi_L$ and $\psi_R \to e^{i\alpha}\psi_R$. We can write the transformation concisely as

$$
\psi \to e^{i\alpha\gamma_5}\psi.
$$
(22.60)

Under this transformation, the kinetic term and QED interaction are invariant,

$$
\bar{\psi}\slashed{D}\psi \to \psi^\dagger e^{-i\alpha\gamma_5^\dagger}\gamma_0 \slashed{D} e^{i\alpha\gamma_5}\psi = \bar{\psi}\slashed{D}\psi,
$$
(22.61)

since $\gamma_5^\dagger = \gamma_5$ and $[\gamma_5, \gamma_0\gamma_\mu] = 0$. However, the mass term is not:

$$
m_e\bar{\psi}\psi \to m_e\bar{\psi}e^{2i\alpha\gamma_5}\psi \neq m_e\bar{\psi}\psi.
$$
(22.62)

Thus, the mass term breaks the chiral symmetry. This is consistent with the expansion in terms of Weyl fermions:

$$
\bar{\psi}\slashed{D}\psi + m_e\bar{\psi}\psi = \bar{\psi}_R\slashed{D}\psi_R + \bar{\psi}_L\slashed{D}\psi_L + m_e\bar{\psi}_L\psi_R + m_e\bar{\psi}_R\psi_L,
$$
(22.63)

which shows that only the mass term couples fields with different charges under the chiral symmetry.

The chiral symmetry is exact in the limit $m_e \to 0$. That means that if $m_e = 0$ then, because of the exact symmetry, m_e will stay 0 to all orders in perturbation theory. For $m_e \neq 0$, if we treat the mass as an interaction rather than a kinetic term, then every diagram that violates the chiral symmetry, including a correction to the mass itself, must be proportional to m_e. We call the symmetry *custodial* because it acts like a custodian and protects the mass from large corrections, even if the symmetry is not exact. We also say sometimes that setting $m_e = 0$ is *technically natural* ['t Hooft, 1979] (See Box 22.1).

It is technically natural for a parameter to be small if quantum corrections to the parameter are proportional to the parameter itself. This happens if the theory has an enhanced symmetry when the parameter is zero.

For another example, consider a vector boson mass. A photon mass term

$$\mathcal{L} = \cdots + m_\gamma^2 A_\mu^2 \tag{22.64}$$

breaks gauge invariance. In the limit $m_\gamma = 0$, gauge invariance is exact, and thus gauge invariance is a custodial symmetry. Thus, any contribution to the photon mass will be proportional to m_γ. For $m_\gamma = 0$, the photon will not get any corrections to any order in perturbation theory. Keep in mind that this only works if the *only* term that breaks gauge invariance is the mass term. If there are other interactions breaking gauge invariance, the mass correction can be proportional to them as well. For example, in the theory of weak interactions, the W and Z bosons have masses that get corrections proportional not only to m_W and m_Z respectively, but also to fermion masses, since these masses are forbidden by the $\text{SU}(2)_{\text{weak}}$ gauge symmetry, which is spontaneously broken in the Standard Model (see Chapter 29).

An important example of a custodial symmetry not related to anything being massless is custodial isospin, which will be defined in Section 31.2.

22.7 Super-renormalizable theories

In four dimensions there are not many options for Lagrangian terms with coefficients of positive mass dimension. The possibilities are mass terms, which we already discussed, a constant term, terms linear in fields, such as $\Lambda^3\phi$, and cubic couplings among bosons, such as $g\phi^3$ or $g\phi A_\mu^2$. That exhausts the possibilities in four dimensions. We have already discussed masses, so now we will quickly go through the other possibilities.

22.7.1 Cosmological constant and tadpoles

The only possible term of mass dimension 4 is a constant:

$$\mathcal{L} = \cdots + \rho. \tag{22.65}$$

This constant ρ has a name: the **cosmological constant**. By itself, this term does nothing. It couples to nothing and in fact it can just be pulled out of the path integral. The reason it is dangerous is because when one couples to gravity and expands $g_{\mu\nu} = \eta_{\mu\nu} + \frac{1}{M_{\text{Pl}}}h_{\mu\nu}$, the Lagrangian becomes

$$\mathcal{L} = \sqrt{g}\,(R + \rho) = \frac{1}{2}\frac{1}{M_{\text{Pl}}}h_{\mu\mu}\rho + \frac{1}{M_{\text{Pl}}^2}h_{\mu\nu}^2\rho + \cdots + \sqrt{g}R. \tag{22.66}$$

The first term generates a vacuum expectation value for $h_{\mu\nu}$, $\langle\Omega|h_{\mu\mu}|\Omega\rangle \neq 0$ (see Section 18.1 and Chapters 28 and 34), which indicates that we are expanding around the wrong vacuum. By redefining $h_{\mu\nu} \to h_{\mu\nu}^0 + h_{\mu\nu}$ for some non-dynamical x-dependent field $h_{\mu\nu}^0(x)$ with $R\left[h_{\mu\nu}^0\right] = \rho$, we can remove this term (we know $h_{\mu\nu}^0$ has to be x-dependent because all the terms in the expansion of R have derivatives, so R will vanish on any space-time-independent $h_{\mu\nu}^0$). Since the renormalized value of the cosmological constant is experimentally quite small, $\rho \sim 10^{-122} M_{\text{Pl}}^4 \sim \left(10^{-3}\,\text{eV}\right)^4$ (and positive – we live in **de Sitter space**), we can ignore it for terrestrial experiments. To account for a non-zero cosmological constant in quantum field theory requires field theory in curved space, a topic beyond the scope of this text.

Terms with coefficients of mass dimension 3 are linear in fields. For example,

$$\mathcal{L} = -\frac{1}{2}\phi(\Box + m^2)\phi + \Lambda^3\phi, \tag{22.67}$$

where Λ is some number with dimensions of mass. The linear term generates a tadpole diagram that gives a vacuum expectation value to ϕ: $\langle\Omega|\phi|\Omega\rangle \neq 0$. Tadpoles were discussed briefly in Section 18.1 and will be studied in detail in the context of spontaneous symmetry breaking in Chapter 28.

22.7.2 Relevant interactions

Next, we consider radiative corrections in a theory with a $g\phi^3$ coupling, that is, with Lagrangian

$$\mathcal{L} = -\frac{1}{2}\phi(\Box + m^2)\phi + \frac{g}{3!}\phi^3. \tag{22.68}$$

We will consider the 3-point function of three scalars, which illustrates a number of interesting features of this theory. At tree-level, the 3-point function is just

$$\mathcal{M}(p, q_1, q_2) = g. \tag{22.69}$$

Here we are allowing the particles to be off-shell, for example, if this vertex were embedded in a larger diagram.

Now consider a radiative correction from loops of ϕ:

$$i\mathcal{M} = \tag{22.70}$$

$$= -g^3 \int \frac{d^4k}{(2\pi)^4} \frac{1}{(k^2 - m^2)\left[(k - q_1)^2 - m^2\right]\left[(k + q_2)^2 - m^2\right]}. \tag{22.71}$$

This integral scales as $\int \frac{d^4k}{k^6}$ at large k and is therefore UV finite. The mass cuts off the IR divergences, and therefore for generic momenta and masses the loop is finite. While there

The entire functional form of this potential is phenomenologically important, especially at low energies, where we saw it gives the Uehling potential and contributes to the Lamb shift. However, when $p \gg m$, the mass drops out and the potential simplifies to

$$\tilde{V}(p^2) = \frac{e_R^2}{p^2}\left(1 + \frac{e_R^2}{12\pi^2}\ln\frac{p^2}{p_0^2}\right). \tag{23.5}$$

In this limit, we can see clearly the problem of **large logarithms**, which the RG will solve. Normally, one would expect that, since the correction is proportional to $\frac{e_R^2}{12\pi^2} \sim 10^{-3}$, higher-order terms would be proportional to the square, cube, etc. of this term and therefore would be negligible. However, there exist scales $p^2 \gg p_0^2$ where $\ln\frac{p^2}{p_0^2} > 10^3$ so that this correction is of order 1. When these logarithms are this large, terms of the form $\left(\frac{e_R^2}{12\pi^2}\ln\frac{p^2}{p_0^2}\right)^2$, which would appear at the next order in perturbation theory, will also be order 1 and so perturbation theory breaks down.

The running coupling was also introduced in Chapter 16, where we saw that we could sum additional 1PI insertions into the photon propagator,

$$\cdots, \tag{23.6}$$

to get

$$\tilde{V}(p^2) = \frac{e_R^2}{p^2}\left[1 + \frac{e_R^2}{12\pi^2}\ln\frac{p^2}{p_0^2} + \left(\frac{e_R^2}{12\pi^2}\ln\frac{p^2}{p_0^2}\right)^2 + \cdots\right] = \frac{1}{p^2}\left[\frac{e_R^2}{1 - \frac{e_R^2}{12\pi^2}\ln\frac{p^2}{p_0^2}}\right]. \tag{23.7}$$

We then defined the effective coupling through the potential by $e_{\text{eff}}^2(p^2) \equiv p^2\tilde{V}(p^2)$, so that

$$e_{\text{eff}}^2(p^2) = \frac{e_R^2}{1 - \frac{e_R^2}{12\pi^2}\ln\frac{p^2}{p_0^2}}. \tag{23.8}$$

This is the effective coupling including the 1-loop 1PI graphs, This is called **leading-logarithmic resummation**.

Once all of these 1PI 1-loop contributions are included, the next terms we are missing should be subleading in some expansion. The terms included in the effective charge are of the form $e_R^2\left(e_R^2\ln\frac{p^2}{p_0^2}\right)^n$ for $n \geq 0$. For the 2-loop 1PI contributions to be subleading, they should be of the form $e_R^4\left(e_R^2\ln\frac{p^2}{p_0^2}\right)^n$. However, it is not obvious at this point that there cannot be a contribution of the form $e_R^6\ln^2\frac{p_0^2}{p^2}$ from a 2-loop 1PI graph. To check, we would need to perform the full $\mathcal{O}(e_R^4)$ calculation, including graphs with loops and counterterms. As you might imagine, trying to resum large logarithms beyond the leading-logarithmic level diagrammatically is extremely impractical. The RG provides a shortcut to systematic resummation beyond the leading-logarithmic level.

The key to systematizing the above QED calculation is to first observe that the problem we are trying to solve is one of large logarithms. If there were no large logarithms, we would not need the RG – fixed-order perturbation theory would be fine. For the Coulomb potential, the large logarithms related the physical scale p^2 where the potential was to be measured to an arbitrary scale p_0^2 where the coupling was defined. The **renormalization group equation** (RGE) then comes from requiring that the potential is independent of p_0^2:

$$p_0^2 \frac{d}{dp_0^2} \tilde{V}(p^2) = 0. \tag{23.9}$$

$\tilde{V}(p^2)$ has both explicit p_0^2 dependence, as in Eq. (23.5), and implicit p_0^2 dependence, through the scale where e_R is defined. In fact, recalling that e_R was defined so that $p_0^2 \tilde{V}(p_0^2) = e_R^2$ exactly, and that the effective charge is defined by $e_{\text{eff}}^2(p^2) \equiv p^2 \tilde{V}(p^2)$, we can make the p_0^2 dependence of $\tilde{V}(p^2)$ explicit by replacing e_R by $e_{\text{eff}}(p_0^2)$.

So, Eq. (23.5) becomes

$$\tilde{V}(p^2) = \frac{e_{\text{eff}}^2(p_0^2)}{p^2} \left(1 - \frac{e_{\text{eff}}^2(p_0^2)}{12\pi^2} \ln \frac{p_0^2}{p^2} \right) + \cdots . \tag{23.10}$$

Then at 1-loop the RGE is

$$0 = p_0^2 \frac{d}{dp_0^2} \tilde{V}(p^2) = \frac{1}{p^2} \left(p_0^2 \frac{de_{\text{eff}}}{dp_0^2} 2e_{\text{eff}} - \frac{e_{\text{eff}}^4}{12\pi^2} - p_0^2 \frac{de_{\text{eff}}}{dp_0^2} \frac{e_{\text{eff}}^3}{3\pi^2} \ln \frac{p_0^2}{p^2} + \cdots \right). \tag{23.11}$$

To solve this equation perturbatively, we note that $\frac{de_{\text{eff}}}{dp_0^2}$ must scale as e_{eff}^3 and so the third term inside the brackets is subleading. Thus, the 1-loop RGE is

$$p_0^2 \frac{de_{\text{eff}}}{dp_0^2} = \frac{e_{\text{eff}}^3}{24\pi^2}. \tag{23.12}$$

Solving this differential equation with boundary condition $e_{\text{eff}}(p_0) = e_R$ gives

$$e_{\text{eff}}^2(p^2) = \frac{e_R^2}{1 - \frac{e_R^2}{12\pi^2} \ln \frac{p^2}{p_0^2}}, \tag{23.13}$$

which is the same effective charge that we got above by summing 1PI diagrams.

Note, however, that we did not need to talk about the geometric series or 1PI diagrams at all to arrive at Eq. (23.13); we only used the 1-loop graph. In this way, the RG efficiently encodes information about some higher-order Feynman diagrams without having to be explicit about which diagrams are included. This improvement in efficiency is extremely helpful, especially in problems with multiple couplings, or beyond 1-loop.

23.1.2 Universality of large logarithms

Before getting to the systematics of the RG, let us think about the large logarithms in a little more detail. Large logarithms arise when one scale is much bigger or much smaller than every other relevant scale. In the vacuum polarization calculation, we considered the limit where the off-shellness p^2 of the photon was much larger than the electron mass m^2. In the limit where one scale is much larger than all the other scales, we can set all the other physical scales to zero to first approximation. If we do this in the vacuum polarization diagram we find from Eq. (23.2) that the full vacuum polarization function $\Pi(p^2) = e_R^2 \Pi_2(p^2) + \delta_3$ at order e_R^2 is

$$\Pi(p^2) = \frac{e_R^2}{12\pi^2} \left[\frac{2}{\varepsilon} + \ln\left(\frac{\mu^2}{-p^2}\right) + \text{const.} \right] + \delta_3 \quad (\text{DR}), \qquad (23.14)$$

where "const." is independent of p.

The equivalent result using a regulator with a dimensional UV cutoff, such as Pauli–Villars, is

$$\Pi(p^2) = \frac{e_R^2}{12\pi^2} \left[\ln\left(\frac{\Lambda^2}{-p^2}\right) + \text{const.} \right] + \delta_3 \quad (\text{PV}). \qquad (23.15)$$

As was discussed in Chapters 21 and 22, the logarithmic, non-analytic dependence on momentum is characteristic of a loop effect and a true quantum prediction. The RG focuses in on these logarithmic terms, which give the dominant quantum effects in certain limits.

If the only physical scale is p^2, the logarithm of p^2 must be compensated by a logarithm of some other *unphysical* scale, in this case, the cutoff Λ^2 (or μ^2 in dimensional regularization). If we renormalize the theory at some scale p_0 by defining $\delta_3 = -\frac{e_R^2}{12\pi^2} \ln\frac{\Lambda^2}{-p_0^2}$, then this becomes

$$\Pi(p^2) = \frac{e_R^2}{12\pi^2} \left[\ln\left(\frac{p_0^2}{p^2}\right) + \text{const.} \right] \qquad (\text{PV}). \qquad (23.16)$$

In dimensional regularization, the $\overline{\text{MS}}$ prescription is that $\delta_3 = \frac{e_R^2}{12\pi^2}\left(-\frac{2}{\varepsilon}\right)$, so that

$$\Pi(p^2) = \frac{e_R^2}{12\pi^2} \left[\ln\left(\frac{\mu^2}{p^2}\right) + \text{const.} \right] \qquad (\text{DR}). \qquad (23.17)$$

In Eqs. (23.14) to (23.17), the logarithmic dependence on the unphysical scales Λ^2, p_0^2 or μ^2 uniquely determines the logarithmic dependence of the amplitude on the physical scale p^2. The Wilsonian RG extracts physics from the $\ln\Lambda^2$ dependence (see Section 23.6), while the continuum RG uses p_0^2 or μ^2.

In practical applications of the RG, dimensional regularization is almost exclusively used. It is therefore important to understand the roles of $\varepsilon = 4-d$, the arbitrary scale μ^2 and scales such as p_0^2 where couplings are defined. Ultraviolet divergences show up as poles of the form $\frac{1}{\varepsilon}$. Do not confuse the scale μ, which was added to make quantities dimensionally correct, with a UV cutoff! Removing the cutoff is taking $\varepsilon \to 0$, *not* $\mu \to \infty$. In minimal subtraction, renormalized amplitudes depend on μ. In observables, such as the difference $p_1^2 \tilde{V}(p_1^2) - p_2^2 \tilde{V}(p_2^2)$, μ necessarily drops out. However, one can imagine choosing

$$\delta_3 = \frac{e_R^2}{12\pi^2} \left[-\frac{2}{\varepsilon} - \ln\frac{\mu^2}{p_0^2} \right] \qquad (23.18)$$

in dimensional regularization so that Eq. (23.14) turns into Eq. (23.16). This is equivalent to *choosing* $\mu^2 = p_0^2$ in Eq. (23.14) and minimally subtracting the $\frac{1}{\varepsilon}$ term. Thus, we usually think of μ as a physical scale where amplitudes are renormalized and μ is often called the **renormalization scale**.

Although we choose μ to be a physical scale, observables should be independent of μ. At fixed order in perturbation theory, verifying μ independence can be a strong theoretical cross check on calculations in dimensional regularization. As we will see by generalizing the vacuum polarization discussion above, the μ independence of physical amplitudes comes from a cancellation between μ dependence of loops and μ dependence of couplings. Since μ is the renormalization point, the effective coupling becomes $e_{\text{eff}}(\mu)$ and the RGE in Eq. (23.12) becomes

$$\mu \frac{de_{\text{eff}}(\mu)}{d\mu} = \frac{e_{\text{eff}}^3(\mu)}{12\pi^2}, \tag{23.19}$$

and we never have to talk about the physical scale p_0 explicitly.

Although μ is a physical, low-energy scale, not taken to ∞, the dependence of amplitudes on μ is closely connected with the dependences on $\frac{1}{\varepsilon}$. For example, in the vacuum polarization calculation, the $\ln \mu^2$ dependence came from the expansion

$$\mu^\varepsilon \left(\frac{2}{\varepsilon} - \ln p^2 + \cdots \right) = \frac{2}{\varepsilon} + \ln \frac{\mu^2}{p^2} + \cdots . \tag{23.20}$$

The $\frac{1}{\varepsilon}$ pole and the $\ln \mu^2$ in unrenormalized amplitudes are inseparable – in four dimensions, there is no ε and no μ. In particular, the numerical coefficient of $\frac{2}{\varepsilon}$ is the same as the coefficient of $\ln \frac{\mu^2}{p^2}$. Thus, even in dimensional regularization, the large logarithms of the physical scale p^2 are connected to UV divergences, as they would be in a theory with a UV regulator Λ. Since the large logarithms correspond to UV divergences, it is possible to resum them entirely from the ε dependence of the counterterms. This leads to the more efficient, but more abstract, derivation of the continuum RGE, as we now show.

23.2 Renormalization group from counterterms

We have seen how large logarithms of the form $\ln \frac{p_1^2}{p_2^2}$ can be resummed through a differential equation which establishes that physical quantities are independent of the scale p_0^2 where the renormalized coupling is defined. Dealing directly with physical renormalization conditions for general amplitudes is extremely tedious. In this section, we will develop the continuum RG with dimensional regularization, exploiting the observations made in the previous section: the large logarithms are associated with UV divergences, which determine the μ dependence of amplitudes; μ^2 can be used as a proxy for the (arbitrary) physical renormalization scale p_0^2; the RGE will then come from μ independence of physical quantities.

Let us first recall where the factors of μ come from. Recall our bare Lagrangian for QED:

$$\mathcal{L} = -\frac{1}{4} F_{\mu\nu}^2 + \bar{\psi}^0 (i \slashed{\partial} - e^0 \gamma^\mu A_\mu^0 - m^0) \psi^0 . \tag{23.21}$$

The quantities appearing here are infinite, or if we are in $d = 4 - \varepsilon$ dimensions they are finite but scale as inverse powers of ε. The dimensions of the fields can be read off from the Lagrangian:

At 1-loop, $Z_m = 1 - \frac{3e_R^2}{8\pi^2\varepsilon}$ and to leading non-vanishing order $\mu\frac{de_R}{d\mu} = \beta(e_R) = -\frac{\varepsilon}{2}e_R$, so

$$\gamma_m = -\frac{1}{1+\delta_m}\left(\frac{2}{e_R}\delta_m\right)\left(-\frac{\varepsilon}{2}e_R\right) = \delta_m\varepsilon = -\frac{3e_R^2}{8\pi^2}. \tag{23.36}$$

We will give a physical interpretation of a running mass in Section 23.5.

23.3 Renormalization group equation for the 4-Fermi theory

We have seen that the RGE for the electric charge allows us to resum large logarithms of kinematic scales, for example in Coulomb scattering. In that case, the logarithms were resummed through the running electric charge. Large logarithms can also appear in pretty much any scattering process, with any Lagrangian, whether renormalizable or not. In fact, non-renormalizable theories, with their infinite number of operators, provide a great arena for understanding the variety of possible RGEs. We will begin with a concrete example: large logarithmic corrections to the muon decay rate from QED. Then we discuss the generalization for renormalizing operators in the Lagrangian and external operators inserted into Green's functions.

The muon decays into an electron and two neutrinos through an intermediate off-shell W boson. In the Standard Model, the decay rate comes from the following tree-level diagram, which leads to

$$\Gamma\left(\mu^- \to \nu_\mu e^- \bar{\nu}_e\right) = \frac{1}{2m_\mu}\int d\Pi_{\text{LIPS}} \left| \begin{array}{c} \text{[diagram]} \end{array} \right|^2 = \left(\frac{\sqrt{2}g^2}{8m_W^2}\right)^2 \frac{m_\mu^5}{192\pi^3} \tag{23.37}$$

plus corrections suppressed by additional factors of $\frac{m_\mu}{m_W}$ or $\frac{m_e}{m_\mu}$, with $g = 0.64$ the weak coupling constant and $m_W = 80.4$ GeV (see Chapter 29 for more details). A photon loop gives a correction to this decay rate of the form

$$\Gamma\left(\mu^- \to \nu_\mu e^- \bar{\nu}_e\right) = \frac{1}{2m_\mu}\int d\Pi_{\text{LIPS}} \left| \begin{array}{c} \text{[diagram]} \end{array} + \begin{array}{c} \text{[diagram]} \end{array} \right|^2$$

$$= \left(\frac{\sqrt{2}g^2}{8m_W^2}\right)^2 \frac{m_\mu^5}{192\pi^3}\left(1 + A\frac{\alpha}{4\pi}\ln\frac{m_W}{m_\mu} + \cdots\right). \tag{23.38}$$

We have only shown the term in this correction that dominates for $m_\mu \ll m_W$, which is a large logarithm. To extract the coefficient A of this logarithm we would need to evaluate the diagram, which is both difficult and unnecessary. At higher order in perturbation theory, there will be additional large logarithms, proportional to $\left(A\frac{\alpha}{4\pi}\ln\frac{m_W}{m_\mu}\right)^n$. While we could attempt to isolate the series of diagrams that contributes these logarithms (as we isolated the geometric series of 1PI corrections to the Coulomb potential in Section 23.1) such an approach is not nearly as straightforward in this case – there are many relevant diagrams with no obvious relation between them. Instead, we resum the logarithms using the RG.

In order to use the RG to resum logarithms besides those in the effective charge, we need another parameter to renormalize besides e_R. To see what we can renormalize, we first expand in the limit that the W is very heavy, so that we can replace $\frac{i}{p^2-m_W^2}\to-\frac{i}{m_W^2}$ for $p^2 \ll m_W^2$. Graphically, this means

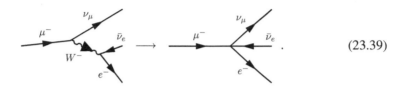

$$\text{(23.39)}$$

This approximation leads to the 4-Fermi theory, discussed briefly in Section 22.2 and to be discussed in more detail here and extensively in Part IV. The 4-Fermi theory replaces the W boson with a set of effective interactions involving four fermions. The relevant Lagrangian interaction in this case is

$$\mathcal{L}_{4F} = \frac{G_F}{\sqrt{2}}\bar{\psi}_\mu\gamma^\mu P_L\psi_{\nu_\mu}\bar{\psi}_e\gamma^\mu P_L\psi_{\nu_e} + h.c., \tag{23.40}$$

where $P_L = \frac{1-\gamma_5}{2}$ projects onto left-handed fermions and $G_F = \frac{\sqrt{2}g^2}{8m_W^2} = 1.166 \times 10^{-5}\,\text{GeV}^{-2}$ (see Section 29.4 for the derivation of Eq. (23.40)). This leads to a decay rate of $\Gamma(\mu^- \to \nu_\mu e^- \bar{\nu}_e) = G_F^2\frac{m_\mu^5}{192\pi^3}$, which agrees with Eq. (23.37). The point of doing this is twofold: first, the 4-Fermi theory is simpler than the theory with the full propagating W boson; second, we can use the RG to compute the RG evolution of G_F that will reproduce the large logarithms in Eq. (23.38) and let us resum them to all orders in α.

It is not hard to go from the RGE for the electric charge to the RGE for a general operator. Indeed, the electric charge can be thought of as the coefficient of the operator $\mathcal{O}_e = \bar{\psi}\slashed{A}\psi$ in the QED Lagrangian. The RGE was determined by the renormalization factor $Z_e = \frac{Z_1}{Z_2\sqrt{Z_3}}$, which was calculated from the radiative correction to the $\bar{\psi}\slashed{A}\psi$ vertex (this gave Z_1), and then subtracting off the field strength renormalizations that came from the electron self-energy graph and vacuum polarization graphs (giving Z_2 and Z_3, respectively).

Unfortunately for the pedagogical purposes of this example, in the actual weak theory, the coefficient A of the large logarithm in Eq. (23.38) is 0 (see Problem 23.2). This fact is closely related to the non-renormalization of the QED current (see Section 23.4.1 below) and is somewhat of an accident. For example, a similar process for the weak decays of quarks does have a non-zero coefficient of the large logarithm, proportional to the strong coupling constant α_s (see Section 31.3). To get something non-zero, let us pretend that the weak interaction is generated by the exchange of a neutral scalar instead, so that the 4-Fermi interaction is

$$\mathcal{L}_4 = \frac{G}{\sqrt{2}} \left(\bar{\psi}_\mu \psi_e \right) \left(\bar{\psi}_{\nu_e} \psi_{\nu_\mu} \right) + h.c. \tag{23.41}$$

In this case, we will get a non-zero coefficient of the large logarithm.

To calculate the renormalization factor for G, we must compute the 1-loop correction to this 4-Fermi interaction. There is only one diagram,

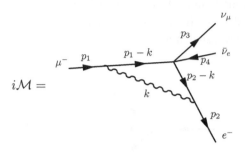

$$= \frac{G}{\sqrt{2}} e_R^2 \mu^{4-d} \int \frac{d^d k}{(2\pi)^d} \frac{\bar{u}(p_2)\, \gamma^\mu \left(\not{p}_2 - \not{k} + m_e \right) \left(\not{p}_1 - \not{k} + m_\mu \right) \gamma^\mu u(p_1)\, \bar{u}(p_3)\, v(p_4)}{\left[(p_1 - k)^2 - m_\mu^2 \right] \left[(p_2 - k)^2 - m_e^2 \right] k^2}. \tag{23.42}$$

To get at the RGE, we just need the counterterm, which comes from the coefficient of the UV divergence of this amplitude. To that end, we can set all the external momenta and masses to zero. Thus,

$$\mathcal{M} = \mathcal{M}_0 \left(-i e_R^2 \mu^{4-d} \int \frac{d^d k}{(2\pi)^d} \frac{d}{k^4} \right) + \text{finite}, \tag{23.43}$$

with the d coming from $\gamma^\mu \not{k} \not{k} \gamma^\mu = d k^2$ and

$$\mathcal{M}_0 = \frac{G}{\sqrt{2}} \bar{u}(p_2)\, u(p_1)\, \bar{u}(p_3)\, v(p_4) \tag{23.44}$$

is the tree-level matrix element from $\mathcal{L}_4$. Extracting the pole gives

$$\mathcal{M} = \mathcal{M}_0 \left(\frac{e_R^2}{2\pi^2} \mu^\varepsilon \frac{1}{\varepsilon} \right) + \text{finite}, \tag{23.45}$$

which is all we will need for the RG analysis.

To remove this divergence, we have to renormalize G. We do so by writing $G = G_R Z_G$, giving

$$\mathcal{L} = \frac{G_R}{\sqrt{2}} Z_G \left(\bar{\psi}_\mu \psi_e \right) \left(\bar{\psi}_{\nu_e} \psi_{\nu_\mu} \right) + h.c. \tag{23.46}$$

To extract the counterterm, we expand $Z_G = 1 + \delta_G$. The counterterm then contributes $\mathcal{M}_0 \delta_G$. To remove the divergence we therefore need to take

$$\delta_G = -\frac{e_R^2}{16\pi^2} \frac{8}{\varepsilon}. \tag{23.47}$$

Now that we know the counterterm, we can calculate the RGE, just as for the electric charge. Expressing the 4-Fermi term in terms of bare fields, we find

$$\frac{G_R}{\sqrt{2}} Z_G (\bar{\psi}_\mu \psi_e)(\bar{\psi}_{\nu_\mu} \psi_{\nu_e}) = \frac{G_R}{\sqrt{2}} \frac{Z_G}{\sqrt{Z_{2\mu} Z_{2e} Z_{2\nu_\mu} Z_{2\nu_e}}} (\bar{\psi}_\mu^{(0)} \psi_e^{(0)})(\bar{\psi}_{\nu_e}^{(0)} \psi_{\nu_\mu}^{(0)}) . \quad (23.48)$$

The coefficient of the bare operator must be independent of μ, since there is no μ in the bare Lagrangian. So, setting $Z_{2\nu} = 1$ since the neutrino is neutral and therefore not renormalized until higher order in e_R, and using $Z_{2\mu} = Z_{2e} = Z_2$ since the muon and electron have identical QED interactions, we find

$$0 = \mu \frac{d}{d\mu} \left(\frac{G_R Z_G}{Z_2} \right) = \frac{G_R Z_G}{Z_2} \left[\frac{\mu}{G_R} \frac{dG_R}{d\mu} + \frac{1}{Z_G} \frac{\partial Z_G}{\partial e_R} \mu \frac{de_R}{d\mu} - \frac{1}{Z_2} \frac{\partial Z_2}{\partial e_R} \mu \frac{de_R}{d\mu} \right],$$
$$(23.49)$$

where we have used that Z_G and Z_2 only depend on μ through e_R in the last step. Using the 1-loop results, $Z_G = 1 - \frac{e_R^2}{16\pi^2} \frac{8}{\varepsilon}$ and $Z_2 = 1 - \frac{e_R^2}{16\pi^2} \frac{2}{\varepsilon}$, and keeping only the leading terms, we have

$$\gamma_G \equiv \frac{\mu}{G_R} \frac{dG_R}{d\mu} = \left(-\frac{\partial Z_G}{\partial e_R} + \frac{\partial Z_2}{\partial e_R} \right) \beta(e_R) = \frac{3e_R}{4\varepsilon\pi^2} \left(-\frac{\varepsilon}{2} e_R \right) = -\frac{3e_R^2}{8\pi^2} = -\frac{3\alpha}{2\pi},$$
$$(23.50)$$

where γ_G is the anomalous dimension for $\mathcal{O}_G = Z_G (\bar{\psi}_\mu \psi_e)(\bar{\psi}_{\nu_\mu} \psi_{\nu_e})$.

Using $\mu \frac{d\alpha}{d\mu} = \beta(\alpha)$, the solution to this differential equation is

$$G_R(\mu) = G_R(\mu_0) \exp \left[\int_{\alpha(\mu_0)}^{\alpha(\mu)} \frac{\gamma_G(\alpha)}{\beta(\alpha)} d\alpha \right] . \quad (23.51)$$

In particular, with $\beta(\alpha) = -\frac{\alpha^2}{2\pi} \beta_0 = \frac{2\alpha^2}{3\pi}$ at leading order we find

$$G_R(\mu) = G_R(\mu_0) \exp \left[-\frac{9}{4} \int_{\alpha(\mu_0)}^{\alpha(\mu)} \frac{d\alpha}{\alpha} \right] = G_R(\mu_0) \left(\frac{\alpha(\mu)}{\alpha(\mu_0)} \right)^{-\frac{9}{4}} . \quad (23.52)$$

Now, we are assuming that we know the value for G at the scale $\mu_0 = m_W$ where the W boson (or its equivalent in our toy model) is integrated out, and we would like to know the value of G at the scale relevant for muon decay, $\mu = m_\mu$. Using Eq. (23.32), we find $\alpha(m_\mu) = 0.007\,36$ and $\alpha(m_W) = 0.007\,43$ so that

$$G_R(m_\mu) = 1.022 \times G_R(m_W), \quad (23.53)$$

which would have given a 4.8% correction to the muon decay rate if the muon decay were mediated by a neutral scalar. In the actual weak theory, where muon decay is mediated by a vector boson coupled to left-handed spinors, the anomalous dimension for the operator in Eq. (23.40) is zero and so G_F does not run in QED.

23.4 Renormalization group equation for general interactions

In the muon decay example, we calculated the running of G, defined as the coefficient of the local operator $\mathcal{O}_G = Z_G (\bar{\psi}_\mu \psi_e)(\bar{\psi}_{\nu_e} \psi_{\nu_\mu})$ in a 4-Fermi Lagrangian. More generally, we can consider adding additional operators to QED, with an effective Lagrangian of the form

$$\mathcal{L} = -\frac{1}{4} Z_3 F_{\mu\nu}^2 + Z_2^i \bar{\psi}_i i \not{\partial} \psi_i - Z_2^i Z_m^i m_i^R \bar{\psi}_i \psi_i + Z_e Z_2^i \sqrt{Z_3} Q_i e_R \bar{\psi}_i \not{A} \psi_i + \sum_j C_j \mathcal{O}_j(x).$$

(23.54)

These operators, $\mathcal{O}_j = Z_j \partial^n \gamma^m A_\mu(x) \cdots A_\nu(x) \bar{\psi}_{i_1}(x) \cdots \psi_{j_n}(x)$, are **composite local operators**, with all fields evaluated at the same space-time point. They can have any number of photons, fermions, γ-matrices, factors of the metric, etc. and analytic (power-law) dependence on derivatives. Keep in mind that the fields A_μ and ψ_j are the renormalized fields. The C_j are known as **Wilson coefficients**. Note that in this convention each Z_j is grouped with its corresponding operator, which is composed of renormalized fields; the Z_j is not included in the Wilson coefficient so that the Wilson coefficient will be a finite number at any given scale. Since the Lagrangian is independent of μ, if we assume no mixing, the RGEs take the form

$$\mu \frac{d}{d\mu}(C_j \mathcal{O}_j) = 0 \quad (\text{no sum on } j).$$

(23.55)

These equations (one for each j) let us extract the RG evolution of Wilson coefficients from the μ dependence of matrix elements of operators. In general, there can be mixing among the operators (see Section 23.5.2 and Section 31.3), in which case this equation must be generalized to $\mu \frac{d}{d\mu} \left(\sum_j C_j \mathcal{O}_j \right) = 0$. One can also have mixing between the operators and the other terms in the Lagrangian in Eq. (23.54), in which case the RGE is just $\mu \frac{d}{d\mu} \mathcal{L} = 0$.

The way these effective Lagrangians are used is that the C_j are first either calculated or measured at some scale μ_0. We can calculate them if we have a (full) theory that is equivalent to this (effective) theory at a particular scale. For example, we found G_F by designing the 4-Fermi theory to reproduce the muon decay rate from the full electroweak theory, to leading order in $\frac{1}{m_W^2}$ at the scale $\mu_0 = m_W$. This is known as **matching**. Alternatively, if a full theory to which our effective Lagrangian can be matched is not known (or is not perturbative), one can simply measure the C_j at some scale μ_0. For example, in the Chiral Lagrangian (describing the low-energy theory of pions) one could in principle match to the theory of strong interactions (QCD), but in practice it is easier just to measure the Wilson coefficients. In either case, once the values of the C_j are set at some scale, we can solve the RGE to resum large logarithms. In the toy muon-decay example, we evolved G_R to the scale $\mu = m_\mu$ in order to incorporate large logarithmic corrections of the form $\alpha \ln \frac{m_\mu}{m_W}$ into the rate calculation.

23.4.1 External operators

Equation (23.55) implies that the RG evolution of Wilson coefficients is exactly compensated for by the RG evolution of the operators. Operator running provides a useful language in which to consider physical implications of the RG. An important example is the running of the current $J_\mu(x) = Z_J \bar{\psi}(x) \gamma^\mu \psi(x)$, which we will now explore. Rather than thinking of J^μ as the coefficient of A_μ in the QED interaction, we will treat $J_\mu(x)$ as an **external**

operator: an operator that is not part of the Lagrangian, but which can be inserted into Green's functions.

The running of J_μ is determined by the μ dependence of Z_J and of the renormalized fields $\bar\psi(x)$ and $\psi(x)$ appearing in the operator. To find Z_J, we can calculate any Green's function involving J^μ. The simplest non-vanishing one is the 3-point function with the current and two fields, whose Fourier transform we already discussed in the context of the Ward–Takahashi identity in Section 14.8 and the proof of $Z_1 = Z_2$ in Section 19.5. We define

$$\langle\Omega|\, T\{J^\mu(x)\psi(x_1)\bar\psi(x_2)\}\,|\Omega\rangle = \int \frac{d^4p}{(2\pi)^4}\frac{d^4q_1}{(2\pi)^4}\frac{d^4q_2}{(2\pi)^4} e^{-ipx}e^{-iq_1x_1}e^{iq_2x_2}$$
$$\times\, i\mathcal{M}^\mu(p,q_1,q_2)(2\pi)^4\delta^4(p+q_1-q_2), \quad (23.56)$$

so that $\mathcal{M}^\mu$ is given by Feynman diagrams without truncating the external lines or adding external spinors. At tree-level,

$$i\mathcal{M}^\mu_{\text{tree}}(p,q_1,q_2) = \frac{i}{\slashed{q}_1 - m}\gamma^\mu\frac{i}{\slashed{q}_2 - m}. \quad (23.57)$$

At next-to-leading order, there is a 1PI loop contribution and a counterterm:

$$\quad (23.58)$$

Here the $\otimes$ indicates an insertion of the current and the $\bigstar$ indicates the counterterm for the current, both with incoming momentum p^μ. The counterterm contribution to the Green's function comes from expanding $Z_J = 1 + \delta_J$ directly in the Green's function (we have not added J^μ to the Lagrangian). These two graphs give, in Feynman gauge,

$$i\mathcal{M}^\mu_{1\text{-loop}} = \frac{i}{\slashed{q}_1 - m}\left[(-ie_R)^2\mu^{4-d}\int\frac{d^dk}{(2\pi)^d}\frac{i\gamma^\nu(\slashed{q}_1 - \slashed{k} + m)}{(q_1 - k)^2 - m^2}\right.$$
$$\left.\times\gamma^\mu\frac{i(\slashed{q}_2 - \slashed{k} + m)\gamma_\nu}{(q_2 - k)^2 - m^2}\frac{-i}{k^2} + \gamma^\mu\delta_J\right]\frac{i}{\slashed{q}_2 - m}. \quad (23.59)$$

Since we are just interested in the counterterm we take $k \gg q_1, q_2$. Then this reduces to

$$i\mathcal{M}^\mu_{1\text{-loop}} = \frac{i}{\slashed{q}_1 - m}\gamma^\mu\frac{i}{\slashed{q}_2 - m}\left[-ie_R^2\mu^{4-d}\frac{(2-d)^2}{d}\int\frac{d^dk}{(2\pi)^d}\frac{1}{k^4} + \delta_J\right]$$
$$= i\mathcal{M}^\mu_{\text{tree}}\left\{\frac{e_R^2}{16\pi^2}\left[\frac{2}{\varepsilon}\right] + \delta_J\right\}. \quad (23.60)$$

Thus, $\delta_J = \frac{e_R^2}{16\pi^2}\left[-\frac{2}{\varepsilon}\right]$, which also happens to equal δ_2 and δ_1. Thus $Z_2 = Z_J$ at 1-loop.

Now that we know Z_J we can calculate the renormalization of the current. The bare current is independent of μ. Since $J^\mu_{\text{bare}}(x) = \bar\psi_0\gamma^\mu\psi_0 = \frac{1}{Z_J}Z_2 J^\mu(x)$, then

$$0 = \mu\frac{d}{d\mu}J^\mu_{\text{bare}} = \mu\frac{d}{d\mu}\left(\frac{Z_2}{Z_J}J^\mu(x)\right) = \mu\frac{d}{d\mu}J^\mu(x). \quad (23.61)$$

$$0 = \mu \frac{d}{d\mu} G_{n,m}^{(0)}$$

$$= Z_3^{\frac{n}{2}} Z_2^{\frac{m}{2}} \left(\mu \frac{\partial}{\partial \mu} + \frac{n}{2} \frac{\mu}{Z_3} \frac{dZ_3}{d\mu} + \frac{m}{2} \frac{\mu}{Z_2} \frac{dZ_2}{d\mu} + \mu \frac{\partial e_R}{\partial \mu} \frac{\partial}{\partial e_R} + \mu \frac{\partial m_R}{\partial \mu} \frac{\partial}{\partial m_R} \right) G_{n,m}. \tag{23.74}$$

Defining

$$\gamma_3 = \frac{\mu}{Z_3} \frac{dZ_3}{d\mu}, \quad \gamma_2 = \frac{\mu}{Z_2} \frac{dZ_2}{d\mu}, \tag{23.75}$$

this reduces to

$$\left(\mu \frac{\partial}{\partial \mu} + \frac{n}{2} \gamma_3 + \frac{m}{2} \gamma_2 + \beta \frac{\partial}{\partial e_R} + \gamma_m m_R \frac{\partial}{\partial m_R} \right) G_{n,m} = 0. \tag{23.76}$$

This equation is known variously as the **Callan–Symanzik equation**, the **Gell-Mann–Low equation**, the **'t Hooft–Weinberg equation** and the **Georgi–Politzer equation**. (The differences refer to different schemes, such as $\overline{\text{MS}}$ or the on-shell physical renormalization scheme.)

One can also calculate Green's functions with external operators inserted, such as $\langle \Omega | T\{ J^\mu(x) \psi_1(x_1) \bar{\psi}_2(x_2) \} | \Omega \rangle$ considered in Section 23.4.1. For a general operator, we define

$$\mu \frac{d}{d\mu} \mathcal{O} = \gamma_\mathcal{O} \mathcal{O}. \tag{23.77}$$

Then a Green's function with an operator $\mathcal{O}$ in it satisfies

$$\left(\mu \frac{\partial}{\partial \mu} + \frac{n}{2} \gamma_3 + \frac{m}{2} \gamma_2 + \beta \frac{\partial}{\partial e_R} + \gamma_m m_R \frac{\partial}{\partial m_R} + \gamma_\mathcal{O} \right) G = 0. \tag{23.78}$$

If there are more operators, there will be more $\gamma_\mathcal{O}$ terms.

23.4.4 Anomalous dimensions

Now let us discuss the term "anomalous dimension". We have talked about the mass dimension of a field many times. For example, in four dimensions, $[\phi] = M^1$, $[m] = M^1$, $[\psi] = M^{3/2}$ and so on. These numbers just tell us what happens if we change units. To be more precise, consider the action for ϕ^4:

$$\mathcal{S} = \int d^4 x \left[-\frac{1}{2} \phi(\Box + m^2)\phi + g\phi^4 \right]. \tag{23.79}$$

This has a symmetry under $x^\mu \to \frac{1}{\lambda} x^\mu$, $\partial_\mu \to \lambda \partial_\mu$, $m \to \lambda m$, $g \to g$ and $\phi \to \lambda \phi$. This operation is called **dilatation** and denoted by $\mathcal{D}$. Thus,

$$\mathcal{D} : \phi \to \lambda^{d_0} \phi. \tag{23.80}$$

The d_0 are called the **classical** or **canonical scaling dimensions** of the various fields and couplings in the theory.

Now consider a correlation function

$$G_n = \langle \Omega | T \{ \phi_1(x_1) \cdots \phi_n(x_n) \} | \Omega \rangle. \tag{23.81}$$

In a classical theory, this Green's function can only depend on the various quantities in the Lagrangian raised to various powers. We might find

$$G_n(x, g, m) = m^a g^b x_1^{c_1} \cdots x_n^{c_n}. \tag{23.82}$$

By dimensional analysis, we must have $a - c_1 - \cdots - c_n = n$. Thus we expect that $\mathcal{D} : G_n \to \lambda^n G_n$.

In the quantum theory, G_n can also depend on the scale where the theory is renormalized, μ. So we could have

$$G_n(x, g, m, \mu) = m^a g^b x_1^{c_1} \cdots x_n^{c_n} \mu^\gamma, \tag{23.83}$$

where now $a - c_1 - \cdots - c_n = n - \gamma$. Note that μ does not transform under $\mathcal{D}$ since it does not appear in the Lagrangian – it is the subtraction point used to connect to experiment. So when we act with $\mathcal{D}$, only the x and m terms change; thus, we find $\mathcal{D} : G_n \to \lambda^{n-\gamma} G_n$. Thus, G_n does not have the canonical scaling dimension. In particular,

$$\mu \frac{d}{d\mu} G_n = \gamma G_n, \tag{23.84}$$

which is how we have been defining anomalous dimensions. Thus, the anomalous dimensions tell us about deviations from the classical scaling behavior.

23.5 Scalar masses and renormalization group flows

In this section we will examine the RG evolution of a super-renormalizable operator, namely a scalar mass term $m^2 \phi^2$. To extract physics from running masses, we have to think of masses more generally than just the location of the renormalized physical pole in an S-matrix, since by definition the pole mass is independent of scale. Rather, we should think of them as a term in a potential, like a ϕ^4 interaction would be. This language is very natural in condensed matter physics. As we will now see, in an off-shell scheme (such as $\overline{\text{MS}}$) masses can have scale dependence. This scale dependence can induce phase transitions and signal spontaneous symmetry breaking (see Chapters 28 and 34).

23.5.1 Yukawa potential correction

Recall that the exchange of a massive particle generates a Yukawa potential, with the mass giving the characteristic scale of the interactions. Just as the Coulomb potential let us understand the physics of a running coupling, the Yukawa potential will help us understand running scalar masses. For example, consider the Lagrangian

$$\mathcal{L} = -\frac{1}{2} \phi (\Box + m^2) \phi - \frac{1}{4!} \lambda \phi^4 + g \phi J, \tag{23.85}$$

which has the scalar field interacting with some external current J. The current–current interaction at leading order comes from an exchange of ϕ, which generates the Yukawa potential. For the static potential, we can drop time derivatives and then Fourier transform the propagator, giving

$$V(r) = -\int \frac{d^3k}{(2\pi)^3} \frac{1}{\vec{k}^2 + m^2} e^{i\vec{k}\cdot\vec{x}} = -\frac{1}{4\pi r} e^{-mr}. \tag{23.86}$$

In the language of condensed matter physics, this correlation function has a correlation length ξ given by the inverse mass, $\xi = \frac{1}{m}$. In this language, we can easily give a physical interpretation to a running mass: the Yukawa potential will be modified by $m \to m(r)$ with calculable logarithmic dependence on r.

To calculate $m(r)$ we will solve the RG evolution induced by the $\lambda\phi^4$ interaction. The first step to studying the RGE for this theory is to renormalize it at 1-loop, for which we need to introduce the various Z-factors into the Lagrangian. In terms of renormalized fields,

$$\mathcal{L} = -\frac{1}{2}Z_\phi \phi \Box \phi - \frac{1}{2}Z_m Z_\phi m_R^2 \phi^2 - \mu^{4-d}\frac{\lambda_R}{4!}Z_\lambda Z_\phi^2 \phi^4. \tag{23.87}$$

Since ϕ has mass dimension $\frac{d-2}{2}$, an extra factor of μ^{4-d} has been added to keep λ_R dimensionless, as was done for the electric charge in QED. The RGE for the mass comes from the μ independence of the bare mass, $m^2 = m_R^2 Z_m$:

$$0 = \mu\frac{d}{d\mu}\left(m^2\right) = \mu\frac{d}{d\mu}\left(m_R^2 Z_m\right) = m_R^2 Z_m\left(\frac{1}{m_R^2}\mu\frac{d}{d\mu}m_R^2 + \frac{1}{Z_m}\mu\frac{d}{d\mu}\delta_m\right). \tag{23.88}$$

Since the only μ dependence in the Lagrangian comes from the ϕ^4 interaction, we need to compute the dependence of δ_m on λ_R and the dependence of λ_R on μ.

We can extract Z_m (and Z_ϕ) from corrections to the scalar propagator. The leading graph is

$$i\Sigma_2(p^2) = \quad\underset{p \qquad\qquad p}{\overset{k}{\bigcirc}}\quad = \frac{-i\lambda_R}{2}\mu^{4-d}\int\frac{d^d k}{(2\pi)^d}\frac{i}{k^2 - m_R^2}$$

$$= \frac{-i\lambda_R\mu^{4-d}}{2(4\pi)^{d/2}}\left(\frac{1}{m_R^2}\right)^{1-\frac{d}{2}}\Gamma\left(1 - \frac{d}{2}\right). \tag{23.89}$$

The quadratic divergence in this integral shows up in dimensional regularization as a pole at $d = 2$ but is hidden if one expands near $d = 4$. Nevertheless, since quadratic divergences are just absorbed into the counterterms, we can safely ignore them and focus on the logarithmic divergences. After all, it is the non-analytic logarithmic momentum dependence that we will resum using the renormalization group.

Expanding in $d = 4 - \varepsilon$ dimensions, $\Sigma_2(p^2) = \frac{\lambda_R m_R^2}{16\pi^2}\frac{1}{\varepsilon} + \cdots$. The counterterms from $Z_\phi = 1 + \delta_\phi$ and $Z_m = 1 + \delta_m$ give a contribution

$$i\Sigma_{\text{c.t.}}(p^2) = \quad\underset{p \qquad\qquad p}{\longrightarrow\!\!\bigstar\!\!\longrightarrow}\quad = i\delta_\phi\left(p^2 - m_R^2\right) - i\delta_m m_R^2. \tag{23.90}$$

So, to order λ_R, $\delta_\phi = 0$ and $\delta_m = \frac{\lambda_R}{16\pi^2}\frac{1}{\varepsilon}$.

An alternative way to extract these counterterms is to use the propagator of the massless theory and to treat $m_R^2 \phi^2$ as a perturbation. This does not change the physics, since the massive propagator is reproduced by summing the usual geometric series of 1PI insertions of the mass:

$$\frac{i}{p^2} + \frac{i}{p^2}\left(-im_R^2\right)\frac{i}{p^2} + \frac{i}{p^2}\left(-im_R^2\right)\frac{i}{p^2}\left(-im_R^2\right)\frac{i}{p^2} + \cdots = \frac{i}{p^2 - m_R^2}. \qquad (23.91)$$

However, one can look at just the first mass insertion to calculate the counterterms. The leading graph with an insertion of the mass and the coupling λ_R is

$$i\Sigma_2(p^2) = \qquad \underset{p \qquad\quad p}{\underset{k \qquad\qquad k}{\bigcirc}} \qquad = \left(-im_R^2\right)\frac{-i\lambda_R}{2}\mu^{4-d}\int\frac{d^dk}{(2\pi)^d}\frac{i}{k^2}\frac{i}{k^2}. \qquad (23.92)$$

This is now only logarithmically divergent. Extracting the UV divergence with the usual trick gives $\Sigma_2(p^2) = \frac{\lambda_R m_R^2}{16\pi^2}\frac{1}{\varepsilon} + \cdots$ and so $\delta_m = \frac{\lambda_R}{16\pi^2}\frac{1}{\varepsilon}$, which is the same result we got from the quadratically divergent integral.

Next, we need the dependence of λ_R on μ. The RGE for λ_R is derived by using that the bare coupling, $\lambda_0 = \mu^{4-d}\lambda_R Z_\lambda$, is μ independent, so

$$0 - \mu\frac{d}{d\mu}(\lambda_0) = \mu\frac{d}{d\mu}\left(\mu^{4-d}\lambda_R Z_\lambda\right) = \mu^\varepsilon\lambda_R Z_\lambda\left(\varepsilon + \frac{\mu}{\lambda_R}\frac{d}{d\mu}\lambda_R + \frac{\mu}{Z_\lambda}\frac{d}{d\mu}\delta_\lambda\right). \qquad (23.93)$$

Then, since δ_λ starts at order λ_R we have $\mu\frac{d}{d\mu}\lambda_R = -\varepsilon\lambda_R + \mathcal{O}(\lambda_R^2)$. Although not necessary for the running of m_R, it is not hard to calculate δ_λ at 1-loop. We can extract it from the radiative correction to the 4-point function. With zero external momenta, the loop gives

$$(-i\lambda_R)^2\frac{3}{2}\mu^{2(4-d)}\int\frac{d^dk}{(2\pi)^d}\frac{i}{k^2}\frac{i}{k^2} = \mu^{2(4-d)}\frac{3\lambda_R^2}{16\pi^2}\frac{i}{\varepsilon}, \qquad (23.94)$$

so that $\delta_\lambda = \frac{3\lambda_R}{16\pi^2}\frac{1}{\varepsilon}$ and then the β-function to order λ_R^2 is

$$\beta(\lambda_R) \equiv \mu\frac{d}{d\mu}\lambda_R(\mu) = -\varepsilon\lambda_R - \frac{3\lambda_R^2}{16\pi^2}\frac{1}{\varepsilon}(-\varepsilon) = -\varepsilon\lambda_R + \frac{3\lambda_R^2}{16\pi^2}. \qquad (23.95)$$

Using the RGE for the mass, Eq. (23.88), $\mu\frac{d}{d\mu}\lambda_R = -\varepsilon\lambda_R$ and $\delta_m = \frac{\lambda_R}{16\pi^2}\frac{1}{\varepsilon}$, we find

$$\gamma_m \equiv \frac{\mu}{m_R^2}\frac{d}{d\mu}m_R^2 = -\frac{1}{Z_m}\frac{\partial\delta_m}{\partial\lambda_R}\mu\frac{d\lambda_R}{d\mu} = \frac{\lambda_R}{16\pi^2} + \mathcal{O}(\lambda_R^3). \qquad (23.96)$$

The solution, treating γ_m as constant, is

$$m_R^2(\mu) = m_R^2(\mu_0)\left(\frac{\mu}{\mu_0}\right)^{\gamma_m}. \qquad (23.97)$$

You can check in Problem 23.3 that the more general solution (including the μ dependence of λ_R following Eq. (23.51)) reduces to Eq. (23.97) for small λ_R.

Now let us return to the Yukawa potential. Since μ just represents an arbitrary scale with dimensions of mass, we can equally well write the solution to the RGE in position space as

$$m^2(r) = m_0^2\left(\frac{r}{r_0}\right)^{-\gamma_m}, \qquad (23.98)$$

where $m_0 = m(r = r_0)$. This leads to a corrected Yukawa potential:

$$V(r) = -\frac{g^2}{4\pi r} \exp[-rm(r)] = -\frac{g^2}{4\pi r} \exp\left[-r^{1-\frac{\gamma_m}{2}} r_0^{\frac{\gamma_m}{2}} m_0\right], \quad (23.99)$$

which is in principle measurable. The final form has been written in a suggestive way to connect to what we will discuss below. Indeed, extracting a correlation length by dimensional analysis, we find

$$V(r) = -\frac{g^2}{4\pi r} \exp\left[-(r/\xi)^{1-\frac{\gamma_m}{2}}\right], \quad \xi = r_0^{1-2\nu} m_0^{-2\nu}, \quad \nu = \frac{1}{2-\gamma_m}. \quad (23.100)$$

In the free theory, ξ scales as m_0^{-1}, by dimensional analysis. With interactions we see that it scales as m_0 to a different power of the mass, determined by ν. This quantity ν is known as a **critical exponent**. Dimensional transmutation has given us another scale with dimensions of mass, r_0^{-1}, which has changed the scaling relation predicted by dimensional analysis. These critical exponents have been measured in a number of situations. We next discuss how to compare the result of our RG calculation to experimental results.

23.5.2 Wilson–Fisher fixed point

It is a remarkable experimental fact that very different physical systems exhibit very similar scaling behaviors in the vicinity of second-order phase transitions. For example, for many materials there is a critical point in the phase diagram when the liquid–gas phase transition becomes second order. In water, this critical point is at a critical temperature $T_C = 173\,°C$ and a critical pressure $p_C = 217$ atm. One can measure correlation functions in water (for example by scattering light off it) and extract from those functions a characteristic scale ξ called the **correlation length**. For example, measuring the intensity of light as a function of momentum, one might find $I(q) = I_0(1 + q^2\xi^2)^{-1}$. In water, near its critical point, the correlation length is found to scale with temperature as $\xi \sim (T - T_C)^{-0.63}$. This 0.63 is an example of a **critical exponent**. This particular critical exponent is called ν and conventionally defined by $\xi \sim (T - T_C)^{-\nu}$. Remarkably, this scaling behavior with the same exponent $\nu = 0.63$, can be seen in thousands of other systems, with very different microscopic descriptions, near their critical points (see [Pelissetto and Vicari, 2002] for a review). A very important example is the 3D Ising model (defined on a rectangular lattice with nearest-neighbor spin–spin interactions). The set of systems that share this scaling behavior near their critical points are said to be in the **3D Ising model universality class**. The universality of the critical exponent ν suggests that it should be calculable without detailed knowledge of the microscopic system. In fact it can. Moreover, the universality can be understood with the RG.

The starting point for a calculation of ν from field theory is to represent the Ising model system with a single scalar field. For water, this field, $\phi(x)$, might be the density, but it does not actually matter what the field is. All that matters is that the effective description shares the symmetries of the microscopic theory (in the case of Ising model systems, there is no symmetry and so a single scalar field will do). The effective description of a field theory near a second-order phase transition can be described by a Ginzburg–Landau model defined by the Lagrangian

$$\mathcal{L}_{\text{eff}} = \mathcal{L}_{\text{kin}} - \frac{1}{2}(T - T_C)\phi^2 - \frac{1}{4!}\lambda\phi^4 + \cdots. \tag{23.101}$$

The $T - T_C$ factor in this Lagrangian is a well-motivated guess. First of all, one expects some kind of temperature dependence in the effective Lagrangian. For $T \sim T_C$, we can then Taylor expand this Lagangian. Thus, if nothing special forces the linear term to vanish, the leading term should be linear in $T - T_C$. The $-\frac{1}{2}(T - T_C)\phi^2$ term gives ϕ a mass $m = \sqrt{T - T_C}$. For $T > T_C$, m^2 is positive and there is a finite correlation length to the system. When T goes below T_C, then m^2 becomes negative, signaling spontaneous symmetry breaking into a different phase (see Chapter 28 for more details). Moreover, the transition is smooth across T_C, as required for a second-order phase transition. Thus, this form of the temperature dependence is a natural guess for an effective description for $T \sim T_C$.

As a quick check, we already know that the 2-point function in such a scalar theory should behave like a Yukawa potential

$$\langle\Omega|\phi(r)\phi(0)|\Omega\rangle \sim \frac{1}{r}e^{-rm} = \frac{1}{r}\exp\left(-r\,(T - T_C)^{1/2}\right). \tag{23.102}$$

Thus, the classical theory predicts $\nu = \frac{1}{2}$, which is not far from the observed universal value ($\nu = 0.63$). To calculate corrections to this classical value, we can use Eq. (23.100):

$$\nu = \frac{1}{2 - \gamma_m}. \tag{23.103}$$

Thus, corrections to the critical exponent are given by an anomalous dimension calculable (analytically or numerically) in quantum field theory.

To calculate γ_m in perturbation theory, it looks like we can use Eqs. (23.95) and (23.96):

$$\mu\frac{d}{d\mu}m_R^2 = \frac{\lambda_R}{16\pi^2}m_R^2 + \mathcal{O}(\lambda_R^2), \tag{23.104}$$

$$\mu\frac{d}{d\mu}\lambda_R = -\varepsilon\lambda_R + \frac{3\lambda_R^2}{16\pi^2} + \mathcal{O}(\lambda_R^3). \tag{23.105}$$

But, what do we take for μ and what do we take for λ_R? Here we arrive at the key reason for universality of critical exponents: although m_R and λ_R are in general scale dependent, for certain values of m_R and λ_R we may find that the right-hand sides of Eqs. (23.104) and (23.105) vanish. It is precisely at these values, which are **fixed points** of the RG evolution equations, that systems become universal. A simple example of a fixed point is where $\lambda_R = m_R = 0$, or more generally when all couplings and masses vanish. Such a solution, for which all the RGEs are trivial, is known as a **Gaussian fixed point** (since at this point the Lagrangian is a free theory of a massless scalar field and the path integral is an exact Gaussian). To calculate ν we want to find a non-trivial fixed point.

In condensed matter physics we are interested in the macroscopic, long-distance behavior of a system. In particle physics we are interested usually in the low-energy limit of a system, which is most accessible experimentally. So, in either case we would like to know what happens as we *lower* μ. The behavior of a system as μ is lowered gives the RG trajectory or **RG flow** of the couplings in a system. For example, suppose we start near (but not on) the Gaussian fixed point. Then the RGE for λ_R at leading order is $\frac{d\ln\lambda_R}{d\ln\mu} = -\varepsilon$, which implies that if $d > 4$ ($\varepsilon < 0$), the system will flow back towards the fixed point as μ decreases, while for $d < 4$ ($\varepsilon > 0$), the system will flow away from the fixed point. The liquid–gas phase transitions for water and the 3D Ising model take place in $d = 3$ (one can ignore time in these non-relativistic systems). For $d = 3$, the flow is away from the fixed point. Thus, the natural question is, where do the couplings flow to? As $\mu \to 0$, they can either blow up, go to zero, or go to some non-trivial fixed point.

Instead of going all the way to $d = 3$, let us explore what happens in $d = 4 - \varepsilon$ dimensions. From Eq. (23.105), we can see that for $0 < \varepsilon \ll 1$, there exist values of λ_R and m_R for which $\frac{d}{d\mu}\lambda_R = \frac{d}{d\mu}m_R = 0$, namely

$$\lambda_\star = \frac{16\pi^2\varepsilon}{3}, \qquad m_\star^2 = 0. \tag{23.106}$$

This is the location of the **Wilson–Fisher** fixed point to order ε (using dimensional regularization). At this fixed point, $\gamma_m = \frac{\varepsilon}{3}$ from Eq. (23.96) and so, from Eq. (23.103),

$$\nu = \frac{3}{6 - \varepsilon}. \tag{23.107}$$

Although the values of $m_\star$ and $\lambda_\star$ are scheme dependent and therefore unphysical, the critical exponents are scheme independent. Indeed, they must be, since they can be measured. You can explore the scheme dependence of the Wilson–Fisher fixed point in Problem 23.6. See [Wilson and Kogut, 1974; Pelissetto and Vicari, 2002] or [Sachdev, 2011, Chapter 4] for more information.

For $\varepsilon = 1$ corresponding to three dimensions, $\nu = 0.6$ at this point, which is quite close to the observed value of 0.63. This (somewhat questionable) practice of expanding around $d = 4$ to get results in $d = 3$ is known as the **epsilon expansion**. You can compute the 2-loop value of ν in Problem 23.5. Currently, ν is known to 5-loops in the epsilon expansion [Kleinert *et al.*, 1991] and has been computed many other ways (with Monte-Carlo methods, high- or low-temperature expansions, Borel resummed perturbation theory, etc.). See [Pelissetto and Vicari, 2002] for a review.

Regardless of whether the epsilon expansion can be justified, we can at least trust the qualitative observation of Wilson and Fisher, that there is a **non-trivial fixed point** (couplings do not all vanish) in this effective theory for $d < 4$. As ε increases, the fixed point will move away from the $\lambda_\star$, due to large ε^2 corrections. This justifies the universality of the critical exponents in three-dimensional systems – even if we cannot calculate the anomalous dimension, we expect that for $d < 3$ it should still exist and should be separate from the Gaussian fixed point.

Fixed points are interesting places. Exactly on the fixed point, the theory is scale invariant, since $\mu\frac{d}{d\mu}m_R^2 = \mu\frac{d}{d\mu}\lambda_R = 0$. While there are many classical theories that are scale invariant (such as QED with massless fermions), theories that are scale invariant at the quantum level are much rarer. Such theories are known as **conformal field theories**. In a conformal theory, the Poincaré group is enhanced to a larger group called the **conformal group**. Recall that the Poincaré group acting on functions of space-time is generated by translations, $P_\mu = -i\partial_\mu$, and Lorentz transformations, $\Lambda_{\mu\nu} = i(x_\mu\partial_\nu - x_\nu\partial_\mu)$. In the conformal group, these are supplemented with a generator for scale transformations, $D = -ix_\mu\partial_\mu$, and four generators for special-conformal transformations, $K_\mu = i(x^2\partial_\mu - 2x_\mu x_\nu\partial_\nu)$. Invariance under the conformal group is so restrictive that

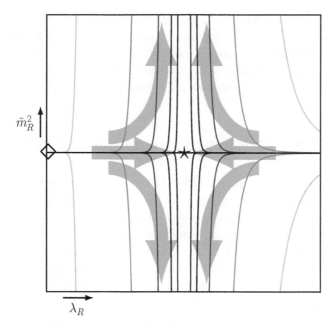

Renormalization group flow in the Wilson–Fisher theory. The Wilson–Fisher fixed point is indicated by the $\star$ at $\tilde{m}_\star = 0$ and $\lambda = \lambda_\star$. The Gaussian fixed point at $m = \lambda = 0$ is indicated by the $\diamond$. The arrows denote flow as the length scale is increased, or equivalently, as μ is decreased. Although the location of the Wilson–Fisher fixed point is scheme dependent, the trajectories near the fixed point can be used to extract scheme-independent information about the conformal field theory living on the fixed point (such as critical exponents).

Fig. 23.1

correlation functions in conformal field theories are strongly constrained. On the other hand, conformal field theories do not have massive particles. In fact, they do not have particles at all. That is, there is no sensible way to define asymptotic single-particle states in such a theory. Thus, they do not have an S-matrix.

One way to find conformal field theories is by looking for fixed points of RG flows in non-conformal field theories, as in the Wilson–Fisher example. Since conformal field theories have no inherent scales, dimensional parameters such as m_R in the Wilson–Fisher theory become dimensionless. To see how the fixed point is approached, it is natural to rescale away any classical scaling dimension of the various couplings. In the Wilson–Fisher case, we do this by defining $\tilde{m}_R(\mu) \equiv \frac{1}{\mu} m_R(\mu)$ so that $\tilde{m}_R$ is dimensionless. Then the RG equations become

$$\mu \frac{d}{d\mu} \tilde{m}_R^2 = \left(-2 + \frac{\lambda_R}{16\pi^2}\right) \tilde{m}_R^2, \tag{23.108}$$

$$\mu \frac{d}{d\mu} \lambda_R = -\varepsilon \lambda_R + \frac{3\lambda_R^2}{16\pi^2}. \tag{23.109}$$

The fixed point is at the same place, $\lambda_\star = \frac{16\pi^2 \varepsilon}{3}$ and $m_R^2 = 0$. The RG flow for $\tilde{m}_R^2$ is shown in Figure 23.1.

The different trajectories in an RG flow diagram represent different values of m_R^2 and λ_R that might correspond to different microscopic systems. For example, changing the temperature of a system moves it from one trajectory to another. The temperature for which $m_R = 0$ is the critical temperature where the theory intersects the non-trivial fixed point. To get close to the non-trivial fixed point, one would have to be very close to the $m_R = 0$ trajectory.

23.5.3 Varieties of asymptotic behavior

One can easily imagine more complicated RG flows than those described by the Wilson–Fisher theory. With just one coupling, such as in QED or in QCD, the RG flow is determined by the β-function $\beta(\alpha) = \mu\frac{d}{d\mu}\alpha$. When the coupling is small, the theory is perturbative, and then the coupling must either increase or decrease with scale. If the coupling increases with μ, as in QED, it goes to zero at long distances. In this case it is said to be **infrared free**. If it decreases with μ (as the strong coupling in QCD does, as we will show in Chapter 26), it goes to zero at short distances and the theory is said to be **asymptotically free**. The third possibility in a perturbative theory is that $\beta(\alpha) = 0$ exactly, in which case the theory is scale invariant. If the coupling is non-perturbative, one can still define a coupling through the value of a Green's function. Then, as long as $\beta(\alpha) > 0$ at one α and $\beta(\alpha) < 0$ at a larger α, there is guaranteed to be an intermediate value where $\beta(\alpha_\star) = 0$. With multiple couplings there are other possibilities for solutions to the RGEs. For example, one could imagine a situation in which couplings circle around each other. It is certainly easy to write down coupled differential equations with bizarre solutions; whether such equations correspond to anything in nature or in a laboratory is another question.

There are not many known examples of perturbative conformal field theories in four dimensions. One is called $\mathcal{N} = 4$ super Yang–Mills theory. Another possibility is if the leading β-function coefficient is small, for example if $\beta(\alpha) = \beta_0\alpha^2 + \beta_1\alpha^3 + \cdots$, where β_0 happens to be of order α. Then there could be a cancellation between β_0 and β_1 and a non-trivial fixed point at some finite value of α. That this might happen in a non-Abelian gauge theory with a large enough number of matter fields was conjectured by Banks and Zaks [Banks and Zaks, 1982] and is known as the **Banks–Zaks theory**.

23.6 Wilsonian renormalization group equation

So far we have been discussing the RGE as an invariance of physical quantities under changes of the scale μ, where the renormalization conditions are imposed. This is the continuum RG, where all comparisons are made after the UV regulator has been completely removed. The Wilsonian picture instead supposes that there is an actual physical cutoff Λ, as there would be in a metal (the atomic spacing) or string theory (the string scale). Then all loops are finite and the theory is well defined. In this case, one can (in principle) integrate over a shell of momentum in the path integral $\Lambda' < p < \Lambda$ and change the couplings of the theory so that low-energy physics is the same. The Wilsonian RGE describes the resulting flow of coupling constants under infinitesimal changes in Λ. The reason we focused on the continuum RG first is that it is easier to connect to observables, which coupling constants are not. However, the Wilsonian RGE helps explain why renormalizable theories play such an important role in physics.

You have perhaps heard people say mysterious phrases such as "a dimension 6 operator, such as $\bar\psi\psi\bar\psi\psi$ is *irrelevant* since it *should* have a coefficient $\frac{1}{\Lambda^2}$, where Λ is an arbitrarily large cutoff." You may also have wondered how the word "should" earned a place in scientific discourse. There is indeed something very odd about this language, since if $\Lambda = 10^{19}$ GeV the operator $\frac{1}{\Lambda^2}\bar\psi\psi\bar\psi\psi$ can safely be ignored at low energy, but if Λ is lowered to 1 GeV this operator becomes extremely important. This language, although imprecise, actually is logical. It originates from the Wilsonian RG, as we will now explain.

To begin, imagine that you have a theory with a physical short-distance cutoff Λ_H, which is described by a Lagrangian with a finite or infinite set of operators $\mathcal{O}_r$ of various mass dimensions r. For example, in a metal with atomic spacing ξ the physical cutoff would be $\Lambda_H \sim \xi^{-1}$ and the operators might include $\frac{1}{\Lambda_H^2}\bar\psi\psi\bar\psi\psi$, where ψ correspond to atoms. Let us write a general Lagrangian with cutoff Λ_H as $\mathcal{L}(\Lambda_H) = \sum C_r(\Lambda_H)\,\Lambda_H^{4-r}\mathcal{O}_r$ with $C_r(\Lambda_H)$ some dimensionless numbers. These numbers can be large and are probably impossible to compute. In principle they could all be measured, but we would need an infinite number of renormalization conditions for all the $C_r(\Lambda_H)$ to completely specify the theory. The key point, however, as we will show, is that not all the $C_r(\Lambda_H)$ are important for long-distance physics.

At low energies, we do not need to take Λ to be as large as ξ^{-1}. As long as Λ is much larger than any energy scale of interest, we can perform loops as if $\Lambda = \infty$ and cutoff-dependent effects will be suppressed by powers of $\frac{E}{\Lambda}$. (For example, for observables with $E \sim 100$ GeV, you do not need $\Lambda = 10^{19}$ GeV; $\Lambda \sim 10^{10}$ GeV works just as well.) So let us compute a different Lagrangian, $\mathcal{L}(\Lambda) = \sum C_r(\Lambda)\,\Lambda^{4-r}\mathcal{O}_r$, with a cutoff $\Lambda < \Lambda_H$, by demanding that physical quantities computed with the two Lagrangians be the same. With $\Lambda = \Lambda_L \ll \Lambda_H$, the coefficients $C_r(\Lambda_L)$ will be some other dimensionless numbers, which may be big or small, and which are (in principle) computable in terms of $C_r(\Lambda_H)$.

Now, if we are making large-distance measurements only, we should be able to work with $\mathcal{L}(\Lambda_L)$ just as well as with $\mathcal{L}(\Lambda_H)$. So we might as well measure $C_r(\Lambda_L)$ to connect our theory to experiment. The important point, which follows from the Wilsonian RG, is that $C_r(\Lambda_L)$ is *independent* of $C_r(\Lambda_H)$ if $r > 4$. Since there will only be a finite number of operators in a given theory with mass dimension $r \leq 4$, if we measure $C_{r\leq 4}(\Lambda_L)$ for these operators (as renormalization conditions), we can then calculate $C_{r>4}(\Lambda_L)$ for all the other operators as functions of the $C_{r\leq 4}(\Lambda_L)$. An explicit example is given below.

This result motivates the definition of **relevant** operators as those with $r < 4$ and **irrelevant** operators as those with $r > 4$. Operators with $r = 4$ are called **marginal**. We only need to specify renormalization conditions for the relevant and marginal operators, of which there are always a finite number. The Wilson coefficients for the irrelevant operators can be computed with very weak dependence on any boundary condition related to short-distance physics, that is, on the values of $C_r(\Lambda_H)$.

Thus, it is true that with $\Lambda = \Lambda_H$ or $\Lambda = \Lambda_L$ the Lagrangian *should* have operators with coefficients determined by Λ to some power. Therefore, irrelevant operators *do* get more important as the cutoff is lowered. However, the important point is not the size of these operators, but that their Wilson coefficients are computable. In other words:

Values of couplings when the cutoff is low are insensitive to the boundary conditions associated with *irrelevant* operators when the cutoff is high.

If we take the high cutoff to infinity then the irrelevant operators are precisely those for which there is zero effect on the low-cutoff Lagrangian. Only relevant operators remain when the cutoff is removed. So:

The space of renormalizable field theories is the space for which the limit $\Lambda_H \to \infty$ exists, holding the couplings fixed when the cutoff is Λ_L.

Another important point is that in the Wilsonian picture one does not want to take Λ_L down to physical scales of interest. One wants to lower Λ enough so that the irrelevant operators become insensitive to boundary conditions, but then to leave it high enough so one can perform loop integrals as if $\Lambda = \infty$. That is:

The Wilsonian cutoff Λ should always be much larger than all relevant physical scales. This is in contrast to the μ in the continuum picture, which should be taken equal to a relevant physical scale.

For example, in the electroweak theory, one can imagine taking $\Lambda = 100\,\text{TeV}$, not $\Lambda = 10^{19}\,\text{GeV}$ and not $\Lambda = 100\,\text{GeV}$.

23.6.1 Wilson–Polchinski renormalization group equation

To prove the above statements, we need to sort out what is being held fixed and what is changing. Since the theory is supposed to be finite with UV cutoff Λ, the path integral is finite (at least to a physicist), and all the physics is contained in the generating functional $Z[J]$. The RGE is then simply $\Lambda \frac{d}{d\Lambda} Z[J] = 0$. If we change the cutoff Λ, then the coupling constants in the Lagrangian must change to hold $Z[J]$ constant. For example, in a scalar theory, we might have

$$Z[J] = \int^{\Lambda_H} \mathcal{D}\phi \exp\left\{ i \int d^4x \left(-\frac{1}{2}\phi(\Box + m^2)\phi + \frac{g_3}{3!}\phi^3 + \frac{g_4}{4!}\phi^4 + \frac{g_6}{6!}\phi^6 \cdots + \phi J \right) \right\}$$

(23.110)

for some cutoff Λ_H on the momenta of the fields in the path integral. All the couplings, m, g_3, g_4 etc., are finite. If we change the cutoff to Λ then the couplings change to m', g_3', g_4' etc., so that $Z[J]$ is the same.

Unfortunately, actually performing the path integral over a Λ-shell is extremely difficult to do in practice. A more efficient way to phrase the Wilsonian RGE in field theory was developed by Polchinski [Polchinski, 1984]. Polchinski's idea was first to cut off the path integral more smoothly by writing

$$Z[J] = \int \mathcal{D}\phi \, e^{iS+\phi J}$$

$$= \int \mathcal{D}\phi \exp\left\{ i \int d^4x \left(-\frac{1}{2}\phi(\Box + m^2)e^{\frac{\Box}{\Lambda^2}}\phi + \frac{g_3}{3!}\phi^3 + \frac{g_4}{4!}\phi^4 + \cdots + \phi J \right) \right\}.$$

(23.111)

The $e^{\Box/\Lambda^2}$ factor makes the propagator go as $e^{-p^2/\Lambda^2} \to 0$ at high energy. You can get away with this only in a scalar theory in Euclidean space, but we will not let such technical details prevent us from making very general conclusions. It is easiest to proceed in momentum space, where $\phi(x)^2 \to \phi(p)\phi(-p)$. Then,

$$Z[J] = \int \mathcal{D}\phi \, e^{iS+\phi J}$$

$$= \int \mathcal{D}\phi \exp\left\{ i \int \frac{d^4p}{(2\pi)^4} \left(\frac{1}{2}\phi(p)(p^2 - m^2)e^{-\frac{p^2}{\Lambda^2}}\phi(-p) + \mathcal{L}_{\text{int}}(\phi) + \phi J \right) \right\}.$$

(23.112)

Taking $\frac{d}{d\Lambda}$ on both sides gives

$$\Lambda\frac{d}{d\Lambda}Z[J] = i \int \mathcal{D}\phi \int \frac{d^4p}{(2\pi)^4} \left(\phi(p)(p^2 - m^2)\phi(-p)\frac{p^2}{\Lambda^2}e^{-\frac{p^2}{\Lambda^2}} + \Lambda\frac{d}{d\Lambda}\mathcal{L}_{\text{int}}(\phi) \right) e^{iS+\phi J}.$$

(23.113)

Since $\frac{p^2}{\Lambda^2}e^{-\frac{p^2}{\Lambda^2}}$ only has support near $p^2 \sim \Lambda^2$, the change in $\mathcal{L}_{\text{int}}$ comes from that momentum region. Therefore, the RGE will be local in Λ. This is a general result, independent of the precise way the cutoff is imposed. It can also be used to define a functional differential equation known as the **exact renormalization group** (see Problem 23.7), which we will not make use of here.

As a concrete example, consider a theory with a dimension-4 operator (with dimensionless coupling g_4) and a dimension-6 operator (with coupling g_6 with mass dimension -2). Then the RGE $\Lambda\frac{d}{d\Lambda}Z[J] = 0$ would imply some equations that we can write as

$$\Lambda\frac{d}{d\Lambda}g_4 = \beta_4\big(g_4, \Lambda^2 g_6\big),$$

(23.114)

$$\Lambda\frac{d}{d\Lambda}g_6 = \frac{1}{\Lambda^2}\beta_6\big(g_4, \Lambda^2 g_6\big),$$

(23.115)

where β_4 and β_6 are some general, complicated functions. The factors of Λ have all been inserted by dimensional analysis since, as we just showed, no other scale can appear in $\Lambda\frac{d}{d\Lambda}Z[J]$. To make these equations more homogeneous, let us define dimensionless couplings $\lambda_4 = g_4$ and $\lambda_6 = \Lambda^2 g_6$. Then,

$$\Lambda\frac{d}{d\Lambda}\lambda_4 = \beta_4(\lambda_4, \lambda_6),$$

(23.116)

$$\Lambda\frac{d}{d\Lambda}\lambda_6 - 2\lambda_6 = \beta_6(\lambda_4, \lambda_6).$$

(23.117)

The $-2\lambda_6$ term implies that if β_6 is small, then $\lambda_6(\Lambda) = \lambda_6(\Lambda_H)\left(\frac{\Lambda}{\Lambda_H}\right)^2$ is a solution. We would like this to mean that as Λ is taken small, $\Lambda \ll \Lambda_H$, the higher-dimension operators die away. However, the actual coupling of the operator for this solution is just $g_6(\Lambda) = \frac{1}{\Lambda_H^2}\lambda_6(\Lambda_H) = g_6(\Lambda_H)$, which does not die off (it does not run since we have set $\beta = 0$), so things are not quite that simple. We clearly need to work beyond zeroth order.

It is not hard to solve the RGEs explicitly in the case when β_4 and β_6 are small. Actually, one does not need the β_i to be small; rather, one can start with an exact solution to the full RGEs and then expand perturbatively around the solution. For simplicity, we will just assume that the β_i can be expanded in their arguments. To linear order, we can write

$$\Lambda\frac{d}{d\Lambda}\lambda_4 = a\lambda_4 + b\lambda_6, \tag{23.118}$$

$$\Lambda\frac{d}{d\Lambda}\lambda_6 = c\lambda_4 + (2+d)\lambda_6, \tag{23.119}$$

and we assume a, b, c, d are small real numbers, so that the anomalous dimension does not overwhelm the classical dimension (otherwise perturbation theory would not be valid). It is now easy to solve this vector of homogeneous linear differential equations by changing to a diagonal basis:

$$\widetilde{\lambda_4} = -\frac{c}{\Delta}\lambda_4 - \frac{2+d-a-\Delta}{2\Delta}\lambda_6, \qquad \widetilde{\lambda_6} = \frac{c}{\Delta}\lambda_4 + \frac{2+d-a+\Delta}{2\Delta}\lambda_6, \tag{23.120}$$

where $\Delta = \sqrt{4bc + (d-a+2)^2}$. The RGEs are easy to solve now:

$$\widetilde{\lambda_4}(\Lambda) = \left(\frac{\Lambda}{\Lambda_0}\right)^{\frac{d+2+a-\Delta}{2}}\widetilde{\lambda_4}(\Lambda_0), \qquad \widetilde{\lambda_6}(\Lambda) = \left(\frac{\Lambda}{\Lambda_0}\right)^{\frac{d+2+a+\Delta}{2}}\widetilde{\lambda_6}(\Lambda_0). \tag{23.121}$$

Back in terms of the original basis, we then have

$$\lambda_4(\Lambda) = \left(\frac{\Lambda}{\Lambda_0}\right)^{\frac{d+2+a-\Delta}{2}}\left[\left(\frac{2+d-a+\Delta}{2\Delta}\right)\lambda_4(\Lambda_0) - \frac{b}{\Delta}\lambda_6(\Lambda_0)\right]$$
$$+ \left(\frac{\Lambda}{\Lambda_0}\right)^{\frac{d+2+a+\Delta}{2}}\left[-\left(\frac{2+d-a-\Delta}{2\Delta}\right)\lambda_4(\Lambda_0) + \frac{b}{\Delta}\lambda_6(\Lambda_0)\right]\Bigg\} \tag{23.122}$$

and

$$\lambda_6(\Lambda) = \left(\frac{\Lambda}{\Lambda_0}\right)^{\frac{d+2+a-\Delta}{2}}\left[-\frac{c}{\Delta}\lambda_4(\Lambda_0) - \left(\frac{2+d-a-\Delta}{2\Delta}\right)\lambda_6(\Lambda_0)\right]$$
$$+ \left(\frac{\Lambda}{\Lambda_0}\right)^{\frac{d+2+a+\Delta}{2}}\left[\frac{c}{\Delta}\lambda_4(\Lambda_0) + \left(\frac{2+d-a+\Delta}{2\Delta}\right)\lambda_6(\Lambda_0)\right]\Bigg\}, \tag{23.123}$$

which is an exact solution to Eqs. (23.118) and (23.119). In these solutions, $\lambda_4(\Lambda_0)$ and $\lambda_6(\Lambda_0)$ are free parameters to be set by boundary conditions.

What we would like to know is the sensitivity of λ_6 at some low scale Λ_L to its initial condition at some high scale Λ_H for fixed, renormalized, values of $\lambda_4(\Lambda_L)$. For simplicity,

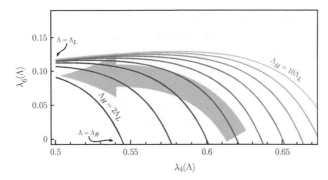

Solutions of the Wilsonian RGEs with $a = 0.1, b = 0.2, c = -0.5$ and $d = 0.3$. We fix $\lambda_4(\Lambda_L) = 0.5$ and look at how the value of $\lambda_6(\Lambda_L)$ depends on $\lambda_6(\Lambda_H)$ for some $\Lambda_H > \Lambda_L$. As $\Lambda_H \to \infty$ the value of $\lambda_6(\Lambda_L)$ goes to a constant value, entirely set by $\lambda_4(\Lambda_L)$ and the anomalous dimensions. Arrow denotes RG flow of decreasing Λ. Note the convergence is extremely quick.

Fig. 23.2

let us take $\lambda_6(\Lambda_H) = 0$ (any other boundary value would do just as well, but the solution is messier). Then, Eqs. (23.122) and (23.123) can be combined into

$$\lambda_6(\Lambda) = \frac{2c\left[\left(\frac{\Lambda}{\Lambda_H}\right)^\Delta - 1\right]}{(2+d-a+\Delta) - (2+d-a-\Delta)\left(\frac{\Lambda}{\Lambda_H}\right)^\Delta}\lambda_4(\Lambda). \qquad (23.124)$$

Setting $\Lambda - \Lambda_L \ll \Lambda_H$ and assuming $a, b, c, d \ll 2$, so that $\Delta \approx 2$, we find

$$\lambda_6(\Lambda_L) = -\frac{c}{2}\left(1 - \frac{\Lambda_L^2}{\Lambda_H^2}\right)\lambda_4(\Lambda_L). \qquad (23.125)$$

In particular, the limit $\Lambda_H \to \infty$ exists. Back in terms of g_4 and g_6 we have fixed $g_4(\Lambda_L)$ and set $g_6(\Lambda_H) = 0$. Thus, as $\Lambda_H \to \infty$ we have $g_6(\Lambda_L) = -\frac{c}{2}\frac{1}{\Lambda_L^2}g_4(\Lambda_L)$. That is, the boundary condition at large Λ_H is totally irrelevant to the value of g_6 at the low scale. That is why operators with dimension greater than 4 are called **irrelevant**. This result is shown in Figure 23.2.

To relate all this rather abstract manipulation to physics, recall the calculation of the electron magnetic moment from Chapter 17. We found that the moment was $g = 2$ at tree-level and $g = 2 + \frac{\alpha}{\pi}$ at 1-loop. If we had added to the QED Lagrangian an operator of the form $\mathcal{O}_\sigma = \frac{e}{4}\bar\psi\sigma^{\mu\nu}\psi F_{\mu\nu}$ with some coefficient C_σ, this would have given $g = 2 + \frac{\alpha}{\pi} + C_\sigma$. Since the measured value of g is in excellent agreement with the calculation ignoring C_σ, we need an explanation of why $\mathcal{O}_\sigma$ should be absent or have a small coefficient. The answer is given by the above calculation, with g_4 representing α and g_6 representing the coefficient of $\mathcal{O}_\sigma$. Say we do add $\mathcal{O}_\sigma$ to the QED Lagrangian with even a very large coefficient, but with the cutoff set to some very high scale, say $\Lambda_H \sim M_{\rm Pl} \sim 10^{19}$ GeV. Then, when the cutoff is lowered, even a little bit (say to 10^{15} GeV), whatever you set your coefficient to at $M_{\rm Pl}$ would be totally irrelevant: the coefficient of $\mathcal{O}_\sigma$ would now be determined completely in terms of α, like g_6 is determined by g_4. Hence g becomes a calculable function of α. The operator $\mathcal{O}_\sigma$ is *irrelevant* to the $g - 2$ calculation.

Note that if we lowered the cutoff down to say 1 MeV, then $\mathcal{O}_\sigma$ would indeed give a contribution to g, but a contribution calculable entirely in terms of α. With such a low cutoff, there would be cutoff dependence in the 1-loop calculation of $g-2$ as well (which is tremendously difficult to actually calculate). Indeed, these two contributions must precisely cancel, since the theory is independent of cutoff. That is why one does not want to take the cutoff Λ_L down to scales near physics of interest in the Wilsonian picture. To repeat, in the continuum picture, μ is of the order of physical scales, but in the Wilsonian picture, Λ is always much higher than all of the relevant physical scales.

Returning to our toy RGEs, suppose we set $\lambda_4(\Lambda_H) = 0$. Then we would have found

$$\lambda_4(\Lambda) = \frac{2b\left[1 - \left(\frac{\Lambda}{\Lambda_H}\right)^\Delta\right]}{2 + d - a - \Delta - (2 + d - a + \Delta)\left(\frac{\Lambda}{\Lambda_H}\right)^\Delta}\lambda_6(\Lambda) . \qquad (23.126)$$

Expanding this for $a, b, c, d \ll 2$ gives

$$\lambda_4(\Lambda_L) = \frac{b}{2}\left(1 - \frac{\Lambda_H^2}{\Lambda_L^2}\right)\lambda_6(\Lambda_L), \qquad (23.127)$$

which diverges as $\Lambda_H \rightarrow \infty$! Thus, we cannot self-consistently hold the irrelevant couplings fixed at low energy and take the high-energy cutoff to infinity.

The same would be true if we had a dimension 4 coupling (such as a gauge coupling) and a dimension-2 parameter, such as m^2 for a scalar. Then, we would have found an extraordinary sensitivity of $m^2(\Lambda_L)$ to the boundary condition $m^2(\Lambda_H)$ if $g(\Lambda_L)$ is held fixed. Of course, like any renormalizable coupling, one should fix $m^2(\Lambda_L)$ through a low-energy experiment, for example measuring the Higgs mass. The Wilsonian RG simply implies that if there is a short-distance theory with cutoff Λ_H in which m is calculable, then $m^2(\Lambda_H)$ should have a very peculiar looking value. For example, suppose $m^2(\Lambda_L) = (100\,\text{GeV})^2$ when $\Lambda_L = 10^5\,\text{GeV}$. Then, there is some value for $m^2(\Lambda_H)$ with $\Lambda_H = 10^{19}\,\text{GeV}$. If there were a different short-distance theory for which $m^2(\Lambda_H)$ were different by a factor of order $\frac{\Lambda_L^2}{\Lambda_H^2} = 10^{-28}$, then $m^2(\Lambda_L)$ would differ by a factor of order 1 (see Problem 23.8). This is the **fine-tuning problem**. It is a sensitivity of long-distance measurements to small deformations of a theory defined at some short-distance scale. The general result is that relevant operators, such as scalar masses, are **UV sensitive** (unless they are protected by a custodial symmetry; see Section 22.6).

23.6.2 Generalization and discussion

The generalization of the above 2-operator example is a theory with an arbitrary set of operators $\mathcal{O}_n$. To match onto the Wilson operator language (this is, after all, the Wilsonian RGE), let us write

$$Z[J] = \int^\Lambda \mathcal{D}\phi \exp\left\{i\int d^4x \sum_n C_n \mathcal{O}_n(\phi)\right\} . \qquad (23.128)$$

Since there is a cutoff, all couplings (Wilson coefficients C_n) in the theory are finite. The RGE in the Wilsonian picture is $\Lambda \frac{d}{d\Lambda} Z[J] = 0$, which forces

$$\Lambda \frac{d}{d\Lambda} C_n = \beta_n(\{C_m\}, \Lambda) \qquad (23.129)$$

for some β_n. In the continuum picture, the RGE we used was

$$\mu \frac{d}{d\mu} C_n = \gamma_{nm} C_m, \qquad (23.130)$$

which looks a lot like the linear approximation to the Wilsonian RGE. In fact, we can linearize the Wilsonian RGE, not necessarily by requiring that all the couplings be small, but simply by expanding around a fixed point, which is a solution of Eq. (23.129) for which $\beta_n = 0$.

In the continuum language, although the cutoff is removed, the anomalous dimensions γ_{mn} are still determined by the UV divergences. So these two equations are very closely related. However, there is one very important difference: in the continuum picture quadratic and higher-order power-law divergences are exactly removed by counterterms. In the continuum picture of renormalization, the only UV divergences corresponding to physically observable effects are logarithmic ones (examples were given in various non-renormalizable theories in Chapter 22). With a finite cutoff, one simply has Λ^2 terms in the RGE. This Λ^2 dependence was critical for the analysis of g_4 and g_6 in the previous subsection.

For a theory with general, possibly non-perturbative β_n, consider a given subset $\mathcal{S}$ of the operators and its complement $\overline{\mathcal{S}}$. Choose coefficients for the operators in $\mathcal{S}$ to be fixed at a scale Λ_L and set the coefficients for the operators in $\overline{\mathcal{S}}$ to 0 at a scale Λ_H. If it is possible to take the limit $\Lambda_H \to \infty$ so that *all* operators have finite coefficients at Λ_L, the theory restricted to the set $\mathcal{S}$ is called a **renormalizable** theory. Actually, one does not have to set all the operators in $\overline{\mathcal{S}}$ to 0 at Λ_H; if there is *any way* to choose their coefficients as a function of Λ_H so that the theory at Λ_L is finite, then the theory is still considered renormalizable.

It is not hard to see that this definition coincides with the one we have been using all along. As you might imagine, generalizing the g_4/g_6 example above, any operator with dimension greater than 4 will be non-renormalizable and **irrelevant**. Operators with dimension less than 4 are **super-renormalizable** and **relevant**. **Marginal** operators have dimension equal to 4; however, if the operator has any anomalous dimension at all it will become marginally relevant or marginally irrelevant. From the Wilsonian point of view, marginally irrelevant operators are the same as irrelevant ones – one cannot keep their couplings fixed at low energy and remove the cutoff.

Technically, the terms relevant and irrelevant should be applied only to operators corresponding to eigenvectors of the RG. Otherwise there is operator mixing. So, let us diagonalize the matrix γ_{mn} and consider its eigenvalues. Any eigenvalue λ_n of γ_{mn} with $\lambda_n > 0$ will will cause the couplings C_n to decrease as μ is lowered. Thus, these operators decrease in importance at long distances. They are the irrelevant operators. Relevant operators have $\lambda_n < 0$. These operators increase in importance as μ is lowered. If we try to take the long-distance limit, the relevant operators blow up. It is sometimes helpful to think of all possible couplings in the theory as a large multi-dimensional surface. An RG fixed point therefore lies on the subsurface of irrelevant operators. Any point on this surface will be attracted to the fixed point, while any point off the surface will be repelled away from it.

In practice, we do not normally work in a basis of operators that are eigenstates of the RG. In a perturbative theory (near a Gaussian fixed point), operators are usually classified by their classical scaling dimension d_n. The coefficient of such an operator (in four dimensions) has classical dimension $[C_n] = 4 - d_n$. If we rescale $C_n \to C_n \mu^{d_n - 4}$ to make the coefficient dimensionless, then the γ_{nn} component in the matrix Eq. (23.130) becomes $\gamma_{nn} = d_n - 4$. Thus, at leading order, irrelevant operators are those with $d_n > 4$. In the quantum theory, loops induce non-diagonal components in γ_{mn}. If a marginal or relevant operator mixes into an irrelevant one, this mixing completely dominates the RG evolution of C_n at low energy. In this way, an operator that is classified as irrelevant based on its scaling dimension can become more important at large distances. However, the value of its coefficient quickly becomes a calculable function of coupling constants corresponding to more relevant operators. We saw this through direct calculation.

Problems

23.1 Consider the operator $\mathcal{O} = \bar{\psi}\slashed{\partial}\psi\bar{\psi}\slashed{\partial}\psi$ in QED.
(a) Evaluate the anomalous dimension of $\mathcal{O}_{\mu\nu}$ at 1-loop.
(b) If the coefficient for this operator is $C = 1$ at 1 TeV, what is C at 1 GeV?

23.2 Show that $A = 0$ in Eq. (23.38) by evaluating the anomalous dimension of G_F from Eq. (23.40) in QED. At an intermediate stage, you may want to use the Fierz identity:

$$\left(\bar{\psi}_1 P_L \gamma^\mu \gamma^\alpha \gamma^\beta \psi_2\right)\left(\bar{\psi}_3 P_L \gamma^\mu \gamma^\alpha \gamma^\beta \psi_4\right) = 16\left(\bar{\psi}_1 P_L \gamma^\mu \psi_2\right)\left(\bar{\psi}_3 P_L \gamma^\mu \psi_4\right),$$
(23.131)

which you derived in Problem 11.8.

23.3 Show that Eq. (23.97) follows from the small λ_R limit of the general solution to $m_R(\mu)$.

23.4 Consider a theory with N real scalar fields ϕ_i with Lagrangian

$$\mathcal{L} = -\frac{1}{2}\phi_i(\Box + m^2)\phi_i - \frac{\lambda}{4}(\phi_i)^2(\phi_j)^2.$$
(23.132)

This effective Lagrangian can describe systems with multiple degrees of freedom near critical points (for example, the superfluid transition in $^4\mathrm{He}$ corresponds to $N = 2$).
(a) Calculate γ_m and $\beta(\lambda_R)$ in this theory. Check that for $N = 1$ you reproduce Eqs. (23.96) and (23.95). (Note that the normalizations of λ in Eqs. (23.85) and (23.132) are different.)
(b) Where is the location of the Wilson–Fisher fixed point in this theory in $4 - \varepsilon$ dimensions?
(c) What is the value of the critical exponent ν is this theory in $d = 3$ in the epsilon expansion?

23.5 Compute the value of the critical exponent ν in the Wilson–Fisher theory (with $N = 1$, as in Section 23.5.2) to order ε^2.

23.6 Scheme dependence in the Wilson–Fisher theory.

(a) Compute the 1-loop RGEs in scalar ϕ^4 theory (with Lagrangian Eq. (23.85)) using a hard cutoff. Show that you get non-zero values for λ and m at the fixed point, but the critical exponent ν is the same as computed in Section 23.5.2.

(b) Plot the RG flow trajectories using the RGEs you just computed with a fixed cutoff. What is different about these trajectories from those in Figure 23.1?

(c) Compute the 1-loop RGEs in the Wilsonian picture by literally integrating over a shell in momentum from $b\Lambda$ to Λ. Show that you get the same value for ν.

(d) Show that the critical exponent ν is independent of regulator and subtraction scheme at 1-loop. Can you choose a scheme so that $\lambda_\star$ and $m_\star$ are whatever you want?

23.7 Derive

$$\Lambda \frac{d}{d\Lambda} \mathcal{L}_{\text{int}}(\phi) = \int d^4p \, \frac{(2\pi)^4}{p^2 + m^2} \frac{p^2}{\Lambda^2} e^{\frac{p^2}{\Lambda^2}} \left[\frac{\delta\mathcal{L}_{\text{int}}}{\delta\phi(p)} \frac{\delta\mathcal{L}_{\text{int}}}{\delta\phi(-p)} - \frac{\delta^2\mathcal{L}_{\text{int}}}{\delta\phi(p)\delta\phi(-p)} \right]$$

$$(23.133)$$

using the Wilson–Polchinski RGE. Show that the first term corresponds to integrating out the tree-level diagram and the second from loops.

23.8 Consider a theory with a dimension-2 mass parameter m^2 and a dimensionless coupling g.

(a) Write down and solve generic Wilsonian RGEs for this theory, as in Eqs. (23.118) and (23.119).

(b) Fix $g(\Lambda_L) = 0.1$ for concreteness with $\Lambda_L = 10^5$ GeV. What value of $m^2(\Lambda_H)$ would lead to $m^2(\Lambda_L) = (100 \, \text{GeV})^2$?

(c) What would $m^2(\Lambda_L)$ be if you changed $m^2(\Lambda_H)$ by 1 part in 10^{20}?

(d) Sketch the RG flows for this theory.

We have discussed the concept of unitarity a number of times now. Informally, unitarity means conservation of probability: something cannot be created from nothing, nor can something just disappear. Our insistence on unitarity constrains the states in the Hilbert space to transform in unitary representations of the Poincaré group. As we will see, this aspect of unitarity provides powerful constraints even if the set of states is not known exactly. (For example, we do not need to know the spectrum of bound states.) Unitarity also constrains the form that interactions can have, since the S-matrix must be unitary.

In Chapter 8, we argued that particles should be identified with states in the Hilbert space that transform in unitary irreducible representations of the Poincaré group. Single- and multi-particle states are eigenstates of the momentum operator $\hat{P}_\mu$, with $\hat{P}_\mu|X\rangle = p_\mu|X\rangle$ for a set of real numbers p_μ with $p_0 > 0$ and $p^2 \geq 0$, which transform in the 4-vector representation of the Lorentz group. The corresponding adjoint states $\langle X|$ satisfy $\langle X|\hat{P}_\mu = \langle X|p_\mu$ for the same p_μ. Single-particle states $|X\rangle$ transform under irreducible unitary representations of the Lorentz group as well, as $|X\rangle \rightarrow \exp(i\theta_{\mu\nu}S^{\mu\nu})|X\rangle$ where $\theta_{\mu\nu}$ are the boost and rotation angles and $S^{\mu\nu}$ are the generators of the Lorentz group in the representation of that particle. The transformations of a multi-particle state are induced by the transformations of the particles in that state. The vacuum $|\Omega\rangle$ is assumed to be Lorentz invariant and to have zero momentum: $\hat{P}|\Omega\rangle = 0$.

An important feature of the Hilbert space is that it is complete, in the sense that

$$\mathbb{1} = \sum_X \int d\Pi_X \, |X\rangle \langle X| \,, \tag{24.1}$$

where the sum is over single- and multi-particle states $|X\rangle$ and

$$d\Pi_X \equiv \prod_{j \in X} \frac{d^3 p_j}{(2\pi)^3} \frac{1}{2E_j}. \tag{24.2}$$

Up to an overall δ-function, this is the Lorentz-invariant phase space of the particles in state X, $d\Pi_{\mathrm{LIPS}} = (2\pi)^4 \, \delta^4\left(\Sigma p\right) d\Pi_X$. We verified the normalization of this completeness relation for one-particle states in Eq. (2.74); Eq. (24.1) is the natural generalization to multi-particle states. For the completeness relation to hold, all possible independent states in the theory must be included. As we will see, there is a close connection between unitarity of the S-matrix and having all the states included in the theory.

We begin the discussion of implications of unitarity in Section 24.1 with the optical theorem. The optical theorem gives a powerful, non-perturbative relationship between cross sections and the imaginary part of scattering amplitudes. In perturbation theory, the optical theorem relates loop amplitudes to tree-level cross sections. To the extent that trees represent classical physics and loops represent quantum effects, the optical theorem implies that the quantum theory is uniquely determined by the classical theory because of unitarity. The relation between loops and trees can be verified in perturbation theory if we have a Lagrangian; however, the optical theorem lets us make statements beyond perturbation theory.

Section 24.2 discusses additional non-perturbative results for general field theories. We show that one-particle states will always give poles in Green's functions. From this, a non-perturbative version of the LSZ reduction formula follows as a special case. Although having states in a theory corresponding to every pole in Green's functions is a requirement of unitarity, unitary theories are not necessarily described by local Lagrangians. Some connections between locality and unitarity are discussed in Section 24.4.

24.1 The optical theorem

Unitarity is a fancy way of saying probabilities add up to 1. Conservation of probability in a quantum theory implies that, in the Schrödinger picture, the norm of a state $|\Psi; t\rangle$ is the same at any time t. For example,

$$\langle \Psi; t | \Psi; t \rangle = \langle \Psi; 0 | \Psi; 0 \rangle. \tag{24.3}$$

Now, since

$$|\Psi; t\rangle = e^{-iHt} |\Psi; 0\rangle, \tag{24.4}$$

unitarity means the Hamiltonian should be Hermitian, $H^\dagger = H$. Then, since the S-matrix is $S = e^{-iHt}$, unitarity implies

$$S^\dagger S = 1. \tag{24.5}$$

That is, the S-matrix is a unitary matrix. Despite its apparent simplicity, this equation has remarkable consequences.

One of the most important implications of unitarity is a relationship between scattering amplitudes and cross sections called (for historical reasons) the optical theorem. To derive the optical theorem, first recall from Chapter 5 that the S-matrix elements that we have been calculating with Feynman graphs were defined by

$$\langle f | T | i \rangle = (2\pi)^4 \delta^4(p_i - p_f) \mathcal{M}(i \to f), \tag{24.6}$$

where the transfer matrix T is the non-trivial part of the S-matrix:

$$S = 1 + iT. \tag{24.7}$$

The matrix T is not Hermitian. In fact, unitarity implies $1 = S^\dagger S = (1 - iT^\dagger)(1 + iT)$ and so

$$i\left(T^\dagger - T\right) = T^\dagger T. \tag{24.8}$$

Sandwiching the left-hand side between $\langle f|$ and $|i\rangle$ gives

$$\langle f|i\left(T^\dagger - T\right)|i\rangle = i\langle i|T|f\rangle^\star - i\langle f|T|i\rangle$$
$$= i(2\pi)^4\delta^4(p_i - p_f)\left(\mathcal{M}^\star(f\to i) - \mathcal{M}(i\to f)\right). \tag{24.9}$$

Using the completeness relation in Eq. (24.1), we get

$$\langle f|T^\dagger T|i\rangle = \sum_X \int d\Pi_X\,\langle f|T^\dagger|X\rangle\,\langle X|T|i\rangle$$
$$= \sum_X \int d\Pi_X(2\pi)^4\delta^4(p_f - p_X)\,(2\pi)^4\delta^4(p_i - p_X)\,\mathcal{M}(i\to X)\,\mathcal{M}^\star(f\to X). \tag{24.10}$$

Thus, unitarity implies:

Box 24.1   **The generalized optical theorem**

$$\mathcal{M}(i\to f) - \mathcal{M}^\star(f\to i) = i\sum_X \int d\Pi_X(2\pi)^4\delta^4(p_i - p_X)\mathcal{M}(i\to X)\mathcal{M}^\star(f\to X).$$

This *generalized optical theorem* must hold order-by-order in perturbation theory. But while its left-hand side has matrix elements, the right-hand side has matrix elements squared. This means that at order λ^2 in some coupling the left-hand side must be a loop to match a tree-level calculation on the right-hand side. Thus, the imaginary parts of loop amplitudes are determined by tree-level amplitudes. In particular, we must have loops – an interacting classical theory by itself, without loops, violates unitarity.

An important special case of the generalized optical theorem is when $|i\rangle = |f\rangle = |A\rangle$ for some state A. Then,

$$2i\,\mathrm{Im}\,\mathcal{M}(A\to A) = i\sum_X \int d\Pi_X(2\pi)^4\delta^4(p_A - p_X)|\mathcal{M}(A\to X)|^2. \tag{24.11}$$

In particular, when $|A\rangle$ is a one-particle state, the decay rate is

$$\Gamma(A\to X) = \frac{1}{2m_A}\int d\Pi_X(2\pi)^4\delta^4(p_A - p_X)|\mathcal{M}(A\to X)|^2. \tag{24.12}$$

So,

$$\mathrm{Im}\mathcal{M}(A\to A) = m_A\sum_X \Gamma(A\to X) = m_A\Gamma_{\text{tot}}, \tag{24.13}$$

where Γ_{tot} is the total decay rate of a particle, equal to its inverse lifetime. This says that the imaginary part of the amplitude associated with the exact propagator is equal to mass times the total decay rate.

If $|A\rangle$ is a two-particle state, then the total cross section is (see Chapter 5)

$$\sigma_{\text{tot}}(A \to X) = \frac{1}{(2E_1)(2E_2)|\vec{v}_1 - \vec{v}_2|} \int d\Pi_X (2\pi)^4 \delta^4(p_A - p_X)|\mathcal{M}(A \to X)|^2.$$

$$(24.14)$$

So,

The optical theorem   Box 24.2

$$\sigma_{\text{tot}}(A \to X) = \frac{1}{(2E_1)(2E_2)|\vec{v}_1 - \vec{v}_2|} \times 2\text{Im}\mathcal{M}(A \to A)$$

This special case is often called the **optical theorem**. It says that the total cross section can be computed from the imaginary part of the forward scattering amplitude.

24.1.1 Decay rates

To see the implications of Eq. (24.13), let us take as an example a simple theory with two scalar fields ϕ and π and Lagrangian

$$\mathcal{L} = -\frac{1}{2}\phi(\Box + M^2)\phi - \frac{1}{2}\pi(\Box + m^2)\pi + \frac{\lambda}{2}\phi\pi^2. \qquad (24.15)$$

If $M > 2m$ then ϕ can decay into $\pi\pi$. Then Eq. (24.13) implies

$$\text{Im}\mathcal{M}(\phi \to \phi) = M\Gamma(\phi \to \pi\pi) + \text{other decay modes}. \qquad (24.16)$$

We will now verify this at order λ^2.

The 1-loop amplitude was evaluated in Chapter 16 (see Eq. (16.10)):

$$i\mathcal{M}_{\text{loop}}(p^2) = \quad \cdots \cdots \overset{p}{\bigcirc} \cdots \cdots \overset{p}{} \quad = -\frac{i\lambda^2}{32\pi^2}\int_0^1 dx \ln\left(\frac{m^2 - i\varepsilon - p^2 x(1-x)}{\Lambda^2}\right),$$

$$(24.17)$$

where we have included the $i\varepsilon$ from the virtual scalar propagator $\left(k^2 - m^2 + i\varepsilon\right)^{-1}$ by $m^2 \to m^2 - i\varepsilon$ to move off the branch cut. For a $1 \to 1$ S-matrix element, we need to put ϕ on-shell by setting $p^2 = M^2$. This gives

$$\mathcal{M}_{\text{loop}}(M^2) = -\frac{\lambda^2}{32\pi^2}\int_0^1 dx \ln\left(\frac{m^2 - M^2 x(1-x) - i\varepsilon}{\Lambda^2}\right). \qquad (24.18)$$

Now, $x(1-x) \le \frac{1}{4}$, so for $M < 2m$ this expression is real, and therefore $\text{Im}\mathcal{M}_{\text{loop}} = 0$. In this regime the decay rate is also zero, so Eq. (24.16) holds for $M < 2m$.

For $M > 2m$ we use

$$\ln(-A - i\varepsilon) = \ln A - i\pi. \qquad (24.19)$$

Then,

$$\text{Im}\mathcal{M}_{\text{loop}} = \frac{\lambda^2}{32\pi}\int_0^1 dx\, \theta(M^2 x(1-x) - m^2)$$

$$= \frac{\lambda^2}{32\pi}\sqrt{1 - 4\frac{m^2}{M^2}}\,\theta(M - 2m). \qquad (24.20)$$

The two-body decay rate (see Chapter 5), including the $\frac{1}{2}$ for identical particles, is

$$\Gamma_{\text{tot}} = \Gamma(\phi \to \pi\pi) = \frac{1}{2} \int \frac{1}{2M} |\mathcal{M}|^2 \frac{|\vec{p}_f|}{M} \frac{d\Omega}{16\pi^2} \theta(M - 2m). \qquad (24.21)$$

With $\vec{p}_f^2 = \left(\frac{M}{2}\right)^2 - m^2$ and $\mathcal{M} = i\lambda$ the total rate is

$$\Gamma_{\text{tot}} = \frac{\lambda^2}{32\pi M} \sqrt{1 - 4\frac{m^2}{M^2}} \theta(M - 2m). \qquad (24.22)$$

So Eq. (24.16) holds and the optical theorem is verified in this case to order λ^2.

24.1.2 Cutting rules

To dissect the calculation we just did, it is helpful to think about the real and imaginary parts of a Feynman propagator. To evaluate the imaginary part of a propagator, note that

$$\text{Im} \frac{1}{p^2 - m^2 + i\varepsilon} = \frac{1}{2i} \left(\frac{1}{p^2 - m^2 + i\varepsilon} - \frac{1}{p^2 - m^2 - i\varepsilon} \right) = \frac{-\varepsilon}{(p^2 - m^2)^2 + \varepsilon^2}. \qquad (24.23)$$

This vanishes as $\varepsilon \to 0$, except near $p^2 = m^2$. If we integrate over p^2, we find

$$\int_0^\infty dp^2 \frac{-\varepsilon}{(p^2 - m^2)^2 + \varepsilon^2} = -\pi, \qquad (24.24)$$

implying that

$$\text{Im} \frac{1}{p^2 - m^2 + i\varepsilon} = -\pi \delta(p^2 - m^2). \qquad (24.25)$$

This is a useful formula. It says that *the propagator is real except for when the particle goes on-shell*. More generally:

Imaginary parts of loop amplitudes come from intermediate particles going on-shell.

Similarly,

$$\frac{1}{k_0 - \omega_k + i\varepsilon} - \frac{1}{k_0 - \omega_k - i\varepsilon} = -2\pi i \delta(k_0 - \omega_k), \qquad (24.26)$$

where $\omega_k = \sqrt{\vec{k}^2 + m^2}$. This lets us write the Feynman propagator as

$$\Pi_F(k) \equiv \frac{i}{k^2 - m^2 + i\varepsilon} = \frac{i}{2\omega_k}\left[\frac{1}{k_0 - \omega_k + i\varepsilon} - \frac{1}{k_0 + \omega_k - i\varepsilon}\right]$$

$$= \Pi_A(k) + \frac{\pi}{\omega_k}\delta(k_0 - \omega_k), \tag{24.27}$$

where the advanced propagator is

$$\Pi_A(k) = \frac{i}{2\omega_k}\left[\frac{1}{k_0 - \omega_k - i\varepsilon} - \frac{1}{k_0 + \omega_k - i\varepsilon}\right]. \tag{24.28}$$

An important point is that while $\Pi_F(k)$ has poles at $k_0 = \pm\omega_k \mp i\varepsilon$, which lie above and below the real k_0 axis, $\Pi_A(k)$ only has poles above the real axis, at $k_0 = \pm\omega_k + i\varepsilon$.

Now consider our loop integral:

$$= \frac{(i\lambda)^2}{2}\int \frac{d^4k}{(2\pi)^4}\frac{i}{(k-p)^2 - m^2 + i\varepsilon}\frac{i}{k^2 - m^2 + i\varepsilon}$$

$$= -\frac{\lambda^2}{2}\int \frac{d^4k}{(2\pi)^4}\left[\Pi_A(k-p) + \frac{\pi}{\omega_{k-p}}\delta(k_0 - p_0 - \omega_{k-p})\right]\left[\Pi_A(k) + \frac{\pi}{\omega_k}\delta(k_0 - \omega_k)\right]. \tag{24.29}$$

The term with $\Pi_A(k-p)\Pi_A(k)$ in it only has poles above the real k_0 axis. Thus, we can close the k_0 integration contour in the lower half-plane and the integral gives zero. Also, the two δ-functions can never be simultaneously satisfied (this is easiest to see in the frame where $\vec{p} = 0$ so that $p_0 = M$ and $\omega_{k-p} = \omega_k$). Dropping such terms, we can use Eq. (24.27) again to write

$$i\mathcal{M}_{\text{loop}}(p^2) = \frac{\lambda^2}{2}\int \frac{d^4k}{(2\pi)^4}\left[i\Pi_F(k-p)\frac{\pi}{\omega_k}\delta(k_0 - \omega_k)\right.$$

$$\left. + i\Pi_F(k)\frac{\pi}{\omega_{k-p}}\delta(k_0 - p_0 - \omega_{k-p})\right]. \tag{24.30}$$

Now, since the δ-functions are real, an imaginary part can only come from i times the Feynman propagators. Thus, to calculate $\text{Im}\mathcal{M}_{\text{loop}}(p)$ we can use Eq. (24.25) to get

$$\text{Im}\mathcal{M}_{\text{loop}}(p^2) = \frac{\lambda^2}{2}\int \frac{d^4k}{(2\pi)^4}\left[\pi\delta\big((k-p)^2 - m^2\big)\frac{\pi}{\omega_k}\delta(k_0 - \omega_k)\right.$$

$$\left. + \pi\delta\big(k^2 - m^2\big)\frac{\pi}{\omega_{k-p}}\delta(k_0 - p_0 - \omega_{k-p})\right]. \tag{24.31}$$

The term on the second line vanishes (as before, this is easiest to see in the p^μ rest frame). Then we use

$$\frac{1}{2\omega_k}\delta(k_0 - \omega_k) = \delta\big(k^2 - m^2\big) - \frac{1}{2\omega_k}\delta(k_0 + \omega_k). \tag{24.32}$$

Box 24.3 **Cutting rules**

1. Cut through the diagram in any way that can put all of the cut propagators on-shell without violating momentum conservation.
2. For each cut, replace $\frac{1}{p^2-m^2+i\varepsilon} \rightarrow -2i\pi\delta(p^2-m^2)\theta(p^0)$ and complex-conjugate everything to the right of the cut.
3. Sum over all cuts.
4. The result is the discontinuity: $\mathrm{Disc}(i\mathcal{M}) = -2\mathrm{Im}\mathcal{M}$.

Since $\int dk_0\delta\big((p-k)^2-m^2\big)\,\delta(k_0+\omega_k)=0$ we find a final simple form

$$2\mathrm{Im}\mathcal{M}_{\mathrm{loop}}(p^2) = -\frac{\lambda^2}{2}\int\frac{d^4k}{(2\pi)^4}(-2\pi i)\,\delta\big((p-k)^2-m^2\big)(-2\pi i)\,\delta\big(k^2-m^2\big).$$

$$(24.33)$$

This equation indicates that the imaginary part of the amplitude can be calculated by putting intermediate particles on-shell.

It turns out the above manipulations can be performed for *any* amplitude. The generalization of Eq. (24.33) is an efficient shortcut to calculating imaginary parts of loop amplitudes known as the **cutting rules**. These rules are given in Box 24.3. Each way of putting intermediate states in a loop amplitude on-shell is known as a **cut**. Cut diagrams are often drawn as

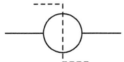

with the dashed line indicating that the particles in the loop intersecting the cut are to be put on-shell. Cuts are directional, in the sense that cut particles should have positive energy when flowing from the left to the right side of the diagrams. You can explore another way to derive the cutting rules in Problem 24.1. An excellent discussion of the cutting rules can be found in [Veltman, 1994].

As an example, one can use the cutting rules to directly confirm the optical theorem. Changing variables in Eq. (24.33) to $k = q_2$ and $p - k = q_1$ and inserting a factor of $1 = \int d^4q_1\delta^4(p-q_1-q_2)$, we get

$$2\mathrm{Im}\mathcal{M}_{\mathrm{loop}} = \frac{\lambda^2}{2}\int\frac{d^4q_1}{(2\pi)^4}\int\frac{d^4q_2}{(2\pi)^4}(2\pi)^6\delta(q_1^2-m^2)\delta(q_2^2-m^2)\delta^4(p-q_1-q_2). \quad (24.34)$$

Since $p^0 > 0$, these δ-functions only have support if $q_1^0 > 0$ and $q_2^0 > 0$ as well. Then we can use

$$\int\frac{d^4q}{(2\pi)^4}2\pi\delta(q^2-m^2)\theta(q_0) = \int\frac{d^3q}{(2\pi)^3}\frac{1}{2\omega_q} \quad\quad (24.35)$$

to find

$$\mathrm{Im}\mathcal{M}_{\mathrm{loop}} = \frac{1}{2}\left(\frac{\lambda^2}{2}\right)\int d\Pi_{\mathrm{LIPS}} = M\Gamma(\phi\rightarrow\pi\pi), \quad\quad (24.36)$$

in agreement with Eq. (24.16).

The *discontinuity* of an amplitude considered as a complex function of momenta is given by the cutting rules [Cutkosky, 1960]. The discontinuity of an amplitude means the difference between the amplitude when the energies are given small positive imaginary parts or small negative imaginary parts. That is,

$$\text{Disc } i\mathcal{M}(p^0) \equiv i\mathcal{M}(p^0 + i\varepsilon) - i\mathcal{M}(p^0 - i\varepsilon) = -2\text{Im}\mathcal{M}(p^0). \qquad (24.37)$$

Amusingly, the word *cut* refers simultaneously to the procedure of slicing open loops to form trees, to branch cut singularities associated with particle thresholds producing the discontinuity, and to Cutkosky's name. The analytic structure of the S-matrix in the complex plane is a fascinating and important subject (see for example [Eden *et al.*, 1966]).

By the way, you may have noticed in Eq. (24.30) that the entire loop amplitude was given by a sum of terms with δ-functions, not just its imaginary part. In fact, one can perform similar substitutions for any loop amplitude, replacing all the propagators with $\Pi_F = \Pi_A + \delta$ and dropping all the terms with only Π_A. For the remaining terms, one can substitute back in $\Pi_A = \Pi_F - \delta$ to produce a set of terms with Feynman propagators, each one of which has at least one δ-function. In this way, loops can be decomposed into tree amplitudes. That this can always be done is known as the **Feynman tree theorem** [Feynman, 1972]. Essentially, the Feynman tree theorem reduces Lorentz-covariant time-dependent perturbation theory to old-fashioned perturbation theory (see Chapter 4), which is formulated in terms of on-shell intermediate states from the start. In fact, one of the simplest ways to prove the generalized optical theorem for a given theory, and hence that the theory is unitary, is using old-fashioned perturbation theory (see, for example [Sterman 1993, Section 9.6]).

24.1.3 Propagators and polarization sums

The optical theorem and the cutting rules work for particles of any spin. For particles with spin, one must sum over final state spins in the decay rate. In fact, the optical theorem efficiently connects propagators to spin sums, as we now explain.

For example, take Yukawa theory with Lagrangian

$$\mathcal{L} = -\frac{1}{2}\phi(\Box + M^2)\phi + \bar{\psi}(i\partial\!\!\!/ - m)\psi + \lambda\phi\bar{\psi}\psi. \qquad (24.38)$$

For the decay of ϕ into $\bar{\psi}\psi$, we find

$$\Gamma(\phi \to \bar{\psi}\psi) = \sum_{s,s'} \frac{\lambda^2}{2M} \int \frac{d^3q_1}{(2\pi)^3} \frac{1}{2\omega_{q_1}} \int \frac{d^3q_2}{(2\pi)^3} \frac{1}{2\omega_{q_2}}$$
$$\times (2\pi)^4 \delta^4(p - q_1 - q_2)\bar{v}_{s'}(q_1)u_s(q_2)\bar{u}_s(q_2)v_{s'}(q_1). \qquad (24.39)$$

Using Eq. (24.35) we can write this as

$$\Gamma = \frac{\lambda^2}{2M} \int \frac{d^4q_2}{(2\pi)^4} \int \frac{d^4q_1}{(2\pi)^4} (2\pi)^4 \delta^4(p - q_1 - q_2)$$
$$\times 2\pi\delta(q_1^2 - m^2)2\pi\delta(q_2^2 - m^2)\text{Tr}[(q\!\!\!/_2 + m)(q\!\!\!/_1 - m)]. \qquad (24.40)$$

The loop is

$$i\mathcal{M}_{\text{loop}} = \quad \text{-----}\bigcirc\text{- - - -}$$

$$= \lambda^2 \int \frac{d^4 q_2}{(2\pi)^4} \int \frac{d^4 q_1}{(2\pi)^4} \frac{\text{Tr}[(\slashed{q}_2 + m)(\slashed{q}_1 - m)]}{[q_1^2 - m^2 + i\varepsilon][q_2^2 - m^2 + i\varepsilon]} (2\pi)^4 \delta^4(p - q_1 - q_2).$$

(24.41)

For the imaginary part of $\mathcal{M}_{\text{loop}}$, we have to put the intermediate states on-shell. This replaces the propagators by $-2\pi i$ times δ-functions, just as for the scalar case. The numerator factor is unaffected, and stays as $\text{Tr}[(\slashed{q}_1 - m)(\slashed{q}_2 + m)]$. Thus, the cut loop amplitude gives $2M$ times the decay rate, which is twice the imaginary part, as expected.

Note, however, that the $\text{Tr}[(\slashed{q}_1 - m)(\slashed{q}_2 + m)]$ factor in the decay rate came from a sum over physical on-shell final states, while this factor in the loop came from the numerators of the propagators. Thus, for the optical theorem to hold in general:

> The numerator of a propagator must be equal to the sum over physical spin states.

This is a consequence of unitarity.

As a check, for a massive spin-1 field, the spin sum is (see Problem 8.5)

$$\sum_{i=1}^{3} \epsilon_i^\mu \epsilon_i^{\nu\star} = -g^{\mu\nu} + \frac{p^\mu p^\nu}{m^2}$$

(24.42)

and the propagator is

$$i\Pi^{\mu\nu}(p^2) = -i\frac{g^{\mu\nu} - \frac{p^\mu p^\nu}{m^2}}{p^2 - m^2 + i\varepsilon}.$$

(24.43)

So the numerator is again given by the sum over physical spin states and the optical theorem holds.

What about a massless spin-1 field? There, the spin sum includes only transverse polarizations. There is no way to write the sum in a Lorentz-invariant way, but we can write it as

$$\sum_{i=1}^{2} \epsilon_i^\mu \epsilon_i^{\nu\star} = -g^{\mu\nu} + \frac{1}{2E^2}(p^\mu \bar{p}^\nu + \bar{p}^\mu p^\nu),$$

(24.44)

where $\bar{p}^\mu = (E, -\vec{p})$ (see Section 13.5.1). The photon propagator is

$$i\Pi^{\mu\nu}(p^2) = -i\frac{g^{\mu\nu} - (1 - \xi)\frac{p^\mu p^\nu}{p^2}}{p^2 + i\varepsilon}.$$

(24.45)

So the numerator of the propagator is *not* just the sum over physical polarizations. However, because of gauge invariance (for the propagator) and the Ward identity (for the decay rate), all the p^μ terms drop out in physical calculations. Thus we see that gauge invariance and the Ward identity are tied together and, moreover, both are required for a unitary theory of a massless spin-1 particle. That is:

Unitarity for massless spin-1 fields requires gauge invariance.

The same analysis can be made for massive and massless spin-2, although it is not terribly illuminating. The result is that we can *always* write the propagator for any spin particle in the form

$$\Pi_s(p) = \frac{\sum_j \epsilon_j \epsilon_j^\star}{p^2 - m^2 + i\varepsilon},\tag{24.46}$$

where ϵ_j are a basis of *physical* polarizations for a particle of given spin.

24.1.4 Unstable particles

In Chapter 18 we showed that after summing all the 1PI insertions the full propagator in the interacting theory becomes (Eq. (18.37) for a scalar)

$$iG(p^2) = \frac{i}{p^2 - m_R^2 + \Sigma(p^2) + i\varepsilon},\tag{24.47}$$

where $i\Sigma(p^2)$ is defined as the sum of 1PI self-energy graphs and m_R is whatever renormalized mass appears in the Lagrangian (e.g. m_R is the $\overline{\text{MS}}$ mass). The pole mass m_P was defined so that $G(p^2)$ has a pole at $p^2 = m_P^2$, which led to $m_P^2 - m_R^2 + \Sigma(m_P^2) = 0$.

If the particle is unstable, then $\Sigma(p^2)$ will in general have an imaginary part, and the definition of pole mass needs to be modified. To see this, recall that by the optical theorem,

$$\Gamma_{\text{tot}} = \frac{1}{m_P}\text{Im}\left(-\!\!\bullet\!\!-\right)$$

$$= \frac{1}{m_P}\text{Im}\left(-\!\!\boxed{\text{1PI}}\!\!- + \cdots\right)$$

$$= \frac{1}{m_P}\text{Im}\Sigma(m_P^2) + \cdots,\tag{24.48}$$

where the $\cdots$ refer to non-1PI diagrams. If we assume that $\Gamma_{\text{tot}} \ll m_P$, as in a weakly coupled theory, then these additional contributions will be suppressed by additional factors of some couplings and can be ignored. Thus, $\text{Im}\Sigma(m_P^2) = m_P\Gamma_{\text{tot}}$, which is non-zero for unstable particles.

A natural way to keep the mass real is to modify the definition of pole mass so that

$$m_P^2 - m_R^2 + \text{Re}\Sigma(m_P^2) = 0.\tag{24.49}$$

This new definition is sometimes called the **real pole mass** or the **Breit–Wigner mass**. By Eq. (24.48), near the pole the propagator has the form

$$iG(p^2) = \frac{i}{p^2 - m_P^2 + im_P\Gamma_{\text{tot}}}.\tag{24.50}$$

This expression is valid for $\Gamma_{\text{tot}} \ll m_P$.

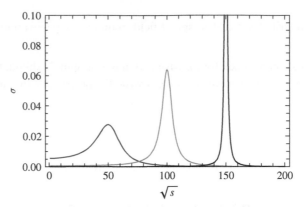

From left to right, the Breit–Wigner distributions for $\Gamma/m_P = 50\%$, 10% and 1%.

For example, consider an s-channel diagram involving this modified propagator:

$$\sigma \propto \left| \bowtie \right|^2 = g^4 \left| \frac{i}{p^2 - m_P^2 + i m_P \Gamma_{\text{tot}}} \right|^2 = g^4 \frac{1}{(p^2 - m_P^2)^2 + (m_P \Gamma_{\text{tot}})^2}.$$

$$(24.51)$$

This is known as a **Breit–Wigner distribution**. It is the characteristic shape of a resonance. Examples are shown in Figure 24.1. The full-width at half-maximum of the Breit–Wigner distribution is $2 m_P \Gamma_{\text{tot}}$. This is why we use the words *width* and *decay rate* interchangeably.

Note also that we can justify treating $\Sigma(p^2)$ as constant when $\Gamma_{\text{tot}} \ll m_P$, since then the cross section only has support for $p^2 \sim m_P^2$. In the $\Gamma_{\text{tot}} \to 0$ limit, we can treat the cross section as a δ-function with coefficient given by the integral over the Breit–Wigner distribution:

$$g^4 \left| \frac{i}{p^2 - m_P^2 + i m_P \Gamma_{\text{tot}}} \right|^2 \approx g^4 \frac{\pi}{m_P \Gamma_{\text{tot}}} \delta(p^2 - m_P^2), \quad \Gamma_{\text{tot}} \ll m_P. \qquad (24.52)$$

This is called the **narrow-width approximation**. It says that near a resonance we can treat the resonant particle as being on-shell. In the narrow-width approximation, the production and decay of the resonance can be treated separately – there can be no interference between production and decay. For example,

$$\bowtie \quad \text{does not interfere with} \quad \bowtie \quad \text{near resonance.} \qquad (24.53)$$

This follows simply because the resonance cannot be on-shell at the same phase space point in the two diagrams. Factorization when intermediate particles go on-shell is a general consequence of unitarity with other important implications, to be discussed further in Section 24.3.

Another implication of the narrow-width approximation is that cross sections can be calculated as production rates. For example, consider the process $e^+e^- \to Z \to \bar{\nu}\nu$ in a simplified model where the Z is a vector boson of mass m_Z that couples only to the electron, e^-, and the neutrino, ν, with strength g. At center-of-mass energies $E_{\rm CM} \ll m_Z$, the total cross section for this process is proportional to g^4. However, for $E_{\rm CM} \sim m_Z$, there is resonant enhancement and σ is proportional only to g^2. Indeed, the total decay rate Γ of the Z is proportional to g^2 (since $\Gamma \sim {\rm Im}(i\Sigma) \sim g^2$) and thus a factor of g^2 cancels near resonance, $\sigma \sim g^4 \frac{\pi}{m\Gamma}\delta(p^2 - m^2) \sim \frac{g^2}{m}\delta(p^2 - m^2)$. To exploit this resonance enhancement, from 1989 to 1996 the Large Electron-Positron (LEP) collider at CERN collided electrons at the Z-pole ($E_{\rm CM} = 91$ GeV). Running at the Z-pole greatly enhanced the production rate of Z's and allowed for precision tests of the Standard Model. To compare this LEP data to theory predictions, the narrow-width approximation works excellently, and one can completely ignore Z/γ interference. At higher center-of-mass energies, at which LEP ran from 1998 to 2000, Z/γ interference is important and must be included.

When $\Gamma_{\rm tot} \gtrsim m_P$, there is no natural definition for the mass of a particle. For example, in a strongly coupled theory the decay rate becomes large as do both the real and imaginary parts of $\Sigma(p^2)$. A particle decaying very fast relative to its mass cannot be reliably identified as a particle. Examples include certain bound states in pure QCD called glueballs. These decay as fast as they are formed and do not form sharp resonances. Identifying a resonance with a particle only makes sense when $\Gamma_{\rm tot} < m_P$.

There are alternatives to the real pole mass. An obvious one is the **complex pole mass**, m_C, defined by $m_C^2 - m_R^2 + \Sigma(m_C^2) = 0$. A much more important mass definition is the $\overline{\rm MS}$ **mass**, m_R, discussed in Section 18.4. The $\overline{\rm MS}$ mass is not defined by any pole prescription. Instead, it is a renormalized quantity which must be extracted from scattering processes that depend on it. Recall from Chapter 18 that a mass definition is equivalent to a subtraction scheme that is a prescription for determining the finite parts of counterterms. For the $\overline{\rm MS}$ mass, one simply sets the finite parts of the counterterms to zero. The $\overline{\rm MS}$ mass can be converted to the pole mass using Eq. (24.49). $\overline{\rm MS}$ masses are particularly useful for particles that do not form asymptotic states and cannot be identified as resonances, such as quarks. For example, there is no way to extract the bottom-quark mass from a Breit–Wigner distribution. $\overline{\rm MS}$ masses are also important for precision physics, as we will see in Chapter 31.

24.1.5 Partial wave unitarity bounds

Another important implication of the optical theorem is that scattering amplitudes cannot be arbitrarily large. That unitarity bounds should exist follows from conservation of probability: what comes out should not be more than what goes in. Roughly speaking, the optical theorem says that $\text{Im}\mathcal{M} \geq |\mathcal{M}|^2$, which implies $|\mathcal{M}| \leq 1$. There are a number of ways to make this more precise. An important example is the **Froissart bound**, which says that total cross sections cannot grow faster than $\ln^2 E_{\text{CM}}$ at high energy [Froissart, 1961]. In this section, we will discuss a different bound, called the partial wave unitarity bound.

Consider $2 \rightarrow 2$ elastic scattering of two particles A and B in the center-of-mass frame: $A(p_1) + B(p_2) \rightarrow A(p_3) + B(p_4)$. The total cross section for this process in the center-of-mass frame is (integrating the general formula in Eq. (5.32) over $d\phi$)

$$\sigma_{\text{tot}}(AB \rightarrow AB) = \frac{1}{32\pi E_{\text{CM}}^2} \int d\cos\theta |\mathcal{M}(\theta)|^2. \tag{24.54}$$

To derive a useful bound, it is helpful to decompose the amplitude into partial waves. We can always write

$$\mathcal{M}(\theta) = 16\pi \sum_{j=0}^{\infty} a_j (2j+1) P_j(\cos\theta), \tag{24.55}$$

where $P_j(\cos\theta)$ are the Legendre polynomials that satisfy $P_j(1) = 1$ and

$$\int_{-1}^{1} P_j(\cos\theta) P_k(\cos\theta) d\cos\theta = \frac{2}{2j+1}\delta_{jk}. \tag{24.56}$$

Thus, we can perform the $\cos\theta$ integral in Eq. (24.54) to get

$$\sigma_{\text{tot}} = \frac{16\pi}{E_{\text{CM}}^2} \sum_{j=0}^{\infty} (2j+1) |a_j|^2. \tag{24.57}$$

Now, the optical theorem relates the imaginary part of the forward scattering amplitude, at $\theta = 0$, to the total cross section:

$$\text{Im}\mathcal{M}(AB \rightarrow AB \text{ at } \theta = 0) = 2E_{\text{CM}}|\vec{p}_i| \sum_X \sigma_{\text{tot}}(AB \rightarrow X)$$

$$\geq 2E_{\text{CM}}|\vec{p}_i|\sigma_{\text{tot}}(AB \rightarrow AB), \tag{24.58}$$

and therefore

$$\sum_{j=0}^{\infty} (2j+1)\text{Im}(a_j) \geq \frac{2|\vec{p}_i|}{E_{\text{CM}}} \sum_{j=0}^{\infty} (2j+1) |a_j|^2. \tag{24.59}$$

Since $|a_j| \geq \text{Im}(a_j)$, this equation says that $|a_j|$ cannot be arbitrarily large. This is an example of a **partial wave unitarity bound**. The sum on j can in fact be dropped by considering scattering of angular momentum eigenstates rather than plane waves (see [Itzykson and Zuber, 1980, Section 5.3]).

To get a cleaner-looking bound, consider the case when the total cross section is well approximated by the elastic cross section; that is, when the only relevant final state is the same as the initial one, in which case the inequality becomes an equality. Moreover, let us take the high-energy limit, $E_{\text{CM}} \gg m_A, m_B$, so that masses can be neglected and $|\vec{p}_i| = \frac{1}{2}E_{\text{CM}}$. Then Eq. (24.59) becomes

$$\text{Im}(a_j) = |a_j|^2. \tag{24.60}$$

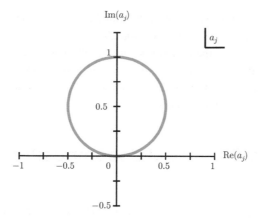

Argand diagram for the condition $\text{Im}\,(a_j) = |a_j|^2$ which corresponds to a circle in the complex plane.

Fig. 24.2

This equation is solved by a circle in the complex plane, as in Figure 24.2. It implies

$$|a_j| \le 1, \quad 0 \le \text{Im}(a_j) \le 1, \quad \text{and} \quad |\text{Re}(a_j)| \le \frac{1}{2} \tag{24.61}$$

for all j. These bounds actually follow more generally, without having to assume the elastic scattering cross section dominates, but the complete derivation is more involved, requiring angular momentum conservation of the S-matrix, which depends on the spins of the particles involved (see Problem 24.3).

The partial wave unitarity bound provides extremely important limitations on the behavior of scattering amplitudes. For example, suppose we have a theory with a dimension-5 interaction, such as

$$\mathcal{L} = \frac{1}{2}\phi\Box\phi + \frac{1}{\Lambda}\psi(\partial_\mu\phi)^2. \tag{24.62}$$

An s-channel exchange diagram gives

$$\mathcal{M}(\phi\phi \to \phi\phi) = \;\;\; \sim \frac{p^2}{\Lambda}\frac{1}{p^2}\frac{p^2}{\Lambda} \sim \frac{s}{\Lambda^2}. \tag{24.63}$$

This has no angular dependence, so $|a_0| = \frac{s}{16\pi\Lambda^2}$ and $a_j = 0$ for $j > 0$. Thus, this amplitude violates the unitarity bound for $E_{\text{CM}} > \sqrt{16\pi}\Lambda$ (including the t- and u-channels does not change this bound by much). That does not mean this theory is not unitary, but that this diagram cannot represent the physics of this process for $E_{\text{CM}} > \sqrt{16\pi}\Lambda$. Of course, we already knew that because this is a non-renormalizable theory loops *should* become important around the scale Λ. The perturbative unitarity bound implies that loops *must* be important around the scale Λ.

For a more physical example, the perturbative unitarity bound would be violated by the scattering of longitudinal W bosons in the Standard Model if there were no Higgs boson. Due to the $\frac{E}{m_W}$ dependence of the longitudinal polarization of the W boson, the amplitude for W boson scattering violates the unitarity bound at ~ 1 TeV. The Higgs boson restores perturbative unitarity, as we will see in Section 29.2.

An important point is that the bound does *not* imply that above some scale unitarity is violated. It says only that unitarity *would be violated* if we could trust perturbation theory, which we cannot. The standard resolution is to introduce new particles or to look for a UV completion above the scale where perturbativity is lost.

24.2 Spectral decomposition

Fields are a crucial ingredient of quantum field theory. In a free theory (or an interacting theory at any fixed time) we have constructed a set of fields out of creation and annihilation operators that add or remove particles from states in the Hilbert space. Constructing fields out of creation and annihilation operators has a number of advantages: it smoothly connects classical field theory and quantum mechanics; it leads naturally to a well-defined perturbation expansion; and it guarantees that the cluster decomposition principle holds.[1] However, one can also consider a generalized notion of fields that is not necessarily connected to creation and annihilation operators.

A field $\phi(x)$ is an operator acting on the Hilbert space which is a function of space-time. Certain fields are associated with particles, meaning they have non-zero matrix elements with some single-particle states:

$$\langle\Omega|\phi(x)|\mathbf{p}\rangle = Ne^{-ipx}, \tag{24.64}$$

where $|\mathbf{p}\rangle$ is some one-particle state with momentum p^μ and $|\Omega\rangle$ is the vacuum. The normalization N is a number. A special case is fields $\phi(x)$ that are the renormalized interacting fields constructed out of creation and annihilation operators for which $N = 1$ by construction. Another special case is the bare fields $\phi_0(x)$ appearing in a bare Lagrangian, related to the renormalized fields by $\phi_0(x) = \sqrt{Z}\phi(x)$, where Z is the field strength renormalization. For these, $N = \sqrt{Z}$. Another example, which will play an important role in Chapter 28, is the pions, which are composite particles of mass $m_\pi \sim 140$ MeV. The neutral pion state $|\pi^0\rangle$ has a non-zero matrix element with the current $J^{\mu 5}(x) = \bar\psi(x)\gamma^\mu\gamma^5\psi(x)$. Explicitly, $\langle\Omega|J_\mu^5(x)|\pi^0(p)\rangle = ie^{ipx}p_\mu F_\pi$ with $F_\pi = 92$ MeV (up to some isospin factors that we are ignoring).

Equation (24.64) does not care if the fields are **elementary**, meaning they appear in a Lagrangian, or **composite**, like the pions which are made of quarks. Indeed, going back-and-forth between elementary and composite notation is the idea behind effective field theory, a powerful technique which will play an important role in Parts IV and V. In this section, we show how one can understand the existence of particles as poles in Green's functions without using creation and annihilation operators.

[1] Recall from Section 7.3.2 that cluster decomposition requires there be no δ-function singularities in the connected part of the S-matrix. Since connected Feynman diagrams only have at most poles or branch cuts, cluster decomposition is automatic in perturbation theory. One can also define the connected part of the S-matrix without Feynman diagrams (see [Eden *et al.*, 1966]).

The general fields $\phi(x)$ are Heisenberg picture operators acting on the Hilbert space. They can therefore be translated to the origin using $e^{-i\hat{P}x}\phi(x)e^{i\hat{P}x} = \phi(0)$. If we have a state $|X\rangle$ with momentum p^μ, so $\hat{P}^\mu|X\rangle = p^\mu|X\rangle$, then we have

$$\langle\Omega|\phi(x)|X\rangle = \langle\Omega|e^{i\hat{P}x}e^{-i\hat{P}x}\phi(x)e^{i\hat{P}x}e^{-i\hat{P}x}|X\rangle = e^{-ipx}\langle\Omega|\phi(0)|X\rangle, \qquad (24.65)$$

where $\langle\Omega|\hat{P} = 0$ has been used, since the vacuum has zero momentum. Similarly, $\langle X|\phi(x)|\Omega\rangle = e^{ipx}\langle\Omega|\phi(0)|X\rangle$. This kind of algebraic trick will let us produce some non-trivial constraints on Green's functions.

24.2.1 Two-point functions

Consider the two-point function $\langle\Omega|\phi(x)\phi(y)|\Omega\rangle$ (no time-ordering). Recalling the completeness relation in Eq. (24.1) we can use Eq. (24.65) to write

$$\langle\Omega|\phi(x)\phi(y)|\Omega\rangle = \sum_X \int d\Pi_X e^{-ip_X(x-y)}\langle\Omega|e^{-i\hat{P}x}\phi(x)e^{i\hat{P}x}|X\rangle\langle X|e^{-i\hat{P}y}\phi(y)e^{i\hat{P}y}|\Omega\rangle$$

$$= \sum_X \int d\Pi_X e^{-ip_X(x-y)}|\langle\Omega|\phi(0)|X\rangle|^2$$

$$= \int \frac{d^4p}{(2\pi)^4}e^{-ip(x-y)}\left\{\sum_X \int d\Pi_X (2\pi)^4\delta^4(p-p_X)|\langle\Omega|\phi(0)|X\rangle|^2\right\},$$
$$(24.66)$$

where a δ-function has been inserted in the last line. Now, the quantity in brackets in Eq. (24.66) is a Lorentz scalar, so it can only depend on p^2. Since the states $|X\rangle$ are physical, on-shell states in the Hilbert space, they all have momentum p_X^μ with $p_X^2 \geq 0$ and positive energy. Thus $p^2 \geq 0$ and $p^0 > 0$ as well. Therefore, we can write

$$\sum_X \int d\Pi_X (2\pi)^4 \delta^4(p-p_X)|\langle\Omega|\phi(0)|X\rangle|^2 = 2\pi\theta(p^0)\rho(p^2), \qquad (24.67)$$

where $\rho(p^2)$ is known as a **spectral density**. From this equation, it follows that $\rho(p^2)$ is real and that $\rho(p^2) \geq 0$ for all $p^2 > 0$ and that $\rho(p^2) = 0$ if $p^2 \leq 0$. That the spectral function is non-negative has important implications, as we will see.

The two-point function can then be written as

$$\langle\Omega|\phi(x)\phi(y)|\Omega\rangle = \int \frac{d^4p}{(2\pi)^3}e^{-ip(x-y)}\theta(p^0)\rho(p^2). \qquad (24.68)$$

To simplify this further we define

$$D(x, y, m^2) \equiv \int \frac{d^3 p}{(2\pi)^3} \frac{1}{2\omega_p} e^{-ip(x-y)}, \quad \omega_p = \sqrt{\vec{p}^2 + m^2}$$

$$= \int \frac{d^4 p}{(2\pi)^3} e^{-ip(x-y)} \theta(p_0) \delta(p^2 - m^2), \tag{24.69}$$

which lets us write

$$\langle \Omega | \phi(x)\phi(y) | \Omega \rangle = \int_0^\infty dq^2 \rho(q^2) D(x, y, q^2). \tag{24.70}$$

For a free scalar field, $D(x, y, m^2) = \langle \Omega | \phi_0(x)\phi_0(y) | \Omega \rangle$ and therefore $\rho(q^2) = \delta(q^2 - m^2)$. However, Eq. (24.70) makes no assumption about expanding around the free theory; $D(x, y, m^2)$ is just the mathematical expression given by Eq. (24.69) and so Eq. (24.70) holds for an arbitrary interacting theory.

To connect to S-matrix elements, we need to relate the spectral function to time-ordered products. This is easy to do:

$$\langle \Omega | T\{\phi(x)\phi(y)\} | \Omega \rangle = \langle \Omega | \phi(x)\phi(y) | \Omega \rangle \, \theta(x^0 - y^0) + \langle \Omega | \phi(y)\phi(x) | \Omega \rangle \, \theta(y^0 - x^0)$$

$$= \int_0^\infty dq^2 \rho(q^2) \left[D(x, y, q^2) \theta(x^0 - y^0) + D(y, x, q^2) \theta(y^0 - x^0) \right]. \tag{24.71}$$

Now, the calculation of the Feynman propagator in Section 6.2 involved the mathematical identity

$$D(x, y, q^2)\theta(x^0 - y^0) + D(y, x, q^2)\theta(y^0 - x^0) = \int \frac{d^4 p}{(2\pi)^4} \frac{i}{p^2 - q^2 + i\varepsilon} e^{ip(x-y)}. \tag{24.72}$$

We therefore find

$$\langle \Omega | T\{\phi(x)\phi(y)\} | \Omega \rangle = \int \frac{d^4 p}{(2\pi)^4} e^{ip(x-y)} i\Pi(p^2), \tag{24.73}$$

where

$$\Pi(p^2) \equiv \int_0^\infty dq^2 \frac{\rho(q^2)}{p^2 - q^2 + i\varepsilon} \tag{24.74}$$

is known as the **spectral representation** or **Källén–Lehmann representation** of the two-point function.

To be clear, we have derived an expression for the Fourier transform of the exact non-perturbative two-point function in terms of a spectral density – no dynamics has been used, and we are not expanding around the free theory in any way. In fact, we have hardly used quantum field theory at all: no mention of creation and annihilation operators went into Eq. (24.73). One can do the same analysis for a fermion or gauge boson two-point function without any unusual complications; however, we stick to the scalar case here for simplicity (see Problem 24.2).

The spectral density has a lot of information in it. Basically, it tells us about all the on-shell intermediate states in the theory. It is observable (in principle) since it is just based on an (in principle) observable Green's function, $\langle \Omega | T \{\phi(x)\phi(y)\} | \Omega \rangle$. For a free theory

$$\Pi(p^2) = \frac{1}{p^2 - m^2 + i\varepsilon} \qquad (24.75)$$

and $\rho(q^2) = \delta(q^2 - m^2)$. For an interacting theory, the spectral function will have singularities at locations of physical, renormalized particle masses and other physical thresholds. Since $\rho(q^2)$ is real and $\rho(q^2) > 0$, we can calculate it from the 2-point function by taking the imaginary part of $\Pi(q^2)$ using Eq. (24.25):

$$\rho(p^2) = -\frac{1}{\pi} \text{Im}\left[\Pi(p^2)\right]. \qquad (24.76)$$

As we have already observed, in a unitary theory $\Pi(p^2)$ can have an imaginary part only when cuts can put intermediate particles on-shell. Thus, the spectral density contains information about the particles in the theory. In particular, it can tell us about these particles regardless of whether there are fundamental fields corresponding to them in the Lagrangian.

As an example, recall the Lagrangian in Eq. (24.15), which describes a scalar ϕ of mass M interacting with a scalar π of mass m, with interaction $\frac{\lambda}{2}\phi\pi^2$. In this case, $\Pi(p^2)$ has an imaginary part at $p^2 = M^2$ (from Eq. (24.25)). This is an isolated pole. For $p^2 > 4m^2$ (above the $\phi \to \pi\pi$ threshold) there is an additional imaginary part. To be explicit, using Eqs. (24.20) and (24.47) we have

$$\rho(q^2) = -\frac{1}{\pi} \text{Im}\left[\Pi(q^2)\right]$$

$$= -\frac{1}{\pi} \text{Im}\left[q^2 - M^2 + i\varepsilon + i\frac{\lambda^2}{32\pi}\sqrt{\frac{q^2 - 4m^2}{q^2}}\theta(q^2 - 4m^2) + \cdots\right]^{-1}$$

$$= \delta(q^2 - M^2) + \theta(q^2 - 4m^2)\frac{\lambda^2}{32\pi^2}\frac{1}{(q^2 - M^2)^2}\sqrt{\frac{q^2 - 4m^2}{q^2}} + \cdots. \qquad (24.77)$$

This is typical of spectral functions: it has a pole at one-particle states (and possible bound states) and then a branch-cut singularity above the multi-particle threshold. Note that the coefficient of $\delta(q^2 - M^2)$ is 1 only when we use on-shell renormalization. Otherwise, it is given by the residue of the pole in the $\Pi(p^2)$ at the pole mass $p^2 = m_P^2$, which is subtraction-scheme dependent.

The spectral representation gives powerful non-perturbative constraints. For example, suppose we tried to define a UV-finite quantum field theory by writing the Lagrangian for a scalar field as

$$\mathcal{L} = -\frac{1}{2}\phi\left(\Box + c\frac{\Box^2}{\Lambda^2} + m^2\right)\phi + \mathcal{L}_{\text{int}}(\phi). \qquad (24.78)$$

This would lead to a propagator with $\Pi(p^2) = \frac{1}{p^2 - m^2 - c\frac{p^4}{\Lambda^2}}$. This propagator would have the appealing feature that $\Pi(p^2) \to -\frac{1}{c}\frac{\Lambda^2}{p^4}$ as $p^2 \to \infty$ so that loops involving this scalar would be much more convergent than in a theory without the $\frac{\Box^2}{\Lambda^2}$ term. More generically, let us consider deformations of $\Pi(p^2)$ to make it vanish as $p^2 \to \infty$. In a Feynman diagram, we would Wick rotate $p^0 \to ip^0$ to evaluate the loop. Then $p^2 \to -p_E^2$, so we would like $\Pi(-p_E^2)$ to go to zero as $p_E^2 \to \infty$ faster than $\frac{1}{p_E^2}$. Unfortunately, any such behavior is forbidden by unitarity. As $p_E^2 \to \infty$ the spectral decomposition implies

$$\left|\Pi(-p_E^2)\right| = \left|\int_0^\infty dq^2 \frac{\rho(q^2)}{p_E^2 + q^2}\right| \geq \left|\int_0^{q_0^2} dq^2 \frac{\rho(q^2)}{p_E^2 + q_0^2}\right| \tag{24.79}$$

for any q_0^2. In taking the limit $p_E^2 \to \infty$, eventually we must have $p_E^2 > q_0^2$. Then

$$\lim_{p_E^2 \to \infty} p_E^2 \left|\Pi(-p_E^2)\right| \geq \lim_{p_E^2 \to \infty} p_E^2 \left|\int_0^{q_0^2} dq^2 \frac{\rho(q^2)}{2p_E^2}\right| = \frac{A}{2} \tag{24.80}$$

for some finite positive number $A = \int_0^{q_0^2} \rho(q^2)\, dq^2$. A propagator such as $\Pi(p^2) = \frac{1}{p^2 - m^2 - c\frac{p^4}{\Lambda^2}}$ would violate this bound for any c and A at large enough p_E^2. Note that the positivity of $\rho(q^2)$, which follows from Eq. (24.67), was critical for this bound. The conclusion is:

Propagators cannot decrease faster than $\frac{1}{p^2}$ at large p^2.

This is a very powerful, general non-perturbative result.

24.2.2 Spectral decomposition for bare fields

Up to this point, $\phi(x)$ has been referring to the renormalized field. However, all the derivations in this section work equally well for a bare field $\phi_0(x)$, since all we have used is that the fields have unitary transformations under the Poincaré group. For a general quantum field theory, the bare fields $\phi_0(x)$ are infinite and meaningless. However, once the theory has been regulated (or if it is finite or conformal), then we can legitimately talk about correlation functions of bare fields, calculated from some bare Lagrangian.

For bare fields, let us write the spectral decomposition as

$$\langle \Omega | T\{\phi_0(x)\phi_0(y)\} | \Omega \rangle = \int \frac{d^4p}{(2\pi)^4} e^{-ip(x-y)} \int_0^\infty dq^2 \frac{i}{p^2 - q^2 + i\varepsilon} \rho_0(q^2). \tag{24.81}$$

One can derive an important normalization condition on the spectral function for bare fields:

$$\int_0^\infty dq^2 \rho_0(q^2) = 1. \tag{24.82}$$

To derive this, first observe that by taking a time derivative of Eq. (24.69) we find

$$2\frac{\partial}{\partial t} D(x, 0, \mu^2)\bigg|_{t=0} = -2\frac{\partial}{\partial t} D(0, x, \mu^2)\bigg|_{t=0} = -i\delta^3(\vec{x}), \tag{24.83}$$

where $x^\mu = (t, \vec{x})$. Next recall the canonical commutation relation among the bare free fields:

$$[\phi_0(\vec{x}', t'), \partial_t \phi_0(\vec{x}, t)]_{t=t'} = i\delta^3(\vec{x} - \vec{x}'). \qquad (24.84)$$

We derived this relation for the free theory in Section 2.3.3, and used it as a specification of the dynamics for interacting theories in Section 7.1. Setting $\vec{x}' = 0$ and $t' = 0$, this relation implies that

$$-i\delta^3(\vec{x}) = \partial_t \langle \Omega | [\phi_0(x), \phi_0(0)] | \Omega \rangle \Big|_{t=0} = \partial_t \int_0^\infty dq^2 \rho_0(q^2) \left[D(x, 0, q^2) - D(0, x, q^2) \right] \Big|_{t=0}$$

$$= -i\delta^3(x) \int_0^\infty dq^2 \rho_0(q^2), \qquad (24.85)$$

from which Eq. (24.82) follows.

The importance of Eq. (24.82) is that it constrains the form of the divergences that can appear. For example, recall that the bare fields are related to the renormalized fields by $\phi_0(x) = \sqrt{Z}\phi(x)$. In the on-shell scheme,

$$\int d^4x\, e^{ipx} \langle \Omega | T\{\phi_0(x)\phi_0(0)\} | \Omega \rangle \xrightarrow{p^2 \to m_P^2} \frac{iZ}{p^2 - m_P^2 + i\varepsilon}. \qquad (24.86)$$

Thus $\rho_0(p^2) = Z\delta(p^2 - m_P^2) + \hat{\rho}_0(p^2)$ as in Eq. (24.77), where $\hat{\rho}_0(p^2)$ is everything beyond the pole and, like $\rho_0(p^2)$, is positive. Thus, the normalization condition implies

$$Z = 1 - \int dp^2\, \hat{\rho}_0(p^2), \qquad (24.87)$$

which then implies $0 \le Z \le 1$.

As an example, we computed $Z = Z_2$ for QED in Chapter 18, finding

$$Z_2 = 1 - \frac{\alpha}{2\pi} \left(\frac{1}{2} \ln\frac{\Lambda^2}{m_P^2} + \frac{9}{4} + \ln\frac{m_\gamma^2}{m_P^2} \right), \qquad (24.88)$$

where Λ is the Pauli–Villars mass and m_γ is an IR regulator. Clearly, Z_2 is not between 0 and 1 as $m_\gamma \to 0$. Unfortunately, the only conclusion we can really draw from this is that Z_2 cannot be computed in perturbation theory, even in a finite theory. This is, of course, not a problem, since Z_2 is not measurable.

24.3 Polology

The spectral decomposition also has non-trivial implications for arbitrary scattering amplitudes. In particular, it will let us associate poles in the S-matrix with on-shell intermediate states. This proof follows [Weinberg, 1995, Section 10.2].

Consider the momentum space Green's function:

$$G_n(p_1, \ldots, p_n) = \int d^4x_1 e^{ip_1x_1} \cdots \int d^4x_n e^{-ip_nx_n} \langle \Omega | T\{\phi(x_1) \cdots \phi(x_n)\} | \Omega \rangle. \qquad (24.89)$$

We will now prove that if $p^\mu = p_1^\mu + \cdots + p_r^\mu = p_{r+1}^\mu + \cdots + p_n^\mu$ for some subset of the momenta and if there is a one-particle state $|\Psi\rangle$ of mass m for which $\langle\Psi|\phi(x_1)\cdots\phi(x_r)|\Omega\rangle \neq 0$ then G will have a pole at $p^2 = m^2$ and the Green's function will factorize near the pole.

To prove this, we first write

$$\langle\Omega|T\{\phi(x_1)\cdots\phi(x_n)\}|\Omega\rangle$$
$$= \Theta_{1r}\langle\Omega|T\{\phi(x_1)\cdots\phi(x_r)\}T\{\phi(x_{r+1})\cdots\phi(x_n)\}|\Omega\rangle + \text{extra}, \quad (24.90)$$

where

$$\Theta_{1r} \equiv \theta(\min(t_1,\ldots,t_r) - \max(t_{r+1},\ldots,t_n)) \quad (24.91)$$

puts the two subsets in time order and "extra" refers to the other time orderings. We have dropped the subscripts on the fields for conciseness.

Next, we insert a complete set of states. The sum time-ordered product can then be written as

$$\langle\Omega|T\{\phi(x_1)\cdots\phi(x_n)\}|\Omega\rangle = \int \frac{d^3 p_\Psi}{(2\pi)^3} \frac{1}{2E_\Psi} \Theta_{1r}$$
$$\times \langle\Omega|T\{\phi(x_1)\cdots\phi(x_r)\}|\Psi\rangle\langle\Psi|T\{\phi(x_{r+1})\cdots\phi(x_n)\}|\Omega\rangle + \text{extra}. \quad (24.92)$$

The complete set of states sums over all one- and multi-particle states, but we are only exhibiting one term from this sum – the one involving the one-particle state $|\Psi\rangle$ of mass m. Other one-particle states and all the multi-particle states in the sum are in the "extra" part.

Now, inserting momentum operators, as in Eq. (24.66), we can write

$$\langle\Omega|T\{\phi(x_1)\cdots\phi(x_r)\}|\Psi\rangle$$
$$= e^{-ip_\Psi x_1}\langle\Omega|T\left\{e^{-i\hat{P}x_1}\phi(x_1)e^{i\hat{P}x_1}e^{-i\hat{P}x_1}\phi(x_2)\cdots\phi(x_r)e^{i\hat{P}x_1}\right\}|\Psi\rangle$$
$$= e^{-ip_\Psi x_1}\langle\Omega|T\{\phi(0)\phi(x_2-x_1)\cdots\phi(x_r-x_1)\}|\Psi\rangle$$
$$= e^{-ip_\Psi x_1}\langle\Omega|T\{\phi(0)\phi(x_2')\cdots\phi(x_r')\}|\Psi\rangle, \quad (24.93)$$

where we have defined $x_j' \equiv x_j - x_1$ for $j \leq r$. Similarly,

$$\langle\Psi|T\{\phi(x_{r+1})\cdots\phi(x_n)\}|\Omega\rangle = e^{ip_\Psi x_{r+1}}\langle\Psi|T\{\phi(0)\phi(x_{r+2}')\cdots\phi(x_n')\}|\Omega\rangle, \quad (24.94)$$

with $x_j' = x_j - x_{r+1}$ for $j > r$. Then, changing variables on all but x_1 and x_{r+1}, we have

$$\int d^4 x_1 e^{ip_1 x_1} \cdots \int d^4 x_n e^{-ip_n x_n}$$
$$= \int d^4 x_1 e^{ip_1 x_1} \cdots \int d^4 x_n' e^{-ip_n x_n'} e^{i(p_2+\cdots+p_r)x_1} e^{-i(p_{r+2}+\cdots+p_n)x_{r+1}}. \quad (24.95)$$

Also,

$$\min(t_1,\ldots,t_r) - \max(t_{r+1},\ldots,t_n)$$
$$= t_1 - t_{r+1} + \min(0, t_2',\ldots,t_r') - \max(0, t_{r+2}',\ldots,t_n'), \quad (24.96)$$

which we will include using the following representation of the θ-function:

$$\theta(x) = \int_{-\infty}^{\infty} \frac{d\omega}{2\pi} \frac{i}{\omega + i\varepsilon} e^{-i\omega x}. \qquad (24.97)$$

Then we have

$$G_n(p_1, \ldots, p_n) = \int \frac{d^3 p_\Psi}{(2\pi)^3} \frac{1}{2E_\Psi} \int d^4 x_1 e^{ip_1 x_1} \cdots \int d^4 x'_n e^{-ip_n x'_n} \int \frac{d\omega}{2\pi} \frac{i}{\omega + i\varepsilon}$$

$$\times e^{-ip_\Psi(x_1 - x_{r+1})} e^{i(p_2 + \cdots + p_r)x_1} e^{-i(p_{r+2} + \cdots + p_n)x_{r+1}}$$

$$\times e^{-i\omega(t_1 - t_{r+1})} e^{-i\omega(\min(0, t'_2, \ldots, t'_r) - \max(0, t'_{r+2} - t'_n))}$$

$$\times \langle \Omega | T\{\phi(0)\phi(x'_2) \cdots \phi(x'_r)\} | \Psi \rangle \langle \Psi | T\{\phi(0)\phi(x'_{r+2}) \cdots \phi(x'_n)\} | \Omega \rangle + \text{extra}. \qquad (24.98)$$

Next, performing the $d^4 x_1$ integral over the exponentials containing x_1 or t_1 gives

$$(2\pi)^3 \delta^3(\vec{p}_1 + \cdots + \vec{p}_r - \vec{p}_\Psi)(2\pi)\delta(E_1 + \cdots + E_r - E_\Psi - \omega), \qquad (24.99)$$

where $E_\Psi = \sqrt{\vec{p}_\Psi^2 + m_\Psi^2}$ since $|\Psi\rangle$ is an on-shell, one-particle state. Similarly, the $d^4 x_{r+1}$ integral gives

$$(2\pi)^3 \delta^3(\vec{p}_{r+1} + \cdots + \vec{p}_n - \vec{p}_\Psi)(2\pi)\delta(E_{r+1} + \cdots + E_n - E_\Psi - \omega). \qquad (24.100)$$

Performing the $d^3 p_\Psi$ integral next over the δ^3-function sets $E_\Psi = \sqrt{(\vec{p}_1 + \cdots + \vec{p}_r)^2 + m_\Psi^2}$ and leads to

$$G_n(p_1, \ldots, p_n)$$

$$= \frac{1}{2E_\Psi} \int d^4 x'_2 e^{ip_2 x'_2} \cdots \int d^4 x'_n e^{-ip_n x'_n} \int \frac{d\omega}{2\pi} \frac{i}{\omega + i\varepsilon} e^{-i\omega(\min(\cdots) - \max(\cdots))}$$

$$\times (2\pi)^5 \delta^4(p_1 + \cdots + p_r - p_{r+1} - \cdots - p_n)\delta(E_1 + \cdots + E_r - E_\Psi - \omega)$$

$$\times \langle \Omega | T\{\phi(0)\phi(x'_2) \cdots \phi(x'_r)\} | \Psi \rangle \langle \Psi | T\{\phi(0)\phi(x'_{r+2}) \cdots \phi(x'_n)\} | \Omega \rangle + \text{extra}. \qquad (24.101)$$

By assumption, the matrix elements on the last line are non-zero. Then, this expression has a pole at $\omega = 0$, which is the pole we were looking for. Near this pole, we can drop the $e^{-i\omega(\min(\cdots) - \max(\cdots))}$ term and perform the ω integral over the δ-function to give

$$G_n(p_1, \ldots, p_n) = \int d^4 x'_2 e^{ip_2 x'_2} \cdots \int d^4 x'_n e^{-ip_n x'_n} \frac{1}{2E_\Psi} \frac{i}{E_1 + \cdots + E_r - E_\Psi + i\varepsilon}$$

$$\times (2\pi)^4 \delta^4(p_1 + \cdots - p_n)\langle \Omega | T\{\phi(0)\phi(x'_2) \cdots \phi(x'_r)\} | \Psi \rangle$$

$$\times \langle \Psi | T\{\phi(0)\phi(x'_{r+2}) \cdots \phi(x'_n)\} | \Omega \rangle + \text{extra}. \qquad (24.102)$$

The factors of E_j can be simplified. Write $p^\mu = p_1^\mu + \cdots + p_r^\mu$, then

$$\frac{1}{E_1 + \cdots + E_r - E_\Psi + i\varepsilon} = \frac{1}{p_0 - \sqrt{\vec{p}^2 + m_\Psi^2} + i\varepsilon}$$

$$= \frac{p_0 + \sqrt{\vec{p}^2 + m_\Psi^2}}{p_0^2 - (\vec{p}^2 + m_\Psi^2) + i\varepsilon}$$

$$= \frac{2E_\Psi}{p^2 - m_\Psi^2 + i\varepsilon}, \tag{24.103}$$

where the last equality holds near the pole, where $E_\Psi = p_0$, and we have used the fact that ε is infinitesimal.

Now the matrix element $\mathcal{M}_\Psi^{1,r}$ for $\phi_1 \cdots \phi_r \to \Psi$ is given by

$$(2\pi)^4 \delta^4(p_1 + \cdots + p_r - p_\Psi) \mathcal{M}_\Psi^{1,r}$$

$$= \int d^4x_1 \cdots d^4x_r e^{ip_1 x_1} \cdots e^{ip_r x_r} \langle \Omega | T\{\phi(x_1) \cdots \phi(x_r)\} | \Psi \rangle$$

$$= (2\pi)^4 \delta^4(p_1 + \cdots + p_r - p_\Psi) \int d^4x_2' \cdots d^4x_r' e^{ip_2 x_2'} \cdots e^{ip_r x_r'}$$

$$\times \langle \Omega | T\{\phi(0)\phi(x_2') \cdots \phi(x_r')\} | \Psi \rangle, \tag{24.104}$$

where Eq. (24.93) has been used to get to the second line.

Thus, for p_ψ^2 near m_ψ^2,

$$G_n(p_1, \ldots, p_n) = (2\pi)^4 \delta^4(\Sigma p) \frac{i}{p_\Psi^2 - m_\Psi^2 + i\varepsilon} \mathcal{M}_\Psi^{1,r} \mathcal{M}_\Psi^{r+1,n\dagger} + \text{extra}, \tag{24.105}$$

where "extra" refers to anything else that contributes. This equation says that Green's functions *always* have poles when on-shell intermediate particles can be produced. For example, positronium (an e^+e^- bound state) would appear as a pole in a Green's function corresponding to e^+e^- scattering.

In deriving Eq. (24.105), the only thing we used was that the state $|\Psi\rangle$ is a one-particle state with overlap with the state with r fields $\phi_1 \cdots \phi_r$. We never needed to associate Ψ with a field in a Lagrangian. This formula does not distinguish elementary particles (those with corresponding fields in a Lagrangian) from composite particles. All that is needed is that the particle transforms in an irreducible representation of the Poincaré group, so that it has some on-shell momentum p^μ with $p^2 = m^2$.

In fact, we never even used the fact that the fields $\phi(x)$ each have non-vanishing matrix elements in one-particle states. Equation (24.105) holds even if the fields $\phi_i(x)$ are generic operators $\mathcal{O}_i(x)$, as long as the product $\mathcal{O}_1(x_1) \cdots \mathcal{O}_r(x_r)$ still has a non-zero matrix element between the vacuum $\langle \Omega |$ and some one-particle state $|\Psi\rangle$. Now suppose that the $\phi_i(x)$ do correspond to elementary fields. In fact, suppose they are the renormalized fields that satisfy

$$\langle \Omega | \phi(x) | \mathbf{p} \rangle = e^{-ipx}, \tag{24.106}$$

where $|\mathbf{p}\rangle$ is the one-particle state corresponding to the field ϕ. Then there will be a pole in the Green's function even for $r = 1$. For $r = 1$, the pole occurs when the momentum

p_1^μ in the Fourier transform of the Green's function Eq. (24.89) goes on-shell: $p_1^2 \to m^2$, where m is the mass of the one-particle state corresponding to the field ϕ. Since there was nothing special about the first field in the time-ordered product, a generic Green's function constructed from elementary fields will have poles when all of the external momenta go on-shell. This is exactly what we expect from the LSZ reduction formula, but now the existence of single-particle poles has been proven non-perturbatively.

Another important implication of Eq. (24.105) is that massless spin-1 particles that interact with each other must transform in the adjoint representation of a non-Abelian gauge group. If this were not true, that is, if the couplings among the particles did not satisfy the Jacobi identity, Eq. (24.105) would be violated. We prove this in Chapter 27. (see also Problem 9.3).

24.4 Locality

We have seen that we do not need to have fields in the Lagrangian corresponding to every particle. Green's functions will always have poles at the mass of any particle that has non-zero overlap with some subset of the fields in the Green's functions. However, if one wants to calculate S-matrix elements involving some particle, it is extremely helpful to have an associated field. In fact, it is often extremely useful to go from one description in which a pole is emergent as a bound state to a description in which that bound state has a corresponding field. For example, we go from a theory (QCD) in which a pion is a pole in a Green's function to a theory (the Chiral Lagrangian) with a field corresponding to the pion. The two descriptions have their own Lagrangians. The QCD Lagrangian is useful for calculating the pion mass, while the Chiral Lagrangian is useful if one wants to calculate the $\pi\pi \to \pi\pi$ cross section, taking the pion mass from data. A great virtue of quantum field theory is its flexibility: one can use different Lagrangians for different processes. A number of examples of effective field theories, such as the Chiral Lagrangian, were given in Chapter 22. More will be discussed in Parts IV and V.

There is an interesting connection between the emergence of particles as poles in Green's functions and locality. Informally, locality means that physics over here is independent of physics over there – we do not have to have the wavefunction of the universe to see what happens in our lab. However, defining locality in terms of observables is not straightforward – there are a number of different definitions we can give. For example, we could identify locality with the cluster decomposition principle (mentioned in Section 7.3.2), which requires the connected S-matrix not to be more singular than having poles or branch cuts (see [Weinberg, 1995, Chapter 4]). Alternatively, we could associate locality with commutators vanishing outside the light cone (a property we called causality in Chapter 12). There are many related ways to define locality.

To be concrete, we will define **locality** in terms of a Lagrangian. We take locality to mean that the Lagrangian is an integral over a Lagrangian density that is a functional of fields and their derivatives evaluated at the same space-time point. For example, a Lagrangian term such as $\phi \Box \phi$ is local by this definition, but a term such as $\phi \frac{1}{\Box} \phi$ is not. To be clear, this definition is mathematical, not physical: it is a property of our calculational framework, not of observables. Nevertheless, it has interesting consequences.

To understand the connection between this definition of locality and unitarity, consider *integrating out* a field. We will integrate out particles (at both the classical and quantum levels) in a number of different ways in later chapters. For now, we use the classical meaning, which is to set a field equal to its classical expectation value, given by the solution to its equations of motion. For example, start with the local Lagrangian in Eq. (24.15) . The equations of motion for ϕ are

$$- (\Box + M^2)\phi + \frac{\lambda}{2}\pi^2 = 0. \qquad (24.107)$$

Integrating out ϕ therefore gives

$$\mathcal{L}_{\text{non-local}} = -\frac{1}{2}\pi(\Box + m^2)\pi + \frac{\lambda^2}{8}\pi^2\frac{1}{\Box + M^2}\pi^2, \qquad (24.108)$$

which now appears non-local. If we expand this Lagrangian for $\Box \ll M^2$ we get a local theory

$$\mathcal{L}_{\text{local}} = -\frac{1}{2}\pi(\Box + m^2)\pi + \frac{\lambda^2}{8}\left(\frac{1}{M^2}\pi^4 - \pi^2\frac{\Box}{M^4}\pi^2 + \cdots\right). \qquad (24.109)$$

This is a very similar procedure to how we integrate out the W and Z bosons to derive the 4-Fermi theory (discussed already in Chapter 22 and an important theme for Part IV). Now, both $\mathcal{L}_{\text{non-local}}$ and $\mathcal{L}_{\text{local}}$ appear to describe exactly the same theory, but one appears non-local and the other local. Thus, our definition of locality, no negative powers of derivatives in the Lagrangian, already appears ambiguous.

What goes wrong with the apparently local (but really non-local) Lagrangian, $\mathcal{L}_{\text{local}}$? At energies $p^2 \sim M^2$ we will see the apparent pole where the ϕ particle should have been, but had been integrated out. If the particle ϕ has really been removed from the Hilbert space when we integrated it out, unitarity would be violated. Indeed, the pole would give a non-vanishing imaginary part to an appropriate amplitude, but there would be no corresponding on-shell state so the optical theorem would be violated. Thus, the non-local theory suggests that one should use a different effective description for energies greater than M in which the particle in the Hilbert space corresponding to the pole (present even in $\mathcal{L}_{\text{local}}$) is given its own field.

Another example is the theory of a massive vector boson, with Lagrangian

$$\mathcal{L} = -\frac{1}{4}F_{\mu\nu}^2 + \frac{1}{2}m^2 A_{\mu}^2. \qquad (24.110)$$

This theory has no gauge invariance and three polarizations. Thus, there are three states with poles at $p^2 = m^2$ in the S-matrix. Now let us integrate in a Stueckelberg field $\pi(x)$ via $A_{\mu} \to A_{\mu} + \partial_{\mu}\pi$, as we did in Section 8.7. This restores gauge invariance, with $\pi \to \pi - \alpha$ and $A_{\mu} \to A_{\mu} + \partial_{\mu}\alpha$. The Lagrangian is now

$$\mathcal{L} = -\frac{1}{4}F_{\mu\nu}^2 + \frac{1}{2}m^2\left(A_{\mu}^2 - 2(\partial_{\mu}A_{\mu})\pi - \pi\Box\pi\right), \qquad (24.111)$$

where we have integrated by parts. The equations of motion for π are $\pi = -\frac{1}{\Box}\partial_\mu A_\mu$. Then integrating π back out gives

$$\mathcal{L} = -\frac{1}{4}F_{\mu\nu}^2 + \frac{1}{2}m^2\left(A_\mu^2 + \partial_\mu A_\mu \frac{1}{\Box}\partial_\nu A_\nu\right) = -\frac{1}{4}F_{\mu\nu}^2 - \frac{1}{4}F_{\mu\nu}\frac{m^2}{\Box}F_{\mu\nu}. \qquad (24.112)$$

This theory is now gauge invariant, but apparently non-local. Because of gauge invariance, there are only two polarizations for the photon, instead of three for the massive vector boson. The non-locality tells us that an on-shell state is missing.

Problems

24.1 In this problem you will show how the cutting rules can be obtained directly from contour integration.

(a) Where are the poles in the integrand in Eq. (24.29) in the complex k^0 plane?

(b) Close the contour upward and write the result as the sum of two residues. Show that one of these residues cannot contribute to the imaginary part of $\mathcal{M}$.

(c) Evaluate the imaginary part of the amplitude by using the other pole. Show that you reproduce Eq. (24.33).

(d) Now consider a more complicated $2 \to 3$ process:

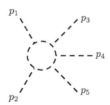

Explore the pole structure of this amplitude in the complex plane and show that the imaginary part of this amplitude is given by the cutting rules.

24.2 Derive the spectral representation for a Dirac spinor.

24.3 Derive the partial wave unitarity bound for elastic scattering for a theory with scalars only.

24.4 LSZ reduction formula in $\overline{\text{MS}}$.

(a) In a subtraction scheme other than on-shell subtraction, $\sqrt{Z} \equiv \langle\Omega|\phi(0)|p\rangle \neq 1$. Use Eq. (24.105) to derive a relation between the 2-point function G_2 and Z.

(b) Calculate Z in $\overline{\text{MS}}$ in QED, including the counterterm contribution. You should find that $Z \neq 1$, and for a massless electron, that Z is UV finite, IR divergent and differs from the $\overline{\text{MS}}$ field-strength renormalzation factor $Z_2 = 1 - 2\frac{e_R^2}{16\pi^2}\frac{1}{\varepsilon_{\text{UV}}}$.

(c) Use Eq. (24.105) again to relate the S-matrix element $S_4 = \langle p_1 p_2|p_3 p_4\rangle$ to the Green's function G_4.

(d) Derive a relation between S-matrix elements and amputated Green's functions valid in $\overline{\text{MS}}$.

(d) What changes in the calculation of $e^+e^- \to \mu^+\mu^-(+\gamma)$ in Section 20.A if $\overline{\text{MS}}$ rather than the on-shell scheme is used?

PART IV

THE STANDARD MODEL

So far, the only massless spin-1 particle we have considered is the photon of QED. Yang–Mills theories are a generalization of QED with multiple massless spin-1 particles that can interact among themselves. Just as the Lagrangian description of QED is strongly constrained by gauge invariance, Lagrangians for Yang–Mills theories are strongly constrained by a generalization called non-Abelian gauge invariance. You already derived a number of these constraints by considering the soft limit in Problem 9.3. In this chapter we begin a systematic study of Yang–Mills theories.

To begin, we review how the QED Lagrangian was determined. In Chapter 8 we saw that to write down a local Lagrangian for a massless spin-1 particle, whose irreducible representation of the Poincaré group has two degrees of freedom, we had to embed the particle in a vector field $A_\mu(x)$, which has four degrees of freedom. The two extra degrees of freedom in $A_\mu(x)$ are removed in quantum field theory through gauge invariance. The gauge symmetry $A_\mu(x) \to A_\mu(x) + \frac{1}{e}\partial_\mu \alpha(x)$ identifies the photon with an equivalence class of vector fields. The kinetic Lagrangian invariant under this symmetry is unique: $\mathcal{L} = -\frac{1}{4}F_{\mu\nu}^2$ with $F_{\mu\nu} = \partial_\mu A_\nu - \partial_\nu A_\mu$. This kinetic Lagrangian propagates two degrees of freedom, as required for an irreducible unitary representation of a massless spin-1 particle. To have the photon interact with matter, the interactions have to preserve the gauge symmetry. We found that an easy way to determine gauge-invariant interactions is with the covariant derivative $D_\mu \psi = (\partial_\mu - iQeA_\mu)\psi$. For example, replacing $\partial_\mu \to D_\mu$ in the fermionic kinetic term $\bar{\psi}\gamma^\mu \partial_\mu \psi$ gives $\bar{\psi}\gamma^\mu D_\mu \psi$, which is gauge invariant under the transformation $\psi(x) \to e^{iQ\alpha(x)}\psi(x)$. In fact, $\bar{\psi}\gamma^\mu D_\mu \psi$ contains the unique renormalizable interaction we can write down in QED. Yang–Mills theories are the unique generalizations of QED in which renormalizable self-interactions among massless spin-1 particles are possible.

We begin our study of Yang–Mills theories with an example. Suppose we have two fields ϕ_1 and ϕ_2. Then the kinetic Lagrangian

$$\mathcal{L}_{\text{kin}} = (\partial_\mu \phi_1^\star)(\partial_\mu \phi_1) + (\partial_\mu \phi_2^\star)(\partial_\mu \phi_2) = (\partial_\mu \vec{\phi})^\dagger(\partial_\mu \vec{\phi}), \qquad (25.1)$$

where $\vec{\phi} = (\phi_1, \phi_2)^T$, is invariant under a global SU(2) symmetry, $\vec{\phi} \to U\vec{\phi}$, with U a special unitary 2×2 matrix.[1] In general, such a U can always be written as

---

[1] This Lagrangian is actually invariant under a larger $U(2) = U(1) \times SU(2)$ symmetry. But, as we will come to understand, there is no point in considering non-simple groups such as $U(N)$ in quantum field theory. For example, in a gauge theory the coupling constants for the $U(1)$ and $SU(2)$ subgroups will in general be different; even if we set them equal at one scale, they will run differently. Moreover, the $U(1)$ symmetry in Lagrangians such as Eq. (25.1) will often be violated by a quantum effect called an anomaly, to be discussed in Chapter 30. Thus, we will restrict attention to the simple $SU(N)$ subgroups.

$$U = \exp\left[i(\alpha_1 \tau_1 + \alpha_2 \tau_2 + \alpha_3 \tau_3)\right] = \exp(i\alpha^a \tau^a), \tag{25.2}$$

where $\tau^a = \frac{1}{2}\sigma^a$ and σ^a are the Pauli sigma matrices (see Eq. (10.3)) and α^a are real numbers. The normalization of the τ^a matrices is chosen so that $[\tau^a, \tau^b] = i\varepsilon^{abc}\tau^c$. Here, ε^{abc} is the **Levi-Civita tensor** (the totally antisymmetric tensor with $\varepsilon^{123} = 1$). Infinitesimally,

$$\vec{\phi} \to \vec{\phi} + i\alpha^a \tau^a \vec{\phi}. \tag{25.3}$$

We can promote the global SU(2) symmetry to a local symmetry by elevating the real numbers α^a to real functions of space-time $\alpha^a(x)$. To make the kinetic terms invariant under the local symmetry, we can elevate the ordinary derivatives to covariant derivatives defined by

$$D_\mu \vec{\phi} = \partial_\mu \vec{\phi} - ig A_\mu^a \tau^a \vec{\phi}, \tag{25.4}$$

where g is a number (the strength of the force) and A_μ^a are a set of three gauge bosons, which transform as

$$A_\mu^a(x) \to A_\mu^a(x) + \frac{1}{g}\partial_\mu \alpha^a(x) - f^{abc}\alpha^b(x)A_\mu^c(x), \tag{25.5}$$

where $f^{abc} = \varepsilon^{abc}$ are the structure constants for SU(2). The unique gauge-invariant kinetic term for the A_μ^a is

$$\mathcal{L}_{\text{YM}} = -\frac{1}{4}\sum_a \left(\partial_\mu A_\nu^a - \partial_\nu A_\mu^a + g f^{abc} A_\mu^b A_\nu^c\right)^2. \tag{25.6}$$

You should check (Problem 25.1) that $\mathcal{L}_{\text{kin}}$ with $\partial_\mu \to D_\mu$ and $\mathcal{L}_{\text{YM}}$ are gauge invariant. This gauge symmetry is called **non-Abelian** because the group generators τ^a do not commute with each other. Yang–Mills theories are also known as **non-Abelian gauge theories**. In Section 25.2 we will see why the form of Eq. (25.6) is natural from a geometric point of view.

Note that the kinetic term in Eq. (25.6) includes renormalizable interactions among the three gauge bosons for SU(2). These interactions are very important. For example, as we will see in Chapter 26, virtual gauge bosons produce a vacuum polarization effect with the opposite sign from virtual spinors or scalars. Thus, in contrast to QED where the fine-structure constant was logarithmically *weaker* at larger distances, coupling constants in Yang–Mills theories can get logarithmically *stronger* at larger distances. This property of Yang–Mills theories explains qualitative features of the strong force, such as why quarks act as essentially free within a nucleus yet can never escape. In the next few chapters, we will study the fascinating physics of Yang–Mills theories.

25.1 Lie groups

We have already seen a few examples of Lie groups: the 3D rotation group SO(3), the Pauli spin group SU(2) and the Lorentz group O(1,3). The Lie group associated with QED, whose elements are phases $e^{i\alpha}$ with $0 \leq \alpha < 2\pi$, is called U(1). This section provides a summary of some of the relevant mathematics of group theory (see also the discussion in Section 10.1).

Lie groups are groups with infinite numbers of elements that are also differentiable manifolds. All groups have an identity element $\mathbb{1}$. Any group element continuously connected to the identity can be written as

$$U = \exp(i\theta^a T^a) \cdot \mathbb{1}, \tag{25.7}$$

where θ^a are numbers parametrizing the group elements and T^a are called the **group generators**. Given any explicit form of the elements U of a Lie group, you can always figure out what the T^a are by expanding in a small neighborhood of $\mathbb{1}$. We performed this exercise for the Lorentz group, $O(1,3)$, in Chapter 10.

The generators of a Lie group T^a form a **Lie algebra**. The Lie algebra is defined through its commutation relations:

$$\left[T^a, T^b\right] = i f^{abc} T^c, \tag{25.8}$$

where f^{abc} are known as **structure constants**. A Lie group is **Abelian** if $f^{abc} = 0$ and **non-Abelian** otherwise. For example, the algebra su(2) associated with the non-Abelian group SU(2) has $f^{abc} = \varepsilon^{abc}$.

Note that we are calling Eq. (25.8) a commutation relation, but really it is just a mapping $\mathcal{G} \times \mathcal{G} \to \mathcal{G}$. This mapping is more generally called a **Lie bracket**. By calling it a commutator, we are implying that it can be represented as

$$[A, B] = AB - BA. \tag{25.9}$$

Such notation implies, in addition to the Lie bracket mapping, that products of elements are well defined. When this holds, then $[A, [B, C]] = ABC - ACB - BCA + CBA$ and it automatically follows that

$$[A, [B, C]] + [B, [C, A]] + [C, [A, B]] = 0. \tag{25.10}$$

This last equation is known as the **Jacobi identity**. In terms of the structure constants, the Jacobi identity can be written as

$$f^{abd} f^{dce} + f^{bcd} f^{dae} + f^{cad} f^{dbe} = 0. \tag{25.11}$$

The formal definition of a Lie algebra does not require that we write $[A, B] = AB - BA$, but it does require that the Jacobi identity holds. The Jacobi identity is formally defined only using the Lie bracket, and not through products. This is really just a technical mathematical point – in all the cases with physics applications, the generators are embedded into matrices and the Lie bracket can be defined as a commutator, so the Jacobi identity is automatically satisfied.

An **ideal** is a subalgebra $\mathcal{I} \subset \mathcal{G}$ satisfying $[g, i] \subset \mathcal{I}$ for any $g \in \mathcal{G}$ and $i \in \mathcal{I}$. A **simple** Lie algebra has no non-trivial **ideals**. Important simple Lie algebras are su(N) and so(N). The Standard Model is based on the gauge group SU(3) $\otimes$ SU(2) $\otimes$ U(1) whose Lie algebra is su(3) $\oplus$ su(2) $\oplus$ u(1). The Standard Model Lie algebra is **semisimple**, meaning it is the direct sum of simple Lie algebras. A theorem that explains the importance of semisimple Lie algebras in physics states that all finite-dimensional representations of semisimple algebras are Hermitian (see Problem 25.3). Hence, one can construct unitary theories based on semisimple algebras. There can be an infinite or finite number of generators T^a for the Lie algebra. If there are a finite number, the algebra and the group it generates are said to be **finite dimensional**.

Unitary groups can be defined as preserving a complex inner product:

$$\langle \vec{\psi} | \vec{\chi} \rangle = \langle \vec{\psi} | U^\dagger U | \vec{\chi} \rangle. \tag{25.12}$$

That is, $U^\dagger U = \mathbb{1}$. Elements of special unitary groups also have $\det(U) = 1$. The group $\mathrm{SU}(N)$ is defined by its action on N-dimensional vector spaces. In the defining representation, group elements can be written as $U = \exp(i\theta^a T^a)$, where T^a is a Hermitian matrix. There are $N^2 - 1$ generators for $\mathrm{SU}(N)$, so we say the dimension of the group $d(G) = N^2 - 1$ for $G = \mathrm{SU}(N)$.

The orthogonal groups preserve a real inner product:

$$V \cdot W = V \cdot O^T \cdot O \cdot W. \tag{25.13}$$

So, $O^T O = \mathbb{1}$. For these $d(\mathrm{O}(N)) = \frac{1}{2}N(N-1)$. Every orthogonal matrix has determinant ± 1. Those with determinant 1 are elements of the special orthogonal group. The dimensions of $\mathrm{O}(N)$ and $\mathrm{SO}(N)$ are the same.

Other finite-dimensional simple Lie groups include the **symplectic** groups, $\mathrm{Sp}(N)$, which are the next step in the generalization from a real to a complex inner product: they preserve a quaternionic inner product. An equivalent definition is that they satisfy $\Omega S = -S^T \Omega$, with $\Omega = \begin{pmatrix} 0 & \mathbb{1} \\ -\mathbb{1} & 0 \end{pmatrix}$. Finally, there are five exceptional simple Lie groups, $\mathrm{G}_2, \mathrm{F}_4, \mathrm{E}_6, \mathrm{E}_7$ and E_8. The algebras for $\mathrm{SU}(N)$, $\mathrm{SO}(N)$, $\mathrm{Sp}(N)$ and the exceptional groups are the only finite-dimensional simple Lie algebras [Cartan, 1894].

25.1.1 Representations

We will now discuss representations of the $\mathrm{SU}(N)$ groups. These groups play an essential role in quantum field theory due to the simple observation that the free theory of N complex fields is automatically invariant under $\mathrm{U}(1) \times \mathrm{SU}(N)$. The $\mathrm{SU}(N)$ groups are simply connected (see Section 10.5.1), meaning that they are topologically trivial. Thus, representations of the $\mathrm{SU}(N)$ groups are in one-to-one correspondence with representations of the $\mathrm{su}(N)$ algebra.

Recall from Section 10.1 that **representations** of a Lie algebra can be constructed by embedding the generators into matrices. The two most important representations are the defining (or fundamental) representation and the adjoint representation. The **fundamental** representation is the smallest non-trivial representation of the algebra. For $\mathrm{SU}(N)$, the fundamental representation is generated by the set of the $N \times N$ Hermitian matrices with trace 0. A set of N fields ϕ_i transforming in the fundamental representation, transform under infinitesimal group transformations as

$$\phi_i \rightarrow \phi_i + i\alpha^a (T^a_{\mathrm{fund}})_{ij} \, \phi_j \tag{25.14}$$

for real numbers α^a. The complex conjugate fields transform in the **anti-fundamental representation** for which $T^a_{\text{anti-fund}} = -\left(T^a_{\text{fund}}\right)^\star$, thus

$$\phi^\star_i \to \phi^\star_i + i\alpha^a \left(T^a_{\text{anti-fund}}\right)_{ij} \phi^\star_j = \phi^\star_i - i\alpha^a \phi^\star_j \left(T^a_{\text{fund}}\right)_{ji}, \tag{25.15}$$

where we have used that T^a_{fund} is Hermitian for $\mathrm{SU}(N)$ in the last step. In this way, we can always replace anti-fundamental generators with fundamental ones.

Our default representation will be the fundamental one, so we write T^a (with no subscript) for T^a_{fund}. Generators in a general representation will be denoted T^a_R. It will occasionally be useful to write explicitly the row and column indices i and j as in T^a_{ij}. We use mid-alphabet Latin letters such as i and j to index the color (for $\mathrm{SU}(3)$ of the strong interactions) or flavor (as in up or down quark for $\mathrm{SU}(2)_{\text{isospin}}$), hence these are sometimes called **color indices** or **flavor indices**. We use early-alphabet Latin letters such as a and b to index different generators in the algebra.

The algebra can be determined by expanding a basis of group elements around $\mathbb{1}$. For $\mathrm{SU}(2)$ the generators in the fundamental representation are the Pauli matrices σ^a conventionally normalized by dividing by 2:

$$T^a = \tau^a \equiv \frac{\sigma^a}{2}. \tag{25.16}$$

These satisfy $\left[T^a, T^b\right] = i\varepsilon^{abc}T^c$. For $\mathrm{SU}(3)$ the generators are often written in a standard basis $T^a = \frac{1}{2}\lambda^a$, with λ^3 and λ^8 diagonal (the **Gell-Mann matrices**):

$$\lambda^1 = \begin{pmatrix} 0 & 1 & \\ 1 & 0 & \\ & & 0 \end{pmatrix}, \quad \lambda^2 = \begin{pmatrix} 0 & -i & \\ i & 0 & \\ & & 0 \end{pmatrix}, \quad \lambda^3 = \begin{pmatrix} 1 & & \\ & -1 & \\ & & 0 \end{pmatrix}, \quad \lambda^4 = \begin{pmatrix} & & 1 \\ & 0 & \\ 1 & & \end{pmatrix},$$

$$\lambda^5 = \begin{pmatrix} 0 & & -i \\ & 0 & \\ i & & 0 \end{pmatrix}, \quad \lambda^6 = \begin{pmatrix} 0 & & \\ & 0 & 1 \\ & 1 & 0 \end{pmatrix}, \quad \lambda^7 = \begin{pmatrix} 0 & & \\ & 0 & -i \\ & i & 0 \end{pmatrix}, \quad \lambda^8 = \frac{1}{\sqrt{3}}\begin{pmatrix} 1 & & \\ & 1 & \\ & & -2 \end{pmatrix}. \tag{25.17}$$

The normalization of the generators is arbitrary and a convention must be chosen. A common convention in physics is to normalize the structure constants by

$$\sum_{c,d} f^{acd} f^{bcd} = N\delta^{ab}. \tag{25.18}$$

(In mathematics, the convention $\sum_{c,d} f^{acd} f^{bcd} = \delta^{ab}$ is often used instead.) Once the normalization of the structure constants is fixed, the normalization of the generators in any representation is also fixed. Indeed, $\left[T^a_R, T^b_R\right] = if^{abc}T^c_R$, which must hold for any representation with the same f^{abc}, is not invariant under rescaling of the T^a_R. Equation (25.18) implies that the generators for $\mathrm{SU}(N)$ in the fundamental representation are normalized so that

$$\mathrm{tr}\left(T^a T^b\right) = \frac{1}{2}\delta^{ab}, \tag{25.19}$$

which you can easily check for $\mathrm{SU}(2)$ or $\mathrm{SU}(3)$ using the explicit generators above.

In a generic Lie algebra, the commutator of generators $[T^a, T^b]$ is well defined but the product $T^a T^b$ is not. In the fundamental representation of $\mathrm{SU}(N)$, the generators are matrices that *can* be multiplied. We write

$$T^a T^b = \frac{1}{2N}\delta^{ab} + \frac{1}{2}d^{abc}T^c + \frac{1}{2}if^{abc}T^c. \tag{25.20}$$

The constants $d^{abc} = 2\,\mathrm{tr}\left[T^a\left\{T^b, T^c\right\}\right]$ provide a totally symmetric group invariant. For $\mathrm{SU}(N)$, there is a unique such invariant up to an overall constant. For $\mathrm{SU}(2)$, $d^{abc} = 0$. One can also show that (see Problem 25.2)

$$\mathrm{tr}\left[T^a T^b T^c\right] = \frac{1}{4}\left(d^{abc} + if^{abc}\right), \tag{25.21}$$

$$\mathrm{tr}\left[T^a T^b T^c T^d\right] = \frac{1}{4N}\delta^{ab}\delta^{cd} + \frac{1}{8}\left(d^{abe} + if^{abe}\right)\left(d^{cde} + if^{cde}\right) \tag{25.22}$$

and so on.

The next important representation is the **adjoint representation**, which acts on the vector space spanned by the generators themselves. For $\mathrm{SU}(N)$ there are $N^2 - 1$ generators, so this is an $N^2 - 1$-dimensional representation. Matrices describing the adjoint representation are given by $(T^a_{\mathrm{adj}})^{bc} = -if^{abc}$. For $\mathrm{SU}(2)$ these are 3×3 matrices:

$$T^1_{\mathrm{adj}} = \begin{pmatrix} 0 & & \\ & 0 & -i \\ & i & 0 \end{pmatrix}, \quad T^2_{\mathrm{adj}} = \begin{pmatrix} 0 & & i \\ & 0 & \\ -i & & 0 \end{pmatrix}, \quad T^3_{\mathrm{adj}} = \begin{pmatrix} 0 & -i & \\ i & 0 & \\ & & 0 \end{pmatrix}. \tag{25.23}$$

For $\mathrm{SU}(3)$ they are 8×8 matrices. It is easy to check that both the adjoint and fundamental representations satisfy $[T^a_{\mathrm{adj}}, T^b_{\mathrm{adj}}] = if^{abc}T^c_{\mathrm{adj}}$ with $f^{abc} = \varepsilon^{abc}$ for $\mathrm{SU}(2)$. As we will soon see, gauge fields must transform in the adjoint representation. There are lots of other representations as well, but the fundamental and adjoint representations are by far the most important for physics.

It will be extremely useful to have basis-independent ways to characterize representations. These are known as **Casimir operators** or **Casimirs**. For example, for $\mathrm{SU}(2)$ we know $\vec{J}^2 = \sum_a T^a_R T^a_R$ is a Casimir operator with eigenvalue $j(j+1)$; j labels the representation and is given the special name *spin*. More generally, we define the **quadratic Casimir** $C_2(R)$ by

$$T^a_R T^a_R = C_2(R)\mathbb{1}, \tag{25.24}$$

where the sum over a is implicit. That this operator will always be proportional to the identity follows from **Schur's lemma**: a group element that commutes with all other group elements in any irreducible representation must be proportional to $\mathbb{1}$. In this case, it is enough to show that our operator commutes with all the generators:

$$[T^a_R T^a_R, T^b_R] = (if^{abc}T^c_R)T^a_R + T^a_R(if^{abc}T^c_R) = if^{abc}\{T^c_R, T^a_R\} = 0. \tag{25.25}$$

We have used the antisymmetry of f^{abc} in the last step. Therefore Eq. (25.24) holds for some $C_2(R)$ by Schur's lemma.

To evaluate the quadratic Casimir, it is helpful to first define an inner product on the generators. In any representation the generators can be chosen so that

$$\text{tr}\left[T_R^a T_R^b\right] = T(R)\delta^{ab}, \tag{25.26}$$

where $T(R)$ is a number known as the **index** of the representation. Sometimes $C(R)$ is written instead of $T(R)$ for the index. For the fundamental representation, our convention in Eq. (25.18) implies

$$T(\text{fund}) = T_F = \frac{1}{2}, \tag{25.27}$$

that is, $T_{ji}^a T_{ij}^b = \frac{1}{2}\delta^{ab}$. For the adjoint representation,

$$T(\text{adj}) = T_A = N, \tag{25.28}$$

that is, $f^{acd} f^{bcd} = N\delta^{ab}$.

Setting $a = b$ in Eq. (25.26) and summing over a gives

$$d(R)\, C_2(R) = T(R)\, d(G), \tag{25.29}$$

where $d(R)$ is the dimension of the representation ($d(\text{fund}) = N$ and $d(\text{adj}) = N^2 - 1$) and $d(G)$ is the dimension of the group (number of group generators: $d(\text{SU}(N)) = N^2 - 1$). Equation (25.29) implies that for $\text{SU}(N)$ the quadratic Casimir for the fundamental representation is

$$C_F \equiv C_2(\text{fund}) = \frac{N^2 - 1}{2N}, \tag{25.30}$$

that is, $(T^a T^a)_{ij} - C_F \delta_{ij}$. In particular, $C_F = \frac{3}{4}$ for SU(2) and $C_F = \frac{4}{3}$ for SU(3). For the adjoint representation,

$$C_A \equiv C_2(\text{adj}) = N, \tag{25.31}$$

that is, $f^{acd} f^{bcd} = C_A \delta^{ab}$. For the adjoint representation the index and quadratic Casimir are the same. Almost every calculation in Yang–Mills theories will have factors of C_F or C_A in it.

Since, for any representation,

$$\text{tr}\left([T_R^a, T_R^b]\, T_R^c\right) = i f^{abd}\text{tr}\left(T_R^d T_R^c\right) = i f^{abc} T(R), \tag{25.32}$$

we can write

$$f^{abc} \equiv -\frac{i}{T_F}\text{tr}\left([T^a, T^b]\, T^c\right), \tag{25.33}$$

where T^a are the fundamental generators. Thus, we can always replace the structure constants with commutators and products of fundamental group generators. This is extremely handy when one tries to compute complicated gluon scattering amplitudes.

In $\text{SU}(N)$ one also has a Fierz identity of the form

$$\sum_a T_{ij}^a T_{kl}^a = \frac{1}{2}\left(\delta_{il}\delta_{kj} - \frac{1}{N}\delta_{ij}\delta_{kl}\right). \tag{25.34}$$

You can check that, since the generators in $\mathrm{SU}(N)$ are traceless, summing over δ^{ij} or δ^{kl} gives zero. This is a useful relation, since it implies

$$\mathrm{tr}[T^a A]\,\mathrm{tr}[T^a B] = \frac{1}{2}\left[\mathrm{tr}(AB) - \frac{1}{N}\mathrm{tr}(A)\,\mathrm{tr}(B)\right] \tag{25.35}$$

for any A and B, which lets us reduce products of traces to single traces.

Another invariant that characterizes $\mathrm{SU}(N)$ representations is the **anomaly coefficient** $A(R)$ defined by

$$\mathrm{tr}\left[T_R^a\{T_R^b, T_R^c\}\right] = \frac{1}{2}A(R)d^{abc} = A(R)\mathrm{tr}\left[T^a\{T^b, T^c\}\right], \tag{25.36}$$

where d^{abc} are as in Eq. (25.20), or equivalently by $A(\mathrm{fund}) = 1$. These anomaly coefficients will be used in the study of anomalies in Chapter 30. Some relations among them are explored in Problem 25.4.

In summary, the main relations we will use often for $\mathrm{SU}(N)$ are

$$\mathrm{tr}\left(T^a T^b\right) = T_{ji}^a T_{ij}^b = T_F \delta^{ab}, \tag{25.37}$$

$$\sum_a \left(T^a T^a\right)_{ij} = C_F \delta_{ij}, \tag{25.38}$$

$$f^{acd} f^{bcd} = C_A \delta^{ab}, \tag{25.39}$$

with $T_F = \frac{1}{2}$, $C_A = N$ and $C_F = \frac{N^2-1}{2N}$. These relations are used in almost every QCD calculation.

25.2 Gauge invariance and Wilson lines

Now that we understand the mathematics of Lie groups, we will develop a more geometric way to think about gauge theories. This is not strictly necessary, and if you just want to know the rules for computation, you can safely skip this section (and in fact the remainder of this chapter; the Feynman rules for non-Abelian gauge theories are given in Section 26.1).

25.2.1 Abelian case

Consider first a complex scalar field $\phi(x)$. The phase of this field is just a convention. Thus, a theory of such a field should be invariant under redefinitions $\phi(x) \to e^{i\alpha}\phi(x)$ (as if $Q = 1$). Now suppose we want to examine the field at two points x^μ and y^μ very far away from each other. In a local theory, the convention that we choose at x^μ should be independent of the convention we choose at y^μ. But then how can we tell if $\phi(x) = \phi(y)$? If we changed conventions we would have

$$\phi(y) - \phi(x) \to e^{i\alpha(y)}\phi(y) - e^{i\alpha(x)}\phi(x). \tag{25.40}$$

So, for example, $|\phi(y) - \phi(x)|$ depends on our choice of local phases. In fact, it is impossible to come up with a convention-independent way to compare these fields at different points. Moreover, it is also impossible to compute $\partial_\mu \phi(x)$, since the derivative is a difference, and the difference depends on the phase choices.

To make comparisons of field values at different points well defined, we need another ingredient. This motivates defining a new field $W(x, y)$ called a Wilson line. It is a kind of bi-local field that depends on two points. We want it to transform as

$$W(x, y) \to e^{i\alpha(x)} W(x, y) e^{-i\alpha(y)} \tag{25.41}$$

so that

$$W(x, y)\,\phi(y) - \phi(x) \to e^{i\alpha(x)} W(x, y) e^{-i\alpha(y)} e^{i\alpha(y)} \phi(y) - e^{i\alpha(x)} \phi(x)$$
$$= e^{i\alpha(x)} [W(x, y)\,\phi(y) - \phi(x)]\,. \tag{25.42}$$

The point of this is that now $|W(x, y)\,\phi(y) - \phi(x)|$ is independent of our choice of a local phase convention.

Taking $y^\mu = x^\mu + \delta x^\mu$, dividing by δx^μ, and letting $\delta x^\mu \to 0$ lets us turn this difference into a derivative:

$$D_\mu \phi(x) \equiv \lim_{\delta x^\mu \to 0} \frac{W(x, x + \delta x)\phi(x + \delta x) - \phi(x)}{\delta x^\mu}\,. \tag{25.43}$$

Then

$$D_\mu \phi(x) \to e^{i\alpha(x)} D_\mu \phi(x), \tag{25.44}$$

which holds from Eq. (25.42) even if δx^μ in (25.43) is not small.

We naturally want $W(x, x) = 1$. So if δx^μ is small, then we should be able to expand

$$W(x, x + \delta x) = 1 - ie\delta x^\mu A_\mu(x) + \mathcal{O}(\delta x^2)\,, \tag{25.45}$$

where e is arbitrary. It then follows from the transformation of $W(x, y)$ in Eq. (25.41) that

$$A_\mu(x) \to A_\mu(x) + \frac{1}{e} \partial_\mu \alpha(x) \tag{25.46}$$

and then, from Eq. (25.43), $D_\mu \phi = \partial_\mu \phi - ieA_\mu \phi$. In this way, the gauge field is introduced as a **connection**, allowing us to compare field values at different points, despite their arbitrary phases. Another example of a connection that you might be familiar with from general relativity is the Christoffel connection, which allows us to compare field values at different points, despite their different local coordinate systems.

It is possible to write a closed-form expression for $W(x, y)$:

$$W_P(x, y) = \exp\left(ie \int_y^x A_\mu(z) dz^\mu \right). \tag{25.47}$$

This functional of $A_\mu(x)$ is known as a **Wilson line**. The integral is a line integral along the path P from y^μ to x^μ. More precisely, the path P is a function $z^\mu(\lambda)$ with $0 \le \lambda \le 1$ with $z^\mu(0) = y^\mu$ and $z^\mu(1) = x^\mu$ and so

$$W_P(x, y) = \exp\left(ie \int_0^1 \frac{dz^\mu(\lambda)}{d\lambda} A_\mu(z(\lambda))\, d\lambda \right). \tag{25.48}$$

where we have used that $T^{a\dagger} = T^a$ for SU(N).

In the non-Abelian case, it is often convenient to represent the gauge field as a Lie-algebra-valued field by writing

$$\mathbf{A}_\mu \equiv A_\mu^a T^a. \tag{25.60}$$

Then, $W_P(x, y) = P\left\{\exp\left(ig \int_y^x \mathbf{A}_\mu(z)\, dz^\mu\right)\right\}$, which looks a lot like the Abelian case. (Technically, A_μ^a are the components of a Lie-algebra-valued one-form $\mathbb{A} = \mathbf{A}_\mu dx^\mu$.)

The infinitesimal expansion of the Wilson line is

$$W(x^\mu, x^\mu + \delta x^\mu) = \mathbb{1} - ig\mathbf{A}_\mu \delta x^\mu. \tag{25.61}$$

The local transformations can be expressed in terms of $U(x) = e^{i\alpha^a(x)T^a} \in$ SU(N), which is the group element for the transformation at point x. They are

$$\vec{\psi}(x) \to U(x) \cdot \vec{\psi}(x) \tag{25.62}$$

and

$$W(x, y) \to U(x)W(x, y)U^\dagger(y), \tag{25.63}$$

where $U^\dagger(y) = U^{-1}(y)$ in SU(N).

To determine how A_μ^a transforms, we could just expand the transformation of W. A more efficient way to derive the transformation law is to use that the covariant derivative must transform like the field $D_\mu\vec{\psi} \to U \cdot D_\mu\vec{\psi}$ and therefore

$$\left(\partial_\mu - ig\mathbf{A}_\mu'\right) \cdot U \cdot \vec{\psi} = U \cdot \left(\partial_\mu - ig\mathbf{A}_\mu\right) \cdot \vec{\psi}, \tag{25.64}$$

where $\mathbf{A}_\mu'$ is the transformed version of $\mathbf{A}_\mu$. Thus,

$$\partial_\mu U - ig\mathbf{A}_\mu' U = -igU\mathbf{A}_\mu, \tag{25.65}$$

which implies

$$\mathbf{A}_\mu' = U\mathbf{A}_\mu U^{-1} - \frac{i}{g}\left(\partial_\mu U\right)U^{-1}. \tag{25.66}$$

In terms of components, the infinitesimal version is

$$A_\mu^a(x) \to A_\mu^a(x) + \frac{1}{g}\partial_\mu \alpha^a(x) - f^{abc}\alpha^b(x)A_\mu^c(x) \tag{25.67}$$

plus terms higher order in α.

Finally, let us look at the commutator of covariant derivatives as before. We now find

$$[D_\mu, D_\nu]\psi(x) = \left(-ig(\partial_\mu\mathbf{A}_\nu - \partial_\nu\mathbf{A}_\mu) - g^2[\mathbf{A}_\mu, \mathbf{A}_\nu]\right)\psi(x). \tag{25.68}$$

As in the Abelian case, there are no derivatives acting on $\psi(x)$ in this expansion. We now see that the natural field strength in the non-Abelian case is

$$\mathbf{F}_{\mu\nu} \equiv \frac{i}{g}[D_\mu, D_\nu] = (\partial_\mu\mathbf{A}_\nu - \partial_\nu\mathbf{A}_\mu) - ig[\mathbf{A}_\mu, \mathbf{A}_\nu]. \tag{25.69}$$

Or in components, $\mathbf{F}_{\mu\nu} = F_{\mu\nu}^a T^a$, where

$$F_{\mu\nu}^a = \partial_\mu A_\nu^a - \partial_\nu A_\mu^a + g f^{abc} A_\mu^b A_\nu^c. \tag{25.70}$$

In the Abelian case $f^{abc} = 0$ and $F_{\mu\nu}^a$ reduces to the electromagnetic field strength. Note that, as in the Abelian case, $F_{\mu\nu}^a$ is antisymmetric: $F_{\mu\nu}^a = -F_{\nu\mu}^a$. The transformation law for $F_{\mu\nu}^a$ is

$$F_{\mu\nu}^a \to F_{\mu\nu}^a - f^{abc} \alpha^b F_{\mu\nu}^c, \tag{25.71}$$

which is the same for a constant α or a local $\alpha(x)$. Thus, although initially $\mathbf{F}_{\mu\nu} = F_{\mu\nu}^a T^a$ was defined with generators in the fundamental representation, the kinetic term just depends on the $F_{\mu\nu}^a$ fields, which naturally transform in the adjoint representation.

We can now write down a locally $\mathrm{SU}(N)$ invariant Lagrangian:

$$\mathcal{L} = -\frac{1}{4}\left(F_{\mu\nu}^a\right)^2 + \sum_{i,j=1}^N \bar\psi_i\left(\delta_{ij} i\slashed\partial + g\slashed A^a T_{ij}^a - m\delta_{ij}\right)\psi_j. \tag{25.72}$$

The first term is exactly the kinetic term in Eq. (25.6). The constant g is the analog of the QED strength e.

There is one more renormalizable term we could add consistent with gauge invariance:

$$\mathcal{L}_\theta = \theta\varepsilon^{\mu\nu\alpha\beta} F_{\mu\nu}^a F_{\alpha\beta}^a = 2\theta\partial_\mu\left(\varepsilon^{\mu\nu\alpha\beta} A_\nu^a F_{\alpha\beta}^a\right), \tag{25.73}$$

where θ is a number. Since this term is a total derivative it does not contribute at any order in perturbation theory (see Section 7.4.2). However, it can contribute due to non-perturbative effects, as will be discussed in Sections 29.5 and 30.5. For example, in QCD, θ is called the strong CP phase. If θ were non-zero, the neutron would have an electric dipole moment proportional to θ. The absence of such a moment experimentally is a mystery known as the strong CP problem.

25.3 Conserved currents

If we expand out the Lagrangian in Eq. (25.72) we find

$$\mathcal{L} = -\frac{1}{4}\left(\partial_\mu A_\nu^a - \partial_\nu A_\mu^a + g f^{abc} A_\mu^b A_\nu^c\right)^2 + \bar\psi_i\left(i\delta_{ij}\gamma^\mu \partial_\mu + g\gamma^\mu A_\mu^a T_{ij}^a - m\delta_{ij}\right)\psi_j, \tag{25.74}$$

where indices that appear twice are summed over. The equations of motion are

$$\partial_\mu F_{\mu\nu}^a + g f^{abc} A_\mu^b F_{\mu\nu}^c = -g\bar\psi_i\gamma_\nu T_{ij}^a\psi_j \tag{25.75}$$

for the gauge fields and

$$\left(i\slashed\partial - m\right)\psi_i = -g\slashed A^a T_{ij}^a\psi_j \tag{25.76}$$

for the spinors.

Because the Lagrangian has a gauge symmetry, it has a global symmetry, under which

$$\psi_i \to \psi_i + i\alpha^a T_{ij}^a\psi_j \tag{25.77}$$

and

$$A_\mu^a \to A_\mu^a - f^{abc}\alpha^b A_\mu^c \tag{25.78}$$

for infinitesimal α. In Section 3.3 we proved Noether's theorem, that a global symmetry implies a conserved current given by

$$J_\mu = \sum_n \frac{\partial \mathcal{L}}{\partial(\partial_\mu \phi_n)} \frac{\delta \phi_n}{\delta \alpha}. \tag{25.79}$$

In the non-Abelian case, there will be $N^2 - 1$ currents, one for each symmetry direction α^a. Summing over both matter fields $\phi_n = \psi_i$ and gauge fields $\phi_n = A_\mu^a$ gives

$$J_\mu^a = -\bar\psi_i \gamma^\mu T_{ij}^a \psi_j + f^{abc} A_\nu^b F_{\mu\nu}^c. \tag{25.80}$$

It is not hard to check that the current is conserved on the equation of motion, $\partial_\mu J_\mu^a = 0$, which we leave for an exercise (Problem 25.5).

In contrast to the QED current, the Noether current associated with a global non-Abelian symmetry in a theory with gauge bosons is not gauge invariant (or even gauge covariant). Thus, it is not physical and there is not a well defined charge that one can measure. Although it is true that the charges

$$Q^a = \int d^3x \, J_0^a \tag{25.81}$$

are conserved, that is $\partial_t Q^a = 0$, these charges depend on our choice of gauge. Thus, in a non-Abelian gauge theory such as QCD there is no such thing as a classical current, like a wire with quarks in it instead of electrons. There is no simple analog of Gauss's law either; the gauge fields are bound up with the matter fields in an intricate and nonlinear way.

One can define a **matter current** constructed only out of fermions as

$$j_\mu^a = -\bar\psi_i \gamma^\mu T_{ij}^a \psi_j, \tag{25.82}$$

which is gauge covariant. However, this current satisfies

$$D_\mu j_\mu^a = 0, \tag{25.83}$$

where $D_\mu j_\nu^a = \partial_\mu j_\nu^a + g f^{abc} A_\mu^b j_\nu^c$ is the covariant derivative in the adjoint representation. Thus, the matter current is not conserved, $\partial_\mu j_\mu^a \neq 0$, and there is no associated conserved charge.

Our observations about currents follow from a very general theorem known as the **Weinberg–Witten theorem** [Weinberg and Witten, 1980]:

Box 25.1   The Weinberg–Witten theorem (for spin 1)

A theory with a global non-Abelian symmetry under which massless spin-1 particles are charged does not admit a gauge-invariant conserved current.

Another way to phrase the theorem without reference to gauge invariance (which is unphysical) is that there cannot be a conserved Lorentz-covariant current in a theory with massless spin-1 particles with non-vanishing values of the charge associated with that current.

Lorentz covariance replaces gauge invariance because the physical polarizations of a massless spin-1 particle transform non-covariantly as $\epsilon_\mu(p) \to \epsilon_\mu(p) + p_\mu$ under certain Lorentz transformations. The connection between these non-covariant transformations and charge conservation was discussed in Sections 8.4.2 and 9.5.

The Weinberg–Witten theorem for spin 2 implies:

A theory with a conserved and Lorentz-covariant energy-momentum tensor can never have a massless particle of spin 2.

In this case also, Lorentz covariance is equivalent to saying that there cannot be a gauge field associated with the local symmetry. For the energy-momentum tensor, this local symmetry is local translations (see Section 3.3.1) and the gauge field is the graviton (see Section 8.7.2).

The Weinberg–Witten theorem does not say anything useful about the Standard Model, since it has non-conserved currents under which non-Abelian gauge fields and gravity are charged and conserved currents that are not gauge invariant. But it does say that if you started with a theory without gravity, say only with scalars, spinors and gauge fields, which *does* have a conserved energy-momentum tensor, you would never have some kind of phase transition that gives you a massless graviton, since the same energy-momentum tensor could no longer exist. String theory and the anti de Sitter/conformal field theory (AdS/CFT) correspondence get around this by having gravity emerge in a different space-time – ten dimensions for string theory from a two-dimensional world sheet, and four dimensions for the CFT from a ten-dimensional string theory. The Weinberg–Witten theorem assumes the space-time dimension is fixed.

25.4 Gluon propagator

The next step is to derive the gluon propagator. For simplicity, we call the massless spin-1 particles gluons and the theory QCD, although the derivation that follows applies for any gauge group. We will compute the gluon propagator in the covariant R_ξ gauges, as we did for the photon propagator, but we will find a new feature: Faddeev–Popov ghosts. These ghosts are unphysical virtual states. They are an artifact of insisting on Lorentz invariance (through the covariant R_ξ gauges) from which reemerges the conflict between unitarity for massless spin-1 particles and manifest Lorentz invariance (this conflict was the subject of Chapter 8). In some non-covariant gauges, such as axial gauges, discussed below, ghosts are absent. However, covariant gauges are vastly simpler for most calculations despite the required ghosts, thus we will learn to work with ghosts as the lesser of two evils.

25.4.1 Faddeev–Popov procedure

Recall our derivation of the photon propagator in QED. We first observed that the equations of motion for a photon coupled to an external current,

$$(g_{\mu\nu}\Box - \partial_\mu\partial_\nu) A_\mu = J_\nu, \qquad (25.84)$$

were not invertible: the operator $k_\mu k_\nu - k^2 g_{\mu\nu}$ has an eigenvector k_ν with eigenvalue 0. We made them invertible by modifying the Lagrangian with a Lagrange multiplier $\frac{1}{2\xi}(\partial_\mu A_\mu)^2$. This modification led to a one-parameter family of propagators; we had to carefully check that our modification would not violate gauge invariance through a dependence of physical quantities on ξ.

A more systematic way of calculating the photon propagator came with the introduction of path integrals in Chapter 14. In Section 14.5 we observed that

$$f(\xi) = \int \mathcal{D}\pi e^{-i\int d^4x \frac{1}{2\xi}(\Box\pi)^2} = \int \mathcal{D}\pi e^{-i\int d^4x \frac{1}{2\xi}(\Box\pi - \partial_\mu A_\mu)^2} \qquad (25.85)$$

was independent of A_μ, since the last step is a simple shift $\pi \to \pi - \frac{1}{\Box}\partial_\mu A_\mu$. We then saw that

$$\int \mathcal{D}A_\mu \mathcal{D}\phi_i e^{i\int d^4x \mathcal{L}[A,\phi_i]} = \frac{1}{f(\xi)} \int \mathcal{D}\pi \mathcal{D}A_\mu \mathcal{D}\phi_i e^{i\int d^4x \left[\mathcal{L}[A,\phi_i] - \frac{1}{2\xi}(\Box\pi - \partial_\mu A_\mu)^2\right]}$$

$$= \left[\frac{1}{f(\xi)} \int \mathcal{D}\pi\right] \int \mathcal{D}A_\mu \mathcal{D}\phi_i e^{i\int d^4x \left[\mathcal{L}[A,\phi_i] - \frac{1}{2\xi}(\partial_\mu A_\mu)^2\right]},$$

$$(25.86)$$

implying that (up to an overall normalization factor $\frac{1}{f(\xi)}\int \mathcal{D}\pi$, which drops out of physical quantities) the un-gauge-fixed Lagrangian will give the same result as the gauge-fixed one. The interpretation of the normalization factor is that it describes the path integral over gauge orbits, as can be seen in Eq. (25.85), on which physical quantities do not depend. Removing this normalization leaves a path integral over only the physical degrees of freedom for a massless spin-1 particle. If any of these steps are not familiar, please review the derivation in Section 14.5.

For the non-Abelian theory, we can do the same trick, but there are some subtleties. To start, we will need $N^2 - 1$ fields π^a. The gauge transformation is more complicated in the non-Abelian case. For infinitesimal transformations parametrized by π^a, we now have

$$A_\mu^a \to A_\mu^a + \frac{1}{g}\partial_\mu\pi^a + f^{abc}A_\mu^b\pi^c. \qquad (25.87)$$

Since π^a is in the adjoint representation, this can be written more concisely as

$$A_\mu^a \to A_\mu^a + \frac{1}{g}D_\mu\pi^a, \qquad (25.88)$$

where $D_\mu\pi^a = \partial_\mu\pi^a + gf^{abc}A_\mu^b\pi^c$ is the way a covariant derivative acts on a field transforming in the adjoint representation. Note that D_μ mixes different π^a fields, thus we might more accurately write $D_\mu^{ab}\pi^b$; instead $D_\mu\pi^a$ is used for simplicity. Now let us multiply and divide our path integral by

$$f[A] = \int \mathcal{D}\pi \exp\left[-i\int d^4x \frac{1}{2\xi}(\partial_\mu D_\mu\pi^a)^2\right]. \qquad (25.89)$$

Unlike in the Abelian case, f is not just a number, but is now a functional of A_μ. Nevertheless, we can still define $\alpha^a[A]$ as the gauge-transformation parameters that take a given A_μ^a

configuration to Lorenz gauge. That is, $\partial_\mu A_\mu^a = -\frac{1}{g}\partial_\mu D_\mu \alpha^a[A]$ has a solution. Shifting π^a by $\frac{1}{g}\alpha^a[A]$ then gives

$$f[A] = \int \mathcal{D}\pi \exp\left[-i \int d^4x \frac{1}{2\xi}\left(\partial_\mu A_\mu^a - \partial_\mu D_\mu \pi^a\right)^2\right], \qquad (25.90)$$

so that

$$\int \mathcal{D}A_\mu \mathcal{D}\phi_i e^{i \int d^4x \mathcal{L}[A,\phi_i]}$$

$$= \int \mathcal{D}\pi \mathcal{D}A_\mu \mathcal{D}\phi_i \frac{1}{f[A]} \exp\left\{i \int d^4x \left[\mathcal{L}[A,\phi_i] - \frac{1}{2\xi}\left(\partial_\mu A_\mu^a - \partial_\mu D_\mu \pi^a\right)^2\right]\right\}$$

$$= \left[\int \mathcal{D}\pi\right]\int \mathcal{D}A_\mu \mathcal{D}\phi_i \frac{1}{f[A]} \exp\left\{i \int d^4x \left[\mathcal{L}[A,\phi_i] - \frac{1}{2\xi}\left(\partial_\mu A_\mu^a\right)^2\right]\right\}, \quad (25.91)$$

where we have redefined $A_\mu^a \to A_\mu^a + D_\mu \pi^a$ in the last step. This redefinition leaves $\mathcal{L}[A,\phi_i]$ unaffected trivially and $f[A]$ unaffected since it already involves a complete integral over gauge orbits. Since this redefinition removes π from the Lagrangian, the π integral just gives an unphysical constant. The result is almost identical to the Abelian case, except now $f[A]$ depends on the gauge fields.

Before evaluating this integral, let us pause and think about what is going on. When we gauge-fix the path integral, we are no longer guaranteed that only the physical transverse modes of A_μ^a propagate. Indeed, there are additional modes π^a that have 4-derivative kinetic terms, and are therefore ghost-like (see Section 8.7). However, the path integral also tells us we have to divide out by the diagrams involving π^a, just as we divide out by the vacuum bubbles in calculating the connected Green's functions. We could just calculate this way. But it is easier to rewrite $f[A]$ in such a way that we can add Feynman diagrams instead of subtracting them.

To simplify $f[A]$, observe that in the form of Eq. (25.89), despite its dependence on A_μ, $f[A]$ is still quadratic in π, so we can perform the Gaussian integral as a functional of A_μ. We find

$$f = \sqrt{\frac{1}{\det(\partial_\mu D_\mu)^2}} \times \text{const.}, \qquad (25.92)$$

so that

$$Z[0] = \text{const.} \times \int \mathcal{D}A_\mu \mathcal{D}\phi_i [\det(\partial_\mu D_\mu)] \exp\left\{i \int d^4x \left[\mathcal{L}[A,\phi_i] - \frac{1}{2\xi}\left(\partial_\mu A_\mu^a\right)^2\right]\right\}, \qquad (25.93)$$

with the determinant in the numerator.

Now recall from Section 14.6 that a determinant can be written as a path integral over Grassmann numbers:

$$\det(\mathcal{O}) = \int \mathcal{D}\bar\psi \mathcal{D}\psi \exp\left(-i \int d^4x\, \bar\psi \mathcal{O}\psi\right). \qquad (25.94)$$

up to an infinite multiplicative constant. Thus, we can write

$$\det(\partial^\mu D_\mu) = \int \mathcal{D}\bar{c}\mathcal{D}c \exp\left(i \int d^4x\, \bar{c}(-\partial^\mu D_\mu)c\right) \qquad (25.95)$$

for Grassmann-valued fields c and $\bar{c}$. Thus, we finally have the gauge-fixed path integral for a non-Abelian gauge theory:

$$Z[0] = \text{const.} \times \int \mathcal{D}A_\mu \mathcal{D}\phi_i \mathcal{D}\bar{c}\mathcal{D}c \exp\left\{ i\int d^4x \left[\mathcal{L}[A,\phi_i] - \frac{1}{2\xi}(\partial_\mu A_\mu^a)^2 - \bar{c}^a \partial^\mu D_\mu c^a \right]\right\}. \tag{25.96}$$

Here c^a and $\bar{c}^a$ are anticommuting Lorentz scalars, called **Faddeev–Popov ghosts** and **anti-ghosts** respectively. There is one ghost and one anti-ghost for each gauge field. The sector of this gauge-fixed Lagrangian involving just the non-Abelian gauge bosons is

$$\mathcal{L}_{R_\xi} = -\frac{1}{4}(F_{\mu\nu}^a)^2 - \frac{1}{2\xi}(\partial_\mu A_\mu^a)^2 + (\partial_\mu \bar{c}^a)(\delta^{ac}\partial_\mu + gf^{abc}A_\mu^b)c^c. \tag{25.97}$$

This is the **Faddeev–Popov Lagrangian**. The resulting propagator is

$$\nu;b \;\underset{p}{\text{\textcircled{000000000}}}\; \mu;a \;=\; i\frac{-g^{\mu\nu} + (1-\xi)\frac{p^\mu p^\nu}{p^2}}{p^2 + i\varepsilon}\delta^{ab}, \tag{25.98}$$

which is the same as the photon propagator up to the δ^{ab} factor. The ghost propagator and the interaction vertices for Yang–Mills theory are given in Section 26.1.

Ghosts are unphysical since they violate the spin-statistics theorem. As we showed in Chapter 12, there cannot be physical states that anticommute and transform like scalars under the Lorentz group. However, nothing prevents ghosts from appearing in the path integral. As with physical fields, one can expand the path integral in perturbation theory, generating Feynman diagrams involving these ghosts. For S-matrix elements of physical states, the ghosts will appear in internal lines.

One way to understand why ghosts have to be fermionic is so that they can cancel unphysical degrees of freedom of the gluons in loops. When we take the gluons off-shell, we are no longer guaranteed to have the right number of physical degrees of freedom. The ghosts are fermionic because they need a -1 in loops to cancel the $+1$ from the unphysical polarizations.

One can generalize the above argument to allow for integrals along arbitrary gauge orbits to be factored out. Begin by picking a gauge, that is, some element of the equivalence class of gauge fields, and call it $\hat{A}_\mu^a$. Fields in this gauge satisfy some constraint $G[\hat{A}] = 0$, where G is a functional. For example, in Lorenz gauge we could take $G[A] = \partial_\mu A_\mu^a$. Any gauge field can be written as $A_\mu^a = \hat{A}_\mu^a + D_\mu \pi^a$ for some π^a. In this way, we should be able to split the path integral into an integral over $\hat{A}_\mu^a$ and an integral over π^a. To do so, we observe that

$$1 = \int \mathcal{D}\pi\, \delta\Big(G\big[A_\mu^a + D_\mu \pi^a\big]\Big) \det\left(\frac{\delta G\big[A_\mu^a + D_\mu \pi^a\big]}{\delta \pi^b}\right). \tag{25.99}$$

For example, with $G[A_\mu^a] = \partial_\mu A_\mu^a$ we find

$$\det\left(\frac{\delta G\big[A_\mu^a + D_\mu \pi^a\big]}{\delta \pi^b}\right) = \det(\partial^\mu D_\mu)$$

$$= \int \mathcal{D}\bar{c}\mathcal{D}c \exp\left(i\int d^4x\, \bar{c}^a(-\partial^\mu D_\mu)c^a \right) \tag{25.100}$$

as above. Folding the 1 in Eq. (25.99) into the path integral gives

$$Z[0] = \text{const.} \times \int \mathcal{D}\pi \int \mathcal{D}A_\mu \mathcal{D}\phi_i \delta\Big(G\big[A_\mu^a + D_\mu \pi^a\big]\Big) \det\left(\frac{\delta G\big[A_\mu^a + D_\mu \pi^a\big]}{\delta \pi^b}\right)$$
$$\times \exp\left(i \int d^4x \, \mathcal{L}[A, \phi_i]\right). \quad (25.101)$$

Now we can shift $A_\mu^a \to A_\mu^a - D_\mu \pi^a$. This is a linear shift, accompanied by a global transformation, so the measure does not change. Assuming the determinant is independent of π, we then have

$$Z[0] = \text{const.} \times \left[\int \mathcal{D}\pi\right] \int \mathcal{D}A_\mu \mathcal{D}\phi_i \delta\Big(G\big[A_\mu^a\big]\Big) \det\left(\frac{\delta G\big[A_\mu^a + D_\mu \pi^a\big]}{\delta \pi^b}\right)\Bigg|_{\pi \to 0}$$
$$\times \exp\left(i \int d^4x \, \mathcal{L}[A, \phi_i]\right). \quad (25.102)$$

The π integral is now just an (infinite) constant. Now we note that if we shift G by a constant the determinant does not change. So we can average over a Gaussian-weighted selection of shifts using

$$\int \mathcal{D}\chi \exp\left(-i \int d^4x \frac{\chi^2}{2\xi}\right) \delta\Big(G\big[A_\mu^a\big] - \chi\Big) = \exp\left(-i \int d^4x \frac{1}{2\xi} G\big[A_\mu^a\big]^2\right). \quad (25.103)$$

Thus,

$$Z[0] = \text{const.} \times \int \mathcal{D}A_\mu \mathcal{D}\phi_i \det\left(\frac{\delta G\big[A_\mu^a + D_\mu \pi^a\big]}{\delta \pi^b}\right)\Bigg|_{\pi \to 0}$$
$$\times \exp\left[i \int d^4x \left(\mathcal{L}[A, \phi_i] - \frac{1}{2\xi} G\big[A_\mu^a\big]^2\right)\right]. \quad (25.104)$$

For $G\big[A_\mu^a\big] = \partial_\mu A_\mu$ this reduces to the Lagrangian for the covariant gauges discussed above.

25.4.2 BRST invariance

Since Faddeev–Popov ghosts are so strange, it is worth considering why they must be there from another perspective. Recall that to be able to renormalize a theory, we need to include every operator consistent with symmetries, or else there may be infinities for which we do not have appropriate counterterms. In QED, although gauge invariance was broken by the $\frac{1}{2\xi}(\partial_\mu A_\mu)^2$ term, we still used gauge invariance to forbid additional gauge-violating terms. We were able to get away with this in QED because A_μ coupled to a conserved current, so modifying only its kinetic term had no effect. In QCD, the gauge fields do not couple to a conserved current because of self-interactions of the gauge fields, so the $\frac{1}{2\xi}(\partial_\mu A_\mu^a)^2$ term, with its associated Faddeev–Popov ghosts, is not so clearly an innocuous deformation. Remarkably, when ghosts are included, the QCD Lagrangian retains an exact global symmetry called BRST invariance (named after Becchi, Rouet, Stora and Tyutin).

BRST invariance is therefore critical in proving renormalizability of non-Abelian gauge theories.

BRST invariance can be seen even in QED, where it is a little simpler. Taking the Abelian limit of the Faddeev–Popov Lagrangian with scalar matter fields ϕ_i included, we find

$$\mathcal{L} = -\frac{1}{4}F_{\mu\nu}^2 + (D_\mu\phi_i^\star)(D_\mu\phi_i) - m^2\phi_i^\star\phi_i - \frac{1}{2\xi}(\partial_\mu A_\mu)^2 - \bar{c}\,\Box\, c. \qquad (25.105)$$

The term $\frac{1}{2\xi}(\partial_\mu A_\mu)^2$ breaks the gauge symmetry down to a residual symmetry under which $A_\mu \to A_\mu + \frac{1}{e}\partial_\mu\alpha$ for fields $\alpha(x)$ satisfying $\Box\alpha = 0$. This is a residual symmetry of the entire Lagrangian. Now note that the equations of motion for c and $\bar{c}$ are $\Box c = \Box\bar{c} = 0$. Thus, instead of gauge transforming with a parameter α, we can gauge transform with $\alpha(x) = \theta c(x)$ for any Grassmann number θ. In other words, the Lagrangian is invariant under

$$\phi_i(x) \to e^{i\theta c(x)}\phi_i(x) = \phi_i(x) + i\theta c(x)\phi_i(x), \qquad (25.106)$$

$$A_\mu(x) \to A_\mu(x) + \frac{1}{e}\theta\partial_\mu c(x), \qquad (25.107)$$

as long as the equations of motion $\Box c = \Box\bar{c} = 0$ can be used. If we do not use the equations of motion, we find the first three terms in Eq. (25.105) are invariant; however,

$$(\partial_\mu A_\mu)^2 \to (\partial_\mu A_\mu)^2 + \frac{2}{e}(\partial_\mu A_\mu)(\theta\Box c) + \frac{1}{e^2}(\theta\Box c)(\theta\Box c). \qquad (25.108)$$

Since θ is Grassmann, $\theta^2 = 0$ and the last term vanishes. Thus, if we also take

$$\bar{c}(x) \to \bar{c}(x) - \frac{1}{e}\theta\frac{1}{\xi}\partial_\mu A_\mu(x), \qquad (25.109)$$

then the Lagrangian is invariant without using the equations of motion. This is an example of **BRST invariance**. Since $\theta c(x)$ acts like $\alpha(x)$ for the A_μ and ϕ_i transformations, BRST is a generalization of gauge invariance that holds despite the gauge-breaking $\frac{1}{2\xi}(\partial_\mu A_\mu)^2$ term.

In the non-Abelian case, the Lagrangian is

$$\mathcal{L}_{\text{FP}} = \mathcal{L}\big[A_\mu^a, \phi_i\big] - \frac{1}{2\xi}\big(\partial_\mu A_\mu^a\big)^2 + (\partial_\mu\bar{c}^a)(D_\mu c^a), \qquad (25.110)$$

where $D_\mu c^a = \partial_\mu c^a + gf^{abc}A_\mu^b c^c$. Thus, we can proceed as in the Abelian case, defining the transformations as

$$\phi_i \to \phi_i + i\theta c^a T_{ij}^a\phi_j, \qquad (25.111)$$

$$A_\mu^a \to A_\mu^a + \frac{1}{g}\theta D_\mu c^a, \qquad (25.112)$$

$$\bar{c}^a \to \bar{c}^a - \frac{1}{g}\theta\frac{1}{\xi}\partial_\mu A_\mu^a. \qquad (25.113)$$

As in the Abelian case, these are just gauge transformations with $\alpha^a(x) = \theta c^a(x)$ when acting on ϕ_i or A_μ^a; thus, $\mathcal{L}\big[A_\mu^a, \phi_i\big]$ is invariant. Also the transformation of $(\partial_\mu\bar{c}^a)$ is designed to exactly cancel the transformation of $-\frac{1}{2\xi}\left(\partial_\mu A_\mu^a\right)^2$. However, unlike in the Abelian case, the $D_\mu c^a$ term is not invariant, because of the A_μ^a hidden in $D_\mu c^a$:

$$D_\mu c^a \to D_\mu c^a + \theta f^{abc}\left(D_\mu c^b\right)c^c. \tag{25.114}$$

To make this covariant derivative invariant, we will need to transform c^a as well. This can be done by defining the BRST transformation for c^a as

$$c^a \to c^a - \frac{1}{2}\theta f^{abc}c^b c^c. \tag{25.115}$$

Note that nowhere did we use that c^a and $\bar{c}^a$ were related in any way (as ψ and $\bar{\psi}$ are related by charge-conjugation invariance); thus, we are free to give them different transformation laws. To check this, we compute

$$D_\mu c^a \to D_\mu c^a + \theta f^{abc}\left(D_\mu c^b\right)c^c - \theta f^{abc}\left[\frac{1}{2}\left(\partial_\mu c^b\right)c^c + \frac{1}{2}c^b(\partial_\mu c^c) + \frac{g}{2}A_\mu^b f^{cde}c^d c^e\right]. \tag{25.116}$$

The first two terms in brackets are equal, since $\left(\partial_\mu c^b\right)c^c = -c^c\left(\partial_\mu c^b\right)$ and $f^{abc} = -f^{acb}$. For the last term, we note that by the Jacobi identity in Eq. (25.11),

$$
\begin{aligned}
f^{abc}f^{cde}A_\mu^b c^d c^e &= -f^{bdc}f^{cae}A_\mu^b c^d c^e - f^{dac}f^{cbe}A_\mu^b c^d c^e \\
&= 2f^{abc}f^{bed}A_\mu^e c^d c^c,
\end{aligned} \tag{25.117}
$$

where we have used antisymmetry of f^{abc} and a fair amount of index relabeling to get to the second line. We therefore have that

$$D_\mu c^a \to D_\mu c^a + \theta f^{abc}\left(D_\mu c^b\right)c^c - \theta f^{abc}\left[\left(\partial_\mu c^b\right)c^c + g f^{bed}A_\mu^e c^d c^c\right] = D_\mu c^a \tag{25.118}$$

and that $\mathcal{L}_{\text{FP}}$ is invariant.

We conclude that the Faddeev–Popov Lagrangian is BRST invariant. BRST invariance is a global symmetry parametrized by a Grassmann number θ under which fields transform as in Eqs. (25.111) to (25.115).

One implication of BRST invariance is for renormalization. Since BRST invariance is an exact symmetry of the Lagrangian, it will be preserved in loops. Thus, one will not need any counterterms that violate BRST invariance. Since the Faddeev–Popov ghosts and anti-ghosts are critical in establishing BRST invariance of the gauge-fixed non-Abelian Lagrangian, this strongly constrains how they can appear at higher orders. The proof of renormalizability for non-Abelian theories shows that all the infinities are canceled with the finite number of counterterms corresponding to terms in the most general BRST-invariant Lagrangian.

By the way, BRST invariance has a sophisticated mathematical foundation and many formal applications. For example, if one writes the transformations as $\phi_i \to \phi_i + \theta\Delta\phi_i$, $A_\mu^a \to A_\mu^a + \theta\Delta A_\mu^a$ and so on, the operator Δ turns out to be **nilpotent**, $\Delta^2 = 0$. You can check this in Problem 25.7. Δ is sometimes called the **Slavnov operator**. Thus, all states that are **exact**, $|X\rangle = \Delta|Y\rangle$, for some $|Y\rangle$ are **closed**, $\Delta|X\rangle = 0$. This establishes a **cohomology**: there is a well-defined equivalence class of states that are closed but not exact. It turns out that one can identify physical states with the cohomology of Δ. Shifting an element of this class by an exact state does not change the physical state. This is a precise mathematical way of saying statements such as the electric and magnetic fields $\vec{E}$ and $\vec{B}$ correspond to an equivalence class of potentials A_μ for which $F_{\mu\nu} = \partial_\mu A_\nu - \partial_\nu A_\mu$. A physical but heuristic discussion can be found in [Peskin and Schroeder, 1995, Section 16.4] and a more formal treatment in [Weinberg, 1996].

25.4.3 Axial gauges

The whole discussion of ghosts and BRST in the previous sections makes it seem like these are crucial things. Ghosts are in fact crucial, in the sense that you have to include them to get the right answer in perturbation theory, at least in covariant gauges. But ghosts are also unphysical. They arise as an artifact of insisting on a gauge in which the gluon propagator is Lorentz covariant. If we never tried to embed the two physical polarizations of a massless spin-1 particle into a field $A_\mu(x)$ we would never have had to deal with ghosts. Or, if we restricted to gauge-invariant objects, such as the field strength $F_{\mu\nu}$ (as is done on the lattice), we also would not have to deal with ghosts.

An alternative to dealing with ghosts is to choose a gauge in which the ghosts decouple from the physical particles, and hence can be ignored. All such gauges are explicitly Lorentz violating. The most important class of non-covariant gauges are the **axial gauges**. The axial-gauge gauge-fixing and ghost terms are

$$\mathcal{L}_{\text{gauge-fixing}} + \mathcal{L}_{\text{ghost}} = -\frac{1}{2\lambda}(r^\mu A_\mu^a)^2 + \bar{c}^a r^\mu (\delta^{ac}\partial_\mu + gf^{abc}A_\mu^b)c^c \qquad (25.119)$$

where there are now two parameters, λ (a number) and r^μ (a 4-vector). For example, $r^\mu = (1,0,0,0)$ would put $-\frac{1}{2\lambda}A_0^2$ in the Lagrangian; then taking the limit $\lambda \to 0$ forces $A_0 = 0$, which is axial gauge in electromagnetism.

The propagator following from this modification is

$$i\Pi_{\text{axial}}^{\mu\nu ab} = \frac{i}{p^2 + i\varepsilon}\left[-g^{\mu\nu} + \frac{r^\mu p^\nu + r^\nu p^\mu}{rp} - \frac{(r^2 + \lambda p^2)p^\mu p^\nu}{(rp)^2} \right]\delta^{ab}. \qquad (25.120)$$

It satisfies $p^\mu \Pi_{\text{axial}}^{\mu\nu ab} = 0$ when p^μ is on-shell ($p^2 = 0$). In addition, for $\lambda = 0$, the axial propagator satisfies $r^\mu \Pi_{\text{axial}}^{\mu\nu ab} = 0$. Then, since the ghost-antighost-gluon vertex is proportional to r^μ, it will vanish when contracted with gluon propagator. Thus, for $\lambda = 0$ and any r^μ, the ghosts decouple.

A special case known as **lightcone gauge** has $r^2 = 0$ (r^μ is light-like) and $\lambda = 0$. Then,

$$i\Pi_{\text{lightcone}}^{\mu\nu ab} = \frac{i}{p^2 + i\varepsilon}\left[-g^{\mu\nu} + \frac{r^\mu p^\nu + p^\mu r^\nu}{rp} \right]\delta^{ab}. \qquad (25.121)$$

In lightcone gauge, there are only two physical polarizations: those transverse to the $r-p$ plane, summed over in the numerator of this propagator. That is, the numerator is the polarization sum of transverse modes in a particular basis. Since only two polarizations are being propagated, we do not need ghosts to cancel the unphysical polarizations, which explains why they decouple. In contrast, in the Feynman-gauge propagator ($\xi = 1$), the numerator $-g^{\mu\nu}$ is the sum over four polarizations and so ghosts *are* needed to cancel the unphysical modes.

Axial gauges make it clear that ghosts are not strictly necessary to describe non-Abelian gauge theories. In practice, unless you are in a situation where there is some natural direction r^μ, axial gauges are very unwieldy. Ghosts are a formal annoyance, but from a practical point of view they are not that bad. On the other hand, for external polarizations, having the freedom to choose r^μ separately for each gluon (corresponding to a basis of transverse polarizations) can be extremely useful. We will show how this freedom can be exploited to great practical advantage in Chapter 27 (on the spinor-helicity formalism). Further discussion of non-covariant gauges can be found in [Liebbrandt, 1987].

25.5 Lattice gauge theories

Before going on to perturbative calculations in non-Abelian gauge theories, it is worth discussing the only systematically improvable method for computing non-perturbative quantities in gauge theories: the lattice. Lattice simulations are useful when gauge fields are strongly interacting, as is the case for QCD at low energies. There are enormous practical difficulties with lattice simulations, and many open theoretical questions (such as how to simulate chiral gauge theories). Here we only superficially summarize the approach to lattice QCD pioneered by Wilson. This discussion is adapted from [Gattringer and Lang, 2010].

Let us define a four-dimensional lattice with n_{sites} sites in each dimension spaced a distance a apart. We denote the lattice sites by n. On the lattice, quantum field theory reduces to quantum mechanics with n_{sites}^4 fields. Matter fields $\vec{\phi}(n)$ naturally reside on the lattice sites. We denote by $\hat{\mu}$ a vector of unit length a in the μ direction, so $\vec{\phi}(n + \hat{\mu})$ and $\vec{\phi}(n)$ are the field values on nearest-neighbor sites. Gauge transformations are also discrete, so we can rotate fields by group elements

$$\vec{\phi}(n) \to U(n) \cdot \vec{\phi}(n), \qquad (25.122)$$

where $U(n) = \exp(i\alpha^a(n)T^a)$ defined separately on each site. To be able to compare field values at different sites in a gauge-invariant way, we need the discrete version of the Wilson line discussed in Section 25.2. We therefore define new fields $W_\mu(n)$ transforming as

$$W_\mu(n) \to U(n)W_\mu(n)U^\dagger(n + \hat{\mu}). \qquad (25.123)$$

Then,

$$\vec{\phi}^\dagger(n)W_\mu(n)\vec{\phi}(n + \hat{\mu}) \to \vec{\phi}^\dagger(n)U^\dagger(n)\,U(n)W_\mu(n)U^\dagger(n + \hat{\mu})\,U(n + \hat{\mu})\vec{\phi}(n + \hat{\mu})$$
$$= \vec{\phi}^\dagger(n)W_\mu(n)\vec{\phi}(n + \hat{\mu}). \qquad (25.124)$$

Products of fields on distant lattice sites can be multiplied in a gauge-invariant way by multiplying together $W_\mu(n_i)$ factors along any path between the sites. The $W_\mu(n)$ can be thought to live between neighboring sites; thus they are called **link fields**. To connect any two sites, it is enough to have one link between every neighboring pair. For convenience,

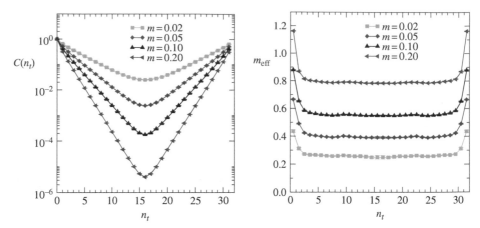

Fig. 25.3   The pion mass can be extracted from the scaling behavior of the correlation function $C(n_t) = \langle \mathcal{O}(0)\mathcal{O}(n_t) \rangle \sim \exp(-m_{\text{eff}} n_t)$, where n_t is the number of sites and m_{eff} is the effective mass in lattice units. The left figure shows $\ln C(n_t)$ and the right plot its slope. One learns from these plots, for example, that as the quark mass is quadrupled from 0.05 to 0.2, the pion mass doubles. That $m_\pi^2 \propto m_q$ will be derived with chiral pertubation theory in Chapter 28. (Figure from [Gattringer and Lang, 2010].)

Problems

25.1 Check that the Yang–Mills Lagrangian in Eq. (25.6) is gauge invariant by explicitly inserting the transformation in Eq. (25.67).

25.2 Derive Eqs. (25.20) to (25.22).

25.3 Semisimplicity.

    (a) The key reason that semisimple Lie algebras are of interest in physics is that all finite-dimensional representations of semisimple algebras are Hermitian. Prove this fact.

    (b) Prove that the Lorentz algebra $so(1,3)$ is *not* semisimple, but its complexification $su(2) \oplus su(2)$ is.

    (c) Show that the unitary group $U(N)$ is neither simple nor semisimple.

    (d) An important algebra that is not semisimple is the **Heisenberg algebra**. It has three generators p, x and $\hbar$ satisfying $[x,p] = i\hbar$ and $[x,\hbar] = [p,\hbar] = 0$. Find a three-dimensional matrix representation of this algebra. Is the algebra simple? Is it semisimple?

25.4 Anomaly coefficients.

    (a) Show that anomaly coefficients for complex conjugate representations are equal with opposite sign: $A(R) = -A(\bar{R})$. Conclude that the anomaly coefficient for a real representation is zero.

    (b) Show that for reducible representations $A(R_1 \oplus R_2) = A(R_1) + A(R_2)$.

    (c) Show that for tensor product representations $A(R_1 \otimes R_2) = A(R_1)d(R_2) + d(R_1)A(R_2)$.

(d) What is $A(\mathbf{10})$ for $SU(4)$? You can use that $\mathbf{4} \otimes \mathbf{4} = \mathbf{6} + \mathbf{10}$, with $\mathbf{6}$ being a real representation.

25.5 Check that the Noether current in Eq. (25.80), $J^a_\mu = -\bar{\psi}_i \gamma^\mu T^a_{ij} \psi_j + f^{abc} A^b_\nu F^c_{\mu\nu}$, is conserved on the equations of motion.

25.6 Show that the path ordering is necessary in the definition of the non-Abelian Wilson line, Eq. (25.57), for the transformation property in Eq. (25.63) to be satisfied.

(a) First show gauge invariance to leading non-trivial order in perturbation theory. That is, show that the θ-functions are necessary in Eq. (25.58).

(b) Show that the Wilson line with path ordering satisfies Eq. (25.63) exactly.

25.7 Check that the Slavnov operator Δ, defined so that $\phi \to \phi + \theta \Delta \phi$ for the various fields under BRST transformations, is nilpotent $\Delta^2 = 0$.

26 Quantum Yang–Mills theory

In the previous chapter, we introduced Yang–Mills theory as the natural generalization of electrodynamics to systems with many fields. If we have N fields ϕ_i, then the Lagrangian $\mathcal{L} = -\phi_i^\star \Box \phi_i$ is invariant under a global $SU(N)$ symmetry, under which $\phi_i \to U_{ij}\phi_j$ for $U \in SU(N)$. In Yang–Mills theory there are massless spin-1 particles which transform in the adjoint representation of $SU(N)$. Since $SU(N)$ is a non-Abelian group, Yang–Mills theories are often called non-Abelian gauge theories. It is perhaps worth emphasizing that the important feature of these theories is not gauge invariance (which is an unphysical feature of Lagrangians) but the existence of massless spin-1 particles that are charged, that is, they carry quantum numbers. In the next chapter we will discuss a method for performing S-matrix calculations in Yang–Mills theories that sidesteps gauge invariance altogether. These caveats aside, introducing a local Lagrangian for Yang–Mills theory with gauge invariance is by far the most powerful and general method for studying these theories. Thus, we focus in this chapter on perturbative calculations in non-Abelian gauge theories.

In Chapter 25, gauge invariance was motivated as allowing us to choose a different $SU(N)$ convention at different points. We saw that we could compare field values at different points in a convention-independent way if we used Wilson lines $W(x, y)$ defined so that $W(x, y)\, \phi(y)$ transforms as $\phi(x)$. Expanding such a Wilson line out for small deviations led to $W(x, x + \delta x) = 1 - ig\delta x^\mu A_\mu^a T^a$, where T^a are the generators of $SU(N)$ in the fundamental representation. In this way, we found that a local theory needs one gauge field A_μ^a for each generator, and thus the gauge fields A_μ^a transform in the adjoint representation of $SU(N)$.

Next, we found that, in computing the propagator for the gauge boson, we had to gauge-fix, as in QED. But in the non-Abelian case, the covariant gauge-fixing (R_ξ gauges), when done properly through the path integral, generated new particles called Faddeev–Popov ghosts, which have spin 0 but fermionic statistics. These particles never appear as external states but must be included in internal lines for consistency. That we need these ghosts is a horrible consequence of the Lagrangian formulation of field theory. There is no observable consequence of ghosts, we just need them to describe an interacting theory of massless spin-1 particles using a local manifestly Lorentz-invariant Lagrangian. Alternative formulations (such as the lattice) do not require ghosts. Perturbative gauge theories in certain non-covariant gauges, such as lightcone gauge, are also ghost free. However, to maintain manifest Lorentz invariance in a perturbative gauge theory, it seems ghosts are unavoidable.

In this chapter we will perform some perturbative calculations in the non-Abelian theory. This will allow us to understand both the theory of the strong interactions, QCD, which is a non-Abelian gauge theory with gauge group $SU(3)$, and the theory of the weak interactions, which is based on $SU(2)$. For simplicity, we will refer to the non-Abelian gauge theory as QCD, and the gauge bosons as gluons. Our results will be more general than this, but it is helpful to talk about QCD for concreteness.

We will discuss some tree-level and 1-loop results, including probably the most important calculation in QCD – vacuum polarization. We will find that the QCD gauge coupling runs in the opposite direction from QED: it gets larger at larger distances. This makes the phenomenology of QCD completely different from the phenomenology of QED. In the next chapter, we will return to tree-level graphs through the spinor-helicity formalism.

Due to the many possible contractions in each vertex, calculations in non-Abelian gauge theories quickly get intractably complicated. For example, the process $gg \to ggg$ even at tree-level contains around $10\,000$ terms. Part of the reason things are so complicated is because there is a huge redundancy when we sum over off-shell intermediate states. In the next chapter, we will see that there is a smarter way to organize the tree-level structure. In this chapter we concentrate on processes with few gluons so that the number of terms is manageable and we can perform the calculations using traditional Feynman rules.

26.1 Feynman rules

The first step in performing perturbative calculations in a non-Abelian gauge theory is to work out the Feynman rules. The $SU(N)$-invariant Lagrangian for a set of N fermions and N scalars interacting with non-Abelian gauge fields is

$$
\begin{aligned}
\mathcal{L} = {}&-\frac{1}{4}\left(F_{\mu\nu}^a\right)^2 - \frac{1}{2\xi}\left(\partial_\mu A_\mu^a\right)^2 + \left(\partial_\mu \bar{c}^a\right)\left(\delta^{ac}\partial_\mu + g f^{abc} A_\mu^b\right) c^c \\
&+ \bar{\psi}_i\left(\delta_{ij} i\slashed{\partial} + g\slashed{A}^a T_{ij}^a - m\delta_{ij}\right)\psi_j \\
&+ \left[\left(\delta_{ki}\partial_\mu - ig A_\mu^a T_{ki}^a\right)\phi_i\right]^*\left[\left(\delta_{kj}\partial_\mu - ig A_\mu^a T_{kj}^a\right)\phi_j\right] - M^2 \phi_i^\star \phi_i,
\end{aligned}
\tag{26.1}
$$

where c^a and $\bar{c}^a$ are the Faddeev–Popov ghosts and anti-ghosts respectively and

$$
F_{\mu\nu}^a = \partial_\mu A_\nu^a - \partial_\nu A_\mu^a + g f^{abc} A_\mu^b A_\nu^c.
\tag{26.2}
$$

We have included scalars in this Lagrangian for generality, even though we have observed no scalar states in nature that are colored (charged under QCD). Many theories, such as supersymmetric QCD, do have colored scalars. The Higgs doublet in the Standard Model is an example of a scalar field charged under the weak gauge group $SU(2)$.

The kinetic terms from the QCD Lagrangian are

$$
\mathcal{L}_{\text{kin}} = -\frac{1}{4}\left(\partial_\mu A_\nu^a - \partial_\nu A_\mu^a\right)^2 - \frac{1}{2\xi}\left(\partial_\mu A_\mu^a\right)^2 + \bar{\psi}_i\left(i\slashed{\partial} - m\right)\psi_i - \phi_i^\star\left(\Box + M^2\right)\phi_i - \bar{c}^a \Box c^a.
\tag{26.3}
$$

Since the kinetic term for the gauge bosons is just the sum over $N^2 - 1$ free gauge bosons, the propagator for each should be just the same as the propagator for a photon. Since we chose the basis of group generators to be orthogonal there is no kinetic mixing between gluons. So the propagator is

$$\underset{p}{\underset{\nu;b}{\text{\it\small 00000000}}}\,{}^{\mu;a} \quad = \quad i\frac{-g^{\mu\nu} + (1 - \xi)\frac{p^\mu p^\nu}{p^2}}{p^2 + i\varepsilon}\delta^{ab}. \tag{26.4}$$

The gluon propagator is the photon propagator with an extra δ^{ab} factor. When gluons appear as intermediate states, one must sum over all possible gluons, which gives a sum over a and b.

The ghost propagator is

$$b \underset{p}{\cdots\blacktriangleright\cdots} a \quad = \quad \frac{i\delta^{ab}}{p^2 + i\varepsilon}. \tag{26.5}$$

The propagators for colored fermions,

$$j \underset{p}{\xrightarrow{\hspace{2cm}}} i \quad = \quad \frac{i\delta^{ij}}{\not{p} - m + i\varepsilon}, \tag{26.6}$$

and for colored scalars,

$$j \underset{p}{\dashrightarrow} i \quad = \quad \frac{i\delta^{ij}}{p^2 - M^2 + i\varepsilon}, \tag{26.7}$$

are the same as in QED but with δ_{ij} factors, where i, j refer to fundamental color indices. These δ^{ab} and δ^{ij} factors just say that the color that comes in is the same as the color that comes out – color is conserved. As with the gluon, we must sum over colors when these propagators appear as intermediate states.

The interactions are

$$\mathcal{L}_{\text{int}} = -gf^{abc}(\partial_\mu A_\nu^a)A_\mu^b A_\nu^c - \frac{1}{4}g^2\left(f^{eab}A_\mu^a A_\nu^b\right)\left(f^{ecd}A_\mu^c A_\nu^d\right) + gf^{abc}(\partial_\mu \bar{c}^a)A_\mu^b c^c$$
$$+ gA_\mu^a \bar{\psi}_i \gamma^\mu T_{ij}^a \psi_j + igA_\mu^a T_{ij}^a(\phi_i^\star \partial_\mu \phi_j - \phi_j \partial_\mu \phi_i^\star) + g^2\phi_i^\star A_\mu^a T_{ik}^a T_{kj}^b A_\mu^b \phi_j, \tag{26.8}$$

where we have used that $\left(T_{ij}^a\right)^\star = T_{ji}^a$ for $\text{SU}(N)$. For the triple-gluon vertex, the derivative can act on any of the gluons, giving the Feynman rule

$$= gf^{abc}[g^{\mu\nu}(k - p)^\rho + g^{\nu\rho}(p - q)^\mu + g^{\rho\mu}(q - k)^\nu]. \tag{26.9}$$

Note that we take all the momentum incoming, so $p+k+q=0$. This is different from the convention we used for QED, where all momenta were going to the right. The four-gluon vertex gives

$$= -ig^2 \times \left[f^{abe} f^{cde} \left(g^{\mu\rho}g^{\nu\sigma} - g^{\mu\sigma}g^{\nu\rho} \right) \right.$$
$$+ f^{ace} f^{bde} \left(g^{\mu\nu}g^{\rho\sigma} - g^{\mu\sigma}g^{\nu\rho} \right)$$
$$\left. + f^{ade} f^{bce} \left(g^{\mu\nu}g^{\rho\sigma} - g^{\mu\rho}g^{\nu\sigma} \right) \right]. \tag{26.10}$$

The ghost vertex Feynman rule is

$$= -g f^{abc} p^\mu. \tag{26.11}$$

Note that there is only one contraction (since ghosts and anti-ghosts are different), in contrast to the scalar QED vertex.

There is one vertex for interaction with a fermion, which gives

$$= ig\gamma^\mu T^a_{ij}. \tag{26.12}$$

As in QED, the orientation of the vertex in a Feynman diagram does not matter. The vertex gets a factor of $ig\gamma^\mu T^a_{ij}$, with i the color of the quark with the arrow pointing away from the vertex and j the other color.

Finally, there are two vertices for the scalar, just as in scalar QED. These are

$$= ig(k^\mu + q^\mu) T^a_{ij} \tag{26.13}$$

and

$$= ig^2 T^a_{ik} T^b_{kj} g^{\mu\nu}. \tag{26.14}$$

Quark	down	up	strange	charm	bottom	top
$\overline{\text{MS}}$ mass (MeV)	4.70	2.15	93.5	1270	4180	163 000
Charge	$-1/3$	$+2/3$	$-1/3$	$+2/3$	$-1/3$	$+2/3$

Table 26.1 Quark masses and charges in the $\overline{\text{MS}}$ scheme [Particle Data Group (Beringer *et al.*), 2012].

acts just like a massive photon with its own set of charges. The photon couples to anything charged, such as the quarks. There are six flavors of quarks each transforming under the fundamental representation of SU(3) whose masses in the $\overline{\text{MS}}$ scheme and charges are shown in Table 26.1. Note that in the first generation, the charge $\frac{2}{3}$ quark (the up) is lighter than the charge $-\frac{1}{3}$ quark (the down), while in the second and third generations the opposite is true. There are many subtleties with quark-mass definitions, since quarks do not appear as asymptotic states and therefore do not have well-defined pole masses.

In Chapter 20 we calculated that the total cross section for unpolarized $e^+e^- \to \gamma^* \to \mu^+\mu^-$ scattering at tree-level is

$$\sigma\left(e^+e^- \to \mu^+\mu^-\right) = \frac{4\pi\alpha_e^2}{3E_{\text{CM}}^2} \equiv \sigma_0. \tag{26.21}$$

The Feynman diagram for quark production is identical, except now we must factor in the charges of the quarks and sum over quark colors. Only color singlet pairs such as red/anti-red can be produced. And thus we get a factor of $N = 3$ from the color sum. Thus, at tree-level,

$$\sigma\left(e^+e^- \to \bar{q}q\right) = 3\sigma_0\left(\left(\frac{2}{3}\right)^2 + \left(-\frac{1}{3}\right)^2 + \left(-\frac{1}{3}\right)^2 + \left(\frac{2}{3}\right)^2 + \left(-\frac{1}{3}\right)^2 + \left(\frac{2}{3}\right)^2\right). \tag{26.22}$$

The center-of-mass energies at LEP (an e^+e^- collider at CERN), which ranged from 90 to 205 GeV, were above the bottom-quark pair-production threshold ($\sim$9 GeV) but below the top-quark pair-production threshold ($\sim$350 GeV), and so only five quarks should be summed over to compare theory to LEP data. The theory prediction for LEP is therefore

$$R_{\text{had}}^\gamma \equiv \frac{\sigma(e^+e^- \to \gamma \to \text{hadrons})}{\sigma(e^+e^- \to \gamma \to \mu^+\mu^-)} = R_{\text{had}}^{\gamma 0} + \mathcal{O}(\alpha_s), \tag{26.23}$$

where

$$R_{\text{had}}^{\gamma 0} = \sum_{\text{colors}} \sum_{q=u}^{b} Q_q^2 = 3.67. \tag{26.24}$$

The equivalent ratio including also an intermediate Z boson (see Chapter 29) is $R_{\text{had}}^0 = 20.09$. We can compare this directly to the measured value of

$$R_{\text{had}} \equiv \frac{\sigma(e^+e^- \to \text{hadrons})}{\sigma(e^+e^- \to \mu^+\mu^-)}. \tag{26.25}$$

The measured value at LEP 1, which ran at $E_{\text{CM}} = M_Z$, was $R_{\text{had}} = 20.79 \pm 0.04$, which is close to R_{had}^0 but about 3.5% higher. Nonetheless, this comparison is only consistent with there being three colors of quarks (not four or two) and five flavors. The correction at the small percentage level is what one expects from loop corrections and can be used to extract α_s from data, as we will see shortly.

By the way it is very convenient, and non-trivial, that we can sum over quarks in the theory calculation and compare to a measurement made on hadrons. The reason this works is that the quarks are produced at short distance and hadronization occurs at long distance. Because the long-distance physics is too slow to affect the short-distance physics, the total rate gets frozen-in well before hadronization, a process known as **factorization**. Factorization is one of the most profound, important, and subtle concepts in QCD. It will be discussed in more detail in Chapters 32 and 36.

26.3.2 $e^+e^- \to$ hadrons at next-to-leading order

Now let us consider the radiative corrections to the total $e^+e^- \to$ hadrons rate. Again, we will be able to steal the results for the radiative corrections to $e^+e^- \to \mu^+\mu^-$, which we computed in Chapter 20, modifying them only with the appropriate color factors when necessary.

There are two real-emission contributions at next-to-leading order given by the diagrams

$$+ \qquad . \tag{26.26}$$

These are identical to the $e^+e^- \to \mu^+\mu^-\gamma$ diagrams from Chapter 20, up to the replacement $e \to -g_s$ and the addition of a color matrix T_{ij}^a, where a is the color of the gluon and i and j are the colors of the quarks. When we square these diagrams to get the cross section for fixed external colors, we get $\mathcal{M}\mathcal{M}^\dagger \sim T_{ij}^a T_{ji}^a$ with no sum over a. We then sum over a, i and j to get a factor of $\text{Tr}[T^a T^a] = N C_F$. Integrating over phase space gives the same thing as in QED with the replacements $\alpha_e \to \alpha_s$, $\sigma_0 \to R_{\text{had}}^{\gamma 0}\sigma_0$ from Eq. (26.24) and an extra C_F factor. So we have, using Eq. (20.A.116) of Chapter 20,

$$\sigma_R = R_{\text{had}}^{\gamma 0}\sigma_0 \left(\frac{4\alpha_s}{\pi}\right) C_F \left(\frac{\tilde{\mu}^2}{Q^2}\right)^{4-d} \left(\frac{1}{\varepsilon^2} + \frac{13}{12\varepsilon} - \frac{5\pi^2}{24} + \frac{259}{144} + \mathcal{O}(\varepsilon)\right). \tag{26.27}$$

The virtual graph

$$\tag{26.28}$$

is also the same as in QED up to $e \to -g_s$, the color matrices, and the factor of $R_{\text{had}}^{\gamma 0}$. In this case, the color matrices are $T_{ik}^a T_{kj}^a$ summed over a, since the gluon propagator contains a δ^{ab} factor. The tree-level graph which contributes to the same final state has only a δ_{ij} factor. Thus, the interference between these two gives $T_{ik}^a T_{kj}^a$. Summing over final-state colors gives the same $\text{Tr}[T^aT^a] = C_F$ factor as for the real emission graphs. Thus, the virtual contribution, using the result in Eq. (20.A.102), is

$$\sigma_V = R^{\gamma 0}_{\text{had}}\sigma_0\left(\frac{4\alpha_s}{\pi}\right)\left(\frac{\tilde{\mu}^2}{Q^2}\right)^{4-d}C_F\left(-\frac{1}{\varepsilon^2}-\frac{13}{12\varepsilon}+\frac{5\pi^2}{24}-\frac{29}{18}+\mathcal{O}(\varepsilon)\right). \quad (26.29)$$

As expected, the IR divergences cancel when we sum these graphs, giving a result

$$\sigma_{\text{NLO}} = \sigma_0 + \sigma_R + \sigma_V = R^{\gamma 0}_{\text{had}}\sigma_0\left(1+\frac{3\alpha_s}{4\pi}C_F\right). \quad (26.30)$$

The Z boson couples like the photon with different charges (and different charges for left- and right-handed quarks, as we will see in Chapter 29). However, QCD corrections are the same for left- or right-handed quarks, since QCD is non-chiral. Thus we find

$$R_{\text{had}} = R^0_{\text{had}}\left(1+\frac{\alpha_s}{\pi}+\mathcal{O}(\alpha_s^2)\right). \quad (26.31)$$

Thus, to explain the 3.5% discrepancy from the LEP 1 data, we require $\frac{\alpha_s}{\pi} = 0.035$ or $\alpha_s = 0.11$. For comparison, the fine-structure constant at LEP 1 energies is $\alpha_e(m_Z) \approx \frac{1}{129} = 0.0077$.

There are many other ways to measure α_s, such as from the hadronic decay rate of the τ lepton, from deep inelastic scattering, lattice calculations, multijet rates, event shapes, etc. In each of these measurements, α_s is extracted from physical quantities. However, α_s is only defined within some regularization and subtraction scheme, so some convention must be chosen to make comparisons between these extractions useful. In particular, since α_s is scale dependent (see next section), one also needs to evolve α_s to a common scale. It is conventional to present results for α_s defined in dimensional regularization with modified minimal subtraction ($\overline{\text{MS}}$) at the scale $\mu = m_Z$. A comparison of various values of α_s extracted at different scales using different methods is shown in Figure 26.1. As of

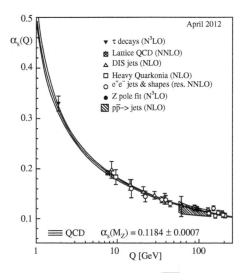

Fig. 26.1 Running coupling and data. The best fit value for the $\overline{\text{MS}}$ strong coupling constant is $\alpha_s(m_Z) = 0.1184 \pm 0.0007$. Image from [Particle Data Group (Beringer *et al.*), 2012].

this writing, the current world average is $\alpha_s = 0.1184 \pm 0.0007$. In the next section, we calculate the QCD β-function, which allows us to evolve α_s between the different scales.

26.4 Vacuum polarization

Now we turn to vacuum polarization and the QCD β-function. Unlike in QED, where only the electron loop contributed at 1-loop order, in QCD there are five contributions:

$$\mathcal{M}^{ab\mu\nu} = \mathcal{M}_F^{ab\mu\nu} + \mathcal{M}_3^{ab\mu\nu} + \mathcal{M}_4^{ab\mu\nu} + \mathcal{M}_{\mathrm{gh}}^{ab\mu\nu} + \mathcal{M}_{\mathrm{c.t.}}^{ab\mu\nu}, \qquad (26.32)$$

given by the graphs

$$(26.33)$$

The first is the fermion (or scalar) loop, the next two are gauge boson loops, the fourth is the ghost loop and the fifth is the counterterm. We will use dimensional regularization to compute these loops, since it preserves gauge invariance. We will also do the calculation in Feynman gauge, $\xi = 1$, for which the propagator is

$$i\Pi^{\mu\nu} = \delta^{ab} \frac{-ig^{\mu\nu}}{p^2 + i\varepsilon}. \qquad (26.34)$$

Results for arbitrary ξ are summarized at the end of this section (you can check them as an exercise). We will also express answers in terms of $\mathrm{SU}(N)$ Casimirs, so they will be valid for any N.

26.4.1 Fermion bubble

The fermionic loop is almost identical to the QED case. The integral is

$$= -\mathrm{tr}[T^a T^b](ig)^2 \int \frac{d^4 k}{(2\pi)^4} \frac{i}{(p-k)^2 - m^2} \frac{i}{k^2 - m^2} \mathrm{Tr}[\gamma^\mu(\slashed{k} - \slashed{p} + m)\gamma^\nu(\slashed{k} + m)]. \qquad (26.35)$$

This is exactly the same as in QED but with a color factor $\mathrm{tr}[T^a T^b] = T_F \delta^{ab}$ out front, with $T_F = \frac{1}{2}$. The result, as in the QED case, manifestly preserves gauge invariance. The result has the form

$$\mathcal{M}_F^{ab\mu\nu} = -g^2(g^{\mu\nu}p^2 - p^\mu p^\nu)\delta^{ab}\Pi_2(p^2). \qquad (26.36)$$

So, taking the loop amplitude from Eq. (16.47), we expand as in Eq. (16.45):

$$
\mathcal{M}_F^{ab\mu\nu} = -\delta^{ab} T_F \frac{g^2}{2\pi^2} \left(p^2 g^{\mu\nu} - p^\mu p^\nu \right)
$$

$$
\times \int_0^1 dx\, x(1-x) \left[\frac{2}{\varepsilon} + \ln\left(\frac{\tilde{\mu}^2}{m^2 - p^2 x(1-x)} \right) + \mathcal{O}(\varepsilon) \right]. \qquad (26.37)
$$

The pole and coefficient of $\ln \tilde{\mu}^2$ are independent of the quark mass, as expected. For massless quarks, this reduces to

$$
\mathcal{M}_F^{ab\mu\nu} = \delta^{ab} T_F \left(\frac{g^2}{16\pi^2} \right) \left(p^2 g^{\mu\nu} - p^\mu p^\nu \right) \left[-\frac{8}{3} \frac{1}{\varepsilon} - \frac{20}{9} - \frac{4}{3} \ln \frac{\tilde{\mu}^2}{-p^2} \right]. \qquad (26.38)
$$

26.4.2 Gluon bubble

For the Ward identity to be satisfied in Yang–Mills theory, the contribution from the gluon and ghost graphs should be proportional to $g^{\mu\nu} p^2 - p^\mu p^\nu$. The gluon bubble is

$\displaystyle i\mathcal{M}_3^{ab\mu\nu} = \quad = \frac{g^2}{2} \int \frac{d^4k}{(2\pi)^4} \frac{-i}{k^2} \frac{-i}{(k-p)^2} f^{ace} f^{bdf} \delta^{cf} \delta^{ed} N^{\mu\nu}.$

$$(26.39)$$

The overall factor of $\frac{1}{2}$ is a symmetry factor that is required since gluons are their own antiparticles (unlike quarks). The numerator is

$$
N^{\mu\nu} = \left[g^{\mu\alpha} (p+k)^\rho + g^{\alpha\rho}(p-2k)^\mu + g^{\rho\mu}(k-2p)^\alpha \right] g^{\alpha\beta} g^{\rho\sigma}
$$

$$
\times \left[g^{\nu\beta} (p+k)^\sigma - g^{\beta\sigma}(2k-p)^\nu - g^{\sigma\nu}(2p-k)^\beta \right]. \qquad (26.40)
$$

We next introduce Feynman parameters, so

$$
\frac{1}{k^2} \frac{1}{(p-k)^2} = \int_0^1 dx \frac{1}{\left[(1-x) k^2 + x (p-k)^2 \right]^2}. \qquad (26.41)
$$

We then complete the square by $k \rightarrow k + xp$. This leads to

$$
i\mathcal{M}_3^{ab\mu\nu} = \frac{g^2}{2} \int_0^1 dx \int \frac{d^4k}{(2\pi)^4} \frac{1}{(k^2 - \Delta)^2} f^{acd} f^{bcd} N^{\mu\nu}, \qquad (26.42)
$$

with $\Delta = x\,(x-1)\,p^2$, and now (keeping in mind that $g^{\mu\mu} = d$ in d dimensions)

$$N^{\mu\nu} = 2k^2 g^{\mu\nu} - (6 - 4d)k^\mu k^\nu$$
$$- \left[6\big(x^2 - x + 1\big) - d(1 - 2x)^2\right]p^\mu p^\nu + (2x^2 - 2x + 5)p^2 g^{\mu\nu}$$
$$- (2 - 4x)g^{\mu\nu}(k \cdot p) + (2d - 3)(2x - 1)(k^\mu p^\nu + k^\nu p^\mu). \quad (26.43)$$

As usual, the $k^\mu p^\nu$ and $k \cdot p$ terms give zero since they are odd in $k \to -k$, so terms in the third line do not contribute. In dimensional regularization, we can replace $k^\mu k^\nu \to \frac{1}{d}k^2 g^{\mu\nu}$. Then the integrals are all straightforward. The result is

$$\mathcal{M}_3^{ab\mu\nu} = -\frac{g^2}{2}\frac{\mu^{4-d}}{(4\pi)^{d/2}}\delta^{ab}C_A \int_0^1 dx \left(\frac{1}{\Delta}\right)^{2-\frac{d}{2}} \times \left\{ g^{\mu\nu}3(d - 1)\Gamma\left(1 - \frac{d}{2}\right)\Delta \right.$$
$$+ p^\mu p^\nu \left[6\big(x^2 - x + 1\big) - d(1 - 2x)^2\right]\Gamma\left(2 - \frac{d}{2}\right)$$
$$\left. + g^{\mu\nu}p^2\left[\big(-2x^2 + 2x - 5\big)\Gamma\left(2 - \frac{d}{2}\right)\right]\right\}. \quad (26.44)$$

Before analyzing this further, let us work out the other graphs. As in QED, it is expected that only the sum of all the relevant graphs will satisfy the Ward identity.

26.4.3 Four-point gluon bubble

The other gluon bubble is the seagull graph:

$$i\mathcal{M}_4^{ab\mu\nu} = \quad \overset{k}{\text{(diagram)}} \quad \sim \int \frac{d^4k}{(2\pi)^4}\frac{1}{k^2} = 0. \quad (26.45)$$

This is a scaleless integral and formally vanishes in dimensional regularization. In a different regulator, such as Pauli–Villars, this would be quadratically divergent. As was discussed in the context of scalar QED in Section 16.2.1, the quadratic divergence shows up as a pole at $d = 2$ in dimensional regularization. This pole is canceled by the $d = 2$ pole in the gluon bubble graph. Although here they add up to zero trivially ($0 + 0 = 0$), it is important to understand that the cancellation of the pole requires that the coupling constants be equal for the two graphs.

Putting in all the factors, the diagram is

$$i\mathcal{M}_4^{ab\mu\nu} = -\frac{ig^2}{2}\mu^{4-d}\int\frac{d^dk}{(2\pi)^d}\frac{-ig^{\rho\sigma}\delta^{cd}}{k^2 + i\varepsilon}$$
$$\times \left[f^{abe}f^{cde}(g^{\mu\rho}g^{\nu\sigma} - g^{\mu\sigma}g^{\nu\rho}) + f^{ace}f^{bde}(g^{\mu\nu}g^{\rho\sigma} - g^{\mu\sigma}g^{\nu\rho}) \right.$$
$$\left. + f^{ade}f^{bce}(g^{\mu\nu}g^{\rho\sigma} - g^{\mu\rho}g^{\nu\sigma})\right]$$
$$= -g^2\delta^{ab}g^{\mu\nu}C_A(d - 1)\mu^{4-d}\int\frac{d^dk}{(2\pi)^d}\frac{1}{k^2 + i\varepsilon}, \quad (26.46)$$

where, as in Eq. (19.46)

$$F_2^{(2A)}(p^2) = \frac{g^2}{4\pi^2}m^2 \int_0^1 dx\, dy\, dz\, \delta(x+y+z-1)\frac{z(1-z)}{(1-z)^2m^2 - xyp^2},\qquad (26.69)$$

which is finite, and as in Eq. (19.54)

$$F_1^{(2A)}(p^2) = \frac{g^2}{8\pi^2}\left(\frac{1}{\varepsilon} - 1\right.$$
$$\left. + \int_0^1 dx\, dy\, dz\, \delta(x+y+z-1)\left[\frac{p^2(1-x)(1-y) + m^2(1-4z+z^2)}{\Delta} + \ln\frac{\tilde{\mu}^2}{\Delta}\right]\right),$$
$$(26.70)$$

with $\Delta = (1-z)^2m^2 - xyp^2$. To extract the divergences, we take $p^2 \gg m^2$, which gives

$$= ig\left(C_F - \frac{1}{2}C_A\right)T_{ij}^a\gamma^\mu\left(\frac{g^2}{16\pi^2}\right)\left(\frac{2}{\varepsilon} + \ln\frac{\tilde{\mu}^2}{-p^2} + \text{finite}\right).$$

$$(26.71)$$

The next graph is new:

$$= (ig)f^{abc}\left(T^cT^b\right)_{ij}\Gamma_{(2B)}^\mu,\qquad (26.72)$$

where $\Gamma_{(2B)}^\mu$ can be written in terms of the same form factors as the other graph, so it only depends on p^2. The color factor is

$$T^cT^bf^{abc} = \frac{1}{2}f^{abc}\left[T^c, T^b\right] = -\frac{i}{2}f^{abc}f^{dbc}T^d = -\frac{i}{2}C_A T^a.\qquad (26.73)$$

The loop is (with $m = 0$ for simplicity)

$$(ig)\Gamma_{(2B)}^\mu(p^2) = (ig)^2 g\int\frac{d^4k}{(2\pi)^4}\gamma^\rho\frac{i\slashed{k}}{k^2}\gamma^\nu\frac{-i}{(q_1+k)^2 + i\varepsilon}\frac{-i}{(q_2-k)^2 + i\varepsilon}$$
$$\times\left[g^{\mu\nu}(2q_1 + q_2 + k)^\rho + g^{\nu\rho}(-q_1 + q_2 - 2k)^\mu + g^{\rho\mu}(k - 2q_2 - q_1)^\nu\right].\quad (26.74)$$

This integral is the same as the vertex correction in QED, up to the numerator structure. For our purposes, we would just like to know the structure associated with the UV divergence. To extract this, let us set all the external momenta to zero. Then

$$\Gamma_{(2B)}^\mu(0) = g^2\int\frac{d^4k}{(2\pi)^4}\frac{\gamma^\rho\slashed{k}\gamma^\nu}{k^6}\left[g^{\mu\nu}k^\rho - 2g^{\nu\rho}k^\mu + g^{\rho\mu}k^\nu\right]$$
$$= g^2\int\frac{d^4k}{(2\pi)^4}\frac{1}{k^6}\left[2k^2\gamma^\mu - 2\gamma^\rho\slashed{k}\gamma^\rho k^\mu\right].\qquad (26.75)$$

Going to d dimensions, replacing $k^\mu k^\nu \to \frac{k^2}{d} g^{\mu\nu}$ and $\gamma^\rho \gamma^\nu \gamma^\rho = (2-d)\gamma^\nu$, we have

$$\Gamma^\mu_{(2B)}(0) = \left(4 - \frac{4}{d}\right)\gamma^\mu g^2 \mu^{4-d} \int \frac{d^d k}{(2\pi)^d} \frac{1}{k^4}$$

$$= i\gamma^\mu \frac{g^2}{8\pi^2}\left(\frac{3}{\varepsilon} + \frac{3}{2}\ln\tilde{\mu}^2 + \cdots\right). \qquad (26.76)$$

Now we know that $\Gamma^\mu_{(2B)}(p^2)$ only depends on p^2 so we can restore the leading non-analytic p^2 dependence by dimensional analysis:

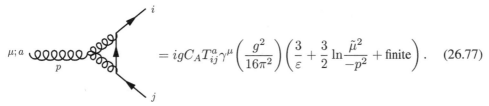

$$= igC_A T^a_{ij} \gamma^\mu \left(\frac{g^2}{16\pi^2}\right)\left(\frac{3}{\varepsilon} + \frac{3}{2}\ln\frac{\tilde{\mu}^2}{-p^2} + \text{finite}\right). \qquad (26.77)$$

Finally, the counterterm gives

$$= igT^a_{ij}\gamma^\mu \delta_1. \qquad (26.78)$$

For this to cancel the UV divergences in the 1-loop graphs, the counterterm must be

$$\delta_1 = \frac{1}{\varepsilon}\left(\frac{g^2}{16\pi^2}\right)[-2C_F - 2C_A]. \qquad (26.79)$$

One can continue this for the gluon 3-point function, 4-point function, and a 3-point function involving the ghost–gluon vertex to find the remaining counterterms, δ_{A^3}, δ_{A^4} and δ_{1c} at 1-loop. The explicit calculations make a useful exercise (Problem 26.2). However, due to gauge invariance, these counterterms are in fact determined by the counterterms we have already computed (see Eqs. (26.80)–(26.87) below).

26.5.3 Summary

For reference, we summarize the results for all the counterterms in QCD at 1-loop, for an arbitrary R_ξ gauge:

$$\delta_1 = \frac{1}{\varepsilon}\left(\frac{g^2}{16\pi^2}\right)\left[-2C_F - 2C_A + 2(1-\xi)C_F + \frac{1}{2}(1-\xi)C_A\right], \qquad (26.80)$$

$$\delta_2 = \frac{1}{\varepsilon}\left(\frac{g^2}{16\pi^2}\right)[-2C_F + 2(1-\xi)C_F], \qquad (26.81)$$

$$\delta_m = \frac{1}{\varepsilon}\left(\frac{g^2}{16\pi^2}\right)[-6C_F], \qquad (26.82)$$

$$\delta_3 = \frac{1}{\varepsilon}\left(\frac{g^2}{16\pi^2}\right)\left[\frac{10}{3}C_A - \frac{8}{3}n_f T_F + (1-\xi)C_A\right], \tag{26.83}$$

$$\delta_{3c} = \frac{1}{\varepsilon}\left(\frac{g^2}{16\pi^2}\right)\left[C_A + \frac{1}{2}(1-\xi)C_A\right], \tag{26.84}$$

$$\delta_{A^3} = \frac{1}{\varepsilon}\left(\frac{g^2}{16\pi^2}\right)\left[\frac{4}{3}C_A - \frac{8}{3}n_f T_F + \frac{3}{2}(1-\xi)C_A\right], \tag{26.85}$$

$$\delta_{A^4} = \frac{1}{\varepsilon}\left(\frac{g^2}{16\pi^2}\right)\left[-\frac{2}{3}C_A - \frac{8}{3}n_f T_F + 2(1-\xi)C_A\right], \tag{26.86}$$

$$\delta_{1c} = \frac{1}{\varepsilon}\left(\frac{g^2}{16\pi^2}\right)[-C_A + (1-\xi)C_A]. \tag{26.87}$$

The answers have been written so the Feynman gauge results with $\xi = 1$ can be easily read off.

26.6 Running coupling

With the results for these 1-loop counterterms, we can now calculate the β function for non-Abelian gauge theories. As discussed in Chapter 23, the renormalization group equation is determined by demanding that observables be independent of, variously, the UV cutoff, the subtraction point where the theory is renormalized, or the arbitrary scale μ in dimensional regularization. In practice, using $\overline{\text{MS}}$ subtraction, one usually sets μ equal to the subtraction point and then uses μ independence to find $\alpha(\mu)$ as a solution to the RGE.

26.6.1 β-function calculation

The fermion–gauge boson interaction in the Lagrangian for a non-Abelian gauge theory is

$$\mathcal{L} = \mu^{\frac{4-d}{2}}g_R Z_1 A_\mu^a \bar{\psi}_i \gamma^\mu T_{ij}^a \psi_j = \mu^{\frac{4-d}{2}}g_R \frac{Z_1}{Z_2\sqrt{Z_3}}A_\mu^{a(0)}\bar{\psi}_i^{(0)}\gamma^\mu T_{ij}^a \psi_j^{(0)}, \tag{26.88}$$

where we have put the $\mu^{\frac{4-d}{2}}$ factors that appear in the loops explicitly in the Lagrangian. So we identify the bare charge as

$$g_0 = g_R \frac{Z_1}{Z_2\sqrt{Z_3}}\mu^{\frac{4-d}{2}}. \tag{26.89}$$

This must be independent of μ, since there is no μ in the bare Lagrangian. So

$$0 = \mu\frac{d}{d\mu}g_0 = \mu\frac{d}{d\mu}\left[g_R\frac{Z_1}{Z_2\sqrt{Z_3}}\mu^{\frac{4-d}{2}}\right]. \tag{26.90}$$

Expanding perturbatively, counting the δ_i as $\mathcal{O}(g_R^2)$, this gives

$$\beta(g_R) = \mu\frac{d}{d\mu}g_R = g_R\left[\left(-\frac{\varepsilon}{2}\right) - \mu\frac{d}{d\mu}\left(\delta_1 - \delta_2 - \frac{1}{2}\delta_3\right)\right] + \cdots. \tag{26.91}$$

Since each δ only depends on μ through g_R, we solve this perturbatively, giving

$$\beta(g_R) = -\frac{\varepsilon}{2}g_R + \frac{\varepsilon}{2}g_R^2\frac{\partial}{\partial g_R}\left(\delta_1 - \delta_2 - \frac{1}{2}\delta_3\right). \qquad (26.92)$$

Using the 1-loop values for the counterterms, we find

$$\beta(g_R) = -\frac{\varepsilon}{2}g_R - \frac{g_R^3}{16\pi^2}\left[\frac{11}{3}C_A - \frac{4}{3}n_f T_F\right]. \qquad (26.93)$$

Note the very important fact that the ξ dependence completely cancels in $\beta(g_R)$.

We could equally well have computed the β-function for the running of the charge in the A^3 interaction. Then we would have computed β from

$$\beta(g_R) = -\frac{\varepsilon}{2}g_R + \frac{\varepsilon}{2}g_R^2\frac{\partial}{\partial g_R}\left(\delta_{A^3} - \frac{3}{2}\delta_3\right). \qquad (26.94)$$

That this gives the same answer as using the coupling to fermions is due to gauge invariance, as discussed in Section 26.6.3 below.

Specializing to QCD now, we take $N = 3$, so $C_A = 3$, and we write $\alpha_s = \frac{g_s^2}{4\pi}$. Then, also using $T_F = \frac{1}{2}$, the RGE at 1-loop (at $\varepsilon = 0$) is $\mu\frac{d}{d\mu}\alpha_s = -\frac{\alpha_s^2}{2\pi}\beta_0$, with $\beta_0 = 11 - \frac{2n_f}{3}$. So as long as there are fewer than 17 flavors of quarks (there are six in nature), $\beta_0 > 0$ and hence $\alpha(\mu)$ decreases with increasing μ. The solution to the 1-loop RGE can be written as

$$\alpha_s(\mu) = \frac{2\pi}{\beta_0}\frac{1}{\ln\frac{\mu}{\Lambda_{\mathrm{QCD}}}}, \qquad (26.95)$$

where Λ_{QCD} is the location of the Landau pole of QCD. In contrast to QED, since $\alpha_s(\mu)$ increases at smaller μ, this equation is valid for $\mu > \Lambda_{\mathrm{QCD}}$. As discussed in Section 23.2, the scale Λ_{QCD} appears through dimensional transmutation as a boundary condition set by a renormalization condition at a particular scale. Measuring α_s at any scale fixes Λ_{QCD}.

That the coupling constant gets weaker at high energy is called **asymptotic freedom**. Asymptotic freedom explains a number of important qualitative features of the strong interactions, such as how QCD can be strong but also short-ranged and why free quarks have never been seen.

26.6.2 Higher-order β-function

The expansion of the QCD β-function, $\beta(\alpha_s) \equiv \mu\frac{d}{d\mu}\alpha_s$, in powers of α_s is

$$\beta(\alpha_s) = -\varepsilon\alpha_s - 2\alpha_s\left[\left(\frac{\alpha_s}{4\pi}\right)\beta_0 + \left(\frac{\alpha_s}{4\pi}\right)^2\beta_1 + \left(\frac{\alpha_s}{4\pi}\right)^3\beta_2 + \left(\frac{\alpha_s}{4\pi}\right)^3\beta_3 + \mathcal{O}(\alpha_s^4)\right]. \qquad (26.96)$$

The ε term is only useful for calculating RGEs for other quantities; when solving this differential equation for $\alpha_s(\mu)$ one can set $\varepsilon = 0$. The QCD β-function is currently known

$T \to \infty$, transient fluctuations drop out and

$$e^{-iET} = \langle \Omega | e^{-iHT} | \Omega \rangle = \frac{\int \mathcal{D}A \exp\left[i \int d^4x \left(-\frac{1}{4} F_{\mu\nu}^2 + e A_\mu J^\mu\right)\right]}{\int \mathcal{D}A \exp\left[i \int d^4x \left(-\frac{1}{4} F_{\mu\nu}^2\right)\right]}. \qquad (26.113)$$

If we identify $E = V(r)$ as the energy of the two charges separated by R, then we have already justified Eq. (26.111) with the Abelian version of Eq. (26.112).

As a cross-check, let us evaluate this path integral explicitly. Since the path integral is quadratic in fields for QED, we can solve it exactly:

$$\exp(-iET) = \exp\left\{ i \int d^4x \int d^4y \frac{e^2}{2} J^\mu(x) D_{\mu\nu}(x,y) J^\nu(y) \right\}, \qquad (26.114)$$

where $i D_{\mu\nu}(x,y)$ is the gauge boson position-space Feynman propagator. In Feynman gauge,

$$i D^{\mu\nu}(x,y) = \langle \Omega | T\{A^\mu(x) A^\nu(y)\} | \Omega \rangle = \frac{1}{4\pi^2} \frac{g^{\mu\nu}}{(x-y)^2 - i\varepsilon} \qquad (26.115)$$

(see Problem 6.1 or Section 33.2). The integrals over x and y will be divergent when both currents are at $z = R$ or both at $z = 0$. However, these contributions will have no R dependence. The only R-dependent part comes from x and y on opposite sides of the loop, which gives

$$-iET = -2 \frac{e^2}{8\pi^2} \int_{-\frac{T}{2}}^{\frac{T}{2}} dx^0 \int_{-\infty}^{\infty} dy^0 \frac{1}{(x_0 - y_0)^2 - R^2 - i\varepsilon} = i\frac{e^2}{4\pi R} \int_{-\frac{T}{2}}^{\frac{T}{2}} dx^0 = i\frac{e^2 T}{4\pi R}. \qquad (26.116)$$

T has been taken to ∞ in the y^0 integral to extract the leading T behavior. This confirms that $E = -\frac{e^2}{4\pi R} = V(R)$. In QED, this result is exact since the path integral is Gaussian.

We conclude that Eqs. (26.108) and (26.109) provide a gauge-invariant definition of a potential that reduces to the expected answer in the QED case. For QCD, the leading-order calculation is identical to QED. At next-to-leading order, the calculation gives, with $n_f = 0$ [Susskind, 1977; Fischler, 1977],

$$E = \tilde{V}(\vec{q}) = -C_F \frac{g_s^2}{\vec{q}^2} \left(1 + \frac{g_s^2}{16\pi^2} \left(\frac{11}{3} C_A \ln \frac{\vec{q}^2}{\mu^2} + \vec{q} - \text{independent} \right) + \mathcal{O}(g_s^4) \right). \qquad (26.117)$$

This expression is gauge invariant as desired. Thus, the expectation value of a Wilson loop can be used to give an exact definition of the potential and therefore of the running coupling.[1]

One motivation for defining a potential in terms of the expectation value of a Wilson loop is in the hope that it could help prove confinement in QCD. If the non-perturbative QCD potential grew linearly with distance, it would take an infinite amount of energy to separate quarks asymptotically. This would explain why free quarks have never been seen and explain **confinement**. Wilson proposed to address this question on the lattice by evaluating the expectation value of a Wilson loop. Indeed, as we saw in Section 25.5,

---

[1] This definition is actually not quite well defined. There is a subtlety at 3-loops where IR divergences in $\langle \Omega | W_{\text{loop}}^P | \Omega \rangle$ appear [Appelquist *et al.*, 1977].

expectation values of Wilson loops are very natural things to evaluate in lattice QCD. Wilson's idea was that if the potential grew with distance it should act like $\ln\langle W_{\text{loop}}\rangle \sim TR$ rather than $\ln\langle W_{\text{loop}}\rangle \sim \frac{T}{R}$. That is, the expectation value would be proportional to the *area* of the Wilson loop.

In his paper [Wilson, 1974] Wilson was able to show analytically that on the lattice $\ln\langle W_{\text{loop}}\rangle$ scales as the area of the loop at strong coupling. His argument was that, as $g_s \to \infty$, contributions that have links not compensated by links in the opposite direction vanish. Thus, the leading contribution comes from configurations in which the entire loop is tiled with plaquettes, as in Figure 26.2. This has been confirmed by numerical simulation [Gattringer and Lang, 2010]. Unfortunately, Wilson's argument holds equally well in any gauge theory, including QED. The challenge with this approach is to show that confinement persists in the continuum limit, that is, after the lattice spacing is removed. This remains an open question in QCD.

By the way, there is indirect experimental evidence for the linear growth of the energy with separation. In the 1970s, by carefully examining the spectrum of various hadrons, people found the interesting relation that the square of the mass of hadrons was proportional to their spin, $m^2 \sim J$. This is known as **Regge behavior**. Such a spectrum is exactly what one would expect from a spinning string. Moreover, a string at constant tension also has energy that grows linearly with the length of the string. One can think of the string as a tube of chromoelectric flux with constant energy density between two quarks. This led people to postulate strings as a fundamental explanation of the strong force before QCD was established and understood. Now we know that the linear growth with distance is explained by QCD, so fundamental strings are not needed. In the 2000s, string theory had a resurgence as a theory of strong interactions when it was found that it could quantitatively explain features of strongly coupled QCD through the AdS/CFT duality [Maldacena, 1998].

Problems

26.1 Calculate δ_{3c} at 1-loop in dimensional regularization by evaluating the ghost 2-point function.

26.2 Work out the remaining counterterms in QCD in Feynman gauge.

26.3 Colored scalars.

   (a) Compute the contribution of a color triplet scalar to δ_3.

   (b) Compute the contribution of a color triplet scalar to δ_{A3}.

   (c) Compute the contribution of a color triplet scalar to the QCD β-function at 1-loop.

   (d) Can you find some number of scalars and/or spinors for which the 1-loop QCD β-function vanishes at 1-loop?

Gluon scattering and the spinor-helicity formalism

Matrix element and cross section calculations in QCD increase in complexity extremely fast. For example, consider the process $gg \to gg$. At tree-level $gg \to gg$ gets contributions from Feynman diagrams with gluons being exchanged in the s, t and u channels, and from diagrams with the 4-point vertex. The s-channel diagram gives (in Feynman gauge)

$$
iM_s(p_1 p_2 \to p_3 p_4) =
$$

$$
= -i\frac{g_s^2}{s} f^{abe} f^{cde} [(\epsilon_1 \cdot \epsilon_2)(p_1 - p_2)^\mu + \epsilon_2^\mu (p_2 + q) \cdot \epsilon_1 + \epsilon_1^\mu (-q - p_1) \cdot \epsilon_2]
$$

$$
\times \left[(\epsilon_4^* \cdot \epsilon_3^*)(p_4 - p_3)^\mu + \epsilon_3^{*\mu}(p_3 + q) \cdot \epsilon_4^* + \epsilon_4^{*\mu}(-q - p_4) \cdot \epsilon_3^* \right], \quad (27.1)
$$

where $q = p_1 + p_2 = p_3 + p_4$. We can simplify this a little, using transversality of the gluons, $p_i \cdot \epsilon_i = 0$, but not much. The answer is still a mess:

$$
M_s(p_1 p_2 \to p_3 p_4) = -\frac{g_s^2}{s} f^{abe} f^{cde}
$$

$$
\times \big\{ -4\epsilon_1 \cdot \epsilon_3^* \epsilon_2 \cdot p_1 p_3 \cdot \epsilon_4^* + 2\epsilon_1 \cdot \epsilon_2 \epsilon_3^* \cdot p_1 \epsilon_4^* \cdot p_3 - 2\epsilon_1 \cdot p_4 \epsilon_2 \cdot p_1 \epsilon_3^* \cdot \epsilon_4^* + \epsilon_1 \cdot \epsilon_2 p_4 \cdot p_1 \epsilon_3^* \cdot \epsilon_4^*
$$

$$
+ 4\epsilon_1 \cdot \epsilon_4^* \epsilon_2 \cdot p_1 \epsilon_3^* \cdot p_4 - 2\epsilon_1 \cdot \epsilon_2 \epsilon_3^* \cdot p_4 \epsilon_4^* \cdot p_1 - 2\epsilon_1 \cdot p_2 \epsilon_2 \cdot p_3 \epsilon_3^* \cdot \epsilon_4^* + \epsilon_1 \cdot \epsilon_2 \epsilon_3^* \cdot \epsilon_4^* p_2 \cdot p_3
$$

$$
+ 4\epsilon_1 \cdot p_2 \epsilon_2 \cdot \epsilon_3^* \epsilon_4^* \cdot p_3 - 2\epsilon_1 \cdot \epsilon_2 \epsilon_3 \cdot p_2 \epsilon_4^* \cdot p_3 + 2\epsilon_1 \cdot p_2 \epsilon_2 \cdot p_4 \epsilon_3^* \cdot \epsilon_4^* - \epsilon_1 \cdot \epsilon_2 \epsilon_3^* \cdot \epsilon_4^* p_4 \cdot p_2
$$

$$
- 4\epsilon_1 \cdot p_2 \epsilon_2 \cdot \epsilon_4^* \epsilon_3^* \cdot p_4 + 2\epsilon_1 \cdot \epsilon_2 \epsilon_3^* \cdot p_4 \epsilon_4^* \cdot p_2 + 2\epsilon_1 \cdot p_3 \epsilon_2 \cdot p_1 \epsilon_3^* \cdot \epsilon_4^* - \epsilon_1 \cdot \epsilon_2 \epsilon_3^* \cdot \epsilon_4^* p_1 \cdot p_3 \big\}.
$$

$$
(27.2)
$$

To get the cross section, you would also need to compute the crossed diagrams, add the 4-point vertex, square the amplitude, sum over polarizations and simplify the color factor. If you managed to do all that, adding all 1000 or so terms, summing over final states and averaging over initial states you would find

$$
\frac{1}{256} \sum_{\substack{\text{pols.} \\ \text{colors}}} |\mathcal{M}|^2 = g_s^4 \frac{9}{2} \left(3 - \frac{tu}{s^2} - \frac{su}{t^2} - \frac{st}{u^2} \right), \quad (27.3)
$$

which is remarkably simple.

Why are the matrix elements for gluon scattering such a mess and the final answer so simple? The root of the problem is our insistence on manifest locality. In fact, the entire formalism of quantum field theory that we have developed so far is based on describing interactions among particles in terms of local Lagrangians. In a local Lagrangian, interactions involve non-negative powers of derivatives, such as $\partial^k \phi_{i_1}(x) \cdots \phi_{i_n}(x)$. While the local Lagrangian description has its advantages, such as manifest Lorentz invariance, it also has disadvantages. In Chapter 8, we encountered subtleties in trying to write a Lagrangian for a massless spin-1 particle that would only propagate the two physical degrees of freedom. We needed to have a redundancy of description, called gauge invariance, that established an equivalence among different components of the vector field $A_\mu(x)$ in which these two polarizations were embedded. We also saw that we could integrate out this redundancy directly at the level of the path integral, which, in the covariant R_ξ gauges, led to an additional complication, Faddeev–Popov ghosts. Even if we work in a gauge without ghosts, such as lightcone gauge, there is still an enormous redundancy built into the entire Feynman-diagram approach. The $A^2 \partial A$ interaction allows for multiple contractions, generating six terms in the Feynman rule, and the A^4 vertex generates another six. That is why even the $gg \to gg$ process above has so many pieces. For five gluon scattering, such as $gg \to ggg$, there are of order $10\,000$ terms in the matrix element. For a cross section, the number of terms is unmanageable without a computer. With just a few more gluons in the final state, even a numerical approach becomes unrealistic.

In this chapter, we describe an alternative approach to constructing amplitudes, using only physical on-shell external states. This approach exploits the spinor-helicity formalism. This formalism is based on the simple observation that spin-1 fields transform in the $\left(\frac{1}{2}, \frac{1}{2}\right)$ representation of the Lorentz group, so that they are naturally represented as bispinors, $\epsilon_{\alpha\dot{\alpha}} = \sigma^\mu_{\alpha\dot{\alpha}} \epsilon_\mu$ (recall, $\sigma^\mu = (\mathbb{1}, \vec{\sigma})$ from Eq. (10.56)). In this way, the redundancy of embedding a massless spin-1 particle into a vector field $A_\mu(x)$ can be avoided. It will take a bit of patience to get used to the notation (as it did for Dirac spinors). Once that is done, we will see some remarkable simplifications. For example, we will find that for $gg \to gg$ there are only two non-vanishing amplitudes, which are

$$\widetilde{\mathcal{M}}(1^- 2^- 3^+ 4^+) = \frac{\langle 12 \rangle^4}{\langle 12 \rangle \langle 23 \rangle \langle 34 \rangle \langle 41 \rangle}, \quad \widetilde{\mathcal{M}}(1^- 2^+ 3^- 4^+) = \frac{\langle 13 \rangle^4}{\langle 12 \rangle \langle 23 \rangle \langle 34 \rangle \langle 41 \rangle}.$$

(27.4)

Adding the appropriate prefactor, squaring and summing over spins and colors then leads to Eq. (27.3) almost effortlessly.

Besides simplifying calculations, the spinor-helicity approach has led to a number of insights into gauge theories, some of which we will discuss (such as their uniqueness), and others (such as dual conformal invariance, or the sense in which gravity $=$ $(\text{gauge theory})^2$) that are still not completely understood. We make some comments on the outlook for this approach in Section 27.7.

27.1 Spinor-helicity formalism

Since momenta transform in the $\left(\frac{1}{2}, \frac{1}{2}\right)$ representation of the Lorentz group, in a sense they are more naturally described as bispinors, $P_{\alpha\dot{\alpha}}$, than as 4-vectors, $P_\mu(x)$. To understand

That is, $\epsilon_\mu \to \epsilon_\mu + \frac{\sqrt{2}}{[pr]} p_\mu$. Since the reference vector is arbitrary, any physical amplitude must be invariant under this transformation. Thus, the Ward identity will be automatically satisfied. Moreover, changing r to any other r is just a gauge transformation, and the polarizations are unchanged.

We will often take r^μ to be the momentum of another gluon in the problem. If the gluons are all labeled by i, then we can write $\epsilon_i(j)$ for the polarization of the gluon with momentum p_i^μ with reference momentum $r^\mu = p_j^\mu$. In this way, any gluon scattering amplitude (or more generally, scattering amplitudes for massless particles of any spin) can be expressed in terms of $[ij]$ and $\langle ij \rangle$ with the i corresponding to momenta in the problem.

With this notation, it is worth working out once and for all the various Lorentz contractions that can appear in scattering amplitudes. We have, using 1 and 2 for the particles and i and j for the reference momenta,

$$\epsilon_1^-(i) \cdot \epsilon_2^-(j) = \frac{1}{2} \mathrm{tr}\left(2 \frac{i]\langle 1\ 2\rangle[j}{[1i]\,[2j]} \right) = \frac{\langle 12\rangle[ji]}{[1i][2j]}. \tag{27.32}$$

Also,

$$\epsilon_1^-(i) \cdot \epsilon_2^+(j) = \frac{\langle 1j\rangle[2i]}{[1i]\langle j2\rangle}, \quad \epsilon_1^+(i) \cdot \epsilon_2^+(j) = \frac{\langle ij\rangle[21]}{\langle i1\rangle\langle j2\rangle} \tag{27.33}$$

and

$$\epsilon_1^-(i) \cdot p_3 = \frac{1}{\sqrt{2}} \frac{\langle 13\rangle[3i]}{[1i]}, \quad \epsilon_1^+(i) \cdot p_3 = \frac{1}{\sqrt{2}} \frac{[13]\langle 3i\rangle}{\langle i1\rangle}, \tag{27.34}$$

and finally $p_1 \cdot p_2 = \frac{1}{2}\langle 21\rangle[12]$ as above. As a check on these, note that parity conjugation flips $+$ to $-$ and $\langle \cdots \rangle$ to $[\cdots]$.

Finally, recall from Chapter 8 that Lorentz transformations which hold a particular momentum fixed are called **little-group** transformations. In terms of helicity spinors, the entire set of transformations that preserve the momentum $p^{\alpha\dot\alpha} = p\rangle[p$ are rescalings:

$$p\rangle \to z\,p\rangle, \quad [p \to \frac{1}{z}\,[p, \tag{27.35}$$

which can also be seen in the explicit decompositions in Eq. (27.17). Thus, little-group transformations must be rescalings of this form. There is a separate little-group transformation associated with each momentum.

If we have a gluon with momentum p^μ then its polarizations transform under the little group associated with p as

$$\epsilon_p^-(r) = \sqrt{2} \frac{p\rangle[r}{[pr]} \to z^2 \epsilon_p^-(r), \quad \epsilon_p^+(r) = \sqrt{2} \frac{r\rangle[p}{\langle rp\rangle} \to z^{-2}\epsilon_p^+(r). \tag{27.36}$$

Note that the polarizations are independent of rescalings of spinors associated with the reference momentum. Since any gluon scattering amplitude can be written entirely in terms of inner products of spinors associated with the momenta in the problem, and since momenta and reference vectors are little-group invariant, the little-group scaling of any amplitude is determined solely by the external polarizations. This strongly constrains the form that a scattering amplitude can have, to all orders in perturbation theory.

Explicitly, the number of factors of $i\rangle$ and $\langle i$ minus the number of factors of $i]$ and $[i$ in the amplitude must be equal to 2 for a negative helicity gluon and -2 for a positive helicity gluon. For example, consider the scattering of two positive and two negative helicity gluons. The result might be

$$\widetilde{\mathcal{M}}\left(1^-, 2^-, 3^+, 4^+\right) = \frac{\langle 21\rangle [34]^2}{[21][14]\langle 41\rangle} \quad \text{or} \quad \frac{\langle 12\rangle^3}{\langle 23\rangle\langle 34\rangle\langle 41\rangle}, \tag{27.37}$$

but it could not be something like $\langle 12\rangle\langle 34\rangle$ since that would scale incorrectly under the little group.

27.1.3 Dirac spinors

Dirac spinors can also be handled smoothly with helicity spinors (although we will not be using them much in this chapter). Recall that Dirac spinors can be either left- or right-handed. Of course, a physical state can only be left- *or* right-handed. Thus we can write left- and right-handed Dirac spinors (in the Weyl basis) as

$$p\rangle = \begin{pmatrix} \lambda^\alpha \\ 0 \end{pmatrix}, \quad p] = \begin{pmatrix} 0 \\ \tilde{\lambda}_{\dot\alpha} \end{pmatrix}, \quad [p = \begin{pmatrix} 0 & \tilde{\lambda}^{\dot\alpha} \end{pmatrix}, \quad \langle p = \begin{pmatrix} \lambda_\alpha & 0 \end{pmatrix}. \tag{27.38}$$

Note that, for massless fermions, particles and antiparticles are represented by the same spin states (cf. Eqs. (11.22) and (11.23)). That is, connecting to our usual Dirac spinor notation, $|p\rangle = P_L u(p)$ and $p] = P_R u(p)$ (for particles) or $|p\rangle = P_L v(p)$ and $p] = P_R v(p)$ (for antiparticles). We see that, using helicity spinors, $p]$ and $p\rangle$ can be seamlessly treated as either Weyl or Dirac.

The γ-matrices in the Weyl basis are

$$\gamma^\mu_{\alpha\dot\alpha} = \begin{pmatrix} 0 & \sigma^{\mu\alpha\dot\alpha} \\ \bar\sigma^\mu_{\dot\alpha\alpha} & 0 \end{pmatrix}. \tag{27.39}$$

We see immediately that

$$[p\gamma^\mu q] = \langle p\gamma^\mu q\rangle = 0. \tag{27.40}$$

Also,

$$\langle p\gamma^\mu q] = \langle p\sigma^\mu q] = [q\bar\sigma^\mu p\rangle = [q\gamma^\mu p\rangle, \tag{27.41}$$

where Eq. (27.8) has been used.

With helicity spinors, Dirac algebra becomes very easy. For example,

$$\langle p\gamma^\mu q]\langle r\gamma_\mu s] = g_{\mu\nu}\langle p\sigma^\mu q]\langle r\sigma^\nu s] = 2\langle pr\rangle [sq], \tag{27.42}$$

where Eq. (27.7) has been used. Similarly, we find

$$\langle p\slashed{k} q] = \langle pk\rangle [kq]. \tag{27.43}$$

For a concrete application, consider unpolarized $e^+ e^- \to \mu^+ \mu^-$ scattering in QED in the high-energy limit. If the electron is right-handed, we denote it as $1]$. Since $[2\gamma^\mu 1] = 0$, the positron must be left-handed. Similarly, take the muon to be $\langle 3$, which forces the antimuon to be $4]$. For these helicities, the amplitude is

$$
i\mathcal{M}(1^-2^+3^-4^+) = \quad = (-ie)^2\,\langle 2\gamma^\mu 1]\frac{-ig^{\mu\nu}}{s}\langle 3\gamma_\nu 4] = 2\frac{ie^2}{s}[41]\langle 23\rangle .
$$

$$(27.44)$$

Squaring this amplitude gives

$$
\left|\mathcal{M}(1^-2^+3^-4^+)\right|^2 = 4e^4\frac{[41]\langle 14\rangle\langle 23\rangle [32]}{s^2} = 16e^4\frac{p_1\cdot p_4 p_2\cdot p_3}{s^2} = 4e^4\frac{u^2}{s^2}. \quad (27.45)
$$

The $1^+2^-3^+4^-$ amplitude is identical (by parity). The other two non-vanishing amplitudes give the same thing with $1 \leftrightarrow 2$, namely $|\mathcal{M}(1^-2^+3^+4^-)|^2 = 4e^4\frac{t^2}{s^2}$. Thus,

$$
\frac{1}{4}\sum_{\text{spins}}|\mathcal{M}|^2 = 2e^4\frac{t^2+u^2}{s^2}, \quad (27.46)
$$

in agreement with Eq. (13.68) when $m_e = m_\mu = 0$.

27.2 Gluon scattering amplitudes

With all this algebra taken care of, we can now start to see some results. Consider first the $2 \to 2$ scattering of gluons, all of which have positive helicity (with incoming momenta). Choose all the polarizations to have the same reference vector r^μ, which can be any random lightlike direction not aligned with any of the p_i^μ. With this choice, it follows from Eq. (27.33) that

$$
\epsilon_i^+(r)\cdot\epsilon_j^+(r) = \frac{\langle rr\rangle [ji]}{\langle ri\rangle\langle rj\rangle} = 0, \quad (27.47)
$$

so that all the polarizations are orthogonal: $\epsilon_i^+\cdot\epsilon_j^+ = 0$. However, every term in the s-channel amplitude has some $\epsilon_i\cdot\epsilon_j$ factor, as can been seen immediately from the explicit expression in Eq. (27.2). Therefore $\mathcal{M}_s(1^+,2^+,3^+,4^+) = 0$. It is easy to see in the same way that all terms in the t-channel, u-channel and 4-point vertex-channel have at least one pair of polarization vectors contracted. We conclude that the tree-level amplitude for $+++ +$ scattering vanishes identically.

This result is actually quite general:

Amplitudes with all positive (or all negative) helicities vanish at tree-level in QCD, for any number of legs.

To see why, again choose r^μ to be different from all the momenta so that $\epsilon_i^+\cdot\epsilon_j^+ = 0$. The only thing a polarization can get contracted with besides another polarization is a momentum. But at tree-level, each vertex can contribute at most one factor of momentum (none for the 4-point vertex). Since there are always fewer vertices than external lines, there must be a polarization contraction in each term in the answer, and thus the amplitude must vanish.

What about having one negative helicity? Call the momentum of the negative helicity gluon p_1^μ. Now choose the reference vector for the $p_{i\neq1}^\mu$ polarizations to be p_1^μ. In this case, we still have $\epsilon_i^+ \cdot \epsilon_j^+ = 0$ for $i, j \neq 1$, but we also now have

$$\epsilon_i^+(1) \cdot \epsilon_1^-(r) = \frac{[ir]\langle 11\rangle}{\langle 1i\rangle[1r]} = 0, \qquad (27.48)$$

so every possible polarization contraction still must vanish. This works for any number of gluons greater than three. Remember that the reference momentum could not have $p_\mu r^\mu = 0$. But for three gluons, $p_1 \cdot p_3 = \frac{1}{2}(p_1 + p_3)^2 = \frac{1}{2}p_2^2 = 0$, so this trick does not work. Of course, for three gluons, you cannot have non-trivial scattering anyway (at least with real momenta; with complex momenta the three-gluon scattering amplitude does not automatically vanish, as we will discuss below).

In summary, we have found:

Amplitudes with all but one positive (or all but one negative) helicity vanish at tree-level for any number of external legs greater than three.

Beyond this, there is no general rule, and indeed amplitudes generally do not vanish.

Finally, QCD is parity invariant, so amplitudes are the same if we flip all the helicities. Therefore:

Amplitudes are invariant under parity, which flips all the helicities $h_i \to -h_i$.

Thus, the leading non-vanishing amplitudes will have at least two negative and two positive helicities. Those with exactly two negative or exactly two positive helicities are called **maximum helicity violating** (MHV) amplitudes.

27.2.1 Color factors

To get the full answer for gluon scattering amplitudes we need to deal with color. First recall from Eq. (25.33) that the structure constants for $SU(N)$ are related to the generators in the fundamental representation by

$$f^{abc} = -2i\mathrm{tr}\left([T^a, T^b]T^c\right). \qquad (27.49)$$

This equation lets us reduce products of f^{abc} factors to traces over products of matrices.

Another important equation from Chapter 25 is Eq. (25.34):

$$\sum_a T_{ij}^a T_{kl}^a = \frac{1}{2}\left(\delta_{il}\delta_{kj} - \frac{1}{N}\delta_{ij}\delta_{kl}\right). \qquad (27.50)$$

so that

$$\sum_{\text{colors}} \left| \mathcal{M}(1^-2^-3^+4^+) \right|^2 = 4g_s^4 N^2 (N^2 - 1) \left\{ \frac{s^2}{t^2} + \frac{s^4}{t^2 u^2} + \frac{s^3}{t^2 u} \right\}$$

$$= 4g_s^4 N^2 (N^2 - 1) \left(\frac{s^4}{t^2 u^2} - \frac{s^2}{tu} \right), \qquad (27.72)$$

where $s + t + u = 0$ has been used to get to a form that is manifestly symmetric in $t \leftrightarrow u$.

With this answer, it is not hard to complete the full cross section calculation. Since only the MHV channels do not vanish, and each one is gauge invariant by itself, they will all be given by some crossing of this result. For example, $\mathcal{M}(1^-2^+3^-4^+)$ is given by $\mathcal{M}(1^-2^-3^+4^+)$ with $s \leftrightarrow u$. The six non-vanishing amplitudes correspond to the six permutations of s, t, u. Summing all of these permutations gives

$$\sum_{\substack{\text{pols.} \\ \text{colors}}} |\mathcal{M}|^2 = 4g_s^4 N^2 (N^2 - 1) \left\{ \left(\frac{s^4}{t^2 u^2} - \frac{s^2}{tu} \right) + \text{perms of } s, t, u \right\}$$

$$= 4g_s^4 N^2 (N^2 - 1) \frac{(s^2 + t^2 + u^2)(s^4 + t^4 + u^4)}{s^2 u^2 t^2}. \qquad (27.73)$$

Averaging over the number of initial states, which is $4 \times (N^2 - 1)^2$ for the spins and colors, taking $N = 3$, and simplifying with $s + t + u = 0$ gives

$$\frac{1}{256} \sum_{\substack{\text{pols.} \\ \text{colors}}} |\mathcal{M}|^2 = \frac{9}{2} g_s^4 \left(3 - \frac{su}{t^2} - \frac{ut}{s^2} - \frac{st}{u^2} \right). \qquad (27.74)$$

This final form is the standard way $gg \to gg$ is presented for QCD.

27.4 Color ordering

As we have seen in the $gg \to gg$ example, crossing relations can be extremely helpful in gluon scattering. For multi-gluon amplitudes, with $n > 4$ gluons, crossings can be complicated, so it is worth understanding how crossings work in general. The first step is to separate the color from the kinematics.

Define a **color-stripped amplitude** as the part of the amplitude with the color factor stripped off. The Feynman rules for computing color-stripped amplitudes are the same as the regular QCD Feynman rules, but without a $\sqrt{2} i g_s f^{abc}$ factor. For example, for the four-gluon amplitude, the color-stripped s-channel amplitude is

$$\widetilde{\mathcal{M}}_s(1234) = \frac{1}{2s} \left[(\epsilon_1 \cdot \epsilon_2)(p_1 - p_2)^\mu + 2\epsilon_2^\mu (p_2 \cdot \epsilon_1) - 2\epsilon_1^\mu (p_1 \cdot \epsilon_2) \right]$$

$$\times \left[(\epsilon_3 \cdot \epsilon_4)(p_3 - p_4)^\mu + 2\epsilon_4^\mu (p_4 \cdot \epsilon_3) - 2\epsilon_3^\mu (p_3 \cdot \epsilon_4) \right]. \qquad (27.75)$$

Here the numbers 1234 have implicit helicities associated with each gluon. Note that $\widetilde{\mathcal{M}}_s$ is antisymmetric under interchange of $1 \leftrightarrow 2$ or $3 \leftrightarrow 4$, so

$$\widetilde{\mathcal{M}}_s(1234) = -\widetilde{\mathcal{M}}_s(2134) = -\widetilde{\mathcal{M}}_s(1243) = \widetilde{\mathcal{M}}_s(2143). \qquad (27.76)$$

Also, $\widetilde{\mathcal{M}}_s(1234) = \widetilde{\mathcal{M}}_s(3412)$.

The color factor for the s-channel diagram can be written in terms of single traces, using the SU(N) tricks from Section 27.2.1:

$$f^{12a} f^{34a} = -2\mathrm{tr}\Big\{ [1,2][3,4] \Big\} = -2\Big[\mathrm{tr}\{1234\} - \mathrm{tr}\{2134\} - \mathrm{tr}\{1243\} + \mathrm{tr}\{2143\} \Big]. \qquad (27.77)$$

This is a sum of four terms that is antisymmetric under $1 \leftrightarrow 2$ or $3 \leftrightarrow 4$. Thus, the full s-channel amplitude for $\mathcal{M}_s(1234)$ can be written as a sum of terms that have the gluons ordered the same way in the color factor and the color-stripped amplitude:

$$\mathcal{M}_s(1234) = 4g_s^2 \Big[\mathrm{tr}\{1234\}\widetilde{\mathcal{M}}_s(1234) + \mathrm{tr}\{2134\}\widetilde{\mathcal{M}}_s(2134)$$
$$+ \mathrm{tr}\{1243\}\widetilde{\mathcal{M}}_s(1243) + \mathrm{tr}\{2143\}\widetilde{\mathcal{M}}_s(2143) \Big]. \qquad (27.78)$$

Note that all the terms in the sum have the same sign.

The t-channel color-stripped amplitude is just the $2 \leftrightarrow 4$ cross of the s-channel one:

$$\widetilde{\mathcal{M}}_t(1234) = \widetilde{\mathcal{M}}_s(1432) . \qquad (27.79)$$

Similarly, the u-channel is a $2 \leftrightarrow 3$ cross:

$$\widetilde{\mathcal{M}}_u(1234) = \widetilde{\mathcal{M}}_s(1324) . \qquad (27.80)$$

Keep in mind, in these crossings, the polarizations stick with the momenta. For example, $\widetilde{\mathcal{M}}_s(1^-2^-3^+4^+) = \widetilde{\mathcal{M}}_t(1^-4^+3^+2^-) \neq \widetilde{\mathcal{M}}_t(1^-4^-3^+2^+)$. Both t- and u-channels also have four terms in the color trace with appropriate signs, so the full amplitude can be written as a sum of single trace color factors and color-stripped amplitudes with positive signs.

The result is that the full amplitude $\mathcal{M}(1234) = \mathcal{M}_s(1234) + \mathcal{M}_t(1234) + \mathcal{M}_u(1234)$ has twelve terms, four each from the s, t, u channels, all of which can be written as $\mathrm{tr}\{ijkl\}\widetilde{\mathcal{M}}_s(ijkl)$. The sum can be simplified further, since not all the color factors are independent due to the cyclic property of the trace $\mathrm{tr}\{1234\} = \mathrm{tr}\{2341\}$. It is helpful to pair up terms, so that

$$\mathrm{tr}\{1234\}\Big[\widetilde{\mathcal{M}}_s(1234) + \widetilde{\mathcal{M}}_s(1432)\Big] = \mathrm{tr}\{1234\}\Big[\widetilde{\mathcal{M}}_s(1234) + \widetilde{\mathcal{M}}_t(1234)\Big]$$
$$= \mathrm{tr}\{1234\}\widetilde{\mathcal{M}}(1234), \qquad (27.81)$$

where

$$\widetilde{\mathcal{M}}(ijkl) \equiv \widetilde{\mathcal{M}}_s(ijkl) + \widetilde{\mathcal{M}}_t(ijkl) = \qquad \qquad \qquad \qquad (27.82)$$

is known as the **color-ordered partial amplitude**. We can then write the four-gluon scattering amplitude as

$$\mathcal{M}(1234) = 4g_s^2 \sum_{\sigma \in S_3} \text{tr}\left\{1\sigma(2)\,\sigma(3)\sigma(4)\right\} \widetilde{\mathcal{M}}\left(1\sigma(2)\sigma(3)\,\sigma(4)\right), \tag{27.83}$$

where S_3 is the permutation group of $\{2, 3, 4\}$. Sometimes this group is written as $S_3 = S_4/Z_4$, with Z_4 referring to the cyclic permutations.

Note that the two diagrams that contribute to the color-ordered partial amplitude are the planar ones. In the double-line notation, the diagrams that are non-planar are suppressed by factors of $\frac{1}{N}$ and drop out of tree-level amplitudes for SU(N). Thus, at tree-level, we will always be able to express gluon scattering in terms of sums of planar diagrams. In fact, the decomposition into partial amplitudes and single traces works for any number of gluons, at tree-level. The generalized formula is simply

$$\mathcal{M}(12\ldots n) = -2\left(\sqrt{2}ig_s\right)^{n-2} \sum_{\sigma \in S_n/\mathbb{Z}_n} \text{tr}\left\{\sigma(1)\sigma(2)\ldots\sigma(n)\right\} \widetilde{\mathcal{M}}(\sigma(1)\sigma(2)\ldots\sigma(n)).$$

$$\tag{27.84}$$

The general definition of $\widetilde{\mathcal{M}}(12\ldots n)$ is the sum over all *planar* color-stripped graphs with a given ordering of the external momenta. This equation can be derived by using the cyclic property of the trace to uncross all the crossed diagrams (see Problem 27.3). Although it should not be obvious at this point why one would want to express an amplitude in terms of $\widetilde{\mathcal{M}}(12\ldots n)$, it turns out that $\widetilde{\mathcal{M}}(12\ldots n)$ can be remarkably simple.

For example, consider the MHV partial amplitude $\widetilde{\mathcal{M}}(1^-2^-3^+4^+)$ for $gg \to gg$. Plugging Eqs. (27.61) and (27.63) into Eq. (27.82), this partial amplitude is

$$\widetilde{\mathcal{M}}\left(1^-2^-3^+4^+\right) = \frac{\langle 12\rangle^4}{\langle 12\rangle\langle 23\rangle\langle 34\rangle\langle 41\rangle} \tag{27.85}$$

because $\widetilde{\mathcal{M}}_t(1^-2^-3^+4^+) = 0$. We can also compute

$$\widetilde{\mathcal{M}}\left(1^-2^+3^-4^+\right) = \widetilde{\mathcal{M}}_s\left(1^-2^+3^-4^+\right) + \widetilde{\mathcal{M}}_t\left(1^-2^+3^-4^+\right)$$
$$= \widetilde{\mathcal{M}}_u\left(1^-3^-2^+4^+\right) + \widetilde{\mathcal{M}}_t\left(1^-3^-2^+4^+\right). \tag{27.86}$$

Again $\widetilde{\mathcal{M}}_t(1^-3^-2^+4^+)$ vanishes, and we computed $\widetilde{\mathcal{M}}_u(1^-2^-3^+4^+)$ in Eq. (27.66). So we have

$$\widetilde{\mathcal{M}}\left(1^-2^+3^-4^+\right) = \widetilde{\mathcal{M}}_u\left(1^-3^-2^+4^+\right) = \frac{\langle 13\rangle^4}{\langle 12\rangle\langle 23\rangle\langle 34\rangle\langle 41\rangle}, \tag{27.87}$$

which is remarkably similar to $\widetilde{\mathcal{M}}(1^-2^-3^+4^+)$. In fact, an amazing feature of gluon scattering is that the color-ordered MHV amplitude for any number of gluons is

$$\widetilde{\mathcal{M}}(1^+2^+\cdots j^-\cdots k^-\cdots n^+) = \frac{\langle jk\rangle^4}{\langle 12\rangle\,\langle 23\rangle\,\langle 34\rangle\cdots\langle n1\rangle}, \tag{27.88}$$

where j and k are the two negative helicity gluons. This is known as the **Parke–Taylor formula**. It is an amazing result that shows that scattering amplitudes in QCD have a lot more symmetry to them than you might guess from looking at the Feynman rules. You are encouraged to verify that the Parke–Taylor formula reproduces the full $gg \to gg$ scattering amplitude at tree-level in Problem 27.2.

As a highly non-trivial example, it is now quite easy to calculate the five-gluon scattering cross section (Problem 27.6). For five gluons, everything but the MHV amplitudes vanish, so as with four gluons there is only one independent amplitude to compute, and it is given by the Parke–Taylor formula. If you tried to do five-gluon scattering with polarization vectors and momenta, it would have 10 000 terms. Using the spinor-helicity formalism, the calculation can be done by hand.

27.5 Complex momenta

We have seen that helicity spinors can be used to simplify Feynman diagrams. But so far, we have only used spinors for the external momenta and polarizations. We still have to compute the Feynman diagrams using the vertices from the Lagrangian. Of course, the spinor-helicity formalism is still an enormous help, but it would be nice to be able to apply the helicity formalism to internal lines too. This is not so simple, since we needed $p^2 = 0$ to write the momentum in terms of spinors, but $p^2 \neq 0$ in general on an internal line. In fact, there is a procedure, not using Feynman diagrams, that uses only on-shell internal states for which the helicities are also $+$ or $-$. For this to work, we need to consider complex momenta. With complex momenta, the 3-point vertex will not identically vanish if the three momenta are on-shell. As we will see, the 4-point and higher-order amplitudes can be built up from the 3-point amplitude, and then the limit of real momenta can be taken.

27.5.1 3-point amplitude

Rather than compute the 3-point amplitude from the Feynman rules, let us just figure out what the most general possible amplitude could be:

$$= \quad ? \tag{27.89}$$

It must depend on the three polarization vectors ϵ_i and the three momenta p_i, or equivalently on the spinors $[1, [2, [3 \text{ and } \langle 1, \langle 2, \langle 3.$ Momentum conservation is

$$1\rangle[1 + 2\rangle[2 + 3\rangle[3 = 0 . \tag{27.90}$$

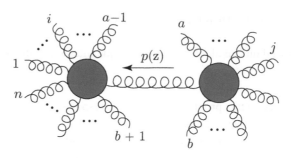

Fig. 27.2   Momentum routing for BCFW recursion.

gluon. Let gluon i be on the left and gluon j be on the right, as shown in Figure 27.2. Then the momentum of the intermediate gluon is

$$\hat{P}(z) = \sum_{k=a}^{b} k\rangle [k - zi\rangle [j. \tag{27.113}$$

So, the pole at $\hat{P}^2(z^\star_{a,b}) = 0$ implies

$$0 = (p_a + \cdots + p_b)^2 - z^\star_{a,b} \sum_{k=a}^{b} \langle ik\rangle \, [kj] + \frac{(z^\star_{a,b})^2}{2} \langle ii\rangle [jj], \tag{27.114}$$

with the last term vanishing. Then,

$$z^\star_{a,b} = \frac{(p_a + \cdots + p_b)^2}{\langle ia\rangle [aj] + \cdots + \langle ib\rangle [bj]}. \tag{27.115}$$

We will get one such $z^\star_{a,b}$ for each partition of the diagram by a, b. For each, we can use

$$-\frac{1}{z^\star_{a,b}} \operatorname*{Res}_{z \to z^\star_{a,b}} \left(\mathcal{M}_1(z) \frac{1}{(p_a + \cdots + p_b)^2 - z \sum \langle ik\rangle \, [kj]} \mathcal{M}_2(z) \right)$$

$$= \mathcal{M}_1(z^\star_{a,b}) \frac{1}{(p_a + \cdots + p_b)^2} \mathcal{M}_2(z^\star_{a,b}), \tag{27.116}$$

where $\mathcal{M}_1$ and $\mathcal{M}_2$ are the diagrams on either side of the partition.

Finally, plugging into Eq. (27.112) we find

$$\mathcal{M}(1\ldots n) = \sum_{a,b,h} \mathcal{M}(1,\ldots,a-1,b+1,\ldots,n \to \hat{P}^h)$$

$$\times \frac{1}{(p_a + \cdots + p_b)^2} \mathcal{M}(\hat{P}^{-h} \to a,\ldots,b), \tag{27.117}$$

where the matrix elements on the right side are to be evaluated with their momentum shifted by $z = z^\star_{a,b}$. This is the **BCFW recursion formula** (Britto–Cachazo–Feng–Witten). The matrix elements on the left and right sides have fewer than n gluons. This formula lets us recursively build up arbitrary tree-level matrix elements algebraically. The helicity h of the internal now on-shell particle with momentum $\hat{P}^\mu$ must be summed over. Note that, to be consistent with our convention that momenta are always incoming, h must flip from the left to the right.

The BCFW formula requires the $z \to \infty$ limit to be well behaved. This is almost always true, except for some choices of i and j. It is easiest to check if we already know the answer. For example, recall the MHV color-ordered partial amplitude for $gg \to gg$:

$$\widetilde{\mathcal{M}}(1^-2^-3^+4^+) = \frac{\langle 12 \rangle^3}{\langle 23 \rangle \langle 34 \rangle \langle 41 \rangle}. \tag{27.118}$$

Let us try $i = 1$ and $j = 2$. Then the only angle shift is $2\rangle \to 2\rangle - z1\rangle$. So, $\langle 12 \rangle \to \langle 12 \rangle$, $\langle 23 \rangle \to \langle 23 \rangle - z\langle 13 \rangle$ and $\langle 41 \rangle \to \langle 41 \rangle$, and at large z this amplitude vanishes as $\frac{1}{z}$ as desired. For the amplitude not to vanish as $z \to \infty$, $\langle 12 \rangle$ would have to shift, which we could only get with $i = 3$ or $i = 4$ and $j = 1$ or $j = 2$. For $i = 3$ and $j = 2$, we find $\langle 12 \rangle \to \langle 12 \rangle - z\langle 13 \rangle$, $\langle 23 \rangle \to \langle 23 \rangle$, $\langle 34 \rangle \to \langle 34 \rangle$ and $\langle 41 \rangle \to \langle 41 \rangle$ so the amplitude blows up as z^3. The general rule for $2 \to 2$ is that the helicity combinations $(i, j) = (+, +), (-, -)$ or $(-, +)$ are good, while $(+, -)$ is bad.

Intriguingly, while BCFW works for gauge theories, it does not work for scalar field theories. For example, in a simple scalar field theory, such as ϕ^4 theory, there are tree-level amplitudes that are just constants. If the amplitude is momentum independent, shifting the momentum introduces no z dependence, and therefore amplitudes will not vanish at $z = \infty$. Thus, BCFW implies that gauge theories are in a way simpler than scalar theories because they can be constructed from sewing together lower point amplitudes. Amplitudes for the exchange of spin-2 particles vanish even faster as $z \to \infty$ than for gauge theories (for certain helicity choices). Thus, in a way, gravity is the simplest theory of them all.

27.6.1 Example

As an example, let us work out $\widetilde{\mathcal{M}}(1^-2^-3^+4^+)$ using BCFW. There are still two diagrams contributing to this partial amplitude, s- and t-channel, but now we will get the answer from the 3-point vertex without using the Lagrangian. We take $i = 1$ and $j = 4$, which is a $(-, +)$ combination and so has good behavior as $z \to \infty$. For there to be a pole these have to be on opposite sides of the internal line. So the t-channel diagram has no poles and does not contribute. The s-channel diagram has $\hat{P}^\mu = -\hat{p}_1^\mu - \hat{p}_2^\mu$, so

$$z^\star_{3,4} = \frac{s}{\langle 13 \rangle [34] + \langle 14 \rangle [44]} = \frac{\langle 34 \rangle}{\langle 31 \rangle}. \tag{27.119}$$

Thus,

$$\widetilde{\mathcal{M}}(1^-2^-3^+4^+) = \sum_h \widetilde{\mathcal{M}}(\hat{1}^-2^-\hat{P}^h)\frac{1}{\langle 12 \rangle [21]}\widetilde{\mathcal{M}}([-\hat{P}^{-h}]3^+\hat{4}^+) \tag{27.120}$$

with $[\hat{1} = [1 + z^\star_{3,4}[4$ and $\hat{4}\rangle = 4\rangle - z^\star_{3,4}1\rangle$. Since $\widetilde{\mathcal{M}}(\hat{1}^-2^-\hat{P}^-)$ vanishes, we must have $h = +$. Then,

$$\widetilde{\mathcal{M}}(1^-2^-3^+4^+) = -\frac{\langle \hat{1}2 \rangle^3}{\langle \hat{1}\hat{P} \rangle \langle \hat{P}2 \rangle}\frac{1}{\langle 12 \rangle [21]}\frac{[3\hat{4}]^3}{[3\hat{P}][\hat{P}4]}. \tag{27.121}$$

Now,

$$\hat{P}\rangle[\hat{P} = 3\rangle[3 + 4\rangle[4 - \frac{\langle 34\rangle}{\langle 31\rangle}1\rangle[4. \tag{27.122}$$

Substituting this in for $\langle \hat{1}\hat{P}\rangle[\hat{P}\hat{4}]$ and $\langle 2\hat{P}\rangle[\hat{P}3]$, we find, after some simplification,

$$\widetilde{\mathcal{M}}(1^-2^-3^+4^+) = \frac{\langle 12\rangle^4}{\langle 12\rangle\langle 23\rangle\langle 34\rangle\langle 41\rangle}. \tag{27.123}$$

This is identical to the MHV amplitude we computed in Section 27.3. Here we computed it without Feynman rules, just using the 3-point MHV amplitude, which is fixed by symmetry (little-group scaling) and sewing things together with scalar propagators, $(p_a + \cdots + p_b)^{-2}$.

One reason BCFW is so efficient is that there is often only one diagram for each step in the recursion. This is always true for MHV amplitudes. For example, for the 7-point MHV amplitude, let us take $i = 1$ and $j = 7$, so that $[\hat{1} = [1 + z[7$ and $\hat{7}\rangle = 7\rangle - z1\rangle$. Then,

$$\begin{aligned}
\widetilde{\mathcal{M}}(1^-2^-3^+4^+5^+6^+7^+) &= \frac{\langle 16\rangle}{\langle 17\rangle\langle 76\rangle}\widetilde{\mathcal{M}}(1^-2^-3^+4^+5^+6^+) \\
&= \frac{\langle 16\rangle}{\langle 17\rangle\langle 76\rangle}\frac{\langle 15\rangle}{\langle 16\rangle\langle 65\rangle}\widetilde{\mathcal{M}}(1^-2^-3^+4^+5^+) \\
&= \frac{\langle 15\rangle}{\langle 17\rangle\langle 76\rangle\langle 65\rangle}\frac{\langle 14\rangle}{\langle 15\rangle\langle 54\rangle}\widetilde{\mathcal{M}}(1^-2^-3^+4^+) \\
&= \frac{\langle 14\rangle}{\langle 17\rangle\langle 76\rangle\langle 65\rangle\langle 54\rangle}\frac{\langle 13\rangle}{\langle 14\rangle\langle 43\rangle}\widetilde{\mathcal{M}}(1^-2^-3^+) \\
&= \frac{\langle 12\rangle^4}{\langle 71\rangle\langle 12\rangle\langle 23\rangle\langle 34\rangle\langle 45\rangle\langle 56\rangle\langle 67\rangle},
\end{aligned} \tag{27.124}$$

with only one non-vanishing amplitude present in each step. In this way, one can use BCFW to prove the Parke–Taylor formula for tree-level MHV amplitudes (see Problem 27.7).

27.7 Outlook

The use of the spinor-helicity formalism and related ideas may provide an entirely new way to calculate amplitudes in quantum field theory. We have already seen that it simplifies gluon scattering at tree-level. These methods also generalize to loop computations, although it seems that the most efficient way to perform loops, using spinors or otherwise, is still not known.

As mentioned in the introduction, part of the reason helicity spinors work so well is because they reduce the amount of extra baggage associated with embedding two helicities into polarization vectors ϵ_μ. This is even more true for higher-spin fields. Indeed, massless fields of arbitrary spin are described by two polarizations, so they can be described by one λ and one $\tilde{\lambda}$, just as for spin 1. Of course, we cannot have interacting theories with massless fields of spin > 2, but you can study their representations this way anyway. For spin 2 (i.e. for gravity) the polarization tensor notation is extremely tedious – one introduces 16 elements of a tensor $h_{\mu\nu}$, then has to impose tracelessness and transversality by hand. Having two spinors makes things much easier.

Little-group scaling for spin 2 implies that the 3-point amplitude must be

$$\mathcal{M}\left(1^{a+}2^{b+}3^{c-}\right) \propto \left(\frac{[12]^4}{[12]\,[23]\,[31]}\right)^2. \qquad (27.125)$$

This equation makes graviton scattering amplitudes appear to be the square of the corresponding gauge-theory amplitudes. This actually seems to be true in a certain sense quite generally, which is a very profound result that is not quite understood.

As an additional bonus, some symmetries become clear from the description of an amplitude in terms of spinors instead of through a Lagrangian. The most well-known one is called **dual conformal invariance**. Dual conformal invariance is a symmetry of amplitudes in certain very symmetric theories when momenta are replaced by momenta differences $x_i^{\mu} = p_i^{\mu} - p_{i+1}^{\mu}$. It is part of a larger infinite-dimensional symmetry called Yangian invariance, which includes conformal invariance and dual conformal invariance.

Due to the accumulation of surprising *theoretical data* (like the Parke–Taylor formula, Eq. (27.88), or dual-conformal invariance) on the remarkable properties of scattering amplitudes, it is reasonable to expect that the simplest way to describe fundamental physics may not be with quantum field theory. For example, we may need to move away from the formulation of a theory in terms of a local Lagrangian, $\mathcal{L}(x)$, to one where locality is rather an emergent property. Of course, quantum field theory is likely to remain the most efficient tool for calculating scattering amplitudes with few final-state particles at low-loop order, much like Newtonian mechanics is still the tool of choice for computing the effect of macroscopic forces on macroscopic objects. However, quantum field theory may well be a certain limit of a more general theory, as classical mechanics is a limit of quantum mechanics. Formulating such a general theory based on purely theoretical data (as opposed to experimental data, as was the case for quantum mechanics) is a formidable but perhaps not insurmountable challenge.

Problems

27.1 What are the explicit polarization vectors $\epsilon_{\pm}^{\mu} = \frac{1}{2}\sigma_{\alpha\dot\alpha}^{\mu}\epsilon_{\pm}^{\alpha\dot\alpha}$ when $p^{\mu} = (E,0,0,E)$ and $r^{\mu} = (1,0,0,1)$? What would you choose r^{μ} to be so that $\epsilon^{\mu} = (0,1,0,0)$ when $p^{\mu} = (E,0,0,E)$?

27.2 Verify that the color-stripped amplitudes and Parke–Taylor formula reproduce the $gg \to gg$ scattering cross section by using Eqs. (27.84) and (27.88) and adding the appropriate color factors.

27.3 Prove the general formula for the matrix element in terms of color-ordered partial amplitudes, Eq. (27.84).

27.4 Compute the Compton scattering cross section, $\gamma e^- \to \gamma e^-$, in the high-energy limit using helicity spinors. Check that you reproduce Eq. (13.141).

27.5 Calculate $|\mathcal{M}|^2$ summed over spins and colors for the remaining $2 \to 2$ processes in QCD. Fill out the following table:

so that Q generates the symmetry transformation. Also, since the charge is conserved, it commutes with the Hamiltonian: $[H, Q] = i\partial_t Q = 0$.

The operator Q corresponds to a conserved charge no matter what vacuum we expand around. Spontaneous symmetry breaking occurs, by definition, if the symmetric vacuum, with $Q|\Omega\rangle_{\text{sym}} = 0$, is unstable and the true (stable) vacuum is charged, $Q|\Omega\rangle \neq 0$. If the vacuum has energy E_0, that is $H|\Omega\rangle = E_0|\Omega\rangle$, then

$$HQ|\Omega\rangle = [H, Q]|\Omega\rangle + QH|\Omega\rangle = E_0 Q|\Omega\rangle \tag{28.7}$$

and therefore the state $Q|\Omega\rangle$ is degenerate with the ground state.

Now, we can always construct states of 3-momentum $\vec{p}$ from the vacuum via

$$|\pi(\vec{p})\rangle = \frac{-2i}{F} \int d^3x\, e^{-i\vec{p}\cdot\vec{x}} J_0(x)|\Omega\rangle, \tag{28.8}$$

which have energy $E(\vec{p}) + E_0$. Here, F is a constant with dimension of mass and the $-2i$ factor has been added for later convenience. Since $|\pi(\vec{0})\rangle = \frac{-2i}{F} Q|\Omega\rangle$ has energy E_0, we can conclude that $E(\vec{p}) \to 0$ as $\vec{p} \to 0$ for these states. Therefore, the states $|\pi\rangle$ must satisfy a massless dispersion relation. This is **Goldstone's theorem**:

Box 28.1   **Goldstone's theorem**

Spontaneous breaking of continuous global symmetries implies the existence of massless particles.

The states $|\pi(\vec{p})\rangle$ are known as **Goldstone bosons**. Goldstone's theorem is very general. Sometimes it is useful to construct the Goldstone bosons from the vacuum using a Noether current. Often it is easier just to locate the Goldstone bosons in the broken phase of the theory through some other means.

Multiplying Eq. (28.8) by $\langle\pi(\vec{q})|$ and integrating over $\int \frac{d^3\vec{p}}{(2\pi)^3} e^{i\vec{p}\cdot\vec{y}}$ gives

$$\langle\pi(\vec{q})|J_0(y)|\Omega\rangle = i\omega_q F e^{i\vec{q}\cdot\vec{y}}, \tag{28.9}$$

where the normalization of one-particle states $\langle\pi(\vec{q})|\pi(\vec{p})\rangle = 2\omega_p (2\pi)^3 \delta^3(\vec{q} - \vec{p})$ has been used. The Lorentz-invariant version of this equation, $\langle\pi(\vec{q})|J_\mu(y)|\Omega\rangle = iq_\mu F e^{i\vec{q}\cdot\vec{y}}$, is a useful way to identify a particle in the spectrum as the Goldstone boson, as we will see below.

28.2.1 Linear sigma model

The simplest relativistic theory with spontaneous symmetry breaking of a continuous global symmetry has a complex scalar field with Lagrangian

$$\mathcal{L} = (\partial_\mu\phi^\star)(\partial_\mu\phi) + m^2\phi\phi^\star - \frac{\lambda}{4}\phi^2\phi^{\star 2}. \tag{28.10}$$

Note that the terms here are canonically normalized for a complex field. This theory has a global U(1) symmetry $\phi(x) \to e^{i\alpha}\phi(x)$ for constant α. For $m^2 > 0$ the theory is unstable around $\phi = 0$. The potential $V(\phi) = -m^2|\phi|^2 + \frac{\lambda}{4}|\phi|^4$ is minimized when $|\phi|^2 = \frac{2m^2}{\lambda}$.

So now there are an infinite number of equivalent vacua $|\Omega_\theta\rangle$ with $\langle\Omega_\theta|\phi|\Omega_\theta\rangle = \sqrt{\frac{2m^2}{\lambda}}e^{i\theta}$ for any constant θ.

All the vacua are equivalent (by symmetry) so we can pick any convenient parametrization. It is conventional to pick $|\Omega\rangle$ so that $\langle\Omega|\phi|\Omega\rangle$ is real. Then $\langle\Omega|\phi|\Omega\rangle = v = \sqrt{\frac{2m^2}{\lambda}}$. Instead of writing $\phi(x) = v + \tilde{\phi}(x)$, with $\tilde{\phi}(x)$ a complex field, it is often more convenient to expand around v by parametrizing $\phi(x)$ in terms of two real fields $\sigma(x)$ and $\pi(x)$ as

$$\phi(x) = \left(\sqrt{\frac{2m^2}{\lambda}} + \frac{1}{\sqrt{2}}\sigma(x)\right)e^{i\frac{\pi(x)}{F_\pi}}, \tag{28.11}$$

with F_π a real number. Then $V(\phi)$ depends only on σ, and not on π. Expanding the Lagrangian around the minimum we find

$$\mathcal{L} = \frac{1}{2}(\partial_\mu\sigma)^2 + \left(\sqrt{\frac{2m^2}{\lambda}} + \frac{1}{\sqrt{2}}\sigma(x)\right)^2 \frac{1}{F_\pi^2}(\partial_\mu\pi)^2$$
$$- \left(-\frac{m^4}{\lambda} + m^2\sigma^2 + \frac{1}{2}\sqrt{\lambda}m\sigma^3 + \frac{1}{16}\lambda\sigma^4\right). \tag{28.12}$$

Choosing $F_\pi = \frac{2m}{\sqrt{\lambda}} = \sqrt{2}v$ then makes the π kinetic term canonically normalized. This theory is called a **linear sigma model**. The π field is massless and is the Goldstone boson. π is often called a **pion** because, as we will see, it is closely related to the real-world hadrons $\pi^\pm$ and π^0.

The Lagrangian in Eq. (28.12) describes a massless particle π, as well as a massive particle σ. Massless Goldstone bosons such as π will appear in any theory with spontaneous symmetry breaking (by Goldstone's theorem), with one massless particle for each broken symmetry. Note that having a massless particle has nothing to do with how we parametrize ϕ; if we wrote $\phi(x) = \frac{2m}{\sqrt{\lambda}} + \tilde{\phi}(x)$ there would be a mass matrix for the two real components of $\tilde{\phi}$ which has a zero eigenvalue. Diagonalizing this matrix would lead back to our sigma model (see Problem 28.1). In the linear sigma model, the σ field has mass $m_\sigma = \sqrt{2}m$. The σ field can be visualized as radial excitations of the potential shown in Figure 28.1, which is commonly called a Mexican hat potential.

Goldstone bosons are naturally associated with shift symmetries. Recall that the broken symmetry was $\phi(x) \to e^{i\theta}\phi(x)$. The vacuum $\langle\phi\rangle = \sqrt{\frac{2m^2}{\lambda}}$ certainly breaks the symmetry. However, the symmetry is still realized as

$$\pi(x) \to \pi(x) + F_\pi\theta \tag{28.13}$$

with σ invariant. This is a symmetry of the sigma-model Lagrangian, Eq. (28.12). That a phase rotation of ϕ amounts to a shift in π can be seen transparently in Eq. (28.11). The symmetry can be used to strongly constrain the sigma model, even if the full theory that is spontaneously broken is not known. In particular, the shift symmetry *forbids a mass term for $\pi(x)$*. In fact, there is a close connection between Goldstone's theorem, which requires a massless mode, and shift symmetries of the Goldstone bosons, corresponding to movement around the flat direction of the potential, as in Figure 28.1.

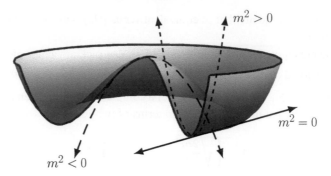

Fig. 28.1 Mexican hat potential. The masses squared of particles are given by the second derivatives of the potential. Expanding around the origin, there are two tachyonic (negative mass-squared) modes (long-dashed line). Expanding around a minimum, there is one mode with positive mass-squared (small-dashed line), corresponding to excitations along the radial direction, and one massless mode (solid line), corresponding to excitations along the symmetry direction where the potential is flat.

To distinguish the Goldstone bosons, whose interactions are determined by symmetry, from the radial modes, such as σ, which are model dependent and invariant under the symmetry, we can take the limit $m \to \infty$ and $\lambda \to \infty$ keeping $F_\pi = \frac{2m}{\sqrt{\lambda}}$ fixed. Then the Lagrangian reduces to

$$\mathcal{L} = \frac{1}{2}(\partial_\mu \pi)^2 \,, \tag{28.14}$$

which is a theory of a free pion. This **decoupling limit** is much more interesting in theories where the pions do not become free particles, like the ones we are about to discuss. The Lagrangian (28.14) is an example of a **nonlinear sigma model**, which is the linear sigma model in which the σ field has been decoupled.

To see that π is the Goldstone boson in Eq. (28.8), we calculate the Noether current in the decoupling limit from Eq. (28.14) using the symmetry transformation in Eq. (28.13). We find

$$J^\mu = \frac{\partial \mathcal{L}}{\partial(\partial_\mu \pi)} \frac{\delta \pi}{\delta \theta} = F_\pi \partial_\mu \pi. \tag{28.15}$$

Thus, defining $|\pi\rangle$ as the state created and annihilated by the π field, we have

$$\langle \Omega | J^\mu(x) | \pi(p) \rangle = ip^\mu F_\pi e^{-ipx}. \tag{28.16}$$

Comparing to Eq. (28.9) we see $|\pi\rangle$ is the Goldstone boson.

28.2.2 SU(2) × SU(2)

Now let us study a more interesting case. The QCD Lagrangian including only the up and down quarks is

$$\mathcal{L} = -\frac{1}{4}\left(F^a_{\mu\nu}\right)^2 + i\bar{u}\slashed{D}u + i\bar{d}\slashed{D}d - m_u\bar{u}u - m_d\bar{d}d. \tag{28.17}$$

If the quark masses were equal, this theory would have a global $SU(2)$ symmetry that rotates the up and down quarks into each other. In reality, the masses of the up and down quarks are close but not equal; more importantly, they are very small compared to Λ_{QCD} (which is the relevant scale as we will see). So let us just set the masses to zero for now. With $m_u = m_d = 0$, the theory actually has two independent $SU(2)$ symmetries, since the left-handed quarks and the right-handed quarks are completely decoupled. Indeed, writing the right- and left-handed spinors as $\psi_q^{R/L} = \frac{1}{2}(1 \pm \gamma_5)\,\psi_q$, the Lagrangian is

$$\mathcal{L} = -\frac{1}{4}\left(F_{\mu\nu}^a\right)^2 + i\bar{u}^L \slashed{D} u^L + i\bar{u}^R \slashed{D} u^R + i\bar{d}^L \slashed{D} d^L + i\bar{d}^R \slashed{D} d^R. \tag{28.18}$$

This is invariant under separate rotations:

$$\begin{pmatrix} u^L \\ d^L \end{pmatrix} \to g_L \begin{pmatrix} u^L \\ d^L \end{pmatrix}, \quad \begin{pmatrix} u^R \\ d^R \end{pmatrix} \to g_R \begin{pmatrix} u^R \\ d^R \end{pmatrix}, \tag{28.19}$$

where $g_L \in SU(2)_L$ and $g_R \in SU(2)_R$. Equivalently, the symmetry can be written as $q \to e^{i(\theta_a \tau^a + \gamma_5 \beta_a \tau^a)} q$ where $q = \begin{pmatrix} u \\ d \end{pmatrix}$ is a flavor doublet of the Dirac spinors u and d. The set of transformations parametrized by θ_a with $\beta_a = 0$ is the diagonal subgroup, called **isospin**. The set of transformations parametrized by β_a with $\theta_a = 0$ are the **axial rotations**.

The $SU(2)_L \times SU(2)_R$ symmetry of QCD is called a **chiral symmetry**, since it acts differently on left- and right-handed fields. Actually, the Lagrangian in Eq. (28.18) is invariant under $U(2) \times U(2) = SU(2)_L \times SU(2)_R \times U(1)_V \times U(1)_A$, with the two $U(1)$ symmetries called **vector** and **axial**. The Noether currents associated with these symmetries are (up to a sign)

$$J_\mu^a = \bar{q} \tau^a \gamma^\mu q, \quad J_\mu^{5a} = \bar{q} \tau^a \gamma^\mu \gamma^5 q, \quad J_\mu^V = \bar{q} \gamma^\mu q, \quad J_\mu^A = \bar{q} \gamma^\mu \gamma^5 q. \tag{28.20}$$

We will see in Chapter 30 that the axial $U(1)$, under which $q \to e^{i\theta \gamma_5} q$, is not an exact symmetry of QCD with massless quarks since it is broken by quantum effects called anomalies. The vector $U(1)$ symmetry, under which $q \to e^{i\theta} q$, is a symmetry even when quark masses are included, as in Eq. (28.17). It corresponds to baryon number conservation (or quark number conservation: quarks contribute $\frac{1}{3}$ to baryon number and antiquarks $-\frac{1}{3}$). In the full Standard Model, including weak interactions, baryon number is also anomalous. However, the difference between baryon number and lepton number, $B - L$, is non-anomalous. Because of these anomalies, we will postpone the discussion of the $U(1)$ symmetries until Chapter 30 and concentrate on the spontaneous breaking of $SU(2) \times SU(2)$.

Spontaneous symmetry breaking of $SU(2) \times SU(2)$ happened 14 billion years ago, when the temperature of the universe cooled below $T_C \sim \Lambda_{QCD}$. Below that scale, the thermal energy of quarks dropped below their binding energy and, instead of a big quark–gluon plasma, hadrons appeared. Although it has not been proven from QCD itself, the ground state of QCD apparently has a non-zero expectation value for the quark bilinears $\bar{u}u$ and $\bar{d}d$:

$$\langle \bar{u}u \rangle = \langle \bar{d}d \rangle = V^3. \tag{28.21}$$

We will confirm this by checking that it implies a spectrum of hadrons consistent with nature. One may have imagined that $\langle \bar{u}u \rangle$ and $\langle \bar{d}d \rangle$ could have had different expectation

Lagrangian. Indeed, the leading $SU(2) \times SU(2)$ invariant term we can add to our nonlinear sigma model is

$$\mathcal{L}_M = \frac{V^3}{2} \mathrm{tr}\left(MU + M^\dagger U^\dagger\right)$$

$$= V^3(m_u + m_d) - \frac{V^3}{2F_\pi^2}(m_u + m_d)\left(\pi_0^2 + \pi_1^2 + \pi_2^2\right) + \mathcal{O}\left(\pi^3\right). \qquad (28.36)$$

The prefactor $\frac{V^3}{2}$ is fixed so that the vacuum energy contributed by $\mathcal{L}_M$ matches the vacuum energy in $\mathcal{L}_m$. Indeed, when $\langle \bar{u}u \rangle = \langle \bar{d}d \rangle = V^3$, we find $\mathcal{L}_m = V^3(m_u + m_d)$, which matches the expansion in Eq. (28.36). We can now read off the pion masses:

$$m_\pi^2 = \frac{V^3}{F_\pi^2}(m_u + m_d). \qquad (28.37)$$

This is known as the **Gell-Mann–Oakes–Renner relation**. It says that the square of the pion mass scales linearly with the quark masses. For example, with $V \sim \Lambda_{\mathrm{QCD}} \sim 250\,\mathrm{MeV}$, $F_\pi = 92\,\mathrm{MeV}$, and $m_\pi = 140\,\mathrm{MeV}$, this relation gives $m_u + m_d \sim 11\,\mathrm{MeV}$. The Gell-Mann–Oakes–Renner relation has been confirmed with lattice QCD (see Figure 25.3). Keep in mind that these quark masses correspond to whatever renormalized masses appear in Eq. (28.35), which are not necessarily pole masses or $\overline{\mathrm{MS}}$ masses.

Thus, using only the pattern of symmetry breaking, we were able to extract the pion decay constant F_π, relate pion masses to quark masses, and calculate quantum effects such as pion scattering. The symmetries also constrain the pion interactions with baryons, such as the proton and neutron. Indeed, it was the modeling of the strong interactions among protons and neutrons through Yukawa forces mediated by pion exchange that elucidated the symmetry principles we have so concisely encoded in the Chiral Lagrangian.

By the way, note that if we contract Eq. (28.30) with p_μ we find, if the current J_μ^{5a} is conserved, that $p^2 F_\pi^2 = m_\pi^2 F_\pi^2 = 0$. This connects the chiral symmetry, with its corresponding conserved current, to masslessness of the Goldstone bosons. If the current is not exactly conserved, as in the real world because of quark masses, then $\partial_\mu J_\mu^{5a} = m_q \bar{q} \gamma^5 \tau^a q \neq 0$, in which case the pion picks up a mass proportional to m_q.

28.2.3 SU(3) × SU(3)

It is only a coincidence (as far as we know) that $SU(2)_{\mathrm{weak}}$ and $SU(2)_L$ relate the same two quarks. To the extent that three quarks can be treated as light, the discussion in Section 28.2.2 can be extended to $SU(3)_L \times SU(3)_R$ in a straightforward way. The third lightest quark is the strange quark, whose mass, $m_s \sim 100\,\mathrm{MeV}$, is not particularly small with respect to $\Lambda_{\mathrm{QCD}} \sim 300\,\mathrm{MeV}$. Nevertheless, the spontaneous breaking of $SU(3) \times SU(3) \to SU(3)$ provides an excellent description of additional strange mesons. The relatively large strange quark masses can be added as a perturbation to this picture, as the up and down quark masses were, and the resulting effective theory seems to work very well phenomenologically.

When $SU(3)_L \times SU(3)_R \to SU(3)_V$ through $\langle \bar{u}u \rangle = \langle \bar{d}d \rangle = \langle \bar{s}s \rangle = V^3$, the 16 symmetries are reduced to 8, leaving $16 - 8 = 8$ pseudo-Goldstone bosons. These are three pions, four kaons and an eta particle, which are embedded into the nonlinear sigma model field $U(x)$ as

$$U(x) \equiv \exp\left[2i\frac{\pi^a T^a}{F_\pi}\right] = \exp\left[\frac{\sqrt{2}i}{F_\pi}\begin{pmatrix} \frac{1}{\sqrt{2}}\pi^0 + \frac{1}{\sqrt{6}}\eta^0 & \pi^+ & K^+ \\ \pi^- & -\frac{1}{\sqrt{2}}\pi^0 + \frac{1}{\sqrt{6}}\eta^0 & K^0 \\ \bar{K}^- & \bar{K}^0 & -\sqrt{\frac{2}{3}}\eta^0 \end{pmatrix}\right].$$
(28.38)

Chiral symmetry relates many properties of these mesons. For details see [Georgi, 1984; Donoghue *et al.*, 1992].

Besides mesons, chiral symmetry breaking also describes baryons, which are bound states of three quarks. Three colored quarks can be combined into a color singlet with the totally antisymmetric tensor as $B = \epsilon_{ijk}q^i q^j q^k$. We need a little-group theory to see how they transform under the unbroken SU(3). The product of three triplets gives (see e.g. [Georgi, 1982])

$$3 \otimes 3 \otimes 3 = (6 \oplus \bar{3}) \otimes 3 = (6 \otimes 3) \oplus (\bar{3} \otimes 3) = 10 \oplus 8 \oplus 8 \oplus 1. \tag{28.39}$$

So there is a decuplet (the 10), two octets (called just 8 since the 8 is the adjoint representation which is real) and a singlet. The proton and neutron sit in one octet:

$$B_8 = \begin{pmatrix} \frac{1}{\sqrt{2}}\Sigma^0 + \frac{1}{\sqrt{6}}\Lambda & \Sigma^+ & P^+ \\ \Sigma^- & -\frac{1}{\sqrt{2}}\Sigma^0 + \frac{1}{\sqrt{6}}\Lambda & N \\ \Xi^- & \Xi^0 & -\frac{2}{\sqrt{6}}\Lambda \end{pmatrix}. \tag{28.40}$$

The meson and baryon octets in Eqs. (28.38) and (28.40) were given the enlightened moniker of **the eightfold way** by Murray Gell-Mann.

Another way to represent the octet or the decuplet is by their quantum numbers. Such diagrams are shown in Figure 28.2. Gell-Mann worked out these representations in 1962, when everything but the Ω^- had been seen. He was therefore able to predict that the Ω^- should exist, and, using symmetry, its mass and quantum numbers. The Ω^- was discovered in 1964 with exactly the properties Gell-Mann predicted. The Ω^- was historically important as a true theoretical prediction and helped people believe in quarks.

28.2.4 Discussion

In summary, we have seen that spontaneous symmetry breaking of chiral SU(2) × SU(2) leads to a triplet of pions (or the meson octet of pseudo-Goldstone bosons for the three-flavor case). The pions can be studied through a nonlinear sigma model with a field $U(x) = \exp(2i\pi^a \tau^a / F_\pi)$. The Lagrangian written in terms of $U(x)$ must be invariant under the full SU(2) × SU(2) symmetry. This strongly constrains the terms that can be written down. In fact, the transformation properties $U(x) \to g_L U g_R^\dagger$, under which the pions themselves transform nonlinearly, determine almost everything about pion couplings. This approach to determining pion couplings was pioneered by Callan, Coleman, Wess and Zumino (CCWZ) in 1969 [Callan *et al.*, 1969; Coleman *et al.*, 1969].

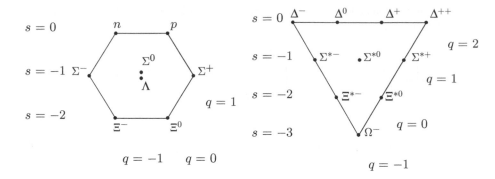

Fig. 28.2 Baryon octet and decuplet organized by quantum numbers. Diagonal lines have the same charge and horizontal lines have the same strangeness (number of strange quarks minus strange antiquarks in the hadron).

The effective theory is extremely predictive even at the quantum level, despite being non-renormalizable. Predictions were discussed in Chapter 22 on non-renormalizable theories. We have actually used the CCWZ trick a couple of times already: one was in building up the Lagrangians for massless spin-1 and spin-2 particles, in Section 8.7, and the other was in the Faddeev–Popov procedure, in Sections 14.5 and 25.4.

More generally, consider a continuous global symmetry G spontaneously broken down to a subgroup H. The vacuum is then invariant under H, but not under the remaining elements of G, which are denoted as a **coset** and written as G/H. The coset is not a subgroup of G (for example, it does not contain the identity element). We have seen that the Goldstone bosons transform in a linear representation of the unbroken subgroup H (e.g. the pions are a triplet of isospin) but nonlinearly under G/H.

An important point is that the nonlinear transformations, under which the Goldstone bosons shift, are transformations of fields, such as $\pi^a(x)$, but not of states appearing as excitations around the same vacuum in a Hilbert space. In that sense, nonlinear transformations are like gauge transformations, which are a concept derived from the Lagrangian description. In contrast, linearly realized global symmetries, as for the unbroken group H, act on states. These are symmetries with associated conserved charges which can be measured. There is no conserved charge for a broken symmetry, despite the fact that it can be restored in a Lagrangian with a nonlinear transformation. Since the vacuum is not invariant, the broken symmetry relates different ground states, and relates excitations around one ground state to excitations around another.

Finally, consider the case when the phase transition under which a symmetry group G is broken is smooth (i.e. second order). Above the symmetry-breaking scale there should be states transforming linearly under the full group. Thus, at the transition scale, since the transition is smooth, it must be possible to describe the system either with Goldstone bosons or with a linear multiplet. Thus, it must be possible to embed the Goldstone bosons into a linear multiplet. Moreover, the whole linear multiplet must be massless at the transition point since the Goldstone bosons are massless and the transition is smooth. The linear multiplet into which the Goldstone bosons are embedded is unique and therefore provides a precise definition of the order parameter. A more detailed discussion of this point is given in [Weinberg, 1996, Section 19.6].

28.3 The Higgs mechanism

We have seen that a spontaneously broken continuous global symmetry generates Goldstone bosons transforming as elements of the coset G/H, where G is the original symmetry group and H is the symmetry group of the vacuum. Now we consider what happens if there is a gauge boson associated with the broken symmetry. As we will see, this causes the Goldstone boson to disappear from the spectrum and the gauge boson to become massive through a procedure known as the Higgs mechanism.

The Higgs mechanism is not quite fairly named, since the same idea was discovered and understood by many people in different contexts, including Anderson (who proposed it first in a non-relativistic context in 1962), as well as Brout, Englert, Ginzburg, Guralnik, Hagan, Kibble, Landau and, of course, Higgs.

We will first discuss one physical example, type-II superconductors, which can be understood through an Abelian Higgs model. Then we will discuss non-Abelian theories, leading up to the Glashow–Weinberg–Salam model of electroweak symmetry, which is the subject of the next chapter.

28.3.1 Abelian Higgs model

Let us return to the linear sigma model from Section 28.2.1 and gauge the U(1) symmetry. The Lagrangian is then

$$\mathcal{L} = -\frac{1}{4}F_{\mu\nu}^2 + (\partial_\mu\phi^\star - ieA_\mu\phi^\star)(\partial_\mu\phi + ieA_\mu\phi) + m^2|\phi|^2 - \frac{\lambda}{4}|\phi|^4, \qquad (28.41)$$

which is known as the **Abelian Higgs model**. As before, the wrong-sign mass term for the scalar indicates that the ground state has $|\langle\phi\rangle| = \frac{v}{\sqrt{2}} = \sqrt{\frac{2m^2}{\lambda}}$. To see what happens to them, we write, as in Eq. (28.11),

$$\phi(x) = \left(\frac{v + \sigma(x)}{\sqrt{2}}\right) e^{i\frac{\pi(x)}{F_\pi}}. \qquad (28.42)$$

Plugging this in, our Lagrangian becomes

$$\mathcal{L} = -\frac{1}{4}F_{\mu\nu}^2 + \left(\frac{v+\sigma}{\sqrt{2}}\right)^2 \left[-i\frac{\partial_\mu\pi}{F_\pi} + \frac{\partial_\mu\sigma}{v+\sigma} - ieA_\mu\right]\left[i\frac{\partial_\mu\pi}{F_\pi} + \frac{\partial_\mu\sigma}{v+\sigma} + ieA_\mu\right]$$
$$- \left(-\frac{m^4}{\lambda} + m^2\sigma^2 + \frac{1}{2}\sqrt{\lambda}m\sigma^3 + \frac{1}{16}\lambda\sigma^4\right). \qquad (28.43)$$

First, look at the terms involving only A_μ:

$$\mathcal{L} = -\frac{1}{4}F_{\mu\nu}^2 + \frac{1}{2}e^2v^2A_\mu^2 + \cdots. \qquad (28.44)$$

This suggests that the gauge boson has picked up a mass:

$$m_A = ev. \tag{28.45}$$

Similarly, the σ field has mass $m_\sigma = \sqrt{2}m$ and π is massless. Unfortunately, because there are bilinear terms mixing π, σ and A_μ, extracting the spectrum is not quite that simple.

We can simplify things by decoupling σ through the limit $m, \lambda \to \infty$ with v fixed. We used this decoupling limit in Section 28.2.1. Taking this limit projects out the nonlinear sigma model, which is constrained by symmetries, from the linear sigma model, which has additional modes such as σ, about which we cannot say much. In the decoupling limit, the Lagrangian in Eq. (28.43) simplifies to

$$\mathcal{L} = -\frac{1}{4}F_{\mu\nu}^2 + \frac{1}{2}m_A^2\left(A_\mu + \frac{1}{eF_\pi}\partial_\mu\pi\right)^2. \tag{28.46}$$

This implies that we should set $F_\pi = v$ so that $\pi(x)$ has canonical normalization. This Lagrangian has a gauge boson mass term, a kinetic term for π, as well as an $A_\mu\partial_\mu\pi$ cross term indicating kinetic mixing between π and A_μ. The kinetic mixing makes interpreting the physical spectrum tricky; however, it can be removed through gauge-fixing, as we will now see.

The gauge symmetry in the Lagrangian in Eq. (28.46) is

$$A_\mu(x) \to A_\mu(x) + \frac{1}{e}\partial_\mu\alpha(x), \qquad \pi(x) \to \pi(x) - F_\pi\alpha(x). \tag{28.47}$$

Note that the π transformation is not the transformation law for a scalar field in a linear representation, but it is a gauge transformation nonetheless. Now we can remove the kinetic mixing by choosing a gauge. One gauge, called **unitary gauge**, just uses the shift to set $\pi(x) = 0$. In this gauge the Lagrangian becomes simply that of a massive gauge boson. Another convenient gauge is Lorenz gauge, $\partial_\mu A_\mu = 0$. In this gauge the cross term vanishes (after integration by parts), and the field π is massless with a normal kinetic term with the correct sign. In this gauge, the constrained gauge field has two degrees of freedom and the pion has one degree of freedom, which are the same three degrees of freedom of the unconstrained massive gauge boson in unitary gauge. Thus, we say that, in unitary gauge, the *gauge boson eats the Goldstone boson* through the **Higgs mechanism**.

In the case of broken local symmetries (in contrast to global symmetries), the low-energy theory has no memory that the symmetry was spontaneously broken instead of explicitly broken. Indeed, the Lagrangian in Eq. (28.44), with explicit symmetry breaking, can be turned into the nonlinear sigma model in Eq. (28.46) by integrating in a pion, that is, by performing a field redefinition $A_\mu \to A_\mu + \frac{1}{eF_\pi}\partial_\mu\pi$ (we performed this exercise in Section 8.7). This introduces a gauge invariance, Eq. (28.47). Using this gauge invariance to set $\pi = 0$ reverts to the theory with the massive gauge boson. In fact, introducing pions in this way turns out to be an efficient way to study the high-energy properties of a theory with a massive gauge boson, since scalars are easier to compute with than longitudinal modes of gauge bosons. That the pions and the longitudinal modes are equivalent is a result known as the Goldstone boson equivalence theorem, to be discussed more in Section 29.2.

Although in the low-energy theory massive gauge bosons will not reveal if the origin of their masses is from spontaneous symmetry breaking or not, spontaneously broken theories are renormalizable while explicitly broken ones are not. How is this possible if they are indistinguishable? The difference is the σ field, also known as the **Higgs boson**, present in the spontaneously broken theory in Eq. (28.43), but not in the explicitly broken one, Eq. (28.44). The Higgs boson plays a crucial role in the renormalizability of spontaneously broken gauge theories. This is easy to see from simply looking at the Lagrangian: the linear sigma model, including the full ϕ, has no terms with mass dimension greater than 4. In contrast, a nonlinear sigma model, with just the π fields, is generally non-renormalizable, as in Eq. (28.28). The Abelian Higgs model is a special case that happens to be renormalizable without the σ field because a photon has no self-interactions, or equivalently, because the π field has no interactions in the nonlinear sigma model.

28.3.2 Superconductors

The Abelian Higgs model is realized in nature in superconductors. The Ginzburg–Landau model of superconductivity simply postulates that the Lagrangian in Eq. (28.41) describes superconductors near the critical temperature T_C with ϕ the order parameter. So the effective Lagrangian is

$$\mathcal{L} = -\frac{1}{4}F_{\mu\nu}^2 + |D_\mu \phi|^2 - m^2 |\phi|^2 - \frac{1}{4}\lambda |\phi|^4, \qquad (28.48)$$

with $m^2 \sim T - T_C$ and D_μ the covariant derivative of QED. Below the critical temperature, the mass-squared for ϕ becomes negative and the $U(1)_{\text{QED}}$ is spontaneously broken. Thus, the photon picks up a mass, m_A. The effective low-energy Lagrangian in unitary gauge in the decoupling limit is

$$\mathcal{L} = -\frac{1}{4}F_{\mu\nu}^2 + \frac{1}{2}m_A^2 A_\mu^2. \qquad (28.49)$$

One immediate consequence of this effective description is that the photon mass term makes it energetically unfavorable to have magnetic fields. Indeed, a constant magnetic field would come from a linearly growing A_μ, giving an enormous contribution to the energy. Thus, magnetic fields must not be able to exist inside superconductors. The screening of magnetic fields inside superconductors is known as the **Meissner effect**. Another way to connect the Meissner effect to a photon mass is to recall that a massive photon generates a Yukawa potential with length scale $R = \frac{1}{m_A}$, known as the **penetration depth**. This is the characteristic scale with which magnetic fields can persist in a superconductor.

What happens if we crank up the magnetic field B? At some point, the field energy would be larger than the energy saved by having $\langle \phi \rangle \neq 0$, so we would lose superconductivity. Of course $\langle \phi \rangle$ does not have to be 0 everywhere or v everywhere; it can have finite-size domains where superconductivity is lost. These domains will have a characteristic size $\xi = \frac{1}{m}$, known as the **correlation length**. The two length scales are set by the two parameters in the Ginzburg–Landau model: m_A and m, with $m_A = ev = e\frac{2m}{\sqrt{\lambda}}$. In the case that $\xi < R$, so-called **type-II superconductors**, the system is unstable to formation of flux tubes of cross-sectional area $\pi \xi^2$ within the superconductor. These are known as

Abrikosov vortices. For $\xi > R$, the **type-I superconductors**, the vortex size is larger than the penetration depth, and so vortices will not spontaneously form.

To connect this model to superconductivity, we note that the flux in the vortices is quantized in units of $B\pi\xi^2$, so the vortices cannot dissipate smoothly to zero. At the microscopic level, these vortices have to be formed by electrons swirling around within the material. If there were any resistance to this motion, the electrons would slow down and the flux would change. Thus, this system must be superconducting. Thus, the Ginzburg–Landau model gives a direct connection between the Meissner effect and superconductivity (see e.g. [Weinberg, 1995] or [Altland and Simons, 2010] for a less hand-waving explanation).

The Ginzburg–Landau effective field theory description corresponds to a beautiful microscopic picture due to Bardeen, Cooper and Schrieffer. There, the phase transition is understood as due to attractive interactions between electrons through phonon exchange. So ϕ is identified with pairs of electrons, $\phi \sim e^- e^-$, which are known as **Cooper pairs** or **BCS pairs**. When $\langle\phi\rangle \neq 0$, the ground state has a non-zero charge, which explains why the symmetry of QED is broken.

28.3.3 Non-Abelian gauge theories

Next, consider the spontaneous breaking of a gauged non-Abelian symmetry. The procedure is almost identical to the Abelian case, but now there will be one massive gauge boson for each broken-symmetry generator. So the number of massive and massless bosons in the low-energy theory will depend on the representation of the group in which the order parameter transforms.

For example, consider an SO(3) gauge theory. We introduce three real scalars ϕ_i and the Lagrangian

$$\mathcal{L} = -\frac{1}{4}\left(F^a_{\mu\nu}\right)^2 + \frac{1}{2}\left(\partial_\mu\phi_i - igA^a_\mu\tau^a_{ij}\phi_j\right)^2 + \frac{1}{2}m^2\phi_i^2 - \frac{\lambda}{4!}\left(\phi_i^2\right)^2. \tag{28.50}$$

These scalars transform in the fundamental representation of SO(3). The potential is minimized for $|\langle\vec{\phi}\rangle| = v = \sqrt{\frac{6m^2}{\lambda}}$. By an SO(3) transformation, we can pick the direction and phase so that $\langle\phi_3\rangle = v$ and $\langle\phi_1\rangle = \langle\phi_2\rangle = 0$. That is, without loss of generality, we take

$$\left\langle\begin{pmatrix}\phi_1\\\phi_2\\\phi_3\end{pmatrix}\right\rangle = \begin{pmatrix}0\\0\\v\end{pmatrix}. \tag{28.51}$$

This vacuum is invariant under $H = \mathrm{SO}(2) \subset G = \mathrm{SO}(3)$, which rotates ϕ_1 and ϕ_2. Since SO(2) has one generator and SO(3) has three, there will be two Goldstone bosons that are eaten to form two massive gauge bosons. To see this explicitly, we can expand the Lagrangian in unitary gauge (that is, the nonlinear sigma model with $\pi = 0$). We find

$$\mathcal{L} = -\frac{1}{4}\left(F^a_{\mu\nu}\right)^2 + \frac{g^2}{4}A^a_\mu A^b_\mu \vec{v}^T\left\{\tau^a,\tau^b\right\}\vec{v}, \tag{28.52}$$

where $\vec{v} = \langle \vec{\phi} \rangle = \begin{pmatrix} 0 \\ 0 \\ v \end{pmatrix}$. We have symmetrized the $\tau^a \tau^b$ using $[A_\mu^a, A_\mu^b] = 0$. Plugging in the SO(3) generators (cf. Eq. (10.14)):

$$\tau^1 = i \begin{pmatrix} 0 & 0 & 0 \\ 0 & 0 & -1 \\ 0 & 1 & 0 \end{pmatrix}, \quad \tau^2 = i \begin{pmatrix} 0 & 0 & 1 \\ 0 & 0 & 0 \\ -1 & 0 & 0 \end{pmatrix}, \quad \tau^3 = i \begin{pmatrix} 0 & -1 & 0 \\ 1 & 0 & 0 \\ 0 & 0 & 0 \end{pmatrix}, \quad (28.53)$$

we see by explicit calculation that $\vec{v}^T \{\tau^a, \tau^b\} \vec{v}$ is only non-zero for $a = b = 1$ or $a = b = 2$. Thus,

$$\mathcal{L} = -\frac{1}{4}\left(F_{\mu\nu}^a\right)^2 + \frac{1}{2}m_A^2\left(A_\mu^1 A_\mu^1 + A_\mu^2 A_\mu^2\right), \quad (28.54)$$

with $m_A^2 = g^2 v^2$, which describes two massive gauge bosons and one massless one, as expected.

As a second example, consider an SU(5) gauge theory where the order parameter is a set of scalar fields Φ^a. This is called the **Georgi–Glashow model**. For SU(5) there are $5^2 - 1 = 24$ generators, which we call τ_{adj}^a in the adjoint representation. These are 24 traceless Hermitian matrices. One can write down a potential for $\Phi = \Phi^a \tau^a$ (see Problem 28.5) so that its expectation value is

$$\langle \Phi \rangle = \mathbf{v} = v \begin{pmatrix} 2 & & & & \\ & 2 & & & \\ & & 2 & & \\ & & & -3 & \\ & & & & -3 \end{pmatrix}. \quad (28.55)$$

The number of massless gauge bosons in the broken phase is determined by the subgroup of SU(5) that is unbroken by this vacuum expectation value. Clearly, there will be an SU(3) subgroup, rotating the top-left 3×3 block, and an SU(2) subgroup, rotating the bottom-right 2×2 block, which are unbroken and commute with each other. More generally, the mass term for A_μ^a is given by $\text{tr}([\mathbf{v}, \tau^a][\mathbf{v}, \tau^a])$, so the unbroken subgroup is generated by the generators of SU(5) that commute with $\mathbf{v}$. In addition to the block-diagonal SU(3) and SU(2) generators, there is also the generator proportional to $\mathbf{v}$ itself, which obviously commutes with $\mathbf{v}$. This generates a U(1) subgroup. So this vacuum expectation value breaks SU(5) $\rightarrow$ SU(3) $\times$ SU(2) $\times$ U(1). That is, it breaks SU(5) to the Standard Model gauge group. This suggests that the Standard Model gauge group might actually be just the unbroken subgroup of a larger SU(5). There are two amazing things about this type of **grand unification**: the gauge coupling constants are related and must unify (which they appear to do, more-or-less), and the quantum numbers of the quarks and leptons are explained from SU(5) representations.

A third example is the spontaneous breaking of SU(2) $\times$ U(1) $\rightarrow$ U(1), corresponding to the breaking of the electroweak symmetry down to the U(1) symmetry of electromagnetism. We will study this example in detail in Chapter 29.

28.4 Quantization of spontaneously broken gauge theories

To derive the Feynman rules for a spontaneously broken gauge theory, we have to work out the propagators for the Goldstone bosons as well as the massive gauge bosons. First of all, as we already observed, a broken gauge theory at low energy is the same whether it is explicitly or spontaneously broken; the difference is in the extra fields, such as σ in the linear sigma model, which are in general heavy. While fields such as σ are relevant to the UV completion of the theory with massive vector bosons, they are also model-dependent (that is, how many there are and their quantum numbers depend on the details of spontaneous symmetry breaking, in contrast to the Goldstone bosons and massive vector bosons). Since the Goldstone bosons can be completely removed in unitary gauge, we can choose this gauge, and then the only field left is the massive vector boson we discussed in Chapter 8. In unitary gauge, taking all the gauge boson masses equal for simplicity, the Lagrangian is

$$\mathcal{L} = -\frac{1}{4}\left(F^a_{\mu\nu}\right)^2 + \frac{1}{2}m_A^2\left(A^a_\mu\right)^2. \tag{28.56}$$

The propagator for the massive vector boson is then

$$\text{~~~~} \quad = \quad \frac{i}{p^2 - m_A^2}\left(-g^{\mu\nu} + \frac{p^\mu p^\nu}{m_A^2}\right). \tag{28.57}$$

This is the unitary-gauge propagator for the vector field. In many circumstances, it is preferable to be able to use other gauges, in which case the Goldstone bosons will also be propagating degrees of freedom.

The easiest way to derive the Lagrangian for the Goldstone bosons and the vector fields after spontaneous symmetry breaking, that is, the gauged nonlinear sigma model, is the CCWZ method discussed in Section 28.2.4. Starting with the Lagrangian for the massive vector bosons, the Goldstone bosons are introduced to restore the broken symmetry. That is, we replace $A^a_\mu \tau^a \to U A^a_\mu \tau^a U^{-1} - \frac{i}{g}\left(\partial_\mu U\right) U^{-1}$ as in Eq. (25.66) with $U = \exp\left(2i\frac{1}{F_\pi}\pi^a \tau^a\right)$. This leads to

$$A^a_\mu \to A^a_\mu + \frac{2}{gF_\pi}\partial_\mu \pi^a - \frac{2}{F_\pi}f^{abc}\pi^b A^c_\mu + \cdots, \tag{28.58}$$

as in Eq. (25.67). With this substitution, the massive vector Lagrangian, Eq. (28.56), becomes

$$\mathcal{L} = -\frac{1}{4}\left(F^a_{\mu\nu}\right)^2 + \frac{1}{2}m_A^2\left(A^a_\mu + \frac{2}{gF_\pi}\partial_\mu \pi^a + \cdots\right)^2. \tag{28.59}$$

As before, we must take $gF_\pi = 2m_A$ to give the pions canonically normalized kinetic terms. It is easy to check that this Lagrangian is gauge invariant under $A^a_\mu \to A^a_\mu + \frac{1}{g}\partial_\mu \alpha^a + \cdots$ and $\pi^a \to \pi^a - \frac{1}{2}F_\pi \alpha^a + \cdots$.

As it stands, this Lagrangian is not terribly convenient, since we still have the $A^a_\mu \partial_\mu \pi^a$ cross term, which mixes the two particles. The kinetic terms are also no longer invertible since we have introduced a redundancy (gauge invariance). To remedy these problems, we can break the gauge invariance we just introduced by adding a gauge-fixing term. In this case, we would like to introduce something that also removes the kinetic mixing. A natural choice, called R_ξ gauges, is[2]

$$\mathcal{L} = -\frac{1}{4}\left(F^a_{\mu\nu}\right)^2 + \frac{1}{2}m_A^2\left(A^a_\mu + \frac{1}{m_A}\partial_\mu\pi^a + \cdots\right)^2 - \frac{1}{2\xi}\left(\partial_\mu A^a_\mu - \xi m_A\pi^a\right)^2. \quad (28.60)$$

The new term removes the kinetic mixing and lets us invert the kinetic terms to find propagators. For the Abelian case, this completes the gauge-fixing. In the non-Abelian case, we have to gauge-fix carefully using the Faddeev–Popov procedure. We take as our gauge-fixing functional

$$G[A^a, \pi^a] = \partial_\mu A^a_\mu - \xi m_A\pi^a. \quad (28.61)$$

Then (changing the notation from Section 25.4 from π to α), we have

$$\det\left(\frac{\delta G\left[A^a_\mu + D_\mu\alpha^a, \pi^a - m_A\alpha^a\right]}{\delta\alpha}\right) = \det\left(\partial^\mu D_\mu + \xi m_A^2\right)$$

$$= \int \mathcal{D}c\,\mathcal{D}\bar{c}\exp\left(-i\int d^4x\,\bar{c}^a\left(\partial^\mu D_\mu + \xi m_A^2\right)c^a\right). \quad (28.62)$$

So, the gauge-fixed Lagrangian with ghosts is now

$$\mathcal{L} = -\frac{1}{4}\left(F^a_{\mu\nu}\right)^2 + \frac{1}{2}m_A^2\left(A^a_\mu + \frac{1}{m_A}\partial_\mu\pi^u\right)^2 - \frac{1}{2\xi}\left(\partial_\mu A^a_\mu - \xi m_A\pi^a\right)^2$$

$$- \bar{c}^a\left(\partial^\mu D_\mu + \xi m_A^2\right)c^a + \cdots. \quad (28.63)$$

The kinetic terms are

$$\mathcal{L}_{\text{kin}} = -\frac{1}{2}A^a_\mu\left(-g^{\mu\nu}\Box + \left(1 - \frac{1}{\xi}\right)\partial^\mu\partial^\nu - m_A^2 g^{\mu\nu}\right)A^a_\nu$$

$$- \frac{1}{2}\pi^a\left(\Box + \xi m_A^2\right)\pi^a - \bar{c}^a\left(\Box + \xi m_A^2\right)c^a. \quad (28.64)$$

Now we can simply calculate the propagators by inverting the kinetic terms. We find for the gauge fields:

$$\nu;b \;\text{〰〰〰}\; \mu;a \quad = \quad \frac{i}{p^2 - m_A^2}\left(-g^{\mu\nu} + \frac{p^\mu p^\nu}{p^2 - \xi m_A^2}(1 - \xi)\right)\delta^{ab}, \quad (28.65)$$

for the Goldstone bosons:

$$b \;\bullet\bullet\bullet\bullet\bullet\bullet\bullet\bullet\bullet\bullet\bullet\bullet\; a \quad = \quad \frac{i}{p^2 - \xi m_A^2}\delta^{ab}, \quad (28.66)$$

and for the ghosts:

$$b \;\bullet\bullet\bullet\bullet\bullet\bullet\blacktriangleright\bullet\bullet\bullet\bullet\bullet\; a \quad = \quad \frac{i}{p^2 - \xi m_A^2}\delta^{ab}. \quad (28.67)$$

---

[2] An alternative choice, with $\Delta\mathcal{L} = -\frac{1}{2\xi}(\partial_\mu A^a_\mu)^2$ are sometimes called *Fermi gauges*.

These are the covariant R_ξ gauge propagators for a spontaneously broken gauge theory.

For $\xi = \infty$ we are back in unitary gauge. Here,

$$i\Pi_{A_\mu^a A_\nu^b}(p) = \frac{i\left(-g^{\mu\nu} + \frac{p^\mu p^\nu}{m_A^2}\right)}{p^2 - m_A^2}\delta^{ab}, \qquad \Pi_{\pi\pi}(p) = 0, \quad \Pi_{c\bar{c}}(p) = 0. \qquad (28.68)$$

The numerator in the vector boson propagator sums over the three physical polarizations. Thus there is no need for anything else. This is called unitary gauge because only physical modes propagate. Thus, if you cut through a diagram, you will find only states that can go on-shell; thus unitarity will be satisfied trivially in the sense of the optical theorem.

For $\xi = 1$, Feynman–'t Hooft gauge,

$$i\Pi_{A_\mu^a A_\nu^b}(p) = \frac{-ig^{\mu\nu}}{p^2 - m_A^2}\delta^{ab}, \quad i\Pi_{\pi^a \pi^b} = \frac{i}{p^2 - m_A^2}\delta^{ab}, \quad i\Pi_{c^a \bar{c}^b}(p) = \frac{i}{p^2 - m_A^2}\delta^{ab}.$$
$$(28.69)$$

In this gauge, all the propagators have the same mass and simple numerators. You can think of the $g^{\mu\nu}$ in the vector boson propagator as summing over all four polarizations with the ghosts removing the longitudinal and timelike polarizations (as in the massless case), but now the Goldstone bosons put the longitudinal polarizations back into the physical spectrum.

Note that in any gauge with finite ξ all the propagators scale as $\frac{1}{p^2}$ at large p, so the theory will be renormalizable in the power-counting sense. And, of course, it can be renormalized. But it is still non-renormalizable in the sense that an infinite number of counterterms are needed. No problems are solved by these fancy gauges – the theory will still be strongly coupled at the scale $F_\pi = \frac{m_A}{g}$, as we discussed in Section 22.2 and will elaborate upon in Chapter 29.

The amazing thing about spontaneously broken gauge theories is that when they come from linear sigma models they are in fact renormalizable – you only need a finite number of counterterms. For renormalizability, the extra scalar field σ in the linear sigma model plays a crucial role. In the Standard Model, this σ is the Higgs boson. Since σ is not charged under the broken or unbroken symmetry, its kinetic terms have nothing to do with π or A_μ. Indeed, its Lagrangian is just that of Eq. (28.12):

$$\mathcal{L} = -\frac{1}{2}\sigma(\Box + 2m^2)\sigma - \frac{1}{2}\sqrt{\lambda}m\sigma^3 - \frac{1}{16}\lambda\sigma^4, \qquad (28.70)$$

where $m_\sigma = \sqrt{2}m$ is a totally separate free parameter from m_A. And so

$$i\Pi_{\sigma\sigma}(p) = \frac{i}{p^2 - m_\sigma^2}. \qquad (28.71)$$

At high energy this propagator scales as $\frac{1}{p^2}$. σ comes into the theory in exactly the right way to cancel the extra divergences of the massive vector theory. This is not at all obvious in any of the R_ξ gauges, but it is obvious physically: at high energy, $E \gg m$, the mass can be neglected and spontaneous symmetry breaking becomes a small effect. Thus, at high energy, the linear sigma model is just a gauge theory coupled to a linearly transforming matter field, and is renormalizable for the same reason that non-Abelian gauge theories are renormalizable.

Problems

28.1 Show that writing $\phi(x) = \sqrt{\frac{2m^2}{\lambda}} + \tilde{\phi}(x)$ for the linear sigma model in Section 28.2.1 leads to a mass matrix with zero eigenvalue. Show that when a linear combination of the two real fields in the complex field $\tilde{\phi}$ is chosen to diagonalize the mass matrix, the expansion in Eq. (28.12) results.

28.2 Work out the transformations to order π^2 and θ^2 in Eq. (28.25) using the Baker–Campbell–Hausdorff lemma:

$$\exp(A)\exp(B)$$
$$= \exp\left(A + B + \frac{1}{2}[A, B] + \frac{1}{12}[A, [A, B]] - \frac{1}{12}[B, [A, B]] + \cdots\right). \quad (28.72)$$

Show that pions transform in the adjoint representation under isospin.

28.3 Work out the interaction terms of order π^3 in the gauged nonlinear sigma model in Eq. (28.59).

28.4 Consider a theory with n real scalar fields and Lagrangian $\mathcal{L} = -\frac{1}{2}\phi_i(\Box - m^2)\phi_i - \frac{\lambda}{4}(\phi_i\phi_i)^2$.

(a) What are the global symmetries of this theory?

(b) What are all the possible vacua of this theory? Are all the vacua equivalent?

(c) Write down the Lagrangian for small excitations around one of the vacua. How many Goldstone bosons are there?

28.5 For grand unification based on $SU(5)$ to work, there must be a potential for the 24 scalar fields Φ^a such that $\Phi = \Phi^a \tau^a$ has a minimum in the form of Eq. (28.55). Consider the most general $SU(5)$-invariant potential for Φ:

$$V = -m^2\text{tr}(\Phi^2) + a\,\text{tr}(\Phi^4) + b\left[\text{tr}(\Phi^2)\right]^2. \quad (28.73)$$

One can always choose a basis where $\langle\Phi\rangle = v\,\text{diag}(a_1, a_2, a_3, a_4, a_5)$ with $\sum_i a_i = 0$.

(a) For what values of m^2, a and b is $\langle\Phi\rangle = v\,\text{diag}(2, 2, 2, -3, -3)$ an extremum?

(b) Show that excitations around $\langle\Phi\rangle = v\,\text{diag}(2, 2, 2, -3, -3)$ all have non-negative mass-squared.

(c) Find all possible minima for this potential. This is easiest if you impose the tracelessness condition with a Lagrange multiplier.

(d) For the minimum of the form $\langle\Phi\rangle = v\,\text{diag}(1, 1, 1, 1, -4)$, what are the masses of the massive gauge bosons, and what is the unbroken gauge group?

be arbitrarily large. For elastic scattering, the bound was that the partial wave amplitudes $|a_j| \le 1$. In this case, the fact that $t = -\frac{1}{2}E_{\text{CM}}^2(1 - \cos\theta)$ at high energy gives $a_0 = \frac{E_{\text{CM}}^2}{32\pi v^2}$, $a_1 = \frac{E_{\text{CM}}^2}{96\pi v^2}$ and $a_j = 0$ for $j > 1$. Thus, perturbative unitarity is violated at $E_{\text{CM}} \approx \sqrt{32\pi}v \approx 2.5\,\text{TeV}$. Considering other channels, the bound can be tightened to show that perturbative unitarity is violated for $E_{\text{CM}} \gtrsim 800$ GeV. That is, it *would* be violated, if there were no Higgs boson.

To be clear, this bound does not imply that a theory without a Higgs is not unitary, only that scattering amplitudes cannot be calculated reliably in perturbation theory. Since $\mathcal{M} \approx 1$ at $E_{\text{CM}} \approx 800\,\text{GeV}$ at tree-level, it is logical that loop amplitudes could be ~ 1 as well. When loop and tree amplitudes are the same size, perturbation theory breaks down. Of course, we already knew that the theory with massive vector bosons was non-renormalizable, so that loops could become important. The unitarity bound says that loops *must* become important at this scale and perturbation theory *must* break down.

Now, let us see how the physical Higgs boson comes to the rescue. There is only one Higgs exchange diagram, in the t-channel. It gives

$$\mathcal{M}_h = \quad\text{(diagram)}\quad = -\frac{e^2}{\sin^2\theta_w \cos^2\theta_w}\epsilon_1^\mu \epsilon_2^\nu \epsilon_3^{\star\alpha}\epsilon_4^{\star\beta}\left(g^{\alpha\mu}g^{\beta\nu}\right)\frac{m_W^2}{t - m_h^2}$$

$$= -\frac{e^2}{4m_Z^2 \sin^2\theta_w \cos^2\theta_w}\frac{t^2\left(t - 4m_W^2\right)\left(t - 4m_Z^2\right)}{(t - m_h^2)(t - 2m_W^2)(t - 2m_Z^2)}$$

$$= -\frac{t}{v^2} + \mathcal{O}(1). \tag{29.26}$$

Now we can see clearly that the high-energy behavior of $\mathcal{M}_{\text{tot}}$ is precisely canceled, for any m_h. On the other hand, if m_h is too large, then the perturbative unitarity bound would be violated before the Higgs contributions could kick in. This is the reason that we knew the Higgs boson (or something else serving its function) had to be found at the Large Hadron Collider. The precise bound from partial wave unitarity, including all the relevant channels, is

$$m_h \le \sqrt{\frac{16\pi}{3}}v \approx 1\,\text{TeV}. \tag{29.27}$$

This is called the **Lee–Quigg–Thacker bound** [Lee *et al.*, 1977].

29.2.1 Goldstone boson equivalence theorem

The above calculation of longitudinal W–Z scattering was done in unitary gauge ($\xi = \infty$), where there are no ghosts and the Goldstone bosons are eaten by gauge bosons. One could also do the calculation in any gauge, and the answer would, of course, be the same. It is illustrative in fact to consider the same calculation in Lorenz gauge ($\xi = 0$). There, the longitudinal modes are given by the Goldstone bosons, which have massless propagators. Therefore, what we want to calculate is Goldstone boson scattering. The Goldstone boson interactions are given by replacing the gauge fields by Goldstone bosons via $W_\mu^a \to \frac{1}{F_\pi}\partial_\mu \pi^a + \cdots$, where $F_\pi = v = \frac{m}{g}$. So, scattering Goldstone bosons should be very similar to scattering longitudinal modes with polarization vectors $\epsilon_\mu = p_\mu$, which is just what we calculated above.

To derive the interactions, we can use the linear sigma model:

$$\mathcal{L} = (D_\mu H)(D_\mu H^\dagger) - \frac{m_h^2}{2v^2}\left(H^\dagger H - \frac{1}{2}v^2\right)^2. \tag{29.28}$$

This is exactly what would result from taking $g \to 0$ and $m_W \to 0$ holding $v = 2\frac{m_W}{g}$ fixed in the full Lagrangian. Indeed, since the longitudinal modes/Goldstone bosons have a characteristic interaction strength of F_π, while the transverse modes interact with g, this limit will completely isolate the longitudinal components.

Next, substitute H as in Eq. (29.3):

$$H = \exp\left[i\frac{\sqrt{2}}{v}\left(\begin{matrix} \frac{z}{\sqrt{2}} & w^- \\ w^+ & -\frac{z}{\sqrt{2}} \end{matrix}\right)\right]\left(\begin{matrix} 0 \\ \frac{v+h}{\sqrt{2}} \end{matrix}\right), \tag{29.29}$$

giving

$$\mathcal{L} = \left[(\partial_\mu w_-)(\partial_\mu w_+) + \frac{1}{2}(\partial_\mu z)^2\right]\left(1 + \frac{2h}{v} + \frac{h^2}{v^2}\right) + \frac{1}{2}(\partial_\mu h)^2 - \frac{1}{2}m_h^2 h^2 + \cdots$$
$$+ \frac{1}{6v^2}\left[w_+^2(\partial_\mu w_-)^2 + w_-^2(\partial_\mu w_+)^2 - 2w_- w_+(\partial_\mu w_-)(\partial_\mu w_+)\right] + \cdots$$
$$- \frac{1}{3v^2}[z\,(\partial_\mu w_-) - w_-(\partial_\mu z)]\,[z(\partial_\mu w_+) - w_+(\partial_\mu z)] + \cdots. \tag{29.30}$$

Here, $w^\pm$ and z are the longitudinal modes for $W_\mu^\pm$ and Z.

By the way, the interactions among the Goldstone bosons here are identical to the pion interactions in the Chiral Lagrangian. Indeed, for the Goldstone bosons, one only needs a nonlinear sigma model, which can be derived using the CCWZ method discussed in Sections 8.7 and 28.2.4: start with the unitary gauge Lagrangian and replace $\mathbf{W}_\mu \to U^{-1}\mathbf{W}_\mu U$ ꟷ $U^{-1}D_\mu U$, where $U = \exp(2i\pi^a\tau^a/v)$. Then, taking $g \to 0$ holding v fixed brings us to the nonlinear sigma model ($h = 0$ above). Thus, the vacuum expectation value $v = \langle h \rangle$ plays the role that $F_\pi \approx \langle \bar{q}q \rangle^{1/3} \approx \Lambda_{\text{QCD}}$ plays in the Chiral Lagrangian.

For $W_L Z_L \to W_L Z_L$, all we need is the contact interaction on the last line. This looks just like a current–current interaction in scalar QED. There are $\binom{4}{2} = 6$ possible contractions, giving

$$\mathcal{M}_4(w^+ z \to w^+ z) = -\frac{1}{3v^2}\left[(p_1 - p_2)^\mu(p_3 - p_4)^\mu + (p_1 + p_4)^\mu(p_2 + p_3)^\mu\right] = \frac{t}{v^2}, \tag{29.31}$$

which agrees with the direct calculation of longitudinal gauge boson scattering above in the high-energy limit.

The Higgs exchange, which occurs only in the t-channel, gives

$$\mathcal{M}_h(w^+ z \to w^+ z) = -\frac{4}{v^2}\frac{(p_1 \cdot p_3)(p_2 \cdot p_4)}{t - m_h^2} = -\frac{t^2}{v^2(t - m_h^2)} + \mathcal{O}(t^0), \tag{29.32}$$

which agrees exactly with Eq. (29.26) in the high-energy limit.

This is known as the **Wolfenstein parametrization**. The angle θ_{12} is known as the **Cabibbo angle**. The Cabibbo angle is the rotation angle between the first two generations, and the only parameter relevant to hadronic physics involving light (u, d, s) quarks, so historically it was very important.

By the way, if there are only two generations, then the counting is as follows. A unitary 2×2 complex matrix has four real degrees of freedom. There is one rotation angle, for $SO(2)$, and three phases. But there is now a $U(1)^4$ chiral symmetry which can remove three phases, so there is in the end only one parameter, the Cabibbo angle. In particular, the CKM matrix can be taken real. As we will see in Section 29.5 below, if the CKM matrix is real, there can be no CP violation. Historically, CP violation was observed in the kaon system well before the third generation was discovered (even before charm was discovered), and a third generation was predicted as necessary for CP violation. We discuss this more below.

29.3.3 The unitarity triangle

In the Standard Model, the CKM matrix is unitary by construction. However, if there were a fourth generation, the restriction of the CKM matrix to the three-generation subsector would not be unitary. Thus, testing the CKM matrix for unitarity assuming three generations is a way to indirectly look for physics beyond the Standard Model. In practice, we try to measure all the CKM elements separately to check whether unitarity in fact holds. The current best measured values are [Particle Data Group (Beringer *et al.*), 2012]

$$
\begin{pmatrix} |V_{ud}| & |V_{us}| & |V_{ub}| \\ |V_{cd}| & |V_{cs}| & |V_{cb}| \\ |V_{td}| & |V_{ts}| & |V_{tb}| \end{pmatrix} = \begin{pmatrix} 0.97 \pm 0.0001 & 0.22 \pm 0.001 & 0.0039 \pm 0.0004 \\ 0.23 \pm 0.01 & 1.02 \pm 0.04 & 0.0041 \pm 0.001 \\ 0.0084 \pm 0.0006 & 0.039 \pm 0.002 & 0.88 \pm 0.07 \end{pmatrix}
$$

(29.60)

and we can see that the matrix is in fact unitary to within current uncertainties. We can also test whether there is a single phase (see Section 29.5 below).

The CKM element magnitudes in this table represent an aggregate compiled by the Particle Data Group. But what if we want to know how a new measurement fits in with this picture? A convenient way to see if the CKM elements associated with a particular measurement are consistent with the CKM matrix being unitary is to represent unitarity graphically with something called a **unitarity triangle**.

Unitarity implies that the rows of the CKM matrix are orthonormal, as are the columns. That is, $\sum_i V_{ij} V_{ik}^\star = \delta_{jk}$ for any i and k. For example, $V_{ud} V_{ub}^\star + V_{cd} V_{cb}^\star + V_{td} V_{tb}^\star = 0$. This equation says that three complex numbers add up to zero (there are five other such equations, but this one is a standard choice). Dividing by the best measured of these quantities, $V_{cd} V_{cb}^\star$, leads to

$$
\frac{V_{ud} V_{ub}^\star}{V_{cd} V_{cb}^\star} + \frac{V_{td} V_{tb}^\star}{V_{cd} V_{cb}^\star} + 1 = 0.
$$

(29.61)

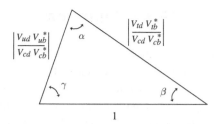

The unitarity triangle gives a graphical representation of CKM elements. Different measurements constrain its angles and side lengths.

Fig. 29.1

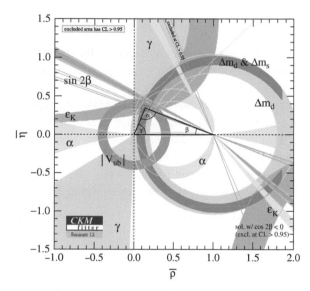

Precision flavor measurements mapped to the unitarity triangle [CKM fitter group (Charles *et al.*), 2012]. Length of the bottom edge has been normalized to 1, as compared to Fig. 29.1, by dividing all edge lengths by $V_{cd}V_{cb}^*$.

Fig. 29.2

This unitarity constraint can be represented as a closed triangle in the complex plane, as shown in Figure 29.1.

The lengths of the sides of the unitarity triangle measure flavor mixing and the angles of the triangle are sensitive to CP violation. Indeed, if all the CKM elements were real, the triangle would collapse to a line. Thus, we define a quantity J as twice the area of the (non-rescaled) triangle:

$$J \equiv 2(\text{area}) = \text{Im}(V_{ud}V_{tb}V_{td}^\star V_{ub}^\star) = (2.96 \pm 0.20) \times 10^{-5}, \qquad (29.62)$$

where J is known as the Jarlskog invariant (see Section 29.5). In practice, data are combined into a global fit for the unitarity triangle. There are public numerical programs for doing these fits, such as the CKMfitter package. A sample output from one of these programs is shown in Figure 29.2.

29.3.4 Neutrinos

Although neutrinos are very light, they are in fact massive. Neutrinos carry no electric charge (hence their name). If we assume that both left- and right-handed neutrinos exist, then neutrino masses can be generated after electroweak symmetry breaking from interactions of the form $Y^\nu_{ij} \bar{L}^i \widetilde{H} \nu^j_R$ (see Eq. (29.48)). Since L^i and $\widetilde{H}$ have the same weak and hypercharge quantum numbers, ν_R must be uncharged under both the weak and electromagnetic force, as in Table 29.1. We thus sometimes refer to the right-handed neutrinos as **sterile neutrinos**. The most general renormalizable mass terms in the lepton sector are

$$\mathcal{L}_{\text{mass}} = -Y^e_{ij} \bar{L}^i H e^j_R - Y^\nu_{ij} \bar{L}^i \widetilde{H} \nu^j_R - i M_{ij} (\nu^i_R)^c \nu^j_R + h.c. \qquad (29.63)$$

The last term in Eq. (29.63) denotes Majorana masses for the neutrinos, which are not forbidden by electroweak symmetry. In this term, $\nu^c_R = \nu^T_R \sigma_2$ is the charge conjugate Weyl spinor (see Section 11.3).

If neutrinos have any quantum numbers at all, then Majorana mass terms are forbidden. The most natural quantum number for right-handed neutrinos to have is lepton number (see Section 30.5.1). That is, if right-handed neutrinos carry lepton number, then Majorana masses are forbidden and the masses must be Dirac.

With neutrinos, we often go back and forth between Dirac spinor notation and Weyl spinor notation. Normally (as for the electron) we construct Dirac spinors out of independent left- and right-handed Weyl spinors, $\psi = \begin{pmatrix} \psi_L \\ \psi_R \end{pmatrix}$. As discussed in Section 11.3, we can also construct Dirac spinors out of single Weyl spinors as $\psi_R = \begin{pmatrix} -i\sigma_2 \nu^\star_R \\ \nu_R \end{pmatrix}$ or $\psi_L = \begin{pmatrix} \nu_L \\ i\sigma_2 \nu^\star_L \end{pmatrix}$. Then, Dirac and Majorana mass terms can be written in a uniform notation as (focusing on one generation for simplicity)

$$\mathcal{L}_{\nu,\text{mass}} = -m \bar{\psi}_L \psi_R - \frac{M}{2} \bar{\psi}_R \psi_R. \qquad (29.64)$$

In this notation, ψ_L and ψ_R can mix. Thus, the mass eigenstates are linear combinations that diagonalize the matrix $\begin{pmatrix} 0 & m \\ m & M \end{pmatrix}$. As you showed in Problem 11.9, the physical masses are $\sqrt{m^2 + \frac{1}{4}M^2} \pm \frac{1}{2}M$. In the limit that $M \gg m$, one mass is $m_{\text{heavy}} \approx M$ and the other is $m_{\text{light}} \approx \frac{m^2}{M} \ll m_{\text{heavy}}$. In particular, if one takes the Dirac masses to be electroweak scale, $m \approx 100$ GeV, and the Majorana masses to be very high, $M \approx M_{\text{Pl}} \approx 10^{19}$ GeV, then one finds $m_{\text{light}} \approx 10^{-6}$ eV. This explanation of the lightness of neutrino masses is called the **see-saw mechanism**: as M goes up, m goes down.

Why should the Majorana masses M_{ij} be so large? On the one hand, the Majorana mass terms are dimension 3 and hence super-renormalizable. So, following the Wilsonian RG picture (Section 23.6) one expects them to be UV sensitive. On the other hand, in the limit that $M_{ij} = 0$, the Lagrangian has a custodial symmetry, lepton number (or its non-anomalous cousin $B - L$, see Section 30.5.1). Thus, radiative corrections to the Majorana masses will be proportional to the Majorana masses themselves. In other words, in a theory with right-handed neutrinos, it is technically natural (see Box 22.1) for the Majorana masses to be small.

To understand the largeness of the see-saw scale, an important observation is that one does not need right-handed neutrinos at all to give neutrinos mass. If we allow non-renormalizable terms in the Lagrangian, then neutrino masses can be produced from a dimension-5 term:

$$\mathcal{L}_{\text{dim-5}} = -\widetilde{M}_{ij}\,(\bar{L}^i\widetilde{H})(\widetilde{H}L^j)^\dagger. \tag{29.65}$$

Such a term is in fact generated if we integrate out the right-handed neutrinos in Eq. (29.63) (see Problem 29.8). If the mass-eigenstate sterile right-handed neutrinos are very heavy, a dimension-3 mass term, like that in Eq. (29.64) is indistinguishable from a dimension-5 mass term, like that in Eq. (29.65). Since right-handed neutrinos have never been observed, a model without them is in a sense simpler. In addition, there is no custodial symmetry when these dimension-5 terms are turned off. Thus, one expects these terms to be generated at least by quantum gravity at the Planck scale. In other words, in a theory without right-handed neutrinos, the left-handed neutrinos naturally have masses parametrically smaller than the weak scale due to the see-saw mechanism.

Regardless of whether neutrinos are Majorana or Dirac, or whether the masses come from operators of dimension 3, 4 or 5, the only neutrinos we can ever measure are left-handed. Since left-handed neutrinos only interact via the weak force, it is more natural to work in the flavor basis than in the mass basis. We denote by ν_{Le}, $\nu_{L\mu}$ and $\nu_{L\tau}$ the left-handed electron, muon and tauon neutrinos respectively. In the flavor basis, the couplings to the W boson are diagonal (but the masses are not):

$$\mathcal{L}_{\nu W} = -\frac{g}{\sqrt{2}}\big(\bar{e}_L \slashed{W}\nu_{Le} + \bar{\mu}_L \slashed{W}\nu_{L\mu} + \bar{\tau}_L \slashed{W}\nu_{L\tau} + h.c.\big). \tag{29.66}$$

The mass eigenstates are related to these by a unitary transformation. We write ν_{L1}, ν_{L2} and ν_{L3} for the mass eigenstates. Then

$$\mathcal{L}_{\nu W} = -\frac{g}{\sqrt{2}}U^{ij}\big(\bar{e}_{Li}\slashed{W}\nu_{Lj} + h.c.\big), \tag{29.67}$$

where $\nu_{Le} = U^{e1}\nu_{L1} + U^{e2}\nu_{L2} + U^{e3}\nu_{L3}$ and so on. The matrix U is called the **Pontecorvo–Maki–Nakagawa–Sakata (PMNS) matrix**. It is the lepton analog of the CKM matrix. It can be written with an almost identical parametrization to Eq. (29.58):

$$U = \begin{pmatrix} c_{12}c_{13} & s_{12}c_{13} & s_{13}e^{-i\delta} \\ -s_{12}c_{23}-c_{12}s_{23}s_{13}e^{i\delta} & c_{12}c_{23}-s_{12}s_{23}s_{13}e^{i\delta} & s_{23}c_{13} \\ s_{12}s_{23}-c_{12}c_{23}s_{13}e^{i\delta} & -c_{12}s_{23}-s_{12}c_{23}s_{13}e^{i\delta} & c_{23}c_{13} \end{pmatrix}\begin{pmatrix} 1 & & \\ & e^{i\frac{\alpha_{12}}{2}} & \\ & & e^{i\frac{\alpha_{31}}{2}} \end{pmatrix}. \tag{29.68}$$

Note that $1,2,3$ refer to mass eigenstates defined in terms of flavor eigenstates by this matrix. We do not assume that $m_1 < m_2 < m_3$.

As with the CKM matrix, the PNMS matrix contains three mixing angles, θ_{12}, θ_{13} and θ_{23} (note that although we use the same notation, these angles are different from the CKM mixing angles). The phase δ is the **Dirac phase**. If neutrino masses are Dirac, there is only this phase. However, if there is a Majorana component to the neutrino masses, then two additional phases, α_{12} and α_{31}, are possible. You can show why exactly two extra phases occur in Problem 29.6. Thus there are three masses, three mixing angles and one or three phases in the neutrino sector.

It is very difficult to measure masses and mixing angles in neutrinos. Neutrino masses were first observed indirectly using **solar neutrinos** (neutrinos coming from the Sun). Practically all of the neutrinos produced by the Sun should be produced as electron (flavor eigenstate) neutrinos. However, the number of electron neutrinos observed on Earth that came from the Sun was found to be only around one-third of the number expected. This was the **solar neutrino problem**. The resolution is that (flavor eigenstate) neutrinos oscillate as they propagate through space. Indeed, it is only in the mass basis that the neutrino propagators are diagonal (see Problem 29.7). The solar neutrino problem was finally convincingly resolved by the Sudbury Neutrino Observatory (SNO) in 2001, which found that 35% of the solar neutrinos were ν_e and 65% were ν_μ or ν_τ. This confirmed not only that neutrinos oscillate, but that the solar models which predicted their production rate were correct.

Atmospheric neutrinos are those produced by cosmic rays. Cosmic rays (mostly protons) hit nuclei in the atmosphere, producing pions, which decay as $\pi^- \to \mu^- \bar{\nu}_\mu \to (e^- \bar{\nu}_e \nu_\mu)\bar{\nu}_\mu$. Thus, one expects a 2:1 ratio of muon to electron neutrinos coming from the atmosphere. Deviations from this ratio constrain other neutrino mixing angles and masses. Neutrino oscillations are also measured using **reactor neutrinos** (produced by nuclear reactors; mostly $\bar{\nu}_e$) and **accelerator** neutrinos (produced by particle accelerators; mostly ν_μ).

Neutrino oscillations are only sensitive to differences in squares of neutrino masses. Solar oscillations give $\Delta m_{21}^2 = m_2^2 - m_1^2 = (7.50 \pm 0.20) \times 10^{-5}$ eV^2, while atmospheric oscillations give $|\Delta m_{32}^2| = |m_3^2 - m_2^2| = 0.002\,32 \pm 0.000\,12$ eV^2. These differences are consistent with either $m_3 > m_2 > m_1$ (**normal hierarchy**) or $m_2 > m_1 > m_3$ (**inverted hierarchy**). The mixing angles are $\sin^2(2\theta_{12}) = 0.857 \pm 0.024$, $\sin^2(2\theta_{23}) > 0.95$ and $\sin^2(2\theta_{13}) = 0.098 \pm 0.013$. The Dirac CP phase δ has not been measured as of this writing (but may be soon), nor have the Majorana CP phases. To measure the Majorana CP phases, one would first have to measure that neutrinos are Majorana, which is extraodinarily challenging on its own. Majorana neutrinos would imply lepton number violation, for example in neutrino-less double β-decay (see Problem 11.9).

29.4 The 4-Fermi theory

Well before the electroweak unification was understood, its effective low-energy description, the 4-Fermi theory, was proven to give a very accurate phenomenological description of the weak interactions. Precision measurements at low energy gave indications of how heavy the W and Z bosons should be. They also indicated that the theory should involve vector currents (V) such as $\bar{\psi}\gamma^\mu\psi$ and axial vector currents (A) such as $\bar{\psi}\gamma^\mu\gamma_5\psi$. In fact, the structure of the electroweak theory was deduced from the $V - A$ (pronounced "V minus A") structure of the 4-Fermi theory. Writing $V - A = \gamma^\mu - \gamma^\mu \gamma^5 = 2\gamma^\mu P_L$, with $P_L = \frac{1}{2}(1 - \gamma^5)$, we see that the $V - A$ structure in the low-energy theory corresponds to a chiral theory in which weak interactions involve only left-handed fermions.[1]

The $W^\pm$ couple to the left-handed currents $J_\mu^{L\pm}$ as $\mathcal{L} = \frac{e}{\sqrt{2}\sin\theta_w}\left(W_+^\mu J_\mu^+ + W_-^\mu J_\mu^-\right)$ where

$$J_\mu^+ = \bar{\nu}_{eL}\gamma^\mu e_L + \bar{\nu}_{\mu L}\gamma^\mu \mu_L + \bar{\nu}_{\tau L}\gamma^\mu \tau_L + V_{ij}\bar{u}_L^i \gamma^\mu d_L^j, \tag{29.69}$$

$$J_\mu^- = \bar{e}_L\gamma^\mu \nu_{eL} + \bar{\mu}_L\gamma^\mu \nu_{\mu L} + \bar{\tau}_L\gamma^\mu \nu_{\tau L} + V_{ij}^\dagger \bar{d}_L^i \gamma^\mu u_L^j. \tag{29.70}$$

To derive the 4-Fermi theory, let us start with the lepton sector treating the neutrinos as massless (so we can ignore mixing angles). At tree-level, the interactions among the electron, muon and their neutrinos are

$$= \left(\frac{ie}{\sqrt{2}\sin\theta}\right)^2 (\bar{e}_L\gamma^\mu \nu_{eL} + \bar{\mu}_L\gamma^\mu \nu_{\mu L})$$

$$\times \frac{-i\left(g^{\mu\nu} - \frac{p^\mu p^\nu}{m_W^2}\right)}{p^2 - m_W^2} (\bar{\nu}_{eL}\gamma^\nu e_L + \bar{\nu}_{\mu L}\gamma^\nu \mu_L). \tag{29.71}$$

We call these **charged-current interactions**. At low energy, $p^2 \ll m_W^2$, and we can approximate these exchanges with a local 4-Fermi interaction:

$$\mathcal{L}_{4F} = -\frac{4G_F}{\sqrt{2}}\left(\bar{e}\gamma^\mu\left(\frac{1-\gamma^5}{2}\right)\nu_e + \bar{\mu}\gamma^\mu\left(\frac{1-\gamma^5}{2}\right)\nu_\mu\right)$$

$$\times \left(\bar{\nu}_e\gamma^\mu\left(\frac{1-\gamma^5}{2}\right)e + \bar{\nu}_\mu\gamma^\mu\left(\frac{1-\gamma^5}{2}\right)\mu\right), \tag{29.72}$$

where we have put in the γ_5 matrices using $P_L = \frac{1-\gamma_5}{2}$ so we can use Dirac spinors to describe the fermions, and

$$\frac{4G_F}{\sqrt{2}} \equiv \frac{e^2}{2m_W^2 \sin^2\theta_w} = \frac{g^2}{2m_W^2} = \frac{2}{v^2}. \tag{29.73}$$

Using the 4-Fermi Lagrangian gives a current–current interaction amplitude that is identical to Eq. (29.72) for $p^2 \ll m_W^2$. Thus, at low energy, the weak theory reduces to a set of 4-Fermi interactions among leptons (and quarks) with a universal strength given by G_F.

In particular, the muon decay rate is easy to calculate from the 4-Fermi theory. In the limit $m_\mu \gg m_e$, we find (cf. Problem 5.3)

$$\Gamma(\mu \to e\bar{\nu}\nu) = G_F^2 \frac{m_\mu^5}{192\pi^3}. \tag{29.74}$$

---

[1] Actually, there were some confusing indications through the 1950s that also scalar currents (S), such as $\bar{\psi}\psi$, or tensor currents (T), such as $\bar{\psi}\sigma^{\mu\nu}\psi$, were involved. Only the vector and axial vector currents can be easily embedded in a spontaneously broken renormalizable gauge theory; thus, careful measurements of spin and angular momentum in low-energy experiments were important inspirations for the electroweak theory.

From the measured muon lifetime, $\tau_\mu = 2.197\,\mu$s; this lets us deduce $G_F = 1.166 \times 10^{-5}\,\mathrm{GeV}^{-2}$. This determines that the electroweak vev is

$$v = 247\,\mathrm{GeV}, \tag{29.75}$$

and constrains one combination of $\sin\theta_w$ and m_W. Note that, since α_e is known and $\sin\theta_w < 1$, we also know that $m_W = \frac{v}{2}\frac{e}{\sin\theta_w} > 37.4\,\mathrm{GeV}$ and $m_Z = \frac{m_W}{\cos\theta_w} > m_W$. Thus, simply from the muon lifetime, we already knew in the 1960s that the W and Z must be quite heavy. Having an idea where to look helped motivate the design of the Super Proton Synchrotron (SPS) at CERN, with which the W and Z bosons were discovered in 1983.

Quarks can be studied with charged-current interactions in the 4-Fermi theory, much like leptons. The only complication is that now flavor mixing is an issue. Including the first two generations, the weak currents are expanded to

$$J_\mu^+ = \cdots + (\bar{u}_L \cos\theta_c - \bar{c}_L \sin\theta_c)\gamma^\mu d_L + (\bar{c}_L \cos\theta_c + \bar{u}_L \sin\theta_c)\gamma^\mu s_L, \tag{29.76}$$

where the $\cdots$ are the terms in Eq. (29.69), and similarly for J_μ^-. These mediate processes such as β-decay, for example $n \to p^+ e^- \bar{\nu}_e$. From precision measurements of radioactive decays and from rates for kaon decay, such as $K^+ \to \pi^0 e^+ \nu_e$ (here, $K^+ = \bar{s}u$), it was deduced that G_F in these processes is consistent with the leptonic measurements, and that $\sin\theta_c = 0.22$.

The **neutral-current interactions**, mediated by Z-boson exchange, are much harder to measure directly in the lepton sector. The first observation was in 1973 when $\nu_\mu e^-$ elastic scattering was observed. This was a great test of the electroweak theory, consistent with a Z boson, but it only gave a very poor measure of m_Z and θ_w. It was not until the mid 1990s that θ_w could be measured from this process directly. We now have very precise measurements: $m_W = 80\,\mathrm{GeV}$, $m_Z = 91.2\,\mathrm{GeV}$ and $\sin^2\theta_w = 0.21$. Moreover, measuring these quantities in multiple ways has provided important tests of the Standard Model and constraints on beyond-the-Standard-Model physics (see Chapter 31).

The Z boson couples to linear combinations of the J_μ^3 and QED currents. The interactions are

$$\mathcal{L}_{\mathrm{int}} = \frac{e}{\sqrt{2}\sin\theta_w}\left(W_+^\mu J_\mu^+ + W_-^\mu J_\mu^-\right) + \frac{e}{\sin\theta_w}Z_\mu J_\mu^Z + eA_\mu J_{\mathrm{EM}}^\mu, \tag{29.77}$$

where, from Eq. (29.40),

$$J_\mu^Z = \frac{1}{\cos\theta_w}J_\mu^3 - \frac{\sin^2\theta_w}{\cos\theta_w}J_\mu^{\mathrm{EM}} = \sum_i\left[\frac{1}{\cos\theta_w}\bar{\psi}_i^L\gamma^\mu T^3\psi_i^L - \frac{\sin^2\theta_w}{\cos\theta_w}Q_i\bar{\psi}_i\gamma^\mu\psi_i\right], \tag{29.78}$$

with ψ_i including quarks and leptons and T^3 being the SU(2) generator in the appropriate representation. Note that J_μ^Z only couples fermions to fermions of the same flavor. The full 4-Fermi theory can then be written as

$$\mathcal{L}_{\mathrm{4F}} = -\frac{4G_F}{\sqrt{2}}\left[J_\mu^+ J_\mu^- + \cos^2\theta_w\left(J_\mu^Z\right)^2\right]. \tag{29.79}$$

There is no $J_{\mathrm{EM}}^\mu J_{\mathrm{EM}}^\mu$ 4-Fermi interaction since the photon is massless and so, unlike the W and Z bosons, its propagator can never be approximated by a constant.

One immediate prediction of $\mathcal{L}_{4F}$ is that, since the neutral current is flavor diagonal, there will be no **flavor-changing neutral current** (FCNC) processes, such as $s \to de^+e^-$. This is an obvious result the way we have set things up, but it is not at all obvious without an electroweak theory. Indeed, historically, in the 1960s it was not at all clear why there were no FCNCs. In the 1960s the only hadrons known were made up of u, d and s quarks. The charm quark was then *predicted* to exist based on the absence of neutral currents, as we will now explain. When charm was discovered in 1974 the electroweak theory was spectacularly confirmed.

To see why charm is required to avoid FCNCs, let us forget about leptons and consider a theory with only two generations of quarks. Then there is only one mixing angle, θ_c, so we can choose a basis so that u and c quarks are flavor and mass eigenstates, while the left-handed d and s quarks are mixed. Then the two left-handed doublets are

$$Q^1 = \begin{pmatrix} u_L \\ \cos\theta_c d_L + \sin\theta_c s_L \end{pmatrix}, \quad Q^2 = \begin{pmatrix} c_L \\ \cos\theta_c s_L - \sin\theta_c d_L \end{pmatrix}. \tag{29.80}$$

The electromagnetic current is flavor diagonal for any number of quarks, so we will ignore it. The neutral current coming from weak interactions is

$$\begin{aligned} J^3_\mu &= \frac{1}{2}\bar{u}\gamma^\mu u - \frac{1}{2}(\cos\theta_c \bar{d} + \sin\theta_c \bar{s})\gamma^\mu(\cos\theta_c d + \sin\theta_c s) \\ &\quad + \frac{1}{2}\bar{c}\gamma^\mu c - \frac{1}{2}\left(\cos\theta_c \bar{s} - \sin\theta_c \bar{d}\right)\gamma^\mu(\cos\theta_c s - \sin\theta_c d) \\ &= \frac{1}{2}\bar{u}\gamma^\mu u - \frac{1}{2}\bar{c}\gamma^\mu c + \frac{1}{2}\bar{d}\gamma^\mu d - \frac{1}{2}\bar{s}\gamma^\mu s, \end{aligned} \tag{29.81}$$

where we have dropped the L subscripts for readability. This current is flavor diagonal, as expected. Now, suppose there were no charm quark. Then there would be no Q^2 and the neutral current would have a non-vanishing cross term $\cos\theta_c \sin\theta_c \bar{d}\gamma^\mu s$, implying $\bar{d}s \to \mu^+\mu^-$ and $K^0 \to \mu^+\mu^-$. So Glashow, Iliopoulos and Maiani (GIM) predicted that there must be a charm quark so that the flavor-changing process would cancel. The absence of FCNCs works for any number of generations, and is known as the **GIM mechanism**. It is a general consequence of the T^3 generator of SU(2) commuting with flavor rotations.

29.5 *CP* violation

That parity is violated in the weak theory is obvious: the left-handed fields couple differently from the right-handed fields. Parity violation is manifest in nuclear β-decay, which always produces left-handed electrons. However, one might imagine that, while the universe is not invariant under reflection in a mirror, it might still be invariant under that reflection accompanied by the interchange of particles and antiparticles. This is CP invariance. We now know that CP invariance is violated by rare processes involving hadrons. We call this weak CP violation. There is another possible form of CP violation, called strong CP violation, which is expected but has not been observed. The non-observation is known as the strong CP problem. We will now discuss both of these aspects of CP physics.

29.5.1 Weak *CP* violation

We derived how C and P act on fields and spinor bilinears in Chapter 11. Under the combination CP, we found:

$$\bar{\psi}_i \psi_j(t, \vec{x}) \to +\bar{\psi}_j \psi_i(t, -\vec{x}), \qquad \bar{\psi}_i \gamma^5 \psi_j(t, \vec{x}) \to -\bar{\psi}_j \gamma^5 \psi_i(t, -\vec{x}), \tag{29.82}$$

$$\bar{\psi}_i \slashed{A} \psi_j(t, \vec{x}) \to +\bar{\psi}_j \slashed{A} \psi_i(t, -\vec{x}), \quad \bar{\psi}_i \slashed{A} \gamma^5 \psi_j(t, \vec{x}) \to \bar{\psi}_j \slashed{A} \gamma^5 \psi_i(t, -\vec{x}), \tag{29.83}$$

which we can use to check which terms in the Standard Model Lagrangian can violate CP.

We showed above that one can perform chiral rotations on the left- and right-handed fermions of the Standard Model so that the quark masses are diagonal and the mixing is moved to the CKM matrix V. The relevant part of the electroweak Lagrangian is

$$\mathcal{L}_{\mathrm{mix}} = \frac{e}{\sqrt{2}\sin\theta_w} \left[\bar{u}_L V \slashed{W}^+ d_L \; + \; \bar{d}_L V^\dagger \slashed{W}^- u_L \right]$$

$$= \frac{e}{\sqrt{2}\sin\theta_w} \left[W_\mu^+ \bar{u} V \gamma^\mu \left(\frac{1-\gamma_5}{2} \right) d \; + W_\mu^- \bar{d} V^\dagger \gamma^\mu \left(\frac{1-\gamma_5}{2} \right) u \right], \tag{29.84}$$

where $\psi_{L/R} = \frac{1}{2}(1 \pm \gamma_5)\psi$ has been used to remove the projectors on the second line. Under CP, W^+ and W^- switch places since they are each other's antiparticles. So,

$$CP: \quad \mathcal{L}_{\mathrm{mix}} \to \frac{e}{\sqrt{2}\sin\theta_w} \left[W_\mu^+ \bar{u} \, (V^\dagger)^T \gamma^\mu \left(\frac{1-\gamma_5}{2} \right) d + W_\mu^- \bar{d} V^T \gamma^\mu \left(\frac{1-\gamma_5}{2} \right) u \right]. \tag{29.85}$$

Thus, the Standard Model Lagrangian is invariant under CP if $V^\star = V$, that is, if V is real. Thus:

> A non-zero phase in the CKM matrix implies CP violation.

There is an easier way to see that complex numbers imply CP violation. We know that any term in any local Lagrangian must be *CPT* invariant, which is true with real or complex coefficients. Since T sends $i \to -i$ in addition to whatever it does on fields, if a term is T invariant for real coefficients, it must be T violating for imaginary coefficients. By *CPT* invariance, we conclude that imaginary coefficients imply CP violation.

Recall that, in the flavor basis, all the flavor structure is in the Yukawa matrices. Consider the up-type quark (uct) mass terms:

$$\mathcal{L}_{\mathrm{Yuk}} = -\frac{v}{\sqrt{2}} \left[\bar{u}_L Y_u u_R + \bar{u}_R Y_u^\dagger u_L \right] = -\frac{v}{2\sqrt{2}} \left[\bar{u}(Y_u + Y_u^\dagger)u + \bar{u}\left(Y_u - Y_u^\dagger\right) \gamma^5 u \right]. \tag{29.86}$$

Under CP, $\bar{u}_i u_j \to \bar{u}_j u_i$ and $\bar{u}_i \gamma_5 u_j \to -\bar{u}_j \gamma_5 u_i$ (along with $\vec{x} \to -\vec{x}$), so

$$\mathcal{L}_{\mathrm{Yuk}} \to -\frac{v}{2\sqrt{2}} \left[\bar{u}(Y_u + Y_u^\dagger)^T u - \bar{u}\left(Y_u - Y_u^\dagger\right)^T \gamma^5 u \right]$$

$$= -\frac{v}{2\sqrt{2}} \left[\bar{u} \left(Y_u^\star + Y_u^{\dagger\star}\right) u + \bar{u}\left(Y_u^\star - Y_u^{\dagger\star}\right) \gamma^5 u \right]. \tag{29.87}$$

Thus, again we see that the Lagrangian is CP invariant if the coefficients are real.

Whether or not a matrix is real is not a basis-invariant statement. Indeed, in the flavor basis where the W interactions are flavor diagonal and the mass matrix is complex, $V = \mathbb{1}$, there is still CP violation. Conversely, even if the mass matrix were diagonal, and V were complex, there might still be no CP violation if some residual chiral rotation could remove the phase. For example, if one of the quarks is massless, this is always true. So it would be useful to have a basis-independent measure of CP violation.

Now recall that we relate the Yukawa couplings to the diagonal mass matrices via

$$Y_d = U_d M_d K_d^\dagger, \qquad Y_u = U_u M_u K_u^\dagger, \tag{29.88}$$

where $M_d = \frac{\sqrt{2}}{v}\text{diag}(m_d, m_s, m_b)$, $M_u = \frac{\sqrt{2}}{v}\text{diag}(m_u, m_c, m_t)$ and $V = U_u^\dagger U_d$. Thus, if $U_u = U_d$, then $V = \mathbb{1}$ with no flavor or CP violation. Before, we used the freedom to rotate right-handed fields without changing the weak interactions to remove K_d and K_u. We could equally well have rotated $d_R \to K_d U_d^\dagger d_R$ and $u_R \to K_u U_u^\dagger u_R$ so that $Y_d = U_d M_d U_d^\dagger$ and $Y_u = U_u M_u U_u^\dagger$, which makes the Yukawa matrices Hermitian. So let us assume Y_u and Y_d are Hermitian, without loss of generality. If Y_u and Y_d could be simultaneously diagonalized, then $V = \mathbb{1}$ and there is no CP violation. Thus, CP violation is all encoded in the commutator

$$-iC = [Y_u, Y_d] = \left[U_u M_u U_u^\dagger, U_d M_d U_d^\dagger \right] = U_u \left[M_u, V M_d V^\dagger \right] U_u^\dagger. \tag{29.89}$$

The matrix C is traceless and Hermitian because Y_u and Y_d are Hermitian. Thus, it is natural to look at its determinant as the obvious basis-invariant quantity:

$$\det C = \frac{16}{v^6}(m_t - m_c)(m_t - m_u)(m_c - m_u)(m_b - m_s)(m_b - m_d)(m_s - m_d)J, \tag{29.90}$$

where, for any i, j, k and l,

$$\text{Im}\left(V_{ij} V_{kl} V_{il}^\star V_{kj}^\star\right) = J \sum_{m,n} \varepsilon_{ikm}\varepsilon_{jln}, \tag{29.91}$$

where ε_{ijk} is the antisymmetric 3-index tensor. This is a fancy way of saying

$$J = \text{Im}(V_{11} V_{22} V_{12}^\star V_{21}^\star) = -\text{Im}(V_{11} V_{32} V_{12}^\star V_{31}^\star) = \text{Im}(V_{22} V_{33} V_{23}^\star V_{32}^\star) = \dots, \tag{29.92}$$

where these products are all equal due to the unitarity of the CKM matrix. J is known as the **Jarlskog invariant**. In terms of the standard parametrization,

$$J = s_{12} s_{23} s_{31} c_{12} c_{23} c_{31}^2 \sin \delta. \tag{29.93}$$

J has a nice geometric interpretation as well: it is twice the area of the unitarity triangle, as in Eq. (29.62).

The important point about the Jarlskog invariant is that it vanishes if and only if there is no CP violation. That is,

All weak CP violation in the Standard Model is proportional to $\text{Im}\,\det[Y_u, Y_d]$.

We have already seen that if V is real there is no CP violation. If V is real then $J = 0$ and so $\det C = 0$. Also, we note that if either two up-type or two down-type quarks are degenerate then $\det C = 0$ as well. For degenerate masses we get an extra phase rotation to remove the CP phase.

Note that since $\det C$ has many factors of masses $m_i \ll v$, it is in general quite small. Thus, even if the CP phase is large, the physical manifestations of CP violation are bound to be small. Another way to see this is to observe that if there were only two generations, then one could remove all the phases completely. Thus, any CP-violating effect in the Standard Model *must involve all three generations*. Consider, for example, the observed CP violation in kaon decays such as $K^+ \to \pi^+\pi^-$. One might imagine that, at the quark level, this is just $s \to \bar{u}du$ through a W exchange. However, such a Feynman diagram only involves the first two generations, and thus cannot explain the observed CP violation. Instead, it must be a loop-induced process. But the CKM elements coupling either of the first two generations to the third are small, thus the amount of observed CP violation is going to be suppressed by products of small CKM elements.

29.5.2 Measurements of weak *CP* violation

There are lots of ways to measure the one CP phase in the Standard Model. That all these measurements are consistent is an important check on the CKM matrix and often provides stringent constraints on beyond-the-Standard-Model physics. We will give only a brief summary of these measurements.

Historically, the first measurement of CP violation was through decays of neutral kaons. Kaons were discovered in 1946 through cosmic rays, and were "strange" because they had long lifetimes – they can only decay through strangeness-violating weak interactions. Their quark content is $K^0 = \bar{s}d$ and $\bar{K}^0 = \bar{d}s$, which are flavor eigenstates, but CP conjugates of each other. The CP eigenstates are

$$K_1 = \frac{K^0 + \bar{K}^0}{\sqrt{2}}, \qquad K_2 = \frac{K^0 - \bar{K}^0}{\sqrt{2}}, \tag{29.94}$$

with K_1 CP-even and K_2 CP-odd. Thus, to the extent that CP is a good symmetry, only K_1 can decay to $\pi\pi$, which is a CP-even final state, while K_2 must decay to $\pi\pi\pi$. This makes K_2 live much longer (52 ns) than K_1 (0.089 ns). What Christenson, Cronin, Fitch and Turlay famously found in 1964 was that the long-lived kaon sometimes *did* decay to $\pi\pi$, about 0.2% of the time, indicating CP violation. If the Hamiltonian commuted with CP, K_1 and K_2 would be the mass eigenstates, but since CP is violated, these states can mix with each other. The mass eigenstates in the K_1/K_2 system can be written as

$$K_S = K_1 + \epsilon K_2, \qquad K_L = K_2 - \epsilon K_1, \tag{29.95}$$

with $\epsilon = 0$ if CP is conserved. Christenson *et al.* found that $\epsilon \sim 2 \times 10^{-3}$. The most precise value today is $|\epsilon| = (2.228 \pm 0.011) \times 10^{-3}$.

The kaon system is actually a little more complicated, since it is also possible that the CP eigenstate K_2 could decay to $\pi\pi$ directly. To be more precise, if all the CP violation were due to mixing between K_1 and K_2 (this is called **indirect CP violation** or **CP violation from mixing**), then

$$\frac{\Gamma(K_L \to \pi^+\pi^-)}{\Gamma(K_L \to \pi^0\pi^0)} = \frac{\Gamma(K_1 \to \pi^+\pi^-)}{\Gamma(K_1 \to \pi^0\pi^0)} = \frac{\Gamma(K_S \to \pi^+\pi^-)}{\Gamma(K_S \to \pi^0\pi^0)}. \tag{29.96}$$

In addition, there can be **direct *CP* violation**, also called ***CP* violation from decay**, for which we introduce a new parameter ϵ' with $\mathcal{M}(K_2 \to \pi\pi) \propto \epsilon'$. Arguments that exploit the approximate isospin invariance of the meson system (see Chapter 28) show that

$$\eta_{+-} \equiv \frac{\mathcal{M}(K_L \to \pi^+\pi^-)}{\mathcal{M}(K_S \to \pi^+\pi^-)} = \epsilon + \epsilon', \qquad \eta_{00} \equiv \frac{\mathcal{M}(K_L \to \pi^0\pi^0)}{\mathcal{M}(K_S \to \pi^0\pi^0)} = \epsilon - 2\epsilon'. \tag{29.97}$$

Experimentally, it is found that $\left|\frac{\eta_{+-}}{\eta_{00}}\right| = 0.9951 \pm 0.0008$ so that $\mathrm{Re}\left(\frac{\epsilon'}{\epsilon}\right) = (1.65 \pm 0.26) \times 10^{-3}$. It is also possible to measure a third type of CP violation, from the **interference between mixing and decay**, which would show up in $\mathrm{Im}(\epsilon)$. Current measurements give $\mathrm{Im}(\epsilon) = (1.57 \pm 0.02) \times 10^{-3}$.

It is not possible to calculate theoretically ϵ or ϵ' due to the non-perturbative QCD effects in the required matrix elements. But it is also not hard to see if the measurements are roughly consistent with theory. Since CP violation requires three generations, at the perturbative level, there must be loops involving top or bottom quarks involved in the decays. For example, we could have a W loop and an intermediate top quark for the $s \to \bar{u}du$ decay. This would be suppressed by $V_{td} \sim 0.084$. The mixing can be estimated from box diagrams. The result is that the sizes of ϵ and ϵ' are apparently consistent with the CKM paradigm.

For many years CP violation had only been measured in kaon decays and mixing (including also additional modes, such as $K_L \to \mu^+\nu_\mu\pi^-$). The advent of B physics opened up a whole new world of CP-violating observables and has provided important checks on the CKM framework and strong constraints on new physics. CP violation has been observed in decays, first in $B^0 \to K^+\pi^-$ then in other modes, such as $B^0 \to \pi^+\pi^-$, $B^0 \to \eta K^{0\star}$, $B^+ \to \rho^0 K^+$, and also in interference $B \to J/\psi K_S$, $B \to \eta' K_S$, etc. So far, to the extent that we can connect these measurements to the CKM matrix (there are sometimes large theory uncertainties), everything seems perfectly consistent with a single CP phase. However, beyond-the-Standard-Model physics in CP violation could be just around the corner!

29.5.3 Strong *CP* violation

There is one more possible source of CP violation in the Standard Model. Sometimes global chiral symmetries, such as $\psi \to e^{i\gamma_5\theta}\psi$, that are symmetries of a classical Lagrangian are not symmetries of a quantum theory. When this happens we say the symmetries are **anomalous**. As we will discuss in the next chapter, anomalies can be understood as arising in situations in which a classical action is invariant under a symmetry transformation, but the path integral measure is not. For example, if we perform a chiral transformation on a quark, we find

$$\int \mathcal{D}\bar{\psi}\mathcal{D}\psi \to \int \mathcal{D}\bar{\psi}\mathcal{D}\psi \exp\left(i\theta \int \frac{g^2}{32\pi^2}\varepsilon^{\mu\nu\alpha\beta}F^a_{\mu\nu}F^a_{\alpha\beta}\right), \tag{29.98}$$

where $F_{\mu\nu}^a$ is the field strength for anything under which quarks are charged, g is the corresponding charge, and $\varepsilon^{\mu\nu\alpha\beta}$ is the totally antisymmetric tensor. For multiple generations, rotating by $\psi_R^i \to R^{ij}\psi_R^j$ and $\psi_L^i \to L^{ij}\psi_L^j$, the angle will be given by $\det(R^\dagger L) = re^{i\theta}$ for some $r \in \mathbb{R}$ (see Problem 29.9). Note that $\theta = \arg\det(R^\dagger L) = 0$ if the rotation is non-chiral.

The term $\varepsilon^{\mu\nu\alpha\beta}F_{\mu\nu}^a F_{\alpha\beta}^a$ is C-conserving but violates P, T and CP. To see this, recall from Chapter 11 that under CP,

$$A_0^a(t,\vec{x}) \to -A_0^a(t,-\vec{x}), \qquad A_i^a(t,\vec{x}) \to A_i^a(t,-\vec{x}), \qquad \partial_0 \to \partial_0, \qquad \partial_i \to -\partial_i.$$
(29.99)

If CP and P are both violated, then the terms

$$\mathcal{L}_{CPV} = \theta_{\rm QCD}\frac{g_s^2}{32\pi^2}\varepsilon^{\mu\nu\alpha\beta}F_{\mu\nu}^a F_{\alpha\beta}^a + \theta_2\frac{g^2}{32\pi^2}\varepsilon^{\mu\nu\alpha\beta}W_{\mu\nu}^a W_{\alpha\beta}^a + \theta_1\frac{g'^2}{16\pi^2}\varepsilon^{\mu\nu\alpha\beta}B_{\mu\nu}B_{\alpha\beta}$$
(29.100)

are allowed. Here $F_{\mu\nu}^a$, $W_{\mu\nu}^a$ and $B_{\mu\nu}$ are the $\mathrm{SU}(3),\mathrm{SU}(2)$ and $\mathrm{U}(1)$ field strengths, respectively. In fact, not only are these terms allowed, but they *must* be included since they may be generated through UV-divergent loop corrections and thus the θ_i are needed to renormalize the divergences. On the other hand, since the θ_i change if we perform chiral rotations, it is not clear whether they can have observable consequences, since observables must be independent of our chiral phase conventions.

To see whether the θ_i have observable consequences, let us revisit the Yukawa matrices, which we saw can be written as

$$Y_d = U_d M_d U_d^\dagger K_d^\dagger, \qquad Y_u = U_u M_u U_u^\dagger K_u^\dagger.$$
(29.101)

Here, extra factors of $U_d^\dagger$ and $U_u^\dagger$ have been inserted, without loss of generality. Then we can first perform chiral rotations on only the right-handed fields to remove K_u and K_d, and then perform non-chiral rotations to remove U_d and U_u. The phase induced by the K_d and K_u chiral rotations is given by (see Problem 29.9)

$$\arg\det(K_d K_u) = -\arg[\det(M_d M_u)\det(Y_d Y_u)] = -\arg\det(Y_d Y_u),$$
(29.102)

since $\det(M_d M_u) \in \mathbb{R}$. Thus, the CP violating term becomes, after this rotation,

$$\mathcal{L}_\theta = \bar{\theta}\frac{g_s^2}{32\pi^2}\varepsilon^{\mu\nu\alpha\beta}F_{\mu\nu}^a F_{\alpha\beta}^a, \qquad \bar{\theta} \equiv \theta_{\rm QCD} - \theta_F,$$
(29.103)

where

$$\theta_F \equiv \arg\det(Y_d Y_u).$$
(29.104)

A chiral rotation moves the phase back and forth between $\theta_{\rm QCD}$ and $\theta_{\rm F}$ leaving $\bar{\theta}$ unchanged. Thus, $\bar{\theta}$ is a basis-independent measure of CP violation, and can be physical. $\bar{\theta}$ is known as the **strong CP phase**. However, if $\det(M_d M_u) = 0$, that is, if any of the quark masses vanish, then $\bar{\theta}$ is again unphysical.

Before discussing the strong CP phase further, we note that the $\mathrm{SU}(2)$ and $\mathrm{U}(1)$ angles can be removed completely by chiral rotations. We saw that rotating only the right-handed fields can make the Yukawa couplings real, but θ_2 is unchanged by these rotations since right-handed fields are uncharged. Thus, we can rotate the left-handed fields only to put θ_2 into the Yukawa couplings then rotate the right-handed fields to remove it. Therefore, there is no basis-independent measure of CP violation for SU(2) and θ_2 is unphysical. Similarly, since neutrinos are uncharged, we can rotate them to show that the U(1) phase is unphysical. Thus, neither θ_2 nor θ_1 can have any physical consequences.

We have seen that $\bar{\theta}$ is basis independent, and if none of the quark masses vanish, then it can potentially be measured. But how will it show up? Not in perturbation theory! To see this, note that we can write

$$\varepsilon^{\mu\nu\alpha\beta}F_{\mu\nu}^a F_{\alpha\beta}^a = \partial^\mu K_\mu, \qquad K_\mu = \varepsilon_{\mu\nu\alpha\beta}\left(A_\nu^a F_{\alpha\beta}^a - \frac{g}{3}f^{abc}A_\nu^a A_\alpha^b A_\beta^c\right), \qquad (29.105)$$

showing that $\varepsilon^{\mu\nu\alpha\beta}F_{\mu\nu}^a F_{\alpha\beta}^a$ is a total derivative. K_μ is known as a **Chern–Simons current**. Total derivatives never contribute in perturbation theory – the Feynman rule would have a factor of the sum of all momenta going into the vertex minus the momenta going out, which gives a factor of zero. Thus, $\bar{\theta}$ can only have physical consequences through non-perturbative effects.

Although we cannot calculate the effect of $\bar{\theta}$ directly in QCD, we can actually make precise quantitative predictions using the Chiral Lagrangian, discussed in Chapter 28. Recall that the Chiral Lagrangian is a nonlinear sigma model in which the pions are embedded in a composite field $U(x) = \exp(2i\pi^a(x)\tau^a/F_\pi)$. Including the mass term, the Chiral Lagrangian is

$$\mathcal{L} = \frac{F_\pi^2}{4}\text{tr}\left[(D_\mu U)(D_\mu U^\dagger)\right] + \frac{V^3}{2}\text{tr}\left[MU + M^\dagger U^\dagger\right], \qquad (29.106)$$

where $V^3 = \langle \bar{u}u \rangle = \langle \bar{d}d \rangle$ and M is the quark mass matrix in QCD. As we saw in Section 28.2.2, the second term leads to the Gell-Mann–Oakes–Renner relation, $F_\pi^2 m_\pi^2 = V^3(m_u + m_d)$. To see the dependence on $\bar{\theta}$ we first use our chiral rotation to remove the phase from the $\varepsilon^{\mu\nu\alpha\beta}F_{\mu\nu}^a F_{\alpha\beta}^a$ term in the QCD Lagrangian completely, putting it back in the Yukawa couplings. This leads to complex quark masses. That is, now $M = \begin{pmatrix} m_u & \\ & m_d \end{pmatrix}e^{i\bar{\theta}}$. One immediate consequence is that the vacuum energy is now $\bar{\theta}$ dependent:

$$E(\bar{\theta}) = -V^3(m_u + m_d)\cos\bar{\theta} = -F_\pi^2 m_\pi^2 \cos\bar{\theta}. \qquad (29.107)$$

Unfortunately, this is not directly measurable.

By the way, the dependence of the energy on $\bar{\theta}$ can be understood a different way, by studying instantons. Briefly, there a countable set of gauge field configurations $|n\rangle$ that are locally gauge equivalent to 0, but cannot be gauged away globally due to a topological obstruction. The system can tunnel from one of these states to another through gauge-field configurations known as **instantons**. The stable energy eigenstates are Bloch-wave-like linear combinations of these $|\theta\rangle = \sum_n e^{in\theta}|n\rangle$, known as the **$\theta$-vacua**. They have energies $E(\theta) \sim \cos\theta$. The physical value of $\bar{\theta}$ is actually a sum of θ_{QCD}, θ_F and the value of θ which specifies the θ-vacuum of the system.

A more important consequence is that the neutron picks up an electric dipole moment proportional to $\bar{\theta}$. The calculation is not trivial, so we will only sketch it. The neutron and the proton form an isospin doublet, so their couplings to the pion have to be of the form

$$\mathcal{L}_{\pi NN} = \pi^a \bar{\Psi} \left(i\gamma^5 g_{\pi NN} + \bar{g}_{\pi NN} \right) \tau^a \Psi, \tag{29.108}$$

where Ψ is the proton–neutron isospin doublet. The first term is the ordinary Yukawa coupling to the pseudoscalar pions, which gives rise to the Yukawa potential describing the strong nuclear force among nucleons. The second term is CP-violating and must be proportional to $\bar{\theta}$. Upgrading isospin to SU(3) and using baryon mass relations one can show that [Crewther et al., 1979]

$$\bar{g}_{\pi NN} = -\frac{m_u m_d}{f_\pi (m_u + m_d)(2m_s - m_u - m_d)} (M_\Xi - M_N)\bar{\theta} \approx -0.04\,\bar{\theta}, \tag{29.109}$$

which can be compared to $g_{\pi NN} = 13.4$. Loops of pions such as

$$\tag{29.110}$$

(with the CP violation coming in at the $\bar{g}_{\pi NN}$ vertex) generate a neutron electric dipole moment. These loops are UV divergent. Cutting off the UV divergences at m_N gives

$$d_N = \frac{m_N}{4\pi^2} g_{\pi NN} \bar{g}_{\pi NN} \ln \frac{m_N}{m_\pi} = \left(5.2 \times 10^{-16} e \cdot \text{cm} \right) \bar{\theta}. \tag{29.111}$$

The current bound on the neutron EDM is $|d_N| < 2.9 \times 10^{-26} e \cdot \text{cm}$, so that

$$\bar{\theta} < 10^{-10}. \tag{29.112}$$

The smallness of $\bar{\theta}$ despite the large amount of CP violation in the weak sector is known as the **strong CP problem**.

Possible solutions to the strong CP problem include:

- One of the quarks is massless, $m_u = 0$. Unfortunately there is no symmetry protecting $m_u = 0$, since the chiral symmetry is anomalous. So m_u would just have to be tuned to be small instead of tuning $\bar{\theta}$ to be small. Thus, the $m_u = 0$ solution just moves the fine-tuning problem around.
- **Axions.** The idea behind axions is to add fields to the Standard Model so that there is a new anomalous U(1) symmetry. This symmetry is known after its authors as a **Peccei–Quinn symmetry**. If this $\mathrm{U(1)_{PQ}}$ is spontaneously broken, it will generate a new Goldstone boson, a. Then a chiral rotation can move the Goldstone boson into the $\bar{\theta}$ parameter, modifying the energy in Eq. (29.107) to

$$E\left(\bar{\theta}, a\right) = F_\pi^2 m_\pi^2 \cos\left(\bar{\theta} - \frac{a(x)}{f_a} \right), \tag{29.113}$$

where f_a is the **axion decay constant**. Then $\langle a \rangle = f_a \bar{\theta}$ and the ground state has no effective $\bar{\theta}$. The excitations around this vacuum are known as axions, and additionally provide a viable dark-matter candidate. Expanding Eq. (29.113), one finds $m_a = \frac{m_\pi f_\pi}{f_a}$, so that the axion decay constant is inversely proportional to the axion mass. Astrophysical bounds (for example, axion emissions from red giants) require $f_a > 10^{10}$ GeV, while cosmological bounds (too many axions would overclose the universe) require $f_a < 10^{12}$ GeV. Thus, the axion should be very weakly coupled with a mass 10^{-4} eV $< m_a < 10^{-2}$ eV. It is of course possible to evade these bounds with clever model building.

One concern about the axion solution to the strong CP problem is that the $\mathrm{U}(1)_{\mathrm{PQ}}$ symmetry must be very special. For example, let ϕ denote the field whose expectation value breaks $\mathrm{U}(1)_{\mathrm{PQ}}$. Since quantum gravity is non-renormalizable, we should generically include dimension n operators such as $c_n \frac{1}{M_{\mathrm{Pl}}^{n-4}} \phi^n + h.c.$ in the Lagrangian. After spontaneous breaking of $\mathrm{U}(1)_{\mathrm{PQ}}$, these will contribute to the potential $E(\bar{\theta}, a)$ terms such as $|c_n| \frac{f_a^n}{M_{\mathrm{Pl}}^{n-4}} \cos(na + \arg(c_n))$ which make $\langle a \rangle \neq \bar{\theta}$. For $\bar{\theta}$ to be consistent with current bounds on the neutron EDM requires operators with $n > 10$ be forbidden (or have exponentially small coefficients). See [Kamionkowski and March-Russell, 1992] for more information. There are of course ways to build models that forbid dangerous operators.

- Spontaneous CP violation. Here one supposes that, at some high scale, CP is an exact symmetry of nature, and is then spontaneously broken. When CP is a symmetry, the θ term is forbidden. Thus, all the CP violation appears in the Yukawa matrices. One can then connect the generation of a large weak CP phase and a small strong CP phase to the generation of mass and mixing angles. There are many ways to do this, but no overwhelmingly compelling model at this point.

29.5.4 Summary of *CP* violation

We have seen that the Standard Model contains two types of CP violation: weak and strong. To date, only weak CP violation has been observed. In the Standard Model, one can describe the weak CP phase in a basis-invariant way in terms of the Jarlskog invariant:

$$J = \mathrm{Im}(V_{11}V_{22}V_{12}^\star V_{21}^\star) = (2.96 \pm 0.20) \times 10^{-5}. \qquad (29.114)$$

As an angle, we can also write

$$\theta_{\mathrm{weak}} = \arg \det[Y_u Y_d - Y_d Y_u]. \qquad (29.115)$$

Or, one can identify the CP phase with the parameter δ in the CKM parametrization in Eq. (29.58). This phase has been experimentally measured to be $\delta = 69° \pm 5°$. One can measure weak CP violation many ways: in decays, in mixing, or in interference between decays and mixing. Historically, CP violation was measured first in the $K \to 2\pi$ decays, but now has been much more thoroughly tested using B mesons.

The strong CP phase has two components. One is the θ_{QCD} angle associated with

$$\mathcal{L}_{CP} = \theta_{\mathrm{QCD}} \frac{g_s^2}{32\pi^2} \varepsilon^{\mu\nu\alpha\beta} F_{\mu\nu}^a F_{\alpha\beta}^a. \qquad (29.116)$$

The other is $\theta_F = \arg \det [Y_u Y_d]$. These two angles rotate into each other under global chiral transformations of the Standard Model quarks. Only the combination $\bar{\theta} = \theta_{QCD} - \theta_F$ is possibly physical. Moving $\bar{\theta}$ into θ_{QCD}, we see that it has no effect to any order in perturbation theory, since $\varepsilon^{\mu\nu\alpha\beta} F^a_{\mu\nu} F^a_{\alpha\beta}$ is a total derivative. But it does have an important effect at low energy, where non-perturbative dynamics of QCD translate it into a CP-violating coupling between pions and nucleons. This should lead to an electric dipole moment for the neutron of order $(5.2 \times 10^{-16}) \, e \cdot \mathrm{cm} \, \bar{\theta}$. Current bounds then force $\bar{\theta} \leq 10^{-10}$.

One of the great mysteries of the Standard Model is why weak CP violation is nearly maximal ($\delta \sim \frac{\pi}{2}$) while strong CP violation is so small ($\bar{\theta} < 10^{-10}$). Another important fact about CP violation is that it is also necessary to explain the abundance of matter over antimatter in the universe. It turns out that there is not enough CP violation in the Standard Model to explain this abundance. Thus, there is good reason to think that there is more to be learned about CP violation.

Problems

29.1 The dominant production mechanism for Higgs bosons at LEP was $e^+ e^- \to ZH$. Calculate the total cross section for this process at tree-level in the Standard Model. How many 100 GeV Higgs bosons would there have been when LEP ran at 206 GeV?

29.2 $e + e^- \to$ hadrons in the Standard Model.
  (a) Calculate the rate for the total cross section $\sigma_{\mathrm{tot}}(e + e^- \to$ hadrons$)$ in the Standard Model at tree-level including both Z-boson and photon contributions and their interference. The contribution using photons alone was calculated in Section 26.3.
  (b) Calculate σ_{tot} at 1-loop.
  (c) Plot the total cross section as a function of center-of-mass energy showing separately the photon contribution, the Z-boson contribution, and their sum. Plot also the sum ignoring interference between the Z-boson and photon contributions. When can interference be ignored?

29.3 Higgs decays.
  (a) Calculate the rate for $H \to b\bar{b}$ in the Standard Model.
  (b) Calculate the rate for $H \to gg$ in the Standard Model. The dominant contribution to this comes from a triangle loop diagram involving top quarks.
  (c) Calculate the rate for $H \to \gamma\gamma$ in the Standard Model. Include contributions both from top loops and from loops of W bosons.
  (d) Plot the branching ratios for $H \to b\bar{b}$, $H \to gg$ and $H \to \gamma\gamma$ as a function of Higgs mass.

29.4 Partial wave unitarity.
  (a) Calculate the matrix element for longitudinal $W_L^+ W_L^- \to W_L^+ W_L^-$ scattering in the Standard Model.
  (b) Show that the high-energy behavior of this matrix element is reproduced using the Goldstone boson equivalence theorem.

(c) Does this give a stronger unitarity constraint than the one using $W_L^+ Z_L \to$ $W_L^+ Z_L$ scattering?

29.5 Figure 29.2 includes a number of experimental constraints on the CKM matrix.

(a) The parameter ε_K is what we were calling ϵ in Section 29.5.2. Why do the curves marked ε_K have the shape they do? That is, what combination of CKM elements is ε_K sensitive to?

(b) What could you measure to produce the curves marked Δm_d or $|V_{ub}|$?

29.6 Show that with general Dirac and Majorana mass matrices, there are three phases in the PNMS matrix, while if the mass matrix is purely Dirac, there is only one. How many phases are there if the masses are purely Majorana?

29.7 Neutrino oscillations.

(a) Neutrinos are produced in the Sun predominantly through the reaction $p + p + e^- \to d + \nu_e$. What is the momentum of the neutrinos produced this way?

(b) Consider a two-neutrino flavor system. The mass eigenstates evolve in time as

$$|\nu_1\rangle = e^{-iE_1 t}\Big(\cos\theta |\nu_e\rangle + \sin\theta |\nu_\mu\rangle \Big), \tag{29.117}$$

$$|\nu_2\rangle = e^{-iE_2 t}\Big(-\sin\theta |\nu_e\rangle + \cos\theta |\nu_\mu\rangle \Big), \tag{29.118}$$

where θ is the mixing angle. Show that the probability of finding a solar neutrino as an electron neutrino after a time T is given by

$$P = 1 - \sin^2(2\theta)\sin^2 \frac{(E_2 - E_1)T}{2}. \tag{29.119}$$

(c) Take the high-energy limit $E \gg m_\nu$ to show that the probability of finding a solar neutrino with energy E as an electron neutrino at a distance L is given by

$$P = 1 - \sin^2(2\theta)\sin^2 \frac{\Delta m^2 L}{4E}. \tag{29.120}$$

(d) How far should you put your detector from a reactor producing $\sim 4\,\mathrm{MeV}$ neutrinos assuming $\Delta m^2 = 7.5 \times 10^{-5}\,\mathrm{eV}^2$ to see the largest effect?

29.8 Show that when you integrate out the right-handed neutrinos in Eq. (29.63), a dimension-5 operator like that in Eq. (29.65) results. What is the exact relationship between M_{ij} and $\widetilde{M}_{ij}$?

29.9 Show that when multiple generations are rotated, then the θ angle shifts by $\arg\det\left(R^\dagger L\right)$.

30 Anomalies

Most of the time, a symmetry of a classical theory is also a symmetry of the quantum theory based on the same Lagrangian. When it is not, the symmetry is said to be **anomalous**. Since symmetries are extremely important for determining the structure of a theory, anomalies are also extremely important. In fact, anomalies have already been mentioned in two important contexts: in Chapter 28 they were invoked to justify why the Chiral Lagrangian was based on $\mathrm{SU}(2) \times \mathrm{SU}(2) \to \mathrm{SU}(2)$ and not $\mathrm{U}(2) \times \mathrm{U}(2)$, and in Chapter 29 they were used to explain the strong CP problem. These results will be reviewed and properly justified in Section 30.5.

Recall from Section 3.3 that continuous global symmetries imply conserved currents, through Noether's theorem. If a symmetry is anomalous then it is not actually a symmetry and the associated current will not be conserved. Such a situation has dire consequences for theories in which the current couples to a massless spin-1 particle, such as QED or Yang–Mills theory. If the current to which a massless spin-1 particle couples is not conserved, the Ward identity will be violated, unphysical longitudinal polarizations can be produced, and unitarity will be violated. Thus, in a unitary quantum theory, gauged symmetries (those with associated massless spin-1 particles) must be **anomaly free**. It turns out that this is a strong requirement for a consistent quantum theory. For example, in the Standard Model, it forces electric charge to be quantized, and the quark and lepton charges to be related, as we will see in Section 30.4.

Anomalies of symmetries associated with gauge bosons are called **gauge anomalies**. If a symmetry is not gauged, nothing goes terribly wrong if it is anomalous. That is, **global anomalies** do not lead to inconsistencies (the phrase *anomaly free* refers to the absence of gauge anomalies only). There are actually many global anomalies in the Standard Model. For example, baryon number conservation, that is, the symmetry that prevents quarks from turning into antiquarks, with associated Noether current $J^{\mu}_{\mathrm{baryon}} = \sum_i \bar{q}_i \gamma^{\mu} q_i$, is anomalous. This anomaly is allowed because there is no massless spin-1 particle in the Standard Model that couples to $J^{\mu}_{\mathrm{baryon}}$. In fact, baryon number violation is a necessary condition to explain the preponderance of matter over antimatter in the universe. Global anomalies also help explain why the η' meson is so heavy (the $\mathrm{U}(1)$ problem) and generate one of the greatest mysteries of the Standard Model: the strong CP problem, discussed in Section 29.5.3. These topics are all discussed in Section 30.5.

An important fact about anomalies is that they are infrared effects, from having massless particles in the spectrum. This leads to the idea of **anomaly matching**: the spectrum of massless particles in a theory below a phase transition is strongly constrained by the spectrum above the transition. For example, we will show in Section 30.6 that anomaly matching implies that the $\mathrm{SU}(3)_L \times \mathrm{SU}(3)_R$ flavor symmetry of QCD *must* be spontaneously broken, a fact that we had to assume in our study of the Chiral Lagrangian in Chapter 28. Anomaly matching provides strong constraints on the spectrum of bound states in strongly coupled theories.

Another type of anomaly, one we have already seen, is that of scale invariance. QCD (in the absence of quark masses) is scale invariant as a classical theory, but the quantum theory is certainly not scale invariant. In this case, the anomaly is called the **trace anomaly** and is proportional to the β-function. Conformal field theories are trace-anomaly free. The study of conformal field theories is a fascinating subject, but beyond our scope. In this chapter we will focus entirely on chiral anomalies, that is, anomalies which arise in theories that treat left-handed and right-handed fermions differently.

As in previous chapters we use the abbreviation $\langle \cdots \rangle \equiv \langle \Omega | T \{ \cdots \} | \Omega \rangle$.

30.1 Pseudoscalars decaying to photons

The way anomalies were first understood was through Feynman diagrams. This is not the easiest way to understand them, but it is important to show that they can be understood using methods you already know. We will start with the case in which a massive fermion runs around the loop. This avoids the ambiguities associated with massless fermions, which are discussed in Section 30.2. It also lets us calculate the rate for the decay $\pi^0 \to \gamma\gamma$, which, as we will see, provides an important way to measure the number of colors of quarks.

30.1.1 Triangle diagrams for massive fermions

To begin, forget about symmetries and just consider the QED Lagrangian with a Yukawa coupling between a fermion ψ and a pseudoscalar π:

$$\mathcal{L} = -\frac{1}{4}F_{\mu\nu}^2 - \frac{1}{2}\pi(\Box + m_\pi^2)\pi + \bar{\psi}(i\partial\!\!\!/ - eA\!\!\!/ - m)\psi + i\lambda\pi\bar{\psi}\gamma^5\psi. \tag{30.1}$$

You can think of π as the neutral pion, ψ as the proton, and the Yukawa coupling as $\lambda = \frac{m_p}{F_\pi}$ if you want (identifications we justify below), but the calculation we will do applies for any π, ψ and λ.

There are two 1-loop diagrams that contribute to $\pi \to \gamma\gamma$:

The sum of these diagrams is

$$iM = -1(-\lambda)(-ie)^2 \epsilon_\mu^{1\star} \epsilon_\nu^{2\star} M^{\mu\nu}(q_1, q_2), \tag{30.3}$$

where

$$M^{\mu\nu} = \int \frac{d^4k}{(2\pi)^4} \mathrm{Tr}\left[\gamma^\mu \frac{i}{\slashed{k} - m} \gamma^\nu \frac{i}{\slashed{k} + \slashed{q}_2 - m} \gamma^5 \frac{i}{\slashed{k} - \slashed{q}_1 - m}\right.$$
$$\left. +\gamma^\nu \frac{i}{\slashed{k} - m} \gamma^\mu \frac{i}{\slashed{k} + \slashed{q}_1 - m} \gamma^5 \frac{i}{\slashed{k} - \slashed{q}_2 - m}\right]. \tag{30.4}$$

Although superficially $M^{\mu\nu} \sim \int \frac{d^4k}{k^3}$ looks linearly divergent, it is easy to see that the result must be UV finite. By Lorentz invariance and symmetry under exchanging $1 \leftrightarrow 2$ and $\mu \leftrightarrow \nu$ (by bosonic statistics), the only two possibilities are that $M^{\mu\nu} \sim q_2^\mu q_1^\nu$ or $M^{\mu\nu} \sim \varepsilon^{\mu\nu\alpha\beta} q_\alpha^1 q_\beta^2$. Either way, by dimensional analysis, we could have, at worst, $M^{\mu\nu} \sim q^2 \int \frac{d^4k}{k^5}$, which is convergent in the UV.

First, we move all the γ-matrices to the numerator to find

$$M^{\mu\nu} = -i \int \frac{d^4k}{(2\pi)^4} \mathrm{Tr}\left[\frac{\gamma^\mu (\slashed{k} + m) \gamma^\nu (\slashed{k} + \slashed{q}_2 + m) \gamma^5 (\slashed{k} - \slashed{q}_1 + m)}{\left[(k - q_1)^2 - m^2\right]\left[(k + q_2)^2 - m^2\right][k^2 - m^2]} + \binom{\mu \leftrightarrow \nu}{1 \leftrightarrow 2}\right]. \tag{30.5}$$

Then we use

$$\mathrm{Tr}\left(\gamma^\mu \gamma^\nu \gamma^\alpha \gamma^\beta \gamma^5\right) = -4i\varepsilon^{\mu\nu\alpha\beta} \tag{30.6}$$

to simplify the numerator as

$$\mathrm{Tr}\left[\gamma^\mu (\slashed{k} + m) \gamma^\nu (\slashed{k} + \slashed{q}_2 + m) \gamma^5 (\slashed{k} - \slashed{q}_1 + m)\right] = 4im\varepsilon^{\mu\nu\alpha\beta} q_1^\alpha q_2^\beta. \tag{30.7}$$

Since this is symmetric under $1 \leftrightarrow 2$ with $\mu \leftrightarrow \nu$, the integral reduces to

$$M^{\mu\nu} = 8m\varepsilon^{\mu\nu\alpha\beta} q_\alpha^1 q_\beta^2 \int \frac{d^4k}{(2\pi)^4} \frac{1}{\left[(k - q_1)^2 - m^2\right]\left[(k + q_2)^2 - m^2\right][k^2 - m^2]}. \tag{30.8}$$

This can be evaluated using Feynman parameters in the usual way. The result is

$$M^{\mu\nu} = 8m\varepsilon^{\mu\nu\alpha\beta} q_\alpha^1 q_\beta^2$$
$$\times \left(\frac{-i}{16\pi^2}\right) \int_0^1 dx \int_0^{1-x} dy \frac{1}{m^2 - x(1-x) q_1^2 - y(1-y) q_2^2 - xy(s - q_1^2 - q_2^2)}, \tag{30.9}$$

where $s = (q_1 + q_2)^2$. We can next set $q_1^2 = q_2^2 = 0$ and $s = m_\pi^2$ since the photons and pion are on-shell. For the purposes of the $\pi^0 \to \gamma\gamma$ decay with the proton in the loop, we take $m_\pi \ll m_p = m$. Then the double integral gives $\frac{1}{2m^2}$. Combining with Eq. (30.3) we get

$$M = \lambda \frac{e^2}{4\pi^2 m} \varepsilon^{\mu\nu\alpha\beta} \epsilon_\mu^{1\star} \epsilon_\nu^{2\star} q_1^\alpha q_2^\beta \tag{30.10}$$

and therefore

$$\Gamma(\pi \to \gamma\gamma) = \frac{\alpha_e^2}{64\pi^3}\lambda^2 \frac{m_\pi^3}{m^2}.\qquad(30.11)$$

Thus, if we know λ and m we can calculate the decay rate to photons. We next discuss how we know λ and m for $\pi^0 \to \gamma\gamma$ and the physical implications of this calculation.

30.1.2 $\pi^0 \to \gamma\gamma$

To relate the result above to the physical pion decay rate, we need to know how the pion couples to charged fermions. These couplings can be extracted by recalling that the pions are Goldstone bosons from the spontaneously broken chiral symmetry of QCD. This interpretation of the pion was explained in Chapter 28, but we will review it here for clarity.

Recall the QCD Lagrangian with two effectively massless flavors ($m_u, m_d \ll \Lambda_{\text{QCD}}$):

$$\mathcal{L}_{\text{QCD}} = i\bar{\psi}_u \slashed{D}\psi_u + i\bar{\psi}_d \slashed{D}\psi_d.\qquad(30.12)$$

This Lagrangian is invariant not only under the global SU(2) symmetry (isospin) under which ψ_u and ψ_d transform as a doublet, but under a larger $\text{SU}(2)_L \times \text{SU}(2)_R$ symmetry under which the left-handed and right-handed quark doublets transform separately. Strong dynamics of QCD induces condensates, $\langle\bar{\psi}_u\psi_u\rangle \approx \langle\bar{\psi}_d\psi_d\rangle \approx \Lambda_{\text{QCD}}^3$, which spontaneously break $\text{SU}(2)_L \times \text{SU}(2)_R$ down to $\text{SU}(2)_{\text{isospin}}$. Thus, in the low-energy theory, particles only form multiplets of $\text{SU}(2)_{\text{isospin}}$. For example, the proton and neutron form an isospin doublet $\Psi = (\psi_p, \psi_n)$. Under elements $g_L \times g_R$ of the chiral symmetry group, this doublet, which can be written as $\Psi = \Psi_L + \Psi_R$, transforms as $\Psi_L \to g_L\Psi_L$ and $\Psi_R \to g_R\Psi_R$. The nucleon mass term, $m_N\bar{\Psi}\Psi = m_N\bar{\Psi}_L\Psi_R + m_N\bar{\Psi}_R\Psi_L$, is only invariant when $g_L = g_R$, that is, under $\text{SU}(2)_{\text{isospin}}$. Since $m_N \sim 1\,\text{GeV}$ is large, in the theory with just the proton and neutron there is little evidence of the original chiral symmetry.

A useful trick is to restore the full chiral symmetry by introducing a triplet of pions, π^a. These transform in the adjoint representation of isospin and nonlinearly under the broken generators of $\text{SU}(2)_L \times \text{SU}(2)_R$. The transformation properties are efficiently encoded by embedding the pions in a field $U = \exp(2i\pi^a\tau^a/F_\pi)$ transforming as $U \to g_L U g_R^\dagger$. This lets us write down a Lagrangian invariant under $\text{SU}(2)_L \times \text{SU}(2)_R$:

$$\mathcal{L} = \frac{F_\pi^2}{4}\text{tr}\left[(\partial_\mu U)(\partial_\mu U)^\dagger\right] + \bar{\Psi}_L i\slashed{\partial}\Psi_L + \bar{\Psi}_R i\slashed{\partial}\Psi_R - m_N\left(\bar{\Psi}_L U\Psi_R + \bar{\Psi}_R U^\dagger\Psi_L\right)$$

$$= \left(-\frac{1}{2}\pi^a\Box\pi^a + \cdots\right) + \bar{\Psi}\left(i\slashed{\partial} - m_N\right)\Psi + i\frac{2m_N}{F_\pi}\pi^a\left(\bar{\Psi}\gamma^5\tau^a\Psi + \cdots\right).\qquad(30.13)$$

To connect to charge-eigenstate fields, recall that the charged pions are $\pi^\pm = \frac{1}{\sqrt{2}}(\pi^1 \pm i\pi^2)$ and the neutral pion is $\pi^0 = \pi^3$. The proton and neutron form an isospin doublet. Using $\tau^3 = \text{diag}(\frac{1}{2}, -\frac{1}{2})$, the interaction involving the π^0 and the proton is then $i\frac{m_N}{F_\pi}\pi^0(\bar{\psi}_p\gamma^5\psi_p)$ with ψ_p the proton. In this way, the coupling of the neutral pion to a charged fermion (the proton) is determined. Thus, we can use Eq. (30.11) with $\lambda = \frac{m_N}{F_\pi}$ and $m = m_N$ to calculate the $\pi^0 \to \gamma\gamma$ decay rate. We find

$$\Gamma(\pi^0 \to \gamma\gamma) = \frac{\alpha_e^2}{64\pi^3}\frac{m_\pi^3}{F_\pi^2} = 7.77\,\text{eV},\qquad(30.14)$$

independent of m_N. The current experimental value is 7.73 ± 0.16 eV. So this is remarkably good!

Although the $\pi^0 \to \gamma\gamma$ rate was originally calculated (correctly) through a proton loop [Steinberger, 1949], as we have done, it was not done using the Chiral Lagrangian. All that is in fact needed is that the neutral pion is one of the Goldstone bosons associated with the spontaneous breaking of $\mathrm{SU}(2)_L \times \mathrm{SU}(2)_R \to \mathrm{SU}(2)$. That is, one just needs to identify $\langle \Omega | J_\mu^{5a}(x) | \pi^a(p) \rangle = i e^{ipx} F_\pi p_\mu$ (as discussed in Chapter 28) and to take $a = 3$ for the neutral pion. In QCD, $J_\mu^{5a} = \bar{q} \tau^a \gamma_\mu \gamma^5 q$ with τ^a the isospin generators. Although the pions are elementary particles in the Chiral Lagrangian but composite particles in QCD, the current-algebra relation does not care: $\langle \Omega | J_\mu^{5a}(x) | \pi^a(p) \rangle = i e^{ipx} F_\pi p_\mu$ holds in either theory. Normally, we cannot calculate anything about pions in perturbative QCD. The decay $\pi^0 \to \gamma\gamma$ is perhaps the unique exception to this rule: it does not get corrections from QCD beyond 1-loop. Although it is not at all obvious at this point, in the limit in which the pion is massless (so it is a Goldstone boson not a pesudo-Goldstone boson), the pion decay rate is exact at 1-loop. Moreover, since the final result is independent of the mass of the particle going around the loop, we do not need to know the quark masses. In other words, we can take Ψ to be either the proton (which is part of an isospin doublet with the neutron) or the up and down quarks (which form an isospin doublet with each other).

When Ψ is the (u, d) quark doublet instead of the (p^+, n) doublet, the factor of e^2 in the amplitude is multiplied by a factor of Q_i^2, where Q_i is the charge of the quark. Using $\tau^3 = \mathrm{diag}\left(\frac{1}{2}, -\frac{1}{2}\right)$ again, we see that the up quark gets the same isospin factor $\frac{1}{2}$ as the proton, but the down quark gets $-\frac{1}{2}$. In addition, we must sum over the number of colors N. Putting these factors together, the rate in Eq. (30.14) is multiplied by

$$ N\left[\left(\frac{2}{3}\right)^2 - \left(\frac{1}{3}\right)^2\right] = \frac{N}{3}. \tag{30.15} $$

Since the rate in Eq. (30.14) is already close to the known experimental value, we conclude that $N = 3$. Historically, this was one of first constraints on QCD [Adler, 1969], and it remains one of the easiest ways to measure the number of colors.[1]

30.1.3 Currents and symmetries

So far, we have just calculated the rate for a pseudoscalar to decay into two photons. We have not yet explained what this has to do with anomalous symmetries. In fact, the connection is not obvious. Indeed, the $\pi^0 \to \gamma\gamma$ rate calculation has a tumultuous history: getting the rate right was one thing, understanding the calculation was another. In the 1940s, when $\pi^0 \to \gamma\gamma$ was of particular interest, neither quantum field theory nor the profound importance of symmetries were well understood. Early attempts at this decay rate were producing non-gauge invariant amplitudes. A gauge-invariant result was finally achieved by Steinberger in 1949, using the recently proposed Pauli-Villars regularization scheme. However, Steinberger's result seemed to depend on the way in which the calculation was done. The

---

[1] We have shamelessly glossed over the fact that the π^0 is massive and its mass is not less than the quark masses (at least the masses defined through the Gell-Mann–Oakes–Renner relation Eq. (28.37), $m_\pi^2 = \frac{V^3}{F_\pi^2}(m_u + m_d)$). A proper treatment of quark masses gives small corrections to our calculation. Details can be found in [Adler, 1969], [Cheng and Li, 1985] or [Donoghue et al., 1992].

puzzle was solved by Schwinger in 1951 who calculated a gauge-invariant rate that was apparently free of ambiguities. (Schwinger's calculation, and his gauge-invariant proper-time formalism, are described in Chapter 33.) The calculation then rested for 20 years, until quantum field theory had matured. It was not until 1969, through the work of Alder, Bell and Jackiw, that the subtleties in the $\pi^0 \to \gamma\gamma$ calculation, and the connection to anomalous symmetries, were finally understood. A discussion of the history of anomalies can be found in [Bastianelli and van Nieuwenhuizen, 2006, Section 5.4].

The relevant symmetries to be considered are present in the QED Lagrangian:

$$
\begin{aligned}
\mathcal{L} &= \bar{\psi}(i\slashed{\partial} - e\slashed{A} - m)\psi \\
&= \bar{\psi}_L(i\slashed{\partial} - e\slashed{A})\psi_L + \bar{\psi}_R(i\slashed{\partial} - e\slashed{A})\psi_R - m\bar{\psi}_L\psi_R - m\bar{\psi}_R\psi_L,
\end{aligned} \tag{30.16}
$$

where the right- and left-handed fields are $\psi_{R/L} = \frac{1}{2}(1 \pm \gamma_5)\psi$ as usual. In the limit $m \to 0$, this Lagrangian is invariant under two independent global symmetries:

$$
\psi \to e^{i\alpha}\psi, \quad \psi \to e^{i\beta\gamma_5}\psi, \tag{30.17}
$$

or equivalently,

$$
\psi_L \to e^{i(\alpha-\beta)}\psi_L, \quad \psi_R \to e^{i(\alpha+\beta)}\psi_R. \tag{30.18}
$$

The symmetries under which the left- and right-handed fields transform the same way are called **vector symmetries**, and the symmetries under which they transform with opposite charge are called **chiral symmetries**. The Noether currents associated with these symmetries are

$$
J^\mu = \bar{\psi}\gamma^\mu\psi, \quad J^{\mu5} = \bar{\psi}\gamma^\mu\gamma^5\psi, \tag{30.19}
$$

which are called the **vector current** and **axial current** respectively. The equations of motion imply

$$
\partial_\mu J^\mu = 0, \quad \partial_\mu J^{\mu5} = 2im\bar{\psi}\gamma^5\psi. \tag{30.20}
$$

Thus, classically the vector symmetry is exactly conserved, which is important since it is the one that couples to QED, while the chiral symmetry is only conserved in the massless limit.

To connect to the $\pi^0 \to \gamma\gamma$ calculation, we first recall that the result of the loop diagram, Eq. (30.10), was that $\mathcal{M} = \lambda\frac{e^2}{4\pi^2 m}\epsilon^{\mu\nu\alpha\beta}\epsilon_\mu^{1\star}\epsilon_\nu^{2\star}q_1^\alpha q_2^\beta$. This loop can be interpreted as saying that the composite operator to which the pion couples, namely $\bar{\psi}\gamma^5\psi$, has a non-zero value in the presence of a background electromagnetic field. More precisely,

$$
\langle A|\bar{\psi}\gamma^5\psi|A\rangle = i\frac{e^2}{32\pi^2}\frac{1}{m}\varepsilon^{\mu\nu\alpha\beta}F_{\mu\nu}F_{\alpha\beta}. \tag{30.21}
$$

This equation will be derived rigorously in Chapter 33 for constant $F_{\mu\nu}$. It is consistent with Eq. (30.20) only if, in the presence of a constant background field $F_{\mu\nu}$, the axial current is not conserved:

$$
\langle A|\partial_\mu J^{\mu5}|A\rangle = -\frac{e^2}{16\pi^2}\varepsilon^{\mu\nu\alpha\beta}F_{\mu\nu}F_{\alpha\beta}. \tag{30.22}
$$

We will derive this result an alternative way in Section 30.3.

An important point is that Eq. (30.22) is independent of the mass m of whatever goes around the loop in the limit when that mass becomes small. Thus, it seems that if $m = 0$ exactly we should still have $\partial_\mu J_\mu^5 \neq 0$. On the other hand, if $m = 0$ the axial current is (classically) exactly conserved: $\partial_\mu J^{\mu 5} = 0$. These two statements are only consistent if the symmetry violation arises due to quantum effects, that is, if the chiral symmetry is anomalous. To clarify the situation we will next attempt to calculate $\partial_\mu J_\mu^5$ directly in the quantum theory with $m = 0$ from the start.

30.2 Triangle diagrams with massless fermions

It is not hard to see that the massless limit of the 1-loop calculation we just did is not going to be smooth: the numerator trace in Eq. (30.7) vanishes as $m \to 0$, since it is proportional to m, and the final result in Eq. (30.10) blows up, since it is proportional to $\frac{1}{m}$. Since what we are really interested in is the symmetry violation, it makes sense to recast the calculation as matrix elements of currents instead of matrix elements of the Goldstone bosons that these currents create from the vacuum.

30.2.1 Current matrix elements

We would like to see if the conservation laws $\partial_\mu J^\mu = \partial_\mu J^{\mu 5} = 0$, which hold in the classical theory with massless fermions, also hold in the quantum theory. Recall from Section 7.1 and Section 14.7 that the difference between classical and quantum theories is encoded in the Schwinger–Dyson equations. These equations describe how the classical equations of motion are modified for quantum fields inside correlation functions. Thus, we consider the correlation function $\langle J^{\alpha 5}(x) J^\mu(y) J^\nu(z) \rangle$, which is closely related to the triangle diagrams computed in the previous section. We would like to know if $\frac{\partial}{\partial x^\mu} \langle J^{\alpha 5}(x) J^\mu(y) J^\nu(z) \rangle = 0$. In this section, we calculate the relevant Feynman diagrams in perturbation theory. In Section 30.3, we use the path integral to rederive and reinterpret our perturbative result with the Schwinger–Dyson equations.

In momentum space, we want to calculate

$$iM_5^{\alpha\mu\nu}(p, q_1, q_2)(2\pi)^4\delta^4(p - q_1 - q_2)$$
$$= \int d^4x\, d^4y\, d^4z\, e^{-ipx}e^{iq_1y}e^{iq_2z}\langle J^{\alpha 5}(x)J^\mu(y)J^\nu(z)\rangle$$
$$= \int d^4x\, d^4y\, d^4z\, e^{-ipx}e^{iq_1y}e^{iq_2z}\langle [\bar\psi(x)\gamma^\alpha\gamma^5\psi(x)][\bar\psi(y)\gamma^\mu\psi(y)][\bar\psi(z)\gamma^\nu\psi(z)]\rangle.$$
(30.23)

Here, the brackets indicate that the spinor indices inside are contracted. This looks like an S-matrix element without the LSZ projection factors putting the external states on-shell. We can evaluate it just as we would any other Green's function, but with the positions of some fields taken at the same point. Indeed, it is not hard to see that the leading diagrams that contribute are the two in Eq. (30.2) without the external pion or photon lines, and without the coupling constants. Thus, at 1-loop the correlation function is

$$iM_5^{\alpha\mu\nu} = -\int \frac{d^4k}{(2\pi)^4} \mathrm{Tr}\left[\gamma^\mu \frac{i}{\not{k}} \gamma^\nu \frac{i}{\not{k}+\not{q_2}} \gamma^\alpha \gamma^5 \frac{i}{\not{k}-\not{q_1}} + \gamma^\nu \frac{i}{\not{k}} \gamma^\mu \frac{i}{\not{k}+\not{q_1}} \gamma^\alpha \gamma^5 \frac{i}{\not{k}-\not{q_2}}\right].$$

$$(30.24)$$

Rather than evaluating $M_5^{\alpha\mu\nu}$ and then contracting it variously with p^α, q_1^μ or q_2^ν, it is simpler to perform the contractions before evaluating the integrals.

Contracting the axial current with its momentum p^α gives

$$p_\alpha M_5^{\alpha\mu\nu} = \int \frac{d^4k}{(2\pi)^4} \left[\frac{\mathrm{Tr}\left[\gamma^\mu \not{k}\gamma^\nu (\not{k}+\not{q_2})\not{p}\gamma^5(\not{k}-\not{q_1})\right]}{k^2(k+q_2)^2(k-q_1)^2} + \left(\begin{array}{c}\mu\leftrightarrow\nu\\1\leftrightarrow 2\end{array}\right)\right].$$

$$(30.25)$$

In this case, the integral is superficially linearly divergent as in the massive case. To simplify the integral, we can use $\{\gamma^5, \gamma^\mu\} = 0$ and $p^\mu = q_1^\mu + q_2^\mu$ so that

$$\not{p}\gamma^5 = (\not{q_1}+\not{q_2})\gamma^5 = \gamma^5(\not{k}-\not{q_1}) + (\not{k}+\not{q_2})\gamma^5.$$

$$(30.26)$$

Then,

$$p_\alpha M_5^{\alpha\mu\nu} - \int \frac{d^4k}{(2\pi)^4} \left[\frac{\mathrm{Tr}\left[\gamma^\mu \not{k}\gamma^\nu (\not{k}+\not{q_2})\gamma^5\right]}{k^2(k+q_2)^2}\right.$$
$$\left. +\frac{\mathrm{Tr}\left[\gamma^\mu \not{k}\gamma^\nu \gamma^5(\not{k}-\not{q_1})\right]}{k^2(k-q_1)^2} + \left(\begin{array}{c}\mu\leftrightarrow\nu\\1\leftrightarrow 2\end{array}\right)\right]$$
$$= 4i\varepsilon^{\mu\nu\rho\sigma}\int \frac{d^4k}{(2\pi)^4} \left[\frac{k^\rho q_2^\sigma}{k^2(k+q_2)^2} + \frac{k^\rho q_1^\sigma}{k^2(k-q_1)^2}\right] + \left(\begin{array}{c}\mu\leftrightarrow\nu\\1\leftrightarrow 2\end{array}\right). \quad (30.27)$$

Each term in this expression has only one type of momentum in it (q_1 or q_2), so by Lorentz invariance the integral of each term has to give either $q_1^\rho q_1^\sigma$ or $q_2^\rho q_2^\sigma$, both of which vanish when contracted with $\varepsilon^{\mu\nu\rho\sigma}$. Thus, $p_\alpha M_5^{\alpha\mu\nu}$ appears to vanish, in contradiction to our expectations.

Before we make any rash conclusions, let us try to evaluate $q_\mu^1 M_5^{\alpha\mu\nu}$, which should be zero by the Ward identity of QED. We find

$$q_\mu^1 M_5^{\alpha\mu\nu} = \int \frac{d^4k}{(2\pi)^4} \left[\frac{\mathrm{Tr}\left[\not{q_1}\not{k}\gamma^\nu (\not{k}+\not{q_2})\gamma^\alpha\gamma^5(\not{k}-\not{q_1})\right]}{k^2(k+q_2)^2(k-q_1)^2}\right.$$
$$\left. +\frac{\mathrm{Tr}\left[\gamma^\nu \not{k}\not{q_1}(\not{k}+\not{q_1})\gamma^\alpha\gamma^5(\not{k}-\not{q_2})\right]}{k^2(k+q_1)^2(k-q_2)^2}\right]. \quad (30.28)$$

We can simplify these terms by writing $\not{q_1} = \not{k} - (\not{k}-\not{q_1})$ in the first term and $\not{q_1} = (\not{k}+\not{q_1}) - \not{k}$ in the second term, to remove terms in the denominator:

$$q_\mu^1 M_5^{\alpha\mu\nu} = \int \frac{d^4k}{(2\pi)^4} \left[\frac{\text{Tr}\left[\gamma^\nu(\slashed{k}+\slashed{q_2})\gamma^\alpha\gamma^5(\slashed{k}-\slashed{q_1})\right]}{(k-q_1)^2(k+q_2)^2} - \frac{\text{Tr}\left[\slashed{k}\gamma^\nu(\slashed{k}+\slashed{q_2})\gamma^\alpha\gamma^5\right]}{k^2(k+q_2)^2} \right.$$
$$\left. + \frac{\text{Tr}\left[\gamma^\nu\slashed{k}\gamma^\alpha\gamma^5(\slashed{k}-\slashed{q_2})\right]}{k^2(k-q_2)^2} - \frac{\text{Tr}\left[\gamma^\nu(\slashed{k}+\slashed{q_1})\gamma^\alpha\gamma^5(\slashed{k}-\slashed{q_2})\right]}{(k+q_1)^2(k-q_2)^2} \right]. \tag{30.29}$$

Evaluating the traces then gives

$$q_\mu^1 M_5^{\alpha\mu\nu} = -4i\varepsilon^{\alpha\nu\rho\sigma} \int \frac{d^4k}{(2\pi)^4} \left[\frac{(k^\rho-q_1^\rho)(k^\sigma+q_2^\sigma)}{(k-q_1)^2(k+q_2)^2} - \frac{k^\rho q_2^\sigma}{k^2(k+q_2)^2} \right.$$
$$\left. + \frac{k^\rho q_2^\sigma}{k^2(k-q_2)^2} - \frac{(k^\rho-q_2^\rho)(k^\sigma+q_1^\sigma)}{(k-q_2)^2(k+q_1)^2} \right]. \tag{30.30}$$

Now, if we were cavalier about the divergent integrals, we would shift $k \to k + q_1$ in the first integrand and $k \to k + q_2$ in the fourth integrand to see that the whole thing must vanish. Unfortunately, this is incorrect: one cannot shift the integration variable in a linearly divergent integral.

The mistake is very subtle. In fact, one of the reasons Schwinger set up his manifestly gauge-invariant proper-time formalism (Chapter 33) was to resolve confusions in the literature about this type of integral. The most obvious way to make a divergent integral well-defined is to introduce a regulator. Unfortunately, none of our favorite regulators will work without careful thought. For example, to use dimensional regularization we have to know how to handle chiral fermions and γ^5 in non-integer dimensions. With Pauli–Villars regularizatoin, the ghost-fermion mass explicitly breaks the chiral symmetry we are trying to verify. Instead, we proceed by trying to make sense of the linearly divergent integrals directly.

30.2.2 Linearly divergent integrals

Consider the one-dimensional integral

$$\Delta(a) = \int_{-\infty}^{\infty} dx[f(x+a) - f(x)], \tag{30.31}$$

where the function $f(x)$ goes to a constant at $x = +\infty$ and a different constant at $x = -\infty$. Then each term is linearly divergent, and we would like to know if the difference is finite or infinite. If we are allowed to shift $x \to x - a$ on just the first term, then $\Delta(a)$ vanishes at the level of the integrand. On the other hand, if we Taylor expand around $a = 0$ we find

$$\Delta(a) = \int_{-\infty}^{\infty} dx \left[af'(x) + \frac{a^2}{2}f''(x) + \cdots \right] = a\left[f(\infty) - f(-\infty)\right], \tag{30.32}$$

where the higher-derivative terms do not contribute since $f(\pm\infty) = \text{const}$. Thus, the difference between a linearly divergent integral and its shifted value has a linear dependence on the shift.

In four dimensions, we can do the same thing. In this case, we will need to evaluate integrals such as

$$\Delta^\alpha(a^\mu) = \int \frac{d^4k}{(2\pi)^4}(F^\alpha[k+a] - F^\alpha[k]). \tag{30.33}$$

Wick rotating, this is

$$\Delta^\alpha(a^\mu) = i \int \frac{d^4 k_E}{(2\pi)^4} \left(F^\alpha[k_E + a] - F^\alpha[k_E] \right). \tag{30.34}$$

Taylor expanding the integrand around $a = 0$, as in the one-dimensional case, we get

$$\Delta^\alpha(a^\mu) = i \int \frac{d^4 k_E}{(2\pi)^4} \left[a^\mu \frac{\partial}{\partial k_E^\mu} (F^\alpha[k_E]) + \frac{1}{2} a^\mu a^\nu \frac{\partial}{\partial k_E^\mu} \frac{\partial}{\partial k_E^\nu} (F^\alpha[k_E]) + \cdots \right]. \tag{30.35}$$

These derivative terms can then be integrated using Gauss's theorem. Since the integral is supposed to be linearly divergent, at large k_E our function must scale as

$$\lim_{k_E \to \infty} F^\alpha(k_E) = A \frac{k_E^\alpha}{k_E^4}. \tag{30.36}$$

Therefore, everything but the term with one derivative vanishes too fast at infinity to contribute. To evaluate the one-derivative term, we write it as a surface integral:

$$\Delta^\alpha(a^\mu) = ia^\mu \int \frac{d^4 k_E}{(2\pi)^4} \frac{\partial}{\partial k_E^\mu} (F^\alpha[k_E]) = ia^\mu \int \frac{d^3 S_\mu}{(2\pi)^4} F^\alpha[k_E]. \tag{30.37}$$

The surface element $d^3 S_\mu$ is normal to the surface of a 4-sphere at $|k_E| = \infty$. So it can be written as $d^3 S_\mu = k^2 k_\mu d\Omega_4$, where we drop the E subscript for clarity. Thus,

$$\Delta^\alpha(a^\mu) = ia^\mu \lim_{|k| \to \infty} \int \frac{d\Omega_4}{(2\pi)^4} A \frac{k^\mu k^\alpha}{k^2}. \tag{30.38}$$

Finally, we use $k^\alpha k^\mu \to \frac{1}{4} k^2 \delta^{\mu\alpha}$ and $\Omega_4 = 2\pi^2$ to get

$$\Delta^\alpha(a^\mu) = \frac{i}{32\pi^2} A a^\alpha. \tag{30.39}$$

This is a general result: linearly divergent integrals that would vanish if we could shift are finite, with the result proportional to the necessary shift.

30.2.3 Vector current conservation, continued

We can now evaluate the integral in Eq. (30.30). Dropping the second and third terms (they depend only on q_2 and must vanish by Lorentz invariance), we have

$$q_\mu^1 M_5^{\alpha\mu\nu} = -4i\varepsilon^{\alpha\nu\rho\sigma} \int \frac{d^4 k}{(2\pi)^4} \left[\frac{(k^\rho - q_1^\rho)(k^\sigma + q_2^\sigma)}{(k - q_1)^2 (k + q_2)^2} - \frac{(k^\rho - q_2^\rho)(k^\sigma + q_1^\sigma)}{(k - q_2)^2 (k + q_1)^2} \right]. \tag{30.40}$$

Part of this integrand is quadratically divergent, but vanishes because $\varepsilon^{\alpha\nu\rho\sigma} k^\rho k^\sigma = 0$. Thus, we have a linear divergence. The first term has k shifted from the second by $a^\sigma = q_2^\sigma - q_1^\sigma$. The linear divergence in the second term has the form

$$F^{\alpha\nu}(k) = -4i\varepsilon^{\alpha\nu\rho\sigma} \frac{(q_1^\rho + q_2^\rho) k^\sigma}{(k + q_1)^2 (k - q_2)^2} \xrightarrow{k \to \infty} -4i\varepsilon^{\alpha\nu\rho\sigma} (q_1^\rho + q_2^\rho) \frac{k^\sigma}{k^4}. \tag{30.41}$$

So $A^{\alpha\nu\sigma} = -4i\varepsilon^{\alpha\nu\rho\sigma}(q_1^\rho + q_2^\rho)$ and we get

$$q_\mu^1 M_5^{\alpha\mu\nu} = \frac{i}{32\pi^2} A^{\alpha\nu\sigma} a^\sigma = \frac{1}{4\pi^2} \varepsilon^{\alpha\nu\rho\sigma} q_1^\rho q_2^\sigma \neq 0. \tag{30.42}$$

Thus, it seems the Ward identity is violated for the vector current, but not the axial current.

The resolution to this mystery is that, although the integral was finite, it depended on the shift of k between the two integrals. But the choice of k as a loop momentum was arbitrary to begin with. The only constraint is that once we make a choice for k, we have to evaluate $M_5^{\alpha\mu\nu}$ once and for all – we cannot change our convention if we want to contract $M_5^{\alpha\mu\nu}$ with a different momentum. So let us take the most general possibility. We change k to

$$k^\mu \to k^\mu + b_1 q_1^\mu + b_2 q_2^\mu \tag{30.43}$$

in the first graph. Since we want to maintain Bose symmetry for the photons, we should take $k^\mu \to k^\mu + b_2 q_1^\mu + b_1 q_2^\mu$ in the second graph. This will change the result to (note that all the terms in Eq. (30.30) now contribute)

$$q_\mu^1 M_5^{\alpha\mu\nu} = \frac{1}{8\pi^2} \varepsilon^{\alpha\nu\rho\sigma} \left[(q_1^\rho + q_2^\rho)(1 - b_1 + b_2)(q_2^\sigma - q_1^\sigma) + (b_1 - b_2) q_1^\rho q_2^\sigma \right]$$

$$= \frac{1}{8\pi^2} \varepsilon^{\alpha\nu\rho\sigma} q_1^\rho q_2^\sigma (2 - b_1 + b_2). \tag{30.44}$$

Similarly, we find

$$p_\alpha M_5^{\alpha\mu\nu} = \frac{1}{4\pi^2} \varepsilon^{\mu\nu\rho\sigma} q_1^\rho q_2^\sigma (b_1 - b_2). \tag{30.45}$$

Thus, if we take $b_1 = b_2$ then

$$p_\alpha M_5^{\alpha\mu\nu} = 0, \quad q_\mu^1 M_5^{\alpha\mu\nu} = \frac{1}{4\pi^2} \varepsilon^{\alpha\nu\rho\sigma} q_1^\rho q_2^\sigma, \tag{30.46}$$

so that the axial current is conserved but the vector is not conserved. Alternatively, we can take $b_1 - b_2 = 2$, in which case

$$p_\alpha M_5^{\alpha\mu\nu} = \frac{1}{2\pi^2} \varepsilon^{\mu\nu\rho\sigma} q_1^\rho q_2^\sigma, \quad q_\mu^1 M_5^{\alpha\mu\nu} = 0, \tag{30.47}$$

so that the vector current is conserved but the axial current is not. This second choice agrees with what we found in the massive case, Eq. (30.22), as it should. Indeed, when the electron has a mass, there is no longer an ambiguity – the chiral symmetry is already broken, so only the vector symmetry could possibly be conserved.

30.2.4 Discussion

We have found that the choice of momentum routing in the loops can affect the symmetry properties of the final result. You can think of this as a choice of regulator, although it is not really a regulator but rather a different type of ambiguity inherent in divergences of individual Feynman diagrams. If one insists on preserving gauge invariance, then for QED with a single Dirac fermion, we showed that $\partial_\mu \langle J^{\alpha 5} J^\mu J^\nu \rangle = \partial_\nu \langle J^{\alpha 5} J^\mu J^\nu \rangle = 0$ so that the Ward identity is satisfied, but $\partial_\alpha \langle J^{\alpha 5} J^\mu J^\nu \rangle \neq 0$ so that the axial current is not conserved in the quantum theory. Moreover, only this choice of momentum routing is consistent with the massless limit of having a massive Dirac fermion in the loop.

Is it always possible to choose a momentum routing that preserves gauge invariance? In QED with any number of Dirac fermions the answer is yes. There, the photon couples to the vector current $J^\mu = \sum_i Q_i \bar\psi_i \gamma^\mu \psi_i$. Let us denote the matrix element corresponding to the 3-point function $\langle J^\alpha J^\mu J^\nu \rangle$ as $M_V^{\alpha\mu\nu}$. Then $M_V^{\alpha\mu\nu}$ vanishes when contracted with *any* momentum. You can check this yourself, but it follows simply from charge-conjugation invariance of QED (a special case of Furry's theorem, see Problem 14.2).

If we had only a Weyl fermion, however, there would be a problem. Then the Lagrangian is

$$\mathcal{L} = -\frac{1}{4}F_{\mu\nu}^2 + \bar{\psi}\left(i\slashed{\partial} - e\slashed{A}\right)P_L\psi, \tag{30.48}$$

where $P_L = \frac{1}{2}(1-\gamma_5)$ as usual. Here, we have explicitly broken charge-conjugation invariance, so Furry's theorem does not apply. In this case, the photon couples to a current $J_L^\mu = \bar{\psi}\gamma^\mu P_L\psi$. Let us denote the matrix element for $\langle J_L^\alpha J_L^\mu J_L^\nu \rangle$ as $M_L^{\mu\nu\alpha}$. Then,

$$M_L^{\alpha\mu\nu} = \int \frac{d^4k}{(2\pi)^4} \left[\frac{\mathrm{Tr}\left[\gamma^\mu P_L \slashed{k} \gamma^\nu P_L (\slashed{k}+\slashed{q}_2)\gamma^\alpha P_L(\slashed{k}-\slashed{q}_1)\right]}{k^2(k+q_2)^2(k-q_1)^2} + \left(\begin{array}{c} \mu \leftrightarrow \nu \\ 1 \leftrightarrow 2 \end{array}\right) \right], \tag{30.49}$$

as in Eq. (30.25) with a slightly different numerator. We can move the factors of P_L past various γ-matrices so that there is only one P_L left. Then we expand $P_L = \frac{1}{2}(1-\gamma_5)$ into two terms. The term without γ_5 is just $M_V^{\alpha\mu\nu}$, corresponding to the 3-point function with all vector currents $\langle J^\alpha J^\mu J^\nu \rangle$. The other has a single γ_5, which gives the quantity $\langle J^{\alpha 5} J^\mu J^\nu \rangle \sim M^{\alpha\mu\nu}$ we calculated above. Thus

$$M_L^{\alpha\mu\nu} = \frac{1}{2}\left(M_V^{\alpha\mu\nu} - M_5^{\alpha\mu\nu}\right). \tag{30.50}$$

We showed above that either $p_\alpha M_5^{\alpha\mu\nu} \neq 0$ or $q_\mu^1 M_5^{\alpha\mu\nu} \neq 0$. Since $p_\alpha M_V^{\alpha\mu\nu} = q_\mu^1 M_V^{\alpha\mu\nu} = 0$, we must therefore have that either $p_\alpha M_L^{\alpha\mu\nu} \neq 0$ or $q_\mu^1 M_L^{\alpha\mu\nu} \neq 0$. In other words, either $\partial_\alpha \langle J_L^\alpha J_L^\mu J_L^\nu \rangle \neq 0$ or $\partial_\mu \langle J_L^\alpha J_L^\mu J_L^\nu \rangle \neq 0$. Thus, the Ward identity cannot be satisfied. The same conclusion obviously holds for a theory with only a single right-handed fermion. In either case, the Ward identity must be violated and

> **QED with a single Weyl fermion is inconsistent.**

What if we had a left- and a right-handed fermions with different charges Q_L and Q_R?

$$\mathcal{L} = -\frac{1}{4}F_{\mu\nu}^2 + \bar{\psi}\left(i\slashed{\partial} + Q_L e\slashed{A}\right)P_L\psi + \bar{\psi}\left(i\slashed{\partial} + Q_R e\slashed{A}\right)P_R\psi. \tag{30.51}$$

In this case, the gauge boson A_μ couples to

$$J_{\text{mix}}^\mu = Q_L \bar{\psi}\gamma^\mu P_L\psi + Q_R \bar{\psi}\gamma^\mu P_R\psi. \tag{30.52}$$

In this case, there is a contribution to $\langle J_{\text{mix}}^\alpha J_{\text{mix}}^\mu J_{\text{mix}}^\nu \rangle$ with either fermion in the loop. There is no source of mixing between left- and right-handed fermions, thus

$$M_{\text{mix}}^{\alpha\mu\nu} = Q_L^3 M_L^{\alpha\mu\nu} + Q_R^3 M_R^{\alpha\mu\nu} = \frac{1}{2}\left(Q_R^3 - Q_L^3\right)M_5^{\alpha\mu\nu} + \frac{1}{2}\left(Q_R^3 + Q_L^3\right)M_V^{\alpha\mu\nu}. \tag{30.53}$$

Therefore, the only way a theory with a gauge boson that couples to a single left-handed and a single right-handed fermion can be consistent is if $Q_L = Q_R$, as in QED.

This leaves us with an obvious follow-up question: Are the weak interactions anomalous? Since the $SU(2)_{\text{weak}}$ gauge group of the Standard Model only couples to left-handed fields, it seems very dangerous. To answer this question, we need the generalization of the above results to non-Abelian currents. But first we repeat the chiral anomaly calculation using a different technique.

30.3 Chiral anomaly from the integral measure

In the previous section, we calculated the chiral anomaly through Feynman diagrams. In the massless case, this calculation was very subtle and involved a careful choice of momentum in a loop integral. A more direct connection between the anomaly and the violation of a symmetry uses the path integral. The intuitive idea, due to Kazuo Fujikawa, is that anomalies arise when there are symmetries of the action that are not symmetries of the functional measure in the path integral.

To begin, we quickly review the path-integral proof of current conservation in the quantum theory from Section 14.5. We start with

$$\langle \mathcal{O}(x_1,\ldots,x_n)\rangle = \frac{1}{Z[0]}\int \mathcal{D}\bar\psi\,\mathcal{D}\psi \exp\left[i\int d^4x\, i\bar\psi\,\partial\!\!\!/\psi\right]\mathcal{O}(x_1,\ldots,x_n),\qquad(30.54)$$

where $\mathcal{O}(x_1,\ldots,x_n)$ is some gauge-invariant operator. For example, you can think of $\mathcal{O} = J^\mu(y)J^\nu(z)$. This action is invariant under the global symmetries $\psi\to e^{i\alpha}\psi$ and $\psi\to e^{i\beta\gamma_5}\psi$. To derive current conservation for the vector symmetry, we redefine $\psi(x)\to e^{i\alpha(x)}\psi(x)$, with α now a function of x. The measure is invariant under this change of variables (we will confirm this in a moment) and $\mathcal{O}(x_1,\ldots,x_n)$ is invariant, but $\bar\psi\,\partial\!\!\!/\psi\to\bar\psi\,\partial\!\!\!/\psi+i\bar\psi\gamma^\mu\psi\partial_\mu\alpha$. Since the path integral integrates over all field configurations, it is invariant under any field redefinition, thus the remaining term proportional to α must vanish. Expanding to first order in α and integrating by parts, we find

$$0 = \frac{1}{Z[0]}\int d^4z\,\alpha(z)\int \mathcal{D}\bar\psi\,\mathcal{D}\psi\exp\left[i\int d^4x\, i\bar\psi\,\partial\!\!\!/\psi\right]\frac{\partial}{\partial z^\mu}\left[\bar\psi(z)\gamma^\mu\psi(z)\right]\mathcal{O}(x_1,\ldots,x_n).\qquad(30.55)$$

Since this holds for all $\alpha(z)$, we must have

$$\partial_\mu\langle J^\mu(x)\mathcal{O}(x_1,\ldots,x_n)\rangle = 0.\qquad(30.56)$$

The only part of the above derivation that changes when we consider an axial rotation $\psi\to e^{i\beta(x)\gamma_5}\psi$ is that the path integral measure is no longer invariant.

To see how the measure changes, consider a general linear transformation $\psi(x)\to\Delta(x)\psi(x)$ and $\bar\psi(x)\to\bar\psi(x)\Delta_c(x)$ which generates a Jacobian factor:

$$\mathcal{D}\bar\psi\,\mathcal{D}\psi\to[\mathcal{J}_c\,\mathcal{J}]^{-1}\mathcal{D}\bar\psi\,\mathcal{D}\psi.\qquad(30.57)$$

The Jacobians $\mathcal{J}=\det\Delta$ and $\mathcal{J}_c=\det\Delta_c$ appear to negative powers because the transformed variables are fermionic (see Section 14.6). To make sense out of $\mathcal{J}$ we write

$$\mathcal{J}=\det\Delta=\exp\operatorname{tr}\ln\Delta=\exp\left[\int d^4x\,\langle x|\operatorname{Tr}\ln\Delta(x)|x\rangle\right],\qquad(30.58)$$

where the tr sums over all the eigenvalues of Δ and the Tr is a Dirac trace. For the sum over positions, we have introduced a one-particle Hilbert space $\{|x\rangle\}$.[2] For example, consider a non-chiral transformation with $\Delta(x) = e^{i\alpha(x)}$ and $\Delta_c(x) = e^{-i\alpha(x)}$. Then,

$$\mathcal{J} = \mathcal{J}_c^\dagger = \exp\left[4i \int d^4x \delta^4(x - x)\,\alpha(x)\right]. \tag{30.59}$$

where $\langle x|y\rangle = \delta^4(x - y)$ has been used. Despite the infinite $\delta^4(0)$ factor, $\mathcal{J}_c\,\mathcal{J} = 1$ since the integrand is real and so the measure is invariant. In contrast, for the axial transformation $\Delta(x) = \Delta_c(x) = e^{i\beta(x)\gamma^5}$ and

$$\mathcal{J} = \mathcal{J}_c = \exp\left[i \int d^4x \delta^4(x - x)\,\beta(x)\mathrm{Tr}[\gamma_5]\right], \tag{30.60}$$

Since $\delta^4(0)\mathrm{Tr}[\gamma_5]$ gives infinity times zero, $\mathcal{J}_c\,\mathcal{J}$ is now undefined.

In full QED, the situation is similar. The QED path integral is

$$\int \mathcal{D}\bar\psi\,\mathcal{D}\psi\,\mathcal{D}A\,\exp\left[i \int d^4x \left(-\frac{1}{4}F_{\mu\nu}^2 + i\bar\psi\slashed{D}\psi\right)\right]. \tag{30.61}$$

The action is still invariant under the global symmetries $\psi \to e^{i\alpha}\psi$ and $\psi \to e^{i\beta\gamma_5}\psi$ with A_μ unchanged. Under the local axial transformation, A_μ is invariant, so its transformation does not contribute to the Jacobian.

To regulate the undefined product in Eq. (30.60), let us first write $\mathcal{J}$ as:

$$\mathcal{J} = \exp\left(i \int d^4x\,\mathrm{Tr}[\langle x|\beta(\hat{x})\gamma_5|x\rangle]\right), \tag{30.62}$$

Now, we regulate the divergence in a gauge-invariant manner by introducing an exponential regulator of the form $\exp(-\slashed{M}^2/\Lambda^2)$, where $\slashed{M} = \slashed{\hat{p}} - e\slashed{A}(\hat{x})$, Λ is some UV cutoff and $\hat{p}$ is the operator conjugate to $\hat{x}$ in the one-particle Hilbert space. The relation $\slashed{D}^2 = D_\mu^2 + \frac{e}{2}F_{\mu\nu}\sigma^{\mu\nu}$, from Eq. (10.106), implies

$$\slashed{M}^2 = \hat{\Pi}^2 - \frac{e}{2}\sigma_{\mu\nu}F^{\mu\nu}(\hat{x}), \tag{30.63}$$

so that

$$\mathrm{Tr}[\langle x|\beta(\hat{x})\gamma^5|x\rangle] = \lim_{\Lambda\to\infty}\mathrm{Tr}\left[\langle x|\beta(\hat{x})\gamma^5 e^{\slashed{M}^2/\Lambda^2}|x\rangle\right]$$

$$= \lim_{\Lambda\to\infty}\beta(x)\langle x|\mathrm{Tr}\left[\gamma^5\exp\left(\frac{(\hat{p} - eA(\hat{x}))^2 - \frac{e}{2}\sigma_{\mu\nu}F^{\mu\nu}}{\Lambda^2}\right)\right]|x\rangle. \tag{30.64}$$

Now, the trace of a product of γ-matrices with one γ^5 vanishes unless there are at least four γ-matrices in the product. Thus, the leading term in the expansion of the exponential

---

[2] To interpret this expression, we do not need a physical interpretation of the one-particle Hilbert space – we just want to use the mathematical tricks we learned in quantum mechanics to write the sum over positions in a suggestive form. There is in fact a beautiful interpretation of one-particle Hilbert spaces like this in quantum field theory, to which much of Chapter 33 is devoted.

is of order $\frac{1}{\Lambda^4}$. Using the identity $\frac{1}{2}\{\sigma^{\mu\nu}, \sigma^{\alpha\beta}\} = g^{\mu\alpha}g^{\nu\beta}\mathbb{1} - g^{\nu\alpha}g^{\mu\beta}\mathbb{1} + i\gamma^5\varepsilon^{\mu\nu\alpha\beta}$, where $\mathbb{1}$ is the identity matrix with Dirac indices, we can derive that

$$(\sigma_{\mu\nu}F^{\mu\nu})^2 = 2F_{\mu\nu}^2\mathbb{1} + i\gamma^5\varepsilon^{\mu\nu\alpha\beta}F_{\mu\nu}F_{\alpha\beta}, \tag{30.65}$$

which leads to

$$\text{Tr}[\langle x|i\beta(\hat{x})\gamma^5|x\rangle]$$
$$= -\frac{e^2}{2}\beta(x)\varepsilon^{\mu\nu\alpha\beta}F_{\mu\nu}(x)F_{\alpha\beta}(x)\lim_{\Lambda\to\infty}\left[\frac{1}{\Lambda^4}\langle x|e^{(\hat{p}-eA)^2/\Lambda^2}|x\rangle + \mathcal{O}\left(\frac{1}{\Lambda^5}\right)\right]. \tag{30.66}$$

To extract the contribution leading in e, we can set $A = 0$ in the exponent. Next insert $\mathbb{1} = (2\pi)^{-4}\int d^4k|k\rangle\langle k|$ with $\hat{p}|k\rangle = k|k\rangle$ to get

$$\frac{1}{\Lambda^4}\langle x|e^{\hat{p}^2/\Lambda^2}|x\rangle = \frac{1}{\Lambda^4}\int\frac{d^4k}{(2\pi)^4}e^{k^2/\Lambda^2} = \frac{i}{\Lambda^4}\int\frac{d^4k_E}{(2\pi)^4}e^{-k_E^2/\Lambda^2} = \frac{i}{16\pi^2}. \tag{30.67}$$

Thus, we find a finite answer as $\Lambda \to \infty$:

$$\mathcal{J} = \exp\left[-i\int d^4x\left(\beta(x)\frac{e^2}{32\pi^2}\varepsilon^{\mu\nu\alpha\beta}F_{\mu\nu}(x)F_{\alpha\beta}(x)\right)\right]. \tag{30.68}$$

Note that if we had used $e^{-\hat{p}^2/\Lambda^2}$ or $e^{-\Pi^2/\Lambda^2}$ the singularity would not have been regulated and the Jacobian would still be undefined.

The result is that under an axial transformation

$$\int \mathcal{D}\bar{\psi}\,\mathcal{D}\psi\,\mathcal{D}A\exp\left[i\int d^4x\,\mathcal{L}_{\text{QED}}\right]$$
$$\to \int\mathcal{D}\bar{\psi}\,\mathcal{D}\psi\,\mathcal{D}A\exp\left[i\int d^4x\left(\mathcal{L}_{\text{QED}} - J_\mu^5\partial_\mu\beta + \beta\frac{e^2}{16\pi^2}\varepsilon^{\mu\nu\alpha\beta}F_{\mu\nu}F_{\alpha\beta}\right)\right]. \tag{30.69}$$

Thus, the Schwinger–Dyson equation in Eq. (30.56) becomes

$$\partial_\mu\langle J^{5\mu}(x)\mathcal{O}(x_1,\ldots,x_n)\rangle = -\frac{e^2}{16\pi^2}\langle\varepsilon^{\mu\nu\alpha\beta}F_{\mu\nu}(x)F_{\alpha\beta}(x)\mathcal{O}(x_1,\ldots,x_n)\rangle. \tag{30.70}$$

We often abbreviate this with

$$\partial_\mu J_\mu^5 = -\frac{e^2}{16\pi^2}\varepsilon^{\mu\nu\alpha\beta}F_{\mu\nu}F_{\alpha\beta}, \tag{30.71}$$

which agrees with Eq. (30.22). This equation confirms the interpretation of the chiral anomaly as due to non-invariance of the path integral measure.

The path integral approach is valuable because it allows to prove that there are no higher-order corrections to the anomaly equation, Eq. (30.71). That is,

> The chiral anomaly is 1-loop exact.

Note that we did not show this. In fact, the path integral transformation amounts to a 1-loop computation, as can be seen from Eq. (30.67) or by restoring factors of $\hbar$ (the correspondence between functional determinants and loops will be explored in Chapters 33 and 34). Nevertheless, *because* the anomaly is exact at 1-loop, the measure transformation at leading order gives the complete answer. The 1-loop exactness of the chiral anomaly was first proposed by Adler and Bell using diagrammatic arguments. Its most satisfying proof uses topological arguments (see for example [Nakahara, 2003] or [Weinberg, 1996] for details).

30.4 Gauge anomalies in the Standard Model

In this section, we will check that the currents associated with the $SU(3)_{QCD} \times SU(2)_{weak} \times U(1)_Y$ gauge symmetries of the Standard Model are non-anomalous. If we write these three currents as J_μ^{QCD}, J_μ^{weak} and J_μ^Y, then we have to show that $\partial_\mu \langle J_\mu^j J_\alpha^k J_\nu^l \rangle = 0$ for j, k, l any of the forces. This is easiest to do by reading charges or anomaly coefficients from the triangle diagrams.

When all three currents involved are associated with $U(1)_Y$, we call the putative anomaly the $U(1)_Y^3$ anomaly. It is easy to check that this vanishes. As we saw in Eq. (30.53), left-handed Weyl fermions and right-handed Weyl fermions contribute to the anomaly with opposite signs. Therefore, we have

$$\partial_\mu J_Y^\mu = \left(\sum_{\text{left}} Y_l^3 - \sum_{\text{right}} Y_r^3 \right) \frac{g'^2}{32\pi^2} \varepsilon^{\mu\nu\alpha\beta} B_{\mu\nu} B_{\alpha\beta}, \tag{30.72}$$

where $B_{\mu\nu}$ is the field strength for $U(1)_Y$. The vanishing of the $U(1)_Y^3$ anomaly requires

$$0 = \left(2Y_L^3 - Y_e^3 - Y_\nu^3 \right) + 3 \left(2Y_Q^3 - Y_u^3 - Y_d^3 \right). \tag{30.73}$$

Here, $Y_L, Y_e, Y_\nu, Y_Q, Y_u$ and Y_d are the hypercharges for the left-handed leptons, the right-handed electrons (or muon or tauon), the right-handed neutrinos (assuming they exist), the left-handed quarks, the right-handed up-type quarks and the right-handed down-type quarks, respectively. As derived in Chapter 29, these charges are (see Table 29.1)

$$Y_L = -\frac{1}{2}, \quad Y_e = -1, \quad Y_\nu = 0, \quad Y_Q = \frac{1}{6}, \quad Y_u = \frac{2}{3}, \quad Y_d = -\frac{1}{3}. \tag{30.74}$$

Plugging in to Eq. (30.73), we find that the anomaly in fact vanishes. Note that the anomaly would vanish for any number of generations, but that it does not vanish for the quarks or leptons alone.

By the way, one can also trivially check that the $U(1)_{EM}^3$ anomaly vanishes in QED. In QED, all the left- and right-handed charged particles are Dirac, and hence have the same charges (QED is non-chiral). Thus, in QED, $\sum_{\text{left}} Q_L^3 = \sum_{\text{right}} Q_R^3$. That the $U(1)_{EM}^3$ anomaly vanishes also follows from the vanishing of anomalies in the electroweak theory, which we have nearly shown.

For non-Abelian gauge theories, the currents associated with the gauge fields are of the form

$$J_\mu^a = \sum_\psi \bar{\psi}_i T_{ij}^a \gamma^\mu \psi_j, \tag{30.75}$$

where T^a_{ij} are the group generators which could be in an arbitrary representation. The triangle diagrams then pick up factors of T^a at the vertices. The two momentum routings give

$$i\mathcal{M} = \text{tr}\left[T^a T^b T^c\right] \times \quad + \quad \text{tr}\left[T^a T^c T^b\right] \times \qquad . \tag{30.76}$$

Now, we can always write the group trace as a sum of symmetric and antisymmetric tensors as (see Eq. (25.20))

$$\text{tr}\left[T^a T^b T^c\right] = \frac{1}{2}\text{tr}\left[\left[T^a, T^b\right]T^c\right] + \frac{1}{2}\text{tr}\left[\{T^a, T^b\}T^c\right] = i\frac{1}{2}T_R f^{abc} + \frac{1}{4}d_R^{abc}. \tag{30.77}$$

The contribution proportional to the f^{abc} gives the difference between the two loops. This difference is UV divergent. However, since it is proportional to f^{abc}, it can be removed through renormalization without violating gauge invariance. Indeed, it contributes to the renormalization of the $f^{abc}A^a_\mu A^b_\nu \partial_\mu A^c_\nu$ vertex in the Yang–Mills Lagrangian.

The contribution proportional to d_R^{abc} is what we are after; d_R^{abc} is a totally a symmetric tensor given by

$$d_R^{abc} = 2\text{tr}\left[T^a_R \{T^b_R, T^c_R\}\right]. \tag{30.78}$$

As mentioned in Section 25.1, for $\text{SU}(N)$ there is a unique totally symmetric three-index tensor up to a constant. Thus for any representation,

$$2\text{tr}\left[T^a_R\{T^b_R, T^c_R\}\right] \equiv A(R)d^{abc}, \tag{30.79}$$

with $A(R)$ the **anomaly coefficient** and d^{abc} (without a subscript) defined using the fundamental representation. Thus, $A(\text{fund}) = 1$.

The contribution proportional to the anomaly constant d^{abc} sums the two triangle diagrams. It is therefore proportional to the result from summing the diagrams in the $\text{U}(1)$ case. We thus find

$$\partial_\alpha J^a_\alpha(x) = \left(\sum_{\text{left}} A(R_l) - \sum_{\text{right}} A(R_r)\right)\frac{g^2}{128\pi^2}d^{abc}\varepsilon^{\mu\nu\alpha\beta}F^b_{\mu\nu}F^c_{\alpha\beta}, \tag{30.80}$$

where the "left" sum is over left-handed particles, with $A(R_l)$ the anomaly coefficients associated with their representations R_l, and similarly for the "right" sum. We can check the normalization using the $\text{U}(1)^3_Y$ anomaly. For a $\text{U}(1)$, $T^a = 1$, $d^{abc} = 4$ and so Eq. (30.80) reduces to Eq. (30.72). Note that Eq. (30.80) can vanish either if the anomaly coefficients cancel in the sum, or if $d^{abc} = 0$.

Now we would like check whether anomalies cancel in the Standard Model. For $\text{SU}(2)$, we can use $\{\tau^a, \tau^b\} = \frac{1}{2}\delta^{ab}\mathbb{1}$. Then $d^{abc} = \delta^{bc}\text{tr}\{\tau^a\} = 0$. Thus, there can never be $\text{SU}(2)^3$ anomalies in any theory. There could in principle be an $\text{SU}(3)^3$ anomaly in some theory, but since QCD is non-chiral, there are no $\text{SU}(3)^3_{\text{QCD}}$ anomalies in the Standard Model. Next, consider mixed anomalies. An $\text{SU}(N)\text{U}(1)^2$ anomaly would be proportional to $2\text{tr}[T^a\{1,1\}] = 4\text{tr}[T^a] = 0$. Hence $\text{SU}(N)\text{U}(1)^2$ anomalies always vanish. In the

| Table 30.1 | Anomaly constraints on the hypercharges of Standard Model particles. |

Anomaly	Constraint
$U(1)_Y^3$	$(2Y_L^3 - Y_e^3 - Y_\nu^3) + 3(2Y_Q^3 - Y_u^3 - Y_d^3) = 0$
$SU(3)^2U(1)_Y$	$2Y_Q - Y_u - Y_d = 0$
$SU(2)^2U(1)_Y$	$Y_L + 3Y_Q = 0$
$\text{grav}^2U(1)_Y$	$(2Y_L - Y_e - Y_\nu) + 3(2Y_Q - Y_u - Y_d) = 0$

same way, any anomaly with exactly one factor of $SU(2)$ or $SU(3)$ vanishes. The only possible anomalies are therefore $SU(3)^2U(1)$ and $SU(2)^2U(1)$.

The $SU(3)_{\text{QCD}}^2U(1)$ anomaly gets contributions only from quarks. Using $\text{tr}\{T^aT^b\} = \frac{1}{2}\delta^{ab}$, which holds for any $SU(N)$, we find that this anomaly is proportional to

$$2\text{tr}[T^a\{T^b, Y\}] = 2\delta^{ab}\left(\sum_{\substack{\text{left} \\ \text{colored}}} Y_l - \sum_{\substack{\text{right} \\ \text{colored}}} Y_r\right) = 2\delta^{ab}(6Y_Q - 3Y_u - 3Y_d). \quad (30.81)$$

Plugging in the values in Eq. (30.74), this vanishes. The $SU(2)^2U(1)$ anomaly only gets contributions from left-handed fields, and so

$$2\text{tr}[\tau^a\{\tau^b, Y\}] = 2\delta^{ab}\sum_{\text{left}} Y_i = 2\delta^{ab}(2Y_L + 6Y_Q). \quad (30.82)$$

For this anomaly to cancel, left-handed leptons must have -3 times the hypercharge of left-handed quarks, as they do. Thus, all possible anomalies associated with the $SU(3)_{\text{QCD}} \times SU(2)_{\text{weak}} \times U(1)_Y$ of the Standard Model exactly vanish.

There is one more type of gauge boson in the Standard Model whose anomalies must cancel: the graviton. The calculation of the anomaly with one gauge boson and two external gravitons produces

$$\partial_\alpha J_\alpha^a(x) \propto \text{Tr}[T_R^a]\,\varepsilon^{\mu\nu\alpha\beta}R_{\mu\nu\gamma\delta}R_{\alpha\beta\gamma\delta}, \quad (30.83)$$

where $R_{\mu\nu\alpha\beta}$ is the Riemann tensor. Since the $SU(N)$ generators are traceless, there are no $\text{grav}^2SU(2)$ or $\text{grav}^2SU(3)$ anomalies. The only thing we have to worry about is $\text{grav}^2U(1)_Y$. Since all fermions couple to gravity, we must have

$$0 = \sum_{\text{left}} Y_l - \sum_{\text{right}} Y_r = (2Y_L - Y_e - Y_\nu) + 3(2Y_Q - Y_u - Y_d). \quad (30.84)$$

This also holds in the Standard Model.

The four nonlinear equations that the six hypercharges must satisfy are summarized in Table 30.1. The general solution to these equations (up to redefining $u_R \leftrightarrow d_R$ or $e_R \leftrightarrow \nu_R$ which the hypercharge constraints do not care about) is either

$$Y_L = -\frac{a}{2} - b, \ \ Y_e = -a - b, \ \ Y_\nu = -b, \ \ Y_Q = \frac{a}{6} + \frac{b}{3}, \ \ Y_u = \frac{2a}{3} + \frac{b}{3}, \ \ Y_d = -\frac{a}{3} + \frac{b}{3}, \quad (30.85)$$

for any a and b, or

$$Y_Q = Y_L = 0, \quad Y_u = c, \quad , Y_d = -c, \quad Y_e = d, \quad Y_\nu = -d \tag{30.86}$$

for any c and d. The Standard Model hypercharge assignments satisfy Eq. (30.85) with $a = 1$ and $b = 0$. Note that we can always rescale the hypercharges (or equivalently redefine the coupling g'), thus these are two one-parameter families of solutions. Suppose we also know that the right-handed neutrino has $Y_\nu = 0$, either because it does not exist, because it is Majorana (in which case it is its own antiparticle and cannot have any quantum numbers, including hypercharge), or for some other reason. That implies, if we take the first solution, that $b = 0$. Then we can set $a = 1$ by rescaling g', and so the Standard Model hypercharges are uniquely determined. The second solution, Eq. (30.86), is not realized in nature.

Notice that any solution of Eq. (30.85) or Eq. (30.86) has $Y_L + 3Y_Q = 0$ exactly. As a consequence, the electron must have *exactly* the same electric charge as the proton. Without anomaly considerations, one might have imagined that the electron could have had say 3.0001 times the quark charge, giving a small residual charge to the atom. Anomaly cancellation says this cannot be true. Charge is quantized!

Another question we can ask is: Can there be another $U(1)$ force acting on the Standard Model particles that we do not know about? Let us call this force $U(1)'_Y$ and the charges under this new group Y'_i. For anomalies to cancel, all the conditions in Table 30.1 must hold with $Y_i \to Y'_i$. In addition, $U(1)^2_Y U(1)_{Y'}$ and $U(1)_Y U(1)^2_{Y'}$ anomalies must cancel. As you can easily check, the only possibility is that Y'_i satisfy Eq. (30.85) with $Y_i \to Y'_i$. Taking $a = 1$ and $b = 0$ sets Y'_i equal to the Standard Model hypercharges. The orthogonal possibility is $a = 0$, $b = 1$, which gives

$$Y'_L = Y'_e = Y'_\nu = -1, \quad Y'_Q = Y'_u = Y'_d = \frac{1}{3}. \tag{30.87}$$

These charges are -1 for the leptons and $\frac{1}{3}$ for quarks, or equivalently $+1$ for baryons. We call this new group $U(1)_{B-L}$ and will discuss it more in the next section.

30.5 Global anomalies in the Standard Model

We have argued that anomalies must vanish for symmetries associated with gauge fields. If this were not true, the Ward identity would be violated and we could no longer guarantee that only the two physical polarizations of a massless spin-1 particle would propagate. On the other hand, if the symmetry is a global symmetry not associated with a gauge field, it can be anomalous. For example, the $\pi^0 \to \gamma\gamma$ decay is due to an anomalous axial current as we discussed in Section 30.1. If G is a global symmetry, then G^3 anomalies have no physical effect. The simplest way to see this is that, for a global symmetry, there is no associated $\varepsilon^{\mu\nu\alpha\beta} F_{\mu\nu} F_{\mu\nu}$ term for a current to diverge to. Thus, the global anomalies of interest are the GH^2 anomalies, where H is one of the Standard Model gauge groups.

30.5.1 Baryogenesis

An important example of a global symmetry of the Standard Model Lagrangian is **baryon number**, for which all quarks have $B = \frac{1}{3}$ and leptons have $B = 0$. That is $u \to e^{i\frac{1}{3}\alpha}u, d \to e^{i\frac{1}{3}\alpha}d,\ e \to e, \nu_e \to \nu_e$, etc. Another example is **lepton number**, for which quarks have $L = 0$ and leptons have $L = 1$. Substituting these quantum numbers into the anomaly constraints in Table 30.1 we see that all of the mixed anomalies vanish except for $\mathrm{SU}(2)^2\mathrm{U}(1)_B$ and $\mathrm{SU}(2)^2\mathrm{U}(1)_L$. For these,

$$\partial_\mu J_\mu^B = \partial_\mu J_\mu^L = n_g \frac{g^2}{32\pi^2} \varepsilon^{\mu\nu\alpha\beta} W_{\mu\nu}^a W_{\alpha\beta}^a, \tag{30.88}$$

where $W_{\mu\nu}^a$ is the $\mathrm{SU}(2)$ field strength and n_g is the number of generations ($n_g = 3$ in the Standard Model). So B and L are anomalous.

On the other hand, this equation implies that the global symmetry $B - L$, where quarks have $B - L = \frac{1}{3}$ and leptons have $B - L = -1$, is non-anomalous (as we saw in Eq. (30.87)). Thus, while it is not possible to have a gauge boson associated with B or L, it is possible to have one associated with $B - L$. In fact, such gauge bosons are common in grand unified theories and are actively searched for at colliders.

Returning to the Standard Model, it is natural to ask what physical effect the anomaly $\partial_\mu J_\mu^B \neq 0$ can have. Recall from Eq. (29.105) that the anomaly term is a total derivative, $\varepsilon^{\mu\nu\alpha\beta} W_{\mu\nu}^a W_{\alpha\beta}^a = \partial_\mu K^\mu$, so it cannot contribute at any order in perturbation theory (any Feynman diagram with this vertex would have a factor of $\sum p_\mu = 0$). However, it could possibly contribute to the path integral through field configurations that are locally gauge equivalent to 0, but are topologically stable. A class of such configurations is the **sphalerons**, which violate B and L but preserve $B - L$. Sphalerons can mediate baryon number violation into leptons.

Sphalerons are static configurations of the $\mathrm{SU}(2)$ gauge fields that can be locally gauged away. For these configurations, $\int d^4x\, \varepsilon^{\mu\nu\alpha\beta} W_{\mu\nu}^a W_{\alpha\beta}^a \sim 3(16\pi^2) \neq 0$. The results of sphaleron calculations imply that the rate per unit volume for the transfer from baryon number to lepton number violation at zero temperature should be roughly $\Gamma/V \sim m_W^4 \exp\left(-\frac{8\pi^2}{g^2}\right) \sim m_W^4 10^{-180}$, which is exceedingly tiny. At temperatures of order m_W, the rate can actually be much higher.

One of the reasons baryon number violation is interesting is because of the preponderance of matter over antimatter in the universe. In order to establish such an asymmetry, Andrei Sakharov showed in 1967 that three conditions must be met [Sakharov, 1967]:

Sakharov conditions to produce a matter–antimatter asymmetry	Box 30.1

1. Baryon number must be violated.
2. *C* and *CP* must be violated.
3. There must have been some departure from thermal equilibrium.

If any of these do not hold, the matter–antimatter asymmetry would have been washed out by thermal fluctuations. For example, by *CPT* invariance, the rate for any conversion of matter into antimatter must be the same as the rate for conversion of antimatter into matter, hence the need for non-equilibrium dynamics. Intriguingly, all of these conditions are in fact satisfied in the Standard Model: baryon number is violated by the anomaly, *CP* is violated because there are three generations and hence a phase in the CKM matrix, and as the universe cools it is out of equilibrium. In particular, as it cools through the electroweak phase transition, a matter–antimatter asymmetry can be produced. Unfortunately, to explain the matter–antimatter asymmetry quantitatively, we need *more* baryon number violation, *more CP* violation, and a phase transition that is *not as smooth* as in the Standard Model (it should be strongly first order). That baryogenesis cannot be explained in the Standard Model remains an important motivation for beyond-the-Standard-Model physics.

30.5.2 The *U*(1) and strong *CP* problems

Another important application of global anomalies is to the strong *CP* problem. This was discussed in Section 29.5. There, we started from the Standard Model with Yukawa couplings to the Higgs doublet, spontaneously broke electroweak symmetry, then performed chiral rotations on the left-handed and right-handed quarks to move all *CP* violation into the CKM matrix. The CKM matrix could be taken real up to a single phase, known as the weak *CP* phase. However, in doing the chiral rotations, since the measure is not invariant, we generate a term

$$\mathcal{L} = \theta_{\mathrm{QCD}} \frac{g_s^2}{32\pi^2} \varepsilon^{\mu\nu\alpha\beta} F_{\mu\nu}^a F_{\alpha\beta}^a, \tag{30.89}$$

where F^a is the QCD field strength (one also generates $\varepsilon^{\mu\nu\alpha\beta} F_{\mu\nu}^a F_{\alpha\beta}^a$ terms for the weak and electromagnetic fields this way, but those phases can be removed with additional rotations of just the right-handed fields). There was therefore an additional chiral-rotation-invariant phase, called the strong *CP* phase, given by $\bar{\theta} = \theta_{\mathrm{QCD}} + \arg\det(Y_d Y_u)$. We argued that the neutron picks up an electric dipole moment proportional to $\bar{\theta}$, and current experimental bounds require $\bar{\theta} < 10^{-10}$. The strong *CP* problem is: Why is this phase so small? Possible solutions were discussed in Chapter 29.

Another example of a global anomaly is the chiral symmetry of QCD. Consider QCD in the limit that the three lightest quark flavors (up, down and strange) can be treated as massless. Then the Lagrangian is just

$$\mathcal{L} = -\frac{1}{4}\left(F_{\mu\nu}^a\right)^2 + i\bar{q}_L^j \slashed{D} q_L^j + i\bar{q}_R^j \slashed{D} q_R^j, \tag{30.90}$$

where L and R refer to the left- and right-handed quarks. This Lagrangian has a global $\mathrm{U}(3)_L \times \mathrm{U}(3)_R$ symmetry. The QCD vacuum has $\langle \bar{q}_L q_R \rangle \approx V^3 \approx \Lambda_{\mathrm{QCD}}^3 \neq 0$, spontaneously breaking $\mathrm{U}(3)_L \times \mathrm{U}(3)_R \to \mathrm{U}(3)_{\mathrm{diagonal}}$. Thus there should be nine massless Goldstone bosons, conveniently written as a matrix when multiplying the $\mathrm{SU}(3)$ generators (see Eq. (28.38)):

$$\pi^a T^a = \frac{1}{\sqrt{2}} \begin{pmatrix} \frac{1}{\sqrt{2}}\pi^0 + \frac{1}{\sqrt{6}}\eta^0 + \frac{1}{\sqrt{3}}\eta' & \pi^+ & K^+ \\ \pi^- & -\frac{1}{\sqrt{2}}\pi^0 + \frac{1}{\sqrt{6}}\eta^0 + \frac{1}{\sqrt{3}}\eta' & K^0 \\ \bar{K}^- & \bar{K}^0 & -\sqrt{\frac{2}{3}}\eta^0 + \frac{1}{\sqrt{3}}\eta' \end{pmatrix}. \tag{30.91}$$

In reality, quarks do have masses, and so the pseudoscalar mesons (the Goldstone bosons) pick up mass (becoming pseudo-Goldstone bosons) according to the Gell-Mann–Oakes–Renner relation, Eq. (28.37): $m_\pi^2 F_\pi^2 \approx V^3 m_q$.

Now, consider neutral mesons. Experimentally, the lightest two neutral mesons are the π^0 (135 MeV) and the η (549 MeV). After that, the next lightest has mass 957 MeV, which we would like to identify with the η'. Unfortunately, if you work out the group theory factors for the Goldstone masses, you find that this is impossible. The mass of the diagonal Goldstone boson, the η', must satisfy $m_{\eta'} < \sqrt{3}m_{\pi^0}$ [Weinberg, 1975]. Why the η' is so heavy is known as the **U(1) problem**. It is called that because the Goldstone boson corresponding to the axial diagonal U(1) is apparently missing.

The solution to the U(1) problem should now be apparent: the symmetry of the QCD Lagrangian is not in fact $U(3)_L \times U(3)_R = U(1)_L \times U(1)_R \times SU(3)_L \times SU(3)_R$ because the $U(1)_A$ under which $q_L \to e^{i\theta}q_L$ and $q_R \to e^{-i\theta}q_R$ is anomalous. Under this $U(1)_A$ all quarks have charge ± 1, so the anomaly corresponds to $SU(3)^2_{QCD}U(1)_A$ triangle diagrams. Since the symmetry is anomalous, it is not a symmetry. If there is no symmetry, it cannot be spontaneously broken and there can be no Goldstone boson. Note that the $SU(3)_L$ and $SU(3)_R$ do not have an $SU(3)^2_{color} \times SU(3)_L$ anomaly, since the $SU(3)$ generators are traceless.

The U(1) problem and the strong *CP* problem are actually closely related. The same chiral rotations that move the *CP* phase between the quark mass matrix and θ_{QCD} are those corresponding to the anomalous $U(1)_A$. In both cases, the anomaly is from $SU(3)^2_{QCD}U(1)_A$. Under the $U(1)_A$ rotation, the measure changes and the Lagrangian shifts to

$$\mathcal{L} \to \mathcal{L} + \theta_{QCD}\frac{g_s^2}{32\pi}\varepsilon^{\mu\nu\alpha\beta}F_{\mu\nu}^a F_{\alpha\beta}^a. \tag{30.92}$$

Thus, the physics of the anomaly for both the strong *CP* and U(1) problems must come from topologically non-trivial gauge configurations. One might have tried to define the path integral excluding these configurations to solve the strong *CP* problem. But then the U(1) problem would not be solved. Thus, the heavy η' tells us that non-perturbative configurations must be important.

It is challenging to calculate the η' mass in QCD, since non-perturbative methods are needed. One such method is the lattice, which has in fact been able to calculate the η' mass purely within QCD to within around 10%. Analytically, one can approach the problem by summing over topological configurations, in this case **instantons**, but the result is only an order of magnitude estimate. Another approach is the large N limit of QCD, which relates the η' mass to the topological susceptibility, defined by

$$\chi_t \equiv \frac{1}{4}\langle (\varepsilon^{\mu\nu\alpha\beta}F_{\mu\nu}^a F_{\alpha\beta}^a)(\varepsilon^{\rho\sigma\kappa\lambda}F_{\rho\sigma}^b F_{\kappa\lambda}^b) \rangle. \tag{30.93}$$

The **Witten–Veneziano relation** is $\chi_t = \frac{F_\pi^2}{12}\left(m_\eta^2 + m_{\eta'}^2 - 2m_K^2\right)$. So if $\varepsilon^{\mu\nu\alpha\beta} F_{\mu\nu}^a F_{\alpha\beta}^a$ had no effect then $\chi_t = 0$ and the η' mass would be small. Solving the Witten–Veneziano relation for χ_t gives $\chi_t = (171\,\text{MeV})^4 \approx \Lambda_{\text{QCD}}^4$, which is roughly what one would expect by dimensional analysis [Witten, 1979; Veneziano, 1979].

30.6 Anomaly matching

An important use of anomalies is in **anomaly matching**, which relates the spectrum of a theory above and below a phase transition ['t Hooft *et al.*, 1980]. Consider QCD with three flavors and its global $G = \text{SU}(3)_L \times \text{SU}(3)_R \times \text{U}(1)_V$ symmetry. In pure QCD, this symmetry is not anomalous, or more precisely there are no $\text{SU}(3)_{\text{QCD}}^2 G$ anomalies. There are however G^3 anomalies, but these have no physical effect since there is no associated $\varepsilon^{\mu\nu\alpha\beta} F_{\mu\nu} F_{\mu\nu}$.

Now let us gauge the whole symmetry group G by introducing gauge bosons, but take their gauge couplings arbitrarily small so that the gauge bosons do not affect the physics. The anomalies, such as an $\text{SU}(3)_L^3$ anomaly, will have physical effects. However, we can cancel these anomalies by introducing a bunch of left- or right-handed **spectator fermions**. It is not hard to choose their quantum numbers so that all the anomalies cancel, and in fact there are many solutions. Since the gauge couplings are infinitesimal, these fermions will also not affect the physics.

Now consider the low-energy theory where the quarks are confined. Then the spectrum comprises not quarks but mesons and baryons which are all color singlets. We have not proven confinement, but it is apparently true, so let us just assume it happens. Indeed it is helpful at this point to have in the back of your mind a more general theory with N colors and n_f flavors, where we do not know if confinement happens or not. In the general case, mesons are still $\bar{q}q$ pairs, but baryons are bound states of N quarks or N antiquarks, which are fermions for N odd.

Since anomalies are determined by massless particles, they are long-distance effects. Thus, they cannot change by a phase transition that happens at a finite scale, such as Λ_{QCD}. This implies that, since the theory above the phase transition was anomaly free, the theory below the transition must also be anomaly free. Another way to see this is that a gauge anomaly would imply an inconsistency of the gauge theory, such as unitarity violation. Such a drastic change from a consistent field theory to an inconsistent one cannot happen just due to a phase transition. But below the transition the massless quarks are no longer around to cancel the anomalies of the spectators, so how can this happen? There must be other massless particles in the spectrum. There are two possibilities: a symmetry can be spontaneously broken, in which case there will be massless Goldstone bosons, or else there might be massless baryons.

Consider first the real world, where $\text{SU}(3)_L \times \text{SU}(3)_R$ is spontaneously broken, generating a triplet of pions, π^a. Let us focus on the π^0. This π^0 is associated with a particular axial U(1) symmetry under which $u \to e^{i\theta\gamma_5} u$ and $d \to e^{-i\theta\gamma_5} d$. Let us call this $\text{U}(1)_{\pi^0}$. Note that this is a *different* $\text{U}(1)_A$ from the one associated with the η'. That one had $q_i \to e^{i\theta\gamma_5} q_i$

and was anomalous to begin with, even without our fictitious gauge bosons. Before symmetry breaking, the theory was anomaly free. But after symmetry breaking, when the quarks are confined, it seems there is a $U(1)^2_{QED} U(1)_{\pi^0}$ anomaly. This must somehow be compensated for by the only relevant massless particle, the π^0. To see how, recall that the pion transforms under the broken symmetry as a shift $\pi^0 \to \pi^0 + \theta$ (this shift is what forbids a mass term for the pion, among other things). Therefore, we can compensate for the anomaly that rotates the coefficient of $\varepsilon^{\mu\nu\alpha\beta} F_{\mu\nu} F_{\alpha\beta}$ by adding a term

$$\mathcal{L} = N \frac{e^2}{16\pi^2} \pi^0 \varepsilon^{\mu\nu\alpha\beta} F_{\mu\nu} F_{\alpha\beta} \tag{30.94}$$

to the Chiral Lagrangian. In fact, this is the unique term whose chiral rotation $\pi^0 \to \pi^0 + \theta$ can exactly compensate the chiral rotation of the spectators. The factor of N comes from the N spectators that compensate for the N colors of quarks in the high-energy theory. It is in this way that the $\pi^0 \to \gamma\gamma$ rate is completely fixed by the anomaly and can be computed in perturbation theory, despite the fact that pions are composite objects. In fact, this was one of the early ways in which the number of colors N was cleanly measured.

Now let us suppose instead that $SU(3)_L \times SU(3)_R$ were not spontaneously broken. Then there would be no Goldstone bosons whose transformations could compensate the anomaly of the spectator. Consider the $SU(3)^3_L$ anomalies. These cancel if and only if the sum of the anomaly coefficients $\sum_i A(R_i) = 0$, where the sum is over all left-handed fermions in the theory. In QCD the quarks transform in the fundamental representation with $A(\text{fund}) = 1$. Including the three colors, the anomaly coefficient in QCD is then 3, thus the spectators contribute -3, by construction.

For the anomalies to be the same in the confined phase, color singlet fermions constructed out of quarks must be able to provide $\sum_i A(R_i) = 3$ to cancel the spectators. Since QCD has $N = 3$, color singlet fermions must be baryons comprising three quarks. To see what the contributions of the baryons could be, we have to decompose the tensor product of three fundamental representations of $SU(3)_L$ into irreducible representations of $SU(3)_L$. The decomposition is [Georgi, 1982]

$$\mathbf{3} \otimes \mathbf{3} \otimes \mathbf{3} = \left(\mathbf{6} \oplus \overline{\mathbf{3}}\right) \otimes \mathbf{3} = (\mathbf{6} \otimes \mathbf{3}) \oplus (\overline{\mathbf{3}} \otimes \mathbf{3}) = \mathbf{10} \oplus \mathbf{8} \oplus \mathbf{8} \oplus \mathbf{1}. \tag{30.95}$$

These are the decuplet, two octets and one singlet. (These are the same decuplet and octet that were shown in Section 28.2.3 in the context of Gell-Mann's eightfold way.) Of these, the $\mathbf{8}$ and $\mathbf{1}$ are real representations so they give $A(R_i) = 0$. To find $A(\mathbf{10})$ we use the identities $A(R_1 \oplus R_2) = A(R_1) + A(R_2)$ and $A(R_1 \otimes R_2) = A(R_1) d(R_2) + d(R_1) A(R_2)$, which you proved in Problem 25.4. First, we find

$$A(\mathbf{6}) = A(\mathbf{3} \otimes \mathbf{3}) - A\left(\overline{\mathbf{3}}\right) = 3A(\mathbf{3}) + 3A(\mathbf{3}) - A\left(\overline{\mathbf{3}}\right) = 7. \tag{30.96}$$

Then we find

$$A(\mathbf{10}) = A(\mathbf{6} \otimes \mathbf{3}) - A(\mathbf{8}) = 3A(\mathbf{6}) + 6A(\mathbf{3}) - A(\mathbf{8}) = 27. \tag{30.97}$$

If there are n decuplets of baryons, they will contribute $27n$, which cannot possibly cancel the -3 from the spectators (we would need $n = \frac{1}{9}$ decuplets!). We conclude that the chiral symmetry $SU(3)_L \times SU(3)_R$ of QCD *must be spontaneously broken*. Note that this argument does not work for $SU(2)_L \times SU(2)_R$ since $d^{abc} = 0$ for $SU(2)$, so there can never be any $SU(2)^3$ anomalies.

Another application of anomaly matching is in **Seiberg dualities** in supersymmetric gauge theories. The starting point is a supersymmetric gauge theory with N colors and n_f flavors. In the regime $\frac{3}{2}N > n_f > N$, this theory seems to flow towards a conformal fixed point in the infrared, but becomes strongly coupled. The duality postulates that this conformal fixed point is the same as one coming from a theory with $n_f - N$ colors and n_f flavors. Away from the fixed point, the two theories have very different particle content. Yet, if the theories agree at the fixed point, the spectators one adds to cancel anomalies at the fixed point should also cancel the anomalies in the two theories separately. As a highly non-trivial check on this duality, the anomalies associated with the global $SU(n_f)_L \times SU(n_f)_R$, $U(1)_{\text{baryon}}$ and an additional $U(1)_R$ symmetry all agree. That the anomalies are identical in the two theories, despite their radically different particle content, is strong evidence for the conjectured duality. See [Terning, 2006] for a more in-depth discussion. Anomaly matching is one the few tools we have for making concrete statements about non-perturbative theories.

Problems

30.1 Baryon number has an anomaly so that $\partial_\mu J_\mu^B \neq 0$ as in Eq. (30.88). Since the right-hand side of Eq. (30.88) has more than two gauge fields, it implies that diagrams such as

with the $\otimes$ indicating $J_\mu^B(x)$, should also give non-zero answers when contracted with ∂_μ. Evaluate this diagram and any other that contributes at the same order to show that the W^3 terms in Eq. (30.88) are correctly reproduced.

30.2 For which types of neutrino masses (Majorana, Dirac or both) is lepton number anomalous? For which types of masses is $B - L$ anomalous?

30.3 Suppose that QCD were based on the gauge group SU(5). Let us assume that the proton still exists as a five-quark bound state with charge $+1$, so that quarks now have five colors and electric charges in $\mathbb{Z}/5$. What values for the $SU(5) \times SU(2)_{\text{weak}} \times SU(1)_Y$ quantum numbers of the Standard Model fields would make this universe anomaly free?

30.4 Can anomaly matching arguments determine if $SU(4)_L \times SU(4)_R$ is spontaneously broken in QCD?

We now have discussed the complete Standard Model of particle physics. The model is based on the gauge group $SU(3)_{QCD} \times SU(2)_{weak} \times U(1)_{hypercharge}$, which is spontaneously broken down to $SU(3)_{QCD} \times U(1)_{EM}$ at a scale $v = 247$ GeV. Assuming Dirac neutrino masses, the Standard Model has 27 parameters: three coupling constants g, g' and g_s; six quark, three charged lepton, and three neutrino masses; three mixing angles and one phase among quarks; three mixing angles and one phase among leptons; the Higgs mass m_h and vev v; the QCD vacuum angle $\bar{\theta}$; and the cosmological constant Λ. While 27 parameters might seem like a lot, there are an *infinite* number of measurements that could conceivably be done. Since the Standard Model is renormalizable, the result of any of these infinite number of measurements can, in principle, be expressed as a function of these 27 parameters. Thus, the Standard Model is an overconstrained system – we can test it by making enough measurements with enough precision. In this chapter, we discuss two ways in which quantum field theory at loop level is required to connect measurements to the parameters of the Standard Model.

First we will discuss constraints on the gauge sector of the electroweak theory. At tree-level, many observables depend only on the three parameters g, g' and v (or equivalently α_e, $\sin^2\theta_w$ and the Fermi constant G_F). The dominant radiative corrections to many of these observables are from virtual top-quark- and Higgs-boson-loop contributions to the W-boson, Z-boson and photon propagators. Corrections to the gauge boson propagators are called *oblique corrections*. Oblique corrections provided important indirect information about the mass of the top quark and Higgs boson before these particles were seen directly, and they continue to provide important constraints on beyond-the-Standard-Model physics. Electroweak precision constraints are often expressed in terms of the S, T, U and ρ parameters, which will be defined and discussed in Section 31.2.

Another area where loops play an important role in connecting observables to parameters of the Standard Model is in the arena of flavor physics. Recall from Chapter 29 that the CKM matrix is unitary in the Standard Model. If enough CKM elements are measured, this unitarity can be directly tested. The sensitivity of such tests to beyond-the-Standard-Model physics is only limited by the level of precision with which theory and experiment can be compared. In Section 31.3, we discuss important loop corrections from QCD. In particular, we will show how virtual gluons modify the relation between CKM elements extracted at low energy (such as V_{cb}, which can be measured from $B^+ \rightarrow \bar{D}^0\pi^+$ decays) and CKM elements at the weak scale. The calculation we perform involves renormalization group evolution with operator mixing, a beautiful subject in its own right.

31.1 Electroweak precision tests

In this section we discuss **precision electroweak physics**, which is concerned (mainly) with observables constructed out of leptons and electroweak gauge bosons. We discuss quark-based observables in Section 31.3 and in Chapters 32, 35 and 36.

There are a few quantities that are basically only sensitive to electroweak physics and have been measured extremely well. We will focus on five of them:

1. The electron magnetic dipole moment $\frac{1}{2}g_e = 1.001\,159\,652\,180\,73 \pm 2.8 \times 10^{-13}$.
2. The lifetime of the muon: $\tau_\mu = (2.196\,981\,1 \pm 0.000\,002\,2) \times 10^{-6}$s. In GeV, the decay rate is $\tau_\mu^{-1} = \Gamma\left(\mu^- \to \nu_\mu e^- \bar{\nu}_e\right) = 2.995\,98 \times 10^{-19}$ GeV.
3. The Z-boson pole mass: $m_{Z,\text{pole}} = 91.1876 \pm 0.0021$ GeV.
4. The W-boson pole mass: $m_{W,\text{pole}} = 80.385 \pm 0.015$ GeV.
5. The polarization asymmetry in Z-boson production:

$$A_e = \frac{\sigma_L - \sigma_R}{\sigma_L + \sigma_R} = \frac{\sigma\left(e_L^- e_L^+ \to Z\right) - \sigma\left(e_R^- e_R^+ \to Z\right)}{\sigma\left(e_L^- e_L^+ \to Z\right) + \sigma\left(e_R^- e_R^+ \to Z\right)} = 0.1515 \pm 0.0019. \quad (31.1)$$

This asymmetry, which can be measured using polarized electron beams, would vanish in a non-chiral theory. Another important observable is the decay rate of the Z boson into electrons $\Gamma\left(Z \to e^+ e^-\right)$, which you can explore in Problem 31.2.

In the Standard Model, at leading order in perturbation theory, each one of these five observables depends only on three electroweak parameters: the strength of the QED coupling e (or equivalently the fine-structure constant $\alpha_e \equiv \frac{e^2}{4\pi}$), the Higgs vev v (or equivalently the Fermi constant $G_F \equiv \frac{1}{\sqrt{2}v^2}$) and the weak mixing angle $s = \sin\theta_w$.

The tree-level dependences of the Z and W masses on e, v and s are

$$m_Z = \frac{ev}{2sc}, \qquad m_W = \frac{ev}{2s}, \quad (31.2)$$

where $c \equiv \cos\theta_w = \sqrt{1 - s^2}$. The muon decay rate involves a virtual W boson. Including the full m_e and m_μ dependence, the rate computed at tree-level is

$$\tau_\mu^{-1} = \Gamma\left(\mu^- \to \nu_\mu e^- \bar{\nu}_e\right) = G_F^2 \frac{m_\mu^5}{192\pi^3}\left(1 - 8r + 8r^3 - r^4 - 12r^2 \ln r\right), \quad r = \frac{m_e^2}{m_\mu^2}. \quad (31.3)$$

The polarization asymmetry A_e is non-zero because the Z boson has different couplings to left- and right-handed fermions. Recalling from Chapter 29 that the Z-boson couplings to the electron can be written as

$$\mathcal{L}_Z = -\frac{e}{sc} Z_\mu \left[\left(\frac{1}{2} - s^2\right)\bar{e}_L \gamma^\mu e_L - s^2 \bar{e}_R \gamma^\mu e_R\right], \quad (31.4)$$

we find that

$$A_e = \frac{\sigma_L - \sigma_R}{\sigma_L + \sigma_R} = \frac{\left(\frac{1}{2} - s^2\right)^2 - s^4}{\left(\frac{1}{2} - s^2\right)^2 + s^4}. \quad (31.5)$$

Now we would like to know whether all the measured values for these observables are consistent with the Standard Model.

To begin, we have to come up with a clean definition of the three parameters e, G_F and $\sin^2\theta_w$ based on experiments. That is, we need to define renormalization conditions for them. We will denote the values of these couplings extracted from the first three measurements above with a circumflex, as $\hat{G}_F$, $\hat{e}$ and $\hat{s}^2$. We also define

$$\hat{m}_Z \equiv m_{Z,\text{pole}}. \tag{31.6}$$

Any other quantity related to these three by tree-level algebraic relations will also be denoted with a circumflex. For example, $\hat{v} = \sqrt{\frac{1}{\sqrt{2}\hat{G}_F}}$ or

$$\hat{m}_W = \frac{\hat{e}\hat{v}}{2\hat{s}}. \tag{31.7}$$

This $\hat{m}_W$ is *not* equal to $m_{W,\text{pole}}$. We compute the difference in this chapter.

Since g_e is known extremely well, we use it to define $\hat{e}$. We worked out that $g_e - 2 = \frac{\alpha_e}{\pi}$ at 1-loop in Chapter 17, but actually the calculation is known to very high orders, competing with the experimental precision. This high-order calculation and the precise g_e measurement give

$$\hat{\alpha}_e(0) = (137.035\,999\,074 \pm 0.000\,000\,044)^{-1}, \tag{31.8}$$

with $\hat{\alpha}_e = \frac{\hat{e}^2}{4\pi}$. The 0 argument of $\hat{\alpha}_e$ refers to this being a long-distance ($p^2 \approx 0$) measurement. That is, this value of the fine-structure constant corresponds to the on-shell coupling $\hat{e}$ defined through the 3-point function in Section 19.3. For precision electroweak physics, it is more useful to work with $\hat{\alpha}_e(m_Z)$ which is [Particle Data Group (Beringer *et al.*), 2012]

$$\hat{\alpha}_e(m_Z)^{-1} = 127.944 \pm 0.014. \tag{31.9}$$

The running of α_e has been discussed elsewhere (Chapters 16 and 23), thus we simply take this value as input.

By the way, we will always evaluate running couplings and running masses with μ set equal to an $\overline{\text{MS}}$ mass. Technically, $\hat{\alpha}_e(m_Z)$ and $\hat{\alpha}_e(\hat{m}_Z)$ differ by corrections that begin at 2-loop and beyond, so which scale we choose is beyond the order we are working in this chapter. However, since the RGEs are calculated in the $\overline{\text{MS}}$ scheme, it makes sense to choose μ to be an $\overline{\text{MS}}$ mass. An example where the choice of scale is important is for the top mass, as discussed around Eq. (31.62) below.

Next, since τ_μ is extremely well measured, we use it to define $\hat{G}_F = \frac{1}{\sqrt{2}\hat{v}^2}$. Using the measured values $m_\mu = 105.658\,371\,5$ MeV and $m_e = 0.510\,998\,910$ MeV in our tree-level decay formula we get

$$\hat{G}_F = 1.163\,93 \times 10^{-5}\,\text{GeV}^{-2}, \tag{31.10}$$

which gives $\hat{v} = \sqrt{\frac{1}{\sqrt{2}\hat{G}_F}} = 246.48$ GeV.

Finally, for $\sin^2\theta_w$, there are many reasonable definitions. For example, one could define $\sin^2\theta_w \equiv 1 - \frac{m_W^2}{m_Z^2}$, or one could define it from A_e. In the $\overline{\text{MS}}$ scheme, one could define it from the renormalized coupling constants as $\tan\theta_w \equiv \frac{g_R'}{g_R}$. For precision tests, a logical choice is to base it on the next-best-measured quantity in this list, $\hat{m}_Z = m_{Z,\text{pole}}$. We call this value $\hat{s}$. It satisfies the relation

$$\hat{s}^2(1 - \hat{s}^2) \equiv \frac{\pi \hat{\alpha}_e(m_Z)}{\sqrt{2}\hat{G}_F \hat{m}_Z^2}. \tag{31.11}$$

Plugging in the numbers (and using that $\hat{s}^2 \approx 1 - \frac{m_W^2}{m_Z^2}$ to determine which root) we find

$$\hat{s}^2 = 0.234\,289. \tag{31.12}$$

This is one possible definition of $\sin^2 \theta_w$.

Plugging these numbers in to Eqs. (31.2) and (31.5) predicts (at tree-level)

$$m_{W,\text{pole}}^{(\text{tree})} = \hat{m}_W \equiv \frac{\hat{e}\hat{v}}{2\hat{s}} = 79.794\,\text{GeV}, \tag{31.13}$$

$$A_e^{(\text{tree})} = \hat{A}_e \equiv \frac{\left(\frac{1}{2} - \hat{s}^2\right)^2 - \hat{s}^4}{\left(\frac{1}{2} - \hat{s}^2\right)^2 + \hat{s}^4} = 0.1252. \tag{31.14}$$

These are well outside the experimental bounds – by nearly 40 standard deviations in the m_W case! This does not mean we have a contradiction within the Standard Model. We cannot make such a conclusion until we include loop corrections and carefully renormalize.

31.1.1 Oblique corrections

In order to test the electroweak sector, we will proceed in four steps:

1. Express our fiducial quantities $\hat{\alpha}_e$, $\hat{G}_F$ and $\hat{m}_Z$ in terms of $\overline{\text{MS}}$ Lagrangian parameters e, m_Z and $\sin^2 \theta_w$.
2. Solve for the Lagrangian parameters in terms of the fiducial quantities.
3. Express any other quantity we want to compute ($m_{W,\text{pole}}$ and A_e) in terms of Lagrangian parameters.
4. Substitute in the measured quantities to get our predictions.

To be clear, in the notation we use for this chapter e and m_Z mean the $\overline{\text{MS}}$ renormalized electric charge and Z-boson $\overline{\text{MS}}$ mass, which are in general different from the charge $\hat{e}$ and pole mass $\hat{m}_Z$ in the on-shell scheme. Quantities with circumflexes (such as $\hat{s}$, $\hat{c}$ and $\hat{m}_W$) are related to the fiducial quantities $\hat{\alpha}_e(\hat{m}_Z)$, $\hat{m}_Z$ and $\hat{G}_F$ by tree-level relations.

There are many loops that can contribute to radiative corrections of the observables listed above. However, since the observables are given at tree-level by gauge boson exchange, the largest contributions will come from loops affecting the gauge boson propagators. For historical reasons, these are called **oblique corrections**. An advantage of these observables is that, since the Standard Model would have the same structure with any number of generations, the oblique corrections from each generation will be gauge invariant and finite. We will therefore focus on the largest corrections, which come from loops of the third generation quarks (t, b) and from the Higgs boson.

In the $\overline{\text{MS}}$ scheme, the tree-level propagators are determined from renormalized values of the masses in the Lagrangian. The Z-boson 2-point function in the free theory is given by

$$iG^{\mu\nu}_{Z(\text{tree})}(p) = i\Pi^{\mu\nu}_Z(p) = \frac{-i\left(g^{\mu\nu} - \frac{p^\mu p^\nu}{m_Z^2}\right)}{p^2 - m_Z^2}, \qquad (31.15)$$

with m_Z the renormalized $\overline{\text{MS}}$ mass parameter in the Lagrangian. Here, we use a shorthand notation defined by

$$\langle\Omega|T\{Z^\mu(x)Z^\nu(y)\}|\Omega\rangle \equiv \int \frac{d^4 p}{(2\pi)^4} e^{ip(x-y)} iG^{\mu\nu}_Z(p), \qquad (31.16)$$

and similarly for other 2-point functions.

The Z-boson propagator gets radiative corrections from loops, which correct both the $g^{\mu\nu}$ and $p^\mu p^\nu$ terms. By Lorentz invariance, we must find

$= i\Pi_{ZZ}g^{\mu\nu} + i\Pi^{pp}_{ZZ}p^\mu p^\nu. \qquad (31.17)$

However, since all the observables in which we are interested have the gauge bosons coupling to essentially massless fermions (which provide conserved currents), the $p^\mu p^\nu$ terms will not contribute. Thus, we can simply write that the corrections will give

$$iG^{\mu\nu}_Z(p) = \frac{-ig^{\mu\nu}}{p^2 - m_Z^2}\left(1 + i\Pi_{ZZ}\frac{-i}{p^2 - m_Z^2} + \cdots\right) + p^\mu p^\nu \text{ terms.} \qquad (31.18)$$

Summing all the one-particle-irreducible contributions Π_{ZZ} leads to

$$iG^{\mu\nu}_Z(p) = \frac{-ig^{\mu\nu}}{p^2 - m_Z^2 - \Pi_{ZZ}(p^2)} + p^\mu p^\nu \text{ terms,} \qquad (31.19)$$

so that the pole mass will be related to the renormalized Lagrangian mass at 1-loop as

$$\hat{m}_Z^2 = m_{Z,\text{pole}}^2 = m_Z^2 + \text{Re}\left[\Pi_{ZZ}(m_Z^2)\right]. \qquad (31.20)$$

The real part of Π_{ZZ} is taken in Eq. (31.20) because the Z boson is unstable.[1] Note that if we use Π_{ZZ} to 1-loop order, it does not matter which m_Z^2 is used in the argument of Π_{ZZ} – the difference is higher order. This is our first equation relating an observable ($\hat{m}_Z$) to an $\overline{\text{MS}}$ quantity (m_Z).

[1] Recall from Section 24.1.4 that 2-point functions for unstable particles have imaginary parts proportional to their decay widths. Since the width of the Z boson ($\Gamma_Z = 2.5$ GeV) is much less than its mass ($m_Z = 91.2$ GeV), the relation $\text{Im}\left[\Pi_{ZZ}\right] = -m_{Z,\text{pole}}\Gamma_Z$ applies. For p^2 near m_Z^2, the 2-point function then becomes

$$iG^{\mu\nu}_Z(p) = \frac{-i\left(g^{\mu\nu} - \frac{p^\mu p^\nu}{m_Z^2}\right)}{p^2 - m_{Z,\text{pole}}^2 + im_{Z,\text{pole}}\Gamma_Z}. \qquad (31.21)$$

This generates a Breit–Wigner line shape which is fit to data to determine the real pole mass $m_{Z,\text{pole}}$ and Γ_Z from data.

Similarly, for the W mass

$$m_{W,\text{pole}}^2 = m_W^2 + \text{Re}\big[\Pi_{WW}\big(m_W^2\big)\big]. \tag{31.22}$$

For the remainder of this chapter, we will not write $\text{Re}\,[\,]$ explicitly. We will simply evaluate the real part of the various self-energy functions at the end of the calculation.

For corrections to the photon propagator, recall that the photon is massless to all orders in perturbation theory, since its mass is forbidden by gauge invariance. Thus we have

$$iG_\gamma^{\mu\nu}(p) = \frac{-ig^{\mu\nu}}{p^2 - \Pi_{\gamma\gamma}(p^2)} + p^\mu p^\nu \text{ terms}, \tag{31.23}$$

where $\Pi_{\gamma\gamma}$ are the 1PI vacuum polarization graphs. Comparing to the notation $\Pi(p^2)$ from Section 19.2, we find $\Pi_{\gamma\gamma} = -p^2 \Pi(p^2)$.

Now, let us relate the renormalized electric charge e (the $\overline{\text{MS}}$ parameter in the Lagrangian) to the value $\hat{e}^2(m_Z) = 4\pi\hat{\alpha}_e(m_Z)$ in Eq. (31.9) (which comes from a physical measurement). One way to define $\hat{e}^2(Q)$ is as the value of the effective charge relevant for s-channel photon exchange at a scale Q. Then, the total cross section for $e^+e^- \to \mu^+\mu^-$ at $s = m_Z^2$ from photon exchange is

$$\sigma_{\text{tot}} = \frac{\hat{e}^4(m_Z)}{12\pi m_Z^2}. \tag{31.24}$$

As explained in Section 20.3.1, the running coupling is defined so that the large logarithms in the vacuum polarization graphs are included in a tree-level graph using $\hat{e}^4(m_Z)$ instead of $\hat{e}^4(0)$. To compare to the $\overline{\text{MS}}$ parameter in the Lagrangian, we note that, had we used the full vacuum polarization contributions, we would have found

$$\sigma_{\text{tot}} = \frac{e^4}{12\pi m_Z^2}\left(\frac{m_Z^2}{m_Z^2 - \Pi_{\gamma\gamma}(m_Z^2)}\right)^2, \tag{31.25}$$

where the factor in brackets comes from replacing p^2 by $p^2 + \Pi_{\gamma\gamma}(p^2)$ in the photon propagator and evaluating at $p^2 = m_Z^2$. Thus,

$$\hat{e}^2(m_Z) = e^2 \frac{1}{1 - \frac{1}{m_Z^2}\Pi_{\gamma\gamma}(m_Z^2)} = e^2\left[1 + \frac{\Pi_{\gamma\gamma}(m_Z^2)}{m_Z^2} + \cdots\right]. \tag{31.26}$$

This equation relates the value of $\hat{e}^2(m_Z)$ extracted from $g-2$ and evolved to m_Z in Eq. (31.9) to the renormalized $\overline{\text{MS}}$ parameters e and m_Z in the Lagrangian.

Finally, we want to relate the muon lifetime $\hat{\tau}_\mu$ (or equivalently $\hat{G}_F$) to the renormalized parameter $s^2 = \sin^2\theta_w$ in the $\overline{\text{MS}}$ Lagrangian. Since muon decay proceeds through a charged-current interaction, the oblique corrections to the decay rate come from Π_{WW}. As $\hat{G}_F$ comes from the low-energy limit of the tree-level W propagator, we have

$$\frac{\hat{G}_F}{\sqrt{2}} = -\frac{g^2}{8}\frac{1}{p^2 - m_W^2 - \Pi_{WW}(p^2)}\bigg|_{p^2=0} = \frac{e^2}{8s^2c^2m_Z^2}\left(1 - \frac{\Pi_{WW}(0)}{m_W^2} + \cdots\right), \tag{31.27}$$

where we have replaced $\frac{g^2}{8m_W^2} \to \frac{1}{2v^2} \to \frac{e^2}{8s^2c^2m_Z^2}$ (note that m_Z and not $\hat{m}_Z$ appears in this expression, since Π_{ZZ} does not contribute to a correction to the muon decay rate at this order). $\hat{G}_F$ on the left-hand side is the measured value, extracted from the muon lifetime in Eq. (31.10), while all the quantities on the right-hand side are $\overline{\text{MS}}$ quantities.

The final observable, the Z-boson production asymmetry A_e, is determined by how the Z boson couples to electrons. In the $\overline{\text{MS}}$ Lagrangian the Z and photon couplings are

$$\mathcal{L}_{\gamma Z} = -\frac{e}{sc} Z_\mu \left[\left(\frac{1}{2} - s^2 \right) \bar{e}_L \gamma^\mu e_L - s^2 \bar{e}_R \gamma^\mu e_R \right] - e A_\mu [\bar{e}_L \gamma^\mu e_L + \bar{e}_R \gamma^\mu e_R] \,, \quad (31.28)$$

where $c = \sqrt{1 - s^2}$. Here, $s = \sin\theta_w$ is the renormalized $\overline{\text{MS}}$ value for the sine of the weak mixing angle. At tree-level, this Lagrangian gives $A_e = \frac{\left(\frac{1}{2} - s^2\right)^2 - s^4}{\left(\frac{1}{2} - s^2\right)^2 + s^4}$, as in Eq. (31.5).

At 1-loop, there is a contribution to Z-boson production from Π_{ZZ} vacuum polarization graphs and from vacuum polarization graphs that mix the Z boson with the photon:

$$\text{(31.29)}$$

The first graph tells us that we should use the corrected propagator with a pole at $m_{Z,\text{pole}}$ rather than at m_Z. Since we evaluate these graphs with momentum $p^2 = m_Z^2$ going through the boson lines, we can account for the Π_{ZZ} correction at 1-loop by simply replacing m_Z by $\hat{m}_Z$ in the tree-level result. Since the tree-level result for A_e has no m_Z dependence, the effect of Π_{ZZ} is higher order.

The second graph gives a factor of $\frac{\Pi_{\gamma Z}(p^2)}{p^2}$ with $p^2 = m_Z^2$ and the photon charge rather than the Z boson charge. That is, it says the effective Z-boson couplings are

$$\mathcal{L}_Z^{\text{eff}} = -\frac{e}{sc} Z_\mu \left[\left(\frac{1}{2} - s^2 \right) \bar{e}_L \gamma^\mu e_L - s^2 \bar{e}_R \gamma^\mu e_R \right] - e \frac{\Pi_{\gamma Z}(m_Z^2)}{m_Z^2} Z_\mu [\bar{e}_L \gamma^\mu e_L + \bar{e}_R \gamma^\mu e_R]$$

$$- \frac{e}{sc} Z_\mu \left[\left(\frac{1}{2} - s_{\text{eff}}^2 \right) \bar{e}_L \gamma^\mu e_L - s_{\text{eff}}^2 \bar{e}_R \gamma^\mu e_R \right], \quad (31.30)$$

where

$$s_{\text{eff}}^2 \equiv s^2 - sc \frac{\Pi_{\gamma Z}(m_Z^2)}{m_Z^2}. \quad (31.31)$$

This leads to a simple formula for the asymmetry,

$$A_e = \frac{\left(\frac{1}{2} - s_{\text{eff}}^2\right)^2 - s_{\text{eff}}^4}{\left(\frac{1}{2} - s_{\text{eff}}^2\right)^2 + s_{\text{eff}}^4}, \quad (31.32)$$

which is valid at 1-loop.

Now let us turn to our four advertised steps from the introduction to this section. **Step 1** is to express the three fiducial measured quantities in terms of Lagrangian parameters e, m_Z and s:

$$\hat{e}^2(m_Z) = e^2 \left[1 + \frac{\Pi_{\gamma\gamma}(m_Z^2)}{m_Z^2} \right], \tag{31.33}$$

$$\hat{G}_F = \sqrt{2}\frac{e^2}{8s^2c^2m_Z^2}\left(1 - \frac{\Pi_{WW}(0)}{m_W^2}\right), \tag{31.34}$$

$$\hat{m}_Z^2 = m_Z^2 + \Pi_{ZZ}(m_Z^2), \tag{31.35}$$

with the real part of Π_{ZZ} implicit in the last equation. The left-hand sides of these three equations are the measured values while the right-hand sides are formal expressions in terms of renormalized $\overline{\text{MS}}$ Lagrangian parameters.

Step 2 is to invert these equations (to leading order in $\hat{\alpha}_e$) giving e, m_Z and s in terms of $\hat{e}$, $\hat{m}_Z$ and $\hat{G}_F$:

$$e^2 = \hat{e}^2 \left[1 - \frac{\Pi_{\gamma\gamma}(\hat{m}_Z^2)}{\hat{m}_Z^2} \right], \tag{31.36}$$

$$m_Z^2 = \hat{m}_Z^2 \left[1 - \frac{\Pi_{ZZ}(\hat{m}_Z^2)}{\hat{m}_Z^2} \right], \tag{31.37}$$

and

$$s^2 c^2 = \sqrt{2}\frac{\hat{e}^2}{8\hat{G}_F \hat{m}_Z^2}\left(1 + \Pi_R\right), \tag{31.38}$$

where

$$\Pi_R \equiv -\frac{\Pi_{\gamma\gamma}(\hat{m}_Z^2)}{\hat{m}_Z^2} + \frac{\Pi_{ZZ}(\hat{m}_Z^2)}{\hat{m}_Z^2} - \frac{\Pi_{WW}(0)}{\hat{m}_W^2}. \tag{31.39}$$

Since these vacuum polarization graphs are already 1-loop, we can use either $\hat{m}_Z$ and $\hat{m}_W$ or m_Z and m_W as the arguments of these vacuum polarization graphs. The difference is formally beyond the order we are working.

It is not hard to get an expression for s^2 instead of s^2c^2 using trigonometric identities. In terms of $\hat{s}^2$, defined in Eq. (31.11) as the value of $\sin^2\theta_w$ extracted directly from our fiducial observables, we find

$$s^2 = \hat{s}^2 \left(1 + \frac{\hat{c}^2}{\hat{c}^2 - \hat{s}^2}\Pi_R \right), \qquad c^2 = \hat{c}^2 \left(1 - \frac{\hat{s}^2}{\hat{c}^2 - \hat{s}^2}\Pi_R \right). \tag{31.40}$$

Step 3 is to express the other observables first in terms of Lagrangian parameters. For A_e we have already done this in Eq. (31.32). For m_W, using $m_W^2 = c^2 m_Z^2$ and Eq. (31.22) we get

$$m_{W,\text{pole}}^2 = c^2 m_Z^2 + \Pi_{WW}(m_W^2). \tag{31.41}$$

Then, **Step 4**, we express $m_{W,\text{pole}}$ and A_e in terms of the measured quantities

$$m_{W,\text{pole}}^2 = \hat{c}^2 \hat{m}_Z^2 \left(1 - \frac{\hat{s}^2}{\hat{c}^2 - \hat{s}^2}\Pi_R - \frac{\Pi_{ZZ}(\hat{m}_Z^2)}{\hat{m}_Z^2} + \frac{\Pi_{WW}(\hat{m}_W^2)}{\hat{c}^2 \hat{m}_Z^2} \right). \tag{31.42}$$

Similarly, Eq. (31.32) is $A_e = \frac{\left(\frac{1}{2}-\hat{s}_{\text{eff}}^2\right)^2-\hat{s}_{\text{eff}}^4}{\left(\frac{1}{2}-\hat{s}_{\text{eff}}^2\right)^2+\hat{s}_{\text{eff}}^4}$, which gives an expression for A_e in terms of s_{eff} and is defined in Eqs. (31.31) in terms of $\overline{\text{MS}}$ quantities. Writing s_{eff} instead in terms of observables, using Eq. (31.40) gives

$$s_{\text{eff}}^2 = \hat{s}^2 + \frac{\hat{s}^2\hat{c}^2}{\hat{c}^2 - \hat{s}^2}\Pi_R - \hat{s}\hat{c}\frac{\Pi_{\gamma Z}\left(\hat{m}_Z^2\right)}{\hat{m}_Z^2}. \tag{31.43}$$

Next, we need to evaluate the various vacuum polarization amplitudes, which we can then plug in to get our experimental predictions.

31.1.2 Electroweak vacuum polarization loops

Now let us evaluate all the vacuum polarization graphs. We will focus on the contributions from the top quark, which gives the largest effect $\left(\text{proportional to } \frac{m_t^2}{m_Z^2}\right)$, and from the bottom quark, which is required by $SU(2)$ invariance. The Higgs boson contributions are also important, but we leave their computation as an exercise (Problem 31.3). To compute Π_{ij}, we need to perform loops with left- or right-handed insertions. We will do this for general masses and couplings and then insert the appropriate masses and couplings for the appropriate self-energy function.

The fermion contributions to the vacuum polarization functions come from loops such as

$$i\Pi^{\mu\nu} = \tag{31.44}$$

where the two masses m_1 and m_2 can be different (for example in Π_{WW}). The LL or RR amplitudes at 1-loop are

$$i\Pi_{LL}^{\mu\nu} = i\Pi_{RR}^{\mu\nu} = (-1)e^2\mu^{4-d}\int\frac{d^dk}{(2\pi)^d}\frac{\text{Tr}\left[(i\gamma^\mu)\,P_L i(\slashed{k}+m_1)(i\gamma^\nu)\,P_L i(\slashed{k}+\slashed{p}+m_2)\right]}{[k^2-m_1^2]\left[(k+p)^2-m_2^2\right]}$$

$$= ig^{\mu\nu}\frac{e^2}{(4\pi)^{d/2}}\mu^{4-d}\int_0^1 dx\frac{\Gamma\left(2-\frac{d}{2}\right)}{\Delta^{2-d/2}}\left[2xm_2^2 + 2(1-x)m_1^2 - 4x(1-x)p^2\right] + p^\mu p^\nu \text{ terms},$$

$$\tag{31.45}$$

where

$$\Delta = xm_2^2 + (1-x)m_1^2 - x(1-x)p^2. \tag{31.46}$$

There is also a $p^\mu p^\nu$ term, which we will just drop from now on since it does not contribute to the observables due to Ward identities. Stripping off the $ig^{\mu\nu}$ as in Eq. (31.17) and expanding with $d = 4 - \varepsilon$ we get

$$\Pi_{LL} = \Pi_{RR} = \frac{e^2}{4\pi^2}\left\{\frac{m_1^2 + m_2^2 - \frac{2}{3}p^2}{2\varepsilon} - \frac{1}{2}\int_0^1 dx\left[x(1-x)p^2 - \Delta\right]\ln\frac{\tilde{\mu}^2}{\Delta}\right\}. \tag{31.47}$$

The LR integral requires a mass insertion to turn $R \leftrightarrow L$ so it must be odd in the masses. By dimensional analysis we therefore expect it will be proportional to $m_1 m_2$. The exact result is

$$i\Pi^{\mu\nu}_{LR} = i\Pi^{\mu\nu}_{RL} = (-1)e^2\mu^{4-d} \int \frac{d^d k}{(2\pi)^d} \frac{\text{Tr}\left[(i\gamma^\mu) P_L i(\slashed{k} + m_1)(i\gamma^\nu) P_R i(\slashed{k} + \slashed{p} + m_2)\right]}{[k^2 - m_1^2]\left[(k+p)^2 - m_2^2\right]}$$

$$= -ig^{\mu\nu} \frac{e^2}{(4\pi)^{d/2}} \mu^{4-d} \int_0^1 dx \frac{\Gamma\left(2 - \frac{d}{2}\right)}{\Delta^{2-d/2}} 2m_1 m_2 + p^\mu p^\nu \text{ terms}, \tag{31.48}$$

so that

$$\Pi_{LR} + \Pi_{RL} = -\frac{e^2}{4\pi^2} m_1 m_2 \left\{ \frac{1}{\varepsilon} + \frac{1}{2} \int_0^1 dx \ln\frac{\tilde{\mu}^2}{\Delta} \right\}. \tag{31.49}$$

As a check, the above calculation with $m_1 = m_2 = m_e$ should reproduce the QED vacuum polarization amplitude (vector–vector or Π_{VV}). We find

$$\Pi_{VV} = \Pi_{LL} + \Pi_{LR} + \Pi_{RL} + \Pi_{RR}$$

$$= -\frac{e^2}{(4\pi)^{d/2}} \mu^{4-d} p^2 \int_0^1 dx \frac{\Gamma\left(2 - \frac{d}{2}\right)}{[m_e^2 - p^2 x(1-x)]^{2-d/2}} 8x(1-x)$$

$$= -\frac{e^2}{2\pi^2} p^2 \left[\frac{1}{3\varepsilon} + \int_0^1 dx \, x(1-x) \ln\left(\frac{\tilde{\mu}^2}{m_e^2 - p^2 x(1-x)} \right) \right]. \tag{31.50}$$

This is proportional to p^2 and agrees with the result for the vacuum polarization graph in Eq. (16.45), since $\Pi_2^{\mu\nu} = g^{\mu\nu}\Pi_{VV}$.

With these amplitudes, it is straightforward to plug in the charges and work out the vacuum polarization amplitudes for the $\gamma/W/Z$ fields. Π_{WW} is proportional to Π_{LL}, since it only involves left-handed fields, and $\Pi_{\gamma\gamma}$ is proportional to Π_{VV}. For Π_{ZZ} and $\Pi_{Z\gamma}$ we can use that the Z boson couples to $T^3 - s^2 Q$ with strength $-\frac{e}{sc}$ (see Eq. (31.4)) to write everything in terms of vector and left-handed amplitudes. For a single (t, b) doublet, we get

$$\Pi_{\gamma\gamma}(p^2) = N \sum_{i=t,b} Q_i^2 \Pi_{VV}(\Delta_{ii}), \tag{31.51}$$

$$\Pi_{\gamma Z}(p^2) = \frac{1}{sc} N \sum_{i=t,b} \left(T_i^3 Q_i \frac{1}{2}\Pi_{VV}(\Delta_{ii}) - s^2 Q_i^2 \Pi_{VV}(\Delta_{ii}) \right), \tag{31.52}$$

$$\Pi_{WW}(p^2) = |V_{tb}|^2 \frac{1}{s^2} N \frac{1}{2}\Pi_{LL}(\Delta_{tb}), \tag{31.53}$$

$$\Pi_{ZZ}(p^2)$$
$$= \frac{1}{s^2 c^2} N \sum_{i=t,b} \left((T_i^3)^2 \Pi_{LL}(\Delta_{ii}) - 2s^2 T_i^3 Q_i \frac{1}{2}\Pi_{VV}(\Delta_{ii}) + s^4 Q_i^2 \Pi_{VV}(\Delta_{ii}) \right). \tag{31.54}$$

Here Δ_{ij} means Δ with $m_1 = m_i$ and $m_2 = m_j$. The factor of $N = 3$ comes from the three colors of quarks; the $\frac{1}{2}$ in Π_{WW} from the normalization of the $W_\pm$ generators; the $\frac{1}{2}\Pi_{VV}$ comes from the T^3/hypercharge mixing, $\Pi_{3Y} \propto \Pi_{LR} + \Pi_{RR} = \frac{1}{2}\Pi_{VV}$. Note that, since the Π_{ij} start at 1-loop, it does not matter which definitions of s and e we use in these expressions. One can easily substitute the charges $Q_t = \frac{2}{3}$, $Q_b = -\frac{1}{3}$, $T_t^3 = \frac{1}{2}$ and $T_b^3 = -\frac{1}{2}$ to simplify these formulas, but the more general formulas help illustrate where some cancellations come from.

With these results, we can now evaluate how the top and bottom quarks affect our predictions. For example, expanding out Eq. (31.43) gives

$$s_{\text{eff}}^2 = \hat{s}^2 + \frac{\hat{s}^2 \hat{c}^2}{\hat{c}^2 - \hat{s}^2}\left(-\frac{\Pi_{\gamma\gamma}(\hat{m}_Z^2)}{\hat{m}_Z^2} + \frac{\Pi_{ZZ}(\hat{m}_Z^2)}{\hat{m}_Z^2} - \frac{\Pi_{WW}(0)}{\hat{m}_W^2}\right) - \hat{s}\hat{c}\frac{\Pi_{\gamma Z}(\hat{m}_Z^2)}{\hat{m}_Z^2}. \quad (31.55)$$

Let us first check that the divergent parts (and hence also the μ-dependent parts) of this and $m_{W,\text{pole}}^2$ in Eq. (31.42) are zero. Noting that

$$\Pi_{LL}(\Delta_{ij}) = e^2 \frac{m_i^2 + m_j^2 - \frac{2}{3}p^2}{8\pi^2\varepsilon} + \mathcal{O}(\varepsilon^0), \qquad \Pi_{VV}(\Delta_{ii}) = -e^2\frac{p^2}{6\pi^2\varepsilon} + \mathcal{O}(\varepsilon^0), \quad (31.56)$$

the divergent parts are

$$\Pi_{\gamma\gamma}(m_Z^2) = -\frac{e^2 m_Z^2}{2\pi^2\varepsilon}\left(Q_t^2 + Q_b^2\right), \quad (31.57)$$

$$\Pi_{WW}(m_W^2) = |V_{tb}|^2 \frac{3e^2}{16\pi^2 s^2\varepsilon}\left(m_b^2 + m_t^2 - \frac{2}{3}m_W^2\right), \quad (31.58)$$

$$\Pi_{\gamma Z}(m_Z^2) = \frac{e^2 m_Z^2}{8\pi^2 sc\varepsilon}\left(Q_b - Q_t + 4s^2(Q_t^2 + Q_b^2)\right), \quad (31.59)$$

$$\Pi_{ZZ}(m_Z^2) = \frac{e^2}{16s^2c^2\pi^2\varepsilon}\left[3m_b^2 + 3m_t^2 - 2m_Z^2\left(1 + 2(Q_b - Q_t)s^2 + 4(Q_b^2 + Q_t^2)s^4\right)\right]. \quad (31.60)$$

Using only $Q_t - Q_b = 1 = |Q_{W^\pm}|$ we find that s_{eff}^2 and $m_{W,\text{pole}}^2$ would be finite if $|V_{tb}|^2 = 1$. For $|V_{tb}| \neq 1$, one must include all the other quark loops to see the finiteness. Doing so, we would find the divergent part of Π_{WW} is

$$\begin{aligned}\Pi_{WW}(m_W^2) &= \frac{3e^2}{16\pi^2 s^2\varepsilon}\left[\left(|V_{tb}|^2 + |V_{ts}|^2 + |V_{td}|^2\right)m_t^2\right.\\ &\qquad \left. + \left(|V_{tb}|^2 + |V_{cb}|^2 + |V_{ub}|^2\right)m_b^2 + \cdots\right]\\ &= \frac{3e^2}{16\pi^2 s^2\varepsilon}\left[m_t^2 + m_b^2 + m_c^2 + m_s^2 + m_u^2 + m_d^2 - 4m_W^2\right], \quad (31.61)\end{aligned}$$

where unitarity of the CKM matrix has been used. Thus, the CKM matrix elements drop out of the divergent parts of the vacuum polarization graphs. Since the finite parts of loops involving light quarks are proportional to the light-quark masses, we can neglect their corrections. Including the m_t^2 contributions from the top-bottom, top-strange and top-down graphs is therefore equivalent to including just the m_t^2 contribution from one of these graphs with $V_{tb} = 1$. Thus, we set $V_{tb} = 1$ and include just the top-bottom loop.

Since the divergent contributions to our predictions for $m_{W,\text{pole}}^2$ and s_{eff}^2 (and hence A_e) cancel, we can now evaluate the 1-loop corrections to these observables. To do so, the only remaining issue is what value to take for the electric charge and top mass.

For the electric charge, we must first convert the on-shell value from Eq. (31.8) to the $\overline{\text{MS}}$ scheme at $\mu = 0$ and then run up to m_Z. The leading-order vacuum polarization diagram $\Pi_{\gamma\gamma}$ determines the leading-order running coupling. Since we have to run over a large range of energy to get an accurate value of $e(m_Z)$ we should include all charged particles and subleading-loop running. The running has in fact been calculated up to 4-loops, and the current best $\overline{\text{MS}}$ value of the fine-structure constant at m_Z is given in Eq. (31.9). We will simply use that value, since renormalization group evolution has already been covered in Chapter 23.

The current experimental value of the top mass is $m_{t,\text{pole}} = 173.5 \pm 1.0$ GeV. This is the value of a parameter in Monte-Carlo simulations which produce distributions with the best fit to the observed line shape. Since this shape is approximately Breit–Wigner, we conclude that this value corresponds most closely to the real pole mass defined in Section 24.1.4. Of course, working only at 1-loop, it does not matter whether we use the top-quark pole mass or the $\overline{\text{MS}}$ mass in the oblique corrections, since differences are higher order. However, because of the strong (quadratic) dependence of the oblique corrections on the top mass, subleading (2-loop and higher) effects can be large. Large logarithms in these higher-order amplitudes can be minimized by using the scale-dependent top-quark mass $m_t(\mu)$ in the $\overline{\text{MS}}$ scheme rather than the pole mass $m_{t,\text{pole}}$. Converting to the $\overline{\text{MS}}$ mass using 3-loop QCD corrections gives [Melnikov and van Ritbergen, 2000]

$$m_t(m_t) = m_{t,\text{pole}} \left[1 - \frac{4}{3} \frac{\alpha_s(m_t)}{\pi} - 9.125 \left(\frac{\alpha_s(m_t)}{\pi} \right)^2 - 80.405 \left(\frac{\alpha_s(m_t)}{\pi} \right)^3 \right]$$

$$= 163.0 \text{ GeV}, \tag{31.62}$$

where $\alpha_s(m_t) = 0.1088$ has been used. Note that there is a 10.5 GeV difference between the pole and the $\overline{\text{MS}}$ top-quark masses, so this is a fairly large effect. The W-boson and Z-boson masses should technically also be used in the $\overline{\text{MS}}$ scheme. However, since these particles are colorless, the scheme dependence is small and the difference can be neglected.[2]

Now we can compute the 1-loop corrections to A_e and $m_{W,\text{pole}}$. Using the values discussed above, $m_t(m_t) = 163.0$ GeV, $\alpha_e(m_Z) = 0.007816$, $\hat{m}_Z = 91.1876$ GeV, $m_b = 4.18$ GeV and $\hat{s}^2 = 0.234\,289$, we get

$$m_{W,\text{pole}} = 80.368 \text{ GeV}. \tag{31.63}$$

Comparing to the experimental value $m_W^{\text{exp}} = 80.399 \pm 0.023$ GeV we now find good agreement. We also get $s_{\text{eff}}^2 = 0.2313$, giving a prediction

$$A_e = 0.1491 \tag{31.64}$$

to be compared to the experimental value $A_e^{\text{exp}} = 0.1514 \pm 0.0019$. Both of these predictions are now within uncertainties of the data. In case you are curious, if we had used the top pole mass instead of $\overline{\text{MS}}$ mass, the 1-loop values would have been $m_{W,\text{pole}} = 80.440$ GeV and $A_e = 0.1522$.

---

[2] Actually, as discussed in Section 22.6.1, for the Higgs mass, $m_h(m_h)^2 - m_{h,\text{pole}}^2$ is actually quadratically sensitive to the top mass. This sensitivity is related to $m_h = 0$ not being technically natural and the hierarchy problem. In the Standard Model, it turns out that $m_{h,\text{pole}} = 125$ GeV gives $m_h(m_h) = 124$ GeV, so the difference happens to be numerically small.

Table 31.1 Standard Model predictions for electroweak observables. Inputs are $\hat{\alpha}_e(m_Z) = 0.007\,757\,5$, $\hat{\tau}_\mu = 2.196\,981\,1 \times 10^{-6}$ s, $m_{Z,\text{pole}} = 91.1876$ GeV and $m_{t,\text{pole}} = 173.5$ GeV. The rightmost column includes a 125 GeV Higgs.

Observable	Exp. value	Tree-level	1-loop (t, b)	1-loop (t, b, h)
m_W (GeV)	80.399 ± 0.023	79.794	80.368	80.333
A_e	0.1514 ± 0.0019	0.1252	0.1491	0.1470

Sometimes it is helpful to have approximate analytic formulas for the oblique corrections. For example, taking $m_b \to 0$ then $m_Z \ll m_t$ we get, from Eq. (31.42),

$$m_{W,\text{pole}}^2 \approx \hat{c}^2 \hat{m}_Z^2 \left[1 + \frac{3\hat{\alpha}_e}{16\pi \hat{s}^2(\hat{c}^2 - \hat{s}^2)} \frac{m_t^2}{\hat{m}_Z^2} \right], \tag{31.65}$$

and from Eq. (31.55),

$$s_{\text{eff}}^2 \approx \hat{s}^2 \left[1 - \frac{3\hat{\alpha}_e}{16\pi \hat{s}^2(\hat{c}^2 - \hat{s}^2)} \frac{m_t^2}{\hat{m}_Z^2} \right]. \tag{31.66}$$

These approximations give $m_{W,\text{pole}} = 80.285$ and $s_{\text{eff}}^2 = 0.2314$, which leads to $A_e = 0.1480$, in close agreement with the exact 1-loop results listed above.

In addition to the top/bottom contribution, the other reasonably sized correction is from the Higgs boson. We use the value

$$m_{H,\text{pole}} = 125\,\text{GeV} \tag{31.67}$$

Calculating the appropriate vacuum polarization graphs, the leading m_h dependence shifts the predictions for s_{eff}^2 and $m_{W,\text{pole}}^2$ as (see Problem 31.3)

$$m_{W,\text{pole}}^2 \to m_{W,\text{pole}}^2 - \frac{5\alpha_e}{24\pi} \frac{\hat{c}^2 \hat{m}_Z^2}{\hat{c}^2 - \hat{s}^2} \ln \frac{m_h^2}{m_W^2} \tag{31.68}$$

and

$$s_{\text{eff}}^2 \to s_{\text{eff}}^2 + \frac{\alpha_e(1 + 9\hat{s}^2)}{48\pi(\hat{c}^2 - \hat{s}^2)} \ln \frac{m_h^2}{m_W^2}. \tag{31.69}$$

Note that the oblique corrections depend quadratically on the top-quark mass but only logarithmically on the Higgs mass. Taking $m_h = 125$ GeV this leads to $m_{W,\text{pole}} = 80.351$ GeV and $s_{\text{eff}}^2 = 0.2314$, which gives $A_e = 0.1477$. These values are summarized in Table 31.1.

31.2 Custodial SU(2), ρ, S, T and U

In the Standard Model, the W-boson and Z-boson masses have a ratio determined by the relative strengths of the weak and electromagnetic gauge couplings. That is,

$$\frac{m_W^2}{m_Z^2} = \frac{g^2}{g^2 + g'^2} = \cos^2\theta_w. \tag{31.70}$$

This is a consequence of the way the $SU(2) \times U(1)$ symmetry is spontaneously broken through the Higgs mechanism with a single $SU(2)$ doublet. If the Higgs sector were more complicated, there might be deviations from this, even at tree-level. It is therefore useful to define something called the ρ-**parameter**, defined as

$$\rho \equiv \frac{m_W^2}{m_Z^2 \cos^2\theta_w}. \tag{31.71}$$

Denoting the tree-level value of ρ as ρ_0, we see that $\rho_0 = 1$ in the Standard Model.

Since the W-boson and Z-boson masses and the gauge couplings g and g' have nothing to do with the linear-sigma-model field h (the Higgs), it is natural to wonder what exactly guarantees that $\rho_0 = 1$. That is, can we see that $\rho_0 = 1$ purely from the low-energy effective theory, the nonlinear sigma model? The answer is yes; $\rho_0 = 1$ is guaranteed by a symmetry. To see this symmetry, recall that our original Higgs doublet H transformed as a doublet under $SU(2)$. Writing $H = \frac{1}{\sqrt{2}} \begin{pmatrix} h_3 + ih_4 \\ h_1 + ih_2 \end{pmatrix}$, we see that the potential

$$V(H) = \lambda \left(H^\dagger H - \frac{v^2}{2} \right)^2 = \frac{\lambda}{2} \left(h_1^2 + h_2^2 + h_3^2 + h_4^2 - v^2 \right)^2 \tag{31.72}$$

is actually invariant under a larger $SO(4)$ symmetry, under which the quadruplet (h_1, h_2, h_3, h_4) transforms in the fundamental representation. Note that $SO(4)$ has six generators, which is twice as many as $SU(2)$. When H gets a vev (such as with $h_1 = v$ and $h_2 = h_3 = h_4 = 0$) the $SO(4)$ symmetry is broken down to $SO(3)$. Thus there are actually three unbroken (global) symmetry directions in the Higgs sector of the Standard Model. In other words, there is a residual global $SU(2)$ symmetry after electroweak symmetry breaking. This is known as **custodial isospin** or **custodial SU(2)**.

Despite the fact that we have introduced this symmetry as acting on H, it is not hard to see that it actually just acts on the Goldstone bosons. Thus, it should be present in the low-energy theory. In fact, it is even present in the 4-Fermi theory. The charged-current and neutral-current 4-Fermi interactions, coming from W and Z exchange respectively, are

$$\mathcal{L} = -\left(\frac{e}{s}\right)^2 \left(J_\mu^1 + iJ_\mu^2\right) \frac{g^{\mu\nu}}{m_W^2} \left(J_\nu^1 - iJ_\nu^2\right) - \left(\frac{e}{sc}\right)^2 \left(J_\mu^3 - s^2 J_\mu^{\text{EM}}\right) \frac{g^{\mu\nu}}{m_Z^2} \left(J_\nu^3 - s^2 J_\nu^{\text{EM}}\right). \tag{31.73}$$

We conventionally define $G_F = \frac{\sqrt{2}e^2}{8s^2 m_W^2}$ so that this can be rewritten as

$$\mathcal{L} = -\frac{8}{\sqrt{2}} G_F \left[\left| J_\mu^1 + iJ_\mu^2 \right|^2 + \rho \left(J_\mu^3 - s^2 J_\mu^{\text{EM}} \right)^2 \right]. \tag{31.74}$$

So we see that the custodial symmetry forcing $\rho_0 = 1$ is just the symmetry that relates the strength of the weak part of the neutral-current interactions to the strength of the charged-current interaction. (If we restore the $SU(2)_L$ symmetry with Goldstone bosons, this equality of coupling strengths would translate to the equality $F_\pi^3 = F_\pi^{1,2}$, which is guaranteed by the custodial $SU(2)$.)

As an example, consider electroweak symmetry breaking by QCD. Recall that even if we did not have a Higgs sector at all, we would still have $SU(2) \times U(1) \to U(1)$ by the QCD $\langle \bar{q}q \rangle$ condensate. If this were the only source of electroweak symmetry breaking, would we still find $\rho_0 = 1$? The answer is yes, because QCD does have a custodial SU(2) symmetry. In the massless quark limit, QCD has a full $SU(2)_L \times SU(2)_R$ symmetry, since the strong interactions treat the left- and right-handed fields identically. After symmetry breaking, since only $SU(2)_L$ has associated gauge bosons, the breaking would be $SU(2)_L \times SU(2)_R \to SU(2)_V$, where this $SU(2)_V$ symmetry is precisely the custodial symmetry that relates the W-boson and Z-boson masses to the gauge charges. That chiral symmetry is broken in QCD is non-trivial but can be proved (at least in the 3-flavor case) using anomaly-matching arguments discussed in Section 30.6. There are many theories without custodial SU(2). For example, a theory with Higgs triplets instead of Higgs doublets generically does not have the symmetry.

The custodial SU(2) symmetry relates $SU(2)_{\text{weak}}$ partners, such as the up and down quarks, or top and bottom quarks. The Yukawa couplings in the Standard Model generally do not respect custodial SU(2). Mostly, the breaking is a small effect, since most of the Yukawa couplings are small. The exception is the top quark, whose Yukawa coupling is close to 1. Thus, the dominant contribution to $\Delta\rho = \rho - 1$ in the Standard Model is from the top quarks.

To see how the top quark affects the ρ parameter, we first need a better definition; Eq. (31.71) depends on which version of $\sin^2\theta_w$ we use. Traditionally, ρ is defined to measure the difference between the charged-current and neutral-current interaction strengths. The charged-current strength G_F^{charged} can be measured from muon decay. The neutral-current strength can, in principle, be measured from pure neutrino–neutrino scattering. Following Eq. (31.27), the neutral current strength is

$$\frac{G_F^{\text{neutral}}}{\sqrt{2}} = \frac{e^2}{8s^2 m_Z^2}\left(1 - \frac{\Pi_{ZZ}(0)}{m_Z^2} + \cdots\right) \tag{31.75}$$

so that using Eq. (31.27)

$$\Delta\rho \equiv \frac{G_F^{\text{neutral}}}{G_F^{\text{charged}}} - 1 = \frac{\Pi_{WW}(0)}{m_W^2} - \frac{\Pi_{ZZ}(0)}{m_Z^2}. \tag{31.76}$$

One of the most straightforward ways to measure $\Delta\rho$ is by scattering neutrinos off hadrons, in which case there is an extra term $2\frac{s}{c}\frac{\Pi_{Z\gamma}(0)}{m_Z^2}$ in $\Delta\rho$. For an SU(2) doublet with masses m_1 and m_2 (such as the top/bottom quark doublets for which Π_{WW} and Π_{ZZ} were calculated above) we find the contribution to $\Delta\rho$ is

$$\Delta\rho = N\frac{\alpha_e}{16\pi s^2 c^2 m_Z^2}\left(m_1^2 + m_2^2 - \frac{2m_1^2 m_2^2}{m_1^2 - m_2^2}\ln\frac{m_1^2}{m_2^2}\right). \tag{31.77}$$

Note that this vanishes in the limit $m_1 \to m_2$. For the top quark, where $m_t \gg m_b$, this simplifies to

$$\Delta\rho^t = \frac{3\alpha_e}{16\pi s^2 c^2}\frac{m_t^2}{m_Z^2} = 0.008. \tag{31.78}$$

It is convenient to absorb corrections like this, from Standard Model particles, into the definition of ρ. We can do this by defining ρ in terms of $\overline{\text{MS}}$ parameters as $\rho \equiv \frac{m_W^2}{m_Z^2 c^2}$. This combination is by definition 1 in the Standard Model. The current experimental value is $\rho = 1.0004$.

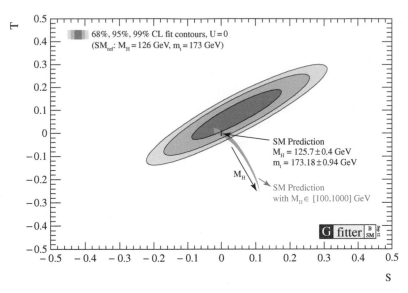

Fig. 31.1 The allowed region in the $S-T$ plane assuming $U = 0$. The small vertical line in the middle uses measured values for m_t and m_h as indicated. If experimental data on the Higgs mass are not included, the central value moves as indicated as m_h is varied between 100 GeV and 1000 GeV. From [Gfitter Group (Baak *et al.*), 2012].

In the same way that looking for deviations of ρ from 1 can tell us about custodial-SU(2)-violating interactions, it is useful to have some additional ways to constrain and characterize new physics. To this end, the **Peskin–Takeuchi parameters** S, T and U are often used [Peskin and Takeuchi, 1992]. These are defined as

$$T \equiv \frac{1}{\alpha_e}\left(\frac{\Pi_{WW}^{\text{new}}(0)}{m_W^2} - \frac{\Pi_{ZZ}^{\text{new}}(0)}{m_Z^2}\right) = \frac{\rho - 1}{\alpha_e},\tag{31.79}$$

$$S \equiv \frac{4c^2 s^2}{\alpha_e}\left[\frac{\Pi_{ZZ}^{\text{new}}(m_Z^2) - \Pi_{ZZ}^{\text{new}}(0)}{m_Z^2} - \frac{c^2 - s^2}{cs}\frac{\Pi_{Z\gamma}^{\text{new}}(m_Z^2)}{m_Z^2} - \frac{\Pi_{\gamma\gamma}^{\text{new}}(m_Z^2)}{m_Z^2}\right],\tag{31.80}$$

$$U \equiv \frac{4s^2}{\alpha_e}\left[\frac{\Pi_{WW}^{\text{new}}(m_W^2) - \Pi_{WW}^{\text{new}}(0)}{m_W^2} - \frac{c}{s}\frac{\Pi_{Z\gamma}^{\text{new}}(m_Z^2)}{m_Z^2} - \frac{\Pi_{\gamma\gamma}^{\text{new}}(m_Z^2)}{m_Z^2}\right] - S,\tag{31.81}$$

where α_e is $\alpha_e(m_Z)$. Here *new* means that S, T and U are normalized by subtracting off the Standard Model prediction. $S = T = U = 0$ is defined with $m_t = 173$ GeV and $m_h = 126$ GeV. Current experimental measurements give $S = 0.03 \pm 0.10$, $T = 0.05 \pm 0.12$ and $U = 0.03 \pm 0.10$. The actual allowed region is an ellipse, as shown in Figure 31.1. Thus, if you propose a model of physics beyond the Standard Model, you can calculate S and T as a shortcut to comparing with electroweak precision data.

In practice, S and T tend to give stronger constraints on beyond-the-Standard-Model physics than U. T measures custodial isospin violation, since it is equivalent to ρ. S would get a contribution, for example, from a new generation of fermions, even if custodial isospin were preserved. For a new doublet, we would have

$$S = \frac{N}{6\pi} \sum_i \left[1 - Y_i \ln \frac{m_1^2}{m_2^2} \right],$$ (31.82)

$$T = \frac{N}{16\pi s^2 c^2 m_Z^2} \left(m_1^2 + m_2^2 - \frac{2m_1^2 m_2^2}{m_1^2 - m_2^2} \ln \frac{m_1^2}{m_2^2} \right),$$ (31.83)

where $m_{1,2}$ are the fermion masses and Y_i are the hypercharges. The mass splitting violates isospin and is strongly constrained by T. Even for one new multiplet with degenerate masses S is in conflict with experiment.

An important application is that S strongly constrains models of new physics that replace the Higgs with QCD-like dynamics. As long as custodial isospin is preserved in these **technicolor** theories, T will be OK, but S will in general get contributions proportional to the number of techniquarks. For a single doublet, with $N_C = 4$ for technicolor, we might find $S = \frac{4}{3\pi} = 0.45$, which is severely ruled out.

It is also often useful to think about S and T as coming from higher-dimension operators. For example, suppose the Standard Model were augmented with the following operators:

$$\mathcal{O}_S = H^\dagger \sigma^i H W_{\mu\nu}^i B^{\mu\nu}, \qquad \mathcal{O}_T = \left| H^\dagger D_\mu H \right|^2.$$ (31.84)

At tree-level, we would get contributions to S and T proportional to the Wilson coefficients for these operators. In practice, one can take one's favorite model of new physics, for example supersymmetry, integrate out the new particles *before* breaking electroweak symmetry, and then look at the coefficients C_S and C_T of the operators $\mathcal{O}_S$ and $\mathcal{O}_T$ that are generated by integrating out the new particles. Then $S = \frac{4sc}{\alpha} v^2 C_S$ and $T = -\frac{1}{2\alpha} v^2 C_T$. It is often easier to use this shortcut than to compute the contributions of new physics to the vacuum polarization graphs and electroweak precision observables directly.

31.3 Large logarithms in flavor physics

So far in this chapter we have studied electroweak precision tests. These exploit the fact that the renormalizability of the Standard Model overconstrained the gauge sector. The Standard Model is also overconstrained in the flavor sector. As we saw in Chapter 29, the CKM matrix, based on three generations of quarks, must be unitary. Unitarity constrains various combinations of the CKM elements, such as

$$V_{ud} V_{ub}^\star + V_{cd} V_{cb}^\star + V_{td} V_{tb}^\star = 0.$$ (31.85)

One way to visualize this constraint is the unitarity triangle, discussed in Section 29.3.3. To test if Eq. (31.85) is satisfied, we must be able to extract the CKM elements from data. To do so at high accuracy requires precision experimental and theoretical physics.

Consider, for example, the extraction of the CKM elements V_{cb}. This element characterizes the strength of the $\bar{c}W b$ coupling in the Standard Model Lagrangian. Thus, it shows up in quark-level $b \to c$ transitions. Of course, quarks are not directly observed, so one can only measure this transition rate indirectly through the decay rates of the various hadrons. An important class of measurements from which CKM elements are extracted are the $B \to D$ decays, where B is a meson containing a bottom quark and D is a meson containing a charm quark. For example, the process $\bar{B}^0 \to D^+\pi^-$, where $\bar{B}^0 = \bar{d}b$ and $D^+ = \bar{d}c$ and $\pi^- = \bar{u}d$, is driven by the quark-level transition $b \to c\bar{u}d$ which proceeds through a highly off-shell W boson. The rate for $\bar{B}^0 \to D^+\pi^-$ is directly proportional to $|V_{cb}|^2$. Hadronic $B \to D$ decays are also important for measuring CP violation and constraining the angles in the unitarity triangle.

Unfortunately, the process $\bar{B}^0 \to D^+\pi^-$ is in fact much more complicated than the underlying quark-level process $b \to c\bar{u}d$. Due to poorly known non-perturbative hadronic matrix elements, there is a huge uncertainty in the prediction for the meson decay even with an accurate calculation of the quark decay rate (one approach to constraining the non-perturbative part using perturbative physics in the heavy-quark limit is discussed in Chapter 35). To understand the contribution from perturbative Standard Model physics, the subject of this chapter, let us for simplicity assume that the relationship between the $\bar{B}^0 \to D^+\pi^-$ decay rate and V_{cb} is known.

What we will consider here is how radiative corrections from QCD affect the relationship between V_{cb} at the scale of the mass of the B hadron (~ 5 GeV) and V_{cb} at the scale of electroweak symmetry breaking (~ 100 GeV), where unitarity of the CKM matrix should hold. As you can easily imagine, the $b \to c\bar{u}d$ decay rate when calculated to 1-loop in QCD will give a large logarithmic correction of the form $\alpha_s \ln \frac{m_W}{m_b}$. This logarithm is large and can have an important effect on the decay rate and hence on the extraction of the correct V_{cb}. The goal of this section is to calculate this large logarithm and similar logarithms to all orders in α_s.

31.3.1 Matching to the 4-Fermi theory

The $b \to c\bar{u}d$ decay is well-described by the 4-Fermi theory. We formally introduced the 4-Fermi theory in Chapter 29, where we observed that the W- and Z-boson propagators can be effectively replaced by $\frac{ig^{\mu\nu}}{m_W^2}$ and $\frac{ig^{\mu\nu}}{m_Z^2}$ when the typical energies are much lower than m_W and m_Z. Thus, the Lagrangian with the full weak interactions would be equivalent to a simpler Lagrangian with just current–current interactions among the fermions. We will now make this correspondence precise beyond leading order.

What we want to have is some non-renormalizable effective Lagrangian with no W or Z which reproduces all the physics of the full electroweak theory, up to corrections that are suppressed by powers of E/m_W. We expect our Lagrangian to be

$$\mathcal{L} = -\frac{1}{4}F_{\mu\nu}^2 - \frac{1}{4}\left(G_{\mu\nu}^a\right)^2 + \sum_q \bar{\psi}_q\left(i\not{D} - m_q\right)\psi_q - \sum_n C_n \mathcal{O}_n(x), \qquad (31.86)$$

with $F_{\mu\nu}$ the QED field strength, $G_{\mu\nu}^a$ the QCD field strength and $\mathcal{O}_n(x)$ a set of composite local operators constructed out of fermions, covariant derivatives, and QED or QCD field strengths. The Wilson coefficients C_n are numbers. They can depend on the renormalization group scale μ and on constants such as m_W, but not on momenta. Momentum factors should appear as derivatives in the operators $\mathcal{O}_n$. In the case of the electroweak theory, the only scale that can appear in C_n is $m_W \sim m_Z$. So, by dimensional analysis, the higher the dimension of the operator, the more the matrix elements of that operator will be suppressed at energies $E \ll m_W$. The great thing about an effective Lagrangian such as Eq. (31.86) is that you can compute the Wilson coefficients by matching to the full theory for one process and then use them to compute amplitudes for other processes. That is, the Wilson coefficients are independent of the external state. Although we have not yet proved it, this fact is the essential content of Wilson's operator product expansion (to be discussed in more detail in Chapter 32).

The amplitude for the transition $b \to c\bar{u}d$ in the Standard Model is given at tree-level by W^- exchange:

$$\mathcal{M} = \quad = \left(\frac{ie}{\sqrt{2}\sin\theta_w} \right)^2 \left(V_{cb} \bar{c}_L^i \gamma^\mu b_L^i \right) \frac{-ig_{\mu\nu}}{p^2 - m_W^2} \left(V_{ud}^\star \bar{d}_L^j \gamma^\nu u_L^j \right),$$

(31.87)

where i and j are color indices and $q_L \equiv \frac{1-\gamma_5}{2} q$. For $p^2 \ll m_W^2$, this same amplitude is reproduced by a Lagrangian term $-C_1 \mathcal{O}_1$ with

$$\mathcal{O}_1(x) = \left[\bar{c}_L^i(x) \gamma^\mu b_L^i(x) \right] \left[\bar{d}_L^j(x) \gamma_\mu u_L^j(x) \right],$$

(31.88)

and

$$C_1 = G \equiv \frac{4G_F}{\sqrt{2}} V_{cb} V_{ud}^\star,$$

(31.89)

where $\frac{G_F}{\sqrt{2}} = \frac{e^2}{8m_W^2 \sin^2\theta_w}$. This is the tree-level matching result. Note that we are employing an efficient abbreviation: the same notation is used for the quark fields in Eq. (31.88) and for the external spinors in Eq. (31.87).

31.3.2 One-loop matching

Since $\alpha_s \sim 0.1 \gg \alpha_e \sim 0.01$, electroweak corrections at 1-loop are typically comparable in strength to 2-loop QCD corrections (for processes involving quarks). Thus, we will work to 1-loop in α_s and ignore electroweak effects. To perform the matching, we need to demand that matrix elements of the quarks be the same in both theories to order α_s. If the theories are matched properly, this equivalence should hold for any final state. An obvious choice is to pick on-shell initial and final states, relevant for the $b \to c\bar{u}d$ decay. An alternative approach is to match by considering $\bar{c}b \to \bar{u}d$, with the external momenta all set to zero. This will give an off-shell matrix element involving fewer scales at the cost of possibly introducing IR divergences. Since we will be working in dimensional regularization, having fewer scales makes the calculation much easier than it would be with on-shell external states.

The tree-level diagrams for $\bar{c}b \to \bar{u}d$ in the full weak theory and in the effective theory are

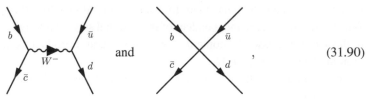

and   ,   (31.90)

which gives $C_1 = G$ as we have just shown. At order α_s there are six 1-loop diagrams and two counterterm diagrams in the full theory:

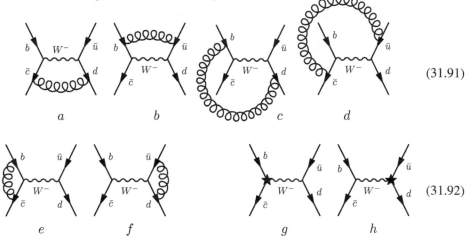

$$a \qquad\qquad b \qquad\qquad c \qquad\qquad d \qquad (31.91)$$

$$e \qquad\qquad f \qquad\qquad g \qquad\qquad h \qquad (31.92)$$

In the effective theory there are six 1-loop diagrams and one counterterm diagram

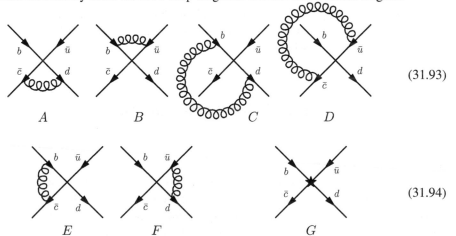

$$A \qquad\qquad B \qquad\qquad C \qquad\qquad D \qquad (31.93)$$

$$E \qquad\qquad F \qquad\qquad\qquad G \qquad\qquad (31.94)$$

Diagram a gives (with all external momenta set to zero)

$$i\mathcal{M}_a = (-m_W^2 G)(ig_s)^2 \mu^{4-d} \int \frac{d^d k}{(2\pi)^d} \frac{-i}{k^2} \frac{-i}{k^2 - m_W^2} \left(\bar{c}_L \gamma^\nu T^a \frac{i\slashed{k}}{k^2} \gamma^\mu b_L \right) \left(\bar{d}_L \gamma_\nu T^a \frac{-i\slashed{k}}{k^2} \gamma_\mu u_L \right).$$

$$(31.95)$$

Here we suppress the color indices and group colored objects together to indicate the color contractions implicitly. For example, $(\bar{c}b)(\bar{d}u) = (\bar{c}^i b^i)(\bar{d}^j u^j)$. Since the integrand must be Lorentz invariant, we can replace two $\not{k}$ terms with a Lorentz contraction $\not{k}\Gamma\not{k} \to \frac{1}{d}k^2\gamma^\alpha\Gamma\gamma_\alpha$. This leaves

$$\mathcal{M}_a = i(m_W^2 G)\frac{g_s^2}{d}\mu^{4-d}\int\frac{d^d k}{(2\pi)^d}\frac{1}{k^4(k^2 - m_W^2)}\left(\bar{c}_L\gamma^\nu T^a\gamma^\alpha\gamma^\mu b_L\right)\left(\bar{d}_L\gamma_\nu T^a\gamma_\alpha\gamma_\mu u_L\right).$$
(31.96)

This integral is IR divergent but UV finite. Thus, the answer is unambiguous in dimensional regularization:

$$\mathcal{M}_a = -\frac{Gg_s^2}{32\pi^2}\left(\frac{1}{\varepsilon_{\text{IR}}} + \frac{3}{4} + \frac{1}{2}\ln\frac{\tilde{\mu}^2}{m_W^2}\right)\left(\bar{c}_L\gamma^\nu T^a\gamma^\alpha\gamma^\mu b_L\right)\left(\bar{d}_L\gamma_\nu T^a\gamma_\alpha\gamma_\mu u_L\right). \quad (31.97)$$

To simplify the spinor part we use the color identity from Eq. (25.34),

$$\sum_a T_{ij}^a T_{kl}^a = \frac{1}{2}\left(\delta_{il}\delta_{jk} - \frac{1}{N}\delta_{ij}\delta_{kl}\right), \quad (31.98)$$

which leads to

$$\mathcal{M}_a = -G\frac{\alpha_s}{16\pi}\left(\frac{1}{\varepsilon_{\text{IR}}} + \frac{1}{2}\ln\frac{\tilde{\mu}^2}{m_W^2} + \frac{3}{4}\right)$$
$$\times\left[\left(\bar{c}_L^i\gamma^\nu\gamma^\alpha\gamma^\mu b_L^j\right)\left(\bar{d}_L^j\gamma^\nu\gamma^\alpha\gamma^\mu u_L^i\right) - \frac{1}{N}\left(\bar{c}_L^i\gamma^\nu\gamma^\alpha\gamma^\mu d_L^i\right)\left(\bar{d}_L^j\gamma^\nu\gamma^\alpha\gamma^\mu u_L^j\right)\right],$$
(31.99)

where i and j are color indices. To simplify these spinor products and match the spinor contractions up with the color contractions, we use Fierz identities such as (see Problem 11.8)

$$\left(\bar{\psi}_{1L}\gamma^\nu\gamma^\alpha\gamma^\mu\psi_{2L}\right)\left(\bar{\psi}_{3L}\gamma^\nu\gamma^\alpha\gamma^\mu\psi_{4L}\right) = 16\left(\bar{\psi}_{1L}\gamma^\mu\psi_{2L}\right)\left(\bar{\psi}_{3L}\gamma^\mu\psi_{4L}\right), \quad (31.100)$$
$$\left(\bar{\psi}_{1L}\gamma^\mu\psi_{2L}\right)\left(\bar{\psi}_{3L}\gamma^\mu\psi_{4L}\right) = \left(\bar{\psi}_{1L}\gamma^\mu\psi_{4L}\right)\left(\bar{\psi}_{3L}\gamma^\mu\psi_{2L}\right), \quad (31.101)$$
$$\left(\bar{\psi}_{1L}\gamma^\nu\gamma^\alpha\gamma^\mu\psi_{2L}\right)\left(\bar{\psi}_{3L}\gamma^\mu\gamma^\alpha\gamma^\nu\psi_{4L}\right) = 4\left(\bar{\psi}_{1L}\gamma^\mu\psi_{2L}\right)\left(\bar{\psi}_{3L}\gamma^\mu\psi_{4L}\right). \quad (31.102)$$

After rearrangement, we find

$$\mathcal{M}_a = -G\frac{\alpha_s}{\pi}\left(\frac{1}{\varepsilon_{\text{IR}}} + \frac{1}{2}\ln\frac{\tilde{\mu}^2}{m_W^2} + \frac{3}{4}\right)\left[(\bar{c}_L\gamma^\mu u_L)(\bar{d}_L\gamma^\mu b_L) - \frac{1}{N}(\bar{c}_L\gamma^\mu b_L)(\bar{d}_L\gamma^\mu u_L)\right],$$
(31.103)

where now our notation indicates that the color contractions are within parentheses (the same as the spinor contractions).

Diagrams b, c and d can be computed similarly (the Fierz identity in Eq. (31.102) is required for diagrams c and d), giving a total result

$$\mathcal{M}_a + \mathcal{M}_b + \mathcal{M}_c + \mathcal{M}_d = -G\frac{3\alpha_s}{2\pi}\left(\frac{1}{\varepsilon_{\text{IR}}} + \frac{1}{2}\ln\frac{\tilde{\mu}^2}{m_W^2} + \frac{3}{4}\right)$$
$$\times\left[(\bar{c}_L\gamma^\mu u_L)(\bar{d}_L\gamma^\mu b_L) - \frac{1}{N}(\bar{c}_L\gamma^\mu b_L)(\bar{d}_L\gamma^\mu u_L)\right]. \quad (31.104)$$

Diagrams e and f just give scaleless integrals which identically vanish in dimensional regularization. Nevertheless, it is helpful to separate out the UV and IR divergences, which gives (see Appendix B)

$$\mathcal{M}_e = \mathcal{M}_f = G C_F \frac{\alpha_s}{2\pi} \left(\frac{1}{\varepsilon_{\mathrm{IR}}} - \frac{1}{\varepsilon_{\mathrm{UV}}} \right) (\bar{c}_L \gamma^\mu b_L)(\bar{d}_L \gamma^\mu u_L). \tag{31.105}$$

The UV divergences must be exactly canceled by the $\overline{\mathrm{MS}}$ counterterms in graphs g and h. Thus we must have

$$\mathcal{M}_g = \mathcal{M}_h = G C_F \frac{\alpha_s}{2\pi} \left(\frac{1}{\varepsilon_{\mathrm{UV}}} \right) (\bar{c}_L \gamma^\mu b_L)(\bar{d}_L \gamma^\mu u_L), \tag{31.106}$$

which can easily be confirmed by direct calculation.

The total result in the full theory, up to one loop with $N = 3$, is then

$$\begin{aligned}
\mathcal{M}_{\text{full}} &= \mathcal{M}_{\text{tree}} + \mathcal{M}_a + \cdots + \mathcal{M}_h \\
&= -G \left[1 - \frac{\alpha_s}{2\pi} \left(\frac{11}{3} \frac{1}{\varepsilon_{\mathrm{IR}}} + \frac{1}{2} \ln \frac{\tilde{\mu}^2}{m_W^2} + \frac{3}{4} \right) \right] (\bar{c}_L \gamma^\mu b_L)(\bar{d}_L \gamma^\mu u_L) \\
&\quad - G \frac{3\alpha_s}{2\pi} \left(\frac{1}{\varepsilon_{\mathrm{IR}}} + \frac{1}{2} \ln \frac{\tilde{\mu}^2}{m_W^2} + \frac{3}{4} \right) (\bar{c}_L \gamma^\mu u_L)(\bar{d}_L \gamma^\mu b_L).
\end{aligned} \tag{31.107}$$

Before going on, we note that the IR divergences are an artifact of setting all external momenta to zero. If we wanted to just calculate the $b \to c\bar{u}d$ rate at 1-loop, we would put all the momenta on-shell, which would make the integrals IR finite. Then there would be no $\frac{1}{\varepsilon_{\mathrm{IR}}}$ term and $\ln \frac{\tilde{\mu}^2}{m_W^2}$ would be replaced by $\ln \frac{p^2}{m_W^2}$ or some other relevant scale. These are the physical large logarithms we are trying to understand through effective field theory.

To match onto the full theory with the effective theory, it is clear from Eq. (31.107) that we are going to need two operators

$$\mathcal{O}_1 = (\bar{c}_L \gamma^\mu b_L)(\bar{d}_L \gamma^\mu u_L), \qquad \mathcal{O}_2 = (\bar{c}_L \gamma^\mu u_L)(\bar{d}_L \gamma^\mu b_L). \tag{31.108}$$

Even with these operators, we cannot just set $C_1 = G \left[1 - \frac{\alpha_s}{2\pi} \left(\frac{11}{3} \frac{1}{\varepsilon_{\mathrm{IR}}} + \frac{1}{2} \ln \frac{\tilde{\mu}^2}{m_W^2} + \frac{3}{4} \right) \right]$, since this is a divergent quantity. Even if we replaced the divergence with a physical scale, we could not set $C_1 = G - G \frac{\alpha_s}{4\pi} \frac{11}{6} \ln \frac{p^2}{m_W^2}$ since then C_1 would be momentum-dependent and our effective Lagrangian would be non-local. To properly do the matching, we have to now compute the 1-loop corrections in the effective theory.

In the 4-Fermi theory, the 1-loop graphs have no W propagator at all, and we get the leading coefficient $C_1 = G$ in front of the whole diagram. Without the W propagator, the diagrams are now UV divergent. For each diagram $\mathcal{M}_A$ through $\mathcal{M}_G$, there are contributions from both $\mathcal{O}_1$ and $\mathcal{O}_2$. For example,

$$\begin{aligned}
i\mathcal{M}_A &= C_1 \frac{g_s^2}{d} \mu^{4-d} \int \frac{d^d k}{(2\pi)^d} \frac{1}{k^4} (\bar{c}_L \gamma^\nu T^a \gamma^\alpha \gamma^\mu b_L)(\bar{d}_L \gamma^\nu T^a \gamma^\alpha \gamma^\mu u_L) \\
&\quad + C_2 \frac{g_s^2}{d} \mu^{4-d} \int \frac{d^d k}{(2\pi)^d} \frac{1}{k^4} (\bar{c}_L \gamma^\nu T^a \gamma^\alpha \gamma^\mu u_L)(\bar{d}_L \gamma^\nu T^a \gamma^\alpha \gamma^\mu b_L).
\end{aligned} \tag{31.109}$$

All the diagrams are scaleless integrals which vanish in dimensional regularization. It is helpful nevertheless to separate out the UV and IR divergences. This leads to

$$
\mathcal{M}_A = \frac{\alpha_s}{\pi}\left(\frac{1}{\varepsilon_{\text{UV}}} - \frac{1}{\varepsilon_{\text{IR}}}\right)C_1\left[(\bar{c}_L\gamma^\mu u_L)(\bar{d}_L\gamma^\mu b_L) - \frac{1}{3}(\bar{c}_L\gamma^\mu b_L)(\bar{d}_L\gamma^\mu u_L)\right]
$$
$$
+ \frac{\alpha_s}{2\pi}\left(\frac{1}{\varepsilon_{\text{UV}}} - \frac{1}{\varepsilon_{\text{IR}}}\right)C_2\left[(\bar{c}_L\gamma^\mu b_L)(\bar{d}_L\gamma^\mu u_L) - \frac{1}{3}(\bar{c}_L\gamma^\mu u_L)(\bar{d}_L\gamma^\mu b_L)\right].
$$
(31.110)

Summing all the diagrams gives

$$
\mathcal{M}_A + \cdots + \mathcal{M}_F = \frac{\alpha_s}{2\pi}\left(\frac{1}{\varepsilon_{\text{UV}}} - \frac{1}{\varepsilon_{\text{IR}}}\right)\left(3C_2 - \frac{11}{3}C_1\right)(\bar{c}_L\gamma^\mu b_L)(\bar{d}_L\gamma^\mu u_L)
$$
$$
+ \frac{\alpha_s}{2\pi}\left(\frac{1}{\varepsilon_{\text{UV}}} - \frac{1}{\varepsilon_{\text{IR}}}\right)\left(3C_1 - \frac{11}{3}C_2\right)(\bar{c}_L\gamma^\mu u_L)(\bar{d}_L\gamma^\mu b_L).
$$
(31.111)

The UV divergences in the effective theory will be removed by counterterms, leaving only the $\frac{1}{\varepsilon_{\text{IR}}}$ and finite terms. Explicitly, the counterterms must give

$$
\mathcal{M}_G = -\frac{\alpha_s}{2\pi}\frac{1}{\varepsilon_{\text{UV}}}\left[\left(3C_2 - \frac{11}{3}C_1\right)(\bar{c}_L\gamma^\mu b_L)(\bar{d}_L\gamma^\mu u_L)\right.
$$
$$
\left.+ \left(3C_1 - \frac{11}{3}C_2\right)(\bar{c}_L\gamma^\mu u_L)(\bar{d}_L\gamma^\mu b_L)\right].
$$
(31.112)

The first spinor product can come from renormalization of $\mathcal{O}_1$, and the second from renormalization of $\mathcal{O}_2$. We will discuss renormalization more after we finish with matching.

Adding all the contributions in the effective theory up to order α_s then gives

$$
\mathcal{M}_{\text{EFT}} = -\left[C_1 + \frac{\alpha_s}{2\pi}\frac{1}{\varepsilon_{\text{IR}}}\left(3C_2 - \frac{11}{3}C_1\right)\right](\bar{c}_L\gamma^\mu b_L)(\bar{d}_L\gamma^\mu u_L)
$$
$$
- \left[C_2 + \frac{\alpha_s}{2\pi}\frac{1}{\varepsilon_{\text{IR}}}\left(3C_1 - \frac{11}{3}C_2\right)\right](\bar{c}_L\gamma^\mu u_L)(\bar{d}_L\gamma^\mu b_L).
$$
(31.113)

Comparing with $\mathcal{M}_{\text{full}}$ in Eq. (31.107) we see that the full theory amplitude can be reproduced up to order α_s if we choose

$$
C_1 = G\left[1 - \frac{\alpha_s}{2\pi}\left(\frac{1}{2}\ln\frac{\tilde{\mu}^2}{m_W^2} + \frac{3}{4}\right)\right], \qquad C_2 = G\left[\frac{3\alpha_s}{2\pi}\left(\frac{1}{2}\ln\frac{\tilde{\mu}^2}{m_W^2} + \frac{3}{4}\right)\right]. \quad (31.114)
$$

Here $\tilde{\mu}$ is the matching scale, which we clearly want to choose to be near m_W to not have large logarithms in the matching coefficients.

Note that these Wilson coefficients are IR finite (they do not depend on ε_{IR}). In other words, the IR-divergent terms from loops in the full theory have been reproduced by loops in the effective theory. The cancellation of IR divergences is a self-consistency check. As long as we are only integrating out heavy particles, such as the W, IR divergences should cancel in the matching. If they did not, it would mean the infrared degrees of freedom are different in the two theories and we have not just integrated out short-distance physics.

31.3.3 Running

At this point, all we have done is construct a theory that agrees with the full weak theory at 1-loop up to corrections of order $\frac{E}{m_W}$. We found that a large logarithm of the form $\frac{\alpha_s}{4\pi}\ln\frac{\tilde{\mu}^2}{m_W^2}$ appears in both theories. In a physical process, the scale $\tilde{\mu}$ should be replaced by a physical scale, such as the B mass. If this were all we could do with the effective theory, it would not be very useful – we might as well use the full theory which gets the E/m_W behavior right too. The real power of the effective theory is that we can now solve the RGEs to resum these logarithms. We saw how this works in Chapter 23. This is a practical application of those methods.

To calculate the RGE, we need the anomalous dimensions of $\mathcal{O}_1$ and $\mathcal{O}_2$. These are determined by the operator renormalizations. We can write our general Lagrangian with these operators as

$$\mathcal{L} = \mathcal{L}_{\text{kin}} - C_1 Z_1 \mathcal{O}_1 - C_2 Z_2 \mathcal{O}_2, \tag{31.115}$$

where $\mathcal{O}_i$ are renormalized operators depending on renormalized fields. From the counterterms above, Eq. (31.112), we see that

$$Z_1 = 1 + \frac{\alpha_s}{2\pi}\frac{1}{\varepsilon}\left(3\frac{C_2}{C_1} - \frac{11}{3}\right), \qquad Z_2 = 1 + \frac{\alpha_s}{2\pi}\frac{1}{\varepsilon}\left(3\frac{C_1}{C_2} - \frac{11}{3}\right). \tag{31.116}$$

These two counterterms cancel all of the 1-loop UV divergences in the effective theory.

The RG evolution is obtained by demanding that the bare Lagrangian be independent of the arbitrary scale μ. First, we write out the operators in terms of bare fields:

$$\mathcal{O}_1 = \left(\bar{c}_L \gamma^\mu b_L\right)\left(\bar{d}_L \gamma^\mu u_L\right) = \frac{1}{Z_{2\psi}^2}\left(\bar{c}_L^{(0)} \gamma^\mu b_L^{(0)}\right)\left(\bar{d}_L^{(0)} \gamma^\mu u_L^{(0)}\right), \tag{31.117}$$

where $Z_{2\psi}$ (normally called Z_2) is the quark field strength renormalization we computed in Chapter 26. In Feynman gauge, from Eq. (26.62),

$$Z_{2\psi} = 1 - \frac{1}{\varepsilon}\frac{2\alpha_s}{3\pi}. \tag{31.118}$$

Using the $\overline{\text{MS}}$ conventions, where all the μ dependence stems from the β-function, with

$$\mu\frac{d}{d\mu}\alpha_s = \beta(\alpha_s) = -\varepsilon\alpha_s - 2\alpha_s\left(\frac{\alpha_s}{4\pi}\right)\beta_0, \qquad \beta_0 = \frac{11}{3}C_A - \frac{4}{3}T_F n_f, \tag{31.119}$$

the RGE $\mu\frac{d}{d\mu}\left(C_1\frac{Z_1}{Z_{2\psi}^2}\right) = 0$ implies, to order α_s,

$$\mu\frac{d}{d\mu}C_1 = \frac{\alpha_s}{2\pi}(-C_1 + 3C_2), \qquad \mu\frac{d}{d\mu}C_2 = \frac{\alpha_s}{2\pi}(3C_1 - C_2). \tag{31.120}$$

That is, the anomalous dimension is a matrix:

$$\mu\frac{d}{d\mu}C_i = \gamma_{ij}C_j = \frac{\alpha_s}{2\pi}\begin{pmatrix} -1 & 3 \\ 3 & -1 \end{pmatrix}_{ij} C_j. \tag{31.121}$$

To solve the RGE, we simply diagonalize the matrix. The eigenoperators are

$$\mathcal{O}_0 = \frac{1}{2}(\mathcal{O}_1 + \mathcal{O}_2), \quad \mathcal{O}_3 = \frac{1}{2}(\mathcal{O}_1 - \mathcal{O}_2). \tag{31.122}$$

The new subscripts reflect a type of isospin quantum number. Since $\mathcal{O}_0$ is symmetric under $d \leftrightarrow c$ it is a singlet while $\mathcal{O}_3$, which is antisymmetric, is a triplet. The symmetry is of course broken by quark masses, but the UV divergences of the theory are independent of these masses. So the matching at $\mu = m_W$ gives

$$C_0(m_W) = G\left[1 + \frac{3}{4\pi}\alpha_s(m_W)\right], \quad C_3(m_W) = G\left[1 - \frac{3}{2\pi}\alpha_s(m_W)\right], \tag{31.123}$$

and these run with

$$\gamma_0 = \frac{\alpha_s}{\pi}, \qquad \gamma_3 = -2\frac{\alpha_s}{\pi}. \tag{31.124}$$

The RGEs can be easily integrated in the diagonal basis using

$$C_i(\mu) = C_i(m_W) \exp\left(\int_{\alpha_s(m_W)}^{\alpha_s(\mu)} \frac{\gamma_i(\alpha')}{\beta(\alpha')} d\alpha'\right). \tag{31.125}$$

Using the 1-loop anomalous dimension and β_0 with $n_f = 5$ (there are five active flavors between m_W and m_b) this gives

$$C_0(m_b) = C_0(m_W)\left(\frac{\alpha_s(m_W)}{\alpha_s(m_b)}\right)^{6/23}, \quad C_3(m_b) = C_3(m_W)\left(\frac{\alpha_s(m_W)}{\alpha_s(m_b)}\right)^{-12/23}. \tag{31.126}$$

Starting with $\alpha_s(m_Z) = 0.1184$, we run α_s at 1-loop to find $\alpha_s(m_W) = 0.121$ and $\alpha_s(m_b) = 0.213$. Plugging in these numbers leads to

$$C_0(m_b) = 0.888G, \qquad C_3(m_b) = 1.27G, \tag{31.127}$$

which implies

$$C_1(m_b) = 1.08G, \qquad C_2(m_b) = -0.189G. \tag{31.128}$$

The root-mean-square value of these, which is relevant for the $b \to cd\bar{u}$ decay rate, is $\sqrt{C_1^2(m_b) + C_2^2(m_b)} = 1.09G$, which is 9% higher than the tree-level value, and 11% higher than the 1-loop value $\sqrt{C_1^2(m_W) + C_2^2(m_W)} = 0.98G$.

Now recall that $G \equiv \frac{4G_F}{\sqrt{2}}V_{cb}V_{ud}^\star$. Suppose we had an accurate way to relate the 4-Fermi theory at the scale m_b to the hadronic $B \to D\pi$ decay rate (for example if hadronic matrix elements were known from the lattice). We could then use the measured rate $\Gamma(B \to D\pi) \propto |V_{cb}|^2$ to extract V_{cb}. If one did not include the loop corrections, since the rate is quadratically sensitive to V_{cb}, the extracted value would come out 18% too low. This could falsely indicate that the CKM matrix is not unitary and give incorrect indications of beyond-the-Standard-Model physics.

Conveniently, we do not have to calculate the running from m_W down to the GeV scale on our own for every possible calculation. Instead, we can just integrate out new physics at m_W, match onto a standard set of operators, and use precomputed results. There is a standard basis, including $\mathcal{O}_1$ and $\mathcal{O}_2$ up to $\mathcal{O}_9$, and additional operators such as $\mathcal{O}_{7\gamma} = \frac{e}{8\pi^2} m_b \bar{s}_i \sigma^{\mu\nu}(1+\gamma_5) b_i F_{\mu\nu}$ that mediate FCNC processes such as $b \to s\gamma$. More information can be found in [Buchalla *et al.*, 1996].

Problems

31.1 Calculate the rate for $\mu^- \to e^- \bar{\nu}_e \nu_\mu$ at tree-level in the 4-Fermi theory and verify Eq. (31.3).

31.2 Another well-measured quantity is the decay rate of the Z boson into leptons, $\Gamma_{e^+e^-} \equiv \Gamma(Z \to e^+e^-)$. At tree-level,

$$\Gamma(Z \to e^+e^-) = \frac{v}{96\pi} \frac{e^3}{s^3 c^3} \left[\frac{1}{4} + \left(2s^2 - \frac{1}{2} \right)^2 \right]. \qquad (31.129)$$

The current experimental value is $\Gamma_{e^+e^-} \equiv \Gamma(Z \to e^+e^-) = 83.99 \pm 0.18\,\text{MeV}$.

(a) Evaluate the tree-level prediction for $\Gamma_{e^+e^-}$. How many standard deviations is the result off from the experimental value?

(b) Derive an expression for $\Gamma_{e^+e^-}$ at 1-loop in terms of $\overline{\text{MS}}$ Lagrangian parameters.

(c) Derive an expression for $\Gamma_{e^+e^-}$ in terms of vacuum polarization graphs.

(d) Evaluate $\Gamma_{e^+e^-}$ numerically at 1-loop. How does your answer compare to the experimental value?

31.3 Calculate the Higgs boson contributions to the various vacuum polarization graphs exactly. Verify the leading behavior in Eqs. (31.68) and (31.69).

31.4 Flavor-changing b decays:

(a) Calculate the rate for $b \to s\gamma$ in the Standard Model. The relevant graphs have the photon coming off a W-boson loop.

(b) Match to an effective theory at tree-level so that the $b \to s\gamma$ rate is reproduced.

(c) Evaluate the order α_s corrections to the effective theory.

(d) Evolve the operator from m_W to m_b. How big are the radiative corrections to this decay rate from QCD?

Quantum chromodynamics and the parton model 32

One of the most remarkable results in all of physics is that the existence and properties of the proton can be explained by a local quantum field theory based on the gauge group SU(3). This result is additionally remarkable because, although we know QCD predicts the proton, we cannot prove it. Despite the powerful tools we have developed for doing perturbative calculations, we only know how to apply QCD to particles that are colored, not color-neutral particles such as hadrons. In this chapter, we will explore the connection between perturbative QCD and hadron physics.

We have discussed two methods for studying hadrons so far. The first, chiral perturbation theory (Section 28.2), takes from QCD only its symmetries. These symmetries are very powerful, and constrain the possible interactions that hadrons (especially the light mesons) can have, allowing for quantitative quantum predictions. Unfortunately, the Chiral Lagrangian is non-renormalizable, so one would need an infinite number of measurements to make an infinite number of predictions. Since the Chiral Lagrangian cannot be matched systematically to QCD within a perturbative framework (in contrast to, say, how the 4-Fermi theory is matched to the electroweak theory), there are some questions it simply cannot answer. The other method is lattice QCD (Section 25.5). Lattice QCD lets us calculate any desired hadronic property, at least in principle. From a practical perspective, lattice calculations are still extremely computationally expensive. Moreover, there are some quantities, in particular scattering amplitudes, that are not well suited to lattice calculations at all. To calculate what happens when we collide two protons, neither of these methods are adequate.

Intuitively, it seems reasonable that perturbative QCD should have some predictive power for high-energy proton scattering. Although hadrons are strongly interacting, the strong force is scale dependent and becomes weak at very short distances (Section 26.6). Thus, one expects a collision between hadrons at very high energy to be dominated by interactions among essentially free quarks or gluons, and perturbative QCD to be applicable. What is not obvious is whether perturbative calculations can be connected to experimental observations. In fact they can, due to the power of tools such as the operator product expansion and effective field theory. These tools allow us to make precise the factorization of short-distance from long-distance physics.

QCD is an extremely rich subject. It is obviously impossible to cover all the important topics in one chapter. Instead, we will focus here on some aspects of perhaps the most important process in QCD: $e^- p^+$ scattering. We will begin with a historically oriented discussion of how the proton was understood by experiments that bombarded electrons at protons at very high energies. This will lead to the parton model, which was a precursor to QCD. We will then discuss the field theory version of the parton model and the DGLAP evolution equations. This will lead into a discussion of factorization. Another approach to factorization is discussed in Chapter 36 using Soft-Collinear Effective Theory.

As you will see, there are a lot of variables floating around in this chapter. Most of our definitions are standard. Unfortunately, there are different conventions used in the literature for the form factors W_1 and W_2. Our convention is convenient for the QCD analysis. For future reference, the letter q will confusingly refer to both quarks and to a momentum transfer $q^\mu = k^\mu - k'^\mu$, with k^μ and k'^μ the incoming and outgoing electron momenta. P^μ will be the proton momentum and p_i^μ a parton momentum. We define $x \equiv \frac{Q^2}{2P \cdot q}$ and $z \equiv \frac{Q^2}{2p_i \cdot q}$, which are kinematic variables, and use ξ, defined by $p_i^\mu \equiv \xi P^\mu$, as a momentum fraction. Other kinematic variables are $\omega \equiv \frac{1}{x}$, $y \equiv \frac{P \cdot q}{P \cdot k}$ and $\nu \equiv \frac{P \cdot q}{m_p}$. In the context of final-state radiation (Section 32.3), z will refer not to $\frac{Q^2}{2p_i \cdot q}$ but to the ratio of daughter-to-mother energies in a collinear emission, $z = \frac{E_{\text{daughter}}}{E_{\text{mother}}}$.

32.1 Electron–proton scattering

Electron–proton scattering is one of the best ways to study hadrons: it uses an essentially pointlike structureless probe (the electron) to make precision measurements of the proton. This is not dissimilar to the way Rutherford and collaborators discovered the atomic nucleus by slamming α-particles into thin metal sheets. From the resulting distributions, not only were they able to conclude that atoms had a hard center, but they also got a rough estimate of the size of the nucleus.

32.1.1 Rutherford's experiment

Rutherford and his team (Geiger and Marsden) produced α-particles (helium nuclei) from the decay of radon atoms. These α-particles have velocities around 2×10^7 m/s, giving them a kinetic energy of around 8 MeV. When shooting these "bullets" at a very thin sheet (a few atoms thick) of foil they sometimes found scattering angles greater than $90°$. This was totally unexpected, considering the then-popular Thomson model (where the negatively charged electrons are embedded in a positively charged medium, like raisins in a plum pudding). Rutherford famously said, "It was quite the most incredible event that ever happened to me in my life. It was almost as incredible as if you had fired a 15-inch shell at a piece of tissue paper and it came back and hit you." [Andrade, 1964, p.111].

To calculate the expected distribution, Rutherford used a classical model (of course he did, this was 1911!). Assuming a central Coulomb potential, the scattering angle θ is fixed by the energy E and impact parameter b of the collision to be

$$b = \frac{1}{2\pi} \frac{Ze^2}{mv^2} \cot \frac{\theta}{2}, \tag{32.1}$$

where Z is the charge of the target nucleus. Averaging over impact parameters, this leads to a cross section

$$\frac{d\sigma}{d\Omega} = \left(\frac{Ze^2}{4\pi mv^2}\right)^2 \frac{1}{\sin^4\frac{\theta}{2}}. \tag{32.2}$$

Rutherford's group found a distribution consistent with this formula.

Actually, Rutherford was hoping to find deviations from his scattering formula, which would have indicated new interactions of the electron with the nucleus. That he did not find any indicated to him that the α-particles must stop before they hit the nucleus. Using conservation of energy at zero impact parameter, an upper bound $r_{\max}$ on the size of the nucleus is then given by $\frac{1}{2}mv^2 = \frac{2Ze^2}{4\pi r_{\max}}$. Using this formula, Rutherford found $r_{\max} = 4.8 \times 10^{-15}$ m [Rutherford, 1911], which was his estimate for the maximal size of the nucleus. Incidentally, his best estimate came not from his famous gold foil but from lighter aluminum foil, a much less exotic material. To improve on this, one would like to take the smallest nucleus possible (a proton), the smallest probe possible (an electron), and the highest energy possible. This leads to high-energy e^-p^+ collisions. But it was not until 50 years after Rutherford that the nucleus could be unraveled this way.

32.1.2 Elastic e^- p^+ scattering

Suppose the proton were structureless too, like the muon. Then we would expect e^-p^+ scattering to look like $e^-\mu^+$ scattering. In fact, it does, at least at low energy. The leading Feynman diagram is just the t-channel photon exchange diagram, which we have studied many times:

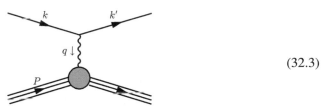

$$\tag{32.3}$$

We call this Coulomb scattering. That the proton is composite is only relevant for photons that have enough energy to see its compositeness – at low energy, the proton is indistinguishable from an elementary fermion such as the muon.

The relativistic cross section for Coulomb scattering of two spin-$\frac{1}{2}$ particles was calculated in Eq. (13.103) of Chapter 13:

$$\left(\frac{d\sigma}{d\Omega}\right)_{\text{lab}} = \frac{\alpha_e^2}{4E^2\sin^4\frac{\theta}{2}}\frac{E'}{E}\left(\cos^2\frac{\theta}{2} - \frac{q^2}{2m_p^2}\sin^2\frac{\theta}{2}\right), \tag{32.4}$$

where E and E' are the electron's initial and final energies and $q^\mu = k^\mu - k'^\mu$ is the **momentum transfer**; θ is the angle between the outgoing and incoming electrons, so $\theta = 0$ is forward scattering. These quantities are related by

$$q^2 = -2k \cdot k' = -\left(4E'E\sin^2\frac{\theta}{2}\right)_{\text{lab}}. \tag{32.5}$$

This formula applies in the lab frame, where the proton is initially at rest. Another useful relation is that in the lab frame $q^2 = 2m_p(E' - E)$. The derivation of Eq. (32.4) used $m_e = 0$, so one cannot take the non-relativistic limit directly. Nevertheless, the non-relativistic limit of $e^- p^+$ scattering in QED does reduce to the Rutherford formula, as explained in Section 13.4.

Equation (32.4) is carefully written in terms of only quantities that can be measured from the initial and final state electron. This is very important, since the early $e^- p^+$ scattering experiments, such as Hofstadter's famous experiments at Stanford in the mid 1950s, collided electron beams with hydrogen gas, and only the outgoing electrons could be measured. The first 4π detector, that is, one that measures all of the final-state particles, including the proton remnants, was not built until 1973 (the MARK I detector at SLAC). As we will see, a tremendous amount can be and was learned about protons by just studying the outgoing electrons.

If we did not know about QCD (as in the 1950s) we might have expected Eq. (32.4) to hold up to arbitrarily small distances. For example, a similar formula does appear to hold up to arbitrarily small distances for electron–muon scattering, which proceeds primarily through QED. Even in QED, Eq. (32.4) gets quantum corrections, as we saw for $e^+ \mu^-$ scattering in Chapter 20. To study these corrections, it is helpful to remove the electron from the problem and think of an off-shell photon with spacelike momentum q^μ as scattering off the proton, as in Chapter 20. In Chapter 17 and Section 19.3, we parametrized the most general type of interaction between an off-shell photon and a spin-$\frac{1}{2}$ particle in terms of two form factors F_1 and F_2. This parametrization did not assume anything about the interactions, and must hold in QCD (or any theory). In this case, as in the QED case, the general vertex can be written as $\bar{u}(p')\,(ie\Gamma^\mu)\,u(p)$, with $\bar{u}(p')$ the outgoing proton spinor and $u(p)$ the incoming proton spinor, both of which we assume to be on-shell. Then the decomposition of Γ^μ into form factors is

$$\Gamma^\mu(q) = F_1(q^2)\gamma^\mu + \frac{i\sigma^{\mu\nu}}{2m_p} q_\nu F_2(q^2). \tag{32.6}$$

Recall that in QED $F_1(q^2)$ gets divergent contributions, and must be renormalized, while $F_2(q^2)$ is finite. The on-shell renormalization condition $F_1(0) = 1$ in this case normalizes the proton charge to $Q = +1$ at large distances. At 1-loop in QED $F_2(0) = \frac{\alpha_e}{2\pi}$, which gives the correction to the electron magnetic moment, usually expressed as its g-factor $g_e = 2 + \frac{\alpha_e}{\pi} + \cdots$.

For the proton, we know its magnetic moment corresponds to a g-factor of $g_p = 5.58$, which is not close to 2. This suggests that the proton is not just a point particle like the electron. (The neutron's g-factor is $g_n = -3.82$, which also seems very strange in perturbation theory, considering that the neutron is neutral.) Repeating the tree-level Coulomb scattering calculation using the $ie\bar{u}\Gamma^\mu u$ vertex (which is not hard, since q^2 is fixed), we get

$$\left(\frac{d\sigma}{d\Omega}\right)_{\text{lab}} = \frac{\alpha_e^2}{4E^2 \sin^4 \frac{\theta}{2}} \frac{E'}{E} \left\{ \left(F_1^2 - \frac{q^2}{4m_p^2}F_2^2\right)\cos^2\frac{\theta}{2} - \frac{q^2}{2m_p^2}\left(F_1 + F_2\right)^2 \sin^2\frac{\theta}{2} \right\}. \tag{32.7}$$

This is known as the **Rosenbluth formula**.

If the proton had only interacted through QED, $F_1(q^2)$ and $F_2(q^2)$ would be calculable and could be compared to data. For example, consider $e^- \tau^+$ scattering. The tauon is a lepton whose mass 1.7 GeV is close to the proton mass. For $e^- \tau^+$ scattering, $F_1(q^2)$

and $F_2(q^2)$ were calculated in QED at 1-loop in Sections 19.3 and 17.2 respectively. For $|q^2| \gg m_\tau^2$, $F_2 \to 0$ and F_1 has logarithmic energy dependence. Comparing $F_1(q^2)$ at two scales, q_1 and q_2, the calculation from Section 19.3 gives

$$F_1(q_1^2) - F_1(q_2^2) \approx -\frac{e^2}{16\pi^2} \ln \frac{q_1^2}{q_2^2}, \quad |q_1^2|, |q_2^2| \gg m_\tau^2, \tag{32.8}$$

which agrees with what is measured. For the proton, very different behavior was observed in the classic scattering experiments from the 1960s. The form factors were found to be well fit by the expressions [Albrecht *et al.*, 1966]

$$F_1(q^2) \sim \frac{1}{\left(1 - \frac{q^2}{0.71 \, \text{GeV}^2}\right)^2}. \tag{32.9}$$

Here a definite scale $0.71 \, \text{GeV}^2$ appears, even in differences such as $F_1(q_1^2) - F_1(q_2^2)$.

Form factors are particularly useful because they correspond to the Fourier transforms of scattering potentials, through the Born approximation (see Section 5.2). Indeed, up to some kinematic factors and normalization, $F_1(q^2) = \int d^3x \, e^{i\vec{q}\cdot\vec{x}} V(x)$, which leads to $V(r) = \frac{m^3}{4\pi} e^{-mr}$ in this case. Thus, the form of the proton is characterized by an exponential shape $\rho(r) \sim e^{-r/r_0}$, with characteristic size $r_0 \sim (0.84 \, \text{GeV})^{-1} \sim 1 \, \text{fm}$.

The conclusion is that the proton has a characteristic size of order $1 \, \text{fm}$. The value of this size is not surprising, since it is of the order of the proton's Compton wavelength. What is surprising is that there is a scale at all! In scattering electrons off tauons, all we would ever see is a form factor with logarithmic dependence on energy. The tauon's size is not of order m_τ^{-1}; if it has a finite size at all, it is much, much smaller than m_τ^{-1}.

To learn more about the proton, experiments had to go to higher energy. At energies $|q^2| > 1 \, \text{GeV}^2$, you might expect $e^- p^+$ to elucidate an even more complicated charge distribution with more and more scales. Instead, somewhat shockingly (from an experimental point of view), they simplify back to the point scattering case. That is, very high energy $e^- p^+$ scattering reveals pointlike constituents *within* the proton, now known as quarks. We will next explain how to see this simplification.

32.1.3 Inelastic $e^- p^+$ scattering

Up until now we have discussed elastic scattering: $e^- p^+ \to e^- p^+$. At center-of-mass energies above m_p, the proton can start to break apart. For example, at high enough energies, the reaction $e^- p^+ \to e^- p^+ \pi^0$ can occur. At very high energies, the proton breaks apart completely, as shown in Figure 32.1. Remarkably, the physics simplifies in this deeply inelastic regime, and we will be able to make precise theoretical predictions.

In deriving the parametrization of the cross section in terms of $F_1(q^2)$ and $F_2(q^2)$, we needed to use the reduction of the photon–proton interactions to terms of the form $\bar{u}(p')\gamma^\mu u(p)$ or $\bar{u}(p')\sigma^{\mu\nu} q_\mu u(p)$. When the proton breaks apart, as in deep inelastic scattering (DIS), this parametrization will no longer do. Instead, we need to parametrize photon–proton–X interactions, where X is anything the proton can break up into. Thus, it makes sense to parametrize the cross section (instead of the vertex) in terms of the momentum transfer q^μ and the proton momentum P^μ.

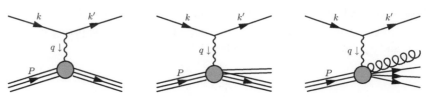

As energy is increased, $e^- p^+$ scattering goes from elastic to slightly inelastic, with $e^- p^+ \pi^0$ in the final state, to deeply inelastic, where the proton breaks apart completely.

In the lab frame, the kinematics are shown in Figure 32.1. We define E and E' as the energies of the incoming and outgoing electron. We also define θ as the angle between $\vec{k}$ and $\vec{k}'$, so $\theta = 0$ is forward scattering. The cross section can be written as

$$\left(\frac{d\sigma}{d\Omega\, dE'} \right)_{\text{lab}} = \frac{\alpha_e^2}{4\pi m_p q^4} \frac{E'}{E} L^{\mu\nu} W_{\mu\nu}, \qquad (32.10)$$

where $L_{\mu\nu}$ is the **leptonic tensor**, which encodes polarization information for the electron or, equivalently, the off-shell photon. We already used a parametrization like this in Chapter 20 while discussing IR divergences. There the $e^+ e^- \to \mu^+ \mu^- (+\gamma)$ cross section simplified using the same lepton tensor. For unpolarized scattering, the lepton tensor is

$$L_{\mu\nu} = \frac{1}{2}\text{Tr}\left[\slashed{k}' \gamma^\mu \slashed{k} \gamma^\nu \right] = 2(k'^\mu k^\nu + k'^\nu k^\mu - k \cdot k' g^{\mu\nu}), \qquad (32.11)$$

where k and k' are the electron's initial and final momentum. The factor of $\frac{1}{2}$ comes from averaging over the initial electron's spin. Note that $L_{\mu\nu} = L_{\nu\mu}$.

The **hadronic tensor** $W^{\mu\nu}$ includes an integral over all the phase space for all final state particles (as did $X^{\mu\nu}$ in Eq. (20.30)). It gives the rate for $\gamma^* p^+ \to$ anything:

$$e^2 \epsilon_\mu \epsilon_\nu^* W^{\mu\nu} = \frac{1}{2} \sum_{X,\text{spins}} \int d\Pi_X (2\pi)^4 \delta^4 (q + P - p_X) |\mathcal{M}\left(\gamma^* p^+ \to X \right)|^2, \qquad (32.12)$$

where ϵ_μ is the polarization of the off-shell photon. Since final states are integrated over, $W_{\mu\nu}$ can depend on P^μ and q^μ only. In unpolarized scattering, it must be symmetric, $W^{\mu\nu} = W^{\nu\mu}$. It also should satisfy $q_\mu W^{\mu\nu} = 0$ by the Ward identity (see Chapter 14), since the interaction is only through a photon. Thus, the most general parametrization is[1]

$$W^{\mu\nu} = W_1 \left(-g^{\mu\nu} + \frac{q^\mu q^\nu}{q^2} \right) + W_2 \left(P^\mu - \frac{P \cdot q}{q^2} q^\mu \right) \left(P^\nu - \frac{P \cdot q}{q^2} q^\nu \right). \qquad (32.13)$$

The Lorentz scalars on which W_1 and W_2 can depend are $P^2 = m_p^2$, q^2 and $P \cdot q$. Natural variables to use are $Q \equiv \sqrt{-q^2} > 0$, which is the energy scale of the collision, and

$$\nu \equiv \frac{P \cdot q}{m_p} = (E - E')_{\text{lab}}, \qquad (32.14)$$

---

[1] A word of caution: there are a number of different conventions for the normalization of W_1 and W_2 in the literature. Ours is convenient for the $Q/m_p \to \infty$ limit.

where ν is a Lorentz-invariant quantity which, in the proton rest frame, reduces to the energy lost by the electron. An alternative to ν is the dimensionless ratio

$$x \equiv \frac{Q^2}{2P \cdot q}, \tag{32.15}$$

which is known as **Bjorken** x and will play an important role in what follows.

Without too much work, one can contract $L^{\mu\nu}$ with $W_{\mu\nu}$ and use Eq. (32.10) to express the result in terms of the scattering angle θ:

$$\left(\frac{d\sigma}{d\Omega \, dE'}\right)_{\text{lab}} = \frac{\alpha_e^2}{8\pi E^2 \sin^4 \frac{\theta}{2}} \left[\frac{m_p}{2} W_2(x, Q) \cos^2 \frac{\theta}{2} + \frac{1}{m_p} W_1(x, Q) \sin^2 \frac{\theta}{2}\right]. \tag{32.16}$$

As in the elastic case, we have set everything up so we only have to know the incoming and outgoing electron momenta, not anything about the final hadronic state X. That is, W_1 and W_2 can be completely determined by measuring only the energy and angular dependence of the outgoing electron.

The defining assumption of the **parton model**, originally due to Feynman, is that some objects called **partons** within the proton are essentially free. When we connect to QCD, we will see that parton refers to not only quarks, but also the gluons and antiquarks in a hadron (and photons and, at least formally, every other particle in the Standard Model too). For now, let us just assume that there exist partons within the proton, some of which are charged. To test the parton model, we need to determine what the form factors W_1 and W_2 would look like if the electron were scattering elastically off partons of mass m_q inside the proton. An elastic parton scattering diagram is

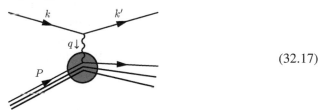

$$\tag{32.17}$$

where the circle represents the proton and the three lines coming in and three lines going out represent partons within the proton, only one of which participates in the interaction with the electron.

This diagram is not that different from the one for electron–muon scattering. To evaluate it, call the scattered parton's initial momentum p_i^μ and its final momentum p_f^μ, so that $p_i^\mu + q^\mu = p_f^\mu$ by momentum conservation. Squaring both sides gives

$$m_q^2 + 2p_i \cdot q + q^2 = m_q^2 \qquad \Rightarrow \qquad \frac{Q^2}{2p_i \cdot q} = 1. \tag{32.18}$$

Unfortunately, the parton momentum is not directly measurable. However, let us just assume it has some fraction ξ of the proton's momentum, $p_i^\mu = \xi P^\mu$. Then $x = \frac{\xi Q^2}{2p_i \cdot q} = \xi$. In particular, if the parton model were valid, then by measuring x we would be measuring the fraction of the proton's momentum involved in the parton-level scattering.

Now let us additionally suppose that the partons are weakly interacting. Then we should be able to calculate $e^- q \rightarrow e^- q$ elastic scattering in perturbation theory. In particular, we expect the form factors to have only weak, logarithmic dependence on Q^2 (just as for

$e^-\mu^- \to e^-\mu^-$ scattering) when the initial partonic momentum is fixed, that is, at fixed x. The cross section's (approximate) independence of Q^2 at fixed x is known as **Bjorken scaling**. We will make this precise in a moment, but you might want to glance ahead at Figure 32.2, which shows Bjorken scaling beautifully confirmed by data.

Another ingredient in the parton model is the classical probabilities $f_i(\xi)d\xi$ of the photon hitting parton species i which has a fraction ξ of the proton momentum. These $f_i(\xi)$ are known as **parton distribution functions (PDFs)**. The physical justification of PDFs is that the momentum sloshes around among proton constituents at time scales $\sim \Lambda_{\mathrm{QCD}}^{-1} \sim m_p^{-1}$. These time scales are much slower than the time scales $\sim Q^{-1}$ that the photon probes. The separation of scales $Q \gg \Lambda_{\mathrm{QCD}}$ allows us to treat the parton wavefunctions within the proton as being decoherent, giving the probabilistic interpretation. To actually prove that this decoherence occurs amounts to a proof of *factorization*. Factorization is discussed in Section 32.4 below.

With PDFs we can be more precise about the predictions of a theory with weakly interacting partons. The parton model assumption is that the cross section for $e^- P^+ \to e^- X$ scattering is given by $e^- p_i \to e^- X$, where p_i is a parton with momentum $p_i^\mu = \xi P^\mu$, integrated over ξ. In equations:

$$\sigma\left(e^- P^+ \to e^- X\right) = \sum_i \int_0^1 d\xi \, f_i(\xi) \hat{\sigma}\left(e^- p_i \to e^- X\right). \tag{32.19}$$

Here we initiate the standard convention that partonic quantities are given circumflexes, for example $\hat{\sigma}$.

Assuming the partons are free except for their QED interactions, the electron can only scatter off the charged particles in the proton which we are calling quarks. For a given quark momentum p_i, the $e^- q \to e^- q$ partonic cross section is just like any pointlike scattering cross section in QED. It is given by the Rosenbluth formula, Eq. (32.7), with $F_1 = 1$ and $F_2 = 0$. Before integrating over final electron energy E', the cross section is

$$\left(\frac{d\hat{\sigma}(e^- q \to e^- q)}{d\Omega \, dE'}\right)_{\mathrm{lab}} = \frac{\alpha_e^2 Q_i^2}{4E^2 \sin^4 \frac{\theta}{2}} \left[\cos^2 \frac{\theta}{2} + \frac{Q^2}{2m_q^2} \sin^2 \frac{\theta}{2}\right] \delta\left(E - E' - \frac{Q^2}{2m_q}\right),$$
$$\tag{32.20}$$

where Q_i is the charge of the quark. You can check that Eq. (32.7) with $F_1 = 1$ and $F_2 = 0$ is reproduced from this if we integrate over the δ-function in light of the constraint in Eq. (32.5). Note that if we did not assume free quarks, there could have been generic form factors $G_1(Q)$ and $G_2(Q)$ in front of the $\sin^2 \frac{\theta}{2}$ and $\cos^2 \frac{\theta}{2}$ terms, as there are for low-energy $e^- p^+$ elastic scattering as in Eq. (32.7). Such form factors would violate Bjorken scaling, and their absence is essentially the content of the parton-model prediction for DIS.

In order to get the DIS cross section from this, we have to integrate over the incoming quark momentum. Since $p_i^\mu = \xi P^\mu$ and in the lab frame the proton is at rest, this implies $m_q = \xi m_p$. We can also use that $E - E' = \nu = \frac{Q^2}{2m_p x}$ from Eqs. (32.14) and (32.15) and therefore

$$\delta\left(E - E' - \frac{Q^2}{2m_q}\right) = \delta\left(\frac{Q^2}{2m_p x} - \frac{Q^2}{2m_p \xi}\right) = \frac{2m_p}{Q^2} x^2 \delta(\xi - x). \tag{32.21}$$

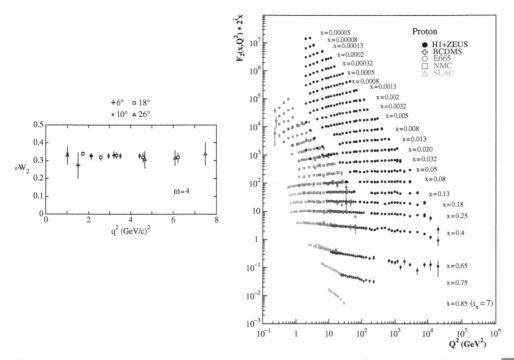

Bjorken scaling is confirmed in deep inelastic scattering data. Left figure is from [Friedman and Kendall, 1972]. Right figure is from [Particle Data Group (Beringer *et al.*), 2012].

Fig. 32.2

And so, using Eq. (32.19), we get

$$\left(\frac{d\sigma(e^- P \to e^- X)}{d\Omega \, dE'}\right)_{lab} = \sum_i f_i(x) \frac{\alpha_e^2 Q_i^2}{4E^2 \sin^4 \frac{\theta}{2}} \left[\frac{2m_p}{Q^2} x^2 \cos^2 \frac{\theta}{2} + \frac{1}{m_p} \sin^2 \frac{\theta}{2}\right].$$
(32.22)

Comparing to Eq. (32.16) we can read off that

$$W_1(x, Q) = 2\pi \sum_i Q_i^2 f_i(x),$$
(32.23)

$$W_2(x, Q) = 8\pi \frac{x^2}{Q^2} \sum_i Q_i^2 f_i(x).$$
(32.24)

Now we have a concrete prediction for Bjorken scaling. The quantities $W_1(x, Q)$ and $Q^2 W_2(x, Q)$ should be independent of Q at fixed x. Remember, although quarks are not observable, the quantity x is, since $x = \frac{Q^2}{2m_p(E-E')}$, where E and E' are the initial and final electron energies in the lab frame. Some early measurements, and some later more accurate ones, demonstrating Bjorken scaling are shown in Figure 32.2.

Another result of the parton model is that $W_1(x, Q) = \frac{Q^2}{4x^2} W_2(x, Q)$ for $Q \gg m_p$, which also follows from Eq. (32.24). This is known as the **Callan–Gross relation**. The proportionality can be traced back to the $\frac{Q^2}{2m_q^2} = \frac{Q^2}{2x^2 m_p^2}$ factor in the $e^- q \to e^- q$ scattering amplitude, which is in turn due to the quarks being free Dirac fermions. Thus the Callan–Gross relation tests that quarks have spin-$\frac{1}{2}$.

For completeness, we point out that the Callan–Gross relation is often given in other forms. We can write it in a Lorentz-invariant way by changing variables to $y = \frac{P \cdot q}{P \cdot k} = \frac{\nu}{E}$ so that $dE' d\Omega = \frac{2m_p E}{E'} \pi y \, dx \, dy$ and then (treating the electron and proton as massless)

$$\frac{d\sigma(e^- P \to e^- X)}{dx \, dy} = \frac{2\pi\alpha^2}{Q^4} s \left(1 + (1 - y)^2\right) \sum_i Q_i^2 x f_i(x), \tag{32.25}$$

with $s = E_{\mathrm{CM}}^2$. This characteristic $1 + (1 - y)^2$ behavior is often identified with the Callan–Gross relation.

Sometimes also dimensionless structure functions are used:

$$\mathcal{F}_1(x) \equiv \frac{1}{4\pi} W_1(x), \qquad \mathcal{F}_2(x) \equiv \frac{Q^2}{8\pi x} W_2(x), \tag{32.26}$$

so that the Callan–Gross relation becomes $\mathcal{F}_1(x) = \frac{1}{2x} \mathcal{F}_2(x) = \frac{1}{2} \sum_i Q_i^2 f_i(x)$. These $\mathcal{F}_i$ should not be confused with the F_i in the original proton form factor, despite their alphabetical similarity. We will follow the standard convention and use these $\mathcal{F}_i$ form factors in the QCD analysis in Section 32.4.

32.1.4 Sum rules

For PDFs to be probabilities, they must satisfy some constraints. For example, if the proton had exactly one down quark, then the down quark must have some momentum, and so $\int d\xi \, f_d(\xi) = 1$. In reality, one can have virtual down–antidown quark pairs within the proton, so there can be more than one down quark. However, since down-quark number is conserved (in QED and QCD) we have

$$\int d\xi \, [f_d(\xi) - f_{\bar{d}}(\xi)] = 1, \tag{32.27}$$

where $f_{\bar{d}}(\xi)$ is the down-antiquark PDF. Similarly, because the proton has up-quark number of 2 and zero strange-quark number:

$$\int d\xi \, [f_u(\xi) - f_{\bar{u}}(\xi)] = 2, \qquad \text{and} \qquad \int d\xi \, [f_s(\xi) - f_{\bar{s}}(\xi)] = 0. \tag{32.28}$$

The strange-quark sum rule also applies for bottom-quark and charm-quark PDFs. There is no conserved gluon number, so f_g has no associated sum rule. In addition,

$$\sum_j \int d\xi \, [\xi f_j(\xi)] = 1. \tag{32.29}$$

This sum rule follows from momentum conservation (see Problem 32.2). Each of these sum rules corresponds to a classically conserved current (up, down, strange number or momentum). Numerically, it turns out that $\int d\xi \xi (f_u(\xi) + f_d(\xi)) \approx 0.38$. Thus, only around 38% of the proton momentum is contained in the **valence quarks** (u and d). The gluon content of the proton, given by $\int d\xi \, \xi f_g(\xi)$, ranges from 35% to 50% depending on the scale (scale dependence of the PDFs will be discussed shortly). The remainder of the proton momentum is in **sea quarks** (meaning s, c or b quarks and $\bar{d}, \bar{u}, \bar{c}, \bar{s}$ or $\bar{b}$ antiquarks).

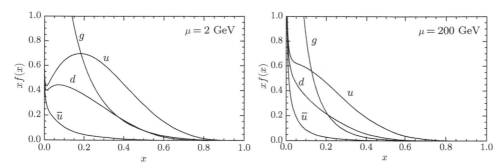

MSTW parton distribution functions [MSTW group (Martin *et al.*), 2009] are shown for various partons. The central values for $xf_i(x, \mu)$ are shown for u, d, g and $\bar{u}$. The factorization scale $\mu = 2\,\text{GeV}$ is used on the left and $\mu = 200\,\text{GeV}$ is used on the right. The sea quark PDFs other than $\bar{u}$ are not shown; they are qualitatively similar to the $\bar{u}$ PDF.

Fig. 32.3

In practice, the PDFs are determined not just from DIS, but from many other high-energy processes, such as $p\bar{p}$ and pp collisions. There are a number of different groups that perform global fits to PDFs. The fits differ by the way they weight different contributions, the order in α_s at which the associated perturbative calculations are performed, and how the PDFs are parametrized. Example parton distributions are shown in Figure 32.3.

32.2 DGLAP equations

We have seen that qualitatively correct features of DIS, such as Bjorken scaling and the Callan–Gross relation, follow from the parton model. However, one can see already in Figure 32.2 that Bjorken scaling does not quite hold – there is some weak (logarithmic) Q^2 dependence visible in the structure function. In this section, we will show how the logarithmic Q^2 dependence can be calculated by combining the parton model with perturbative QCD. Thus, for now, we will continue to *assume* the parton model holds, so that the e^-p^+ cross section is given by a sum of parton-scattering rates, with the initial parton's energy given by classical probability functions $f_i(\xi)$. In the next section, we will discuss to what extent the parton model itself can be proven within QCD.

In Eq. (32.10) we wrote the $e^-p^+ \rightarrow e^-X$ cross section in terms of the leptonic tensor $L^{\mu\nu}$ and the hadronic tensor $W^{\mu\nu}(x, Q)$, with the hadronic tensor given by $|\mathcal{M}(\gamma^\star p^+ \rightarrow X)|^2$ integrated over final states, as in Eq. (32.12). Let us write $\hat{W}^{\mu\nu}(z, Q)$ as the partonic version of $W^{\mu\nu}(x, Q)$, given by $|\mathcal{M}(\gamma^\star q \rightarrow X)|^2$ integrated over final states. Here z is the partonic version of x:

$$z \equiv \frac{Q^2}{2p_i \cdot q}.\tag{32.30}$$

Now we use the parton model assumption that the probability of finding $p_i^\mu = \xi P^\mu$ for some $0 \leq \xi \leq 1$ is given by a PDF $f_i(\xi)$. Thus, $x = z\xi$ and we have to integrate over ξ. This leads to

$$W^{\mu\nu}(x,Q) = \sum_i \int_0^1 dz \int_0^1 d\xi f_i(\xi) \hat{W}^{\mu\nu}(z,Q) \delta(x-z\xi)$$

$$= \sum_i \int_x^1 \frac{d\xi}{\xi} f_i(\xi) \hat{W}^{\mu\nu}\left(\frac{x}{\xi},Q\right). \tag{32.31}$$

Let us check this at leading-order QCD. At order $\mathcal{O}\left(\alpha_s^0\right)$, the only partonic process that contributes to $W^{\mu\nu}$ is $\gamma^* q \to q$. Then, with p_i^μ and $p_f^\mu = p_i^\mu + q^\mu$ the initial and final quark momenta, we have

$$\hat{W}^{\mu\nu}(z,Q) = \frac{Q_i^2}{2} \int \frac{d^3\vec{p}_f}{(2\pi)^3} \frac{1}{2E_f} \text{Tr}\left[\gamma^\mu \not{p}_i \gamma^\nu \not{p}_f\right] (2\pi)^4 \delta^4(p_i + q - p_f)$$

$$= 2\pi Q_i^2 \left[\left(-g^{\mu\nu} + \frac{q^\mu q^\nu}{q^2}\right) + \frac{4z}{Q^2}\left(p_i^\mu - \frac{p_i \cdot q}{q^2} q^\mu\right)\left(p_i^\nu - \frac{p_i \cdot q}{q^2} q^\nu\right)\right] \delta(1-z). \tag{32.32}$$

We find $\hat{W}_1 = 2\pi Q_i^2 \delta(1-z) = \frac{Q^2}{4z}\hat{W}_2$, confirming the Callan–Gross relation at leading order. Plugging this leading-order $\hat{W}^{\mu\nu}(z,Q)$ into Eq. (32.31) reproduces Eq. (32.24), confirming the normalization.

For simplicity, let us consider the form factor $W_0 \equiv -g^{\mu\nu}W_{\mu\nu}$. For the hadronic tensor,

$$W_0(x,Q) \equiv -g^{\mu\nu}W_{\mu\nu} = 3W_1(x,Q) - W_2\left(x,Q\right)\left(m_p^2 + \frac{Q^2}{4x^2}\right). \tag{32.33}$$

For $Q \gg m_p$, this simplifies to $W_0 = 3W_1 - \frac{Q^2}{4x^2}W_2$ so that $W_0 = 2W_1$ at leading order. In particular,

$$W_0(x,Q) = 4\pi \sum_i Q_i^2 f_i(x). \tag{32.34}$$

This equation motivates using W_0 as a *definition* of PDFs, valid beyond leading order. Defining the PDFs in this way lets us calculate the Q dependence of the PDFs, as we will now see. In particular, we can now forget about all those confusing structure functions and focus on W_0, which is basically just the unpolarized cross section for $\gamma^* p^+ \to X$.

At the parton level, at leading order, $\hat{W}_0^{\text{LO}} = 4\pi Q_i^2 \delta(1-z)$. At next-to-leading order in the parton model in QCD there is a virtual $\gamma^* q \to q$ graph and s- and t-channel $\gamma^* q \to qg$ graphs:

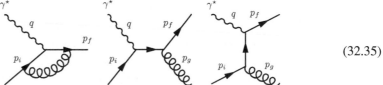

$$\tag{32.35}$$

These diagrams are essentially just crossings of the $\gamma^* \to \mu^+\mu^-(+\gamma)$ diagrams in Chapter 20. We will assume the reader is thoroughly familiar with the calculations in Chapter 20, so that we can just present and discuss the relevant results without repeating similar calculational details.

Using the same techniques described in Chapter 20, we can compute the virtual contributions at NLO (see Eq. (20.A.101)). The interference between the leading-order graph and the loop in Eq. (32.35) in $d = 4 - \varepsilon$ dimensions gives

$$\hat{W}_0^V = 4\pi Q_i^2 \frac{\alpha_s}{2\pi} C_F \left(\frac{4\pi\mu^2}{Q^2}\right)^{\frac{\varepsilon}{2}} \frac{\Gamma\left(1-\frac{\varepsilon}{2}\right)}{\Gamma\left(1-\varepsilon\right)} \left(-\frac{8}{\varepsilon^2} - \frac{6}{\varepsilon} - 8 - \frac{\pi^2}{3}\right) \delta(1-z) \quad (32.36)$$

up to terms that will not contribute when $\varepsilon \to 0$. In this expression, the UV divergence has already been removed with the counterterm, so these ε are all ε_{IR}. For the real emission graphs, the calculation is a bit more strenuous, but also can be done using techniques from Chapter 20. The result is [Altarelli *et al.*, 1979]

$$\hat{W}_0^R = 4\pi Q_i^2 \frac{\alpha_s}{2\pi} C_F \left(\frac{4\pi\mu^2}{Q^2}\right)^{\frac{\varepsilon}{2}} \frac{\Gamma\left(1-\frac{\varepsilon}{2}\right)}{\Gamma\left(1-\varepsilon\right)}$$
$$\times \left\{3z + z^{\frac{\varepsilon}{2}}(1-z)^{-\frac{\varepsilon}{2}} \left(-\frac{2}{\varepsilon}\frac{1+z^2}{1-z} + 3 - z - \frac{3}{2}\frac{1}{1-z} - \frac{7}{4}\frac{\varepsilon}{1-z}\right)\right\}. \quad (32.37)$$

Looking at these results, it appears that $\hat{W}_0^V$ has a $\frac{1}{\varepsilon^2}$ double pole but $\hat{W}_0^R$ does not, so that the poles will not cancel. However, there is in fact a $\frac{1}{\varepsilon^2}$ pole in $\hat{W}_0^R$, coming from the $\frac{1}{1-z}(1-z)^{-\frac{\varepsilon}{2}}$ terms. To see this, we need to use the fact that $(1-z)^{-1-\varepsilon}$ expanded around $\varepsilon = 0$ gives a distribution. The relevant identity is

$$\frac{1}{(1-z)^{1+\varepsilon}} = -\frac{1}{\varepsilon}\delta(1-z) + \frac{1}{[1-z]_+} - \varepsilon \left[\frac{\ln(1-z)}{1-z}\right]_+ + \sum_{n=2}^{\infty} \frac{(-\varepsilon)^n}{n!} \left[\frac{\ln^n(1-z)}{1-z}\right]_+, \quad (32.38)$$

which you can derive in Problem 32.3. Here the plus function is defined so that

$$\int_0^1 dz \frac{f(z)}{[1-z]_+} = \int_0^1 dz \frac{f(z) - f(1)}{1-z} \quad (32.39)$$

and so that $\frac{1}{[1-z]_+} = \frac{1}{1-z}$ for $z \neq 1$. These two conditions uniquely define the distribution for any limits of integration. The other plus functions are defined similarly:

$$\int_0^1 dz\, f(z) \left[\frac{\ln^n(1-z)}{1-z}\right]_+ \equiv \int_0^1 dz (f(z) - f(1)) \frac{\ln^n(1-z)}{1-z}, \quad (32.40)$$

with $\left[\frac{\ln^n(1-z)}{1-z}\right]_+ = \frac{\ln^n(1-z)}{1-z}$ for $z \neq 1$. Then we find

$$\hat{W}_0^R = 4\pi Q_i^2 \frac{\alpha_s}{2\pi} C_F \left(\frac{4\pi\mu^2}{Q^2}\right)^{\frac{\varepsilon}{2}} \frac{\Gamma\left(1-\frac{\varepsilon}{2}\right)}{\Gamma\left(1-\varepsilon\right)} \times \left\{3 + 2z - \frac{1+z^2}{1-z}\ln z\right.$$
$$\left. + \left(\frac{8}{\varepsilon^2} + \frac{3}{\varepsilon} + \frac{7}{2}\right)\delta(1-z) - \left(2\frac{1+z^2}{\varepsilon} + \frac{3}{2}\right)\left[\frac{1}{1-z}\right]_+ + (1+z^2)\left[\frac{\ln(1-z)}{1-z}\right]_+\right\}, \quad (32.41)$$

and therefore, up to next-to-leading order,

$$\hat{W}_0 = \hat{W}_0^{\text{LO}} + \hat{W}_0^V + \hat{W}_0^R = 4\pi Q_i^2 \left\{\left[\delta(1-z) - \frac{1}{\varepsilon}\frac{\alpha_s}{\pi}P_{qq}(z)\left(\frac{4\pi\mu^2}{Q^2}\right)^{\frac{\varepsilon}{2}}\frac{\Gamma\left(1-\frac{\varepsilon}{2}\right)}{\Gamma\left(1-\varepsilon\right)}\right]\right.$$
$$+ \frac{\alpha_s}{2\pi}C_F \left[(1+z^2)\left[\frac{\ln(1-z)}{1-z}\right]_+ - \frac{3}{2}\left[\frac{1}{1-z}\right]_+\right.$$
$$\left.\left. - \frac{1+z^2}{1-z}\ln z + 3 + 2z - \left(\frac{9}{2} + \frac{1}{3}\pi^2\right)\delta(1-z)\right]\right\}, \quad (32.42)$$

where

$$P_{qq}(z) = C_F\left[(1+z^2)\left[\frac{1}{1-z}\right]_+ + \frac{3}{2}\delta(1-z)\right].\tag{32.43}$$

This distribution, $P_{qq}(z)$, is known as a **DGLAP splitting function**, after Dokshitzer, Gribov, Lipatov, Altarelli and Parisi.

At this point, all the double poles have canceled, but there is still a single $\frac{1}{\varepsilon}$ pole in the cross section whose residue is proportional to $P_{qq}(z)$. Having a pole in a parton-level cross section is not a problem, as long as it drops out of physical predictions. Focusing on this pole, we can insert $\hat{W}_0$ into Eq. (32.31) to get

$$W_0(x,Q) = 4\pi\sum_i Q_i^2 \int_x^1 \frac{d\xi}{\xi} f_i(\xi)\left[\delta\left(1-\frac{x}{\xi}\right) - \frac{\alpha_s}{2\pi}P_{qq}\left(\frac{x}{\xi}\right)\left(\frac{2}{\varepsilon} + \ln\frac{\tilde{\mu}^2}{Q^2}\right) + \text{finite}\right].\tag{32.44}$$

Now, using the definition of plus functions, we find that the splitting function in Eq. (32.43) satisfies

$$\int_0^1 P_{qq}(z)dz = 0.\tag{32.45}$$

Thus, if we integrate $W_0(x,Q)$ over x, to get the *total* DIS cross section at a given Q, the $\frac{1}{\varepsilon}$ pole exactly vanishes.

At fixed x the $\frac{1}{\varepsilon}$ pole does not cancel and $W_0(x,Q)$ is divergent. However, as in many other examples (see Chapter 16), we need to take differences of cross sections to find finite answers. The difference in $W_0(x,Q)$ at the same x but different scales Q and Q_0 is

$$W_0(x,Q) - W_0(x,Q_0) = 4\pi\sum_i Q_i^2 \int_x^1 \frac{d\xi}{\xi} f_i(\xi)\left[\frac{\alpha_s}{2\pi}P_{qq}\left(\frac{x}{\xi}\right)\ln\frac{Q^2}{Q_0^2}\right].\tag{32.46}$$

This difference is a finite integral. The finite parts of Eq. (32.42) drop out of such differences, but the $\frac{1}{\varepsilon}$ pole in the parton-level cross section leads to a physical quantum prediction for the logarithmic Q dependence of the hadronic cross section. (The finite parts of Eq. (32.42) do show up in differences of structure functions [Altarelli *et al.*, 1979; Sterman, 1993].)

Why should we have to calculate differences? Should $W_0(x,Q)$ not be observable and hence finite without any new renormalization, since QCD is renormalizable? There are two answers. First, if we did the calculation in full QCD, the IR divergence would be cut off by some physical scale such as a quark mass m_q or Λ_{QCD}. Indeed, the same divergence occurs in Compton scattering in QED, and is cut off by the electron mass. However, this misses the point. Doing the calculation with massive quarks would replace the logarithm by $\ln\frac{m_q}{Q}$, which for $Q \gg m_q$ would be very large. Thus, the second answer is simply that the difference between $W_0(x,Q)$ at two scales is a more practical quantity to calculate: we can get a testable answer in perturbation theory. Indeed, the logarithm in Eq. (32.46) exactly explains the violation of Bjorken scaling seen in Figure 32.2.

As we have seen many times, renormalization lets us replace the calculation of differences with the calculation of observables in terms of renormalized quantities. In this case, we need to define renormalized PDFs. We could do this by saying W_0 is given exactly by Eq. (32.34) at some reference scale Q_0. Since Q_0 is arbitrary, the independence of the cross section of Q_0 should lead to a renormalization group equation. In anticipation of a connection to the RG, we define

$$W_0(x, Q) \equiv 4\pi \sum_i Q_i^2 f_i(x, \mu = Q) \tag{32.47}$$

for *every* scale Q. For this equation to be consistent with Eq. (32.46) we need

$$f_i(x, \mu_1) = f_i(x, \mu) + \frac{\alpha_s}{2\pi} \int_x^1 \frac{d\xi}{\xi} f_i(\xi, \mu_1) P_{qq}\left(\frac{x}{\xi}\right) \ln\frac{\mu_1^2}{\mu^2}, \tag{32.48}$$

which implies

$$\mu \frac{d}{d\mu} f_i(x, \mu) = \frac{\alpha_s}{\pi} \int_x^1 \frac{d\xi}{\xi} f_i(\xi, \mu) P_{qq}\left(\frac{x}{\xi}\right). \tag{32.49}$$

This is known as a **DGLAP evolution equation**. It allows us to resum large logarithms in structure functions.

We can do a quick check on the self-consistency of our results. For $f_q(x)$ to have a probabilistic interpretation, sum rules such as Eq. (32.27) should hold for any μ. Integrating over x in Eq. (32.48) and using Eq. (32.45) we see that $\int f_i(x, \mu)$ is indeed μ independent. In fact, if we assume Eq. (32.27), one can derive the singular part of $P_{qq}(z)$ uniquely by knowing that for $z > 1$ it behaves as $\frac{1+z^2}{1-z}$. This is a shortcut to deriving the splitting functions, discussed more in the next section.

So far we have only considered partonic processes relevant for $e^- p^+ \to e^- X$, such as $\gamma^* q \to q$ and $\gamma^* q \to qg$, which have quarks in the initial state. At next-to-leading order there are also processes such as $\gamma^* g \to q\bar{q}$ with initial state gluons. Since there is a probability of finding antiquarks and gluons in the proton, there are PDFs $f_{\bar{q}}$ and f_g for these partons as well. All of these PDFs mix under RG group evolution. Thus, DGLAP is really a set of coupled integro-differential equations. For quarks and gluons, these can be written in the form

$$\mu \frac{d}{d\mu} \begin{pmatrix} f_i(x, \mu) \\ f_g(x, \mu) \end{pmatrix} = \sum_j \frac{\alpha_s}{\pi} \int_x^1 \frac{d\xi}{\xi} \begin{pmatrix} P_{q_i q_j}(\frac{x}{\xi}) & P_{q_i g}(\frac{x}{\xi}) \\ P_{g q_j}(\frac{x}{\xi}) & P_{gg}(\frac{x}{\xi}) \end{pmatrix} \begin{pmatrix} f_j(\xi, \mu) \\ f_g(\xi, \mu) \end{pmatrix}. \tag{32.50}$$

The various splitting functions can be derived from cross sections for processes such as $g \to gg$ or $g \to \bar{q}q$ as we did for $q \to qg$ above. At leading order, they are

$$P_{qq}(z) = C_F \left[\frac{1+z^2}{[1-z]_+} + \frac{3}{2}\delta(1-z)\right], \tag{32.51}$$

$$P_{qg}(z) = T_F \left[z^2 + (1-z)^2\right], \tag{32.52}$$

$$P_{gq}(z) = C_F \left[\frac{1+(1-z)^2}{z}\right], \tag{32.53}$$

$$P_{gg}(z) = 2C_A \left[\frac{z}{[1-z]_+} + \frac{1-z}{z} + z(1-z) \right] + \frac{\beta_0}{2} \delta(1-z), \qquad (32.54)$$

where $\beta_0 = \frac{11}{3}C_A - \frac{4}{3}T_F n_f$. Derivations of these other splitting functions can be found in numerous references, for example [Peskin and Schroeder, 1995] or [Ellis *et al.*, 1996].

32.3 Parton showers

In the previous section we derived the next-to-leading order prediction for deep inelastic scattering in the parton model. The key result was that the cross section for $\gamma^* q \to qg$ was IR divergent, but that this divergence could be absorbed in renormalized PDFs. In this section, we will trace the origin of the IR divergence, discuss its universality, and show how that universality can be exploited in an important semi-classical approximation called the **parton shower**.

While regulating divergences in $d = 4 - \varepsilon$ dimensions is efficient mathematically, it obscures some of the physics. So let us return to the $\gamma^* q \to qg$ cross section and see what it looks like in four dimensions. Summing over final state spins and colors and averaging over initial state spins and colors, the real emission diagrams in Eq. (32.35) give

$$|\mathcal{M}|^2 = 2e^2 Q_i^2 C_F g_s^2 \left(-\frac{\hat{t}}{\hat{s}} - \frac{\hat{s}}{\hat{t}} + \frac{2\hat{u}Q^2}{\hat{s}\hat{t}} \right), \qquad (32.55)$$

where

$$\hat{s} = (q + p_i)^2 = Q^2 \frac{1-z}{z}, \qquad \hat{t} = (p_g - p_i)^2, \qquad \hat{u} = (p_i - p_f)^2 \qquad (32.56)$$

satisfy $\hat{s} + \hat{t} + \hat{u} = -Q^2$. The physical region has $Q^2 = -q^2 > 0$, $\hat{s} > 0$ and $\hat{t}, \hat{u} < 0$. As usual, we are putting hats on the partonic quantities.

Now, $|\mathcal{M}|^2$ is singular at $\hat{s} = 0$ and at $\hat{t} = 0$. At fixed incoming partonic momenta (fixed z and Q^2), $\hat{s}$ is non-zero; thus, the only relevant singularity for calculating $\hat{\sigma} = \hat{\sigma}(\gamma^* q \to qg)$ is the $\hat{t} = 0$ one. Defining θ as the angle between the gluon and the incoming quark in the partonic center-of-mass frame, we find

$$0 = \hat{t} = (p_g - p_i)^2 = -2p_g \cdot p_i = -4E_g E_i \sin^2 \frac{\theta}{2}, \qquad (32.57)$$

so that the singularity occurs when $\theta \to 0$. That is, it is a collinear singularity. This same collinear singularity occurs in Compton scattering in QED, as discussed in Sections 13.5.4 and 20.3.2.

In the partonic center-of-mass frame, the transverse momentum of the outgoing gluon with respect to the incoming quark can be written as $p_T^2 = \frac{\hat{s}\hat{t}\hat{u}}{(\hat{s}+Q^2)^2}$. The collinear $\hat{t} = 0$ singularity implies $p_T \to 0$. At small p_T, $d\Omega \sim \frac{4\pi}{\hat{s}} dp_T^2$, and the partonic cross section can be written in terms of p_T^2 at fixed z as

$$\frac{d\hat{\sigma}(\gamma^* q \to qg)}{dp_T^2} = \hat{\sigma}_0 \frac{1}{p_T^2} \left[\frac{\alpha_s}{2\pi} C_F \frac{1+z^2}{1-z} + \mathcal{O}\left(\frac{p_T^2}{Q^2} \right) \right], \qquad (32.58)$$

where $\hat{\sigma}_0 = \frac{\pi^2 \alpha_e}{\hat{s}} Q_i^2$. Here we recognize the non-singular part of the splitting function $P_{qq}(z)$ from Eq. (32.43), although in this case the singularity at $z = 1$ is unregulated since we have worked in four dimensions. The dimensionally regularized calculation shows that the residue of the pole at $p_T^2 = 0$ is the full distribution $P_{qq}(z)$. A neat trick to derive the δ-function and distribution part of $P_{qq}(z)$ from the $z < 1$ part is to exploit sum rules such as Eq. (32.27), which, to be consistent with Eq. (32.48), imply that Eq. (32.45) must hold (as discussed in the previous section). The equivalences in this paragraph all require a fair bit of calculation, which we leave to Problem 32.10.

A remarkable fact about QCD is that the residue of $\frac{1}{p_T^2}$ as $p_T \to 0$ is *always* given by $P_{qq}(z)$ for any process in which a final state gluon goes collinear to a quark. This is true both when the quark is in the initial state and when it is in the final state. For example, consider the decay rate of a massive vector boson $\gamma^\star \to \bar{q}qg$ with $\gamma^\star$ having mass Q. The diagrams

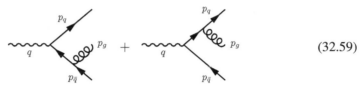

$$ (32.59) $$

were computed in Chapter 20. In four dimensions, the result we found (see Eq. (20.44)) was

$$ \frac{d\Gamma(\gamma^\star \to \bar{q}qg)}{dx_1 dx_2} = \Gamma_0 \frac{\alpha_s}{2\pi} C_F \frac{x_1^2 + x_2^2}{(1 - x_1)(1 - x_2)}, \qquad (32.60) $$

where $\Gamma_0 = Q\alpha_e$, $x_1 = \frac{2E_q}{Q}$, $x_2 = \frac{2E_{\bar{q}}}{Q}$ and $x_y = 2 - x_1 - x_2 = \frac{2E_y}{Q}$. Changing variables to $z = \frac{E_q}{E_g + E_q} = \frac{x_1}{2 - x_2}$ and $m^2 = t = (p_q + p_g)^2 = Q^2(1 - x_2)$, which is the invariant mass of the $q-g$ pair, we find

$$ \frac{d\Gamma(\gamma^\star \to \bar{q}qg)}{dm^2 dz} = Q^2 \Gamma_0 \frac{1}{m^2} \left[\frac{\alpha_s}{2\pi} C_F \frac{1 + z^2}{1 - z} + \mathcal{O}\left(\frac{m^2}{Q^2}\right) \right]. \qquad (32.61) $$

Thus, the residue of $\frac{1}{m^2}$ for this final state radiation case is proportional to the splitting function. In this case z is, by definition, the ratio of the energy carried by the final state quark to the energy of the **mother parton**, that is, the off-shell quark that splits into a quark and gluon. Alternatively, we could write the rate in terms of z and the transverse momentum of the quark with respect to its mother $p_T^2 = \frac{Q^2}{x_1^2}(1 - x_1)(1 - x_g)(1 - x_2)$. In that case, we would also find that the residue of $\frac{1}{p_T^2}$ is proportional to $P_{qq}(z)$.

The general result, for any process in the region of phase space where a gluon is nearly collinear to a quark or antiquark, is that

$$ d\sigma(X \to Y + g) = d\sigma(X \to Y) dt\, dz \frac{1}{t} \left[\frac{\alpha_s}{2\pi} C_F \frac{1 + z^2}{1 - z} + \mathcal{O}\left(\frac{t}{Q^2}\right) \right], \qquad (32.62) $$

where t is any variable, such as m^2 or p_T^2 or the splitting angle θ, that becomes singular in the collinear limit, and Q is any hard scale, that is, any function of momenta that does not vanish in the collinear limit. The variable z is always the fraction of the mother quark's energy carried by the daughter quark. We will prove this in Section 36.4.

One important use of the universality of the collinear limit is that it leads to an efficient semi-classical approximation used in Monte Carlo simulations. One can interpret the splitting functions as probabilities for off-shell partons to branch. These probabilities grow as $\frac{1}{t}$ and are largest for nearly collinear emissions. Since nearly collinear emissions are often not measurable, the simulations work by first picking a momentum for the hardest gluon to be emitted, then picking the next hardest and so on, evolving as a Markov process in a virtuality scale t. One can think of evolution in t as evolution in time from the moment of the collision, or evolution in distance from the collision point.

To be more specific, let us integrate over $z = \frac{E_{\text{daughter}}}{E_{\text{mother}}}$. At fixed small t, in which the collinear approximation is valid, z can be small, but not zero. The lower and upper bounds on z depend on the variable chosen for t (m^2, p_T^2 or θQ^2), but since these all go to zero in the strict collinear limit, the lower bound is $z \gtrsim c\frac{t}{Q^2}$ for some constant c. Thus, for $t \ll Q^2$, the probability of finding *any* gluon at the scale t is approximately

$$R(t) = \frac{\alpha_s}{2\pi}C_F\frac{1}{t}\int_{z_{\min}(t,Q)}^{z_{\max}(t,Q)} dz\frac{1+z^2}{1-z} \approx \frac{\alpha_s}{2\pi}C_F\frac{1}{t}\left(\ln\frac{Q^2}{t}+\mathcal{O}(1)\right). \qquad (32.63)$$

Here, $z_{\min}$ and $z_{\max}$ are the minimum and maximum energies the gluon can have at fixed t. For $t = p_T^2$, $z_{\min}(t,Q) = 1 - z_{\max}(t,Q) = \sqrt{\frac{t}{Q^2}}$ as you can check in Problem 32.6.

We then define the **Sudakov factor** $\Delta(t_0,t)$ as the probability of finding no gluons between the scales t and t_0. To calculate Δ, note that for small shifts,

$$\Delta(t_0,t+\delta t) = \Delta(t_0,t)(1 - \int_t^{t+\delta t} dt'R(t')) = \Delta(t_0,t) - R(t)\,\delta\Delta(t_0,t). \qquad (32.64)$$

This should be consistent with the Taylor expansion $\Delta(t_0,t+\delta t) = \Delta(t_0,t)+\delta t\frac{d}{dt}\Delta(t_0,t)$. Therefore

$$\frac{d}{dt}\Delta(t_0,t) = -R(t)\Delta(t_0,t). \qquad (32.65)$$

The solution to this differential equation with $t_0 = Q^2$ is

$$\Delta(Q,t) = \exp\left(-\int_t^{Q^2} R(t')dt'\right) \approx \exp\left(-\frac{\alpha_s}{4\pi}C_F\ln^2\frac{Q^2}{t}\right). \qquad (32.66)$$

The $\ln^2\frac{Q^2}{t}$ in this expression is the same Sudakov double logarithm characterizing soft-collinear IR divergences we have encountered before (cf. Eq. (20.23)).

And so the cross section for the *hardest* gluon starting from a scale Q is

$$\frac{d\sigma}{dt\,dz} = \Delta(Q,t)\frac{1}{t}\frac{\alpha_s}{2\pi}P_{qq}(z) \approx \exp\left(-\frac{\alpha_s}{4\pi}C_F\ln^2\frac{Q^2}{t}+\cdots\right)\frac{1}{t}\frac{\alpha_s}{2\pi}C_F\frac{1+z^2}{1-z}, \qquad (32.67)$$

with the $\cdots$ subleading at small t. This Sudakov factor is equivalent to performing resummation in QCD at the first non-trivial order (leading logarithmic resummation). It has the important qualitative effect of sending the cross section for producing a gluon at $t = 0$ from $\sigma = \infty$ to $\sigma = 0$: a quark must branch (probability is 1) before it evolves down to $t = 0$. If we take $t = m^2$, then this formula tells us that the rate for the largest invariant mass of a branching, which well approximates the invariant mass of a jet, should not be too small, and not be too large. In other words, Sudakov factors explain the existence of jets.

More details about parton showers can be found in [Sjostrand *et al.*, 2006, Section 10] and [Ellis *et al.*, 1996, Section 5.2].

32.4 Factorization and the parton model from QCD

For practical purposes, the parton model is all one needs to perform perturbative QCD calculations relevant for high-energy scattering involving hadrons. This phenomenological approach assumes **factorization**: that PDFs are universal objects, and any scattering process involving protons can be computed using the same PDFs with a different perturbative calculation. It is remarkable that this procedure works so well, and it is therefore desirable to have a precise derivation of factorization.

Unfortunately, factorization has only been proven in a couple of examples: inclusive deep inelastic scattering (where one measures only the outgoing electron) and the Drell–Yan process (lepton pair production from pp or $p\bar{p}$ collisions). Even in these cases, the proofs are incredibly complicated, with subtlety after subtlety confounding the intuitive picture. The rigorous proofs involve characterizing the infrared singular regions of Feynman diagrams (through pinch surfaces and Landau equations) and are beyond the scope of this text. We will discuss only the classic factorization proof for inclusive deep inelastic scattering using the operator product expansion. This leads to the identification of moments of the PDFs with operator matrix elements. In the next section, an alternative and more generally useful view of the PDFs as lightcone quark matrix elements is given.

The first step to proving factorization is to define what exactly we mean by it. Intuitively, factorization says that the same universal non-perturbative objects (the PDFs), representing the long-distance physics, can be combined with many short-distance calculations in QCD. Roughly, $\sigma = f \otimes H$, where f are the PDFs, H is the perturbative hard calculation, and $\otimes$ denotes a convolution. Such a separation cannot be exactly true: the exact σ must depend on all the brown muck inside the proton. Factorization really means that the calculation done this way is correct up to something small: $\sigma = f \otimes H + \mathcal{O}\!\left(\frac{\Lambda_{\mathrm{QCD}}}{Q}\right)$, where Q is some characteristic high-energy scale in the process. Already, you can see why proofs in cases that are not completely inclusive are so challenging: if there are many measured final states, there can be many scales Q and it is hard to make sure they are all always large in all regions of phase space. For inclusive DIS, we know what Q is, $Q = \sqrt{-(k-k')^2}$, which we take large while holding $x = \frac{Q^2}{2P \cdot q}$ fixed. Thus, there is some hope that we can derive a factorization theorem.

Our approach will first relate the DIS cross section to a product of currents $J^\mu(x) J^\nu(y)$. We then rewrite this product of currents in terms of local operators, $J^\mu(x) J^\nu(y) = \sum_n C_n(x-y)\, \mathcal{O}_n(x)$. The DIS limit $Q^2 \to \infty$ at fixed Bjorken x will correspond to $x^\mu - y^\mu \to 0$ so that we can Taylor expand the Wilson coefficients $C_n(x-y)$ around $x^\mu = y^\mu$, keeping only the leading term. Then, matrix elements of these operators in proton states will give us a definition of the PDFs: $f \sim \langle P | \mathcal{O} | P \rangle$.

32.4.1 The operator product expansion

The operator product expansion is the position-space version of the low-energy expansion used to derive effective Lagrangians. The operators in an effective Lagrangian are **composite operators**, where fields are taken at the same point. For example, recall how the 4-Fermi theory approximates the theory of weak interactions (see Chapters 22, 29 and 31). If we integrate out the W boson at tree-level we end up with a non-local Lagrangian:

$$\mathcal{L}_W \sim g^2 \int d^4x \, d^4y \, \bar{\psi}(x)\gamma^\mu \psi(x) \, D^{\mu\nu}(x,y) \, \bar{\psi}(y)\gamma^\nu \psi(y), \qquad (32.68)$$

where

$$D^{\mu\nu}(x,y) = \int \frac{d^4p}{(2\pi)^4} \frac{-g^{\mu\nu}}{p^2 - m_W^2} e^{ip(x-y)} = \frac{g^{\mu\nu}}{\Box_x + m_W^2} \int \frac{d^4p}{(2\pi)^4} e^{ip(x-y)} \qquad (32.69)$$

is the W-boson propagator. For $\Box \sim p^2 \ll m_W^2$ we expand

$$\frac{g^2}{\Box + m_W^2} = G_F \left(1 - \frac{\Box}{m_W^2} + \left(\frac{\Box}{m_W^2} \right)^2 + \cdots \right) \qquad (32.70)$$

with $G_F \sim \frac{g^2}{m_W^2}$, so that

$$\mathcal{L}_W \sim G_F \int d^4x \left[\bar{\psi}\gamma^\mu \psi \bar{\psi}\gamma^\mu \psi - \bar{\psi}\gamma^\mu \psi \frac{\Box}{m_W^2} \bar{\psi}\gamma^\mu \psi + \bar{\psi}\gamma^\mu \psi \frac{\Box^2}{m_W^4} \bar{\psi}\gamma^\mu \psi + \cdots \right], \qquad (32.71)$$

with all fields at the same point x. This effective Lagrangian is now local.

The **operator product expansion (OPE)** writes products of local operators evaluated at different points, in the limit that the points approach each other, as a sum over composite local operators. Let all possible operators in the theory be denoted by $\mathcal{O}_n$. Then the OPE says that

$$\lim_{x \to y} \mathcal{O}_1(x)\mathcal{O}_2(y) = \sum_n C_n(x-y) \, \mathcal{O}_n(x) \qquad (32.72)$$

for any two operators $\mathcal{O}_1$ and $\mathcal{O}_2$. The reason the OPE is powerful is because *the expansion holds at the level of operators*. That is, the Wilson coefficients C_n are just numbers, independent of the external state. Thus, the C_n can be computed once and for all in perturbation theory and can then be used for any process. Moreover, to compute the C_n one just needs to evaluate any matrix element sensitive to them, then one determines the C_n relevant for *all* matrix elements.

For example, the 4-Fermi theory comes from the expansion of two weak currents $J^\mu(x) = \bar{\psi}(x)\gamma^\mu \psi(x)$ and $J^\mu(y) = \bar{\psi}(y)\gamma^\mu \psi(y)$ approaching each other. We performed the 1-loop OPE through matching to the 4-Fermi theory in Section 31.3. In the 4-Fermi case, as in other perturbative effective field theories, only a finite number of operators are relevant for a given precision. In the case of DIS, we will see that an infinite number of operators are important (the twist-2 operators, defined below) but the OPE will still be useful.

Intuitively, the existence of an OPE makes perfect sense: long-distance physics should be independent of short-distance physics. This is resoundingly true in many other contexts: Newton's laws are independent of quantum mechanics, chemistry is independent of nuclear physics, etc. That is, the OPE should work for the same reason effective field theories work: physics naturally compartmentalizes itself so that all irrelevant scales can be taken to be either 0 or ∞ without strongly affecting the physics in which we are interested. Despite the fact that the OPE is physically sensible, a rigorous mathematical proof is still lacking.

A practical form of the OPE is

$$\int d^4x \, e^{iqx} \mathcal{O}(x)\mathcal{O}(0) = \sum_n C_n(q)\mathcal{O}_n(0), \qquad (32.73)$$

with the Wilson coefficients in momentum space and the operators in position space. We usually calculate the OPE by evaluating $C_n(q)$.

32.4.2 Products of currents

To apply the OPE to DIS we first want to express $W^{\mu\nu}$ in terms of matrix elements of the electromagnetic current constructed from quarks. Treating the quark charge as $Q = 1$ for simplicity, this current is $J^\mu(x) = \bar{\psi}(x)\gamma^\mu\psi(x)$, with $\psi(x)$ the quark field. You may recall from Eq. (14.152) that S-matrix elements for photons, which have the photon propagator amputated by LSZ, are equal to matrix elements of the current J^μ to which the photon couples. This equivalence follows because in pure quark states with spinors $u_1(p)$ and $u_2(p')$ with momentum p and p', the current has matrix element

$$\langle p'|J^\mu(x)|p\rangle = \bar{u}_2(p')\gamma^\mu u_1(p)e^{i(p'-p)x}. \qquad (32.74)$$

To check this equation, simply plug in the expression for $J^\mu(x)$ as a product of the quantum quark fields in terms of creation and annihilation operators. Thus, a shorthand for the spinor product $\bar{u}_2(p')\gamma^\mu u_1(p)$ coming out of a Feynman diagram matrix element calculation is just the current matrix element at $x = 0$: $\langle p'|J^\mu(0)|p\rangle$.

For DIS, $\gamma^\star p^+ \to X$, we need the matrix element of this current (since that is what the photon couples to) at $x = 0$ between an initial proton state $|P\rangle$ and an arbitrary hadronic final state $\langle X|$. That is, we need

$$\mathcal{M}(\gamma^\star p^+ \to X) = e\,\epsilon^\mu \langle X|J_\mu(0)|P\rangle. \qquad (32.75)$$

Comparing to Eq. (32.12) we see that

$$W_{\mu\nu}(\omega, Q) = \sum_X \int d\Pi_X \langle P|J_\mu(0)|X\rangle \langle X|J_\nu(0)|P\rangle (2\pi)^4 \,\delta^4(q^\mu + P^\mu - p_X^\mu)$$

$$= \sum_X \int d\Pi_X \int d^4x \, e^{i(q+P-p_X)x} \langle P|J_\mu(0)|X\rangle \langle X|J_\nu(0)|P\rangle. \qquad (32.76)$$

Here we write $W_{\mu\nu}$ as a function of the Lorentz invariants $\omega = \frac{1}{x} = \frac{2P\cdot q}{Q^2} > 1$ and Q^2 (using ω instead of Bjorken x avoids confusion with position). There is an implicit average over proton spins in $W_{\mu\nu}$.

We next simplify this using

$$\langle P|J_\mu(0)|X\rangle = \langle P|e^{-i\hat{P}\cdot x}J_\mu(x)e^{i\hat{P}\cdot x}|X\rangle = e^{-i(P-p_X)\cdot x}\langle P|J_\mu(x)|X\rangle, \qquad (32.77)$$

where $\hat{P}$ is the momentum operator that generates translations. This gives

$$W_{\mu\nu}(\omega,Q) = \sum_X \int d\Pi_X \int d^4x\, e^{iq\cdot x}\langle P|J_\mu(x)|X\rangle\langle X|J_\nu(0)|P\rangle$$

$$= \int d^4x\, e^{iq\cdot x}\langle P|J_\mu(x)J_\nu(0)|P\rangle. \qquad (32.78)$$

Having performed the sum over $|X\rangle$, we no longer have to think explicitly about what the final states are. Now we can focus on the product of two current operators.

We would now like to use the $Q \to \infty$ limit (at fixed ω) to expand the operator product $J_\mu(x)J_\nu(0)$ around $x^\mu = 0$. Unfortunately, there are two problems with such an expansion. The first problem is that, while we know how to calculate matrix elements of time-ordered products of fields at different points using Feynman rules, we do not know how to calculate products that are not time ordered. The second problem is that large Q^2 implies $x_\mu x^\mu \to 0$ (see Problem 32.7), but it does not imply that $x^\mu \to 0$. In fact, the currents can be separated very far on the lightcone at large Q^2. In momentum space, the problem is that we would like to Taylor expand in Q^{-2}. Since $\omega = \frac{2P\cdot q}{Q^2}$ this limit implies $\omega \to 0$. However, kinematically $P \cdot q > \frac{1}{2}Q^2$, implying $\omega > 1$ (i.e. Bjorken $x < 1$), so a naive large Q^2 expansion will take us out of the physical region. To solve this problem, we need to rearrange things so we can Taylor expand around $\omega = 0$.

To solve the first problem, we use the optical theorem to turn the product of currents into a time-ordered product. The optical theorem says that the total rate for $\gamma^\star p^+ \to X$ is given by the imaginary part of the forward scattering rate $\gamma^\star p^+ \to \gamma^\star p^+$. Using Eqs. (32.12) and (24.11), we can write

$$W^{\mu\nu} = 2\mathrm{Im}T^{\mu\nu}, \qquad (32.79)$$

where

$$e^2\epsilon_\mu\epsilon_\nu^\star T^{\mu\nu}(\omega,Q) = \mathcal{M}(\gamma^\star p^+ \to \gamma^\star p^+). \qquad (32.80)$$

$T_{\mu\nu}$ is called the **forward Compton amplitude**. It is a forward amplitude since the (off-shell) photon and proton have the same momentum in the initial and final states. In terms of currents, we can write $T_{\mu\nu}$ as

$$T_{\mu\nu}(\omega,Q) = i\int d^4x\, e^{iq\cdot x}\langle P|T\{J_\mu(x)J_\nu(0)\}|P\rangle. \qquad (32.81)$$

We have expressed a matrix element squared $W_{\mu\nu} \sim |\mathcal{M}(\gamma^\star p^+ \to X)|^2 \sim |\langle T\{J\}\rangle|^2$ as the imaginary part of a matrix element $T_{\mu\nu} \sim \mathcal{M}(\gamma^\star p^+ \to \gamma^\star p^+) \sim \langle T\{JJ\}\rangle$.

It is conventional to expand $T_{\mu\nu}$ in terms of its own structure functions, as in (32.13), with a slightly different normalization:

$$T^{\mu\nu}(\omega,Q) = T_1\left(-g^{\mu\nu} + \frac{q^\mu q^\nu}{q^2}\right) + \frac{T_2}{P\cdot q}\left(P^\mu - \frac{P\cdot q}{q^2}q^\mu\right)\left(P^\nu - \frac{P\cdot q}{q^2}q^\nu\right). \qquad (32.82)$$

The DIS structure functions are then $W_1 = 2\mathrm{Im}T_1$ and $W_2 = 4\mathrm{Im}\frac{1}{\omega Q^2}T_2$. It is also conventional to use the form factors $\mathcal{F}_{1,2}$ in Eq. (32.26) for the factorization analysis in which $\mathcal{F}_{1,2} = \frac{1}{2\pi}\text{Im}(T_{1,2})$. Thus, we expect $\text{Im}T_2 = 2\pi \sum_i Q_i^2 x f_i(x)$ at leading order, which will allow us to match to the parton model, once T_2 is calculated.

Writing the hadronic tensor in terms of a time-ordered product solves the first problem since it lets us use Feynman rules to calculate the operator product. But it does not change the fact that $Q^2 \to \infty$ does not imply $x^\mu \to 0$, and thus we cannot justify a small x^μ expansion. Fortunately, although we cannot justify a small x^μ expansion in general, we will be able to justify it in certain cases. In particular, we will be able to justify it when we integrate over ω.

To see how an integration over ω works, recall first, from Section 24.1.2, that the imaginary part of $T^{\mu\nu}$ can only come from on-shell intermediate states $|X\rangle$ (these are, of course, the same physical states contributing to $W^{\mu\nu}$). Since ω is real and greater than 1 in the physical region, it is helpful to analytically continue to the complex ω plane at fixed Q^2. At fixed $Q^2 > 0$, $T^{\mu\nu}$ is an analytic function of ω except for when $(P \pm q)^2 = Q^2(1 \pm \omega)$ is the mass of a physical on-shell state $|X\rangle$.[2] Therefore $T^{\mu\nu}(\omega, Q)$ has branch cuts on the real ω axis, with $\omega > 1$ (the physical region) or $\omega < -1$ (an unphysical region[3]). Then we can use that the imaginary part of a function with a cut is given by the discontinuity across the cut:

$$W_{\mu\nu}(\omega, Q) = 2\text{Im}T_{\mu\nu}(\omega, Q) = -iT_{\mu\nu}(\omega + i\varepsilon, Q) + iT_{\mu\nu}(\omega - i\varepsilon, Q) = \text{Disc}(-iT_{\mu\nu}),$$
$$(32.83)$$

with Disc standing for discontinuity (see Section 24.1.2). You should check this equation yourself (Problem 32.8). $W_{\mu\nu}$ is sometimes called the **absorptive part** of $T_{\mu\nu}$.

Now, suppose we integrate over $1 \le \omega < \infty$, which corresponds to integrating Bjorken x from 0 to 1. Such an integral according to Eq. (32.83) can be performed in the complex plane above and below the cut. Since $T_{\mu\nu}$ is analytic away from the real axis, we can deform this contour to be around $\omega = 0$, as shown in Figure 32.4. Thus, we only need to know $T_{\mu\nu}(\omega, Q)$ near $\omega = 0$ and we can justify Taylor expanding at small ω. In other words, we can justify using the OPE of $J^\mu(x)J^\nu(0)$ as $x^\mu \to 0$ to derive results about $W^{\mu\nu}$ as long as we integrate over all ω.

32.4.3 Operator product expansion for DIS

Now let us apply the OPE to DIS. We want to write

$$T\{J^\mu(x)J^\nu(y)\} = \sum_n C_n(x-y)\mathcal{O}_n^{\mu\nu}(x). \qquad (32.84)$$

What we will do is first calculate the OPE for quark external states. Then, since the OPE applies at the level of operators, independent of external states, we will apply the OPE in proton external states to get a definition of the PDFs.

---

[2] While it is true that $T^{\mu\nu}$ is analytic away from the real axis, it is not easy to show. The proof uses that $T^{\mu\nu}$ is a two-point function in an essential way [Sterman, 1993]. One difficulty in proving factorization for processes where the final state is not inclusive over all hadrons is that the analytic structure of general scattering amplitudes can be incredibly complicated.

[3] The on-shell states for $-\infty < \omega < 1$ cut are not physical for DIS. Since $\omega \to -\omega$ corresponds to $P \to -P$, this cut corresponds to deep inelastic scattering of electrons off antiprotons.

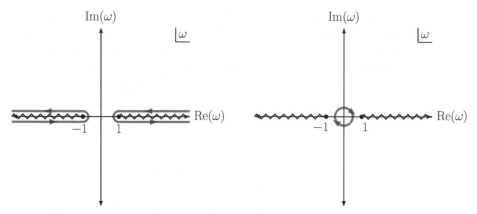

Fig. 32.4 The hadronic tensor $W^{\mu\nu}$ is determined by regions of the forward Compton tensor $T^{\mu\nu}$ along the contours on the left. Integrating over all ω lets us deform the contour and justifies using an operator product expansion derived around $\omega = 0$.

The current–current matrix element in a quark state is the same as the forward scattering matrix element for Compton scattering $\gamma^\star q \to \gamma^\star q$ with the photon off-shell and photon polarizations removed. At leading order in perturbation theory, the result is then

$$i \int d^4x\, e^{iqx} \langle p|T\{J^\mu(x)J^\nu(0)\}|p\rangle$$
$$= -\bar{u}(p) \frac{\gamma^\mu(\not{p}+\not{q})\gamma^\nu}{(p+q)^2 + i\varepsilon} u(p) - \bar{u}(p) \frac{\gamma^\nu(\not{p}-\not{q})\gamma^\mu}{(p-q)^2 + i\varepsilon} u(p). \quad (32.85)$$

Note that this is a forward scattering amplitude, so the quark has the same momentum p^μ in both the initial and the final state.

Let us first concentrate on the $p + q$ term. To calculate the OPE coefficients, at leading order, we expand the denominator in Eq. (32.85) for $Q^2 \gg p^2$. (This is the equivalent of expanding $\frac{1}{p^2 - m_W^2}$ for $m_W^2 \gg p^2$ to generate the 4-Fermi theory.) The expansion of the denominator gives

$$\frac{1}{(p+q)^2} = \frac{1}{-Q^2 + 2q \cdot p + p^2} = -\frac{1}{Q^2} \sum_{n=0}^\infty \left(\frac{2p \cdot q + p^2}{Q^2} \right)^n. \quad (32.86)$$

So,

$$i \int d^4x\, e^{iqx} \langle p|T\{J^\mu(x)J^\nu(0)\}|p\rangle = \frac{1}{Q^2}\bar{u}(p)\gamma^\mu(\not{p}+\not{q})\gamma^\nu u(p) \sum_{n=0}^\infty \left(\frac{2p \cdot q + p^2}{Q^2} \right)^n + \cdots, \quad (32.87)$$

with the $\cdots$ representing the second term in Eq. (32.85).

Whenever we have such a momentum-space expansion, we can read off the Wilson coefficients and operators in the OPE. For the OPE to make sense, all factors of p^μ should come from factors of $i\partial^\mu$ in the operators evaluated on external states (which depend on p^μ). On the other hand, all dependence on the short-distance scale q^μ (and $Q^2 = -q^2$)

must be in the Wilson coefficients. For example, a term $\left(\frac{p^2}{Q^2}\right)^n$ in such an expansion would come from an operator $\mathcal{O}_n = \bar{\psi}\Box^n\psi$ with Wilson coefficient $C_n = \frac{1}{Q^{2n}}$. A term $\left(\frac{2p\cdot q}{Q^2}\right)^3\frac{p^2}{Q^2}$ would come from $\mathcal{O}_n = \bar{\psi}\partial^{\mu_1}\partial^{\mu_2}\partial^{\mu_3}\Box\psi$ with a Wilson coefficient $C_n = \frac{8}{Q^8}q^{\mu_1}q^{\mu_2}q^{\mu_3}$, and so on. For Eq. (32.87), the Wilson operators are messy, and so we will simplify before reading off the OPE.

So far, we have not made any approximations. We want to evaluate the OPE in proton external states, using the operators and Wilson coefficients calculated in quark external states. In the DIS limit, $Q \to \infty$ at fixed ω, we can drop terms in the operators that will give contributions proportional to powers of $\frac{\Lambda_{\text{QCD}}}{Q}$. In the proton, p^μ is replaced by some component of the proton momentum $p^\mu = \xi P^\mu$, so that $p^2 = \xi^2 P^2 = \xi^2 m_p^2 \lesssim \Lambda_{\text{QCD}}^2$. We do not need to know exactly what p^μ is, but we do need to know that it has no access to Q. Terms such as $\Box/Q^2$ in operators give factors of p^2/Q^2 that are small. Thus, we can take the $p^2/Q^2 \to 0$ limit in Eq. (32.87) to extract simplified operators. On the other hand, terms such as ∂^μ/Q in operators then give factors of $q \cdot p/Q^2 \sim \omega$ that are not small (we will be integrating over ω). Thus we only need to keep terms with ∂^μ.

We can also simplify the Dirac structure in Eq. (32.87). Since the final result must be symmetric in $\mu \leftrightarrow \nu$, we can symmetrize and use the relation

$$\gamma^\mu(\slashed{p}+\slashed{q})\gamma^\nu + \gamma^\nu(\slashed{p}+\slashed{q})\gamma^\mu = 2\gamma^\mu(p^\nu + q^\nu) + 2\gamma^\nu(p^\mu + q^\mu) - 2g^{\mu\nu}(\slashed{p}+\slashed{q}). \tag{32.88}$$

The $\slashed{p}$ term acting on quark states gives $m_q = 0$. Acting on proton states using $\slashed{p} = \xi\slashed{P}$ it gives $\xi m_p \ll Q$, so it can be dropped there as well.

The second term in Eq. (32.85) gives the same OPE as the first with $q \to -q$. Therefore, we can drop terms odd in q and double the ones even in q. Thus we can write

$$i\int d^4x\, e^{iqx}\langle p|T\{J^\mu(x)J^\nu(0)\}|p\rangle = \frac{2}{Q^2}\bar{u}(p)(p^\mu\gamma^\nu + \gamma^\mu p^\nu)\sum_{n=0,2,\cdots}^{\infty}\left(\frac{2q\cdot p}{Q^2}\right)^n u(p)$$

$$+ \frac{2}{Q^2}\bar{u}(p)\left(\gamma^\mu q^\nu + q^\mu\gamma^\nu - g^{\mu\nu}\slashed{q}\right)\sum_{n=1,3,\cdots}^{\infty}\left(\frac{2q\cdot p}{Q^2}\right)^n u(p) \tag{32.89}$$

up to terms that give $\frac{\Lambda_{\text{QCD}}}{Q}$-suppressed contributions in proton external states. Note that all of the terms in the series have one γ-matrix in them. All p^μ terms should come from derivatives in operators in the OPE (through the replacement $p_\mu \to i\partial_\mu$) and all q^μ terms should be in the Wilson coefficients. For example, the first line in Eq. (32.89) is reproduced by

$$\mathcal{O}_{n+2}^{\mu\nu\mu_1\cdots\mu_n} = \bar{\psi}_q(x)(i\gamma^\mu\partial^\nu + i\gamma^\nu i\partial^\mu)\, i\partial^{\mu_1}\cdots i\partial^{\mu_n}\psi_q(x) \tag{32.90}$$

with Wilson coefficients

$$C_{n+2}^{\mu_1\cdots\mu_n}(q) = \frac{2}{Q^2}\frac{2^n}{Q^{2n}}q^{\mu_1}\cdots q^{\mu_n}. \tag{32.91}$$

The second line in Eq. (32.91) decomposes similarly.

It is standard to work in a basis of gauge-invariant operators that transform in irreducible representations of the Lorentz group. An operator of spin s will be a symmetric, traceless tensor of rank s. For example, for spin 2,

$$\hat{\mathcal{O}}_{2,0}^{\mu\nu} = \bar{\psi}_q \left(i\gamma^\mu \partial^\nu + i\gamma^\nu \partial^\mu - \frac{1}{2} i g^{\mu\nu} \partial\!\!\!/ \right) \psi_q. \tag{32.92}$$

This has $g_{\mu\nu}\hat{\mathcal{O}}_{2,0}^{\mu\nu} = 0$. It differs from the $\hat{\mathcal{O}}_2^{\mu\nu}$ in Eq. (32.90) by a scalar operator $\bar{\psi}\partial\!\!\!/\psi$. That is,

$$\mathcal{O}_2^{\mu\nu} = \hat{\mathcal{O}}_{2,0}^{\mu\nu} + \frac{1}{2} i g^{\mu\nu} \bar{\psi}_q \, \partial\!\!\!/ \psi_q. \tag{32.93}$$

The basis of spin-s operators is

$$\hat{\mathcal{O}}_{s,r}^{\mu_1\cdots\mu_s} = \bar{\psi}\gamma^{\mu_1} i\partial^{\mu_2} \cdots i\partial^{\mu_s}(-\Box)^r \psi \; + \; \text{symmetrizations of } \mu_i \; - \; \text{traces.} \tag{32.94}$$

These operators have mass dimension $d = 2 + s + 2r$. Knowing the dimension and the spin fixes $t \equiv 2 + 2r = d - s$. This quantity t is known as the **twist** of an operator. That is, twist = dimension $-$ spin. Since the operators with extra $\Box$ factors are suppressed, *the OPE will be dominated by operators with the lowest twist*. These are operators such as $\hat{\mathcal{O}}_{2,0}^{\mu\nu}$ in Eq. (32.92), which is dimension 4 and spin 2 and hence has twist 2. In general, promoting the derivatives to covariant derivatives and adding a label for the quark flavor, we define

$$\hat{\mathcal{O}}_q^{\mu_1\cdots\mu_n}(x) = \bar{\psi}_q(x)\gamma^{\mu_1} iD^{\mu_2} \cdots iD^{\mu_n} \psi_q(x) + \text{symmetrizations of } \mu_i - \text{traces.} \tag{32.95}$$

This is the canonical basis of gauge-invariant twist-2 quark operators.

There are no gauge-invariant operators in QCD with twist less than 2. To see that, first note that gauge-invariant operators must have at least two quark fields or two gluon field strengths $F_{\mu\nu}$. Adding more fields adds to the dimension and hence to the twist. Without derivatives, two quarks have dimension 3 and can only have spin 0 or 1; hence quark operators have at least twist 2. Derivatives add 1 to the dimension and at most 1 to the spin and hence cannot lower twist below 2. For gluons, $F_{\mu\nu}^2$ has dimension 4 and spin at most 2. Thus, gluonic operators also have $t \geq 2$. Explicitly, the twist-2 gluon operators are

$$\hat{\mathcal{O}}_g^{\mu_1\cdots\mu_n} = F^{\mu_1\nu} iD^{\mu_2} \cdots iD^{\mu_{n-1}} F^{\mu_n\nu} + \text{symmetrizations of } \mu_i - \text{traces.} \tag{32.96}$$

The Wilson coefficients for these operators are zero at leading order.

All gauge-invariant operators in QCD with twist higher than 2 are generically called *higher twist*. It is common to think of twist-2 operators as synonymous with the large Q^2, fixed x limit, and higher twist operators as providing power corrections.

In summary, after a bit of algebra and restoring quark charges, the OPE can be written in terms of twist-2 operators as (see [Manohar, 2003, Section 1.8])

$$i \int d^4x \, e^{iq\cdot x} T\{J_\mu(x)J_\nu(0)\}$$

$$= \sum_q Q_q^2 \left\{ \sum_{n=2,4,\cdots}^\infty \frac{(2q^{\mu_1})\cdots(2q^{\mu_n})}{Q^{2n}} \left(-g^{\mu\nu} + \frac{q^\mu q^\nu}{q^2} \right) \hat{\mathcal{O}}_q^{\mu_1\cdots\mu_n} \right.$$

$$\left. + 4 \sum_{n=2,4,\cdots}^\infty \frac{(2q^{\mu_3})\cdots(2q^{\mu_n})}{Q^{2n-2}} \left(g^{\mu\mu_1} - \frac{q^\mu q^{\mu_1}}{q^2} \right) \left(g^{\nu\mu_2} - \frac{q^\nu q^{\mu_2}}{q^2} \right) \hat{\mathcal{O}}_q^{\mu_1\cdots\mu_n} \right\}. \tag{32.97}$$

This OPE is valid to leading power in $\frac{\Lambda_{\text{QCD}}^2}{Q^2}$ and at leading order in α_s.

To use the OPE to calculate the time-ordered hadronic tensor $T^{\mu\nu}$ for DIS, we need to take the matrix element of this OPE in a proton state. By Lorentz invariance, all we can get after summing over proton spins is

$$\sum_{\text{spins}} \langle P | \hat{\mathcal{O}}_q^{\mu_1 \cdots \mu_n} | P \rangle = \mathcal{A}_q^n P^{\mu_1} \cdots P^{\mu_n} - \text{traces}, \qquad (32.98)$$

with $\mathcal{A}_q^n$ functions of Q. This expression is automatically symmetric. The traces give factors of $P^2 = m_p^2 \ll Q^2$ which are subleading compared to contractions of P^{μ_i} with q^μ from the Wilson coefficients, which give factors of $q \cdot P = \frac{1}{2}\omega Q^2$. Thus, we can drop the traces at leading power and we find

$$T^{\mu\nu} = \sum_q Q_q^2 \Bigg\{ \left(-g^{\mu\nu} + \frac{q^\mu q^\nu}{q^2} \right) \sum_{n=2,4,\cdots}^{\infty} \omega^n \mathcal{A}_q^n$$
$$+ \frac{4}{Q^2 \omega^2} \left(P^\mu - \frac{P \cdot q}{q^2} q^\mu \right) \left(P^\nu - \frac{P \cdot q}{q^2} q^\nu \right) \sum_{n=2,4,\cdots}^{\infty} \omega^n \mathcal{A}_q^n \Bigg\}. \qquad (32.99)$$

Comparing to Eq. (32.82) we conclude that

$$T_1 = \frac{\omega}{2} T_2 = \sum_q Q_q^2 \left[\sum_{n=2,4,\cdots} \omega^n \mathcal{A}_q^n \right]. \qquad (32.100)$$

In particular, since $\mathcal{F}_1 = \frac{1}{2\pi} \text{Im} T_1$ and $\mathcal{F}_2 = \frac{1}{2\pi} \text{Im} T_2$ we reproduce the Callan–Gross relation $\mathcal{F}_1 = \frac{\omega}{2} \mathcal{F}_2$. More importantly, since $\mathcal{F}_1 = \frac{1}{2} \sum_q Q_q^2 f_1(x)$ in the parton model, we find

$$f_q(x) = \frac{1}{\pi} \sum_{n=2,4,\cdots} x^{-n} \text{Im} \mathcal{A}_q^n. \qquad (32.101)$$

This gives an operator definition of the PDFs in QCD.

One consequence of this way of defining the PDFs is that it lets us calculate the PDF evolution from the RG evolution of the twist-2 operators. Beyond leading order, amplitudes $\mathcal{A}^n$ are divergent and thus the operators $\mathcal{O}_n$ must be renormalized. The RG evolution of the operators is compensated for by RG evolution of the Wilson coefficients, as discussed in Chapter 23. It is a straightforward exercise to work out the anomalous dimensions for the quark and gluon twist-2 operators. As in the example in Section 31.3, there will be operator mixing. The result of the calculation is that the μ dependence of the PDFs defined through operator matrix elements exactly agrees with the Altarelli–Parisi evolution, as derived in the parton model. The details of the calculation are clearly explained in [Peskin and Schroeder, 1995, Chapter 18].

32.4.4 Moments of the PDFs

With an operator definition of the PDFs, we can now check that the PDFs satisfy the sum rules from Section 32.1.4 that we deduced with physical arguments. For example, $\int dx\, f_q(x)$ should give the total number of valence quarks of a particular species.

The sum rules are generally of the form of integrals of x^m times the PDFs:

$$C_q^m = \int_0^1 dx\, x^{m-1} f_q(x). \tag{32.102}$$

This is known as a **Mellin moment**. Plugging in Eq. (32.101) we find

$$C_q^m = \operatorname{Im} \frac{1}{\pi} \int_1^\infty d\omega \sum_n \omega^{n-m-1} \mathcal{A}_q^n. \tag{32.103}$$

Writing the imaginary part as a discontinuity and deforming the contour to a small circle around the origin, as in Figure 32.4, we have

$$C_q^m = \sum_n \frac{1}{2\pi i} \int \frac{d\omega}{\omega^{m-1}} \omega^{n-2} \mathcal{A}_q^n = \mathcal{A}_q^m. \tag{32.104}$$

Thus, the $\mathcal{A}^n$ are precisely the Mellin moments of the PDFs. It is these moments that are rigorously defined by the OPE in DIS.

Two important special cases are $m = 2$ and $m = 1$. For $m = 2$, the relevant twist-2 operator is

$$\hat{\mathcal{O}}_q^{\mu\nu} = \bar{\psi}_q (\gamma^\mu D^\nu + \gamma^\nu D^\mu) \psi_q. \tag{32.105}$$

This is a symmetrized version of the canonical energy-momentum tensor for a quark, which we derived in Eq. (12.62). The full energy-momentum tensor in QCD is a sum over the quark (and gluon) energy-momentum tensors. Thus, we can evaluate this sum in a proton state to get twice the proton's energy-momentum:

$$\sum_j \langle P | \hat{\mathcal{O}}_j^{\mu\nu} | P \rangle = P^\mu P^\nu, \tag{32.106}$$

where the sum is over all partons (not just quarks). Using Eq. (32.102) with $m = 2$ and Eq. (32.98) we then find

$$\sum_j \int_0^1 dx\, x f_j(x) = 1. \tag{32.107}$$

The operator analysis therefore gives a justification to the interpretation that $\int dx\, x f_j(x) = \langle x \rangle_j$ is the average fraction of momentum carried by species j. Moreover, since the energy-momentum tensor is conserved, this sum rule is independent of μ. Intriguingly, $\langle x \rangle_{\mathrm{up}} \approx 0.3$ and $\langle x \rangle_{\mathrm{down}} \approx 0.1$ and therefore 60% of the proton momentum is carried by gluons and sea quarks. You can explore the $m = 1$ case in Problem 32.9.

32.4.5 Summary

In this section we have given a field theory definition of the parton distribution functions. We wrote the hadronic tensor $W^{\mu\nu}$ for deep inelastic scattering in terms of expectation values of twist-2 operators: $\mathcal{A}^n \sim \langle P|\mathcal{O}^n|P\rangle$. Matching to the parton-model picture, these $\mathcal{A}_n$ can be identified with Mellin moments of the PDFs. This approach allows us to prove certain features of PDFs, such as their sum rules, that can only be justified semi-classically using the parton model. Although the $\mathcal{A}_n$ are non-perturbative, their scale dependence can be calculated in perturbation theory. Thus, one can predict logarithmic Q^2 dependence of the DIS structure functions and calculable corrections to Bjorken scaling. The scaling violation is the same as we found with the DGLAP equations, but with this method, we did not have to assume the parton model.

Although we have defined the PDFs for DIS non-perturbatively in terms of $W^{\mu\nu}$, this definition is not tremendously useful for processes other than DIS. What we would like to do is show that any process involving high-energy scattering of protons can be written as $\sigma = H \otimes f$, with H a calculable hard function and f the same universal PDFs. For the DIS case, we simply defined f in terms of the non-perturbative hadronic form factor $W^{\mu\nu}$. This is called the **DIS PDF scheme**. In this scheme, H is defined to be 1 to all orders, and the only prediction one can make is the scale dependence of the PDFs (or differences between form factors). In global PDF fits, this scheme is not used; instead the $\overline{\text{MS}}$ scheme is used, where only the $\frac{1}{\varepsilon}$ poles are absorbed into the PDFs. Changing schemes of course does not make our calculations any more predictive for DIS.

32.5 Lightcone coordinates

The proof of factorization above using the OPE relied on being able to perform a Taylor expansion at large Q^2 in which we could drop subleading terms. There is another way to set up the DIS calculation so that subleading terms can be dropped, which leads to an alternative way to think about PDFs: as lightcone projections of proton matrix elements. This approach, while somewhat less rigorous than the OPE, is more friendly to more general factorization arguments.

In the parton model, the PDFs $f(\xi)$ are interpreted as the probability to find a parton inside the proton with momentum $p^\mu = \xi P^\mu$ (we use ξ instead of x to avoid confusion with x^μ). We know what probabilities are in quantum mechanics (or quantum field theory): they are matrix elements squared. Thus, we should be able to write

$$f(\xi) = \sum_X |\langle X|\psi|P\rangle|^2 \, \delta(\xi P^\mu - p^\mu) = \langle P|\psi^\dagger\psi|P\rangle\delta(\xi P^\mu - p^\mu), \qquad (32.108)$$

with p^μ the quark momentum as before. This is almost right, since $\psi^\dagger\psi = \bar{\psi}\gamma^0\psi$ is the quark-number density (the zero component of the quark-number current $J^\mu = \bar{\psi}\gamma^\mu\psi$). However, it is not quite right, since the parton's momentum does not have to be exactly proportional to the proton momentum. The momenta only have to be proportional up to some small transverse fluctuations. That is, we expect the component of p^μ in the proton's direction to be ξP^μ, and the other components are small, $\sim \frac{m_P}{Q}$. Since the proton's direction has no meaning in the proton rest frame, it is natural instead to work in a frame where the proton is very energetic. At hadron–hadron colliders, such as the LHC, the center-of-mass frame works fine, but for fixed target experiments (such as typical $e^- p^+$ experiments), it is not. So we will first change frames for DIS, then return to Eq. (32.108).

A useful frame for DIS is one where the off-shell photon $\gamma^\star$ has no energy. We can always go to such a frame since the photon momentum is spacelike, $q^2 = -Q^2 < 0$. This frame is called the **Breit frame**. In the Breit frame, the photon and proton momenta are

$$q_\mu = (0, 0, 0, Q), \qquad P^\mu = (E_P, 0, 0, -P_z). \tag{32.109}$$

where $E_P = \sqrt{m_P^2 + P_z^2}$. Since $q_\mu = k_\mu - k'_\mu$, the incoming and outgoing electron momenta must be

$$k = \left(E_k, k_x, k_y, \frac{Q}{2} \right), \qquad k' = \left(E_k, k_x, k_y, -\frac{Q}{2} \right), \tag{32.110}$$

where $E_k = \sqrt{k_x^2 + k_y^2 + \frac{Q^2}{4}}$. So the electron bounces right off the proton. For this reason, the Breit frame is sometimes referred to as the **brick wall frame**.

Now consider some parton in the proton with momentum p^μ. Its momentum should be collinear with the proton's momentum, $p^\mu = \xi P^\mu$, up to some small transverse component p_T. When we boost from the proton rest frame to the Breit frame, p_T does not change, thus we expect $p_T \sim m_p \ll Q$. A clean way to think about which momentum components are small at large Q is using **lightcone coordinates**. Let n^μ be any lightlike 4-vector, that is, $n^2 = 0$, and normalized such that $n^\mu = (1, \vec{n})$. For DIS, we can take $n^\mu = (1, 0, 0, 1)$, which is backwards to the proton direction. Define the backwards direction to n^μ as $\bar{n}^\mu = (1, -\vec{n})$, so that $n \cdot \bar{n} = 2$. For DIS in the Breit frame, $\bar{n}^\mu = (1, 0, 0, -1)$ is the proton's direction. In general, any momentum k^μ can be decomposed as

$$k^\mu \equiv \frac{1}{2} (\bar{n} \cdot k) \, n^\mu + \frac{1}{2} (n \cdot k) \, \bar{n}^\mu + k_T^\mu \tag{32.111}$$

with $k_T \cdot n = k_T \cdot \bar{n} = 0$. k_T^μ is the part of k^μ in the transverse (x and y) directions. This can be checked by contracting with n^μ or $\bar{n}^\mu$. We also find

$$k^2 = (n \cdot k)(\bar{n} \cdot k) + k_T^2. \tag{32.112}$$

With this notation, we can interpret the momentum fraction ξ of the parton inside the proton to be the component of the momentum in the n direction. That is, $n \cdot p = \xi \, (n \cdot P)$. The $\bar{n} \cdot p$ and p_T components of the parton momentum are much smaller. That is, $\not{n}\psi \approx 0$.

Now that we are in a frame where the proton is very energetic, we can make Eq. (32.108) precise. We write

$$f(\xi) = \sum_X \int d\Pi_X |\langle X | \psi | P \rangle|^2 \delta(\xi n \cdot P - n \cdot p). \tag{32.113}$$

This is the probability of finding a quark within a proton with a given momentum fraction. To be clear, in this equation, there is no scattering. Rather, it describes how the proton momentum splits up into $P^\mu = p^\mu + p_X^\mu$, where p^μ is the momentum of the parton and p_X^μ

is the momentum of everything else in the proton. Inserting a factor of $\int d^4p\, \delta^4(P-p-p_X)$ we find

$$
\begin{aligned}
f(\xi) &= \int \frac{d^4 p_i}{(2\pi)^4} \sum_X \int d\Pi_X (2\pi)^4 \delta^4(P^\mu - p^\mu - p_X^\mu)\delta(\xi n \cdot P - n \cdot p)|\langle X|\psi(0)|P\rangle|^2 \\
&= \int_{-\infty}^{\infty} \frac{dt}{2\pi} \sum_X \int d\Pi_X e^{-itn\cdot(\xi P - P + p_X)}|\langle X|\psi(0)|P\rangle|^2 \\
&= \int_{-\infty}^{\infty} \frac{dt}{2\pi} \sum_X \int d\Pi_X e^{-i\xi tn\cdot P}\langle P|e^{i(n\cdot\hat{P})t}\psi^\dagger(0)e^{-i(n\cdot\hat{P})t}|X\rangle\langle X|\psi(0)|P\rangle \\
&= \int_{-\infty}^{\infty} \frac{dt}{2\pi} e^{-it\xi(n\cdot P)}\langle P|\bar{\psi}(tn^\mu)\gamma^0\psi(0)|P\rangle.
\end{aligned}
\tag{32.114}
$$

We can simplify this further by noting that since the quark is going mostly in the $\bar{n}^\mu$ direction $\not{n}\psi \approx 0$. This implies $\gamma^0 \psi = -(\vec{n}\cdot\vec{\gamma})\,\psi$ and so $2\gamma^0\psi = \not{n}\psi$. Then,

$$
f(\xi) = \int_{-\infty}^{\infty} \frac{dt}{4\pi} e^{-it\xi(n\cdot P)}\langle P|\bar{\psi}(tn^\mu)\not{n}\psi(0)|P\rangle.
\tag{32.115}
$$

To make this gauge invariant, we can insert a Wilson line (see Section 25.2) stretching between the points $x = 0$ and $x^\mu = tn^\mu$ where the quark fields are evaluated:

$$
W_n = P\exp\left\{ ig_s n_\nu \int_0^t ds\, A^\nu(sn^\mu) \right\}.
\tag{32.116}
$$

Thus, we arrive at

$$
f_q(\xi) = \int_{-\infty}^{\infty} \frac{dt}{2\pi} e^{\, it\xi(n\cdot P)}\langle P|\bar{\psi}_q(tn^\mu)\frac{\not{n}}{2}W_n\psi_q(0)|P\rangle.
\tag{32.117}
$$

To be clear, $n^\mu = (1, \vec{n})$ is a lightlike 4-vector pointing *opposite* to the direction of the proton's momentum. You can check in Problem 32.11 that moments of the PDFs defined this way reproduce matrix elements of twist-2 operators from Section 32.4.4.

The advantage of an expression such as Eq. (32.117) is that it appears generically in high-energy processes in a frame where the proton is ultra-relativistic, such as the lab frame in hadron–hadron collisions. Similar analyses can therefore be done for other processes, such as Drell–Yan, direct photon production ($pp \rightarrow \gamma + X$), dijet production, etc. Each of these has a scale Q (the invariant mass of the lepton pair in Drell–Yan, the transverse momentum of the photon, or the invariant mass of the dijet system). When $Q \gg m_p$, by considering how the relevant momenta scale with Q (such as P^μ, p^μ and q^μ for DIS) one can often write down factorization formulas for cross sections using lightcone PDFs. If one is content with scaling arguments as a proof of these factorization formulas, then it is possible to have a tremendous amount of predictive power without an OPE.

Problems

32.1 Derive an expression for the mean charge radius $\langle r^2 \rangle = \int d^3x \, r^2 \rho(x)$ in terms of a form factor $F(q^2)$ by expanding $F(q^2) = \int d^3x \, e^{i\vec{q}\cdot\vec{x}} V(x)$ around $x = 0$. What is the mean charge radius of the proton from Eq. (32.9)?

32.2 Show that the PDFs, as classical probabilities, should satisfy $\sum_j \int dx \, x f_j(x) = 1$, as in Eq. (32.29). [Hint: consider the average momentum for each parton.]

32.3 Derive the expansion in Eq. (32.38). One way to do this is to write

$$\int_0^1 dx \, x^{-1+\varepsilon} f(x) = \int_0^1 dx \, x^{-1+\varepsilon} f(0) + \int_0^1 dx \, x^{-1+\varepsilon} [f(x) - f(0)] \quad (32.118)$$

and to evaluate the first term and Taylor expand the second term.

32.4 Evaluate the relationship between W_1 and W_2 that would result instead of the Callan–Gross relation if quarks were scalars. How could you test this prediction?

32.5 Calculate the $g \to gg$ splitting function by taking the collinear limit of $gg \to gg$ scattering. You can use the cross section calculated in Chapter 27.

32.6 Find the limits of integration on z for $t = p_T^2$ in the process $\gamma^* \to q\bar{q}g$ discussed in Section 32.3. Then calculate $P(t)$ and the Sudakov factor $\Delta(Q, t)$ explicitly. Repeat the exercise for $t = m^2$ and $t = \theta$. Which part of the Sudakov factor is universal?

32.7 In this problem, you will show that $Q \to \infty$ at fixed $\omega = \frac{2P\cdot q}{Q^2}$ or equivalently fixed $\chi \equiv \frac{2m_p}{\omega}$ implies that $J(x^\mu)J(0)$ is dominated by the lightcone, where $x_\mu^2 \to 0$.

(a) In the proton rest frame, show that

$$q^\mu x_\mu = \frac{\omega Q^2}{2m_p}(x^0 - r) - \frac{m_p}{\omega} r + \mathcal{O}\left(\frac{1}{Q^2}\right), \quad (32.119)$$

where $r \equiv \frac{\vec{q}\cdot\vec{x}}{|\vec{q}|}$.

(b) Use the method of stationary phase to show that at fixed ω, $W^{\mu\nu}$, in the form of Eq. (32.78), is dominated by $|x^0 - r| \le \frac{c_1}{Q^2}$ and $r \le c_1$ for two constants c_1 and c_2 as $Q \to \infty$.

(c) Show that $x^2 \le \frac{\text{const}}{Q^2}$ and therefore that $J(x^\mu)J(0)$ is dominated by lightlike separations in the DIS limit.

32.8 *Relating imaginary parts to discontinuities.* The goal of this problem is to verify Eq. (32.83).

(a) By expanding the time ordering in terms of $\theta(t)$ and $\theta(-t)$ show that $T^{\mu\nu}$ as in Eq. (32.81) can be written as

$$T_{\mu\nu}(\omega, Q) = \sum_X \frac{(2\pi)^3 \delta^3(\vec{p}_X - \vec{q} - \vec{P})}{p_X^0 - p^0 - q^0 - i\varepsilon} \langle p^+|J_\mu(0)|X\rangle\langle X|J_\nu(0)|p^+\rangle$$

$$+ \sum_X \frac{(2\pi)^3 \delta^3(\vec{p}_X + \vec{q} - \vec{P})}{p_X^0 - p^0 + q^0 - i\varepsilon} \langle p^+|J_\mu(0)|X\rangle\langle X|J_\nu(0)|p^+\rangle.$$

$$(32.120)$$

You may want to use $\theta(t) = \frac{1}{2\pi i}\int_{-\infty}^{\infty} \frac{ds}{s - i\varepsilon} e^{ist}$.

(b) Use part (a) to show that one of the terms above does not contribute to the discontinuity in the physical region and that $W_{\mu\nu} = -i \operatorname{Disc} T_{\mu\nu}$.

32.9 Show that current conservation implies a sum rule for each flavor in QCD using spin-1 operators in the OPE, as we did for spin 2 in Section 32.4.4.

32.10 Show that $p_T^2 = \frac{\hat{s}\hat{t}\hat{u}}{(\hat{s}+Q^2)^2}$ and verify Eq. (32.58).

32.11 Relate the lightcone PDF definition from Eq. (32.117) to the Mellin moments from Section 32.4.4.

(a) Compute the $m = 1$ moment of the lightcone PDF definition to show that you get the matrix element of the spin-1 operator $\hat{\mathcal{O}}_q^\mu = \bar{\psi}\gamma^\mu\psi$. Be careful with the limits of integration.

(b) Show that you can reproduce the matrix elements of the twist-2 spin-m operators by taking moments.

(c) Can you construct the lightcone PDF definition from the Mellin moments?

PART V

ADVANCED TOPICS

We have mentioned effective actions a few times already. For example, the effective action for the 4-Fermi theory is derived from the Standard Model by *integrating out* the W and Z bosons. It is an *effective* action since it is valid only in some regime, in this case for energies less than m_W. More generally, an effective action is one that gives the same results as a given action but has different degrees of freedom. For the 4-Fermi theory, the effective action does not have the W and Z bosons. In this chapter we will develop powerful tools to calculate effective actions more generally. We will discuss three ways to calculate effective actions: through matching (or the operator product expansion), through field-dependent expectation values using Schwinger proper time, and with functional determinants coming from Feynman path integrals.

The first step is to define what we mean by an effective action. The term **effective action**, denoted by Γ, generally refers to a functional of fields (like any action) defined to give the same Green's functions and S-matrix elements as a given action S, which is often called the action for the **full theory**. We write $\Gamma = \int d^4x\, \mathcal{L}_{\text{eff}}(x)$, where $\mathcal{L}_{\text{eff}}$ is called the **effective Lagrangian**. Differences between Γ and S include that Γ often has fewer fields, is non-renormalizable, and only has a limited range of validity. When a field is in the full theory but not in the effective action, we say it has been **integrated out**.

The advantage of using effective actions over full theory actions is that by focusing only on the relevant degrees of freedom for a given problem calculations are often easier. For example, in Section 31.3 we saw that in the 4-Fermi theory large logarithmic corrections to $b \to c\bar{d}u$ decays of the form $\alpha_s^n \ln^n \frac{m_W}{m_b}$ could be summed to all orders in perturbation theory. The analogous calculation in the full Standard Model would have been a nightmare.

The effective action we will focus on for the majority of this chapter is the one arising from integrating out a fermion of mass m in QED. We can define this effective action $\Gamma[A_\mu]$ by

$$\int \mathcal{D}A \exp(i\Gamma[A_\mu]) \equiv \int \mathcal{D}A\, \mathcal{D}\bar{\psi}\, \mathcal{D}\psi \, \exp\left[i \int d^4x \left(-\frac{1}{4}F_{\mu\nu}^2 + \bar{\psi}(i\slashed{D} - m)\psi \right) \right]. \tag{33.1}$$

When A_μ corresponds to a constant electromagnetic field, $\mathcal{L}_{\text{eff}}[A]$ is called the Euler–Heisenberg Lagrangian. The Euler–Heisenberg Lagrangian is amazing: it gives us the QED β-function, Schwinger pair creation, scalar and pseudoscalar decay rates, the chiral anomaly, and the low-energy limit for scattering n photons, including the light-by-light scattering cross section. As we will see, the Euler–Heisenberg Lagrangian can be calculated to all orders in α_e using techniques from non-relativistic quantum mechanics.

33.1 Effective actions from matching

So far, we have only discussed how effective actions can be calculated through matching. This approach requires that matrix elements of states agree in the full and effective theories. For example, in the 4-Fermi theory, we asked that

$$\langle \Omega | T \{ \bar{\psi}\psi\bar{\psi}\psi \} | \Omega \rangle_S = \langle \Omega | T \{ \bar{\psi}\psi\bar{\psi}\psi \} | \Omega \rangle_\Gamma, \tag{33.2}$$

where the subscript on the correlation function indicates the action used to calculate it. Writing the effective Lagrangian as a sum over operators $\mathcal{L}_{\text{eff}}(x) = \sum C_i \mathcal{O}_i(x)$ we were able to determine the Wilson coefficients C_i by asking that Eq. (33.2) hold order-by-order in perturbation theory. One-loop matching in the 4-Fermi theory was discussed in Section 31.3. Other examples of matching that we considered include the Chiral Lagrangian (Section 28.2.2) and deep inelastic scattering (Section 32.4).

In the 4-Fermi theory and for deep inelastic scattering, we matched by expanding propagators $\frac{1}{p^2 - m_W^2}$ or $\frac{1}{p^2 + Q^2}$ respectively (see Eqs. (32.70) and (32.71)). The reason one can expand propagators to derive an effective Lagrangian is because when a scale such as m_W or Q is taken large, the propagator can only propagate over a small distance. In terms of Feynman diagrams, we expand an exchange graph in a set of local interactions:

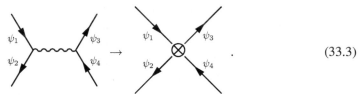

$$\tag{33.3}$$

To see how this works in position space, consider matching a Yukawa theory with a massive scalar,

$$\mathcal{L}_Y = i\bar{\psi}\slashed{\partial}\psi - \frac{1}{2}\phi(\Box + m^2)\phi + \lambda\phi\bar{\psi}\psi, \tag{33.4}$$

to an effective Lagrangian $\mathcal{L}_{\text{eff}}$ which lacks that scalar and is useful for energies much less than m. For large m, fluctuations of ϕ around its classical configuration are highly suppressed. Thus, to leading order we can assume ϕ satisfies its classical equations of motion, $\phi = \frac{\lambda}{\Box + m^2}\bar{\psi}\psi$, and that loops of ϕ are small corrections. Plugging the classical solution back into the Lagrangian gives

$$\mathcal{L}_{\text{eff}} = i\bar{\psi}\slashed{\partial}\psi + \frac{\lambda^2}{2}\bar{\psi}\psi\frac{1}{\Box + m^2}\bar{\psi}\psi. \tag{33.5}$$

In this way $\mathcal{L}_{\text{eff}}$ is guaranteed to give the same correlation functions as $\mathcal{L}_Y$ but has no ϕ field in it. As long as m is larger than typical momentum scales, we can also Taylor expand this non-local effective Lagrangian in a series of local operators:

$$\mathcal{L}_{\text{eff}} = i\bar{\psi}\slashed{\partial}\psi + \frac{\lambda^2}{2m^2}\bar{\psi}\psi\bar{\psi}\psi - \frac{\lambda^2}{2m^4}\bar{\psi}\psi\Box\bar{\psi}\psi + \cdots. \tag{33.6}$$

If ϕ were the W and Z, this would give the 4-Fermi theory supplemented by additional operators that have effects suppressed by powers of $\frac{E^2}{m_W^2}$ at low energy.

Setting ϕ to its classical equations of motion amounts to taking the steepest descent approximation in the path integral. To integrate out ϕ to all orders, we have to perform the path integral exactly. Thus, we can define the effective action as

$$\int \mathcal{D}\bar{\psi}\,\mathcal{D}\psi\,\exp\left(i\int d^4x\,\mathcal{L}_{\text{eff}}[\psi,\bar{\psi}]\right) = \int \mathcal{D}\phi\,\mathcal{D}\bar{\psi}\,\mathcal{D}\psi\,\exp\left(i\int d^4x\,\mathcal{L}_Y[\phi,\psi,\bar{\psi}]\right),$$
(33.7)

which connects back to the definition given in Eq. (33.1).

33.2 Effective actions from Schwinger proper time

The next method we discuss for computing effective actions is through Schwinger proper time. The idea here is to evaluate the propagator for the particle we want to integrate out as a functional of the other fields. Pictorially, we can write this as

$$G_A(x,y) = \quad\longrightarrow\quad + \quad\longrightarrow\quad + \quad\longrightarrow\quad + \cdots. \quad (33.8)$$

Then, when we integrate out the field, we will generate an infinite set of interactions among the other fields.

The key to Schwinger's proper-time formalism is the mathematical identity

$$\frac{i}{A+i\varepsilon} = \int_0^\infty ds\, e^{is(A+i\varepsilon)}, \tag{33.9}$$

which holds for $A \in \mathbb{R}$ and $\varepsilon > 0$ (see Appendix B). This lets us write the Feynman propagator for a scalar as

$$\begin{aligned}
D_F(x,y) &= \int \frac{d^4p}{(2\pi)^4} e^{ip(x-y)} \frac{i}{p^2 - m^2 + i\varepsilon} \\
&= \int \frac{d^4p}{(2\pi)^4} e^{ip(x-y)} \int_0^\infty ds\, e^{is(p^2 - m^2 + i\varepsilon)}.
\end{aligned} \tag{33.10}$$

The integral over d^4p is Gaussian and can be done exactly using Eq. (14.7) with $A = -2isg^{\mu\nu}$, giving

$$D_F(x,y) = \frac{-i}{16\pi^2} \int_0^\infty \frac{ds}{s^2} e^{-i\left[\frac{(x-y)^2}{4s} + sm^2 - i\varepsilon s\right]}, \tag{33.11}$$

which is an occasionally useful representation of the propagator. For $m = 0$ it provides a shortcut to the position-space Feynman propagator $D_F(x,y) = -\frac{1}{4\pi^2}\frac{1}{(x-y)^2 - i\varepsilon}$.

An alternative to performing the integral over p directly is first to introduce a one-particle Hilbert space spanned by $|x\rangle$, as in non-relativistic quantum mechanics. This lets us write $\langle p|x\rangle = e^{ipx}$. Then, from Eq. (33.10) we get

$$D_F(x,y) = \int \frac{d^4p}{(2\pi)^4} \langle y|p\rangle \int_0^\infty ds\, e^{is(p^2 - m^2 + i\varepsilon)} \langle p|x\rangle. \tag{33.12}$$

The analogy with quantum mechanics can be taken even further. Introduce momentum operators $\hat{p}^\mu$ with $\hat{p}^\mu|p\rangle = p^\mu|p\rangle$ and define $\hat{H} = -\hat{p}^2$. Then $e^{isp^2}\langle p|x\rangle = \langle p|e^{-is\hat{H}}|x\rangle$. This lets us use $(2\pi)^{-4}\int d^4p\,|p\rangle\langle p| = 1$ in Eq. (33.12) to get

$$D_F(x,y) = \int_0^\infty ds\, e^{-s\varepsilon}e^{-ism^2}\langle y|e^{-is\hat{H}}|x\rangle \equiv \int_0^\infty ds\, e^{-s\varepsilon}e^{-ism^2}\langle y;0|x;s\rangle, \tag{33.13}$$

where $|x;s\rangle \equiv e^{-is\hat{H}}|x\rangle$. In the second step, we have interpreted $\hat{H}$ as a Hamiltonian and s as a time variable known as **Schwinger proper time**.[1] Schwinger proper time gives an intuitive interpretation of a propagator:

> A propagator is the amplitude for a particle to propagate from x to y in proper time s, integrated over s.

One has to be careful interpreting $\hat{H}$ however, since it conventionally includes only the p dependence and not the m dependence (as $\hat{H} = m^2 - \hat{p}^2$ would).

We can go even further into quantum mechanics by defining the Green's function as an operator matrix element. Define the Green's function operator for a massive scalar as

$$\hat{G} \equiv \frac{i}{\hat{p}^2 - m^2 + i\varepsilon}. \tag{33.14}$$

Then the Feynman propagator is

$$D_F(x,y) = \int \frac{d^4p}{(2\pi)^4} e^{ip(x-y)} \frac{i}{p^2 - m^2 + i\varepsilon} = \int \frac{d^4p}{(2\pi)^4} \langle y|p\rangle\langle p| \frac{i}{\hat{p}^2 - m^2 + i\varepsilon}|x\rangle$$
$$= \langle y|\hat{G}|x\rangle. \tag{33.15}$$

Or we can go directly to proper time, without ever introducing the p integral, through Eq. (33.9):

$$D_F(x,y) = \langle y|\hat{G}|x\rangle = \int_0^\infty ds\, e^{-s\varepsilon}e^{-ism^2}\langle y|e^{-i\hat{H}s}|x\rangle, \tag{33.16}$$

where $\hat{H} = -\hat{p}^2$ as before.

By the way, when you have two propagators, as in a loop, the relevant identity is

$$\frac{1}{AB} = -\int_0^\infty ds \int_0^\infty dt\, e^{isA + itB} \tag{33.17}$$

(the $i\varepsilon$ factors are implicit). If we then write $s = x\tau$ and $t = (1-x)\tau$, so that s and t are the fractions x and $(1-x)$ of the total proper time τ, this becomes

$$\frac{1}{AB} = -\int_0^1 dx \int_0^\infty \tau\, d\tau\, e^{i\tau(xA + (1-x)B)} = \int_0^1 dx \frac{1}{[Ax + B(1-x)]^2}, \tag{33.18}$$

---

[1] To understand why s is called a proper time, recall from relativity that proper time s is defined by the differential $ds^2 = g_{\mu\nu}dx^\mu dx^\nu$. Since $\hat{H} = -g_{\mu\nu}\hat{p}^\mu\hat{p}^\nu$, it naturally generates translations in proper time through $g^{\mu\nu}\frac{\partial}{\partial x^\mu}\frac{\partial}{\partial x^\nu}$.

which is a Feynman parameter integral. Thus, in a loop, each particle has its own proper time, s or t, which denote how long each particle has taken to get around its part of the loop. Then the Feynman parameter $x = \frac{s}{s+t}$ is how far one particle is behind the other one.

33.2.1 Background fields

Now suppose a field ϕ interacts with a photon field, through the usual scalar QED Lagrangian:

$$\mathcal{L} = -\frac{1}{4}F_{\mu\nu}^2 - \phi^\star\left(D^2 + m^2\right)\phi, \tag{33.19}$$

with $D_\mu = \partial_\mu + ieA_\mu$. As a step towards calculating the Euler–Heisenberg Lagrangian, we will need the scalar propagator in the presence of a fixed external A_μ field. We write $\langle A|\cdots|A\rangle$ instead of $\langle\Omega|\cdots|\Omega\rangle$ when matrix elements are taken in the presence of an external field rather than the vacuum. Thus, the propagator in the presence of an external field A_μ is written as

$$G_A(x, y) = \langle A|T\{\phi(y)\phi^\star(x)\}|A\rangle. \tag{33.20}$$

Using operator notation, we use $\partial_\mu \to -i\hat{p}_\mu$ to define

$$\hat{G}_A = \frac{i}{(\hat{p} - eA(\hat{x}))^2 - m^2 + i\varepsilon}. \tag{33.21}$$

This equation illustrates an advantage of the quantum mechanics operator formalism over Feynman diagrams: we can work in position and momentum space at the same time, through operators such as $\hat{p} - eA(\hat{x})$.

Then, as in Eq. (33.15), we have

$$G_A(x, y) = \langle y|\hat{G}_A|x\rangle = \langle y|\frac{i}{(\hat{p} - eA(\hat{x}))^2 - m^2 + i\varepsilon}|x\rangle = \int ds\, e^{-s\varepsilon}e^{-ism^2}\langle y|e^{-i\hat{H}s}|x\rangle, \tag{33.22}$$

where now

$$\hat{H} = -(\hat{p} - eA(\hat{x}))^2. \tag{33.23}$$

So we get the same formula as for the free theory, but with a different Hamiltonian. The interpretation of Eq. (33.22) is that $G_A(x, y)$ describes the evolution of ϕ from x to y in time s, including all possible interactions with a field A_μ over all possible times s. This is shown diagrammatically in Eq. (33.8).

For a spinor, we want to evaluate

$$G_A(x, y) = \langle A|T\{\psi(y)\bar{\psi}(x)\}|A\rangle. \tag{33.24}$$

First, recall from Eq. (10.106) that

$$\slashed{D}^2 = D_\mu^2 + \frac{e}{2}F_{\mu\nu}\sigma^{\mu\nu}. \tag{33.25}$$

We used this identity in Chapter 10 to show that Dirac spinors satisfy the Klein–Gordon equation with an additional magnetic moment term. Here, the $F_{\mu\nu}\sigma^{\mu\nu}$ term will again produce the differences between the scalar and Dirac spinor cases of quantities we calculated. Then, in momentum space, we have

$$(\not{p} - e\not{A}(\hat{x}))^2 = (\hat{p} - eA(\hat{x}))^2 - \frac{e}{2}F_{\mu\nu}(\hat{x})\sigma^{\mu\nu}. \tag{33.26}$$

This identity lets us write the spinor Green's function operator as

$$\begin{aligned}
\hat{G}_A &= \frac{i}{\not{p} - e\not{A}(\hat{x}) - m + i\varepsilon} \\
&= (\not{p} - e\not{A}(\hat{x}) + m)\,\frac{i}{(\hat{p} - eA(\hat{x}))^2 - \frac{e}{2}F_{\mu\nu}(\hat{x})\sigma^{\mu\nu} - m^2 + i\varepsilon},
\end{aligned} \tag{33.27}$$

and so the Dirac propagator is

$$G_A(x,y) = \langle y|\frac{i}{\not{p} - e\not{A} - m + i\varepsilon}|x\rangle = \int_0^\infty ds\, e^{-s\varepsilon} e^{-ism^2} \langle y|(\not{p} - e\not{A}(\hat{x}) + m)e^{-i\hat{H}s}|x\rangle \tag{33.28}$$

as before, but now with

$$\hat{H} = -(\hat{p}^\mu - eA^\mu(\hat{x}))^2 + \frac{e}{2}F_{\mu\nu}(\hat{x})\sigma^{\mu\nu}. \tag{33.29}$$

Note that there is no Dirac trace here, since the Green's function is a matrix in spinor space.

33.2.2 Field-dependent expectation values

To connect to effective actions, recall from Section 33.1 that to integrate out a field at tree-level we set it equal to its equations of motion. Another way to phrase this procedure is that we set the field equal to a configuration for which the Lagrangian has a minimum. Now, classically, we can always expect to find the field at the minimum. So the minimum can be thought of as a classical expectation. The generalization to the quantum theory is to replace a field by its quantum vacuum expectation value:

$$\phi \to \langle\Omega|\phi|\Omega\rangle. \tag{33.30}$$

The classical and quantum expectation values agree at tree-level, but can be different when loops or non-perturbative effects are included. We will consider how the vacuum can be destabilized by quantum effects in Chapter 34. Our focus here is not on the expectation value in the vacuum, but in the presence of a fixed electromagnetic field. Thus, in a background field, we can integrate out ϕ by replacing $\phi \to \langle A|\phi|A\rangle$.

Let us go straight to the fermion case. The Lagrangian is

$$\mathcal{L} = -\frac{1}{4}F_{\mu\nu}^2 + \bar{\psi}(i\not{\partial} - m)\psi - eA_\mu\bar{\psi}\gamma^\mu\psi. \tag{33.31}$$

We now want to replace this by the effective Lagrangian where the current that A_μ couples to is replaced by its expectation value in the given fixed configuration, which we are denoting as A_μ:

$$\mathcal{L}_{\text{eff}} = -\frac{1}{4}F_{\mu\nu}^2 - eA_\mu J_A^\mu, \tag{33.32}$$

where

$$J_A^\mu \equiv \langle A | \bar\psi(x) \gamma^\mu \psi(x) | A \rangle. \tag{33.33}$$

This is not a vacuum matrix element, but a matrix element in the presence of a given state $|A\rangle$.

Now we can calculate J_A^μ using Schwinger proper time. First note that $A = 0$ is the vacuum, so J_0^μ should reduce to the propagator $G(x, y)$ with $x = y$ when the field is turned off. Indeed, being explicit about the spin indices

$$J_0^\mu(x) = \langle \Omega | \bar\psi_{\dot\alpha}(x) \gamma^\mu_{\dot\alpha\alpha} \psi_\alpha(x) | \Omega \rangle = -\mathrm{Tr}\left[\langle \Omega | \psi_\alpha(x) \bar\psi_{\dot\alpha}(x) \gamma^\mu_{\dot\alpha\beta} | \Omega \rangle\right] \equiv -\mathrm{Tr}\,\langle x | \hat G \gamma^\mu | x \rangle. \tag{33.34}$$

The third form is meant to indicate that the trace of the matrix $\left[\psi\bar\psi\gamma^\mu\right]_{\alpha\beta}$ is being taken. In the presence of a non-zero A field, we just have to replace this by the propagator in the A_μ background:

$$J_A^\mu(x) = -\mathrm{Tr}\,\langle x | \hat G_A \gamma^\mu | x \rangle, \tag{33.35}$$

where $\hat G_A$ is the Green's function in Eq. (33.27). So,

$$
\begin{aligned}
J_A^\mu &= -\mathrm{Tr}\left[\int_0^\infty ds\, e^{-s\varepsilon} e^{-ism^2} \langle x | \gamma^\mu (\slashed p - e\slashed A + m) e^{-i\hat H s} | x \rangle\right] \\
&= -\int_0^\infty ds\, e^{-s\varepsilon} e^{-ism^2} \langle x | \mathrm{Tr}\left[\gamma^\mu(\slashed p - e\slashed A) e^{i((\hat p - eA)^2 - \frac{e}{2}\sigma_{\mu\nu}F^{\mu\nu})s}\right] | x \rangle,
\end{aligned} \tag{33.36}
$$

where we have used that Tr of an odd number of γ-matrices is zero. Next, note that the current is itself a variation:

$$J_A^\mu = -\frac{i}{2e}\frac{\partial}{\partial A_\mu}\int_0^\infty \frac{ds}{s} e^{-s\varepsilon} e^{-ism^2} \mathrm{Tr}\left[\langle x | e^{-i\hat H s} | x \rangle\right]. \tag{33.37}$$

Integrating both sides with respect to A_μ and using Eq. (33.32) gives

$$\mathcal{L}_{\mathrm{eff}}(x) = -\frac{1}{4}F_{\mu\nu}^2(x) + \frac{i}{2}\int_0^\infty \frac{ds}{s} e^{-s\varepsilon} e^{-ism^2} \mathrm{Tr}\left[\langle x | e^{-i\hat H s} | x \rangle\right], \tag{33.38}$$

which is only a function of the background field A_μ. For a spinor, $\hat H$ is given in Eq. (33.29). For a complex scalar, the effective Lagrangian has a similar form:

$$\mathcal{L}_{\mathrm{eff}}(x) = -\frac{1}{4}F_{\mu\nu}^2(x) - i\int_0^\infty \frac{ds}{s} e^{-s\varepsilon} e^{-ism^2} \langle x | e^{-i\hat H s} | x \rangle, \tag{33.39}$$

with $\hat H = -(\hat p - eA(\hat x))^2$ as in Eq. (33.23). The scalar case is actually more difficult to derive than the spinor case using Schwinger's method because of the $A_\mu^2 \phi^\star \phi$ term in the scalar QED Lagrangian. We produce this Lagrangian using Feynman path integrals in Eq. (33.52) below.

33.2.3 Interpretation and cross check

Up to an extra factor of $\frac{1}{s}$, the proper-time integral in Eq. (33.39) looks just like $\langle x|\hat{G}_A|y\rangle$ in Eq. (33.22) with $x = y$. This is easy to understand: the effective action sums closed loops, where the particle propagates back to where it started after some proper time s. That is, it is an integral over $\langle x; 0|x; s\rangle$. In terms of Feynman diagrams, the effective action includes all diagrams with any number of external photons and one closed fermion loop:

$$\mathcal{L}_{\text{eff}} = -\frac{1}{4}F_{\mu\nu}^2 + \bigcirc + \bigcirc + \bigcirc + \bigcirc + \cdots .$$

(33.40)

The physical interpretation of the expectation value $\langle x|e^{-i\hat{H}s}|x\rangle = \langle x; 0|x; s\rangle$ in Eq. (33.38) is therefore that it is the amplitude for a particle to go around a loop in proper time s based on evolution with the Hamiltonian $\hat{H}$.

Note that the first diagram in Eq. (33.40) does not involve any photons at all, thus it should represent the vacuum energy of the system. This provides a nice consistency check. Setting $A = 0$, to get just the first diagram, the effective action becomes (in the complex scalar case)

$$\Gamma[0] = -i \int d^4x \int_0^\infty \frac{ds}{s} e^{-s\varepsilon} e^{-ism^2} \langle x|e^{i\hat{p}^2 s}|x\rangle.$$

(33.41)

Inserting $\mathbb{1} = \int \frac{d^4k}{(2\pi)^4} |k\rangle\langle k|$ we find

$$\Gamma[0] = -iVT \int_0^\infty \frac{ds}{s} \int \frac{d^4k}{(2\pi)^4} \exp\left[i(k_0^2 - \vec{k}^2 - m^2 + i\varepsilon)s\right],$$

(33.42)

where VT is the volume of space-time. It is convenient to remove this factor by writing $\Gamma[0] = -(VT)V_{\text{eff}}$ with V_{eff} an effective potential energy density, which in this case is just a constant.

The integral over proper time is divergent from the $s \sim 0$ region, corresponding to where the loop has zero proper length. However, Schwinger proper time conveniently gives us a Lorentz-invariant and gauge-invariant way to regulate such divergences: cut off the integral for $s > s_0$. To evaluate V_{eff}, we Wick rotate $k_0 \to ik_0$ and can integrate over the imaginary axis. This gives

$$V_{\text{eff}} = -\int_{s_0}^\infty \frac{ds}{s} \int \frac{d^3k}{(2\pi)^3} \int \frac{dk^0}{2\pi} \exp\left[-i(k_0^2 + \vec{k}^2 + m^2)s\right]$$

$$= -\frac{1}{2\sqrt{\pi}} \int \frac{d^3k}{(2\pi)^3} \int_{s_0}^\infty \frac{ds}{s^{3/2}} \exp\left[-(\vec{k}^2 + m^2)s\right],$$

(33.43)

where we have replaced $s \to -is$ in the second step. Then we find

$$V_{\text{eff}} = \int \frac{d^3k}{(2\pi)^3} \left(-\frac{1}{\sqrt{\pi s_0}} + \sqrt{\vec{k}^2 + m^2} + \mathcal{O}(\sqrt{s_0})\right).$$

(33.44)

The $-\frac{1}{\sqrt{\pi s_0}}$ is a divergent constant, corresponding to an extrinsic cutoff-dependent vacuum energy. This can be removed with a vacuum energy counterterm. The important term is in the integral over $\sqrt{\vec{k}^2 + m^2} = \omega_k$, which counts the ground-state energies of the modes. It was this sum, not the constant, that led to the Casimir force discussed in Chapter 15.

Note that we get ω_k instead of $\frac{1}{2}\omega_k$ since this is the effective action for a complex scalar that has twice the energy of a real scalar. For a Dirac fermion, the calculation is identical, since $\hat{H} = -\hat{p}^2$ in both cases when $A = 0$. The only difference is that the Dirac trace and $-\frac{1}{2}$ in Eq. (33.38) give a factor of $4(-\frac{1}{2}) = -2$ compared to the scalar case in Eq. (33.39). The minus sign is consistent with a fermion loop and the factor of 2 is consistent with a Dirac spinor having twice the number of degrees of freedom of a complex scalar. These are the same results we found in Section 12.5 by computing the energy density from the energy-momentum tensor. One consequence is that in a theory with a Weyl fermion and a complex scalar of the same mass, such as in theories with supersymmetry, the vacuum energy is zero.

33.3 Effective actions from Feynman path integrals

An alternative approach to calculating the effective action is based on the Feynman path integral. Here we want to integrate over some fields by performing the path integral. For scalar QED, integrating out the scalar means

$$\int \mathcal{D}A \exp(i\Gamma[A]) = \int \mathcal{D}A\, \mathcal{D}\phi\, \mathcal{D}\phi^\star \exp\left[i\int d^4x\left(-\frac{1}{4}F_{\mu\nu}^2 - \phi^\star(D^2 + m^2)\phi\right)\right].$$
(33.45)

In this case, since the original action is quadratic in ϕ, we can evaluate the path integral exactly. We will ignore the $i\varepsilon$ in this section for simplicity.

Recall the general formula from Problem 14.1:

$$\int \mathcal{D}\phi^\star \mathcal{D}\phi \exp\left[i\int d^4x(\phi^\star M\phi + JM)\right] = \mathcal{N}\frac{1}{\det M}\exp(iJM^{-1}J),$$
(33.46)

where $\mathcal{N}$ is some (infinite) normalization constant. Thus, for the scalar QED Lagrangian we find

$$\int \mathcal{D}A \exp(i\Gamma[A]) = \mathcal{N}\int \mathcal{D}A \exp\left[i\int d^4x\left(-\frac{1}{4}F_{\mu\nu}^2\right)\right]\frac{1}{\det(-D^2 - m^2)}.$$
(33.47)

This equation will be satisfied if

$$\exp\left[i\Gamma[A] + i\int d^4x\frac{1}{4}F_{\mu\nu}^2\right] = \mathcal{N}\frac{1}{\det(-D^2 - m^2)}.$$
(33.48)

produces the full Green's function $G(x, y) = \langle y; 0|x; s\rangle$, which is more generally useful than the effective action alone. For $x = y$, which is relevant for the effective action, the differential equation reduces to (cf. Eq. (33.A.150)):

$$i\partial_s \langle x; 0|x; s\rangle = -\text{tr}\left[\frac{i}{2}e\mathbf{F}\coth(es\mathbf{F}) + \frac{e}{2}\sigma\mathbf{F}\right]\langle x; 0|x; s\rangle, \qquad (33.75)$$

where $\mathbf{F} = F_{\mu\nu}$ and $\sigma = \sigma_{\mu\nu}$ are matrices. The solution with appropriate boundary conditions is

$$\langle x; 0|x; s\rangle = \frac{-i}{16\pi^2}\frac{1}{s^2}\exp\left(-\frac{1}{2}\text{tr}\ln\left[\frac{\sinh es\mathbf{F}}{es\mathbf{F}}\right] - i\frac{es}{2}\sigma_{\mu\nu}F^{\mu\nu}\right)$$

$$= -i\frac{e^2}{64\pi^2}\frac{F^{\mu\nu}\tilde{F}_{\mu\nu}}{\text{Im}\cos(esX)}\exp\left(-i\frac{es}{2}\sigma_{\mu\nu}F^{\mu\nu}\right). \qquad (33.76)$$

Again, this can be checked by differentiation. For a constant magnetic field, this is equivalent to Eq. (33.67).

The Euler–Heisenberg Lagrangian was first calculated by Heisenberg and his student Hans Euler by finding exact solutions to the Dirac equation in a constant $F_{\mu\nu}$ background [Euler and Heisenberg, 1936]. Our derivation of it, particularly the one in Appendix 33.A, is due to Schwinger [Schwinger, 1951].

33.4.1 Vacuum polarization

Expanding the unrenormalized Euler–Heisenberg Lagrangian, as in Eq. (33.72), we found two divergent terms which were removed with counterterms in Eq. (33.73). If we do not include these counterterms, the expansion gives

$$\mathcal{L}_{\text{EH}} = -\frac{1}{4}F_{\mu\nu}^2 - \frac{e^2}{8\pi^2}\int_0^\infty \frac{ds}{s}e^{is\varepsilon}e^{-sm^2}\left[\frac{1}{e^2s^2} + \frac{1}{6}F_{\mu\nu}^2\right] + \text{finite}. \qquad (33.77)$$

The first term in brackets is constant. It gives the vacuum energy density, as discussed in Section 33.2.3. The second term looks just like the tree-level QED kinetic term, $-\frac{1}{4}F_{\mu\nu}^2$. Keeping only this term (before renormalization), we have

$$\mathcal{L}_{\text{EH}} = -\frac{1}{4}F_{\mu\nu}^2 - \frac{1}{6}F_{\mu\nu}^2\frac{e^2}{8\pi^2}\int_0^\infty \frac{ds}{s}e^{is\varepsilon}e^{-sm^2}. \qquad (33.78)$$

This is UV divergent, from the $s \sim 0$ region. Regulating with a Lorentz-invariant UV cutoff s_0, we find

$$\mathcal{L}_{\text{EH}} = -\frac{1}{4}F_{\mu\nu}^2\left(1 + \frac{e^2}{12\pi^2}\int_{s_0}^\infty \frac{ds}{s}e^{is\varepsilon}e^{-sm^2}\right)$$

$$= -\frac{1}{4}F_{\mu\nu}^2\left(1 - \frac{e^2}{12\pi^2}\ln(s_0 m^2) + \text{const}\right). \qquad (33.79)$$

This logarithmic dependence on the cutoff is exactly what we found from computing the full vacuum polarization graph in QED. As discussed in Chapter 23, UV divergences determine RGEs, and this one determines the leading order β-function coefficient. We can read off from the coefficient of the logarithm in Eq. (33.79) (as discussed in Chapter 23), that the β-function in QED at 1-loop is

$$\beta(e) = \frac{e^3}{12\pi^2},\qquad(33.80)$$

which agrees with Eq. (16.73) (or Eq. (23.29)).

33.4.2 Light-by-light scattering

The original motivation of Heisenberg and Euler was to calculate the rate for photons to scatter off other photons. This problem was suggested to them by Otto Halpern and is sometimes called Halpern scattering. The relevant Feynman diagram is

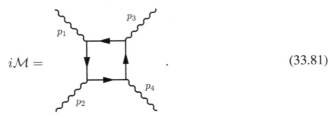

$$i\mathcal{M} = \qquad\qquad\qquad .\qquad(33.81)$$

This is a difficult loop to compute directly, even with today's technology, much less with what Euler and Heisenberg knew in 1936. We can get the answer (in the limit of low-frequency light $\omega \ll m$) directly from the Euler–Heisenberg Lagrangian. The relevant term is the one to fourth order in e, which has the form $\frac{\alpha^2}{90}\frac{1}{m^4}\left[(F^2)^2 + \frac{7}{4}(F\tilde{F})^2\right]$. This term was computed first in a paper by Euler and Kockel [Euler and Kockel, 1935]. Using it for light-by-light scattering corresponds to a tree-level Feynman diagram of the form

$$i\mathcal{M} = \qquad\qquad\qquad .\qquad(33.82)$$

Note that our effective Lagrangian is only valid when $\partial_\mu F_{\alpha\beta} = 0$; thus we will only get the result to leading order in $\frac{p^2}{m^2}$. From the experimental point of view, this is enough, since light-by-light scattering of real on-shell photons has not yet been experimentally observed, at any frequency.

The matrix element is

$$\mathcal{M} = \frac{\alpha^2}{90}\frac{1}{m^4}\Big\{(p_\mu^1\epsilon_\nu^1 - p_\nu^1\epsilon_\mu^1)(p_\mu^2\epsilon_\nu^2 - p_\nu^2\epsilon_\mu^2)(p_\alpha^3\epsilon_\beta^{3\star} - p_\beta^3\epsilon_\alpha^{3\star})(p_\alpha^4\epsilon_\beta^{4\star} - p_\beta^4\epsilon_\alpha^{4\star})$$

$$+\frac{7}{16}\left[\varepsilon^{\mu\nu\alpha\beta}(p_\mu^1\epsilon_\nu^1 - p_\nu^1\epsilon_\mu^1)(p_\alpha^2\epsilon_\beta^2 - p_\beta^2\epsilon_\alpha^2)\right] \times \left[\varepsilon^{\mu\nu\alpha\beta}(p_\mu^3\epsilon_\nu^{3\star} - p_\nu^3\epsilon_\mu^{3\star})(p_\alpha^4\epsilon_\beta^{4\star} - p_\beta^4\epsilon_\alpha^{4\star})\right]$$

$$+\text{permutations}\Big\}.$$

$$(33.83)$$

Summing over final polarizations and averaging over initial polarizations, the result is

$$\frac{1}{4}\sum_{\text{pols.}}\mathcal{M}^2 = \frac{1}{4}\frac{\alpha^4}{90^2}\frac{1}{m^8}2224(s^2t^2 + s^2u^2 + t^2u^2),$$

(33.84)

which leads to a cross section

$$\sigma_{\text{tot}} = \frac{973}{10125\pi}\alpha^4\frac{\omega^6}{m^8}.$$

(33.85)

This is the correct low-energy limit of the exact light-by-light scattering diagram. The exact result from the 1-loop graphs can be found in [Berestetsky *et al.*, 1982].

33.4.3 Schwinger pair production

Notice that the effective Lagrangian in Eq. (33.73) has singularities for certain values of the electromagnetic field. To see where the singularities are, we first consider the case with $\vec{B}$ and $\vec{E}$ parallel. Then,

$$F_{\mu\nu}^2 = 2(\vec{B}^2 - \vec{E}^2) = 2(B^2 - E^2),$$

(33.86)

where $E = |\vec{E}|$ and $B = |\vec{B}|$, and

$$F^{\mu\nu}\tilde{F}_{\mu\nu} = -4\vec{E}\cdot\vec{B} = -4EB,$$

(33.87)

and then, from Eq. (33.70),

$$X^2 = \frac{1}{2}(F_{\mu\nu}^2 - iF^{\mu\nu}\tilde{F}_{\mu\nu}) = (B + iE)^2.$$

(33.88)

Then the Euler–Heisenberg Lagrangian in Eq. (33.73) simplifies to

$$\mathcal{L}_{\text{EH}} = \frac{1}{2}\left(E^2 - B^2\right)$$
$$- \frac{e^2}{8\pi^2}\int_0^\infty \frac{ds}{s}e^{i\varepsilon s}e^{-m^2 s}\left[EB\cot(esE)\coth(esB) - \frac{1}{e^2 s^2} - \frac{B^2 - E^2}{3}\right].$$

(33.89)

Since $\coth(x)$ has no poles for $x > 0$, the singularities are all associated with constant electric fields. Thus, we take the limit $B \to 0$, in which case the fact that we took $\vec{E}$ and $\vec{B}$ parallel is immaterial. From Eq. (33.89) we find

$$\mathcal{L}_{\text{EH}} = \frac{1}{2}E^2 - \frac{1}{8\pi^2}\int_0^\infty \frac{ds}{s^3}e^{i\varepsilon s}e^{-sm^2}\left[eEs\cot(eEs) - 1 + \frac{1}{3}(esE)^2\right].$$

(33.90)

In this form, we can see that the Euler–Heisenberg Lagrangian has poles for real E when s is equal to $s_n = \frac{n\pi}{eE}$ for $n = 1, 2, \dots$ As we will now see, these poles indicate that strong electric fields can create electron–positron pairs, a process known as **Schwinger pair production** (although it was predicted first by Euler and Heisenberg).

How can electrons and positrons be produced from the Euler–Heisenberg Lagrangian, which has no electron field in it? They cannot. However, in a unitary quantum field theory, forward scattering rates are related to the sum over real production rates via the optical theorem. Recall from Section 24.1 that by the optical theorem (see Eq. (24.11))

$$\text{Im}\mathcal{M}(A \to A) = \frac{1}{2} \sum_X d\Pi_{\text{LIPS}}^X |\mathcal{M}(A \to X)|^2. \tag{33.91}$$

We can apply this theorem to QED in the situation where $|A\rangle$ corresponds to a coherent collection of photons describing a large electric field. In QED, the sum over states $|X\rangle$ includes states with on-shell electrons and positrons. Since QED is unitary, the optical theorem holds. In the Euler–Heisenberg Lagrangian the states $|A\rangle$ are the same states as in QED. Thus, if the calculation of $\mathcal{L}_{\text{EH}}$ has been done correctly, the left-hand side of Eq. (33.91) should be unchanged, as one would expect from a matching calculation. The right-hand side of Eq. (33.91), on the other hand, cannot be the same as in full QED, since QED has electrons in it and the Euler–Heisenberg theory does not. Thus, what would be a unitary process in full QED now appears as a non-unitary process in the effective theory. Unfortunately, it is not easy to use Eq. (33.91) to calculate the pair-production rate, since one would have to sum over an infinite number of multi-particle states.

There is a nice shortcut, due to Schwinger, for evaluating the total pair-production rate. If there were no pair production, then the electric field state $|A\rangle$ would be constant in time. Thus $\langle A|S|A\rangle = 1$ where S is the S-matrix. Since in this case the action is constant, $S = e^{i\Gamma}$. Therefore, $|\langle A| e^{i\Gamma} |A\rangle|^2 = |e^{i\Gamma}|^2$ measures the probability for something other than A to be produced. In other words, $|e^{i\Gamma}|^2$ gives the probability that no pairs are produced over the time T and volume V of the experiment. We then have

$$|e^{i\Gamma}|^2 = e^{i\Gamma}e^{-i\Gamma^\star} = e^{i(\Gamma - \Gamma^\star)} = e^{-2\text{Im}[\Gamma]} = e^{-2VT\text{Im}\mathcal{L}_{\text{EH}}}, \tag{33.92}$$

where in the last step we use that, for given background fields, the Euler–Heisenberg Lagrangian is just a number. Thus $2\text{Im}\mathcal{L}_{\text{EH}}$ is the probability, per unit time and volume, that any number of pairs are created. This is the continuum field version of the optical theorem relation $\text{Im}\mathcal{M}(A \to A) = m_A\Gamma_{\text{tot}}$, where Γ_{tot} is the total decay rate of a single particle of mass m_A.

In order to calculate $\text{Im}\mathcal{L}_{\text{EH}}$ we note that the integrand in Eq. (33.71) has poles at $s_n = \frac{\pi}{eE}n$. There is no pole at $s = 0$, as can be seen from expanding the integrand at small s. The imaginary part of this expression can be calculated using contour integration (Problem 33.3). The result is that[2]

$$2\text{Im}(\mathcal{L}_{\text{eff}}) = \frac{1}{4\pi} \sum_{n=1}^\infty \frac{1}{s_n^2} e^{-m^2 s_n} = \frac{\alpha E^2}{\pi^2} \sum_{n=1}^\infty \frac{1}{n^2} \exp\left(\frac{-n\pi m^2}{eE}\right). \tag{33.93}$$

Performing this sum, we find

$$\Gamma\left(E \to e^+e^- \text{ pairs}\right) = \frac{\alpha E^2}{\pi^2} \text{Li}_2\left(e^{-\frac{\pi m^2}{eE}}\right), \tag{33.94}$$

with $\text{Li}_2(x)$ the dilogarithm function. This is the rate for Schwinger pair production in an external electric field.

---

[2] This sum also has an interpretation as a sum over instantons (see for example [Kim and Page, 2002]).

The rate for pair production is negligible until $E \gtrsim E_{\text{critical}} = \frac{m_e^2}{e} \approx 10^{18}$ volts/meter, which is an enormous field. As of this writing, Schwinger pair production in QED has still not been observed, since it is extremely difficult to get such fields in the lab. One might imagine, however, that such strong fields might be produced close to a particle with a very large charge, such as an atomic nucleus. The field around a nucleus is $E \sim \frac{e}{4\pi r^2} Z$. Now, the Euler–Heisenberg Lagrangian is only valid for fields that have wavelengths greater than $\frac{1}{m_e}$, so the best we can say is that pair production would begin for Z large enough that $E_{\text{critical}} \sim \frac{e}{4\pi \left(m_e^{-2}\right)} Z$, which gives $Z = \frac{4\pi^2}{e^2} = \frac{1}{\alpha} \sim 137$. This result is sometimes invoked to explain why the periodic table has less than 137 elements![3]

33.4.4 Connection to perturbation theory

It is informative to consider which of the predictions we have derived from $\mathcal{L}_{\text{EH}}$ are equivalent to perturbative calculations in QED, and which are not.

We found that the Schwinger pair-production rate depended on $\exp(-\frac{\pi m^2}{eE})$. This dependence on e indicates that pair production is a non-perturbative effect – you would never see pair production from constant electric fields at any fixed order in perturbative QED. Of course, you can get pair production in perturbation theory. But this would involve photon modes of frequencies larger than m. More precisely, one can show that [Itzykson and Zuber, 1980]

$$\Gamma(E \to e^+ e^-) = \frac{\alpha}{3} \int d^4 q \, \theta(q^2 - 4m^2) \left[\vec{E}(q^2)\right]^2 \sqrt{1 - \frac{4m^2}{q^2}} \left(1 + \frac{2m^2}{q^2}\right), \quad (33.95)$$

which vanishes when $\vec{E}$ is constant. The Schwinger pair-production rate is one of the very few analytic non-perturbative calculations in quantum field theory that give physical predictions.

Other results, such as the rate for light-by-light scattering, could be calculated in perturbative QED. Nevertheless, the Euler–Heisenberg Lagrangian efficiently encodes the result of many loop calculations all at once. It is worth discussing exactly what graphs are included in the Euler–Heisenberg Lagrangian, since this understanding will apply to similar effective actions in other contexts.

Recall our expression for the effective Lagrangian where the fermion is integrated out, Eq. (33.38),

$$\mathcal{L}_{\text{eff}}[A] = -\frac{1}{4} F_{\mu\nu}^2 + \frac{i}{2} \int \frac{ds}{s} e^{-ism^2} \langle x | e^{-i \slashed{D}^2 s} | x \rangle. \quad (33.96)$$

We have not assumed $F_{\mu\nu}$ is constant at this point, and in fact this effective action is *exact*. That is, since the Lagrangian was quadratic in ψ, this is a formal expression for the result of evaluating the path integral of ψ completely. It does, however, correspond to only 1-loop

---

[3] This result actually follows more simply from dimensional analysis. The ground state of a hydrogen-like atom has energy $E_0 \sim -Z^2 \alpha^2 m_e$. To get pair production, a nucleus has to be able to capture an electron from the vacuum, emitting a positron into the continuum, so $E_0 \lesssim -m_e$ giving $Z \gtrsim \frac{1}{\alpha}$, up to order 1 factors, which we cannot get by dimensional analysis.

graphs, those in Eq. (33.40), since there is only a single propagator going from x back to x in proper time s. But how can this expression be exact if it does not include higher loops? Are graphs such as

$$\text{(33.97)}$$

which have internal photon and/or fermion loops, included or not?

To answer this question, first recall that in the calculation of the effective action, and in the formal exact expression Eq. (33.38), the photon propagator plays no role. In fact, if we dropped the photon kinetic term from the original action, the only change in the effective action would be that the $-\frac{1}{4}F_{\mu\nu}^2$ term would be missing. Thus, neither of the graphs above are included in the effective action calculation, since both involve the photon propagator. On the other hand, since nothing is thrown out (assuming the effective action $\Gamma[A]$ is known exactly), any physical effect associated with these graphs must be reproducible within the effective theory. For example, these graphs in full QED contribute to the QED β-function, which has physical effects. The way the effective theory reproduces the physics of these loops is with its own loops involving effective vertices. Basically, the fermion loops are computed first, treating the photon lines as external, which generates new vertices. Then the photon lines coming off these vertices are sewn together in a loop amplitude using the photon propagator in the effective theory.

For example, to reproduce the physics of the first graph in Eq. (33.97), the relevant effective vertex can be determined by cutting through the intermediate photon and then contracting the fermion loop to a point:

$$\text{(33.98)}$$

The second graph in Eq. (33.97) involves this vertex, associated with the inner fermion loop, and a 6-point vertex associated with the outer fermion loop. The physics of the diagrams in Eq. (33.97) are then reproduced by connecting the legs in these effective vertices:

$$\text{(33.99)}$$

These graphs would reproduce the complete result from the graphs in Eq. (33.97), but we need the full $\mathcal{L}_{\text{eff}}[A]$ to compute them.

In the Euler–Heisenberg Lagrangian, we took $F_{\mu\nu}$ constant. Thus, the full physics of the loops in Eq. (33.97) is not reproduced by the Euler–Heisenberg Lagrangian alone. Only if we had the full effective Lagrangian, by evaluating $\Gamma[A]$ exactly, which would supplement the Euler–Heisenberg Lagrangian with additional terms depending on $\partial_\mu F_{\alpha\beta}$ (and give corrections at higher order in α to the terms without derivatives), would the full theory be reproduced. This exact $\Gamma[A]$ is not known.

Even at energies above m_e, the exact effective Lagrangian can be used. The electron still shows up as a pole in the scattering amplitude, as is clear already from Schwinger pair production in the constant $F_{\mu\nu}$ approximation. Thus, one can treat the electron like a bound state and calculate S-matrix elements for it. Of course, this is a terribly inefficient way to calculate electron production and scattering, since we already know the full theory. It is more efficient to use the UV completion of Γ, namely QED, which has a Lagrangian that is local and real.

33.5 Coupling to other currents

The effective action from integrating out ψ can be generalized to the case where ψ couples to other things besides A_μ. In this way, we can calculate things such as the $\pi^0 \to \gamma\gamma$ rate, where π^0 is the neutral pion from QCD (see Chapter 28).

When ψ couples to things other than A_μ, the effective Lagrangian has more terms. Say we had

$$\mathcal{L} = \bar{\psi}(i\slashed{\partial} - m)\psi - \frac{1}{2}\phi(\Box + m_\phi^2)\phi - \frac{1}{2}\pi(\Box + m_\pi^2)\pi - eA_\mu\bar{\psi}\gamma^\mu\psi + \lambda\phi\bar{\psi}\psi + ig\pi\bar{\psi}\gamma^5\psi,$$
(33.100)

which has a scalar ϕ and a pseudoscalar π in addition to the external field A_μ. When we integrate out ψ, the effective Lagrangian (without ψ) will just contain the other fields coupled to the expectation value of the various ψ bilinears in the background electromagnetic field, as in Section 33.2.2. That is,

$$\mathcal{L}_{\text{eff}}[A, \phi, \pi] = -\frac{1}{2}\phi(\Box + m_\phi^2)\phi - \frac{1}{2}\pi(\Box + m_\pi^2)\pi - eA_\mu J_A^\mu + \lambda\phi J_\phi + ig\pi J_\pi, \quad (33.101)$$

where

$$J_A^\mu = \langle A|\bar{\psi}\gamma^\mu\psi|A\rangle, \quad J_\phi = \langle A|\bar{\psi}\psi|A\rangle, \quad J_\pi = \langle A|\bar{\psi}\gamma^5\psi|A\rangle. \quad (33.102)$$

We sometimes call these field-dependent expectation values **classical currents**, since they are just classical functionals of background $A_\mu(x)$ fields. The calculation of these classical currents corresponds to the evaluation of Feynman diagrams such as

$$J_\phi = \quad \otimes \quad + \quad \otimes \quad + \quad \otimes \quad + \quad \otimes \quad + \cdots. \quad (33.103)$$

Here, the $\otimes$ refers to insertions of the external current in the original theory, corresponding to an interaction with the scalar. The photon lines are the background electromagnetic fields.

For the scalar current,

$$
\begin{aligned}
J_\phi &= \langle A|\bar\psi(x)\psi(x)|A\rangle = -\mathrm{Tr}\left[\langle x|\hat G_A|x\rangle\right] \\
&= -\mathrm{Tr}\left[\int_0^\infty ds\, e^{-ism^2}\langle x|(\slashed p - e\slashed A + m)e^{i(\slashed p - e\slashed A)^2 s}|x\rangle\right] \\
&= -4m\int_0^\infty ds\, e^{-ism^2}\langle x|e^{-i\hat H s}|x\rangle.
\end{aligned}
\tag{33.104}
$$

You may notice that $J_\phi = -\frac{\partial}{\partial m}\mathcal L_{\mathrm{eff}}[A]$, with $\mathcal L_{\mathrm{eff}}[A]$ in Eq. (33.38), a result that is useful and not surprising, since the $\phi\bar\psi\psi$ interaction and the mass term $m\bar\psi\psi$ have the same form.

For the pseudoscalar current,

$$
\begin{aligned}
J_\pi &= \langle A|\bar\psi(x)\gamma^5\psi(x)|A\rangle = -\mathrm{Tr}\left[\langle x|\hat G_A\gamma^5|x\rangle\right] \\
&= -\mathrm{Tr}\left[\int_0^\infty ds\, e^{-ism^2}\langle x|(\slashed p - e\slashed A + m)e^{i(\slashed p - e\slashed A)^2 s}\gamma^5|x\rangle\right] \\
&= -m\int_0^\infty ds\, e^{-ism^2}\mathrm{Tr}\left[\langle x|\gamma^5 e^{-i\hat H s}|x\rangle\right].
\end{aligned}
\tag{33.105}
$$

This current does not have a simple relation to $\mathcal L_{\mathrm{eff}}[A]$, but as we will see, is not hard to compute.

33.5.1 Currents at low energy

Since the scalar current is $J_\phi = -\frac{\partial}{\partial m}\mathcal L_{\mathrm{eff}}[A]$, for the case of constant electromagnetic fields, we can read the answer from the Euler–Heisenberg Lagrangian, although additional counterterms may be required. We find (hiding the counterterms)

$$
\begin{aligned}
J_\phi &= -\frac{e^2}{32\pi^2}\frac{\partial}{\partial m}\int_0^\infty \frac{ds}{s}e^{-m^2 s}\frac{\mathrm{Re}\cosh(esX)}{\mathrm{Im}\cosh(esX)}F^{\mu\nu}\tilde F_{\mu\nu} \\
&= \frac{e^2}{8\pi^2}\frac{\partial}{\partial m}\int_0^\infty \frac{ds}{s}e^{-m^2 s}\left[\frac{1}{e^2 s^2}+\frac{1}{6}F_{\mu\nu}^2 + \cdots\right] \\
&= -\frac{e^2}{4\pi^2}m\int_0^\infty ds\, e^{-m^2 s}\left[\frac{1}{e^2 s^2}+\frac{1}{6}F_{\mu\nu}^2 + \cdots\right].
\end{aligned}
\tag{33.106}
$$

The first term is infinite and can be removed with a renormalization of the bare term $\Lambda^3\phi$ in the Lagrangian. The second term is finite and gives

$$
J_\phi = -\frac{\alpha}{6\pi}\frac{1}{m}\left(F_{\mu\nu}^2 + \cdots\right),
\tag{33.107}
$$

where the $\cdots$ are higher order in e.

For the pseudoscalar, we need

$$
J_\pi = -m\int_0^\infty ds\, e^{-ism^2}\mathrm{Tr}[\gamma_5\langle x|e^{-i\hat H s}|x\rangle].
\tag{33.108}
$$

Now, from Eq. (33.76),

$$\langle x|e^{-i\hat{H}s}|x\rangle = \langle x; 0|x; s\rangle = -i\frac{e^2}{64\pi^2}\frac{F^{\mu\nu}\tilde{F}_{\mu\nu}}{\text{Im}\cos(esX)}\exp\left(-i\frac{es}{2}\sigma_{\mu\nu}F^{\mu\nu}\right). \quad (33.109)$$

and so

$$J_\pi = \frac{ie^2 m}{64\pi^2}\int_0^\infty ds\, e^{-ism^2}\frac{F^{\mu\nu}\tilde{F}_{\mu\nu}}{\text{Im}\cos(esX)}\text{Tr}[\gamma_5 e^{-i\frac{e}{2}\sigma_{\mu\nu}F^{\mu\nu}s}]. \quad (33.110)$$

Since $\text{Tr}[\gamma_5] = \text{Tr}[\sigma_{\mu\nu}\gamma_5] = 0$, only terms with $\sigma_{\mu\nu}$ to an even power will survive. Using $(\sigma_{\mu\nu}F^{\mu\nu})^2 = 2F_{\mu\nu}^2 + 2i\gamma_5 F^{\mu\nu}\tilde{F}_{\mu\nu}$ we get

$$\text{Tr}[\gamma_5 e^{-i\frac{e}{2}\sigma_{\mu\nu}F^{\mu\nu}s}] = -4i\text{Im}\cos(esX). \quad (33.111)$$

And thus,

$$J_\pi = \frac{e^2 m}{16\pi^2}\int_0^\infty ds\, e^{-ism^2}F^{\mu\nu}\tilde{F}_{\mu\nu} = -i\frac{\alpha}{4\pi m}F^{\mu\nu}\tilde{F}_{\mu\nu}. \quad (33.112)$$

Plugging J_ϕ and J_π and the Euler–Heisenberg Lagrangian into Eq. (33.101) gives

$$\mathcal{L}_{\text{eff}}[A, \phi, \pi] = \mathcal{L}_{\text{EH}}[A] - \frac{1}{2}\phi(\Box + m_\phi^2)\phi + \frac{\lambda}{m}\phi\left(-\frac{\alpha}{6\pi}F_{\mu\nu}^2 + \cdots\right)$$
$$- \frac{1}{2}\pi(\Box + m_\pi^2)\pi + \frac{\alpha}{4\pi}\frac{g}{m}\pi F^{\mu\nu}\tilde{F}_{\mu\nu}. \quad (33.113)$$

Note that the π coupling has just one term. The decay rates predicted from this effective Lagrangian are

$$\Gamma(\phi \to \gamma\gamma) = \frac{\alpha^2}{144\pi^3}\lambda^2\frac{m_\phi^3}{m^2}, \quad (33.114)$$

$$\Gamma(\pi \to \gamma\gamma) = \frac{\alpha^2}{64\pi^3}g^2\frac{m_\pi^3}{m^2}. \quad (33.115)$$

Not surprisingly, the pseudoscalar rate agrees exactly with Eq. (30.11). In this method of calculation, however, we gain additional insight into the associated anomaly.

33.5.2 Chiral anomaly

Connecting the $\pi \to \gamma\gamma$ rate to an anomalous symmetry is straightforward in the effective action language. Recall that the QED Lagrangian,

$$\mathcal{L} = \bar{\psi}(i\slashed{\partial} - e\slashed{A})\psi - m\bar{\psi}\psi, \quad (33.116)$$

is invariant under a vector symmetry, $\psi \to e^{i\alpha}\psi$, and, in the limit $m \to 0$, under a chiral symmetry, $\psi \to e^{i\gamma_5}\psi$. The associated Noether currents are $J^\mu = \bar{\psi}\gamma^\mu\psi$ and $J^{\mu 5} = \bar{\psi}\gamma^\mu\gamma^5\psi$. By the equations of motion, the axial current satisfies

$$\partial_\mu J^{\mu 5} = 2im\bar{\psi}\gamma^5\psi. \quad (33.117)$$

So the amount by which the axial current is not conserved is proportional to the fermion mass.

Now, we already calculated the expectation value of $\bar{\psi}\gamma^5\psi$ in the background electromagnetic field. In Eq. (33.112) we found $\langle A|\bar{\psi}\gamma^5\psi|A\rangle = i\frac{\alpha}{4\pi m}F^{\mu\nu}\tilde{F}_{\mu\nu}$. This is consistent with Eq. (33.117) only if

$$\langle A|\partial_\mu J^{\mu 5}|A\rangle = -\frac{\alpha}{2\pi}F^{\mu\nu}\tilde{F}_{\mu\nu}, \tag{33.118}$$

which agrees with Eq. (30.22).

33.6 Semi-classical and non-relativistic limits

The Schwinger proper-time method is not only useful for calculating loops using quantum mechanics, it also gives a new perspective on the semi-classical and non-relativistic limits of quantum field theory. In particular, it illustrates where the particles are hiding in the path integral. As we will see, Schwinger proper time lets us derive one-particle quantum mechanics as the low-energy limit of quantum field theory.

To begin, we return to the expression for the Green's function we derived above for a scalar particle in a background electromagnetic field, Eq. (33.22):

$$G_A(x,y) = \langle A|T\{\phi(x)\phi(y)\}|A\rangle = \int_0^\infty ds\, e^{-ism^2}\langle y|e^{-i\hat{H}s}|x\rangle, \tag{33.119}$$

with $\hat{H} = -(\hat{p} - eA(\hat{x}))^2$. This operator $\hat{H}$ is the Hamiltonian in a one-particle quantum mechanical system that generates translations in Schwinger proper time s. The function $G_A(x,y)$ is computed for constant electromagnetic fields in Appendix 33.A. In this section, we rewrite $G_A(x,y)$ in terms of a quantum mechanical path integral.

In quantum mechanics, the path integral gives the amplitude for a particle to propagate from x^μ to y^μ in time s (see Section 14.2.2):

$$\langle y|e^{-i\hat{H}s}|x\rangle = \int_{z(0)=x}^{z(s)=y} \mathcal{D}z(\tau)\exp(i\int d\tau\, \mathcal{L}(z,\dot{z})), \tag{33.120}$$

where $\mathcal{L} = \hat{p}\dot{\hat{x}} - \hat{H}$ is the Legendre transform of the Hamiltonian. We would like to work out this Lagrangian in the case of a scalar in an electromagnetic field.

To simplify things, we first write $\hat{H} = -\hat{\Pi}^2$, where $\hat{\Pi}^\mu = \hat{p}^\mu - eA^\mu(\hat{x})$. The Heisenberg equations of motion for translation in s are

$$\dot{\hat{x}}^\mu \equiv \frac{d\hat{x}^\mu}{ds} = i[\hat{H}, \hat{x}^\mu] = i[-\hat{\Pi}^2, \hat{x}^\mu] = 2\hat{\Pi}^\mu, \tag{33.121}$$

where $[\hat{\Pi}^\mu, \hat{x}^\nu] = [\hat{p}^\mu, \hat{x}^\nu] = ig^{\mu\nu}$ has been used in the last step. So,

$$\mathcal{L} = \hat{p}^\mu\frac{\partial\hat{H}}{\partial p^\mu} - \hat{H} = -\hat{\Pi}^2 - 2eA^\mu\Pi^\mu = -\left(\frac{d\hat{x}^\mu}{2ds}\right)^2 - eA^\mu\frac{d\hat{x}^\mu}{ds}, \tag{33.122}$$

giving

$$\langle y|e^{-i\hat{H}s}|x\rangle = \int_{z(0)=x}^{z(s)=y} \mathcal{D}z(\tau)\exp\left(-i\int_0^s d\tau\left(\frac{dz^\mu}{2d\tau}\right)^2 - ie\int A_\mu(z)dz^\mu\right), \tag{33.123}$$

with the integral over A_μ a line integral along the path $z(s)$. So the Green's function is

$$G_A(x, y) = \int_0^\infty ds\, e^{-ism^2} \int_{z(0)=x}^{z(s)=y} \mathcal{D}z(\tau) \exp\left(-i\int_0^s d\tau \left(\frac{dz^\mu}{2d\tau}\right)^2 - ie\int A_\mu(z)dz^\mu\right).$$

(33.124)

This is an exact formal expression, only useful to the extent that we can solve for $z(\tau)$.

This world-line formulation was derived by a different method by Feynman [Feynman, 1950], although it had little application for many years. Interest in this approach was revived by Polyakov [Polyakov, 1981] in the context of string theory, and by Bern and Kosower [Bern and Kosower, 1992] who used it to develop an efficient way to compute loop diagrams in QCD.

33.6.1 Semi-classical limit

In the limit that a particle is very massive, loops involving that particle are suppressed. Thus, it should be possible to treat a massive particle classically and the radiation it produces quantum mechanically.

To take the large mass limit, we first rescale $s \to \frac{s}{m^2}$ and $\tau \to \frac{\tau}{m^2}$. This gives

$$G_A(x, y) = \frac{1}{m^2} \int_0^\infty ds\, e^{-is} \int_{z(0)=x}^{z\left(\frac{s}{m^2}\right)=y} \mathcal{D}z(\tau)$$
$$\times \exp\left(-i\int_0^s d\tau \left[m^2\left(\frac{dz^\mu}{2d\tau}\right)^2\right] - ie\int A_\mu(z)dz^\mu\right). \quad (33.125)$$

Now we see that, for large m, the $m^2\left(\frac{dz^\mu}{2d\tau}\right)^2$ term completely dominates the path integral. Moreover, as $m \to \infty$, the action is dominated by the point of stationary phase, which is also the classical free-particle solution:

$$z^\mu(\tau) = x^\mu + v^\mu\tau, \quad (33.126)$$

where $v^\mu = \frac{y^\mu - x^\mu}{s}$ is the particle's velocity. So we get, rescaling $s \to sm^2$ back again, and plugging in the stationary phase solution,

$$G_A(x, y) = \int_0^\infty ds \exp\left(-i\left[sm^2 + \frac{(y-x)^2}{4s} + ev^\mu\int_0^s d\tau A_\mu(z(\tau))\right]\right). \quad (33.127)$$

The first two terms in the exponent are independent of e and represent propagation of a free particle, similar to Eq. (33.11). The next term is equivalent to adding a term to the Lagrangian $\mathcal{L} = -eA_\mu J_c^\mu$, where J_c^μ is the source current from a classical massive particle moving at constant velocity:

$$J_c^\mu(x) = v^\mu\delta(x - v\tau). \quad (33.128)$$

In words, a heavy particle produces a gauge potential A_μ as if it is moving at a constant velocity.

This is the **semi-classical** limit. When a particle is heavy, the quantum field theory can be approximated by treating that particle as a classical source, but treating everything else quantum mechanically. You can study the fermion case in Problem 33.4.

33.6.2 Non-relativistic limit

In the non-relativistic limit, not only is the particle's mass assumed to be larger than the energy of typical photons, but the particle's velocity is also assumed to be much less than the speed of light. Define $\Delta t = y^0 - x^0$ and $\Delta x = |\vec{y} - \vec{x}|$. A particle moving slowly from x^μ to y^μ has $\Delta t \gg \Delta x$.

Separating out the time component, the 2-point function in Eq. (33.124) becomes

$$
G_A(x, y) = \int_0^\infty ds \int_{z(0)=x}^{z(s)=y} \mathcal{D}z^0(\tau)\mathcal{D}\vec{z}(\tau)
$$
$$
\times \exp\left(-i\int_0^s d\tau\left[\left(\frac{dz^0}{2d\tau}\right)^2 - \left(\frac{d\vec{z}}{2d\tau}\right)^2 + m^2\right] - ie\int A_\mu(z)dz^\mu\right).
$$
(33.129)

The classical path that minimizes the action, from the large m limit, has

$$
z^0(\tau) = x^0 + \frac{\Delta t}{s}\tau.
$$
(33.130)

We want to treat this time evolution classically, and leave the rest of the field fluctuations quantum mechanical. However, we can see that since both $(\frac{dz^0}{2d\tau})^2$ and m^2 are large, the stationary phase will have $\frac{\Delta t}{2s} \sim m$ and so $s \sim \frac{\Delta t}{2m}$. That is, the integral is dominated by the region near $z^0 = x^0 + 2m\tau$ and $s = \frac{\Delta t}{2m}$. To leading order in the expansion of s and z^0 around their stationary-phase points, we then find

$$
G_A(x, y) = \int_{z(0)=x}^{z(\frac{\Delta t}{2m})=y} \mathcal{D}\vec{z}(\tau) \exp\left(i\int_0^{\frac{\Delta t}{2m}} d\tau\left[\left(\frac{d\vec{z}}{2d\tau}\right)^2 - 2m^2\right] - ie\int A_\mu(z)dz^\mu\right).
$$
(33.131)

Now we change variables to $\tau = \frac{t}{2m}$ to find

$$
G_A(x, y) = \int_{z(0)=x}^{z(\Delta t)=y} \mathcal{D}\vec{z}(t) \exp\left(i\int_0^{\Delta t} dt\left[\frac{1}{2}m\left(\frac{d\vec{z}}{dt}\right)^2 - m\right] - ie\int A_\mu(z)dz^\mu\right).
$$
(33.132)

This result is exactly the path integral expression in non-relativistic, first-quantized quantum mechanics with a potential $V = m$. We have just derived that the non-relativistic limit of quantum field theory is quantum mechanics!

33.A Schwinger's method

In this appendix, we explicitly calculate the 1-loop effective action for constant background electromagnetic fields $F_{\mu\nu}$ using Schwinger's original method [Schwinger, 1951]. This is an alternative way to calculate the Euler–Heisenberg Lagrangian than the sum over Landau levels method discussed in Section 33.4. This method, although a bit longer, is appealing because it avoids having to regulate the system in a box. It also produces a general expression for the propagator $G_A(x,y)$ of a particle in a constant background electromagnetic field.

Our starting point is the formula for the effective action in Eq. (33.38):

$$\mathcal{L}_{\text{eff}}(x) = -\frac{1}{4}F_{\mu\nu}^2(x) + \frac{i}{2}\int_0^\infty \frac{ds}{s}e^{-ism^2}\text{Tr}\left[\langle x|e^{-i\hat{H}s}|x\rangle\right], \qquad (33.\text{A}.133)$$

with $\hat{H} = -(\hat{p}^\mu - eA^\mu(\hat{x}))^2 + \frac{e}{2}F_{\mu\nu}(\hat{x})\sigma^{\mu\nu}$. We have dropped the ε term, since we will not need it with this method. Here $A_\mu(\hat{x})$ is to be thought of as a classical gauge field configuration with position replaced by the operator $\hat{x}$. We would like to calculate $\mathcal{L}_{\text{eff}}(x)$ when $F_{\mu\nu}(\hat{x}) = (\partial_\mu A_\nu - \partial_\nu A_\mu)(\hat{x})$ is constant. We begin by calculating $\langle y|e^{-i\hat{H}s}|x\rangle$. Once this is known, we will set $y = x$ and integrate over s to get $\mathcal{L}_{\text{eff}}$.

33.A.1 Proper-time propagation

States such as $|x\rangle$ are eigenstates of an operator $\hat{x}^\mu$ in a first-quantized Hilbert space. The operators $\hat{x}^\mu$ are Schrödinger-picture operators. They are related to Heisenberg-picture operators by $\hat{x}^\mu(s) = e^{i\hat{H}s}\hat{x}^\mu e^{-i\hat{H}s}$. Using the definition $|x;s\rangle \equiv e^{-i\hat{H}s}|x\rangle$ we find

$$i\partial_s\langle y;0|x;s\rangle = i\partial_s\langle y|e^{-i\hat{H}s}|x\rangle = \langle y|e^{-i\hat{H}s}\hat{H}|x\rangle. \qquad (33.\text{A}.134)$$

Now,

$$\langle y|e^{-i\hat{H}s}\hat{x}^\mu(s) = \langle y|\hat{x}^\mu e^{-i\hat{H}s} = y^\mu\langle y|e^{-i\hat{H}s}, \qquad (33.\text{A}.135)$$

and

$$\hat{x}^\mu(0)|x;0\rangle = \hat{x}^\mu|x;0\rangle = x^\mu|x;0\rangle. \qquad (33.\text{A}.136)$$

Thus, if we can write $\hat{H}$ in terms of $\hat{x}(0)$ and $\hat{x}(s)$ we can turn Eq. (33.A.134) into an ordinary differential equation whose solution gives $\langle y;0|x;s\rangle$.

In quantum mechanics, the position and momentum operators satisfy $[\hat{x},\hat{p}] = i$. In our 4D first-quantized setup we generalize this to

$$[\hat{x}^\mu(s), \hat{p}^\nu(s)] = -ig^{\mu\nu}, \qquad (33.\text{A}.137)$$

with the commutation applying at the same proper time s. To simplify the form of the Hamiltonian, we introduce the operator $\hat{\Pi}^\mu = \hat{p}^\mu - eA^\mu(\hat{x})$. Then, assuming $F_{\mu\nu}$ is constant, we get

$$\left[\hat{x}^\mu(s), \hat{\Pi}^\nu(s)\right] = -ig^{\mu\nu}, \qquad (33.\text{A}.138)$$

$$[\hat{\Pi}^\mu(s), \hat{\Pi}^\nu(s)] = -ieF^{\mu\nu}. \tag{33.A.139}$$

In terms of $\hat{\Pi}^\mu$, the Hamiltonian is

$$\hat{H}(s) = -\hat{\slashed{\Pi}}^2 = -\hat{\Pi}_\mu(s)\hat{\Pi}^\mu(s) + \frac{e}{2}F_{\mu\nu}\sigma^{\mu\nu}. \tag{33.A.140}$$

For simplicity, we will drop circumflexes on operators from now on. As a notational convenience, we will also replace μ and ν indices with boldface type. So the vectors x^μ and Π^μ are written as $\mathbf{x}$ and $\mathbf{\Pi}$, respectively, and the matrices $F^{\mu\nu}$ and $\sigma^{\mu\nu}$ are written as $\mathbf{F}$ and $\boldsymbol{\sigma}$ respectively. Then $\text{tr}(\boldsymbol{\sigma}\mathbf{F}) = \sigma_{\nu\mu}F^{\mu\nu} = -\sigma_{\mu\nu}F^{\mu\nu}$, with $\text{tr}(\cdots)$ referring to a trace over μ and ν indices in this context.

In this notation, the evolution of $\Pi^\mu(s)$ generated by the Hamiltonian $H(s)$ through the Heisenberg equations of motion becomes

$$\frac{d\mathbf{\Pi}}{ds} = i[\hat{H}, \mathbf{\Pi}] = 2e\mathbf{F}\cdot\mathbf{\Pi}, \tag{33.A.141}$$

where we have used that since $\mathbf{F}$ is constant it commutes with all operators, including $\mathbf{\Pi}$. This equation is solved by $\mathbf{\Pi}(s) = e^{2es\mathbf{F}}\mathbf{\Pi}(0)$. Similarly,

$$\frac{d\mathbf{x}}{ds} = i[\hat{H}, \mathbf{x}] = 2\mathbf{\Pi}, \tag{33.A.142}$$

which gives

$$\mathbf{x}(s) = \mathbf{x}(0) + 2se^{es\mathbf{F}}\frac{\sinh(es\mathbf{F})}{se\mathbf{F}}\cdot\mathbf{\Pi}(0). \tag{33.A.143}$$

This solution is easy to check by differentiating. In the limit $\mathbf{A} \to 0$, $\mathbf{\Pi} \to \mathbf{p}$ and this becomes $\mathbf{x}(s) = \mathbf{x}(0) + 2s\mathbf{p}(0)$, which is consistent with the eigenstates of $\mathbf{x}(s)$ being those which evolve into position x^μ after a time s.

Thus we have

$$\mathbf{\Pi}(0) = e^{-es\mathbf{F}}\frac{e\mathbf{F}}{2\sinh(es\mathbf{F})}\cdot[\mathbf{x}(s) - \mathbf{x}(0)], \tag{33.A.144}$$

$$\mathbf{\Pi}(s) = e^{es\mathbf{F}}\frac{e\mathbf{F}}{2\sinh(es\mathbf{F})}\cdot[\mathbf{x}(s) - \mathbf{x}(0)]. \tag{33.A.145}$$

The Hamiltonian then becomes

$$\hat{H} = -\mathbf{\Pi}(s)\cdot\mathbf{\Pi}(s) - \frac{e}{2}\text{tr}(\boldsymbol{\sigma}\mathbf{F}) = -[\mathbf{x}(s) - \mathbf{x}(0)]\mathbf{K}[\mathbf{x}(s) - \mathbf{x}(0)] - \frac{e}{2}\text{tr}(\boldsymbol{\sigma}\mathbf{F}), \tag{33.A.146}$$

with $\mathbf{K} \equiv \frac{e^2\mathbf{F}^2}{4\sinh^2(e\mathbf{F}s)}$. Note that $K_{\mu\nu} = K_{\nu\mu}$.

To evaluate $\langle y|e^{-i\hat{H}s}\hat{H}|x\rangle$ in Eq. (33.A.134) using $\hat{H}$, it is helpful first to rewrite $\hat{H}$ so that $\mathbf{x}(s)$ is on the left and $\mathbf{x}(0)$ is on the right. This is not hard:

$$\mathbf{\Pi}(s)\cdot\mathbf{\Pi}(s) = \mathbf{x}(s)\mathbf{K}\mathbf{x}(s) - 2\mathbf{x}(s)\mathbf{K}\mathbf{x}(0) + \mathbf{x}(0)\mathbf{K}\mathbf{x}(0) + K_{\mu\nu}[x^\mu(s), x^\nu(0)]. \tag{33.A.147}$$

Now,

$$K_{\mu\nu}[x^\mu(s), x^\nu(0)] = -\text{tr}\left\{\mathbf{K}\left[\mathbf{x}(0), \mathbf{x}(0) + 2e^{es\mathbf{F}}\frac{\sinh(es\mathbf{F})}{e\mathbf{F}}\cdot\mathbf{\Pi}(0)\right]\right\}$$

$$= \frac{i}{2}\text{tr}[e\mathbf{F} + e\mathbf{F}\coth(es\mathbf{F})]. \tag{33.A.148}$$

So, since $\operatorname{tr}[\mathbf{F}] = 0$, we have

$$\hat{H} = -\mathbf{x}(s)\mathbf{K}\mathbf{x}(s) + 2\mathbf{x}(s)\mathbf{K}\mathbf{x}(0) - \mathbf{x}(0)\mathbf{K}\mathbf{x}(0) - \frac{i}{2}\operatorname{tr}[e\mathbf{F}\coth(es\mathbf{F})] - \frac{e}{2}\operatorname{tr}(\boldsymbol{\sigma}\mathbf{F}) . \tag{33.A.149}$$

In this canonical form, $\hat{H}$ can be evaluated in position eigenstates.

Equation (33.A.134) becomes

$$i\partial_s \langle y; 0|x; s\rangle = -\left\{ (\mathbf{y} - \mathbf{x})\frac{e^2\mathbf{F}^2}{4\sinh^2(es\mathbf{F})}(\mathbf{y} - \mathbf{x}) \right.$$
$$\left. + \frac{i}{2}\operatorname{tr}[e\mathbf{F}\coth(es\mathbf{F})] + \frac{e}{2}\operatorname{tr}(\boldsymbol{\sigma}\mathbf{F}) \right\} \langle y; 0|x; s\rangle, \tag{33.A.150}$$

where $\mathbf{x} = x^\mu$ and $\mathbf{y} = y^\mu$ are position vectors, not operators anymore. This is just a differential equation. The general solution is

$$\langle y; 0|x; s\rangle = C(x, y)\exp\left\{ -i(\mathbf{y} - \mathbf{x})\frac{e\mathbf{F}}{4}\coth(es\mathbf{F})(\mathbf{y} - \mathbf{x}) \right.$$
$$\left. - \frac{1}{2}\operatorname{tr}\ln\left[\frac{\sinh(es\mathbf{F})}{e\mathbf{F}}\right] + i\frac{es}{2}\operatorname{tr}(\boldsymbol{\sigma}\mathbf{F}) \right\} \tag{33.A.151}$$

This can be checked by differentiation and holds for any $C(x, y)$.

To determine $C(x, y)$, we use the additional information that

$$\left(i\frac{\partial}{\partial\mathbf{x}} - e\mathbf{A}\right)\langle y; 0|x; s\rangle = \langle y; 0|e^{-i\hat{H}s}\mathbf{\Pi}(0)|x; s\rangle$$
$$= e^{-es\mathbf{F}}\frac{e\mathbf{F}}{2\sinh(es\mathbf{F})}(\mathbf{y} - \mathbf{x})\langle y; 0|x; s\rangle, \tag{33.A.152}$$

and similarly

$$\left(-i\frac{\partial}{\partial\mathbf{y}} - e\mathbf{A}\right)\langle y; 0|x; s\rangle = e^{es\mathbf{F}}\frac{e\mathbf{F}}{2\sinh(es\mathbf{F})}(\mathbf{y} - \mathbf{x})\langle y; 0|x; s\rangle . \tag{33.A.153}$$

Plugging in our general solution, we find

$$\left[i\frac{\partial}{\partial\mathbf{x}} - e\mathbf{A} - \frac{e}{2}\mathbf{F}(\mathbf{x} - \mathbf{y})\right]C(x, y) = 0, \tag{33.A.154}$$

and

$$\left[-i\frac{\partial}{\partial\mathbf{y}} - e\mathbf{A} - \frac{e}{2}\mathbf{F}(\mathbf{x} - \mathbf{y})\right]C(x, y) = 0. \tag{33.A.155}$$

The solution is

$$C(x, y) = C\exp\left[ie\int_x^y dz^\mu\left(A_\mu(z) + \frac{1}{2}F_{\mu\nu}(z^\nu - y^\nu)\right)\right] . \tag{33.A.156}$$

This line integral is independent of path since the integrand has zero curl. The constant C can be fixed by demanding that the result reduce to the free theory as $A \to 0$. The final result is

$$\langle y; 0|x; s\rangle = \frac{-i}{16\pi^2 s^2} \exp\left[ie \int_x^y dz^\mu \left(A_\mu(z) + \frac{1}{2}F_{\mu\nu}(z^\nu - y^\nu)\right)\right]$$

$$\times \exp\left[-i(\mathbf{y} - \mathbf{x})\frac{e\mathbf{F}}{4}\coth(es\mathbf{F})(\mathbf{y} - \mathbf{x}) + i\frac{es}{2}\mathrm{tr}(\boldsymbol{\sigma}\mathbf{F}) - \frac{1}{2}\mathrm{tr}\ln\left[\frac{\sinh(es\mathbf{F})}{es\mathbf{F}}\right]\right],$$

(33.A.157)

which is manifestly gauge invariant. Taking $A \to 0$ reproduces Eq. (33.11), which confirms the normalization.

Equation (33.A.157) is more generally useful than just for the calculation of the Euler–Heisenberg Lagrangian. The special case when $x = y$ is quoted in Eq. (33.76) and used for the calculation of the $\pi^0 \to \gamma\gamma$ rate in Section 33.5.1.

33.A.2 Effective Lagrangian

Now that we have the proper-time Hamiltonian, we are a small step away from the Euler–Heisenberg Lagrangian. We need to calculate

$$\mathcal{L}_{\mathrm{EH}}(x) = -\frac{1}{4}F_{\mu\nu}^2(x) + \frac{i}{2}\int_0^\infty \frac{ds}{s} e^{-ism^2}\mathrm{Tr}\left\{\langle x|e^{-i\hat{H}s}|x\rangle\right\}$$

$$= -\frac{1}{4}F_{\mu\nu}^2(x) + \frac{1}{32\pi^2}\mathrm{Tr}\left\{\int_0^\infty ds\frac{1}{s^3}\exp\left[-ism^2 + i\frac{es}{2}\mathrm{tr}(\boldsymbol{\sigma}\mathbf{F}) - \frac{1}{2}\mathrm{tr}\ln\left[\frac{\sinh(es\mathbf{F})}{es\mathbf{F}}\right]\right]\right\},$$

(33.A.158)

where Tr is the Dirac trace and tr contracts μ and ν as above.

Now, recall from Eq. (30.65) that

$$[\mathrm{tr}(\boldsymbol{\sigma}\mathbf{F})]^2 = -2\mathrm{tr}(\mathbf{F}^2) - 2i\gamma_5\mathrm{tr}(\mathbf{F}\tilde{\mathbf{F}}) = 8(\mathcal{F} - i\gamma_5\mathcal{G}),$$

(33.A.159)

where $\tilde{F}^{\mu\nu} \equiv \frac{1}{2}\varepsilon^{\mu\nu\alpha\beta}F_{\alpha\beta}$ and

$$\mathcal{F} \equiv \frac{1}{4}F_{\mu\nu}^2 = \frac{1}{2}(\vec{B}^2 - \vec{E}^2),$$

(33.A.160)

$$\mathcal{G} \equiv -\frac{1}{4}F^{\mu\nu}\tilde{F}_{\mu\nu} = \vec{E} \cdot \vec{B}.$$

(33.A.161)

Then, since γ_5 has eigenvalues ± 1, the Dirac eigenvalues of $\mathrm{Tr}(\boldsymbol{\sigma}\mathbf{F})$ are

$$\lambda_i^{\boldsymbol{\sigma}\mathbf{F}} = \pm\sqrt{8(\mathcal{F} \pm i\mathcal{G})},$$

(33.A.162)

with all four sign combinations possible. So,

$$\mathrm{Tr}\left[e^{i\frac{es}{2}\mathrm{tr}(\boldsymbol{\sigma}\mathbf{F})}\right] = 2\cos\left[es\sqrt{2(\mathcal{F} + i\mathcal{G})}\right] + 2\cos\left[es\sqrt{2(\mathcal{F} - i\mathcal{G})}\right]$$

$$= 4\mathrm{Re}\cos[esX],$$

(33.A.163)

where

$$X \equiv \sqrt{\frac{1}{2}F_{\mu\nu}^2 - \frac{i}{2}F^{\mu\nu}\tilde{F}_{\mu\nu}} = \sqrt{2(\mathcal{F} + i\mathcal{G})} = \sqrt{(\vec{B} + i\vec{E})^2}.$$

(33.A.164)

Next we need

$$\frac{1}{2}\operatorname{tr}\ln\left[\frac{\sinh(e\mathbf{F}s)}{es\mathbf{F}}\right]=\ln\sqrt{\lambda_1\lambda_2\lambda_3\lambda_4}, \tag{33.A.165}$$

where λ_i are the four eigenvalues of $\frac{\sinh(e\mathbf{F}s)}{es\mathbf{F}}$. These eigenvalues are determined from the eigenvalues of a constant $F_{\mu\nu}$, which are (see Problem 33.5)

$$\lambda_i^{\mathbf{F}}=\pm\frac{i}{\sqrt{2}}\left[\sqrt{\mathcal{F}+i\mathcal{G}}\pm\sqrt{\mathcal{F}-i\mathcal{G}}\right], \tag{33.A.166}$$

with all four possible sign choices. After some simplification the result is

$$\exp\left\{-\frac{1}{2}\operatorname{tr}\ln\left[\frac{\sinh(e\mathbf{F}s)}{es\mathbf{F}}\right]\right\}=-\frac{(es)^2\mathcal{G}}{\operatorname{Im}\cos(esX)}. \tag{33.A.167}$$

Putting everything together, we find

$$\mathcal{L}_{\mathrm{EH}}(x)=-\frac{1}{4}F_{\mu\nu}^2+\frac{e^2}{32\pi^2}\int_0^\infty ds\frac{1}{s}e^{-im^2s}\frac{\operatorname{Re}\cos(esX)}{\operatorname{Im}\cos(esX)}F^{\mu\nu}\tilde{F}_{\mu\nu}, \tag{33.A.168}$$

which is the final answer for the unrenormalized Euler–Heisenberg effective Lagrangian, in agreement with Eq. (33.71).

Problems

33.1 Complete the calculation of the Euler–Heisenberg Lagrangian using Landau levels in an arbitrary $F_{\mu\nu}$. Show that for an electric field $B\to iE$ is justified. Also show that the result for a general electromagnetic field is given by Eq. (33.71).

33.2 Calculate light-by-light scattering using helicity spinors.

33.3 Calculate the contour integral to derive the pair-production rate Eq. (33.94) from Eq. (33.93). It is helpful to first expand the integration limits to $\int_{-\infty}^{\infty}ds$, then deform the contour to pick up the poles.

33.4 Repeat the analysis in Section 33.6.1 for a fermion. Show that in the non-relativistic limit, the spin is irrelevant.

33.5 Show that the eigenvalues of $F_{\mu\nu}$ are given by Eq. (33.A.166).

Background fields

34

In Chapter 33, we explored how fields could be integrated out exactly leaving an effective action. The concrete example we considered was integrating out the electron from QED. Then we defined the effective action $\Gamma[A] = \int d^4x\, \mathcal{L}_{\text{eff}}$ by the relation

$$\int \mathcal{D}A \exp\left(i \int d^4x\, \mathcal{L}_{\text{eff}}[A]\right) = \int \mathcal{D}A\, \mathcal{D}\bar{\psi}\, \mathcal{D}\psi \exp\left(i \int d^4x\, \mathcal{L}[A, \psi, \bar{\psi}]\right). \quad (34.1)$$

In the special case where we are not concerned with the dynamics of A_μ we can treat A_μ as a classical background and drop the $\int \mathcal{D}A$ on both sides. For example, in Chapter 33 we were able to do the integral over the electron field $\bar{\psi}$ and ψ exactly if we assumed that $F_{\mu\nu}$ was a space-time-independent classical background field, leading to the Euler–Heisenberg effective Lagrangian. The Euler–Heisenberg Lagrangian differs from the exact $\mathcal{L}_{\text{eff}}[A]$ in that it does not contain terms with derivatives acting on $F_{\mu\nu}$ such as $\frac{1}{m_e^2}(\partial_\alpha F_{\mu\nu})^2$ (since these terms vanish for constant $F_{\mu\nu}$). For predictions at low energy, terms with extra derivatives have effects suppressed by factors of $\frac{E^2}{m_e^2}$, and we can ignore them to a first approximation.

A fantastic feature of the Euler–Heisenberg example was that all the work was done in calculating $\mathcal{L}_{\text{eff}}[A]$. Once $\mathcal{L}_{\text{eff}}$ was known, it was used to make a number of quantitative physical predictions with little additional effort: Schwinger pair production, light-by-light scattering, the chiral anomaly, etc. These predictions were made by using $\mathcal{L}_{\text{eff}}$ as a classical Lagrangian generating only tree-level Feynman diagrams. Of course, for the Euler–Heisenberg case, this was only an approximation since we were just ignoring the dynamics of A_μ. But imagine how powerful the effective action would be if it *were* exact. An action $\Gamma[A_\mu, \bar{\psi}, \psi]$ which when used classically (at tree-level) reproduces all of the physics of a full quantum theory is called a **1PI effective action** (for reasons that will soon become clear). The only difference between 1PI effective actions and effective actions like those discussed in Chapter 33 is that those actions had only some subset of the fields integrated out; for a 1PI effective action, all the fields are integrated out.

One can compute a 1PI effective action by matching, that is by evaluating loops in the full theory and demanding that the effective action, when used at tree-level, agrees order-by-order in perturbation theory with the full theory. As we show in Section 34.1.1, terms in the effective Lagrangian computed in this way are easily seen to correspond to one-particle irreducible diagrams in the full theory. An alternative approach (Section 34.1.2) is to identify the effective action as a Legendre transform of the **generating functional of connected diagrams** $W[J]$. ($W[J]$ is related to the generating functional $Z[J]$ from Section 14.3 as $W[J] = -i \ln Z[J]$.) This lets us identify the minimum of $\Gamma[\phi]$ as the expectation of ϕ in the vacuum: $\Gamma'[\langle\phi\rangle] = 0$. As we will see, the Legendre transform approach also leads to an alternative way to compute the effective action by shifting fields (Section 34.1.4): write $S[\phi_b + \phi]$ and integrate out ϕ leading to $\Gamma[\phi_b]$. In this approach, called the **background-field method**, ϕ represents the quantum fluctuations around a classical background ϕ_b. If we assume the background field ϕ_b is constant then the 1PI effective action can be written as $\Gamma[\phi_b] = -VT V_{\text{eff}}(\phi_b)$ with $V_{\text{eff}}(\phi_b)$ known as the **effective potential**.

An example where the details of the effective potential are very important is in the case of spontaneous symmetry breaking. Consider the case of a single scalar field with Lagrangian

$$\mathcal{L} = -\frac{1}{2}\phi\Box\phi - \frac{1}{2}m^2\phi^2 - \frac{\lambda}{4!}\phi^4. \tag{34.2}$$

As we saw in Chapter 28, if $m^2 > 0$, the system is stable, but if $m^2 < 0$ the system is unstable and spontaneous symmetry breaking occurs. But what happens if $m = 0$? To find out if the system with $m = 0$ is stable or not, we need to include quantum corrections. More generally, we would like to know how big the quantum corrections are, since they could conceivably flip the sign of the mass term. Quantum corrections are efficiently encoded in the effective potential. In this case, the effective potential is known as the **Coleman–Weinberg potential**. Its minimum tells us the true quantum value of ϕ. This potential has important implications for the Higgs potential in the Standard Model, as we discuss in Section 34.2.3.

Another important application of the background-field method is in non-Abelian gauge theories. If we replace $A_\mu^a \rightarrow \widetilde{A}_\mu^a + A_\mu^a$ and integrate out A_μ^a, we get a 1PI effective action $\Gamma[\widetilde{A}_\mu^a]$. We can integrate out A_μ^a order-by-order in perturbation theory by computing Feynman diagrams with fixed background fields $\widetilde{A}_\mu^a$. An advantage of this approach over ordinary perturbation theory is that since only the quantum field A_μ^a propagates, only it has to be gauge-fixed. Thus, one can choose a gauge-fixing functional that respects an exact gauge invariance associated with the background field $\widetilde{A}_\mu^a$. In this **background-field gauge**, which really is a family of gauges parametrized by a number $\widetilde{\xi}$, the 1PI effective action is guaranteed to be gauge invariant. This is in contrast to the approach in Chapter 26, where in covariant R_ξ gauges, the gauge-fixing parameter ξ appeared all over the place: in Green's functions, in counterterms, etc. In background-field gauge, the renormalization constants will be $\widetilde{\xi}$ independent. This will provide a quick way to produce the QCD β-function, requiring only 1-loop corrections to the gluon 2-point function be computed, not any 3-point functions (in the ordinary method, we had to compute some 1-loop 3-point graphs).

In general, the 1PI effective action is not guaranteed to be gauge invariant, although physical predictions coming from it must be. Two examples of how this works are discussed: the effective potential in scalar QED is discussed in Section 34.2.4 and the $\Gamma\left[\widetilde{A}_\mu^a\right]$ in non-covariant gauges are discussed in Section 34.3.3.

Here, we leave time ordering and the vacuum implicit, so $\langle \cdots \rangle \equiv \langle \Omega | T\{\cdots\} | \Omega \rangle$. Also $\langle J | \cdots | J \rangle \equiv \langle \Omega | T\{\cdots\} | \Omega \rangle_J$ refers to Green's functions evaluated in the presence of a given classical background configuration $J(x)$ in a Lagrangian with terms such as $J\phi$ added (as in Chapter 33). We will also commonly refer to the 1PI effective action as simply the effective action.

34.1 1PI effective action

In this section we discuss how to compute a 1PI effective action, by matching (Section 34.1.1), through a Legendre transform (Section 34.1.2), or through a background-field calculation (Section 34.1.4).

34.1.1 Matching

Our goal is to compute a 1PI effective action Γ defined so that if used classically (at tree-level), it reproduces Green's functions in the full quantum theory. For example, in QED, we saw in Chapter 18 that the 2-point Green's function for the electron, including all quantum corrections, could be written as

$$G(x, y) = \langle \psi(x)\bar{\psi}(y) \rangle = \int \frac{d^4p}{(2\pi)^4} e^{-ip(x-y)} \frac{i}{\not{p} - m + \Sigma(\not{p})}, \tag{34.3}$$

where $\Sigma(\not{p})$ is the sum of all 1PI contributions to the electron self-energy graph. For this to come out of a tree-level calculation, we must take the kinetic terms for ψ in the effective Lagrangian to have the form $\mathcal{L}_{\text{eff}}^{\text{kin}} = \bar{\psi}[i\not{\partial} - m + \Sigma(i\not{\partial})]\psi$.

In fact, in QED we already know what a number of the terms should look like:

$$\Gamma = \int d^4x \Big\{ \bar{\psi}[i\not{\partial} - m + \Sigma(i\not{\partial})]\psi$$
$$- e\bar{\psi}\Gamma^\mu(i\not{\partial})A_\mu\psi - \frac{1}{2}A_\mu(\partial^\mu\partial^\nu - \Box g^{\mu\nu})(1 + \Pi(-\Box))A_\nu + \cdots \Big\}, \tag{34.4}$$

where $\Sigma(\not{p})$ is the sum of all 1PI contributions to the electron self-energy graph, $-e\Gamma^\mu(\not{p})$ is the sum of all 1PI vertex corrections (with on-shell spinors), and $\Pi(p^2)$ is the 1PI contributions to the vacuum polarization function. Each of these was computed at 1-loop in Part III (see Chapter 19). The $\cdots$ represent higher-dimension operators, such as $(F_{\mu\nu}^2)^2$, which should generically be present. To compute these terms, one would need the set of 1PI contributions to 4-point and higher-point functions. For example, an additional set of terms (those with no derivatives acting on $F_{\mu\nu}$) was computed at 1-loop in the Euler–Heisenberg Lagrangian. In general, each term in the effective action contains all the 1PI contributions to the n-point function with the fields in that term. That is why we call Γ the 1PI effective action. (For the kinetic term, we conventionally define $\Sigma(i\not{\partial})$ to be all the 1PI graphs *except* for the free propagator.)

Taking derivatives of the effective Lagrangian with respect to the fields generates 1PI Green's functions. Two derivatives give the inverse of the 1PI 2-point function:

$$\langle \psi(x)\bar{\psi}(y) \rangle_{1\text{PI}} = \left\{ \frac{\partial}{\partial\psi(x)} \frac{\partial}{\partial\bar{\psi}(y)} (-i)\Gamma[\bar{\psi}, \psi, A] \right\}^{-1}_{\psi=\bar{\psi}=A=0}. \tag{34.5}$$

Applying Eq. (34.5) to Eq. (34.4) reproduces Eq. (34.3). This should be reminiscent of how Green's functions come from derivatives of the generating functional $Z[J]$, a connection that will be exploited in Section 34.1.2. In the same way, the 1PI contributions to the 3-point function are

$$\langle \psi(x) A_\mu(z) \bar{\psi}(y) \rangle_{\text{1PI}} = \frac{\partial}{\partial \psi(x)} \frac{\partial}{\partial \bar{\psi}(y)} \frac{\partial}{\partial A_\mu(z)} i\Gamma[\bar{\psi}, \psi, A] \Big|_{\psi=\bar{\psi}=A=0}. \tag{34.6}$$

For 4-point functions we find

$$\langle \psi(x_1) \bar{\psi}(x_2) \psi(x_3) \bar{\psi}(x_4) \rangle_{\text{1PI}} = \frac{\partial}{\partial \psi(x_1)} \frac{\partial}{\partial \bar{\psi}(x_2)} \frac{\partial}{\partial \psi(x_3)} \frac{\partial}{\partial \bar{\psi}(x_4)} i\Gamma[\bar{\psi}, \psi, A] \Big|_{\psi=\bar{\psi}=A=0}. \tag{34.7}$$

and so on.

To compute complete Green's functions, we can string together 1PI diagrams. For the 2-point function,

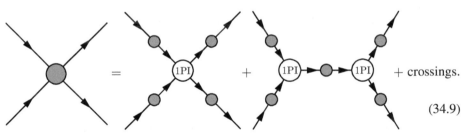

$$\tag{34.8}$$

The 4-point function gets tree-level contributions from the effective action from its 1PI vertex as well as connected contributions to lower-point amplitudes:

$$\tag{34.9}$$

The 4-point function also gets a contribution from disconnected diagrams given by the product of two connected 2-point functions. Although disconnected diagrams can be computed with the effective action, they are not of much interest since they do not contribute to S-matrix elements. Proceeding in this way, Green's functions in the full quantum theory can be constructed from tree-level diagrams using the 1PI effective action.

By the way, you might be wondering, if we can get the results of the full quantum theory with classical physics, why bother with loops at all? That is, why not just *start* from the effective action? The answer is that effective actions are in general highly non-local and hopelessly unconstrained. We have written $\Gamma = \int d^4 x \, \mathcal{L}_{\text{eff}}$ for notational simplicity, but not all effective actions can be written this way. For example, we might have found a contribution to a $\bar{\psi}(q_1)\psi(q_2)\psi(q_3)\bar{\psi}(q_4)$ 1PI diagram of the form $\frac{(q_2+q_3)^2}{(q_1+q_2)^2}$. This could come from a term in an effective action such as

$$\Gamma = \int d^4 x_1 \, d^4 x_2 \, d^4 x_3 \, d^4 x_4 \frac{(\partial_2 + \partial_3)^2}{(\partial_1 + \partial_2)^2} \bar{\psi}(x_1)\psi(x_2)\psi(x_3)\bar{\psi}(x_4), \tag{34.10}$$

which is very manifestly non-local. If we start with a local classical action and then perform the loops, things such as locality, Lorentz invariance and causality are easier to check and confirm. Since the effective action contains basically the same information as the full S-matrix, a construction of an effective action from first principles (unitarity, analyticity etc.) is essentially equivalent to the S-matrix program of the 1950s and 1960s. This approach may eventually prove predictive, but at this point only effective actions derived from local classical Lagrangians are known to be consistent.

On the other hand, if one expands the effective action at energies well below some physical scale, one should be able to write it as a series of local terms. In fact, this is the approach to most effective field theories, such as the 4-Fermi theory or the Chiral Lagrangian (although in these cases one leaves some fields as dynamical). In a low-energy limit, thinking of the effective action as being a series of terms can be useful. Since the effective action should obey the same symmetries as the classical action, the effective action should contain all terms consistent with those symmetries. Thus, a quick-and-dirty approach is to write down an effective action with all possible terms respecting the symmetries of the classical action, with coefficients (representing all the 1PI graphs) estimated by dimensional analysis. We usually assume that anything that can happen from such an effective action will happen – if something does not happen, there should be a symmetry reason for it. This approach is particularly useful in strongly coupled theories, such as low-energy QCD, and explains the success of the Chiral Lagrangian.

34.1.2 Legendre transform

At this point, we have defined a 1PI effective action so that the vertices correspond to the sum of all 1PI contributions to Green's functions. Although one can compute these vertices by simply evaluating the 1PI graphs, one can also compute the effective action through a Legendre transform. This method is very powerful and gives us new intuition for how to think about the effective action, for example, showing that $\Gamma'[\langle\phi\rangle] = 0$. It also lets us justify the background-field method, which is the subject of most of this chapter.

The key result, which we will now derive, is that $\Gamma[\phi]$ is the Legendre transform of a functional $W[J] \equiv -i \ln Z[J]$:

$$\Gamma[\phi] = W[J_\phi] - \int d^4x \, J_\phi(x) \, \phi(x). \tag{34.11}$$

In this equation, J_ϕ is an implicit functional of ϕ defined as the solution to

$$\left.\frac{\partial W[J]}{\partial J(x)}\right|_{J=J_\phi} = \phi(x). \tag{34.12}$$

These are the analogs of the Hamiltonian being the Legendre transform of the Lagrangian: $H[\pi] = L[\dot{x}_\pi] - \dot{x}_\pi \pi$ with $\dot{x}_\pi$ an implicit function of π defined so that $\left.\frac{\partial L[\dot{x}]}{\partial \dot{x}}\right|_{\dot{x}=\dot{x}_\pi} = \pi$. The conjugate relation comes from varying Eq. (34.11) with respect to ϕ:

$$\frac{\partial \Gamma[\phi]}{\partial \phi(x)} = \int d^4y \left[\frac{\partial J_\phi(y)}{\partial \phi(x)} \frac{\partial W[J_\phi]}{\partial J_\phi(y)} - \frac{\partial J_\phi(y)}{\partial \phi(x)}\phi(y)\right] - J_\phi(x) = -J_\phi(x). \tag{34.13}$$

This lets us define ϕ_J as an implicit functional of J satisfying

$$\frac{\partial \Gamma[\phi]}{\partial \phi}\bigg|_{\phi=\phi_J} = -J, \qquad (34.14)$$

which gives the inverse Legendre transform

$$W[J] = \Gamma[\phi_J] + \int d^4x\, J(x)\phi_J(x). \qquad (34.15)$$

Equations (34.14) and (34.15) are the analogs of $L[\dot{x}] = H[\pi] + \pi_{\dot{x}}\dot{x}$ with $\pi_{\dot{x}}$ an implicit function of $\dot{x}$ defined so that $\frac{\partial H[\pi]}{\partial \pi}\big|_{\pi=\pi_{\dot{x}}} = -\dot{x}$. Varying Eq. (34.15) with respect to J gives

$$\frac{\partial W[J]}{\partial J(x)} = \int d^4y \left[\frac{\partial \phi_J(y)}{\partial J(x)} \frac{\partial \Gamma[\phi_J]}{\partial \phi_J(y)} + J(y)\frac{\partial \phi_J(y)}{\partial J(x)} \right] + \phi_J(x) = \phi_J(x). \qquad (34.16)$$

and brings us full circle back to Eq. (34.12). We will now take Eq. (34.11) as a new definition of $\Gamma[\phi]$ and show that it agrees with our previous definition, as the functional whose vertices are the sums of 1PI graphs.

The first task is to get to know $W[J]$. It is defined by

$$e^{\frac{i}{\hbar}W[J]} = Z[J] = \int \mathcal{D}\phi \exp\left\{ \frac{i}{\hbar}\left[S[\phi] + \int d^4x\, J\phi \right] \right\}, \qquad (34.17)$$

where the $\hbar$ factors are restored here for later reference. Now, recall from Section 14.3 that $Z[J]$ generates Green's functions via

$$\langle J|\phi(x_1)\cdots\phi(x_n)|J\rangle = (-i\hbar)^n \frac{1}{Z[J]} \frac{\partial^n Z}{\partial J(x_1)\cdots\partial J(x_n)}. \qquad (34.18)$$

Usually, we set $J = 0$ after taking the derivatives, turning the left-hand side of this equation into a vacuum matrix element. With $J \neq 0$, $Z[J]$ generates Green's functions for ϕ in a background given by a classical current $J(x)$ (see Chapter 33 for an example). A helpful way to think about $W[J]$ is that it generates all *connected* diagrams:

$$(-i\hbar)^n \frac{\partial^n W[J]}{\partial J(x_1)\cdots\partial J(x_n)} = -i\hbar\langle J|\phi(x_1)\cdots\phi(x_n)|J\rangle_{\text{connected}}. \qquad (34.19)$$

For example, taking $n = 2$ we find using $W[J] = -i\hbar \ln Z[J]$ that

$$\begin{aligned}
(-i\hbar)^2 \frac{\partial^2 W}{\partial J_1\, \partial J_2} &= (-i\hbar)^3 \frac{\partial}{\partial J_1}\left(\frac{1}{Z}\frac{\partial Z}{\partial J_2} \right) \\
&= (-i\hbar)^3 \frac{1}{Z}\frac{\partial^2 Z}{\partial J_1 \partial J_2} - (-i\hbar)^3 \left(\frac{1}{Z}\frac{\partial Z}{\partial J_1} \right)\left(\frac{1}{Z}\frac{\partial Z}{\partial J_2} \right) \\
&= -i\hbar\Big[\langle J|\phi(x_1)\phi(x_2)|J\rangle - \langle J|\phi(x_1)|J\rangle\langle J|\phi(x_2)|J\rangle \Big],
\end{aligned} \qquad (34.20)$$

where $J_i = J(x_i)$ is the same shorthand used in Section 14.3.2. The first term on the second line is the full Green's function, in the presence of the source, including connected and disconnected pieces. The second term is the disconnected pieces, which are subtracted off. You can try other examples and prove this interpretation of $W[J]$ in Problem 34.1.

Table 34.1 Scaling with $\hbar$ of some representative Feynman diagrams. Connected tree-level diagrams scale as $\hbar^{-1}$ and connected loops as $\hbar^{\#\text{loops}-1}$. Disconnected diagrams violate these rules.

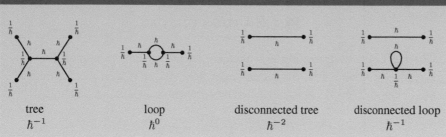

For a 1-point function (which is always connected), we find

$$\frac{\partial W[J]}{\partial J(x)} = -i\hbar \frac{1}{Z} \frac{\partial Z[J]}{\partial J(x)} = \langle J|\phi(x)|J\rangle. \tag{34.21}$$

This gives us a physical interpretation of the Legendre relation in Eq. (34.12): for a given (classical) field configuration $\phi_c(x)$, the current $J_{\phi_c}(x)$ is precisely the current in whose presence the expectation value of ϕ is ϕ_c: $\langle J_{\phi_c}|\phi|J_{\phi_c}\rangle = \phi_c$. Conversely, since $\phi_J = \frac{\partial W[J]}{\partial J(x)}$, from Eq. (34.16) we see that ϕ_J can be identified as the expectation value of ϕ in the presence of a given current J: $\phi_J = \langle J|\phi|J\rangle$. Thus, $\phi_{J_\phi} = \phi$ and $J_{\phi_J} = J$, as one would expect.

Now we are ready to show that $\Gamma[\phi]$ in Eq. (34.11) is the same effective action whose vertices contain all the 1PI graphs. Such an effective action, when used at tree-level, should give the same Green's functions as the action $S[\phi]$ would, including loops. Since tree-level contributions are classical, we only need to isolate the leading contribution as $\hbar \to 0$. Since each term in $\frac{1}{\hbar}S[\phi]$ has a factor of $\frac{1}{\hbar}$, vertices and external states come with factors of $\frac{1}{\hbar}$ while propagators come with factors of $\hbar$. The overall scaling of some sample diagrams is shown in Table 34.1. We see that connected tree-level diagrams scale as $\hbar^{-1}$, and each loop adds a factor of $\hbar$. Disconnected diagrams can violate this scaling, but do not contribute to the S-matrix (see Problem 34.1).

Now we can relate the effective action to $W[J]$. Since all connected diagrams can either be computed with $W[J] = -i\hbar \ln Z[J]$ using $S[\phi]$ to all orders in $\hbar$, or they can be computed with the equivalent of $W[J]$ constructed using $\Gamma[\phi]$ instead of $S[\phi]$ in the $\hbar \to 0$ limit, we must have

$$W[J] = \lim_{\hbar \to 0}(-i\hbar) \ln\left[\int \mathcal{D}\phi \exp\left\{\frac{i}{\hbar}\left[\Gamma[\phi] + \int d^4x\, J\phi\right]\right\}\right]. \tag{34.22}$$

Taking $\hbar \to 0$ isolates exactly the tree-level diagrams computed using $\Gamma[\phi]$. Taking $\hbar \to 0$ also forces the field configurations contributing to the path integral to be exactly those that extremize the action. Thus, the ϕ integral just replaces ϕ by ϕ_J defined by $\frac{\partial \Gamma[\phi_J]}{\partial \phi_J} = -J$ as in Eq. (34.14). This leads to

$$W[J] = \Gamma[\phi_J] + \int d^4x\, J(x)\,\phi_J(x), \tag{34.23}$$

which is the (inverse) Legendre transform in Eq. (34.15).

In conclusion, the functional $\Gamma[\phi]$, defined as the Legendre transform of $W[J]$, is the 1PI effective action $\Gamma[\phi]$ described in Section 34.1.1, whose vertices are sums of 1PI diagrams in the full quantum theory.

34.1.3 Cross checks

To make the Legendre transform less abstract, let us try some examples. First, we will calculate the effective action for a free scalar field. We start with the Lagrangian $\mathcal{L} = -\frac{1}{2}\phi(\Box + m^2)\phi$. Then $W[J]$ is calculated by performing a Gaussian path integral:

$$
\begin{aligned}
W[J] &= -i\ln \int \mathcal{D}\phi \exp\left\{i\int d^4x\left[-\frac{1}{2}\phi(\Box + m^2)\phi + J\phi\right]\right\} \\
&= \int d^4x \frac{1}{2}J\frac{1}{\Box + m^2}J + \text{const.}
\end{aligned}
\tag{34.24}
$$

To Legendre transform, we need J_ϕ, which is defined to solve

$$
\phi(x) = \left.\frac{\partial W[J]}{\partial J(x)}\right|_{J=J_\phi} = \frac{1}{\Box + m^2}J_\phi(x).
\tag{34.25}
$$

Thus, we find $J_\phi = (\Box + m^2)\phi$ and

$$
\Gamma[\phi] = W[J_\phi] - \int d^4x\, J_\phi(x)\,\phi(x) = \int d^4x\left[-\frac{1}{2}\phi(\Box + m^2)\phi\right] + \text{const.}
\tag{34.26}
$$

So, the effective action is identical to the classical action for a free-field theory (up to a constant), as we should expect.

As another check, let us verify that, in an interacting theory, derivatives of $\Gamma[\phi]$ with respect to ϕ do generate full 1PI Green's functions, as in Section 34.1.1. First, we take one derivative and reproduce Eq. (34.13):

$$
\frac{\partial \Gamma[\phi]}{\partial \phi} = -J_\phi.
\tag{34.27}
$$

This equation implies that the field configuration ϕ_0 that satisfies this equation for $J_\phi = 0$ satisfies the classical equations of motion using the action $\Gamma[\phi]$. The same field configuration also has $\phi_0 = \left.\frac{\partial W[J]}{\partial J}\right|_{J=0} = \langle\phi\rangle$ by Eq. (34.21). In other words, the effective action is minimized by the field configuration given by the expectation value of the quantum field ϕ in the true vacuum of the theory.

Taking two derivatives should give the inverse of the 2-point function, as in Eq. (34.5). To check this, we note that

$$
G(x,y) = \langle\phi(x)\phi(y)\rangle = -i\left.\frac{\partial^2 W}{\partial J(x)\,\partial J(y)}\right|_{J=0} = -i\left.\frac{\partial \phi_J(y)}{\partial J(x)}\right|_{J=0},
\tag{34.28}
$$

where $\frac{\partial W[J]}{\partial J(x)} = \phi_J(x)$ from Eq. (34.16) has been used. Thus we find

$$
\left.\frac{\partial^2 \Gamma[\phi]}{\partial \phi(x)\partial \phi(y)}\right|_{\phi=\phi_0} = -\left.\frac{\partial J(y)}{\partial \phi(x)}\right|_{J=0} = iG(x,y)^{-1},
\tag{34.29}
$$

where Eq. (34.27) has been used. This agrees with the equivalent of Eq. (34.5) for a scalar theory, except instead of evaluating $\phi = 0$ we must evaluate ϕ on its expectation value

$\phi_0 = \langle \phi \rangle$ (in QED, these are the same since $\psi, \bar{\psi}$ and A_μ have vanishing expectation values).

34.1.4 Background fields

At this point we have discussed some properties of the functional $\Gamma[\phi]$:

- It can be used at tree-level to generate Green's functions of ϕ.
- Its vertices correspond to the sum of 1PI contributions to Green's functions with a given number of external states.
- It is formally given by the Legendre transform of $W[J] = -i \ln Z[J]$.
- Its minimum is at the value $\phi_0 = \langle \phi \rangle$ in the true vacuum of the theory.

We can calculate $\Gamma[\phi]$ by matching – each term is just the 1PI Green's function for certain external states, as in Eq. (34.4). We also know how to calculate an approximation to $\Gamma[\phi]$ by integrating out other fields besides ϕ, as in Eq. (34.1). What we would like to show next is how to include fluctuations of ϕ itself in the computation of $\Gamma[\phi]$.

To begin, suppose we have an action $S[\phi]$. From $S[\phi]$ we can compute its 1PI effective action $\Gamma[\phi]$ through a Legendre transform. Now shift $\phi \rightarrow \phi_b + \phi$ for a non-dynamical, but arbitrary, background-field configuration $\phi_b(x)$. Non-dynamical in this context means that ϕ_b is not integrated over in the path integral. For example, if $S[\phi] = \int d^4x \left(-\frac{1}{2} \phi \Box \phi + \frac{g}{3!} \phi^3 \right)$ we would find

$$S_b[\phi_b, \phi] = S[\phi + \phi_b] = \int d^4x \left(-\frac{1}{2} \phi \Box \phi - \phi \Box \phi_b - \frac{1}{2} \phi_b \Box \phi_b \right.$$
$$\left. + \frac{g}{3!} \phi^3 + \frac{g}{2} \phi^2 \phi_b + \frac{g}{2} \phi_b^2 \phi + \frac{g}{3!} \phi_b^3 \right) \quad (34.30)$$

This new action $S_b[\phi_b, \phi]$ leads to a new effective action $\Gamma_b[\phi_b, \phi]$ which depends on ϕ_b.

We can compute $\Gamma_b[\phi_b, \phi]$ by matching, designing it so that its vertices reproduce the 1PI graphs with external ϕ fields. In doing this, we need to use the new vertices such as $\frac{g}{2} \phi^2 \phi_b$, which give new interactions. For example, the 2-point function might get contributions from

$$G_b(x,y) = \left\{ \frac{-i \partial \Gamma_b[\phi_b, \phi]}{\partial \phi(x) \, \partial \phi(y)} \bigg|_{\phi=0} \right\}^{-1}$$

$$= \underline{\qquad} + \underline{\quad} \quad + \quad \underline{\quad} \quad + \cdots , \quad (34.31)$$

where the single lines represent ϕ fields and the grounded lines represent ϕ_b fields. Terms such as $\frac{1}{2} \phi_b \Box \phi_b$ or $\frac{g}{3!} \phi_b^3$ with no ϕ fields can be pulled right out of the path integral, since ϕ_b is not dynamical. Also terms such as $\frac{g}{2} \phi_b^2 \phi$ can be ignored, since they will never contribute to a 1PI graph. Thus, the computation is equivalent to one performed with

$$S_{\text{trunc}}[\phi_b, \phi] \equiv S[\phi + \phi_b] - S'[\phi_b]\phi. \quad (34.32)$$

Reducing the Lagrangian in this way is not necessary, but simplifies some computations.

What we will show next is that $\Gamma_b[\phi_b, 0] = \Gamma_b[0, \phi_b] = \Gamma[\phi_b]$, where $\Gamma[\phi]$ is the effective action containing the 1PI graphs in $S[\phi]$. To prove this, start by defining

$$\exp(iW_b[\phi_b, J]) = \int \mathcal{D}\phi \exp\left\{ iS[\phi_b + \phi] + i \int d^4x \, J\phi \right\}. \tag{34.33}$$

The analog of Eq. (34.16) is then $\frac{\partial W_b[\phi_b, J]}{\partial J} = \phi_{J;b}$. Now shift $\phi \to \phi - \phi_b$ in the path integral to give

$$W_b[\phi_b, J] = W[J] - \int d^4x \, J\phi_b. \tag{34.34}$$

Differentiating with respect to J gives $\frac{\partial W_b[\phi_b, J]}{\partial J} = \frac{\partial W[J]}{\partial J} - \phi_b$, which implies

$$\phi_{J;b} = \phi_J - \phi_b. \tag{34.35}$$

This is quite a natural relation. It says that when we replace $\phi \to \phi_b + \phi$ in the path integral, the expectation value ϕ of a field in a given current shifts by $-\phi_b$.

Now, $\Gamma_b[\phi_b, \phi]$ can be computed through the Legendre transform, as in Eq. (34.12) but with additional dependence on the background field:

$$\Gamma_b[\phi_b, \phi] = W_b[\phi_b, J_{\phi;b}] - \int d^4x \, J_{\phi;b}\phi. \tag{34.36}$$

Then using Eq. (34.34), we find

$$\Gamma_b[\phi_b, \phi] = W[J_{\phi;b}] - \int d^4x \, J_{\phi;b}(\phi_b + \phi). \tag{34.37}$$

Now we take $\phi = \phi_{J;b}$, use Eq. (34.35), and that $J_{\phi_J} = J$ to get

$$\Gamma_b[\phi_b, \phi_{J;b}] = W[J] - \int d^4x \, J\phi_J = \Gamma[\phi_J] = \Gamma[\phi_{J;b} + \phi_b]. \tag{34.38}$$

Since this holds for any J, we find $\Gamma_b[\phi_b, \phi] = \Gamma[\phi + \phi_b]$ and therefore $\Gamma[\phi_b] = \Gamma_b[\phi_b, 0]$ as desired. It is also true that $\Gamma[\phi] = \Gamma_b[0, \phi]$, but setting $\phi_b = 0$ does not get us anywhere.

Computing $\Gamma[\phi]$ with the relation $\Gamma[\phi_b] = \Gamma_b[\phi_b, 0]$ means computing graphs with no ϕ_b particles in loops and no external ϕ fields from $S[\phi + \phi_b]$. At zeroth order, $\Gamma[\phi_b] = S[\phi_b]$. The leading-order contribution comes from vacuum bubbles (0-point functions) in a constant background field:

$$\Gamma_b[\phi_b, 0] = S[\phi_b] \ + \ \bigcirc \ + \ \bigcirc \ + \ \bigcirc \ + \ \oslash \ + \cdots. \tag{34.39}$$

Calculating the effective action this way is known as the **background-field method**.

An occasionally useful shorthand for $\Gamma[\phi_b] = \Gamma_b[\phi_b, 0]$ is

$$e^{i\Gamma[\phi_b]} = \int_{\text{restr.}} \mathcal{D}\phi \, e^{iS[\phi + \phi_b]}, \tag{34.40}$$

where the subscript "restr." means that only a restricted set of field configurations can be integrated over. Without some kind of restriction, we could just shift $\phi \to \phi - \phi_b$ and $\Gamma[\phi_b]$ would be independent of ϕ_b. In perturbation theory, this restricted set is the 1PI diagrams. More generally, whatever field configurations are necessary to produce the effective action must be included.

34.1.5 Summary

In Chapter 33, we discussed effective actions coming from integrating out a subset of fields, such as $\mathcal{L}_{\text{eff}}[A]$ from integrating out $\bar{\psi}$ and ψ in QED. In this chapter, we have introduced 1PI effective actions $\Gamma[\phi]$ (or $\Gamma[A, \bar{\psi}, \psi]$ for QED), in which quantum fluctuations of *all* fields have been integrated out. The vertices in this action correspond to 1PI diagrams. Thus, the 1PI effective action can be used at tree-level to give Green's functions including all quantum corrections. In general, the 1PI effective action is highly non-local.

We then saw that the 1PI effective action could be written concisely as a Legendre transform $\Gamma[\phi] = W[J_\phi] - \int d^4x\, \phi J_\phi$, where J_ϕ is an implicit functional of ϕ defined so that $\Gamma'[\phi] = -J_\phi$ and $W[J] = -i \ln Z[J]$. This leads to a physical picture of the effective action: the minimum of $\Gamma[\phi]$ is at $\phi_c = \langle \phi \rangle$, the expectation value of ϕ in the true vacuum of the theory. The value of $\Gamma[\phi]$ away from its minimum corresponds to the action for the quantum system in the presence of an external current J_ϕ. Conversely, with a non-zero current J, the minimum of the action shifts to $\langle J|\phi|J \rangle$. Thus, the effective action maps out how the minimum moves when the theory is modified. To repeat: only the true ground state, with $J = 0$, describes the solution to the quantum theory with a given classical action $S[\phi]$. Values of $\Gamma[\phi]$ for $\phi \neq \langle \phi \rangle$ correspond to solutions to a different quantum theory, one where an external current is present.

Finally, we found a convenient relation: that $\Gamma[\phi_b]$ could be computed by evaluating 1PI graphs in a theory using $S[\phi + \phi_b]$ instead of $S[\phi]$. Since only 1PI graphs contribute, we can equally well use $S[\phi + \phi_b] - S'[\phi_b]\phi$, which removes all the tadpole terms from the Lagrangian. It is worth emphasizing that removing the tadpoles is very important – it is the main reason we had to go through this whole rigmarole with the Legendre transform. Away from the true minimum of $\Gamma[\phi]$ there is a non-zero current J_ϕ. For example, suppose the action $S[\phi]$ has a minimum at $\phi = 0$ but the effective action $\Gamma[\phi]$ does not; then $J_0 \neq 0$. In computing the effective action, we really want to integrate out fluctuations about the true minimum, not around $\phi = 0$. That is where the current comes in: the $J\phi$ term exactly compensates the tadpole for *any* value of ϕ. Thus, we can do the $\Gamma[\phi]$ calculation for a general J_ϕ and then simply set $J = 0$ to find the true minimum.

34.2 Background scalar fields

In this section we give some examples of effective action calculations with background scalar fields. We begin with a simple ϕ^4 theory, where one can ask the interesting question of whether spontaneous symmetry breaking occurs for $\mathcal{L} = -\frac{1}{2}\phi \Box \phi - \frac{\lambda}{4!}\phi^4$. The effective potential in this case is called the Coleman–Weinberg potential. The Coleman–Weinberg potential has important implications for the Standard Model: it can tell us whether very large values of the Higgs field, $h \gg v$, can give a lower-energy state than $h = v$. We also discuss the contribution of gauge fields to an effective action, through a scalar QED example. Although the effective potential becomes gauge dependent when gauge fields are included, physical predictions will be gauge independent. We give an example of such a physical prediction.

34.2.1 The Coleman–Weinberg potential

Consider a general theory of a single scalar field ϕ, with Lagrangian

$$\mathcal{L} = -\frac{1}{2}\phi\Box\phi - V[\phi], \tag{34.41}$$

where $V[\phi]$ is some potential. For example, $V = \frac{1}{2}m^2\phi^2 + \frac{1}{4!}\lambda\phi^4$. With these signs, the classical minimum is at $\phi = 0$, which preserves the Z_2 symmetry under $\phi \to -\phi$ of this Lagrangian. A natural question is whether quantum corrections change this. In other words, is $\langle\phi\rangle = 0$ in the full theory? If not, the Z_2 symmetry under $\phi \to -\phi$ of the classical Lagrangian is spontaneously broken. The question is particularly intriguing in the case when $m^2 = 0$, where an infinitesimal shift that makes $m^2 < 0$ would destabilize the system. The 1PI effective action is exactly what we need to find out what happens at the quantum level: since $\Gamma'[\langle\phi\rangle] = 0$, we just need to see if $\Gamma'[0] = 0$.

Following the general method outlined in the previous section, we shift $\phi \to \phi_b + \phi$ and drop the tadpoles to get

$$e^{i\Gamma[\phi_b]} = e^{i\int d^4x \left(-\frac{1}{2}\phi_b\Box\phi_b - V[\phi_b]\right)} \int_{\mathrm{restr.}} \mathcal{D}\phi\, e^{i\int d^4x\left(-\frac{1}{2}\phi\Box\phi - \frac{1}{2}\phi^2 V''(\phi_b) - \frac{1}{3!}\phi^3 V'''(\phi_b) + \cdots\right)}$$
$$\tag{34.42}$$

For a general potential, we will never be able to evaluate this path integral. However, we can easily evaluate it at 1-loop order. 1-loop means one ϕ-loop (since ϕ_b does not propagate) with an arbitrary number of external ϕ_b fields, as in Eq. (34.39). In the language of Chapter 33, we want to compute the ϕ propagator in terms of a background ϕ_b and close the ends of the propagator together to form a loop.

Technically, 1-loop means next-to-leading order in an expansion in $\hbar$. Since $\hbar = 1$, this loop expansion generically makes no sense. However, when there is a small coupling, then 2-loop graphs will be suppressed by some coupling compared to 1-loop graphs with the same number of external ϕ_b legs. Thus, the loop expansion in background-field calculations is not any less justified than in ordinary perturbation theory. Nevertheless, we will have to check for self-consistency to see that the quantum corrections are not large.

Since none of the ϕ^3 or higher vertices can contribute at 1-loop, we can truncate to quadratic order in ϕ giving

$$e^{i\Gamma[\phi_b]} = e^{i\int d^4x\left(-\frac{1}{2}\phi_b\Box\phi_b - V[\phi_b]\right)} \int \mathcal{D}\phi\, \exp\left[i\int d^4x\left(-\frac{1}{2}\phi\Box\phi - \frac{1}{2}\phi^2 V''(\phi_b)\right)\right],$$
$$\tag{34.43}$$

where the "restr." subscript has been dropped since the only diagrams left have one closed loop and are 1PI. Now we have reduced our problem to a Gaussian integral, which we can do exactly:

$$e^{i\Gamma[\phi_b]} = \text{const.} \times e^{i\int d^4x\left(-\frac{1}{2}\phi_b\Box\phi_b - V[\phi_b]\right)} \frac{1}{\sqrt{\det(\Box + V''[\phi_b])}}. \tag{34.44}$$

So the calculation is reduced to evaluating this functional determinant. Using the standard tricks (see Chapter 33), we have $\Gamma[\phi] = \int d^4x\left(-\frac{1}{2}\phi_b\Box\phi_b - V[\phi_b]\right) + \Delta\Gamma[\phi_b]$, where

$$i\Delta\Gamma[\phi_b] = \ln\frac{1}{\sqrt{\det(\Box + V''(\phi_b))}} + \text{const.} = -\frac{1}{2}\text{tr}\ln(\Box + V''(\phi_b)) + \text{const.} \tag{34.45}$$

Pulling out a ϕ_b-independent trace over $\ln\Box$ we can make the logarithm dimensionless. Then, writing the trace as a d^4x integral (see Chapters 30 or 33), we have

$$i\Delta\Gamma[\phi_b] = -\frac{1}{2}\int d^4x\langle x|\ln\left(1 + \frac{V''[\phi_b]}{\Box}\right)|x\rangle + \text{const.} \tag{34.46}$$

Unfortunately, unless ϕ_b is constant, this integral is very hard to evaluate. So let us assume ϕ_b is constant. Then $V''(\phi_b)$ becomes just a function rather than a functional. For example, if $V(\phi) = \frac{1}{2}m^2\phi^2 + \frac{\lambda}{4!}\phi^4$ then $V''(\phi_b) = m^2 + \frac{\lambda}{2}\phi_b^2$. This motivates us to think of $V''(\phi_b)$ as an effective mass-squared, and thus we define $m_{\text{eff}}^2(\phi_b) \equiv V''(\phi_b)$, often leaving the dependence of m_{eff} on ϕ_b implicit.

Inserting a complete set of momentum states, we find

$$i\Delta\Gamma[\phi_b] = -\frac{1}{2}\int d^4x \int \frac{d^4k}{(2\pi)^4}\ln\left(1 - \frac{m_{\text{eff}}^2}{k^2}\right) + \text{const.} \tag{34.47}$$

The $\int d^4x$ just gives VT, which allows us to write $\Delta\Gamma[\phi_b] = -VT\Delta V_{\text{eff}}(\phi_b)$, with $V_{\text{eff}}(\phi_b)$ what we call the effective potential. The d^4k integral is very badly divergent. Since this is just a scalar field theory, nothing goes wrong if we Wick rotate $k^2 \to -k_E^2$ and impose a hard cutoff $k_E < \Lambda$. Then we get

$$\Delta\Gamma[\phi_b] = -VT\frac{2\pi^2}{2(2\pi)^4}\int_0^\Lambda dk_E\, k_E^3\ln\left(1 + \frac{m_{\text{eff}}^2}{k_E^2}\right) + \text{const.}$$

$$= -\frac{VT}{128\pi^2}\left(2m_{\text{eff}}^2\Lambda^2 + 2m_{\text{eff}}^4\ln\frac{m_{\text{eff}}^2}{\Lambda^2} + \text{const.}\right). \tag{34.48}$$

In the second line, $\Lambda \gg m_{\text{eff}}$ has been used to replace $\ln(1 + \frac{\Lambda^2}{m_{\text{eff}}^2})$ by $-\ln\frac{m_{\text{eff}}^2}{\Lambda^2}$. The full effective potential defined by $\Gamma[\phi_b] = -VTV_{\text{eff}}(\phi_b)$ with ϕ_b assumed constant is then

$$V_{\text{eff}}(\phi_b) = V(\phi_b) + c_1 + c_2 m_{\text{eff}}^2(\phi_b) + \frac{1}{64\pi^2}m_{\text{eff}}^4\ln\frac{m_{\text{eff}}^2(\phi_b)}{c_3}, \tag{34.49}$$

with c_1, c_2 and c_3 some uninteresting, regulator-dependent but ϕ_b-independent, divergent constants (e.g. $c_2 = \frac{1}{64\pi^2}\Lambda^2$ with the hard cutoff used above).

As usual, the various divergences can be removed through renormalization with the physics content residing in the logarithmic term. Adding counterterms in the usual way lets us write

$$V(\phi) = \frac{1}{2}m_R^2(1 + \delta_m)\phi^2 + \frac{\lambda_R}{4!}(1 + \delta_\lambda)\phi^4 + \Lambda_R(1 + \delta_\Lambda), \tag{34.50}$$

with δ_m, δ_Λ and δ_λ assumed to start at $\mathcal{O}(\lambda_R^2)$. Then $m_{\text{eff}}^2(\phi) = V''(\phi) = m_R^2 + \frac{\lambda_R}{2}\phi^2 + \mathcal{O}(\lambda_R^2)$ and the effective action is

$$V_{\text{eff}}(\phi) = \Lambda_R(1 + \delta_\Lambda) + \frac{1}{2}m_R^2(1 + \delta_m)\phi^2 + \frac{\lambda_R}{4!}(1 + \delta_\lambda)\phi^4$$
$$+ c_1 + c_2(m_R^2 + \frac{\lambda_R}{2}\phi^2) + \frac{\left(m_R^2 + \frac{\lambda_R}{2}\phi^2\right)^2}{64\pi^2}\ln\frac{m_R^2 + \frac{\lambda_R}{2}\phi^2}{c_3}. \quad (34.51)$$

Now we need renormalization conditions to fix δ_Λ, δ_m and δ_λ.

To address the question of whether the Z_2 symmetry of the classical Lagrangian is spontaneously broken for $m_R = 0$ we need to define $m_R = 0$ carefully. Normally we might define m_R as the mass of ϕ. Such a definition is problematic in the current case since it presupposes that ϕ is a physical degree of freedom with positive m_R^2, which is what we are trying to check. For a classical potential, $V(\phi) = \Lambda + \frac{m^2}{2}\phi^2 + \frac{\lambda}{4!}\phi^4$, the mass can be determined by $m^2 = V''(0)$. Thus, an alternative to asking that a mass vanish is that $V_{\text{eff}}''(0) = 0$. We therefore take $m_R^2 = V_{\text{eff}}''(0) = 0$ as one renormalization condition. Similarly, $\Lambda = 0$ classically can be written as $V(0) = 0$, so we set $\Lambda_R = V_{\text{eff}}(0) = 0$, which sets a second renormalization condition. For λ_R, the analogous renormalization condition $V_{\text{eff}}''''(0) = \lambda_R$ does not work, since the effective potential is singular at $\phi = 0$ (when $m_R = 0$). Instead, we can set $V_{\text{eff}}''''(\phi_R) = \lambda_R$ for some fixed (but arbitrary) scale ϕ_R.

Using $m_R^2 = V_{\text{eff}}''(0) = \Lambda_R = V_{\text{eff}}(0)$ and $V_{\text{eff}}''''(\phi_R) = \lambda_R$, solving for the counterterms, and plugging in gives

$$V_{\text{eff}}(\phi) = \frac{1}{4!}\phi^4\left\{\lambda_R + \frac{3\lambda_R^2}{32\pi^2}\left[\ln\left(\frac{\phi^2}{\phi_R^2}\right) - \frac{25}{6}\right]\right\}, \quad (34.52)$$

which is known as the **Coleman–Weinberg potential** [Coleman and Weinberg, 1973].

Now let us return to our original question: Is $\langle\phi\rangle = 0$ or not? It seems from Eq. (34.52) that the minimum is now not at zero, but $V'(\langle\phi\rangle) = 0$ when

$$\lambda_R \ln\frac{\langle\phi\rangle^2}{\phi_R^2} = \frac{11}{3}\lambda_R - \frac{32}{3}\pi^2. \quad (34.53)$$

Unfortunately, since $\frac{32}{3}\pi^2 \approx 105$ this is a large logarithm: $\left|\lambda_R \ln\frac{\langle\phi\rangle^2}{\phi_R^2}\right| \gg 1$. Thus, one expects higher-order terms in the background-field calculation (such as a $\phi^4\lambda_R^3\ln^2\frac{\phi^2}{\phi_R^2}$ correction to $V(\phi)$) to be at least as important as the term we calculated. We cannot proceed further without resumming these large logarithms.

34.2.2 Resummation of large logarithms

In order to resum the large logarithms in the effective potential, it is useful to think about other equivalent ways that the potential can be calculated. First, it is possible, and almost as easy, to compute the effective potential by summing the relevant Feynman diagrams. For simplicity, let us specialize to the massless theory with $\mathcal{L} = -\frac{1}{2}\phi\Box\phi - \frac{\lambda_R}{4!}\phi^4$. Since a constant background field carries zero momentum, a 1-loop diagram with n background fields is the same as a scalar loop with no external fields multiplied by a factor of $\left(-\frac{i}{2}\lambda_R\phi_b^2\right)^n$. For example, with 10 external ϕ_b, the loop is

$$\int \frac{d^4k}{(2\pi)^4} \frac{1}{10} \left(\frac{\frac{1}{2}\lambda\phi_b^2}{k^2 + i\varepsilon}\right)^5, \qquad (34.54)$$

with the $\frac{1}{10}$ a symmetry factor (rotation and reflection). This diagram is badly IR divergent; and diagrams with more external fields are even more IR divergent. However, summing all the diagrams at the integrand level, we find

$$i\Delta\Gamma = VT \int \frac{d^4k}{(2\pi)^4} \sum_{n=1}^{\infty} \frac{1}{2n} \left(\frac{\frac{1}{2}\lambda\phi_b^2}{k^2 + i\varepsilon}\right)^n = -VT\frac{1}{2} \int \frac{d^4k}{(2\pi)^4} \ln\left(1 + \frac{\lambda\phi_b^2}{2(-k^2 - i\varepsilon)}\right), \qquad (34.55)$$

which is now IR finite. Rotating to Euclidean space and integrating up to $k_E < \Lambda$ as before,

$$\Delta\Gamma = -VT\frac{2\pi^2}{2(2\pi)^4} \int_0^\Lambda dk_E\, k_E^3 \ln\left(1 + \frac{\lambda\phi_b^2}{2k_E^2}\right), \qquad (34.56)$$

we reproduce Eq. (34.48) with $m_{\text{eff}}^2 = \frac{\lambda}{2}\phi_b^2$. This approach, which is how Coleman and Weinberg originally calculated their potential, illustrates that ϕ_b acts as an IR cutoff on diagrams such as Eq. (34.54).

Next we observe that the entire logarithmic term in the Coleman–Weinberg potential could have been extracted from the 4-point interaction alone. If we calculate the 4-point amplitude for non-zero external momenta, we find

$$i\mathcal{M}_4 = $$

$$= -i\lambda_R + i\frac{3\lambda_R^2}{32\pi^2}\left(\frac{2}{\varepsilon} + \ln\frac{\mu^2}{(stu)^{1/3}} + \cdots\right) - i\delta_\lambda + \mathcal{O}(\lambda_R^3), \qquad (34.57)$$

where the $\cdots$ are non-logarithmic terms which are subdominant when the logarithm is large and which we will ignore. In $\overline{\text{MS}}$, δ_λ is chosen to remove the $\frac{1}{\varepsilon}$ pole and μ is taken to be some scale near $Q \equiv (stu)^{1/6}$ to minimize the logarithms. Then,

$$\mathcal{M}_4(Q) = -\left[\lambda_R + \frac{3\lambda_R^2}{32\pi^2}\ln\frac{Q^2}{\mu^2}\right] + \mathcal{O}(\lambda_R^3), \qquad (34.58)$$

with $\mathcal{M}_4(\mu) = -\lambda_R$. Comparing to Eq. (34.52), we see that the large logs are reproduced by

$$V_{\text{eff}}(\phi) = -\frac{\phi^4}{4!}\mathcal{M}_4(\phi), \qquad (34.59)$$

with ϕ_R replacing μ. This implies that we can resum the large logarithms in $V_{\text{eff}}(\phi)$ by resumming the large logs of μ in $\mathcal{M}_4$, which we can do by running λ with the renormalization group.

As discussed in Chapter 23, the renormalization group works by making the scale μ at which λ_R is defined arbitrary. Thus, we replace $\lambda_R \to \lambda(\mu)$ to get

$$\mathcal{M}_4(Q) = -\lambda(\mu) - \frac{3\lambda(\mu)^2}{32\pi^2}\ln\frac{Q^2}{\mu^2} + \mathcal{O}(\lambda^3). \qquad (34.60)$$

Since $\mathcal{M}_4(Q)$ is independent of μ, we have

$$\beta(\lambda) \equiv \mu\frac{d}{d\mu}\lambda(\mu) = \frac{3}{16\pi^2}\lambda^2 + \mathcal{O}(\lambda^3), \qquad (34.61)$$

which is the 1-loop RGE for λ. We can solve this equation to find a result much like the 1-loop running electric charge (see Chapters 16 and 23):

$$\lambda(\mu) = \frac{\lambda_R}{1 - \frac{3\lambda_R}{32\pi^2}\ln\frac{\mu^2}{\mu_R^2}} = \lambda_R + \frac{3\lambda_R^2}{32\pi^2}\ln\frac{\mu^2}{\mu_R^2} + \mathcal{O}(\lambda_R^3). \qquad (34.62)$$

This implies that the resummed Coleman–Weinberg potential in the leading-log approximation is

$$V_{\text{eff}}(\phi) = \frac{1}{4!}\phi^4\lambda(\phi) = \frac{1}{4!}\phi^4\frac{\lambda_R}{1 - \frac{3\lambda_R}{32\pi^2}\ln\frac{\phi^2}{\phi_R^2}}. \qquad (34.63)$$

Expanding out to order λ_R^2 reproduces the large logarithm of Eq. (34.52).

Note that more generally we would also need to include the wavefunction renormalization factor γ in the running of $\lambda(\mu)$. The Callan–Symanzik equation, Eq. (23.76), implies

$$\left(\mu\frac{\partial}{\partial\mu} + \beta\frac{\partial}{\partial\lambda} + 2\gamma\right)\mathcal{M}_4 = 0. \qquad (34.64)$$

For $\mathcal{L} = -\frac{1}{2}\phi\Box\phi - \frac{\lambda}{4!}\phi^4$ we find $\gamma = \mathcal{O}(\lambda^4)$, and so Eq. (34.64) with Eq. (34.57) gives Eq. (34.61) up to terms of higher order in λ. At 2-loops and higher in this theory, the resummed effective potential is given by $V_{\text{eff}}(\phi) = \frac{1}{4!}\phi(\mu)^4\lambda(\mu)$ with $\mu = \phi$, where the running of ϕ must be included as well.

34.2.3 Higgs effective potential

An important application of the Coleman–Weinberg approach is to the effective potential for the Standard Model Higgs boson. It will let us answer the very important question: Is the Standard Model vacuum stable? If not, there must be physics beyond the standard model coming in to make it stable.

The most important contribution to the Higgs effective potential is from the top quark, since its Yukawa coupling is close to one. Including just the Higgs and the top quark, the relevant part of the electroweak Lagrangian from Chapter 29 is

$$\mathcal{L} = |D_\mu H|^2 + m^2|H|^2 - \lambda|H|^4 + i\bar{Q}\slashed{\partial}Q + i\bar{t}_R\slashed{\partial}t_R + (Y_t\bar{Q}\tilde{H}t_R + \text{h.c.}), \quad (34.65)$$

where H is the Higgs doublet, Q is the lefthanded top/bottom quark doublet, and t_R is the righthanded top quark. Expanding $H = \dfrac{1}{\sqrt{2}}\begin{pmatrix} 0 \\ v+h \end{pmatrix}$ with $v = \frac{m}{\sqrt{\lambda}} = 247$ GeV generates a canonically normalized physical Higgs scalar h with mass $m_h = \sqrt{2}m$. In this normalization we can read off that $Y_t = \sqrt{2}\frac{m_t}{v} \approx 0.93$.

You can show in Problem 34.2 that the contribution to the effective potential from a Dirac fermion with Yukawa coupling $Y\phi\bar{\psi}\psi$ is $\Delta V_{\text{eff}}(\phi) = -\frac{1}{16\pi^2}Y^4\phi^4 \ln\frac{\phi^2}{\phi_R^2}$. More generally, a useful formula for general contributions is

$$V_{\text{eff}}(\phi) = V(\phi) + \sum_i (-1)^{2s_i}\frac{n_d^i}{64\pi^2}m_{i,\text{eff}}^4(\phi)\ln\frac{m_{i,\text{eff}}^2(\phi)}{\phi_R^2}, \quad (34.66)$$

where $(-1)^{2s_i}$ is -1 for a fermion and 1 for a boson, n_d^i is the number of degrees of freedom (e.g. 12 for a colored Dirac spinor and 1 for a neutral scalar) and $m_{i,\text{eff}}(\phi)$ is the ϕ-dependent mass, e.g. $m_{i,\text{eff}} = Y\phi$ for a fermion or $m_{i,\text{eff}}^2 = V''(\phi)$ for the scalar itself. You can prove this general formula in Problem 34.2.

Using Eq. (34.66) for the top-quark contribution to the effective potential, the effective mass is $m_{t,\text{eff}} = \frac{1}{\sqrt{2}}Y_t h$, while for the Higgs, $m_{h,\text{eff}}^2 = -m^2 + 3\lambda h^2$. Therefore,

$$V_{\text{eff}}(h) = -m^2 h^2 + \frac{\lambda}{4}h^4$$
$$+ \frac{1}{64\pi^2}\left(-m^2 + 3\lambda h^2\right)^2 \ln\frac{-m^2+3\lambda h^2}{v^2} - \frac{12}{64\pi^2}\left(\frac{1}{2}Y_t^2 h^2\right)^2 \ln\frac{\frac{1}{2}Y_t^2 h^2}{v^2}. \quad (34.67)$$

The factor of 12 in the top contribution comes from the 3 colors times 4 components of a Dirac spinor. We have chosen $h_R = v$ since this is the scale where we know the Higgs potential should be well approximated by its classical form.

Now let us explore some of the physics of this potential. As long as $h \sim v \sim m$ the logarithmic terms have little effect. However, at large values of h, the coefficient of h^4 in the potential can get significant corrections. Taking the $h \gg v$ limit, we get

$$V_{\text{eff}}(h) = \frac{\lambda}{4}h^4 + \left(\frac{9}{64\pi^2}\lambda^2 - \frac{3}{64\pi^2}Y_t^4\right)h^4 \ln\frac{h^2}{v^2}. \quad (34.68)$$

For our vacuum to be absolutely stable, this potential should be positive for all h, which means the coefficient of the logarithm should be positive. Using $\lambda = \frac{m_h^2}{2v^2}$, $Y_t = \sqrt{2}\frac{m_t}{v}$ and the $\overline{\text{MS}}$ mass for the top quark, $m_t = 163$ GeV, we find absolute stability holds if $m_h > 247.7$ GeV. This bound assumes the potential can be trusted for all h; however, for $m_h = 247.7$ GeV, the potential only goes negative at a value well above the Planck scale where quantum gravity becomes strong. Asking that the potential be positive up to $M_{\text{Pl}} \approx 10^{19}$ GeV gives a weaker bound, $m_h > 221$ GeV. For $h \lesssim M_{\text{Pl}}$, we do not have to worry about quantum gravity; however, we do have to worry about large logarithms.

Indeed, $\frac{1}{64\pi^2} \ln \frac{M_{\rm Pl}^2}{v^2} = 0.12$, which is comparable to $\frac{\lambda}{4} \approx 0.10$ (for $m_h \approx 221$ GeV) and so we cannot trust our bound without resumming the effective potential.

To get a more accurate instability bound, one should also include contributions from gauge bosons, which are smaller than the top contribution, but not unimportant. Including the full 2-loop effective potential in the Standard Model and performing resummation at 3-loop order, the absolute stability bound becomes [Degrassi *et al.*, 2012]

$$m_h > 129.4 \text{ GeV} \pm 1.8 \text{ GeV}. \tag{34.69}$$

In other words, if $m_h < 125.8$ GeV we are 95% confident that our patch of the universe will eventually tunnel into a more stable vacuum. To refine this bound, one can ask not that our vacuum be absolutely stable, but that it be stable only for a Hubble time ($\approx 10^{10}$ years). This weakens the bound by a few GeV. That the real Higgs boson mass is close to the bound of instability is an intriguing and unexplained fact.

34.2.4 Scalar QED

As a final example with background scalar fields, consider the effective potential in scalar QED. In this case we will find that the effective potential is not gauge invariant although some simple physical predictions coming from it are.

We start with the Lagrangian

$$\mathcal{L} = -\frac{1}{4}F_{\mu\nu}^2 - \frac{1}{2\xi}(\partial_\mu A_\mu)^2 + \frac{1}{2}|D_\mu \phi|^2 - \frac{1}{2}m^2|\phi|^2 - \frac{\lambda}{4!}|\phi|^4, \tag{34.70}$$

with $D_\mu \phi = \partial_\mu \phi + ieA_\mu \phi$. For $m^2 < 0$, this is the Abelian Higgs model, studied in Section 28.3.1. Note that we have chosen normalization conventions so that expanding $\phi = \phi_1 + i\phi_2$ provides canonical normalization for the real fields ϕ_1 and ϕ_2. In this way, the effective potential which is only a function of $\phi = \sqrt{\phi_1^2 + \phi_2^2}$ will have the canonical normalization for a real scalar field.

Now, we want to ask whether $m_R = 0$, meaning $V_{\rm eff}''(0) = 0$, leads to $\langle \phi \rangle \neq 0$ in the quantum theory. We leave most of the details of the calculation in this case to Problem 34.3. The result of 1-loop graphs involving ϕ or A_μ is an effective potential

$$V_{\rm eff}(\phi) = \frac{1}{4!}\phi^4 \left\{ \lambda_R + \frac{1}{8\pi^2}\left(\frac{5\lambda_R^2}{6} + 9e_R^4 - \xi e_R^2 \lambda_R \right)\left[\ln\left(\frac{\phi^2}{\phi_R^2} \right) - \frac{25}{6} \right] \right\}. \tag{34.71}$$

Here we have chosen the same renormalization conditions as in Section 34.2.1: $V(0) = 0$, $V''(0) = 0$ and $V''''(\phi_R) = \lambda_R$ (how e_R is defined is irrelevant at this order).

The first important thing to notice is that the potential in this case is not gauge invariant. One way to see why this is so is to recall that only 1PI graphs are included in an effective potential. Thus, wavefunction renormalization graphs such as

$$\tag{34.72}$$

which would be included in an S-matrix calculation are simply discarded [Jackiw, 1974]. Another way to see the origin of the gauge dependence is to recall that through the Legendre-transform derivation of the effective action, we learned that the effective potential away from $\phi = \langle\phi\rangle$ is the potential in the presence of a non-zero external current. However, if this current $J(x)$ is gauge invariant, then the interaction $-J(x)\phi(x)$ used to perform the Legendre transform is *not* gauge invariant, since ϕ transforms. Despite the gauge dependence of the effective potential, we expect physical quantities computed from V_{eff} should be gauge invariant. We will now see how this can happen.

The vacuum expectation value of ϕ, given by $V'(\langle\phi\rangle) = 0$, is where

$$\ln\frac{\langle\phi\rangle^2}{\phi_R^2} = \frac{11}{3} - \frac{48\pi^2\lambda_R}{5\lambda_R^2 + 54e_R^4 - 6e_R^2\lambda_R\xi}. \tag{34.73}$$

The second term on the right looks generically like a large number. However, in the situation where $e_R^4 \approx \lambda_R \ll 1$ we find simply

$$\ln\frac{\langle\phi\rangle^2}{\phi_R^2} \approx \frac{11}{3} - \frac{8\pi^2\lambda_R}{9e_R^4}. \tag{34.74}$$

In particular, if we choose $\phi_R = \langle\phi\rangle$, meaning we define $\lambda_R \equiv V''''(\langle\phi\rangle)$, then $\lambda_R = \frac{33}{8\pi^2}e_R^4$. Therefore, our assumption $e_R^4 \approx \lambda_R \ll 1$ is proven self-consistent and there are no large logarithms. With the choice $\lambda_R = \frac{33}{8\pi^2}e_R^4$ we find

$$V_{\text{eff}}(\phi) = \frac{3e_R^4}{64\pi^2}\phi^4\left(\ln\frac{\phi^2}{\langle\phi\rangle^2} - \frac{1}{2}\right). \tag{34.75}$$

By the way, in defining λ_R in terms of $\langle\phi\rangle$ we are trading a dimensionless parameter for a dimensionful one. In fact, the term *dimensional transmutation*, which we have used to describe running couplings in Chapter 23, originated from the paper of Coleman and Weinberg where the relation $\lambda_R = \frac{33}{8\pi^2}e_R^4$ was first derived [Coleman and Weinberg, 1973].

Although we have shown that our calculation is self-consistent, we have yet to make any physical predictions. To do so, consider the spectrum in the spontaneously broken theory. When $\langle\phi\rangle \neq 0$, the U(1) of the classical Lagrangian is broken. As in the Abelian Higgs model from Section 28.3.1, one of the real components of ϕ remains in the spectrum, with mass $m_S^2 = V''(\langle\phi\rangle) = \frac{3e_R^4}{8\pi^2}\langle\phi\rangle^2$. The other component of ϕ is eaten by the photon. The photon's mass is given by the relation $m_V^2 = e_R^2\langle\phi\rangle^2$ as in Eq. (28.45). Thus, we find a prediction

$$\frac{m_S^2}{m_V^2} = \frac{3e_R^2}{8\pi^2} \tag{34.76}$$

independent of the only dimensionful scale $\langle\phi\rangle$ (or equivalently, independent of λ_R).

This scalar to vector mass ratio is a gauge-invariant prediction, despite the gauge dependence of the effective potential. In the regime where predictions can be trusted, the ξ-dependent part of the potential only had effects at higher order in perturbation theory. More specifically, since $\lambda_R = \frac{33}{8\pi^2}e_R^4$, the $\xi e_R^2\lambda_R$ term in Eq. (34.71) could be canceled by something such as ξe_R^6 appearing at 2-loops. The 2-loop calculation has been done [Kang, 1974]. Indeed, terms such as ξe_R^6 do appear and the appropriate cancellations do happen to leave physical quantities gauge invariant. For example, the scalar-to-vector mass ratio at 2-loops is $\frac{m_S^2}{m_V^2} = \frac{3e_R^2}{8\pi^2} - \frac{61e_R^4}{768\pi^2}$, independent of ξ.

The lesson is that, although the effective potential itself is gauge dependent, physical predictions made using the 1PI effective action formalism should still be gauge independent. We will see another example of how this can happen in Section 34.3.3 below.

34.3 Background gauge fields

An important application of effective actions and background fields is to non-Abelian gauge theories. In this case, we want to integrate out the fluctuations A_μ^a for a fixed background field $\widetilde{A}_\mu^a$ (we put a tilde on the background field rather than a b subscript as above to avoid confusion with $\mathrm{SU}(N)$ indices). In this case, we will not assume the background field is constant, so that the vertices in $\Gamma[A_\mu^b]$ will represent the full 1PI graphs. To be able to calculate anything, we will have to work to fixed order in the coupling constant g (in contrast to the Coleman–Weinberg calculation, which assumed constant background fields and was valid to all orders in coupling constants but to fixed order in $\hbar$). Effectively, we will just be doing normal perturbation theory in a new language. As we will see, this approach makes some calculations simpler, such as the calculation of the QCD β-function.

Substituting $A_\mu^a \to \widetilde{A}_\mu^a + A_\mu^a$ in the Yang–Mills field strength leads to

$$F_{\mu\nu}^a \to \partial_\mu A_\nu^a - \partial_\nu A_\mu^a + \partial_\mu \widetilde{A}_\nu^a - \partial_\nu \widetilde{A}_\mu^a + g f^{abc} \left(A_\mu^b A_\nu^c + \widetilde{A}_\mu^b A_\nu^c + A_\mu^b \widetilde{A}_\nu^c + \widetilde{A}_\mu^b \widetilde{A}_\nu^c \right)$$
$$= \widetilde{F}_{\mu\nu}^a + \widetilde{D}_\mu A_\nu^a - \widetilde{D}_\nu A_\mu^a + g f^{abc} A_\mu^b A_\nu^c, \tag{34.77}$$

where $\widetilde{F}_{\mu\nu}^a = \partial_\mu \widetilde{A}_\nu^a - \partial_\nu \widetilde{A}_\mu^a + g f^{abc} \widetilde{A}_\mu^b \widetilde{A}_\nu^c$ is the field strength for the background field and

$$\widetilde{D}_\mu A_\nu^a = \partial_\mu A_\nu^a + g f^{abc} \widetilde{A}_\mu^b A_\nu^c \tag{34.78}$$

is a derivative that is covariant with respect to the background-field gauge transformations. Then the Lagrangian

$$\mathcal{L}_{\mathrm{BF}} = -\frac{1}{4} \left(\widetilde{F}_{\mu\nu}^a + \widetilde{D}_\mu A_\nu^a - \widetilde{D}_\nu A_\mu^a + g f^{abc} A_\mu^b A_\nu^c \right)^2 \tag{34.79}$$

is invariant under

$$\widetilde{A}_\mu^a \to \widetilde{A}_\mu^a + \frac{1}{g} \partial_\mu \alpha^a - f^{abc} \alpha^b \widetilde{A}_\mu^c, \quad A_\mu^a \to A_\mu^a - f^{abc} \alpha^b A_\mu^c. \tag{34.80}$$

Since $\mathcal{L}_{\mathrm{BF}}$ is symmetric in $A_\mu^a \leftrightarrow \widetilde{A}_\mu^a$, it is also invariant under

$$A_\mu^a \to A_\mu^a + \frac{1}{g} \partial_\mu \beta^a - f^{abc} \beta^b A_\mu^c, \quad \widetilde{A}_\mu^a \to \widetilde{A}_\mu^a - f^{abc} \beta^b \widetilde{A}_\mu^c. \tag{34.81}$$

Unfortunately, we cannot keep both symmetries manifest at the same time. The advantage of keeping gauge invariance with respect to the background field manifest is that, since $\widetilde{A}_\mu$

is not dynamical, we do not need its propagator and do not have to gauge-fix. That is, we can keep background-field gauge invariance manifest throughout the calculation.

Now let us gauge-fix A_μ through the Faddeev–Popov procedure. We can of course go to the usual one-parameter family of covariant gauges by adding

$$\mathcal{L}_{\text{fix}} = -\frac{1}{2\xi}\left(\partial_\mu A_\mu^a\right)^2 + (\partial_\mu \bar{c}^a)(\partial_\mu c^a) + g f^{abc}(\partial_\mu \bar{c}^a) A_\mu^b c^c \qquad (34.82)$$

to $\mathcal{L}_{\text{BF}}$. This set of gauges is not ideal for background-field calculations, since it breaks all of the gauge symmetry. Instead, we will choose a different family of gauges corresponding to the condition $\widetilde{D}_\mu A_\mu = 0$ (instead of $\partial_\mu A_\mu = 0$). Such a gauge-fixing breaks gauge invariance for the propagating A_μ, but keeps manifest the gauge symmetry with respect to $\widetilde{A}_\mu$. Working out the Faddeev–Popov procedure, as in Section 25.4, the result is what one would expect, Eq. (34.82) with $\partial_\mu \to \widetilde{D}_\mu$. The Lagrangian is

$$\mathcal{L}_{\text{BFG}} = \mathcal{L}_{\text{BF}} - \frac{1}{2\widetilde{\xi}}(\widetilde{D}_\mu A_\mu^a)^2 + (\widetilde{D}_\mu \bar{c}^a)(\widetilde{D}_\mu c^a) + g f^{abc}(\widetilde{D}_\mu \bar{c}^a) A_\mu^b c^c. \qquad (34.83)$$

Here the ghosts c^a, anti-ghosts $\bar{c}^a$ and A_μ^c all transform as adjoints under the gauge symmetry with respect to $\widetilde{A}_\mu$ and so the background gauge invariance is still exact. This type of gauge-fixing is called **background-field gauge**. It is really a family of gauges parametrized by $\widetilde{\xi}$. Other choices of gauge are possible, as we discuss in Section 34.3.3.

34.3.1 Renormalization

A compelling virtue of the background-field method is that renormalization is drastically simpler than in regular non-Abelian gauge theories. For example, since the quantum fields A_μ and c only appear in loops, it is useless to renormalize them: their renormalization factors would always cancel between vertices and propagators. We do need to renormalize the background gauge field $\widetilde{A}_\mu$, which we do with $\widetilde{A}_\mu = \sqrt{\widetilde{Z}_3}\widetilde{A}_\mu^R$, and also its self-interactions. Another way to see why only the background fields need renormalization is that the divergences from loops of A_μ will appear as divergences in the effective action, which is a functional of only the background fields. Thus, these divergences can only be removed by renormalizing the background fields and their interactions.

Renormalization is simplest in background-field gauge where the gauge invariance associated with the background field is unbroken. Thus, if the regulator respects gauge invariance, the effective action must be gauge invariant as well, and there are even fewer counterterms required. For example, at tree-level, $\mathcal{L}_{\text{eff}}[\widetilde{A}] = -\frac{1}{4}(\widetilde{F}_{\mu\nu}^a)^2$. At 1-loop, possible divergences can only be removed by renormalizing fields in the tree-level effective Lagrangian. Since

$$(\widetilde{F}_{\mu\nu}^a)^2 = \widetilde{Z}_3\Big[(\partial_\mu \widetilde{A}_\nu^a - \partial_\nu \widetilde{A}_\mu^a)^2 + 4\frac{\widetilde{Z}_{A^3}}{\widetilde{Z}_3} g_R f^{abc}(\partial_\mu \widetilde{A}_\nu^a)\widetilde{A}_\mu^b \widetilde{A}_\nu^c$$

$$+ g_R^2 \frac{\widetilde{Z}_{A^4}}{\widetilde{Z}_3}(f^{eab}\widetilde{A}_\mu^a \widetilde{A}_\nu^b)(f^{ecd}\widetilde{A}_\mu^c \widetilde{A}_\nu^d)\Big], \qquad (34.84)$$

the only way for gauge invariance to be preserved is if $\widetilde{Z}_{A^3} = \widetilde{Z}_3$ and $\widetilde{Z}_{A^4} = \widetilde{Z}_3$. This is in contrast to the ordinary renormalization of non-Abelian gauge theories discussed in Chapter 26, where we could only show that $\widetilde{Z}_{A^3}/\widetilde{Z}_3 = \sqrt{\widetilde{Z}_{A^4}/\widetilde{Z}_3}$. Indeed, this ratio was *not* 1, but equal to $1 - \frac{1}{\varepsilon}\frac{g_R^2}{32\pi^2}C_A(\xi + 3)$ at 1-loop.

To see why $\widetilde{Z}_{A^3} = \widetilde{Z}_3$ and $\widetilde{Z}_{A^4} = \widetilde{Z}_3$ in another way, consider that at 1-loop in dimensional regularization the bare 1-loop effective Lagrangian must have the form

$$\mathcal{L}_{\text{eff}}[\widetilde{A}] = -\frac{1}{4}(\widetilde{F}_{\mu\nu}^a)^2 + \frac{c}{\varepsilon}(\widetilde{F}_{\mu\nu}^a)^2 + \mathcal{O}(\varepsilon^0) \tag{34.85}$$

for some number c. If divergences appeared in any other form, the theory would not be gauge invariant or not be renormalizable. Therefore, we must be able to remove the divergence by renormalizing $\widetilde{F}_{\mu\nu}^a$ through the renormalization of $\widetilde{A}_\mu$ only and there cannot be any divergence in $\widetilde{Z}_{A^3}/\widetilde{Z}_3$ or $\widetilde{Z}_{A^4}/\widetilde{Z}_3$.

So in background-field gauge, $\widetilde{\delta}_3 \equiv \widetilde{Z}_3 - 1$ must equal $\widetilde{\delta}_{A^3} = \widetilde{Z}_{A^3} - 1$. Thus, the formula for the 1-loop β-function from Eq. (26.94) simplifies to

$$\beta(g_R) = \frac{\varepsilon}{2}g_R^2\frac{\partial}{\partial g_R}\left(\delta_{A^3} - \frac{3}{2}\delta_3\right) = \frac{\varepsilon}{2}g_R^2\frac{\partial}{\partial g_R}\left(\widetilde{\delta}_{A^3} - \frac{3}{2}\widetilde{\delta}_3\right) = \frac{\varepsilon}{2}g_R^2\frac{\partial}{\partial g_R}\left(-\frac{1}{2}\widetilde{\delta}_3\right). \tag{34.86}$$

We can therefore extract the β-function from the gluon 2-point function alone – we will not have to compute any 3-point loops. Moreover, since the result should be the same β-function as we computed in Chapter 26, $\widetilde{\delta}_3$ cannot be $\widetilde{\xi}$ dependent. Recall from Section 26.5.3 that $\delta_3 = \frac{1}{\varepsilon}(\frac{g^2}{16\pi^2})(\frac{13}{3} - \xi)C_A$ and $\delta_{A^3} = \frac{1}{\varepsilon}(\frac{g^2}{16\pi^2})(\frac{17}{6} - \frac{3}{2}\xi)C_A$. In background-field gauge, we therefore expect $\widetilde{\delta}_3 = \frac{1}{\varepsilon}(\frac{g^2}{16\pi^2})(\frac{22}{3})C_A$. In particular, $\widetilde{\delta}_3$ must itself be independent of the parameter $\widetilde{\xi}$ of the background-field gauges. We will now verify this explicitly.

34.3.2 Background-field Feynman graphs

One approach to performing background-field calculations is using path integrals and functional determinants. This method was discussed in Section 34.2.1. To see how it works for background-gauge fields, see [Peskin and Schroeder, 1995, Section 16.6]. For the β-function calculation, it is perhaps more enlightening to see how the relevant 1-loop graphs differ from what we computed in Chapter 26. Thus, we proceed following [Abbott, 1982], deriving the Feynman rules and computing the loops explicitly.

The Feynman rules are derived from Eq. (34.83). The quantum-field propagator to zeroth order in the background field is just the ordinary R_ξ gauge propagator with $\widetilde{\xi}$ replacing ξ:

$$\nu;b \;\; \text{\small 0000000} \;\; \mu;a \;\; = \;\; i\frac{-g^{\mu\nu} + (1 - \widetilde{\xi})\frac{p^\mu p^\nu}{p^2}}{p^2 + i\varepsilon}\delta^{ab}. \tag{34.87}$$

The vertices involving quantum fields are all identical to those in non-Abelian gauge theories (see Section 26.1). The new vertices all have background fields. The important ones for the β-function include the $\widetilde{A}AA$ vertex:

$$= g f^{abc} \left[g^{\mu\nu} \left(k - p + \frac{1}{\tilde{\xi}} q \right)^{\rho} + g^{\nu\rho} \left(p - q - \frac{1}{\tilde{\xi}} k \right)^{\mu} + g^{\rho\mu} (q-k)^{\nu} \right].$$

(34.88)

Except for the $\tilde{\xi}$ term, this is identical to the gauge theory vertex in Eq. (26.9). The $\bar{c}c\tilde{A}$ vertex is

$$= -g f^{abc} (p^{\mu} + q^{\mu}).$$

(34.89)

This also differs from the gauge theory vertex, which would be just $-g f^{abc} p^{\mu}$. The other new vertices are

$$= -i g^2 f^{ade} f^{ecb} g^{\mu\nu},$$

(34.90)

$$= -i g^2 \left(f^{ace} f^{edb} + f^{ade} f^{ecb} \right) g^{\mu\nu}$$

(34.91)

and

$$
\begin{aligned}
= -i g^2 \times \Big[& f^{abe} f^{cde} (g^{\mu\rho} g^{\nu\sigma} - g^{\mu\sigma} g^{\nu\rho} - \tfrac{1}{\tilde{\xi}} g^{\mu\nu} g^{\rho\sigma}) \\
& + f^{ace} f^{bde} (g^{\mu\nu} g^{\rho\sigma} - g^{\mu\sigma} g^{\nu\rho} - \tfrac{1}{\tilde{\xi}} g^{\mu\rho} g^{\nu\sigma}) \\
& + f^{ade} f^{bce} (g^{\mu\nu} g^{\rho\sigma} - g^{\mu\rho} g^{\nu\sigma}) \Big].
\end{aligned}
$$

(34.92)

The only other new vertex has one background gluon, but this is identical to the 4-gluon vertex in ordinary non-Abelian gauge theories:

$$
\begin{aligned}
= -i g^2 \times \Big[& f^{abe} f^{cde} (g^{\mu\rho} g^{\nu\sigma} - g^{\mu\sigma} g^{\nu\rho}) \\
& + f^{ace} f^{bde} (g^{\mu\nu} g^{\rho\sigma} - g^{\mu\sigma} g^{\nu\rho}) \\
& + f^{ade} f^{bce} (g^{\mu\nu} g^{\rho\sigma} - g^{\mu\rho} g^{\nu\sigma}) \Big]
\end{aligned}
=
$$

(34.93)

Vertices with one regular gauge field and two or three background gauge fields cannot contribute to 1PI diagrams, so they can be ignored.

To extract the QCD β-function, we will only need the divergent parts of the 1-loop graphs. The ghost loop is gauge invariant by itself, and very similar to the graph in Eq. (26.48) with a slightly different numerator:

$$i\mathcal{M}_A^{ab\mu\nu} = \left\|\!\left| \text{◉◉◉◉◉} \right.\!\!\!\!\!\raisebox{1em}{$k-p$}\!\!\!\!\! \left.\text{◉◉◉◉◉} \right|\!\right\| = g^2 C_A \delta^{ab} \mu^{4-d} \int \frac{d^d k}{(2\pi)^d} \frac{i(2k-p)^\mu}{(p-k)^2+i\varepsilon} \frac{i(2k-p)^\nu}{k^2+i\varepsilon}.$$

$$(34.94)$$

This graph is quadratically divergent. As in the QED vacuum polarization graph, there is also a diagram with the 4-point vertex:

$$i\mathcal{M}_B^{ab\mu\nu} = \left\|\!\left| \text{◉◉◉◉◉◉◉◉◉} \right|\!\right\| = 2g^2 C_A \delta^{ab} \mu^{4-d} \int \frac{d^d k}{(2\pi)^d} \frac{g^{\mu\nu}}{k^2+i\varepsilon}. \qquad (34.95)$$

The sum of these two graphs has only logarithmic UV divergences. Evaluating the graphs and expanding in $d = 4 - \varepsilon$ we find

$$\mathcal{M}_A^{ab\mu\nu} + \mathcal{M}_B^{ab\mu\nu} = \frac{g^2}{16\pi^2} \frac{C_A}{3} \left(\frac{2}{\varepsilon} - \ln \frac{p^2}{\tilde{\mu}^2} \right) \delta^{ab} \left(g^{\mu\nu} p^2 - p^\mu p^\nu \right) + \mathcal{O}(\varepsilon^0). \quad (34.96)$$

Note that this is different from the equivalent ghost loop in full QCD.

The vacuum polarization graphs with virtual gluons are

$$i\mathcal{M}_C^{ab\mu\nu} = \left\|\!\left| \text{◉◉◉◉◉} \right.\!\!\!\!\!\raisebox{1em}{$k-p$}\!\!\!\!\! \left.\text{◉◉◉◉◉} \right|\!\right\| = \frac{g^2}{2} C_A \delta^{ab} \mu^{4-d} \int \frac{d^d k}{(2\pi)^d} \frac{-i}{k^2+i\varepsilon} \frac{-i}{(k-p)^2+i\varepsilon} N_C^{\mu\nu}$$

$$(34.97)$$

with

$$N_C^{\mu\nu} = \left[g^{\mu\alpha} \left(p + k + \frac{p-k}{\tilde{\xi}} \right)^\rho + g^{\alpha\rho}(p-2k)^\mu + g^{\rho\mu} \left(k - 2p - \frac{k}{\tilde{\xi}} \right)^\alpha \right]$$

$$\times \left[g^{\rho\sigma} + (1-\tilde{\xi}) \frac{(p-k)^\rho (p-k)^\sigma}{(p-k)^2} \right] \left[g^{\alpha\beta} + (1-\tilde{\xi}) \frac{k^\alpha k^\beta}{k^2} \right]$$

$$\times \left[-g^{\nu\beta} \left(p + k + \frac{p-k}{\tilde{\xi}} \right)^\sigma + g^{\beta\sigma}(2k-p)^\nu + g^{\sigma\nu} \left(2p - k + \frac{k}{\tilde{\xi}} \right)^\beta \right]$$

$$(34.98)$$

and

$$i\mathcal{M}_D^{ab\mu\nu} = \left\|\!\left| \text{◉◉◉◉◉◉◉◉◉} \right|\!\right\| = \frac{g^2}{2} \delta^{ab} C_A \mu^{4-d} \int \frac{d^4 k}{(2\pi)^4} \frac{1}{k^2+i\varepsilon} N_D^{\mu\nu}, \quad (34.99)$$

where

$$N_D^{\mu\nu} = \left[\left(1 - \frac{1}{\widetilde{\xi}}\right) g^{\mu\rho} g^{\nu\sigma} + \left(1 - \frac{1}{\widetilde{\xi}}\right) g^{\mu\nu} g^{\rho\sigma} - 2g^{\mu\sigma} g^{\nu\rho}\right] \left[g^{\rho\nu} - (1 - \widetilde{\xi}) \frac{k^\rho k^\nu}{k^2}\right]. \tag{34.100}$$

After a straightforward calculation, we find that the quadratic divergence in these two loops also exactly cancels. The divergent part of these graphs gives

$$\mathcal{M}_C^{ab\mu\nu} + \mathcal{M}_D^{ab\mu\nu} = \frac{g^2}{16\pi^2} \frac{10C_A}{3} \left(\frac{2}{\varepsilon} - \ln\frac{p^2}{\widetilde{\mu}^2}\right) \delta^{ab}\left(g^{\mu\nu} p^2 - p^\mu p^\nu\right) + \mathcal{O}(\varepsilon^0). \tag{34.101}$$

The $\widetilde{\xi}$ dependence from the modified interactions has exactly canceled the $\widetilde{\xi}$ dependence from the propagators and the coefficient of the $\frac{1}{\varepsilon}$ pole is gauge invariant. We leave the $\mathcal{O}(\varepsilon^0)$ parts of these graphs to Problem 34.8.

We have found that the UV divergences coming from both the ghost and gauge boson contributions are separately gauge invariant. Adding the result of all four graphs, we find that the bare effective Lagrangian at 1-loop,

$$\mathcal{L}_{\text{eff}}^{\text{BFG}} = -\frac{1}{4}(\widetilde{F}_{\mu\nu}^a)^2 \left[1 + \frac{g^2}{16\pi^2} \frac{22C_A}{3}\left(\frac{2}{\varepsilon} - \ln\frac{p^2}{\widetilde{\mu}^2}\right) + \mathcal{O}(\varepsilon^0)\right], \tag{34.102}$$

is canceled with a 1-loop $\overline{\text{MS}}$ counterterm $\widetilde{\delta}_3 = \frac{g^2}{16\pi^2} C_A \frac{22}{3\varepsilon}$. Using Eq. (34.86), this leads to the 1-loop QCD β-function:

$$\beta(g) = \frac{\varepsilon}{2} g^2 \frac{\partial}{\partial g}\left(-\frac{1}{2}\widetilde{\delta}_3\right) = -\frac{g^2}{16\pi^2} \frac{11}{3} C_A, \tag{34.103}$$

in agreement with what we found in Chapter 26 (the C_A term in Eq. (26.93)).

34.3.3 Gauge dependence

The calculation of the QCD β-function using the background-field method is clearly easier than the calculation we did in Chapter 26 since it involves fewer diagrams. In the traditional way of doing the calculation, we needed not just the vacuum polarization graphs, but also corrections to a vertex. For example, in Chapter 26, we used δ_2, δ_3 and δ_1 (from the quark and gluon field strength renormalizations and the $\bar{q}\slashed{A}q$ vertex), or we could have used δ_3 and δ_{A^3}, with δ_{A^3} coming from gluon 3-point function (or we could have used the ghost vertex or the 4-point QCD vertex). Since 3- and 4-point Feynman diagrams are more complicated than 2-point diagrams, not only does the background-field method require fewer graphs, but the most complicated graphs are avoided.

While the background-field advantage may not seem so important for the 1-loop β-function, consider the 2-loop or 3-loop β-function. There, reducing the number and complexity of the graphs required is enormously beneficial. Or consider a more complicated theory, such as quantum gravity. In perturbative quantum gravity, there are an enormous number of interactions and graphs that need to be considered even for the 1-loop running of G_N. The background-field method, which keeps a copy of general coordinate invariance manifest, makes the study of the renormalization in this theory much simpler ['t Hooft and Veltman, 1974].

The background-field method is important for conceptual reasons as well. One important application is to renormalizability. To show non-Abelian gauge theories are renormalizable, we need to show that all the infinities can be reabsorbed into coupling and field strength renormalizations of terms present in the original action. The reason the proof is difficult is because gauge invariance has to be broken to compute the propagator, and therefore non-gauge-invariant divergences cannot be forbidden on gauge-symmetry grounds alone (one needs tools such as BRST invariance for the proof). For example, a term such as $\left(f^{abc} A^b_\mu A^c_\nu\right)^2$ is not forbidden. With the background-field method, since background-field gauge invariance is manifest at every step, and since the regulator respects gauge invariance, it is impossible for a non-gauge-invariant term to be generated in the effective action.

Background-field gauge is a natural gauge to pick since it preserves background gauge invariance. However, physical predictions should come out the same even if we choose a less natural gauge for which background-field gauge invariance is explicitly broken. It is instructive to see how this actually happens. Recall that background-field gauge corresponds to using a gauge-fixing term of the form $\mathcal{L}^{\text{BFG}}_{\text{fix}} = -\frac{1}{2\xi}(\widetilde{D}_\mu A^a_\mu)^2 +$ ghosts, as in Eq. (34.83). From this, we found the 1-loop effective Lagrangian in Eq. (34.102). What would happen if we used $\mathcal{L}_{\text{fix}} = (\partial_\mu A^a_\mu)^2 +$ ghosts as in Eq. (34.82) instead? This gauge-fixing is independent of $\widetilde{A}_\mu$ and explicitly breaks background gauge invariance. In this case, the divergent part of the 1-loop effective action is [Grisaru et al., 1975]

$$\mathcal{L}_{\text{eff}} = -\frac{1}{4}(\widetilde{F}^a_{\mu\nu})^2 + \frac{g^2}{16\pi^2} C_A \frac{2}{\varepsilon} \left[\frac{10}{3}(\widetilde{F}^a_{\mu\nu})^2 - 4g f^{abc} \widetilde{F}^a_{\mu\nu} \widetilde{A}^b_\mu \widetilde{A}^c_\nu \right]. \qquad (34.104)$$

At first glance, this seems troubling, since the coupling must now be renormalized differently in the 4-point vertex and 3-point vertices to cancel the $\frac{1}{\varepsilon}$ poles. To see in what way this effective Lagrangian is equivalent to Eq. (34.102), recall that the effective action is only to be used for classical computations. In particular, for S-matrix elements, the background field states are on-shell. Then, substituting the identity

$$- g f^{abc} \widetilde{F}^a_{\mu\nu} \widetilde{A}^b_\mu \widetilde{A}^c_\nu = (\widetilde{F}^a_{\mu\nu})^2 + 2\widetilde{A}^a_\nu D_\mu \widetilde{F}^a_{\mu\nu} - 2\partial_\mu(\widetilde{A}^a_\nu \widetilde{F}^a_{\mu\nu}) \qquad (34.105)$$

into Eq. (34.104), dropping the total derivative term and the term that vanishes on-shell (using the equations of motion $\widetilde{D}_\mu F^a_{\mu\nu} = 0$), Eq. (34.102) is reproduced. This conclusion reinforces what we found in Section 34.2.4: although the effective action itself is gauge dependent and unphysical, physical predictions coming from the effective action can still be gauge invariant.

Problems

34.1 $W[J]$ as the generating functional of connected diagrams.
   (a) Take the third variational derivative of $W[J]$ to show that it gives only the connected contributions to the 3-point function.
   (b) Show that $W[J]$ does generate all the connected diagrams for any n-point function.

34.2 General scalar effective potential.
  (a) Calculate the contribution of a fermion to the scalar potential starting with the Lagrangian $\mathcal{L} = -\frac{1}{2}\phi\Box\phi - V(\phi) + i\bar{\psi}\partial\!\!\!/\psi - Y\phi\bar{\psi}\psi$.
  (b) Show that the general 1-loop effective potential is given by

$$V_{\text{eff}}(\phi) = V(\phi) + \sum_i (-1)^{2s_i} \frac{n_d^i}{64\pi^2} m_i^4(\phi) \ln \frac{m_i^2(\phi)}{\phi_R^2}, \qquad (34.106)$$

  as in Eq. (34.66), where s_i is the spin and n_d^i is the real number of degrees of freedom on-shell for a given particle.

34.3 Calculate the Coleman–Weinberg potential in scalar QED and verify Eq. (34.71).

34.4 Calculate the W- and Z-boson contributions to the Higgs effective potential.

34.5 Improve the Higgs stability bound in the Standard Model.
  (a) Show that including the $SU(2) \times U(1)$ gauge fields, you get

$$V_{\text{eff}}(h) = -m^2 h^2 + \frac{\lambda}{4} h^4$$
$$+ \left(\frac{9}{64\pi^2} \lambda^2 - \frac{3}{64\pi^2} Y_t^4 + \frac{3}{8} g^4 + \frac{3}{16} (g^2 + g'^2) \right) h^4 \ln \frac{h^2}{v^2}. \qquad (34.107)$$

  (b) Plug in the Standard Model values for g and g' and see how the lower bound on the Higgs mass changes.
  (c) Calculate $\beta_\lambda = \mu \frac{d}{d\mu} \lambda$ and $\gamma_2 = \mu \frac{d}{d\mu} h$ including top and Higgs correction in the Standard Model.
  (d) Solve the RGEs from the previous part to get an RG-improved effective potential.
  (e) What is the lower bound on the Higgs mass for absolute stability using this RG-improved potential?

34.6 Calculate the coefficient of the A^4 vertex in the 1PI effective action using the background-field method.

34.7 Calculate the fermion contributions to the QCD β-function using the background-field method.

34.8 Background-field effective action.
  (a) Calculate the finite parts of the vacuum polarization loops from Section 34.3.2 in background-field gauge. You should find that the finite parts are in fact $\tilde{\xi}$ dependent. For example, the contribution at order $\tilde{\xi}^2$ comes only from the graph in Eq. (34.97)
  (b) Why is it OK for the finite parts to have $\tilde{\xi}$ dependence, but not the divergent parts?

There are only six quarks. Three of them (up, down and strange) are light with masses $m_q \lesssim \Lambda_{\text{QCD}}$. Hadrons containing these quarks only, for example the pions and kaons, can be studied by expanding around the $m_q = 0$ limit. Expanding around $m_q = 0$ leads to the Chiral Lagrangian (Chapter 28), which is a low-energy effective theory, perturbative when $\frac{E}{4\pi F_\pi}$ and $\frac{m_q}{4\pi F_\pi}$ are small, with E a typical energy scale and $4\pi F_\pi \sim 1200\,\text{MeV}$. The heaviest quark, the top, does not hadronize. Since $m_t \gg \Lambda_{\text{QCD}}$, one can make accurate predictions about top physics using perturbation theory in α_s (which is small at scales $\mu \sim m_t$). The remaining two quarks, the charm and bottom, with masses $m_c \sim 1275\,\text{MeV}$ and $m_b \sim 4180\,\text{MeV}$, are unstable but do form metastable hadrons (such as the D and B mesons). Is there any way to study charm and bottom physics in perturbation theory? Yes, by expanding in $\frac{\Lambda_{\text{QCD}}}{m_b}$ or $\frac{\Lambda_{\text{QCD}}}{m_c}$.

The heavy-quark limit presents a picture of heavy mesons qualitatively similar to the hydrogen atom. Consider, for example, a B meson that comprises a heavy b quark and a light valence antiquark ($\bar{u}$ or $\bar{d}$). This is similar to how a hydrogen atom comprises a heavy proton and a light electron. Just as the proton is a static source of Coulomb potential in the Schrödinger equation, the b quark acts as a static source for gluons. Unfortunately, because QCD is strongly coupled at low energies, the Coulomb potential is a bad approximation to full QCD. Thus, we cannot just solve the Schrödinger equation to study the spectrum of the $b\bar{u}$ system. Nevertheless, as we will see, the b quark acts as a classical source to leading order in $\frac{1}{m_b}$, which gives us a handle to perform some useful calculations. A useful qualitative picture is to think of a B meson as being like a proton but with the electron cloud replaced by non-perturbative **brown muck**: $|B\rangle = |b;\,\text{muck}\rangle$.[1]

For example, the spin states of a heavy–light meson, such as a $b\bar{u}$ bound state, are $\frac{1}{2} \otimes \frac{1}{2} = 0 \oplus 1$, with the spin-0 state called the B meson and the spin-1 triplet called the B^*. The mass difference between these is analogous to the energy splitting between the S and P states of the hydrogen atom: it is 0 at leading order. In the hydrogen atom, the splitting between S and P is due to magnetic moment interactions. If the proton is at rest (as it is in the $m_p \to \infty$ limit), it only produces an electric field. Therefore, the S/P splitting must be suppressed by at least a factor of E/m_p with $E \sim 10$ eV the binding energy. To leading order in Λ_{QCD}/m_b the dynamics of a B meson is similarly independent of spin, which is why B and B^* are degenerate to leading order. This is known as **heavy-quark spin symmetry**. In more detail, the splitting should come from the $\mu \vec{S} \cdot \vec{B}$ interaction between the spin $\vec{S}$ and the magnetic field $\vec{B}$, where μ is the magnetic moment of the heavy quark

---

[1] We owe the delightful phrase "brown muck" to Nathan Isgur.

which scales as m_b^{-1} by dimensional analysis. Thus we can write

$$m_B = m_b + \bar{\Lambda} - \frac{\lambda_1}{2m_b} - c_1 \frac{\lambda_2}{2m_b} + \mathcal{O}(m_b^{-2}), \tag{35.1}$$

$$m_{B^*} = m_b + \bar{\Lambda} - \frac{\lambda_1}{2m_b} - c_3 \frac{\lambda_2}{2m_b} + \mathcal{O}(m_b^{-2}), \tag{35.2}$$

where $\bar{\Lambda} \sim \Lambda_{\mathrm{QCD}}$, $\lambda_1 \sim \Lambda_{\mathrm{QCD}}^2$, $\lambda_2 \sim \Lambda_{\mathrm{QCD}}^2$, and c_1 and c_3 are coefficients related to the spin, which we explain in Section 35.4. So we expect (using $\Lambda_{\mathrm{QCD}} \approx 300$ MeV) that

$$m_{B^*} - m_B = \frac{\lambda_2}{2m_b}(c_1 - c_3) \sim \frac{\Lambda_{\mathrm{QCD}}^2}{m_b} \sim 20 \text{ MeV}. \tag{35.3}$$

Experimentally, $m_B^* = 5325$ MeV and $m_B = 5279$ MeV and their difference 44 MeV is consistent with this expectation.

In addition to the spin symmetry, bottom and charm quark physics also simplifies due to a flavor symmetry. This is the analog of the fact that the nucleus of a hydrogen atom or a deuteron look the same to the electron. In the rest frame of the heavy quark, the hadron can be thought of as just the quark, sitting there. If $m_Q = \infty$ the quark cannot move. In fact, in this limit, the quark just acts as a source for gluons. This leads to a **heavy-quark flavor symmetry**: the dynamics is independent of the flavor of the quark, to leading order in m_Q^{-1}. This symmetry provides very strong constraints on the physics of heavy hadrons. For example, the D mesons should satisfy the same parametrization as in Eqs. (35.1) and (35.2) with $m_b \to m_c$:

$$m_D = m_c + \bar{\Lambda} - \frac{\lambda_1}{2m_c} - c_1 \frac{\lambda_2}{m_c} + \mathcal{O}(m_c^{-2}), \tag{35.4}$$

$$m_{D^*} = m_c + \bar{\Lambda} - \frac{\lambda_1}{2m_c} - c_3 \frac{\lambda_2}{m_c} + \mathcal{O}(m_c^{-2}). \tag{35.5}$$

This implies that

$$m_{B^*}^2 - m_B^2 = m_{D^*}^2 - m_D^2 + \mathcal{O}\left(\frac{\Lambda_{\mathrm{QCD}}^3}{m_Q}\right). \tag{35.6}$$

So now we get a prediction for the masses-squared that is accurate up to m_Q^{-1} corrections, a stronger result than that in Eq. (35.3). In particular, using m_B, m_{B^*} and $m_D = 1869$ MeV, this equation predicts that $m_{D^*} = 1993$ MeV. The experimental value is $m_D^* = 2010$ MeV, so the heavy-quark limit prediction is off by only 0.8%.

The momentum of a hadron containing a heavy quark can be written as

$$p^\mu = m_Q v^\mu + k^\mu, \tag{35.7}$$

where v^μ is the hadron's 4-velocity, normalized to $v^2 = 1$, and $k^\mu \ll m_Q$. The key to understanding the heavy-quark flavor symmetry is that the brown muck has energies of order Λ_{QCD}. Therefore, fluctuations in the muck do not have enough energy to reorient the heavy-quark velocity v^μ – the muck can only change k^μ. In this chapter, we discuss an effective theory for heavy quarks in which v^μ is promoted to a conserved quantum number of the heavy-quark field. This leads to Heavy-Quark Effective Theory (HQET), a beautiful and predictive framework for studying bottom and charmed hadrons. Before introducing

HQET, we will describe some more consequences of the heavy flavor and spin symmetries, which can be understood without even introducing the effective Lagrangian. Our presentation here will be somewhat brief, emphasizing important results and conceptual points. More details can be found in the classic review [Georgi, 1990] and the texts [Manohar and Wise, 2000] and [Grozin, 2004].

35.1 Heavy-meson decays

In this section we discuss how heavy-quark flavor and spin symmetries constrain decay rates of heavy mesons. We use the notation m_Q to refer the mass of a heavy quark (b or c) and m_q to refer to the mass of a light quark (u, d or s).

35.1.1 Leptonic decays

First, consider the weak decays of the pseudoscalar mesons $B^- = (\bar{u}b) \to \tau^- \bar{\nu}$ and $D^+ = (\bar{u}c) \to \mu^+ \nu$. As with pions, we define decay constants f_B and f_D through the matrix element of an axial current (see Chapter 28):

$$\langle 0|\bar{u}\gamma^\mu\gamma^5 b|B^-\rangle = -if_B p^\mu, \quad \langle 0|\bar{c}\gamma^\mu\gamma^5 u|D\rangle = -if_D p^\mu. \tag{35.8}$$

These definitions correspond to the conventional relativistic normalization, in which

$$\langle B(p')|B(p)\rangle = 2p^0(2\pi)^3 \delta^3(\vec{p}-\vec{p}') = \langle D(p')|D(p)\rangle, \tag{35.9}$$

and lead to the decay rate

$$\Gamma\left(B^- \to \tau^- \bar{\nu}\right) = \frac{G_F^2|V_{ub}|^2}{8\pi} f_B^2 m_\tau^2 m_B \left(1 - \frac{m_\tau^2}{m_B^2}\right)^2, \tag{35.10}$$

and similarly for other leptonic modes. Since $m_\tau = 1776\,\text{MeV} \gg m_\mu = 105\,\text{MeV}$, the branching ratio to tauons dominates. The formula for leptonic D^+ decays is identical, with m_B replaced by m_D. For D^+ decays, the branching ratio to $\mu^+\nu$ dominates due to the limited phase space for $D^+ \to \tau^+\nu$.

The relativistic normalization is not useful to extract scaling behavior as $m_Q \to \infty$ since $p^0 \to \infty$. Instead, we should use non-relativistic normalization, with

$$_\text{nr}\langle B(p')|B(p)\rangle_\text{nr} = (2\pi)^3 \delta^3(\vec{p}-\vec{p}') = {}_\text{nr}\langle D(p')|D(p)\rangle_\text{nr}. \tag{35.11}$$

You can think of the B or D decay as the b or c quark within the meson annihilating with the brown muck, which has the quantum numbers of the light quark. The important point is that, in the heavy-quark limit, the muck has no knowledge of the heavy-quark mass. Thus, the matrix elements should be mass independent in the heavy-quark limit. Therefore we should have

$$-iav^\mu = \langle 0|\bar{u}\gamma^\mu\gamma^5 b|B^-\rangle_\text{nr} = \frac{1}{\sqrt{2m_B}}\langle 0|\bar{u}\gamma^\mu\gamma^5 b|B^-\rangle = \frac{-if_B m_B v^\mu}{\sqrt{2m_B}}, \tag{35.12}$$

where a is some constant related to the brown muck, and v^μ is the velocity. Similarly,

$$- iav^\mu = \langle 0|\bar{c}\gamma^\mu\gamma^5 u|D\rangle_{\rm nr} = \frac{1}{\sqrt{2m_D}}\langle 0|\bar{c}\gamma^\mu\gamma^5 u|D\rangle = \frac{-if_D m_D v^\mu}{\sqrt{2m_D}} \qquad (35.13)$$

with *the same* a. Therefore, we predict that

$$\frac{f_B}{f_D} = \sqrt{\frac{m_D}{m_B}} \qquad (35.14)$$

up to $\frac{\Lambda_{\rm QCD}}{m_Q} \sim 10\%$ corrections.

Is Eq. (35.14) actually satisfied to 10%? We can use the measured rate $\Gamma(D^+ \to \mu^+\nu)$ $= 2.42 \times 10^{-13}\,\text{MeV}$ to predict the $B \to \tau\nu$ rate. Using the masses $m_{B^-} = 5279\,\text{MeV}$ and $m_{D^+} = 1869\,\text{MeV}$, Eq. (35.14) predicts $\Gamma(B^- \to \tau^-\bar{\nu}) = 1.55 \times 10^{-14}\,\text{MeV}$. The current best-measured value is $(6.7\pm1.7)\times10^{-14}\,\text{MeV}$. Thus Eq. (35.14) is off by a factor of 3! So there is not fantastic agreement with current data, to say the least. Another way to phrase this is that the values of the decay constants extracted from the decay rates are $f_D = 202\,\text{MeV}$ and $f_B = 253\,\text{MeV}$. Their ratio is 1.25, compared to $\sqrt{\frac{m_D}{m_B}} = 0.595$, so again the heavy-quark-limit prediction is off by a large factor. This indicates that there must be an unusually large power correction; that is, the $\frac{\Lambda_{\rm QCD}}{m_D}$ term must have a coefficient of order 10 or so. Intriguingly, lattice calculations give $f_D = 197\,\text{MeV}$ and $f_B = 193\,\text{MeV}$ [Particle Data Group (Beringer *et al.*), 2012], whose ratio is only a factor of 2 off from the heavy-quark limit prediction. The lattice also seems to confirm that there is a large $\frac{1}{m_Q}$ correction to the decay constants.

35.1.2 Exclusive semi-leptonic decays

We can develop a more general view of how the brown muck wavefunctions factorize out of the heavy-quark wavefunctions. Let us continue using the decomposition $p^\mu = m_Q v^\mu + k^\mu$ with $k^\mu \sim \Lambda_{\rm QCD}$. Then the brown muck in the B or D meson (recall $|B\rangle = |b;\,\text{muck}\rangle$), with its fluctuations of order $\Lambda_{\rm QCD}$, cannot affect the velocity v^μ or the spin s_Q of the heavy quark. Thus, a general heavy-meson state, for example for a B, can be written as

$$|B\rangle = |Bvs_Q s_q\rangle = |b;\,vs_b\rangle|\,\text{muck};\,vs_q\rangle, \qquad (35.15)$$

where s_b is the b quark spin, and s_q is the spin of the light quark. Note that the light-quark spin is a good quantum number because the B and heavy-quark spins are good quantum numbers. Although the muck cannot change v^μ, the muck wavefunction can depend on v^μ.

This factorization has immediate and important implications, such as the leptonic decay rates of B^+ and D^- discussed above. More generally, to measure properties of heavy mesons, we look at their current matrix elements, as we did above for weak decays. We are generally interested in couplings to the W bosons, through $J_L^\mu = \overline{Q}\gamma^\mu P_L q$, with $P_L = \frac{1}{2}(1 - \gamma_5)$, or to photons, which interact through $J_V^\mu = \overline{Q}\gamma^\mu Q$. We are interested in these quark currents, since the interaction strength of the W-boson and photon to these currents is related to the interaction strength of the equivalent leptonic currents by electroweak symmetry. By writing hadronic matrix elements in terms of currents we can factorize off the calculable electroweak part of the decay and effectively exploit the above factorization. For example, an interaction of a B meson with a photon would be determined by

$$\langle B'|J_V^\mu|B\rangle = \langle b; v's_b'|\bar{b}\gamma^\mu b|b; vs_b\rangle \langle \text{muck}; v's_q'|\text{muck}; s_q'v\rangle. \qquad (35.16)$$

In particular, in the limit that B and B' are both at rest with the same spins, then the vector current, which is conserved, just picks up the number of b quarks and we get

$$\langle B'|J_V^\mu|B\rangle = 2m_B v^\mu, \qquad (35.17)$$

with the $2m_B$ coming from the relativistic normalization.

When the velocities are not the same, we need to evaluate $\langle\,\text{muck}; v's_q'|\,\text{muck}; s_q v\rangle$. We can always write

$$\langle\,\text{muck}; v's_q'|\,\text{muck}; s_q v\rangle = \xi_{s_q s_q'}(v, v')\,. \qquad (35.18)$$

First of all, there are only two possible spins for the pseudoscalar meson matrix elements, so $s_q, s_q' = \pm\frac{1}{2}$, and the amplitudes must be the same for both by parity. More importantly, by Lorentz invariance, ξ can only depend on the combination $w = v \cdot v'$. Quite generally, since the muck is independent of the spin, we have

$$\langle B|\bar{b}\Gamma c|D\rangle = \langle b; v's_b'|\bar{b}\Gamma c|c; vs_c\rangle \xi(w), \qquad (35.19)$$

where Γ can be any tensor structure. The function $\xi(w)$ is known as an **Isgur–Wise function** and is a universal non-perturbative object. Since the Isgur–Wise function just depends on the muck wavefunctions, it is the same if we swap out one heavy quark for another and if we change the current. In particular, using the non-relativistic normalization, Eq. (35.11), and the vector current matrix element (35.17) we find the boundary condition $\xi(1) = 1$.

As an application, consider the extraction of the CKM element V_{cb} from data. There are a number of ways of measuring V_{cb} but one of the cleanest is from exclusive decays, such as $B \to D^* l\nu$. The rate for such decays can be measured as a function of the velocities v and v' or the B and D^* mesons. Working out all the phase space factors, the result is

$$\Gamma\left(\bar{B} \to D^* e\bar{\nu}\right) = \frac{G_F^2 |V_{cb}|^2 m_B^2}{48\pi^3}$$
$$\times \sqrt{w^2 - 1}(w+1)^2 r^3 (1-r)^2 \left[1 + \frac{4w}{w+1}\frac{1 - 2wr + r^2}{(1-r)^2}\right] F_{D^*}(w)^2, \qquad (35.20)$$

with $r = \frac{m_{D^*}}{m_B}$ and $F_{D^*}(w)$ a form factor. The prediction at leading order in the heavy-quark limit is that $F_{D^*}(w)$, and the analogous form factor $F_D(w)$ for $\bar{B} \to De\bar{\nu}$, should be a universal Isgur–Wise function $F_D(w) = F_{D^*}(w) = \xi(w)$. Since $\xi(1) = 1$ in the heavy-quark limit, all one has to do to extract V_{cb} is to measure the decay rate at zero recoil, that is, where $w = v \cdot v' = 1$. An example of using Eq. (35.20) to extract $|V_{cb}|$ from data by extrapolating to $w = 1$ is shown in Figure 35.1.

In reality, $F_D(1) \neq F_{D^*}(1) \neq 1$, due to both perturbative and non-perturbative corrections. Since $\xi(1) = 1$ exactly in the heavy-quark limit, the perturbative corrections can only come from differences between $\alpha_s(m_b)$ and $\alpha_s(m_c)$. We give an example of how such corrections can be computed using heavy-quark effective theory in the next section. Up to

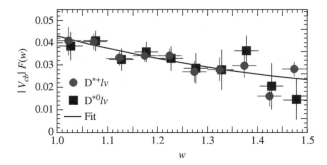

Extraction of V_{cb} from two different $B \to D^*$ decays [CLEO Collaboration (Briere *et al.*) 2002].

Fig. 35.1

order α_s^2, these give $F_{D^*}(1) \approx 0.96$ [Manohar and Wise, 2000]. The non-perturbative corrections could, a priori, give a correction of order $\frac{\Lambda_{QCD}}{m_Q} \approx 0.21$. However, as it turns out, the leading power correction to F_{D^*} actually starts at order m_Q^{-2}, due to a general result known as Luke's theorem. Since $\frac{\Lambda_{QCD}^2}{m_Q^2} \approx 0.04$, we then expect that $F_{D^*}(1) \gtrsim 0.92$ or so. Estimates from the lattice give $F_{D^*}(1) \approx 0.9$ [Bailey *et al.*, 2010], which is reasonably close to the value predicted in the heavy-quark limit. The resulting value of $|V_{cb}|$ extracted from exclusive semi-leptonic $B \to D$ decays combined with other measurements is

$$|V_{cb}| = (40.9 \pm 1.1) \times 10^{-3}. \tag{35.21}$$

35.2 Heavy-quark effective theory

We have seen a number of leading-order predictions from the heavy-quark limit. To make systematic improvements on these predictions, it is helpful to have an effective field theory where the heavy-quark symmetries are exact.

To derive this effective theory, we begin with the decomposition as in Eq. (35.7):

$$p^\mu = m_Q v^\mu + k^\mu. \tag{35.22}$$

Here v^μ is the 4-velocity with $v^2 = 1$ and the components of k^μ are assumed to be much smaller than m_Q. This decomposition is not unique, since we can shift $k^\mu \to k^\mu + \Delta k^\mu$ by a small amount and $v^\mu \to v^\mu - \Delta k^\mu / m_Q$. However, to leading order in m_Q^{-1}, v^μ is unique and this decomposition is well defined. In a hadron, the light quarks and gluons can have momenta $k^\mu \sim \Lambda_{QCD}$, but not much larger, so interactions can only change v by $\frac{\Lambda_{QCD}}{m_Q}$. Thus, to order m_Q^{-1}, v^μ is a good quantum number of the heavy quark. Thus, we want to have an effective theory where quarks carry this quantum number, and the conservation of heavy-quark velocity is apparent at the level of the Lagrangian.

Recall from Section 5.2 that to take the non-relativistic limit of a scalar field theory, we rescale the fields by $\phi \to \frac{1}{\sqrt{2m}} e^{-imt} \phi_{\text{nr}}$. The e^{-imt} factor is a plane wave solution for a particle at rest. For moving particles, we generalize this to the replacement $\phi \to \frac{1}{\sqrt{2m_Q}} e^{-im_Q v \cdot x} \chi_v$. This change of variables induces

$$\mathcal{L} = |D_\mu\phi|^2 - m_Q^2|\phi|^2 \to \chi_v^\star iv^\mu D_\mu\chi_v + \frac{1}{2m_Q}|D_\mu\chi_v|^2. \tag{35.23}$$

The $\frac{1}{2m_Q}$ term is subleading as $m_Q \to \infty$ and can be dropped. Thus, the Heavy-Scalar Effective Theory Lagrangian at leading power is simply

$$\mathcal{L}_{\text{HSET}} = \chi_v^\star iv^\mu D_\mu\chi_v. \tag{35.24}$$

To see how to generalize to the spinor case, let us take the heavy-scalar limit a different way. Just as $e^{-im_Q v\cdot x}$ is the plane wave solution for a particle, $e^{im_Q v\cdot x}$ is the plane wave solution for an antiparticle. In the heavy-scalar limit, pair production is suppressed and we should be able to integrate the antiparticle out. To do this, we write $\phi = \frac{1}{\sqrt{2m_Q}}e^{-im_Q v\cdot x}(\chi_v + \tilde{\chi}_v)$, where

$$\chi_v = e^{im_Q v\cdot x}\frac{1}{\sqrt{2m_Q}}(iv\cdot D + m_Q)\phi, \tag{35.25}$$

$$\tilde{\chi}_v = e^{im_Q v\cdot x}\frac{1}{\sqrt{2m_Q}}(-iv\cdot D + m_Q)\phi. \tag{35.26}$$

Then,

$$\mathcal{L} = |D_\mu\phi|^2 - m_Q^2|\phi|^2 = \chi_v^\star iv\cdot D\chi_v - \tilde{\chi}_v(iv\cdot D + 2m_Q)\tilde{\chi}_v + \mathcal{O}\left(\frac{1}{m_Q}\right). \tag{35.27}$$

Thus, in deriving Eq. (35.24) we are removing the antiparticle field from the Lagrangian.

For the spinor case, note that when $p^\mu = m_Q v^\mu$ exactly, the Dirac equation for a heavy quark, $\slashed{p}\psi = m_Q\psi$, implies

$$(1 - \slashed{v})\psi = 0. \tag{35.28}$$

Thus, we decompose the spinor field as

$$\psi(x) = \psi_v(x) + \tilde{\psi}_v(x), \tag{35.29}$$

where

$$\psi_v(x) = \frac{1 + \slashed{v}}{2}\psi(x), \qquad \tilde{\psi}_v(x) = \frac{1 - \slashed{v}}{2}\psi(x). \tag{35.30}$$

In the heavy-quark limit $\tilde{\psi}_v(x) \approx 0$ since Eq. (35.28) holds. Thus, heavy-quark effective theory is defined by integrating the components $\tilde{\psi}_v$ of ψ out of the theory. This can be done systematically in powers of m_Q^{-1}.

Setting $\tilde{\psi}_v = 0$ gives the HQET Lagrangian at leading power. It amounts to replacing

$$\psi(x) \to e^{-im_Q v\cdot x}\frac{1 + \slashed{v}}{2}Q_v(x) \tag{35.31}$$

in analogy with $\phi \to e^{-im_Q v\cdot x}\chi_v$ in the scalar case. Inserting this replacement into the QCD Lagrangian gives

$$\bar{\psi}(i\slashed{D} - m_Q)\psi \to \bar{Q}_v\frac{1 + \slashed{v}}{2}(i\slashed{D} + m_Q\slashed{v} - m_Q)\frac{1 + \slashed{v}}{2}Q_v = \bar{Q}_v\frac{1 + \slashed{v}}{2}i\slashed{D}\frac{1 + \slashed{v}}{2}Q_v. \tag{35.32}$$

We can then anticommute the Dirac matrices to get

$$\bar{\psi}(i\slashed{D} - m_Q)\psi \to i\bar{Q}_v v \cdot D \frac{1+\slashed{v}}{2} Q_v, \tag{35.33}$$

which is independent of m_Q as expected. Including the gluon and light quarks, the full leading-order HQET Lagrangian is then

$$\mathcal{L}_{\text{HQET}} = -\frac{1}{4}\left(F_{\mu\nu}^a\right)^2 + \bar{q}(i\slashed{D} - m_q)q + \sum_v i\bar{Q}_v v^\mu D_\mu \frac{1+\slashed{v}}{2} Q_v + \mathcal{O}\left(\frac{1}{m_Q}\right), \tag{35.34}$$

where q are light quarks and $F_{\mu\nu}^a$ is the gluon field strength. The HQET Lagrangian at subleading power is discussed in Section 35.4.

Note that the field Q_v has a **label** v, which is the velocity of the heavy quark. This velocity is an exactly conserved quantum number in the effective theory, although it is only approximately conserved in full QCD. The sum over velocities can be thought of as a division of the momentum space for the heavy quark into blocks of size Λ_{QCD}. Using $p^\mu = m_Q v^\mu + k^\mu$, every heavy quark then lives in one of the blocks whose center is $m_Q v^\mu$. It is not necessary to indicate precisely how division into blocks is done or to worry about the block boundaries. In fact, the sum over v in $\mathcal{L}_{\text{HQET}}$ is just formal. In practice, one fixes the velocity v based on the observable, such as the cross section for $B_v \to D_{v'} l\nu$ at a given v and v', which is measured. Then only two values of v are relevant and we can avoid giving a precise definition to what the sum actually means.

From the HQET Lagrangian, we can read off that the propagator for the heavy quark is

$$\begin{array}{c}\xrightarrow{\hspace{2cm}} \\ k\end{array} = \frac{i}{v \cdot k + i\varepsilon} \frac{1+\slashed{v}}{2}. \tag{35.35}$$

This is just the heavy-quark limit of the propagator in QCD:

$$i\frac{\slashed{p} + m_Q}{p^2 - m_Q^2 + i\varepsilon} = i\frac{m_Q(1+\slashed{v}) + \slashed{k}}{2m_Q(v \cdot k) + k^2 + i\varepsilon} \sim \frac{i}{k \cdot v + i\varepsilon}\frac{1+\slashed{v}}{2}, \tag{35.36}$$

where $k \ll m_Q$ has been used in the last step. The HQET vertex is

$$\begin{array}{c} \mu; a \\ \\ j \quad\quad i \end{array} = igv^\mu T_{ij}^a \frac{1+\slashed{v}}{2}. \tag{35.37}$$

The v^μ factor can be understood as following from the $\frac{1+\slashed{v}}{2}$ factors in the propagators, since

$$\frac{1+\slashed{v}}{2}\gamma^\mu\frac{1+\slashed{v}}{2} = v^\mu\frac{1+\slashed{v}}{2}. \tag{35.38}$$

Finally, the Feynman rules for gluon self-interactions and gluon interactions with light quarks are the same as in full QCD.

35.3 Loops in HQET

Now let us turn to an application of HQET: calculating radiative corrections to leading-order heavy-quark prediction, $\frac{f_B}{f_D} = \sqrt{\frac{m_D}{m_B}}$, or equivalently, for the relative decay rates, $\Gamma(B \to \tau\nu)/\Gamma(D \to \mu\nu)$. We would like in include loop corrections involving virtual gluons in these rates.

The first step is to match to the effective theory. The B meson states in the full theory have momentum p^μ and relativistic normalization, as in Eq. (35.9):

$$\langle B(p')|B(p)\rangle = 2p^0(2\pi)^3\delta^3(\vec{k} - \vec{k}'). \tag{35.39}$$

In HQET, states have velocities v^μ and residual momenta k^μ, with non-relativistic normalization, as in Eq. (35.11):

$$_\text{nr}\langle B(v', k')|B(v, k)\rangle_\text{nr} = \delta_{vv'}(2\pi)^3\delta^3(\vec{p} - \vec{p}'). \tag{35.40}$$

The relevant current in the full theory is $J^\mu = \bar{b}\Gamma^\mu u$ for some tensor Γ^μ (in the electroweak theory, $\Gamma^\mu = \frac{1}{2}(1 - \gamma_5)\gamma^\mu$, but the heavy-quark system does not care what Γ^μ is). At leading order, this current matches directly onto the equivalent current constructed out of HQET fields:

$$\mathcal{O}_\Gamma^\mu = \bar{Q}_v(x)\Gamma^\mu q(x), \tag{35.41}$$

with Q_v the heavy-quark field for the b, and q representing the light-quark field. The matrix element relevant for the leptonic decay is then

$$\langle\Omega|\bar{Q}_v\Gamma^\mu q|B(v)\rangle = -iav^\mu \tag{35.42}$$

for some constant a determined by the brown muck. This is the same as Eq. (35.12), but written with HQET fields. v^μ is the only vector that can appear in this equation. Note that we have taken the residual momentum $k^\mu = 0$ in $B(v, k)$, which corresponds to defining v^μ through the momentum of the B hadron as $p^\mu = m_B v^\mu$ exactly. The formula $\frac{f_B}{f_D} = \sqrt{\frac{m_D}{m_B}}$ then follows, as in Section 35.1.1.

Now that we have an effective theory which reproduces $\frac{f_B}{f_D} = \sqrt{\frac{m_D}{m_B}}$ at leading order, we can consider perturbative corrections to this prediction. The dominant corrections in the limit where $m_b \gg m_c \gg \Lambda_\text{QCD}$ are large logarithms of the form $(\alpha_s \ln\frac{m_b}{m_c})^n$. These corrections can be resummed in HQET through the renormalization group evolution of $\mathcal{O}^\mu$. In particular, for B decays, this operator should be evaluated at m_B, while for the D decays it should be evaluated at m_D. Note that the equivalent current in full QCD does not run, because it is conserved. So one needs HQET to calculate this radiative effect through the renormalization group.

35.3.1 Renormalization of HQET

To resum large logarithms through the running of $\mathcal{O}^\mu$ we follow the same approach used to resum large logarithms in the 4-Fermi theory in Section 31.3 (see also Chapter 23). The first step is to renormalize the HQET Lagrangian.

The renormalized fields are related to the bare fields as usual:

$$A_\mu = \frac{1}{\sqrt{Z_3}} A_\mu^0, \quad q = \frac{1}{\sqrt{Z_2}} q^0, \quad g_s = \frac{Z_2 \sqrt{Z_3}}{Z_1} \mu^{\frac{d-4}{2}} g^0, \quad Q_v = \frac{1}{\sqrt{Z_h}} Q_v^0. \quad (35.43)$$

In general, the light-quark field strength renormalization Z_2, which is the same as in QCD, could be different from the field strength renormalization Z_h for the heavy quark. Interpreting the original Lagrangian as comprising bare fields, the renormalized Lagrangian is then (ignoring the light-quark masses)

$$\mathcal{L} = -\frac{1}{4} Z_3 F_{\mu\nu}^2 + Z_2 \bar{q} \left(i\slashed{\partial} + \frac{Z_1}{Z_2} \mu^{\frac{4-d}{2}} g_s \slashed{A}^a T^a \right) q$$

$$+ Z_h \bar{Q}_v v^\mu \left(i\partial_\mu + \frac{Z_1}{Z_h} \mu^{\frac{4-d}{2}} g_s A_\mu^a T^a \right) Q_v. \quad (35.44)$$

To order α_s, Z_2 is the same as in pure QCD, since the light-quark–gluon graphs are the same.

It turns out that Z_3 is also the same as in pure QCD. The only possible difference could come from vacuum polarization diagrams involving heavy quarks; however, these vanish. The technical reason is that the heavy-quark propagators give $\frac{i}{v \cdot k + i\varepsilon} \frac{i}{v \cdot (p-k) + i\varepsilon}$, with k the loop momentum. This has only two poles in k^0 (in contrast to the vacuum polarization graph in full QCD, which would have four), both of which are below the real k^0 axis. Thus, the integral over k^0 can be closed in the upper half plane and the loop integral is zero. A more physical explanation is that in the heavy-quark limit, heavy particles and antiparticles are completely different species: one is a fundamental and the other an antifundamental of $SU(3)_{QCD}$. Thus, the field Q_v that annihilates a heavy quark does not create the corresponding antiquark – this is why there is only a single pole in $\frac{i}{v \cdot k + i\varepsilon}$ instead of the usual two. The simplest but most boring explanation is that virtual $Q\bar{Q}$ pairs are suppressed and in fact do not contribute at all in the $m_Q \to \infty$ limit.

The remaining quantity to be computed in the HQET Lagrangian is Z_h, which comes from the heavy-quark self-energy graph. Expanding $Z_h = 1 + \delta_h$, the contribution of the counterterm will be

$$ = i\delta_h (v \cdot k) \frac{1 + \slashed{v}}{2}. \quad (35.45)$$

Thus, we expect the loop graph to have a $\frac{1}{\varepsilon}(v \cdot p)$ divergence.

Using the HQET Feynman rules, the loop is

$$i\mathcal{M} = \quad\underset{p\quad p-k\quad p}{\text{[diagram]}}\quad = -C_F g_s^2 \mu^{4-d} \int \frac{d^d k}{(2\pi)^d} \frac{1}{k^2 v \cdot (p-k)} \frac{1+\not{v}}{2}. \quad (35.46)$$

This graph is IR divergent, as was the electron self-energy graph we computed in Chapter 18. Since we only want the UV divergence, to extract the anomalous dimension, we will simply use the same trick we have used in many places (e.g. Section 26.4) to extract the pole from a scaleless integral in dimensional regularization (cf. Eq. (B.49)). Since the graph has mass dimension 1, the UV divergence can only be $\frac{1}{\varepsilon} v \cdot p \frac{1+\not{v}}{2}$, as expected from the form of the counterterm, and all we need is the coefficient of this term.

Taking the derivative with respect to $v \cdot p$ and then setting $p = 0$ (since the divergence is now p independent) gives

$$\frac{d\mathcal{M}}{d(v \cdot p)} = -iC_F g_s^2 \mu^{4-d} \int \frac{d^d k}{(2\pi)^d} \frac{1}{k^2 (v \cdot k)^2} \frac{1+\not{v}}{2}. \quad (35.47)$$

The denominators in this graph (and in HQET graphs in general) are not of the form $(k+X)^2$ and therefore it will not help to combine them using Feynman parameters. Instead, we use Schwinger parameters through the identity

$$\frac{1}{AB^2} = 8 \int_0^\infty ds \frac{s}{(A + 2sB)^3}, \quad (35.48)$$

with $A = k^2$ and $B = v \cdot k$, so that

$$\frac{d\mathcal{M}}{d(v \cdot p)} = -\frac{1+\not{v}}{2} 8iC_F g_s^2 \mu^{4-d} \int_0^\infty ds \int \frac{d^d k}{(2\pi)^d} \frac{s}{(k^2 + 2s\, v \cdot k)^3}. \quad (35.49)$$

Now shift $k \to k - sv$, use $v^2 = 1$ and rescale $k \to \frac{k}{s}$, giving

$$\frac{d\mathcal{M}}{d(v \cdot p)} = -\frac{1+\not{v}}{2} 8iC_F g_s^2 \mu^{4-d} \int \frac{d^d k}{(2\pi)^d} \int_0^\infty ds \frac{s}{(k^2 - s^2)^3}$$
$$= \frac{1+\not{v}}{2} 2iC_F g_s^2 \mu^{4-d} \int \frac{d^d k}{(2\pi)^d} \frac{1}{k^4}. \quad (35.50)$$

This is the ordinary scaleless, UV- and IR-divergent dimensionally regularized integral we have seen many times before. We can extract the UV divergence using Eq. (B.49). Writing $d = 4 - \varepsilon$ we find

$$\mathcal{M} = -C_F \frac{g_s^2}{4\pi^2} \frac{1}{\varepsilon_{\text{UV}}} (v \cdot p) \frac{1+\not{v}}{2} + \cdots. \quad (35.51)$$

For this divergence to be canceled by the counterterm contribution from Z_h, we must take

$$Z_h = 1 + \frac{1}{\varepsilon} C_F \frac{\alpha_s}{\pi}. \quad (35.52)$$

Note that the heavy-quark renormalization is different from the light-quark renormalization, which was $Z_2 = 1 - \frac{1}{\varepsilon} C_F \frac{\alpha_s}{2\pi}$ in Feynman gauge.

35.3.2 Running of $\mathcal{O}_\Gamma^\mu$

Now that we have the complete 1-loop renormalization factors for the HQET Lagrangian, we can turn to the renormalization of the heavy–light current, which we wrote as $\mathcal{O}_\Gamma^\mu = \bar{Q}_v \Gamma^\mu q$. This is a composite operator and must be renormalized separately from its constituent fields. The bare operator $\mathcal{O}_{\Gamma,\text{bare}}^\mu = \bar{Q}_v^0 \Gamma^\mu q^0$ is related to the renormalized operator by $\mathcal{O}_{\Gamma,\text{bare}}^\mu = Z_\mathcal{O} \Gamma$, so that

$$\mathcal{O}_\Gamma^\mu = \frac{1}{Z_\mathcal{O}} \mathcal{O}_{\Gamma,\text{bare}}^\mu = \frac{1}{Z_\mathcal{O}} \bar{Q}_v^0 \Gamma^\mu q^0 = \frac{\sqrt{Z_h Z_q}}{Z_\mathcal{O}} \bar{Q}_v \Gamma^\mu q. \tag{35.53}$$

To find $Z_\mathcal{O}$, we can evaluate the correlation function $\langle Q | \mathcal{O}_\Gamma^\mu | q \rangle$ at 0-momentum (any momentum would do, since we are interested in the UV divergence).

Writing $Z_\mathcal{O} = 1 + \delta_\mathcal{O}$, the counterterms give

$$\frac{\sqrt{Z_h Z_q}}{Z_\mathcal{O}} = \frac{1}{2}\delta_2 + \frac{1}{2}\delta_h - \delta_\mathcal{O} + \cdots = C_F \frac{\alpha_s}{4\pi}\frac{1}{\varepsilon} - \delta_\mathcal{O} + \cdots. \tag{35.54}$$

The 1-loop graph is

$$iM - \quad\begin{array}{c}\includegraphics\end{array}\quad = C_F(ig_s)^2 \mu^{4-d} \int \frac{d^d k}{(2\pi)^d} i\gamma^\nu \frac{i\slashed{k}}{k^2}\Gamma^\mu \frac{i}{v\cdot k} iv^\rho \frac{-ig^{\rho\nu}}{k^2}$$

$$= -iC_F g_s^2 \mu^{4-d} \int \frac{d^d k}{(2\pi)^d} \frac{\slashed{v}\slashed{k}}{k^4 v\cdot k}\Gamma^\mu. \tag{35.55}$$

We can simplify this by inserting a Schwinger parameter, through Eq. (35.48), as for the self-energy graph. The result is the exact same integral as (35.50):

$$iM = iC_F g_s^2 \mu^{4-d} \int \frac{d^d k}{(2\pi)^d} \int_0^\infty ds \frac{s}{(k^2 - s^2)^3}\Gamma^\mu = C_F \frac{g_s^2}{8\pi^2}\frac{1}{\varepsilon}\Gamma^\mu + \text{finite}. \tag{35.56}$$

The total divergent contribution at order α_s is therefore

$$\left(C_F \frac{\alpha_s}{4\pi}\frac{1}{\varepsilon} + C_F \frac{\alpha_s}{2\pi}\frac{1}{\varepsilon} - \delta_\mathcal{O}\right)\Gamma^\mu, \tag{35.57}$$

and so we find

$$Z_\mathcal{O} = 1 + C_F \frac{3\alpha_s}{4\pi}\frac{1}{\varepsilon}. \tag{35.58}$$

The RGE comes from μ independence of the bare operator $\mathcal{O}^0$. That is,

$$0 = \mu\frac{d}{d\mu}(\mathcal{O}^0) = \mu\frac{d}{d\mu}(Z_\mathcal{O}\mathcal{O}) \tag{35.59}$$

so that

$$\gamma_\mathcal{O} = -\frac{\mu}{Z_\mathcal{O}}\frac{d}{d\mu}Z_\mathcal{O} = -\frac{1}{Z_\mathcal{O}}\frac{\partial Z_\mathcal{O}}{\partial \alpha_s}\beta(\alpha_s). \tag{35.60}$$

Plugging in $Z_\mathcal{O}$ and $\beta(\alpha_s)$ from Eq. (26.96), we then find

$$\gamma_\mathcal{O} = -C_F \frac{3}{4\pi} \frac{1}{\varepsilon} \left(-\varepsilon \alpha_s - \frac{\alpha_s^2}{2\pi} \beta_0 + \mathcal{O}(\alpha_s^3) \right) = C_F \frac{3\alpha_s}{4\pi} + \mathcal{O}(\alpha_s^2). \qquad (35.61)$$

This is the anomalous dimension for the heavy–light quark operator in HQET at 1-loop.

We are interested in the evolution of the Wilson coefficient C for this operator. We matched $C = 1$ at tree-level. Using $\frac{d}{d\mu}(C\mathcal{O}) = 0$, the Wilson coefficient evolves with $-\gamma_\mathcal{O}$. Then, the RGE is solved with

$$
\begin{aligned}
C(\mu) &= C(\mu_0) \exp\left[-\int_{\alpha_s(\mu_0)}^{\alpha_s(\mu)} \frac{\gamma_\mathcal{O}(\alpha)}{\beta(\alpha)} d\alpha \right] \\
&= C(\mu_0) \exp\left[\frac{\frac{3}{4}C_F}{\frac{1}{2}\beta_0} \ln\frac{\alpha_s(\mu)}{\alpha_s(\mu_0)} \right] \\
&= C(\mu_0) \left(\frac{\alpha_s(\mu)}{\alpha_s(\mu_0)} \right)^{\frac{3}{2}C_F\beta_0^{-1}}.
\end{aligned}
\qquad (35.62)
$$

For the f_B/f_D comparison, we are interested in the renormalization group effects between m_M and m_D. Including four flavors, $\beta_0 = \frac{11}{3}C_A - \frac{4}{3}T_F n_f = \frac{25}{3}$ and so, with $\mu_0 = m_b$, $\mu = m_c$, $\alpha_s(m_b) = 0.22$ and $\alpha_s(m_c) = 0.35$ we get

$$\frac{f_B\sqrt{m_B}}{f_D\sqrt{m_D}} = \left[\frac{\alpha_s(m_b)}{\alpha_s(m_c)} \right]^{-\frac{6}{25}} = 1.12. \qquad (35.63)$$

So there is a 12% correction from this calculation. This does not explain the factor $\sim 200\%$ by which the ratio is off in the real world. This large correction could be explained by power corrections proportional to $\frac{\Lambda_{\text{QCD}}}{m_D}$, which happens to have a numerically large coefficient. This is unfortunate. On the other hand, it is not the effective theory's fault that the charm quark is so light!

35.4 Power corrections

Much of the predictive power of heavy-quark effective theory comes from the way the expansion of corrections in inverse powers of the heavy-quark mass is organized. At each order in m_Q^{-1} there will only be a finite number of operators that can contribute. These operators have matrix elements that although unknown are universal, such as the $\langle \text{muck}; v's'_q | \text{muck}; s'_q v \rangle$ matrix elements involved in the leading-order predictions. In some cases m_Q^{-1} corrections vanish and therefore we can make predictions accurate to the small percentage level.

To derive the subleading HQET Lagrangian we have to integrate out the small component of the heavy-quark field (as opposed to just setting it to zero as we did in Section 35.2). To begin, we project out the large and small components of the heavy-quark field (cf. Eq. (35.31)):

$$\psi(x) = e^{-im_Q v \cdot x} \left[\frac{1 + \slashed{v}}{2} Q_v(x) + \frac{1 - \slashed{v}}{2} \widetilde{Q}_v(x) \right], \qquad (35.64)$$

where $\frac{1+\slashed{v}}{2} Q_v = Q_v$ and $\frac{1-\slashed{v}}{2} \tilde{Q}_v = \tilde{Q}_v$. We then find

$$\mathcal{L} = \bar{\psi}(i\slashed{D} - m_Q)\psi = i\bar{Q}_v v \cdot DQ_v + \overline{\tilde{Q}}_v(-iv \cdot D - 2m_Q)\tilde{Q}_v + i\bar{Q}_v\slashed{D}\tilde{Q}_v + i\overline{\tilde{Q}}_v\slashed{D}Q_v, \quad (35.65)$$

with the $\frac{1\pm\slashed{v}}{2}$ projectors left implicit. It is helpful to simplify this using

$$D_\perp^\mu \equiv D^\mu - v^\mu(v \cdot D). \quad (35.66)$$

Note that if $v^\mu = (1, \vec{0})$ then $D_\perp^\mu = (0, \vec{D})$ is just the spatial derivatives. Hence,

$$\bar{Q}_v\slashed{D}\tilde{Q}_v = \bar{Q}_v\frac{1+\slashed{v}}{2}\left[\slashed{D}_\perp + \slashed{v}(v \cdot D)\right]\frac{1-\slashed{v}}{2}\tilde{Q}_v = \bar{Q}_v\slashed{D}_\perp\tilde{Q}_v, \quad (35.67)$$

so we can write

$$\mathcal{L} = \bar{\psi}(i\slashed{D} - m_Q)\psi = i\bar{Q}_v v \cdot DQ_v + \overline{\tilde{Q}}_v(-iv \cdot D - 2m_Q)\tilde{Q}_v + i\bar{Q}_v\slashed{D}_\perp\tilde{Q}_v + i\overline{\tilde{Q}}_v\slashed{D}_\perp Q_v. \quad (35.68)$$

The field Q_v can be thought of as describing fluctuations in components of the heavy-quark momentum that leave its velocity fixed. These are massless excitations. The field $\tilde{Q}_v$ apparently has mass $2m_Q$. It describes processes in which heavy-quark–heavy-antiquark pairs are created.

Since $\tilde{Q}_v$ is heavy, it can be integrated out of the Lagrangian. The easiest way to integrate out a field at tree-level is to set it equal to its equations of motion. These are

$$(iv \cdot D + 2m_Q)\tilde{Q}_v = i\slashed{D}_\perp Q_v, \quad (35.69)$$

so that

$$\mathcal{L} = i\bar{Q}_v v \cdot DQ_v + \bar{Q}_v i\slashed{D}_\perp\frac{1}{2m_Q + iv \cdot D}i\slashed{D}_\perp Q_v \quad (35.70)$$

$$= i\bar{Q}_v v \cdot DQ_v + \frac{1}{2m_Q}\sum_{n=0}^\infty \bar{Q}_v i\slashed{D}_\perp\left(-\frac{iv \cdot D}{2m_Q}\right)^n i\slashed{D}_\perp Q_v.$$

The first new term, of order m_Q^{-1}, can be simplified by using the relation (see Eq. 10.106)

$$\bar{Q}_v\slashed{D}_\perp^2 Q_v = \bar{Q}\left[D_\perp^2 + \frac{g_s}{2}\sigma_{\mu\nu}F_{\mu\nu}\right]Q_v, \quad (35.71)$$

so that, including the gluon field strength and light-quark fields,

$$\mathcal{L}_{\text{HQET}} = -\frac{1}{4}(F_{\mu\nu}^a)^2 + \bar{q}(i\slashed{D} - m_q)q + i\bar{Q}_v v \cdot DQ_v$$

$$- \bar{Q}_v\frac{D_\perp^2}{2m_Q}Q_v - \frac{g_s}{4m_Q}\bar{Q}_v\sigma_{\mu\nu}Q_v F^{\mu\nu} + \left(\frac{1}{m_Q^2}\right). \quad (35.72)$$

The $\bar{Q}_v\frac{D_\perp^2}{2m_Q}Q_v$ term is a covariant version of the non-relativistic kinetic energy $\frac{\vec{p}^2}{2m}$ of the heavy-quark field. Because of the $D_\perp$, it contains only spatial components perpendicular to v. The $g_s\bar{Q}_v\sigma_{\mu\nu}Q_v F^{\mu\nu}$ term is the chromomagnetic-moment interaction.

One can use the subleading HQET Lagrangian to prove a number of powerful results about hadronic matrix elements. One example is Luke's theorem, mentioned in Section 35.1.2, that the power corrections to certain form factors at zero recoil do not receive m_Q^{-1} corrections. A discussion of this theorem and its proof can be found in [Manohar and Wise, 2000]. Here we discuss only a simpler application: the parametrization of power corrections to meson masses.

35.4.1 Hadron masses

With the subleading-power HQET Lagrangian we can now parametrize the m_Q^{-1} corrections to hadron masses. To calculate masses we can take the expectation value of the HQET Hamiltonian. Let us write

$$\mathcal{H} = \mathcal{H}_{-1} + \mathcal{H}_0 + \mathcal{H}_1 + \cdots \tag{35.73}$$

with $\mathcal{H}_i \sim m_Q^{-i}$. In the $m_Q \to \infty$ limit, the Hamiltonian is just the heavy-quark rest mass, thus $\mathcal{H}_{-1} = m_Q$. The leading-order HQET Lagrangian, Eq. (35.34), leads to a Hamiltonian (cf. Eq. (12.63))

$$\mathcal{H}_0 = \mathcal{E}_0^{\text{QCD}} + \bar{q}(i\gamma^i \partial_i + m_q)q + \bar{Q}_v i\vec{v} \cdot \vec{D} Q_v, \tag{35.74}$$

where (using Eq. (8.26)) $\mathcal{E}_0^{\text{QCD}} = \frac{1}{2}(\vec{E}^2 + \vec{B}^2) + \cdots$ is the energy density of the gluons, whose precise form we do not need. The m_Q^{-1} HQET Lagrangian has no time derivatives, so the m_Q^{-1} Hamiltonian is just the negative of the m_Q^{-1} Lagrangian:

$$\mathcal{H}_1 = \bar{Q}_v \frac{D_\perp^2}{2m_Q} Q_v + \frac{g_s}{4m_Q} \bar{Q}_v \sigma_{\mu\nu} Q_v F^{\mu\nu}. \tag{35.75}$$

Note that all of the quark-mass dependence is explicit in the m_Q factors; thus, the matrix elements of these operators are heavy-quark-flavor independent.

For example, consider meson states $|H_{\vec{J}}\rangle$ in the same flavor multiplet, where $\vec{J}$ is the spin. As in Eq. (35.15), we write $|H_{\vec{J}}\rangle = |\vec{S}_Q\rangle |\text{muck}; \vec{S}_q\rangle$ where $|\vec{S}_Q\rangle$ refers to the heavy-quark state with a given spin $\vec{S}$ and $|\text{ muck}; \vec{S}_q\rangle$ refers to the gluons and light quarks, with $\vec{S}_q$ the light-quark spin. We can evaluate the masses in the heavy-quark rest frame, where $v = (1, \vec{0})$. Then,

$$\langle H_{\vec{J}} | \mathcal{H}_0 | H_{\vec{J}} \rangle = \bar{\Lambda}, \tag{35.76}$$

where $\bar{\Lambda} \sim \Lambda_{\text{QCD}}$ is a non-perturbative matrix element coming from the light quarks and gluons. The prediction for the masses up to order m_Q^0 is then that the B and B^* masses are degenerate, as are the D and D^* masses. The splitting comes at order m_Q^{-1}.

The m_Q^{-1} corrections contain a kinetic energy term, which is spin independent:

$$\frac{1}{2m_b} \langle H_{\vec{J}} | \bar{Q}_v D_\perp^2 Q_v | H_{\vec{J}} \rangle = -\lambda_1 \frac{1}{2m_Q}. \tag{35.77}$$

Here, $\lambda_1 \sim \Lambda_{\text{QCD}}^2$ is some new non-perturbative parameter. We expect this matrix element to be negative (so $\lambda_1 > 0$) since the kinetic energy $\frac{p^2}{2m_Q}$ should be positive.

Matrix elements of the other m_Q^{-1} term, $\overline{Q}_v \sigma_{\mu\nu} Q_v F^{\mu\nu}$, depend on spin. We have

$$\frac{2}{m_Q} \langle H_{\vec{j}} | \frac{g_s}{8} \overline{Q}_v \sigma_{\mu\nu} Q_v F^{\mu\nu} | H_{\vec{j}} \rangle = \frac{1}{m_Q} \langle \vec{S}_Q | \overline{Q}_v \sigma_{\mu\nu} Q_v | \vec{S}_Q \rangle \langle \text{muck}; \vec{S}_q | \frac{g_s}{4} F^{\mu\nu} | \text{muck}; \vec{S}_q \rangle$$

$$= \frac{\lambda_2}{m_Q} 2 \vec{S}_Q \cdot \vec{S}_q, \qquad (35.78)$$

where $\lambda_2 \sim \Lambda_{\text{QCD}}^2$ is some new flavor- and spin-independent non-perturbative parameter. That the muck matrix element is proportional to the light-quark spin follows from the Wigner–Eckhart theorem: $\vec{A}_q$ is the only vector available. Now, $2\vec{S}_Q \cdot \vec{S}_q = \vec{J}^2 - \vec{S}_Q^2 - \vec{S}_q$, so that $2\vec{S}_Q \cdot \vec{S}_q = -\frac{3}{2}$ for the spin-0 mesons and $2\vec{S}_Q \cdot \vec{S}_q = \frac{1}{2}$ for the spin-1 mesons.

Putting these results together, we get

$$m_B = m_b + \bar{\Lambda} - \frac{\lambda_1}{2m_b} - \frac{3\lambda_2}{4m_b}, \qquad (35.79)$$

$$m_{B^*} = m_b + \bar{\Lambda} - \frac{\lambda_1}{2m_b} + \frac{\lambda_2}{2m_b}. \qquad (35.80)$$

An important result from these equations is that the difference between the squares of meson masses in the same multiplet begins at order m_Q^{-1} and is flavor independent: $m_{B^*}^2 - m_B^2 = m_{D^*}^2 - m_D^2 + \mathcal{O}(m_c^{-1})$. This led to the accurate prediction for m_{D^*} mentioned in the introduction to this chapter.

Although two new non-perturbative quantities, λ_1 and λ_2, have appeared at subleading power, *only* two quantities have appeared. These same quantities contribute to other masses, form factors and inclusive decay rates. Thus, one can measure λ_1 and λ_2 and use those values, along with the computable corrections perturbative in $\alpha_s(m_Q)$, to make many quantitive predictions in HQET.

Problems

35.1 Reparametrization invariance.

    (a) Show that the HQET Lagrangian including the leading m_Q^{-1} corrections, Eq. (35.72), is invariant under

$$v^\mu \to v^\mu + \frac{1}{m_Q} k^\mu, \quad Q_v \to e^{ik \cdot x} \left(1 + \frac{\slashed{k}}{2m_Q} \right) Q_v, \qquad (35.81)$$

    with $v \cdot k = 0$ and $k \ll m_Q$. This transformation is known as **reparametrization invariance**. It corresponds to the arbitrariness in the choice of v^μ.

    (b) Use reparametrization invariance to show that the $\overline{Q}_v \frac{D^2}{2m_Q} Q_v$ term in the HQET Lagrangian cannot be renormalized separately from the $\overline{Q}_v v \cdot D Q_v$ term.

    (c) Confirm through a direct 1-loop calculation that these two terms are indeed renormalized in the same way.

35.2 Calculate the anomalous dimension of the HQET operator $\frac{g}{4m_Q} \overline{Q}_v \sigma_{\mu\nu} Q_v F^{\mu\nu}$ at 1-loop.

36 Jets and effective field theory

Almost every event of interest at high-energy colliders contains collimated collections of particles known as jets. An example event with jets is shown in Figure 36.1. The intuitive picture of how jets form is the semi-classical parton shower discussed in Section 32.3: a hard parton (quark or gluon) is produced at short distance. As the parton moves out from the collision point it radiates gluons; gluons in the radiation field then split into other gluons and quark–antiquark pairs. When the collection has spread out over length scales of order $\Lambda_{\rm QCD}^{-1}$, the quarks and gluons hadronize into color-neutral objects. These hadrons then decay into stable or metastable particles (mostly pions), which the experiments attempt to measure. Since the radiation is dominantly in the direction of the original hard parton, it can be added together to form a jet 4-momentum $p_J^\mu = \sum_{i\in{\rm jet}} p_i^\mu$, which approximates the 4-momentum of the hard parton originally produced. For example, if the two jets are produced from the decay of a W boson ($W \to \bar{q}q$ at parton level), the dijet invariant mass should be close to the W-boson mass $(p_{J_1} + p_{J_2})^2 \approx m_W^2$. Thus, jets provide a window into short-distance physics. Jets are useful both in Standard Model studies and in searches for physics beyond the Standard Model.

The distribution of jets is described quite accurately by perturbative QCD. For example, the $gg \to gg$ cross section (computed in Chapter 27), when convolved with PDFs (discussed in Chapter 32), gives a contribution to the distribution of dijet events at hadron colliders. When all parton channels are included, the theoretical calculations are in excellent agreement with data over a wide range of energies and production angles. The theoretical tools necessary for computing the distribution of jets in perturbative QCD have been explained in Chapters 25, 26, 27 and 32.

Fig. 36.1 Event display for a dijet event at the LHC as observed by the ATLAS experiment.

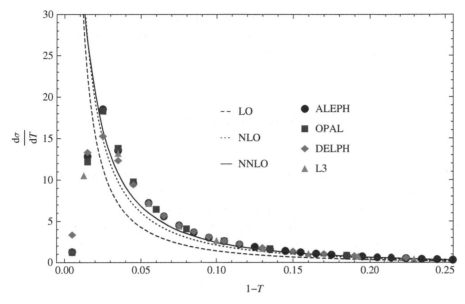

Comparison of thrust data from four experiments at LEP to the calculation in perturbative QCD at up to next-to-next-to-leading order in α_s. The fixed-order calculation has good agreement for $1 - T \gtrsim 0.15$, but fails to describe the peak region even qualitatively.

Fig. 36.2

On the other hand, some properties of jets, such as their mass, are not described well at any fixed order in α_s. For example, Figure 36.2 shows the distribution of thrust at LEP compared to the perturbative calculation at order α_s, α_s^2 and α_s^3. Thrust, which is defined and discussed in Section 36.1, is one way to characterize how dijet-like an event is. Events that produce values of thrust near 1 (the left side of the figure) appear to have two very collimated jets. In fact, near $T = 1$, one can show $1 - T \approx \frac{1}{Q^2}\left(m_{J_1}^2 + m_{J_2}^2\right)$, where m_{J_1} and m_{J_2} are the masses of the two jets and Q the center-of-mass energy. Clearly, the thrust distribution near $T \sim 1$ is not described well in perturbation theory. In fact, the cross section computed in perturbative QCD blows up as $T \to 1$ at any finite order in α_s. One goal of this chapter is to understand the origin of these (unphysical) singularities. To reproduce the experimental fact that the distribution goes to zero as $T \to 1$ requires the resummation of contributions to all orders in α_s. This resummation will generate distributions that turn over, qualitatively reproducing the **Sudakov peak** (the turnover in the data in Figure 36.2), and quantitatively improving the agreement between theory and experiment (cf. Figure 36.3).

The singular terms in observables, such as the thrust distribution, are qualitatively similar to the large logarithms we have resummed with the renormalization group in previous chapters (see Chapters 16, 23, 26, 31 and 35). In previous applications of the renormalization group, the singular terms were of the form $(\alpha \ln x)^n$ with one additional logarithm at each subsequent order in α. With jets, there are often two logarithms. For example, in the cumulant thrust distribution, $R(T) = \frac{1}{\sigma_{\text{tot}}} \int_T^1 \frac{d\sigma}{dT'} dT' \approx 1 - C_F \frac{\alpha_s}{\pi} \ln^2(1-T) + \cdots$, we can see the double logs explicitly. In Section 32.3, we saw how such **Sudakov double logarithms** could be resummed semi-classically with Sudakov factors. Here we will be more systematic about the resummation by developing an effective field theory, called Soft-Collinear Effective Theory. This theory will let us resum the double logarithms systematically using the renormalization group.

36.1 Event shapes

Many applications of jet physics require exclusive jet definitions which isolate the radiation going into jets from the rest of the event. On the other hand, certain properties of events with jets in them can be studied efficiently through inclusive observables called **event shapes**. Event shapes are, by definition, global observables (meaning that all final-state particles contribute) with no free parameters. They have predominantly been useful at e^+e^- colliders.[1]

The most widely studied event shape is called **thrust**. It is defined as

$$T \equiv \max_{\vec{n}} \frac{\sum_j |\vec{p}_j \cdot \vec{n}|}{\sum_j |\vec{p}_j|}, \tag{36.1}$$

where the sum is over the 3-momenta $\vec{p}_j$ of all particles in the event, and the maximum is taken over all 3-vectors $\vec{n}$ with $|\vec{n}| = 1$. The direction that maximizes thrust is called the **thrust axis**. Data for thrust from various experiments at LEP are shown in Figure 36.2.

To develop intuition for thrust, consider the final state of an $e^+e^- \rightarrow$ hadrons event in which two very narrow jets are produced. Such pencil-like jets will have $T \sim 1$ since $|p_j \cdot \vec{n}| \sim |\vec{p}_j|$ if $\vec{n}$ points along the direction of the pencil. For such events, the thrust axis will be close to the jet axis, independently of the jet definition. If an event has particles distributed evenly in all directions then there is no preferred $\vec{n}$ and (very roughly) $|\vec{p}_j \cdot \vec{n}| \sim |\cos\theta \vec{p}_j| \sim \frac{1}{2}|\vec{p}_j|$. Thus, $T \sim \frac{1}{2}$ indicates a spherical event. In this way, thrust is a quantitative measure of how pencil-like or spherical an event is. In the following we will use

$$\tau \equiv 1 - T, \tag{36.2}$$

which goes to 0 in the dijet configuration and goes to $\frac{1}{2}$ for spherical events.

Although thrust is measured on metastable particles coming out of e^+e^- collisions (mostly pions), it can also be computed in perturbation theory using quarks and gluons. Let $Q = E_{\mathrm{CM}}$ be the center-of-mass energy. For $Q \gg \Lambda_{\mathrm{QCD}}$ one expects the shape of the event to be frozen-in on time scales much shorter than the hadronization time. Thus, perturbative QCD should provide a reasonable description of thrust up to corrections suppressed by some power of $\frac{\Lambda_{\mathrm{QCD}}}{Q}$.

Two event shapes closely related to thrust are **heavy jet mass** and **light jet mass**. To compute them, first find the thrust axis for a particular event using Eq. (36.1). Then partition the particles in the event into two hemispheres by the sign of $\vec{p} \cdot \vec{n}$. Call the sum of the

---

[1] At hadron colliders, the beam remnant makes it impractical to include all final-state radiation in an observable. While there are generalizations of e^+e^- event shapes to hadronic event shapes, we will not discuss them.

4-momenta in one hemisphere p_1^μ and the rest p_2^μ. Then, heavy jet mass ρ_H and light jet mass ρ_L are defined by

$$\rho_H \equiv \frac{1}{Q^2} \max\left(p_1^2, p_2^2\right), \quad \rho_L \equiv \frac{1}{Q^2} \min\left(p_1^2, p_2^2\right). \tag{36.3}$$

Thus, ρ_H and ρ_L are really masses-squared. We also define

$$\tau_1 \equiv \frac{1}{Q^2}\left(p_1^2 + p_2^2\right) = \rho_L + \rho_H. \tag{36.4}$$

Other event shapes include jet broadening, sphericity, spherocity, Y_{23} and the C-parameter (see [Ellis $et\ al.$, 1996] for their definitions and some discussion).

36.1.1 Thrust in perturbative QCD

Now we will compute thrust at leading order in perturbation theory in QCD. At zeroth order, the final state consists of two quarks ($e^+e^- \rightarrow q\bar{q}$). These quarks have massless back-to-back 4-vectors, and hence $\tau = 0$. Thus, the zeroth-order distribution is $\frac{d\sigma}{d\tau} = \sigma_0 \delta(\tau)$, where $\sigma_0 = \frac{4\pi\alpha_e^2}{3E_{\mathrm{CM}}^2} R_{\mathrm{had}}$ and $R_{\mathrm{had}} = \sum_{\mathrm{colors}} \sum_{q=u}^{b} Q_q^2 = 3.67$ from Eq. (26.24).

For the $\mathcal{O}(\alpha_s)$ thrust distribution, conventionally called leading order (LO), the partonic process is $e^+e^- \rightarrow \bar{q}qg$. The total cross section at order α_s was calculated in Section 26.3 using the results from the analogous process in QED computed in Chapter 20. There we found that $\sigma_{\mathrm{tot}} = \sigma_0\left(1 + \frac{3}{4\pi}C_F\alpha_s\right)$. To compute thrust at LO define $s = (p_g + p_q)^2$, $t = (p_g + p_{\bar{q}})^2$ and $u = (p_q + p_{\bar{q}})^2$. Since we treat the quarks as massless, $s+t+u = Q^2$. From now on, we set $Q = 1$ for simplicity, so $s+t+u = 1$. The differential cross section at order α_s is (see Section 20.1.2)

$$\frac{1}{\sigma_0}\frac{d\sigma}{ds\,dt} = C_F\frac{\alpha_s}{2\pi}\frac{s^2 + t^2 + 2u}{st}. \tag{36.5}$$

The maximization in the definition of thrust is a minimization over τ. For three massless partons, $\tau = \min(s, t, u) \leq \frac{1}{3}$. The thrust distribution is then

$$\frac{1}{\sigma_0}\frac{d\sigma}{d\tau} = \frac{1}{\sigma_0}\int ds\,dt\frac{d\sigma}{ds\,dt}\Big[\delta(\tau - s)\,\theta(t-s)\,\theta(u-s)$$
$$+ \delta(\tau - t)\,\theta(s-t)\,\theta(u-t) + \delta(\tau - u)\,\theta(t-u)\,\theta(s-u)\Big]$$
$$= \frac{2}{\sigma_0}\int_\tau^{1-2\tau} ds\int dt\frac{d\sigma}{ds\,dt}\delta(\tau - t) + \frac{1}{\sigma_0}\int_\tau^{1-2\tau} ds\int dt\frac{d\sigma}{ds\,dt}\delta(\tau - u)$$
$$= C_F\frac{\alpha_s}{2\pi}\left\{\frac{3(1+\tau)(3\tau - 1)}{\tau} + \frac{[4 + 6\tau(\tau - 1)]\ln\frac{1-2\tau}{\tau}}{\tau(1 - \tau)}\right\}, \tag{36.6}$$

where $u = 1 - s - t$ and the symmetry under $s \leftrightarrow t$ have been used. This result is valid for $\tau > 0$, and shown as the leading-order (LO) curve in Figure 36.2.

At $\tau = 0$ there is an IR divergence. This is canceled by the IR divergence in the virtual contributions, from the 1-loop correction to $e^+e^- \rightarrow \bar{q}q$. The sum of the two is IR finite since thrust is an infrared-safe observable. To see the cancellation one must regulate the virtual graph and the real emission graphs with an IR regulator such as a gluon mass or dimensional regularization (see Chapter 20) and then combine them. Fortunately, we can extract the combined answer from Eq. (36.6) using a trick: the regulated answer must be a distribution whose integral gives the total cross section $\sigma_T = \sigma_0\left(1 + C_F \frac{3\alpha_s}{4\pi}\right)$. Since the virtual graph must be proportional to $\delta(\tau)$ we can deduce that

$$\frac{1}{\sigma_0}\frac{d\sigma}{d\tau} = \delta(\tau) + C_F \frac{\alpha_s}{2\pi}\left\{\delta(\tau)\left(\frac{\pi^2}{3} - 1\right)\right.$$
$$\left. + \left[3(1+\tau)(3\tau - 1) + \frac{[4 + 6\tau(\tau - 1)]\ln(1 - 2\tau)}{1 - \tau}\right]\left[\frac{1}{\tau}\right]_+ - \frac{4 + 6\tau(\tau - 1)}{1 - \tau}\left[\frac{\ln\tau}{\tau}\right]_+\right\}.$$

$$(36.7)$$

Recall that the *plus distribution*, $\left[\frac{\ln^k \tau}{\tau}\right]_+$, defined in Section 32.2, has the property that $\int_0^1 \left[\frac{\ln^k \tau}{\tau}\right]_+ g(\tau) = \int_0^1 d\tau \frac{\ln^k \tau}{\tau}[g(\tau) - g(0)]$ and $\left[\frac{\ln^k \tau}{\tau}\right]_+ = \frac{\ln^k \tau}{\tau}$ for $\tau > 0$. The singular terms in this expression at small τ are

$$\left(\frac{1}{\sigma_0}\frac{d\sigma}{d\tau}\right)_{\text{sing}} = \delta(\tau) + C_F \frac{\alpha_s}{2\pi}\left\{\delta(\tau)\left(\frac{\pi^2}{3} - 1\right) - 3\left[\frac{1}{\tau}\right]_+ - 4\left[\frac{\ln\tau}{\tau}\right]_+\right\}. \qquad (36.8)$$

It is these singular terms that are the main focus of this chapter.

In the region where $\tau \ll 1$, so that the event has pencil-like dijet kinematics, then $\tau \sim \tau_1$. You can prove this in Problem 36.1. An easy check is that at leading order $\tau = \min(s, t, u)$ and s, t and u are the invariant masses of pairs of partons so $\tau = \tau_1 = \rho_H$ and $\rho_L = 0$. We will use the equivalence between τ and τ_1 in the singular limit in Sections 36.5.2 and 36.6.

36.2 Power counting

Our first task is to understand the origin of the singularities in the distribution of jet mass and related jet properties. To calculate jet mass in QCD, or to measure it, one needs a jet definition (for example, Sterman–Weinberg jets, discussed in Section 20.2). For any given jet definition, the distribution of the jet mass can be written in the form

$$\frac{d\sigma}{dm^2} = \left(\frac{d\sigma}{dm^2}\right)_{\text{sing}} + \left(\frac{d\sigma}{dm^2}\right)_{\text{non-sing}}, \qquad (36.9)$$

where "sing" refers to the part of the distribution that is singular as $m^2 \to 0$. The part labeled "non-sing" is regular as $m^2 \to 0$. For example, the singular part of the thrust distribution at leading order is shown in Eq. (36.8). The singular terms dominate the behavior of the distribution at small m^2 (or small τ in the thrust case). Our approach will be to calculate these terms to all orders in α_s using effective field theory methods. We can add in the non-singular part to the resummed singular distribution by matching to perturbative QCD order-by-order in α_s.

To calculate the singular part of $\frac{d\sigma}{dm^2}$ we need an expansion paramater λ (the analog of F_π^{-1} in the Chiral Lagrangian, m_Q^{-1} in HQET or m_W^{-1} in the 4-Fermi theory). A natural choice is the ratio of the jet mass m to scale Q, $\lambda = \frac{m}{Q}$. In practice, it is often easier to use an expansion parameter that is **inclusive**, meaning that it gets a contribution from every observed hadronic particle in an event, rather than **exclusive**, like a jet mass, where only particles within the jet contribute. Examples of observables and inclusive expansion parameters are

- Event shapes at e^+e^- collisions (see Section 36.1). We can take $\lambda = \tau = 1 - T$ for thrust $\lambda = \rho_H$ for heavy jet mass.
- Deep inelastic scattering: $e^-p^+ \to e^-X$. Recall from Chapter 32 that deep inelastic scattering can be thought of as an off-shell photon with spacelike momentum q^μ scattering off a proton with momentum P^μ into a hadronic final state with momentum X^μ. The inclusive observables $Q^2 = -q^2$ and $x = \frac{Q^2}{2P \cdot q}$ can be measured from the outgoing electron only. The interesting kinematical region for jet physics is when the mass of the entire hadronic final state becomes small. The jet mass is $m_J^2 = p_X^2 = (q + P)^2 = Q^2 \frac{1-x}{x} + m_P^2$. Neglecting the proton mass, this is $m_J^2 = Q^2 \frac{1-x}{x}$. Thus, $m_J \to 0$ as $x \to 1$ and the $\lambda = \frac{m_J}{Q} = \sqrt{1-x}$ can be used as an expansion parameter to describe the jet-like limit.
- Heavy-to-light B meson decays. Consider the decay $B \to X_s\gamma$, where X_s is any hadronic final state with strangeness $s = 1$. At the parton level this is $|b; \text{muck}\rangle \to |s; \text{muck}\rangle |\gamma\rangle$ (see Chapter 35). In this case, the energy of the outgoing photon E_γ provides a clean inclusive observable. In the limit that $E_\gamma \to \frac{m_B}{2}$, X_s must be massless, and hence jet-like. Thus $\lambda = 1 - \frac{2E_\gamma}{m_B}$ is an inclusive expansion parameter.

Let us take $B \to X_s\gamma$ for concreteness, where $\lambda = 1 - \frac{E_\gamma}{2m_B}$. In this case the jet is defined to include all hadrons in the final state. Of course, in a typical event, this jet definition does not look jet-like (it can be a single kaon). However, events that have small values of λ are jet-like, in the sense that the invariant mass of the hadronic final state is small. In perturbative QCD, we compute the final-state distribution in terms of quarks and gluons, ignoring hadronization to a first approximation. What collection of final-state quarks and gluons can have a small invariant mass? By momentum conservation, in the B meson rest frame the jet points backwards to the photon $\vec{p}_J = -\vec{p}_\gamma$. Near $\lambda = 0$, $|\vec{p}_J| = |\vec{p}_\gamma| = E_\gamma \sim \frac{m_B}{2}$, so the jet must have large energy and small invariant mass. Since $p_J^2 = (\sum_i p_i^\mu)^2$, for any two particles in the jet with momenta p_i^μ and p_j^μ, we have $p_J^2 > 2p_i \cdot p_j = 2E_iE_j(1 - \cos\theta_{ij})$, where θ_{ij} is the angle between $\vec{p}_i$ and $\vec{p}_j$. So, if any two particles have energies E_i and E_j that are a substantial fraction of E_γ then they must have $\cos\theta_{ij} \sim 1$; that is, they must point in nearly the same direction. Such particles are said to be **collinear**. Alternatively, a particle can have small energy, in which case we say the particle is **soft**.

They key to understanding jet properties in the $\lambda \to 0$ limit is that QCD simplifies in soft and collinear limits. As we will see, the soft radiation depends only on the directions of the various jets or incoming hadrons in the event and their colors; it is independent of how

the collinear radiation is distributed within each jet and of the spins of the collinear particles. The collinear radiation, on the other hand, can be computed for each jet separately,
independently of the distribution and colors of the other jets in the event.

To be precise about soft and collinear limits, lightcone coordinates are useful (see Section 32.5). Suppose we have a jet with 4-momentum p_J^μ, energy Q and invariant mass m.
By assumption, $\lambda = \frac{m}{Q} \ll 1$. If the jet were a single parton, as it is at leading order
in perturbation theory, then (neglecting quark masses) its momentum would be simply
$p_{J,\mathrm{LO}}^\mu = Q n^\mu$, where n^μ is a lightlike 4-vector, $n^2 = 0$. We conventionally normalize
$n^0 = 1$ so that $n^\mu = (1, \vec{n})$. Any 4-vector can be written in lightcone coordinates as

$$p^\mu \equiv \frac{1}{2}(\bar{n} \cdot p) n^\mu + \frac{1}{2}(n \cdot p)\, \bar{n}^\mu + p_\perp^\mu, \qquad (36.10)$$

where $\bar{n}^\mu = (1, -\vec{n})$, which satisfies $n \cdot \bar{n} = 2$ and $p_\perp \cdot n = p_\perp \cdot \bar{n} = 0$. $p_\perp^\mu$ is the part of p^μ
in the transverse directions. In coordinates where $n^\mu = (1, 0, 0, 1)$ then $\bar{n}^\mu = (1, 0, 0 - 1)$
and $p_\perp^\mu = (0, p_x, p_y, 0)$.

The invariant mass of a 4-vector in lightcone coordinates is

$$p^2 = (n \cdot p)(\bar{n} \cdot p) + p_\perp^2. \qquad (36.11)$$

Up to terms subleading in λ, the large component of the jet is its energy $\frac{1}{2}\bar{n} \cdot p = Q$.
Thus, we must have $n \cdot p \sim \lambda^2 Q$ so that $m^2 = (n \cdot p)(\bar{n} \cdot p) + p_\perp^2 = Q^2 \lambda^2$ has the right
scaling. The transverse components can scale at most as λQ. Thus, the jet momentum can
be written as

$$p_J^\mu = \{\bar{n} \cdot p,\ n \cdot p,\ p_\perp\} \sim Q\{\lambda^2, 1, \lambda\}, \qquad (36.12)$$

where $\sim$ indicates λ scaling. This is called **collinear scaling**.

A jet is not a single particle. In perturbation theory, we calculate the cross section for
jet production by computing the cross section for producing a bunch of particles with
momenta p_i^μ and writing $p_J^\mu = \sum_i p_i^\mu$. In order for $p_J^2 \sim Q^2 \lambda^2$, all the p_i^μ in the jet
must have collinear or softer scaling in all their components. Thus, the particle could have
$p_i^\mu = Q(\lambda^2, 1, \lambda)$ like the jet itself, or

$$p_i^\mu \sim \{\bar{n} \cdot p,\ n \cdot p,\ p_\perp\} \sim Q\{\lambda^2, \lambda^2, \lambda^2\}, \qquad (36.13)$$

which is known as **ultrasoft scaling**. We cannot have $p_i^\mu \sim Q\{1, 1, 1\}$ (**hard scaling**) or
$p_i^\mu = Q(\lambda, \lambda, \lambda)$ (**soft scaling**).[2] Since soft and ultrasoft modes will not both be relevant
for a single calculation, we will use the terms soft and ultrasoft interchangeably.

36.3 Soft interactions

In this section we discuss how cross sections for producing gluons simplify when that
radiation has (ultra)soft scaling. The physical argument for simplifications in the soft limit

---

[2] Another possibility is $p_i^\mu \sim Q(\lambda^2, \lambda, \lambda)$ (**Glauber scaling**); however, then $p_i^2 = \lambda^3 - \lambda^2$, which cannot
vanish. Thus, these Glauber modes are purely virtual. Glauber modes play an important role in the rigorous
proof of factorization for Drell–Yan production but can be safely ignored in the applications we consider here.

is similar to the argument that justifies the use of Gauss's law in classical electrodynamics. At large distance from a collection of charges, the electromagnetic field is determined almost completely by the net charge. One can include corrections through a multipole expansion (the dipole moment of the charge distribution gives the first subleading effect), but the leading effect at large distances is determined by Gauss's law. The soft limit of QCD is equivalent to a large-distance limit, where only the net color charge of the various jets is relevant, not the detailed distribution of colored particles within the jets. Leading power in λ in the soft limit corresponds to the leading order in the multipole expansion for a charge distribution.

We saw the usefulness of the soft limit back in Section 9.5, where we used it to connect charge conservation to Lorentz invariance of massless spin-1 particles. In this section, we generalize aspects of that discussion and introduce an efficient way to describe soft radiation patterns using Wilson lines. We begin with the discussion in an Abelian theory, where we show spin independence and the connection to Wilson lines, and then we discuss how things change in QCD. In this section, we work at tree-level and drop all $i\varepsilon$ factors. We also assume photon polarizations are real, so that we can write ϵ_i instead of $\epsilon_i^\star$.

36.3.1 Soft photon emission

Suppose we have some process involving n external states with momenta p_i^μ (which can be incoming or outgoing) and charges Q_i. In this discussion, Q_i will refer to the charge of the particle state $\langle p_i|$ not the field $\psi_i(x)$, so electrons have $Q = -1$ and positrons have $Q = +1$. We are interested in the case where these p_i^μ are all hard and well separated, so that they establish the jet directions. We will then consider how the matrix element in the state of just the particles with momenta p_i^μ is related to a matrix element in a state with additional soft photons. Let us write the matrix element $\mathcal{M}(p_i)$ for the process with just the p_i^μ as

$$\langle p_1 \cdots |\bar{\psi}_1(0) \cdots \psi_n(0)| \cdots p_n \rangle = i\mathcal{M}(p_i). \tag{36.14}$$

In this and the next section, we will abbreviate $\psi_j \equiv \psi_j(0)$, since all fields in matrix elements like this will be evaluated at the same point, which we can take to be $x = 0$.[3] We would like to know how the matrix element changes when m photons with momenta k_i^μ are added to the final state in the limit that all the k_i^μ are soft, meaning $k^\mu \ll p_i^\mu$ for all i. That is, we would like to know how $\langle p_1 \cdots ; k_1 \cdots k_m|\bar{\psi}_1 \cdots \psi_n| \cdots p_n \rangle$ relates to $\mathcal{M}(p_i)$.

Let us first recall the result in scalar QED derived in Section 9.5. There we showed that for the emission of a soft photon with momentum k^μ and polarization ϵ^μ from an outgoing

---

[3] For a physical process, such as $e^+ e^- \to \mu^+ \mu^-$, the matrix element should of course be calculated with the fields at different points, with those points integrated over as in the LSZ formula. However, since we are interested in the case where these p_i are all hard and well separated, we can expand the product of fields at different points in terms of local operators (through the operator product expansion). In momentum space, the difference between a matrix element for $e^+ e^- \to \mu^+ \mu^-$ and a matrix element of $\bar{\psi}\psi\bar{\psi}\psi$ is some calculable function $c(p_i)$. This $c(p_i)$ is a Wilson coefficient for matching onto the local operators that we focus on here. Since $c(p_i)$ is independent of additional soft and collinear radiation, we set it to 1 for simplicity.

electron, the matrix element is modified as

$$\mathcal{M}(p_i) = \quad \longrightarrow \quad = e\frac{p_i \cdot \epsilon}{p_i \cdot k}\mathcal{M}(p_i). \quad (36.15)$$

Emission from incoming positrons gives the same $e\frac{p_i \cdot \epsilon}{p_i \cdot k}$ factor, while outgoing positrons or incoming electrons give $-e\frac{p_i \cdot \epsilon}{p_i \cdot k}$. The full matrix element for soft gluon emission is the sum over these **eikonal factors** $\mp eQ_i\frac{p_i \cdot \epsilon}{p_i \cdot k}$ for all charged particles.

The calculation in spinor QED is similar. Pulling off the spinor for an outgoing electron, we write $\mathcal{M} = \bar{u}(p_i)\,\widetilde{\mathcal{M}}(p_i)$ and then find

$$\bar{u}(p_i)\widetilde{\mathcal{M}}(p_i) \to -ie\bar{u}(p_i)\not\epsilon\frac{i\left[(\not p_i + \not k) + m\right]}{(p_i + k)^2 - m^2}\widetilde{\mathcal{M}}(p_i + k) \cong e\bar{u}(p_i)\not\epsilon\frac{\not p_i + m}{2p_i \cdot k}\widetilde{\mathcal{M}}(p_i).$$
$$(36.16)$$

Using $\bar{u}(p)\not\epsilon\not p = \bar{u}(p)(-m\not\epsilon + 2\epsilon \cdot p)$ we then have

$$\mathcal{M}(p_i) \to e\frac{p_i \cdot \epsilon}{p_i \cdot k}\mathcal{M}(p_i). \quad (36.17)$$

So the same $e\frac{p_i \cdot \epsilon}{p_i \cdot k}$ factor appears in the scalar and the spinor case.

We can understand why the scalar and spinor give the same factor in a different way. The spinor propagator can be written as

$$\frac{\not p + m}{p^2 - m^2} = \sum_{s'}\frac{u_{s'}(p)\bar{u}_{s'}(p)}{p^2 - m^2}. \quad (36.18)$$

Thus, when a photon adds to an external spinor, it produces the shift

$$\bar{u}_s(p_i)\widetilde{\mathcal{M}}(p_i) \to \sum_{s'}\epsilon_\mu\bar{u}_s(p_i)\gamma^\mu u_{s'}(p_i + k)\frac{e}{(p_i + k)^2 - m^2}\bar{u}_{s'}(p_i + k)\widetilde{\mathcal{M}}(p_i + k)$$
$$\cong \sum_{s'}\epsilon_\mu\bar{u}_s(p_i)\gamma^\mu u_{s'}(p_i)\frac{e}{2p_i \cdot k}\bar{u}_{s'}(p_i)\widetilde{\mathcal{M}}(p_i), \quad (36.19)$$

where we have taken the soft limit on the second line. We can then use the identity (see Problem 11.2)

$$\bar{u}_s(p)\gamma^\mu u_{s'}(p) = 2\delta_{ss'}p^\mu \quad (36.20)$$

to see that Eq. (36.17) again results.

That the soft photon interaction is independent of the spin follows from general arguments about Lorentz invariance. The denominator $p_i \cdot q$ follows from there being a pole associated with the emitting particle. By dimensional analysis and the fact that the only 4-vectors available are ϵ^μ and p_i^μ, the form $\frac{p_i \cdot \epsilon}{p_i \cdot q}$ is unique up to a possible factor that might depend on the incoming particle's helicity. However, if the photon were to flip the helicity, then there is no way the Ward identity could be satisfied: the modified amplitude is not even proportional to the original one. In fact, it is obvious physically that soft photons cannot go around flipping helicities of particles, otherwise helicity would not be a very useful concept. More simply, we know charge must be conserved even when charged particles of different spin are scattered. For this to follow from the Ward identity, the form of the interaction in the soft limit must always be $e\frac{p\cdot\epsilon}{p\cdot q}\delta_{hh'}$, where h is the helicity of the particle before the emission and h' its helicity afterwards. For a rigorous proof, see [Weinberg, 1964].

An important point is that the eikonal factors $Q_i e\frac{p_i\cdot\epsilon}{p_i\cdot k}$ are independent of the energy of the charged-particle emitting photons. Writing $p^\mu = E v^\mu$ with v^μ the 4-velocity normalized to $v^0 = 1$, the eikonal factor becomes $Q_i e\frac{v_i\cdot\epsilon}{v_i\cdot k}$. For massless particles, we usually write $p_i^\mu = E n_i^\mu$ with $n^2 = 0$ and $n^0 = 1$; then the eikonal factor is $Q_i e\frac{n_i\cdot\epsilon}{n_i\cdot k}$. So the amplitude for emitting a soft photon depends on the directions that the charged particles are going and their charges, but not their energy.

Now suppose we have two soft photon emissions in QED. If these both come from the same outgoing electron, then an amplitude is modified as

$$= e^2\left[\frac{p_i\cdot\epsilon_1}{p_i\cdot k_1}\frac{p_i\cdot\epsilon_2}{p_i\cdot(k_1+k_2)} + \frac{p_i\cdot\epsilon_1}{p_i\cdot(k_1+k_2)}\frac{p_i\cdot\epsilon_2}{p_i\cdot k_2}\right]\mathcal{M}$$
$$= e^2\left[\frac{(p_i\cdot\epsilon_1)(p_i\cdot\epsilon_2)}{(p_i\cdot k_1)(p_i\cdot k_2)}\right]\mathcal{M}, \tag{36.21}$$

where the second step is just algebra. Actually, this simple algebraic step even has a name; it is called the **eikonal identity**. The result is that the amplitude for two soft photons is given by the square of the one-photon emission amplitude.

If there are multiple charged particles involved, then there are also diagrams where different particles emit the two photons. For these diagrams, the eikonal factors simply multiply. The result is that the sum of all the two-photon emission diagrams gives

$$\mathcal{M}(p_i) \rightarrow \left[\sum_{j=1}^n e\eta_j Q_j\frac{p_j\cdot\epsilon_1}{p_j\cdot k_1}\right]\left[\sum_{j=1}^n e\eta_j Q_j\frac{p_j\cdot\epsilon_2}{p_j\cdot k_2}\right]\mathcal{M}(p_i), \tag{36.22}$$

where $\eta_j = -1$ for an outgoing particle and $\eta_j = 1$ for an incoming particle. One corollary is that if we have two massless particles going in the same direction with momenta $p_1^\mu = E_1 n^\mu$ and $p_2^\mu = E_2 n^\mu$ the sum of the emissions from those particles is $(Q_1 + Q_2)\frac{n\cdot\epsilon}{n\cdot k}$. In other words, the rate for emitting soft photons depends only on the total charge for particles in each direction. This is the reason that soft emissions factorize from collinear emissions: soft radiation is only sensitive to the net charge going in each collinear direction.

The generalization to multiple emissions is straightforward. The amplitude for m photon emissions from the same particle simplifies using the eikonal identity to the product of m one-photon emission amplitudes. For different particles, the eikonal factors simply multiply. Writing $p_j^\mu = E_j n_j^\mu$, in the notation of Eq. (36.14) the result is that

$$\langle p_1 \cdots ; k_1 \cdots k_m | \bar{\psi}_1 \cdots \psi_n | \cdots p_n \rangle$$

$$\xrightarrow{k_i \ll p_i} \prod_{k=1}^{m} \left[\sum_{j=1}^{n} e\eta_j Q_j \frac{n_j \cdot \epsilon_k}{n_j \cdot k_k} \right] \langle p_1 \cdots | \bar{\psi}_1 \cdots \psi_n | \cdots p_n \rangle. \quad (36.23)$$

This equation says that in the soft limit any of the m photons can come from any of the n charged particles. For each emission, the amplitude is corrected by an eikonal factor independent of any other emission. As we will now see, this same amplitude can be reproduced by Wilson lines.

36.3.2 Soft Wilson lines

Recall from Section 25.2 that a Wilson line in QED is the exponential of a line integral over the gauge field. In this case, we want to integrate over the path of the charged particle. Writing $n^\mu = \frac{1}{E} p^\mu$ as the direction of a particle with momentum p^μ, the relevant Wilson line is

$$Y_n^\dagger(x) = \exp\left(ieQ_n n^\mu \int_0^\infty ds\, A_\mu(x^\nu + sn^\nu) e^{-\varepsilon s} \right), \quad (36.24)$$

which goes from the point x out to ∞ along the n^μ direction. We have inserted a convergence factor $e^{-\varepsilon s}$ to the expression in Eq. (25.47) to ensure that the photon field decouples at $t = \infty$. Such decoupling is required for S-matrix calculations that involve asymptotic states (the $e^{-\varepsilon s}$ factor is similar to the one derived in Section 14.4). We write this Wilson line as $Y^\dagger$ instead of Y since the particle is in the final state. For a final-state antiparticle, we would use Y. Q_n can be either the charge of a single particle or the net charge of all particles in the n^μ direction. Indeed, a product of Wilson lines in the same direction is equivalent to a single Wilson line with the sum of the charges.

Now consider the matrix element of this Wilson line in states with photons of momenta k_i: $\langle k_1 \cdots k_m | Y_n^\dagger | \Omega \rangle$. If there is one photon, we need only expand $Y_n^\dagger$ to order e. A photon field at position y will annihilate a photon with momentum k and polarization $\epsilon(k)$ in the external state:

$$\langle k | A_\mu(y) | \Omega \rangle = e^{ik \cdot y} \epsilon_\mu(k) \quad (36.25)$$

We then have

$$\langle k | Y_n^\dagger(0) | \Omega \rangle = ieQ_n n^\mu \langle k | \int_0^\infty ds\, A_\mu(sn^\nu) e^{-\varepsilon s} | \Omega \rangle$$

$$= ieQ_n (n \cdot \epsilon(k)) \int_0^\infty ds\, e^{i(k \cdot n + i\varepsilon)s}$$

$$= -eQ_n \frac{n \cdot \epsilon}{n \cdot k + i\varepsilon}. \quad (36.26)$$

This matches the leading-order eikonal interaction for an outgoing particle of charge Q_n.

For incoming charged particles, the appropriate Wilson line is

$$\overline{Y}(x) = \exp\left(ieQ_n n^\mu \int_{-\infty}^0 ds\, A_\mu(x^\nu + sn^\nu) e^{\varepsilon s} \right), \quad (36.27)$$

which leads to

$$\langle k|\overline{Y}(0)|0\rangle = eQ_n \frac{n \cdot \epsilon}{n \cdot k - i\varepsilon}, \tag{36.28}$$

which also agrees with the soft limit (you can check that the $i\varepsilon$ comes with the correct sign). We will drop these $i\varepsilon$ factors unless they are relevant from now on.

Higher-order terms in the expansion of the Wilson line can be contracted with other external states. The $\frac{1}{n!}$ from the expansion of the Wilson line is exactly what is needed to avoid any extra symmetry factor in the Feynman rules (see Chapter 7). Thus,

$$\langle k_1 \cdots k_m|Y_1^\dagger(0)|\Omega\rangle = \prod_{k=1}^{m}\left[-eQ_n \frac{n \cdot \epsilon_k}{n \cdot k_k}\right]. \tag{36.29}$$

Now consider the matrix element $\langle k_1 \cdots k_m | Y_1^\dagger(0) \cdots \bar{Y}_n(0)|\Omega\rangle$ with multiple Wilson lines in directions n_j with corresponding charges Q_j. Each photon can be contracted with the field from any line. The combinatorics works out perfectly (as you can check) so that

$$\langle k_1 \cdots k_m|Y_1^\dagger(0) \cdots \overline{Y}_n(0)|\Omega\rangle = \prod_{k=1}^{m}\left[\sum_{j=1}^{n} e\eta_j Q_j \frac{n_j \cdot \epsilon_k}{n_j \cdot k_k}\right], \tag{36.30}$$

where η_j is -1 for $Y_j^\dagger$ factors (which correspond to outgoing charged particles) and $\eta_j = +1$ for $\bar{Y}_j$ factors (which correspond to incoming charged particles).

The identity in Eq. (36.30) holds independently of any interactions in the Lagrangian. Indeed, it would hold even with a free $U(1)$ gauge theory with no matter. When we include matter, comparing to Eq. (36.23), we find the tree-level relation

$$\langle p_1 \cdots ; k_1 \cdots k_m|\bar{\psi}_1 \cdots \psi_n| \cdots p_n\rangle_{\mathcal{L}_{QED}}$$
$$\xrightarrow{k_i \text{ soft}} \langle p_1 \cdots ; k_1 \cdots k_m|\bar{\psi}_1 Y_1^\dagger \cdots \overline{Y}_n \psi_n| \cdots p_n\rangle_{\mathcal{L}_{free}}, \tag{36.31}$$

where all the fields are to be evaluated at $x = 0$. Here, $\mathcal{L}_{QED}$ means the matrix element is to be calculated using the interactions in the QED Lagrangian, while $\mathcal{L}_{free}$ implies that the interactions in the Lagrangian are to be set to zero. We have to use the free Lagrangian on the right-hand side to avoid double-counting. In fact, having moved all the photons into the operator rather than the Lagrangian, we now have a simple factorized form for the amplitude:

$$\langle p_1 \cdots ; k_1 \cdots k_m|\bar{\psi}_1 \cdots \psi_n| \cdots p_n\rangle_{\mathcal{L}_{QED}}$$
$$\xrightarrow{k_i \text{soft}} \langle p_1 \cdots|\bar{\psi}_1 \cdots \psi_n| \cdots p_n\rangle\langle k_1 \cdots k_m|Y_1^\dagger \cdots \overline{Y}_n|\Omega\rangle. \tag{36.32}$$

In this form, we no longer need to write $\mathcal{L}_{free}$ since the states $\langle p_1 \cdots|$ and $|\cdots p_n\rangle$ have no photons and the state $|k_1 \cdots k_m\rangle$ has no charged particles.

That the interactions of soft gluons with energetic charged particles can be described completely through Wilson lines; which are pure phase, is reminiscent of the description of interference patterns in geometric optics through the evolution of phase factors called eikonals. This is the reason that the $e^{\frac{n\cdot\epsilon}{n\cdot k}}$ factors are called eikonal factors and the soft limit is sometimes called the eikonal limit. (Wilson lines are also sometimes called eikonal factors as well.)

Keep in mind that there is no restriction on the photon field A_μ appearing in the soft Wilson line; it is the same as a photon field in full QED. The only place the soft approximation is used in the whole derivation above is in saying that the momenta p_i^μ entering the amplitude $\mathcal{M}(p_i)$ are the same before and after the soft photon emission. This is equivalent to the Wilson line $Y(x)$ and the field $\psi(x)$ being evaluated at the same space-time point. In other words, the soft emissions leave the collinear momentum precisely unchanged, to leading power. The position-space language is very natural for soft emissions: a particle just moves along its classical trajectory, casually emitting soft photons. In fact, we already showed in Section 33.6.1 that Wilson lines naturally describe the semi-classical limit of a propagating charged particle.

36.3.3 Soft gluon emission

The above arguments for QED generalize in a straightforward way to QCD. We start with the matrix element for the process just involving quarks:

$$\langle p_1 \cdots |\bar{\psi}_1(x) \cdots \psi_n(x)| \cdots p_n \rangle = i\mathcal{M}e^{ix(p_1 + \cdots + p_n)}. \tag{36.33}$$

You can think of the subscript on the quark fields as a flavor index. We include to make it clear which field corresponds to which state.

To see what happens when a soft gluon is emitted from a quark, we write $\mathcal{M} = \bar{u}_i(p)\widetilde{\mathcal{M}}_i$. Abusing notation slightly, i now denotes the quark color index, and we leave the momentum label implicit. The kinematical factors are the same for emitting a gluon as for emitting a photon, so all that changes is a group factor T_{ij}^a gets added:

$$\tag{36.34}$$

The eikonal factor is now $-gT_{ij}^a \frac{p \cdot \epsilon}{p \cdot k}$. As in QED, this factor is independent of the spin of the colored particle.

Now consider a final state with two soft gluons, one with momentum k_1, polarization ϵ_1 and color a, and the other with k_2, ϵ_2 and b. If these gluons both come from the same quark, there are three graphs:

$$\tag{36.35}$$

Graphs A and B modify the matrix element as

$$\bar{u}_i \widetilde{\mathcal{M}}_i \to (-g_s)^2 \, \bar{u}_i \left[T_{ij}^a T_{jk}^b \frac{p \cdot \epsilon_1}{p \cdot k_1} \frac{p \cdot \epsilon_2}{p \cdot (k_1 + k_2)} + T_{ij}^b T_{jk}^a \frac{p \cdot \epsilon_1}{p \cdot (k_1 + k_2)} \frac{p \cdot \epsilon_2}{p \cdot k_2} \right] \widetilde{\mathcal{M}}_k. \tag{36.36}$$

In the Abelian case, the two-photon emission amplitude simplified with the eikonal identity to a form that was manifestly equal to what came out of the expansion of a Wilson line.

In the non-Abelian case, the eikonal identity does not produce an obvious simplification, since $[T^a, T^b] \neq 0$. Nevertheless, this amplitude *is* reproduced from a Wilson line.

Recall from Section 25.2 that the Wilson line in a non-Abelian theory is path ordered:

$$Y_n^\dagger(x) = P\left\{\exp\left[ig_s T_{ij}^a n^\mu \int_0^\infty ds\, A_\mu^a(x^\nu + sn^\nu)e^{-\varepsilon s}\right]\right\}. \tag{36.37}$$

Path ordering refers to ordering of the T^a matrices such that the ones associated with the gluons closer to $s = 0$ are moved to the right. For an incoming particle, the Wilson line is

$$\overline{Y}_n(x) = P\left\{\exp\left[ig_s T_{ij}^a n^\mu \int_{-\infty}^0 ds\, A_\mu^a(x^\nu + sn^\nu)e^{\varepsilon s}\right]\right\}. \tag{36.38}$$

As in the QED case, emissions from outgoing and incoming antiquarks will be reproduced using Y_n and $\overline{Y}_n^\dagger$ respectively.

We can expand to order g_s^2 to get

$$\langle k_{1a} k_{2b}|Y_n^\dagger(0)|\Omega\rangle = (ig_s)^2\, T^c T^d \int_0^\infty ds \int_0^s dt\, \langle k_{1a} k_{2b}|n \cdot A^c(tn^\mu)\, n \cdot A^d(sn^\nu)|\Omega\rangle. \tag{36.39}$$

We can contract either gluon field with either gluon canceling the factor of 2 in front. These integrals are easy to evaluate, as in Eq. (36.26), with the result

$$\langle k_{1a} k_{2b}|Y_n^\dagger(0)|\Omega\rangle = (-g_s)^2\left[T^a T^b \frac{p \cdot \epsilon_1}{p \cdot k_1}\frac{p \cdot \epsilon_2}{p \cdot(k_1 + k_2)} + T^b T^a \frac{p \cdot \epsilon_1}{p \cdot(k_1 + k_2)}\frac{p \cdot \epsilon_2}{p \cdot k_2}\right], \tag{36.40}$$

in agreement with Eq. (36.36) coming from graphs A and B in Eq. (36.35). In the Abelian case, $T^a = \mathbb{1}$ and the two factors can be combined with the eikonal identity to reproduce the QED result.

The result is that factorization works in QCD just as in QED:

$$\langle p_1 \cdots; k_1 \cdots k_m|\bar{\psi}_1 \cdots \psi_n|\cdots p_n\rangle \xrightarrow{k_i \text{ soft}}$$
$$\langle p_1 \cdots|\bar{\psi}_1 \cdots \psi_n|\cdots p_n\rangle\langle k_1 \cdots k_m|Y_1^\dagger \cdots \overline{Y}_n|\Omega\rangle. \tag{36.41}$$

Note that the Y_i just account for emissions from the hard colored particles. Other graphs, such as graph C in Eq. (36.35), which come from a vertex in the Lagrangian among gluons, are not accounted for in either Eq. (36.36) or Eq. (36.40). Indeed, the three-gluon vertex does not simplify in the soft limit, since when soft gluons interact among themselves, there is no separation of scales to produce a simplification. Thus, once the gluons leave the hard colored particles, they propagate and interact as in full QCD. Therefore, the gluon Lagrangian on the right- and left-hand sides of Eq. (36.41) should be the same as the full QCD Lagrangian: $\mathcal{L} = -\frac{1}{4}(F_{\mu\nu}^a)^2$.

The matrices in the Wilson lines can be in any representation. There is a different Y_n for antiquarks, or gluons. For example, for quarks $Y_n = (Y_n)_{ij}$ where i and j are fundamental color indices. The gluon Wilson line is often denoted $\mathcal{Y}_n$ and the antiquark Wilson line by $Y_n^\dagger$. An often helpful relation is that $YT^aY^\dagger = \mathcal{Y}_{ab}T^b$ or more explicitly

$$(Y)_{ij}(T^a)_{jk}(Y^\dagger)_{kl} = (\mathcal{Y})_{ab}(T^b)_{il}, \tag{36.42}$$

where a and b are adjoint indices and i, j, k and l are fundamental indices. Thus, the soft matrix element (the final term in Eq. (36.41)) for any process with quark, antiquark or gluon jets, can always be written entirely in terms of fundamental Wilson lines Y_n and their adjoints, $Y_n^\dagger$.

36.4 Collinear interactions

In this section we will show why QCD cross sections factorize into matrix elements of jet fields in the limit that all radiation is in some number of collinear directions. To be concrete, consider dijet projection in e^+e^- collisions. At leading order in perturbation theory, the final state consists of two quarks. Let us write the amplitude for this process when the quarks have momenta p_1 and p_2 as $\langle p_1 p_2 | \bar{\psi} \gamma^\mu \psi | \Omega \rangle$. Now add to the final state gluons with momenta $q_{a_1} \cdots q_{b_1}$ collinear to p_1 and momenta $q_{a_2} \cdots q_{b_2}$ collinear to p_2. Then, the matrix element factorizes as

$$\langle p_1 p_2; q_{a_1} \cdots q_{b_2} | \bar{\psi} \gamma^\mu \psi | \Omega \rangle = \gamma_{\alpha\beta}^\mu \langle p_1; q_{a_1} \cdots q_{b_1} | \bar{\chi}_{n_1}^\alpha | \Omega \rangle \langle p_2; q_{a_2} \cdots q_{b_2} | \chi_{n_3}^\beta | \Omega \rangle, \tag{36.43}$$

where α and β are spinor indices. In this expression, the fields χ_n are quark **jet fields**, defined as

$$\chi_n(x) \equiv W_{t_n}^\dagger(x) \psi(x), \tag{36.44}$$

where n^μ is the direction of the jet and where W_{t_n} is a path-ordered QCD Wilson line pointing in some lightlike direction t_n^μ:

$$W_{t_n}^\dagger(x) = P \left\{ \exp \left(i g_s T^a t_n^\mu \int_0^\infty ds \, A_\mu^a(x^\nu + s t_n^\nu) e^{-\varepsilon s} \right) \right. \tag{36.45}$$

It is common to take $t_n^\mu = \bar{n}^\mu$, but in fact the only restriction on t_n^μ is that it is not collinear to the jet direction n^μ. For incoming collinear particles one should use $\overline{W}_{t_n}$, defined analogously to $\overline{Y}_n$ in Eq. (36.38).

As in Section 36.3, we will demonstrate the equivalence of Eq. (36.43) for scalar QED, where all of the essential features of the simplifications can be seen. Adding color and spin is then straightforward. For collinear emissions, gauge invariance plays a much more important role than for soft emissions. In order to understand the gauge dependence efficiently, we will employ the spinor-helicity formalism from Chapter 27. The reader who needs motivation to learn about helicity spinors is encouraged to check Eq. (36.43) using polarization vectors.

36.4.1 Collinear photon emission

Let us begin, as in the soft emission case, with the matrix element for producing some set of charged particles with momenta p_i^μ in scalar QED:

$$\langle p_1 \cdots p_n | \phi_1^\star \cdots \phi_n | \Omega \rangle = i \mathcal{M}(p_i). \tag{36.46}$$

As in the previous section, all fields are implicitly evaluated at $x = 0$. The fields ϕ_i create the states with momentum p_i and charges Q_i. As in the soft case, the Q_i for particles and antiparticles have opposite sign. If the particle is an antiparticle, we use ϕ_i instead of $\phi_i^\star$. Any combination of ϕ and $\phi^\star$ fields is possible as long as the operator is gauge invariant. We simply write $\phi_1^\star \cdots \phi_n$ to avoid cumbersome indices. We also take all the particles to be outgoing, for simplicity. Now we would like to see how $\mathcal{M}$ changes for a final state with additional photons, when each of those photons becomes collinear to one of the p_i.

First, consider one photon with momentum q that is nearly collinear to one of the p_i that which we denote simply as p_1. Let us write $n^\mu = \frac{1}{E} p_1^\mu$ as the normalized lightlike 4-vector in the p_1 direction. In lightcone coordinates, both q and p_1 scale as

$$(n \cdot p_1, \bar{n} \cdot p_1, p_{1\perp}) \sim (n \cdot q, \bar{n} \cdot q, q_\perp) \sim (\lambda^2, 1, \lambda). \tag{36.47}$$

Thus, $q \cdot p_1 \sim \lambda^2$ and $q \cdot p_i \sim 1$ for $i \neq 1$. We want to extract the most dominant term in λ^{-1} in $\langle p_1 \cdots p_n; q | \phi_1^\star \cdots \phi_n | \Omega \rangle$.

The photon with momentum q can be emitted from any of the p_i. Working only at tree-level, but without making any other approximations yet, the scalar QED Feynman rules imply that

$$\mathcal{M}(p_i, q) = \sum_i -eQ_i \frac{p_i \cdot \epsilon}{p_i \cdot q} \mathcal{M}(p_i + q). \tag{36.48}$$

The notation $\mathcal{M}(p_i + q)$ means the $\mathcal{M}(p_i)$ matrix element with p_i changed to $p_i + q$ holding the other momentum fixed. If the $p_i \cdot \epsilon$ terms in the numerator scale uniformly with λ then, since $p_1 \cdot q \sim \lambda^2$, and $p_i \cdot q \sim 1$ for $i \neq 1$, the term with $i = 1$ will dominate this sum. That is, only the diagram where the photon is emitted from the leg to which it is collinear needs to be included at leading power. The $i = 1$ term does in fact dominate in a generic (non-collinear) gauge, as we will shortly see. However, one can choose a gauge where $p_1 \cdot \epsilon = 0$ exactly (this is an axial gauge with $p_1^\mu \partial_\mu A(x) = 0$), in which case the $i = 1$ term vanishes. Thus, to extract the behavior of $\mathcal{M}(p_i, q)$ in the collinear limit we have to be careful with the gauge dependence.

An advantage of helicity spinors (see Chapter 27) is that one can easily choose different gauges for polarizations in different collinear sectors. Gauge dependence for helicity spinors amounts to dependence on the choice of reference vector r^μ to which the polarizations are orthogonal. Recall from Chapter 27 that polarizations satisfy $\epsilon(q) \cdot r = \epsilon(q) \cdot q = 0$ and the only restriction on r^μ is that it cannot be proportional to q^μ.

In the one-photon case, let us take the photon to have negative helicity, so that $[\epsilon^-(r)]^{\alpha\dot\alpha} = \sqrt{2} \frac{q \rangle [r}{[qr]}$. Then each term in the sum in Eq. (36.48) becomes

$$\frac{p_i \cdot \epsilon}{p_i \cdot q} = \sqrt{2} \frac{[ri]\langle iq \rangle}{\langle iq \rangle [qi][qr]} = \sqrt{2} \frac{[ri]}{[qi][qr]}. \tag{36.49}$$

Since $\langle ij \rangle = [ji]$ up to a phase for real momenta and $p_1 \cdot q = \frac{1}{2} \langle 1q \rangle [q1] \sim \lambda^2$ we must have $[q1] \sim \langle q1 \rangle \sim \lambda$. Similarly, since $p_i \cdot q \sim \lambda^0$ for $i > 1$ we have $[qi] \sim \langle qi \rangle \sim \lambda^0$. In a **generic gauge**, where $r \neq p_i$ for any i, then $[ri] \sim \lambda^0$. The term with $i = 1$ in Eq. (36.49) then scales as $\frac{[r1]}{[q1][qr]} \sim \frac{\lambda^0}{\lambda \lambda^0} \sim \lambda^{-1}$. The other terms scale as $\frac{[ri]}{[qi][qr]} \sim \frac{\lambda^0}{\lambda^0 \lambda^0} \sim \lambda^0$. Thus, in a non-collinear gauge, the diagram where the photon is emitted from the leg to which it is collinear does in fact dominate. In a **collinear gauge** where $r = p_1$ then the $i = 1$ term vanishes exactly. However, each of the other terms now scales as $\frac{[1i]}{[qi][q1]} \sim \frac{\lambda^0}{\lambda^0 \lambda} \sim \lambda^{-1}$. Thus, in a collinear gauge, the diagram with the photon coming from the collinear leg is zero and all the other diagrams get enhanced. Moreover, since the amplitude in scalar QED is gauge invariant, the sum of the $i \neq 1$ diagrams in collinear gauge must exactly reproduce the $i = 1$ diagram in the collinear limit.

Now, consider multiple photon emissions. Say we want the amplitude for a final state in which all photons are collinear to some charged particle. Say momenta $q_{a_1} \cdots q_{b_1}$ are collinear to p_1, momenta $q_{a_2} \cdots q_{b_2}$ are collinear to p_2 and so on. In a generic gauge, the matrix element is enhanced by a factor of $\frac{1}{\lambda}$ for each photon only for diagrams in which that photon connects directly to the charged particle collinear to it. Thus, in a generic gauge,

$$\langle p_1 \cdots p_n; q_{a_1} \cdots q_{b_n} | \phi_1^\star \cdots \phi_n | \Omega \rangle \cong \langle p_1; q_{a_1} \cdots q_{b_1} | \phi_1^\star | \Omega \rangle \cdots \langle p_1; q_{a_n} \cdots q_{b_n} | \phi_n | \Omega \rangle.$$
$$(36.50)$$

Note that while the left-hand side is gauge invariant (assuming $\sum Q_i = 0$), the right-hand side is not. A gauge-invariant generalization of Eq. (36.50) is

$$\langle p_1 \cdots p_n; q_{a_1} \cdots q_{b_n} | \phi_1^\star \cdots \phi_n | \Omega \rangle$$
$$\cong \langle p_1; q_{a_1} \cdots q_{b_1} | \phi_1^\star W_1 | \Omega \rangle \cdots \langle p_1; q_{a_n} \cdots q_{b_n} | W_n^\dagger \phi_n | \Omega \rangle, \quad (36.51)$$

where W_i is a Wilson line pointing in some direction t_i^μ that is not collinear to p_i^μ:

$$W_i(x) = \exp\left(ieQ_i t_i^\mu \int_0^\infty ds\, A_\mu^i(x^\nu + st_i^\nu) \right). \quad (36.52)$$

These are the same Wilson lines as in Eq. (36.24) but now pointing in the t_i^μ direction instead of the n^μ direction.

As a first check on Eq. (36.51), note that in a generic gauge the Wilson line gives a factor of $\frac{t \cdot \epsilon}{t \cdot q} = \frac{[rt]}{[qt][qr]}$. Since t^μ and r^μ are not collinear to q^μ, these factors are subdominant to the λ^{-1} contributions coming from Eq. (36.50). Thus, in a generic gauge the Wilson lines give only a power-suppressed contributions to matrix elements and so Eq. (36.51) reduces to Eq. (36.50).

To verify Eq. (36.51) in scalar QED, first consider amplitudes with one photon of momentum q^μ going in the p_1^μ direction. Then the right-hand side of Eq. (36.51) contributes

$$\langle p_1; q/\phi_1^\star W_1 | \Omega \rangle = \sqrt{2} eQ_1 \left(\frac{[r1]}{[rq][q1]} - \frac{[rt]}{[rq][qt]} \right) = \sqrt{2} eQ_1 \frac{[1t]}{[1q][tq]}, \quad (36.53)$$

where $t^\mu = t_1^\mu$ is the direction of the Wilson line and r^μ can be collinear to p_1^μ or not. The $\frac{[r1]}{[rq][q1]}$ term in the middle expression comes from the emission from ϕ_1 while the $\frac{[rt]}{[rq][qt]}$ term comes from $W_1^\dagger$. The final form, which is manifestly gauge invariant, can be derived with the Schouten identity, Eq. (27.27) (or more simply by substituting $[r = \frac{[rt]}{[1t]}[1 + \frac{[r1]}{[1t]}[t$, which is possible since spinors are two-dimensional).

The amplitude for one emission in full scalar QED gets contributions from all lines:

$$\langle p_1 \cdots p_n; q | \phi_1^\star \cdots \phi_n | \Omega \rangle = \sum_i eQ_i \frac{p_i \cdot \epsilon}{p_i \cdot q} = \sqrt{2} \sum_i eQ_i \frac{[ri]}{[rq][qi]}. \quad (36.54)$$

We can separate out the r-dependence and the i-dependence using

$$\frac{[ri]}{[rq][qi]} = \frac{[rt]}{[rq][qt]} + \frac{[it]}{[iq][tq]}. \tag{36.55}$$

Since $\sum Q_i = 0$ the $\frac{[rt]}{[rq][qt]}$ terms do not contribute. We then have

$$\langle p_1 \cdots p_n; q | \phi_1^\star \cdots \phi_n | \Omega \rangle = \sqrt{2} \sum_i e Q_i \frac{[it]}{[iq][tq]}. \tag{36.56}$$

The terms in this remaining sum are all of order λ^0 unless $i = 1$. We thus find

$$\langle p_1 \cdots p_n; q | \phi_1^\star \cdots \phi_n | \Omega \rangle \cong \sqrt{2} e Q_1 \frac{[1t]}{[1q][tq]}, \tag{36.57}$$

in agreement with Eq. (36.53). Thus Eq. (36.51) holds for one emission. For multiple emissions, the proof is almost as simple and we leave it to Problem 36.2.

Collinear factorization in QED is almost identical to scalar QED, although the checks are messier. The equivalent of Eq. (36.51) in QED is

$$\langle p_1 \cdots p_n; q_{a_1} \cdots q_{b_n} | \bar{\psi}_1 \cdots \psi_n | \Omega \rangle$$
$$\cong \langle p_1; q_{a_1} \cdots q_{b_1} | \bar{\psi}_1 W_1 | \Omega \rangle \cdots \langle p_1; q_{a_n} \cdots q_{b_n} | W_n \psi_n^\dagger | \Omega \rangle. \tag{36.58}$$

Both sides of this equation are gauge invariant, so it is enough to check this factorization in a generic gauge. Consider again a one-photon emission in the p_1^μ direction. If this comes off the particle in the 1 direction, it gives

$$- e Q_1 \bar{u}(p_1) \not{\epsilon} \frac{\not{p}_1 + \not{q}}{2 p_1 \cdot q} \mathcal{M} = -e Q_1 \bar{u}(p_1) \left(\frac{p_1 \cdot \epsilon}{p_1 \cdot q} + \frac{\not{\epsilon}\not{q}}{2 p_1 \cdot q} \right) \mathcal{M}. \tag{36.59}$$

In a generic gauge, $p_i \cdot \epsilon \neq 0$ and so this is enhanced by λ^{-1}, as in the scalar case (independently of the $\frac{\not{q}\not{\epsilon}}{2 p_1 \cdot q}$ term, which could only make it more enhanced). This is the dominant contribution and has identical form coming from both sides of Eq. (36.58). On the left-hand side of Eq. (36.58), an emission can also come from particles in the i direction. These give

$$- e Q_i \bar{u}(p_i) \not{\epsilon} \frac{\not{p}_i + \not{q}}{2 p_i \cdot q} \mathcal{M}. \tag{36.60}$$

In a generic gauge, there is no reason anything in this expression should be enhanced as q^μ becomes collinear to p_1^μ. Thus, these $i \neq 1$ emissions scale as λ^0 in a generic gauge and can be ignored compared to the λ^{-1} enhanced emissions in Eq. (36.59). On the right-hand side, emissions from the Wilson lines give the same thing as in the scalar QED case, which also scale as λ^0. Thus, the two sides of Eq. (36.58) agree at leading power in a generic gauge. Since they are both gauge invariant, they therefore agree in any gauge.

Collinear factorization in QCD is almost identical to QED. For example, the factorization formula for a process involving a quark jet and an antiquark jet is given in Eq. (36.43). We can perform the same check on Eq. (36.43) as we did in QED on Eq. (36.58). In a generic gauge, the only diagrams that contribute at leading power in QCD are those in which gluons are emitted from colored particles to which they are collinear. These diagrams are identical when coming from the factorized expression. In the factorized form, the Wilson lines only give power-suppressed contributions in generic gauges. Thus, the two sides agree at leading power. When multiple gluons are emitted, one must also consider contributions in which a collinear gluon splits due to the A^3 or A^4 vertex in the QCD Lagrangian. Although not obvious, these again agree in a generic gauge. You are encouraged to check the equivalence in Problem 36.2.

36.4.2 Splitting functions

One consequence of collinear factorization is the universality of the Altarelli–Parisi splitting functions. Since the amplitude for emitting a collinear gluon from a quark is proportional to $\langle \Omega | \bar{\psi} W | p; q \rangle$, we can calculate the splitting function by squaring this amplitude. The amplitude is

$$\langle p, q | \bar{\psi} W | \Omega \rangle \, \mathcal{M}(p + q) = -g_s \bar{u}_i(p) \left[\frac{\not{\epsilon} (\not{q} + \not{p})}{2p \cdot q} + \frac{t \cdot \epsilon}{t \cdot q} \right] T^a_{ij} \mathcal{M}_j(p + q) \,, \qquad (36.61)$$

where the first term in brackets comes from $\bar{\psi}$ and the other from W. Choosing the spinor to be left-handed, so $\bar{u}_i(p) = \langle p |$, we find for a negative-helicity gluon,

$$\mathcal{M}_- = -\frac{\sqrt{2} g_s T^a}{[pq]} \left(\langle q \mathcal{M} \rangle + \frac{[pr]}{[qr]} \langle p \mathcal{M} \rangle + \frac{[rt][pq]}{[qr][qt]} \langle p \mathcal{M} \rangle \right) , \qquad (36.62)$$

and for a positive-helicity gluon,

$$\mathcal{M}_+ = -\frac{\sqrt{2} g_s T^a}{\langle qp \rangle} \left(\frac{\langle pr \rangle}{\langle qr \rangle} + \frac{\langle tr \rangle \langle qp \rangle}{\langle rq \rangle \langle qt \rangle} \right) \langle p \mathcal{M} \rangle. \qquad (36.63)$$

These amplitudes are both gauge invariant. So let us choose $r^\mu = t^\mu$, in which case the final terms in both amplitudes vanish.

Now let us write $P^\mu = p^\mu + q^\mu$. Since p^μ and q^μ are nearly collinear, $p^\mu = zP^\mu$ and $q^\mu = (1 - z) P^\mu$ at leading power, so $[p = \sqrt{z}[P$ and $[q = \sqrt{1 - z}\,[P$ up to a phase that will drop out of the cross section. We then find

$$\mathcal{M}_- = -\frac{\sqrt{2} g_s T^a}{[pq]} \left(\sqrt{1 - z} + \frac{z}{\sqrt{1 - z}} \right) \langle P \mathcal{M} \rangle \qquad (36.64)$$

and

$$\mathcal{M}_+ = -\frac{\sqrt{2} g_s T^a}{\langle qp \rangle} \left(2 \frac{z}{\sqrt{1 - z}} \right) \langle P \mathcal{M} \rangle, \qquad (36.65)$$

both of which are independent of the Wilson line direction t^μ. Squaring the amplitudes and summing over polarizations and colors gives

$$\sum_{\text{colors}} |\mathcal{M}_+|^2 = g_s^2 C_F \frac{1}{P^2} \frac{1 + z^2}{1 - z} [\mathcal{M} P] \langle P \mathcal{M} \rangle, \qquad (36.66)$$

which we recognize as the DGLAP splitting function. Since we have already proven that collinear emissions for any process are given by matrix elements $\langle p, q | \bar{\psi} W | \Omega \rangle$, we have hereby proven the universality of the DGLAP splitting functions. You can calculate the gluon splitting function in a similar manner in Problem 36.3.

36.5 Soft-Collinear Effective Theory

In the previous sections, we have seen how matrix elements in QCD factorize for processes involving soft or collinear radiation at tree-level. We also saw how soft radiation is only sensitive to the total (color) charge going in each direction, a result familiar from the multipole expansion in classical electromagnetism. It should therefore not surprise you that processes with soft *and* collinear radiation factorize (see also Problem 36.4). That is, at leading power,

$$\langle X_1; \cdots ; X_m; X_s | \bar{\psi}_1 \cdots \psi_m | \Omega \rangle$$
$$\cong \langle X_1 | \bar{\psi}_1 W_1 | \Omega \rangle \cdots \langle X_m | W_m^\dagger \psi_m | \Omega \rangle \langle X_s | Y_1 \cdots Y_m^\dagger | \Omega \rangle, \quad (36.67)$$

where X_j contains gluons going in the direction collinear to the jth jet and X_s contains the soft gluons. As before, all fields are evaluated at $x = 0$ and the subscripts on ψ denote the quark flavor. We showed that this factorized expression holds, at tree-level, if all the final-state particles have momenta that fall into one of these sectors.

The fact that the only relevant interactions at leading power are among particles going in the same direction or among soft gluons is a kind of superselection rule which can be imposed at the level of the Lagrangian. With this insight, we can write Eq. (36.67) as

$$\langle X_1; \cdots ; X_m; X_s | \bar{\psi}_1 \cdots \psi_m | \Omega \rangle_{\mathcal{L}_{\text{QCD}}}$$
$$\cong C(Q) \langle X_1; \cdots ; X_m; X_s | \bar{\psi}_1 W_1 Y_1^\dagger \cdots Y_m W_m^\dagger \psi_m | \Omega \rangle_{\mathcal{L}_{\text{SCET}}}, \quad (36.68)$$

where $\mathcal{L}_{\text{SCET}}$ is a Lagrangian in which all the sectors have been decoupled and $C(Q) = 1$ (at tree-level). More explicitly, let us assign a new quantum number $j = 1 \ldots m$ or "soft" to the states in X_j and X_s respectively. We also introduce fields ψ_j and A_j for each sector that can create and annihilate only particles with those quantum numbers. Then

$$\mathcal{L}_{\text{SCET}} = \mathcal{L}_1 + \cdots + \mathcal{L}_m + \mathcal{L}_{\text{soft}}, \quad (36.69)$$

where $\mathcal{L}_j$ contains quarks and gluons in the jth collinear sector and $\mathcal{L}_{\text{soft}}$ contains the soft quarks and gluons. Each of these $\mathcal{L}_i$ and $\mathcal{L}_{\text{soft}}$ are identical to $\mathcal{L}_{\text{QCD}}$. The Lagrangian $\mathcal{L}_{\text{SCET}}$ is the Lagrangian for **Soft-Collinear Effective Theory** (SCET).[4]

We have only demonstrated Eq. (36.68) at tree-level where $C(Q) = 1$. Loop contributions to the matrix elements on both sides of this equation will generically be both UV and IR divergent. However, since the soft and collinear tree-level matrix elements on both sides agree, the IR divergences in the loops should agree as well. After all, the IR divergences in loops must be able to cancel the IR divergences in phase space integrals over tree-level graphs (see Section 20.3).[5] The UV divergences may be different, but they can be removed with counterterms that also can be different on the two sides. Thus, the difference between

---

[4] There are actually a number of different formulations of SCET, all of which are equivalent at leading power, and equivalent to the formulation we have described. A discussion of power corrections is beyond our scope.

[5] Technically, the IR divergences only agree if the overlapping region between soft and collinear momenta is not double-counted in SCET. Conveniently, for the application discussed in this chapter, this zero bin gives zero in dimensional regularization, so we will ignore it.

the two sides of Eq. (36.68) should not depend on the IR scales or the UV cutoff. We therefore expect to be able to absorb the differences into the short-distance Wilson coefficient $C(Q)$, which may depend on hard scales Q but not on soft or collinear scales. We will not prove this assertion, but we will verify it in explicit examples below.

One of the important applications of SCET is to simplify derivations of factorization formulas. In the traditional approach to factorization, derivations are done using Feynman diagrams. Derivations in SCET are done at the level of fields. Working with fields has the great advantage of making universality manifest: the same objects appear in different factorization theorems. The simplest processes for which factorization can be analyzed are $e^+ e^- \to$ hadrons (which is $e^+ e^- \to \bar{q}q$ at tree-level) or its crossings: deep inelastic scattering ($e^- P \to e^- X$) and Drell–Yan ($PP \to e^+ e^-$). Deep inelastic scattering was studied using full QCD in Chapter 32. Here we discuss Drell–Yan and $e^+ e^- \to$ hadrons.

36.5.1 The Drell-Yan process

The Drell–Yan process refers to the creation of a pair of leptons in the collision of two hadrons, such as in $PP \to \mu^+ \mu^- + X$ [Drell and Yan, 1970]. Let us denote the incoming hadron momenta as P_1^μ and P_2^μ, the outgoing lepton momenta as k_1^μ and k_2^μ, and the hadronic final-state momentum as p_X^μ. Let us also write $q^\mu = k_1^\mu + k_2^\mu$ as the momentum of the off-shell photon decaying to leptons (we ignore the weak interactions for simplicity). Thus $P_1^\mu + P_2^\mu = q^\mu + p_X^\mu$ and $q^2 > 0$.

A rigorous factorization theorem exists for *inclusive* Drell–Yan, meaning only the final state leptons are measured [Collins *et al.*, 1988]. This theorem states that the cross section is given by a convolution among parton distribution functions and a perturbative hard process, up to corrections suppressed by factors of $\frac{\Lambda_{\mathrm{QCD}}}{M}$ where M is the invariant mass of the lepton pair. Since everything we have shown in this chapter so far is based on tree-level matrix elements, we are in no position to derive a rigorous factorization theorem in SCET. On the other hand, while the rigorous factorization theorem justifies performing perturbative calculations, it does not indicate a way to perform these calculations more efficiently than we would if we simply *assumed* factorization holds. Thus, we will simply *assume* that our tree-level results hold to all orders in perturbation theory and apply the effective field theory technology to resum large logarithms.

We will focus on the **threshold region**, where the invariant mass of the lepton pair $M \equiv \sqrt{q^2}$ approaches the center-of-mass energy $\sqrt{s} = \sqrt{(P_1 + P_2)^2}$ of the hadronic collision. The key property of this kinematic region is that the scattering is almost elastic. In the center-of-mass frame, the energy E_X of the hadronic final state must be small. That is, the hadronic final state is soft. Therefore, the process near threshold involves incoming protons (which can be described with collinear fields) and outgoing soft radiation. To be clear, there are two small scales in this problem: $\lambda = 1 - \frac{M^2}{s} \ll 1$ and $\frac{\Lambda_{\mathrm{QCD}}}{M} \ll 1$. We are not interested in resumming logs of $\frac{\Lambda_{\mathrm{QCD}}}{M}$ (beyond what is encoded in α_s). We will treat scales of order Λ_{QCD} (such as the proton mass) as being exactly zero and focus on logarithms of λ.

The setup for the factorization begins by pulling out the leptonic tensor $L_{\mu\nu}$, as in Chapters 20 and 32. The Drell–Yan cross section can be written as

$$d\sigma = \frac{2e^4}{M^4 s} W^{\mu\nu} L_{\mu\nu} d\Pi^{1,2}_{\text{LIPS}}, \tag{36.70}$$

where the leptonic tensor $L^{\mu\nu} = \text{Tr}[\not{k}_1 \gamma^\mu \not{k}_2 \gamma^\nu]$ is the same as for DIS, Eq. (32.11) (up to a factor of 2 from the spin averaging), and $d\Pi^{1,2}_{\text{LIPS}}$ refers to the leptonic phase space (the hadronic phase space is included in $W^{\mu\nu}$). Ignoring the weak interactions, and using just one quark flavor for simplicity, the lepton pair is produced through a neutral current $J^\mu = \bar{\psi}\gamma^\mu\psi$. The hadronic tensor can be expressed in terms of this current (see Chapters 20 and 32) as

$$W^{\mu\nu} = Q_q^2 \sum_X (2\pi)^4 \delta^4(P_1 + P_2 - q - p_X)\langle PP|J^\mu(0)|X\rangle\langle X|J^\nu(0)|PP\rangle$$

$$= Q_q^2 \int d^4x\, e^{-iqx} \langle P_1 P_2|J^\mu(x)J^\nu(0)| P_1 P_2\rangle, \tag{36.71}$$

where Q_q is the quark charge. The second line is derived by inserting factors of $e^{i\hat{\mathcal{P}}x}$ with $\hat{\mathcal{P}}^\mu$ the translation operator, just as in Eq. (32.77). We can sum over lepton spins (cf. Eq. (20.29)) remaining differential in the q^μ to get

$$\frac{d\sigma}{dM^2} = -\frac{2\alpha^2 Q_q^2}{3M^2 s} \int \frac{d^3q}{(2\pi)^3\, 2q_0} \int d^4x\, e^{-iqx} \langle P_1 P_2|J^\mu(x)J_\mu(0)|P_1 P_2\rangle, \tag{36.72}$$

with $q_0 = \sqrt{\vec{q}^2 + M^2}$. So far, everything is exact and applies in any kinematical regime.

Now let us exploit the observation that as $M \to \sqrt{s}$ the only relevant partonic states have either collinear scaling (with respect to the incoming hadrons) or soft scaling if they are in the hadronic final state. Let $n^\mu = \frac{1}{E_1} P_1^\mu$. In the center-of-mass frame, $\bar{n}^\mu = \frac{1}{E_2} P_2^\mu$ points backwards to n^μ. The effective field theory operator we need to match onto at leading order is therefore

$$\mathcal{O}^\mu = \bar{\psi}\overline{W}_1 \overline{Y}_n^\dagger \gamma^\mu \overline{Y}_{\bar{n}} \overline{W}_2^\dagger \psi = \left(\bar{\psi}_\alpha \overline{W}_1\right)_i \gamma^\mu_{\alpha\beta} \left(\overline{W}_2^\dagger \psi_\beta\right)_j \left(\overline{Y}_n^\dagger \overline{Y}_{\bar{n}}\right)_{ij}, \tag{36.73}$$

where $\overline{Y}_n$ and $\overline{Y}_{\bar{n}}$ are as in Eq. (36.38) and $\overline{W}_1$ and $\overline{W}_2$ are as in Eq. (36.45), pointing in directions t_1^μ and t_2^μ, with the integrals going from $-\infty$ to 0 (as the protons are incoming). The only restriction on t_1^μ is that it is not collinear to n^μ, thus we can take $t_1^\mu = \bar{n}^\mu$. Similarly, we can take $t_2^\mu = n^\mu$. The second form in Eq. (36.73) makes the color and spin indices explicit.

Writing $\chi_n = \overline{W}_n^\dagger \psi$ to simplify the notation, we then have

$$\langle P_1 P_2|\mathcal{O}^{\dagger\mu}(x)\mathcal{O}^\mu(0)|P_1 P_2\rangle = -\langle\Omega|\left[\overline{Y}_{\bar{n}}^\dagger \overline{Y}_n(x)\right]_{ij} \left[\overline{Y}_n^\dagger \overline{Y}_{\bar{n}}(0)\right]_{kl}|\Omega\rangle$$

$$\times \gamma^\mu_{\alpha\sigma}\langle P_1|\bar{\chi}^\alpha_{1k}(0)\chi^\beta_{1j}(x)|P_1\rangle \times \gamma^\mu_{\rho\beta}\langle P_2|\bar{\chi}^\rho_{2i}(x)\chi^\sigma_{2l}(0)|P_2\rangle, \tag{36.74}$$

where i, j, k, l are color indices and $\alpha, \beta, \rho, \sigma$ are spinor indices. The factor of -1 comes from anticommuting the spinors to get them into this form.

Since the proton is a color-neutral object, the collinear matrix elements must be diagonal in color space. Thus, we can average over colors to write

$$\langle P_2 | \bar{\chi}_{2i}^\rho(x) \chi_{2l}^\sigma(0) | P_2 \rangle = \frac{\delta_{il}}{N} \langle P_2 | \bar{\chi}_2^\rho(x) \chi_2^\sigma(0) | P_2 \rangle. \tag{36.75}$$

The matrix element on the right has the usual implicit color sum. These δ_{il} factors induce a color trace on the soft Wilson lines. The color-averaged soft matrix element is called a **soft function**. The Drell–Yan soft function is defined by

$$W_{DY}(x) \equiv \frac{1}{N} \text{tr} \langle \Omega | \bar{Y}_{\bar{n}}^\dagger Y_n(x) \bar{Y}_n^\dagger Y_{\bar{n}}(0) | \Omega \rangle, \tag{36.76}$$

with tr denoting a color trace.

The collinear matrix elements are closely related to parton distribution functions. To make the connection precise, we first multipole expand the collinear field:

$$\langle P_2 | \bar{\chi}_2(x) = \langle P_2 | \left[1 + \frac{1}{2}(\bar{n} \cdot x)(n \cdot \partial) + \frac{1}{2}(n \cdot x)(\bar{n} \cdot \partial) + x_\perp \cdot \partial_\perp + \cdots \right] \bar{\chi}_2(0). \tag{36.77}$$

The derivatives can then act as momentum operators on the proton state, pulling out factors of the proton momentum. Now, the proton momentum is $P_2^\mu = E_2 \bar{n}^\mu + \mathcal{O}(\Lambda_{\text{QCD}})$. Thus, at leading power in $\frac{\Lambda_{\text{QCD}}}{E}$, the $\bar{n} \cdot \partial$ and $\partial_\perp$ terms in this expansion can be dropped. Then the series is resummed into

$$\langle P_2 | \bar{\chi}_2(x) \sim \langle P_2 | \bar{\chi}_1(x_-), \tag{36.78}$$

where $x_-^\mu \equiv \frac{1}{2}(\bar{n} \cdot x) n^\mu$. Moreover, we must have

$$\int dx_- e^{-iq_+ x_-} \langle P_2 | \bar{\chi}_2^\rho(x_-) \chi_2^\sigma(0) | P_2 \rangle \propto \bar{\not{n}}^{\sigma \rho}, \tag{36.79}$$

where $dx_- = \frac{1}{\sqrt{2}} d(\bar{n} \cdot x)$ and $q_+^\mu = \frac{1}{2}(n \cdot q) \bar{n}^\mu$. This proportionality follows from Lorentz invariance, since the matrix element can only be proportional to the two 4-vectors around: P_2^μ and q_+^μ, which are both proportional to $\bar{n}^\mu$. To find the proportionality constant, we can contract both sides with $\not{n}^{\rho\sigma}$. This gives

$$\int dx_- e^{-iq_+ x_-} \langle P_2 | \bar{\chi}_2^\rho(x_-) \chi_2^\sigma(0) | P_2 \rangle = \frac{\bar{\not{n}}^{\sigma\rho}}{4} \int dx_- e^{-iq_+ x_-} \langle P_2 | \bar{\chi}_2(x_-) \frac{\not{n}}{2} \chi_2(0) | P_2 \rangle. \tag{36.80}$$

Finally, taking the inverse Fourier transform we can connect to the PDFs:

$$\langle P_2 | \bar{\chi}_2(x_-) \frac{\not{n}}{2} \chi_2(0) | P_2 \rangle = (n \cdot P_2) \int d\xi \, f_q(\xi) e^{i \frac{\xi}{2}(\bar{n} \cdot x)(n \cdot P_2)}, \tag{36.81}$$

where f_q coincides with the lightcone definition of the PDFs from Eq. (32.117):

$$f_q(\xi) = \int_{-\infty}^\infty \frac{dt}{2\pi} e^{-it\xi(n \cdot P_2)} \langle P_2 | \bar{\psi}(tn^\mu) \overline{W}_n(tn^\mu) \frac{\not{n}}{2} \overline{W}_n^\dagger(0) \psi(0) | P_2 \rangle. \tag{36.82}$$

Now we can put everything together. The γ-matrices combine into a Dirac trace: $\text{Tr}[\gamma^\mu \bar{\not{n}} \gamma^\mu \not{n}] = -16$. We also use $\bar{n} \cdot P_1 = n \cdot P_2 = \sqrt{s}$ to find

$$
\frac{d\sigma}{dM^2} = \frac{2\alpha^2 Q_q^2}{3M^2 N} |C|^2 \int \frac{d^3q}{(2\pi)^3 2q_0} \int d\xi_1 \, d\xi_2 \, f_q(\xi_1) \, f_q(\xi_2)
$$
$$
\times \int d^4x \, e^{i\left[\frac{1}{2}\xi_1(\bar{n}\cdot P_1)n^\mu + \frac{1}{2}\xi_2(n\cdot P_2)\bar{n}^\mu - q\right]\cdot x} W_{\mathrm{DY}}(x). \quad (36.83)
$$

Here, C is the Wilson coefficient from matching between J^μ in QCD to $\mathcal{O}^\mu$ in SCET. Our normalization is such that $C = 1$ at leading order. You can calculate C and $W_{\mathrm{DY}}(x)$ at 1-loop in Problem 36.6.

36.5.2 $e^+e^- \to$ dijets

Next we will discuss the factorization formula for $e^+e^- \to$ dijets. This is a crossing of Drell–Yan, so up to some kinematic factors, the starting point is the same. Let k_1^μ and k_2^μ be the electron momenta, and $q^\mu = k_1^\mu + k_2^\mu$ the total momentum. In the center-of-mass frame, $q^\mu = (Q, 0, 0, 0)$ with $Q > 0$. The cross section averaged over the incoming e^+e^- spins is first written in terms of the current $J^\mu = \bar{\psi}\gamma^\mu\psi$ as (cf. Eq. (20.34))

$$
\sigma = \sigma_0 \frac{-2\pi}{NQ^2} \sum_X \int d\Pi_X (2\pi)^4 \, \delta^4(q - p_X) \, \langle \Omega | J^\mu(0) | X \rangle \langle X | J^\mu(0) | \Omega \rangle, \quad (36.84)
$$

where $\sigma_0 = N \frac{4\pi\alpha_e^2}{3E_{\mathrm{CM}}^2} \sum_q Q_q^2$ is the tree-level $e^+e^- \to$ hadrons cross section. Again, we are ignoring the weak interactions for simplicity. The sum over states includes a color and spin sum. To check the normalization, at tree-level $\langle \Omega | J^\mu(0) | X \rangle \langle X | J^\mu(0) | \Omega \rangle = N \mathrm{Tr}[\not{p}_1 \gamma^\mu \not{p}_2 \gamma^\mu] = -4NQ^2$, where p_1^μ and p_2^μ are the momenta of the two outgoing quarks. Also, the inclusive integral over two-body phase space is $\int d\Pi_X (2\pi)^4 \delta^4(q - p_X) = \frac{1}{8\pi}$ (see Eqs. (5.29) or (20.A.85)). Thus we find $\sigma = \sigma_0$ at tree-level.

For dijet production, only certain hadronic final states $|X\rangle$ can contribute to this sum. To be concrete, we consider the cross section when thrust is close to 1, so $\tau = 1 - T \ll 1$. Let n^μ denote the thrust axis. As discussed in Section 36.1, in order to have $\tau \ll 1$, all of the final state particles must either be collinear to n^μ, collinear to $\bar{n}^\mu$ or soft. We denote states in these regions of phase space as $|X_1\rangle$, $|X_2\rangle$ and $|X_s\rangle$ respectively. As shown at tree-level in Sections 36.3 and 36.4, matrix elements with final states in this kinematical regime agree with those from an effective theory with collinear sectors in the n^μ and $\bar{n}^\mu$ directions and a soft sector. The different sectors are completely decoupled from each other. In the effective theory, the cross section is given by

$$
\sigma = \sigma_0 \frac{-2\pi}{NQ^2} \sum_X (2\pi)^4 \, \delta^4(q - p_X) \, \langle \Omega | C^\star \mathcal{O}^{\dagger\mu}(0) | X \rangle \langle X | C\mathcal{O}^\mu(0) | \Omega \rangle, \quad (36.85)
$$

with the same operator as in Drell–Yan, given in Eq. (36.73) and C its Wilson coefficient. In terms of the jet fields $\chi_n = W_n^\dagger \psi_n$ defined in Eq. (36.44), the operator is $\mathcal{O}^\mu = \bar{\chi}_1 Y_n^\dagger \gamma^\mu Y_{\bar{n}} \chi_2$.

Since $|X\rangle = |X_1; X_2; X_s\rangle$, the matrix elements factorize:

$$\sum_X \langle\Omega|\mathcal{O}^{\dagger\mu}|X\rangle\langle X|\mathcal{O}^\mu|\Omega\rangle = \text{Tr}\left[\gamma^\mu \frac{\bar{\slashed{n}}}{8}\gamma^\mu\frac{\slashed{n}}{8}\right]\sum_{X_s X_1 X_2}\text{tr}\left\{\langle\Omega|Y_{\bar{n}}^\dagger Y_n|X_s\rangle\langle X_s|Y_n^\dagger Y_{\bar{n}}|\Omega\rangle\right\}$$

$$\times \frac{1}{N}\text{Tr}\{\langle\Omega|\slashed{n}\chi_1|X_1\rangle\langle X_1|\bar{\chi}_1|\Omega\rangle\} \times \frac{1}{N}\text{Tr}\{\langle\Omega|\bar{\chi}_2|X_2\rangle\langle X_2|\slashed{n}\chi_2|\Omega\rangle\}, \quad (36.86)$$

where Tr is a Dirac trace. To arrive at this form, simplifications have been applied following the Drell–Yan example above: The collinear matrix elements are color diagonal, so we have averaged over color. Also, the collinear scaling of the states $|X_1\rangle$ and $|X_2\rangle$ allowed us to insert the $\slashed{n}$ and $\bar{\slashed{n}}$ factors. Note that we did not need to talk about the scaling of x in this case, since all the operators are evaluated at $x = 0$.

To further simplify the cross section, we use that $q^\mu = p_{X_1}^\mu + p_{X_2}^\mu + p_{X_s}^\mu$. Since $q \sim \lambda^0$ is the hard scale, it fixes the only λ^0 components, which are the large components of the collinear fields: $\bar{n}\cdot p_{X_1} = n\cdot p_{X_2} = Q$. The $\perp$ components of the collinear momenta scale as λ^1. Thus, we must also have $p_{X_1}^\perp = -p_{X_2}^\perp$. Therefore, overall momentum conservation amounts to

$$\delta^4(q - p_X) = 2\delta(Q - \bar{n}\cdot p_{X_1})\,\delta(Q - n\cdot p_{X_2})\,\delta^2(\vec{p}_{X_1}^\perp + \vec{p}_{X_2}^\perp). \quad (36.87)$$

Since the initial states are averaged over, the cross section cannot depend on the dijet axis $\vec{n}$. Let us therefore choose $\vec{n}$ to be at $\theta = \phi = 0$. We then compensate for omitting an angular integral by adding a factor of $4\pi\delta(\theta)\delta(\phi) = \pi Q^2\delta^2(p_{X_1}^\perp)$, where the $\frac{Q^2}{4}$ comes from $|\vec{p}_{X_1}| = \frac{Q}{2}$. Thus, with fixed $\vec{n}$, we substitute

$$\delta^4(q - p_X) \to 2\pi Q^2\delta(Q - \bar{n}\cdot p_{X_1})\,\delta(Q - n\cdot p_{X_2})\,\delta^2(\vec{p}_{X_1}^\perp)\,\delta^2(\vec{p}_{X_2}^\perp). \quad (36.88)$$

Next, we insert

$$1 = \int dr_{1n}\,dr_{2\bar{n}}\,\delta(r_{1n} - n\cdot p_{X_1})\,\delta(r_{2\bar{n}} - \bar{n}\cdot p_{X_2}) \quad (36.89)$$

to get

$$(2\pi)^4\,\delta^4(q - p_X) \to \frac{2\pi Q^2}{(2\pi)^4}\int dr_{1n}\,dr_{2\bar{n}}(2\pi)^4\frac{1}{2}\delta^4(r_1 - p_{X_1})(2\pi)^4\frac{1}{2}\delta^4(r_2 - p_{X_2}), \quad (36.90)$$

where

$$r_1^\mu \equiv \frac{Q}{2}n^\mu + \frac{1}{2}r_{1n}\bar{n}^\mu \quad \text{and} \quad r_2^\mu \equiv \frac{1}{2}Q\bar{n}^\mu + \frac{1}{2}r_{2\bar{n}}n^\mu. \quad (36.91)$$

Noting that $dr_1^2 = (\bar{n}\cdot r_1)dr_{1n}$ and $dr_2^2 = (n\cdot r_2)\,dr_{2\bar{n}}$, we thus have

$$\sigma = \sigma_0\frac{1}{16}H\int dr_1^2\,dr_2^2\sum_{X_s}\frac{1}{N}\text{tr}\{\langle\Omega|Y_{\bar{n}}^\dagger Y_n|X_s\rangle\langle X_s|Y_n^\dagger Y_{\bar{n}}|\Omega\rangle\}$$

$$\times \sum_{X_1}\frac{1}{N}\frac{1}{2\pi(\bar{n}\cdot r_1)}\int d^4x\,e^{i(r_1 - p_{X_1})x}\,\text{tr}\{\langle\Omega|\slashed{n}\chi_1|X_1\rangle\langle X_1|\bar{\chi}_1|\Omega\rangle\}$$

$$\times \sum_{X_2}\frac{1}{N}\frac{1}{2\pi(n\cdot r_2)}\int d^4y\,e^{i(r_2 - p_{X_2})y}\,\text{tr}\{\langle\Omega|\bar{\chi}_2|X_2\rangle\langle X_2|\slashed{n}\chi_2|\Omega\rangle\}, \quad (36.92)$$

with r_1^μ and r_2^μ the 4-vectors given in Eq. (36.91). $H \equiv |C|^2$ in this expression is called the **hard function**.

To progress, we specialize to the calculation of thrust. As discussed in Section 36.1, if $\tau = 1 - T \ll 1$, then $\tau \sim \tau_1$ where $\tau_1 \equiv \frac{1}{Q^2}(p_1^2 + p_2^2)$, with p_1^μ and p_2^μ defined as the sums of the momenta of all particles going into the two hemispheres defined by the thrust axis. All particles in $|X_1\rangle$ go into hemisphere 1, all particles in $|X_2\rangle$ go into hemisphere 2, and soft particles in $|X_s\rangle$ can go either way. Let us write $k_{X_s^1}^\mu$ for the sum of soft momenta that go into hemisphere 1. From the power-counting discussion in Section 36.2, with $\lambda = \frac{1}{Q^2}p_{X_1}^2 \sim \tau$ in this case, the collinear and soft momenta scale as

$$\left(n \cdot p_{X_1}, \bar{n} \cdot p_{X_1}, p_{X_1}^\perp\right) \sim Q(\lambda^2, 1, \lambda), \qquad \left(n \cdot k_{X_s^1}, \bar{n} \cdot k_{X_s^1}, k_{X_s^1}^\perp\right) \sim Q(\lambda^2, \lambda^2, \lambda^2),$$
(36.93)

so that $p_{X_1}^2 \sim \lambda^2$ and $k_1^2 \sim \lambda^4$. Also $p_{X_1} \cdot k_{X_s^1} \sim \frac{1}{2}(\bar{n} \cdot p_{X_1})(n \cdot k_{X_s^1}) \sim \lambda^2$. Thus, the hemisphere-1 mass at leading power is

$$p_{X_1}^2 = (p_{X_1}^\mu + k_{X_s^1}^\mu)^2 \sim p_{X_1}^2 + Q(n \cdot k_{X_s^1}) = r_1^2 + Q(n \cdot k_{X_s^1}).$$
(36.94)

We have used that $r_i = p_{X_i}$ from the δ-functions in Eq. (36.90). Therefore,

$$Q^2\tau \sim p_{X_1}^2 + p_{X_2}^2 \sim r_1^2 + r_2^2 + Q(n \cdot k_{X_s^1}) + Q(\bar{n} \cdot k_{X_s^2}).$$
(36.95)

This equation implies that the observable of interest, τ, when small, reduces to a sum of a contribution from each collinear sector plus a contribution from the soft sector, with no interference.

To calculate the thrust distribution we insert two more integrals and two more δ-functions into our cross section to get

$$\frac{d\sigma}{d\tau} = \sigma_0 H \int dr_1^2 \, dr_2^2 \, dk_{1n} \, dk_{2\bar{n}} \delta\left(Q^2\tau - r_1^2 - r_2^2 - Qk_{1n} - Qk_{2\bar{n}}\right)$$

$$\times \sum_{X_s} \frac{1}{N} \mathrm{tr}\left\{\langle\Omega|Y_{\bar{n}}^\dagger Y_n|X_s\rangle\langle X_s|Y_n^\dagger Y_{\bar{n}}|\Omega\rangle\right\} \delta(k_{1n} - n \cdot k_{X_s^1}) \, \delta(k_{2\bar{n}} - \bar{n} \cdot k_{X_s^2})$$

$$\times \sum_{X_1} \frac{1}{N} \frac{1}{8\pi(\bar{n} \cdot r_1)} \int d^4x \, e^{i(r_1 - p_{X_1})x} \mathrm{tr}\{\langle\Omega|\slashed{\bar{n}}\chi_1|X_1\rangle\langle X_1|\bar{\chi}_1|\Omega\rangle\}$$

$$\times \sum_{X_2} \frac{1}{N} \frac{1}{8\pi(n \cdot r_2)} \int d^4y \, e^{i(r_2 - p_{X_2})y} \mathrm{tr}\{\langle\Omega|\bar{\chi}_2|X_2\rangle\langle X_2|\slashed{n}\chi_2|\Omega\rangle\}.$$
(36.96)

Now, when $\tau_1 \ll 1$, r_{1n} and r_{2n} must be small. Therefore r_1^μ and r_2^μ, as defined in Eq. (36.91), must have collinear scaling. Thus, we can extend the sum over collinear states $|X_1\rangle\langle X_1|$ and $|X_2\rangle\langle X_2|$ to sums over all states. This lets us write the cross section in terms of a universal object called a **jet function**. The jet function in the n^μ direction is defined as

$$J(p^\mu) \equiv \frac{1}{8\pi N(\bar{n} \cdot p)} \sum_X \int d^4x e^{ipx} \mathrm{Tr}\left[\langle\Omega|\bar{\chi}_n(x)|X\rangle\langle X|\slashed{\bar{n}}\chi_n(0)|\Omega\rangle\right].$$
(36.97)

Here, Tr is a Dirac trace and the sum over colors is implicit. The normalization is set so that $J(p^\mu) = \delta(p^2)$ at leading order, as we show below. Since the sum over $|X\rangle$ in the jet function is complete, it can be written as the discontinuity (twice the imaginary part) of a forward matrix element (see Section 24.1.2):

$$J(p^2) = \frac{1}{8\pi N \bar{n} \cdot p} \text{Disc} \left\{ \text{Tr} \, i \int d^4 x \, e^{ipx} \langle \Omega | T \{ \bar{\chi}_n(x) \slashed{\bar{n}} \chi_n(0) \} | \Omega \rangle \right\}. \quad (36.98)$$

By Lorentz invariance and invariance under rescaling of $\bar{n}$, the jet function can only depend on p^2, as we have written. Physically, the jet function gives something close to the probability of finding a jet with invariant mass p^2 (it is not exactly this probability since soft radiation also contributes to jet masses). This same jet function appears in the factorization formulas for many processes (for example, $B \to X_s \gamma$, deep inelastic scattering and direct photon production). Note that the jet function is only useful when evaluated at values of $p^2 \ll Q^2$ for some hard scale Q. Otherwise, extending the sum from collinear states to all states induces uncontrolled subleading power contributions.

We also define the **hemisphere soft function** as

$$S_{\text{hemi}}(k_{1n}, k_{2\bar{n}}) \equiv \sum_{X_s} \frac{1}{N} \text{tr} \left\{ \langle \Omega | Y_{\bar{n}}^\dagger Y_n | X_s \rangle \langle X_s | Y_n^\dagger Y_{\bar{n}} | \Omega \rangle \right\}$$
$$\times \delta(k_{1n} - n \cdot k_{X_s^1}) \delta(k_{2\bar{n}} - \bar{n} \cdot k_{X_s^2}) \quad (36.99)$$

As with the collinear radiation, the scale at which the soft function is to be evaluated is determined by the factorization formula. For $\tau_1 \ll 1$ it implies $k_{1n} \ll 1$ and $k_{2\bar{n}} \ll 1$. Thus, we will extend the sum to include all rather than just soft states. The soft function for thrust is related to the hemisphere soft function by

$$S_T(k) = \int_0^\infty dk_{1n} \, dk_{2\bar{n}} \, S_{\text{hemi}}(k_{1n}, k_{2\bar{n}}) \, \delta(k - k_{1n} - k_{2\bar{n}}). \quad (36.100)$$

Putting everything together, the singular part of the thrust distribution can be calculated in SCET by

$$\frac{1}{\sigma_0} \left(\frac{d^2 \sigma}{d\tau} \right)_{\text{sing}} = H \int dr_1^2 \, dr_1^2 \, dk \, J(r_1^2) \, J(r_2^2) \, S_T(k) \, \delta(Q^2 \tau - r_1^2 - r_2^2 - Qk). \quad (36.101)$$

We will next compute the hard, jet and soft functions to order α_s in perturbation theory using SCET and check that the singular behavior of thrust is reproduced.

36.6 Thrust in SCET

Having set up the factorization formula for thrust in the dijet limit, we can now compute the hard, jet and soft functions in perturbation theory. We will work to order α_s, which allows for leading-log resummation. All our calculations will be done in Feynman gauge.

36.6.1 Hard function

The hard function is defined as $H(Q) = |C(Q)|^2$. We compute C by matching J^μ to $\mathcal{O}^\mu$, which can be done independently of the dijet observable we are interested in. The hard function for dijet production is the same as for Drell–Yan and related to the hard function for deep inelastic scattering by analytic continuation.

An example of matching was worked out in Section 31.3 for the 4-Fermi theory. The procedure here is identical. The Wilson coefficient is computed from the difference between radiative corrections to the current J^μ in QCD and to $\mathcal{O}^\mu$ in SCET. We did the hard work for this loop in Chapter 20 and applied it to QCD in Chapter 26. From Eq. (20.A.101), replacing $e_R \to -g_s$ and adding the QCD color factor (see Section 26.3.1), we have

$$\mathcal{M}_{\text{QCD}} = \quad = C_F \frac{g_s^2}{2\pi^2} \left(\frac{4\pi e^{-\gamma_E} \mu^2}{Q^2} \right)^{\frac{4-d}{2}} \left(-\frac{1}{\varepsilon^2} - \frac{\frac{3}{4} + \frac{i\pi}{2}}{\varepsilon} + \frac{7\pi^2}{48} - 1 - \frac{3\pi i}{8} \right).$$

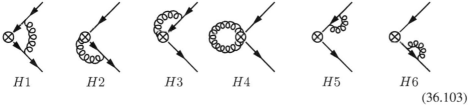

$$(36.102)$$

There are also the wavefunction renormalization graphs and counterterm graphs which do not have to be calculated, as we explain shortly.

In SCET, the loops are the virtual corrections to $\mathcal{O}^\mu = \bar{\psi} W_1 Y_n \gamma^\mu Y_{\bar{n}}^\dagger W_2^\dagger \psi$ in a Lagrangian with decoupled fields $\mathcal{L}_{\text{SCET}} = \mathcal{L}_{\text{soft}} + \mathcal{L}_n + \mathcal{L}_{\bar{n}}$. The virtual diagrams at order α_s have one of the following topologies:

$$H1 \qquad H2 \qquad H3 \qquad H4 \qquad H5 \qquad H6$$

$$(36.103)$$

In diagrams $H2$, $H3$ and $H4$, gluons with an endpoint on the operator vertex correspond to terms coming from the expansion of the Wilson line. For example, expanding Y_n and $Y_{\bar{n}}$ gives factors of $g_s \frac{n \cdot A}{n \cdot k}$ and $g_s \frac{\bar{n} \cdot A}{\bar{n} \cdot k}$ respectively; these gluon fields can then be contracted with a propagator from $\mathcal{L}_{\text{soft}}$ generating diagram $H4$. Diagrams $H5$ and $H6$ are identical to the wavefunction graphs in pure QCD. Thus they will cancel in the matching, which is why we could ignore them in $\mathcal{M}_{\text{QCD}}$.

Since $\mathcal{L}_n$ and $\mathcal{L}_{\bar{n}}$ are completely decoupled, diagram $H1$ does not actually exist. In diagram $H2$, the gluon must be an $\bar{n}$-collinear gluon, and in diagram $H3$, the gluon must be n-collinear. In diagram $H4$, each Wilson line (soft or collinear) gives a factor of $g_s \frac{t \cdot A}{t \cdot k}$ for some lightlike 4-vector t^μ. Since $t \cdot t = 0$ the loop vanishes if the same Wilson line produces both gluons. Since collinear sectors are decoupled, the only contribution to $H4$ can therefore be from soft gluons, with one vertex from $Y_{\bar{n}}$ and the other from $Y_n^\dagger$. Thus we need to compute $H2$ and $H3$ for collinear gluons and $H4$ for soft gluons.

Diagram $H2$ gives

$$iM_{H2} = \quad = 2ig_s^2 C_F \mu^{4-d} \int \frac{d^d k}{(2\pi)^d} \frac{\bar{n}\cdot(p-k)}{(p-k)^2} \frac{1}{\bar{n}\cdot k} \frac{1}{k^2} = 0. \qquad (36.104)$$

The integrand has to produce a Lorentz-invariant quantity of mass dimension $d-4$. The only Lorentz invariants around are p^2, which is zero, and $\bar{n}\cdot p$. However, the integral is also invariant under $\bar{n}^\mu \to \lambda \bar{n}^\mu$ for any λ, thus it cannot be $(\bar{n}\cdot p)^{d-4}$. Thus it must vanish.

The soft graph is

$$iM_{H4} = \quad = -ig_s^2 C_F \mu^{4-d} \int \frac{d^d k}{(2\pi)^d} \frac{n\cdot\bar{n}}{(\bar{n}\cdot k)(n\cdot k)\,k^2} = 0. \qquad (36.105)$$

Now there is simply no quantity with any mass dimension on which the graph could depend. Thus it also must vanish in dimension regularization.

The result is that all of the purely virtual graphs completely vanish in SCET in dimensional regularization. This is a feature of SCET that is incredibly useful. The virtual graphs can also be thought of as converting $\frac{1}{\varepsilon_{IR}}$ poles into $\frac{1}{\varepsilon_{UV}}$ poles. Since the IR singularities of QCD are identical to those in SCET, the $\frac{1}{\varepsilon_{IR}}$ poles must drop out of the matching.[6] The $\frac{1}{\varepsilon_{UV}}$ poles in both SCET and QCD are removed with counterterms in the respective theories. Thus, in dimensional regularization with $\overline{MS}$, we simply drop all virtual graphs and all $\frac{1}{\varepsilon}$ poles of any sort. Thus, the Wilson coefficient can be read off from the virtual graph in QCD. From Eq. (36.102) we find

$$C = 1 + \frac{\alpha_s}{4\pi} C_F \left(-8 + \frac{7\pi^2}{6} - 3\pi i - \ln^2 \frac{Q^2}{\mu^2} + (3 + 2\pi i) \ln \frac{Q^2}{\mu^2} \right) + \mathcal{O}(\alpha_s^2) \qquad (36.106)$$

and

$$H(Q,\mu) = |C|^2 = 1 + \frac{\alpha_s}{4\pi} C_F \left(-16 + \frac{7\pi^2}{3} - 2\ln^2 \frac{Q^2}{\mu^2} + 6\ln \frac{Q^2}{\mu^2} \right) + \mathcal{O}(\alpha_s^2). \qquad (36.107)$$

36.6.2 Jet function

The jet function is defined in Eq. (36.98). Pulling the $\not{\bar{n}}$ out of the integral, it can be written as

$$J(p^2) = \frac{1}{8\pi N \bar{n}\cdot p} \mathrm{Disc}\left\{ i\mathrm{Tr}\left[\not{\bar{n}}_{\alpha\beta} \int d^4 x\, e^{ipx} \langle \Omega | T\{\chi_\beta(0)\bar{\chi}_\alpha(x)\} |\Omega\rangle \right] \right\}. \qquad (36.108)$$

The matrix element in this expression is the quark propagator. At leading order,

$$J(p^2) = -\frac{\mathrm{Tr}[\not{\bar{n}}\not{p}]}{8\bar{n}\cdot p} \frac{1}{\pi} \left\{ 2\mathrm{Im}\left[\frac{1}{p^2 + i\varepsilon} \right] \right\} = \delta(p^2), \qquad (36.109)$$

---

[6] An important check on SCET is provided by using an IR regulator other than dimensional regularization. Then one can see explicitly that the IR divergences of SCET and QCD match up. See for example [Manohar, 2003].

where Eq. (24.25) has been used on the last step.

At order α_s the jet function is easiest to compute with cut diagrams using the optical theorem. There are eight possible cuts. Four cut the gluon and a quark:

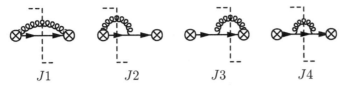

and four cut just a quark

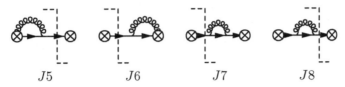

These last four diagrams put the massless quark on shell, so they give scaleless integrals and vanish in dimensional regularization.

One fairly easy way to calculate the jet function is in lightcone gauge, $\bar{n} \cdot A = 0$. In lightcone gauge, the collinear Wilson line is $W = 1$ and so diagrams $J1$, $J2$ and $J3$ vanish. $J4$ is just the self-energy graph in QCD. Thus, the jet function is just the imaginary part of the quark propagator in lightcone gauge. We leave this approach to the calculation to Problem 36.5.

We will instead evaluate the graphs in Feynman gauge. In Feynman gauge, diagram $J1$ is proportional to $\bar{n} \cdot \bar{n}$ and hence vanishes. Diagram $J2$ (before the cut) gives

$$i\mathcal{M}_{J2} = \quad = C_F g_s^2 \mu^\varepsilon \int \frac{d^d k}{(2\pi)^d} \frac{\bar{n}^\mu}{\bar{n}\cdot k} \frac{-i}{k^2 + i\varepsilon} \frac{i(\not{p}-\not{k})}{(p-k)^2 + i\varepsilon} i\gamma^\mu \frac{i\not{p}}{p^2 + i\varepsilon}.$$

(36.110)

Following the cutting rules in Section 24.1.2, we compute the discontinuity by replacing $\frac{1}{p^2+i\varepsilon} \to (-2\pi i)\delta(p^2)$ for the cut lines and summing over spins. After some algebra, this results in

$$\text{Disc}\{i\mathcal{M}_{J2}\} = \frac{\alpha_s}{2} C_F(\bar{n}\cdot p)\frac{1}{p^2}\left(\frac{4\pi\mu^2}{p^2}\right)^\varepsilon \frac{\Gamma(2-\varepsilon)\Gamma(-\varepsilon)}{\Gamma(1-\varepsilon)\Gamma(2-2\varepsilon)}.$$

(36.111)

Diagram $J3$ gives the same answer with $n \leftrightarrow \bar{n}$. Graph $J4$ is computed similarly, giving

$$\text{Disc}\{i\mathcal{M}_{J4}\} = \frac{\alpha_s}{2} C_F(\bar{n}\cdot p)\frac{1}{p^2}\left(\frac{4\pi\mu^2}{p^2}\right)^\varepsilon (1-\varepsilon)\frac{\Gamma(2-\varepsilon)}{\Gamma(3-2\varepsilon)}.$$

(36.112)

Summing diagrams $J1$ to $J4$ and the leading order result gives

$$J_n(p) = \delta(p^2) + \frac{\alpha_s}{2\pi}C_F\frac{1}{p^2}\left(\frac{4\pi\mu^2}{p^2}\right)^\varepsilon \left[2\frac{\Gamma(2-\varepsilon)\Gamma(-\varepsilon)}{\Gamma(1-\varepsilon)\Gamma(2-2\varepsilon)} +(1-\varepsilon)\frac{\Gamma(2-\varepsilon)}{\Gamma(3-2\varepsilon)}\right].$$

(36.113)

To expand this in ε we can use the identity

$$\frac{(\mu^2)^\varepsilon}{(p^2)^{1+\varepsilon}} = -\frac{1}{\varepsilon}\delta(p^2) + \left[\frac{1}{p^2}\right]_\star - \varepsilon\left[\frac{\ln\frac{p^2}{\mu^2}}{p^2}\right]_\star + \cdots, \qquad (36.114)$$

where the $\star$-distribution is a generalization of a $+$-distribution for dimensional variables. $\star$-distributions satisfy

$$\int_0^{\mu^2} dp^2 \left[f(p^2)\right]_\star g(p^2) = \int_0^{\mu^2} dp^2 f(p^2)\left[g(p^2) - g(0)\right] \qquad (36.115)$$

and $\left[f(p^2)\right]_\star = f(p^2)$ for $p^2 > 0$. We then find

$$J(p^2) = \delta(p^2) + C_F\frac{\alpha_s}{4\pi}\left\{\delta(p^2)\left(\frac{4}{\varepsilon^2} + \frac{3}{\varepsilon} + 7 - \pi^2\right) - \left[\frac{3}{p^2}\right]_\star - 4\left[\frac{\frac{1}{\varepsilon} - \ln\frac{p^2}{\mu^2}}{p^2}\right]_\star\right\}.$$
$$(36.116)$$

Since the jet function is an inclusive cross section at fixed p^2, it should be IR finite. Thus, the $\frac{1}{\varepsilon}$ and $\frac{1}{\varepsilon^2}$ IR divergences in Eq. (36.116) must be exactly canceled by the virtual graphs. We have not computed the virtual graphs (diagrams $J5$ through $J8$), since they vanish exactly in dimensional regularization. If one were to separate the UV from IR singularities, these virtual graphs would have to give $\frac{1}{\varepsilon_{\rm IR}^2} - \frac{1}{\varepsilon_{\rm UV}^2}$ terms with coefficients to precisely cancel the IR divergences in Eq. (36.116). Thus, adding the virtual graphs simply converts all $\frac{1}{\varepsilon}$ and $\frac{1}{\varepsilon^2}$ divergences to UV divergences. These UV divergences are then removed with $\overline{\rm MS}$ counterterms, just as in the hard function calculation. The result is that

$$J(p^2) = \delta(p^2) + C_F\frac{\alpha_s}{4\pi}\left\{\delta(p^2)(7 - \pi^2) + \left[\frac{-3 + 4\ln\frac{p^2}{\mu^2}}{p^2}\right]_\star\right\} + \mathcal{O}(\alpha_s^2). \quad (36.117)$$

36.6.3 Soft function

The soft function is $S(k_1, k_2) = \delta(k_1)\,\delta(k_2)$ at zeroth order. This is simply because no radiation is emitted so the total soft momentum going into each hemisphere is zero. At next-to-leading order, the soft function is an integral over real emission graphs summed over gluon polarizations. We write

$$S_{\rm hemi}(k_1, k_2) \sim \int d\Pi_k \left| \otimes\!\!\!\!\!\!\!\!\diagram + \otimes\!\!\!\!\!\!\!\!\diagram \right|^2. \qquad (36.118)$$

The diagrams are meant to indicate emissions from the Y_n and $Y_{\bar{n}}$ Wilson lines (as diagrams $H2, H3$ and $H4$ in Eq. (36.103)). To distinguish which Wilson line the gluons are coming from, we draw the diagrams as we would in full QCD. Using Wilson lines instead of the full QCD Feynman rules is equivalent to taking the soft limit *before* the diagrams are evaluated.

There is only one sector of soft gluons, thus either emission in Eq. (36.118) can go into either hemisphere. In Feynman gauge the terms that come from the square of one diagram are proportional to $n \cdot n = 0$ or $\bar{n} \cdot \bar{n} = 0$. Thus, we only need to evaluate the cross term. We find

$$
S_{\text{hemi}}(k_1, k_2) = -g_s^2 C_F \mu^{4-d} \int \frac{d^{d-1}k}{(2\pi)^{d-1}} \frac{\bar{n} \cdot n}{(n \cdot k)(\bar{n} \cdot k)}
$$
$$
\times \left[\delta(k_1 - n \cdot k)\, \theta(\bar{n} \cdot k - n \cdot k) + (k_2 - \bar{n} \cdot k)\, \theta(n \cdot k - \bar{n} \cdot k) \right]
$$
$$
= C_F \frac{\alpha_s}{\pi} \frac{\mu^{2\varepsilon}}{\varepsilon \Gamma(1-\varepsilon)} \left[\frac{\theta(k_2)}{k_2^{1+2\varepsilon}} \delta(k_1) + \frac{\theta(k_1)}{k_1^{1+2\varepsilon}} \delta(k_2) \right]. \tag{36.119}
$$

We then expand near $\varepsilon = 0$ using Eq. (36.114). Including the leading-order result, the hemisphere soft function to order α_s is

$$
S_{\text{hemi}}(k_1, k_2) = \delta(k_1)\,\delta(k_2) \left[1 + C_F \frac{\alpha_s}{4\pi} \left(\frac{\pi^2}{3} \right) \right]
$$
$$
- 8 C_F \frac{\alpha_s}{4\pi} \left\{ \left[\frac{\ln \frac{k_1}{\mu}}{k_1} \right]_\star \delta(k_2) + \left[\frac{\ln \frac{k_2}{\mu}}{k_2} \right]_\star \delta(k_1) \right\}. \tag{36.120}
$$

The thrust soft function is then

$$
S_T(k) = \int_0^\infty dk'\, S_{\text{hemi}}(k', k - k')
$$
$$
= \delta(k) \left[1 + C_F \frac{\alpha_s}{4\pi} \left(\frac{\pi^2}{3} \right) \right] - 16 C_F \frac{\alpha_s}{4\pi} \left[\frac{\ln \frac{k}{\mu}}{k} \right]_\star + \mathcal{O}(\alpha_s^2). \tag{36.121}
$$

36.6.4 Singular part of thrust

Now let us put everything together to show that SCET reproduces the singular terms in the thrust distribution as $\tau \to 0$. Plugging Eqs. (36.107), (36.117) and (36.121) into Eq. (36.101) we get

$$
\frac{1}{\sigma_0} \left(\frac{d\sigma}{d\tau} \right)_{\text{sing}} = \delta(\tau) + C_F \frac{\alpha_s}{2\pi} \left\{ \delta(\tau) \left(\frac{\pi^2}{3} - 1 \right) - 3 \left[\frac{1}{\tau} \right]_+ - 4 \left[\frac{\ln \tau}{\tau} \right]_+ \right\}, \tag{36.122}
$$

in perfect agreement with Eq. (36.8). Note that the μ dependence exactly drops out of this expression.

36.6.5 Resummed thrust

To resum the singular parts of the thrust distribution, we need to calculate and solve the renormalization group equations for the hard jet and soft functions. These RGEs are easiest to derive by differentiating the fixed-order expressions with respect to μ. Taking the μ-derivative of the hard function in Eq. (36.107) gives

$$
\mu \frac{dH}{d\mu} = \frac{\alpha_s(\mu)}{4\pi} C_F \left(8 \ln \frac{Q^2}{\mu^2} - 12 \right) H + \mathcal{O}(\alpha_s^2). \tag{36.123}
$$

The solution to this RGE is

$$H(Q, \mu) = H(Q, \mu_h) \exp\left(4S(\mu_h, \mu) - 2A_H(\mu_h, \mu) - 2A_\Gamma(\mu_h, \mu) \ln \frac{Q^2}{\mu_h^2}\right), \quad (36.124)$$

where

$$A_H(\nu, \mu) = -\int_{\alpha_s(\nu)}^{\alpha_s(\mu)} d\alpha \frac{\gamma_H(\alpha)}{\beta(\alpha)} \quad (36.125)$$

and

$$S(\nu, \mu) = -C_F \int_{\alpha_s(\nu)}^{\alpha_s(\mu)} d\alpha \frac{\gamma_{\text{cusp}}(\alpha)}{\beta(\alpha)} \int_{\alpha_s(\nu)}^{\alpha} \frac{d\alpha'}{\beta(\alpha')}, \quad (36.126)$$

with $\gamma_H(\alpha) = -6C_F \frac{\alpha}{4\pi} + \mathcal{O}(\alpha^2)$ and $\gamma_{\text{cusp}}(\alpha) = \frac{\alpha}{\pi}$. $A_\Gamma(\nu, \mu)$ is defined as $A_H(\nu, \mu)$ but with $C_F \gamma_{\text{cusp}}(\alpha)$ replacing $\gamma_H(\alpha)$. You can verify that Eq. (36.124) solves Eq. (36.123) and work out closed-form expressions for $S[(\nu, \mu)$ and $A_H(\nu, \mu)$ in Problem 36.8.

The RGEs for the jet and soft functions are non-local, like the RGEs for parton distribution functions. The jet function RGE is

$$\mu \frac{dJ(p^2, \mu)}{d\mu} = \frac{\alpha_s(\mu)}{4\pi} C_F \left[\left(-8 \ln \frac{p^2}{\mu^2} + 6\right) J(p^2, \mu) + 8 \int_0^{p^2} dq^2 \frac{J(p^2, \mu) - J(q^2, \mu)}{p^2 - q^2}\right]. \quad (36.127)$$

One can check by direct substitution that the $\mathcal{O}(\alpha_s)$ jet function in Eq. (36.117) satisfies this RGE. The RGE can be solved through the Laplace transform, as you can explore in Problem 36.7. The result is

$$J(Q, \mu) = e^{-4S(\mu_j, \mu) + 2A_J(\mu_j, \mu)} \tilde{j}(\partial_\eta, \mu_j) \frac{1}{p^2} \left(\frac{p^2}{\mu_j^2}\right)^\eta \frac{e^{-\gamma_E \eta}}{\Gamma(\eta)}\Bigg|_{\eta \to 2A_\Gamma(\mu_j, \mu)}, \quad (36.128)$$

where

$$\tilde{j}(\partial_\eta, \mu_j) = 1 + C_F \frac{\alpha_s(\mu)}{4\pi}\left(2\partial_\eta^2 - 3\partial_\eta + 7 - \frac{2\pi^2}{3}\right) + \mathcal{O}(\alpha_s^2) \quad (36.129)$$

and A_J is defined as in Eq. (36.125) but with $\gamma_J(\alpha) = -3C_F \frac{\alpha}{4\pi} + \mathcal{O}(\alpha^2)$ replacing γ_H.

The thrust soft function satisfies

$$\mu \frac{dS_T(k, \mu)}{d\mu} = \frac{\alpha_s(\mu)}{4\pi} 16 C_F \left[\ln \frac{k}{\mu} S(k, \mu) - \int_0^k dk' \frac{S_T(k, \mu) - S_T(k', \mu)}{k - k'}\right], \quad (36.130)$$

with solution

$$S_T(k, \mu) = e^{4S(\mu_s, \mu) + 2A_S(\mu_s, \mu)} \tilde{s}_T(\partial_\eta, \mu_s) \frac{1}{k}\left(\frac{k}{\mu_s}\right)^\eta \frac{e^{-\gamma_E \eta}}{\Gamma(\eta)}\Bigg|_{\eta = -4A_\Gamma(\mu_s, \mu)}, \quad (36.131)$$

where

$$\tilde{s}_T(\partial_\eta, \mu) = 1 + C_F \frac{\alpha_s(\mu)}{4\pi}\left(-8\partial_\eta^2 - \pi^2\right) + \mathcal{O}(\alpha_s^2). \quad (36.132)$$

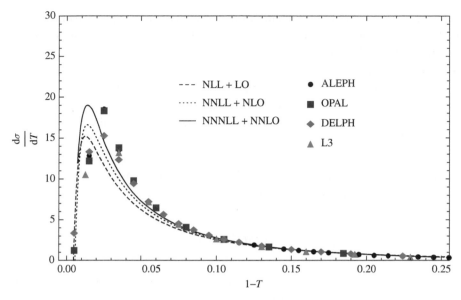

The thrust distribution resummed with SCET compared to data from LEP at $Q = 91.2$ GeV. **Fig. 36.3** Here NNNLL+NNLO means the resummation is performed at the next-to-next-to-next-to-leading logarithmic level and the non-singular distribution is calculated exactly at next-to-next-to-leading order, $\mathcal{O}(\alpha_s^3)$, in perturbative QCD. The agreement with data is excellent for $1 - T > 0.1$ or so. For lower values of $1 - T$, hadronization effects become important.

The resummed hard, jet and soft functions can be combined and simplified to

$$
\frac{1}{\sigma_0}\frac{d\sigma}{d\tau} = \frac{1}{\tau}\exp[4S(\mu_h, \mu_j) + 4S(\mu_s, \mu_j) - 2A_H(\mu_h, \mu_s) + 4A_J(\mu_j, \mu_s)]
$$
$$
\times \left(\frac{Q^2}{\mu_h^2}\right)^{-2A_\Gamma(\mu_h, \mu_j)} H(Q^2, \mu_h)
$$
$$
\times \left[\tilde{j}\left(\ln\frac{\mu_s Q}{\mu_j^2} + \partial_\eta, \mu_j\right)\right]^2 \tilde{s}_T(\partial_\eta, \mu_s)\left[\left(\frac{\tau Q}{\mu_s}\right)^\eta \frac{e^{-\gamma_E \eta}}{\Gamma(\eta)}\right]_{\eta = 4A_\Gamma(\mu_j, \mu_s)}.
$$
$$(36.133)$$

This final expression is manifestly independent of μ. Instead, it depends on μ_h, μ_j and μ_s. These three scales should be chosen as the characteristic scales associated with hard, jet and soft degrees of freedom. More precisely, one can see from the various combinations appearing in this expression that $\mu_h = Q$, $\mu_s = \tau Q$, and $\mu_j = \sqrt{\mu_s Q} = \sqrt{\tau}Q$ are natural choices. Choosing these scales gives

$$
\frac{1}{\sigma_0}\frac{d\sigma}{d\tau} = \frac{1}{\tau}\exp\left[4S(Q, \sqrt{\tau}Q) + 4S(\tau Q, \sqrt{\tau}Q) - 2A_H(Q, \tau Q) + 4A_J(\sqrt{\tau}Q, \tau Q)\right]
$$
$$
\times H(Q^2, Q^2)\left[\tilde{j}(\partial_\eta, \sqrt{\tau}Q)\right]^2 \tilde{s}_T(\partial_\eta, \tau Q)\left.\frac{e^{-\gamma_E \eta}}{\Gamma(\eta)}\right|_{\eta = 4A_\Gamma(\sqrt{\tau}Q, \tau Q)}. \quad (36.134)
$$

To compare to data, one should add to this distribution the non-singular part of the thrust distribution computed at fixed order in perturbative QCD. The non-singular distribution is currently known at $\mathcal{O}(\alpha_s^4)$, called NNLO.

Plots of the thrust distribution computed in SCET, resummed, and supplemented with the non-singular distribution from perturbative QCD are shown in Figure 36.3. The resummation is critical to providing qualitative agreement with the data. For small values of τ the soft scale becomes comparable to hadron masses and then hadronization can no longer be ignored. Since $\mu_s = \tau Q$ this happens for $\tau \lesssim \frac{m_p}{Q} \sim 0.1$. One can see the importance of these hadronization corrections directly in Figure 36.3. For values of $\tau \gtrsim 0.1$ the quantitative agreement with data is excellent. While power corrections can also be treated with effective field theory, they are beyond our scope.

Problems

36.1 Show that in the dijet region $\tau \approx \tau_1$. In particular, show that the singular terms in $\frac{d\sigma}{d\tau}$ are the same as the singular terms in $\frac{d\sigma}{d\tau_1}$ for any number of particles.

36.2 Collinear factorization.

(a) Show that the collinear factorization in Eq. (36.51) holds for multiple emissions in scalar QED.

(b) Show that the collinear factorization in Eq. (36.43) holds for multiple emissions in QCD.

36.3 Calculate the $g \to gg$ splitting function from the matrix element of gluon jet fields following the approach in Section 36.4.2. Average over azimuthal angle, you should find $P_{gg} = 2C_A \left[\frac{z}{1-z} + \frac{1-z}{z} + z(1-z) \right]$, as in Eq. (32.54).

36.4 Show soft-collinear factorization at leading power for two emissions in scalar QED. That is, show that

$$\langle \Omega | \phi_1^\star \phi_2 | p_1 p_2; q; k \rangle \sim \langle \Omega | \phi_1^\star W_1 | p_1; q \rangle \langle \Omega | W_2^\dagger \phi_2 | p_2 \rangle \langle \Omega | Y_1 Y_2^\dagger | k \rangle, \quad (36.135)$$

where p_1^μ and p_2^μ are the momenta of the scalars, k^μ is the momentum of a soft photon and q is the momentum of a photon collinear to p_1.

36.5 Calculate the quark self-energy graph at 1-loop in lightcone gauge. Show that the imaginary part gives the same jet function as computed in Section 36.6.2.

36.6 Threshold Drell–Yan.

(a) Show that near partonic threshold, the Drell–Yan cross section can be written as

$$\frac{d\sigma}{dM^2} = \frac{4\pi\alpha^2 Q_q^2}{3NM^2\sqrt{s}} |C|^2 \int \frac{d\xi_1}{\xi_1} \frac{d\xi_2}{\xi_2} f(\xi_1) f(\xi_2) \hat{W}_{\mathrm{DY}}\left(\sqrt{s}(1-z)\right), \quad (36.136)$$

where

$$\hat{W}_{\mathrm{DY}}(\omega) = \int \frac{dt}{4\pi} e^{\frac{i}{2}\omega x^0} W_{\mathrm{DY}}\left(x^0, \vec{0}\right). \quad (36.137)$$

(b) Compute the Wilson coefficient C for $\mathcal{O}^\mu$ in Eq. (36.73) at order α_s.

(c) Calculate $W_{\mathrm{DY}}(x)$ and $\hat{W}_{\mathrm{DY}}(\omega)$ to 1-loop.

36.7 Laplace transforms are extremely useful for solving RGEs in SCET. We define the Laplace transform of a function $f(\tau)$ as

$$\tilde{f}(\nu) \equiv \int_0^\infty d\tau\, e^{-\nu\tau} f(\tau). \qquad (36.138)$$

(a) Show that the cross section in Eq. (36.101) simplifies to

$$\tilde{\sigma}(\nu) = H\tilde{j}(\nu)^2 \tilde{s}_T(\nu) \qquad (36.139)$$

in Laplace space.

(b) Show that the RGE for the jet function in Eq. (36.127) simplifies to

$$\mu\frac{d}{d\mu}\tilde{j}(\nu,\mu) = \alpha_s(\mu)\left[-2\Gamma_J \ln\frac{Q^2}{e^{\gamma_E}\nu\mu^2} - 2\gamma_J\right]\tilde{j}(\nu,\mu). \qquad (36.140)$$

What are Γ_J and γ_J? Find a similar RGE for the Laplace-transformed soft function.

(c) Solve the RGE for the jet function in Laplace space and show that the result, in position space, is as in Eq. (36.128).

36.8 Sukakov RGEs.

(a) Verify that Eq. (36.124) solves Eq. (36.123).

(b) Show that the function $S(\nu,\mu)$ in Eq. (36.126) has the expansion

$$S(\nu,\mu) = \frac{\pi C_F}{\beta_0^2 \alpha_s(\nu)}\left\{1 - \frac{\alpha_s(\mu)}{\alpha_s(\nu)} - \ln\frac{\alpha_s(\mu)}{\alpha_s(\nu)} + \mathcal{O}(\alpha_s)\right\}. \qquad (36.141)$$

(c) Find a similar expansion for $A_H(\nu,\mu)$.

APPENDICES

Appendix A **Conventions**

A.1 Dimensional analysis

In relativistic quantum field theory, it is standard to set

$$c = 2.998 \times 10^8 \, \text{meters/second} = 1, \tag{A.1}$$

which turns meters into seconds and

$$\hbar = \frac{h}{2\pi} = 1.054\,572 \times 10^{-34} \, \text{joules} \cdot \text{seconds} = 1, \tag{A.2}$$

which turns joules into inverse seconds. This gives all quantities dimensions of energy (or mass, using $E = mc^2$) to some power. Quantities with positive mass dimension (e.g. momentum p) can be thought of as energies, and quantities with negative mass dimension (e.g. position x) can be thought of as lengths.

Sometimes we write the mass dimension of a quantity with brackets, as in $[p] = \left[\frac{1}{x}\right] = 1$, meaning these quantities have mass dimension 1. Other examples are

$$[dx] = [x] = [t] = -1, \tag{A.3}$$

$$[\partial_\mu] = [p_\mu] = 1, \tag{A.4}$$

$$[\text{velocity}] = \left[\frac{x}{t}\right] = [x] - [t] = 0. \tag{A.5}$$

Thus,

$$[d^4 x] = -4. \tag{A.6}$$

The action should be a dimensionless quantity:

$$[S] = \left[\int d^4 x \mathcal{L}\right] = 0. \tag{A.7}$$

So Lagrangians (really, Lagrangian densities) have dimension 4:

$$[\mathcal{L}] = 4. \tag{A.8}$$

For example, a free scalar field has Lagrangian $\mathcal{L} = \frac{1}{2}(\partial_\mu \phi)(\partial^\mu \phi)$ so

$$[\phi] = 1, \tag{A.9}$$

and so on. In general, bosons (whose kinetic terms have two derivatives) have mass dimension 1 and fermions (whose kinetic terms have one derivative) have mass dimension $\frac{3}{2}$.

You can always put the $\hbar$ and c factors back by dimensional analysis. For example, a cross section has units of area, which might be measured in picobarns (pb):[1]

$$1 \text{ picobarn} = 10^{-40} \text{ meters}^2. \tag{A.10}$$

A quantum field theory calculation might produce $\sigma = \frac{1}{m_P^2} \sim \frac{1}{\text{GeV}^2}$, where

$$1 \text{ gigaelectronvolt} = 1.602 \times 10^{-10} \text{ joules}. \tag{A.11}$$

So we need a combination of $\hbar$ and c that converts GeV^{-2} into area. The unique answer is $\hbar^2 c^2 = 9.996 \times 10^{-52} \text{ joules}^2 \cdot \text{meters}^2$. Thus,

$$\frac{1}{\text{GeV}^2} \hbar^2 c^2 = 3.894 \times 10^{-32} \text{ meters}^2 = 3.894 \times 10^8 \text{ picobarns}, \tag{A.12}$$

which is a useful conversion factor.

A.1.1 Factors of 2π

Keeping the factors of 2π straight is important. The origin of all the 2π's is the relation

$$\delta(x) = \int_{-\infty}^{\infty} dp \, e^{\pm 2\pi i p x}. \tag{A.13}$$

This identity holds with either sign; our sign convention for quantum fields is discussed below. To remove the 2π from the exponent, we can rescale either x or p. We rescale p. Then

$$\int_{-\infty}^{\infty} dp \, e^{\pm i p x} = 2\pi \, \delta(x). \tag{A.14}$$

Our convention for the Fourier transform is

$$f(x) = \int \frac{d^4 p}{(2\pi)^4} \tilde{f}(p) e^{-ipx} \quad \leftrightarrow \quad \tilde{f}(p) = \int d^4 x \, f(x) e^{ipx}. \tag{A.15}$$

In general, momentum space integrals will have $\frac{1}{2\pi}$ factors while position space integrals have no 2π factors. Thus, you should get used to writing $\frac{d^4 p}{(2\pi)^4}$ in momentum space integrals. Although physical quantities do not care about our 2π convention, the factors of 2π have important physical effects. Our Fourier transform convention is consistent with

$$p_\mu \leftrightarrow i\partial_\mu, \tag{A.16}$$

which has spatial components $\vec{p} \leftrightarrow -i\vec{\nabla}$, as in quantum mechanics.

---

[1] The origin of the term **barn** comes from the fact that inducing nuclear fission by hitting ^{235}U with neutrons is as easy as hitting the broad side of a barn. The inelastic neutron–^{235}U scattering cross section is around 1 barn $= 10^{-28} \text{m}^2$ at $E \sim 1 \text{ MeV}$.

A.2 Signs

Although the meat of most calculations is independent of the signs, physical results are very dependent on getting the sign right. Here we tabulate some of the signs in important equations.

First, we will never use curved-space backgrounds, so the metric $g_{\mu\nu}$ and the Minkowski metric $\eta^{\mu\nu}$ are interchangeable. The metric we use has sign convention

$$g^{\mu\nu} = \eta^{\mu\nu} = \begin{pmatrix} 1 & & & \\ & -1 & & \\ & & -1 & \\ & & & -1 \end{pmatrix}. \tag{A.17}$$

This convention makes $p^2 = p_0^2 - \vec{p}^2 = m^2 > 0$. The alternative, $g = \mathrm{diag}(-1, 1, 1, 1)$, makes $p^2 < 0$.

The signs of kinetic terms in Lagrangians are set so that the total energy is positive (see Sections 8.2 and 12.5). It is easiest to remember the signs by writing the Lagrangian as $\mathcal{L} = \mathcal{L}_{\mathrm{kin}} - V$, where V is the potential energy, which should be positive in a stable system. For example, for a scalar field, the mass term $\frac{1}{2}m^2\phi^2$ should give positive energy, so $V = \frac{1}{2}m^2\phi^2$ and $\mathcal{L} = -\frac{1}{2}m^2\phi^2$. The kinetic term sign can then be recalled from $p^2 \to -\Box = -\partial_\mu^2$ in Fourier space and $p^2 = m^2$ on-shell, so that the equations of motion should be $(\Box + m^2)\phi = 0$. Therefore, we have

$$\mathcal{L} = -\frac{1}{2}\phi(\Box + m^2)\phi = \frac{1}{2}(\partial_\mu\phi)(\partial^\mu\phi) - \frac{1}{2}m^2\phi^2. \tag{A.18}$$

The factor of $\frac{1}{2}$ makes the kinetic term contribute $(\Box + m^2)\phi$ to the equations of motion (instead of $2(\Box + m^2)\phi$). For a complex scalar, the Lagrangian is

$$\mathcal{L} = -\phi^\star(\Box + m^2)\phi = (\partial^\mu\phi^\star)(\partial_\mu\phi) - m^2\phi^\star\phi \tag{A.19}$$

without the $\frac{1}{2}$, since now variation with respect to $\phi^\star$ will give $(\Box + m^2)\phi$.

For gauge bosons, the Lagrangian is

$$\mathcal{L} = -\frac{1}{4}F_{\mu\nu}^2 = -\frac{1}{2}\partial_\mu A_\nu \partial_\mu A_\nu + \frac{1}{2}\partial_\mu A_\nu \partial_\nu A_\mu = \frac{1}{2}A_\nu \Box A_\nu - \frac{1}{2}A_\mu(\partial_\mu\partial_\nu)A_\nu, \tag{A.20}$$

where $F_{\mu\nu} = \partial_\mu A_\nu - \partial_\nu A_\mu$. In this equation and many others we employ the modern summation convention under which contracted indices can be raised or lowered without ambiguity: $x \cdot p = x^\mu p_\mu = x_\mu p^\mu = x_\mu p_\mu$. All of these contractions are equal to $g^{\mu\nu}x_\mu p_\nu = g_{\mu\nu}x^\mu p^\nu$. The sign and normalization of the $-\frac{1}{4}$ factor in Eq. (A.20) can be understood as follows. In Lorenz gauge $\partial_\mu A_\mu = 0$ the Lagrangian is just $\mathcal{L} = \frac{1}{2}A_\nu \Box A_\nu = \frac{1}{2}A_0\Box A_0 - \frac{1}{2}\vec{A}\Box\vec{A}$. This gives the three spatial components $\vec{A}$, which actually contain the propagating transverse degrees of freedom, the same kinetic terms as for scalars. (That the scalar component A_0 with the wrong sign is not problematic is explained in Section 8.2.)

Dirac fermions are normalized so that

$$\mathcal{L} = \bar{\psi}(i\slashed{\partial} - e\slashed{A} - m)\psi, \tag{A.21}$$

where $\not{\partial} = \gamma^\mu \partial_\mu$ and $\not{A} = \gamma^\mu A_\mu$. As in the scalar case, the $-m\bar\psi\psi$ is fixed so that the corresponding energy density is positive.

The covariant derivative in a non-Abelian gauge theory is

$$D_\mu = \partial_\mu - ig T_R^a A_\mu^a, \tag{A.22}$$

with T_R^a the generators in the appropriate representation. Normalization conventions for these generators are discussed in Section 25.1. We write tr for a sum over group generators or a sum over states, while Tr is used exclusively to denote a Dirac trace. For QED, $D_\mu = \partial_\mu - ieQA_\mu$, where e is the strength of the electromagnetic force ($e = 0.303$ in dimensionless units) and Q is a particle's electric charge (its U(1) quantum number). The electron is defined to have $Q = -1$, which leads to

$$D_\mu \psi_e = (\partial_\mu + ieA_\mu)\psi_e. \tag{A.23}$$

We use this simple form of the covariant derivative throughout Parts II and III.

The Feynman propagators in our conventions are

$$\langle 0|T\{\phi(x)\phi(y)\}|0\rangle = \int \frac{d^4p}{(2\pi)^4} e^{ip(x-y)} \frac{i}{p^2 - m^2 + i\epsilon} \tag{A.24}$$

for a real scalar and

$$\langle 0|T\{A_\mu(x)A_\nu(y)\}|0\rangle = \int \frac{d^4p}{(2\pi)^4} e^{ip(x-y)} \frac{-i(g^{\mu\nu} - (1-\xi)\frac{p^\mu p^\nu}{p^2})}{p^2 + i\epsilon} \tag{A.25}$$

for a massless spin-1 field in covariant gauges. The $-i$ in the photon propagator versus the $+i$ in the scalar propagator is the same sign difference as in $\mathcal{L} = -\frac{1}{2}\phi\Box\phi + \frac{1}{2}A_\nu\Box A_\nu$. The Dirac fermion propagator is

$$\langle 0|T\{\psi(x)\bar\psi(y)\}|0\rangle = \int \frac{d^4p}{(2\pi)^4} e^{-ip(x-y)} \frac{i}{\not{p} - m + i\varepsilon} = \int \frac{d^4p}{(2\pi)^4} e^{-ip(x-y)} \frac{i(\not{p} + m)}{p^2 - m^2 + i\epsilon}. \tag{A.26}$$

It is conventional to write $\psi(x)\bar\psi(y) = \psi(x)_\alpha \bar\psi(y)_\beta$ instead of $\bar\psi(x)\psi(y)$ so one is not tempted to mistake the spinors as being contracted. $\psi(x)\bar\psi(y)$ is a matrix in spinor space, just as $\vec{v}\vec{w}^T$ is a matrix.

When we expand fields in terms of creation and annihilation operators, we write for a single real scalar field

$$\phi(x) = \int \frac{d^3p}{(2\pi)^3} \frac{1}{\sqrt{2\omega_p}} \left[a_p(t)e^{i\vec{p}\vec{x}} + a_p^\dagger(t)e^{-i\vec{p}\vec{x}}\right], \tag{A.27}$$

where $\omega_p \equiv \sqrt{\vec{p}^2 + m^2}$. Including the free-field time dependence and generalizing to the complex case, this becomes

$$\phi(x) = \int \frac{d^3p}{(2\pi)^3} \frac{1}{\sqrt{2\omega_p}} \left(a_p e^{-ipx} + b_p^\dagger e^{ipx}\right), \tag{A.28}$$

$$\phi^\star(x) = \int \frac{d^3p}{(2\pi)^3} \frac{1}{\sqrt{2\omega_p}} \left(a_p^\dagger e^{ipx} + b_p e^{-ipx}\right). \tag{A.29}$$

Similarly, we take

$$\psi(x) = \sum_s \int \frac{d^3p}{(2\pi)^3} \frac{1}{\sqrt{2\omega_p}} \left(a_p^s u_p^s e^{-ipx} + b_p^{s\dagger} v_p^s e^{ipx}\right), \tag{A.30}$$

$$\bar{\psi}(x) = \sum_s \int \frac{d^3p}{(2\pi)^3} \frac{1}{\sqrt{2\omega_p}} \left(a_p^{s\dagger} \bar{u}_p^s e^{ipx} + b_p^s \bar{v}_p^s e^{-ipx}\right). \tag{A.31}$$

The sign of the phases follows from $a(t) = e^{-i\omega t} a(0)$ for annihilation operators by Heisenberg's equations of motion in any simple harmonic oscillator.

A.3 Feynman rules

The conventions for the Feynman rules follow from the sign conventions above. How the rules are derived is described in Chapter 7. The Feynman rules for various theories covered in the text are given in the appropriate chapter.

For scalar QED, the Feynman rules can be found in Section 9.2, for QED in Section 13.1, for QCD in Section 26.1, for the electroweak theory in Section 29.1, for background fields in Section 34.3.2 and for heavy-quark effective theory in Section 35.2. The notation for various symbols appearing in diagrams throughout the book is shown in Table A.1.

Table A.1 Symbols appearing in Feynman diagrams.

Symbol	Meaning	Symbol	Meaning
————	generic particle	——▶——	fermion
– – – – –	scalar	– – ▶ – –	charged scalar
∿∿∿∿	photon or Z boson	∘∘∘∘▶∘∘∘∘	ghost
⌒⌒⌒⌒⌒	gluon	∿∿▶∿∿	W boson
∿∿∿∿	graviton	══▶══	heavy quark
‖⊢————	background field	★	counterterm
⊗	operator or current	⬤	generic amplitude
(1PI)	all one-particle irreducible contributions	▨	alternative generic amplitude

A.4 Dirac algebra

The Dirac matrices satsify $\{\gamma^\mu, \gamma^\nu\} = 2g^{\mu\nu}$. We define

$$\gamma_5 \equiv i\gamma^0\gamma^1\gamma^2\gamma^3, \tag{A.32}$$

which leads to $\{\gamma^5, \gamma^\mu\} = 0$. We also define

$$\sigma^{\mu\nu} = \frac{i}{2}[\gamma^\mu, \gamma^\nu]. \tag{A.33}$$

Some useful identities are

$$g^{\mu\nu}g_{\mu\nu} = 4, \tag{A.34}$$

$$\gamma^\mu\gamma_\mu = 4, \tag{A.35}$$

$$\gamma^\mu\gamma^\nu\gamma_\mu = -2\gamma^\nu, \tag{A.36}$$

$$\gamma^\mu\gamma^\nu\gamma^\rho\gamma_\mu = 4g^{\nu\rho}, \tag{A.37}$$

$$\gamma^\mu\gamma^\nu\gamma^\rho\gamma^\sigma\gamma_\mu = -2\gamma^\sigma\gamma^\rho\gamma^\nu. \tag{A.38}$$

Some useful trace identities are

$$\mathrm{Tr}[\gamma_5] = \mathrm{Tr}[\gamma^\mu] = \mathrm{Tr}[\gamma^\mu\gamma^\alpha\gamma^\nu] = \mathrm{Tr}[\text{odd \# of } \gamma\text{-matrices}] = 0, \tag{A.39}$$

and

$$\mathrm{Tr}[\gamma^\mu\gamma^\nu] = 4g^{\mu\nu}, \tag{A.40}$$

$$\mathrm{Tr}[\gamma^\alpha\gamma^\mu\gamma^\beta\gamma^\nu] = 4(g^{\alpha\mu}g^{\beta\nu} - g^{\alpha\beta}g^{\mu\nu} + g^{\alpha\nu}g^{\mu\beta}), \tag{A.41}$$

$$\mathrm{Tr}[\gamma^\mu\gamma^\nu\gamma^\alpha\gamma^\beta\gamma^5] = -4i\varepsilon^{\mu\nu\alpha\beta}. \tag{A.42}$$

where $\varepsilon^{0123} = -\varepsilon_{0123} = 1$. The projectors are

$$P_L = \frac{1-\gamma_5}{2}, \quad P_R = \frac{1+\gamma_5}{2}, \tag{A.43}$$

so that left-handed fields satisfy $\gamma_5\psi_L = -\psi_L$ and right-handed fields satisfy $\gamma_5\psi_R = \psi_R$. A Dirac spinor in the $\left(\frac{1}{2}, 0\right) \oplus \left(0, \frac{1}{2}\right)$ representation is written with the left-handed spinor on top:

$$\psi = \begin{pmatrix} \psi_L \\ \psi_R \end{pmatrix}. \tag{A.44}$$

Spinor sums are, for particles,

$$\sum_{s=1}^{2} u_s(p)\bar{u}_s(p) = \not{p} + m \tag{A.45}$$

and for antiparticles,

$$\sum_{s=1}^{2} v_s(p)\bar{v}_s(p) = \not{p} - m. \tag{A.46}$$

Also,

$$\bar{u}_\sigma(p)\gamma^\mu u_{\sigma'}(p) = 2\delta_{\sigma\sigma'}p^\mu \tag{A.47}$$

is occasionally useful. Left- and right-handed photon polarizations (circularly polarized light) are

$$\epsilon_L^\mu = \frac{1}{\sqrt{2}}(0,1,-i,0), \quad \epsilon_R^\mu = \frac{1}{\sqrt{2}}(0,1,i,0). \tag{A.48}$$

These polarization vectors are consistent with Eq. (A.43) and the representations of the Lorentz group discussed in Chapter 17.

Some other useful identities are

$$\slashed{D}^2 = D_\mu^2 + \frac{e}{2}F_{\mu\nu}\sigma^{\mu\nu} \tag{A.49}$$

and

$$(\sigma_{\mu\nu}F^{\mu\nu})^2 = 2F_{\mu\nu}^2 + 2i\gamma_5 F_{\mu\nu}\tilde{F}_{\mu\nu}, \tag{A.50}$$

where

$$\tilde{F}^{\mu\nu} \equiv \frac{1}{2}\varepsilon^{\mu\nu\alpha\beta}F_{\alpha\beta}. \tag{A.51}$$

Problems

A.1 Dimensional analysis.

    (a) A photon coupled to a complex scalar field in d dimensions has action

$$S = \int d^d x \left[-\frac{1}{4}F_{\mu\nu}^2 - \phi^\star \Box \phi + gA_\mu \phi^\star \partial_\mu \phi + \lambda\phi^3 + \cdots \right], \tag{A.52}$$

    where $F_{\mu\nu} = (\partial_\mu A_\nu - \partial_\nu A_\mu)$ and $\Box = \partial^\mu \partial_\mu$ as always, but now $\mu = 0, 1, \cdots, d-1$. What are the mass dimensions of A_μ, ϕ, g and λ (as functions of d)?

    (b) An interaction is said to be *renormalizable* if its coupling constant is dimensionless. In what dimension d is the electromagnetic interaction renormalizable? How about the ϕ^3 interaction?

Appendix B **Regularization**

B.1 Integration parameters

To evaluate loop integrals in quantum field theory, it is often helpful to introduce Feynman or Schwinger parameters.

B.1.1 Feynman parameters

Feynman parameters are based on a number of easily verifiable mathematical identities. The simplest is

$$\frac{1}{AB} = \int_0^1 dx \frac{1}{[A + (B - A)x]^2} = \int_0^1 dx\, dy\, \delta(x + y - 1) \frac{1}{[xA + yB]^2}. \tag{B.1}$$

Other useful identities are

$$\frac{1}{AB^n} = \int_0^1 dx\, dy\, \delta(x + y - 1) \frac{ny^{n-1}}{[xA + yB]^{n+1}}, \tag{B.2}$$

$$\frac{1}{ABC} = \int_0^1 dx\, dy\, dz\, \delta(x + y + z - 1) \frac{2}{[xA + yB + zC]^3}. \tag{B.3}$$

These are useful because they let us complete the square in the denominator. For example,

$$\int \frac{d^4k}{(2\pi)^4} \frac{1}{k^2} \frac{1}{(k - p)^2} = \int \frac{d^4k}{(2\pi)^4} \int_0^1 dx \frac{1}{[k^2 + x((k - p)^2 - k^2)]^2}$$

$$= \int_0^1 dx \int \frac{d^4k}{(2\pi)^4} \frac{1}{[(k - xp)^2 - \Delta]^2}, \tag{B.4}$$

where $\Delta = -p^2 x(1 - x)$. Then we can shift $k \to k + xp$ leaving an integral that only depends on k^2.

B.1.2 Schwinger parameters

Another useful set of integration parameters are called Schwinger parameters. They are based on the following mathematical identities, which hold when $\text{Im}(A) > 0$:

$$\frac{i}{A} = \int_0^\infty ds\, e^{isA}, \tag{B.5}$$

$$\left[\frac{i}{A}\right]^2 = \int_0^\infty s\, ds\, e^{isA}. \tag{B.6}$$

You can derive further identities by taking additional derivatives with respect to A. Also, Eq. (B.5) implies

$$\frac{1}{AB} = -\int_0^\infty ds \int_0^\infty dt\, e^{isA+itB} \tag{B.7}$$

when $\text{Im}(A) > 0$ and $\text{Im}(B) > 0$ (i.e. with Feynman propagators). These **Schwinger parameters** s and t have a nice physical interpretation: s and t are the proper times of the particles as they travel along their paths in the Feynman graph. This Schwinger proper-time interpretation is discussed in Chapter 32.

Note that writing $s+t = \tau$ and $x = \frac{t}{s+t}$, or $t = x\tau$ and $s = (1-x)\tau$, Eq.(B.7) becomes

$$\frac{1}{AB} = -\int_0^\infty \tau\, d\tau \int_0^1 dx\, e^{i\tau(A+(B-A)x)}$$

$$= \int_0^1 dx \frac{1}{[A+(B-A)x]^2}. \tag{B.8}$$

So the Feynman parameter x also has an interpretation, as the relative proper time $\frac{s}{s+t}$ of the two particles in the loop.

Other useful related identities are

$$\frac{1}{A^n B^m} = \frac{\Gamma(n+m)}{\Gamma(n)\,\Gamma(m)} \int_0^\infty ds \frac{s^{m-1}}{(A+Bs)^{n+m}}, \tag{B.9}$$

$$\frac{1}{AB} = \int_0^\infty ds \frac{1}{(A+Bs)^2}. \tag{B.10}$$

Schwinger parameters are used in Chapters 34 and 35.

B.2 Wick rotations

After introducing Feynman parameters and completing the square, one is often left with an integral over a loop momentum k^μ in Minkowski space. Once the $i\varepsilon$ factors are put in for Feynman propagators, 1-loop integrals often appear as

$$\int \frac{d^4 k}{(2\pi)^4} \frac{1}{(k^2 - \Delta + i\varepsilon)^n}. \tag{B.11}$$

Assuming $\Delta > 0$ (you can check that Wick rotation still works for $\Delta < 0$ in Problem B.1), this integral has poles at $k_0 = \sqrt{\vec{k}^2 + \Delta} - i\varepsilon$ and $k_0 = -\sqrt{\vec{k}^2 + \Delta} + i\varepsilon$, as shown in Figure B.1. Since the poles are in the top-left and bottom-right quadrants of the k_0 complex plane, the integral over the figure-eight contour shown vanishes. Thus, the integrals over the real axis and the imaginary axis are equal and opposite. Therefore, we can substitute $k_0 \to ik_0$ so that $k^2 \to -k_0^2 - \vec{k}^2 = -k_E^2$, where $k_E^2 = k_0^2 + \vec{k}^2$ is the Euclidean momentum. This

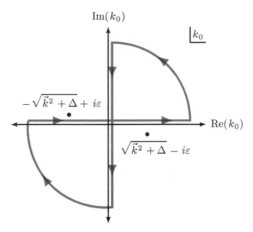

Fig. B.1 Wick rotations. Poles in integrations over Feynman propagators often have poles at at $k_0 = \pm\sqrt{\vec{k}^2 + \Delta} \mp i\varepsilon$. Integrating over the real axis is then equivalent to integrating over the imaginary axis.

is known as a **Wick rotation**. After the Wick rotation, the $i\varepsilon$ will no longer play a role and we can just set $\varepsilon = 0$.

Once Wick-rotated, the integrals are evaluated in a straightforward way. We will need the formula for the surface area of the Euclidean 4-sphere: $\int d\Omega_4 = 2\pi^2$. Using this, we find

$$\int \frac{d^4 k_E}{(2\pi)^4} f(k_E^2) = \frac{1}{16\pi^4} \int d\Omega_4 \int_0^\infty k_E^3 \, dk_E \, f(k_E^2) = \frac{1}{8\pi^2} \int_0^\infty k_E^3 \, dk_E \, f(k_E^2).$$
(B.12)

Then, for example, Eq. (B.11) with $n = 3$ is evaluated as

$$\int \frac{d^4 k}{(2\pi)^4} \frac{1}{(k^2 - \Delta + i\varepsilon)^3} = i \int \frac{d^4 k_E}{(2\pi)^4} \frac{1}{(-k_E^2 - \Delta)^3}$$

$$= (-1)^3 \frac{i}{8\pi^2} \int_0^\infty dk_E \frac{k_E^3}{(k_E^2 + \Delta)^3}$$

$$= \frac{-i}{32\pi^2 \Delta}.$$
(B.13)

Other useful formulas following from Wick rotations are

$$\int \frac{d^4 k}{(2\pi)^4} \frac{k^2}{(k^2 - \Delta + i\varepsilon)^4} = \frac{-i}{48\pi^2} \frac{1}{\Delta},$$
(B.14)

$$\int \frac{d^4 k}{(2\pi)^4} \frac{1}{(k^2 - \Delta + i\varepsilon)^r} = i \frac{(-1)^r}{(4\pi)^2} \frac{1}{(r-1)(r-2)} \frac{1}{\Delta^{(r-2)}}, \quad r > 2,$$
(B.15)

$$\int \frac{d^4 k}{(2\pi)^4} \frac{k^2}{(k^2 - \Delta + i\varepsilon)^r} = i \frac{(-1)^{r-1}}{(4\pi)^2} \frac{2}{(r-1)(r-2)(r-3)} \frac{1}{\Delta^{(r-3)}}, \quad r > 3, \quad (B.16)$$

and so on.

Keep in mind that the Wick rotation is just a trick for evaluating integrals. There is nothing physical about it. In addition, note that the Wick rotation can only be justified if there are no new poles that invalidate the contour rotation. This caveat is only relevant for 2-loop and higher integrals, which we will not encounter.

B.3 Dimensional regularization

The most important regularization scheme for modern applications is dimensional regularization ['t Hooft and Veltman, 1972]. The key observation is that an integral such as

$$\int \frac{d^d k}{(2\pi)^d} \frac{1}{(k^2 - \Delta + i\varepsilon)^2} \tag{B.17}$$

is divergent only if $d \geq 4$. If $d < 4$, then it will converge. If it is convergent we can Wick rotate, and the answer comes from analytically continuing all our formulas above to d dimensions.

B.3.1 Spinor algebra

In d dimensions, the metric is

$$g^{\mu\nu} = \text{diag}(1, -1, -1, \cdots, -1), \tag{B.18}$$

which means that there is exactly one timelike dimension in even non-integer d. This metric satisfies

$$g^{\mu\nu} g_{\mu\nu} = d. \tag{B.19}$$

The Lorentz-invariant phase space is

$$d\Pi_{\text{LIPS}} \equiv (2\pi)^d \prod_{\text{final states } j} \frac{d^{d-1} p_j}{(2\pi)^{d-1}} \frac{1}{2E_{p_j}} \delta^d(\Sigma p). \tag{B.20}$$

We can define spinor algebra to work the same way in $d = 4 - \varepsilon$ dimensions as in $d = 4$. More precisely, we assume there are d four-dimensional γ-matrices satisfying $\{\gamma^\mu, \gamma^\nu\} = 2g^{\mu\nu}$. The identity matrix in spinor space satisfies $\text{Tr}\mathbb{1}_{\alpha\beta} = 4$ as in four dimensions. In theories that involve γ_5 we also assume such a matrix exists satisfying

$$\{\gamma_5, \gamma_\mu\} = 0. \tag{B.21}$$

Theories with anomalies are the only places in which there can be subtleties with such a definition (see Chapter 30). An excellent discussion of spinors in various dimensions can be found in [Polchinski, 1998, Appendix B].

B.3.2 Scalar integrals

We will manipulate the expressions so that they are only functions of the magnitude of k. Then we will use

$$\int d^d k = \int d\Omega_d \int k^{d-1} dk, \qquad (B.22)$$

where $d\Omega_d$ denotes the differential solid angle of the d-dimensional unit sphere. Explicitly,

$$d\Omega_d = \sin^{d-2}(\phi_{d-1}) \sin^{d-3}(\phi_{d-2}) \cdots \sin(\phi_2) \, d\phi_1 \cdots d\phi_{d-1}, \qquad (B.23)$$

where ϕ_i is the angle to the i th axis, with $0 \le \phi_1 < 2\pi$ and $0 \le \phi_i < \pi$ for $i > 1$. For example, $d\Omega_2 = d\phi$. For $d = 3$, we normally write $\phi_1 = \phi$ and $\phi_2 = \theta$ giving

$$d\Omega_3 = d\cos\theta \, d\phi, \qquad (B.24)$$

which is the usual volume element of a two-dimensional surface. Remember, d is the dimension of the solid volume, not the surface, which has dimension $d - 1$. The $(d - 1)$-dimensional surface areas of a ball of radius 1 in integer dimensions are

$$\Omega_2 = \int d\Omega_2 = 2\pi \text{ (circle)}, \int d\Omega_3 = 4\pi \text{ (sphere)}, \int d\Omega_4 = 2\pi^2 \text{ (three-sphere)}, \cdots,$$
$$(B.25)$$

The equivalent volumes are

$$V_d = \Omega_d \int_0^R dr \, r^{d-1} = \Omega_d \frac{1}{d} R^d, \qquad (B.26)$$

which are $V_2 = \pi R^2$, $V_3 = \frac{4}{3}\pi R^3$, $V_4 = \frac{1}{2}\pi^2 R^4$, etc.

For non-integer dimensions, the surface area formula can be derived using the same trick used for Gaussian integrals in Section 14.2.1:

$$(\sqrt{\pi})^d = \left(\int_{-\infty}^{\infty} dx \, e^{-x^2} \right)^d = \int d\Omega_d \int_0^{\infty} dr \, r^{d-1} e^{-r^2} = \frac{1}{2} \Gamma\left(\frac{d}{2}\right) \int d\Omega_d, \quad (B.27)$$

so that

$$\Omega_d = \int d\Omega_d = \frac{2\pi^{d/2}}{\Gamma\left(\frac{d}{2}\right)}. \qquad (B.28)$$

Alternatively, one can just integrate Eq. (B.23):

$$\Omega_d = 2\pi \prod_{n=2}^{d-1} \left(\int_0^{\pi} d\phi_n \sin^{n-1}\phi_n \right) = 2\pi \prod_{n=2}^{d-1} \sqrt{\pi} \left(\frac{\Gamma\left(\frac{n}{2}\right)}{\Gamma\left(\frac{n+1}{2}\right)} \right)$$

$$= 2\pi^{d/2} \frac{\Gamma\left(\frac{2}{2}\right) \Gamma\left(\frac{3}{2}\right)}{\Gamma\left(\frac{3}{2}\right) \Gamma\left(\frac{4}{2}\right)} \cdots \frac{\Gamma\left(\frac{d-1}{2}\right)}{\Gamma\left(\frac{d}{2}\right)} = 2\pi^{d/2} \frac{\Gamma(1)}{\Gamma\left(\frac{d}{2}\right)}. \qquad (B.29)$$

Using $\Gamma(1) = 1$, this reproduces Eq. (B.28).

In these expressions, $\Gamma(x)$ is the **Gamma function**, which is the analytic continuation of the factorial. For integer arguments, it evaluates to

$$\Gamma(1) = 1, \quad \Gamma(2) = 1, \quad \Gamma(3) = 2, \quad \Gamma(x) = (x - 1)! \qquad (B.30)$$

$\Gamma(z)$ has simple poles at 0 and all the negative integers. We will often need to expand $\Gamma(x)$ around the pole at $x = 0$:

$$\Gamma(\epsilon) = \frac{1}{\varepsilon} - \gamma_E + \mathcal{O}(\varepsilon), \tag{B.31}$$

where γ_E is the Euler–Mascheroni constant, $\gamma_E \approx 0.577$. Sometimes relations such as

$$\sin(\pi x) = \frac{\pi(1-x)}{\Gamma(x)\Gamma(2-x)}, \quad \cos(\pi x) = \left(\frac{1-2x}{2x}\right) \frac{\Gamma(1-x)\Gamma(1+x)}{\Gamma(2-2x)\Gamma(2x)}, \tag{B.32}$$

or the Euler β-function

$$\beta(a, b) = \frac{\Gamma(a)\Gamma(b)}{\Gamma(a+b)} = \int_0^1 dx (1-x)^{a-1} x^{b-1} \tag{B.33}$$

allow us to simplify expressions.

The integrals over Euclidean k_E are straightforward:

$$\int dk_E \frac{k_E^a}{(k_E^2 + \Delta)^b} = \Delta^{\frac{a+1}{2}-b} \frac{\Gamma\left(\frac{a+1}{2}\right)\Gamma\left(b - \frac{a+1}{2}\right)}{2\Gamma(b)}. \tag{B.34}$$

Equations (B.22), (B.28) and (B.34) can be combined into a general formula:

$$\int \frac{d^d k}{(2\pi)^d} \frac{k^{2a}}{(k^2 - \Delta)^b} = i(-1)^{a-b} \frac{1}{(4\pi)^{d/2}} \frac{1}{\Delta^{b-a-\frac{d}{2}}} \frac{\Gamma\left(a + \frac{d}{2}\right)\Gamma\left(b - a - \frac{d}{2}\right)}{\Gamma(b)\Gamma\left(\frac{d}{2}\right)}. \tag{B.35}$$

Special cases used in the text are

$$\int \frac{d^d k}{(2\pi)^d} \frac{1}{(k^2 - \Delta + i\varepsilon)^2} = \frac{i}{(4\pi)^{d/2}} \frac{1}{\Delta^{2-\frac{d}{2}}} \Gamma\left(\frac{4-d}{2}\right), \tag{B.36}$$

$$\int \frac{d^d k}{(2\pi)^d} \frac{k^2}{(k^2 - \Delta + i\varepsilon)^2} = -\frac{d}{2} \frac{i}{(4\pi)^{d/2}} \frac{1}{\Delta^{1-\frac{d}{2}}} \Gamma\left(\frac{2-d}{2}\right), \tag{B.37}$$

$$\int \frac{d^d k}{(2\pi)^d} \frac{k^2}{(k^2 - \Delta + i\varepsilon)^3} = \frac{d}{4} \frac{i}{(4\pi)^{d/2}} \frac{1}{\Delta^{2-\frac{d}{2}}} \Gamma\left(\frac{4-d}{2}\right), \tag{B.38}$$

$$\int \frac{d^d k}{(2\pi)^d} \frac{1}{(k^2 - \Delta + i\varepsilon)^3} = \frac{-i}{2(4\pi)^{d/2}} \frac{1}{\Delta^{3-\frac{d}{2}}} \Gamma\left(\frac{6-d}{2}\right). \tag{B.39}$$

This last integral is convergent in $d = 4$; however, the d-dimensional form is important for loops with IR divergences (see Chapter 20).

All dimensionally regulated versions of divergent integrals will have poles at $d = 4$. Therefore, we often expand $d = 4 - \varepsilon$ and drop terms of order ε. Another common convention is $d = 4 - 2\varepsilon$. If you are ever off by a factor of 2 in comparing to someone else's result, check the convention!

B.3.3 Field dimensions

Next, we should calculate the dimensions of all the fields and couplings in the Lagrangian. For the action to be dimensionless, the Lagrangian density should have mass dimension d. For example, in QED, the Lagrangian is

$$\mathcal{L}_{\text{QED}} = -\frac{1}{4}(\partial_\mu A_\nu - \partial_\nu A_\mu)^2 + \bar{\psi}(i\gamma^\mu \partial_\mu - m)\psi - e\bar{\psi}\gamma^\mu\psi A_\mu, \qquad \text{(B.40)}$$

which implies the mass dimensions

$$[A_\nu] = \frac{d-2}{2}, \quad [\psi] = \frac{d-1}{2}, \quad [m] = 1, \qquad \text{(B.41)}$$

and also $[e] = \frac{4-d}{2}$. However, rather than have a non-integer dimensional coupling, it is conventional to take

$$e \to \mu^{\frac{4-d}{2}} e, \qquad \text{(B.42)}$$

where μ is an arbitrary parameter of mass dimension 1. Then e remains dimensionless.

One usually only makes this change for the factors of e (or other gauge couplings) directly participating in a loop. If a loop graph is not one-particle irreducible, there may be other factors of e for which it is often simpler to leave four-dimensional. This is just a convention. If all factors of e are modified as in Eq. (B.42), the answer will still be correct, but may contain awkward logarithms of dimensionful scales when expanded around $d = 4$. These awkward logarithms drop out of physical quantities, of course, but they can be avoided at intermediate steps as well by only adding factors of μ to coupling constants participating in the loop.

The factors of μ coming from Eq. (B.42) modify loop integrals as

$$\int \frac{d^4 k}{(2\pi)^4} \frac{e^2}{(k^2 - \Delta + i\varepsilon)^2} \to \mu^{4-d} \int \frac{d^d k}{(2\pi)^d} \frac{e^2}{(k^2 - \Delta + i\varepsilon)^2}. \qquad \text{(B.43)}$$

Keep in mind that μ is *not* a large scale. It is *not* a UV cutoff. The dimensional regularization is removed when $d \to 4$, not when $\mu \to \infty$. Thus, μ is not like the Pauli–Villars mass M or a generic UV scale Λ. In fact, we will often use μ as a proxy for a physical infrared scale associated with a renormalization group point. Nevertheless, there are two unphysical parameters in dimensional regularization, ε and μ; both must drop out of physical predictions.

Including this factor of μ, the logarithmically divergent integral becomes

$$\int \frac{d^4 k}{(2\pi)^4} \frac{e^2}{(k^2 - \Delta + i\varepsilon)^2} \to \mu^{4-d} \frac{ie^2}{(4\pi)^{d/2}} \Gamma\left(\frac{4-d}{2}\right)\left(\frac{1}{\Delta}\right)^{2-\frac{d}{2}}. \qquad \text{(B.44)}$$

Now letting $d = 4 - \varepsilon$ we expand this around $\varepsilon = 0$ and get

$$\mu^{4-d} \frac{ie^2}{(4\pi)^{d/2}} \Gamma\left(\frac{4-d}{2}\right)\left(\frac{1}{\Delta}\right)^{2-\frac{d}{2}} = \frac{ie^2}{16\pi^2}\left[\frac{2}{\varepsilon} + (-\gamma_E + \ln 4\pi + \ln \mu^2 - \ln \Delta) + \mathcal{O}(\varepsilon)\right]$$

$$= \frac{ie^2}{16\pi^2}\left[\frac{2}{\varepsilon} + \ln\frac{4\pi e^{-\gamma_E}\mu^2}{\Delta} + \mathcal{O}(\varepsilon)\right]. \qquad \text{(B.45)}$$

The γ_E comes from the integral $\int \frac{d^d k}{k^4}$, the 4π comes from the phase space $\frac{1}{(4\pi)^{d/2}}$ and the μ comes from the μ^{4-d}. This combination, $4\pi e^{-\gamma_E}\mu^2$, shows up frequently, so we give it a symbol

$$\tilde{\mu}^2 \equiv 4\pi e^{-\gamma_E}\mu^2 \qquad \text{(B.46)}$$

leading to

$$\int \frac{d^4 k}{(2\pi)^4} \frac{e^2}{(k^2 - \Delta + i\varepsilon)^2} \to \frac{ie^2}{16\pi^2}\left[\frac{2}{\varepsilon} + \ln\frac{\tilde{\mu}^2}{\Delta} + \mathcal{O}(\varepsilon)\right]. \qquad \text{(B.47)}$$

Sometimes we will omit the tilde and just write μ for $\tilde{\mu}$. Note that there is still a divergence in this expression as $\varepsilon \to 0$.

Dimensional regularization characterizes the degree to which integrals diverge at high energy through analytic properties of regulated results, rather than through powers of a cutoff scale. For example, the integral $\int \frac{d^4 k}{(k^2 - \Delta)^2}$ is logarithmically divergent. In d dimensions, the equivalent integral $\int \frac{d^d k}{(k^2 - \Delta)^2} \sim \Gamma(\frac{4-d}{2})$ has a simple pole at $d = 4$, and no other poles for $d < 4$. A quadratically divergent integral, such as $\int \frac{d^4 k}{k^2 - \Delta}$, becomes $\int \frac{d^d k}{k^2 - \Delta} \sim \Gamma(\frac{2-d}{2})$ in d dimensions. Expanding this result around $d = 4$ gives a $\frac{1}{\varepsilon}$ pole as did the expansion of the logarithmically divergent integral. However, this does not mean that power divergences are absent with dimensional regularization. Rather they are hidden, as poles in integer $d < 4$. For example, the quadratic divergence translates to a pole in $\Gamma(\frac{2-d}{2})$ at $d = 2$. Thus, dimensional regularization translates the degree of divergence into the singularity structure of amplitudes in d dimensions.

Dimensional regularization can also be used to regulate IR-divergent integrals. For example, $\int d^d k \frac{1}{(k^2 - m^2)k^4}$ is IR divergent for $d < 4$. We can evaluate this integral in $d = 4 - \varepsilon$ dimensions with $\varepsilon < 0$ instead of $\varepsilon > 0$. A nice feature of dimensional regularization as an IR regulator is that it can be used for both virtual graphs and phase space integrals.

Occasionally when using dimensional regularization we encounter an integral that is both UV and IR divergent; for example, the scaleless integral $\int \frac{d^d k}{k^4}$. This integral is not convergent for any d. Nevertheless, it is useful to be able to do such integrals. To progress, we can introduce an arbitrary scale Λ to divide the UV and IR regions of Euclidean momenta:

$$
\int \frac{d^d k_E}{k_E^4} = \Omega_d \int_0^\Lambda dk_E k_E^{d-5} + \Omega_d \int_\Lambda^\infty dk_E \, k_E^{d-5}
$$
$$
= \Omega_d \left(\ln \Lambda - \frac{1}{\varepsilon_{\text{IR}}} \right) + \Omega_d \left(\frac{1}{\varepsilon_{\text{UV}}} - \ln \Lambda \right), \tag{B.48}
$$

where we have written $d = 4 - \varepsilon_{\text{IR}}$ for the first integral, assuming $\varepsilon_{\text{IR}} < 0$, and $d = 4 - \varepsilon_{\text{UV}}$ for the second integral, assuming $\varepsilon_{\text{UV}} > 0$. Rather than doing this split for every scaleless integral, since we know ε_{IR} and ε_{UV} must vanish from physical quantities, we often just set $\varepsilon_{\text{IR}} = \varepsilon_{\text{UV}} = \varepsilon$. When this is done, the integral is just 0. A simpler justification is that since there is no available quantity with non-zero mass dimension, scaleless integrals such as $\int \frac{d^4 k}{k^4}$ must vanish in d dimensions.

Often we are interested in just the UV divergence of an integral, which can be extracted from a scaleless integral as

$$
\left[\int \frac{d^d k}{(2\pi)^d} \frac{1}{k^4} \right]_{\text{UV-div}} = i \frac{\Omega_d}{(2\pi)^d} \frac{1}{\varepsilon_{\text{UV}}} = i \frac{2}{(2\pi)^d} \frac{\pi^{d/2}}{\Gamma(d/2)} \frac{1}{\varepsilon_{\text{UV}}} = \frac{i}{8\pi^2} \frac{1}{\varepsilon_{\text{UV}}}. \tag{B.49}
$$

This is a very useful shortcut to extracting the UV divergence.

B.3.4 k^μ integrals

We will often have integrals with factors of momenta, such as $k^\mu k^\nu$, in the numerator:

$$F^{\mu\nu}(\Delta) = \int \frac{d^4k}{(2\pi)^4} \frac{k^\mu k^\nu}{(k^2 - \Delta)^n}. \tag{B.50}$$

These can be simplified using a trick. Since the integral is a tensor under Lorentz transformations but only depends on the scalar Δ, it must be proportional to the only tensor around, $g^{\mu\nu}$. Then, just by dimensional analysis, we must get the same thing as in an integral with $k^\mu k^\nu$ replaced by $ck^2 g^{\mu\nu}$ for some number c. Contracting with $g^{\mu\nu}$, we see that $c = \frac{1}{4}$ or more generally $c = \frac{1}{d}$. Therefore,

$$\int \frac{d^dk}{(2\pi)^d} \frac{k^\mu k^\nu}{(k^2 - \Delta)^n} = \frac{1}{d} g^{\mu\nu} \int \frac{d^dk}{(2\pi)^d} \frac{k^2}{(k^2 - \Delta)^n}. \tag{B.51}$$

If there is just one factor of k^μ in the numerator, for example in

$$F(p^2) = \int \frac{d^4k}{(2\pi)^4} \frac{k \cdot p}{(k^2 - p^2)^4}, \tag{B.52}$$

then the integrand is antisymmetric under $k \to -k$. Since we are integrating over all k, the integral must vanish. So we will only need to keep terms with even powers of k in the numerator.

B.4 Other regularization schemes

While dimensional regularization has a number of important advantages (it respects gauge invariance, it can regulate IR or UV divergences, no new fields are needed, etc.), it has the disadvantage of being unphysical. That is, one cannot think of analytical continuation into $4 - \varepsilon$ dimensions as representing some sort of short-distance deformation. A number of regulators that do have short-distance interpretations, such as the hard cutoff regulator or heat-kernel regulator, are discussed in Chapter 15 in the context of the Casimir effect. Those regulators are unfortunately not useful for general field theory calculations. Here we discuss two regulation schemes that do have widespread applicability, the derivative method and Pauli–Villars regularization, and briefly mention a few more.

B.4.1 Derivative method

A quick way to extract the UV divergence of an integral is by taking derivatives. Consider a logarithmically divergent integral, such as

$$\mathcal{I}(\Delta) = \int \frac{d^4k}{(2\pi)^4} \frac{1}{(k^2 - \Delta + i\varepsilon)^2} = \infty. \tag{B.53}$$

If we take the derivative, the integral can be done:

$$\frac{d}{d\Delta}\mathcal{I}(\Delta) = \int \frac{d^4k}{(2\pi)^4} \frac{2}{(k^2 - \Delta + i\varepsilon)^3} = -\frac{i}{16\pi^2\Delta}. \tag{B.54}$$

So,

$$\mathcal{I}(\Delta) = -\frac{i}{16\pi^2} \ln\frac{\Delta}{\Lambda^2}, \tag{B.55}$$

where Λ is an integration constant representing the UV cutoff and is formally infinite. Similarly, for a quadratically divergent integral, one could take the second derivative and then integrate twice to give

$$\int \frac{d^4k}{(2\pi)^4} \frac{k^2}{(k^2 - \Delta + i\varepsilon)^2} = 6 \int d\Delta \int d\Delta \left(\frac{-i}{48\pi^2} \frac{1}{\Delta}\right) = -\frac{i}{8\pi^2}\left(\Delta \ln\frac{\Delta}{\Lambda_1^2} + \Lambda_2^2\right) \tag{B.56}$$

for two integration constants Λ_1 and Λ_2.

The derivative method is not an ideal regulator. Since the cutoff Λ appears as a constant of integration, there is no way to relate Λ from one integral to Λ from another. In particular, cancellations that we expect due to constraints such as gauge invariance are not guaranteed to hold. Nevertheless, the derivative method is a quick way to check the coefficient of the logarithms appearing in any particular integral.

B.4.2 Pauli–Villars regularization

Pauli–Villars regularization requires that for each particle of mass m a new unphysical **ghost** particle of mass Λ be added with either the wrong statistics or the wrong-sign kinetic term. These new particles are designed to cancel exactly loop amplitudes with physical particles at asymptotically large loop momentum. For example, one can write down a Pauli–Villars Lagrangian for QED, which works at the 1-loop level, as

$$\mathcal{L}_{\text{PV}} = -\frac{1}{4}F_{\mu\nu}^2 + \bar{\psi}(i\slashed{\partial} - e\slashed{A} - e\slashed{\tilde{A}} - m)\psi + \frac{1}{4}\widetilde{F}_{\mu\nu}^2 - \frac{1}{2}\Lambda^2\widetilde{A}_\mu^2 + \bar{\widetilde{\psi}}(i\slashed{\partial} - e\slashed{A} - e\slashed{\tilde{A}} - \Lambda)\widetilde{\psi}, \tag{B.57}$$

with $\widetilde{A}_\mu$ the ghost photon and $\widetilde{\psi}$ the ghost electron and $\widetilde{F}_{\mu\nu} = \partial_\mu\widetilde{A}_\nu - \partial_\nu\widetilde{A}_\mu$. We assume that both the ghost photon and ghost electron have bosonic statistics; the ghost photon has a wrong-sign kinetic term.

For example, $\mathcal{L}_{\text{PV}}$ leads to a Feynman-gauge ghost-photon propagator of the form

$$\langle 0|T\{\widetilde{A}_\mu(x)\widetilde{A}_\nu(y)\}|0\rangle = \int \frac{d^4p}{(2\pi)^4} e^{ip(x-y)} \frac{ig^{\mu\nu}}{p^2 - \Lambda^2 + i\varepsilon}. \tag{B.58}$$

Since this has the opposite sign from the photon propagator, it will cancel the photon's contribution, for example, to the electron self-energy loop for loop momenta $k^\mu \gg \Lambda$ (see Chapter 18). The sign of the residue of the propagator is normally dictated by unitarity – a particle whose propagator has the sign in Eq.(B.58) has negative norm, and would generate probabilities greater than 1. So, $\widetilde{A}_\mu$ cannot create or destroy physical on-shell particles. Thus, fields such as $\widetilde{A}_\mu$ are said to be associated with **Pauli–Villars ghosts**. The ghost electron propagator is the same as the regular electron propagator; however, ghost electron loops do not get a factor of -1 (since they are bosonic) and therefore cancel regular electron loops when $k^\mu \gg \Lambda$.

In more detail, an amplitude with Pauli–Villars regularization will sum over the real particle, with mass m, and the ghost particle, with fixed large mass $\Lambda \gg m$:

$$\int \frac{d^4k}{(2\pi)^4} \frac{1}{(k^2 - m^2 + i\varepsilon)^2} \to \int \frac{d^4k}{(2\pi)^4} \left[\frac{1}{(k^2 - m^2 + i\varepsilon)^2} - \frac{1}{(k^2 - \Lambda^2 + i\varepsilon)^2} \right].$$
$$(B.59)$$

For $k \gg \Lambda, m$ both terms in the new integrand scale as $\frac{1}{k^4}$ and so the integrand vanishes at least as $\frac{1}{k^6}$ making the integral convergent. We can now perform this integral by Wick rotation

$$\int \frac{d^4k}{(2\pi)^4} \left[\frac{1}{(k^2 - m^2 + i\varepsilon)^2} - \frac{1}{(k^2 - \Lambda^2 + i\varepsilon)^2} \right]$$
$$= \frac{i}{8\pi^2} (-1)^2 \int_0^\infty dk_E \left[\frac{k_E^3}{(k_E^2 + m^2)^2} - \frac{k_E^3}{(k_E^2 + \Lambda^2)^2} \right]$$
$$= -\frac{i}{16\pi^2} \ln \frac{m^2}{\Lambda^2} \qquad (B.60)$$

so that

$$\int \frac{d^4k}{(2\pi)^4} \frac{1}{(k^2 - m^2 + i\varepsilon)^2} \to \frac{i}{16\pi^2} \ln \frac{\Lambda^2}{m^2}. \qquad (B.61)$$

Note that the coefficient of the logarithm is consistent with what we found using the derivative method, in Eq. (B.55) and with dimensional regularization in Eq. (B.47).

When using Pauli–Villars regularization, the identity

$$\frac{1}{k^2 - m^2} - \frac{1}{k^2 - \Lambda^2} = \int_{m^2}^{\Lambda^2} \frac{-1}{(k^2 - \Xi)^2} d\Xi \qquad (B.62)$$

is often useful. It allows us to evaluate divergent integrals by squaring the propagator and adding an integration parameter Ξ. In fact, due to the identity

$$\int dm^2 \frac{d}{dm^2} \left[\frac{1}{k^2 - m^2} \right] = \int dm^2 \frac{1}{(k^2 - m^2)^2}, \qquad (B.63)$$

Pauli–Villars can be viewed as a systematic implementation of the derivative method.

Pauli–Villars was historically important and serves a useful pedagogical function. Indeed, the introduction of Pauli–Villars ghosts is much more clearly a deformation in the UV, relevant at energy scales of order the Pauli–Villars mass or larger, than analytically continuing to $4 - \varepsilon$ dimensions. However, in modern applications, Pauli–Villars is only occasionally useful. The problem is that complicated multi-loop diagrams necessitate many fictitious particles (one for each real particle will not do it; the Lagrangian $\mathcal{L}_{\text{PV}}$ only works at 1-loop). Thus, Pauli–Villars quickly becomes impractical. In addition, it is not useful in non-Abelian gauge theories, since a massive gauge boson breaks gauge invariance. (Pauli–Villars does work in an Abelian theory, at least at 1-loop, as long as the gauge boson couples to a conserved current.)

B.4.3 Other regulators

There are several other regulators that are sometimes used:

- Hard cutoff: $k_E < \Lambda$. This breaks Lorentz invariance, and usually every symmetry in the theory, but is perhaps the most intuitive regularization procedure.
- Point splitting. Divergences at $k \to \infty$ correspond to two fields approaching each other $x_1 \to x_2$. Point splitting puts a lower bound on this, $|x_1^\mu - x_2^\mu| > |\epsilon^\mu|$. This also breaks translation invariance and is impractical for gauge theories, but is useful in theories with composite operators.
- Lattice regularization. Although a lattice breaks both translation invariance and Lorentz invariance, it is possible to construct a lattice such that translation and Lorentz invariance are restored in the continuum limit (see Section 25.5).

Problems

B.1 Show that the Wick rotation still works if $\Delta < 0$.

References

Abbasi, R. U., *et al.* 2008. First observation of the Greisen–Zatsepin–Kuzmin suppression. *Phys. Rev. Lett.*, **100**, 101101.

Abbott, L. F. 1982. Introduction to the background field method. *Acta Phys. Polon.*, **B13**, 33.

Adler, S. L. 1969. Axial vector vertex in spinor electrodynamics. *Phys. Rev.*, **177**, 2426–2438.

Albrecht, W., Behrend, H. J., Brasse, F. W., *et al.* 1966. Elastic electron–proton scattering at momentum transfers up to $245 \, F^{-2}$. *Phys. Rev. Lett.*, **17**, 1192–1195.

Altarelli, G., Ellis, R. K., and Martinelli, G. 1979. Large perturbative corrections to the Drell–Yan process in QCD. *Nucl. Phys. B*, **157**, 461.

Altland, A., and Simons, B. 2010. *Condensed Matter Field Theory*, 2nd edn. Cambridge: Cambridge University Press.

Andrade, E. N. da C. 1964. *Rutherford and the Nature of the Atom*. New York: Doubleday.

Appelquist, T., Dine, M., and Muzinich, I. J. 1977. The static potential in quantum chromodynamics. *Phys. Lett. B*, **69**, 231.

Arkani-Hamed, N., Georgi, H., and Schwartz, M. D. 2003. Effective field theory for massive gravitons and gravity in theory space. *Ann. Phys.*, **305**, 96–110.

Atlas Collaboration. 2013. Measurements of the properties of the Higgs-like boson in the four lepton decay channel with the ATLAS detector using 25 fb^{-1} of proton-proton collision data. Technical report ATLAS-CONF-2013-013. CERN, Geneva.

Bailey, J. A., *et al.* 2010. $B \to D^* \ell \nu$ at zero recoil: an update. *Proc. Sci.* LATTICE2010, 311.

Banks, T., and Zaks, A. 1982. On the phase structure of vector-like gauge theories with massless fermions. *Nucl. Phys. B*, **196**, 189.

Bastianelli, F., and van Nieuwenhuizen, P. 2006. *Path Integrals and Anomalies in Curved Space*. Cambridge: Cambridge University Press.

Berestetsky, V. B., Lifshitz, E. M., and Pitaevsky, L. P. 1982. *Quantum Electrodynamics*. Oxford: Elsevier.

Bern, Z., and Kosower, D. A. 1992. The computation of loop amplitudes in gauge theories. *Nucl. Phys. B*, **379**, 451–561.

Bethe, H., and Fermi, E. 1932. Über die Wechselwirkung von zwei Elektronen. *Z. Phys.*, **77**, 296–306.

Bhabha, H. J. 1936. The scattering of positrons by electrons with exchange on Dirac's theory of the positron. *Proc. Roy. Soc. London A*, **154**, 195–206.

Bloch, F., and Nordsieck, A. 1937. Note on the radiation field of the electron. *Phys. Rev.*, **52**, 54–59.

Bohr, N. 1938. The causality problem in physics. In *New Theories in Physics*, conference organized in collaboration with the International Union of Physics and the Polish Intellectual Co-Operation Committee, Warsaw, 30 May – 3 June, 1938.

Bohr, N., Kramers, H. A., and Slater, J. C. 1924. Über die Quantentheorie der Strahlung. *Z. Phys.*, **24**, 69–87.

Born, M., Heisenberg, W., and Jordan, P. 1926. Zur Quantenmechanik. II. *Z. Phys.*, **35**, 557–615.

Bressi, G., Carugno, G, Onofrio, R., and Ruoso, G. 2002. Measurement of the Casimir Force between Parallel Metallic Surfaces. *Phys. Rev. Lett.*, **88**, 041804.

Brown, L. S. 1992. *Quantum Field Theory*. Cambridge: Cambridge University Press.

Brown, L. M., and Hoddesdon, L. H. (Eds.). 1984. *The Birth of Particle Physics*. Proceedings, International Symposium, Batavia, USA, May 28–31, 1980. Cambridge: Cambridge University press.

Buchalla, G., Buras, A. J., and Lautenbacher, M. E. 1996. Weak decays beyond leading logarithms. *Rev. Mod. Phys.*, **68**, 1125–1144.

Callan, C. G. 1970. Broken scale invariance in scalar field theory. *Phys. Rev. D*, **2**, 1541–1547.

Callan, C. G., Jr., Coleman, S. R., Wess, J., and Zumino, B. 1969. Structure of phenomenological Lagrangians. 2. *Phys. Rev.*, **177**, 2247–2250.

Cartan, E. 1894. Sur la structure des groupes de transformations finis et continus. Thesis, Paris.

Casimir, H. B. G. 1948. On the attraction between two perfectly conducting plates. *Indag. Math.*, **10**, 261–263.

Cheng, T. P., and Li, L. F. 1985. *Gauge Theory of Elementary Particle Physics*. New York: Oxford University Press.

Chew, F. G. 1961. *S-Matrix Theory of Strong Interactions*. San Francisco, CA: Benjamin.

Christenson, J. H., Cronin, J. W., Fitch, V. L., and Turlay, R. 1964. Evidence for the 2π decay of the K_2^0 meson. *Phys. Rev. Lett.*, **13**, 138–140.

CKM fitter group (Charles *et al.*) 2012. *Eur. Phys. J. C*, **41**, 1–131 (2005). Updated results and plots available at http://ckmfitter.in2p3.fr.

CLEO Collaboration (Briere *et al.*) 2002. Improved measurement of $|V_{cb}|$ using $B \to D^* l_\nu$ decays. *Phys. Rev. Lett.*, **89**, 081803.

Coleman, S. 1985. *Aspects of Symmetry: Selected Erice Lectures*. Cambridge: Cambridge University Press.

Coleman, S. R., and Weinberg, E. J. 1973. Radiative corrections as the origin of spontaneous symmetry breaking. *Phys. Rev. D*, **7**, 1888–1910.

Coleman, S. R., Wess, J., and Zumino, B. 1969. Structure of phenomenological Lagrangians. 1. *Phys. Rev.*, **177**, 2239–2247.

Collins, J. C., Soper, D. E., and Sterman, G. F. 1988. Soft gluons and factorization. *Nucl. Phys. B*, **308**, 833–856.

Compton, A. H. 1923. A quantum theory of the scattering of x-rays by light elements. *Phys. Rev.*, **21**, 483–502.

Crewther, R. J., Di Vecchia, P., Veneziano, G., and Witten, E. 1979. Chiral estimate of the electric dipole moment of the neutron in quantum chromodynamics. *Phys. Lett. B*, **88**, 123–127.

Cutkosky, R. E. 1960. Singularities and discontinuities of Feynman amplitudes. *J. Math. Phys.*, **1**, 429–433.

Degrassi, G., Di Vita, S., Elias-Miro, J., *et al.* 2012. Higgs mass and vacuum stability in the Standard Model at NNLO. *J. High Energy Phys.*, **1208**, 098.

Dirac, P. A. M. 1927. Quantum theory of emission and absorption of radiation. *Proc. Roy. Soc. London A*, **114**, 243–265.

Dirac, P. A. M. 1930. *The Principles of Quantum Mechanics*. Oxford: Clarendon Press.

Dirac, P. A. M. 1936. Does conservation of energy hold in atomic processes? *Nature*, **137**, 298–299.

Donoghue, J. F. 1994. Leading quantum correction to the Newtonian potential. *Phys. Rev. Lett.*, **72**, 2996–2999.

Donoghue, J. F., Golowich, E., and Holstein, B. R. 1992. *Dynamics of the Standard Model*. Cambridge: Cambridge University Press.

Doria, R., Frenkel, J., and Taylor, J. C. 1980. Counter example to nonabelian Bloch–Nordsieck theorem. *Nucl. Phys. B*, **168**, 93.

Drell, S. D., and Yan, T.-M. 1970. Massive lepton-pair production in hadron–hadron collisions at high energies. *Phys. Rev. Lett.*, **25**, 316–320.

Dyson, F. J. 1949. The radiation theories of Tomonaga, Schwinger, and Feynman. *Phys. Rev.*, **75**, 486–502.

Eden, R. J., Landshoff, P. V., Olive, D. I., and Polkinghorne, J. C. 1966. *The Analytic S-Matrix*. Cambridge: Cambridge University Press.

Ellis, R. K., Stirling, W. J., and Webber, B. R. 1996. *QCD and Collider Physics*. Cambridge: Cambridge University Press.

Euler, H., and Heisenberg, W. 1936. Consequences of Dirac's theory of positrons. *Z. Phys.*, **98**, 714.

Euler, H., and Kockel, B. 1935. The scattering of light by light in the Dirac theory. *Naturwiss.*, **23**, 246.

Feynman, R. P. 1950. Mathematical formulation of the quantum theory of electromagnetic interaction. *Phys. Rev.*, **80**, 440–457.

Feynman, R. P. 1972. Closed loop and tree diagrams. *Selected Papers of Richard Feynman*, Brown, L. M. (Ed.), World Scientific, Singapore, 2000, p. 867.

Feynman, R. P., Morinigo, F. B., Wagner, W. G., and Hatfield, B. (Eds.). 1996. *Feynman Lectures on Gravitation*. Reading, MA: Addison-Wesley.

Fierz, M., and Pauli, W. 1939. On relativistic wave equations for particles of arbitrary spin in an electromagnetic field. *Proc. Roy. Soc. London A*, **173**, 211–232.

Fischler, W. 1977. Quark–anti-quark potential in QCD. *Nucl. Phys. B*, **129**, 157–174.

Friedman, J. I., and Kendall, H. W. 1972. Deep inelastic electron scattering. *Annu. Rev. Nucl. Part. Sci.*, **22**, 203–254.

Froissart, M. 1961. Asymptotic behavior and subtractions in the Mandelstam representation. *Phys. Rev.*, **123**, 1053–1057.

Gattringer, C., and Lang, C. B. 2010. Quantum chromodynamics on the lattice. *Lect. Notes Phys.*, **788**, 1–211.

Gell-Mann, M., and Low, F. E. 1954. Quantum electrodynamics at small distances. *Phys. Rev.*, **95**, 1300–1312.

Georgi, H. 1982. Lie algebras in particle physics. *Front. Phys.*, **54**, 1–255.

Georgi, H. 1984. *Weak Interactions and Modern Particle Theory*. San Francisco, CA: Benjamin Cummings.

Georgi, H. 1990. An effective field theory for heavy quarks at low-energies. *Phys. Lett.*, *B*, **240**, 447–450.

Gfitter Group (Baak *et al.*) 2012. The electroweak fit of the Standard Model after the discovery of a new boson at the LHC. *Eur. Phys. J. C*, **72**, 2205.

Glashow, S. L., Iliopoulos, J., and Maiani, L. 1970. Weak interactions with lepton-hadron symmetry. *Phys. Rev. D*, **2**, 1285–1292.

Greisen, K. 1966. End to the cosmic-ray spectrum? *Phys. Rev. Lett.*, **16**, 748–750.

Greiner, W, and Reinhardt, J. 1996. *Field Quantization*. Berlin, Heidelberg: Springer.

Griffiths, D. 2008. *Introduction to Elementary Particles*. Weinheim: Wiley-VCH.

Grisaru, M. T., van Nieuwenhuizen, P., and Wu, C. C. 1975. Background field method versus normal field theory in explicit examples: one loop divergences in S matrix and Green's functions for Yang–Mills and gravitational fields. *Phys. Rev. D*, **12**, 3203.

Grozin, A. G. 2004. *Heavy Quark Effective Theory*. Berlin, Heidelberg: Springer.

Halpern, O. 1933. Scattering processes produced by electrons in negative energy states. *Phys. Rev.*, **44**, 855 856.

Halzen, F., and Martin, A. D. 1984. *Quarks and Leptons: An Introductory Course in Modern Particle Physics*. New York: John Wiley & Sons.

Heisenberg, W., and Euler, H. 1936. Consequences of Dirac's theory of positrons. *Z. Phys.*, **98**, 714–732.

Itzykson, C., and Zuber, J. B. 1980. *Quantum Field Theory*. New York: McGraw-Hill.

Jackiw, R. 1974. Functional evaluation of the effective potential. *Phys. Rev. D*, **9**, 1686.

Kadanoff, L. P. 1966. Scaling laws for Ising models near T_c. *Physics*, **2**, 263–272.

Kamionkowski, M., and March-Russell, J. 1992. Planck-scale physics and the Peccei–Quinn mechanism. *Phys. Lett. B*, **282**, 137–141.

Kang, J. S. 1974. Gauge invariance of the scalar-vector mass ratio in the Coleman–Weinberg model. *Phys. Rev. D*, **10**, 3455–3467.

Kim, Sang Pyo, and Page, D. N. 2002. Schwinger pair production via instantons in a strong electric field. *Phys. Rev. D*, **65**, 105002.

Kinoshita, T. 1962. Mass singularities of Feynman amplitudes. *J. Math. Phys.*, **3**, 650–677.

Klein, O., and Nishina, Y. 1929. On the scattering of radiation by free electrons according to Dirac's new relativistic quantum dynamics. In *The Oskar Klein Memorial Lectures, Vol. 2: Lectures by Hans A. Bethe and Alan H. Guth with Translated Reprints by Oskar Klein*, Gösta Ekspong (Ed.), Singapore: World Scientific, 1994, pp. 113–139.

Kleinert, H., Neu, J., Schulte-Frohlinde, V., Chetyrkin, K. G., and Larin, S. A. 1991. Five-loop renormalization group functions of O(n)-symmetric ϕ^4-theory and ϵ expansions of critical exponents up to ϵ^5. *Phys. Lett. B*, **272**, 39–44.

Lamb, W. E., and Retherford, R. C. 1947. Fine structure of the hydrogen atom by a microwave method. *Phys. Rev.*, **72**, 241–243.

Lamoreaux, S. K. 1997. Demonstration of the Casimir force in the 0.6 to 6 μm range. *Phys. Rev. Lett.*, **78**, 5–8.

Landau, L. D. 1959. On analytic properties of vertex parts in quantum field theory. *Nucl. Phys.*, **13**, 181–192.

Lee, T. D., and Nauenberg, M. 1964. Degenerate systems and mass singularities. *Phys. Rev.*, **133**, B1549.

Lee, T. D., and Yang, C. N. 1956. Question of parity conservation in weak interactions. *Phys. Rev.*, **104**, 254–258.

Lee, B. W., Quigg, C., and Thacker, H. B. 1977. Weak interactions at very high energies: the role of the Higgs-boson mass. *Phys. Rev. D*, **16**, 1519–1531.

Leibbrandt, G. 1987. Introduction to Noncovariant Gauges. *Rev. Mod. Phys.*, **59**, 1067–1119.

Logiurato, F. 2012. Teaching waves with Google Earth. *Phys. Edu.*, **47**(1), 73.

Magnea, L., and Sterman, G. F. 1990. Analytic continuation of the Sudakov form-factor in QCD. *Phys. Rev. D*, **42**, 4222–4227.

Maldacena, J. M. 1998. The large N limit of superconformal field theories and supergravity. *Adv. Theor. Math. Phys.*, **2**, 231–252.

Manohar, A. V. 2003. Deep inelastic scattering as $x \to 1$ using soft collinear effective theory. *Phys. Rev. D*, **68**, 114019.

Manohar, A. V., and Wise, M. B. 2000. *Heavy Quark Physics*. Cambridge: Cambridge University Press.

Mehra, J., and Milton, K. A. 2000. *Climbing the Mountain: The Scientific Biography of Julian Schwinger*. Oxford: Oxford University Press.

Melnikov, K., and van Ritbergen, T. 2000. The three-loop relation between the $\overline{\text{MS}}$ and the pole quark masses. *Phys. Lett. B*, **482**, 99–108.

Møller, C. 1932. *Ann. Phys*, **14**, 531.

MSTW Group (Martin, A. D., Stirling, W. J., Thorne, R. S., and Watt, G.) 2009. Parton distributions for the LHC. *Eur. Phys. J. C*, **63**, 189–285.

Muta, T. 2010. *Foundations of Quantum Chromodynamics: An Introduction to Perturbative Methods in Gauge Theories*. Singapore: World Scientific.

Nakahara, M. 2003. *Geometry, Topology and Physics*. Bristol: Institute of Physics.

Oppenheimer, J. R. 1930. Note on the theory of the interaction of field and matter. *Phys. Rev.*, **35**, 461–477.

Page, L. A. 1950. A measurement of electron-electron scattering. Ph.D thesis, Cornell University.

Pais, A. 1986. *Inward Bound of Matter and Forces in the Physical World*. New York: Oxford University Press.

Particle Data Group (Beringer *et al.*) 2012. Review of particle physics (RPP). *Phys. Rev. D*, **86**, 010001.

Passarino, G., and Veltman, M. J. G. 1979. One loop corrections for e^+e^- annihilation into $\mu^+\mu^-$ in the Weinberg model. *Nucl. Phys. B*, **160**, 151–207.

Pauli, W. 1940. The connection between spin and statistics. *Phys. Rev.*, **58**, 716.

Pelissetto, A., and Vicari, E. 2002. Critical phenomena and renormalization group theory. *Phys. Rep.*, **368**, 549–727.

Peskin, M. E. 1990. Theory of precision electroweak measurements. Seventeenth SLAC Summer Institute, SLAC-PUB-5210.

Peskin, M. E., and Schroeder, D. V. 1995. *An Introduction to Quantum Field Theory*. Reading, MA: Addison-Wesley.

Peskin, M. E., and Takeuchi, T. 1992. Estimation of oblique electroweak corrections. *Phys. Rev. D*, **46**, 381–409.

Planck, M. 1901. On the law of distribution of energy in the normal spectrum. *Ann. Phys.*, **4**, 553.

Polchinski, J. 1984. Renormalization and effective Lagrangians. *Nucl. Phys. B*, **231**, 269–295.

Polchinski, J. 1998. *String Theory. Volume II: Superstring Theory and Beyond*. Cambridge: Cambridge University Press.

Polyakov, A. M. 1981. Quantum geometry of bosonic strings. *Phys. Lett. B*, **103**, 207–210.

Rutherford, E. 1911. The scattering of alpha and beta particles by matter and the structure of the atom. *Phil. Mag.*, **21**, 669–688.

Sachdev, S. 2011. *Quantum Phase Transitions* (2nd edition). Cambridge: Cambridge University Press.

Sakharov, A. D. 1967. Violation of CP invariance, C asymmetry, and baryon asymmetry of the universe. *Pisma Zh. Eksp. Teor. Fiz.*, **5**, 32–35.

Sakurai, J. J. 1993. *Modern Quantum Mechanics* (Revised edition). Reading, MA: Addison-Wesley.

Schumacher, M., Borchert, I., Smend, F. and Rullhusen, P. 1975. Delbruck Scattering of 2.75-MeV photons by lead. *Phys. Lett. B*, **59**, 134–136.

Schweber, S. S. 1994. *QED and the Men Who Made It: Dyson, Feynman, Schwinger, and Tomonaga*. Princeton, NJ: Princeton University Press.

Schwinger, J. S. 1951. On gauge invariance and vacuum polarization. *Phys. Rev.*, **82**, 664–679.

Shankland, R. S. 1936. An apparent failure of the photon theory of scattering. *Phys. Rev.*, **49**, 8–13.

Sjostrand, T., Mrenna, S., and Skands, P. Z. 2006. PYTHIA 6.4 physics and manual. *J. High Energy Phys.*, **0605**, 026.

Srednicki, M. 2007. *Quantum Field Theory*. Cambridge: Cambridge University Press.

Steinberger, J. 1949. On the use of subtraction fields and the lifetimes of some types of meson decay. *Phys. Rev.*, **76**, 1180–1186.

Stelle, K. S. 1978. Classical gravity with higher derivatives. *Gen. Rel. Grav.*, **9**, 353–371.

Sterman, G. F. 1993. *An Introduction to Quantum Field Theory*. Cambridge: Cambridge University Press.

Sterman, G. F., and Weinberg, S. 1977. Jets from quantum chromodynamics. *Phys. Rev. Lett.*, **39**, 1436.

Streater, R. F., and Wightman, A. S. 1989. *PCT, Spin and Statistics, and All That*. Princeton, NJ: Princeton University Press.

Stueckelberg, E. C. G. 1938. Interaction energy in electrodynamics and in the field theory of nuclear forces. *Helv. Phys. Acta*, **11**, 225–244.

Stueckelberg, E. C. G., and Petermann, A. 1953. Normalization of constants in the quanta theory. *Helv. Phys. Acta*, **26**, 499–520.

Susskind, L. 1977. Dynamics of spontaneous symmetry breaking in the Weinberg Salam theory. In Balian, R. and Llewellyn Smith, C. H. (Eds.), *Weak and Electromagnetic Interactions at High-Energies*. Proceedings 29th Summer School on Theoretical Physics, Les Houches, July 5–August 14, 1976.

Symanzik, K. 1970. Small distance behaviour in field theory and power counting. *Commun. Math. Phys.*, **18**, 227–246.

't Hooft, G. 1971. Renormalizable Lagrangians for massive Yang–Mills fields. *Nucl. Phys. B*, **35**, 167–188.

't Hooft, G. 1974. A two-dimensional model for mesons. *Nucl. Phys. B.*, **75**, 461–477.

't Hooft, G. 1979. In *Recent Developments in Gauge Theories*. Reprinted in *Dynamical Gauge Symmetry Breaking: A Collection of Reprints*. Farhi, E., and Jackiw, R. (Eds.), Singapore: World Scientific, 1982, pp. 345–367.

't Hooft, G., Itzykson, C., Jaffe, A., Lehmann, H., Mitter, P. K., *et al.* (Eds.) 1980. *Recent Developments in Gauge Theories*. Proceedings Nato Advanced Study Institute, Cargese, France, August 26 – September 8, 1979. *NATO Adv. Study Inst. Ser. B Phys.*, **59**, 135.

't Hooft, G., and Veltman, M. J. G. 1972. Regularization and renormalization of gauge fields. *Nucl. Phys. B*, **44**, 189–213.

't Hooft, G., and Veltman, M. J. G. 1974. One loop divergencies in the theory of gravitation. *Ann. Poincare Phys. Theor.*, **A20**, 69–94.

Terning, J. 2006. *Modern Supersymmetry: Dynamics and Duality*. Oxford: Oxford University Press.

Uehling, E. A. 1935. Polarization effects in the positron theory. *Phys. Rev.*, **48**, 55–63.

van Ritbergen, T., Vermaseren, J. A. M., and Larin, S. A. 1997. The four-loop beta-function in quantum chromodynamics. *Phys. Lett. B*, **400**, 379–384.

Veltman, M. J. G. 1994. *Diagrammatica: The Path to Feynman Rules*. Cambridge: Cambridge University Press.

Veneziano, G. 1979. U(1) without instantons. *Nucl. Phys. B*, **159**, 213–224.

Weinberg, S. 1964. Photons and gravitons in S-matrix theory: derivation of charge conservation and equality of gravitational and inertial mass. *Phys. Rev.*, **135**, B1049–B1056.

Weinberg, S. 1975. The U(1) problem. *Phys. Rev. D*, **11**, 3583–3593.

Weinberg, S. 1979. Phenomenological Lagrangians. *Physica A*, **96**, 327.

Weinberg, S. 1995. *The Quantum Theory of Fields. Volume 1: Foundations*. Cambridge: Cambridge University Press.

Weinberg, S. 1996. *The Quantum Theory of Fields. Volume 2: Modern Applications*. Cambridge: Cambridge University Press.

Weinberg, S., and Witten, E. 1980. Limits on massless particles. *Phys. Lett. B*, **96**, 59.

Wess, J., and Bagger, J. 1992. *Supersymmetry and Supergravity*. Princeton, NJ: Princeton University Press.

Wigner, E. P. 1939. On unitary representations of the inhomogeneous Lorentz group. *Ann. Math.*, **40**, 149–204.

Wilson, K. G. 1971. Renormalization group and critical phenomena. 1. Renormalization group and the Kadanoff scaling picture. *Phys. Rev. B*, **4**, 3174–3183.

Wilson, K. G. 1974. Confinement of quarks. *Phys. Rev. D*, **10**, 2445–2459.

Wilson, K. G., and Kogut, J. B. 1974. The renormalization group and the epsilon expansion. *Phys. Rep.*, **12**, 75–200.

Witten, E. 1979. Current algebra theorems for the U(1) Goldstone boson. *Nucl. Phys. B*, **156**, 269.

Yukawa, H. 1935. On the interaction of elementary particles. *Proc. Phys. Math. Soc. Japan*, **17,** 48–57.

Zatsepin, G. T., and Kuz'min, V. A. 1966. Upper limit of the spectrum of cosmic rays. *J. Exp. Theor. Phys. Lett.*, **4**, 78–80.

Zee, A. 2003. *Quantum Field Theory in a Nutshell*. Princeton, NJ: Princeton University Press.

Index